Advanced Vehicle Technology

Second edition

Heinz Heisler MSc., BSc., F.I.M.I., M.S.O.E., M.I.R.T.E., M.C.I.T., M.I.L.T.

Formerly Principal Lecturer and Head of Transport Studies,
College of North West London, Willesden Centre, London, UK

OXFORD AMSTERDAM BOSTON LONDON NEW YORK PARIS
SAN DIEGO SAN FRANCISCO SINGAPORE SYDNEY TOKYO

Butterworth-Heinemann
An imprint of Elsevier Science
Linacre House, Jordan Hill, Oxford OX2 8DP
225 Wildwood Avenue, Woburn, MA 01801-2041

First published by Edward Arnold 1989
Reprinted by Reed Educational and Professional Publishing Ltd 2001

Second edition 2002

Library of Congress Cataloguing in Publication Data
A catalogue record for this book is available from the Library of Congress

ISBN 0 7506 5131 8

Typeset by Integra Software Services Pvt. Ltd, Pondicherry, India
www.integra-india.com
Printed and bound in Great Britain

Contents

Preface to the second edition

It is thirteen years since the first edition of this book was published, in this time there have been great strides in improving the vehicle's ability to react quickly, precisely and positively to changing road conditions, such as varying road surface texture and irregularities, body roll when negotiating bends, retaining traction when accelerating and directional stability when braking under adverse road conditions. Improvements in the technical design of the vehicle have to a great extent been made possible by the development of electric and electronic equipment in conjunction with hydraulic or pneumatic servo systems which are capable of rapidly sensing and responding to variable and sometimes unpredictable driving modes and emergencies. Thus the application of electronics to the transmission, steering, suspension and braking systems now dominates their effectiveness. The demand for better vehicle performance in terms of power, torque, acceleration, fuel economy and simplicity in driving considerably increases the use of the five speed electronic/hydraulic controlled automatic transmissions. The necessity to improve the life expectancy of commercial vehicles clutches and gearboxes and to reduce driver fatigue has brought about the use of electronic/pneumatic controlled semi-automatic transmissions. Improvement in tractive grip when driving hard over a slippery, rough or boggy terrain has been made possible with the use of electronic/hydraulic controlled 'limited slip differentials' and 'traction control systems'. Electronic controlled hydraulic pump driven or electric motor driven power steering systems can now provide a varied degree of servo assistance for different driving conditions, such as for parking in confined spaces, low speed or fast driving. The quality of ride expected in the mid-and upper range of cars now justifies the installation of electronic/hydraulic controlled active suspension to regulate the height, levelling and roll stiffness of the car. The need to improve the quality of ride both for special off/on road cars and for commercial vehicles to protect the passengers or goods being carried, has resulted in a preference to electronic controlled air suspensions on many vehicles. With these air–spring suspensions the operation height can be altered to suit various driving conditions and loading applications.

Advances have been made to improve the braking performance of commercial vehicles by introducing electronic/pneumatic control to the braking system with built-in safeguards to brake failure. Part of the job of transporting goods from door-to-door these days is to keep perishable goods cool or even frozen, thus refrigerated vehicles are necessary and have become common, therefore a working understanding of refrigeration for the mechanic and engineer is desirable if not a must. As the competition between manufactures to sell more cars and trucks has increased, designers have looked towards improving the streamlining of the body shape to enhance the style, to reduce wind resistance, noise and fuel consumption and to improve road holding and vehicle stability at speed, it therefore follows that an appreciation of the fundamentals of vehicle aerodynamics is now essential for students, maintenance mechanics, fleet engineers and design engineers.

Preface to the first edition

Advanced Vehicle Technology has been written with the aim of presenting an up to date, broad-based and in-depth treatise devoted to the design, construction and maintenance of all the essential components of the motor vehicle in a clear and precise manner.

The intention is to introduce and explain fundamental automotive engineering concepts so that the reader can appreciate design considerations and grasp an understanding of the difficulties the manufacturer has in producing components which satisfy the designers' requirements and the production engineer and maintenance assessibility. This will enable the mechanic or technician to develop an enquiring mind to search for and reason out symptoms leading to the diagnosis and rectification of component faults.

Line diagrams are used to illustrate basic principles and construction detail. Photographs of dismantled or assembled components are deliberately excluded because these can be obtained from any manufacturer's repair manual and do not adequately explain the information which is being sought.

This follow on from *Vehicle and Engine Technology* covers levels 2 and 3 of CGLI 383 Repair and Servicing of Road Vehicles and the objective bank for levels 2 and 3 of BTEC Motor Vehicle Engineering. However, in places the content goes beyond the requirements of these syllabuses to provide a better foundation for subsequent studies, and the fullness of information should also make the book a useful reference source for practising mechanics and technicians as well as students.

Acknowledgements
The author would like to express his thanks to the many friends and colleagues who provided suggestions and encouragement in the preparation of this book; to Ruth Cripwell of Edward Arnold (Publishers) Ltd who scrutinised and edited the text; and to his daughter Glenda who typed the manuscript.

1 Vehicle Structure

1.1 Integral body construction

The integral or unitary body structure of a car can be considered to be made in the form of three box compartments; the middle and largest compartment stretching between the front and rear road wheel axles provides the passenger space, the extended front box built over and ahead of the front road wheels enclosing the engine and transmission units and the rear box behind the back axle providing boot space for luggage.

These box compartments are constructed in the form of a framework of ties (tensile) and struts (compressive), pieces (Fig. 1.1(a & b)) made from rolled sheet steel pressed into various shapes such as rectangular, triangular, trapezium, top-hat or a combination of these to form closed box thin gauge sections. These sections are designed to resist direct tensile and compressive or bending and torsional loads, depending upon the positioning of the members within the structure.

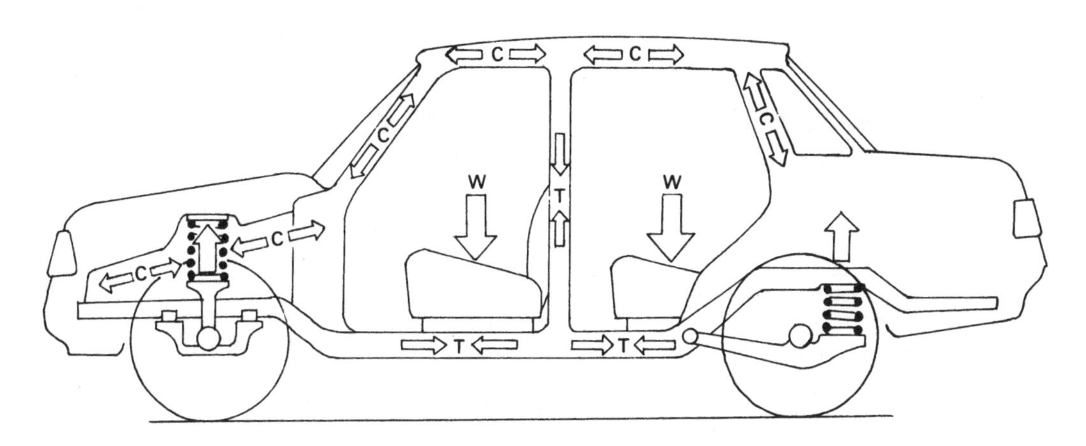

(a) Lengthwise body loading

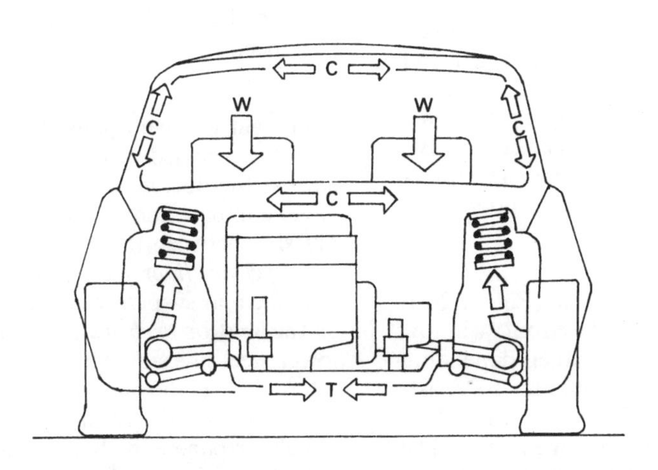

(b) Transverse body loading

Fig. 1.1 (a and b) Structural tensile and compressive loading of car body

1.1.1 Description and function of body components (Fig. 1.2)

The major individual components comprising the body shell will now be described separately under the following subheadings:

1 Window and door pillars
2 Windscreen and rear window rails
3 Cantrails
4 Roof structure
5 Upper quarter panel or window
6 Floor seat and boot pans
7 Central tunnel
8 Sills
9 Bulkhead
10 Scuttle
11 Front longitudinals
12 Front valance
13 Rear valance
14 Toe board
15 Heel board

Window and door pillars (Fig. 1.2(3, 5, 6, and 8)) Windowscreen and door pillars are identified by a letter coding; the front windscreen to door pillars are referred to as *A post*, the centre side door pillars as *BC post* and the rear door to quarter panel as *D post*. These are illustrated in Fig. 1.2.

These pillars form the part of the body structure which supports the roof. The short form A pillar and rear D pillar enclose the windscreen and quarter windows and provide the glazing side channels, whilst the centre BC pillar extends the full height of the passenger compartment from roof to floor and supports the rear side door hinges. The front and rear pillars act as struts (compressive members) which transfer a proportion of the bending effect, due to underbody sag of the wheelbase, to each end of the cantrails which thereby become reactive struts, opposing horizontal bending of the passenger compartment at floor level. The central BC pillar however acts as ties (tensile members), transferring some degree of support from the mid-span of the cantrails to the floor structure.

Windscreen and rear window rails (Fig. 1.2(2)) These box-section rails span the front window pillars and rear pillars or quarter panels depending upon design, so that they contribute to the resistance opposing transverse sag between the wheel track by acting as compressive members. The other function is to support the front and rear ends of the roof panel. The undersides of the rails also include the glazing channels.

Cantrails (Fig. 1.2(4)) Cantrails are the horizontal members which interconnect the top ends of the vertical A and BC or BC and D door pillars (posts). These rails form the side members which make up the rectangular roof framework and as such are subjected to compressive loads. Therefore, they are formed in various box-sections which offer the greatest compressive resistance with the minimum of weight and blend in with the roofing. A drip rail (Fig. 1.2(4)) is positioned in between the overlapping roof panel and the cantrails, the joins being secured by spot welds.

Roof structure (Fig. 1.2) The roof is constructed basically from four channel sections which form the outer rim of the slightly dished roof panel. The rectangular outer roof frame acts as the compressive load bearing members. Torsional rigidity to resist twist is maximized by welding the four corners of the channel-sections together. The slight curvature of the roof panel stiffens it, thus preventing winkling and the collapse of the unsupported centre region of the roof panel. With large cars, additional cross-rail members may be used to provide more roof support and to prevent the roof crushing in should the car roll over.

Upper quarter panel or window (Fig. 1.2(6)) This is the vertical side panel or window which occupies the space between the rear side door and the rear window. Originally the quarter panel formed an important part of the roof support, but improved pillar design and the desire to maximize visibility has either replaced them with quarter windows or reduced their width, and in some car models they have been completely eliminated.

Floor seat and boot pans (Fig. 1.3) These constitute the pressed rolled steel sheeting shape to enclose the bottom of both the passenger and luggage compartments. The horizontal spread-out pressing between the bulkhead and the heel board is called the *floor pan*, whilst the raised platform over the rear suspension and wheel arches is known as the *seat* or *arch pan*. This in turn joins onto a lower steel pressing which supports luggage and is referred to as the *boot pan*.

To increase the local stiffness of these platform panels or pans and their resistance to transmitted vibrations such as drumming and droning, many narrow channels are swaged (pressed) into the steel sheet, because a sectional end-view would show a

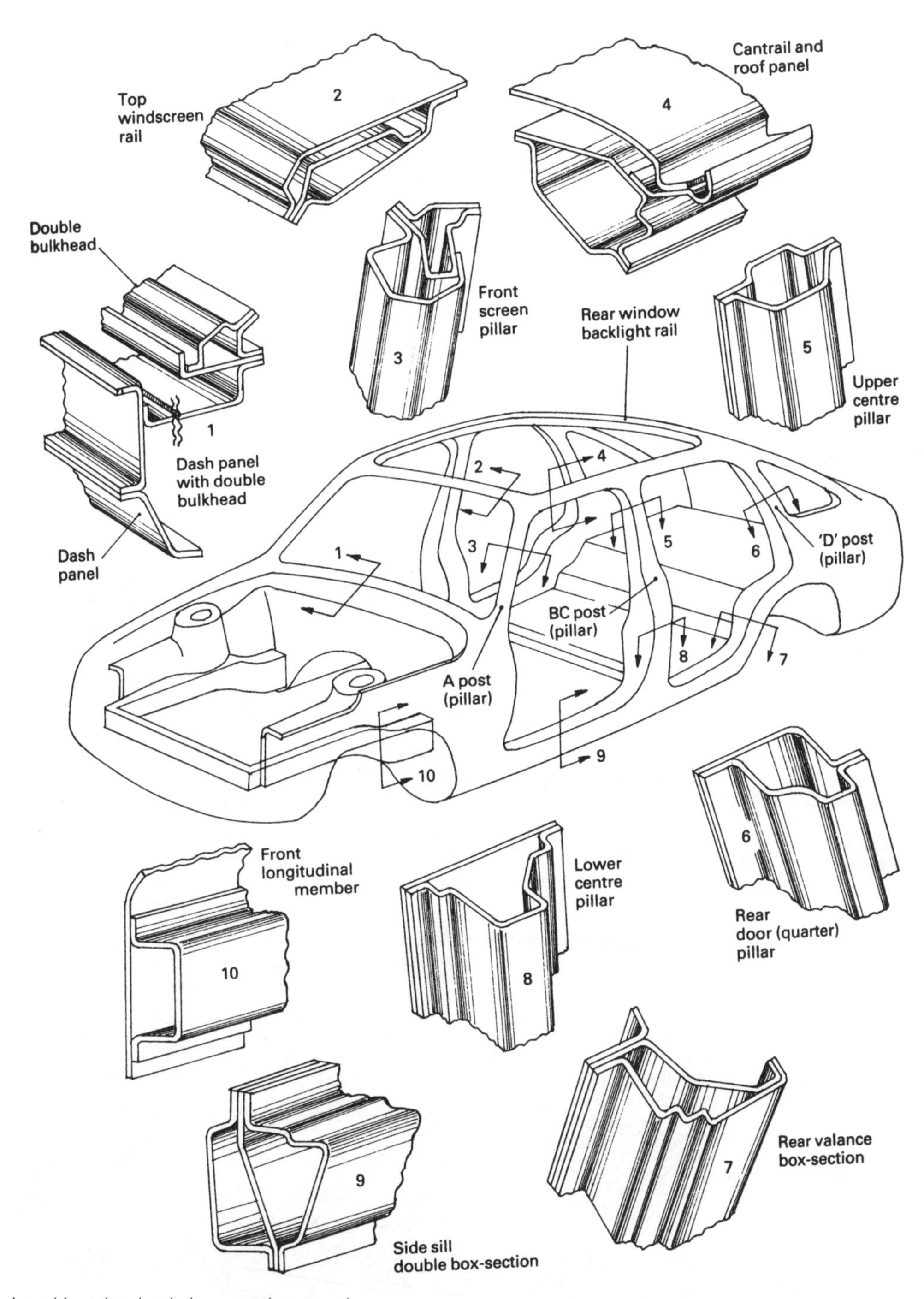

Fig. 1.2 Load bearing body box-section members

semi-corrugated profile (or ribs). These channels provide rows of shallow walls which are both bent and stretched perpendicular to the original flat sheet. In turn they are spaced and held together by the semicircular drawn out channel bottoms. Provided these swages are designed to lay the correct way and are not too long, and the metal is not excessively stretched, they will raise the rigidity

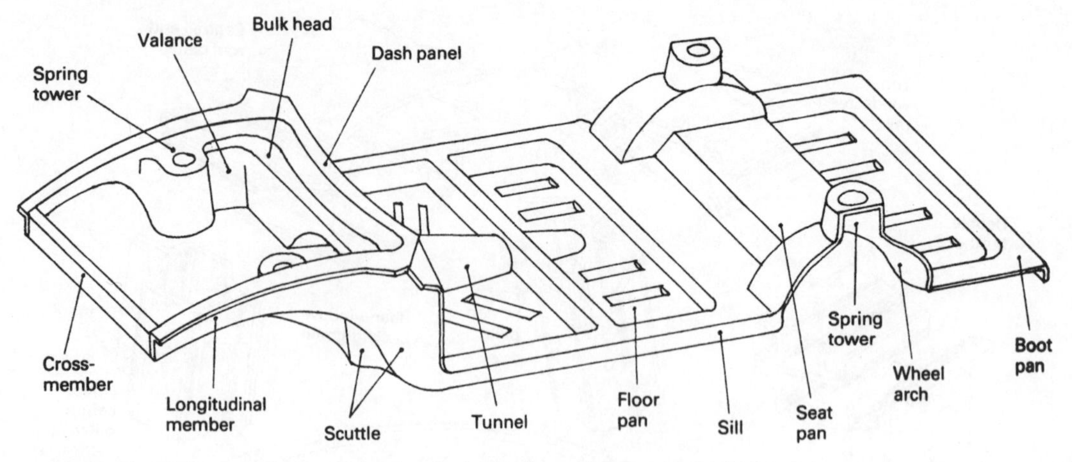

(a) Small central tunnel, sills, front valance, rear wheel arches and all round spring towers

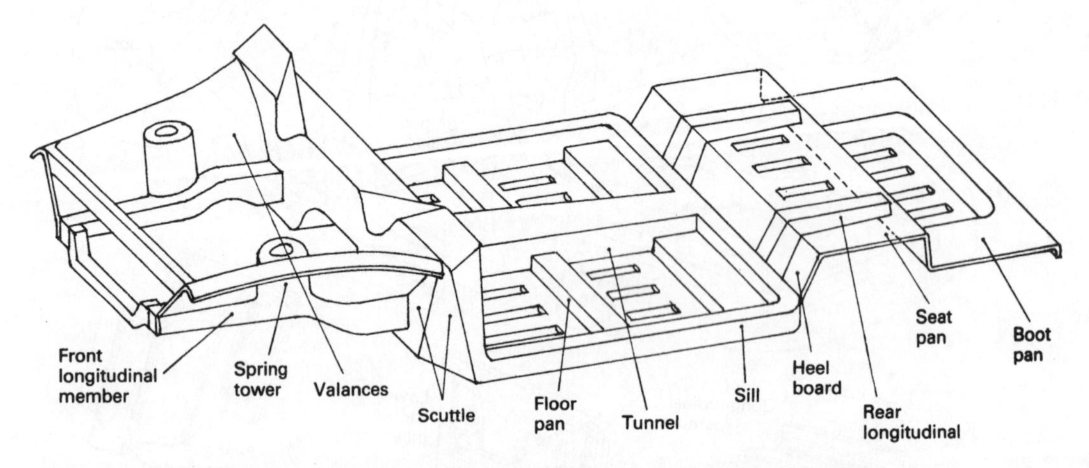

(b) Large central tunnel, sills, front valance and spring tower with rear box-section reinforced seat pan

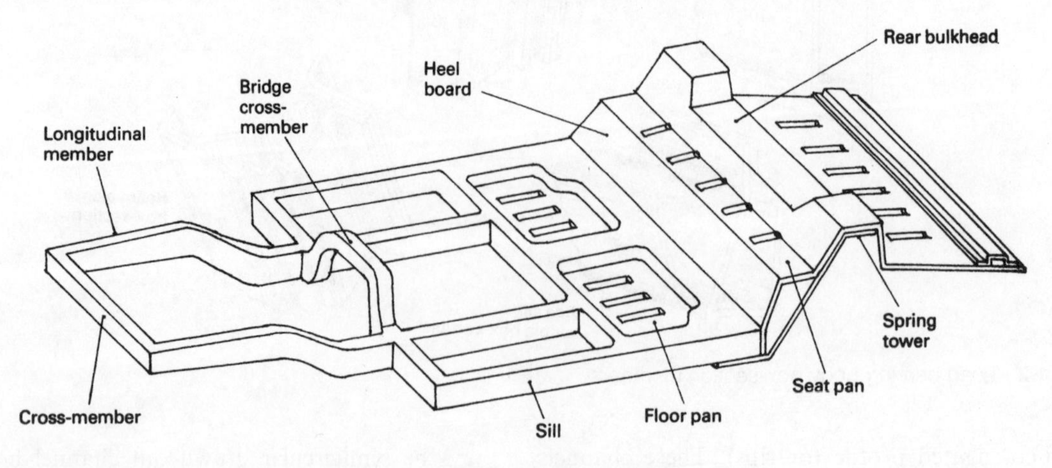

(c) Longitudinal front members, sills, rear seat pan with bulkhead and spring towers

Fig. 1.3 (a–c) Platform chassis

of these panels so that they are equivalent to a sheet which may be several times thicker.

Central tunnel (Fig. 1.3(a and b)) This is the curved or rectangular hump positioned longitudinally along the middle of the floor pan. Originally it was a necessary evil to provide transmission space for the gearbox and propeller shaft for rear wheel drive, front-mounted engine cars, but since the chassis has been replaced by the integral box-section shell, it has been retained with front wheel drive, front-mounted engines as it contributes considerably to the bending rigidity of the floor structure. Its secondary function is now to house the exhaust pipe system and the hand brake cable assembly.

Sills (Figs 1.2(9) and 1.3(a, b and c)) These members form the lower horizontal sides of the car body which spans between the front and rear road-wheel wings or arches. To prevent body sag between the wheelbase of the car and lateral bending of the structure, the outer edges of the floor pan are given support by the side sills. These sills are made in the form of either single or double box-sections (Fig. 1.2(9)). To resist the heavier vertical bending loads they are of relatively deep section.

Open-top cars, such as convertibles, which do not receive structural support from the roof members, usually have extra deep sills to compensate for the increased burden imposed on the underframe.

Bulkhead (Figs 1.2(1) and 1.3(a and b)) This is the upright partition separating the passenger and engine compartments. Its upper half may form part of the dash panel which was originally used to display the driver's instruments. Some body manufacturers refer to the whole partition between engine and passenger compartments as the dash panel. If there is a double partition, the panel next to the engine is generally known as the *bulkhead* and that on the passenger side the *dash board* or *panel*. The scuttle and valance on each side are usually joined onto the box-section of the bulkhead. This braces the vertical structure to withstand torsional distortion and to provide platform bending resistance support. Sometimes a bulkhead is constructed between the rear wheel arches or towers to reinforce the seat pan over the rear axle (Fig. 1.3(c)).

Scuttle (Fig. 1.3(a and b)) This can be considered as the panel formed under the front wings which spans between the rear end of the valance, where it meets the bulkhead, and the door pillar and wing. The lower edge of the scuttle will merge with the floor pan so that in some cases it may form part of the toe board on the passenger compartment side. Usually these panels form inclined sides to the bulkhead, and with the horizontal ledge which spans the full width of the bulkhead, brace the bulkhead wall so that it offers increased rigidity to the structure. The combined bulkhead dash panel and scuttle will thereby have both upright and torsional rigidity.

Front longitudinals (Figs 1.2(10) and 1.3(a and b)) These members are usually upswept box-section members, extending parallel and forward from the bulkhead at floor level. Their purpose is to withstand the engine mount reaction and to support the front suspension or subframe. A common feature of these members is their ability to support vertical loads in conjunction with the valances. However, in the event of a head-on collision, they are designed to collapse and crumble within the engine compartment so that the passenger shell is safeguarded and is not pushed rearwards by any great extent.

Front valance (Figs 1.2 and 1.3(a and b)) These panels project upwards from the front longitudinal members and at the rear join onto the wall of the bulkhead. The purpose of these panels is to transfer the upward reaction of the longitudinal members which support the front suspension to the bulkhead. Simultaneously, the longitudinals are prevented from bending sideways because the valance panels are shaped to slope up and outwards towards the top. The panelling is usually bent over near the edges to form a horizontal flanged upper, thus presenting considerable lateral resistance. Furthermore, the valances are sometimes stepped and wrapped around towards the rear where they meet and are joined to the bulkhead so that additional lengthwise and transverse stiffness is obtained.

If coil spring suspension is incorporated, the valance forms part of a semi-circular tower which houses and provides the load reaction of the spring so that the merging of these shapes compounds the rigidity for both horizontal lengthwise and lateral bending of the forward engine and transmission compartment body structure. Where necessary, double layers of sheet are used in parts of the spring housing and at the rear of the valance where they are attached to the bulkhead to relieve some of the concentrated loads.

Rear valance (Fig. 1.2(7)) This is generally considered as part of the box-section, forming the front half of the rear wheel arch frame and the panel immediately behind which merges with the heel board and seatpan panels. These side inner-side panels position the edges of the seat pan to its designed side profile and thus stiffen the underfloor structure above the rear axle and suspension. When rear independent coil spring suspension is adopted, the valance or wheel arch extends upwards to form a spring tower housing and, because it forms a semi-vertical structure, greatly contributes to the stiffness of the underbody shell between the floor and boot pans.

Toe board The toe board is considered to form the lower regions of the scuttle and dash panel near where they merge with the floor pan. It is this panelling on the passenger compartment side where occupants can place their feet when the car is rapidly retarded.

Heel board (Fig. 1.3(b and c)) The heel board is the upright, but normally shallow, panel spanning beneath and across the front of the rear seats. Its purpose is to provide leg height for the passengers and to form a raised step for the seat pan so that the rear axle has sufficient relative movement clearance.

1.1.2 Platform chassis (Fig. 1.3(a–c))

Most modern car bodies are designed to obtain their rigidity mainly from the platform chassis and to rely less on the upper framework of window and door pillars, quarter panels, windscreen rails and contrails which are becoming progressively slender as the desire for better visibility is encouraged.

The majority of the lengthwise (wheelbase) bending stiffness to resist sagging is derived from both the central tunnel and the side sill box-sections (Fig. 1.3(a and b)). If further strengthening is necessary, longitudinal box-section members may be positioned parallel to, but slightly inwards from, the sills (Fig. 1.3(c)). These lengthwise members may span only part of the wheelbase, or the full length, which is greatly influenced by the design of road wheel suspension chosen for the car, the depth of both central tunnel and side sills, which are built into the platform, and if there are subframes attached fore and aft of the wheelbase (Fig. 1.6 (a and b)).

Torsional rigidity of the platform is usually derived at the front by the bulkhead, dash pan and scuttle (Fig. 1.3(a and b)) at the rear by the heel board, seat pan, wheel arches (Fig. 1.3(a, b and c)), and if independent rear suspension is adopted, by the coil spring towers (Fig. 1.3(a and c)). Between the wheelbase, the floor pan is normally provided with box-section cross-members to stiffen and prevent the platform sagging where the passenger seats are positioned.

1.1.3 Stiffening of platform chassis
(Figs 1.4 and 1.5)

To appreciate the stresses imposed on and the resisting stiffness offered by sheet steel when it is subjected to bending, a small segment of a beam greatly magnified will now be considered (Fig. 1.4(a)). As the beam deforms, the top fibres contract and the bottom fibres elongate. The neutral plane or axis of the beam is defined as the plane whose length remains unchanged during deformation and is normally situated in the centre of a uniform section (Fig. 1.4(a and b)).

The stress distribution from top to bottom within the beam varies from zero along the neutral axis (NA), where there is no change in the length of the fibres, to a maximum compressive stress on the outer top layer and a maximum tensile stress on the outer bottom layer, the distortion of the fibres being greatest at their extremes as shown in Fig. 1.4(b).

It has been found that bending resistance increases roughly with the cube of its distance from the neutral axis (Fig. 1.5(a)). Therefore, bending resistance of a given section can be greatly improved for a given weight of metal by taking metal away from the neutral axis where the metal fibres do not contribute very much to resisting distortion and placing it as far out as possible where the distortion is greatest. Bending resistance may be improved by using longitudinal or cross-member deep box-sections (Fig. 1.5(b)) and tunnel sections (Fig. 1.5(c)) to restrain the platform chassis from buckling and to stiffen the flat horizontal floor seat and boot pans. So that vibration and drumming may be reduced, many swaged ribs are pressed into these sheets (Fig. 1.5(d)).

1.1.4 Body subframes (Fig. 1.6)

Front or rear subframes may be provided to brace the longitudinal side members so that independent suspension on each side of the car receives adequate support for the lower transverse swing arms (wishbone members). Subframes restrain the two halves of the suspension from splaying outwards or the

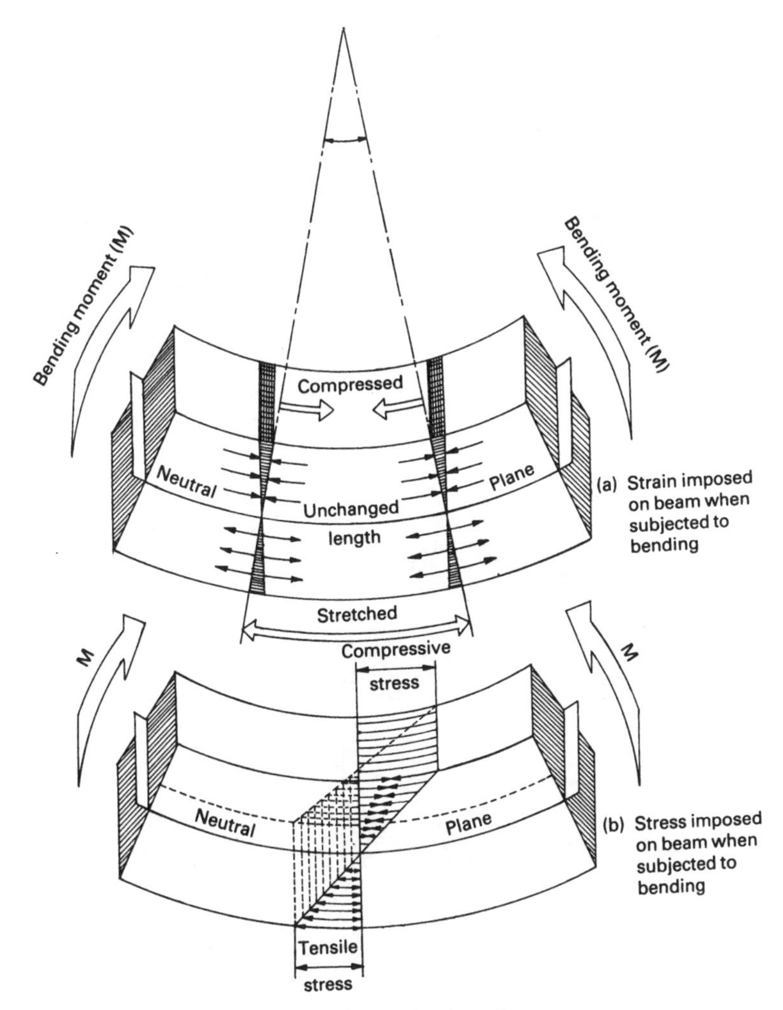

Fig. 1.4 Stress and strain imposed on beam when subjected to bending

longitudinal side members from lozenging as alternative road wheels experience impacts when travelling over the irregularities of a normal road surface.

It is usual to make the top side of the subframe the cradle for the engine or engine and transmission mounting points so that the main body structure itself does not have to be reinforced. This particularly applies where the engine, gearbox and final drive form an integral unit because any torque reaction at the mounting points will be transferred to the subframe and will multiply in proportion to the overall gear reduction. This may be approximately four times as great as that for the front mounted engine with rear wheel drive and will become prominent in the lower gears.

One advantage claimed by using separate subframes attached to the body underframe through the media of rubber mounts is that transmitted vibrations and noise originating from the tyres and road are isolated from the main body shell and therefore do not damage the body structure and are not relayed to the occupants sitting inside.

Cars which have longitudinally positioned engines mounted in the front driven by the rear wheels commonly adopt beam cross-member subframes at the front to stiffen and support the hinged transverse suspension arms (Fig. 1.6(a)). Saloon cars employing independent rear suspension sometimes prefer to use a similar subframe at the rear which provides the pivot points for the semi-trailing arms because this type of suspension requires greater support than most other arrangements (Fig. 1.6(a)).

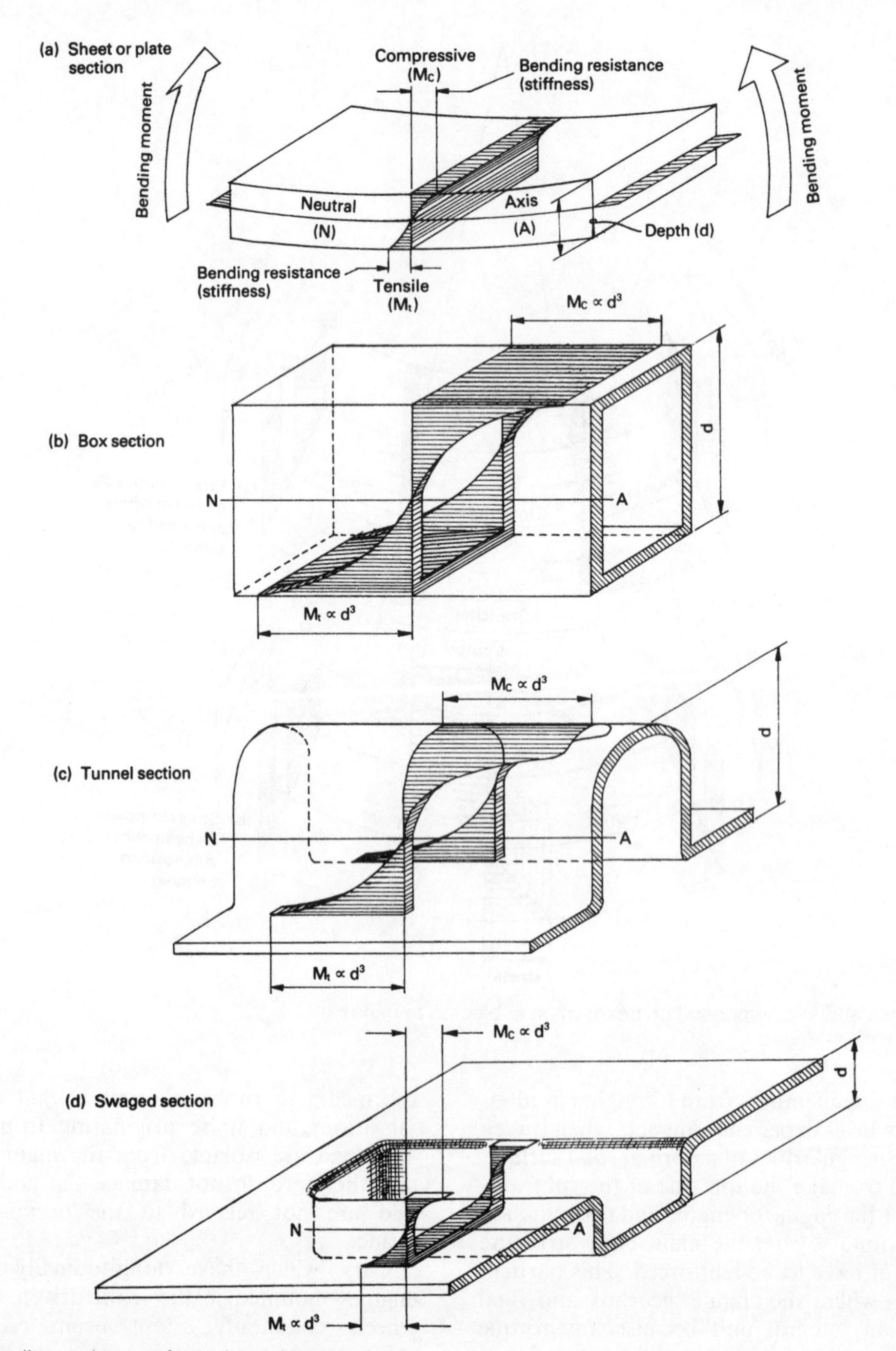

Fig. 1.5 Bending resistance for various sheet sections

When the engine, gearbox and final drive are combined into a single unit, as with the front longitudinally positioned engine driving the front wheels where there is a large weight concentration, a subframe gives extra support to the body longitudinal side members by utilising a horseshoe shaped frame (Fig. 1.6(b)). This layout provides a platform for the entire mounting points for both the swing arm and anti-roll bar which between them make up the lower part of the suspension.

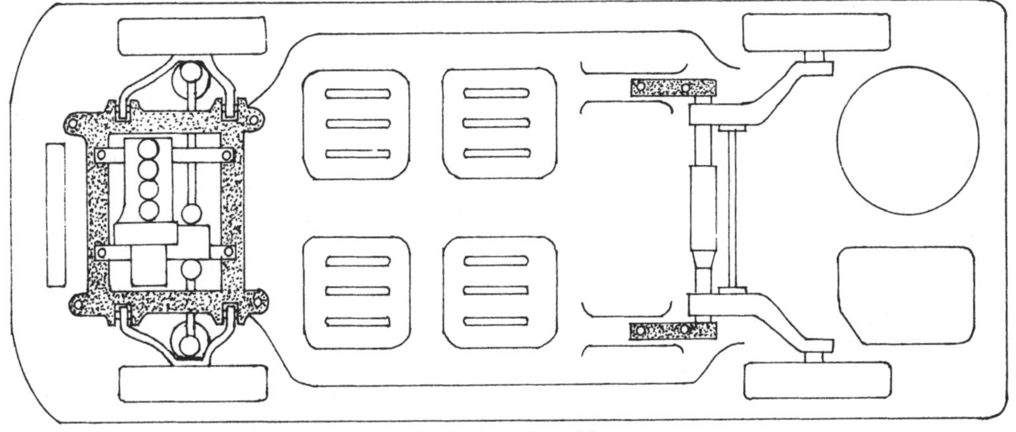

(c) Rectangular subframe

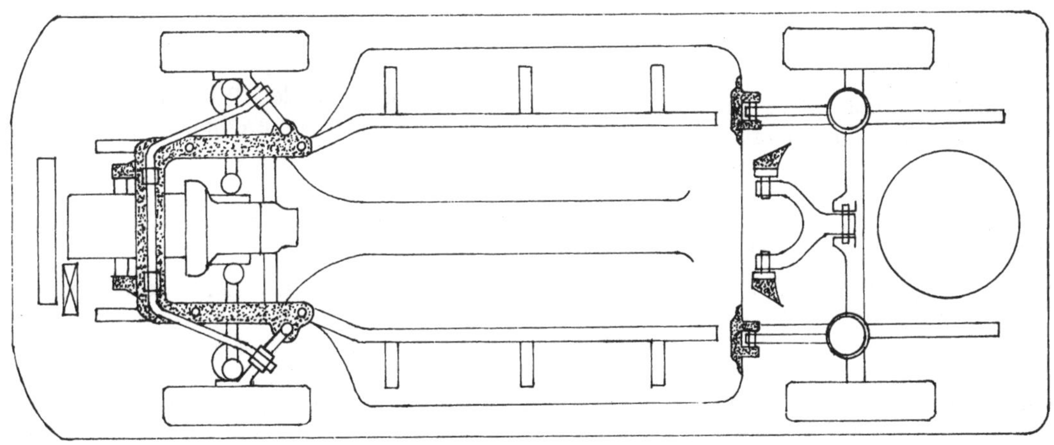

(b) Horseshoe subframe

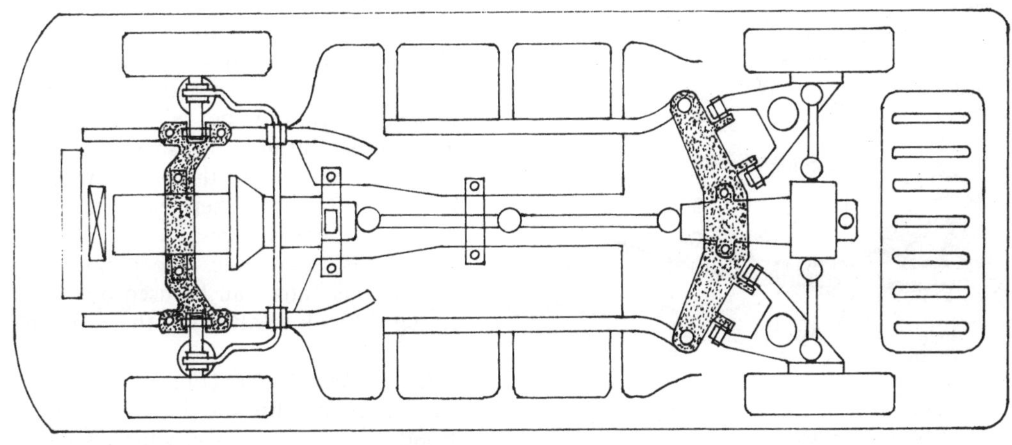

(a) Beam cross-member subframe

Fig. 1.6 (a–c) Body subframe and underfloor structure

Front wheel drive transversely positioned engines with their large mounting point reactions often use a rectangular subframe to spread out both the power and transmission unit's weight and their dynamic reaction forces (Fig. 1.6(c)). This configuration provides substantial torsional rigidity between both halves of the independent suspension without relying too much on the main body structure for support.

Soundproofing the interior of the passenger compartment (Fig. 1.7)
Interior noise originating outside the passenger compartment can be greatly reduced by applying layers of materials having suitable acoustic properties over floor, seat and boot pans, central tunnel, bulkhead, dash panel, toeboard, side panels, inside of doors, and the underside of both roof and bonnet etc. (Fig. 1.7).

Acoustic materials are generally designed for one of three functions:

a) Insulation from noise — This may be created by forming a non-conducting noise barrier between the source of the noises (which may come from the engine, transmission, suspension tyres etc.) and the passenger compartment.
b) Absorption of vibrations — This is the transference of excited vibrations in the body shell to a media which will dissipate their resultant energies and so eliminate or at least greatly reduce the noise.
c) Damping of vibrations — When certain vibrations cannot be eliminated, they may be exposed to some form of material which in some way modifies the magnitude of frequencies of the vibrations so that they are less audible to the passengers.

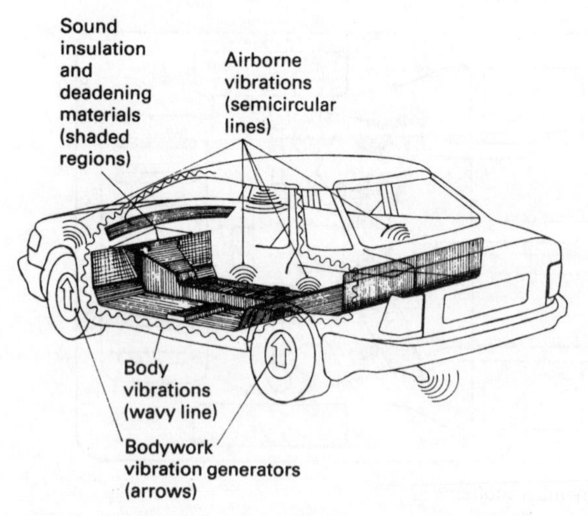

Fig. 1.7 Car body sound generation and its dissipation

The installation of acoustic materials cannot completely eliminate boom, drumming, droning and other noises caused by resonance, but merely reduces the overall noise level.

Insulation Because engines are generally mounted close to the passenger compartment of cars or the cabs of trucks, effective insulation is important. In this case, the function of the material is to reduce the magnitude of vibrations transmitted through the panel and floor walls. To reduce the transmission of noise, a thin steel body panel should be combined with a flexible material of large mass, based on PVC, bitumen or mineral wool. If the insulation material is held some distance from the structural panel, the transmissibility at frequencies above 400 Hz is further reduced. For this type of application the loaded PVC material is bonded to a spacing layer of polyurethane foam or felt, usually about 7 mm thick. At frequencies below 400 Hz, the use of thicker spacing layers or heavier materials can also improve insulation.

Absorption For absorption, urethane foam or lightweight bonded fibre materials can be used. In some cases a vinyl sheet is bonded to the foam to form a roof lining. The required thickness of the absorbent material is determined by the frequencies involved. The minimum useful thickness of polyurethane foam is 13 mm which is effective with vibration frequencies above 1000 Hz.

Damping To damp resonance, pads are bonded to certain panels of many cars and truck cabs. They are particularly suitable for external panels whose resonance cannot be eliminated by structural alterations. Bituminous sheets designed for this purpose are fused to the panels when the paint is baked on the car. Where extremely high damping or light weight is necessary, a PVC base material, which has three times the damping capacity of bituminous pads, can be used but this material is rather difficult to attach to the panelling.

1.1.5 Collision safety (Fig. 1.8)
Car safety may broadly be divided into two kinds: Firstly the *active safety*, which is concerned with the car's road-holding stability while being driven, steered or braked and secondly the *passive safety*,

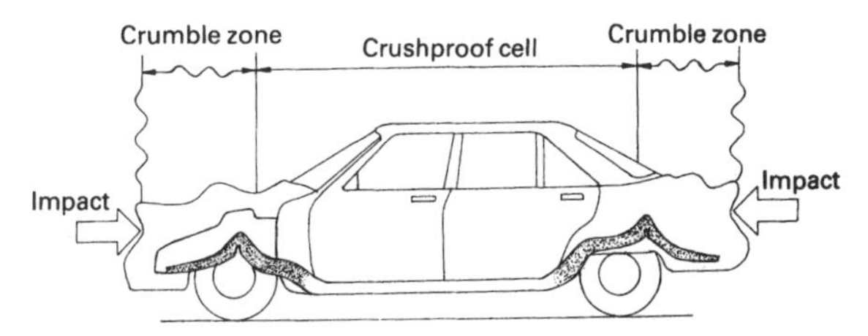

Fig. 1.8 Collision body safety

which depends upon body style and design structure to protect the occupants of the car from serious injury in the event of a collision.

Car bodies can be considered to be made in three parts (Fig. 1.8); a central cell for the passengers of the welded bodywork integral with a rigid platform, acting as a floor pan, and chassis with various box-section cross- and side-members. This type of structure provides a reinforced rigid crushproof construction to resist deformation on impact and to give the interior a high degree of protection. The extension of the engine and boot compartments at the front and rear of the central passenger cell are designed to form zones which collapse and crumble progressively over the short duration of a collision impact. Therefore, the kinetic energy due to the car's initial speed will be absorbed fore and aft primarily by strain and plastic energy within the crumble zones with very little impact energy actually being dissipated by the central body cell.

1.1.6 Body and chassis alignment checks (Fig. 1.9)

Body and chassis alignment checks will be necessary if the vehicle has been involved in a major collision, but overall alignment may also be necessary if the vehicle's steering and ride characteristics do not respond to the expected standard of a similar vehicle when being driven.

Structural misalignment may be caused by all sorts of reasons, for example, if the vehicle has been continuously driven over rough ground at high speed, hitting an obstacle in the road, mounting steep pavements or kerbs, sliding off the road into a ditch or receiving a glancing blow from some other vehicle or obstacle etc. Suspicion that something is wrong with the body or chassis alignment is focused if there is excessively uneven or high tyre wear, the vehicle tends to wander or pull over to one side and yet the track and suspension geometry appears to be correct.

Alignment checks should be made on a level, clear floor with the vehicle's tyres correctly inflated to normal pressure. A plumb bob is required in the form of a stubby cylindrical bar conical shaped at one end, the other end being attached to a length of thin cord. Datum reference points are chosen such as the centre of a spring eye on the chassis mounting point, transverse wishbone and trailing arm pivot centres, which are attachment points to the underframe or chassis, and body cross-member to side-member attachment centres and subframe bolt-on points (Fig. 1.9).

Initially the cord with the plumb bob hanging from its end is lowered from the centre of each reference point to the floor and the plumb bob contact point with the ground is marked with a chalked cross. Transverse and diagonal lines between reference points can be made by chalking the full length of a piece of cord, holding it taut between reference centres on the floor and getting somebody to pluck the centre of the line so that it rebounds and leaves a chalked line on the floor.

A reference longitudinal centre line may be made with a strip of wood baton of length just greater than the width between adjacent reference marks on the floor. A nail is punched through one end and this is placed over one of the reference marks. A piece of chalk is then held at the tip of the free end and the whole wood strip is rotated about the nailed end. The chalk will then scribe an arc between adjacent reference points. This is repeated from the other side. At the points where these two arcs intersect a straight line is made with a plucked, chalked cord running down the middle of the vehicle. This procedure should be followed at each end of the vehicle as shown in Fig. 1.9.

Once all the reference points and transverse and diagonal joining lines have been drawn on the

Table 1.1 Summary of function and application of soundproofing materials

Function	Acoustic materials	Application
Insulation	Loaded PVC, bitumen, with or without foam or fibres base, mineral wool	Floor, bulkhead dash panel
Damping	Bitumen or mineral wool	Doors, side panels, underside of roof
Absorption	Polyurethane foam, mineral wool, or bonded fibres	Side panels, underside of roof, engine compartment, bonnet

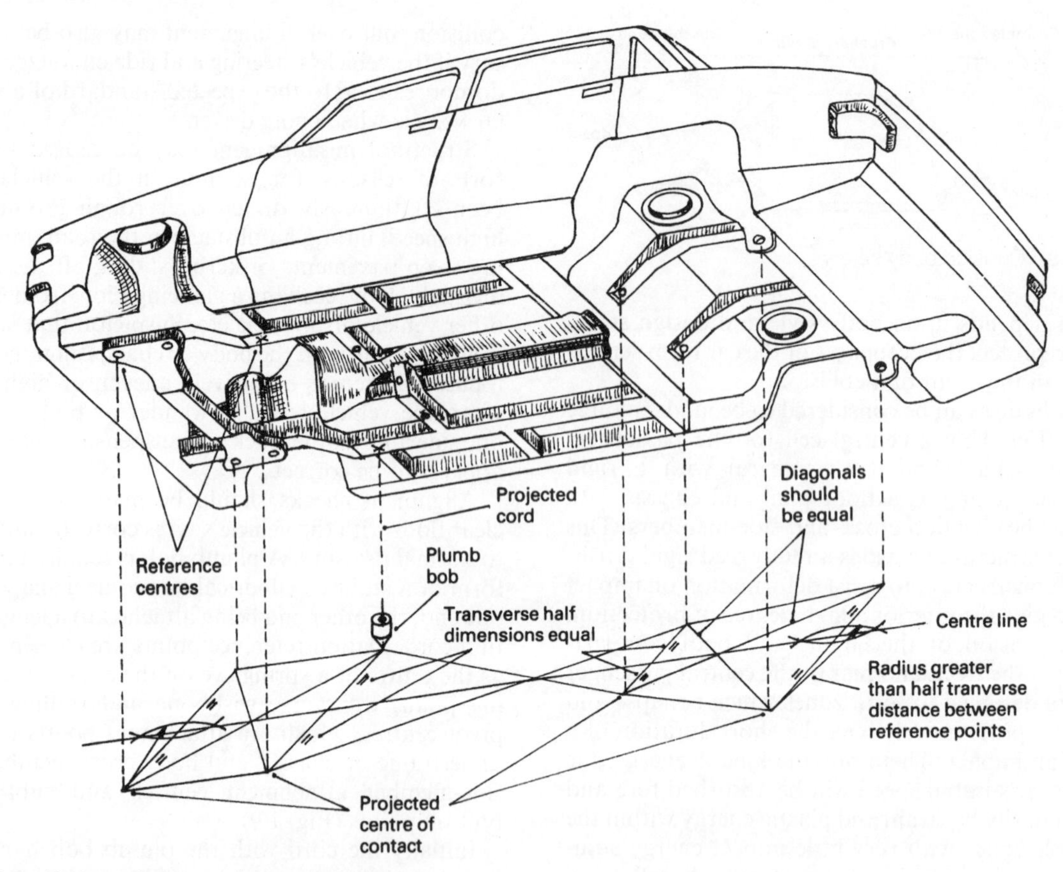

Fig. 1.9 Body underframe alignment checks

floor, a rule or tape is used to measure the distances between centres both transversely and diagonally. These values are then chalked along their respective lines. Misalignment or error is observed when a pair of transverse or diagonal dimensions differ and further investigation will thus be necessary.

Note that transverse and longitudinal dimensions are normally available from the manufacturer's manual and differences between paired diagonals indicates lozenging of the framework due to some form of abnormal impact which has previously occurred.

1.2 Engine, transmission and body structure mountings

1.2.1 Inherent engine vibrations

The vibrations originating within the engine are caused by both the cyclic acceleration of the reciprocating components and the rapidly changing cylinder gas pressure which occurs throughout each cycle of operation.

Both the variations of inertia and gas pressure forces generate three kinds of vibrations which are transferred to the cylinder block:

1 Vertical and/or horizontal shake and rock
2 Fluctuating torque reaction
3 Torsional oscillation of the crankshaft

1.2.2 Reasons for flexible mountings

It is the objective of flexible mounting design to cope with the many requirements, some having conflicting constraints on each other. A list of the duties of these mounts is as follows:

1 To prevent the fatigue failure of the engine and gearbox support points which would occur if they were rigidly attached to the chassis or body structure.
2 To reduce the amplitude of any engine vibration which is being transmitted to the body structure.
3 To reduce noise amplification which would occur if engine vibration were allowed to be transferred directly to the body structure.

4 To reduce human discomfort and fatigue by partially isolating the engine vibrations from the body by means of an elastic media.
5 To accommodate engine block misalignment and to reduce residual stresses imposed on the engine block and mounting brackets due to chassis or body frame distortion.
6 To prevent road wheel shocks when driving over rough ground imparting excessive rebound movement to the engine.
7 To prevent large engine to body relative movement due to torque reaction forces, particularly in low gear, which would cause excessive misalignment and strain on such components as the exhaust pipe and silencer system.
8 To restrict engine movement in the fore and aft direction of the vehicle due to the inertia of the engine acting in opposition to the accelerating and braking forces.

1.2.3 Rubber flexible mountings (Figs 1.10, 1.11 and 1.12)

A rectangular block bonded between two metal plates may be loaded in compression by squeezing the plates together or by applying parallel but opposing forces to each metal plate. On compression, the rubber tends to bulge out centrally from the sides and in shear to form a parallelogram (Fig. 1.10(a)).

To increase the compressive stiffness of the rubber without greatly altering the shear stiffness, an interleaf spacer plate may be bonded in between the top and bottom plate (Fig. 1.10(b)). This interleaf plate prevents the internal outward collapse of the rubber, shown by the large bulge around the sides of the block, when no support is provided, whereas with the interleaf a pair of much smaller bulges are observed.

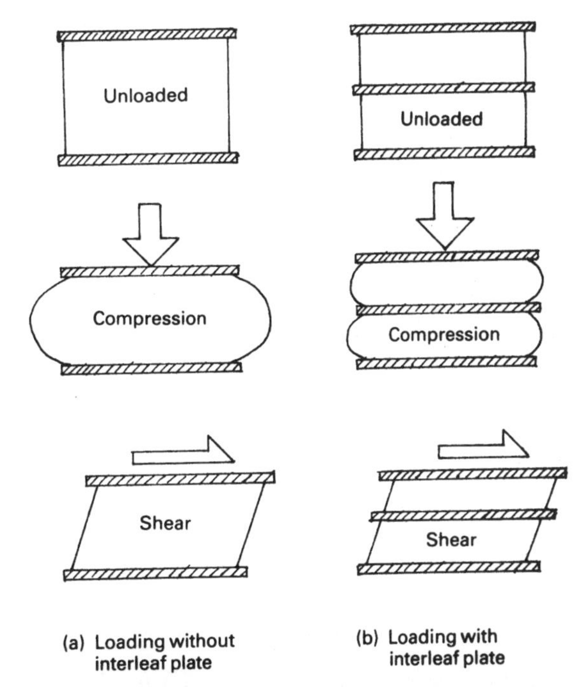

Fig. 1.10 (a and b) Modes of loading rubber blocks

When two rubber blocks are inclined to each other to form a 'V' mounting, see Fig. 1.11, the rubber will be loaded in both compression and shear shown by the triangle of forces. The magnitude of compressive force will be given by W_c and the much smaller shear force by W_S. This produces a resultant reaction force W_R. The larger the wedge angle Θ, the greater the proportion of compressive load relative to the shear load the rubber block absorbs.

The distorted rubber provides support under light vertical static loads approximately equal in both compression and shear modes, but with heavier loads the proportion of compressive stiffness

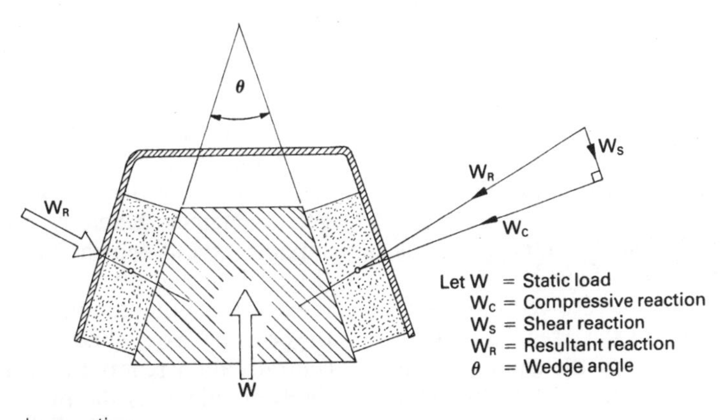

Fig. 1.11 'V' rubber block mounting

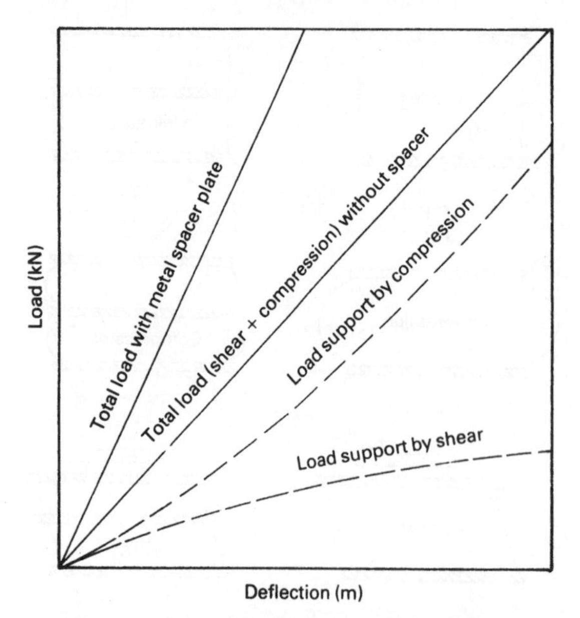

Fig. 1.12 Load–deflection curves for rubber block

to that of shear stiffness increases at a much faster rate (Fig. 1.12). It should also be observed that the combined compressive and shear loading of the rubber increases in direct proportion to the static deflection and hence produces a straight line graph.

1.2.4 Axis of oscillation (Fig. 1.13)

The engine and gearbox must be suspended so that it permits the greatest degree of freedom when oscillating around an imaginary centre of rotation known as the principal axis. This principal axis produces the least resistance to engine and gearbox sway due to their masses being uniformly distributed about this axis. The engine can be considered to oscillate around an axis which passes through the centre of gravity of both the engine and gearbox (Figs 1.13(a, b and c)). This normally produces an axis of oscillation inclined at about 10–20° to the crankshaft axis. To obtain the greatest degree of freedom, the mounts must be arranged so that they offer the least resistance to shear within the rubber mounting.

1.2.5 Six modes of freedom of a suspended body (Fig. 1.14)

If the movement of a flexible mounted engine is completely unrestricted it may have six modes of vibration. Any motion may be resolved into three linear movements parallel to the axes which pass through the centre of gravity of the engine but at right angles to each other and three rotations about these axes (Fig. 1.14).

These modes of movement may be summarized as follows:

Linear motions	**Rotational motions**
1 Horizontal longitudinal	4 Roll
2 Horizontal lateral	5 Pitch
3 Vertical	6 Yaw

1.2.6 Positioning of engine and gearbox mountings (Fig. 1.15)

If the mountings are placed underneath the combined engine and gearbox unit, the centre of gravity is well above the supports so that a lateral (side) force acting through its centre of gravity, such as experienced when driving round a corner, will cause the mass to roll (Fig. 1.15(a)). This condition is undesirable and can be avoided by placing the mounts on brackets so that they are in the same plane as the centre of gravity (Fig. 1.15(b)). Thus the mounts provide flexible opposition to any side force which might exist without creating a roll couple. This is known as a *decoupled* condition.

An alternative method of making the natural modes of oscillation independent or uncoupled is achieved by arranging the supports in an inclined 'V' position (Fig. 1.15(c)). Ideally the aim is to make the compressive axes of the mountings meet at the centre of gravity, but due to the weight of the power unit distorting the rubber springing the inter-section lines would meet slightly below this point. Therefore, the mountings are tilted so that the compressive axes converge at some focal point above the centre of gravity so that the actual lines of action of the mountings, that is, the direction of the resultant forces they exert, converge on the centre of gravity (Fig. 1.15(d)).

The compressive stiffness of the inclined mounts can be increased by inserting interleafs between the rubber blocks and, as can be seen in Fig. 1.15(e), the line of action of the mounts converges at a lower point than mounts which do not have interleaf support.

Engine and gearbox mounting supports are normally of the three or four point configuration. Petrol engines generally adopt the three point support layout which has two forward mounts (Fig. 1.13(a and c)), one inclined on either side of the engine so that their line of action converges on the principal axis, while the rear mount is supported centrally at the rear of the gearbox in approximately the same plane as the principal axis. Large diesel engines tend to prefer the four point support

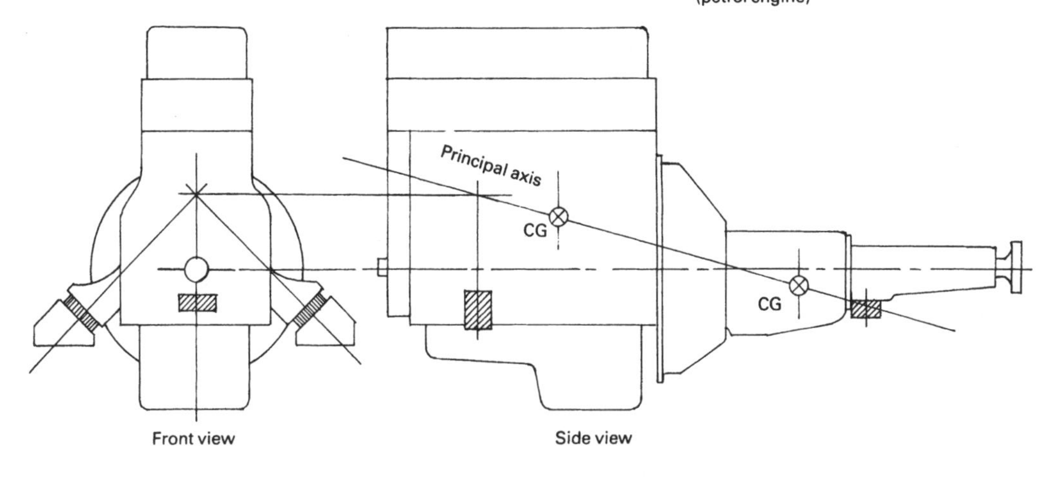

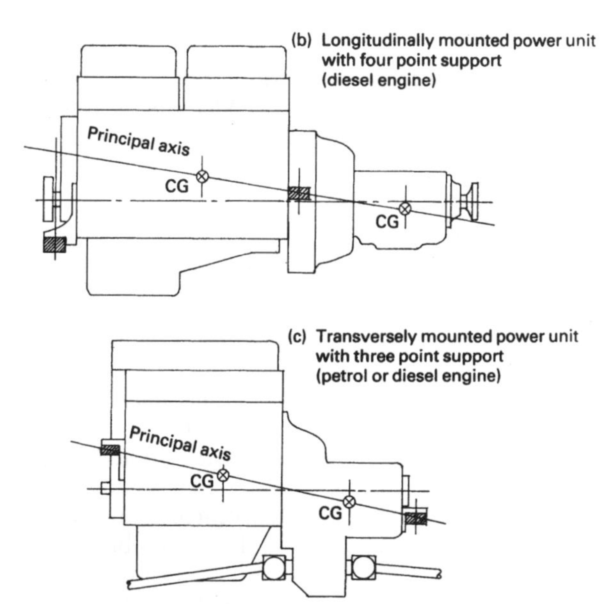

Fig. 1.13 Axis of oscillation and the positioning of the power unit flexible mounts

arrangement where there are two mounts either side of the engine (Fig. 1.13(b)). The two front mounts are inclined so that their lines of action pass through the principal axis, but the rear mounts which are located either side of the clutch bell housing are not inclined since they are already at principal axis level.

1.2.7 Engine and transmission vibrations

Natural frequency of vibration (Fig. 1.16) A sprung body when deflected and released will bounce up and down at a uniform rate. The amplitude of this cyclic movement will progressively decrease and the number of oscillations per minute of the rubber mounting is known as its *natural frequency of vibration.*

There is a relationship between the static deflection imposed on the rubber mount springing by the suspended mass and the rubber's natural frequency of vibration, which may be given by

$$n_0 = \frac{30}{\sqrt{x}}$$

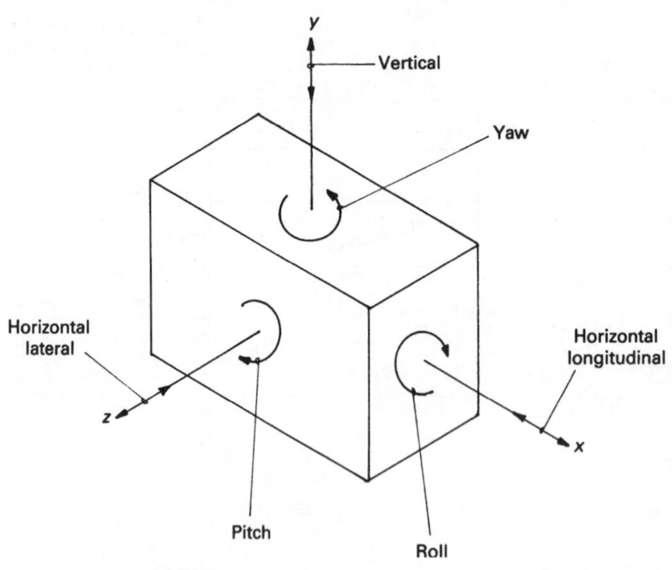

Fig. 1.14 Six modes of freedom for a suspended block

where n_0 = natural frequency of vibration (vib/min)
x = static deflection of the rubber (m)

This relationship between static deflection and natural frequency may be seen in Fig. 1.16.

Resonance Resonance is the unwanted synchronization of the disturbing force frequency imposed by the engine out of balance forces and the fluctuating cylinder gas pressure and the natural frequency of oscillation of the elastic rubber support mounting, i.e. resonance occurs when

$$\frac{n}{n_0} = 1$$

where n = disturbing frequency
n_0 = natural frequency

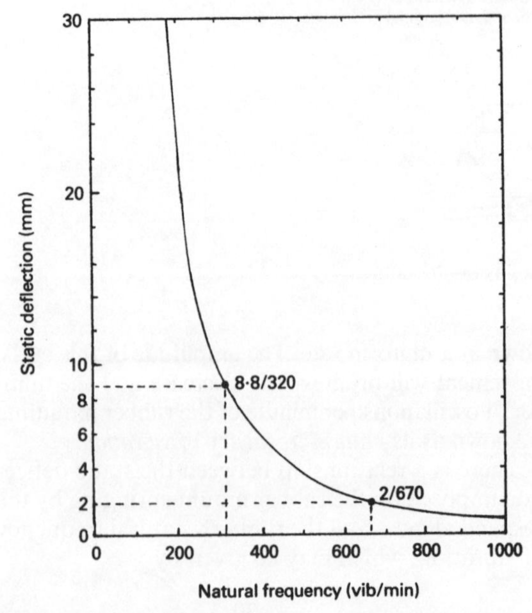

Fig. 1.16 Relationship of static deflection and natural frequency

Transmissibility (Fig. 1.17) When the designer selects the type of flexible mounting the Theory of Transmissibility can be used to estimate critical resonance conditions so that they can be either prevented or at least avoided.

Transmissibility (T) may be defined as the ratio of the transmitted force or amplitude which passes through the rubber mount to the chassis to that of the externally imposed force or amplitude generated by the engine:

$$T = \frac{F_t}{F_d} = \frac{1}{1 - \left(\frac{n}{n_0}\right)^2}$$

where F_t = transmitted force or amplitude
F_d = imposed disturbing force or amplitude

This relationship between transmissibility and the ratio of disturbing frequency and natural frequency may be seen in Fig. 1.17.

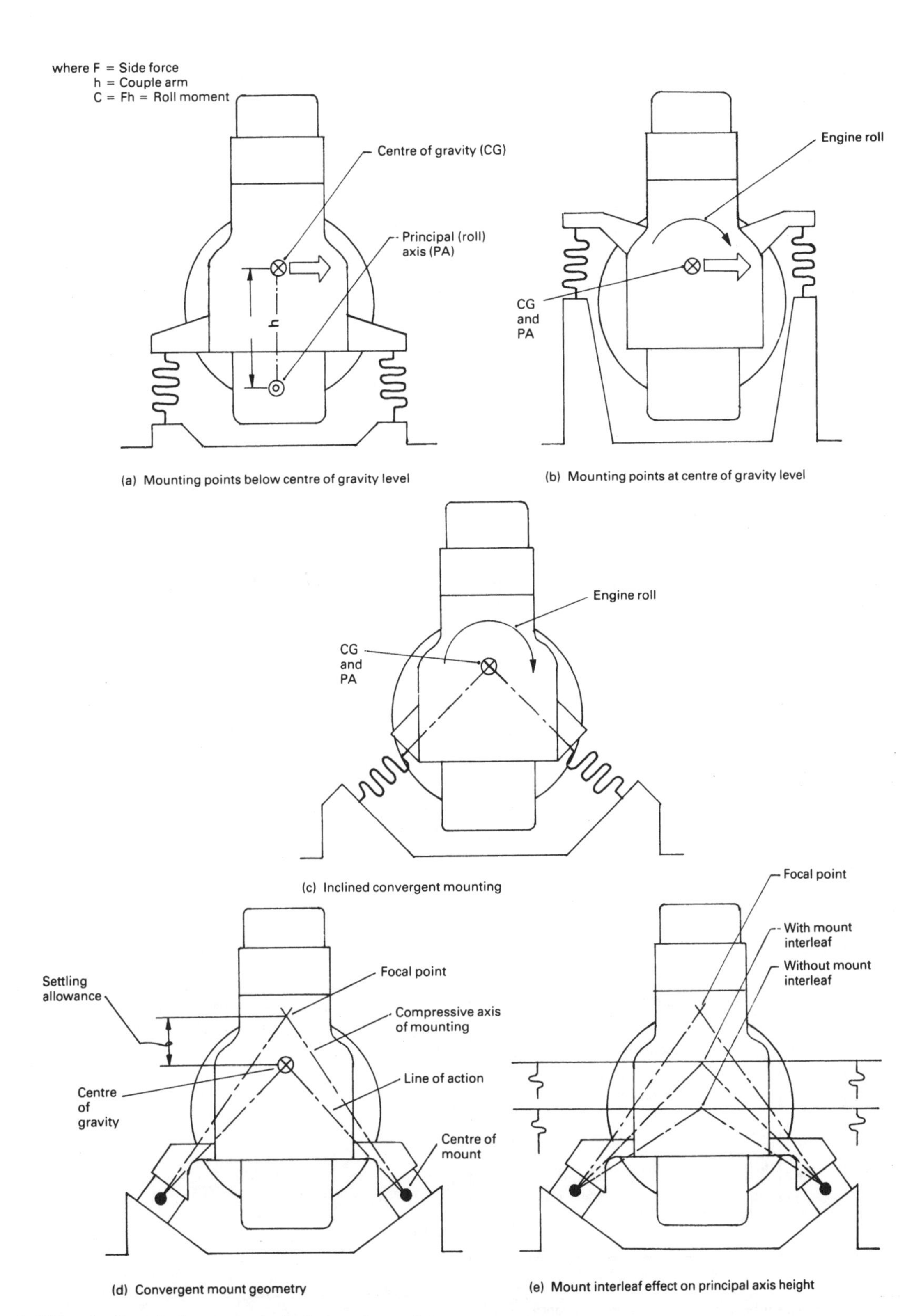

Fig. 1.15 (a–e) Coupled and uncoupled mounting points

The transmissibility to frequency ratio graph (Fig. 1.17) can be considered in three parts as follows:

Range (I) This is the resonance range and should be avoided. It occurs when the disturbing frequency is very near to the natural frequency. If steel mounts are used, a critical vibration at resonance would go to infinity, but natural rubber limits the transmissibility to around 10. If Butyl synthetic rubber is adopted, its damping properties reduce the peak transmissibility to about 2½. Unfortunately, high damping rubber compounds such as Butyl rubber are temperature sensitive to both damping and dynamic stiffness so that during cold weather a noticeably harsher suspension of the engine results.

Damping of the engine suspension mounting is necessary to reduce the excessive movement of a flexible mounting when passing through resonance, but at speeds above resonance more vibration is transmitted to the chassis or body structure than would occur if no damping was provided.

Range (II) This is the recommended working range where the ratio of the disturbing frequency to that of the natural frequency of vibration of the rubber mountings is greater than 1½ and the transmissibility is less than one. Under these conditions off-peak partial resonance vibrations passing to the body structure will be minimized.

Range (III) This is known as the shock reduction range and only occurs when the disturbing frequency is lower than the natural frequency. Generally it is only experienced with very soft rubber mounts and when the engine is initially cranked for starting purposes and so quickly passes through this frequency ratio region.

Example An engine oscillates vertically on its flexible rubber mountings with a frequency of 800 vibrations per minute (vpm). With the information provided answer the following questions:

a) From the static deflection–frequency graph, Fig. 1.16, or by formula, determine the natural frequency of vibration when the static deflection of the engine is 2 mm and then find the disturbing to natural frequency ratio. Comment on these results.
b) If the disturbing to natural frequency ratio is increased to 2.5 determine the natural frequency

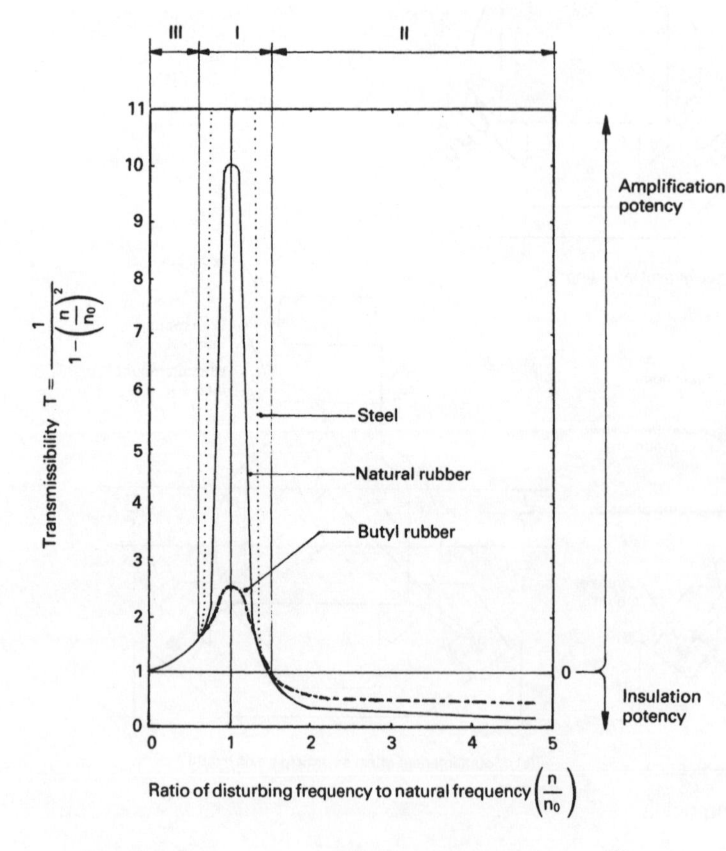

Fig. 1.17 Relationship of transmissibility and the ratio of disturbing and natural frequencies for natural rubber, Butyl rubber and steel

of vibration and the new static deflection of the engine. Comment of these conditions.

a) $n_0 = \dfrac{30}{\sqrt{x}} = \dfrac{30}{\sqrt{0.002}}$

$= \dfrac{30}{0.04472} = 670.84\,\text{vib/min}$

$\therefore \dfrac{n}{n_0} = \dfrac{800}{670.84} = 1.193$

The ratio 1.193 is very near to the resonance condition and should be avoided by using softer mounts.

b) $\dfrac{n}{n_0} = \dfrac{800}{n_0} = 2.5$

$\therefore n_0 = \dfrac{800}{2.5} = 320\,\text{vib/min}$

Now $n_0 = \dfrac{30}{\sqrt{x}}$

thus $\sqrt{x} = \dfrac{30}{n_0}$

$\therefore x = \left(\dfrac{30}{n_0}\right)^2 = \left(\dfrac{30}{320}\right)^2$

$= 0.008789\,\text{m or } 8.789\,\text{mm}$

A low natural frequency of 320 vib/min is well within the insulation range, therefore from either the deflection–frequency graph or by formula the corresponding rubber deflection necessary is 8.789 mm when the engine's static weight bears down on the mounts.

1.2.8 Engine to body/chassis mountings

Engine mountings are normally arranged to provide a degree of flexibility in the horizontal longitudinal, horizontal lateral and vertical axis of rotation. At the same time they must have sufficient stiffness to provide stability under shock loads which may come from the vehicle travelling over rough roads. Rubber sprung mountings suitably positioned fulfil the following functions:

1 Rotational flexibility around the horizontal longitudinal axis which is necessary to allow the impulsive inertia and gas pressure components of the engine torque to be absorbed by rolling of the engine about the centre of gravity.
2 Rotational flexibility around both the horizontal lateral and the vertical axis to accommodate any horizontal and vertical shake and rock caused by unbalanced reciprocating forces and couples.

1.2.9 Subframe to body mountings

(Figs 1.6 and 1.19)

One of many problems with integral body design is the prevention of vibrations induced by the engine, transmission and road wheels from being transmitted through the structure. Some manufacturers adopt a subframe (Fig. 1.6(a, b and c)) attached by resilient mountings (Fig. 1.19(a and b)) to the body to which the suspension assemblies, and in some instances the engine and transmission, are attached. The mass of the subframes alone helps to damp vibrations. It also simplifies production on the assembly line, and facilitates subsequent overhaul or repairs.

In general, the mountings are positioned so that they allow strictly limited movement of the subframe in some directions but provide greater freedom in others. For instance, too much lateral freedom of a subframe for a front suspension assembly would introduce a degree of instability into the steering, whereas some freedom in vertical and longitudinal directions would improve the quality of a ride.

1.2.10 Types of rubber flexible mountings

A survey of typical rubber mountings used for power units, transmissions, cabs and subframes are described and illustrated as follows:

Double shear paired sandwich mounting (Fig. 1.18(a)) Rubber blocks are bonded between the jaws of a 'U' shaped steel plate and a flat interleaf plate so that a double shear elastic reaction takes place when the mount is subjected to vertical loading. This type of shear mounting provides a large degree of flexibility in the upright direction and thus rotational freedom for the engine unit about its principal axis. It has been adopted for both engine and transmission suspension mounting points for medium-sized diesel engines.

Double inclined wedge mounting (Fig. 1.18(b)) The inclined wedge angle pushes the bonded rubber blocks downwards and outwards against the bent-up sides of the lower steel plate when loaded in the vertical plane. The rubber blocks are subjected to both shear and compressive loads and the proportion of compressive to shear load becomes greater with vertical deflection. This form of mounting is suitable for single point gearbox supports.

Inclined interleaf rectangular sandwich mounting (Fig. 1.18(c)) These rectangular blocks are

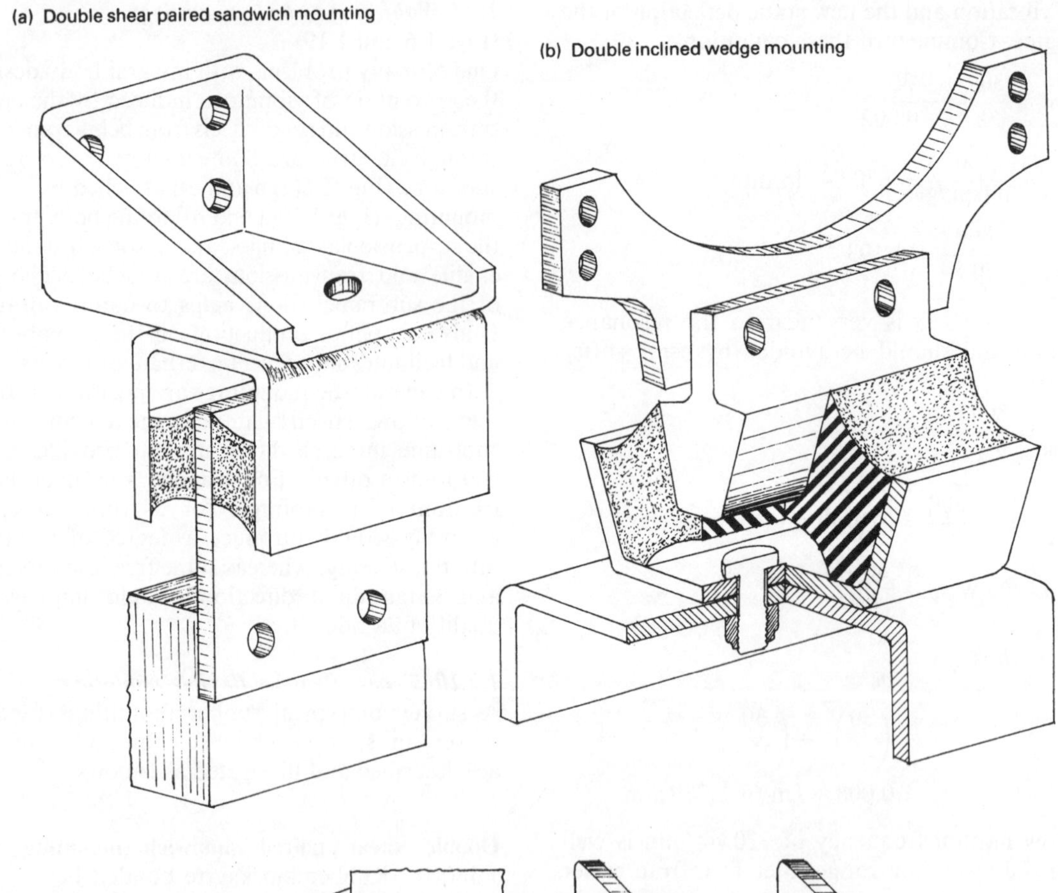

(a) Double shear paired sandwich mounting

(b) Double inclined wedge mounting

Front view

(c) Inclined interleaf rectangular sandwich mounting

Pictorial view

Fig. 1.18 (a–h) Types of rubber flexible mountings

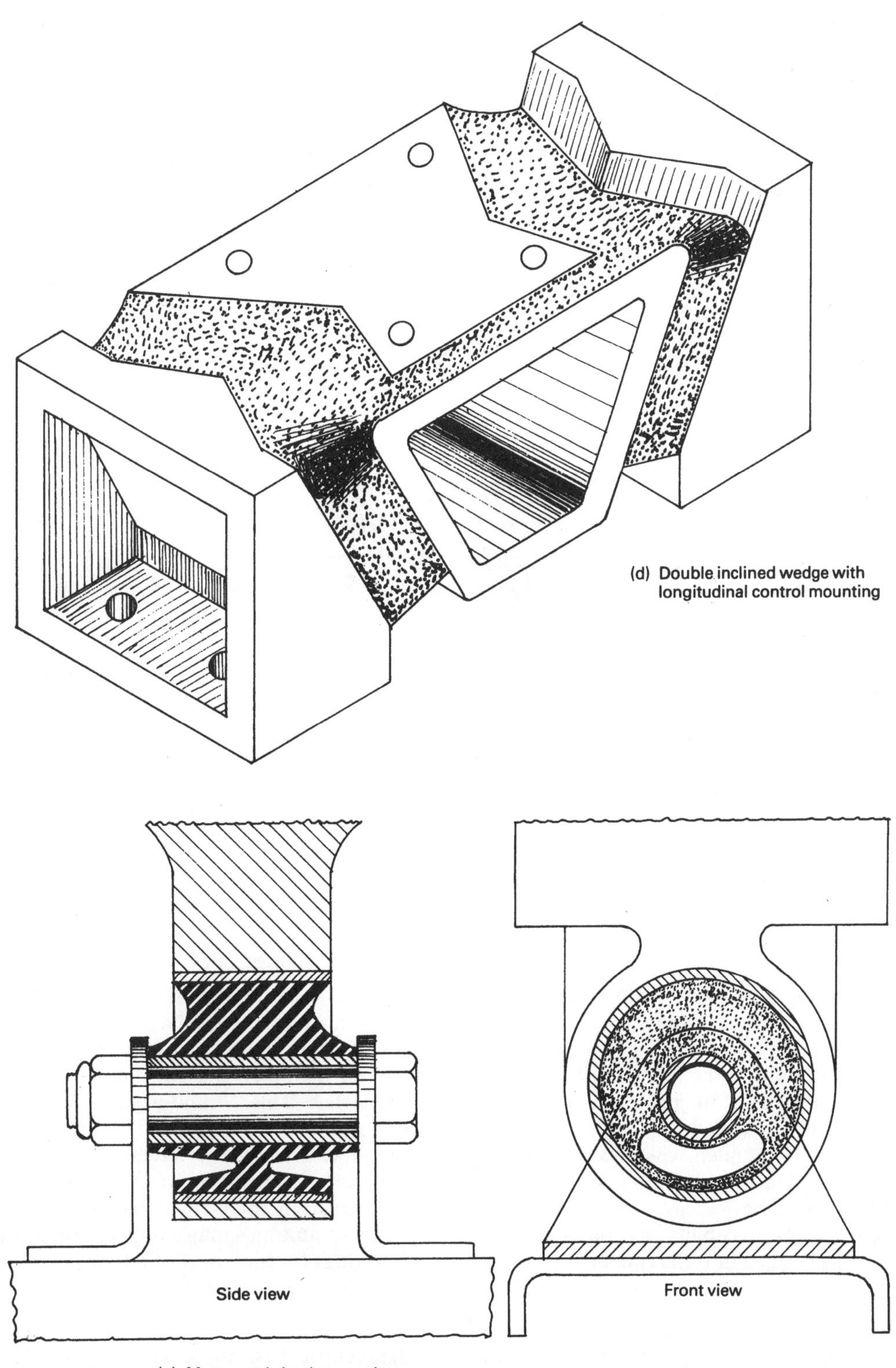

(d) Double inclined wedge with longitudinal control mounting

(e) Metaxentric bush mounting

Fig. 1.18 *contd*

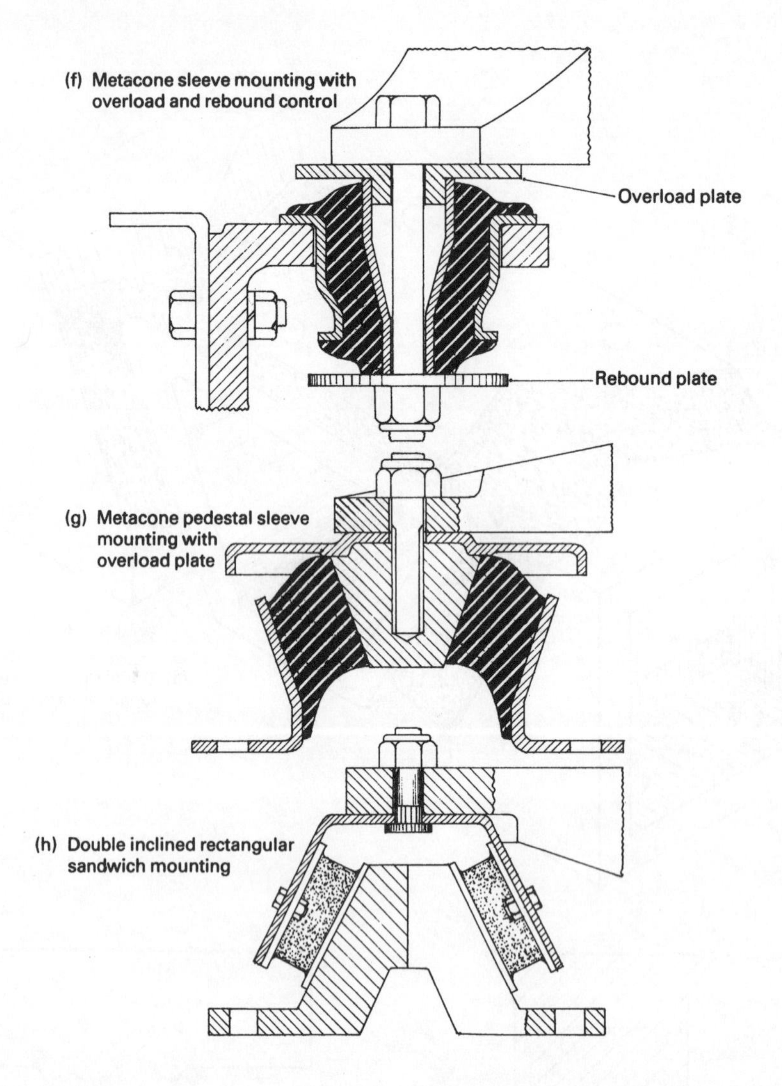

Fig. 1.18 *contd*

designed to be used with convergent 'V' formation engine suspension system where the blocks are inclined on either side of the engine. This configuration enables the rubber to be loaded in both shear and compression with the majority of engine rotational flexibility being carried out in shear. Vertical deflection due to body pitch when accelerating or braking is absorbed mostly in compression. Vertical elastic stiffness may be increased without greatly effecting engine roll flexibility by having metal spacer interleafs bonded into the rubber.

Double inclined wedge with longitudinal control mounting (Fig. 1.18(d)) Where heavy vertical loads and large rotational reactions are to be absorbed, double inclined wedge mounts positioned on either side of the power unit's bell housing at principal axis level may be used. Longitudinal movement is restricted by the double 'V' formed between the inner and two outer members seen in a plan view. This 'V' and wedge configuration provides a combined shear and compressive strain to the rubber when there is a relative fore and aft movement between the engine and chassis, in addition to that created by the vertical loading of the mount.

This mounting's major application is for the rear mountings forming part of a four point suspension for heavy diesel engines.

Metaxentric bush mounting (Fig. 1.18(e)) When the bush is in the unloaded state, the steel inner sleeve is eccentric relative to the outer one so that there is more rubber on one side of it than on the other. Precompression is applied to the rubber expanding the inner sleeve. The bush is set so that the greatest thickness of rubber is in compression in the laden condition. A slot is incorporated in the rubber on either side where the rubber is at its minimum in such a position as to avoid stressing any part of it in tension.

When installed, its stiffness in the fore and aft direction is greater than in the vertical direction, the ratio being about 2.5 : 1. This type of bush provides a large amount of vertical deflection with very little fore and aft movement which makes it suitable for rear gearbox mounts using three point power unit suspension and leaf spring eye shackle pin bushes.

Metacone sleeve mountings (Fig. 1.18(f and g)) These mounts are formed from male and female conical sleeves, the inner male member being centrally positioned by rubber occupying the space between both surfaces (Fig. 1.18(f)). During vertical vibrational deflection, the rubber between the sleeves is subjected to a combined shear and compression which progressively increases the stiffness of the rubber as it moves towards full distortion. The exposed rubber at either end overlaps the flanged outer sleeve and there is an upper and lower plate bolted rigidly to the ends of the inner sleeve. These plates act as both overload (bump) and rebound stops, so that when the inner member deflects up or down towards the end of its movement it rapidly stiffens due to the surplus rubber being squeezed in between. Mounts of this kind are used where stiffness is needed in the horizontal direction with comparative freedom of movement for vertical deflection.

An alternative version of the Metacone mount uses a solid aluminium central cone with a flanged pedestal conical outer steel sleeve which can be bolted directly onto the chassis side member, see Fig. 1.18(g). An overload plate is clamped between the inner cone and mount support arm, but no rebound plate is considered necessary.

These mountings are used for suspension applications such as engine to chassis, cab to chassis, bus body and tanker tanks to chassis.

Double inclined rectangular sandwich mounting (Fig. 1.18(h)) A pair of rectangular sandwich rubber blocks are supported on the slopes of a triangular pedestal. A bridging plate merges the resilience of the inclined rubber blocks so that they provide a combined shear and compressive distortion within the rubber. Under small deflection conditions the shear and compression is almost equal, but as the load and thus deflection increases, the proportion of compression over the shear loading predominates.

These mounts provide very good lateral stability without impairing vertical deflection flexibility and progressive stiffness control. When used for road wheel axle suspension mountings, they offer good insulation against road and other noises.

Flanged sleeve bobbin mounting with rebound control (Fig. 1.19(a and b)) These mountings have the rubber moulded partially around the outer flange sleeve and in between this sleeve and an inner tube. A central bolt attaches the inner tube to the body structure while the outer member is bolted on two sides to the subframe.

When loaded in the vertical downward direction, the rubber between the sleeve and tube walls will be in shear and the rubber on the outside of the flanged sleeve will be in compression.

There is very little relative sideway movement between the flanged sleeve and inner tube due to rubber distortion. An overload plate limits the downward deflection and rebound is controlled by the lower plate and the amount and shape of rubber trapped between it and the underside of the flanged sleeve. A reduction of rubber between the flanged sleeve and lower plate (Fig. 1.19(a)) reduces the rebound, but an increase in depth of rubber increases rebound (Fig. 1.19(b)). The load deflection characteristics are given for both mounts in Fig. 1.19c. These mountings are used extensively for body to subframe and cab to chassis mounting points.

Hydroelastic engine mountings (Figs 1.20(a–c) and 1.21) A flanged steel pressing houses and supports an upper and lower rubber spring diaphragm. The space between both diaphragms is filled and sealed with fluid and is divided in two by a separator plate and small transfer holes interlink the fluid occupying these chambers (Fig. 1.20(a and b)). Under vertical vibratory conditions the fluid will be displaced from one chamber to the other through transfer holes. During downward deflection (Fig. 1.20(b)), both rubber diaphragms are subjected to a combined shear and compressive action and some of the fluid in the upper chamber will be pushed into the lower and back again by way of the transfer holes when the rubber rebounds (Fig. 1.20(a)). For low vertical vibratory frequencies,

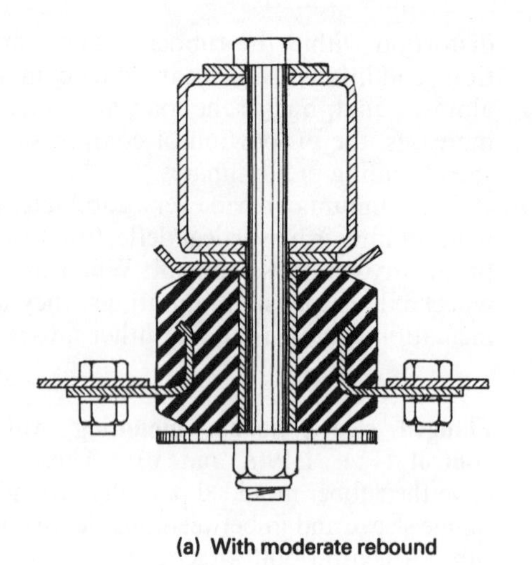
(a) With moderate rebound

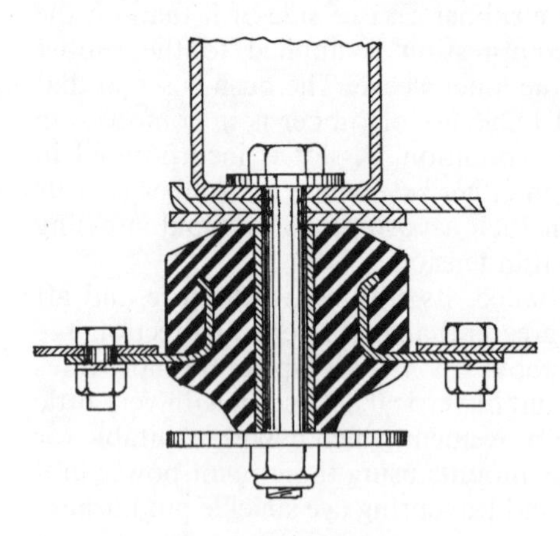
(b) With large rebound

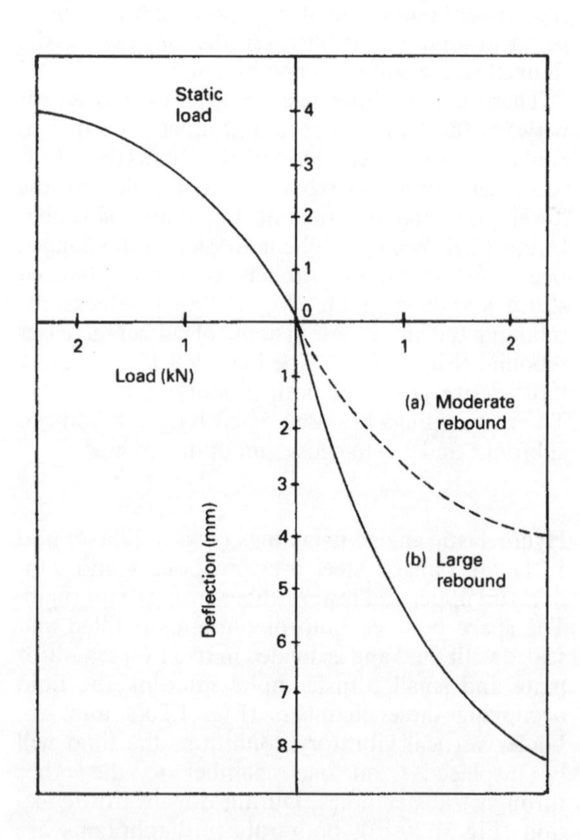

(c) Relation between load and deflection for a subframe to body mounting

Fig. 1.19 (a–c) Flanged sleeve bobbin mounting with rebound control

the movement of fluid between the chambers is unrestricted, but as the vibratory frequencies increase, the transfer holes offer increasing resistance to the flow of fluid and so slow down the up and down motion of the engine support arm. This damps and reduces the amplitude of mountings vertical vibratory movement over a number of cycles. A comparison of conventional rubber and hydroelastic damping resistance over the normal operating frequency range for engine mountings is shown in Fig. 1.20(c).

Instead of adopting a combined rubber mount with integral hydraulic damping, separate diagonally mounted telescopic dampers may be used in conjunction with inclined rubber mounts to reduce both vertical and horizontal vibration (Fig. 1.21).

1.3 Fifth wheel coupling assembly
(Fig. 1.22(a and b))

The fifth wheel coupling attaches the semi-trailer to the tractor unit. This coupling consists of a semicircular table plate with a central hole and a vee section cut-out towards the rear (Fig. 1.22(b)). Attached underneath this plate are a pair of pivoting coupling jaws (Fig. 1.22(a)). The semi-trailer has an upper fifth wheel plate welded or bolted to the underside of its chassis at the front and in the centre of this plate is bolted a kingpin which faces downwards (Fig. 1.22(a)).

When the trailer is coupled to the tractor unit, this upper plate rests and is supported on top of the tractor fifth wheel table plate with the two halves of the coupling jaws engaging the kingpin. To permit

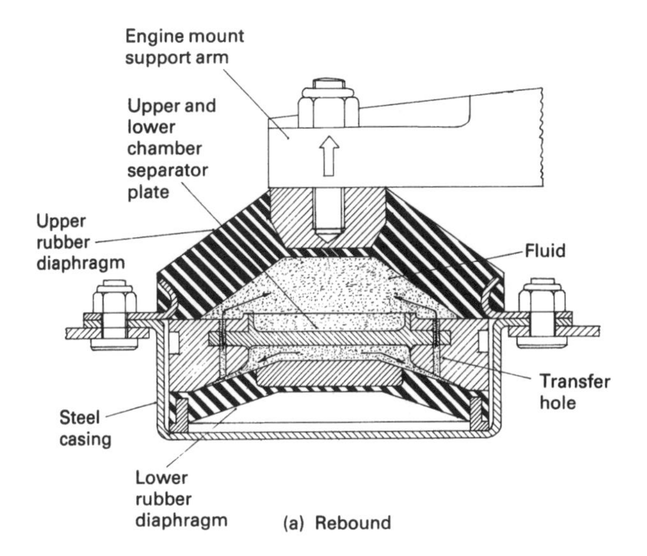

(a) Rebound

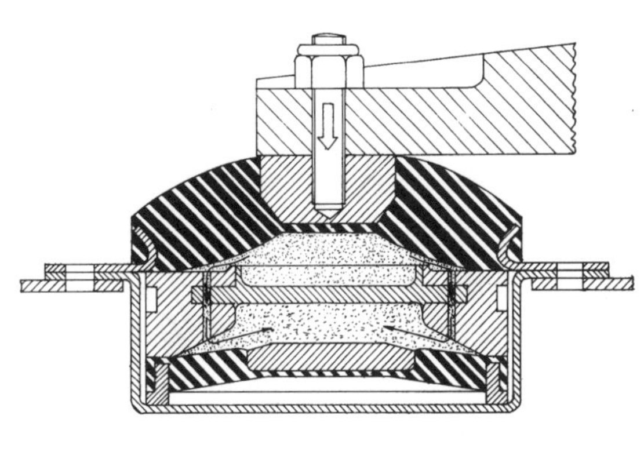

(b) Bump

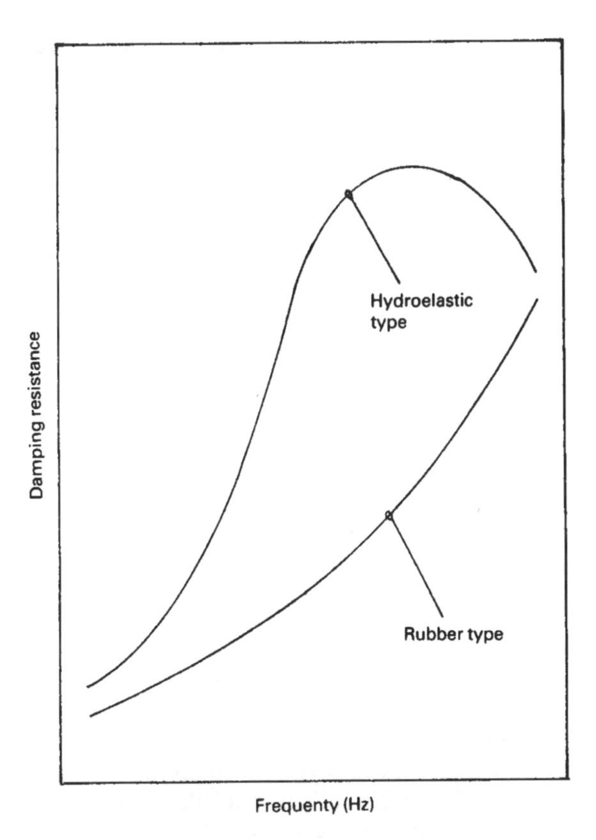

(c) Relationship of frequency of vibration and damping resistance with conventional rubber and hydroelastic type mountings

Fig. 1.20 (a–c) Hydroelastic engine mount

relative swivelling between the kingpin and jaws, the two interfaces of the tractor fifth wheel tables and trailer upper plate should be heavily greased. Thus, although the trailer articulates about the kingpin, its load is carried by the tractor table.

Flexible articulation between the tractor and semi-trailer in the horizontal plane is achieved by permitting the fifth wheel table to pivot on horizontal trunnion bearings that lie in the same vertical plane as the kingpin, but with their axes at right angles to that of the tractor's wheel base (Fig. 1.22(b)). Rubber trunnion rubber bushes normally provide longitudinal oscillations of about $\pm 10°$.

The fifth wheel table assembly is made from either a machined cast or forged steel sections, or from heavy section rolled steel fabrications, and the upper fifth wheel plate is generally hot rolled steel welded to the trailer chassis. The coupling locking system consisting of the jaws, pawl, pivot pins and kingpin is produced from forged high carbon manganese steels and the pressure areas of these components are induction hardened to withstand shock loading and wear.

1.3.1 Operation of twin jaw coupling

(Fig. 1.23(a–d))

With the trailer kingpin uncoupled, the jaws will be in their closed position with the plunger withdrawn from the lock gap between the rear of the jaws, which are maintained in this position by the pawl contacting the hold-off stop (Fig. 1.23(a)). When coupling the tractor to the trailer, the jaws of the

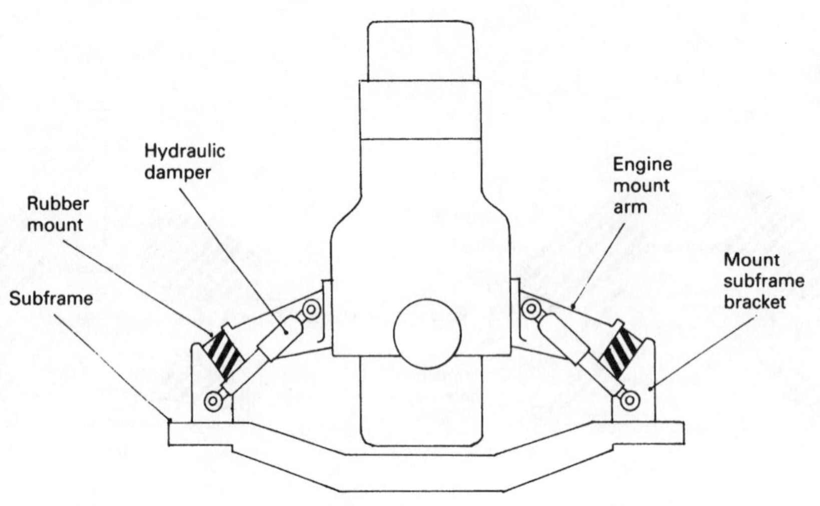

Fig. 1.21 Diagonally mounted hydraulic dampers suppress both vertical and horizontal vibrations

fifth wheel strike the kingpin of the trailer. The jaws are then forced open and the kingpin enters the space between the jaws (Fig. 1.23(b)). The kingpin contacts the rear of the jaws which then automatically pushes them together. At the same time, one of the coupler jaws causes the trip pin to strike the pawl. The pawl turns on its pivot against the force of the spring, releasing the plunger, allowing it to be forced into the jaws' lock gap by its spring (Fig. 1.23(c)). When the tractor is moving, the drag of the kingpin increases the lateral force of the jaws on the plunger.

To disconnect the coupling, the release hand lever is pulled fully back (Fig. 1.23(d)). This draws the plunger clear of the rear of the jaws and, at the same time, allows the pawl to swing round so that it engages a projection hold-off stop situated at the upper end of the plunger, thus jamming the plunger in the fully out position in readiness for uncoupling.

1.3.2 Operation of single jaw and pawl coupling (Fig. 1.24(a–d))

With the trailer kingpin uncoupled, the jaw will be held open by the pawl in readiness for coupling (Fig. 1.24(a)). When coupling the tractor to the trailer, the jaw of the fifth wheel strikes the kingpin of the trailer and swivels the jaw about its pivot pin against the return spring, slightly pushing out the pawl (Fig. 1.24(b)). Further rearward movement of the tractor towards the trailer will swing the jaw round until it traps and encloses the kingpin. The spring load notched pawl will then snap over the jaw projection to lock the kingpin in the coupling position (Fig. 1.24(c)). The securing pin should then be inserted through the pull lever and table eye holes. When the tractor is driving forward, the reaction on the kingpin increases the locking force between the jaw projection and the notched pawl.

To disconnect the coupling, lift out the securing pin and pull the release hand lever fully out (Fig. 1.24(d)). With both the tractor and trailer stationary, the majority of the locking force applied to notched pawl will be removed so that with very little effort, the pawl is able to swing clear of the jaw in readiness for uncoupling, that is, by just driving the tractor away from the trailer. Thus the jaw will simply swivel allowing the kingpin to pull out and away from the jaw.

1.4 Trailer and caravan drawbar couplings

1.4.1 Eye and bolt drawbar coupling for heavy goods trailers (Figs 1.25 and 1.26)

Drawbar trailers are normally hitched to the truck by means of an 'A' frame drawbar which is coupled by means of a towing eye formed on the end of the drawbar (Fig. 1.25). When coupled, the towing eye hole is aligned with the vertical holes in the upper and lower jaws of the truck coupling and an eye bolt passes through both coupling jaws and drawbar eye to complete the attachment (Fig. 1.26). Lateral drawbar swing is permitted owing to the eye bolt pivoting action and the slots between the

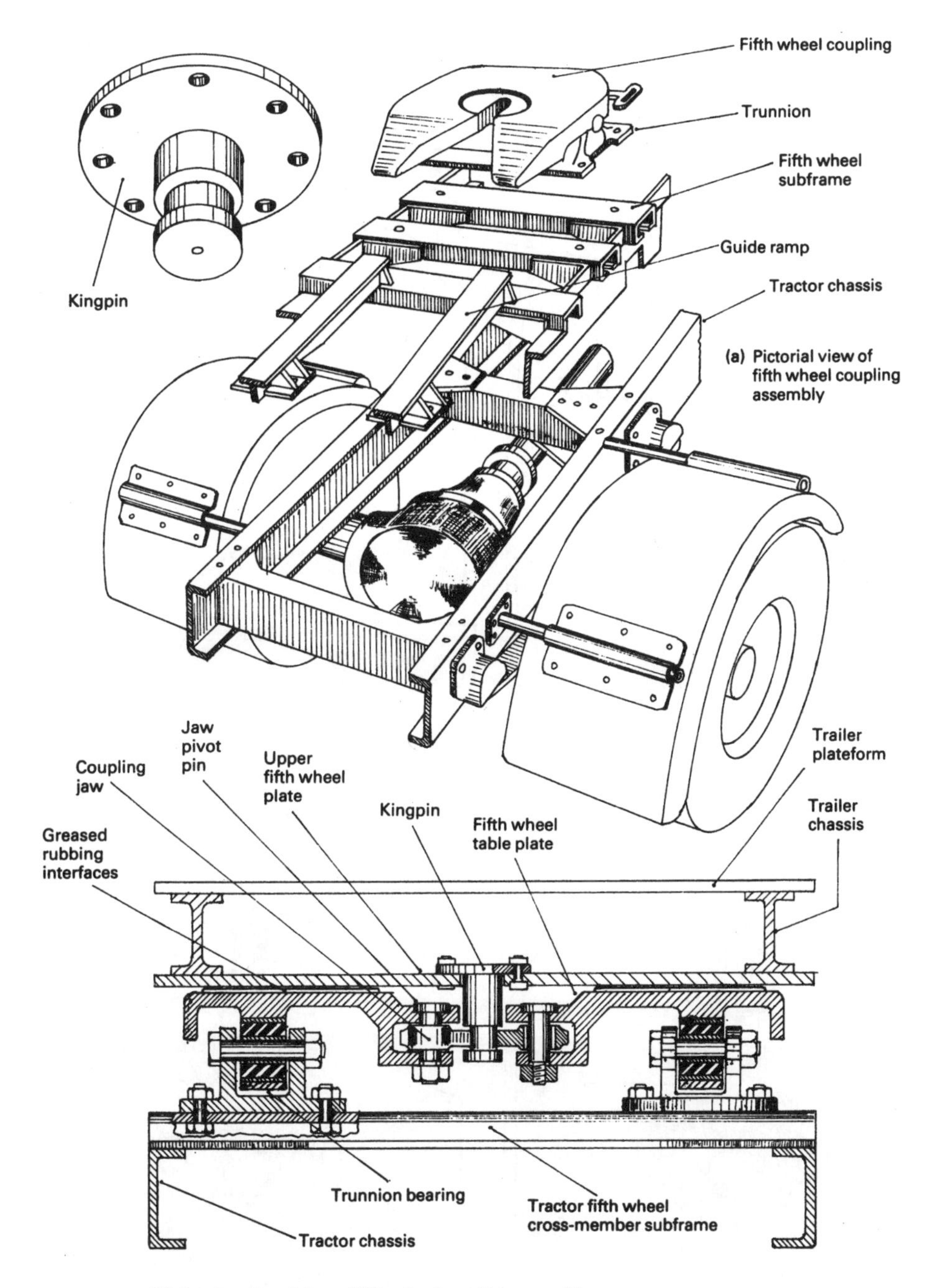

Fig. 1.22 (a and b) Fifth wheel coupling assembly

jaws on either side. Aligning the towing eye to the jaws is made easier by the converging upper and lower lips of the jaws which guide the towing eye as the truck is reversed and the jaws approach the drawbar. Isolating the coupling jaws from the truck draw beam are two rubber blocks which act as a damping media between the towing vehicle and trailer. These rubber blocks also permit additional deflection of the coupling jaw shaft relative to the draw beam under rough abnormal operating conditions, thus preventing over-straining the drawbar and chassis system.

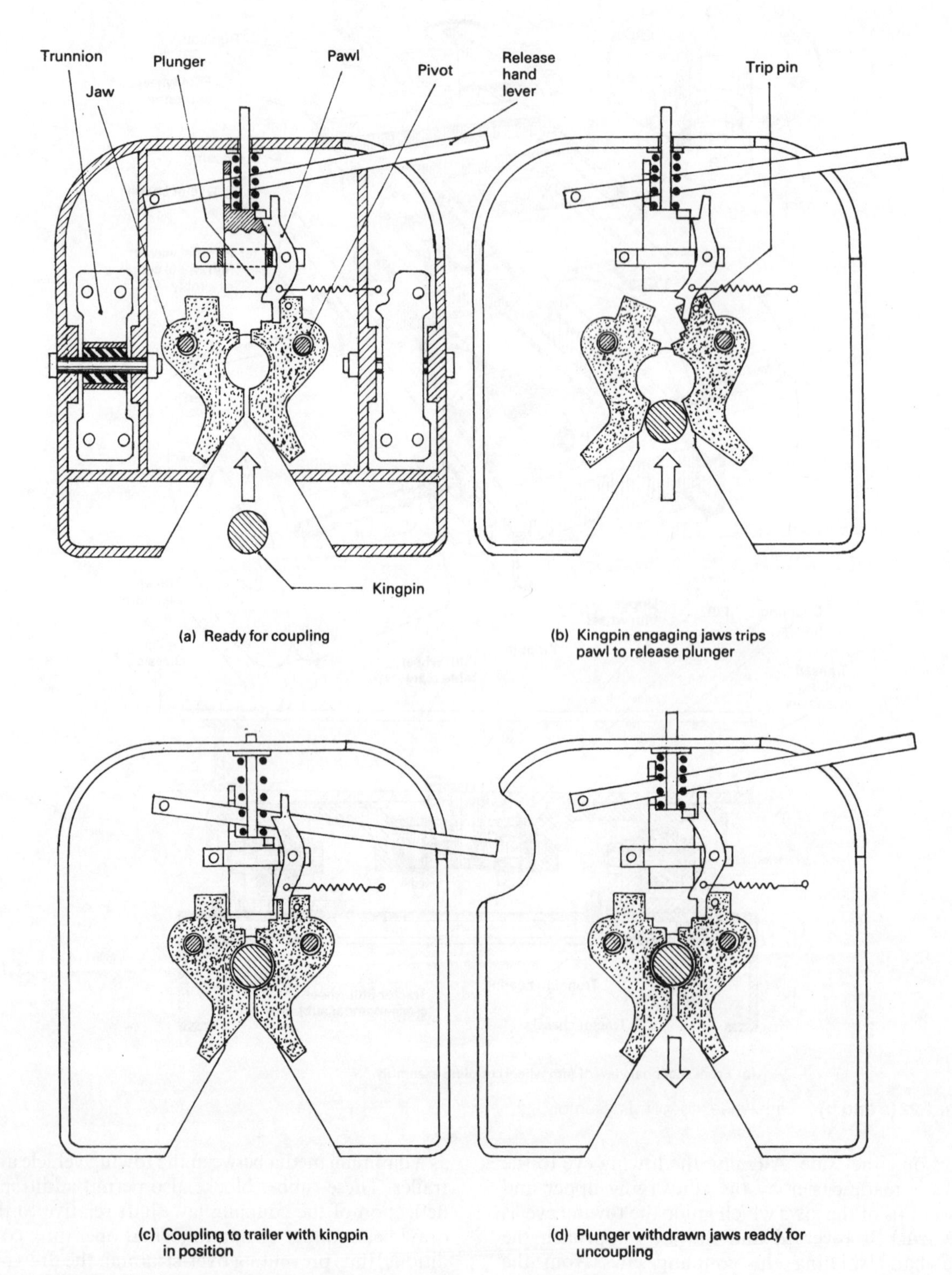

(a) Ready for coupling

(b) Kingpin engaging jaws trips pawl to release plunger

(c) Coupling to trailer with kingpin in position

(d) Plunger withdrawn jaws ready for uncoupling

Fig. 1.23 (a–d) Fifth wheel coupling with twin jaws plunger and pawl

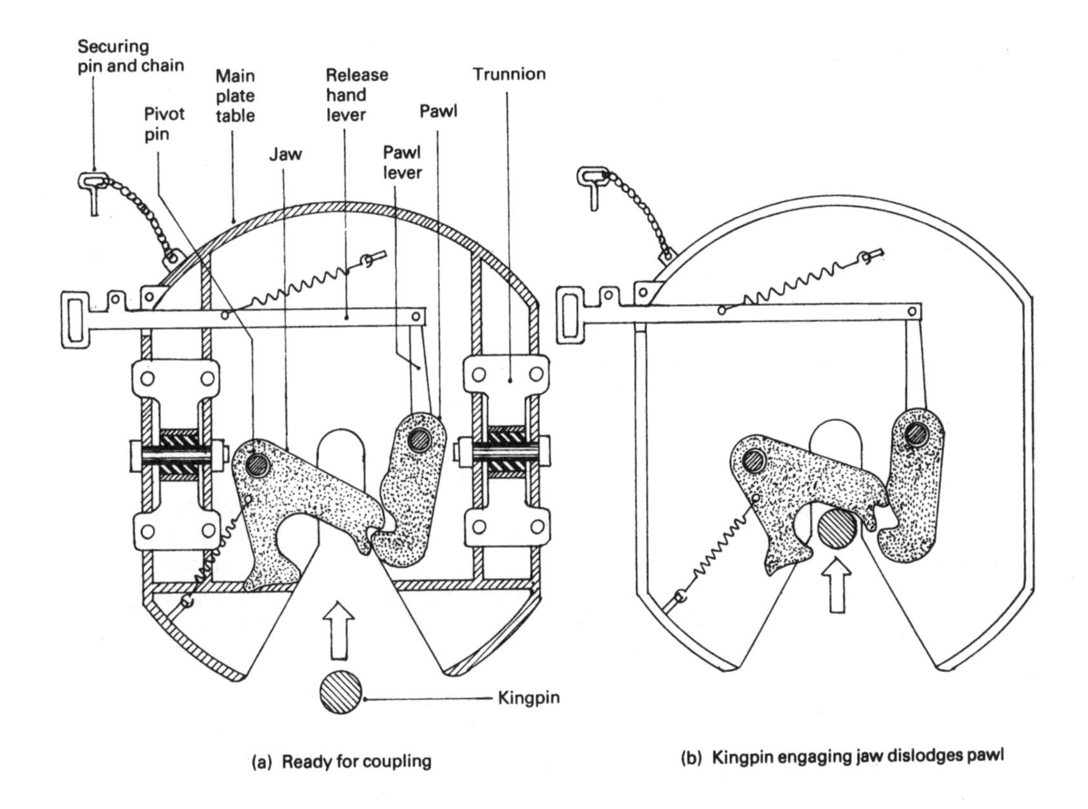

(a) Ready for coupling

(b) Kingpin engaging jaw dislodges pawl

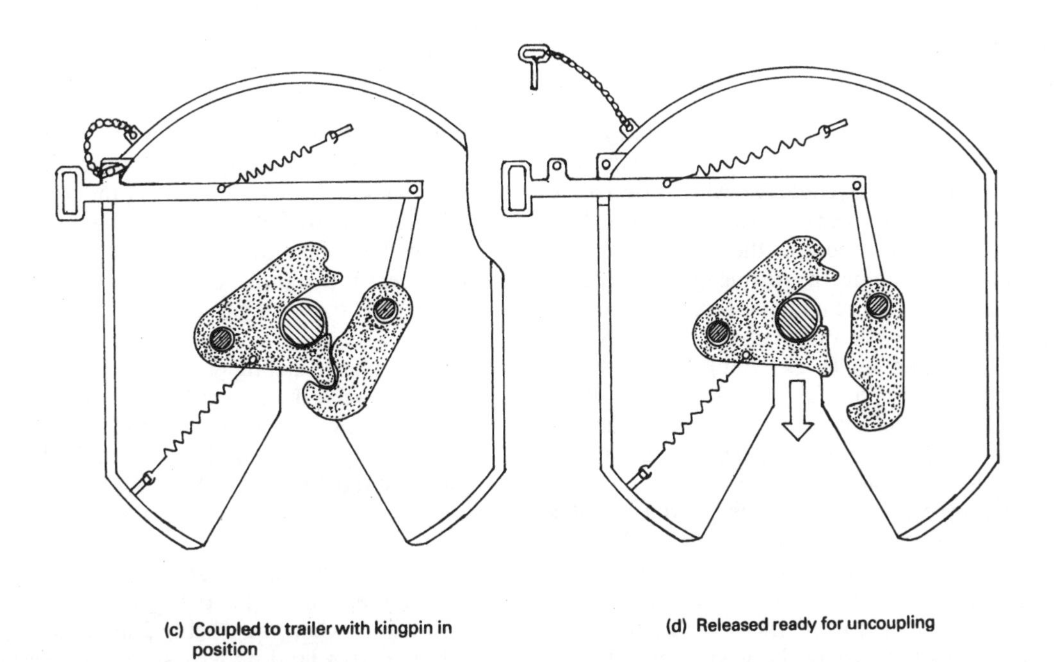

(c) Coupled to trailer with kingpin in position

(d) Released ready for uncoupling

Fig. 1.24 (a–d) Fifth wheel coupling with single jaw and pawl

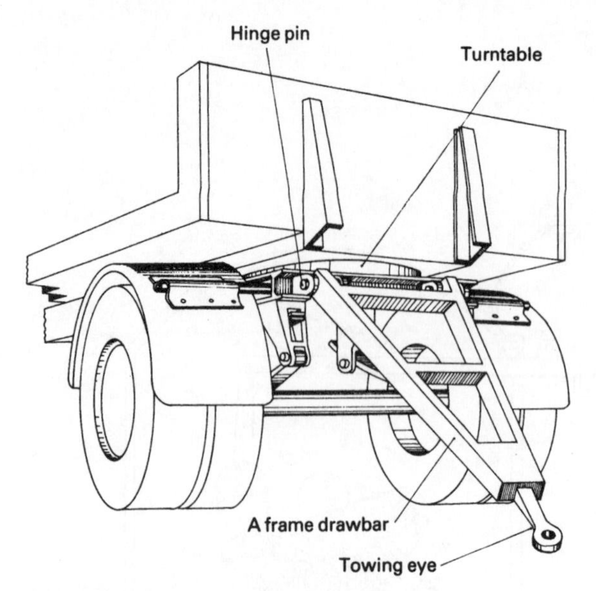

Fig. 1.25 Drawbar trailer

The coupling jaws, eye bolt and towing eye are generally made from forged manganese steel with induction hardened pressure areas to increase the wear resistance.

Operation of the automatic drawbar coupling (Fig. 1.26) In the uncoupled position the eyebolt is held in the open position ready for coupling (Fig. 1.26(a)). When the truck is reversed, the jaws of the coupling slip over the towing eye and in the process strike the conical lower end of the eye bolt (Fig. 1.26(b)). Subsequently, the eye bolt will lift. This trips the spring-loaded wedge lever which now rotates clockwise so that it bears down on the eye bolt. Further inward movement of the eye bolt between the coupling jaws aligns the towing eye with the eye bolt. The spring pressure now acts through the wedge lever to push the eye bolt through the towing eye and the lower coupling jaw (Fig. 1.26(c)). When the eye bolt stop-plate has been fully lowered by the spring tension, the wedge lever will slot into its groove formed in the centre of the eye bolt so that it locks the eye bolt in the coupled position.

To uncouple the drawbar, the handle is pulled upwards against the tension of the coil spring mounted on the wedge level operating shaft (Fig. 1.26(d)). This unlocks the wedge, freeing the eyebolt and then raises the eye bolt to the uncoupled position where the wedge lever jams it in the open position (Fig. 1.26(a)).

1.4.2 Ball and socket towing bar coupling for light caravan/trailers (Fig. 1.27)

Light trailers or caravans are usually attached to the rear of the towing car by means of a ball and socket type coupling. The ball part of the attachment is bolted onto a bracing bracket fitted directly to the boot pan or the towing load may be shared out between two side brackets attached to the rear longitudinal box-section members of the body.

A single channel section or pair of triangularly arranged angle-section arms may be used to form the towbar which both supports and draws the trailer.

Attached to the end of the towbar is the socket housing with an internally formed spherical cavity. This fits over the ball member of the coupling so that it forms a pivot joint which can operate in both the horizontal and vertical plane (Fig. 1.27).

To secure the socket over the ball, a lock device must be incorporated which enables the coupling to be readily connected or disconnected. This lock may take the form of a spring-loaded horizontally positioned wedge with a groove formed across its top face which slips underneath and against the ball. The wedge is held in the closed engaged position by a spring-loaded vertical plunger which has a horizontal groove cut on one side. An uncoupling lever engages the plunger's groove so that when the coupling is disconnected the lever is squeezed to lift and release the plunger from the wedge. At the same time the whole towbar is raised by the handle to clear the socket and from the ball member.

Coupling the tow bar to the car simply reverses the process, the uncoupling lever is again squeezed against the handle to withdraw the plunger and the socket housing is pushed down over the ball member. The wedge moves outwards and allows the ball to enter the socket and immediately the wedge springs back into the engaged position. Releasing the lever and handle completes the coupling by permitting the plunger to enter the wedge lock groove.

Sometimes a strong compression spring is interposed between the socket housing member and the towing (draw) bar to cushion the shock load when the car/trailer combination is initially driven away from a standstill.

1.5 Semi-trailer landing gear (Fig. 1.28)

Landing legs are used to support the front of the semi-trailer when the tractor unit is uncoupled.

Extendable landing legs are bolted vertically to each chassis side-member behind the rear wheels of

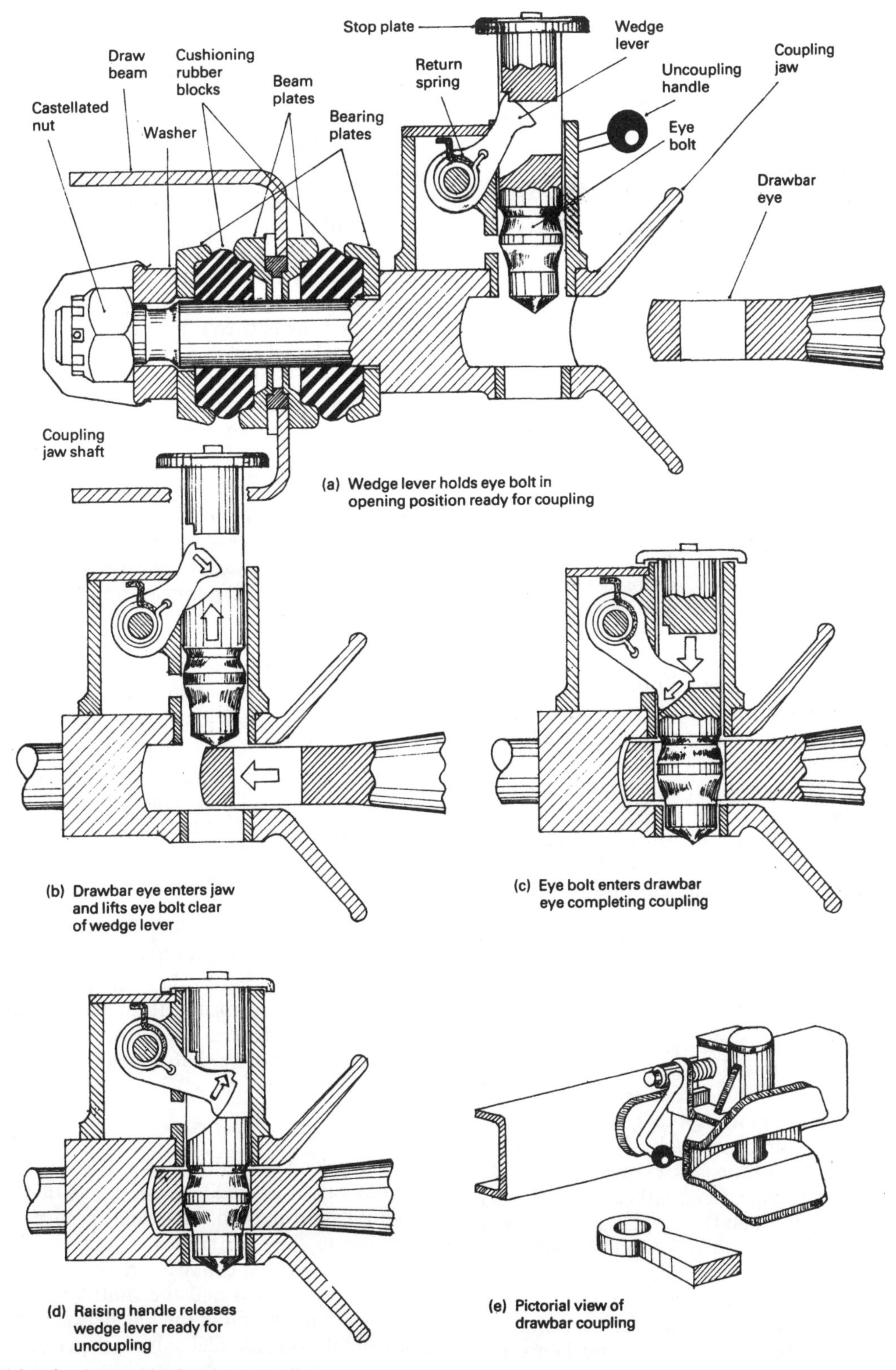

Fig. 1.26 (a–e) Automatic drawbar coupling

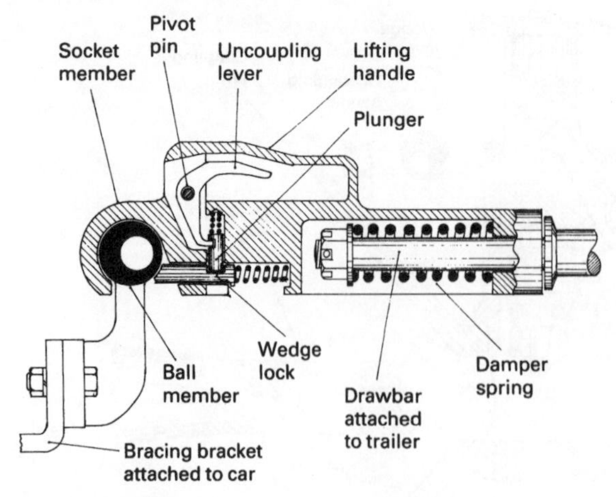

Fig. 1.27 Ball and socket caravan/trailer towing attachment

the tractor unit, just sufficiently back to clear the rear tractor road wheels when the trailer is coupled and the combination is being manoeuvred (Fig. 1.28(a)). To provide additional support for the legs, bracing stays are attached between the legs and from the legs diagonally to the chassis cross-member (Fig. 1.28(b)).

The legs consist of inner and outer high tensile steel tubes of square section. A jackscrew with a bevel wheel attached at its top end supported by the outer leg horizontal plate in a bronze bush bearing. The jawscrew fits into a nut which is mounted at the top of the inner leg and a taper roller bearing race is placed underneath the outer leg horizontal support plate and the upper part of the jackscrew to minimize friction when the screw is rotated (Fig. 1.28(b)). The bottom ends of the inner legs may support either twin wheels, which enable the trailer to be manoeuvred, or simply flat feet. The latter are able to spread the load and so permit greater load capacity.

To extend or retract the inner legs, a winding handle is attached to either the low or high speed shaft protruding from the side of the gearbox. The upper high speed shaft supports a bevel pinion which meshes with a vertically mounted bevel wheel forming part of the jackscrew.

Rotating the upper shaft imparts motion directly to the jackscrew through the bevel gears. If greater leverage is required to raise or lower the front of the trailer, the lower shaft is engaged and rotated. This provides a gear reduction through a compound gear train to the upper shaft which then drives the bevel pinion and wheel and hence the jackscrew.

1.6 Automatic chassis lubrication system

1.6.1 The need for automatic lubrication system (Fig. 1.29)

Owing to the heavy loads they carry commercial vehicles still prefer to use metal to metal joints which are externally lubricated. Such joints are kingpins and bushes, shackle pins and bushes, steering ball joints, fifth wheel coupling, parking brake linkage etc. (Fig. 1.29). These joints require lubricating in proportion to the amount of relative movement and the loads exerted. If lubrication is to be effective in reducing wear between the moving parts, fresh oil must be pumped between the joints frequently. This can best be achieved by incorporating an automatic lubrication system which pumps oil to the bearing's surfaces in accordance to the distance travelled by the vehicle.

1.6.2 Description of airdromic automatic chassis lubrication system (Fig. 1.30)

This lubrication system comprises four major components; a combined pump assembly, a power unit, an oil unloader valve and an air control unit.

Pump assembly (Fig. 1.30) The pump assembly consists of a circular housing containing a ratchet operated drive (cam) shaft upon which are mounted one, two or three single lobe cams (only one cam shown). Each cam operates a row of 20 pumping units disposed radially around the pump casing, the units being connected to the chassis bearings by nylon tubing.

Power unit (Fig. 1.30) This unit comprises a cylinder and spring-loaded air operated piston which is mounted on the front face of the pump assembly housing, the piston rod being connected indirectly to the drive shaft ratchet wheel by way of a ratchet housing and pawl.

Oil unloader valve (Fig. 1.30) This consists of a shuttle valve mounted on the front of the pump assembly housing. The oil unloader valve allows air pressure to flow to the power unit for the power stroke. During the exhaust stroke, however, when air flow is reversed and the shuttle valve is lifted from its seat, any oil in the line between the power unit and the oil unloader valve is then discharged to atmosphere.

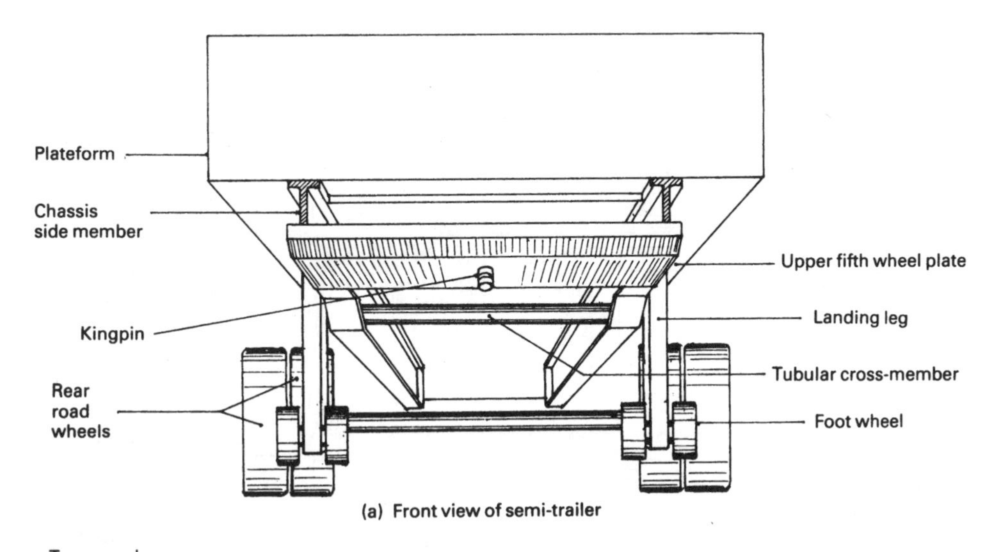

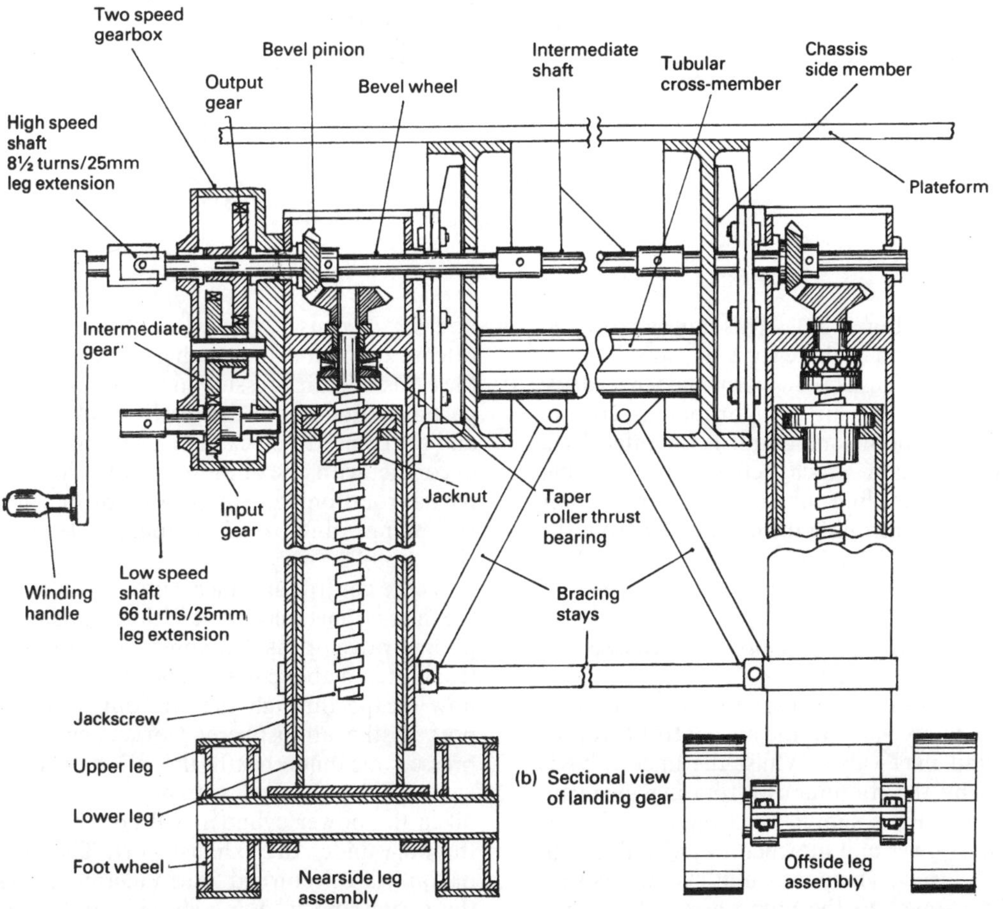

Fig. 1.28 (a and b) Semi-trailer landing gear

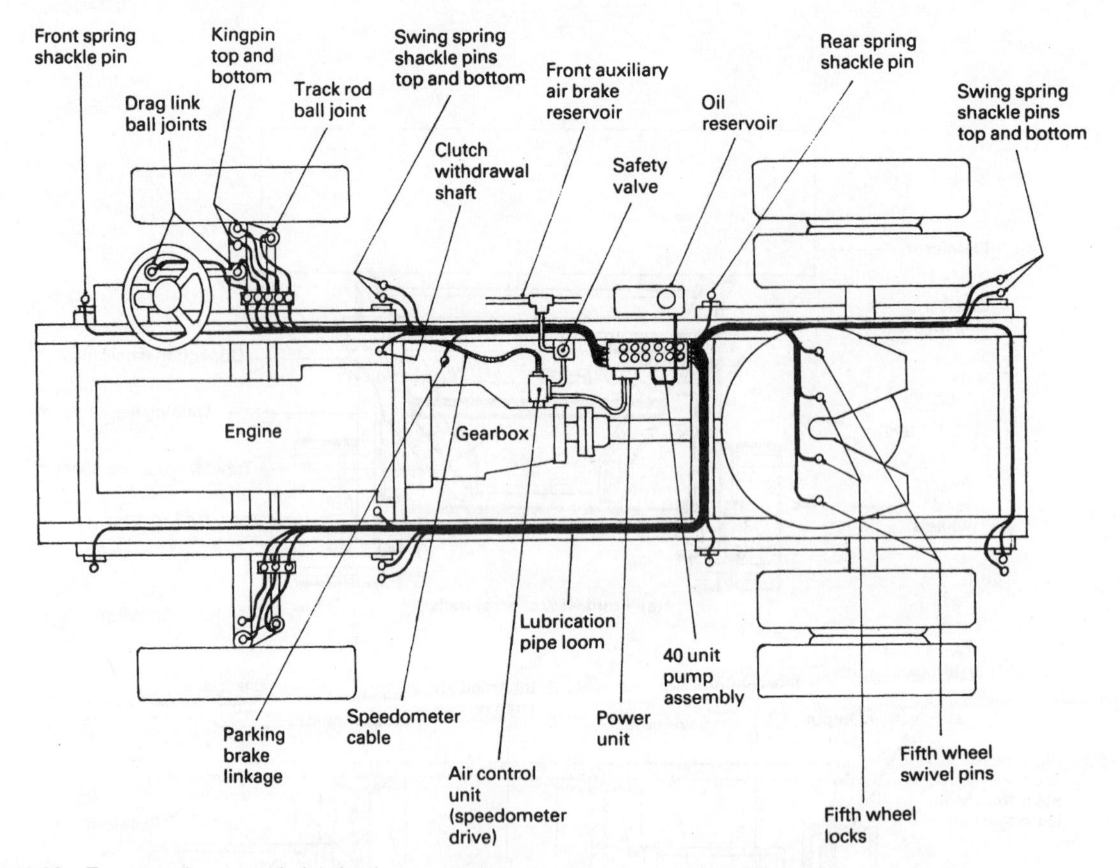

Fig. 1.29 Tractor unit automatic lubrication system

Air control unit (Fig. 1.30) This unit is mounted on the gearbox and is driven via the speedometer take-off point. It consists of a worm and wheel drive which operates an air proportioning control unit. This air proportioning unit is operated by a single lift face cam which actuates two poppet valves, one controlling air supply to the power unit, the other controlling the exhaust air from the power unit.

1.6.3 Operation of airdromic automatic chassis lubrication system (Fig. 1.30)

Air from the air brake auxiliary reservoir passes by way of the safety valve to the air control (proportioning) unit inlet valve. Whilst the inlet valve is held open by the continuously rotating face cam lobe, air pressure is supplied via the oil unloader valve to the power unit attached to the multipump assembly housing. The power unit cylinder is supported by a pivot to the pump assembly casing, whilst the piston is linked to the ratchet and pawl housing. Because the pawl meshes with one of the ratchet teeth and the ratchet wheel forms part of the camshaft, air pressure in the power cylinder will partially rotate both the ratchet and pawl housing and the camshaft clockwise. The cam (or cams) are in contact with one or more pump unit, and so each partial rotation contributes to a proportion of the jerk plunger and barrel pumping cycle of each unit (Fig. 1.30).

As the control unit face cam continues to rotate, the inlet poppet inlet valve is closed and the exhaust poppet valve opens. Compressed air in the air control unit and above the oil control shuttle valve will now escape through the air control unit exhaust port to the atmosphere. Consequently the compressed air underneath the oil unloader shuttle valve will be able to lift it and any trapped air and oil in the power cylinder will now be released via the hole under the exhaust port. The power unit piston will be returned to its innermost position by the spring and in doing so will rotate the ratchet and pawl housing anti-clockwise. The pawl is thus

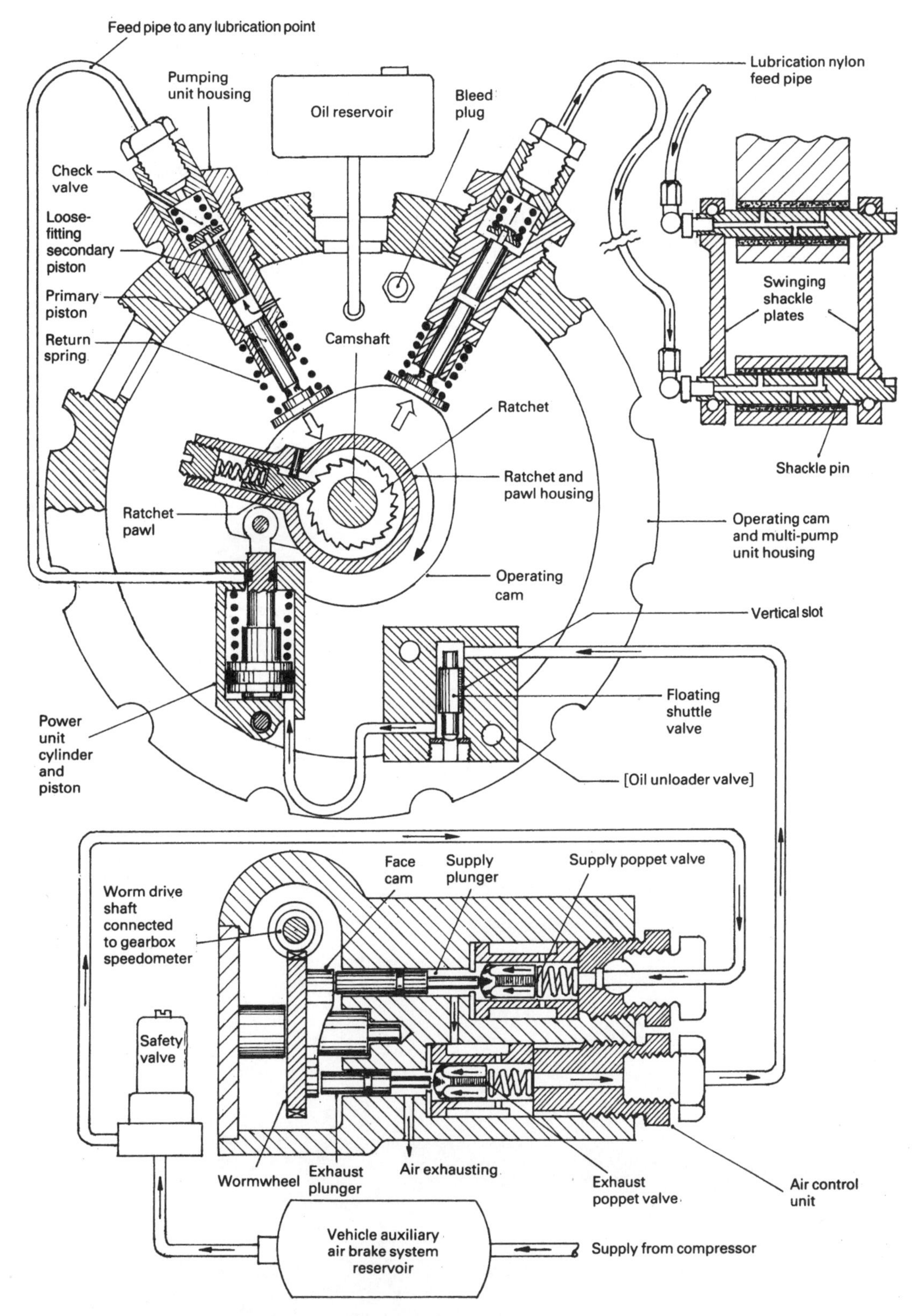

Fig. 1.30 Airdromic automatic chassis lubrication system

able to slip over one or more of the ratchet teeth to take up a new position. The net result of the power cylinder being charged and discharged with compressed air is a slow but progressive rotation of the camshaft (Fig. 1.30).

A typical worm drive shaft to distance travelled relationship is 500 revolutions per 1 km. For 900 worm drive shaft revolutions the pumping cam revolves once. Therefore, every chassis lubrication point will receive one shot of lubricant in this distance.

When the individual lubrication pump unit's primary plunger is in its outermost position, oil surrounding the barrel will enter the inlet port, filling the space between the two plungers. As the cam rotates and the lobe lifts the primary plunger, it cuts off the inlet port. Further plunger rise will partially push out the secondary plunger and so open the check valve. Pressurised oil will then pass between the loose fitting secondary plunger and barrel to lubricate the chassis moving part it services (Fig. 1.30).

2 Friction clutch

2.1 Clutch fundamentals

Clutches are designed to engage and disengage the transmission system from the engine when a vehicle is being driven away from a standstill and when the gearbox gear changes are necessary. The gradual increase in the transfer of engine torque to the transmission must be smooth. Once the vehicle is in motion, separation and take-up of the drive for gear selection must be carried out rapidly without any fierceness, snatch or shock.

2.1.1 Driven plate inertia

To enable the clutch to be operated effectively, the driven plate must be as light as possible so that when the clutch is disengaged, it will have the minimum of spin, i.e. very little flywheel effect. Spin prevention is of the utmost importance if the various pairs of dog teeth of the gearbox gears, be they constant mesh or synchromesh, are to align in the shortest time without causing excessive pressure, wear and noise between the initial chamfer of the dog teeth during the engagement phase.

Smoothness of clutch engagement may be achieved by building into the driven plate some sort of cushioning device, which will be discussed later in the chapter, whilst rapid slowing down of the driven plate is obtained by keeping the diameter, centre of gravity and weight of the driven plate to the minimum for a given torque carrying capacity.

2.1.2 Driven plate transmitted torque capacity

The torque capacity of a friction clutch can be raised by increasing the coefficient of friction of the rubbing materials, the diameter and/or the spring thrust sandwiching the driven plate. The friction lining materials now available limit the coefficient of friction to something of the order of 0.35. There are materials which have higher coefficient of friction values, but these tend to be unstable and to snatch during take-up. Increasing the diameter of the driven plate unfortunately raises its inertia, its tendency to continue spinning when the driven plate is freed while the clutch is in the disengaged position, and there is also a limit to the clamping pressure to which the friction lining material may be subjected if it is to maintain its friction properties over a long period of time.

2.1.3 Multi-pairs of rubbing surfaces (Fig. 2.1)

An alternative approach to raising the transmitted torque capacity of the clutch is to increase the number of pairs of rubbing surfaces. Theoretically the torque capacity of a clutch is directly proportional to the number of pairs of surfaces for a given clamping load. Thus the conventional single driven plate has two pairs of friction faces so that a twin or triple driven plate clutch for the same spring thrust would ideally have twice or three times the torque transmitting capacity respectively of that of the single driven plate unit (Fig. 2.1). However, because it is very difficult to dissipate the extra heat generated in a clutch unit, a larger safety factor is necessary per driven plate so that the torque capacity is generally only of the order 80% per pair of surfaces relative to the single driven plate clutch.

2.1.4 Driven plate wear (Fig. 2.1)

Lining life is also improved by increasing the number of pairs of rubbing surfaces because wear is directly related to the energy dissipation per unit area of contact surface. Ideally, by doubling the surface area as in a twin plate clutch, the energy input per unit lining area will be halved for a given slip time which would result in a 50% decrease in facing wear. In practice, however, this rarely occurs (Fig. 2.1) as the wear rate is also greatly influenced by the peak surface rubbing temperature and the intermediate plate of a twin plate clutch operates at a higher working temperature than either the flywheel or pressure plate which can be more effectively cooled. Thus in a twin plate clutch, half the energy generated whilst slipping must be absorbed by the intermediate plate and only a quarter each by the flywheel and pressure plate. This is usually borne out by the appearance of the intermediate plate and its corresponding lining faces showing evidence of high temperatures and increased wear compared to the linings facing the flywheel and pressure plate. Nevertheless, multiplate clutches do have a life expectancy which is more or less related to the number of pairs of friction faces for a given diameter of clutch.

For heavy duty applications such as those required for large trucks, twin driven plates are used, while for high performance cars where very

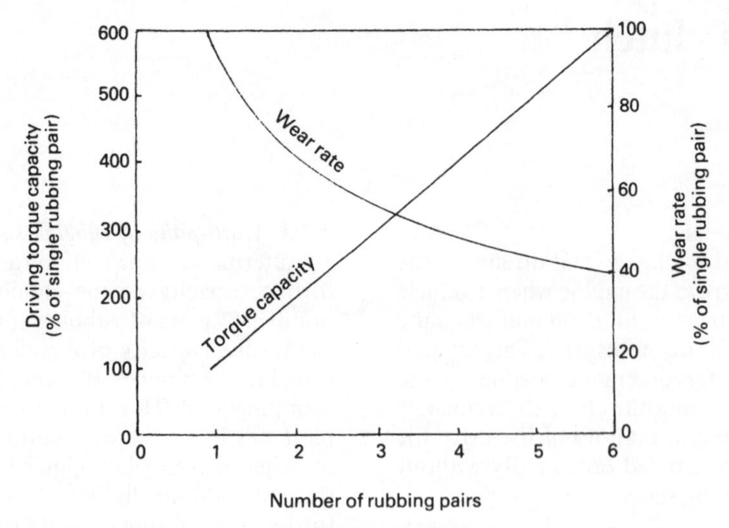

Fig. 2.1 Relationship of torque capacity wear rate and pairs of rubbing faces for multiplate clutch

rapid gear changes are necessary and large amounts of power are to be developed, small diameter multiplate clutches are preferred.

2.2 Angular driven plate cushioning and torsional damping (Figs 2.2–2.8)

2.2.1 Axial driven plate friction lining cushioning (Figs 2.2, 2.3 and 2.4)

In its simplest form the driven plate consists of a central splined hub. Mounted on this hub is a thin steel disc which in turn supports, by means of a ring of rivets, both halves of the annular friction linings (Figs 2.2 and 2.3).

Axial cushioning between the friction lining faces may be achieved by forming a series of evenly spaced 'T' slots around the outer rim of the disc. This then divides the rim into a number of segments (Arcuate) (Fig. 2.4(a)). A horseshoe shape is further punched out of each segment. The central portion or blade of each horseshoe is given a permanent set to one side and consecutive segments have opposite sets so that every second segment is riveted to the same friction lining. The alternative set of these central blades formed by the horseshoe punch-out spreads the two half friction linings apart.

An improved version uses separately attached, very thin spring steel segments (borglite) (Fig. 2.4(b)), positioned end-on around a slightly thicker disc plate. These segments are provided with a wavy 'set' so as to distance the two half annular friction linings.

Both forms of crimped spring steel segments situated between the friction linings provide

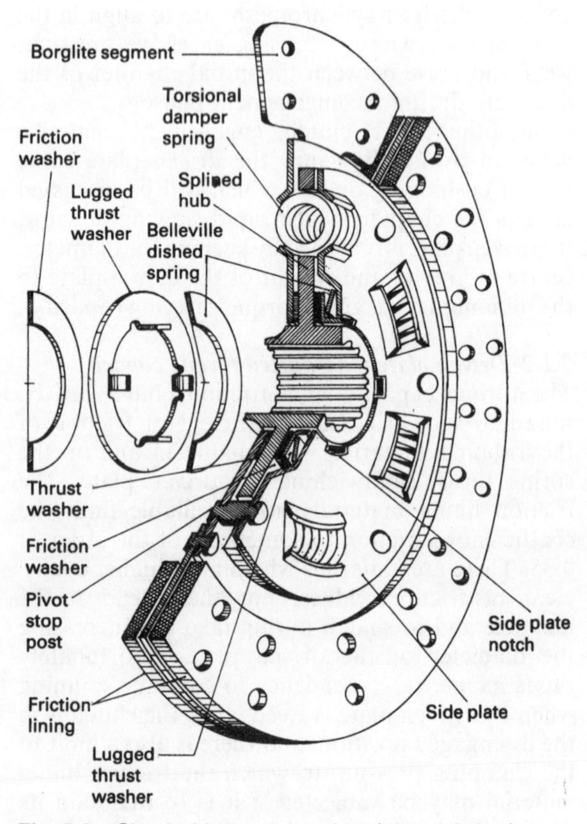

Fig. 2.2 Clutch driven centre plate (pictorial view)

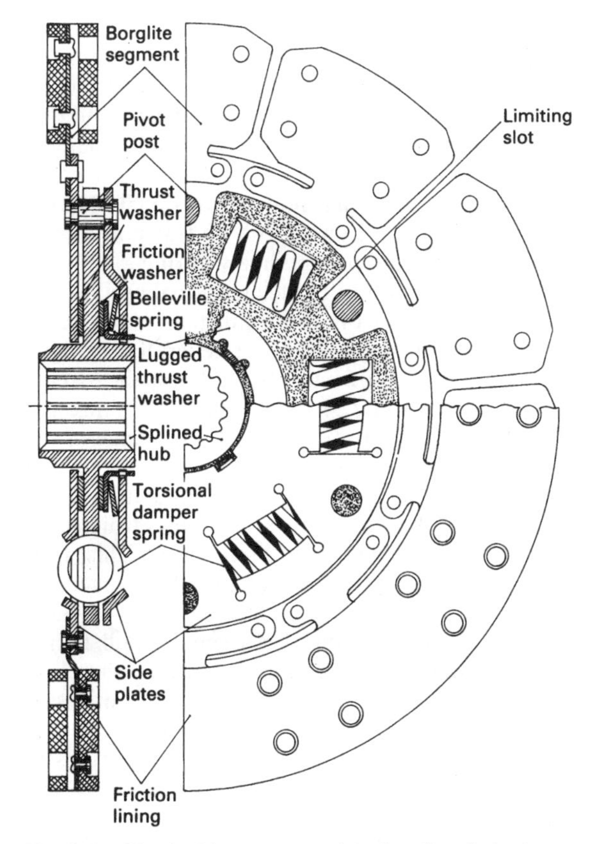

Fig. 2.3 Clutch driven centre plate (sectional view)

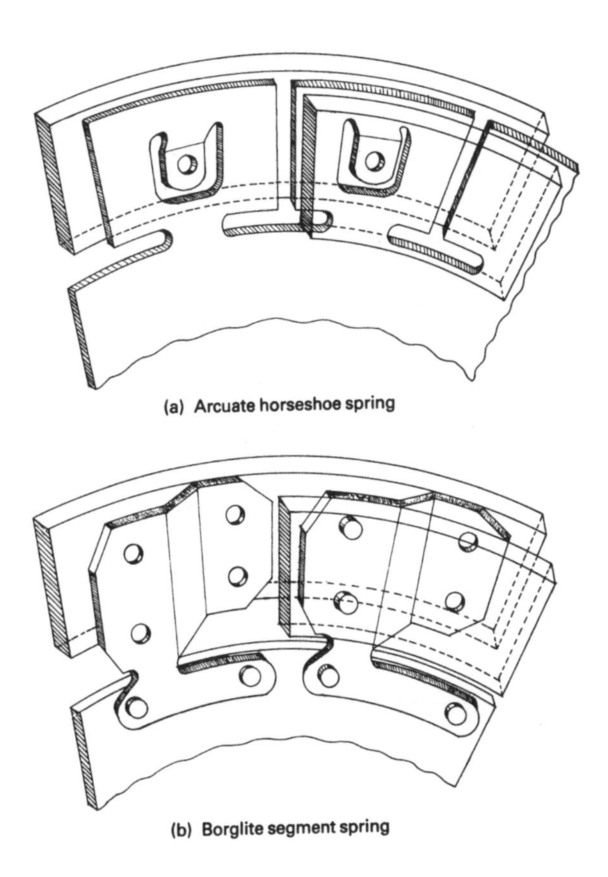

Fig. 2.4 (a and b) Driven plate cushion take-up

progressive take-up over a greater pedal travel and prevent snatch. The separately attached spring segments are thinner than the segments formed out of the single piece driven plate, so that the squeeze take-up is generally softer and the spin inertia of the thinner segments is noticeably reduced.

A further benefit created by the spring segments ensures satisfactory bedding of the facing material and a more even distribution of the work load. In addition, cooling between the friction linings occurs when the clutch is disengaged which helps to stabilise the frictional properties of the face material.

The advantages of axial cushioning of the face linings provide the following:

a) Better clutch engagement control, allowing lower engine speeds to be used at take-up thus prolonging the life of the friction faces.
b) Improved distribution of the friction work over the lining faces reduces peak operating temperatures and prevents lining fade, with the resulting reduction in coefficient of friction and subsequent clutch slip.

The spring take-up characteristics of the driven plate are such that when the clutch is initially engaged, the segments are progressively flattened so that the rate of increase in clamping load is provided by the rate of reaction offered by the spring segments (Fig. 2.5). This first low rate take-up period is followed by a second high rate engagement, caused by the effects of the pressure plate springs exerting their clamping thrust as they are allowed to expand against the pressure plate and so sandwich the friction lining between the flywheel and pressure plate faces.

2.2.2 Torsional damping of driven plate

Crankshaft torsional vibration (Fig. 2.6) Engine crankshafts are subjected to torsional wind-up and vibration at certain speeds due to the power impulses. Superimposed onto some steady mean rotational speed of the crankshaft will be additional fluctuating torques which will accelerate and decelerate the crankshaft, particularly at the front pulley

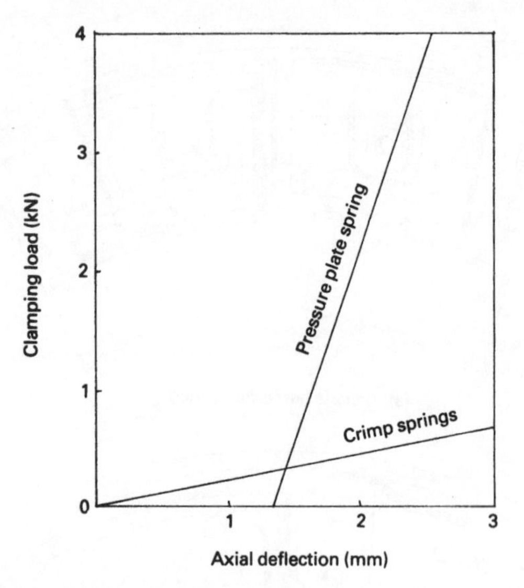

Fig. 2.5 Characteristics of driven plate axial clamping load to deflection take-up

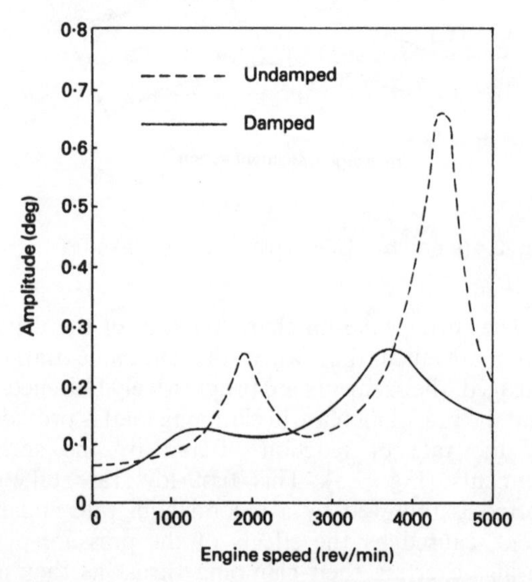

Fig. 2.6 Characteristics of crankshaft torsional vibrations undamped and damped

end and to a lesser extent the rear flywheel end (Fig. 2.6). If the flywheel end of the crankshaft were allowed to twist in one direction and then the other while rotating at certain critical speeds, the oscillating angular movements would take up the backlash between meshing gear teeth in the transmission system. Consequently, the teeth of the driving gears would be moving between the drive (pressure side) and non-drive tooth profiles of the driven gears. This would result in repeated shockloads imposed on the gear teeth, wear, and noise in the form of gear clatter. To overcome the effects of crankshaft torsional vibrations a torsion damping device is normally incorporated within the driven plate hub assembly which will now be described and explained.

Construction and operation of torsional damper springs (Figs 2.2, 2.3 and 2.7) To transmit torque more smoothly and progressively during take-up of normal driving and to reduce torsional oscillations being transmitted from the crankshaft to the transmission, compressed springs are generally arranged circumferentially around the hub of the driven plate (Figs 2.2 and 2.3). These springs are inserted in elongated slots formed in both the flange of the splined hub and the side plates which enclose the hub's flange (Fig. 2.3). These side plates are riveted together by either three or six rivet posts which pass through the flanged hub limit slots. This thus provides a degree of relative angular movement between hub and side plates. The ends of the helical coil springs bear against both central hub flange and the side plates. Engine torque is therefore transmitted from the friction face linings and side plates through the springs to the hub flange, so that any fluctuation of torque will cause the springs to compress and rebound accordingly.

Multistage driven plate torsional spring dampers may be incorporated by using a range of different springs having various stiffnesses and spring location slots of different lengths to produce a variety of parabolic torsional load–deflection characteristics (Fig. 2.7) to suit specific vehicle applications.

The amount of torsional deflection necessary varies for each particular application. For example, with a front mounted engine and rear wheel drive vehicle, a moderate driven plate angular movement is necessary, say six degrees, since the normal transmission elastic wind-up is almost adequate, but with an integral engine, gearbox and final drive arrangement, the short transmission drive length necessitates considerably more relative angular deflection, say twelve degrees, within the driven plate hub assembly to produce the same quality of take-up.

Construction and operation of torsional damper washers (Figs 2.2, 2.3 and 2.8) The torsional energy created by the oscillating crankshaft is partially absorbed and damped by the friction washer clutch situated on either side of the hub flange (Figs 2.2 and 2.3). Axial damping load is achieved by a Belleville dished washer spring mounted between one of the side plates and a four lug thrust washer.

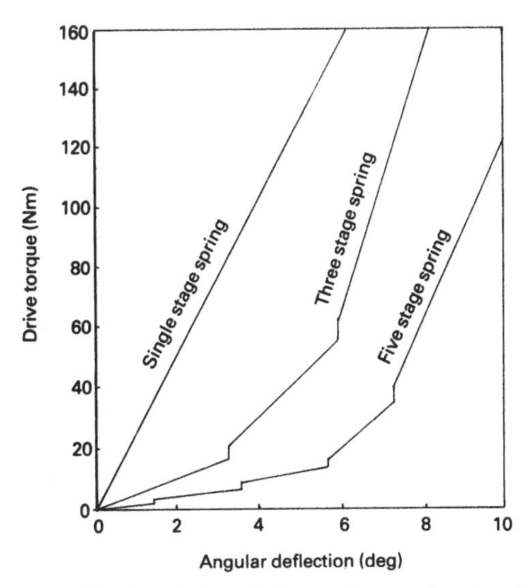

Fig. 2.7 Characteristics of driven plate torsional spring torques to deflection take-up

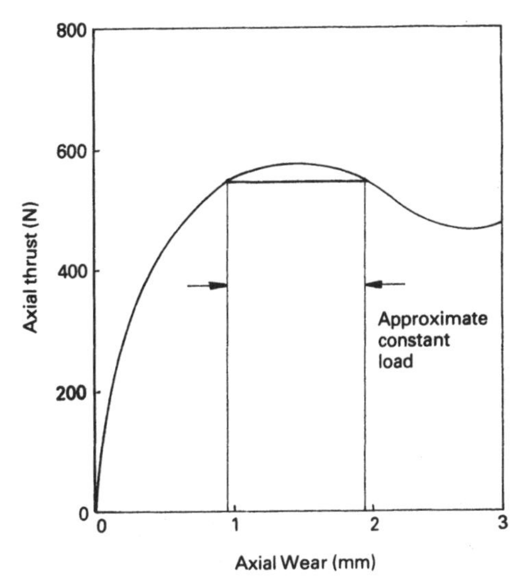

Fig. 2.8 Characteristics of driven plate torsional damper thrust spring

The outer diameter of this dished spring presses against the side plate and the inner diameter pushes onto the lugged thrust washer. In its free state the Belleville spring is conical in shape but when assembled it is compressed almost flat. As the friction washers wear, the dished spring cone angle increases. This exerts a greater axial thrust, but since the distance between the side plate and lugged thrust washer has increased, the resultant clamping thrust remains almost constant (Fig. 2.8).

2.3 Clutch friction materials

Clutch friction linings or buttons are subjected to severe rubbing and generation of heat for relatively short periods. Therefore it is desirable that they have a combination of these properties:

a) Relatively high coefficient of friction under operating conditions,
b) capability of maintaining friction properties over its working life,
c) relatively high energy absorption capacity for short periods,
d) capability of withstanding high pressure plate compressive loads,
e) capability of withstanding bursts of centrifugal force when gear changing,
f) adequate shear strength to transmit engine torque,
g) high level of cyclic working endurance without the deterioration in friction properties,
h) good compatibility with cast iron facings over the normal operating temperature range,
i) a high degree of interface contamination tolerance without affecting its friction take-up and grip characteristics.

2.3.1 Asbestos-based linings (Figs 2.2 and 2.3)
Generally, clutch driven plate asbestos-based linings are of the woven variety. These woven linings are made from asbestos fibre spun around lengths of brass or zinc wire to make lengths of threads which are both heat resistant and strong. The woven cloth can be processed in one of two ways:

a) The fibre wire thread is woven into a cloth and pressed out into discs of the required diameter, followed by stitching several of these discs together to obtain the desired thickness. The resultant disc is then dipped into resin to bond the woven asbestos threads together.
b) The asbestos fibre wire is woven in three dimensions in the form of a disc to obtain in a single stage the desired thickness. It is then pressed into shape and bonded together by again dipping it into a resin solution. Finally, the rigid lining is machined and drilled ready for riveting to the driven plate.

Development in weaving techniques has, in certain cases, eliminated the use of wire coring so that asbestos woven lining may be offered as either non- or semi-metallic to match a variety of working conditions.

Asbestos is a condensate produced by the solidification of rock masses which cool at differential rates. When the moisture content of one layer is transferred to another, fibres are produced on solidification from which, as a result of high compression, these brittle, practically straight and exceptionally fine needle-like threads are made. During processing, these break down further with a diameter of less than 0.003 mm. They exhibit a length/thickness ratio of at least three to one. It is these fine fibres which can readily be inhaled into the lungs which are so dangerous to health.

The normal highest working temperature below which these asbestos linings will operate satisfactorily giving uniform coefficient of friction between 0.32 and 0.38 and a reasonable life span is about 260 °C. Most manufacturers of asbestos-based linings quote a maximum temperature (something like 360 °C) beyond which the lining, if operated continuously or very frequently, will suffer damage, with consequent alteration to its friction characteristics and deterioration in wear resistance.

2.3.2 Asbestos substitute friction material (Figs 2.2 and 2.3)

The DuPont Company has developed a friction material derived from aromatic polyamide fibres belonging to the nylon family of polymers and it has been given the trade name Kevlar aramid.

The operating properties relative to asbestos based linings are as follows:

1 High endurance performance over its normal working pressure and temperature range.
2 It is lighter in weight than asbestos material therefore a reduction in driven plate spin shortens the time required for gear changing.
3 Good take-up characteristics, particularly with vehicles which were in the past prone to grab.
4 Weight for weight Kevlar has five times the tensile strength of steel.
5 Good centrifugal strength to withstand lining disintegration as a result of sudden acceleration which may occur during the changing of gears.
6 Stable rubbing properties at high operating temperatures. It is not until a temperature of 425 °C is reached that it begins to break down and then it does not simply become soft and melt, but steadily changes to carbon, the disintegration process being completed at about 500 °C.

Kevlar exists in two states; firstly as a 0.12 mm thick endless longitudinal fibre, which has a cut length varying between 6 and 100 mm, and secondly in the form of an amorphous structure of crushed and ground fibre known as *pulp*. In either form these fibres are difficult to inhale because of their shape and size.

2.3.3 Metallic friction materials

Metallic and semi-metallic facings have been only moderately successful. The metallic linings are normally made from either sintered iron or copper-based sintered bronze and the semi-metallic facings from a mixture of organic and metallic materials. Metallic lining materials are made from a powder produced by crushing metal or alloy pieces into many small particles. They are then compressed and heated in moulds until sufficient adhesion and densification takes place. This process is referred to as *sintering*. The metallic rings are then ground flat and are then riveted back to back onto the driven plate.

Generally the metallic and semi-metallic linings have a higher coefficient of friction, can operate at higher working temperatures, have greater torque capacity and have extended life compared to that of the organic asbestos based linings. The major disadvantages of metallic materials are their relatively high inertia, making it difficult to obtain rapid gear changes; high quality flywheel and pressure plate. Cast iron must be used to match their friction characteristics and these facings are more expensive than organic materials.

2.3.4 Cerametallic friction materials (Fig. 2.9)

Cerametallic button friction facings are becoming increasingly popular for heavy duty clutches. Instead of a full annular shaped lining, as with organic (asbestos or substitute) friction materials, four or six cerametallic trapezoidal-shaped buttons are evenly spaced on both sides around the driven plate. The cerametallic material is made from a powder consisting mainly of ceramic and copper. It is compressed into buttons and heated so that the copper melts and flows around each particle of solid ceramic. After solidification, the copper forms a strong metal-ceramic interface bond. These buttons are then riveted to the clutch driven plate and then finally ground flat.

The inherent advantages of these cerametallic-lined driven plates are:

1 A very low inertia (about 10% lower than the organic disc and 45% lower than a comparable sintered iron disc). Consequently it will result in quicker gear changes and, in the case of synchronized transmission, will increase synchronizer life.
2 A relatively high and stable coefficient of friction, providing an average value in the region of

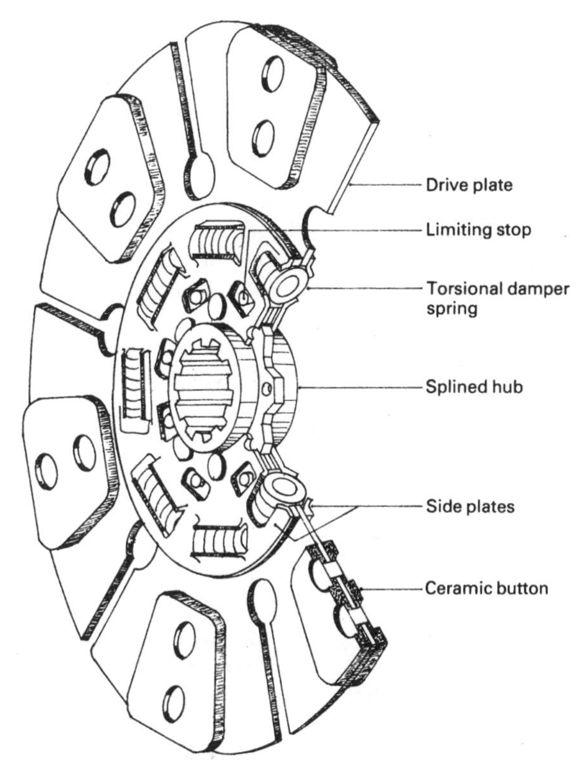

Fig. 2.9 Clutch driven plate with ceramic facings

0.4, which increases the torque capacity of clutches using these driven plates.

3 The capability of operating at high working temperatures of up to 440 °C for relatively long periods without showing signs of fade.
4 Button type driven plates expose more than 50% of the flywheel and pressure plate surfaces to the atmosphere during clutch engagement, so that heat transfer to the surrounding by convection may be improved by as much as 100%.
5 Button type friction pads do not suffer from warpage as do full ring metallic or organic linings and therefore are less prone to distort and cause clutch drag.
6 Button type friction pads permit the dust worn from the friction surfaces to be thrown clear of the clutch areas, thus preventing the possibility of any trapped work-hardened particles from scoring the friction faces.
7 Cerametallic materials are not as sensitive to grease and oil contamination as organic asbestos based linings.
8 The early ceramic-metallic friction buttons had a poor reputation as they tended to wear tracks in flywheel and pressure plate facings. A prolonged development programme has virtually eliminated this problem and has considerably extended the driven plate life span compared to driven plates using organic (asbestos-based) annular disc linings.

2.4 Clutch drive and driven member inspection

This inspection entails the examination of both the driven plate linings and the flywheel and pressure plate facings and will now be considered.

2.4.1 Driven plate lining face inspection

Driven plate friction facings should, after a short period of service, give a polished appearance due to the frequent interface rubbing effect. This smooth and polished condition will provide the greatest friction grip, but it must not be confused with a glazed surface created by the formation of films of grease or oil worked into the rubbing surfaces, heated and oxidized.

A correctly bedded-in friction facing will appear highly polished through which the grain of the friction material can be clearly seen. When in perfect condition, these polished facings are of a grey or mid-brown colour. A very small amount of lubricant on the facings will burn off due to the generated heat. This will only slightly darken the facings, but providing polished facings remain so that the grain of the material can be clearly distinguished, it does not reduce its effectiveness.

Large amounts of lubricant gaining access to the friction surfaces may result in the following:

a) The burning of the grease or oil may leave a carbon deposit and a high glaze, this hides the grain of the material and is likely to cause clutch slip.
b) If the grease or oil is only partially burnt and leaves a resinous deposit on the facings it may result in a fierce clutch and may in addition produce clutch spin caused by the rubbing interfaces sticking.
c) If both carbon and resinous deposits are formed on the linings, clutch judder may develop during clutch take-up.

2.4.2 Flywheel and pressure plate facing inspection

Cast iron flywheel or pressure plate faces should have a smooth polished metallic appearance, but abnormal operating conditions or badly worn driven plate linings may be responsible for the following defects:

a) Overheated clutch friction faces can be identified by colouring of the swept polished tracks. The actual surface temperatures reached can be distinguished broadly by the colours; straw, brown, purple and blue which relate to 240 °C, 260 °C, 280 °C and 320 °C respectively.
b) Severe overheating will create thermal stresses within the cast iron mass of the flywheel and pressure plate, with the subsequent appearance of radial hairline cracks.
c) Excessively worn driven plate linings with exposed rivets and trapped work-hardened dust particles will cause scoring of the rubbing faces in the form of circular grooves.

2.5 Clutch misalignment (Fig. 2.10(a–d))

Clutch faults can sometimes be traced to misalignment of the crankshaft to flywheel flange joint, flywheel housing and bell housing. Therefore, if misalignment exists, the driven plate plane of rotation will always be slightly skewed to that of the restrained hub which is made to rotate about the spigot shaft's axis. Misalignment is generally responsible for the following faults:

1 Rapid wear on the splines of the clutch driven plate hub, this being caused mainly by the tilted hub applying uneven pressure over the interface length of the splines.
2 The driven plate breaking away from the splined hub due to the continuous cyclic flexing of the plate relative to its hub.
3 Excessively worn pressure plate release mechanism, causing rough and uneven clutch engagement.
4 Fierce chattering or dragging clutch resulting in difficult gear changing.

If excessive clutch drag, backlash and poor changes are evident and the faults cannot be corrected, then the only remedy is to remove both gearbox and clutch assembly so that the flywheel housing alignment can be assessed (Fig. 2.10).

2.5.1 Crankshaft end float (Fig. 2.10(a))

Before carrying out engine crankshaft, flywheel or flywheel housing misalignment tests, ensure that the crankshaft end float is within limits. (Otherwise inaccurate run-out readings may be observed.)

To measure the crankshaft end float, mount the magnetic dial gauge base to the back of the flywheel housing and position the indicator pointer against the crankshaft flanged end face. Zero the dial gauge and with the assistance of a suitable lever, force the crankshaft back and forth and, at the same time, observe the reading. Acceptable end float values are normally between 0.08 and 0.30 mm.

2.5.2 Crankshaft flywheel flange runout
(Fig. 2.10(a))

The crankshaft flange flywheel joint face must be perpendicular to its axis of rotation with no permissible runout. To check for any misalignment, keep the dial gauge assembly mounted as for the end float check. Zero gauge the dial and rotate the crankshaft by hand for one complete revolution whilst observing any dial movement. Investigate further if runout exists.

2.5.3 Flywheel friction face and rim face runout
(Fig. 2.10(a and b))

When the flywheel is centred by the crankshaft axis, it is essential that the flywheel friction face and rim rotate perpendicularly to the crankshaft axis.

Mount the dial gauge magnetic base to the engine flywheel housing. First set the indicator pointer against the friction face of the flywheel near the outer edge (Fig. 2.10(a and b)) and set gauge to zero. Turn the flywheel one revolution and observe the amount of variation. Secondly reset indicator pointer against the flywheel rim and repeat the test procedure (Fig. 2.10(b)). Maximum permissible runout in both tests is 0.02 mm per 20 mm of flywheel radius. Thus with a 300 mm diameter clutch fitted, maximum run-out would be 0.15 mm. Repeat both tests 2 or 3 times and compare readings to eliminate test error.

2.5.4 Flywheel housing runout (Fig. 2.10(c))

When the gearbox bell housing is centred by the inside diameter and rear face of the engine flywheel housing, it is essential that the inside diameter and rear face of the housing should be concentric and parallel respectively with the flywheel.

Mount the dial gauge magnetic base to the flywheel friction face and position. Set the indicator pointer against the face of the housing. Make sure that the pointer is not in the path of the fixing holes in the housing face or else incorrect readings may result. Zero the indicator and observe the reading whilst the crankshaft is rotated one complete revolution. Reset the indicator pointer against the internally machined recess of the clutch housing and repeat the test procedure. Maximum permissible runout is 0.20 mm. Repeat both tests two or three times and compare readings to eliminate errors.

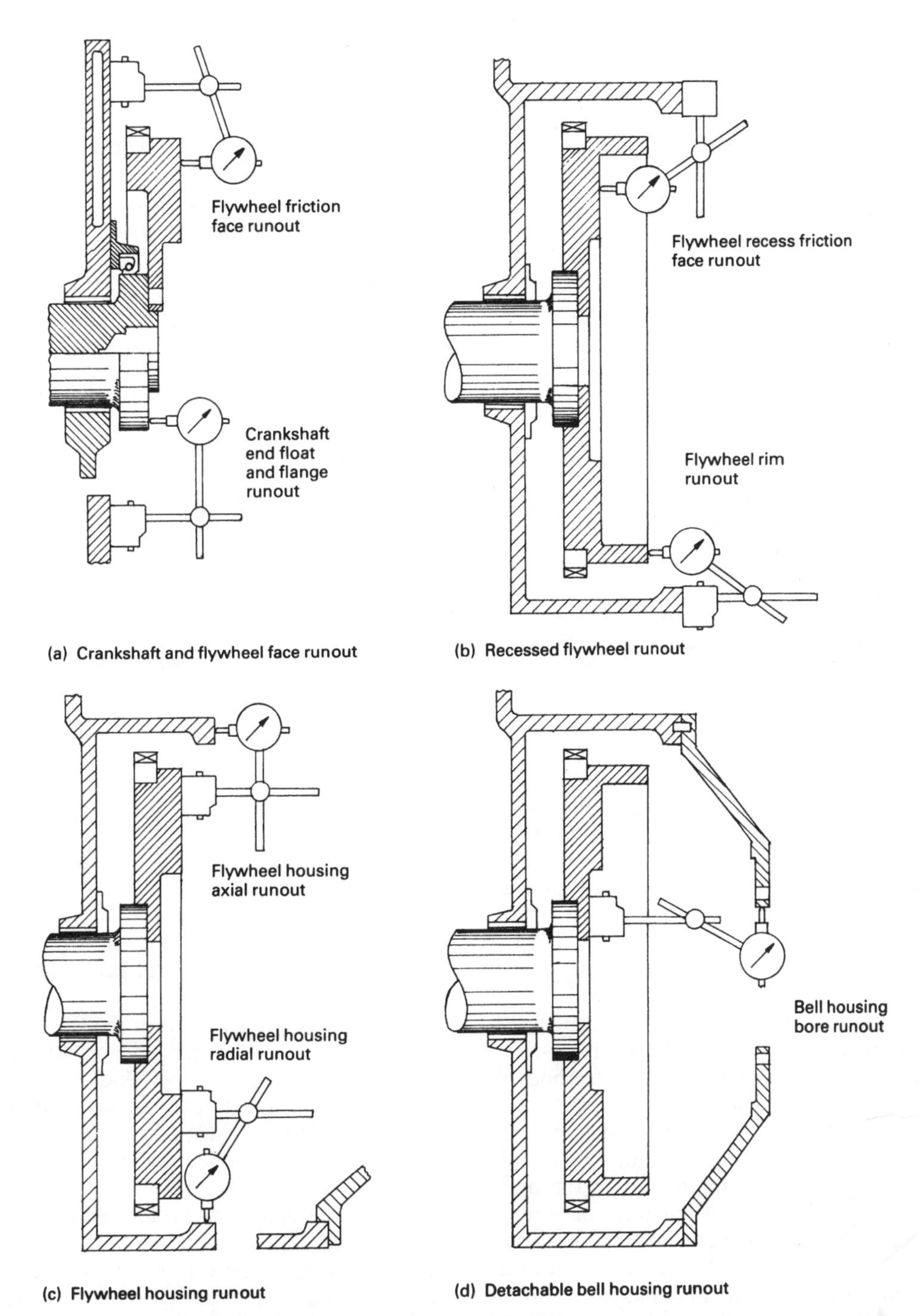

(a) Crankshaft and flywheel face runout

(b) Recessed flywheel runout

(c) Flywheel housing runout

(d) Detachable bell housing runout

Fig. 2.10 (a–d) Crankshaft flywheel and clutch housing alignment

2.5.5 Detachable bell housing runout

(Fig. 2.10(c and d))

When the gearbox bell housing is located by dowel pins instead of the inside diameter of the engine flywheel housing (Fig. 2.10(c)) (shouldered bell housing), it is advisable to separate the clutch bell housing from the gearbox and mount it to the flywheel housing for a concentric check.

Mount the dial gauge magnetic base onto the flywheel friction face and position the indicator pointer against the internal recess of the bell housing gearbox joint bore (Fig. 2.10(d)). Set the gauge to zero and turn the crankshaft by hand one complete revolution. At the same time, observe the dial gauge reading.

Maximum permissible runout should not exceed 0.25 mm.

2.6 Pull type diaphragm clutch (Fig. 2.11)

With this type of diaphragm clutch, the major components of the pressure plate assembly are a cast iron pressure plate, a spring steel diaphragm disc and a low carbon steel cover pressing (Fig. 2.11). To actuate the clutch release, the diaphragm is made to pivot between a pivot ring positioned inside the rear of the cover and a raised circumferential ridge formed on the back of the pressure plate. The diaphragm disc is divided into fingers caused by radial slits originating from the central hole. These fingers act both as leaf springs to provide the pressure plate thrust and as release levers to disengage the driven plate from the drive members.

When the driven and pressure plates are bolted to the flywheel, the diaphragm is distorted into a dished disc which therefore applies an axial thrust between the pressure plate and the cover pressing. This clutch design reverses the normal method of operation by pulling the diaphragm spring outwards to release the driven plate instead of pushing it.

Owing to its configuration, the pull type clutch allows a larger pressure plate and diaphragm spring to be used for a given diameter of clutch. Advantages of this design over a similar push type clutch include lower pedal loads, higher torque capacity, improved take-up and increased durability. This clutch layout allows the ratio of the diaphragm finger release travel to pressure plate movement to be reduced. It is therefore possible to maintain the same pressure plate movement as that offered by a conventional push type clutch, and yet increase the ratio between clamp load and pedal load from 4:1 to 5:1.

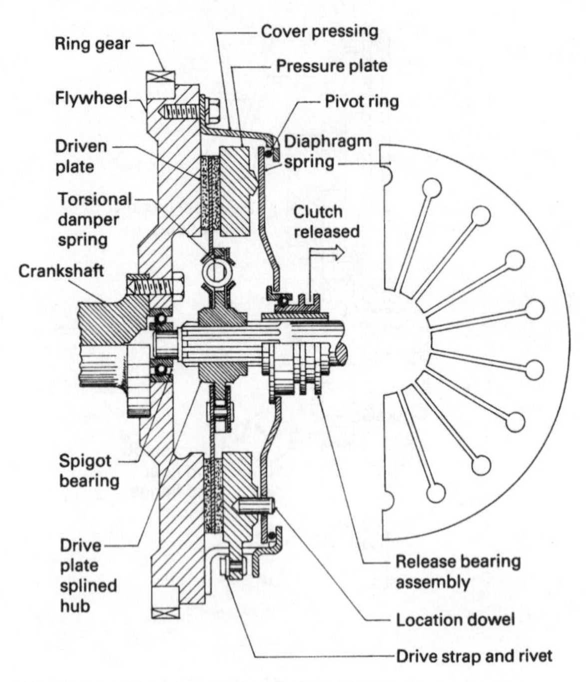

Fig. 2.11 Diaphragm single plate pull type clutch

2.7 Multiplate diaphragm type clutch (Fig. 2.12)

These clutches basically consist of drive and driven plate members. The drive plates are restrained from rotating independently by interlocking lugs and slots which permit axial movement, but not relative rotational spin, whilst the driven plates are attached and supported by internally splined hubs to corresponding splines formed on the gearbox spigot shaft, see Fig. 2.12.

The diaphragm spring is in the form of a dished annular disc. The inner portion of the disc is radially slotted, the outer ends being enlarged with a circular hole to prevent stress concentration when the spring is distorted during disengagement. These radial slots divide the disc into a number of release levers (fingers).

The diaphragm spring is located in position with a shouldered pivot post which is riveted to the cover pressing. These rivets also hold a pair of fulcrum rings in position which are situated either side of the diaphragm.

Whilst in service, the diaphragm cone angle will change continuously as wear occurs and as the clutch is engaged and disengaged during operation. To enable this to happen, the diaphragm pivots and rolls about the fulcrum rings. When the clutch is engaged the diaphragm bears against the outer

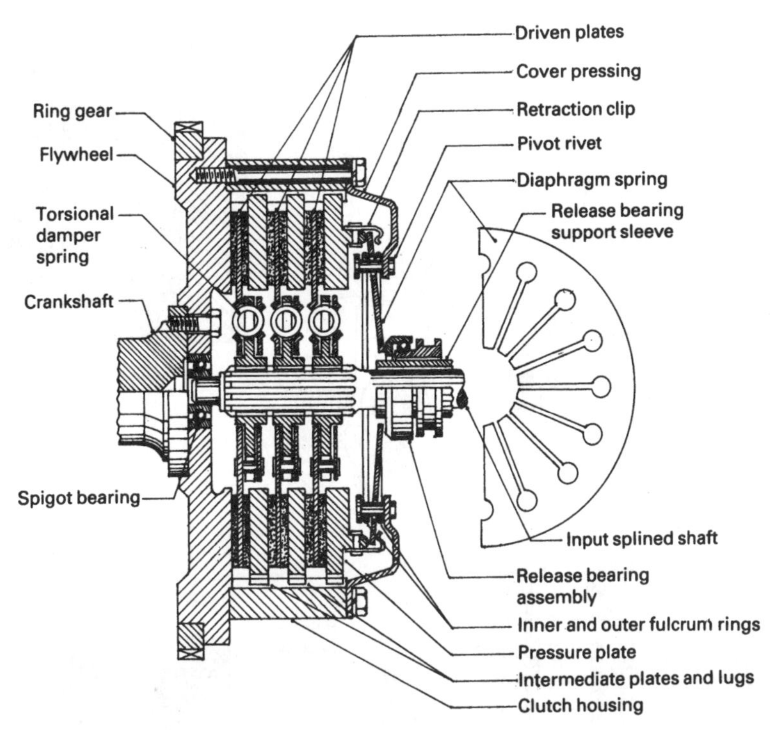

Fig. 2.12 Multiplate diaphragm type clutch

ring, but when disengagement takes place the reaction load is then taken by the inner ring.

As the friction linings wear, the spring diaphragm will become more dished and subsequently will initially exert a larger axial clamping load. It is only when the linings are very worn, so that the distance between the cover pressing and pressure plate become excessive, that the axial thrust will begin to decline.

2.8 Lipe rollway twin driven plate clutch (Fig. 2.13)

These clutches have two circular rows of helical coil springs which act directly between the pressure plate and the cover housing, see Fig. 2.13. The release mechanism is of the pull type in which a central release bearing assembly is made to withdraw (pull out) three release levers to disengage the clutch. The clutch pressure plate assembly is bolted to the flywheel and the driven plate friction linings are sandwiched between the flywheel, intermediate plate and pressure plate facings. The central hub of the driven plates is mounted on a splined gearbox spigot shaft (input shaft). The splined end of the input shaft is supported by a ball race bearing mounted inside the flywheel-crankshaft attachment flange. The other end of this shaft is supported inside the gearbox by either ball or taper roller bearings. There are two types of pressure plate cover housings; one with a deep extended cover rim which bolts onto a flat flywheel facing and the shallow cover type in which the pressure plate casting fits into a recessed flywheel.

The release mechanism is comprised of three lever fingers. The outer end of each lever pivots on a pin and needle race mounted inside each of the adjustable eye bolt supports, which are attached to the cover housing through an internally and externally threaded sleeve which is secured to the cover housing with a lock nut. Inwards from the eye bolt, one-sixth of the release lever length, is a second pin which pivots on a pair of needle-bearing races situated inside the pressure plate lugs formed on either side of each layer.

Release lever adjustment

Initially, setting up of the release levers is achieved by slackening the locknuts and then rotating each sleeve in turn with a two pronged fork adaptor tool which fits into corresponding slots machined out of the adjustment sleeve end. Rotating the sleeve one way or the other will screw the eye bolts in or out until the correct dimension is obtained between the back of the release lever fingers and the outer cover rim edge. This setting dimension is provided by the

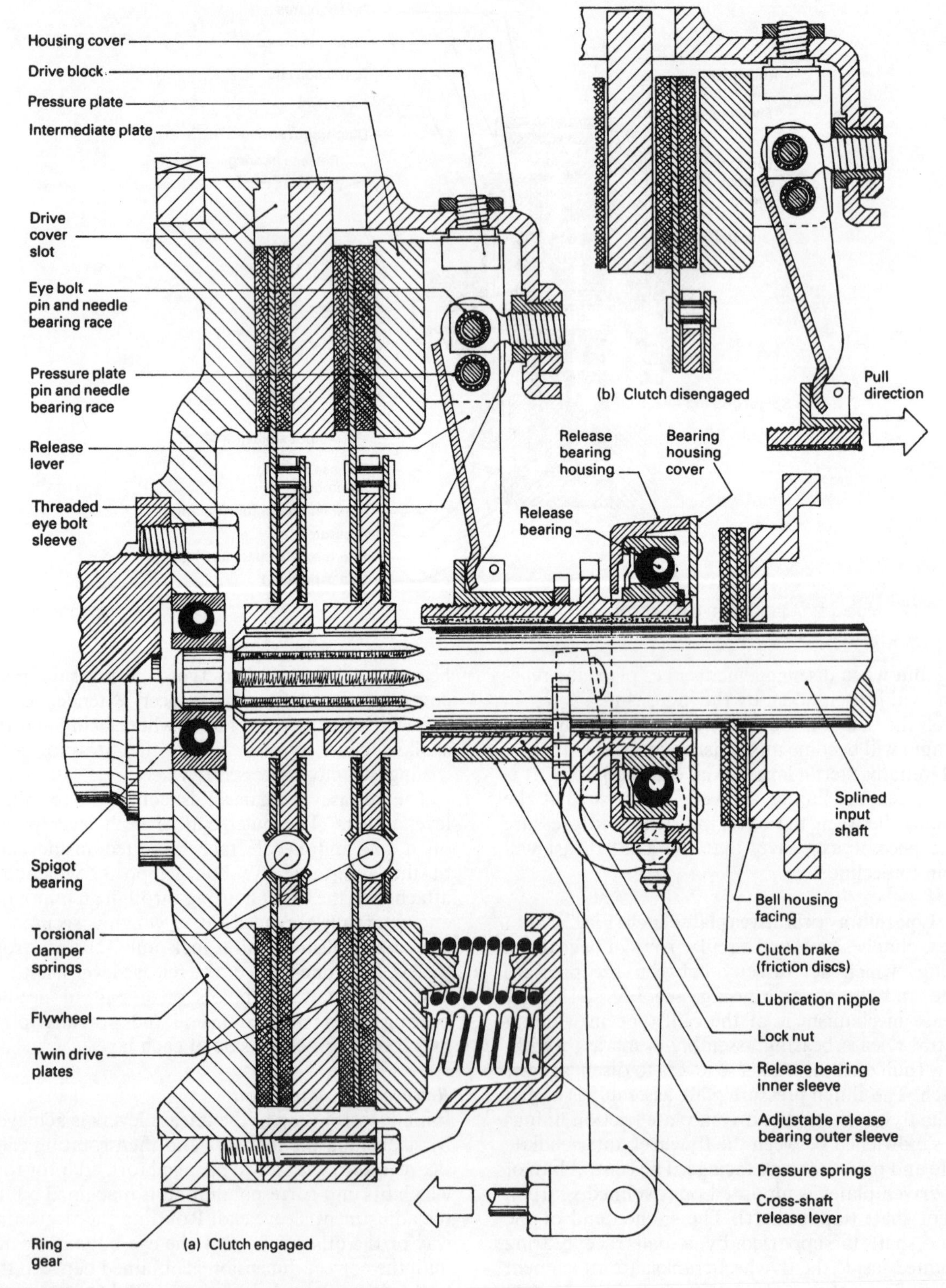

Fig. 2.13 (a–b) Twin driven plate pull type clutch

manufacturers for each clutch model and engine application. Finally, tighten the locknuts of each eye bolt and re-check each lever dimension again.

Release bearing adjustment
Slacken sleeve locknut with a 'C' shaped spanner. Rotate the inner sleeve either way by means of the slotted adjusting nut until the recommended clearance is obtained between the bearing housing cover face and clutch brake.

i.e.	9.5 mm for 355 mm	—	1LP
	13 mm for 355 mm	—	2LP
	13 mm for 294 mm	—	2LP

Finally tighten sleeve locknut and re-check clearance.

2.9 Spicer twin driven plate angle spring pull type clutch (Fig. 2.14)
An interesting clutch engagement and release pressure plate mechanism employs three pairs of coil springs which are inclined to the axial direction of the driven plates. These springs are mounted between the pressure plate cover housing, which takes the spring reaction, and the release lever central hub (Fig. 2.14). The axial clamping thrust is conveyed by the springs to the six to one leverage ratio release levers (six of them) spaced evenly around the release hub. These release levers span between the release hub and a large annular shaped adjustable pivot ring which is screwed inside the pressure plate cover housing. Towards the pivot pin end of the release levers a kink is formed so that it can bear against the pressure plate at one point. The pressure plate and intermediate plate are both prevented from spinning with the driven plates by cast-in drive lugs which fit into slots formed into the cover housing.

In the engaged position, the six springs expand and push the release hub and, subsequently, the release levers towards the pressure plate so that the driven plates are squeezed together to transmit the drive torque.

To release the clutch driven plates, the release bearing assembly is pulled out from the cover housing. This compels the release lever hub to compress and distort the thrust springs to a much greater inclined angle relative to the input shaft axis and so permits the pressure plate to be withdrawn by means of the retraction springs.

Because the spring thrust does not operate directly against the pressure plate, but is relayed through the release levers, the actual spring's stiffness is reduced by a factor of the leverage ratio; in this instance one-sixth of the value if the springs were direct acting.

The operating characteristics of the clutch mechanism are described as follows:

New engaged position (Fig. 2.14(a))
The spring thrust horizontal component of 2.2 kN, multiplied by the lever ratio, provides a pressure plate clamping load of 13.2 kN (Fig. 2.14(a)). The axial thrust horizontal component pushing on the pressure plate does not vary in direct proportion with the spring load exerted between its ends, but is a function of the angle through which the mounted springs operate relative to the splined input shaft.

Worn engagement position (Fig. 2.14(b))
When the driven plate facings wear, the release bearing moves forward to the pressure plate so that the springs elongate. The spring load exerted between the spring ends is thus reduced. Fortunately, the inclined angle of spring axis to that of the thrust bearing axis is reduced so that as the spring load along its axis declines, the horizontal thrust component remains essentially the same. Therefore, the pressure plate clamping load remains practically constant throughout the life of the clutch (Fig. 2.14(b)).

Release position (Fig. 2.14(c))
When the clutch is released, that is when the bearing is pulled rearwards, the springs compress and increase in load, but the spring angle relative to the thrust bearing axis increases so that a greater proportion of the spring load will be acting radially instead of axially. Consequently, the horizontal component of axial release bearing load, caused by the spring thrust, gradually reduces to a value of about 1.7 kN as the bearing moves forwards, which results in the reduced pedal effort. This is shown in Fig. 2.14(c).

Internal manual adjustment
Release bearing adjustment is made by unscrewing the ring lock plate bolt and removing the plate. The clutch pedal is then held down to relieve the release levers and adjusting ring load. The adjusting ring is then rotated to screw it in or out so that it alters the release lever hub axial position.

Turning the adjusting ring clockwise moves the release bearing towards the gearbox (increasing free pedal movement). Conversely, turning the adjusting ring anticlockwise moves the release bearing towards the flywheel (decreasing free pedal movement).

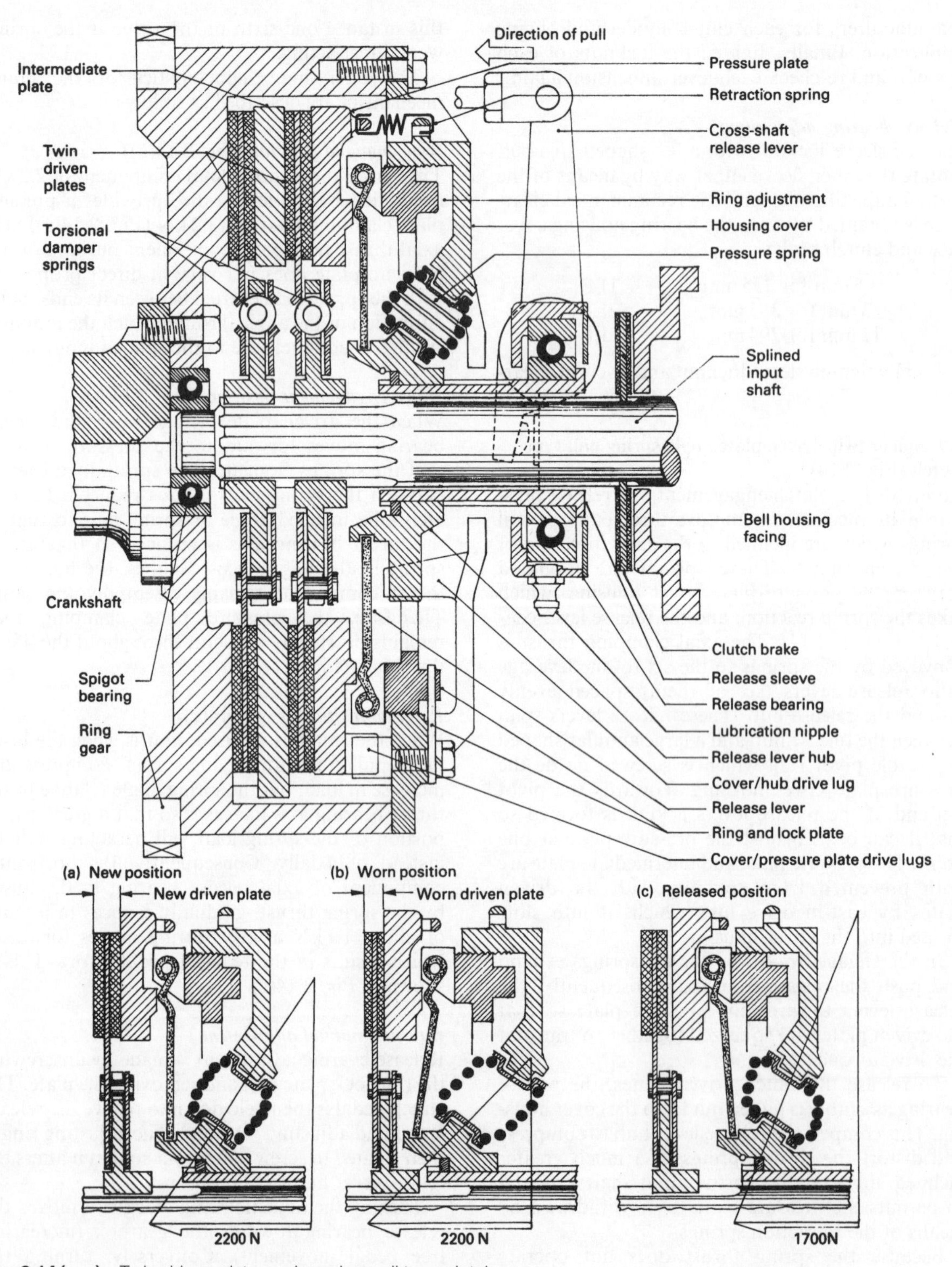

Fig. 2.14 (a–c) Twin driven plate angle spring pull type clutch

The adjusting ring outer face is notched so that it can be levered round with a screwdriver when adjustment is necessary. The distance between each notch represents approximately 0.5 mm. Thus three notches moved means approximately 1.5 mm release bearing movement.

With the pedal released, there should be approximately 13 mm clearance between the release bearing face and clutch brake.

Internal self-adjustment

A clutch self-adjustment version has teeth cut on the inside of the adjusting ring and a small worm and spring self-adjusting device replaces the lock plate. The worm meshes with the adjusting ring. One end of the spring is located in a hole formed in the release lever hub whilst the other end is in contact with the worm. Each time the clutch is engaged and disengaged, the release lever movement will actuate the spring. Consequently, once the driven plates have worn sufficiently, the increased release lever movement will rotate the worm which in turn will partially screw round the adjusting ring to compensate and so reset the position of the release levers.

2.10 Clutch (upshift) brake (Fig. 2.15)

The clutch brake is designed primarily for use with unsynchronized (crash or constant mesh) gearboxes to permit shifting into first and reverse gear without severe dog teeth clash. In addition, the brake will facilitate making unshafts by slowing down the input shaft so that the next higher gear may be engaged without crunching of teeth.

The brake disc assembly consists of a pair of Belleville spring washers which are driven by a hub having internal lugs that engage machined slots in the input shaft. These washers react against the clutch brake cover with facing material positioned between each spring washer and outer cover (Fig. 2.15).

When the clutch pedal is fully depressed, the disc will be squeezed between the clutch release bearing housing and the gearbox bearing housing, causing the input spigot shaft to slow down or stop. The hub and spring washer combination will slip with respect to the cover if the applied torque load exceeds 34 Nm, thus preventing the disc brake being overloaded.

In general, the clutch brake comes into engagement only during the last 25 mm of clutch pedal

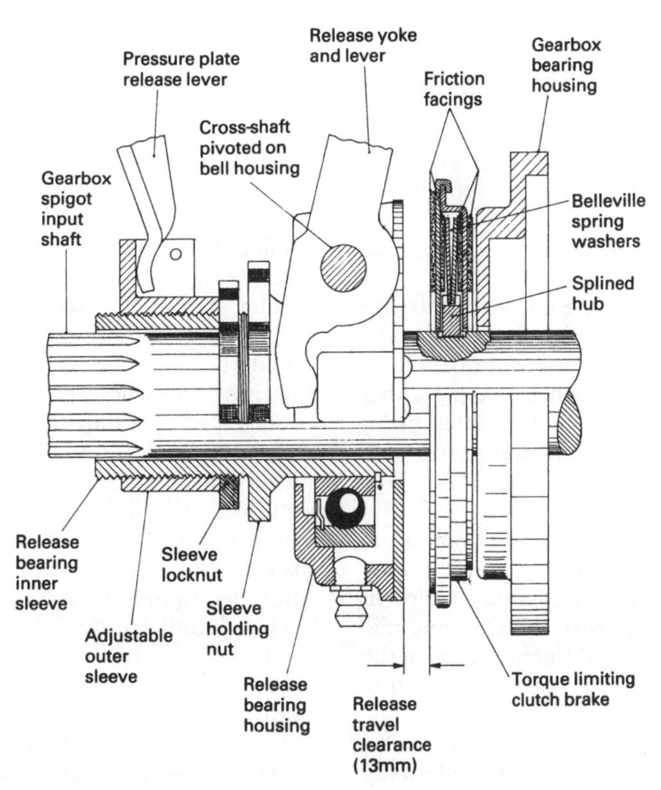

Fig. 2.15 Clutch upshift brake (torque limiting)

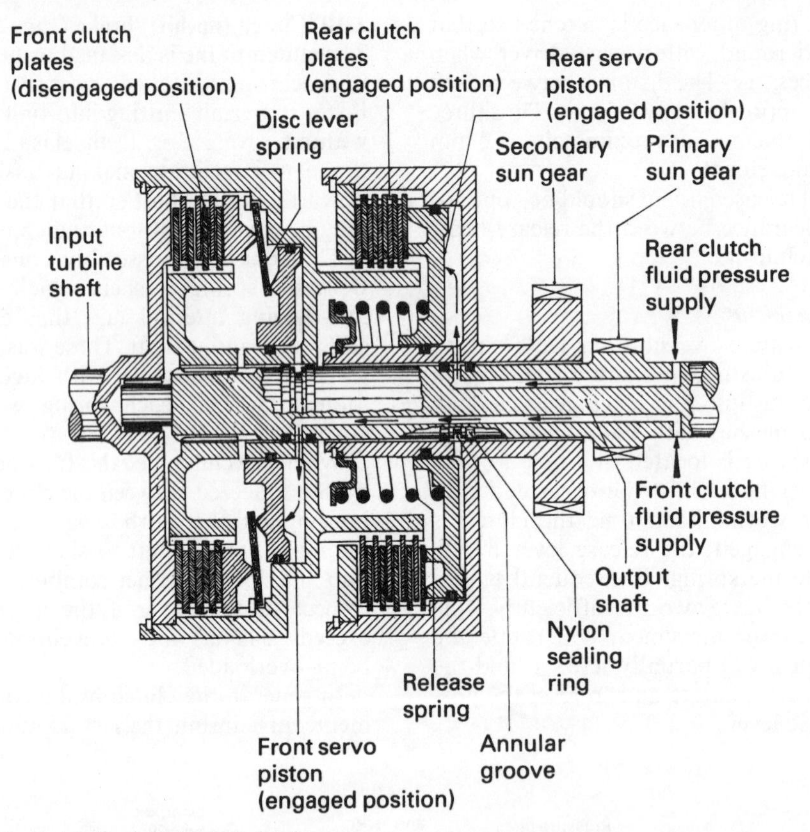

Fig. 2.16 Multiplate hydraulically actuated clutches

travel. Therefore, the pedal must be fully depressed to squeeze the clutch brake. The clutch pedal should never be fully depressed before the gearbox is put into neutral. If the clutch brake is applied with the gearbox still in gear, a reverse load will be put on the gears making it difficult to get the gearbox out of gear. At the same time it will have the effect of trying to stop or decelerate the vehicle with the clutch brake and rapid wear of the friction disc will take place. Never apply the clutch brake when making down shifts, that is do not fully depress the clutch pedal when changing from a higher to a lower gear.

2.11 Multiplate hydraulically operated automatic transmission clutches (Fig. 2.16)

Automatic transmissions use multiplate clutches in addition to band brakes extensively with epicyclic compound gear trains to lock different stages of the gearing or gear carriers together, thereby providing a combination of gear ratios.

These clutches are comprised of a pack of annular discs or plates, alternative plates being internally and externally circumferentially grooved to match up with the input and output splined drive members respectively (Fig. 2.16). When these plates are squeezed together, torque will be transmitted from the input to the output members by way of these splines and grooves and the friction torque generated between pairs of rubbing surfaces. These steel plates are faced with either resinated paper linings or with sintered bronze linings, depending whether moderate or large torques are to be transmitted. Because the whole gear cluster assembly will be submerged in fluid, these linings are designed to operate wet (in fluid). These clutches are hydraulically operated by servo pistons either directly or indirectly through a lever disc spring to multiplate, the clamping load which also acts as a piston return spring. In this example of multiplate clutch utilization hydraulic fluid is supplied under pressure through radial and axial passages drilled in the output shaft. To transmit pressurized fluid from one member to another where there is relative angular movement between components, the output shaft has machined grooves on either side of all the radial supply passages. Square sectioned nylon sealing rings are then pressed into these grooves so that when the shaft is in position, these rings expand and seal lengthwise portions of the shaft with their corresponding bore formed in the outer members.

Front clutch (FC)
When pressurized, fluid is supplied to the front clutch piston chamber. The piston will move over to the right and, through the leverage of the disc spring, will clamp the plates together with considerable thrust. The primary sun gear will now be locked to the input turbine shaft and permit torque to be transmitted from the input turbine shaft to the central output shaft and primary sun gear.

Rear clutch (RC)
When pressurized, fluid is released from the front clutch piston chamber, and is transferred to the rear clutch piston chamber. The servo piston will be forced directly against the end plate of the rear clutch multiplate pack. This compresses the release spring and sandwiches the drive and driven plates together so that the secondary sun gear will now be locked to the input turbine shaft. Torque can now be transmitted from the input turbine shaft to the secondary sun gear.

2.12 Semicentrifugal clutch (Figs 2.17 and 2.18)
With this design of clutch lighter pressure plate springs are used for a given torque carrying capacity, making it easier to engage the clutch in the lower speed range, the necessary extra clamping thrust being supplemented by the centrifugal force at higher speeds.

The release levers are made with offset bob weights at their outer ends, so that they are centrifugally out of balance (Fig. 2.17). The movement due to the centrifugal force about the fixed pivot tends to force the pressure plate against the driven plate, thereby adding to the clamping load. While the thrust due to the clamping springs is constant, the movement due to the centrifugal force varies as the square of the speed (Fig. 2.18). The reserve factor for the thrust spring can be reduced to 1.1 compared to 1.4–1.5 for a conventional helical coil spring clutch unit. Conversely, this clutch design may be used for heavy duty applications where greater torque loads are transmitted.

2.13 Fully automatic centrifugal clutch
(Figs 2.19 and 2.20)
Fully automatic centrifugal clutches separate the engine from the transmission system when the engine is stopped or idling and smoothly take up the drive with a progressive reduction in slip within a narrow rising speed range until sufficient engine power is developed to propel the vehicle directly. Above this speed full clutch engagement is provided.

To facilitate gear changes whilst the vehicle is in motion, a conventional clutch release

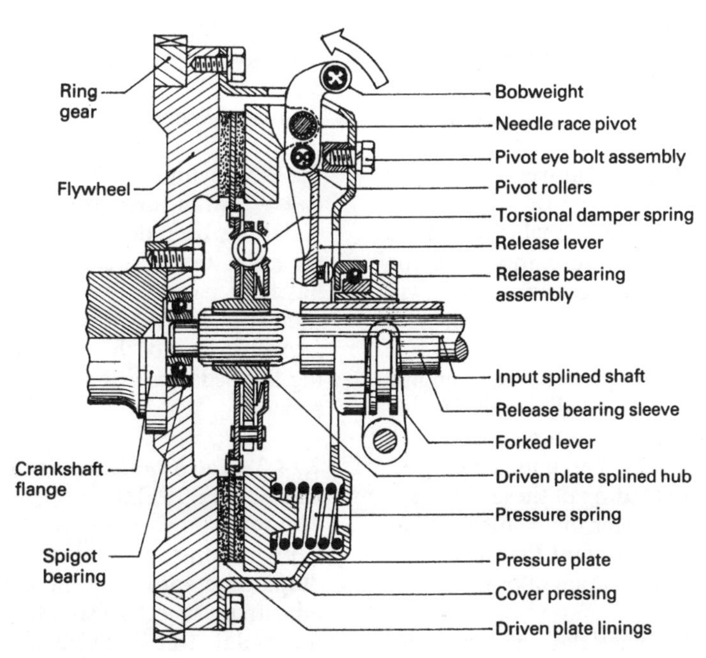

Fig. 2.17 Semicentrifugal clutch

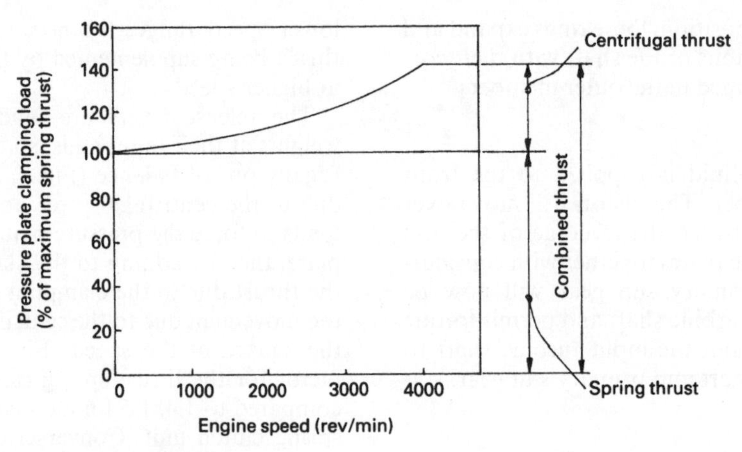

Fig. 2.18 Semicentrifugal clutch characteristics

lever arrangement is additionally provided. This mechanism enables the driver to disengage and engage the clutch independently of the flyweight action so that the drive and driven gearbox member speeds can be rapidly and smoothly unified during the gear selection process.

The automatic centrifugal mechanism consists of a reaction plate situated in between the pressure plate and cover pressing. Mounted on this reaction plate by pivot pins are four equally spaced bobweights (Fig. 2.19). When the engine's speed increases, the bobweight will tend to fly outward. Since the centre of gravity of their masses is to one side of these pins, they will rotate about their pins. This will be partially prevented by short struts offset to the pivot pins which relay this movement and effort to the pressure plate. Simultaneously, the reaction to this axial clamping thrust causes the reaction plate to compress both the reaction and pressure springs so that it moves backwards towards the cover pressing.

The greater the centrifugal force which tends to rotate the bobweights, the more compressed the springs will become and their reaction thrust will be larger, which will increase the pressure plate clamping load.

To obtain the best pressure plate thrust to engine speed characteristics (Fig. 2.20), adjustable reactor springs are incorporated to counteract the main compression spring reaction. The initial compression length and therefore loading of these springs is set up by the adjusting nut after the whole unit has been assembled. Thus the resultant thrust of both lots of springs determine the actual take-up engine speed of the clutch.

Gear changes are made when the clutch is disengaged, which is achieved by moving the release bearing forwards. This movement pulls the reactor plate rearwards by means of the knife-edge link and also withdraws the pressure plate through the retractor springs so as to release the pressure plate clamping load.

2.14 Clutch pedal actuating mechanisms

Some unusual ways of operating a clutch unit will now be described and explained.

2.14.1 Clutch pedal with over-centre spring (Fig. 2.21)

With this clutch pedal arrangement, the over-centre spring supplements the foot pressure applied when disengaging the clutch, right up until the diaphragm spring clutch is fully disengaged (Fig. 2.21). It also holds the clutch pedal in the 'off' position. When the clutch pedal is pressed, the master cylinder piston forces the brake fluid into the slave cylinder. The slave piston moves the push rod, which in turn disengages the clutch. After the pedal has been depressed approximately 25 mm of its travel, the over-centre spring change over point has been passed. The over-centre spring tension is then applied as an assistance to the foot pressure.

Adjustment of the clutch is carried out by adjusting the pedal position on the master cylinder push rod.

2.14.2 Clutch cable linkage with automatic adjuster (Fig. 2.22)

The release bearing is of the ball race type and is kept in constant contact with the fingers of the diaphragm spring by the action of the pedal self-adjustment mechanism. In consequence, there is no pedal free movement adjustment required (Fig. 2.22).

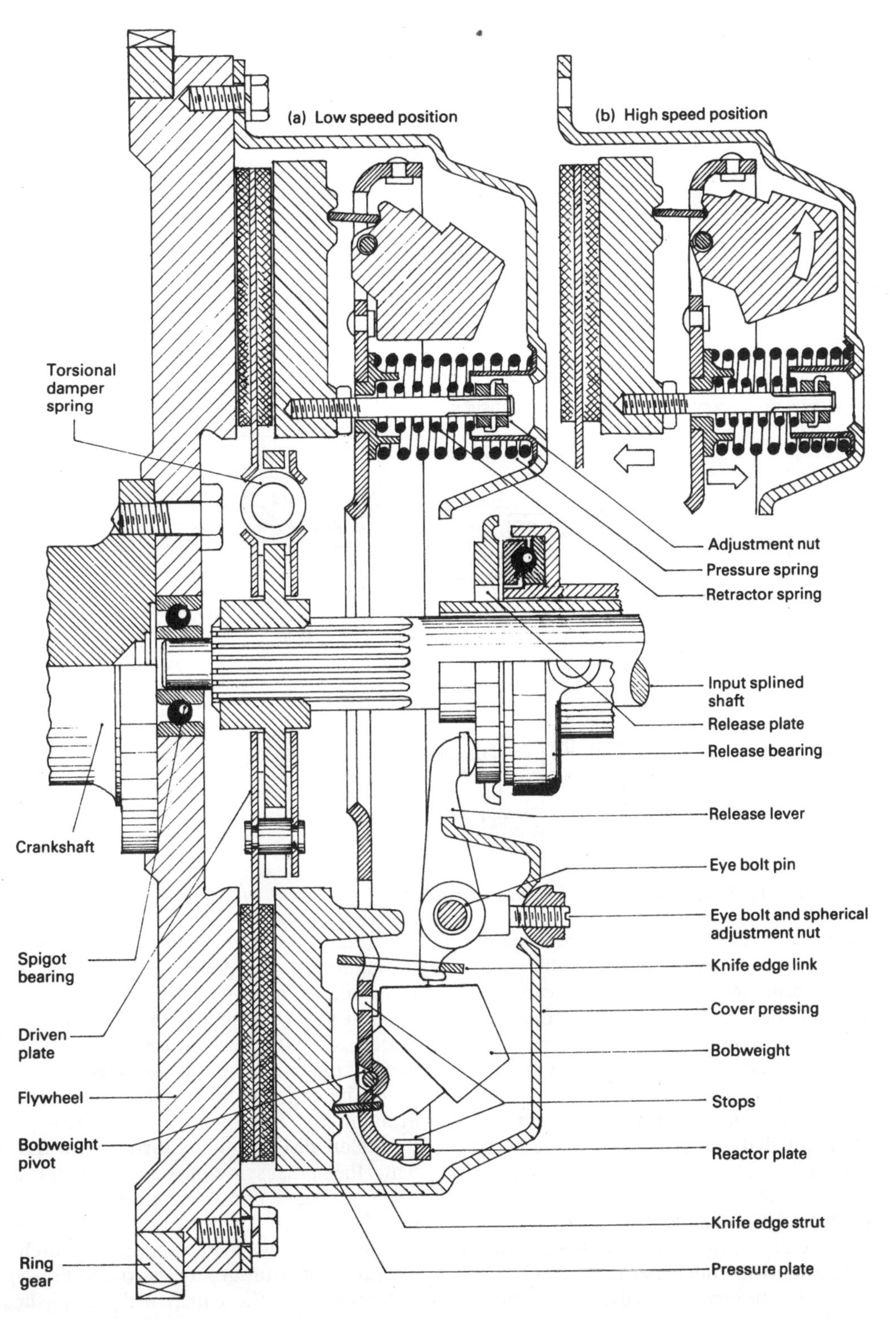

Fig. 2.19 Fully automatic centrifugal clutch

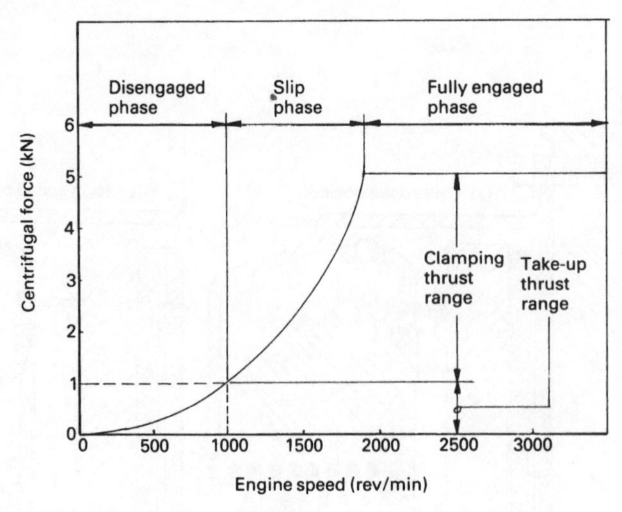

Fig. 2.20 Fully automatic centrifugal clutch characteristics

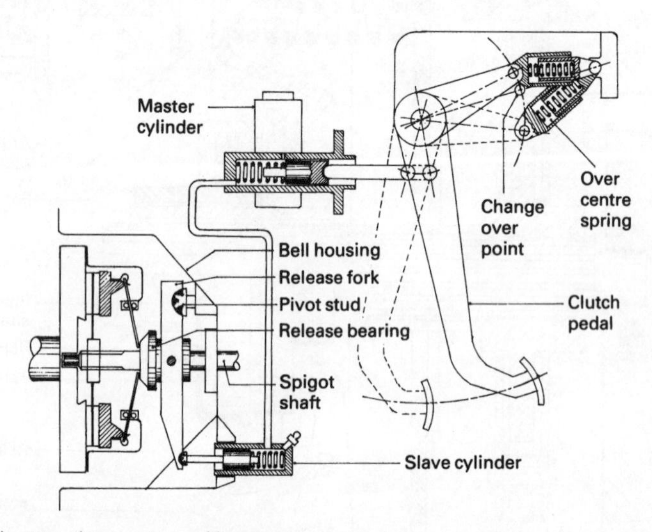

Fig. 2.21 Hydraulic clutch operating system with over-centre spring

When the pedal is released, the adjustment pawl is no longer engaged with the teeth on the pedal quadrant. The cable, however, is tensioned by the spring which is located between the pedal and quadrant. As the pedal is depressed, the pawl engages in the nearest vee between the teeth. The particular tooth engagement position will gradually change as the components move to compensate for wear in the clutch driven plate and stretch in the cable.

2.14.3 Clutch air/hydraulic servo (Fig. 2.23)

In certain applications, to reduce the driver's foot effort in operating the clutch pedal, a clutch air/hydraulic servo may be incorporated into the actuating linkage. This unit provides power assistance whenever the driver depresses the clutch pedal or maintains the pedal in a partially depressed position, as may be necessary under pull-away conditions. Movement of the clutch pedal is immediately relayed by way of the servo to the clutch in proportion to the input pedal travel.

As the clutch's driven plate wears, clutch actuating linkage movement is automatically taken up by the air piston moving further into the cylinder. Thus the actual servo movement when the clutch is being engaged and disengaged remains approximately constant. In the event of any interruption of the air supply to the servo the clutch will still operate, but without any servo assistance.

Immediately the clutch pedal is pushed down, the fluid from the master cylinder is displaced into

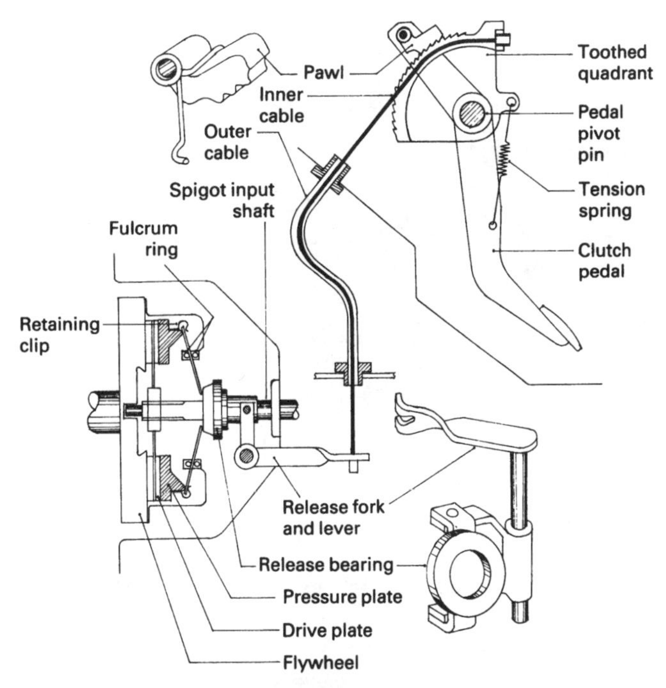

Fig. 2.22 Clutch cable linkage with automatic adjuster

the servo hydraulic cylinder. The pressure created will act on both the hydraulic piston and the reaction plunger. Subsequently, both the hydraulic piston and the reaction plunger move to the right and allow the exhaust valve to close and the inlet valve to open. Compressed air will now pass through the inlet valve port and the passage connecting the reaction plunger chamber to the compressed air piston cylinder. It thereby applies pressure against the air piston. The combination of both hydraulic and air pressure on the hydraulic and air piston assembly causes it to move over, this movement being transferred to the clutch release bearing which moves the clutch operating mechanism to the disengaged position (Fig. 2.23(d)).

When the clutch pedal is held partially depressed, the air acting on the right hand side of the reaction plunger moves it slightly to the left which now closes the inlet valve. In this situation, further air is prevented from entering the air cylinder. Therefore, since no air can move in or out of the servo air cylinder and both valves are in the lapped position (both seated), the push rod will not move unless the clutch pedal is again moved (Fig. 2.23(c)).

When the clutch pedal is released fluid returns from the servo to the master cylinder. This permits the reaction plunger to move completely to the left and so opens the exhaust valve. Compressed air in the air cylinder will now transfer to the reaction plunger chamber. It then passes through the exhaust valve and port where it escapes to the atmosphere. The released compressed air from the cylinder allows the clutch linkage return spring to move the air and hydraulic piston assembly back to its original position in its cylinder and at the same time this movement will be relayed to the clutch release bearing, whereby the clutch operating mechanism moves to the engaged drive position (Fig. 2.23(a)).

2.15 Composite flywheel and integral single plate diaphragm clutch (Fig. 2.24)

This is a compact diaphragm clutch unit built as an integral part of the two piece flywheel. It is designed for transaxle transmission application where space is at a premium and maximum torque transmitting capacity is essential.

The flywheel and clutch drive pressing acts as a support for the annular flywheel mass and functions as the clutch pressure plate drive member. The advantage of having the flywheel as a two piece assembly is that its mass can be concentrated more effectively at its outer periphery so that its overall mass can be reduced for the same cyclic torque and speed fluction which it regulates.

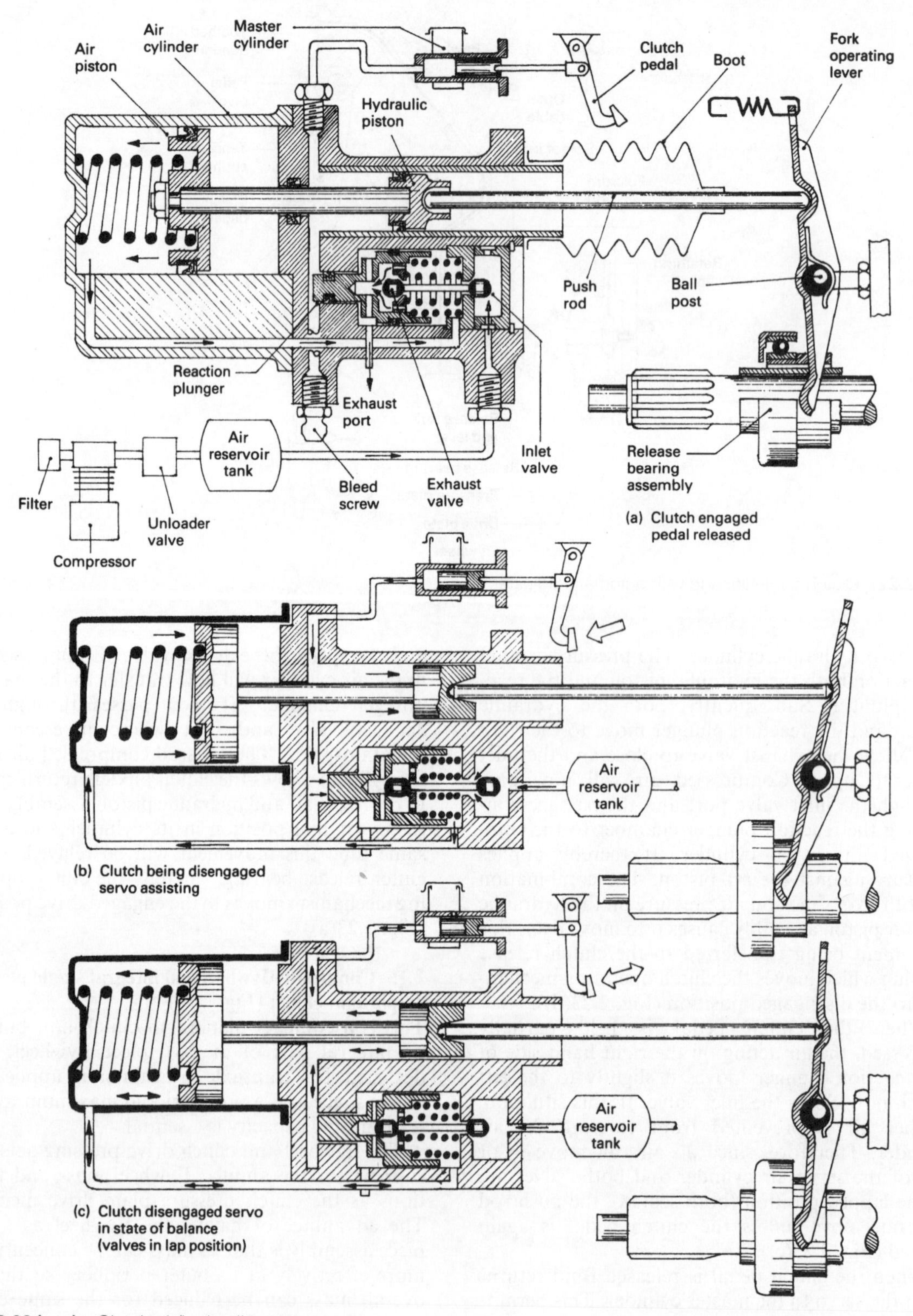

Fig. 2.23 (a–c) Clutch air/hydraulic servo

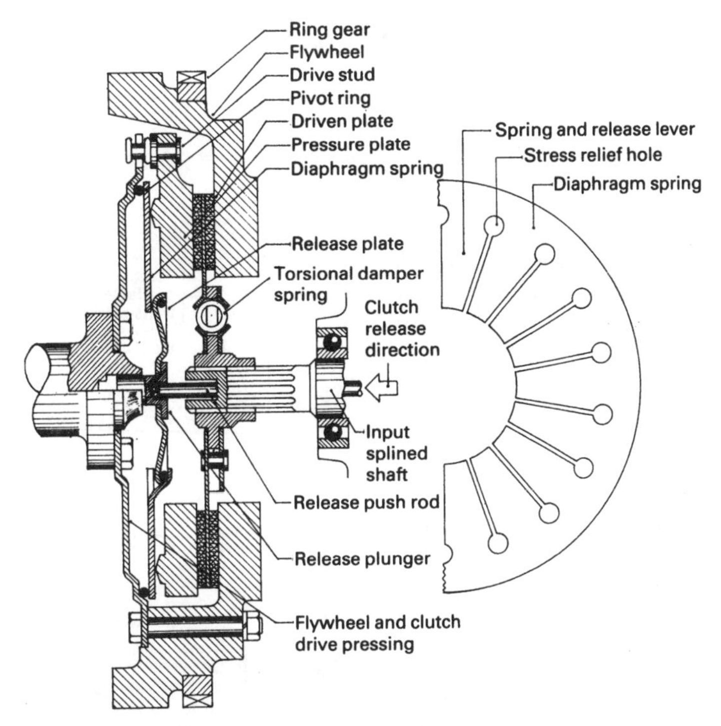

Fig. 2.24 Integral single plate clutch and composite flywheel

The diaphragm spring takes the shape of a dished annular disc. The inner portion of the disc is radially slotted, the outer ends being enlarged with a circular hole to prevent stress concentration when the spring is deflected during disengagement. These radial slots divide the disc into many inwardly pointing fingers which have two functions, firstly to provide the pressure plate with an evenly distributed multileaf spring type thrust, and secondly to act as release levers to separate the driven plate from the sandwiching flywheel and pressure plate friction faces.

To actuate the clutch release, the diaphragm spring is made to pivot between a pivot spring positioned inside the flywheel/clutch drive pressing near its outer periphery and a raised circumferential rim formed on the back of the pressure plate. The engagement and release action of the clutch is similar to the pull type diaphragm clutch where the diaphragm is distorted into a dished disc when assembled and therefore applies on axial thrust between the pressure plate and its adjacent flywheel/clutch drive pressing. With this spring leverage arrangement, a larger pressure plate and diaphragm spring can be utilised for a given overall diameter of clutch assembly. This design therefore has the benefits of lower pedal effort, higher transmitting torque capacity, a highly progressive engagement take-up and increased clutch life compared to the conventional push type diaphragm clutch.

The engagement and release mechanism consists of a push rod which passes through the hollow gearbox input shaft and is made to enter and contact the blind end of a recess formed in the release plunger. The plunger is a sliding fit in the normal spigot bearing hole made in the crankshaft end flange. It therefore guides the push rod and transfers its thrust to the diaphragm spring fingers via the release plate.

3 Manual gearboxes and overdrives

3.1 The necessity for a gearbox

Power from a petrol or diesel reciprocating engine transfers its power in the form of torque and angular speed to the propelling wheels of the vehicle to produce motion. The object of the gearbox is to enable the engine's turning effect and its rotational speed output to be adjusted by choosing a range of under- and overdrive gear ratios so that the vehicle responds to the driver's requirements within the limits of the various road conditions. An insight of the forces opposing vehicle motion and engine performance characteristics which provide the background to the need for a wide range of gearbox designs used for different vehicle applications will now be considered.

3.1.1 Resistance to vehicle motion

To keep a vehicle moving, the engine has to develop sufficient power to overcome the opposing road resistance power, and to pull away from a standstill or to accelerate a reserve of power in addition to that absorbed by the road resistance must be available when required.

Road resistance is expressed as *tractive resistance* (kN). The propelling thrust at the tyre to road interface needed to overcome this resistance is known as *tractive effect* (kN) (Fig. 3.1). For matching engine power output capacity to the opposing road resistance it is sometimes more convenient to express the opposing resistance to motion in terms of *road resistance power*.

The road resistance opposing the motion of the vehicle is made up of three components as follows:

1 Rolling resistance
2 Air resistance
3 Gradient resistance

Rolling resistance (Fig. 3.1) Power has to be expended to overcome the restraining forces caused by the deformation of tyres and road surfaces and the interaction of frictional scrub when tractive effect is applied. Secondary causes of rolling resistance are wheel bearing, oil seal friction and the churning of the oil in the transmission system. It has been found that the flattening distortion of the tyre casing at the road surface interface consumes more energy as the wheel speed increases and therefore the rolling resistance will also rise slightly as shown in Fig. 3.1. Factors which influence the magnitude of the rolling resistance are the laden weight of the vehicle, type of road surface, and the design, construction and materials used in the manufacture of the tyre.

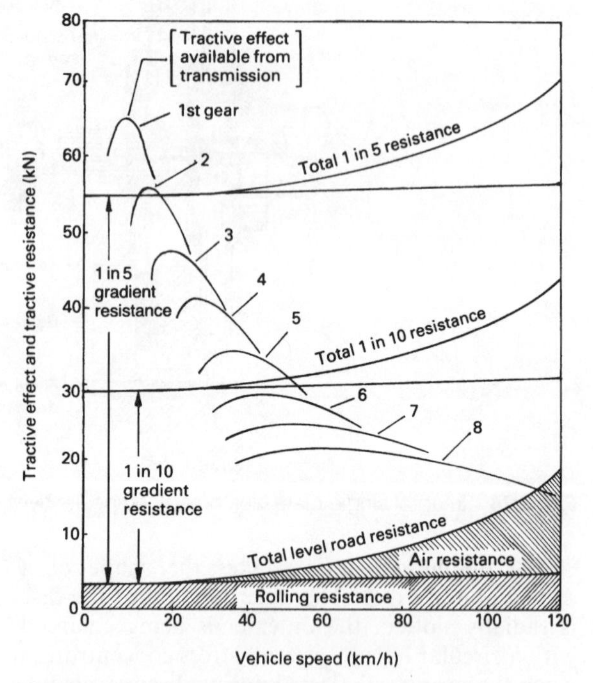

Fig. 3.1 Vehicle tractive resistance and effort performance chart

Air resistance (Fig. 3.1) Power is needed to counteract the tractive resistance created by the vehicle moving through the air. This is caused by air being pushed aside and the formation of turbulence over the contour of the vehicle's body. It has been found that the air rèsistance opposing force and air resistance power increase with the square and cube of the vehicle's speed respectively. Thus at very low vehicle speeds air resistance is insignificant, but it becomes predominant in the upper speed range. Influencing factors which determine the amount of air resistance are frontal area of vehicle, vehicle speed, shape and streamlining of body and the wind speed and direction.

Gradient resistance (Fig. 3.1) Power is required to propel a vehicle and its load not only along a level road but also up any gradient likely to be encountered. Therefore, a reserve of power must be available when climbing to overcome the potential energy produced by the weight of the vehicle as it is progressively lifted. The gradient resistance opposing motion, and therefore the tractive effect or power needed to drive the vehicle forward, is directly proportional to the laden weight of the vehicle and the magnitude of gradient. Thus driving up a slope of 1 in 5 would require twice the reserve of power to that needed to propel the same vehicle up a gradient of 1 in 10 at the same speed (Fig. 3.1).

3.1.2 Power to weight ratio

When choosing the lowest and highest gearbox gear ratios, the most important factor to consider is not just the available engine power but also the weight of the vehicle and any load it is expected to propel. Consequently, the power developed per unit weight of laden vehicle has to be known. This is usually expressed as the *power to weight ratio*.

i.e. $$\text{Power to weight ratio} = \frac{\text{Brake power developed}}{\text{Laden weight of vehicle}}$$

There is a vast difference between the power to weight ratio for cars and commercial vehicles which is shown in the following examples.

Determine the power to weight ratio for the following modes of transport:

a) A car fully laden with passengers and luggage weighs 1.2 tonne and the maximum power produced by the engine amounts to 120 kW.
b) A fully laden articulated truck weighs 38 tonne and a 290 kW engine is used to propel this load.

a) $\text{Power to weight ratio} = \dfrac{120}{1.2} = 100\,\text{kW/tonne}$

b) $\text{Power to weight ratio} = \dfrac{290}{38} = 7.6\,\text{kW/tonne}.$

3.1.3 Ratio span

Another major consideration when selecting gear ratios is deciding upon the steepest gradient the vehicle is expected to climb (this may normally be taken as 20%, that is 1 in 5) and the maximum level road speed the vehicle is expected to reach in top gear with a small surplus of about 0.2% gradeability.

The two extreme operating conditions just described set the highest and lowest gear ratios. To fix these conditions, the ratio of road speed in highest gear to road speed in lowest gear at a given engine speed should be known. This quantity is referred to as the *ratio span*.

i.e. $$\text{Ratio span} = \frac{\text{Road speed in highest gear}}{\text{Road speed in lowest gear}}$$

(both road speeds being achieved at similar engine speed).

Car and light van gearboxes have ratio spans of about 3.5:1 if top gear is direct, but with overdrive this may be increased to about 4.5:1. Large commercial vehicles which have a low power to weight ratio, and therefore have very little surplus power when fully laden, require ratio spans of between 7.5 and 10:1, or even larger for special applications.

An example of the significance of ratio span is shown as follows:

Calculate the ratio span for both a car and heavy commercial vehicle from the data provided.

Type of vehicle	Gear	Ratio	km/h/1000 rev/min
Car	Top	0.7	39
	First	2.9	9.75
Commercial vehicle (CV)	Top	1.0	48
	First	6.35	6

$\text{Car ratio span} = \dfrac{39}{9.75} = 4.0{:}1$

$\text{Commercial vehicle ratio span} = \dfrac{48}{6} = 8.0{:}1$

3.1.4 Engine torque rise and speed operating range (Fig. 3.2)

Commercial vehicle engines used to pull large loads are normally designed to have a positive torque rise curve, that is from maximum speed to peak torque with reducing engine speed the available torque increases (Fig. 3.2). The amount of engine torque rise is normally expressed as a percentage of the peak torque from maximum speed (rated power) back to peak torque.

$$\%\ \text{torque rise} = \frac{\text{Maximum speed torque}}{\text{Peak torque}} \times 100$$

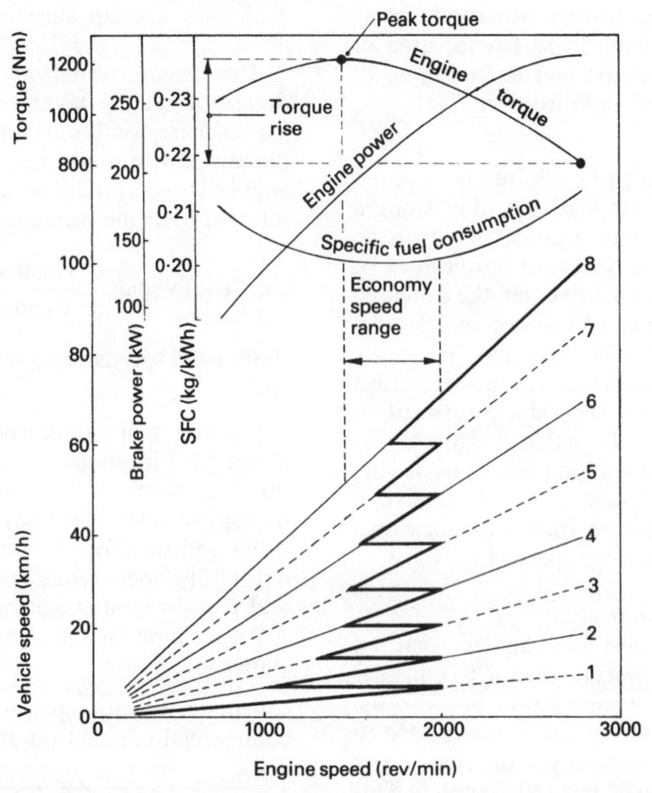

Fig. 3.2 Engine performance and gear split chart for an eight speed gearbox

The torque rise can be shaped depending upon engine design and taking into account such features as naturally aspirated, resonant induction tuned, turbocharged, turbocharged with intercooling and so forth. Torque rises can vary from as little as 5 to as high as 50%, but the most common values for torque rise range from 15 to 30%.

A large torque rise characteristic raises the engine's operating ability to overcome increased loads if the engine's speed is pulled down caused by changes in the road conditions, such as climbing steeper gradients, and so tends to restore the original running conditions. If the torque rise is small it cannot help as a buffer to supplement the high torque demands and the engine speed will rapidly fade. Frequent gear changes therefore become necessary compared to engines operating with high torque rise characteristics. Once the engine speed falls below peak torque, the torque rise becomes negative and the pulling ability of the engine drops off very quickly.

Vehicle driving technique should be such that engines are continuously driven between the speed range of peak torque and governed speed. The driver can either choose to operate the engine's speed in a range varying just below the maximum rated power to achieve maximum performance and journey speed or, to improve fuel economy, wear and noise, within a speed range of between 200 to 400 rev/min on the positive torque rise side of the engine torque curve that is in a narrow speed band just beyond peak torque. Fig. 3.2 shows that the economy speed range operates with the specific fuel consumption at its minimum and that the engine speed band is in the most effective pulling zone.

3.2 Five speed and reverse synchromesh gearboxes

With even wider engine speed ranges (1000 to 6000 rev/min) higher car speeds (160 km/h and more) and high speed motorways, it has become desirable, and in some cases essential, to increase the number of traditional four speed ratios to five, where the fifth gear, and sometimes also the fourth gear, is an overdrive ratio. The advantages of increasing the number of ratio steps are several; not only does the extra gear provide better acceleration response, but it enables the maximum engine rotational speed to be reduced whilst in top gear cruising, fuel

Table 3.1 Typical four and five speed gearbox gear ratios

Five speed box		Four speed box	
Gear	Ratio	Gear	Ratio
top	0.8	top	1.0
4	1.0	3	1.3
3	1.4	2	2.1
2	2.0	1	3.4
1	3.5	R	3.5
R	3.5		

consumption is improved and engine noise and wear are reduced. Typical gear ratios for both four and five speed gearboxes are as shown in Table 3.1.

The construction and operation of four speed gearboxes was dealt with in *Vehicle and Engine Technology*. The next section deals with five speed synchromesh gearboxes utilized for longitudinal and transverse mounted engines.

3.2.1 Five speed and reverse double stage synchromesh gearbox (Fig. 3.3)

With this arrangement of a five speed double stage gearbox, the power input to the first motion shaft passes to the layshaft and gear cluster via the first stage pair of meshing gears. Rotary motion is therefore conveyed to all the second stage layshaft and mainshaft gears (Fig. 3.3). Because each pair of second stage gears has a different size combination, a whole range of gear ratios are provided. Each mainshaft gear (whilst in neutral) revolves on the mainshaft but at some relative speed to it. Therefore, to obtain output powerflow, the selected mainshaft gear has to be locked to the mainshaft. This then completes the flow path from the first motion shaft, first stage gears, second stage gears and finally to the mainshaft.

In this example the fifth gear is an overdrive gear so that to speed up the mainshaft output relative to the input to the first motion shaft, a large layshaft fifth gear wheel is chosen to mesh with a much smaller mainshaft gear.

For heavy duty operations, a forced feed lubrication system is provided by an internal gear crescent type oil pump driven from the rear end of the layshaft (Fig. 3.3). This pump draws oil from the base of the gearbox casing, pressurizes it and then forces it through a passage to the mainshaft. The oil is then transferred to the axial hole along the centre of the mainshaft by way of an annular passage formed between two nylon oil seals. Lubrication to the mainshaft gears is obtained by radial branch holes which feed the rubbing surfaces of both mainshaft and gears.

3.2.2 Five speed and reverse single stage synchromesh gearbox (Fig. 3.4)

This two shaft gearbox has only one gear reduction stage formed between pairs of different sized constant mesh gear wheels to provide a range of gear ratios. Since only one pair of gears mesh, compared to the two pairs necessary for the double stage gearbox, frictional losses are halved.

Power delivered to the input primary shaft can follow five different flow paths to the secondary shaft via first, second, third, fourth and fifth gear wheel pairs, but only one pair is permitted to transfer the drive from one shaft to another at any one time (Fig. 3.4).

The conventional double stage gearbox is designed with an input and output drive at either end of the box but a more convenient and compact arrangement with transaxle units where the final drive is integral to the gearbox is to have the input and output power flow provided at one end only of the gearbox.

In the neutral position, first and second output gear wheels will be driven by the corresponding gear wheels attached to the input primary shaft, but they will only be able to revolve about their own axis relative to the output secondary shaft. Third, fourth and fifth gear wheel pairs are driven by the output second shaft and are free to revolve only relative to the input primary shaft because they are not attached to this shaft but use it only as a supporting axis.

When selecting individual gear ratios, the appropriate synchronizing sliding sleeve is pushed towards and over the dog teeth forming part of the particular gear wheel required. Thus with first and second gear ratios, the power flow passes from the input primary shaft and constant mesh pairs of gears to the output secondary shaft via the first and second drive hub attached to this shaft. Gear engagement is completed by the synchronizing sleeve locking the selected output gear wheel to the output secondary shaft. Third, fourth and fifth gear ratios are selected when the third and fourth or fifth gear drive hub, fixed to the input primary shaft, is locked to the respective gear wheel dog clutch by sliding the synchronizing sleeve in to mesh with it. The power flow path is now transferred from the input primary shaft drive hub and selected pair of constant mesh gears to the output secondary shaft.

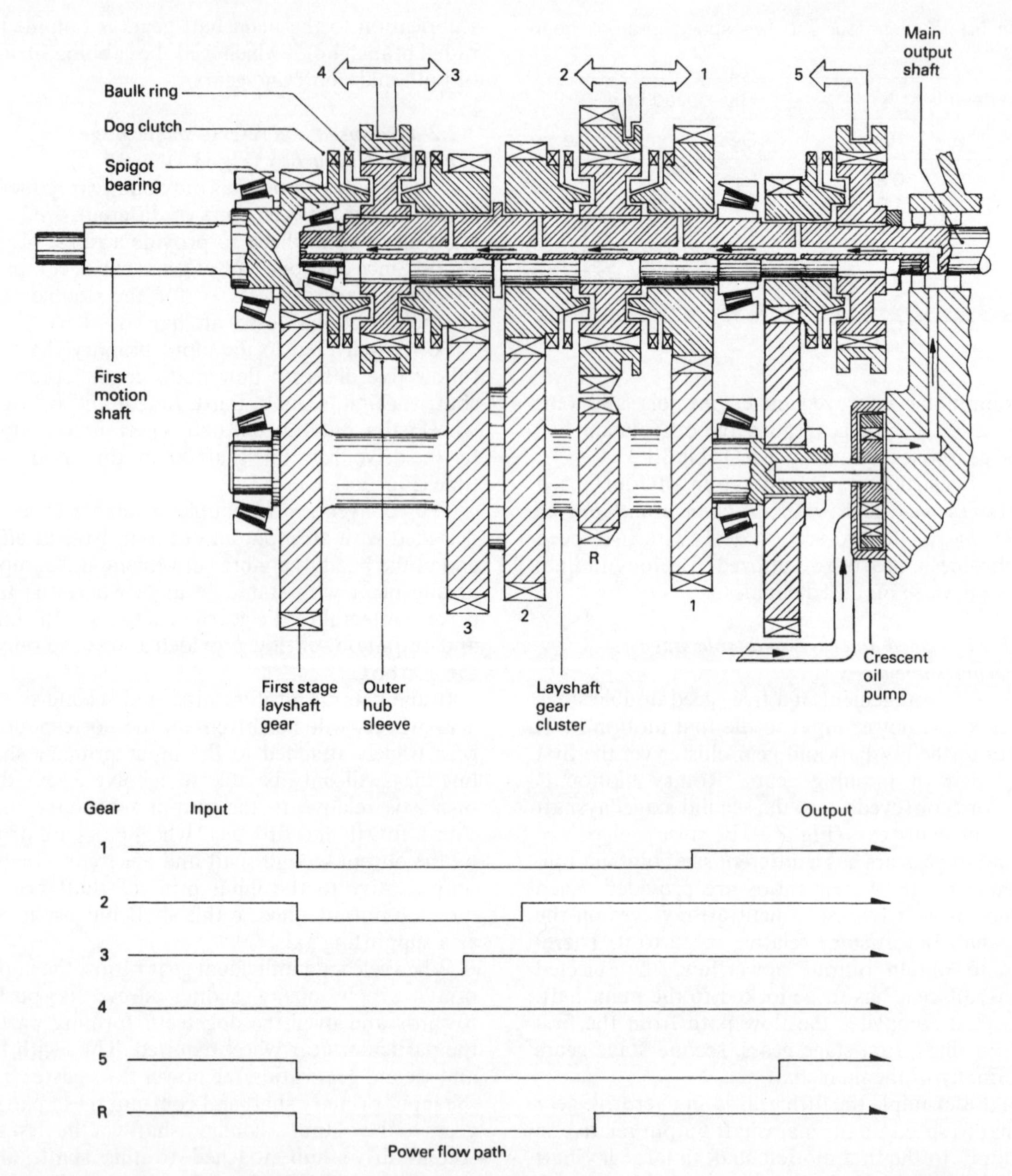

Fig. 3.3 Five speed and reverse double stage synchromesh gearbox

Transference of power from the gearbox output secondary shaft to the differential left and right hand drive shafts is achieved via the final drive pinion and gear wheel which also provide a permanent gear reduction (Fig. 3.4). Power then flows from the differential cage which supports the final drive gear wheel to the cross-pin and planet gears where it then divides between the two side sun gears and accordingly power passes to both stub drive shafts.

3.3 Gear synchronization and engagement

The gearbox basically consists of an input shaft driven by the engine crankshaft by way of the clutch and an output shaft coupled indirectly either

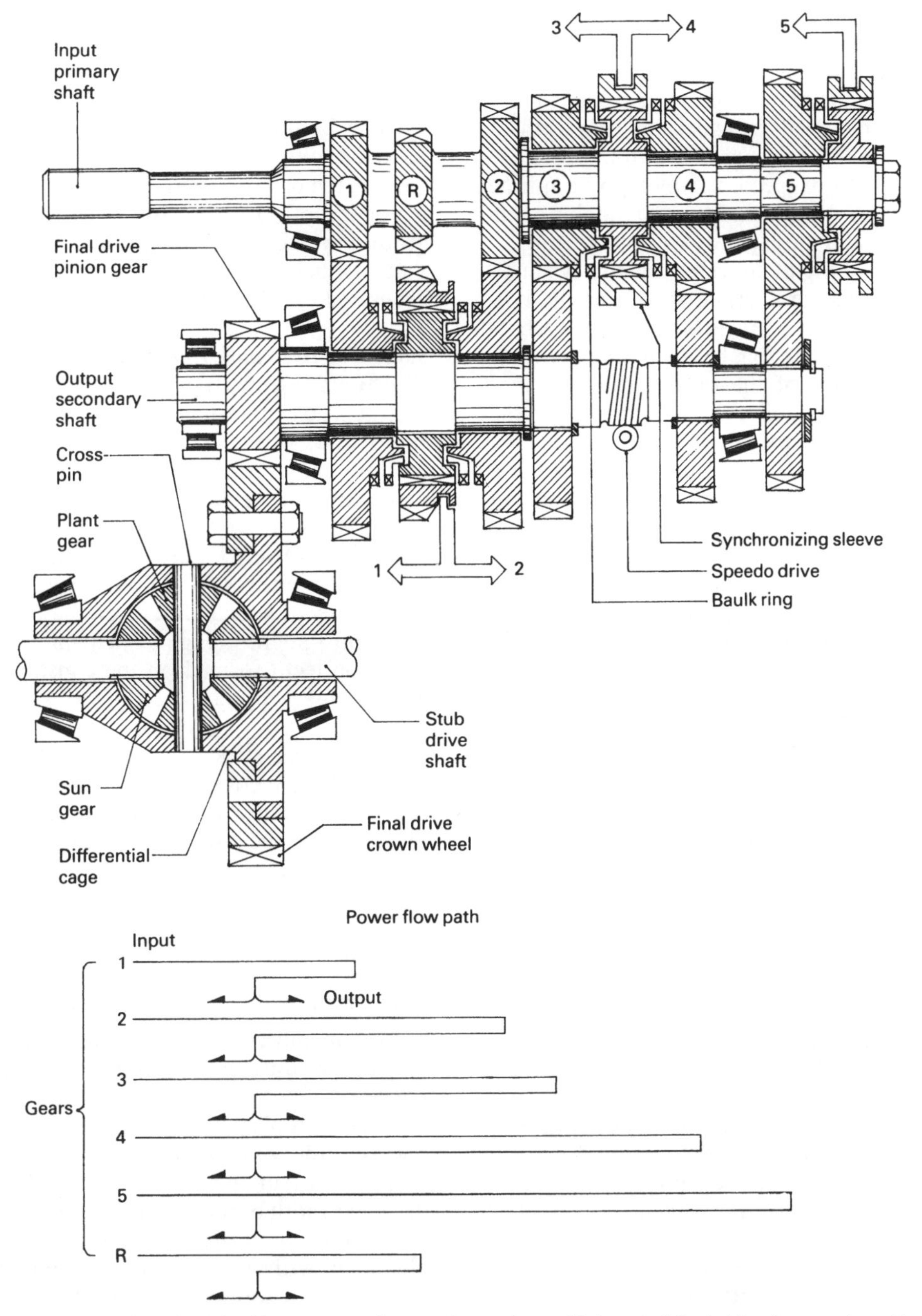

Fig. 3.4 Five speed and reverse single stage synchromesh gearbox with integral final drive (transaxle unit)

through the propellor shaft or intermediate gears to the final drive. Between these two shafts are pairs of gear wheels of different size meshed together.

If the gearbox is in neutral, only one of these pairs of gears is actually attached rigidly to one of these shafts while the other is free to revolve on the second shaft at some speed determined by the existing speeds of the input and output drive shafts.

To engage any gear ratio the input shaft has to be disengaged from the engine crankshaft via the clutch to release the input shaft drive. It is then only the angular momentum of the input shaft, clutch drive plate and gear wheels which keeps them revolving. The technique of good gear changing is to be able to judge the speeds at which the dog teeth of both the gear wheel selected and output shaft are rotating at a uniform speed, at which point in time the dog clutch sleeve is pushed over so that both sets of teeth engage and mesh gently without grating.

Because it is difficult to know exactly when to make the gear change a device known as the *synchromesh* is utilized. Its function is to apply a friction clutch braking action between the engaging gear and drive hub of the output shaft so that their speeds will be unified before permitting the dog teeth of both members to contact.

Synchromesh devices use a multiplate clutch or a conical clutch to equalise the input and output rotating members of the gearbox when the process of gear changing is taking place. Except for special applications, such as in some splitter and range change auxiliary gearboxes, the conical clutch method of synchronization is generally employed.

With the conical clutch method of producing silent gear change, the male and female cone members are brought together to produce a synchronizing frictional torque of sufficient magnitude so that one or both of the input and output members' rotational speed or speeds adjust automatically until they revolve as one. Once this speed uniformity has been achieved, the end thrust applied to the dog clutch sleeve is permitted to nudge the chamfered dog teeth of both members into alignment, thereby enabling the two sets of teeth to slide quietly into engagement.

3.3.1 Non-positive constant load synchromesh unit (Fig. 3.5(a, b and c))

When the gear stick is in the neutral position the spring loaded balls trapped between the inner and outer hub are seated in the circumferential groove formed across the middle of the internal dog teeth (Fig. 3.5(a)). As the driver begins to shift the gear stick into say top gear (towards the left), the outer and inner synchromesh hubs move as one due to the radial spring loading of the balls along the splines formed on the main shaft until the female cone of the outer hub contacts the male cone of the first motion gear (Fig. 3.5(b)). When the pair of conical faces contact, frictional torque will be generated due to the combination of the axial thrust and the difference in relative speed of both input and output shaft members. If sufficient axial thrust is applied to the outer hub, the balls will be depressed inwards against the radial loading of the springs. Immediately the balls are pushed out of their groove, the chamfered edges of the outer hub's internal teeth will then be able to align with the corresponding teeth spacing on the first motion gear. Both sets of teeth will now be able to mesh so that the outer hub can be moved into the fully engaged position (Fig. 3.5(c)).

Note the bronze female cone insert frictional face is not smooth, but consists of a series of tramline grooves which assist in cutting away the oil film so that a much larger synchronizing torque will be generated to speed up the process.

3.3.2 Positive baulk ring synchromesh unit (Fig. 3.6(a, b and c))

The gearbox mainshaft rotates at propellor shaft speed and, with the clutch disengaged, the first motion shaft gear, layshaft cluster gears, and mainshaft gears rotate freely.

Drive torque will be transmitted when a gear wheel is positively locked to the mainshaft. This is achieved by means of the outer synchromesh hub internal teeth which slide over the inner synchromesh hub splines (Fig. 3.6(a)) until they engage with dog teeth formed on the constant mesh gear wheel being selected.

When selecting and engaging a particular gear ratio, the gear stick slides the synchromesh outer hub in the direction of the chosen gear (towards the left). Because the shift plate is held radially outwards by the two energizing ring type springs and the raised middle hump of the plate rests in the groove formed on the inside of the hub, the end of the shift plate contacts the baulking ring and pushes it towards and over the conical surface, forming part of the constant mesh gear wheel (Fig. 3.6(b)).

The frictional grip between the male and female conical members of the gear wheel and baulking ring and the difference in speed will cause the baulking ring to be dragged around relative to the inner hub and shift plate within the limits of the clearance between the shift plate width and that of the recessed slot in the baulking ring. Owing to the designed width of the shift plate slot in the baulking ring, the teeth on the baulking ring are now out of alignment with those on the outer hub by approximately half a tooth width, so that the chamfered faces of the teeth of the baulking ring and outer hub bear upon each other.

As the baulking ring is in contact with the gear cone and the outer hub, the force exerted by the driver on the gear stick presses the baulking ring female cone hard against the male cone of the gear. Frictional torque between the two surfaces will eventually cause these two members to equalize

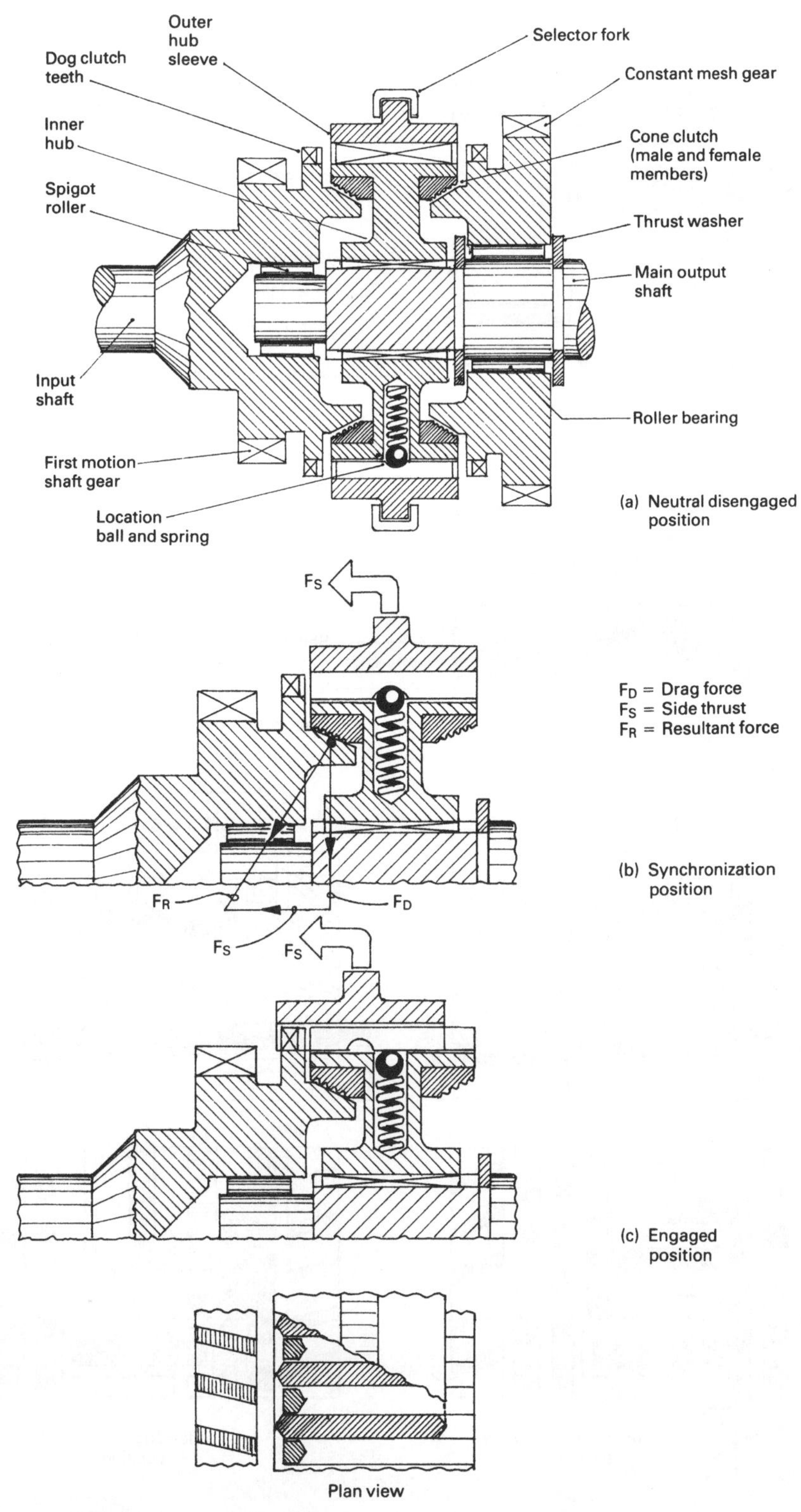

Fig. 3.5 Non-positive constant load synchromesh unit

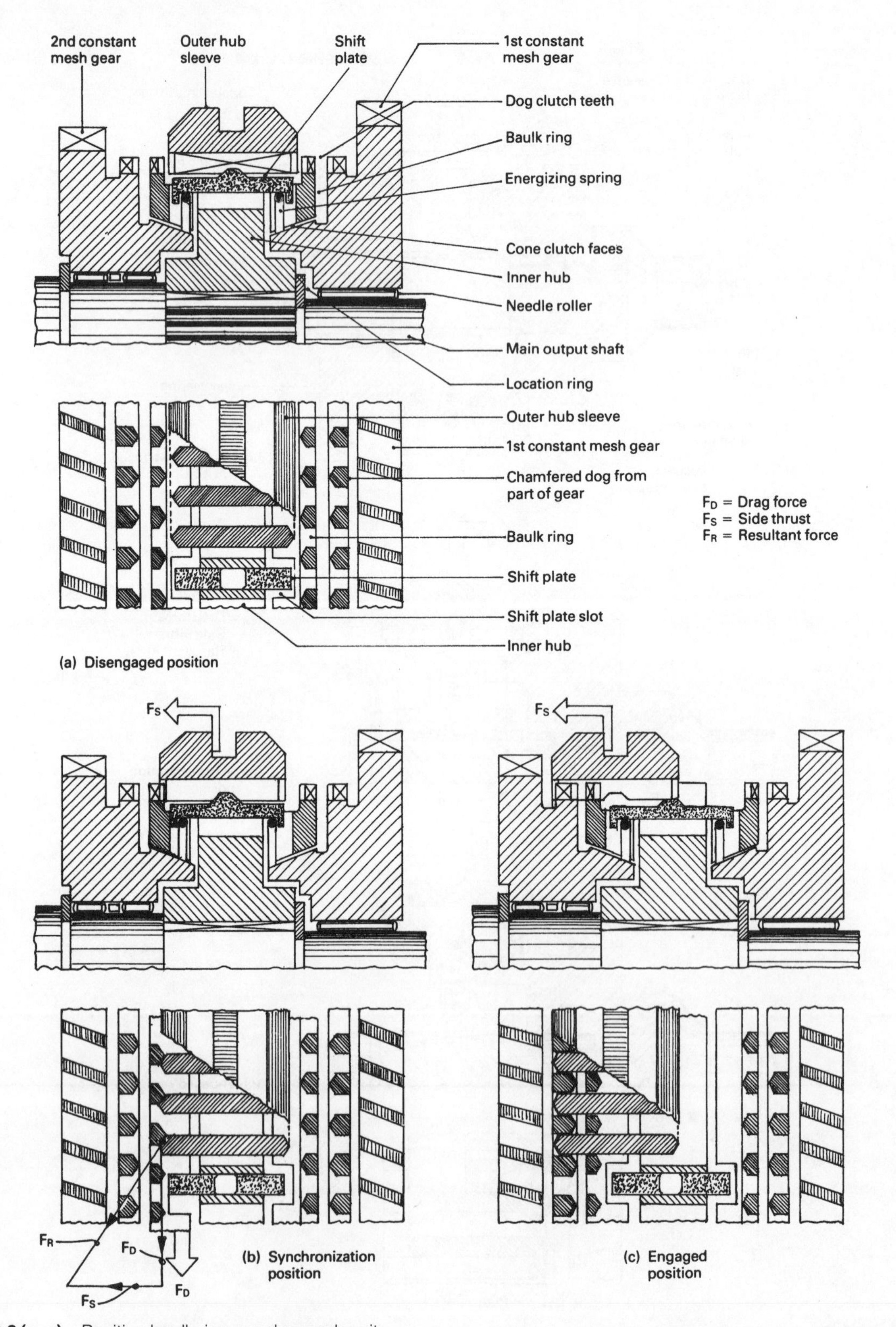

Fig. 3.6 (a–c) Positive baulk ring synchromesh unit

their speeds. Until this takes place, full engagement of the gear and outer hub dog teeth is prevented by the out of alignment position of the baulking ring teeth. When the gear wheel and main shaft have unified their speeds, the synchronizing torque will have fallen to zero so that the baulking ring is no longer dragged out of alignment. Therefore the outer hub can now overcome the baulk and follow through to make a positive engagement between hub and gear (Fig. 3.6(c)). It should be understood that the function of the shift plate and springs is to transmit just sufficient axial load to ensure a rapid bringing together of the mating cones so that the baulking ring dog teeth immediately misalign with their corresponding outer hub teeth. Once the cone faces contact, they generate their own friction torque which is sufficient to flick the baulking ring over, relative to the outer hub. Thus the chamfers of both sets of teeth contact and oppose further outer hub axial movement towards the gear dog teeth.

3.3.3 Positive baulk pin synchromesh unit
(Fig. 3.7(a, b, c and d))

Movement of the selector fork synchronizing sleeve to the left (Fig. 3.7(a and b)) forces the female (internal) cone to move into contact with the male (external) cone on the drive gear. Frictional torque will then synchronize (unify) the input and output speeds. Until speed equalization is achieved, the collars on the three thrust pins (only one shown) will be pressed hard into the enlarged position of the slots (Fig. 3.5(c)) in the synchronizing sleeve owing to the frictional drag when the speeds are dissimilar. Under these conditions, unless extreme pressure is exerted, the dog teeth cannot be crushed by forcing the collars into the narrow portion of the slots. However, when the speeds of the synchromesh hub and drive gear are equal (synchronized) the collars tend to 'float' in the enlarged portion of the slots, there is only the pressure of the spring loaded balls to be overcome. The collars will then slide easily into the narrow portion of the slots (Fig. 3.5(d)) allowing the synchronizer hub dog teeth to shift in to mesh with the dog teeth on the driving gear.

3.3.4 Split baulk pin synchromesh unit
(Fig. 3.8(a, b, c and d))

The synchronizing assembly is composed of two thick bronze synchronizing rings with tapered female conical bores, and situated between them is a hardened steel drive hub internally splined with external dog teeth at each end (Fig. 3.8(a)). Three shouldered pins, each with a groove around its centre, hold the bronze synchronizing cone rings apart. Alternating with the shouldered pins on the same pitch circle are diametrically split pins, the ends of which fit into blind bores machined in the synchronizing cone rings. The pin halves are sprung apart, so that a chamfered groove around the middle of each half pin registers with a chamfered hole in the drive hub.

If the gearbox is in the neutral position, both sets of shouldered and split pins are situated with their grooves aligned with the central drive hub (Fig. 3.8(a and b)).

When an axial load is applied to the drive hub by the gear stick, it moves over (in this case to the left) until the synchronizing ring is forced against the adjacent first motion gear cone. The friction (synchronizing) torque generated between the rubbing tapered surfaces drags the bronze synchronizing ring relative to the mainshaft and drive hub until the grooves in the shouldered pins are wedged against the chamfered edges of the drive hub (Fig. 3.8(c)) so that further axial movement is baulked.

Immediately the input and output shaft speeds are similar, that is, synchronization has been achieved, the springs in the split pins are able to expand and centralize the shouldered pins relative to the chamfered holes in the drive hub. The drive hub can now ride out of the grooves formed around the split pins, thus permitting the drive hub to shift further over until the internal and external dog teeth of both gear wheel hub mesh and fully engage (Fig. 3.8(d)).

3.3.5 Split ring synchromesh unit
(Fig. 3.9(a, b, c and d))

In the neutral position the sliding sleeve sits centrally over the drive hub (Fig. 3.9(a)). This permits the synchronizing ring expander band and thrust block to float within the constraints of the recess machine in the side of the gear facing the drive hub (Fig. 3.9(b)).

For gear engagement to take place, the sliding sleeve is moved towards the gear wheel selected (to the left) until the inside chamfer of the sliding sleeve contacts the bevelled portion of the synchronizing ring. As a result, the synchronizing ring will be slightly compressed and the friction generated between the two members then drags the synchronizing ring round in the direction of whichever member is rotating fastest, be it the gear or driven hub. At the same time, the thrust block is pulled round so that it applies a load to one end of the expander band, whilst the other end is restrained from moving by the anchor block (Fig. 3.9(c)).

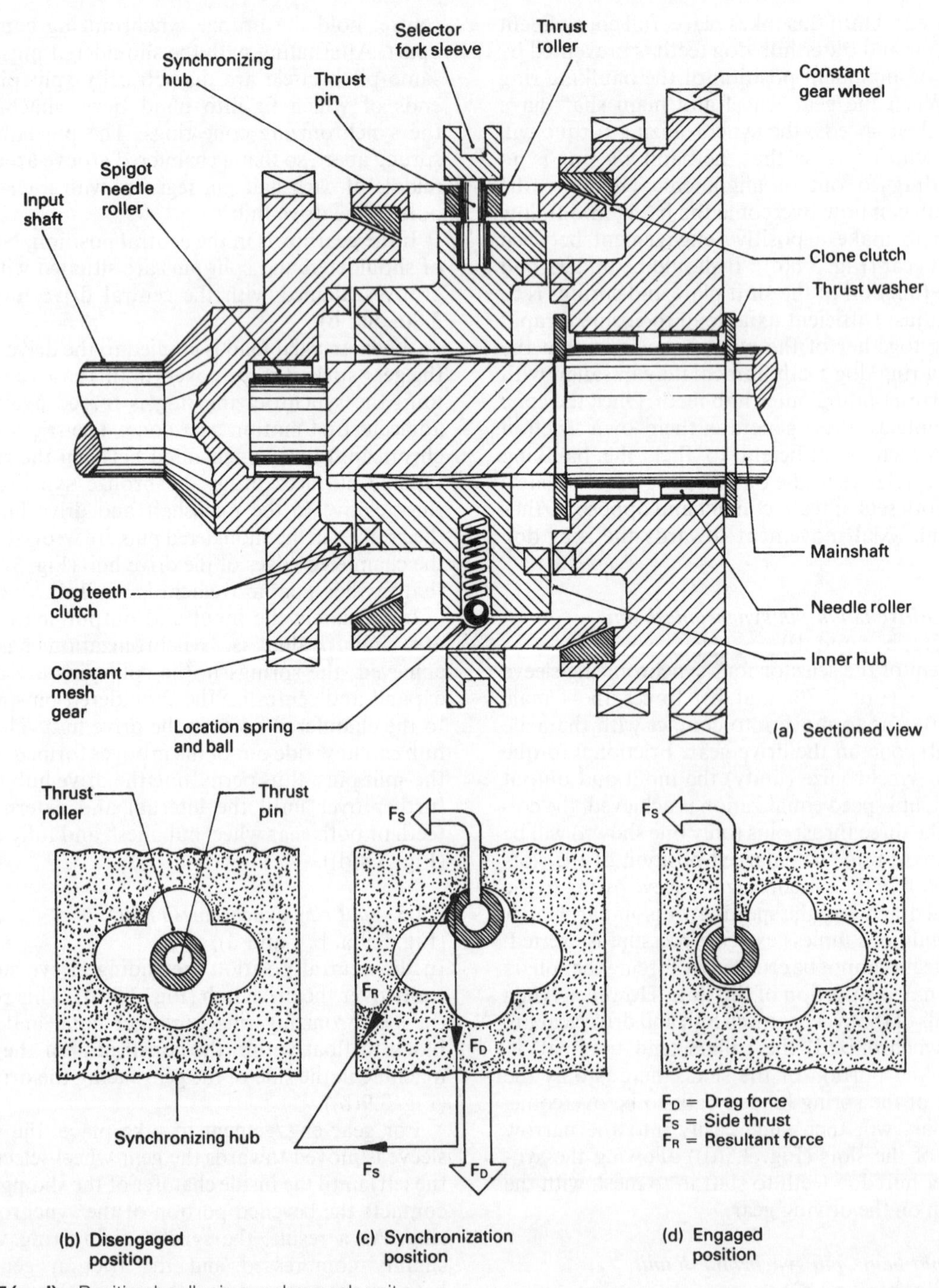

Fig. 3.7 (a–d) Positive baulk pin synchromesh unit

Whilst this is happening the expander is also pushed radially outwards. Consequently, there will be a tendency to expand the synchronizing slit ring, but this will be opposed by the chamfered mouth of the sliding sleeve. This energizing action attempting to expand the synchronizing ring prevents the sliding sleeve from completely moving over and engaging the dog teeth of the selected

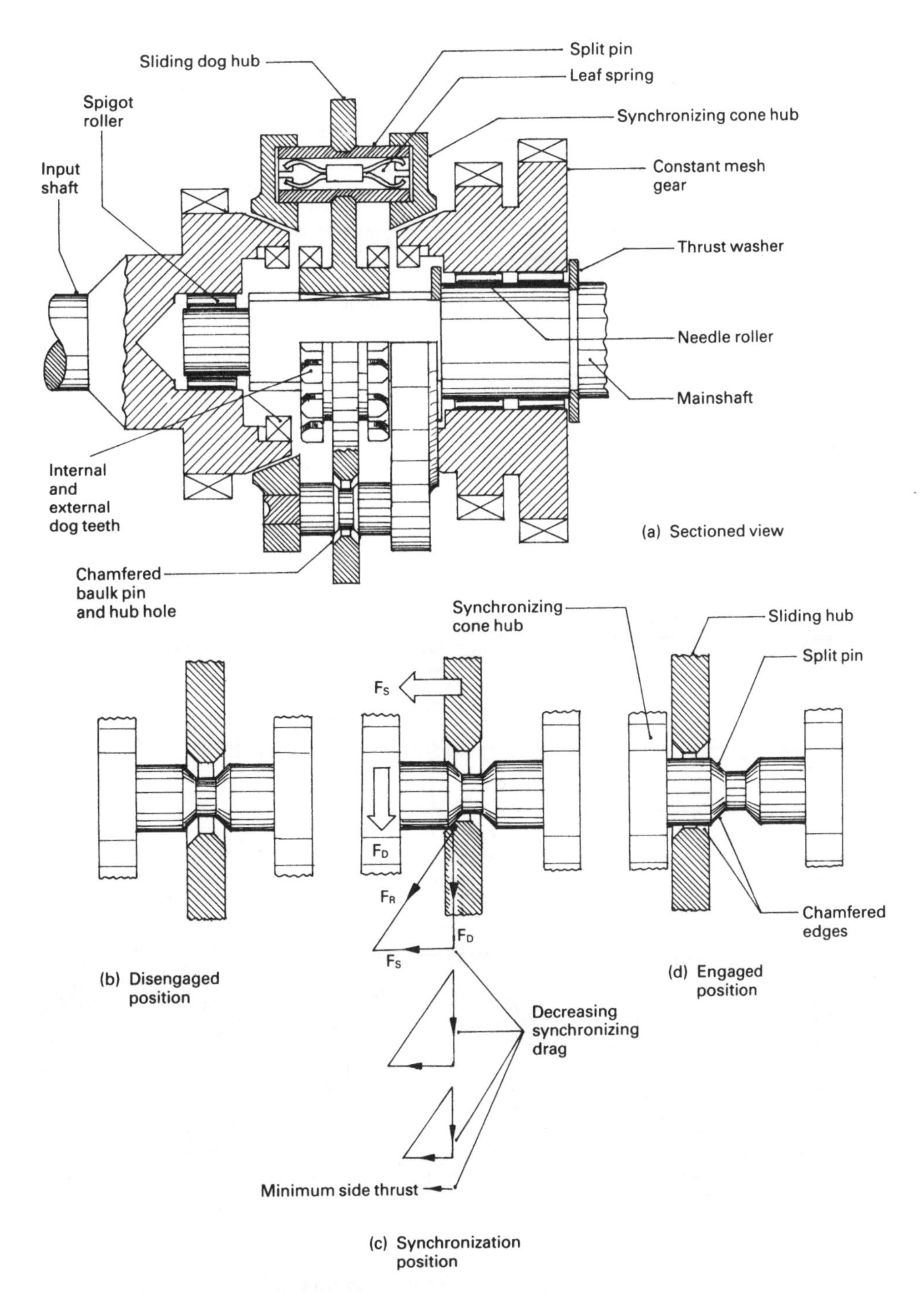

Fig. 3.8 (a–c) Split baulk pin synchromesh unit

gear wheel until both the drive hub and constant mesh gear wheel are revolving at the same speed.

When both input and output members are unified, that is, rotating as one, there cannot be any more friction torque because there is no relative speed to create the frictional drag. Therefore the expander band immediately stops exerting radial force on the inside of the synchronizing ring.

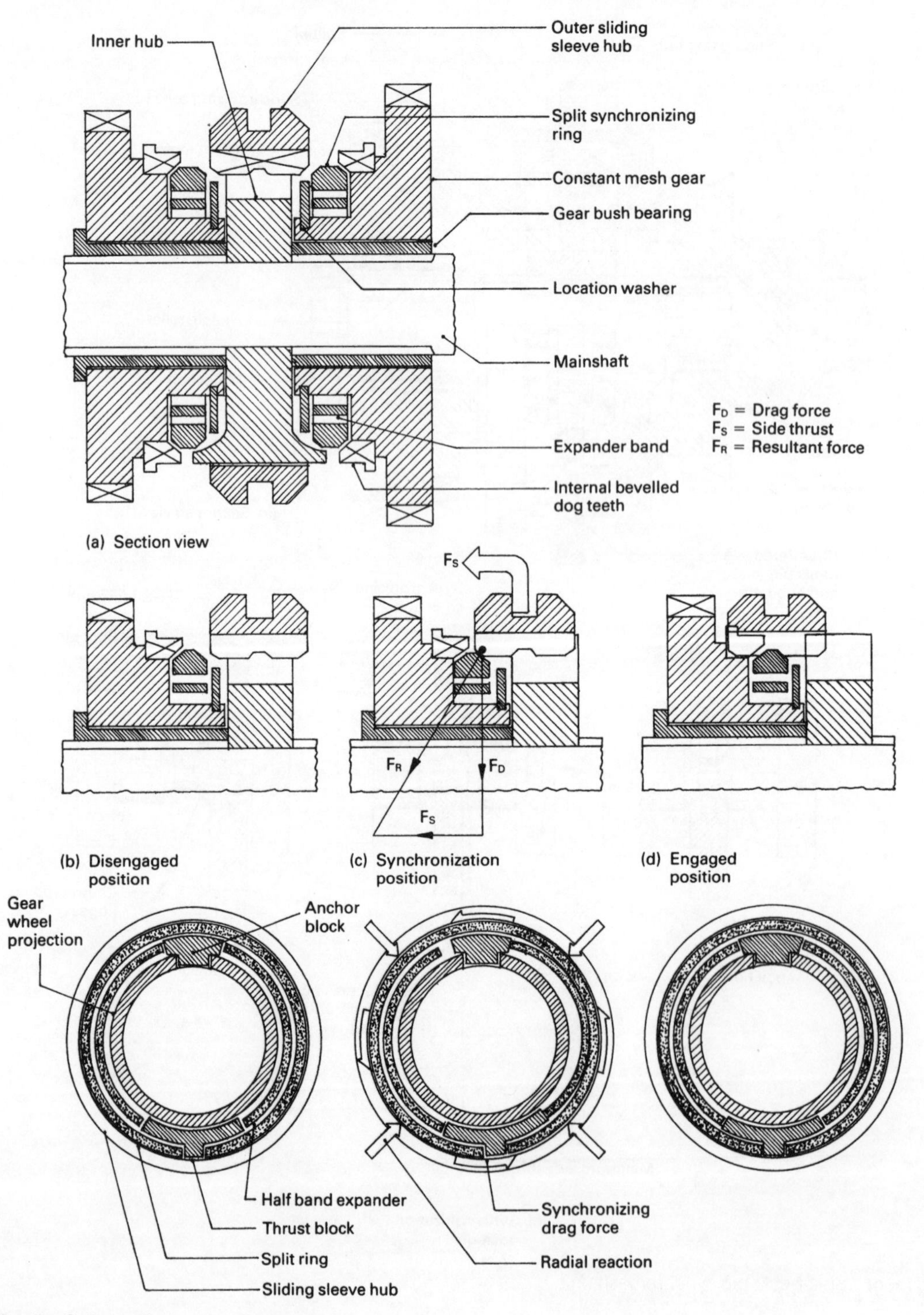

Fig. 3.9 (a–d) Split baulk ring synchromesh unit

The axial thrust applied by the gear stick to the sliding sleeve will now be sufficient to compress the split synchronizing ring and subsequently permits the sleeve to slide over the gear wheel dog teeth for full engagement (Fig. 3.9(d)).

3.4 Remote controlled gear selection and engagement mechanisms

Gear selection and engagement is achieved by two distinct movements:

1 The selection of the required gear shift gate and the positioning of the engagement gate lever.
2 The shifting of the chosen selector gate rod into the engagement gear position.

These two operations are generally performed through the media of the gear shift lever and the remote control tube/rod. Any transverse movement of the gear shift lever by the driver selects the gear shift gate and the engagement of the gate is obtained by longitudinal movement of the gear shift lever.

Movement of the gear shift lever is conveyed to the selection mechanism via the remote control tube. Initially the tube is twisted to select the gate shift gate, followed by either a push or pull movement of the tube to engage the appropriate gear.

For the gear shift control to be effective it must have some sort of flexible linkage between the gear shift lever supported on the floor of the driver's compartment and the engine and transmission integral unit which is suspended on rubber mountings. This is essential to prevent engine and transmission vibrations being transmitted back to the body and floor pan and subsequently causing discomfort to the driver and passengers.

3.4.1 Remote controlled double rod and bell cranked lever gear shift mechanism, suitable for both four and five speed transverse mounted gearbox (Talbot) (Fig. 3.10)

Twisting the remote control tube transfers movement to the first selector link rod. This motion is then redirected at right angles to the transverse gate selector/engagement shaft via the selector relay lever (bell crank) to position the required gear gate (Fig. 3.10). A forward or backward movement of the remote control tube now conveys motion via the first engagement relay lever (bell crank), engagement link rod and second relay lever to rotate the transverse gate selector/engagement shaft. Consequently, this shifts the transverse selector/engagement shaft so that it pushes the synchronizing sliding sleeve into engagement with the selected gear dog teeth.

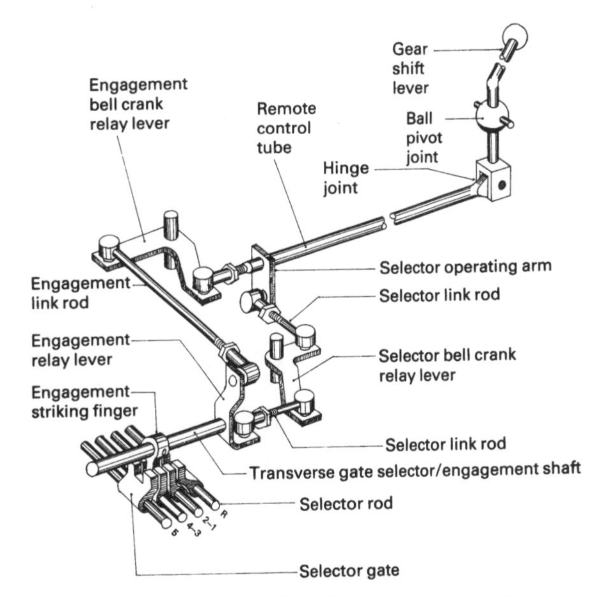

Fig. 3.10 Remote controlled double rod and bell crank lever gearshift mechanism suitable for both four and five speed transversely mounted gearbox

3.4.2 Remote controlled bell cranked lever gear shift mechanism for a four speed transverse mounted gearbox (Ford) (Fig. 3.11)

Gear selection and engagement movement is conveyed from the gear shift lever pivot action to the remote control rod universal joint and to the control shift and relay lever guide (Fig. 3.11). Rocking the gear shift lever transversely rotates the control shaft and relay guide. This tilts the selector relay lever and subsequently the selection relay lever guide and shaft until the striker finger aligns with the chosen selector gate. A further push or pull movement to the gear shift lever by the driver then transfers a forward or backward motion via the remote control rod, control shaft and relay lever guide to the engagement relay lever. Movement is then redirected at right angles to the selector relay guide and shaft. Engagement of the gear required is finally obtained by the selector/engagement shaft forcing the striking finger to shift the gate and selector fork along the single selector rod so that the synchronizing sleeve meshes with the appropriate gear wheel dog clutch.

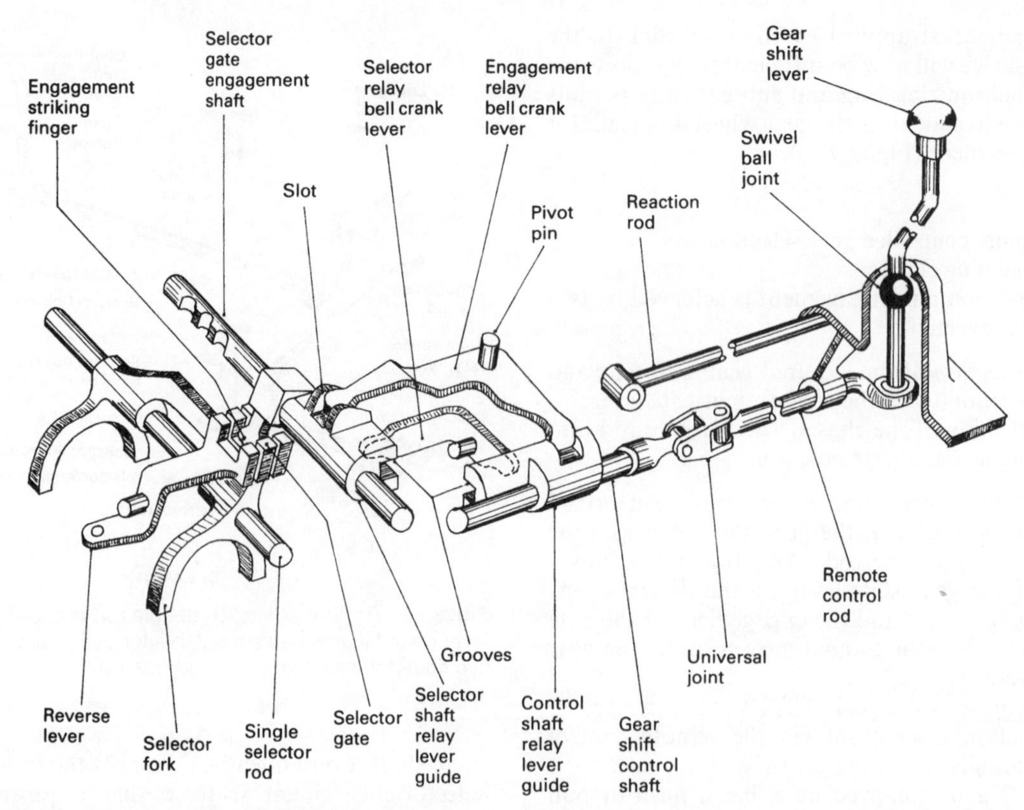

Fig. 3.11 Remote controlled bell crank level gear shift mechanism for a four speed transversely mounted gearbox

3.4.3 Remote controlled sliding ball joint gear shift mechanism suitable for both four and five speed longitudinal or transverse mounted gearbox (VW) (Fig. 3.12)

Selection and engagement of the different gear ratios is achieved with a swivel ball end pivot gear shift lever actuating through a sliding ball relay lever a single remote control rod (Fig. 3.12). The remote control rod transfers both rotary and push-pull movement to the gate selector and engagement shaft. This rod is also restrained in bushes between the gear shift lever mounting and the bulkhead. It thus permits the remote control rod to transfer both rotary (gate selection) and push-pull (select rod engagement shift) movement to the gate selector and engagement shaft. Relative movement between the suspended engine and transmission unit and the car body is compensated by the second sliding ball relay lever. As a result the gate engagement striking finger is able to select and shift into engagement the appropriate selector rod fork.

This single rod sliding ball remote control linkage can be used with either longitudinally or transversely mounted gearboxes, but with the latter an additional relay lever mechanism (not shown) is needed to convey the two distinct movements of selection and engagement through a right angle.

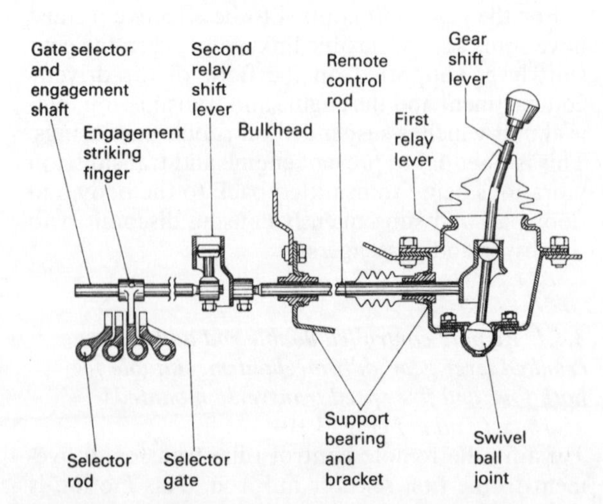

Fig. 3.12 Remote controlled sliding ball joint gear shift mechanism suitable for both four and five speed longitudinally or transversely mounted gearbox

3.4.4 Remote controlled double rod and hinged relay joint gear shift mechanism suitable for both four and five speed longitudinal mounted gearbox (VW) (Fig. 3.13)

With this layout the remote control is provided by a pair of remote control rods, one twists and selects the gear gate when the gear shift lever is given a transverse movement, while the other pushes or pulls when the gear shift lever is moved longitudinally (Fig. 3.13). Twisting movement is thus conveyed to the engagement relay lever which makes the engagement striking finger push the aligned selector gate and rod. Subsequently, the synchronizing sleeve splines mesh with the corresponding dog clutch teeth of the selected gear wheel. Relative movement between the gear shift lever swivel support and rubber mounted gearbox is absorbed by the hinged relay joint and the ball joints at either end of the remote control engagement rod.

3.4.5 Remote controlled single rod with self aligning bearing gear shift mechanism suitable for both five and six speed longitudinal mounted gearbox (Ford) (Fig. 3.14)

A simple and effective method of selecting and engaging the various gear ratios suitable for commercial vehicles where the driver cab is forward of the gearbox is shown in Fig. 3.14.

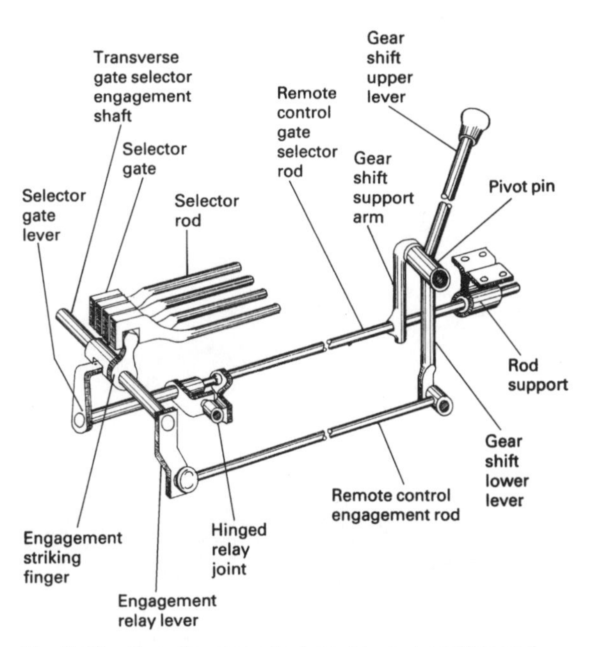

Fig. 3.13 Remote controlled double rod and hinged relay joint gear shift mechanism suitable for both four and five speed longitudinally mounted gearbox

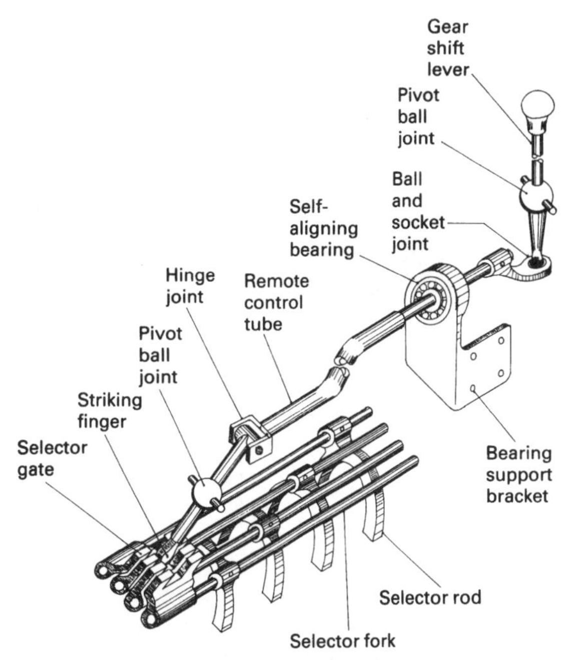

Fig. 3.14 Remote controlled single rod with self-aligning bearing gear shift mechanism suitable for both five and six speed longitudinally mounted gearbox

Movement of the gear shift lever in the usual transverse and longitudinal directions provides both rotation and a push-pull action to the remote control tube. Twisting the remote control tube transversely tilts the relay gear shift lever about its ball joint so that the striking finger at its lower end matches up with the selected gear gate. Gear engagement is then completed by the driver tilting the gear shift lever away or towards himself. This permits the remote control tube to move axially through the mounted self-aligning bearing. As a result, a similar motion will be experienced by the relay gear shift lever, which then pushes the striking finger, selector gate and selector fork into the gear engaged position. It should be observed that the self-aligning bearing allows the remote control tube to slide to and fro. At the same time it permits the inner race member to tilt if any relative movement between the gearbox and chassis takes place.

3.4.6 Remote controlled single rod with swing arm support gear shift mechanism suitable for five and six speed longitudinally mounted gearbox (ZF) (Fig. 3.15)

This arrangement which is used extensively on commercial vehicles employs a universal cross-pin joint to transfer both the gear selection and

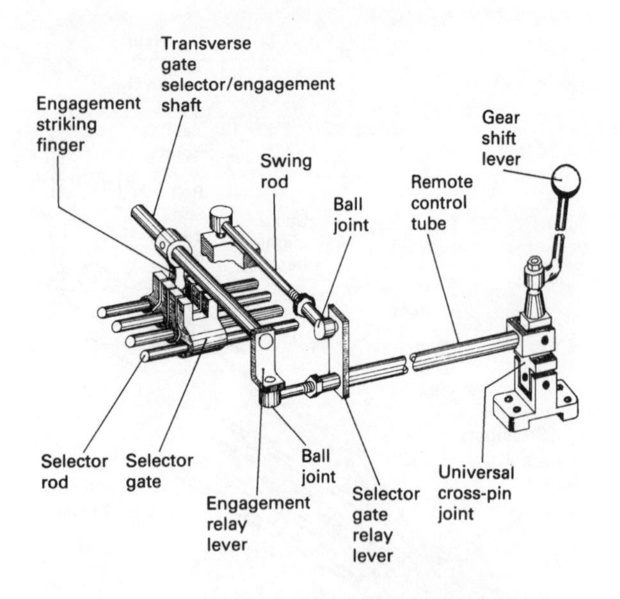

Fig. 3.15 Remote controlled single rod with swing arm support gear shift mechanism suitable for five and six speed longitudinally mounted gearbox

engagement motion to the remote control tube (Fig. 3.15). Twisting this remote control tube by giving the gear shift lever a transverse movement pivots the suspended selector gate relay lever so that the transverse gate selector/engagement shift moves across the selector gates until it aligns with the selected gate. The gear shift lever is then given a to or fro movement. This causes the transverse selector/engagement shaft to rotate, thereby forcing the striking finger to move the selector rod and fork. The synchronizing sleeve will now be able to engage the dog clutch of the appropriate gear wheel. Any misalignment between the gear shift lever support mounting and the gear shift mechanism forming part of the gearbox (caused by engine shake or rock) is thus compensated by the swing rod which provides a degree of float for the selector gate relay lever pivot point.

3.5 Splitter and range change gearboxes

Ideally the tractive effect produced by an engine and transmission system developing a constant power output from rest to its maximum road speed would vary inversely with its speed. This characteristic can be shown as a smooth declining tractive effect curve with rising road speed (Fig. 3.16).

In practice, the transmission has a fixed number of gear ratios so that the ideal smooth tractive effect curve would be interrupted to allow for loss

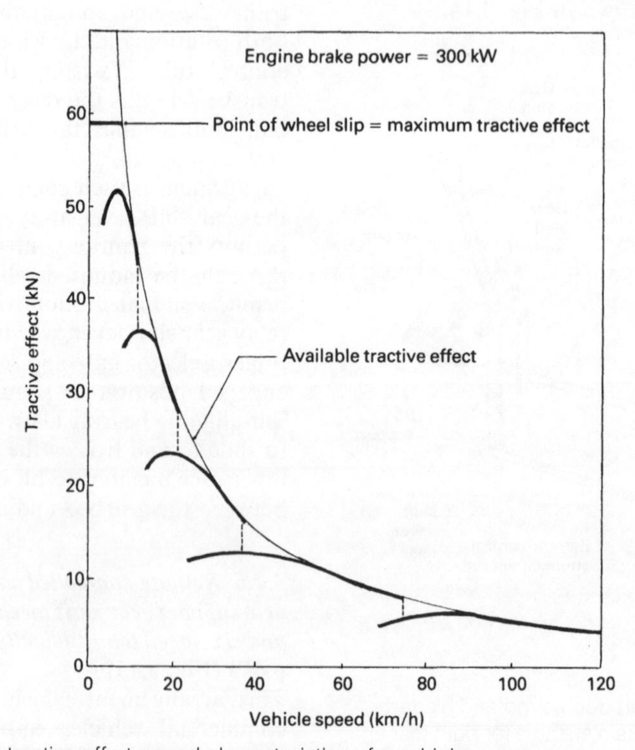

Fig. 3.16 Ideal and actual tractive effort-speed characteristics of a vehicle

of engine speed and power between each gear change (see the thick lines of Fig. 3.16).

For a vehicle such as a saloon car or light van which only weighs about one tonne and has a large power to weight ratio, a four or five speed gearbox is adequate to maintain tractive effect without too much loss in engine speed and vehicle performance between gear changes.

Unfortunately, this is not the situation for heavy goods vehicles where large loads are being hauled so that the power to weight ratio is usually very low. Under such operating conditions if the gear ratio steps are too large the engine speed will drop to such an extent during gear changes that the engine torque recovery will be very sluggish (Fig. 3.17). Therefore, to minimize engine speed fall-off whilst changing gears, smaller gear ratio steps are required, that is, more gear ratios are necessary to respond to the slightest change in vehicle load, road conditions and the driver's requirements. Figs 3.2 and 3.18 show that by doubling the number of gear ratios, the fall in engine speed between gear shifts is much smaller. To cope with moderate payloads, conventional double stage compound gearboxes with up to six forward speeds are manufactured, but these boxes tend to be large and heavy. Therefore, if more gear ratios are essential for very heavy payloads, a far better way of extending the number of gear ratios is to utilize a two speed auxiliary gearbox in series with a three, four, five or six speed conventional compound gearbox. The function of this auxiliary box is to double the number of gear ratios of the conventional gearbox. With a three, four, five or six speed gearbox, the gear ratios are increased to six, eight, ten or twelve respectively (Figs 3.2 and 3.18). For very special applications, a three speed auxiliary gearbox can be incorporated so that the gear ratios are trebled. Usually one of these auxiliary gear ratios provides a range of very low gear ratios known as *crawlers* or *deep gears*. The auxiliary gearbox may be situated either in front or to the rear of the conventional compound gearbox.

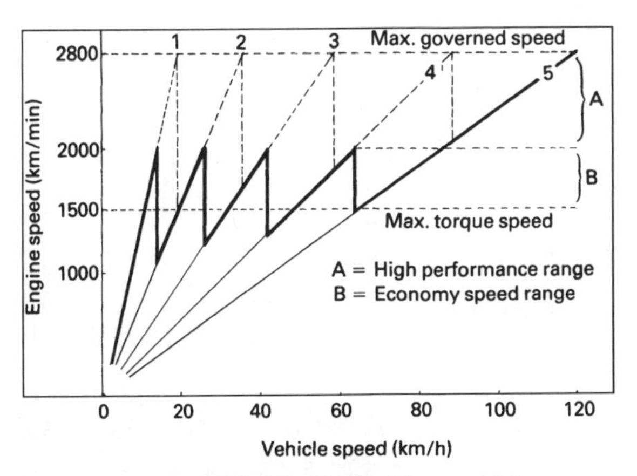

Fig. 3.17 Engine speed ratio chart for a vehicle employing a five speed gearbox

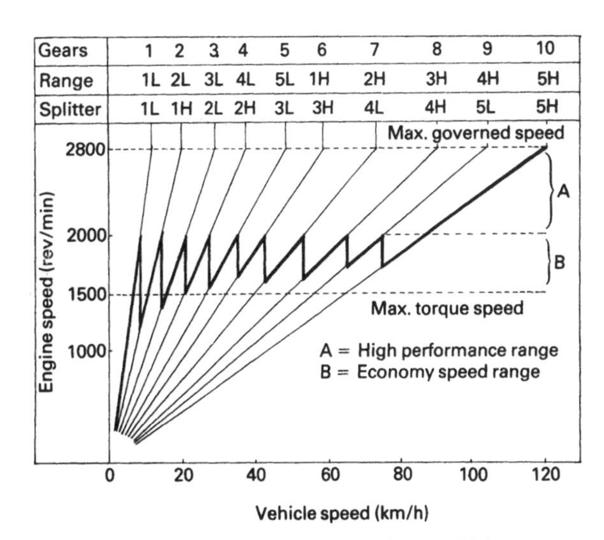

Fig. 3.18 Engine speed ratio chart for a vehicle employing either a ten speed range change or a splitter change gearbox

The combination of the auxiliary gearbox and the main gearbox can be designed to be used either as a splitter gear change or as a range gear change in the following way.

3.5.1 Splitter gear change (Figs 3.19 and 3.20)

With the splitter arrangement, the main gearbox gear ratios are spread out wide between adjacent gears whilst the two speed auxiliary gearbox has one direct gear ratio and a second gear which is either a step up or down ratio (Fig. 3.19). The auxiliary second gear ratio is chosen so that it splits the main gearbox ratio steps in half, hence the name *splitter gear change*. The splitter indirect gear ratio normally is set between 1.2 and 1.4:1. A typical ratio would be 1.25:1.

A normal upchange sequence for an eight speed gearbox (Fig. 3.20), consisting of a main gearbox with four forward gear ratios and one reverse and a two speed auxiliary splitter stage, would be as follows:

Auxiliary splitter low ratio and main gearbox first gear selected results in 'first gear low' (1L); auxiliary splitter switched to high ratio but with the main gearbox still in first gear results in 'first gear high' (1H);

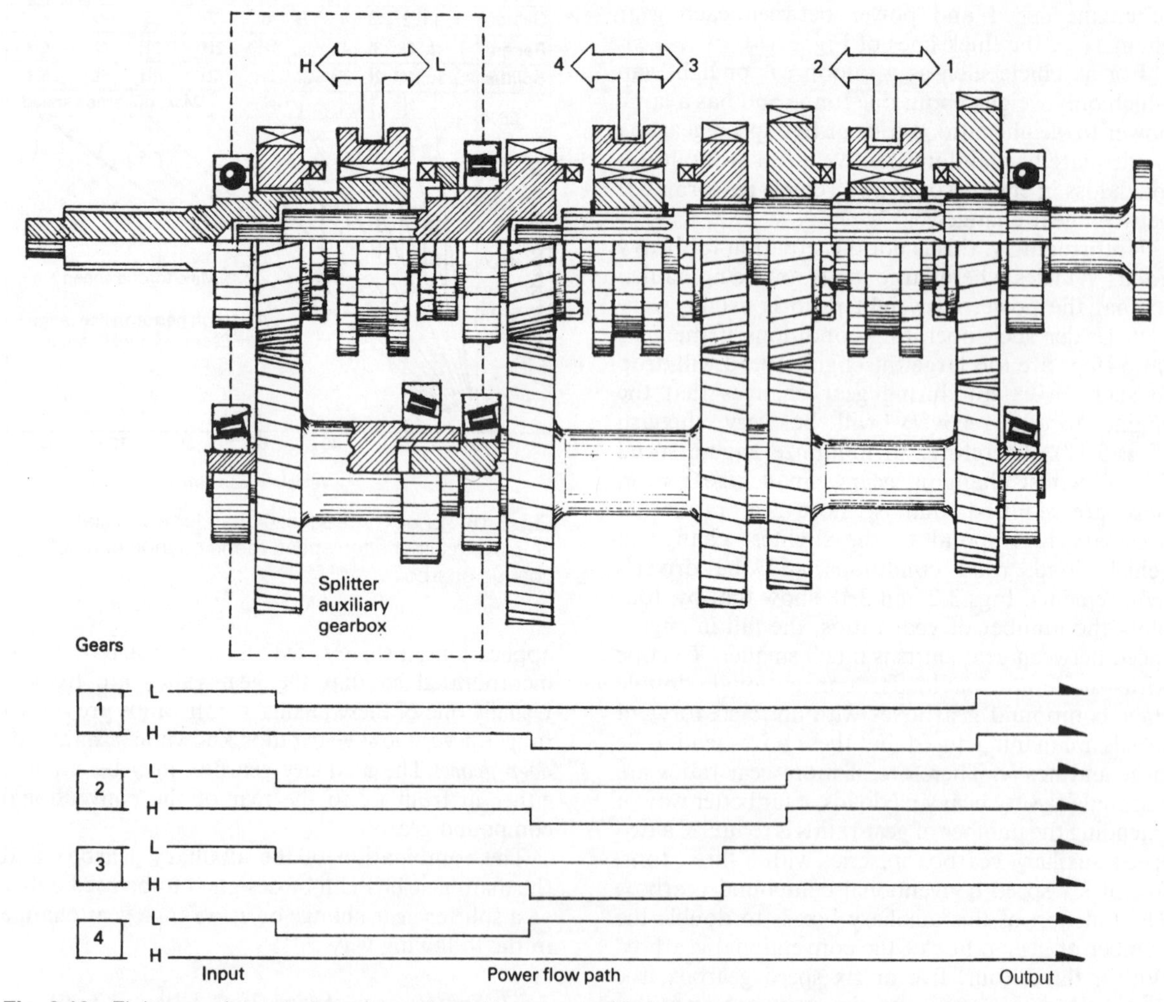

Fig. 3.19 Eight speed constant mesh gearbox with two speed front mounted splitter change

splitter switched again to low ratio and main gearbox second gear selected results in 2L; splitter switched to high ratio, main gearbox gear remaining in second gives 2H; splitter switched to low ratio main gearbox third gear selected gives 3L; splitter switched to high ratio main gearbox still in third gives 3H. This procedure continues throughout the upshift from bottom to top gear ratio. Thus the overall upshift gear ratio change pattern would be:

Gear ratio	1	2	3	4	5	6	7	8	Reverse	
Upshift sequence	1L	1H	2L	2H	3L	3H	4L	4H	RL	RH

It can therefore be predicted that alternate changes involve a simultaneous upchange in the

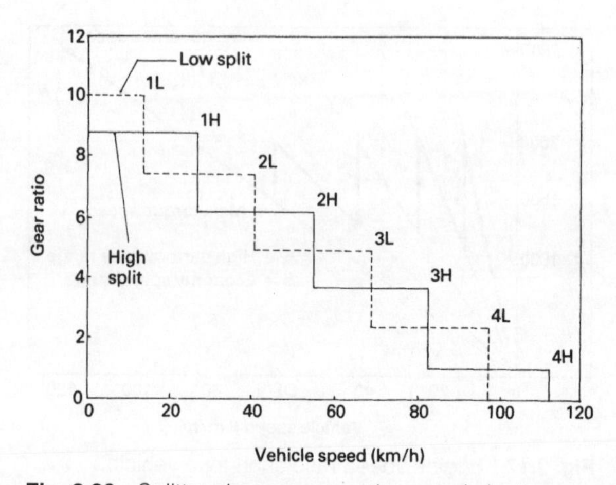

Fig. 3.20 Splitter change gear ratio–speed chart

main gearbox and downchange in the splitter stage, or vice versa.

Referring to the thick lines in Figs 3.2, 3.17 and 3.18, these represent the recommended operating speed range for the engine for best fuel economy, but the broken lines in Fig. 3.17 represent the gear shift technique if maximum road speed is the criteria and fuel consumption, engine wear and noise become secondary considerations.

3.5.2 Range gear change (Figs 3.21 and 3.22)

In contrast to the splitter gear change, the range gear change arrangement (Fig. 3.21) has the gear ratios between adjacent gear ratio steps set close together. The auxiliary two speed gearbox will have one ratio direct drive and the other one normally equal to just over half the gear ratio spread from bottom to top. This is slightly larger than the main gearbox gear ratio spread.

To change from one gear ratio to the next with, for example, an eight speed gearbox comprising four normal forward gears and one reverse and a two speed auxiliary range change, the pattern of gear change would be as shown in Fig. 3.22.

Through the gear ratios from bottom to top 'low gear range' is initially selected, the main gearbox order of upchanges are first, second, third and fourth gears. At this point the range change is moved to 'high gear range' and the sequence of gear upchanges again becomes first, second, third and fourth. Therefore the total number of gear ratios is the sum of both low and high ranges, that is, eight. A tabulated summary of the upshift gear change pattern will be:

Gear ratio	1	2	3	4	5	6	7	8	Reverse	
Upshift sequence	1L	2L	3L	4L	1H	2H	3H	4H	RL	RH

3.5.3 Sixteen speed synchromesh gearbox with range change and integral splitter gears (Fig. 3.23)

This heavy duty commercial gearbox utilizes both a two speed range change and a two speed splitter gear change to enable the four speed gearbox to

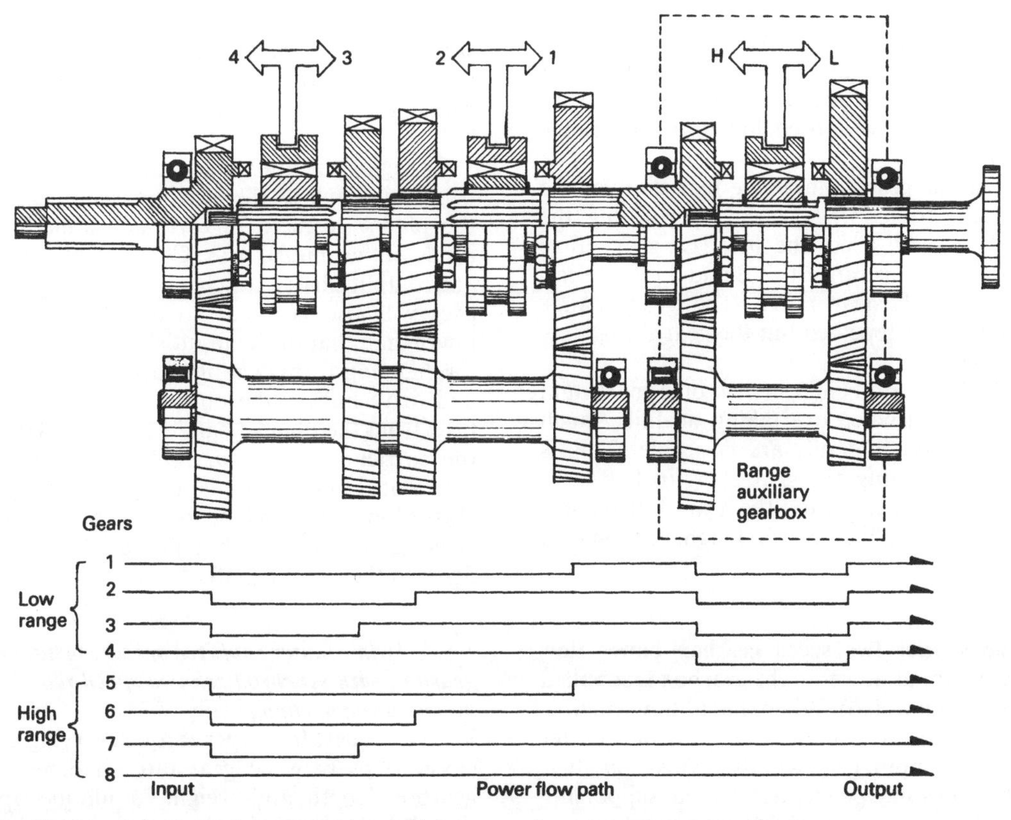

Fig. 3.21 Eight speed constant mesh gearbox with two speed rear mounted range change

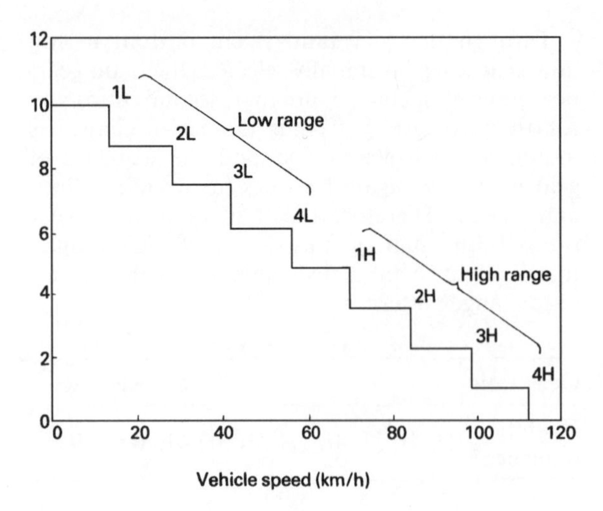

Fig. 3.22 Range change gear ratio–speed chart

extend the gear ratio into eight steps and, when required, to sixteen split (narrow) gear ratio intervals.

The complete gearbox unit can be considered to be divided into three sections; the middle section (which is basically a conventional double stage four speed gearbox), and the first two pairs of gears at the front end which make up the two speed splitter gearbox. Mounted at the rear is an epicyclic gear train providing a two speed low and high range change (Fig. 3.23).

The epicyclic gear train at the rear doubles the ratios of the four speed gearbox permitting the driver to initially select the low (L) gear range driving through this range 1, 2, 3 and 4 then selecting the high (H) gear range. The gear change sequence is again repeated but the gear ratios now become 5, 6, 7 and 8.

If heavy loads are being carried, or if maximum torque is needed when overtaking on hills, much closer gear ratio intervals are desirable. This is provided by splitting the gear steps in half with the two speed splitter gears; the gear shift pattern of 1st low, 1st high, 2nd low, 2nd high, 3rd low and so on is adopted.

Front end splitter two speed gearbox power flow (Fig. 3.23) Input power to the gearbox is supplied to the first motion shaft. When the splitter synchronizing sliding sleeve is in neutral, both the splitter low and high input gear wheels revolve on their needle bearings independently of their supporting first motion shaft and mainshaft respectively.

When low or high splitter gears are engaged, the first motion shaft drive hub conveys power to the first or second pair of splitter gear wheels and hence to the layshaft gear cluster.

Mid-four speed gearbox power flow (Fig. 3.23) Power from the first motion shaft at a reduced speed is transferred to the layshaft cluster of gears and subsequently provides the motion to all the other mainshaft gear wheels which are free to revolve on the mainshaft, but at relatively different speeds when in the neutral gear position.

Engagement of one mid-gearbox gear ratio dog clutch locks the corresponding mainshaft drive hub to the chosen gear so that power is now able to pass from the layshaft to the mainshaft through the selected pair of gear wheels.

Reverse gear is provided via an idler gear which, when meshed between the layshaft and mainshaft, alters the direction of rotation of the mainshaft in the usual manner.

Rear end range two speed gearbox power flow (Fig. 3.23) When the range change is in the neutral position, power passes from the mainshaft and sun gear to the planet gears which then revolve on the output shaft's carrier pin axes and in turn spin round the annular gear and synchronizing drive hub.

Engaging the low range gear locks the synchronizing drive hub to the gearbox casing. This forces the planet gears to revolve and walk round the inside of the annular gear. Consequently, the carrier and output shafts which support the planet gear axes will also be made to rotate but at a speed lower than that of the input shaft.

Changing to high range locks the annular gear and drive hub to the output shaft so that power flow from the planet gears is then divided between the carrier and annular, but since they need to rotate at differing speeds, the power flow forms a closed loop and jams the gearing. As a result, there is no gear reduction but just a straight through drive to the output shaft.

3.5.4 Twin counter shaft ten speed constant mesh gearbox with synchromesh two speed rear mounted range change (Fig. 3.24)

With the quest for larger torque carrying capacity, closer steps between gear ratio changes, reduced gearbox length and weight, a unique approach to fulfil these requirements has been developed

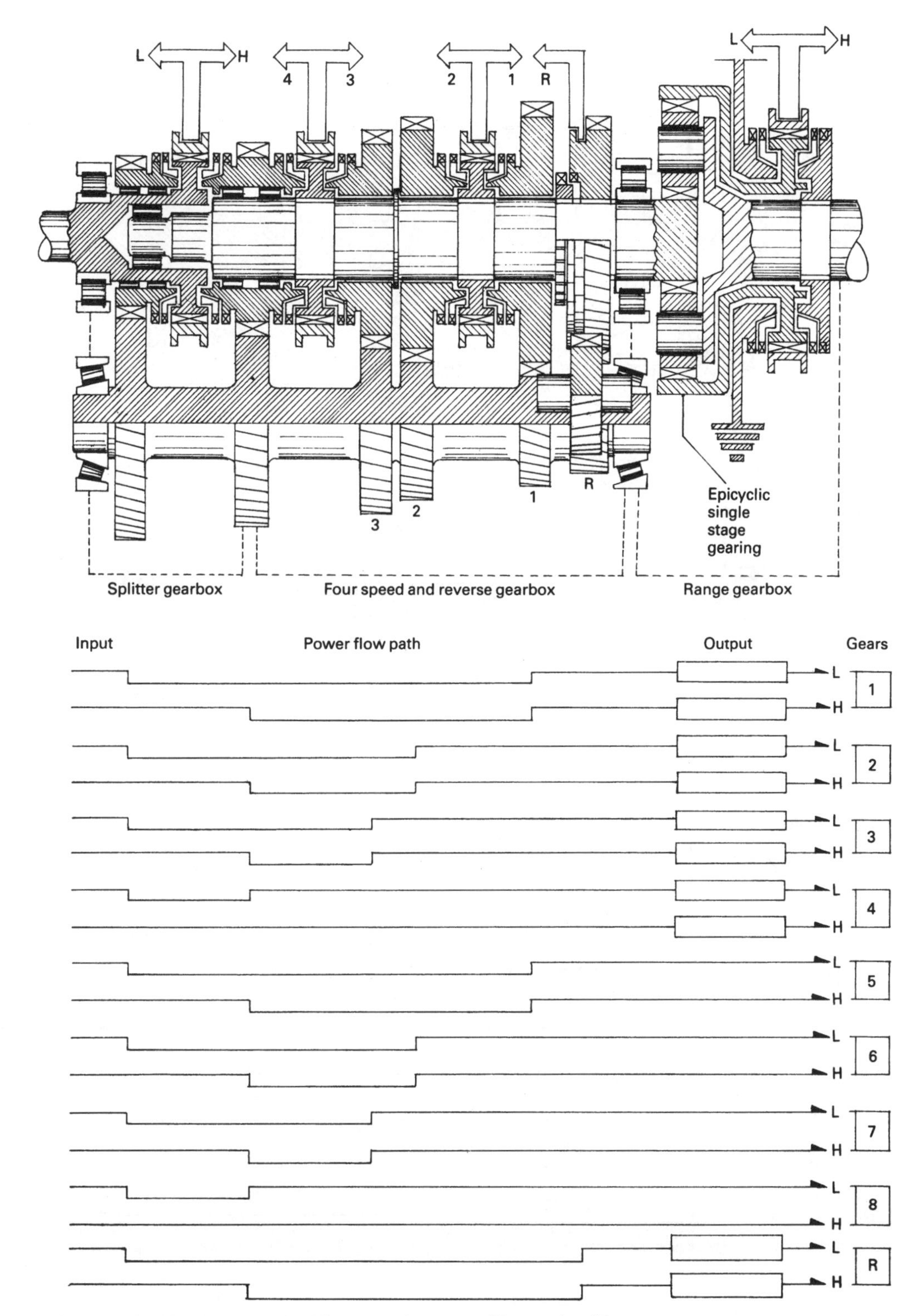

Fig. 3.23 Sixteen speed synchromesh with range change and integral splitter gears

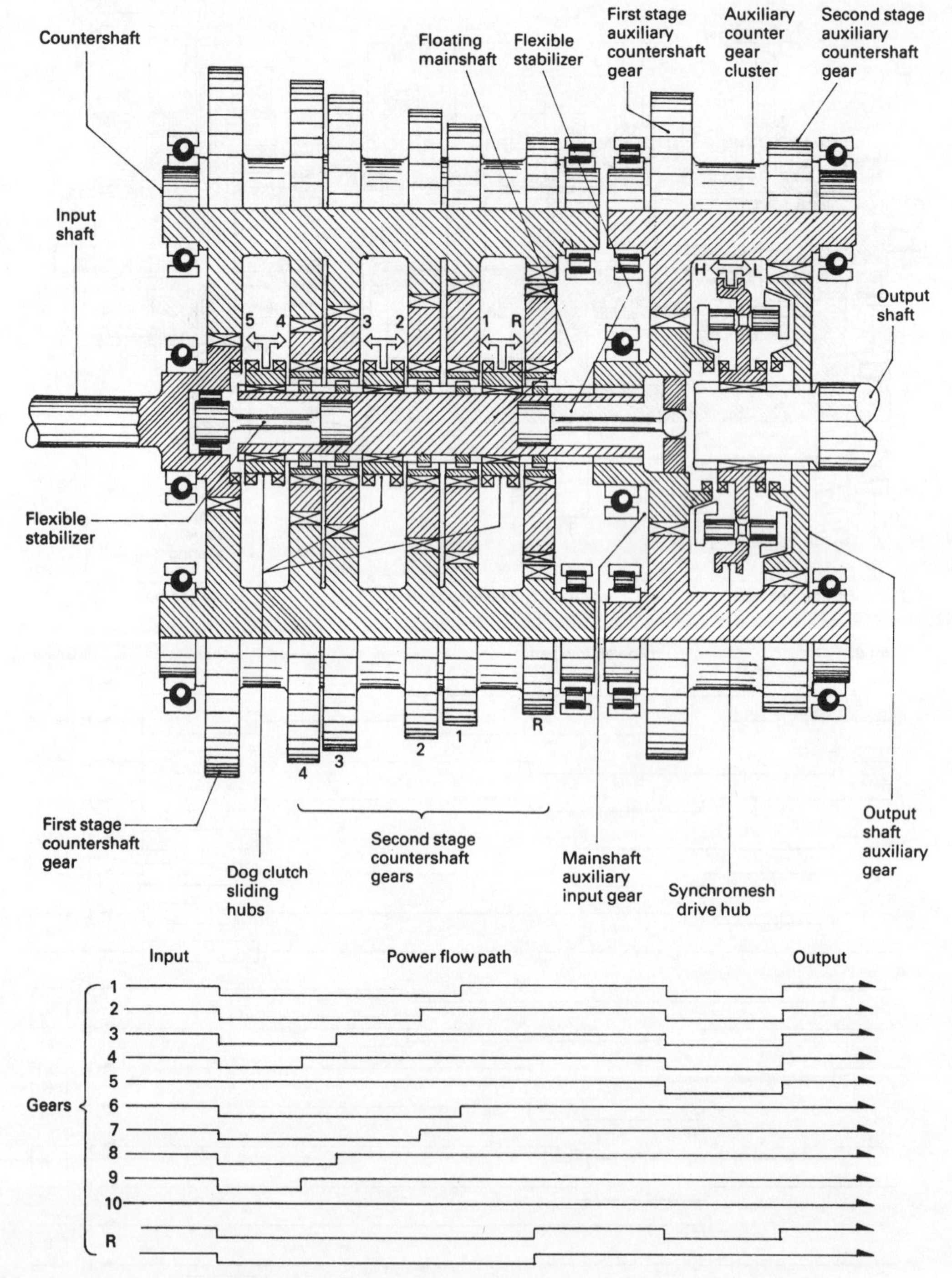

Fig. 3.24 Twin countershaft ten speed constant mesh gearbox with synchromesh two speed range change

adopting the two countershaft constant mesh gearbox which incorporates a synchromesh two speed rear mounted range change (Fig. 3.24).

The main gearbox is in fact a double stage compound conventional gearbox using two countershafts (layshaft) instead of the normal single arrangement.

Design and construction Referring to Fig. 3.24, there is a countershaft either side of the mainshaft and they are all in the same plane. What cannot be seen is that this single plane is inclined laterally at 19° to the horizontal to reduce the overall height of the gearbox.

The mainshaft is hollow and is allowed to float in the following manner: each end is counterbored, and into each counterbore is pressed a stabilizing rod. The front end of this rod projects into the rear of the input shaft which is also counterbored to house a supporting roller bearing for the stabilizer rod. The rear projecting stabilizer rod has a spherically shaped end which rests in a hole in the centre of a steel disc mounted inside the auxiliary drive gear immediately behind the mainshaft. This gear itself is carried by a ball bearing mounted in the gearbox housing. When torque is transmitted through the gearbox, the centrally waisted 11 mm diameter section of both stabilizers deflects until radial loads applied by the two countershaft gears to the mainshaft gear are equalized. By these means, the input torque is divided equally between the two countershafts and two diametrically opposite teeth on the mainshaft gear at any one time. Therefore, the face width of the gear teeth can be reduced by about 40% compared to gearboxes using single countershafts. Another feature of having a mainshaft which is relatively free to float in all radial directions is that it greatly reduces the dynamic loads on the gear teeth caused by small errors of tooth profile during manufacture. A maximum radial mainshaft float of about 0.75 mm has proved to be sufficient to permit the shaft to centralize and distribute the input torque equally between the two countershafts. To minimize end thrust, all the gears have straight spur teeth which run acceptably quietly due to the balanced loading of the gears.

Each of the five forward speeds and reverse are engaged by dog teeth clutches machined on both ends of the drive hubs. The ends of the external teeth on the drive hubs and the internal teeth in the mainshaft gears are chamfered at about 35° to provide some self-synchronizing action before engagement.

Power flow path Power flows into the main gearbox through the input first motion shaft and gear wheel. Here it is divided between the two first stage countershaft gears and is then conveyed via each countershaft gear wheel to the corresponding second stage mainshaft gears. Each of these rotate at relative speeds about the mainshaft. Torque is only transmitted to the mainshaft when the selected dog clutch drive hub is slid in to mesh with the desired gear dog teeth.

The power flow can then pass directly to the output shaft by engaging the synchromesh high range dog teeth. Conversely, a further gear reduction can be made by engaging the low range synchromesh dog teeth so that the power flow from the mainshaft auxiliary gear is split between the two auxiliary countershafts. The additional speed reduction is then obtained when the split power path comes together through the second stage auxiliary output gear. It should be observed that, unlike the mainshaft, the auxiliary gear reduction output shaft has no provision for radial float.

Reverse gear is obtained by incorporating an idle gear between the second stage countershaft reverse gears and the mainshaft reverse gear so that the mainshaft reverse gear is made to rotate in the opposite direction to all the other forward drive mainshaft gears.

3.6 Transfer box power take-off (PTO) (Fig. 3.25)
A power take-off (PTO) provides some shaft drive and coupling to power specialized auxiliary equipment at a specified speed and power output. Power take-offs (PTOs) can be driven directly from the engine's timing gears, but it is more usual and practical to take the drive from some point off the gearbox. Typical power take-off applications are drives for hydraulic pumps, compressors, generators, hoists, derricks, capstain or cable winch platform elevators, extended ladders, hose reels, drain cleaning vehicles, tippers, road sweepers, snow plough blade and throwing operations and any other mechanical mechanism that needs a separate source of power drive output.

The power take-off can be driven either by one of the layshaft cluster gears, so that it is known as a *side mounted transfer box*, or it may be driven from the back end of the layshaft, in which case it is known as a *rear mounted transfer box* (Fig. 3.25).

Transfer boxes can either be single or two speed arrangements depending upon the intended application. The gear ratios of the transfer box are so chosen that output rotational speeds may be anything from 50 to 150% of the layshaft input speed.

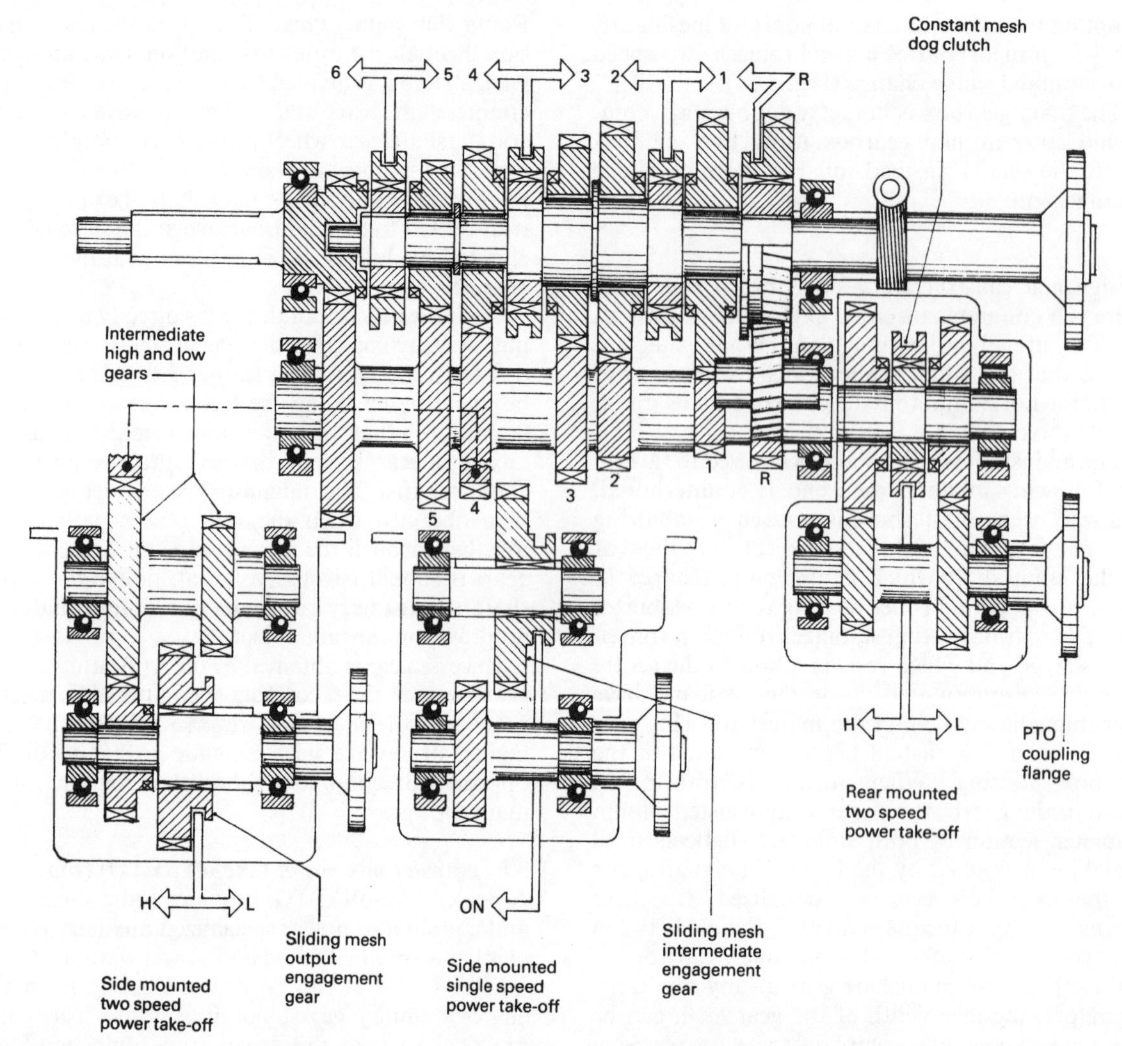

Fig. 3.25 Six speed constant mesh gearbox illustrating different power take-off point arrangement

3.6.1 Side mounted single speed transfer box
(Fig. 3.25)

With the single speed side mounted transfer box, the drive is conveyed to the output gear and shaft by means of an intermediate gear mounted on a splined idler shaft which is itself supported by two spaced out ball bearings (Fig. 3.25). Engagement of the transfer output shaft is obtained by sliding the intermediate straight toothed gear into mesh with both layshaft gear and output shaft gear by a selector fork mounted on a gear shift not shown.

3.6.2 Side mounted two speed transfer box
(Fig. 3.25)

If a more versatile transfer power take-off is required, a two speed transfer box can be incorporated. With this gear train layout, the drive is conveyed to the intermediate shaft by a gear wheel which is in constant mesh with both the layshaft gear wheel and the high speed output gear (Fig. 3.25). The output shaft supports the high speed output gear which is free to revolve relative to it when the transfer drive is in neutral or low gear is engaged. Also attached to this shaft on splines is the low speed output gear.

High transfer gear ratio engagement is obtained by sliding the low speed output gear towards the high speed output gear until its internal splines mesh with the dog teeth on the side of the gear. This then transfers the drive from the layshaft to the output shaft and coupling through a simple single stage gear reduction.

Low transfer gear ratio engagement occurs when the low speed output gear is slid into mesh with the smaller intermediate shaft gear. The power flow then takes place through a double stage (compound) gear reduction.

Rear mounted two speed transfer box (Fig. 3.25) In some gearbox designs, or where the auxiliary equipment requires it, a rear mounted transfer box may be more convenient. This transfer drive arrangement uses either an extended monolayshaft or a short extension shaft attached by splines to the layshaft so that it protrudes out from the rear of the gearbox (Fig. 3.25). The extended layshaft supports a pair of high and low speed gears which are in permanent mesh with corresponding gears mounted on the output shaft.

When the transfer box is in neutral, the gears on the extended layshaft are free to revolve independently on this shaft. Engagement of either high or low gear ratios is achieved by sliding the output drive hub sleeve in to mesh with one or other sets of adjacent dog teeth forming part of the transfer box layshaft constant mesh gears. Thus high gear ratio power flow passes from the layshaft to the constant mesh high range gears to the output shaft and coupling. Conversely, low gear ratio power transmission goes from the layshaft through the low range gears to the output drive.

3.7 Overdrive considerations

Power is essential to propel a vehicle because it is a measure of the rate of doing work, that is, the amount of work being developed by the engine in unit time. With increased vehicle speed, more work has to be done by the engine in a shorter time.

The characteristic power curve over a speed range for a petrol engine initially increases linearly and fairly rapidly. Towards mid-speed the steepness of the power rise decreases until the curve reaches a peak. It then bends over and declines with further speed increase due to the difficulties experienced in breathing at very high engine speeds (Fig. 3.26).

A petrol engined car is usually geared so that in its normal direct top gear on a level road the engine speed exceeds the peak power speed by about 10 to 20% of this speed. Consequently, the falling power curve will intersect the road resistance power curve. The point where both the engine and road resistance power curves coincide fixes the road speed at which all the surplus power has been absorbed. Therefore it sets the maximum possible vehicle speed.

By selecting a 20% overdrive top gear, say, the transmission gear ratios can be so chosen that the engine and road resistance power curves coincide at peak engine power (Fig. 3.26). The undergearing has thus permitted the whole of the engine power curve to be shifted nearer the opposing road resistance power curve so that slightly more engine power is being utilized when the two curves intersect. As a result, a marginally higher maximum vehicle speed is achieved. In other words, the engine will be worked at a lower speed but at a higher load factor whilst in this overdrive top gear.

If the amount of overdrive for top gear is increased to 40%, the engine power curve will be shifted so far over that it intersects the road resistance power curve before peak engine power has been obtained (Fig. 3.26) and therefore the maximum possible vehicle speed cannot be reached.

Contrasting the direct drive 20% and 40% overdrive with direct drive top gear power curves with respect to the road resistance power curve at 70 km/h, as an example, it can be seen (Fig. 3.26) that the reserve of power is 59%, 47% and 38% respectively. This surplus of engine power over the power absorbed by road resistance is a measure of the relative acceleration ability for a particular transmission overall gear ratio setting.

A comparison of the three engine power curves shows that with direct drive top gear the area in the loop made between the developed and opposing power curves is the largest and therefore the engine would respond to the changing driving conditions with the greatest flexibility.

If top gear is overdriven by 20%, as shown in Fig. 3.26, the maximum engine power would be developed at maximum vehicle speed. This then provides the highest possible theoretical speed, but the amount of reserve power over the road resistance power is less, so that acceleration response will not be as rapid as if a direct drive top gear is used. Operating under these conditions, the engine speed would never exceed the peak power speed and so the engine could not 'over-rev', and as a result engine wear and noise would be reduced. Benefits are also gained in fuel consumption as shown in Fig. 3.26. The lowest specific fuel consumption is shifted to a higher cruising speed which is desirable on motorway journeys.

Indulging in an excessive 40% overdrive top gear prevents the engine ever reaching peak power so that not only would maximum vehicle speed be reduced compared to the 20% overdrive gearing, but the much smaller difference in power developed to power dissipated shown on the power curves would

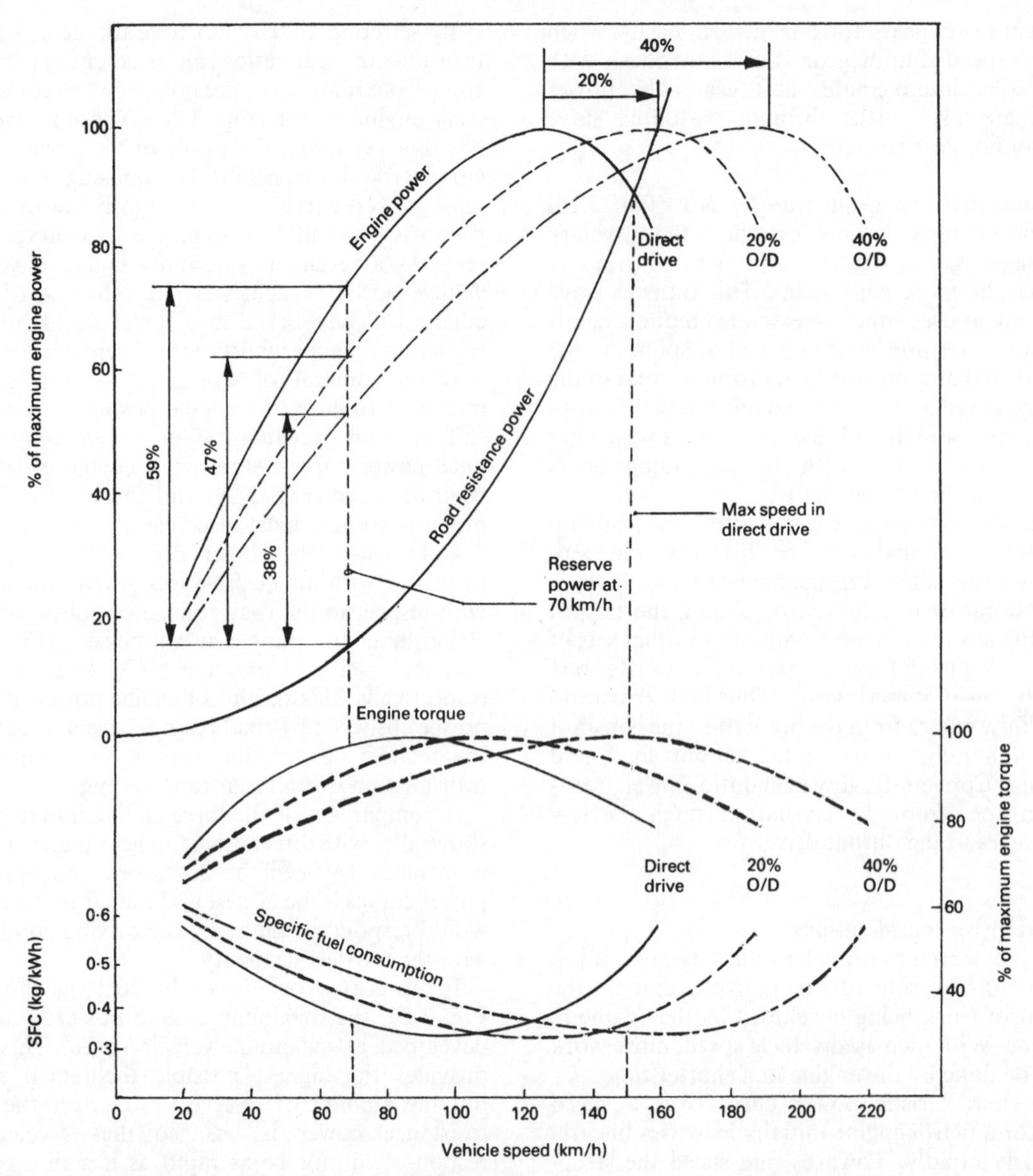

Fig. 3.26 Effect of over and undergearing on vehicle performance

severely reduce the flexibility of driving in this gear. It therefore becomes essential for more frequent down changes with the slightest fall-off in road speed. A further disadvantage with excessive overdrive is that the minimum specific fuel consumption would be shifted theoretically to the engine's upper speed range which in practice could not be reached.

An analysis of matching an engine's performance to suit the driving requirements of a vehicle shows that with a good choice of undergearing in top gear for motorway cruising conditions, benefits of prolonged engine life, reduced noise, better fuel economy and less driver fatigue will be achieved. Another major consideration is the unladen and laden operation of the vehicle, particularly if it is to haul heavy loads. Therefore most top gear overdrive ratios are arrived at as a compromise.

3.7.1 Epicyclic overdrive gearing

Epicyclic gear train overdrives are so arranged that the input shaft drives the pinion carrier while the output shaft is driven by the annular gear ring

(Figs 3.27 and 3.28). The gear train may be either of simple (single stage) or compound (double stage) design and the derived formula for each arrangement is as follows:

Simple gear train (Fig. 3.27)

$$\text{Overdrive gear ratio} = \frac{A}{S+S}$$

also $A = S + 2P$

where A = number of annulus ring gear teeth

S = number of sun gear teeth

P = number of planet gear teeth

Compound gear train (Fig. 3.28)

$$\text{Overdrive gear ratio} = \frac{P_L(P_L + P_S + S)}{P_L(P_L + P_S + S) + P_S S}$$

also $A = P_L + P_S + S$

where A = number of annulus ring gear teeth

S = number of sun gear teeth

P_S = number of small planet gear teeth

P_L = number of large planet gear teeth

The amount of overdrive (undergearing) used for cars, vans, coaches and commercial vehicles varies from as little as 15% to as much as 45%. This corresponds to undergearing ratios of between 0.87:1 and 0.69:1 respectively. Typical overdrive ratios which have been frequently used are 0.82:1 (22%), 0.78:1 (28%) and 0.75:1 (37%).

Example 1 An overdrive simple epicyclic gear train has sun and annulus gears with 21 and 75 teeth respectively. If the input speed from the engine drives the planet carrier at 3000 rev/min, determine

a) the overdrive gear ratio,
b) the number of planet gear teeth,
c) the annulus ring and output shaft speed,
d) the percentage of overdrive.

a) $\text{Overdrive gear ratio} = \frac{A}{A+S} = \frac{75}{75+21}$

$$= \frac{75}{96} = 0.78125$$

b) $A = S + 2P$

Therefore $P = \frac{A-S}{2}$

$$= \frac{75-21}{2}$$

$$= \frac{54}{2} = 27 \text{ teeth}$$

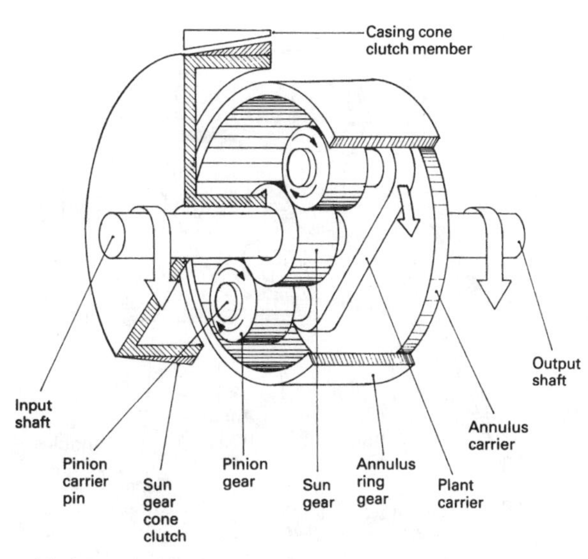

Fig. 3.27 Simple epicycle overdrive gear train

c) Output speed $= \frac{3000}{0.78125} = 3840$ rev/min

d) Percentage of overdrive $= \left(\frac{3840 - 3000}{3000}\right) 100$

$$= \frac{840 \times 100}{3000} = 28\%$$

Example 2 A compound epicyclic gear train overdrive has sun, small planet and large planet gears with 21, 15 and 24 teeth respectively. Determine the following if the engine drives the input planet carrier at 4000 rev/min.

a) The overdrive gear ratio,
b) the number of annulus ring gear teeth,
c) the annulus ring and output shaft speed,
d) the percentage of overdrive.

a) Overdrive gear ratio $= \frac{P_L\,(P_L + P_S + S)}{P_L\,(P_L + P_S + S) + P_S S}$

$$= \frac{24\,(24 + 15 + 21)}{24\,(24 + 15 + 21) + (15 \times 21)}$$

$$= \frac{24 \times 60}{(24 \times 60) + 315} = \frac{1400}{1755}$$

$$= 0.82$$

b) $A = P_L + P_S + S$

$$= 21 + 24 + 15$$

$= 60$ teeth

c) Output speed $= \frac{4000}{0.82} = 4878$ rev/min

d) Percentage of overdrive $= \frac{4878 - 4000}{4000} \times 100$

$$= \frac{878 \times 100}{4000}$$

$$= 21.95\%$$

3.7.2 *Simple epicyclic overdrive gear train* (Fig. 3.27)

If the sun gear is prevented from rotating and the input shaft and planet carrier are rotated, the pinion gears will be forced to revolve around the fixed sun gear and these pinions will revolve simultaneously on their own axes provided by the carrier pins.

As a result, motion will be transferred from the carrier and pinion gears to the annulus ring gear due to the separate rotary movement of both the planet carrier and the revolving planet gears, thus

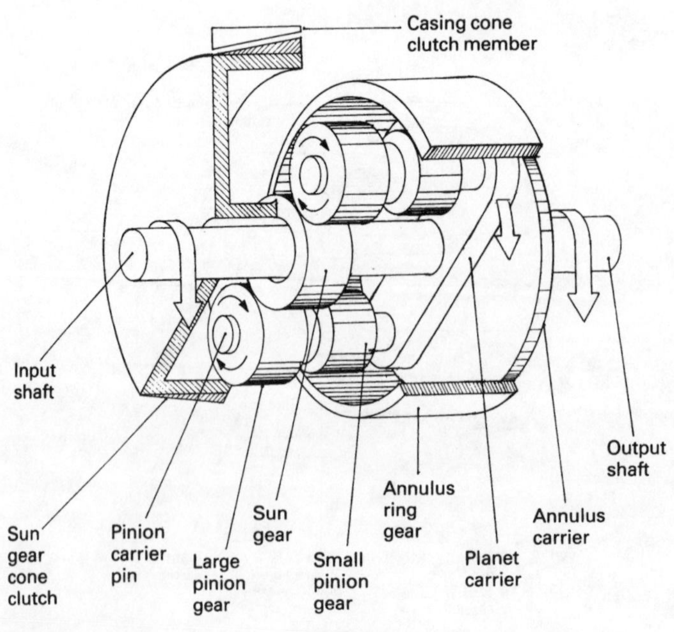

Fig. 3.28 Compound epicycle gear train

the annulus and therefore the output shaft will be compelled to revolve at a slightly faster speed.

3.7.3 Compound epicyclic overdrive gear train (Fig. 3.28)

For only small degrees of overdrive (undergearing), for example 0.82:1 (22%), the simple epicyclic gearing would need a relatively large diameter annulus ring gear; about 175 mm if the dimension of the gear teeth are to provide adequate strength. A way of reducing the diameter of the annulus ring gear for a similar degree of overdrive is to utilize a compound epicyclic gear train which uses double pinion gears on each carrier pin instead of one size of pinion. By this method, the annulus diameter is reduced to about 100 mm and there are only 60 teeth compared to the 96 teeth annulus used with the simple epicyclic gear train.

To transmit power, the sun gear is held still whilst the input shaft and planet carrier are rotated. This compels the large planet gear to roll around the stationary sun gear and at the same time forces each pair of combined pinion gears to revolve about their carrier pin axes.

Consequently the small pinion gear will impart both the pinion carrier orbiting motion and the spinning pinion gear motion to the annulus ring gear so that the output shaft will be driven at a higher speed to that of the input shaft.

3.7.4 Laycock simple gear train overdrive

Description (Fig. 3.29) The overdrive unit is attached to the rear of the gearbox and it consists of a constant mesh helical toothed epicyclic gear train which has a central sunwheel meshing with three planet gears which also mesh with an internally toothed annulus gear. The planet gears are supported on a carrier driven by the input shaft whilst the annulus is attached to the output shaft via a carrier forming an integral part of both members. A double cone clutch selects the different ratios; when engaged one side of the clutch provides direct drive and when the other side is used, overdrive.

Direct drive (Fig. 3.29) Direct drive is obtained when the inner cone clutch engages with the outer cone of the annular gear. Power will then be conveyed via the unidirectional clutch to the output shaft by means of the rollers which are driven up inclined ramps and wedged between the inner and outer clutch members. When the vehicle overruns the engine, the output shaft will try to run faster than the input shaft and so tend to release the unidirectional clutch rollers, but this is prevented by the inner cone clutch locking the sun gear to the annulus, thereby jamming the sun, planet and annular epicyclic gear train so that they cannot revolve relative to each other.

Engagement of the inner cone clutch to the external cone surface of the annulus gear is provided by four stationary thrust springs (only one shown) which are free to exert their axial load against a thrust plate. This in turn transfers thrust by way of a ball bearing to the rotating cone clutch support member splined to the sun gear sleeve. This overrun and reverse torque will be transmitted between the engine and transmission in direct drive.

Owing to the helical cut teeth of the gear wheels, an end thrust exists between the planet gears and the sun gear during overrun and reverse which tends to push the latter rearwards. Therefore, additional clamping load between the cone clutch faces is necessary.

Overdrive (Fig. 3.29) When overdrive is engaged, the cone clutch, which is supported on the splined sleeve of the sun gear, is moved over so that its outer friction facing is in contact with the internal cone brake attached to the casing. Consequently the sun gear is held stationary. With the sun gear held still and the input shaft and planet carrier rotating, the planet gears are forced to rotate about their own axes and at the same time roll around the fixed sun gear, with the result that the annulus gear is driven at a faster speed than the input shaft. This causes the unidirectional clutch outer member (annular carrier) to overrun the inner member (planet carrier) so that the wedged rollers on their ramps are released. Pulling the cone clutch away from the annulus cone and into frictional contact with the brake casing cone against the axial load of the six thrust springs is achieved by means of hydraulic oil pressure. This pressure acts upon two slave pistons (only one shown) when a valve is opened by operating the driver controlled selector switch.

The outward movement of the slave pistons, due to the hydraulic pressure, draws the stationary thrust plate, ball bearing and rotating clutch member away from the annular cone and into engagement with the outer brake cone, thereby locking the sun gear to the casing. Of the helix angle of the gear teeth, the torque reaction tends to push the sun gear forward so that extra end

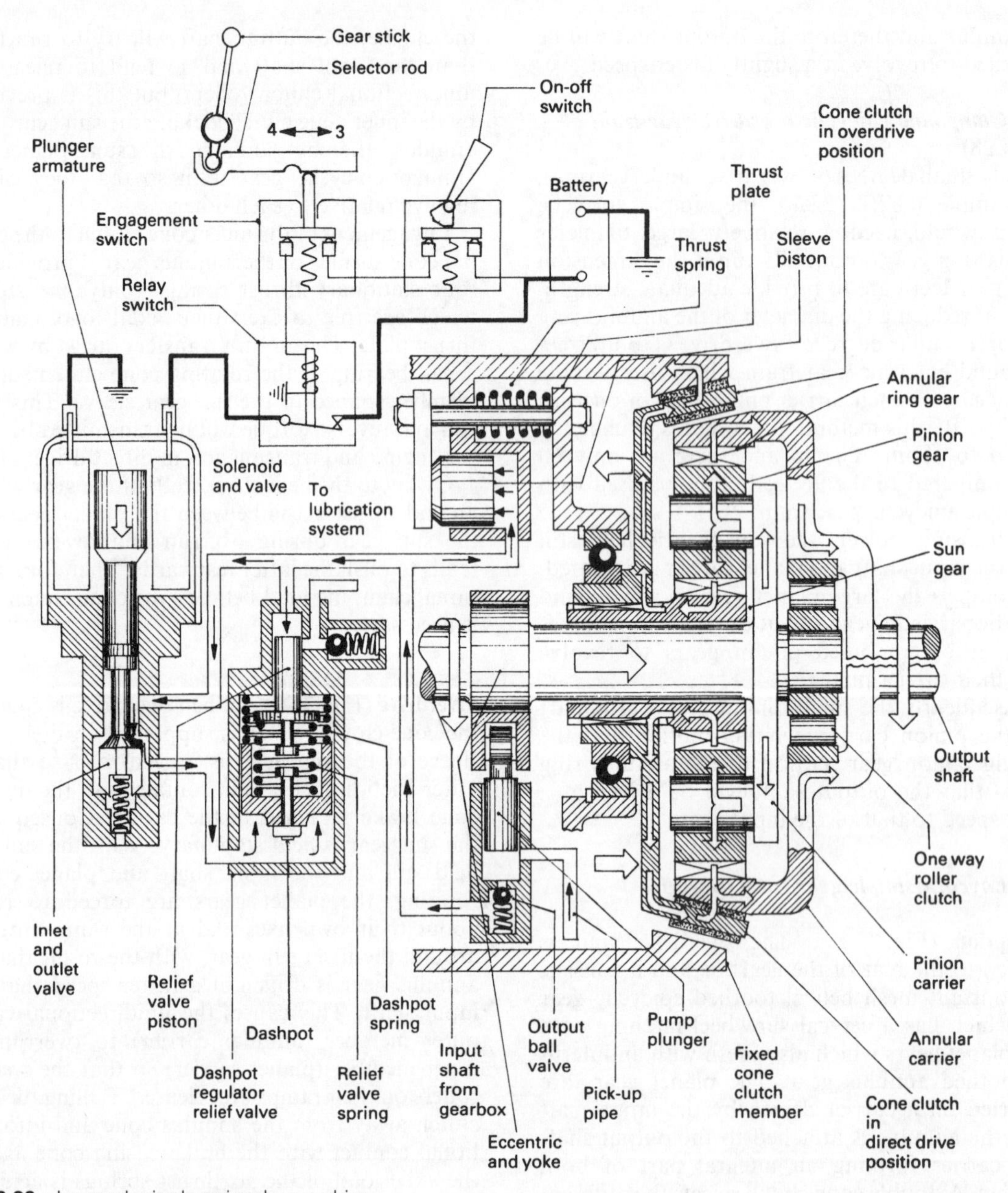

Fig. 3.29 Laycock single epicycle overdrive

thrust is necessary to maintain sufficient clamping thrust between the frictional faces of the cones in the brake position.

Direct and overdrive controlled gear change action (Fig. 3.29) When direct drive is selected, hydraulic pressure is steadily increased and this gradually releases the double-sided cone clutch member from the cone brake fixed to the casing. The release of the cone clutch frees the sun gear and removes the load from the engine. The engine speed increases immediately until it catches up with the output shaft, at which point the unidirectional clutch rollers climb up their respective ramps and jam. The input shaft's power coming from the engine is now permitted to drive the output shaft, which in turn transmits drive to the propellor shaft. At the same time the double-sided cone clutch completes its movement and engages the annular ring cone.

Overdrive is engaged when the double-sided cone clutch moves away from the annulus gear cone and makes contact with the stationary cone brake, thus bringing the sliding cone clutch member and sun gear to rest. As a result of the sun gear being held stationary, the gears now operate as an epicyclic step up gear ratio transmission. During the time the double-sided cone clutch member moves from the annulus cone to the brake cone the clutch will slip. This now permits the unidirectional roller clutch to transmit the drive. Whilst the input ramp member rotates as fast as the output ramp member the roller clutch drives. However, as the annular ring gear speed rises above that of the input shaft, the rollers will disengage themselves from their respective ramps thereby diverting the drive to the epicyclic gear train.

Electrical system (Fig. 3.29) Overdrive or direct drive gear ratio selection is controlled by an electrical circuit which includes an overdrive on/off switch, inhibitor switch and a relay switch. An inhibitor switch is incorporated in the circuit to prevent the engagement of overdrive in reverse and some or all of the indirect gears. A relay switch is also included in the circuit so that the overdrive on/off switch current rating may be small compared to the current draw requirements of the control solenoid. The overdrive may be designed to operate only in top gear, but sometimes the overdrive is permitted to be used in third or even second gear. Selection and engagement of overdrive by the driver is obtained by a steering column or fascia panel switch. When the driver selects overdrive in top gear or one of the permitted indirect gears, say third, the on/off switch is closed and the selected gearbox gear ratio selector rod will have pushed the inhibitor switch button into the closed switch position. Current will now flow from the battery to the relay switch, magnetizing the relay winding so that as the relay contacts close, a larger current will immediately energize the solenoid and open the control valve so that overdrive will be engaged.

Hydraulic system (Fig. 3.29) A plunger type pump driven by an eccentric formed on the input shaft supplies the hydraulic pressure to actuate the slave pistons and thereby operates the clutch. The pump draws oil from the sump through a filter (not shown). It is then pressurized by the plunger and delivered through a non-return valve to both slave cylinders (only one shown) and also to the solenoid controlled valve and dashpot regulator relief valve. The dashpot pressure regulator ensures a smooth overdrive engagement and disengagement under differing operating conditions. When in direct drive the pump to slave cylinder's line pressure is determined by the regulator relief valve spring tension which controls the blow-off pressure of the oil escaping to the lubrication system. This line residual pressure in direct drive is normally maintained at about 2.8 bar, but when engaging overdrive it is considerably raised by the action of the dashpot to about 20–40 bar.

Overdrive engagement Energizing the solenoid draws down the armature, thereby opening the inlet valve and closing the outlet valve. Oil at residual line pressure will now pass through the control orifice to the base of the dashpot regulator relief valve causing the dashpot to rise and compress both the dashpot spring and relief valve spring. Consequently, the pump to slave cylinder pressure circuit will gradually build up as the dashpot spring shortens and increases in stiffness until the dashpot piston has reached its stop, at which point the operating pressure will be at a maximum. It is this gradual increase in line pressure which provides the progressive compression of the clutch thrust springs and the engagement of the cone clutch with the fixed cone brake.

Direct drive engagement De-energizing the solenoid closes the inlet valve and opens the outlet valve. This prevents fresh oil entering the dashpot cylinder and allows the existing oil under the dashpot to exhaust by way of the control orifice and the outlet valve back to the sump. The control orifice restricts the flow of escaping oil so that the pressure drop is progressive. This enables the clutch thrust springs to shift the cone clutch very gradually into contact with the annulus cone.

3.7.5 Laycock compound gear train overdrive (Fig. 3.30)

Overdrive When overdrive is selected, the double-sided cone clutch contacts the brake cone which forms part of the casing. This brings the sun gear which is attached to the sliding clutch member to a standstill.

The input drive passes from the pinion carrier to the annulus ring and hence to the output shaft through the small planet gear. At the same time, the

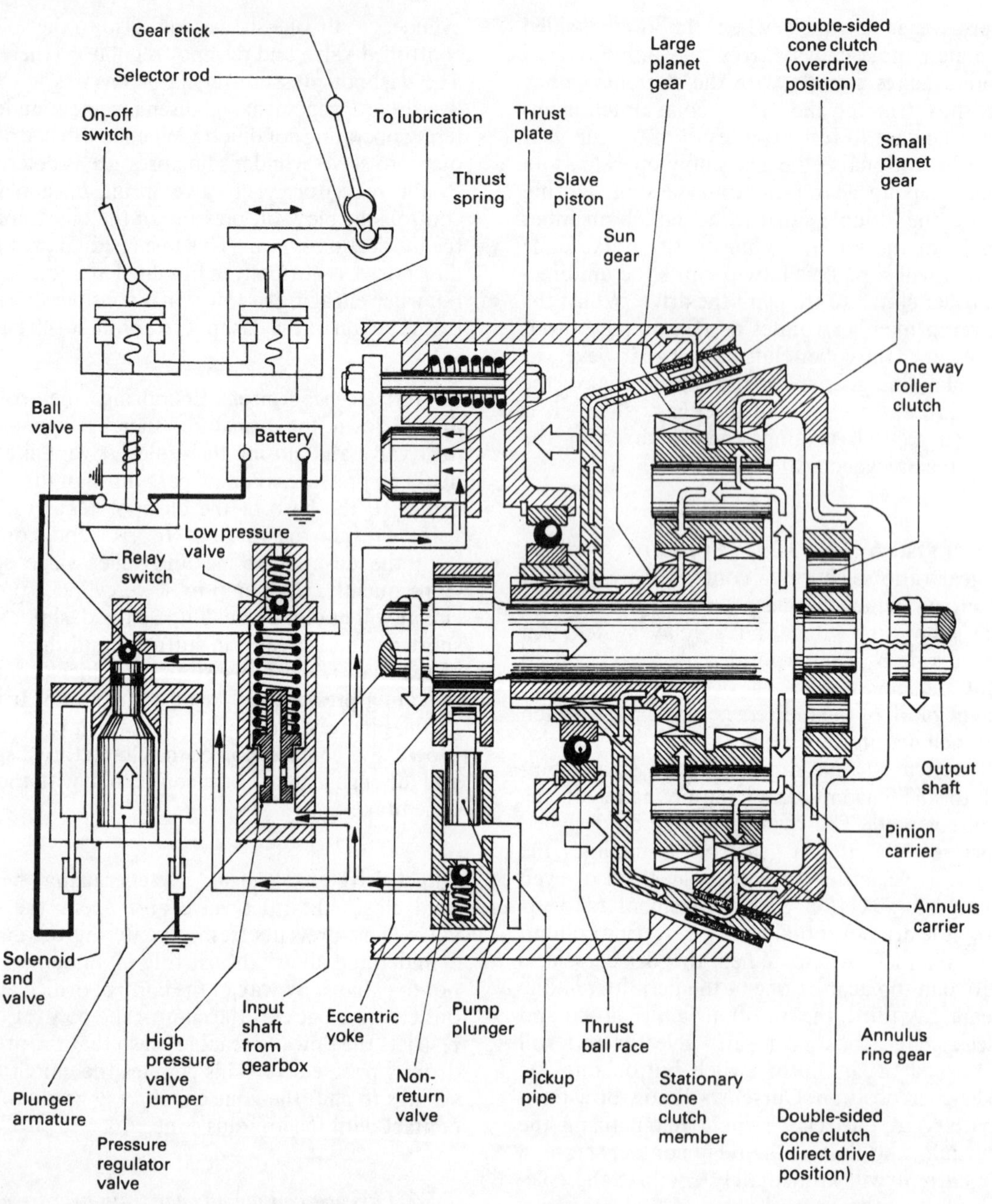

Fig. 3.30 Laycock double epicycle overdrive

large planet gear absorbs the driving torque reaction and in the process is made to revolve around the braked sun gear. The overdrive condition is created by the large planet gears being forced to roll 'walk' about the sun gear, while at the same time revolving on their own axes. As a result, the small planet gears, also revolving on the same carrier pins as the large planet gears, drive forward the annular ring gear at a faster speed relative to that of the input.

The overall gear ratio step up is achieved by having two stages of meshing gear teeth; one between the large pinion and sun gear and the other between the small pinion and annulus ring gear. By using this compound epicyclic gear train, a relatively large step up gear ratio can be obtained for a given diameter of annulus ring gear compared to a single stage epicyclic gear train.

Direct drive (Fig. 3.30) Direct drive is attained by releasing the double-sided cone clutch member from the stationary conical brake and shifting it over so that it contacts and engages the conical frictional surface of the annulus ring gear. The power flow from the input shaft and planet carrier now divides into two paths — the small planet gear to annulus ring gear route and the large planet gear, sun gear and double-sided clutch member route, again finishing up at the annulus ring gear. With such a closed loop power flow arrangement, where the gears cannot revolve independently to each other, the gears jam so that the whole gear train combination rotates as one about the input to output shaft axes. It thereby provides a straight through direct drive. It should be observed that the action of the unidirectional roller clutch is similar to that described for the single stage epicyclic overdrive.

Clutch operating (Fig. 3.30) Engagement of direct drive and overdrive is achieved in a similar manner to that explained under single stage epicyclic overdrive unit.

Direct drive is provided by four powerful springs holding the double-sided conical clutch member in frictional contact with the annulus ring gear. Conversely, overdrive is obtained by a pair of hydraulic slave pistons which overcome and compress the clutch thrust springs, pulling the floating conical clutch member away from the annulus and into engagement with the stationary conical brake.

Hydraulic system (Fig. 3.30) Pressure supplied by the hydraulic plunger type pump draws oil from the sump and forces it past the non-return ball valve to both the slave cylinders and to the solenoid valve and the relief valve.

Direct drive engagement When direct drive is engaged, the solenoid valve opens due to the solenoid being de-energized. Oil therefore flows not only to the slave cylinders but also through the solenoid ball valve to the overdrive lubrication system where it then spills and returns to the sump. A relatively low residual pressure will now be maintained within the hydraulic system. Should the oil pressure rise due to high engine speed or blockage, the low pressure ball valve will open and relieve the excess pressure. Under these conditions the axial load exerted by the clutch thrust springs will clamp the double-sided floating conical clutch member to the external conical shaped annulus ring gear.

Overdrive engagement To select overdrive the solenoid is energized. This closes the solenoid ball valve, preventing oil escaping via the lubrication system back to the sump. Oil pressure will now build up to about 26–30 bar, depending on vehicle application, until sufficient thrust acts on both slave pistons to compress the clutch thrust springs, thereby permitting the double-sided clutch member to shift over and engage the conical surface of the stationary brake. To enable the engagement action to overdrive to progress smoothly and to limit the maximum hydraulic pressure, a high pressure valve jumper is made to be pushed back and progressively open. This controls and relieves the pressure rise which would otherwise cause a rough, and possibly sudden, clutch engagement.

3.8 Setting gear ratios

Matching the engine's performance characteristics to suit a vehicle's operating requirements is provided by choosing a final drive gear reduction and then selecting a range of gear ratios for maximum performance in terms of the ability to climb gradients, achievement of good acceleration through the gears and ability to reach some predetermined maximum speed on a level road.

3.8.1 Setting top gear

To determine the maximum vehicle speed, the engine brake power curve is superimposed onto the power requirement curve which can be plotted from the sum of both the rolling (R_r) and air (R_a) resistance covering the entire vehicle's speed range (Fig. 3.31).

The total resistance R opposing motion at any speed is given by:

$$R = R_r + R_a$$
$$= 10C_r W + C_D A V^2$$

where C_r = coefficient of rolling resistance
W = gross vehicle weight (kg)
C_D = coefficient of aerodynamic resistance (drag)
A = projected frontal area of vehicle (m^2)
V = speed of vehicle (km/h)

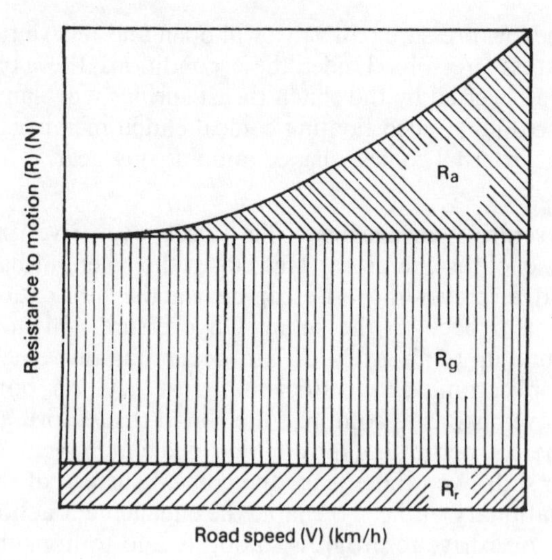

Fig. 3.31 Forces opposing vehicle motion over its speed range

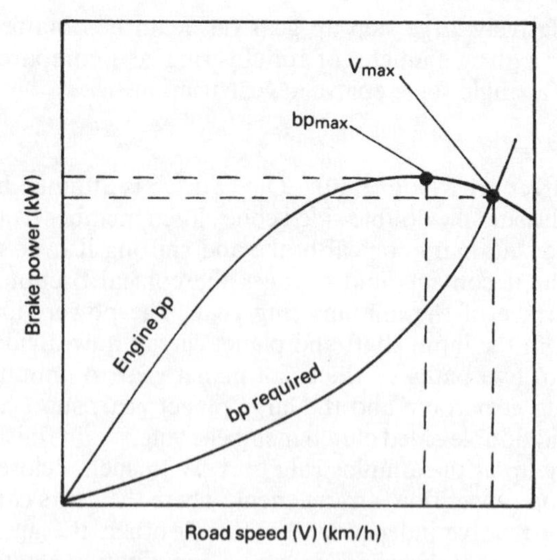

Fig. 3.32 Relationship of power developed and road power required over the vehicle's speed range

The top gear ratio is chosen so that the maximum road speed corresponds to the engine speed at which maximum brake power is obtained (or just beyond) (Fig. 3.32).

Gearing is necessary to ensure that the vehicle speed is at a maximum when the engine is developing approximately peak power.

Thus Linear wheel speed = Linear road speed

$$\frac{\pi dN}{G_F} = \frac{1000}{60} V \text{ (m/min)}$$

$$\therefore \text{ Final drive gear ratio } G_F = \frac{60\ \pi dN}{100\ V}$$

$$= 0.06\ \frac{\pi dN}{V}$$

where G_F = final drive gear ratio
N = engine speed (rev/min)
d = effective wheel diameter (m)
V = road speed at which peak power is developed (km/h)

Example A vehicle is to have a maximum road speed of 150 km/h. If the engine develops its peak power at 6000 rev/min and the effective road wheel diameter is 0.54 m, determine the final drive gear ratio.

$$G_F = \frac{0.06\ \pi dN}{V}$$

$$= \frac{0.06 \times 3.142 \times 0.54 \times 6000}{150}$$

$$= 4.07:1$$

3.8.2 Setting bottom gear

The maximum payload and gradient the vehicle is expected to haul and climb determines the necessary tractive effort, and hence the required overall gear ratio. The greatest gradient that is likely to be encountered is decided by the terrain the vehicle is to operate over. This normally means a maximum gradient of 5 to 1 and in the extreme 4 to 1. The minimum tractive effort necessary to propel a vehicle up the steepest slope may be assumed to be approximately equivalent to the sum of both the rolling and gradient resistances opposing motion (Fig. 3.31).

The rolling resistance opposing motion may be determined by the formula:

$$R_r = 10 C_r W$$

where R_r = rolling resistance (N)
C_r = coefficient of rolling resistance
W = gross vehicle weight (kg)

Average values for the coefficient of rolling resistance for different types of vehicles travelling at very slow speed over various surfaces have been determined and are shown in Table 3.2.

Likewise, the gradient resistance (Fig. 3.33) opposing motion may be determined by the formula:

$$R_g = \frac{10W}{G} \quad \text{or} \quad 10W \sin\theta$$

where R_g = gradient resistance (N)
W = gross vehicle weight ($10W$ kg $= W$N)
G = gradient (1 in x) $= \sin\theta$

Table 3.2 Average values of coefficient of rolling resistance

Vehicle type	Coefficient of rolling resistance (C_r)		
	Concrete	Medium hard soil	Sand
Passenger Car	0.015	0.08	0.30
Trucks	0.012	0.06	0.25
Tractors	0.02	0.04	0.20

Note The coefficient of rolling resistance is the ratio of the rolling resistance to the normal load on the tyre.

i.e. $C_r = \frac{R_r}{W}$

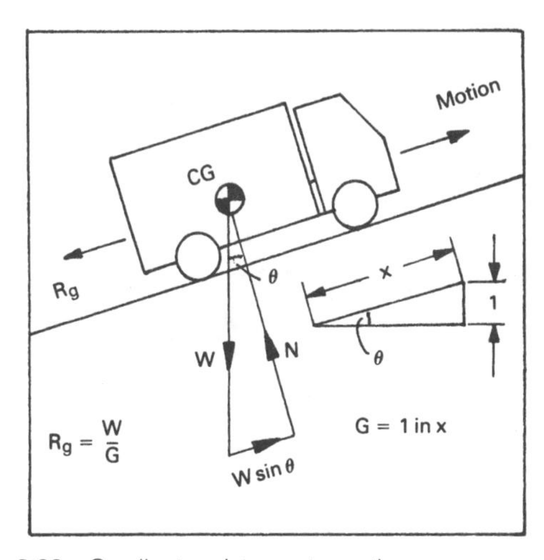

Fig. 3.33 Gradient resistance to motion

Tractive effort = Resisting forces opposing motion

$$E = R$$
$$= R_r + R_g \text{ (N)}$$

where E = tractive effort (N)

R = resisting forces (N)

Once the minimum tractive effort has been calculated, the bottom gear ratio can be derived in the following way:

Driving torque = Available torque

$$ER = T\ G_B G_F \eta_M$$

$$\therefore \text{ Bottom gear ratio } G_B = \frac{ER}{TG_F\eta_M}$$

where G_F = final drive gear ratio

G_B = bottom gear ratio

η_M = mechanical efficiency

E = tractive effort (N)

T = maximum engine torque (Nm)

R = effective road wheel radius (m)

Example A vehicle weighing 1500 kg has a coefficient of rolling resistance of 0.015. The transmission has a final drive ratio 4.07:1 and an overall mechanical efficiency of 85%.

If the engine develops a maximum torque of 100 Nm (Fig. 3.34) and the effective road wheel radius is 0.27 m, determine the gearbox bottom gear ratio.

Assume the steepest gradient to be encountered is a one in four.

$$R_r = 10C_rW$$
$$= 10 \times 0.015 \times 150 = 225\text{N}$$

$$R_g = \frac{10W}{G}$$
$$= \frac{10 \times 1500}{4} = 3750\text{N}$$

$$E = R_r + R_g$$
$$= 3750 + 225 = 3975\text{N}$$

$$G_B = \frac{eR}{TG_F\eta_M}$$
$$= \frac{3975 \times 0.27}{100 \times 4.07 \times 0.85} = 3.1{:}1$$

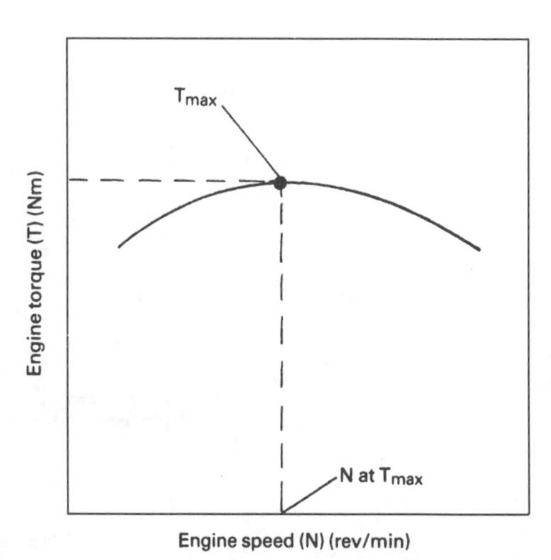

Fig. 3.34 Engine torque to speed characteristics

3.8.3 Setting intermediate gear ratios

Ratios between top and bottom gears should be spaced in such a way that they will provide the tractive effort–speed characteristics as close to the ideal as possible. Intermediate ratios can be best selected as a first approximation by using a geometric progression. This method of obtaining the gear ratios requires the engine to operate within the same speed range in each gear, which is normally selected to provide the best fuel economy.

Consider the engine to vehicle speed characteristics for each gear ratio as shown (Fig. 3.35). When changing gear the engine speed will drop from the highest N_H to the lowest N_L without any change in road speed, i.e. V_1, V_2, V_3 etc.

Let G_1 = 1st overall gear ratio
G_2 = 2nd overall gear ratio
G_3 = 3rd overall gear ratio
G_4 = 4th overall gear ratio
G_5 = 5th overall gear ratio

where $$\text{Overall gear ratio} = \frac{\text{Engine speed (rev/min)}}{\text{Road wheel speed (rev/min)}}$$

Wheel speed when engine is on the high limit N_H in first gear $G_1 = \frac{N_H}{G_1}$ (rev/min)

Wheel speed when engine is on the low limit N_L in second gear $G_2 = \frac{N_L}{G_2}$ (rev/min)

These wheel speeds must be equal for true rolling

Hence $$\frac{N_H}{G_1} = \frac{N_L}{G_2}$$

$$\therefore G_2 = G_1 \frac{N_L}{N_H}$$

Also $$\frac{N_H}{G_2} = \frac{N_L}{G_3}$$

$$\therefore G_3 = G_2 \frac{N_L}{N_H}$$

and $$\frac{N_H}{G_3} = \frac{N_L}{G_4}$$

$$\therefore G_4 = G_3 \frac{N_L}{N_H}$$

$$\frac{N_H}{G_4} = \frac{N_L}{G_5}$$

$$\therefore G_5 = G_4 \frac{N_L}{N_H}$$

The ratio $\frac{N_L}{N_H}$ is known as the minimum to maximum speed range ratio K for a given engine.

Now, gear $G_2 = G_1 \frac{N_L}{N_H} = G_1 K$,

since $\frac{N_L}{N_H} = k$ (a constant)

$$\text{gear } G_3 = G_2 \frac{N_L}{N_H} = G_2\text{K} = (G_1 K)K = G_1\text{K}^2$$

$$\text{gear } G_4 = G_3 \frac{N_L}{N_H} = G_3\text{K} = (G_1 K^2)K = G_1\text{K}^3$$

$$\text{gear } G_5 = G_4 \frac{N_L}{N_H} = G_4\text{K} = (G_1 K^3)K = G_1\text{K}^4.$$

Hence the ratios form a geometric progression.

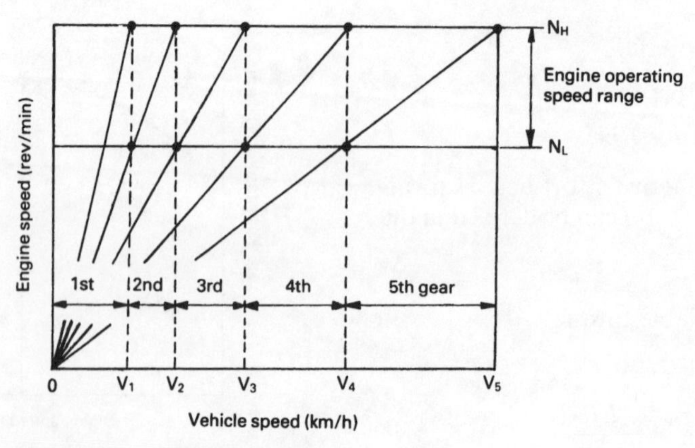

Fig. 3.35 Gear ratios selected on geometric progression

The following relationship will also apply for a five speed gearbox:

$$\frac{G_2}{G_1} = \frac{G_3}{G_2} = \frac{G_4}{G_3} = \frac{G_5}{G_4} = \frac{N_L}{N_H} = K$$

and $G_5 = G_1 K^4$ or $K^4 = \dfrac{G_5}{G_1}$

Hence $K = \left[\dfrac{G_5}{G_1}\right]^{\frac{1}{4}}$ or $\sqrt[4]{\dfrac{G_5}{G_1}}$

In general, if the ratio of the highest gear (G_T) and that of the lowest gear (G_B) have been determined, and the number of speeds (gear ratios) of the gearbox n_G is known, the constant K can be determined by:

$$K = \left[\frac{G_T}{G_B}\right]^{\frac{1}{n\mathrm{G}^{-1}}}$$

So $$\frac{G_T}{G_B} = K^{n\mathrm{G}^{+1}}$$

$\therefore$ $$G_T = G_B K^{n\mathrm{G}^{-1}}$$

For commercial vehicles, the gear ratios in the gearbox are often arranged in geometric progression. For passenger cars, to suit the changing traffic conditions, the step between the ratios of the upper two gears is often closer than that based on geometric progression. As a result, this will affect the selection of the lower gears to some extent.

Example A transmission system for a vehicle is to have an overall bottom and top gear ratio of 20:1 and 4.8 respectively. If the minimum to maximum speeds at each gear changes are 2100 and 3000 rev/min respectively, determine the following:

a) the intermediate overall gear ratios
b) the intermediate gearbox and top gear ratios.

$$K = \frac{N_L}{N_H} = \frac{2100}{3000} = 0.7$$

a) 1st gear ratio $G_1 = 20.0{:}1$
2nd gear ratio $G_2 = G_1 K = 20 \times 0.7 = 14.0{:}1$
3rd gear ratio $G_3 = G_1 K^2 = 20 \times 0.7^2 = 9.8{:}1$
4th gear ratio $G_4 = G_1 K^3 = 20 \times 0.7^3 = 6.86{:}1$
5th gear ratio $G_5 = G_1 K^4 = 20 \times 0.7^4 = 4.8{:}1$

b) $G_1 = \dfrac{20.0}{4.8} = 4.166{:}1$

$G_2 = \dfrac{14.0}{4.8} = 2.916{:}1$

$G_3 = \dfrac{9.8}{4.8} = 2.042{:}1$

$G_4 = \dfrac{6.86}{4.8} = 1.429{:}1$

Top gear $G_5 = \dfrac{4.8}{4.8} = 1.0{:}1$

4 Hydrokinetic fluid couplings and torque converters

A fluid drive uses hydrokinetic energy as a means of transferring power from the engine to the transmission in such a way as to automatically match the vehicle's speed, load and acceleration requirement characteristics. These drives may be of a simple two element type which takes up the drive smoothly without providing increased torque or they may be of a three or more element unit which not only conveys the power as required from the engine to the transmission, but also multiplies the output torque in the process.

4.1 Hydrokinetic fluid couplings

(Figs 4.1 and 4.2)

The hydrokinetic coupling, sometimes referred to as a fluid flywheel, consists of two saucer-shaped discs, an input impeller (pump) and an output turbine (runner) which are cast with a number of flat radial vanes (blades) for directing the flow path of the fluid (Fig. 4.1).

Owing to the inherent principle of the hydrokinetic coupling, there must be relative slip between the input and output member cells exposed to each

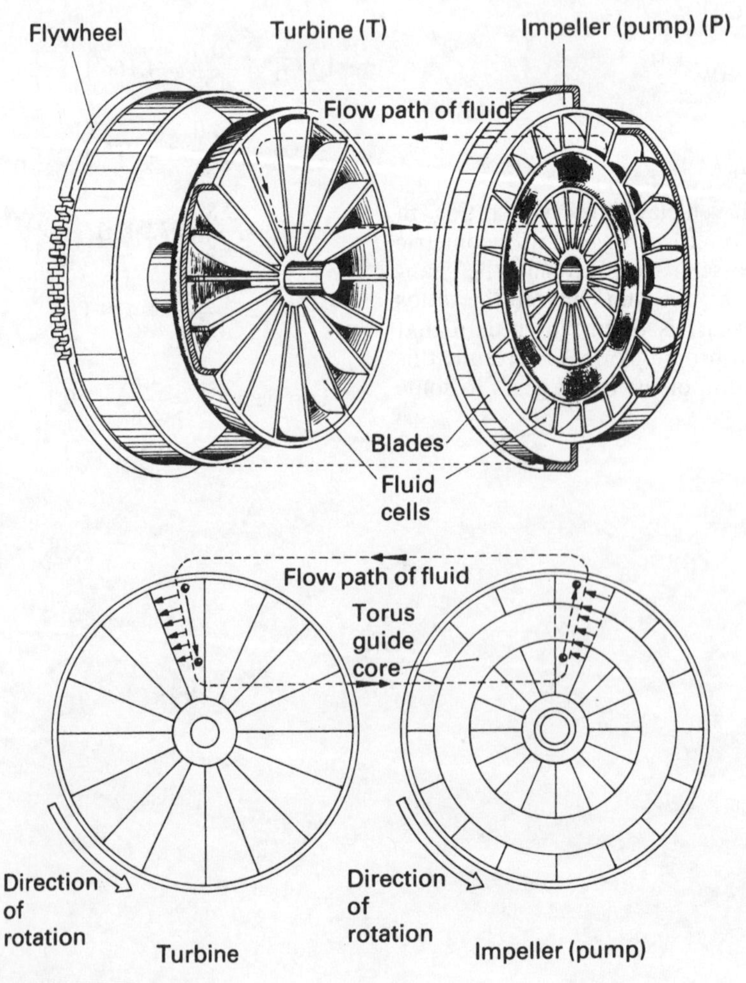

Fig. 4.1 Fluid coupling action

other, and the vortex flow path created by pairs of adjacent cells will be continuously aligned and misaligned with different cells.

With equal numbers of cells in the two half members, the relative cell alignment of all the cells occurs together. Consequently, this would cause a jerky transfer of torque from the input to the output drive. By having differing numbers of cells within the impeller and turbine, the alignment of each pair of cells at any one instant will be slightly different so that the impingement of fluid from one member to the other will take place in various stages of circulation, with the result that the coupling torque transfer will be progressive and relatively smooth.

The two half-members are put together so that the fluid can rotate as a vortex. Originally it was common practice to insert at the centre of rotation a hollow core or guide ring (sometimes referred to as the *torus*) within both half-members to assist in establishing fluid circulation at the earliest moment of relative rotation of the members. These couplings had the disadvantage that they produced considerable drag torque whilst idling, this being due mainly to the effectiveness of the core guide in circulating fluid at low speeds. As coupling development progressed, it was found that turbine drag was reduced at low speeds by using only a core guide on the impeller member (Fig. 4.2). With the latest design

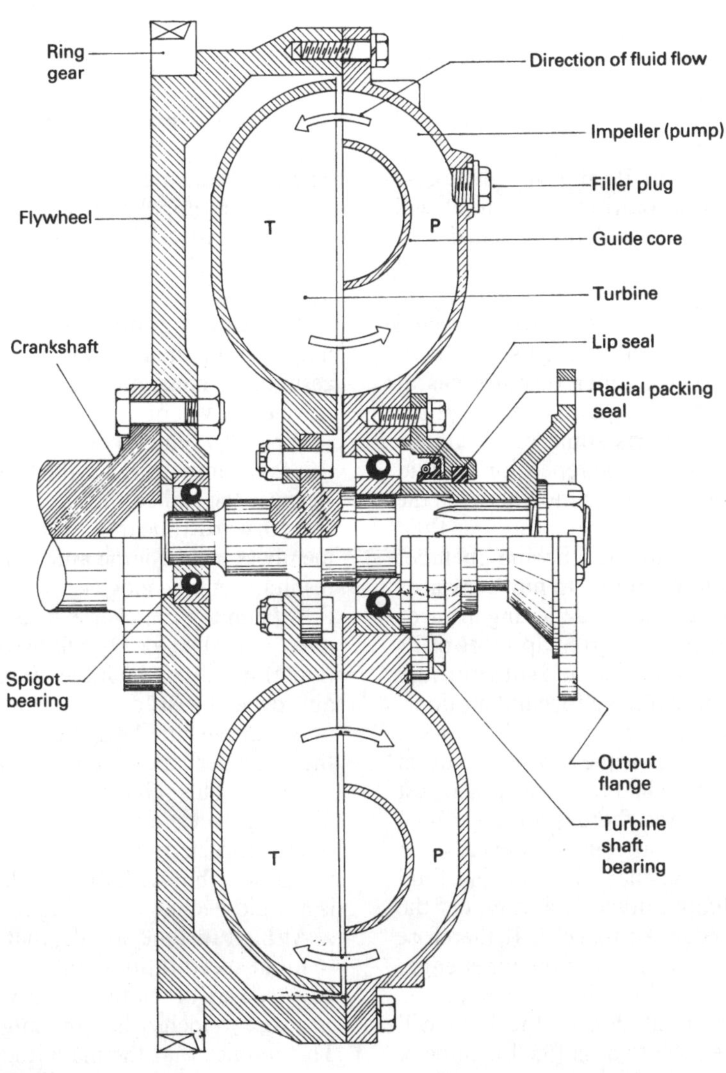

Fig. 4.2 Fluid coupling

these cores are eliminated altogether as this also reduces fluid interference in the higher speed range and consequently reduces the degree of slip for a given amount of transmitted torque (Fig. 4.6).

4.1.1 Hydrokinetic fluid coupling principle of operation (Figs 4.1 and 4.3)

When the engine is started, the rotation of the *impeller* (pump) causes the working fluid trapped in its cells to rotate with it. Accordingly, the fluid is subjected to centrifugal force and is pressurized so that it flows radially outwards.

To understand the principle of the hydrokinetic coupling it is best to consider a small particle of fluid circulating between one set of impeller and turbine vanes at various points A, B, C and D as shown in Figs 4.1 and 4.3.

Initially a particle of fluid at point A, when the engine is started and the impeller is rotated, will experience a centrifugal force due to its mass and radius of rotation, r. It will also have acquired some kinetic energy. This particle of fluid will be forced to move outwards to point B, and in the process of increasing its radius of rotation from r to R, will now be subjected to considerably more centrifugal force and it will also possess a greater amount of kinetic energy. The magnitude of the kinetic energy at this outermost position forces it to be ejected from the mouth of the impeller cell, its flow path making it enter one of the outer turbine cells at point C. In doing so it reacts against one side of the turbine vanes and so imparts some of its kinetic energy to the turbine wheel. The repetition of fluid particles being flung across the junction between the impeller and turbine cells will force the first fluid particle in the slower moving turbine member (having reduced centrifugal force) to move inwards to point D. Hence in the process of moving inwards from R to r, the fluid particle gives up most of its kinetic energy to the turbine wheel and subsequently this is converted into propelling effort and motion.

The creation and conversion of the kinetic energy of fluid into driving torque can be visualized in the following manner: when the vehicle is at rest the turbine is stationary and there is no centrifugal force acting on the fluid in its cells. However, when the engine rotates the impeller, the working fluid in its cells flows radially outwards and enters the turbine at the outer edges of its cells. It therefore causes a displacement of fluid from the inner edges of the turbine cells into the inner edges of the impeller cells, thus a circulation of the fluid will be established between the two half cell members. The fluid has two motions; firstly it is circulated by the impeller around its axis and secondly it circulates round the cells in a vortex motion.

This circulation of fluid only continues as long as there is a difference in the angular speeds of the impeller and turbine, because only then is the centrifugal force experienced by the fluid in the faster moving impeller greater than the counter centrifugal force acting on the fluid in the slower moving turbine member. The velocity of the fluid around the couplings' axis of rotation increases while it flows radially outwards in the impeller cells due to the increased distance it has moved from the centre of rotation. Conversely, the fluid velocity decreases when it flows inwards in the turbine cells. It therefore follows that the fluid is given kinetic energy by the impeller and gives up its kinetic energy to the turbine. Hence there is a transference of energy from the input impeller to the output turbine, but there is no torque multiplication in the process.

4.1.2 Hydrokinetic fluid coupling velocity diagrams (Fig. 4.3)

The resultant magnitude of direction of the fluid leaving the impeller vane cells, V_R, is dependent upon the exit velocity, V_E, this being a measure of the vortex circulation flow rate and the relative linear velocity between the impeller and turbine, V_L.

The working principle of the fluid coupling may be explained for various operating conditions assuming a constant circulation flow rate by means of velocity vector diagrams (Fig. 4.3).

When the vehicle is about to pull away, the engine drives the impeller with the turbine held stationary. Because the stalled turbine has no motion, the relative forward (linear) velocity V_L between the two members will be large and consequently so will the resultant entry velocity V_R. The direction of fluid flow from the impeller exit to turbine entrance will make a small angle Θ_1, relative to the forward direction of motion, which therefore produces considerable drive thrust to the turbine vanes.

As the turbine begins to rotate and catch up to the impeller speed the relative linear speed is reduced. This changes the resultant fluid flow direction to Θ_2 and decreases its velocity. The net output thrust, and hence torque carrying capacity, will be less, but with the vehicle gaining speed there is a rapid decline in driving torque requirements.

At high turbine speeds, that is, when the output to input speed ratio is approaching unity, there will be only a small relative linear velocity and resultant entrance velocity, but the angle Θ_3 will be large. This implies that the magnitude of the fluid thrust will be very small and its direction ineffective in

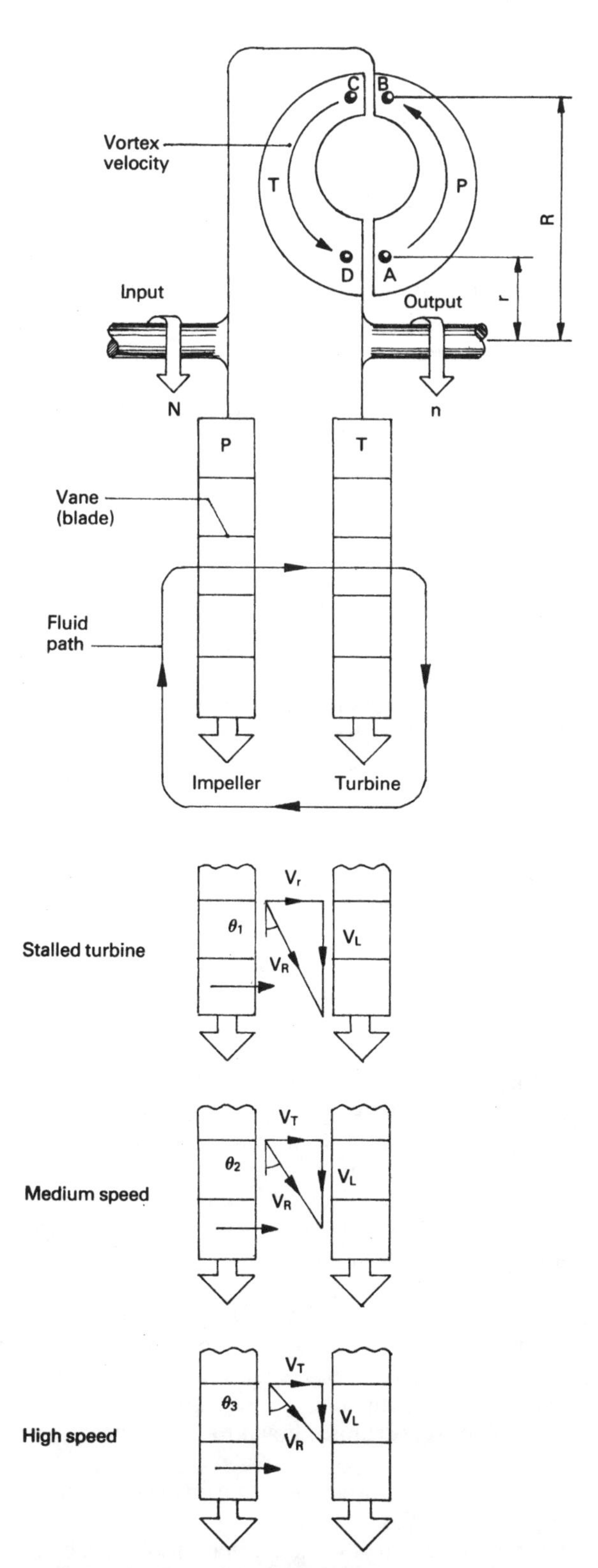

Fig. 4.3 Principle of the fluid coupling

V_E = Impeller exit velocity
V_L = Relative linear velocity of both pump and turbine
V_R = Resultant effective velocity and direction of fluid

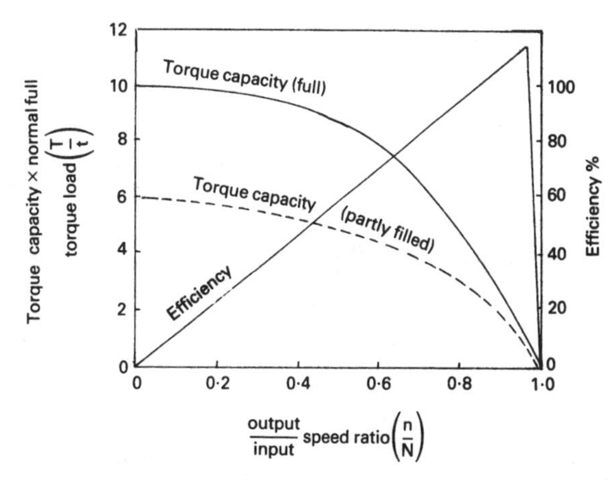

Fig. 4.4 Relationship of torque capacity efficiency and speed ratio for fluid couplings

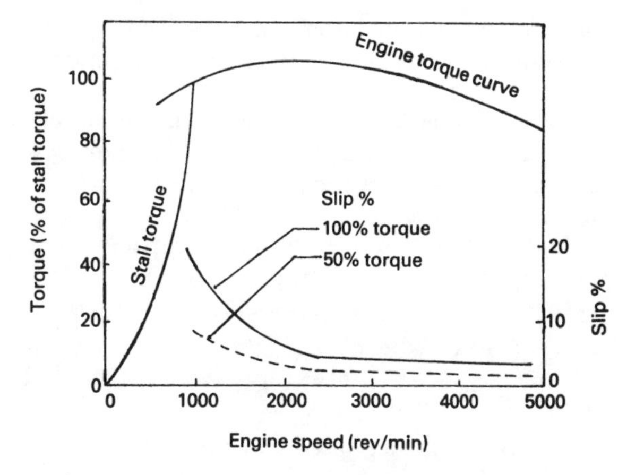

Fig. 4.5 Relationship of engine speed, torque and slip for a fluid coupling

rotating the turbine. Thus the output member will slip until sufficient circulating fluid flow imparts enough energy to the turbine again.

It can be seen that at high rotational speeds the cycle of events is a continuous process of output speed almost, but never quite, catching up to input speed, the exception being when the drive changes from engine driven to overrun transmission driven when the operating conditions will be reversed.

4.2 Hydrokinetic fluid coupling efficiency and torque capacity (Figs 4.4 and 4.5)

Coupling efficiency is the ratio of the power available at the turbine to the amount of power supplied to the impeller. The difference between input and output power, besides the power lost by fluid shock, friction and heat, is due mainly to the relative slip between the two members (Fig. 4.4). A more useful term is the *percentage slip*, which is defined as the ratio of the difference in input and output speeds divided by the input speed and multiplied by 100.

i.e. $$\% \text{ slip} = \left(\frac{N-n}{N}\right) \times 100$$

The percentage slip will be greatly influenced by the engine speed and output turbine load conditions (Fig. 4.5). A percentage of slip must always exist to create a sufficient rate of vortex circulation which is essential to impart energy from the impeller to the turbine. The coupling efficiency is at best about 98% under light load high rotational speed conditions, but this will be considerably reduced as turbine output load is increased or impeller speed is lowered. If the output torque demand increases, more slip will occur and this will increase the vortex circulation velocity which will correspondingly impart more kinetic energy to the output turbine member, thus raising the torque capacity of the coupling. An additional feature of such couplings is that if the engine should tend to stall due to overloading when the vehicle is accelerated from rest, the vortex circulation will immediately slow down, preventing further torque transfer until the engine's speed has recovered.

Fluid coupling torque transmitting capacity for a given slip varies as the fifth power of the impeller internal diameter and as the square of its speed.

i.e. $$T \propto D^5 N^2$$

where D = impeller diameter

N = impeller speed (rev/min)

Thus it can be seen that only a very small increase in impeller diameter, or a slight increase in impeller speed, considerably raises the coupling torque carrying capacity. A further controlling factor which affects the torque transmitted is the quantity of fluid circulating between the impeller and turbine. Raising or lowering the fluid level in the coupling increases or decreases the torque which can be transmitted to the turbine (Fig. 4.4).

4.3 Fluid friction coupling (Figs 4.6 and 4.7)

A fluid coupling has the take-up characteristics which particularly suit the motor vehicle but it suffers from two handicaps that are inherent in the system. Firstly, idling drag tends to make the vehicle creep forwards unless the parking brake is fully applied, and secondly there is always a small amount of slip which is only slight under part load (less than 2%) but becomes greater when transmitting anything near full torque.

These limitations have been overcome for large truck applications by combining a shoe and drum centrifugally operated clutch to provide a positive lock-up at higher output speeds with a smaller coreless fluid coupling than would be necessary if the drive was only to be through a fluid coupling. The reduced size and volume of fluid circulation in the coupling thereby eliminate residual idling drag (Fig. 4.6).

With this construction there is a shoe carrier between the impeller and flywheel attached to the output shaft. Mounted on this carrier are four brake shoes with friction material facings. They are each pivoted (hinged) to the carrier member at one end and a garter spring (coil springs shown on front view to illustrate action) holds the shoes in their retraction position when the output shaft is at rest.

When the engine is accelerated the fluid coupling automatically takes up the drive with maximum smoothness. Towards maximum engine torque speed the friction clutch shoes are thrown outwards by the centrifugal effect until they come into contact with the flywheel drum. The frictional grip will now lock the input and output drives together. Subsequently the fluid vortex circulation stops and the fluid coupling ceases to function (Fig. 4.7).

Relative slip between input and output member in low gear is considerably reduced, due to the automatic friction clutch engagement, and engine braking is effectively retained down to idling speeds.

4.4 Hydrokinetic three element torque converter (Figs 4.8 and 4.9)

A three element torque converter coupling is comprised of an input impeller casing enclosing the

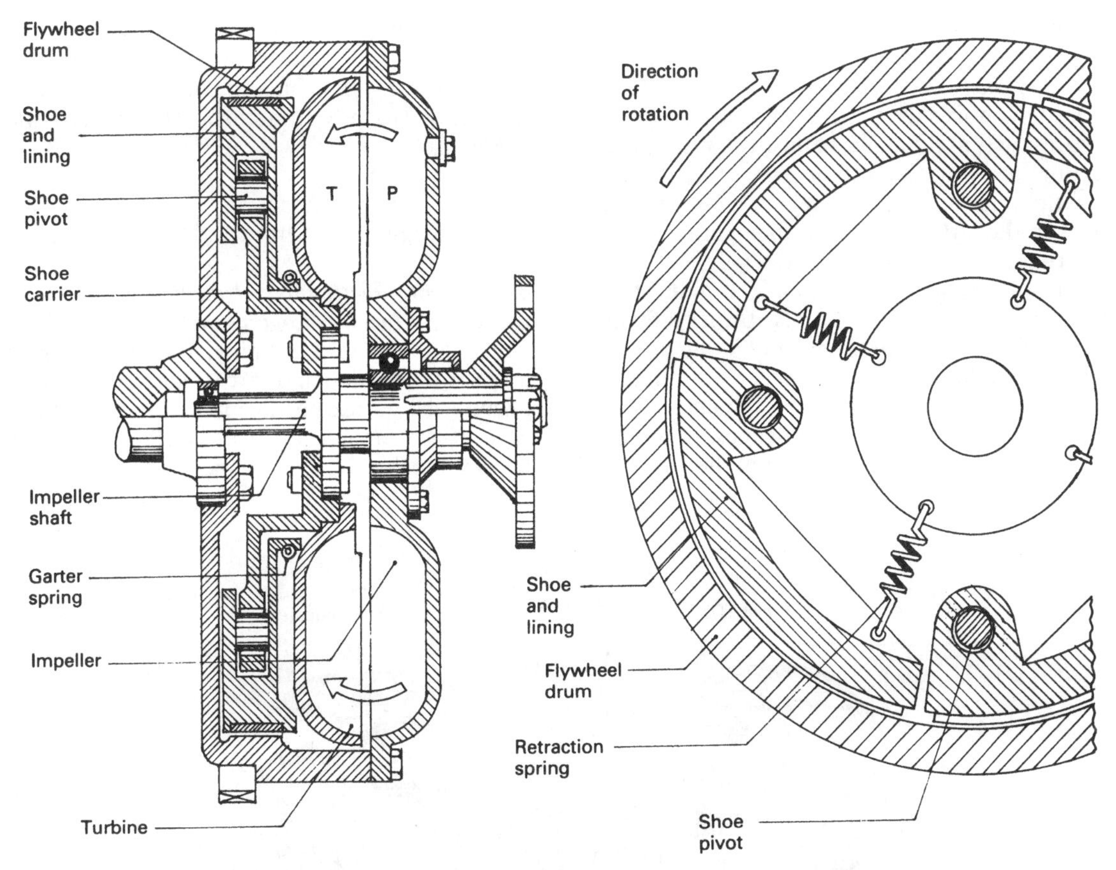

Fig. 4.6 Fluid friction coupling

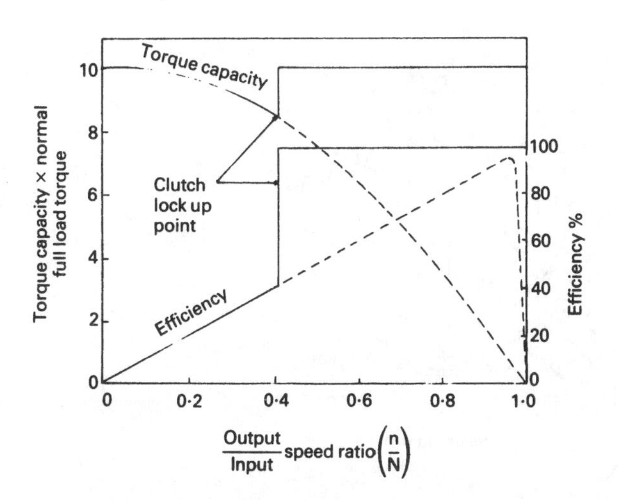

Fig. 4.7 Relationship of torque carrying capacity, efficiency and output speed for a fluid coupling

output turbine wheel. There are about 26 and 23 blades for the impeller and turbine elements respectively. Both of these elements and their blades are fabricated from low carbon steel pressings. The third element of the converter called the *stator* is usually an aluminium alloy casting which may have something in the order of 15 blades (Figs 4.8 and 4.9).

The working fluid within a converter when the engine is operating has two motions:

1 Fluid trapped in the impeller and turbine vane cells revolves bodily with these members about their axis of rotation.
2 Fluid trapped between the impeller and turbine vane cells and their central torus core rotates in a circular path in the section plane, this being known as its *vortex motion*.

When the impeller is rotated by the engine, it acts as a centrifugal pump drawing in fluid near the centre of rotation, forcing it radially outwards through the cell passages formed by the vanes to the impeller peripheral exit. Here it is ejected due to its momentum towards the turbine cell passages and in the process acts at an angle against the vanes, thus imparting torque to the turbine member (Fig. 4.8).

The fluid in the turbine cell passages moves inwards to the turbine exit. It is then compelled to flow between the fixed stator blades (Fig. 4.9). The reaction of the fluid's momentum as it glides over the curved surfaces of the blades is absorbed by the casing to which the stator is held and in the process it is redirected towards the impeller entrance. It enters the passages shaped by the impeller vanes. As it acts on the drive side of the vanes, it imparts a torque equal to the stator reaction in the direction of rotation (Fig. 4.8).

It therefore follows that the engine torque delivered to the impeller and the reaction torque transferred by the fluid to the impeller are both transmitted to the output turbine through the media of the fluid.

i.e. $\text{Engine torque} + \text{Reaction torque} = \text{Output turbine torque}$

4.4.1 Hydrokinetic three element torque converter principle of operation (Fig. 4.8)

When the engine is running, the impeller acts as a centrifugal pump and forces fluid to flow radially around the vortex passage made by the vanes and core of the three element converter. The rotation of the impeller by the engine converts the engine power into hydrokinetic energy which is utilized in

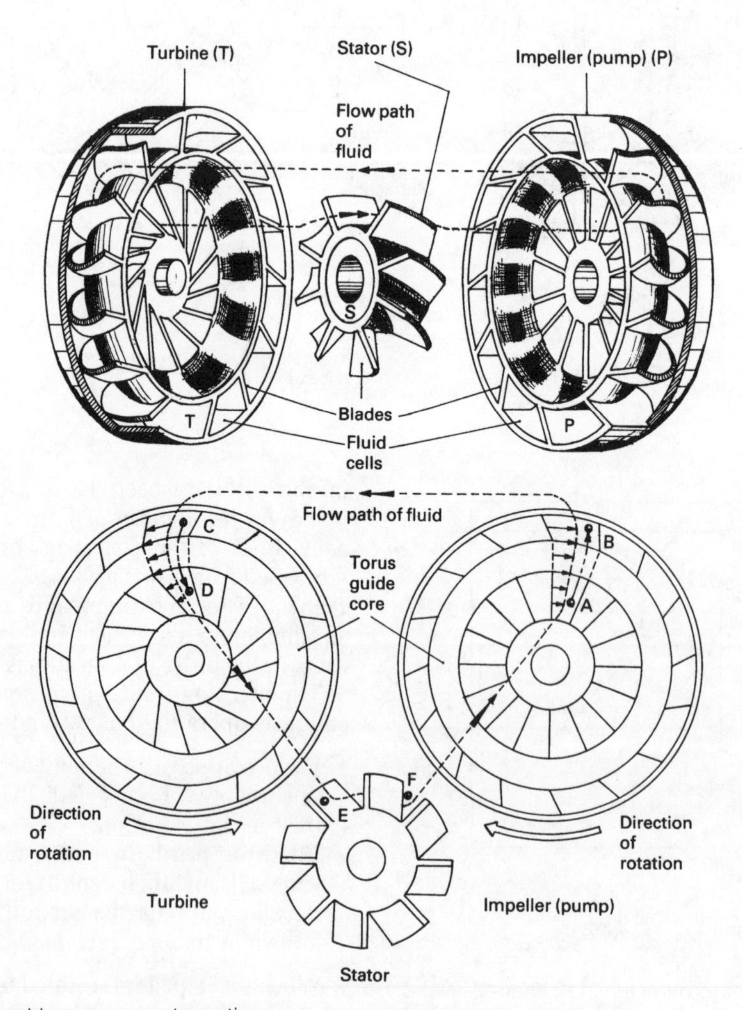

Fig. 4.8 Three element torque converter action

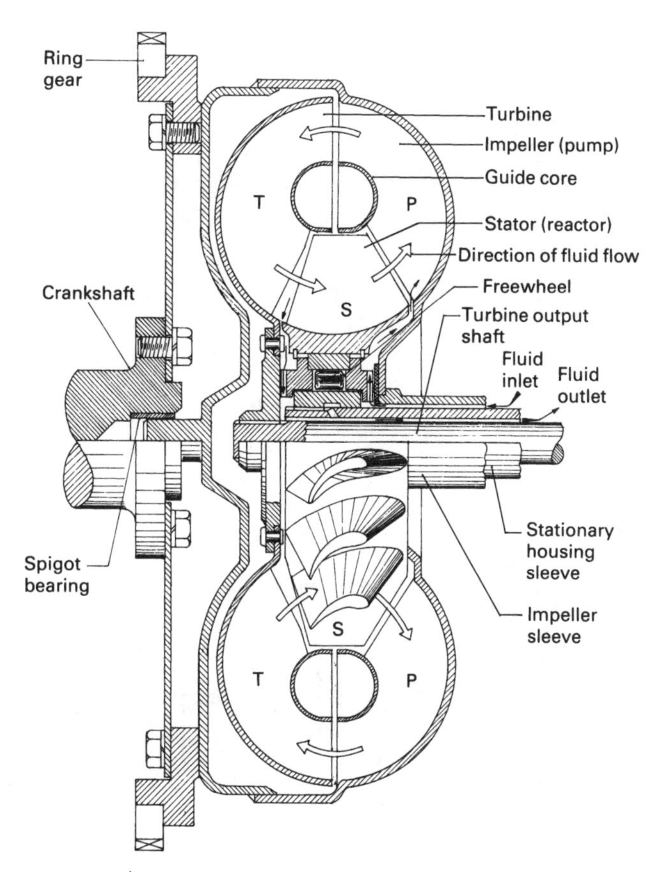

Fig. 4.9 Three element torque converter

providing a smooth engine to transmission take-up and in producing torque multiplication if a third fixed stator member is included.

An appreciation of the principle of the converter can be obtained by following the movement and events of a fluid particle as it circulates the vortex passage (Fig. 4.8).

Consider a fluid particle initially at the small diameter entrance point A in the impeller. As the impeller is rotated by the engine, centrifugal force will push the fluid particle outwards to the impeller's largest exit diameter, point B. Since the particle's circumferential distance moved every revolution will be increased, its linear velocity will be greater and hence it will have gained kinetic energy.

Pressure caused by successive particles arriving at the impeller outermost cell exit will compel the particle to be flung across the impeller–turbine junction where it acts against the side of cell vane it has entered at point C and thereby transfers some of its kinetic energy to the turbine wheel. Because the turbine wheel rotates at a lower speed relative to the impeller, the pressure generated in the impeller will be far greater than in the turbine. Subsequently the fluid particle in the turbine curved passage will be forced inwards to the exit point D and in doing so will give up more of its kinetic energy to the turbine wheel.

The fluid particle, still possessing kinetic energy at the turbine exit, now moves to the stator blade's entrance side to point E. Here it is guided by the curvature of the blades to the exit point F.

From the fixed stator (reactor) blades the fluid path is again directed to the impeller entrance point A where it imparts its hydrokinetic energy to the impeller, this being quite separate to the kinetic energy produced by the engine rotating the impeller. Note that with the fluid coupling, the transfer of fluid from the turbine exit to the impeller entrance is direct. Thus the kinetic energy gained by the input impeller is that lost by the output turbine and there is no additional gain in output turning effort, as is the case when a fixed intermediate stator is incorporated.

4.4.2 *Hydrokinetic three element torque converter velocity diagrams* (Figs 4.9 and 4.10)
The direction of fluid leaving the turbine to enter the stator blades is influenced by the tangential exit velocity which is itself determined by the vortex circulating speed and the linear velocity due to the rotating turbine member (Fig. 4.10).

When the turbine is in the stalled condition and the impeller is being driven by the engine, the direction of the fluid leaving the impeller will be determined entirely by the curvature and shape of the turbine vanes. Under these conditions, the fluid's direction of motion, Θ_1, will make it move deep into the concave side of the stator blades where it reacts and is

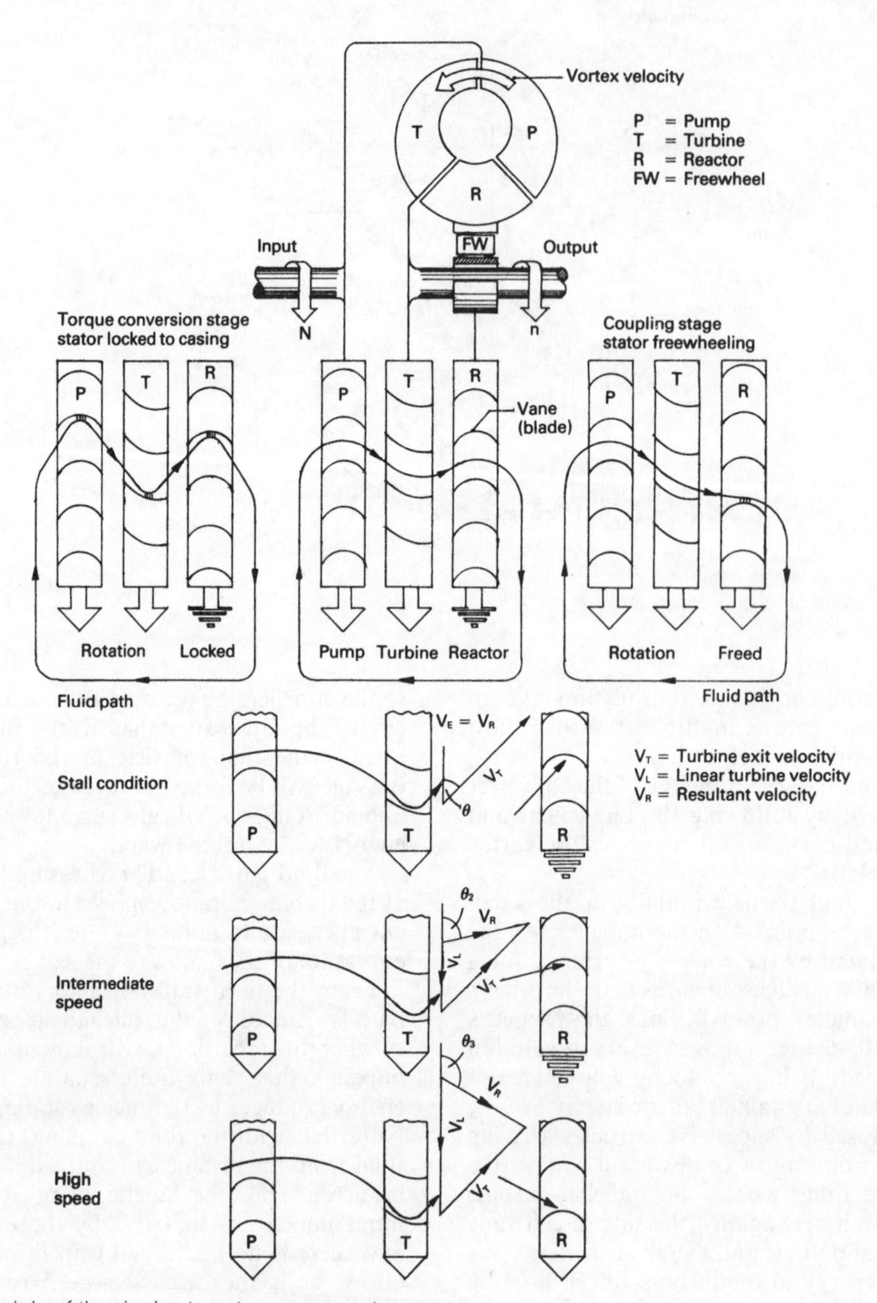

Fig. 4.10 Principle of the single stage torque converter

made to flow towards the entrance of the impeller in a direction which provides the maximum thrust.

Once the turbine begins to rotate, the fluid will acquire a linear velocity so that the resultant effective fluid velocity direction will now be Θ_2. A reduced backward reaction to the stator will be produced so that the direction of the fluid's momentum will not be so effective.

As the turbine speed of rotation rises, the fluid's linear forward velocity will also increase and, assuming that the turbine's tangential exit velocity does not alter, the resultant direction of the fluid will have changed to Θ_3 where it now acts on the convex (back) side of the stator blades.

Above the critical speed, when the fluid's thrust changes from the concave to the convex side of the blades, the stator reaction torque will now act in the opposite sense and redirect the fluid. Thus its resultant direction towards the impeller entry passages will hinder instead of assist the impeller motion. The result of this would be in effect to cancel out some of the engine's input torque with further speed increases.

The inherent speed limitation of a hydrokinetic converter is overcome by building into the stator hub a one way clutch (freewheel) device (Fig. 4.9). Therefore, when the direction of fluid flow changes sufficiently to impinge onto the back of the blades, the stator hub is released, allowing it to spin freely between the input and output members. The freewheeling of the stator causes very little fluid interference, thus the three element converter now becomes a two element coupling. This condition prevents the decrease in torque for high output speeds and produces a sharp rise in efficiency at output speeds above the coupling point.

4.4.3 Hydrokinetic torque converter characteristics (Figs. 4.11 and 4.12)

Maximum torque multiplication occurs when there is the largest speed difference between the impeller and turbine. A torque output to input ratio of about 2:1 normally occurs with a three element converter when the turbine is stationary. Under such conditions, the vortex rate of fluid circulation will be at a peak. Subsequently the maximum hydrokinetic energy transfer from the impeller to turbine then stator to impeller again takes place (Figs 4.11 and 4.12). As the turbine output speed increases relative to the impeller speed, the efficiency rises and the vortex velocity decreases and so does the output to input torque ratio until eventually the circulation rate of fluid is so low that it can only support a 1:1 output to input torque ratio. At this point the reaction torque will be zero. Above this speed the stator is freewheeled. This offers less resistance to the circulating fluid and therefore produces an improvement in coupling efficiency (Figs 4.11 and 4.12).

This description of the operating conditions of the converter coupling shows that if the transmission is suddenly loaded the output turbine speed will automatically drop, causing an increase in fluid circulation and correspondingly a rise in torque multiplication, but conversely a lowering of efficiency due to the increased slip between input and output members. When the output conditions have changed and a reduction in load or an increase in turbine speed follows the reverse happens; the efficiency increases and the output to input torque ratio is reduced.

4.5 Torque converter performance terminology (Figs 4.11 and 4.12)

To understand the performance characteristics of a fluid drive (both coupling and converter), it is essential to identify and relate the following terms used in describing various relationships and conditions.

4.5.1 Fluid drive efficiency (Figs 4.11 and 4.12)

A very convenient method of expressing the energy losses, due mainly to fluid circulation within a fluid drive at some given output speed or speed ratio, is

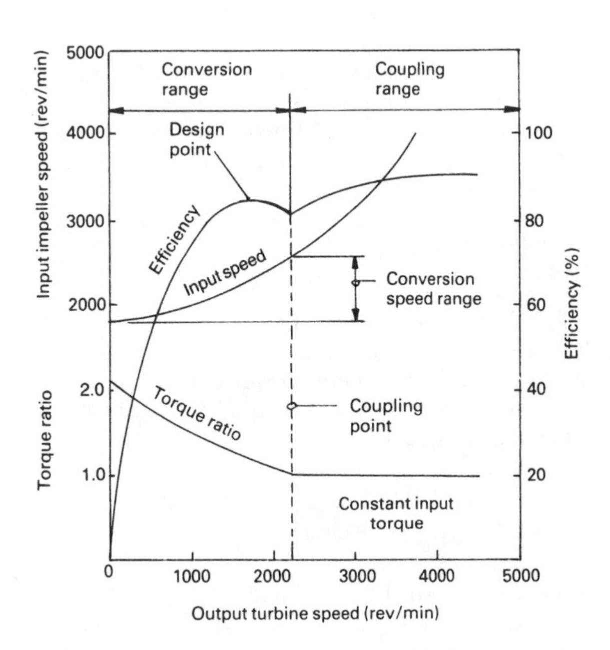

Fig. 4.11 Characteristic performance curves for a three element converted coupling

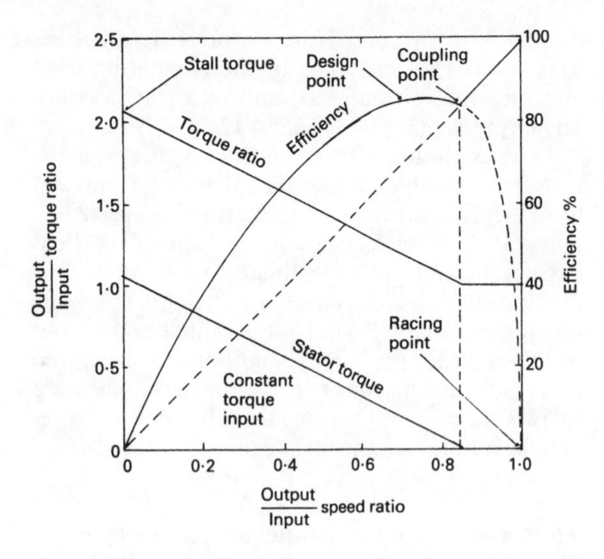

Fig. 4.12 Characteristic performance curves for a converter coupling plotted to a base of output (turbine speed) to input (impeller speed)

to measure its efficiency, that is, the percentage ratio of output to input work done.

$$\text{i.e.}\quad \text{Efficiency} = \frac{\text{Output work done}}{\text{Input work done}} \times 100$$

4.5.2 Speed ratio (Fig. 4.12)

It is frequently necessary to compare the output and input speed differences at which certain events occur. This is normally defined in terms of a speed ratio of output (turbine) speed N_2 to the input (impeller) speed N_1.

$$\text{i.e.}\quad \text{Speed ratio} = \frac{N_2}{N_1}$$

4.5.3 Torque ratio (Fig. 4.12)

The torque multiplication within a fluid drive is more conveniently expressed in terms of a torque ratio of output (turbine) torque T_2 to the input (impeller) torque T_1.

$$\text{i.e.}\quad \text{Torque ratio} = \frac{T_2}{T_1}$$

4.5.4 Stall speed (Figs 4.11 and 4.12)

This is the maximum speed which the engine reaches when the accelerator pedal is fully down, the transmission in drive and the foot brake is fully applied. Under such conditions there is the greatest impeller to turbine speed variation, with the result that the vortex fluid circulation and correspondingly torque conversion are at a maximum, conversely converter efficiency is zero. Whilst these stall conditions prevail, torque conversion loading drags the engine speed down to something like 60–70% of the engine's maximum torque speed, i.e. 1500–2500 rev/min. A converter should only be held in the stall condition for the minimum of time to prevent the fluid being overworked.

4.5.5 Design point (Figs 4.11 and 4.12)

Torque converters are so designed that their internal passages formed by the vanes are shaped so as to make the fluid circulate with the minimum of resistance as it passes from one member to another member at definite impeller to turbine speed ratio, known as the *design point*. A typical value might be 0.8:1.

Above or below this optimum speed ratio, the resultant angle and direction of fluid leaving one member to enter another will alter so that the flow from the exit of one member to the entry of another will no longer be parallel to the surfaces of the vanes, in fact it will strike the sides of the passage vanes entered. When the exit and entry angles of the vanes do not match the effective direction of fluid motion, some of its momentum will be used up in entrance losses and consequently the efficiency declines as the speed ratio moves further away on either side of the design point. Other causes of momentum losses are internal fabrication finish, surface roughness and inter-vane or blade thickness interference. If the design point is shifted to a lower speed ratio, say 0.6, the torque multiplication will be improved at stall and lower speed conditions at the expense of an earlier fall-off in efficiency at the high speed ratio such as 0.8. There will be a reduction in the torque ratio but high efficiency will be maintained in the upper speed ratio region.

4.5.6 Coupling point (Figs 4.11 and 4.12)

As the turbine speed approaches or exceeds that of the impeller, the effective direction of fluid entering the passages between the stator blades changes from pushing against the concave face to being redirected towards the convex (back) side of the blades. At this point, torque conversion due to fluid transfer from the fixed stator to the rotating impeller, ceases. The turbine speed when the direction of the stator reaction is reversed is known as the *coupling point* and is normally between 80 and 90% of the impeller speed. At this point the stator is released by the freewheel device and is then driven in the same direction as the impeller and turbine. At and above this speed the stator blades will spin with the impeller and turbine which then simply act as a fluid coupling, with the benefit of increasing efficiency as the turbine output speed approaches but never reaches the input impeller speed.

4.5.7 Racing or run-away point (Fig. 4.12)

If the converter does not include a stator freewheel device or if the mechanism is jammed, then the direction of fluid leaving the stator would progressively change from transferring fluid energy to assist the impeller rotation to one of opposition as the turbine speed catches up with that of the impeller. Simultaneously, the vortex fluid circulation will be declining so that the resultant torque capacity of the converter rapidly approaches zero. Under these conditions, with the accelerator pedal fully down there is very little load to hold back the engine's speed so that it will tend to race or *run-away*. Theoretically racing or run-away should occur when both the impeller and turbine rotate at the same speed and the vortex circulation ceases, but due to the momentum losses caused by internal fluid resistance, racing will tend to begin slightly before a 1:1 speed ratio (a typical value might be 0.95:1).

4.5.8 Engine braking transmitted through converter or coupling on overrun

Torque converters are designed to maximize their torque multiplication from the impeller to the turbine in the forward direction by adopting backward swept rotating member circulating passage vanes. Unfortunately, in the reverse direction when the turbine is made to drive the impeller on transmission overrun, the exit and entry vane guide angles of the members are unsuitable for hydrokinetic energy transference, so that only a limited amount of engine braking torque can be absorbed by the converter except at high output overrun vehicle speeds. Conversely, a fluid coupling with its flat radial vanes is able to transmit torque in either drive or overrun direction with equal effect.

4.6 Overrun clutches

Various names have been used for these mechanisms such as *freewheel, one way clutch* and *overrun clutch*, each one signifying the nature of the device and is therefore equally appropriate.

A freewheel device is a means whereby torque is transmitted from one stationary or rotating member to another member, provided that input torque (drive speed) is greater than that of the output member. If the conditions are reversed and the output member's applied torque (or speed) becomes greater than that of the input, the output member will overrun the input member (rotate faster). Thus the lock between the two members will be automatically released. Immediately the drive will be discontinued which permits the input and output members to revolve independently to one another.

Overrun clutches can be used for a number of applications, such as starter motor pre-engagement drives, overdrives, torque converter stator release, automatic transmission drives and final differential drives.

Most overrun clutch devices take the form of either the *roller and wedge or sprag lock* to engage and disengage drive.

4.6.1 Overrun clutch with single diameter rollers (Fig. 4.13)

A roller clutch is comprised of an inner and outer ring member and a series of cylindrical rollers spaced between them (see Fig. 4.13). Incorporated between the inner and outer members is a cage which positions the rollers and guides so that they roll up and down their ramps simultaneously. One of the members has a cylindrical surface concentric with its axis, this is usually made the outer member. The other member (inner one) has a separate wedge ramp formed for each roller to react against. The shape of these wedge ramps may be flat or curved depending upon design. In operation each roller provides a line contact with both the outer internal cylindrical track and the external wedge ramp track of the inner member.

When the input wedge member is rotated clockwise and the output cylindrical member is prevented from rotating or rotates anticlockwise in the opposite direction, the rollers revolve and climb up the wedge ramps, and thereby squeeze themselves between the inner and outer member tracks. Eventually the elastic compressive and frictional forces created by the rollers against these tracks prevents further roller rotation. Torque can now transfer from the input inner member to the outer ring member by way of these jammed (locked) rollers.

If the output outer member tries to rotate in the same direction but faster than the inner member, the rollers will tend to rotate and roll down their ramps, thereby releasing (unlocking) the outer member from that of the input drive.

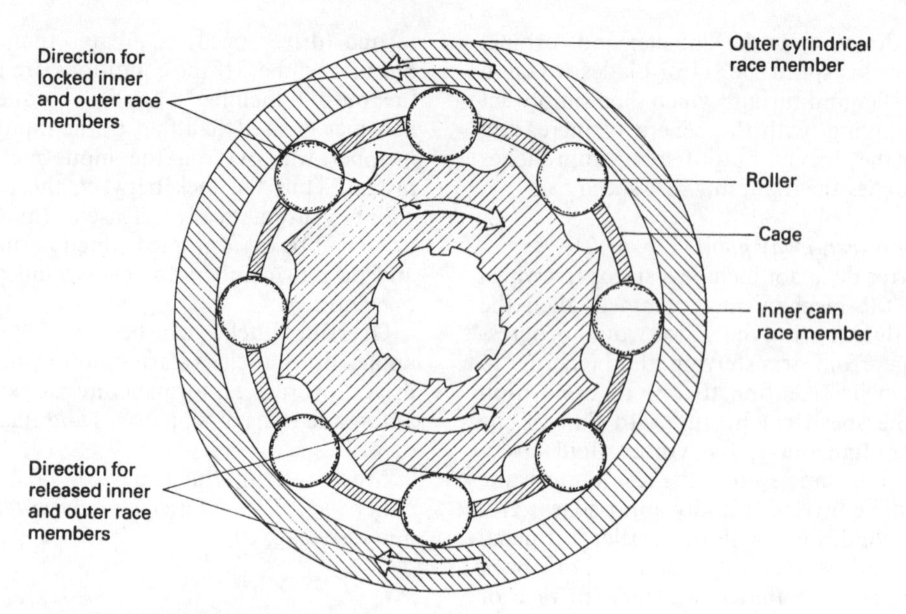

Fig. 4.13 Overrun freewheel single diameter roller type clutch

4.6.2 Overrun clutch with triple diameter rollers (Fig. 4.14)

This is a modification of the single roller clutch in which the output outer member forms an internal cylindrical ring, whereas the input inner member has three identical external inclined plane profiles (see Fig. 4.14). Situated between the inner and outer tracks are groups of three different sized rollers. An anchor block and energizing shoe is arranged, between each group of rollers; the blocks are screwed to the inner member while the shoes (with the assistance of the springs) push the rollers together and against their converging contact tracks. The inclined plane profile required to match the different diameter rollers provides a variable wedge angle for each size of roller. It is claimed that the take-up load of each roller will be progressive and spread more evenly than would be the case if all the rollers were of the same diameter.

When the input inner ring takes up the drive, the rollers revolve until they are wedged between the inclined plane on the inner ring and the cylindrical internal track of the outer member. Consequently the compressive load and the frictional force thus created between the rollers and tracks locks solid the inner and outer members enabling them to transmit torque.

If conditions change and the outer member overruns the inner member, the rollers will be compelled to revolve in the opposite direction to when the drive was established towards the diverging end of the tracks. It thus releases the outer member and creates the freewheel phase.

4.6.3 Sprag overrun clutch (Fig. 4.15)

A very reliable, compact and large torque-carrying capacity overrun clutch is the sprag type clutch. This dispenses with the wedge ramps or inclined plane formed on the inner member which is essential with roller type clutches (see Fig. 4.15). The sprag clutch consists of a pair of inner and outer ring members which have cylindrical external and internal track surfaces respectively. Interlinking the input and output members are circular rows of short struts known as *sprags*. Both ends of the sprags are semicircular with their radius of curvature being offset to each other so that the sprags appear lopsided. In addition a tapered waste is formed in their mid-region. Double cages are incorporated between inner and outer members. These cages have rectangular slots formed to equally space and locate the sprags around the inner and outer tracks. During clutch engagement there will be a slight shift between relative positions of the two cages as the springs tilt, but the spacing will be

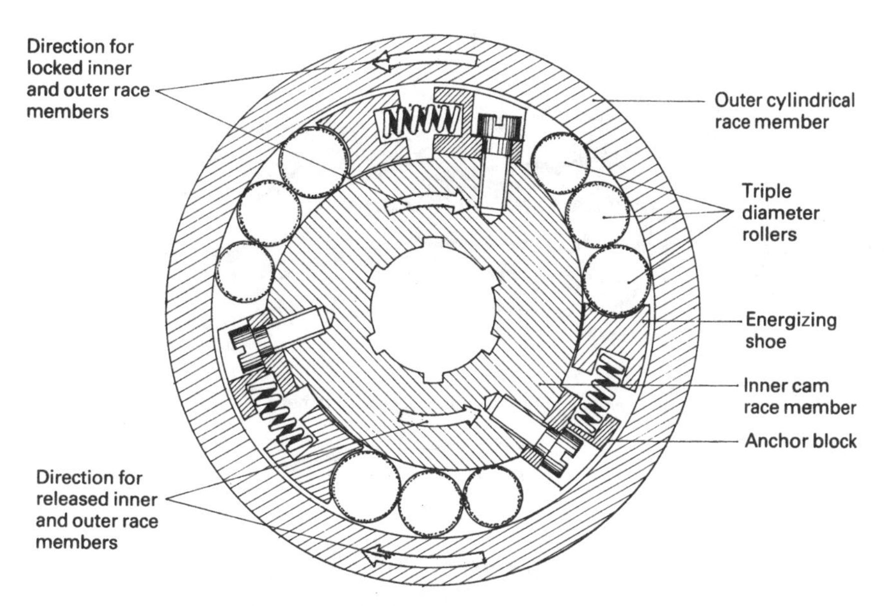

Fig. 4.14 Overrun freewheel triple diameter roller type clutch

accurately kept. This ensures that each sprag equally contributes its share of wedge action under all operating conditions. In between the cages is a ribbon type spring which twists the sprags into light contact with their respective track when the clutch is in the overrun position.

When the inner ring member is rotated clockwise and the outer ring member is held stationary or is rotated anticlockwise, the spring tension lightly presses the sprags against their track. This makes the inner and outer members move in opposite directions. The sprags are thus forced to tilt anticlockwise, consequently wedging their inclined planes hard against the tracks and thereby locking the two drive and driven members together.

As conditions change from drive to overrun and the outer member rotates faster than the inner one, the sprags will rotate clockwise and so release the outer member: a freewheel condition is therefore established.

4.7 Three stage hydrokinetic torque converter

(Figs 4.16, 4.17 and 4.18)

A disadvantage with the popular three element torque converter is that its stall torque ratio is only in the region of 2:1, which is insufficient for some applications, but this torque multiplication can be doubled by increasing the number of turbine and stator members within the converter, so that there are more stages of conversion (Fig. 4.16).

Consider the three stage torque converter. As shown in Fig. 4.17, it is comprised of one impeller, three interlinked output turbines and two fixed stator members.

Tracing the conversion vortex circuit starting from the input rotating member (Fig. 4.18), fluid is pumped from the impeller P by centrifugal force to the two velocity components V_t and V_r, making up the resultant velocity V_p which enters between the first turbine blades T_1 and so imparts some of its hydrokinetic energy to the output. Fluid then passes with a velocity V_{T1} to the first fixed stator, S_1, where it is guided and redirected with a resultant velocity V_{S1}, made up from the radial and tangential velocities V_r and V_t to the second set of turbine blades T_2, so that momentum is given to this member. Fluid is now transferred from the exit of the second turbine T_2 to the entrance of the second stator S_2. Here the reaction of the curved blades deflects the fluid towards the third turbine blades T_3 which also absorb the fluid's thrust. Finally the fluid completes its circulation cycle by again entering the impeller passages.

The limitation of a multistage converter is that there are an increased number of entry and exit junctions between various members which raise

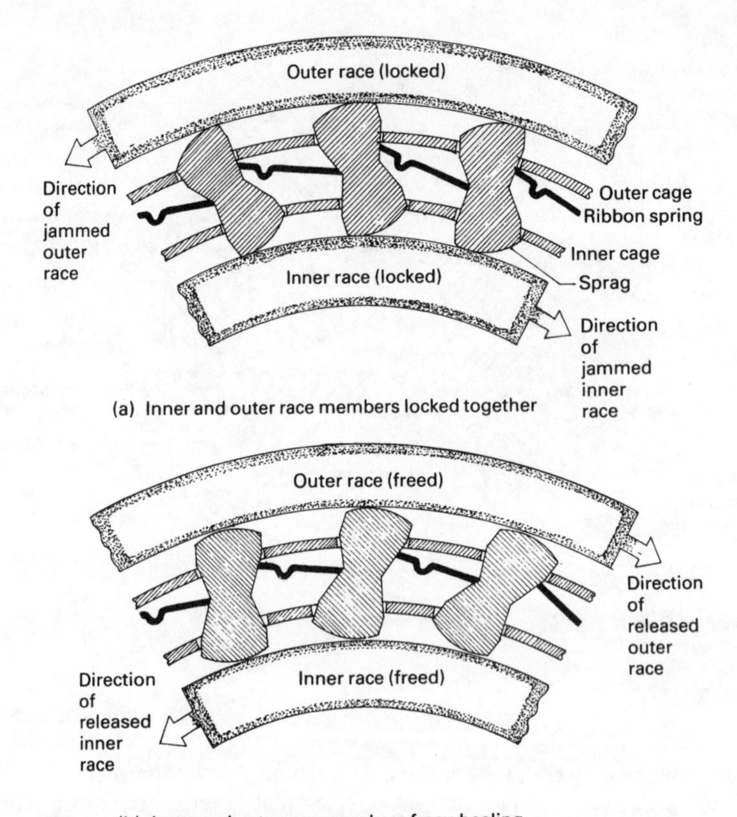

Fig. 4.15 (a and b) Overrun freewheel sprag type clutch

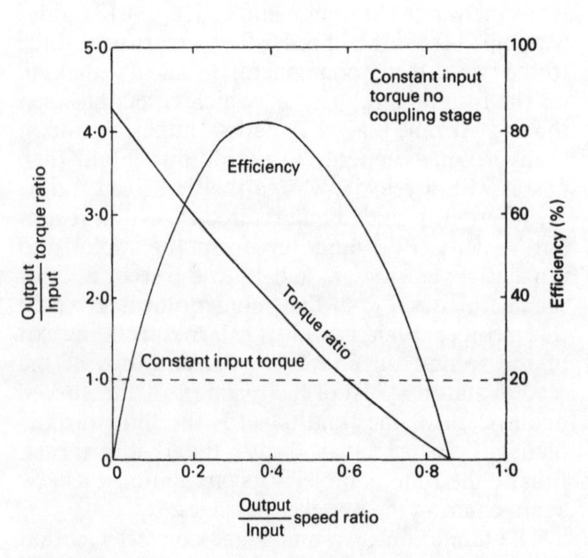

Fig. 4.16 Characteristic performance curves of a three stage converter

the fluid flow resistance around the torus passages. Subsequently, efficiency drops off fairly rapidly with higher speed ratios compared to the three element converter (Fig. 4.16).

4.8 Polyphase hydrokinetic torque converter
(Figs 4.19 and 4.20)

The object of the polyphase converter is to extend the high efficiency speed range (Fig. 4.20) of the simple three element converter by altering the vane or blade shapes of one element. Normally the stator is chosen as the fluid entrance direction changes with increased turbine speed. To achieve this, the stator is divided into a number of separate parts, in this case three, each one being mounted on its own freewheel device built into its hub (Fig. 4.19). The turbine exit and linear velocities V_E and V_L produce an effective resultant velocity V_R which changes its direction of entry between stator blades as the impeller and turbine relative speeds

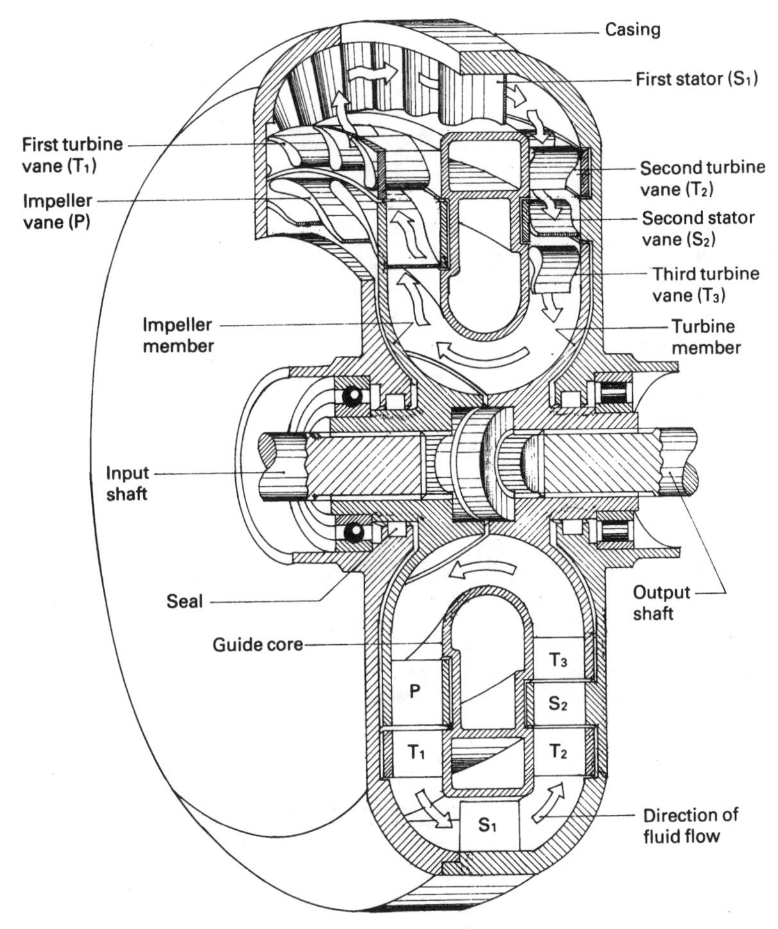

Fig. 4.17 Multistage (six element) torque converter

approach unity. It is this direction of fluid entering between the stator blades which in phases releases the various stator members.

Initial phase
Under stall speed conditions, the fluid flow from the turbine to the stator is such as to be directed onto the concave (rear) side of all three sections of the divided stator blades, thus producing optimum stator reaction for maximum torque multiplication conditions.

Second phase
As the turbine begins to rotate and the vehicle is propelled forwards, the fluid changes its resultant direction of entry to the stator blades so that it impinges against the rear convex side of the first stator blades S_1. The reaction on this member is now reversed so that it is released and is able to spin in the same direction as the input and output elements. The two remaining fixed stators now form the optimum blade curvatures for high efficiency.

Third phase
With higher vehicle and turbine speeds, the fluid's resultant direction of entry to the two remaining held stators changes sufficiently to push from the rear of the second set of stator blades S_2. This section will now be released automatically to enable the third set of stator blades to operate with optimum efficiency.

Coupling phase
Towards unity speed ratio when the turbine speed has almost caught up with the impeller, the fluid entering the third stator blades S_3 will have altered its direction to such an extent that it releases this

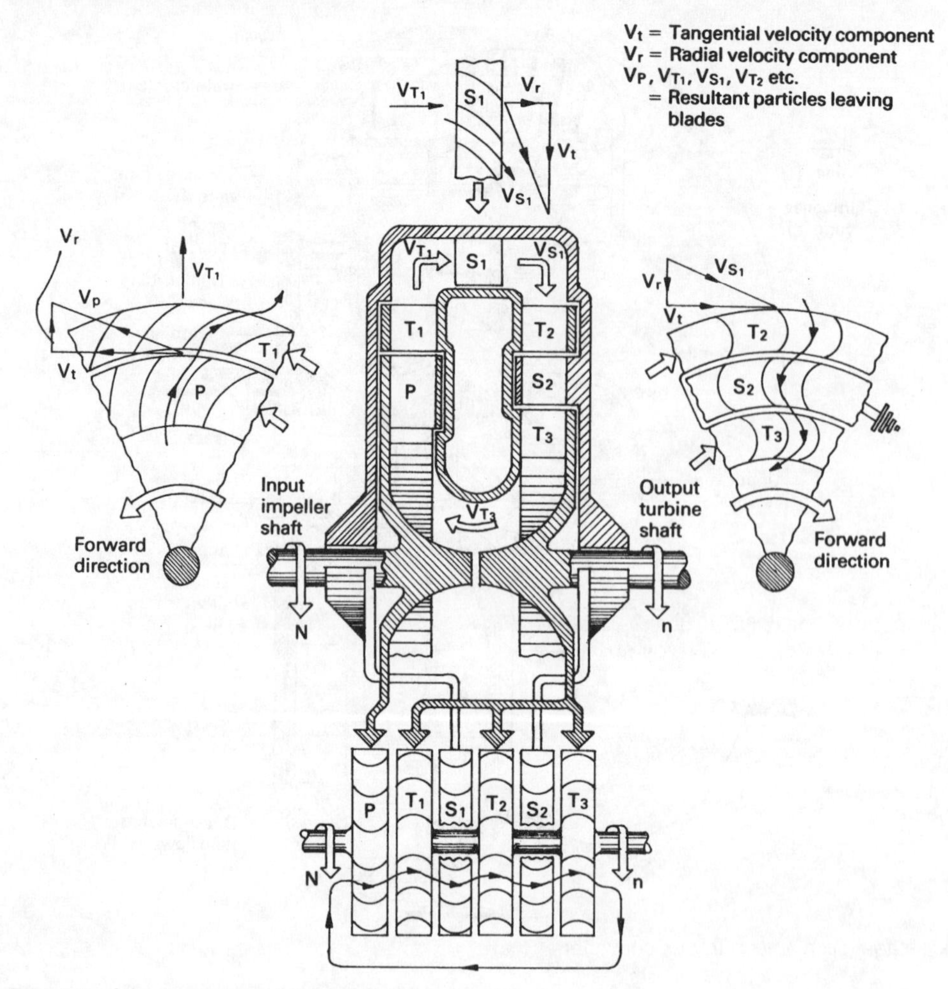

Fig. 4.18 Principle of the three stage torque converter

last fixed set of blades. Since there is no more reaction torque, conversion ceases and the input and output elements act solely as a fluid coupling.

4.9 Torque converter with lock-up and gear change friction clutches (Figs 4.21 and 4.22)

The two major inherent limitations with the torque converter drive are as follows:

Firstly, the rapid efficiency decline once the relative impeller to turbine speed goes beyond the design point, which implies higher input speeds for a given output speed and increased fuel consumption. Secondly, the degree of fluid drag at idle speed which would prevent gear changing with constant mesh and synchromesh gearboxes.

The disadvantage of the early fall in efficiency with rising speed may be overcome by incorporating a friction disc type clutch between the flywheel and converter which is hydraulically actuated by means of a servo piston (Fig. 4.21). This lock-up clutch is designed to couple the flywheel and impeller assembly directly to the output turbine shaft either manually, at some output speed decided by the driver which would depend upon the vehicle load and the road conditions or automatically, at a definite input to output speed ratio normally in the region of the design point here where efficiency is highest (Fig. 4.22).

To overcome the problem of fluid drag between the input and output members of the torque converter when working in conjunction with either

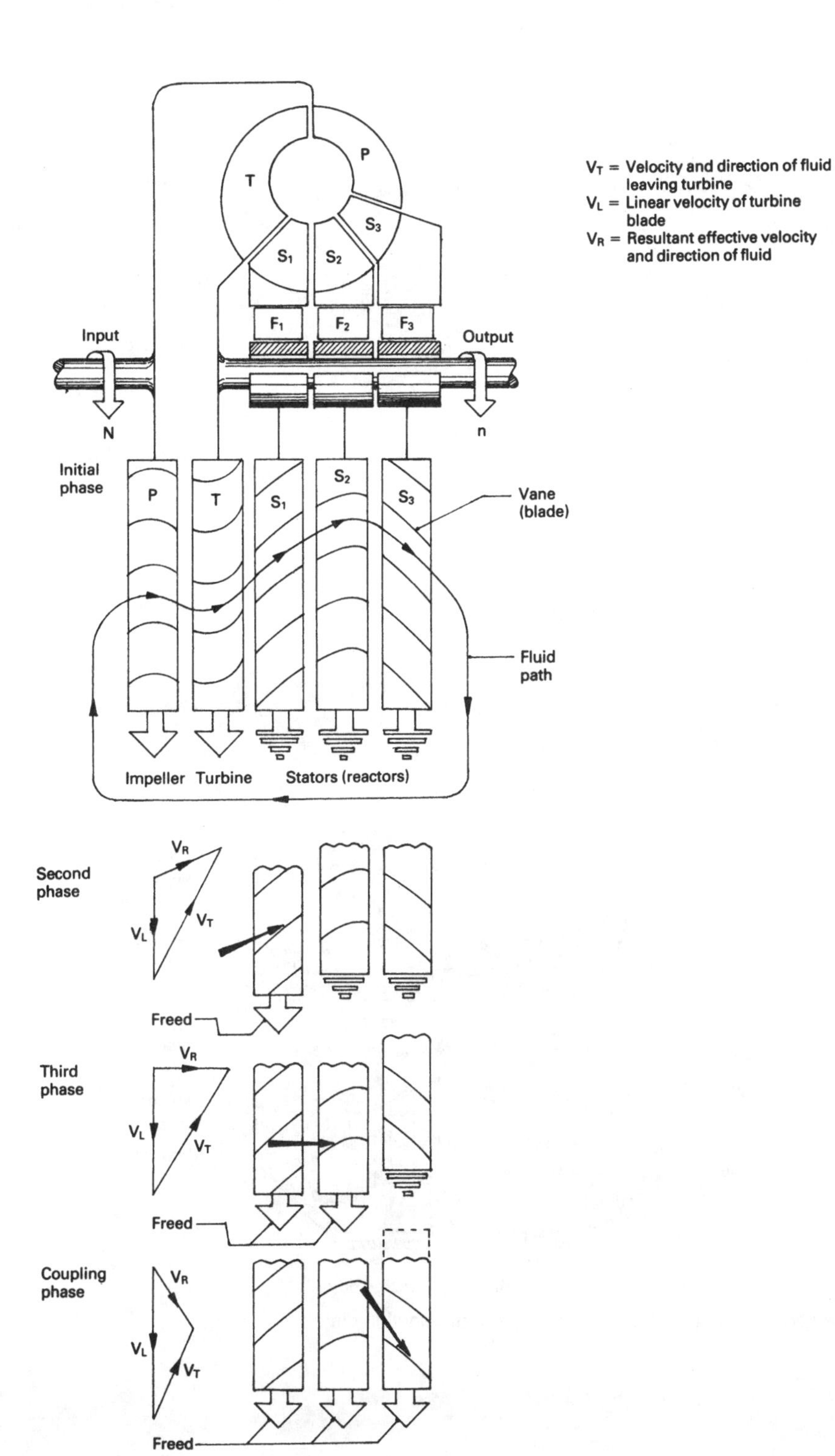

Fig. 4.19 Principle of a polystage torque converter

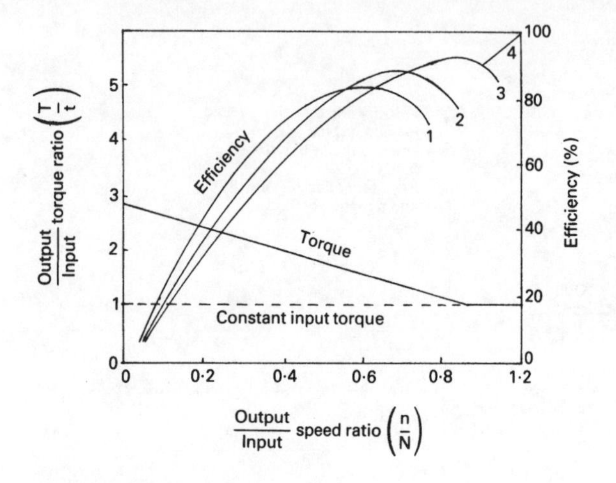

Fig. 4.20 Relationship of speed ratio, torque ratio and efficiency for a polyphase stator torque converter

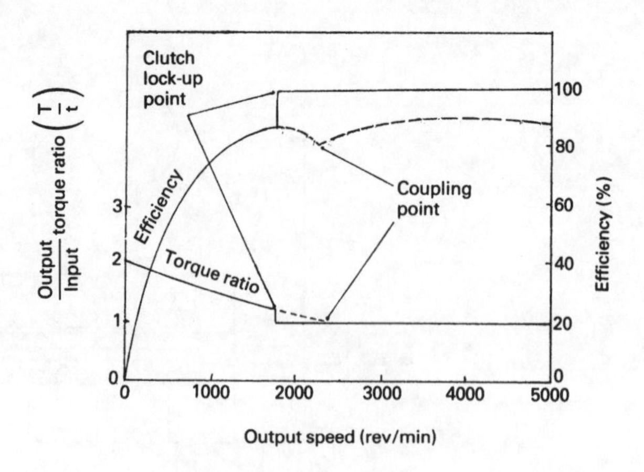

Fig. 4.22 Characteristic performance curves of a three element converter with lock-up clutch

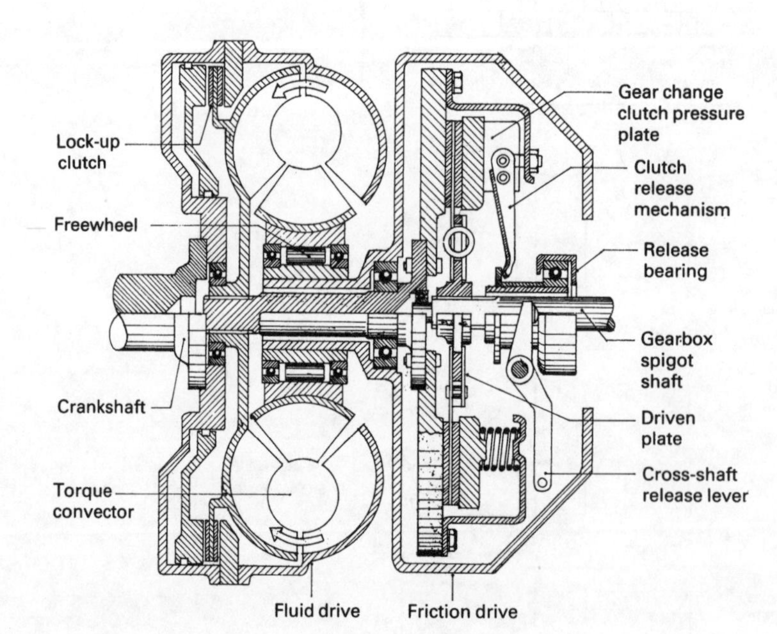

Fig. 4.21 Torque converter with lock-up and gear change function clutches

constant mesh or synchromesh gearboxes, a conventional foot operated friction clutch can be utilized between the converter and the gearbox. When the pedal is depressed and the clutch is in its disengaged position, the gearbox input primary shaft and the output main shaft may be unified, thereby enabling the gear ratio selected to be engaged both smoothly and silently.

5 Semi- and fully automatic transmission

5.1 Automatic transmission considerations

Because it is difficult to achieve silent and smooth gear ratio changes with a conventional constant mesh gear train, automatic transmissions commonly adopt some sort of epicyclic gear arrangement, in which different gear ratios are selected by the application of multiplate clutches and band brakes which either hold or couple various members of the gear train to produce the necessary speed variations. The problem of a gradual torque take-up when moving away from a standstill has also been overcome with the introduction of a torque converter between the engine and transmission gearing so that engine to transmission slip is automatically reduced or increased according to changes in engine speed and road conditions. Torque converter performance characteristics have been discussed in Chapter 3.

The actual speed at which gear ratio changes occur is provided by hydraulic pressure signals supplied by the governor valve and a throttle valve. The former senses vehicle speed whereas the latter senses engine load.

These pressure signals are directed to a hydraulic control block consisting of valves and pistons which compute this information in terms of pressure variations. The fluid pressure supplied by a pressure pump then automatically directs fluid to the various operating pistons causing their respective clutch, clutches or band brakes to be applied. Consequently, gear upshifts and downshifts are performed independently of the driver and are so made that they take into account the condition of the road, the available output of the engine and the requirements of the driver.

5.1.1 The torque converter (Fig. 5.1)

The torque converter provides a smooth automatic drive take-up from a standstill and a torque multiplication in addition to that provided by the normal mechanical gear transmission. The performance characteristics of a hydrokinetic torque converter incorporated between the engine and the gear train is shown in Fig. 5.1 for light throttle and full throttle maximum output conditions over a vehicle speed range. As can be seen, the initial torque multiplication when driving away from rest is considerable and the large gear ratio steps of the conventional transmission are reduced and smoothed out by the converter's response between automatic gear shifts. Studying Fig. 5.1, whilst in first gear, the torque converter provides a maximum torque multiplication at stall pull away conditions which progressively reduces with vehicle speed until the converter coupling point is reached. At this point, the reaction member freewheels. With further speed increase, the converter changes to a simple fluid coupling so that torque multiplication ceases. In second gear the converter starts to operate nearer the coupling point causing it to contribute far less torque multiplication and in third and fourth gear the converter functions entirely beyond the coupling point as a fluid coupling. Consequently, there is no further torque multiplication.

5.2 Four speed and reverse longitudinally mounted automatic transmission mechanical power flow (Fig. 5.2)

(Similar gear trains are adopted by some ZF, Mercedes-Benz and Nissan transmissions)

The epicyclic gear train is comprised of three planetary gear sets, an overdrive gear set, a forward gear set and a reverse gear set. Each gear set consists of an internally toothed outer annular ring gear, a central externally toothed sun gear and a planet carrier which supports three intermediate planet gears. The planet gears are spaced evenly between and around the outer annular gear and the central sun gear.

The input to the planetary gear train is through a torque converter which has a lock-up clutch. Different parts of the gear train can be engaged or released by the application of three multiplate clutches, two band brakes and one first gear one way roller clutch.

Table 5.1 simplifies the clutch and brake engagement sequence for each gear ratio.

A list of key components and abbreviations used are as follows:

1	Manual valve	MV
2	Vacuum throttle valve	VTV
3	Governor valve	GV
4	Pressure regulating valve	PRV
5	Torque converter	TC

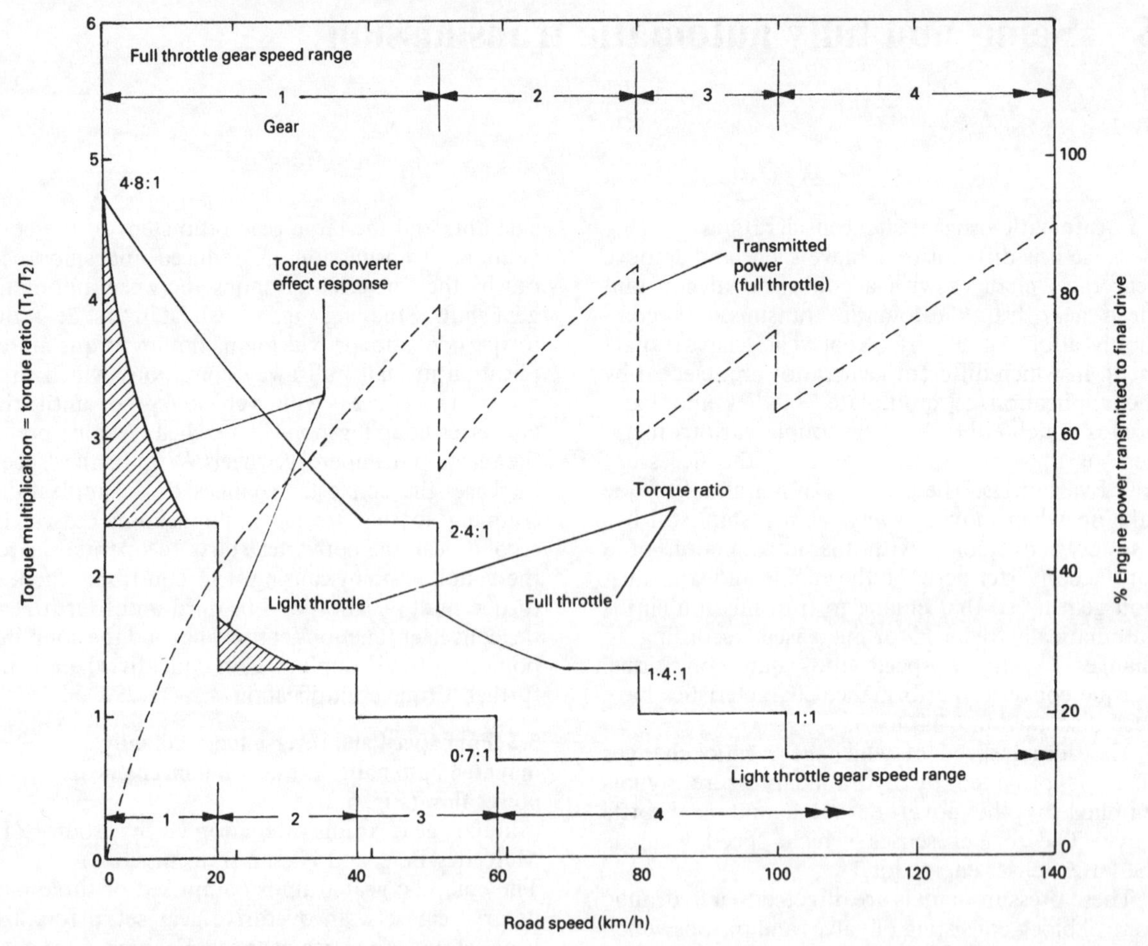

Fig. 5.1 Torque multiplication and transmitted power performance relative to vehicle speed for a typical four speed automatic transmission

6	1–2 shift valve	(1–2)SV
7	2–3 shift valve	(2–3)SV
8	3–4 shift valve	(3–4)SV
9	Converter check valve	CCV
10	Drive clutch	DC
11	High and reverse multiplate clutch	(H + R)C
12	Forward clutch	FC
13	Overdrive band brake	ODB
14	Second gear band brake	2GB
15	Low and reverse multiplate brake	(L + R)B
16	First gear one way roller clutch	OWC
17	Torque converter one way clutch	OWC_R
18	Parking lock	PL

5.2.1 D drive range — first gear

(Figs 5.3(a) and 5.4(a))

With the selector lever in D range, engine torque is transmitted to the overdrive pinion gears via the output shaft and pinion carrier. Torque is then split between the overdrive annular gear and the sun gear, both paths merging due to the engaged direct clutch. Therefore the overdrive pinion gears are prevented from rotating on their axes, causing the overdrive gear set to revolve as a whole without any gear ratio reduction at this stage. Torque is then conveyed from the overdrive annular gear to the intermediate shaft where it passes through the applied forward clutch plates to the annular gear of the forward gear set. The clockwise rotation of the forward annular gear causes the forward planet gears to rotate clockwise, driving the double sun gear counter clockwise. The forward planetary carrier is attached to the output shaft so that the planet gears drive the sun gear instead of walking around the sun gear. This anticlockwise rotation of the sun gear causes the reverse planet gears to rotate

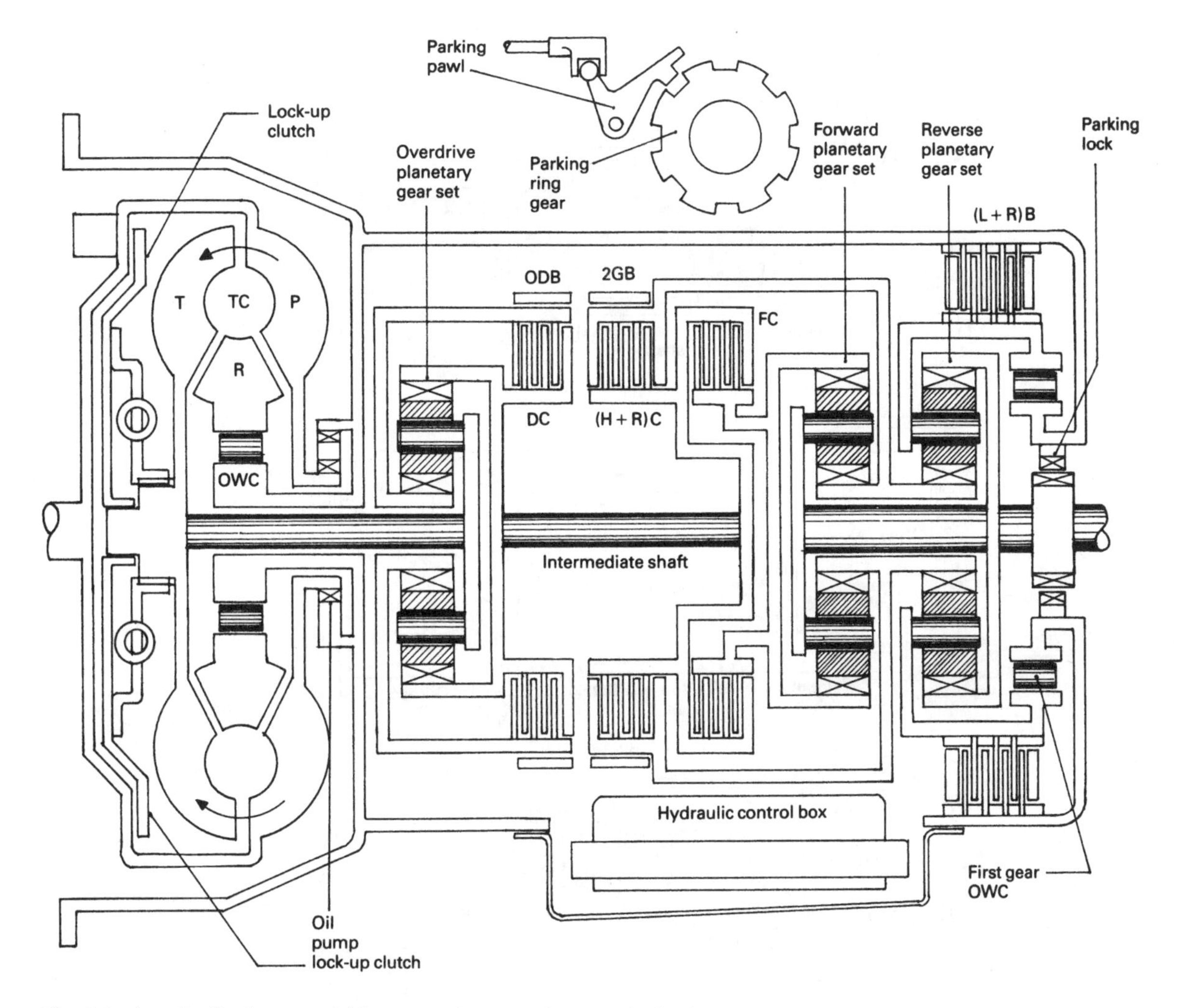

Fig. 5.2 Longitudinally mounted four speed automatic transmission layout

Table 5.1 Clutch and brake engagement sequence

Range	Drive clutch DC	High and reverse clutch (H + R) C	Second gear band brake 2GB	Forward clutch FC	Overdrive brake ODB	Low and reverse brake (L + R)B	One way clutch OWC	Ratio
P and N	–	–	–	–	–	–	–	–
First D	Applied	–	–	Applied	–	–	Applied	2.4:1
Second D	Applied	–	Applied	Applied	–	Applied	–	1.37:1
Third D	Applied	Applied	–	Applied	–	–	–	1:1
Fourth D	–	Applied	–	Applied	Applied	–	–	0.7:1
Reverse R	Applied	Applied	–	–	–	Applied	–	2.83:1

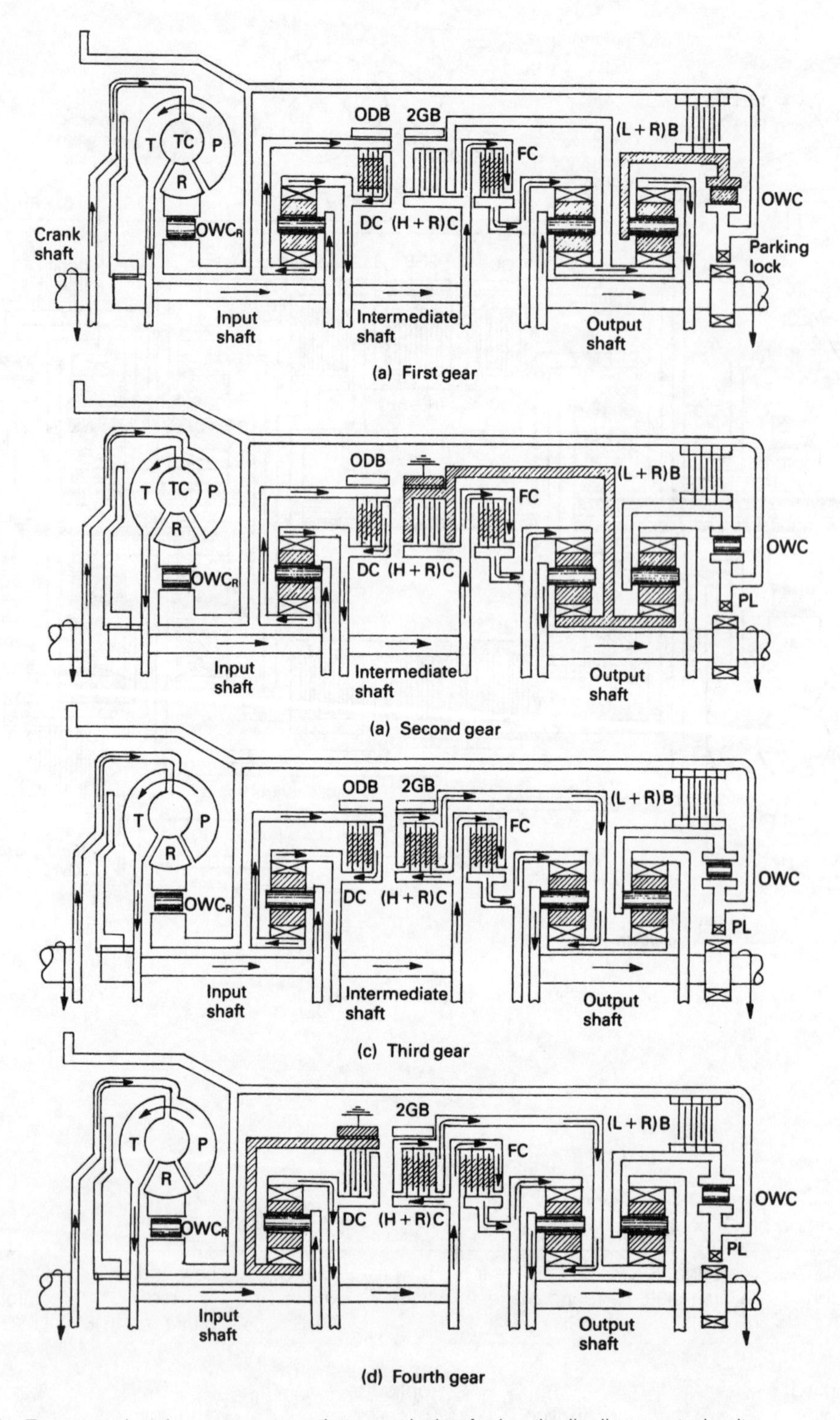

Fig. 5.3 (a–e) Four speed and reverse automatic transmission for longitudinally mounted units

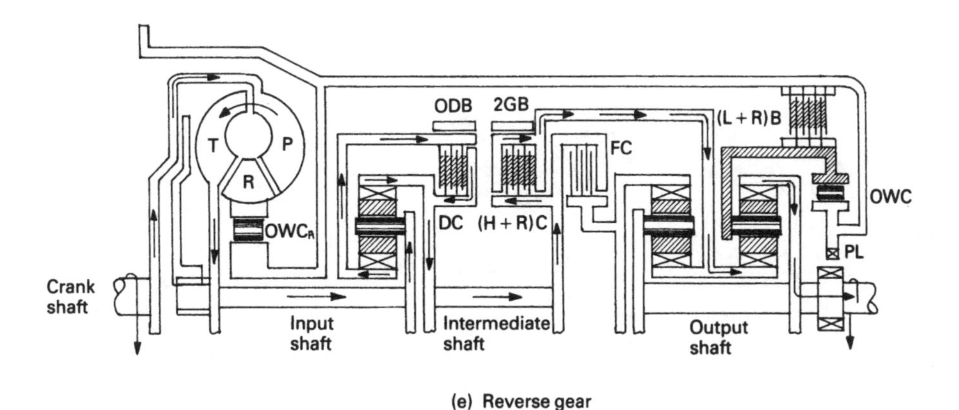

(e) Reverse gear

Fig. 5.3 *contd*

clockwise. With the one way roller clutch holding the reverse planet carrier, the reverse planetary gears turn the reverse annular gear and output shaft clockwise in a low speed ratio of approximately 2.46:1.

5.2.2 D drive range — second gear

(Figs 5.3(b) and 5.4(b))
In D range in second gear, both direct and forward clutches are engaged. At the same time the second gear band brake holds the double sun gear and reverse pinion carrier stationary.

Engine torque is transmitted through the locked overdrive gear set similarly to first gear. It is then conveyed through the applied forward clutch via intermediate shaft to the forward annular gear. With the double sun gear held by the applied second gear band brake, the clockwise rotation of the forward annular gear compels the pinion gears to rotate on their own axes and roll 'walk' around the stationary sun gear in a clockwise direction. Because the forward pinion gear pins are mounted on the pinion carrier, which is itself attached to the output shaft, the output shaft will be driven clockwise at a reduced speed ratio of approximately 1.46.

5.2.3 D drive range — third or top gear

(Figs 5.3(c) and 5.4(c))
With the selector lever in D range, hydraulic line pressure will apply the direct clutch, high and reverse clutch and forward clutch.

As for first and second gear operating conditions, the engine torque is transmitted through the locked overdrive gear set to the high and reverse multiplate clutch and the forward multiplate clutch, both of which are applied. Subsequently, the high and reverse clutch will rotate the double sun gear clockwise and similarly the forward clutch will rotate the forward annular gear clockwise. This causes both external and internal gears on the forward gear set to revolve in the same direction at similar speeds so that the bridging planet gears become locked and the whole gear set therefore revolves together as one. The output shaft drive via the reverse carrier therefore turns clockwise with no relative speed reduction to the input shaft, that is as a direct drive ratio 1:1.

5.2.4 D drive range — fourth or overdrive gear

(Figs 5.3(d) and 5.4(d))
In D range in fourth gear, the overdrive band brake, the high and reverse clutch and the forward clutch are engaged. Under these conditions, torque is conveyed from the input shaft to the overdrive carrier, causing the planet gears to rotate clockwise around the held overdrive sun gear. As a result, the overdrive annular gear will be forced to rotate clockwise but at a higher speed than the input overdrive carrier. Torque is then transmitted via the intermediate shaft to the forward planetary gear set which are then locked together by the engagement of the high and reverse clutch and the forward clutch. Subsequently, the gear set is compelled to rotate bodily as a rigid straight through drive. The torque then passes from the forward planet carrier to the output shaft. Hence there is a gear ratio step up by the overdrive planetary gear set of roughly 30%, that is, the output to input shaft gear ratio is about 0.7:1.

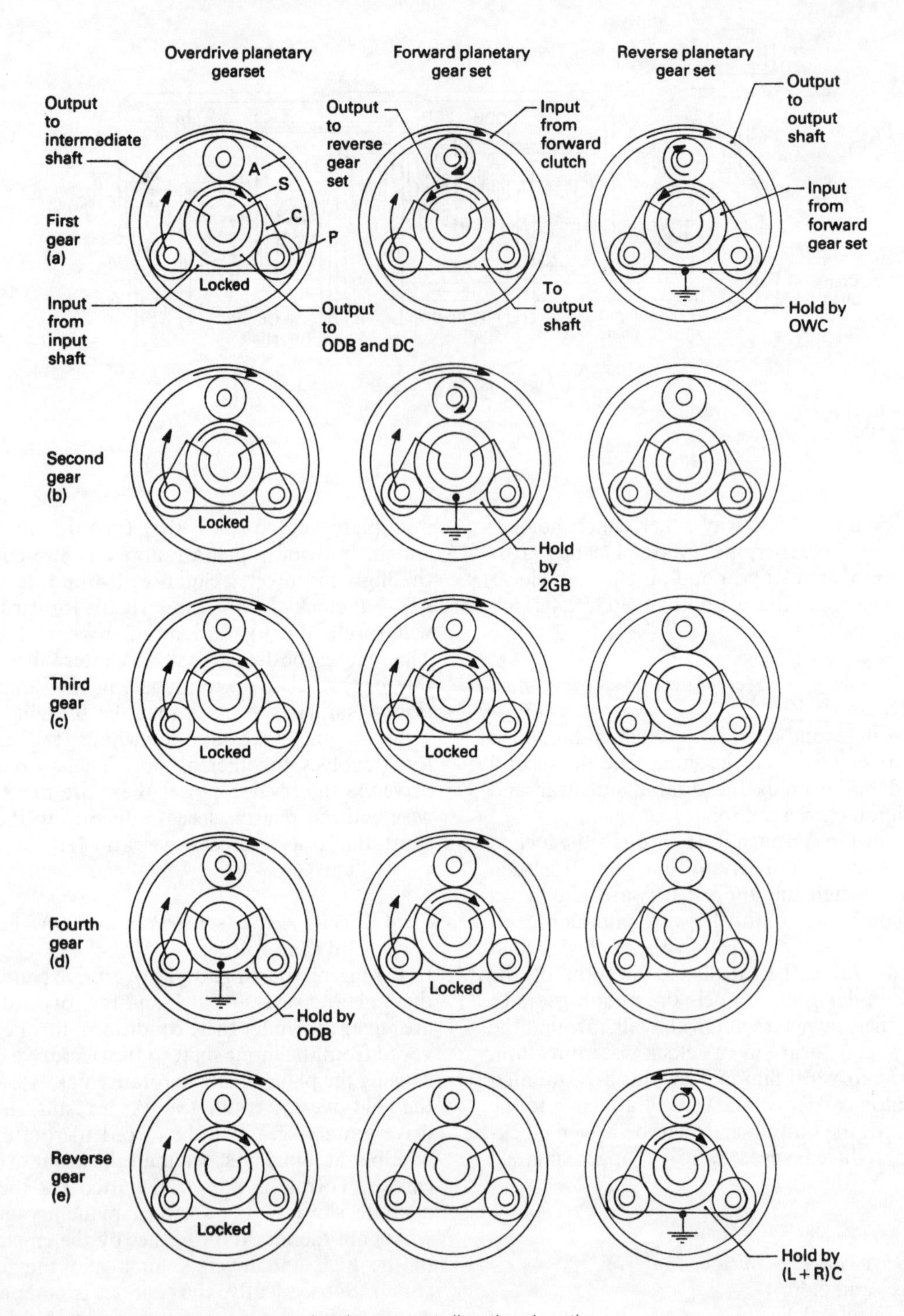

Fig. 5.4 (a–e) Four speed and reverse epicycle gear set directional motion

5.2.5 R range — reverse gear

(Figs 5.3(e) and 5.4(e))

With the selector lever in reverse position all three clutches and the low and reverse multiplate brake are engaged. Subsequently, engine torque will be transmitted from the input shaft through the locked overdrive gear set through the locked forward gear set via the intermediate shaft to the reverse sun gear in a clockwise direction.

Because the reverse planet carrier is held by the low and reverse multiplate brake, the planet gears are forced to rotate counterclockwise on their axes, and in doing so compel the reverse annular gear to also rotate counterclockwise. As a result, the output shaft, which is attached to the reverse annular gear, rotates counterclockwise, that is, in the reverse direction, to the input shaft at a reduction ratio of approximately 2.18:1.

5.3 The fundamentals of a hydraulic control system

The effective operation of an automatic transmission relies upon a hydraulic control circuit to actuate the gear changes relative to the vehicle's road speed and acceleration pedal demands with the engine delivering power. Only a very small proportion of a transmission's operating time is spent in performing gear changes. In fact, the hydraulic system is operational for less than 1% of the driving time. The transition time from one gear ratio to the next takes roughly one second or less and therefore the hydraulic control valves must be designed to direct fluid pressure to the appropriate operating pistons which convert the fluid pressure into mechanical force and movement to energize the respective clutches and band brakes instantly and precisely.

An understanding of a basic hydraulic control system can best be considered under the four headings:

1 Pressure supply and regulating valves
2 Speed and load sensing valves
3 Gear shift valves
4 Clutch and brake coupling and hold devices

5.3.1 Pressure supply and regulating valve

(Fig. 5.5)

The essential input to the hydraulic control system is fluid pressure generated by a pump and driven by the engine. The pump's output pressure will increase roughly in proportion to the engine's speed. However, the pressure necessary to actuate the various valves and to energize the clutch and band servo pistons will vary under different working conditions. Therefore the fluid pressure generated by the pump is unlikely to suit the many operating requirements. To overcome these difficulties, a pressure regulating valve is used which automatically adjusts the pump's output pressure to match the working requirements at any one time. One of the functions of the pressure regulating valve is to raise the line pressure reaching the clutch and brake when the vehicle is driven hard with large throttle opening to prevent the friction surfaces slipping. Conversely under light loads and with a small throttle opening, a much lower line pressure is adequate to clamp the friction plates or bands. By reducing the line pressure, fierce clutch and brake engagements are eliminated which promotes smooth and gentle gear changes. Power consumption, which is needed to drive the hydraulic pump, is also reduced as actuating pressures are lowered. The pressure regulating valve is normally a *spring-loaded spool type valve*, that is, a plunger with one or more reduced diameter sections along its length, positioned in a cylinder which has a number of passages intersecting the cylinder walls.

When the engine speed, and correspondingly pump pressure, is low, fluid flows via the inlet port around the wasted section of the plunger and out unrestricted along a passage leading to the manual valve where it is distributed to the various control valves and operating pistons. As the pump pressure builds up with rising engine speed, line pressure is conveyed to the rear face of the plunger and will progressively move the plunger forward against the control spring, causing the middle land to uncover an exhaust port which feeds back to the pump's intake. Hence as the pump output pressure tends to rise, more fluid is passed back to the suction intake of the pump. It therefore regulates the output fluid pressure, known as line pressure, according to the control spring stiffness. To enable the line pressure to be varied to suit the operating conditions, a throttle pressure is introduced to the spring end of the plunger which opposes the line pressure. Increasing the throttle pressure raises line pressure and vice versa.

In addition to the main pressure regulating valve there is a secondary regulating valve which limits the fluid flowing through to the torque converter. Raising the torque converter's fluid pressure increases its torque transmitting capacity which is desirable when driving in low gear or when the engine is delivering its maximum torque.

5.3.2 Speed and load sensing valves

(Figs 5.5 and 5.6)

For gear changes to take place effectively at the optimum engine and road speed, taking into account the driver's demands expressed in throttle opening, some means of sensing engine load and vehicle road speed must be provided. Engine output torque is simply monitored by a throttle valve which is linked to the accelerator pedal, either directly or indirectly, via a vacuum diaphragm operated linkage which senses the change in induction depression, which is a measure of the engine load. The amount the accelerator pedal or manifold vacuum alters is relayed to the throttle valve which accordingly raises or lowers the output pressure. This is then referred to as *throttle pressure*.

Road speed changes are measured by a centrifugal force-sensitive regulating valve which senses transmission output shaft speed and transmits this information in the form of a fluid pressure, referred to as *governor pressure*, which increases or decreases according to a corresponding variation in road speed. Both throttle pressure and governor pressure are signalled to each gear shift valve so that these may respond to the external operating conditions (i.e. engine torque developed and vehicle speed) by permitting fluid pressure to be either applied or released from the various clutch and brake actuating piston chambers.

5.3.3 Gear shift valves (Fig. 5.5)

Shift valves are of the spool plunger type, taking the form of a cylindrical plunger reduced in diameter in one or more sections so as to divide its length into a number of lands. When operating, these valves shift from side to side and cover or uncover passages leading into the valve body so that different hydraulic circuits are switched on and off under various operating conditions.

The function of a shift valve is to direct the fluid pressure to the various clutch and brake servo pistons to effect gear changes when the appropriate load and speed conditions prevail. Shift valves are controlled by line or throttle pressure, which is introduced into the valve at the spring end, and governor pressure, which is introduced directly to the valve at the opposite end. Generally, the governor valve end is of a larger diameter than the spring end so that there will be a proportionally greater movement response due to governor pressure variation. Sometimes the shift valve plunger at the governor pressure end is referred to as the *governor plug*.

The position of the shift valve at any instant depends upon the state of balance between the opposing end forces acting on the spool valve end faces.

$$\text{Spring load} + \text{Throttle pressure load} = \text{Governor pressure load}$$

$$F_S + P_T A_T = P_G A_G$$

But $PA = F$

hence $F_S + F_T = F_G$

where F_S = Spring load
F_T = Throttle pressure load
P_T = Throttle pressure
F_G = Governor pressure load
A_T = CSA of plunger at throttle pressure end
P_G = Governor pressure
A_G = CSA of plunger at governor pressure end

Thus increasing or decreasing the spring stiffness or enlarging or reducing the diameter of the spool valve at one end considerably alters the condition when the shift valve moves from one end to the other to redirect line pressure to and from the various clutch and brakes and so produce the necessary gear change.

Each shift valve control spring will have a particular stiffness so that different governor pressures, that is, road speeds, are required to cause either a gear upshift or downshift for a given opposing throttle pressure. Conversely, different engine power outputs will produce different throttle pressures and will alter the governor pressure accordingly when a particular gear shift occurs. Large engine loads (high throttle pressure) will delay gear upshifts whereas light engine load demands (low throttle pressure) and high vehicle speeds (high governor pressure) will produce early upshifts and prevent early downshift.

To improve the quality of the time sequence of up or down gear shift, additional valves and components are included to produce a smooth transition from one gear to the next. Some of these extra devices are described in Section 5.6.

5.3.4 Clutch and brake coupling and hold devices

(Figs 5.5 and 2.16)

Silent gear change synchronization is made possible by engaging or locking out various members of the epicyclic gear train gear sets with the engine's power being transmitted continuously. It therefore requires a rapid and accurate gear change which is achieved by utilizing multiplate clutches and band brakes. A gear up- or downshift therefore occurs with the almost simultaneous energizing of one

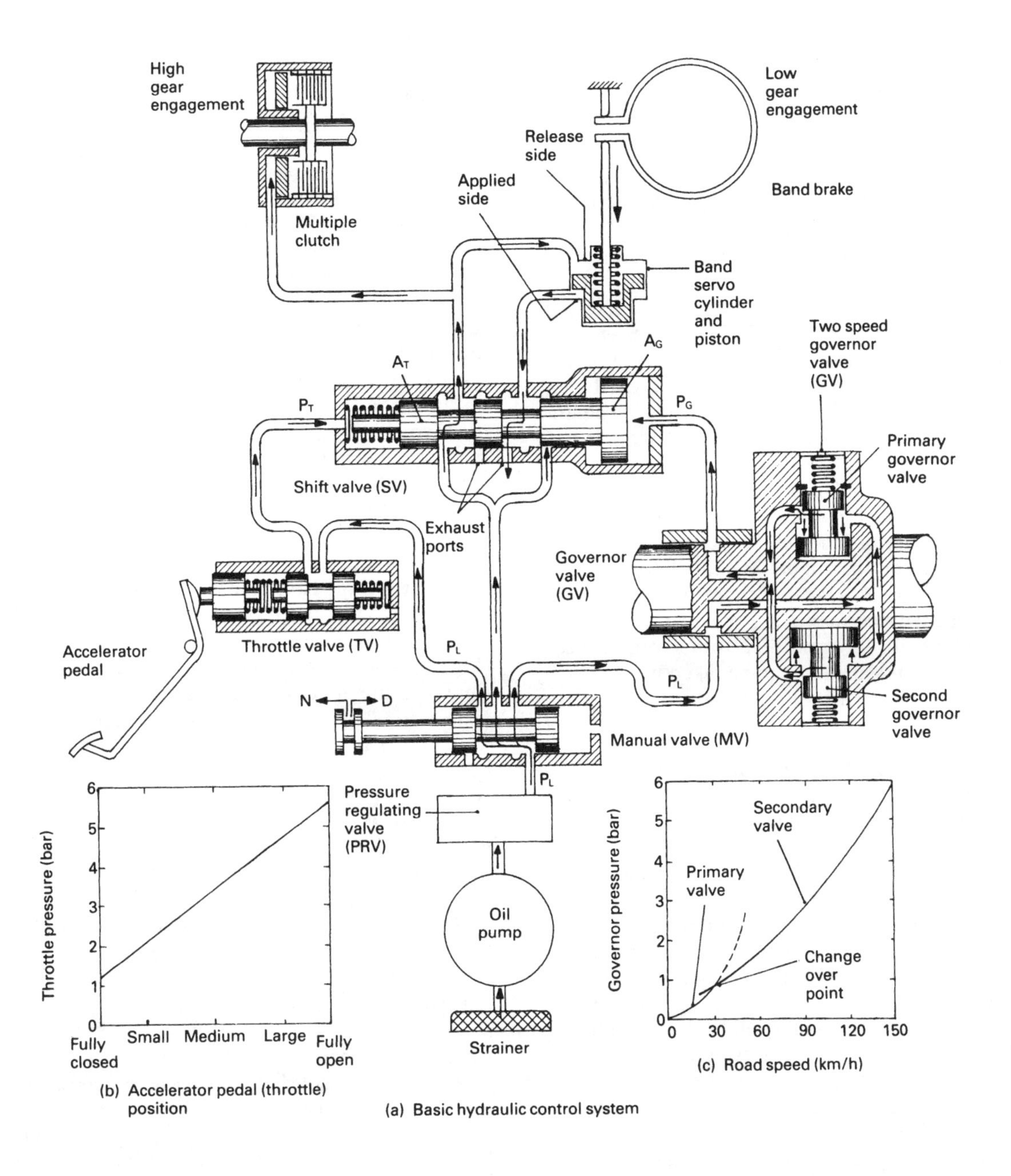

Fig. 5.5 (a and b) Basic multiplate clutch and band brake transmission hydraulic control system

clutch or brake and a corresponding de-energizing of another clutch or brake.

Multiplate clutch (Figs 2.16 and 5.5) Wet multiplate type clutches are very compact for their torque transmitting and heat dissipating capacity. They are used to lock any two members of a planetary gear set together or to transfer drive from one shaft or member to another quickly and smoothly. The rotating and fixed friction plates can be energized by an annular shaped, hydraulically operated piston either directly or indirectly by a dished washer which acts also as a lever to multiply the operating clamping load. Return springs are used to separate the pairs of rubbing faces when the fluid pressure is released. Wear and adjustment of the friction plate pack is automatically compensated by the piston being free to move further forward (see Chapter 2, Fig. 2.16).

Band brake (Fig. 5.5) This form of brake consists of a friction band encompassing an external drum so that when the brake is applied the band contracts, thereby wrapping itself tightly around the drum until the drum holds. The application of the band is achieved through a double acting stepped servo cylinder and piston. Fluid line pressure is introduced to the small diameter end of the piston to energize the band brake. To release the band, similar line pressure is directed to the spring chamber side of the cylinder. Band release is obtained due to the larger piston area side producing a greater force to free the band. This method of applying and releasing the band enables a more prolonged and controllable energizing and de-energizing action to be achieved. This class of brake is capable of absorbing large torque reactions without occupying very much space, which makes the band brake particularly suitable for low gear high torque output gear sets. Band wear slackness can be taken up by externally adjusting the anchor screw.

5.4 Basic principle of a hydraulically controlled gearshift (Fig. 5.5)

Selecting the drive D range positions the manual valve spool so that line pressure from the pressure regulator valve passes through to the shift valve, throttle valve and governor valve (Fig. 5.5(a)).

Throttle pressure will be introduced to the spring end of the shift valve via the throttle valve. Depressing the accelerator pedal allows the spool valve to move outwards. This increases the valve opening so that a high throttle pressure will be delivered to the shift valve. Conversely, depressing the accelerator pedal partially restricts the flow of fluid and therefore reduces the throttle pressure reaching the shift valve (Fig. 5.5(b)).

At the same time, line pressure enters the governor valve, flows between the wasted region of both primary and secondary spool valves and reacts against the difference in the annular adjacent face areas of each spool valve. Both valves are forced inwards, covering up the two exits from the governor valve housing. As the vehicle moves forwards, the rotation of the governor causes a centrifugal force to act through the mass of each governor valve so that it tends to draw the valve spools outwards in opposition to the hydraulic pressure which is pushing each valve inwards (Fig. 5.5(a)).

With rising output shaft speed, the centrifugal force acting through the primary valve is sufficient to overcome the hydraulic line pressure, which is acting against the shouldered groove face area and will therefore progressively move outwards as the rotational speed increases until the valve borders on an end stop. The opening of the governor valve outlet passage now allows fluid to flow out from the governor, where it is then directed to the large diameter end of the shift valve. This output pressure is known as *governor pressure*. With even higher rotational output shaft speed (vehicle speed), greater centrifugal force will be imposed on the secondary valve until it is able to overcome the much larger hydraulic inward load imposed on the large shoulder of this valve. The secondary valve will start to move out from the centre of rotation, uncovering the secondary valve outlet passage so that increased governor pressure passes to the shift valve.

This two stage governor valve action enables the governor to be more sensitive at the very low speeds but not oversensitive at the higher speeds (Fig. 5.5(c)). Sensitivity refers to the amount of fluid pressure increase or decrease for a unit change in rotational speed. If there is a large increase or decrease in governor pressure per unit charge in speed, then the governor is sensitive. If there is very little variation in governor pressure with a change in rotational speed (i.e. vehicle speed), then the governor is insensitive and therefore not suitable for signalling speed changes to the hydraulic control systems.

The reason a single stage governor would not perform satisfactorily over the entire output shift speed range is due to the *centrifugal force square law*: at low speeds the build-up in centrifugal force for a small increase in rotational speed is very small, whereas at higher speeds only a small rise in speed produces a considerable increase in centrifugal force. If the governor has the correct sensitivity at high speed it would be insensitive at low speed or if it has the desired sensitivity at low speed it would be far too responsive to governor pressure changes in the higher speed range.

Once the governor pressure end load ($P_G \times A_G$) equals the spring and throttle pressure load ($F_S + P_T \times A_T$) with rising vehicle speed, any further speed increase will push the shift valve plunger towards the spring end to the position shown in Fig. 5.5(a). The fluid on the applied side of the band brake servo piston will now exhaust (drain) through the shift valve to the inlet side of the oil pump. Simultaneously, line pressure from the manual valve is directed via the shift valve to both the release side of the band servo piston and to the multiplate clutch piston which then energizes the friction plates.

Supply fluid to the spring side of the servo piston (known as the *release side*), provides a more progressive and controllable transition from one gear change to another which is not possible when relying only on the return spring.

When the vehicle's speed is reduced or the throttle pressure is raised sufficiently, the shift valve plunger will move to the governor pressure end of the valve (Fig. 5.5(a)). The line pressure transmitted to the shift valve is immediately blocked and both the multiplate clutch and the band brake hydraulic feed passages are released of fluid pressure by the middle plunger land uncovering the exhaust part. Simultaneously, as the same middle land covers the right hand exhaust port and uncovers the line pressure passage feeding from the manual valve, fluid will flow to the applied side of the band servo piston, causing the band to contract and so energize the brake.

5.5 Basic four speed hydraulic control system

A simplified hydraulic control system for a four speed automatic transmission will now be examined for the reader to obtain an appreciation of the overall function of the hydraulic computer (control) system.

5.5.1 First gear (Fig. 5.6)

With the manual valve in D, drive position, fluid is delivered from the oil pump to the pressure regulating valve. It then divides, some being delivered to the torque converter, the remainder passing out to the manual valve as regulating pressure (more commonly known as *line pressure*). Line pressure from the manual valve is then channelled to the forward clutch, which is energized, and to the overdrive band servo on the applied side. At the same time, line pressure from the pressure regulating valve passes through the 3–4 shift valve where it is directed to energize the drive clutch and to the released side of the overdrive band servo, thus preventing the engagement of the band. Line pressure is also directed to both the governor valve and to the vacuum throttle valve. The reduced pressure output from the governor valve which is known as governor pressure is directed to the end faces of each of the three shaft valves, whereas the output pressure from the throttle valve, known as throttle pressure, is conveyed to the spring end of the 2–3 and 3–4 shift valves. On the other hand, the 1–2 shift valve spring end is subjected to line pressure from the manual valve.

Whilst the transmission is in drive first gear the one way clutch will engage, so preventing the reverse planetary carrier from rotating (not shown in hydraulic system).

5.5.2 Second gear (Fig. 5.7)

With the manual valve still in D, drive position, hydraulic conditions will be similar to first gear, that is, the overdrive and forward clutches are engaged, except that rising vehicle speed increases the governor pressure sufficiently to push the 1–2 shift valve against both spring and line pressure end loads. As a result, the 1–2 shift valve middle land uncovers the line pressure supply passage feeding from the manual valve. Line pressure is now directed to the second gear band servo on the applied side, energizing the second gear brake and causing both the forward and reverse sun gears to hold.

If there is a reduction in vehicle speed or if the engine load is increased sufficiently, the resulting imbalance between the spring and throttle pressure load as opposed to governor pressure acting on the 1–2 shift valve at opposite ends causes the shift valve to move against the governor pressure. Consequently the hydraulic circuitry will switch back to first gear conditions, causing the transmission to shift down from second to first gear again.

5.5.3 Third gear (Fig. 5.8)

At even higher road speeds in D, drive position, the governor pressure will have risen to a point where it is able to overcome the spring and throttle pressure load of the 2–3 shift valve. This causes the spool valve to shift over so that the line pressure passage feed from the manual valve is uncovered. Line pressure will now flow through the 2–3 shift valve where it is directed to the high and reverse clutch to energize the respective fixed and rotating friction plates. At the same time, line pressure passes to the second gear band servo on the release side to disengage the band. Consequently both overdrive and forward planetary gear sets lock-up, permitting the input drive from the torque converter to be transmitted directly through to the transmission's output shaft.

The actual vehicle speed at which the 2–3 shift valve switches over will be influenced by the throttle opening (throttle pressure). A low throttle pressure will cause an early gear upshift whereas a large engine load (high throttle pressure) will raise the upshift speed.

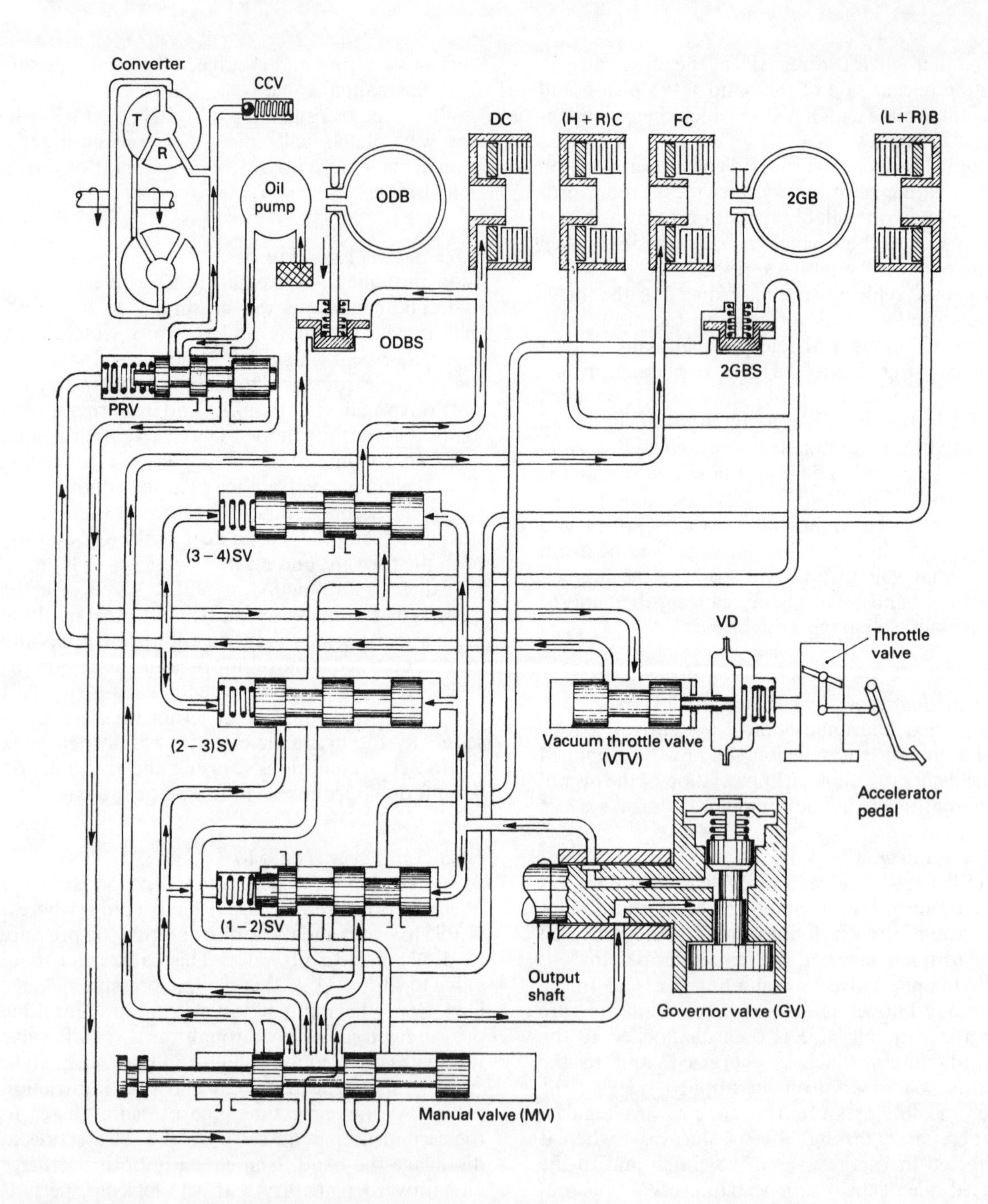

Fig. 5.6 Hydraulic control system (D) range first gear

5.5.4 Fourth gear (Fig. 5.9)

With still higher road speeds in D, drive position, the increased governor pressure will actuate the 3–4 shift valve, forcing it to shift across so that it covers up the line pressure supply passage and at the same time uncovers the exhaust or drain port. As a result, the line pressure exhausts from the release side of the overdrive band servo which then permits the band to be energized. At the same time the drive clutch will be de-energized because of the collapse of line pressure as it is released through the 3–4 shift valve exhaust port.

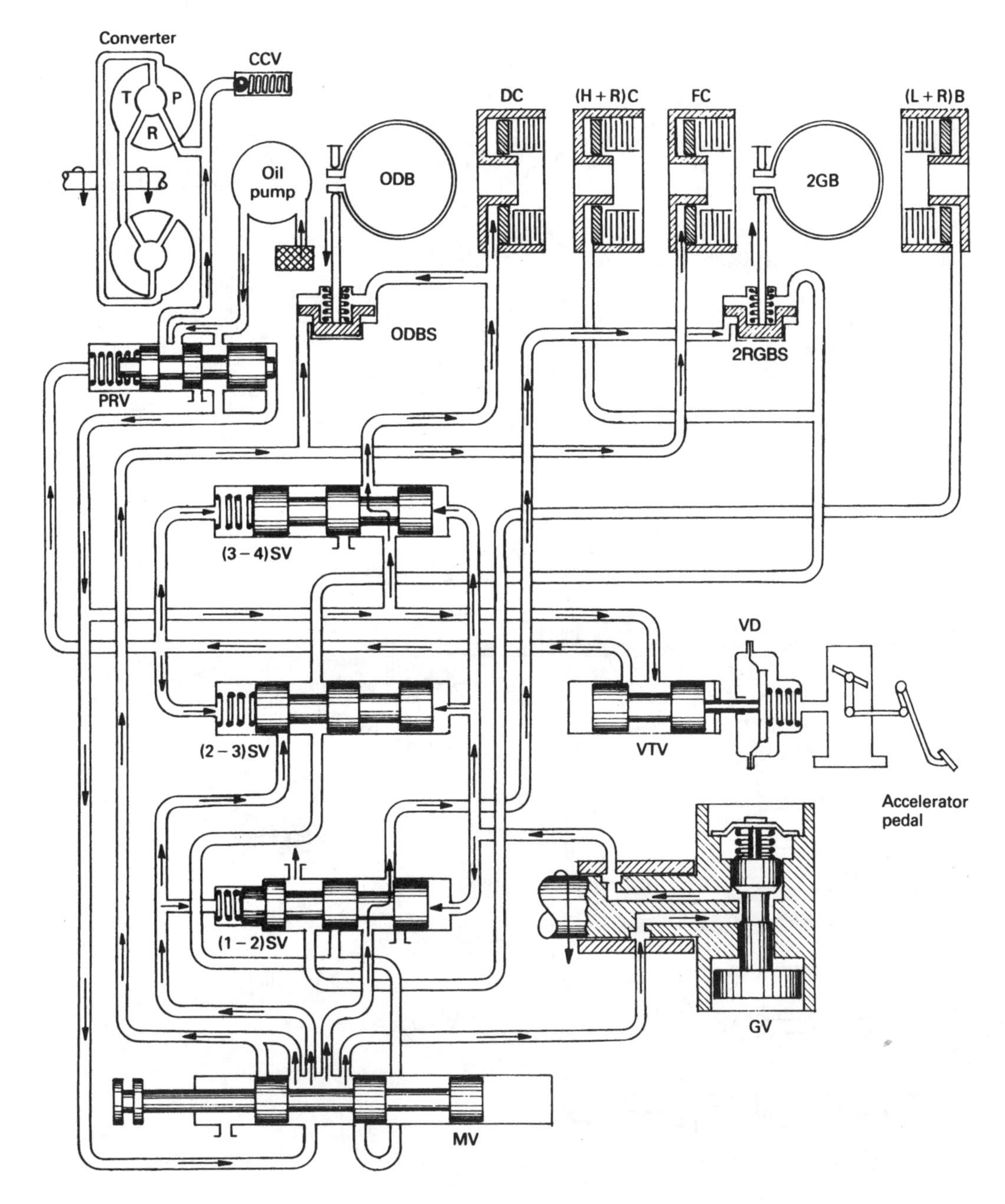

Fig. 5.7 Hydraulic control system (D) range second gear

Under these operating conditions the overdrive shaft planetary gear set reduces the intermediate shift speed and, since the forward clutch is in a state of lock-up only, this speed step up is transmitted through to the output shaft.

5.5.5 ***Reverse gear*** (Fig. 5.10)
With the manual valve in R, reverse position, line pressure from the manual valve is directed via the 2–3 shift valve to the release side of the second gear band servo, causing the band to disengage. At the

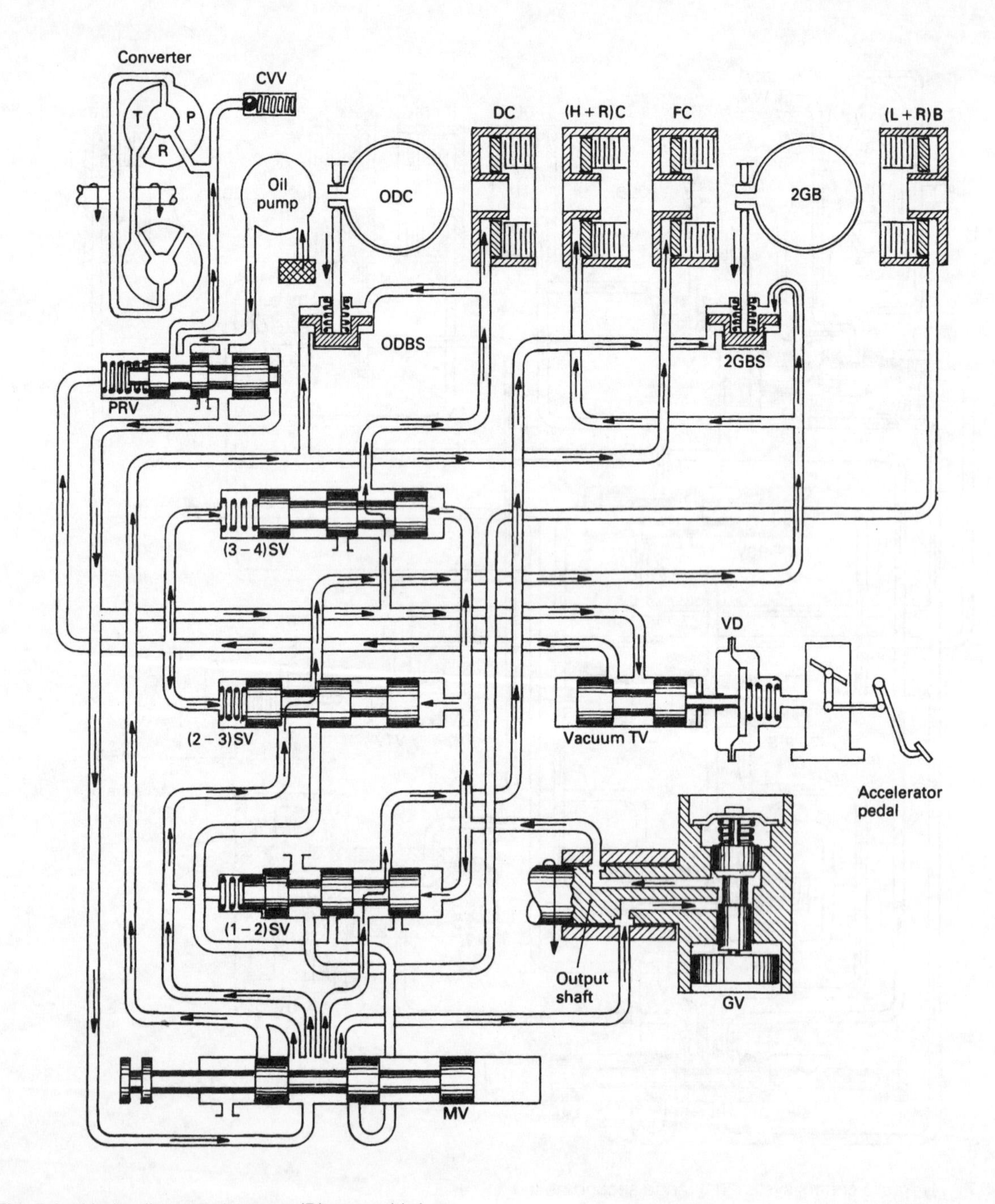

Fig. 5.8 Hydraulic control system (D) range third gear

same time line pressure from the same supply passage engages the high and reverse clutch. The manual valve also supplies line pressure to the low and reverse band brake via the 1–2 shift valve to hold the reverse planetary carrier. In addition, line pressure from the pressure regulating valve output side is directed via the 3–4 shift valve to the release side of the overdrive brake servo to disengage the band and to the drive clutch piston to engage the friction plates. Note that both band brake servos on the applied sides have been exhausted of line pressure and so has the forward clutch piston chamber.

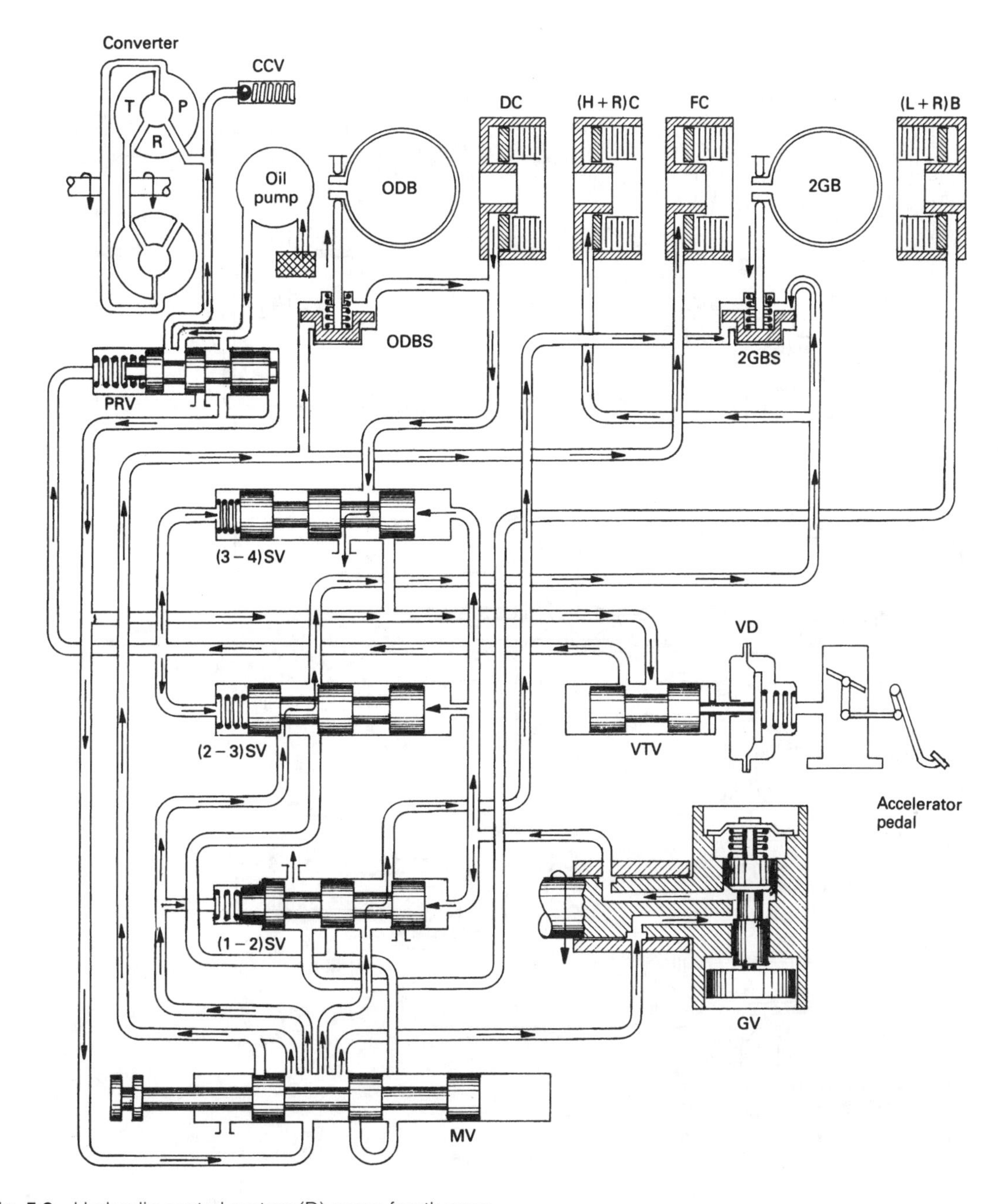

Fig. 5.9 Hydraulic control system (D) range fourth gear

5.5.6 Lock-up torque converter (Fig. 5.11)

Introduction To overcome the inherent relative slip which always occurs between the torque converter's pump impeller and the turbine runner, even driving at moderate speeds under light load conditions, a lock-up friction clutch may be incorporated between the input pump impeller and the turbine output shaft. The benefits of this lock-up can only be realised if the torque converter is allowed to operate when light torque demands are made on the engine and only when the converter is operating above its torque multiplication range that is beyond the coupling point. Consequently, converter

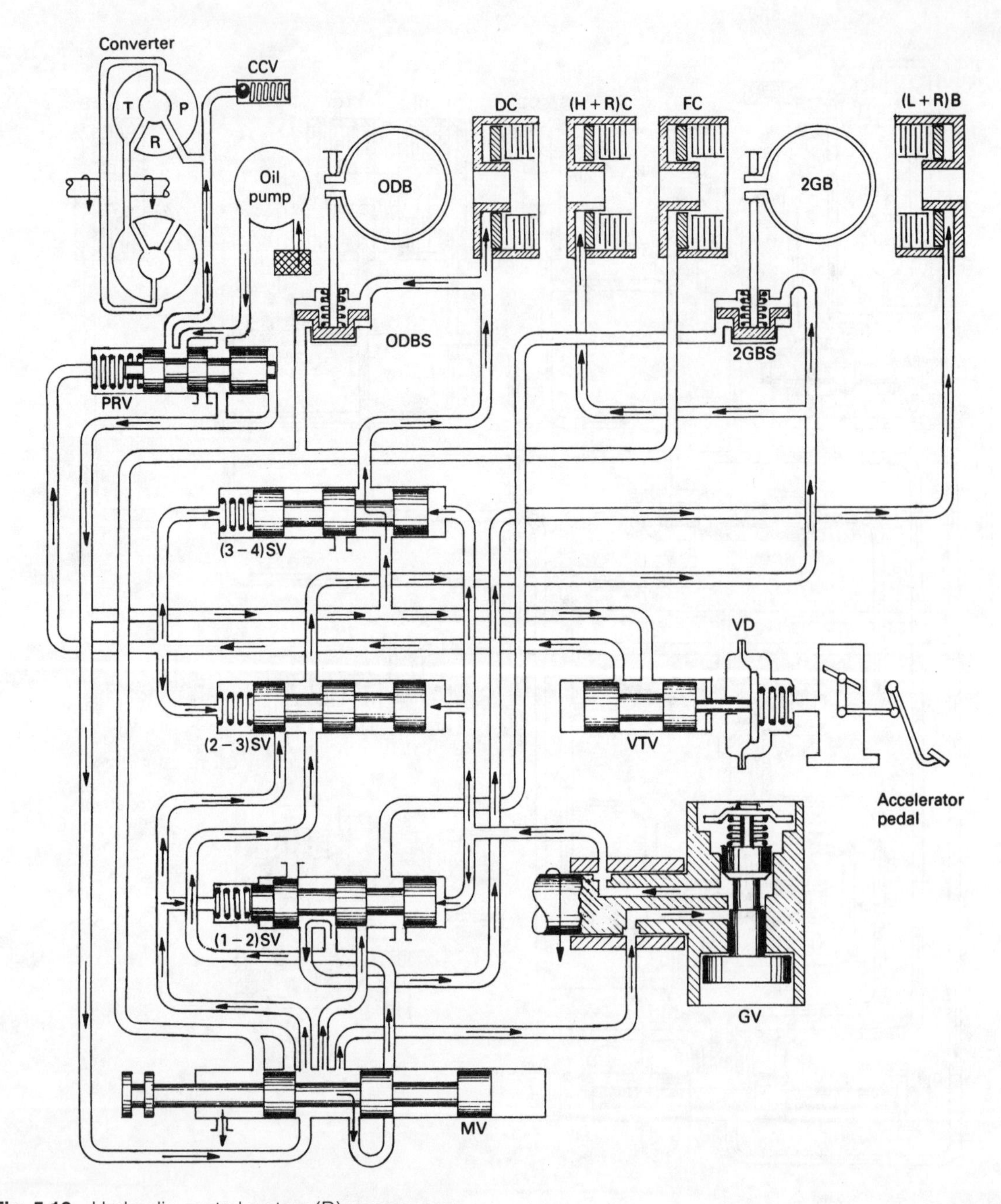

Fig. 5.10 Hydraulic control system (D) reverse gear

lock-up is only permitted to be implemented when the transmission is in either third or fourth gear. The advantages of bypassing the power transfer through the circulating fluid and instead transmitting the engine's output directly to the transmission input shaft eliminates drive slippage, thereby increasing the power actually propelling the vehicle. Due to this net gain in power output, fuel wastage will be reduced.

Lock-up clutch description The lock-up clutch consists of a sliding drive plate which performs two functions; firstly to provide the friction coupling device and secondly to act as a hydraulic con-

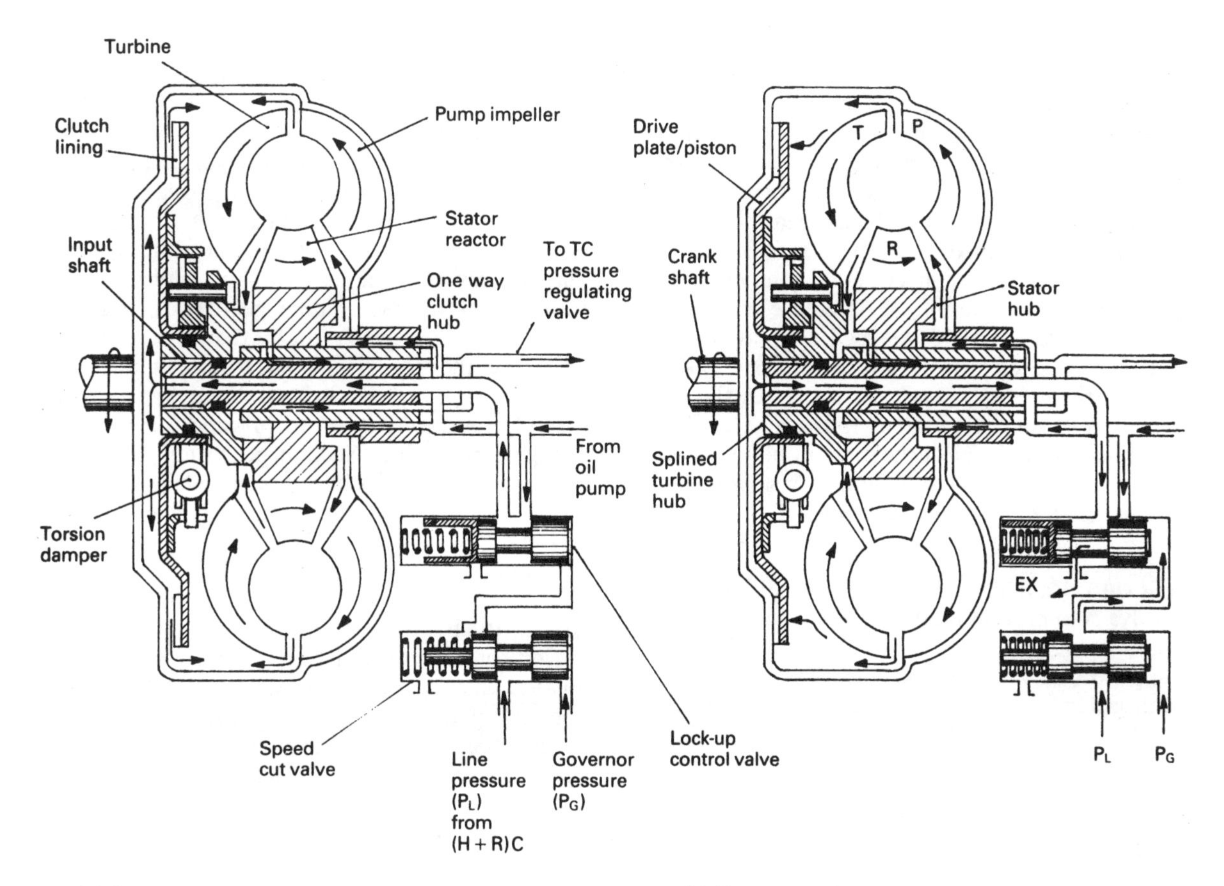

Fig. 5.11 (a and b) Lock-up torque converter

trolled piston to energize or de-energize the clutch engagement facings. The lock-up drive plate/piston is supported by the turbine hub which is itself mounted on the transmission input shaft. A transmission damper device, similar to that used on a conventional clutch drive-plate, is incorporated in the lock-up plate to absorb and damp shock impacts when the lock-up clutch engages.

Lock-up control The automatic operation of the converter lock-up is controlled by a speed cut valve and a lock-up control valve. The function of these valves is to open and close fluid passages which supply and discharge fluid from the space formed between the torque converter casing and the lock-up drive plate/piston.

Lock-up disengaged (Fig. 5.11(a)) With the vehicle driven in either first or second gear at relatively low speeds, low governor pressure permits the speed cut and lock-up control valve return springs to push their respective plunger to the right. Under these conditions, pressurized fluid from the torque converter flows into the space separating the lock-up plate/piston from the turbine. At the same time, fluid from the oil pump is conveyed to the space formed between the torque converter's casing and the lock-up plate/piston via the lock-up control valve and the central axial passage in the turbine input shaft. Consequently, the pressure on both sides of the lock-up plate will be equalized and so the lock-up plate/piston cannot exert an engagement load to energize the friction contact faces.

Lock-up engaged (Fig. 5.11(b)) As the speed of the vehicle rises, increased governor pressure will force the speed cut valve plunger against its spring until it uncovers the line pressure passage leading into the right hand end of the lock-up control valve. Line pressure fed from the high and reverse clutch is directed via the speed cut valve to the right hand end of lock-up control valve, thereby pushing its plunger to the left to uncover the lock-up clutch drain port. Instantly, pressurized fluid from the chamber created between the torque converter casing and lock-up plate/piston escapes via the central input shaft passage through the wasted region of the lock-up control valve plunger back to the inlet side of the oil pump. As a result, the difference of pressure across the two sides of the lock-up plate/piston causes it to slide towards the torque converter casing until the friction faces contact. This closes the exit for the converter fluid so that full converter fluid pressure is exerted against the lock-up plate/piston. Hence the input and output shafts are now locked together and therefore rotate as one.

Speed cut valve function The purpose of the speed cut valve is to prevent fluid draining from the space formed between the converter casing and lock-up plate/piston via the lock-up control valve if there is a high governor pressure but the transmission has not yet changed to third or fourth gear. Under these conditions, there is no line pressure in the high and reverse clutch circuit which is controlled by the shift valve. Therefore when the speed cut valve plunger moves to the left there is no line pressure to actuate the lock-up control valve so that the lock-up plate/piston remains pressurized on both sides in the disengaged position.

5.6 Three speed and reverse transaxle automatic transmission mechanical power flow

(Gear train as adopted by some Austin-Rover, VW and Audi 1.6 litre cars)

The operating principle of the mechanical power or torque flow through a transaxle three speed automatic transmission in each gear ratio will now be considered in some depth, see Fig. 5.12.

The planetary gear train consists of two sun gears, two sets of pinion gears (three in each set), two sets of annular (internal) gears and pinion carriers which support the pinion gears on pins. Helical teeth are used throughout.

For all forward gears, power enters the gear train via the forward annular gear and leaves the gear train by the reverse annular gear. In reverse gear, power enters the gear train by the reverse sun gear and leaves the gear train via the reverse annular gear.

First gear compounds both the forward gear set and the reverse gear set to provide the necessary low gear reduction. Second gear only utilizes the forward planetary gear set to produce the intermediate gear reduction. Third gear is achieved by locking the forward planetary gear set so that a straight through drive is obtained. With planetary gear trains the gears are in constant mesh and gear ratio changes are effected by holding, releasing or rotating certain parts of the gear train by means of a one way clutch, two multiplate clutches, one multiplate brake and one band brake.

The operation of the automatic transmission gear train can best be explained by referring to Table 5.2 which shows which components are engaged in each manual valve selection position.

5.6.1 Selector lever (Table 5.2)

The selector lever has a number of positions marked P R N D 2 1 with definite functions as follows:

P — park When selected, there is no drive through the transmission. A mechanical lock actuated by a linkage merely causes a parking pawl to engage in the slots around a ring gear attached to the output shaft (Fig. 5.2). Thus the parking pawl

Table 5.2 Manual valve selection position

Range	Forward clutch FC	Drive and reverse clutch (D + R)C	First and reverse brake (I + R)B	Second gear band 2GB	One way clutch OWC	Ratio
P and N	–	–	–	–	–	–
D – 1st 2 – 1st	Applied	–	–	–	Applied	2.71:1
1 – 1st	Applied	–	Applied	–	–	2.71:1
D – 2nd 2 – 2nd	Applied	–	–	Applied	–	1.5:1
D – 3rd	Applied	Applied	–	–	–	1.00:1
R	–	Applied	Applied	–	–	2.43:1

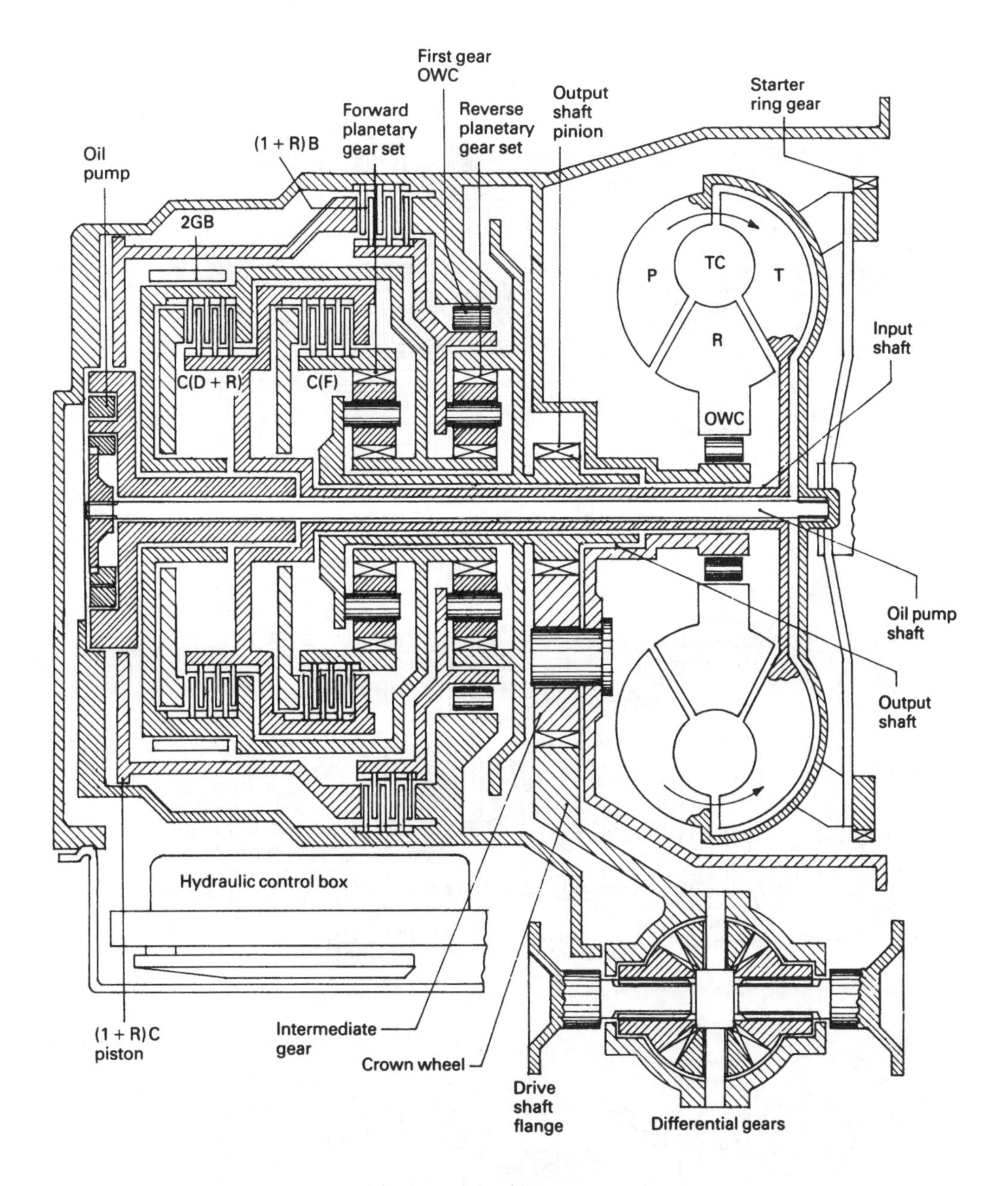

Fig. 5.12 Transaxle three speed automatic transmission layout

locks the output shaft to the transmission casing so that the vehicle cannot roll backwards or forwards. This pawl must not be engaged whilst the vehicle is moving. The engine may be started in this position.

R — reverse When selected, the output shaft from the automatic transmission is made to rotate in the opposite direction to produce a reverse gear drive. The reverse position must only be selected when the vehicle is stationary. The engine will not start in reverse position.

N — neutral When selected, all clutches and band brake are disengaged so that there is no drive through the transmission. The engine may be started in N — neutral range.

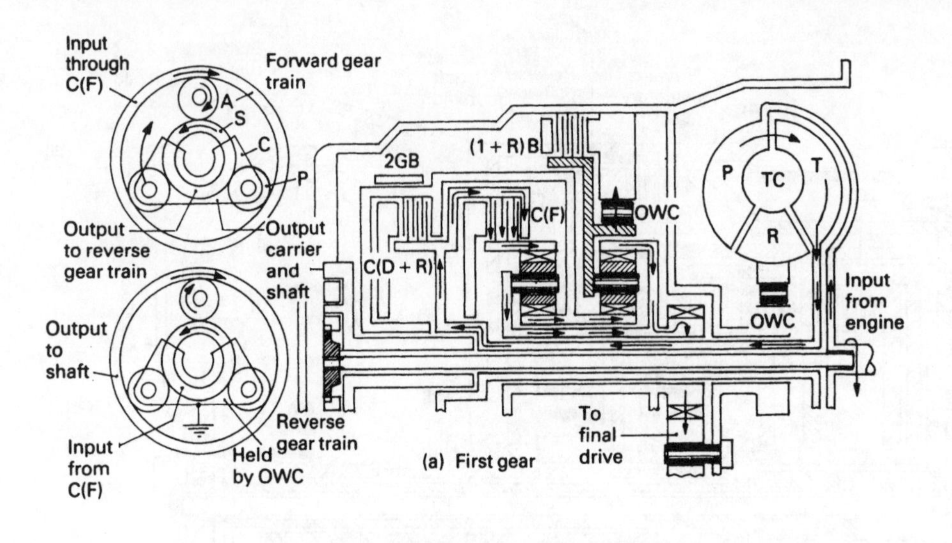

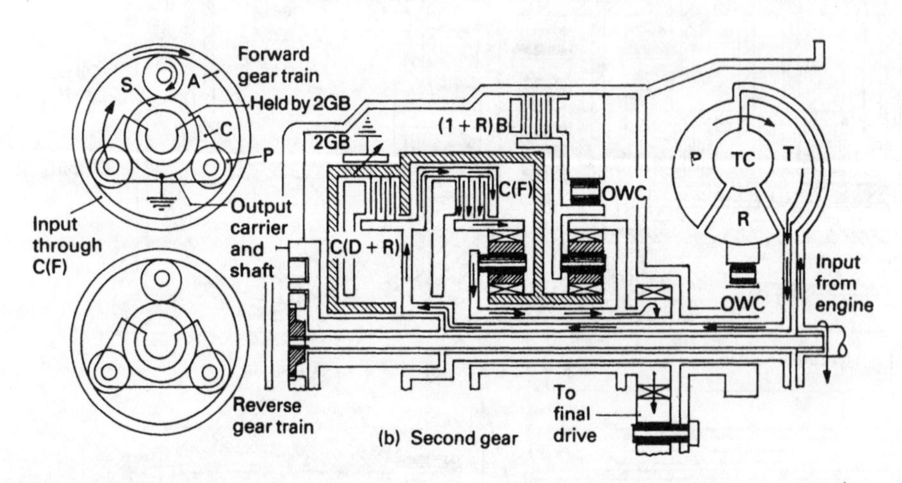

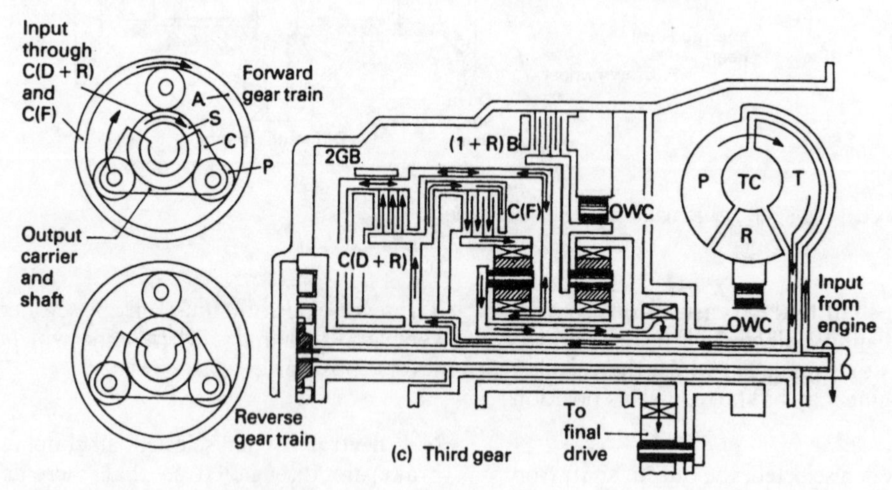

Fig. 5.13 (a–d) Three speed and reverse automatic transmission transaxle units

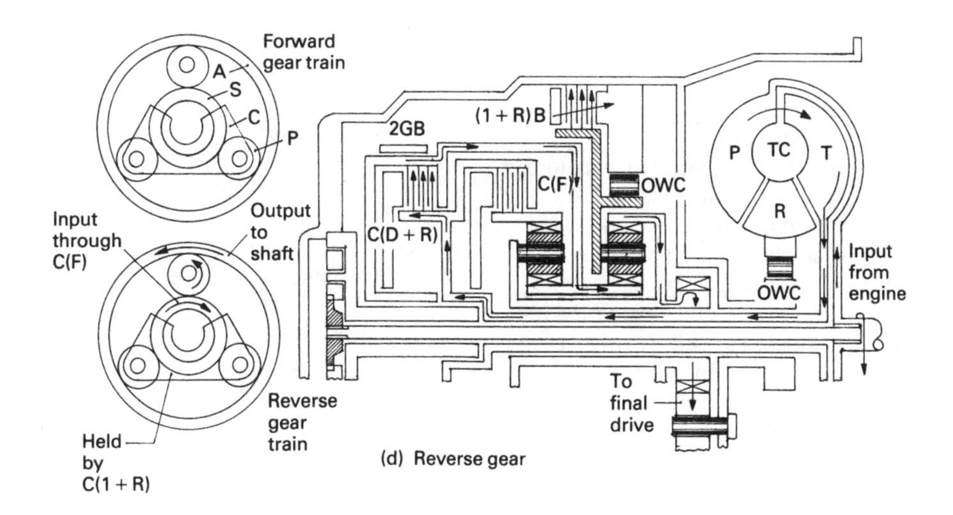

Fig. 5.13 *contd*

D — drive This position is used for all normal driving conditions, automatically producing 1–2, 2–3 upshifts and 3–2, 2–1 downshifts at suitable road speeds or according to the position of the accelerator pedal. The engine will not start in D — drive range.

2 — First and second This position is selected when it is desired to restrict gear changes automatically from 1–2 upshift and 2–1 downshifts only. The selector must not be positioned in 2 range above 100 km/h (70 mph). The engine will not start in this range position.

1 — First gear When this range is selected, the transmission is prevented from shifting into second and third gear. A friction clutch locks out the one way roller clutch so that better control may be obtained when travelling over rough or wet ground or icy roads. Engine braking on overrun is available when descending steep hills.

5.6.2 First gear (D — 1st) (Fig. 5.13(a))
With the manual selector valve in D range, engine torque is transmitted from the converter through the applied forward clutch to the annular gear of the forward planetary gear train. The clockwise rotation of the forward annular gear causes the forward planet gears to rotate clockwise, driving the double (compound) sun gear anticlockwise. The forward planetary carrier is splined to the output shaft. This causes the planet gears to drive the double sun gear instead of rolling 'walking' around the sun gears. This counterclockwise rotation of the sun gears causes the reverse planet gears to rotate clockwise. With the one way clutch holding the reverse planet carrier stationary, the reverse planetary gears turn the reverse annular gear and output shaft clockwise in a reduction ratio of something like 2.71:1.

When first gear is selected in the D range, a very smooth transmission take-up is obtained when the one way clutch locks, but on vehicle overrun the one way clutch is released so that the transmission freewheels.

5.6.3 First gear manual (1 — 1st) (Fig. 5.13(a))
The power flow in first gear manual differs from the D range in that the first and reverse brake are applied to hold the reverse planet carrier stationary. Under these conditions on vehicle overrun, engine braking is provided.

5.6.4 Second gear (D — 2nd) (Fig. 5.13(b))
In D range in second gear, the forward clutch and the second gear band brake are applied. The forward clutch then transmits the engine torque from the input shaft to the forward annular gear and the second gear band brake holds the double sun gear stationary. Thus engine torque is delivered to the annular gear of the forward planetary train in a clockwise rotation. Consequently, the planet gears are compelled to revolve on their axes and roll 'walk' around the stationary sun gear in a clockwise direction. As a result the output shaft, which is splined to the forward planet carrier, is made to turn in a clockwise direction at a slower speed

relative to the input shaft with a reduction ratio of approximately 1.50:1.

5.6.5 Third gear (D — 3rd) (Fig. 5.13(c))

In D range engine torque is transmitted through both forward clutch and drive and reverse clutch. The drive and reverse clutch rotate the sun gear of the forward gear train clockwise and similarly the forward clutch turns the annular gear of the same gear set also clockwise. With both the annular gear and sun gear of the forward gear train revolving in the same direction at the same speed, the planet gear becomes locked in position, causing the forward gear train to revolve as a whole. The output shaft, which is splined to the forward planet carrier, therefore rotates at the same speed as the input shaft, that is as a direct drive ratio 1:1.

5.6.6 Reverse gear (R) (Fig. 5.13(d))

With the manual selector valve in the R position, the drive and reverse multiplate brake is applied to transmit clockwise engine torque to the reverse gear set sun gear. With the first and reverse brake applied, the reverse planet gear carrier is held stationary. The planet gears are compelled to revolve on their own axes, thereby turning the reverse annular gear which is splined to the output shaft in an anticlockwise direction in a reduction ratio of about 2.43:1.

5.7 Hydraulic gear selection control components (Fig. 5.24)

(Three speed and reverse transaxle automatic transmission)

An explanation of how the hydraulic control system is able to receive pressure signals which correspond to vehicle speed, engine load and the driver's requirements, and how this information produces the correct up or down gear shift through the action of the control system's various plunger (spool) valves will now be considered by initially explaining the function of each component making up the control system.

A list of key components and abbreviations used in the description of the hydraulic control system is as follows:

1	Manual valve	MV
2	Kickdown valve	KDV
3	Throttle pressure valve	TPV
4	Valve for first gear manual range	V(1G)MR
5	1–2 shift valve	(1–2)SV
6	1–2 governor plug	(1–2)GP
7	Throttle pressure limiting valve	TPLV
8	Main pressure limiting valve	MPLV
9	Main pressure regulating valve	MPRV
10	Valve for first and reverse gear brake	V(1 + R)GB
11	Converter pressure valve	CPV
12	Soft engagement valve	SEV
13	2–3 shift valve	(2–3)SV
14	2–3 governor plug	(2–3)GP
15	Valve for direct and reverse clutch	V(D + R)C
16	3–2 control valve	(3–2)CV
17	3–2 kickdown valve	(3–2)KDV
18	Governor valve	GV
19	Forward clutch piston	FCP
20	Oil pump	P
21	Converter check valve	CCV
22	Second gear band servo	2GBS
23	Accumulator	A
24	Forward clutch piston	FCP
25	Direct and reverse clutch piston	(D + R)CP
26	First and reverse brake piston	(1 + R)BP
27	One way clutch	OWC

5.7.1 The pressure supply system

This consists of an internal gear crescent oil pump driven by the engine via a shaft splined to the torque converter impeller. The oil pressure generated by the oil pump is directed to the pressure regulating valve. By introducing limited throttle pressure into the regulator valve spring chamber, the thrust acting on the left hand end of the valve is increased during acceleration. This prevents the regulator valve being pushed back and spilling oil into the intake side of the oil pump. As a result, the line pressure will rise as the engine speed increases.

5.7.2 Main pressure regulator valve (MPRV) (Fig. 5.14(a and b))

This valve controls the output pressure which is delivered to the brake band, multiplate brake and clutch servos. Oil pressure from the pump acts on the left hand end of the valve and opposes the return spring. This oil pressure moves the valve to the right, initially permitting oil to pass to the converter pressure valve and its circuit, but with further valve movement oil will be exhausted back to the pump intake passage. The line pressure build-up is also controlled by introducing limited throttle pressure into the regulator spring chamber

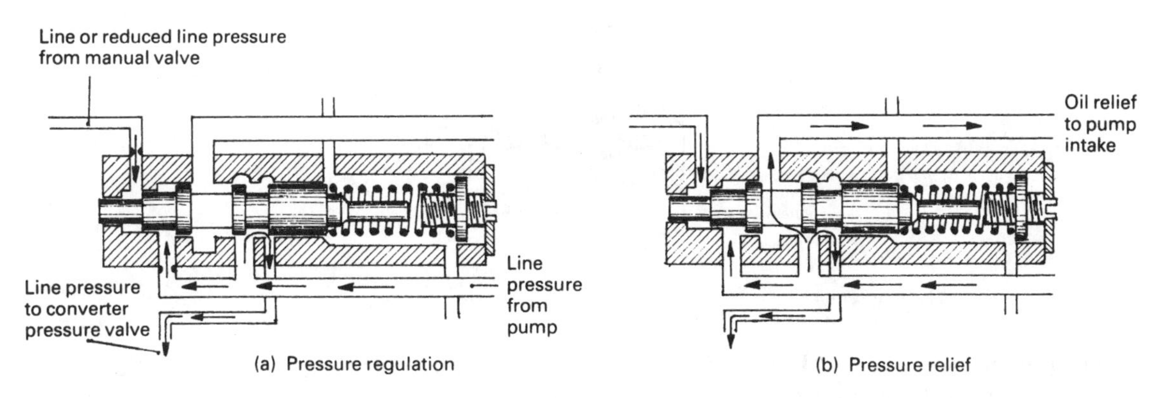

Fig. 5.14 (a and b) Main pressure regulating valve (MPRV)

which assists the spring in opposing the valve moving to the right. In addition, oil pressure from the manual valve passage, indirectly controlled by the governor, is imposed on the left hand end of the regulator valve. This modifies the valve movement to suit the various gear train and road condition requirements.

5.7.3 Throttle pressure valve (TPV)

(Fig. 5.15(a and b))

The throttle pressure valve transmits regulated pressure based on engine throttle position. Opening or closing the engine throttle moves the kickdown valve spool so that the throttle valve spring tension is varied. The amount of intermediate pressure allowed through the throttle pressure valve is determined by the compression of the spring. The reduced pressure on the output side of the throttle valve is then known as throttle pressure. Throttle pressure is directed to the main pressure limiting valve, the *kickdown valve*, and to one end of the shift valves in opposition to governor pressure, which acts on the other end of the shift valves controlling upshift and downshift speeds.

5.7.4 Main pressure limiting valve (MPLV)

(Fig. 5.16)

This valve is designed to limit or even cut off the variable throttle pressure passing through to the main regulating valve and the soft engagement valve. The pressure passing out from the valve to the main pressure regulator valve is known as *limited throttle pressure*. As the pressure passes through the valve it reacts on the left hand end of the main pressure limiting valve so that the valve will progressively move to the right, until at some predetermined pressure the valve will close the throttle pressure port feeding the main pressure regulating valve circuit.

When the throttle pressure port closes, the high pressure in the regulator spring chamber is permitted to return to the throttle pressure circuit via the non-return ball valve.

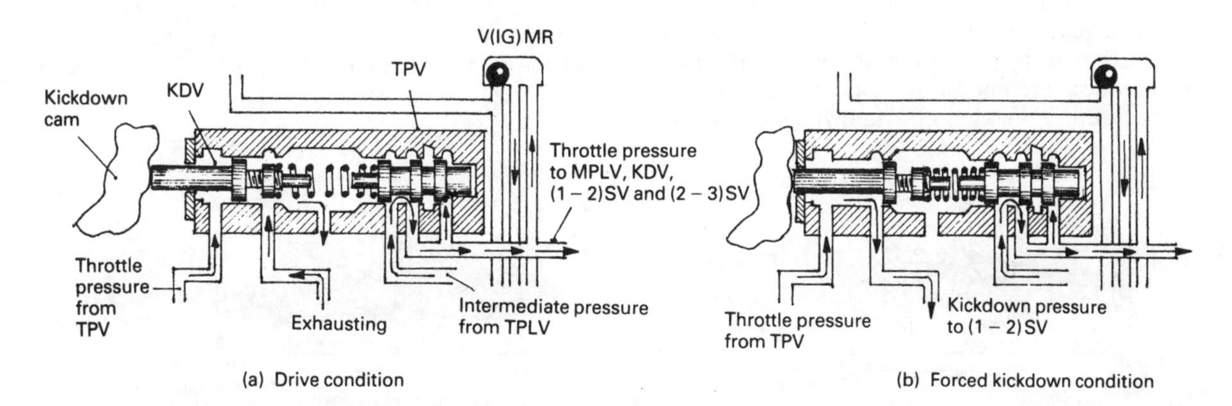

Fig. 5.15 (a and b) Kickdown valve (KDV), throttle pressure valve (TPV) and valve for first gear manual range (1GMR)

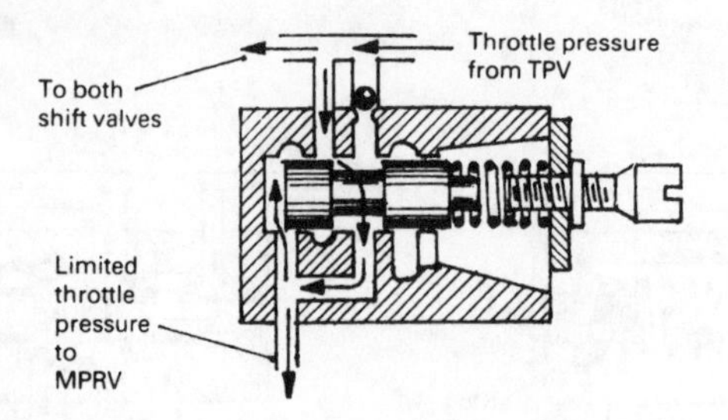

Fig. 5.16 Main pressure limiting valve

5.7.5 ***Converter pressure valve (CPV)*** (Fig. 5.17)
This valve shuts off the oil supply to the torque converter once the delivery pressure reaches 6 bar. Line pressure from the main pressure regulator valve passes through the valve to the torque converter and acts on its right hand end until the preset pressure is reached. At this point the valve is pushed back against its spring, closing off the oil supply to the torque converter until the converter pressure is reduced again.

The force on the output side of the converter pressure valve feeding into the converter is known as converter pressure.

5.7.6 ***Converter check valve (CCV)***
This valve, which is located inside the stator support, prevents the converter oil drainage when the vehicle is stationary with the engine switched off. This valve is not shown in the diagrams.

5.7.7 ***Throttle pressure limiting valve (TPLV)*** (Fig. 5.18)
This valve converts line pressure, supplied by the pump and controlled by the main pressure regulator valve, into intermediate pressure. The pressure reduction is achieved by line pressure initially passing through the diagonal passage in the valve so that it reacts against the left hand end of the valve. Consequently the valve shifts over and partially reduces the line pressure port opening. The reduced output pressure now known as intermediate pressure then passes to the throttle pressure valve.

5.7.8 ***Kickdown valve (KDV)***
(Fig. 5.15(a and b))
This valve permits additional pressure to react on the shift valves and governor plugs when a rapid acceleration (forced throttle) response is required by the driver so that the governor pressure is compelled to rise to a higher value before a gear upshift occurs. When the throttle is forced wide open, the kickdown valve is moved over to the right, thus allowing throttle pressure to pass through the valve. The output pressure is known as kickdown pressure. The kickdown pressure feeds in between both 1–2 and 2–3 shift valves and governor plug combinations. As a result, this kickdown pressure opposes and delays the governor pressure movement of the governor plug and shift valve, thereby preventing a gear upshift occurring until a much higher speed is reached.

5.7.9 ***1–2 Shift valve and governor plug (1–2)SV and (1–2)GP*** (Fig. 5.19)
This valve combination automatically controls and shifts the transmission from first to second or from second to first depending upon governor and throttle pressure. When governor pressure on the right hand governor plug side overcomes throttle pressure on the left hand 1–2 shift valve side, both

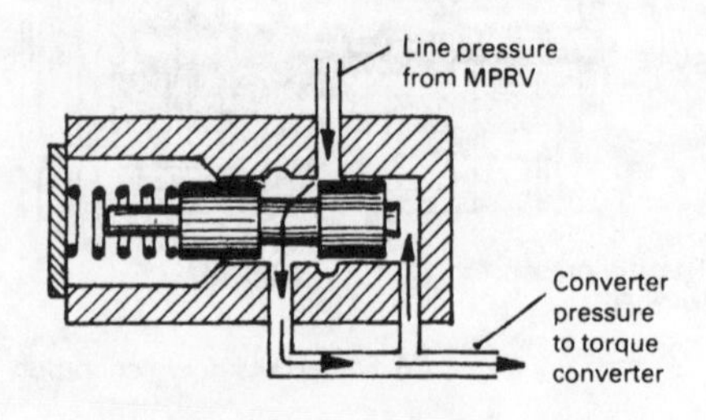

Fig. 5.17 Converter pressure valve (CPV)

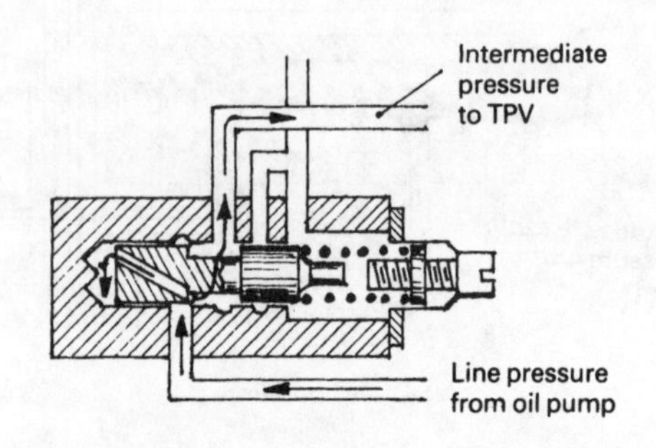

Fig. 5.18 Throttle pressure limiting valve (TPLV)

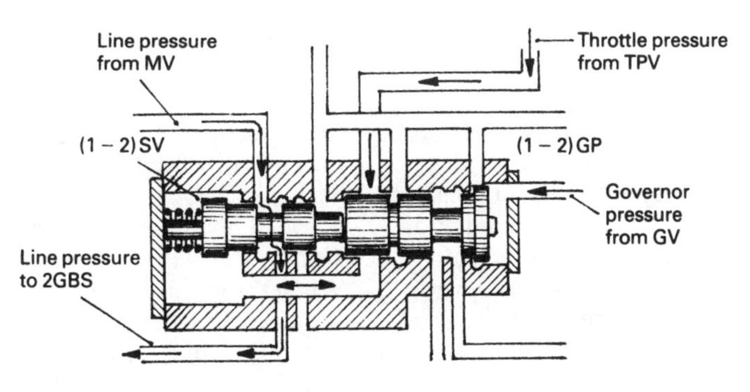

Fig. 5.19 1–2 shift valve (1–2)SV, and 1–2 governor plug (1–2)GP in 1–2 upshift condition

1–2 governor plug and 1–2 shift valve move to the left thereby opening the line pressure port which delivers oil from the pump. Line pressure will now pass unrestricted through the valve to feed into the brake band servo. As a result an upchange occurs. If, in addition to the throttle pressure, kickdown pressure is introduced to the valve combination, gear upshifts will be prolonged. If '1' manual valve is selected, line pressure will be supplied to the governor plug chamber (large piston area) and the throttle spring chamber, preventing a 1–2 upshift. '1' manual position cannot be engaged at speeds above 72 km/h because the 1–2 shift valve cannot move across, due to the governor pressure.

5.7.10 2–3 Shift valve and governor plug (2–3)SV and (2–3)GP (Fig. 5.20(a and b))

The 2–3 shift valve and governor plug control the gear change from second to top gear or from top to second depending upon governor and throttle pressure. As governor pressure exceeds throttle pressure, the shift valve and governor plug are pushed over to the left. This permits line pressure to pass through the valve so that it can supply pressure to the drive and reverse clutch piston, so that an upchange can now take place. When '2' manual valve position is selected, there is no pressure feeding to the shift valve which therefore prevents a 2–3 upshift.

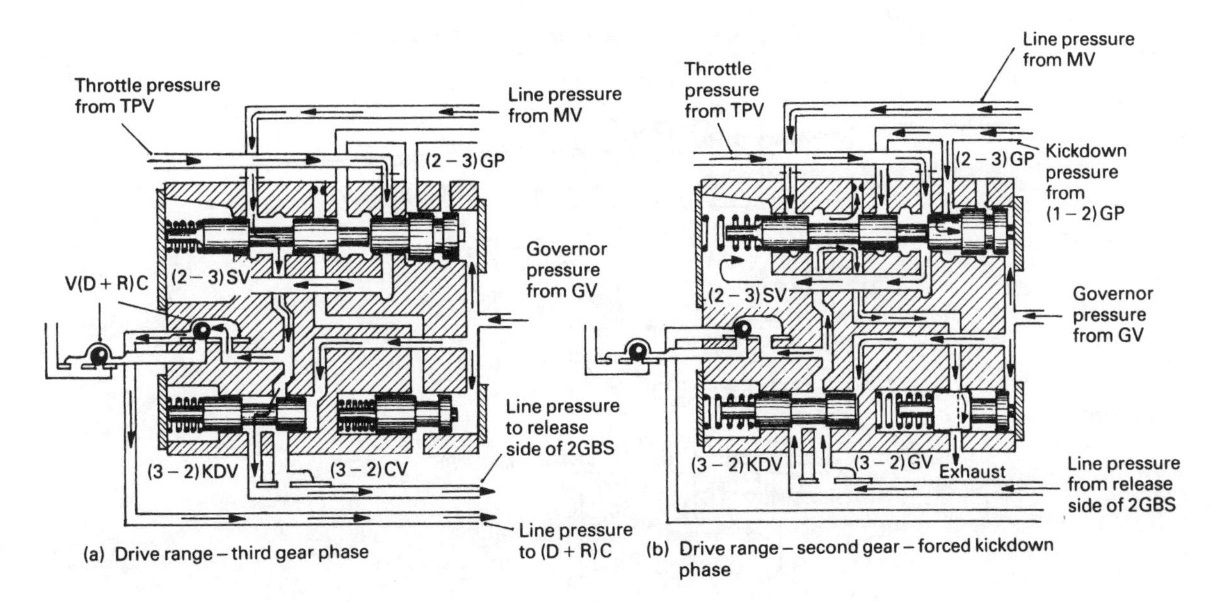

Fig. 5.20 (a and b) 2–3 shift valve (2–3)SV, 2–3 governor plug (2–3)GP, 3–2 control valve (3–2)CV, 3–2 kickdown valve (3–2)KDV and valve for direct and reverse clutch V(D + R)C

5.7.11 3–2 Kickdown valve (3–2)KDV
(Fig. 5.20(a and b))
This valve is provided to prolong the downshift from third to second gear during rapid acceleration from above 90 km/h so that the change takes place relatively smoothly. With rising output shaft speed, the governor pressure acting on the right hand end of the valve moves it to the right, thus practically restricting the oil outflow from the servo spring chamber and therefore extending the second gear band engagement time.

5.7.12 3–2 Control valve (3–2)CV
(Fig. 5.20(a and b))
This valve controls the expulsion of oil from the spring side of the second gear band servo piston at speeds in the region of 60 km/h. The time period for oil to exhaust then depends upon the governor pressure varying the effective exhaust port restriction. Line pressure oil from the spring side of the second gear band servo piston passes through a passage leading to the 3–2 kickdown valve annular groove and from there to the 2–3 shift valve annular groove. Here some oil exhausts out from a fixed restriction while the remainder passes via a passage to the 3–2 control valve. As the vehicle speed approaches 60 km/h the governor pressure rises sufficiently to force back the 3–2 control valve piston, thus causing the wasted (reduced diameter) part of the control valve to complete the exhaustion of oil.

5.7.13 Valves for direct and reverse clutch V(D + R)C (Fig. 5.21(a and b))
When the manual selector valve is moved to reverse position the left hand ball valve drops onto its seat so that line pressure oil from the manual selector is compelled to move through a restriction. At the same time, the right hand ball valve is pushed to the right, immediately closing off the second gear servo piston spring side from the line pressure. The right hand ball valve is dislodged to the left when third gear is selected so that the manual reverse line pressure is prevented feeding the drive and reverse clutch piston. During the time third gear is selected the left hand ball serves no purpose.

5.7.14 Valve for first gear manual range V(1G)MR (Figs 5.15(a and b) and 5.22)
The selection of first gear manual supplies line pressure to the underside passage to the ball valve,

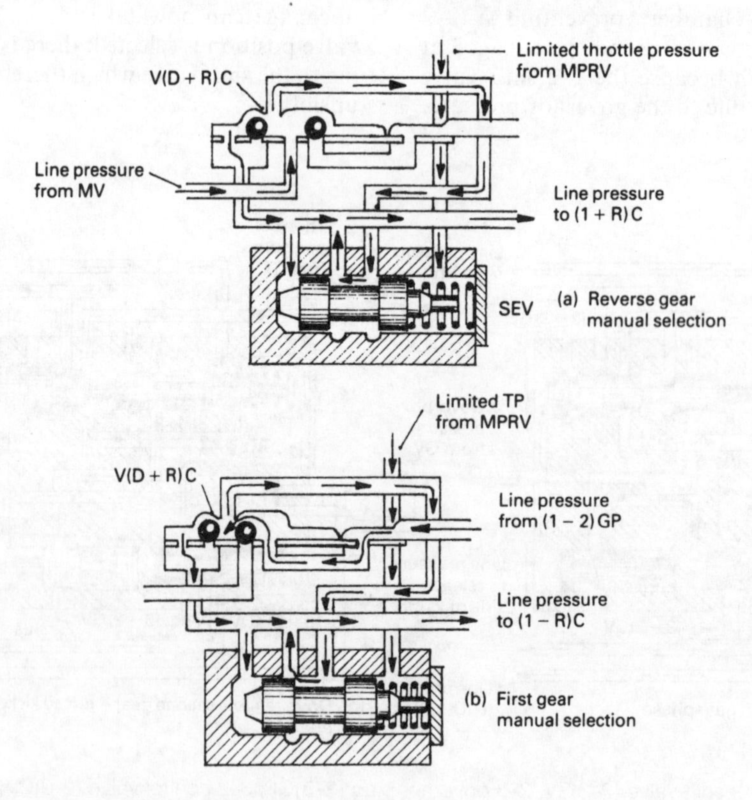

Fig. 5.21 (a and b) Soft engagement valve (SEV), valve for first and reverse clutch, V(1 + R)GC

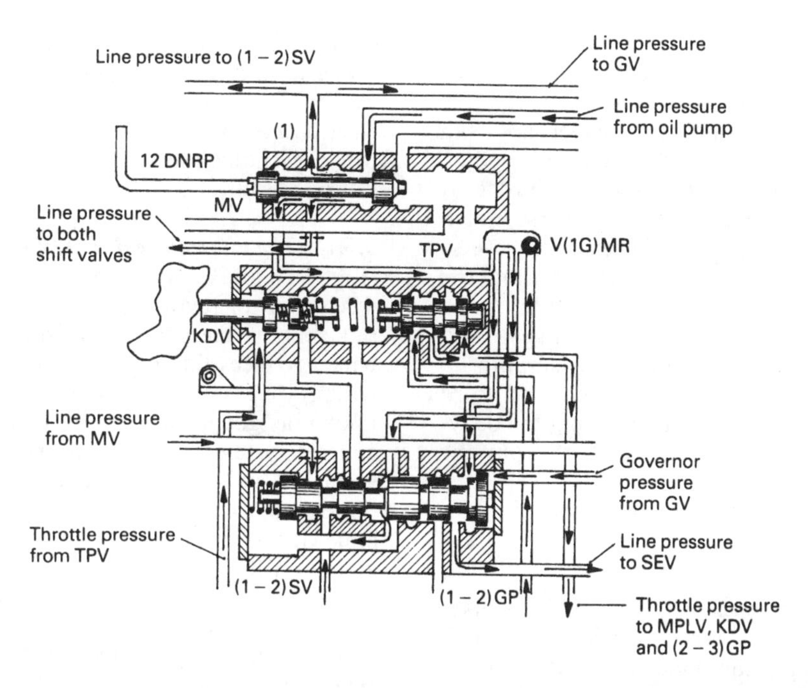

Fig. 5.22 Manual valve (MV), kickdown valve (KDV), throttle pressure valve (TPV), 1–2 shift valve (1–2)SV and 1–2 governor plug (1–2)GP in first gear – manual selection

causing it to move to the right. Line pressure then fills the throttle pressure lines leading to the left hand end of the 1–2 shift valve and therefore a 1–2 upshift is prevented.

5.7.15 Valves for first and reverse gear brake V(1 + R)GB (Fig. 5.20(a and b))

The selection of first gear manual position causes line pressure to dislodge the right hand ball valve to the left, thereby closing the reverse line passage from the selector valve. At the same time the left hand ball valve closes so that line pressure flow for the engagement of the first and reverse multiplate is slowed down.

The selection of reverse gear causes the right hand ball valve to be pushed by line pressure to the right and so the first gear line passage from the selector valve is closed. Similarly the left hand ball valve closes so that line pressure flow to the first and reverse brake is restricted, thus prolonging the clutch engagement period.

5.7.16 Soft engagement valve (SEV) (Fig. 5.21(a and b))

This valve provides a cushioning effect for the engagement of first and reverse gear brake. This effect is achieved by line pressure acting on the left hand valve end pushing the valve to the right against the opposing variable throttle pressure. The result of this movement is to restrict and slow down the pressure build-up on the first and reverse gear brake piston.

5.7.17 Clutches and brakes

Front clutch piston (FCP) (Fig. 5.24) This is an annular shaped piston which directs a clamping load to a multiplate clutch via a diaphragm type spring when line pressure is introduced behind the piston. The engagement of the clutch couples the output shaft from the torque converter turbine to the forward annular gear ring. The forward clutch is applied in all forward drive gear ranges.

First and reverse brake piston (1 + R)CP (Fig. 5.24) Introducing line pressure to the first and reverse clutch piston cylinder engages the multiplate brake which locks the reverse planetary carrier to the transmission casing. The first and reverse brake is applied only in first and reverse range.

Drive and reverse clutch piston (D + R)CP (Fig. 5.24) Directing line pressure behind the drive and reverse clutch piston applies the clutch, thereby transmitting torque from the torque converter turbine output shaft to the forward sun gear. When the forward clutch is also applied, the forward planetary gears (annular, planet and sun gears) are locked together and they rotate bodily, thus producing a straight through 1:1 third gear drive. However, when the first and reverse clutch is applied instead of the forward clutch, the reverse planetary carrier is held stationary causing the reverse gear reduction ratio to be engaged.

The timing of the release of one set of gears and the engagement of another to produce smooth up and down gear shifts between second and third gears is achieved by carefully controlling the delivery and exhaustion of hydraulic fluid from the clutch and band brake servo.

These operating conditions are explained under second gear band servo.

Second gear band servo (2GBS) (Fig. 5.24) This is a double acting piston servo which has a small piston area to apply the band brake and a large piston area which is on the release spring chamber side of the servo.

Directing line pressure to the small piston area chamber of the servo applies the band brake against the resistance of the return spring and thereby holds stationary both sun gears. Introducing line pressure on the large piston area spring chamber side of the servo produces an opposing force which releases the grip on the band brake. The piston returns to the 'off' position and the relaxing of the band brake is made possible by the difference in piston area on each side, both sides being subjected to the same line pressure. The band brake is applied only in the second gear forward speed range.

During upshift from 2–3 it is important that the second gear band brake does not release too quickly relative to the drive and reverse clutch engagement, in order to avoid *run-up* (rapid engine speed surge) during the transition from 2nd to 3rd gear. During downshift it is also important that the second gear band brake does not engage before the drive and rear clutch releases in order to avoid *tie-up* (gear jamming) on the 3–2 shift.

The 3–2 control valve and the 3–2 kickdown valves therefore affect the timing relationship between the second gear band servo and the drive and reverse clutch to provide correct shift changes under all operating conditions.

First gear one way clutch (OWC) (Fig. 5.13(a)) When in drive range, the one way roller type clutch operates in place of the first and reverse multiplate brake to prevent the rotation of the reverse pinion carrier. This one way clutch enables the gear set to freewheel on overrun and to lock-up on drive, therefore preventing a jerky gear ratio in 1–2 upshift and 2–1 downshift.

5.7.18 The governor valve (GV)

(Figs 5.23 and 5.24)
The governor revolving with the transmission output shaft is basically a pressure regulating valve which reduces line pressure to a value that varies with output vehicle speed. This variable pressure is known as governor pressure and is utilized in the control system to effect up and down gear shifts from 1–2 and 2–3 shift valves. Governor pressure opposes shift valve spring force, throttle pressure and kickdown pressure, and the resulting force acting on the governor plug and shift valve determines the vehicle's gear change speeds. The governor drive is achieved through a skew gear meshing with a ring gear mounted on the reverse annular gear carrier which is attached to the output pinion shaft.

The two types of governor valves used for this class of automatic transmission are the ball and pivot flyweight and the plunger and flyweight. These governors are described below.

Plunger and flyweight type governor (Fig. 5.23) Rotation of the governor at low speed causes the governor weight and valve to produce a centrifugal force. This outward force is opposed by an equal and opposite hydraulic force produced by governor pressure acting on the stepped annular area of the governor valve. Because the governor valve is a regulating valve, and will attempt to remain in equilibrium, governor pressure will rise in accordance with the increase in centrifugal force caused by increased rotational speed. As the output shaft speed increases, the governor weight moves outwards (due to the centrifugal force) to a stop in the governor body, when it can move no further. When this occurs, the governor spring located between the weight and the governor valve becomes effective. The force of this spring then combines with centrifugal force of the governor valve to oppose the hydraulic pressure, thus making the pressure less sensitive to output shaft speed variation. Therefore the governor provides two distinct phases of regulation, the first being used for accurate control of the low speed shift points.

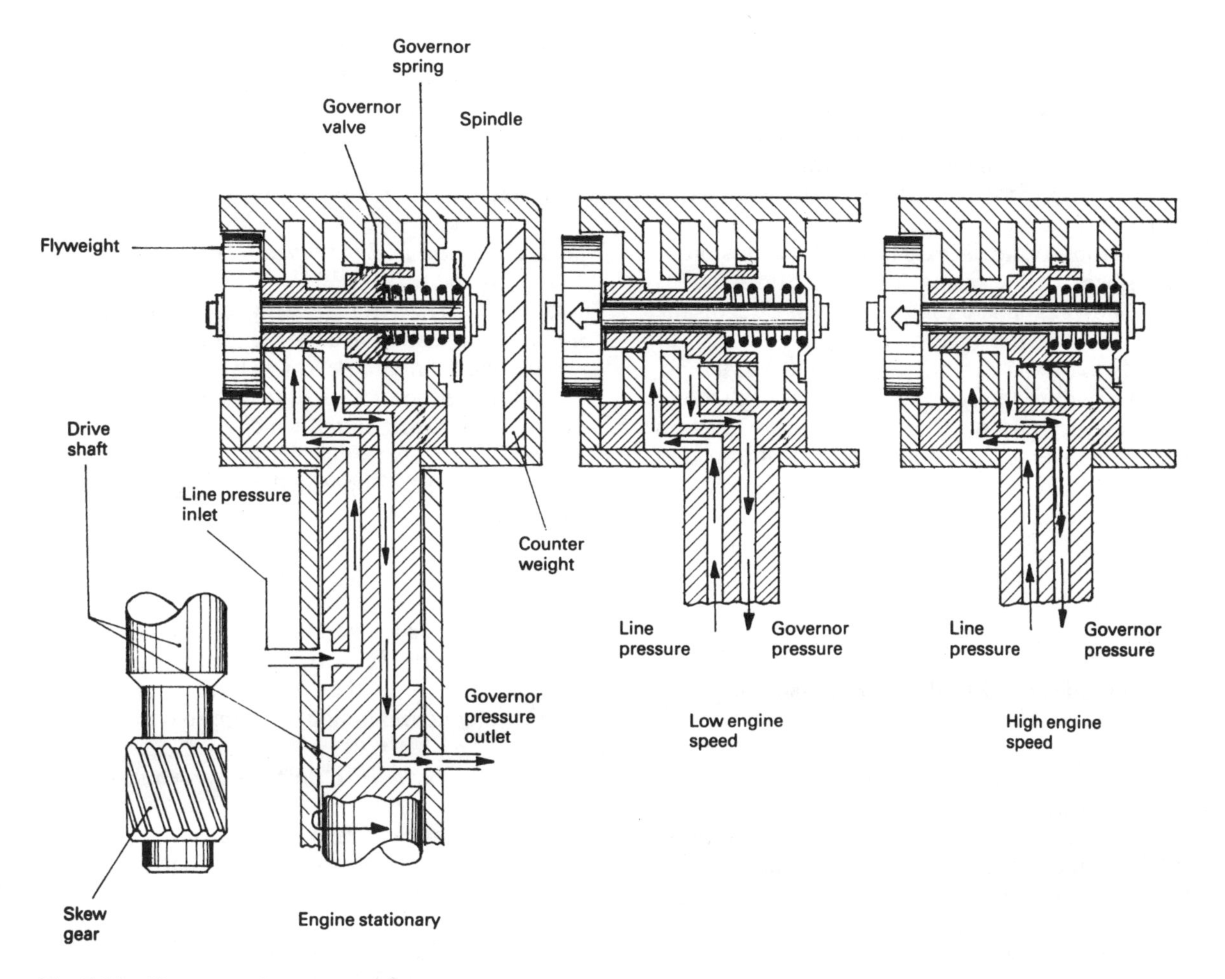

Fig. 5.23 Plunger and governor valve

Ball and pivot flyweight type governor (Fig. 5.24) This type of governor consists of a ball valve controlled by a hinged flyweight and a pressure relief ball valve. Fluid from the oil pump at line pressure is introduced via a restriction into an axial passage formed in the governor drive shaft. When the transmission output shaft stops rotating (vehicle stationary) with the engine idling, fluid pressure forces the governor ball valve off its seat, permitting fluid to escape back to the sump. Rotation of the output shaft as the vehicle accelerates from a standstill causes the flyweight centrifugal force to close the ball valve. Therefore fluid trapped in the governor drive shaft passage, known as governor pressure, has to reach a higher pressure before fluid exhausts through the valve. By these means the line pressure is regulated to a valve that varies with the output shaft and vehicle speed. A pressure relief valve is also included to safeguard the system from excessively high pressure if the governor valve malfunctions.

5.7.19 Hydraulic accumulator (Fig. 5.24)

This is a cylinder and spring loaded piston which is used to store a small amount of pressure energy to enable a rapid flow of fluid under pressure to one of the operating components or to absorb and smooth fluctuating fluid delivery. The piston is pushed back when the fluid pressure exceeds the spring load and fluid enters and fills up the space left behind by the displaced piston.

With the transmission in neutral or park, line pressure from the pressure pump enters the accumulator at the opposite end to the spring, thereby displacing the piston and compressing the spring. When the hydraulic control shifts into the second gear phase, line pressure from the 1–2 shift valve is directed to the second gear band servo applied end and the spring end of the accumulator.

When the accumulator spring is compressed, fluid from the supply can flow rapidly to the applied side of the band servo piston. As soon as the servo piston meets resistance (starts to apply its load), the fluid pressure increases and the accumulator piston spring is extended as the piston is pushed back by the spring. This is because there is equal line pressure acting on either side of the accumulator piston and so the spring is able to apply its load and extend. As a result, the supply of fluid is reduced to the applied side of the second gear band servo piston. The accumulator therefore smooths and times the application of the second gear band brake in order to reduce the risk of shock and a jerky operation. In addition, the extra quantity of fluid in the system due to the accumulator leads to a slow rate of release of the servo piston and band.

5.8 Hydraulic gear selection control operation

5.8.1 Fluid flow in neutral (Fig. 5.24)

Pressurized oil from the pump flows to the main pressure regulating valve. The valve shifts over due to the oil pressure, thus opening a passage supplying the torque converter via the converter pressure valve. Increased pump pressure moves the valve further until it uncovers the exhaust port dumping the oil back into the oil pump suction intake passage.

The oil pressure generated between the pump and main pressure limiting valve is known as line pressure and is directed to the throttle pressure limiting valve, accumulator, manual selection valve and through the latter valve to the 2–3 shift valve.

When the throttle foot released, the intermediate pressure exhausts so that there will be no throttle pressure. Depressing the throttle pedal increases the throttle spring tension and creates a throttle pressure which is then directed to the kickdown valve, main pressure limiting valve, 1–2 shift valve and the 2–3 shift valve.

At the same time, a limited throttle pressure is created between the main pressure limiting valve and main pressure regulating valve. This pressure is also directed to the soft engagement valve.

With the manual gear selector valve in neutral, there is no line pressure to the rear of the main regulating valve and therefore the trapped line pressure will be at a maximum.

5.8.2 Fluid flow in park (Fig. 5.24)

With the manual valve in park position, the hydraulic flow is the same as in neutral, except there is no line pressure to the 2–3 shift valve and the main pressure regulating valve provides maximum increase in line pressure.

5.8.3 Fluid flow in drive range — first gear (Figs 5.24 and 5.19)

Both main pressure regulating valve and throttle pressure valve operate as for neutral and park.

With the manual selector valve in D, line pressure is directed to the 1–2 shift valve, 2–3 shift valve and to the forward clutch which it engages.

The shift valve is subjected to governor pressure at one end which opposes the spring tension and throttle pressure at the opposite end. As the car speed increases governor pressure will overcome throttle pressure causing a 1–2 upshift to take place.

Throughout this period a reduced line pressure reacts against the left hand end of the main pressure regulating valve. The valve movement then allows more oil to pass to the torque converter, thereby causing a reduction in line pressure to occur.

5.8.4 Fluid flow in drive range — second gear (Figs 5.24 and 5.19)

As for drive range — first gear, the main regulator valve and throttle pressure valve function as for neutral and park.

When the manual selector valve is positioned in D, line pressure is directed to the 1–2 shift valve, 2–3 shift valve and to the forward clutch which it applies.

The rising governor pressure imposes itself against one end of the 1–2 governor plug counteracting throttle pressure until at some predetermined pressure difference the valve shifts across. This then permits line pressure to flow to the accumulator and the second gear servo piston thus causing the second gear brake band to be applied.

At the same time, full line pressure is applied to the left hand end of the main pressure regulating valve so that there will be a further decrease in line pressure relative to drive range – first gear operating conditions.

5.8.5 Fluid flow in drive range — third gear (Figs 5.24 and 5.20(a))

As for drive range – first and second gears, the main regulator valve and throttle pressure valve perform as for neutral and park.

The manual selector valve will still be in D position so that line pressure is directed to both 1–2 and

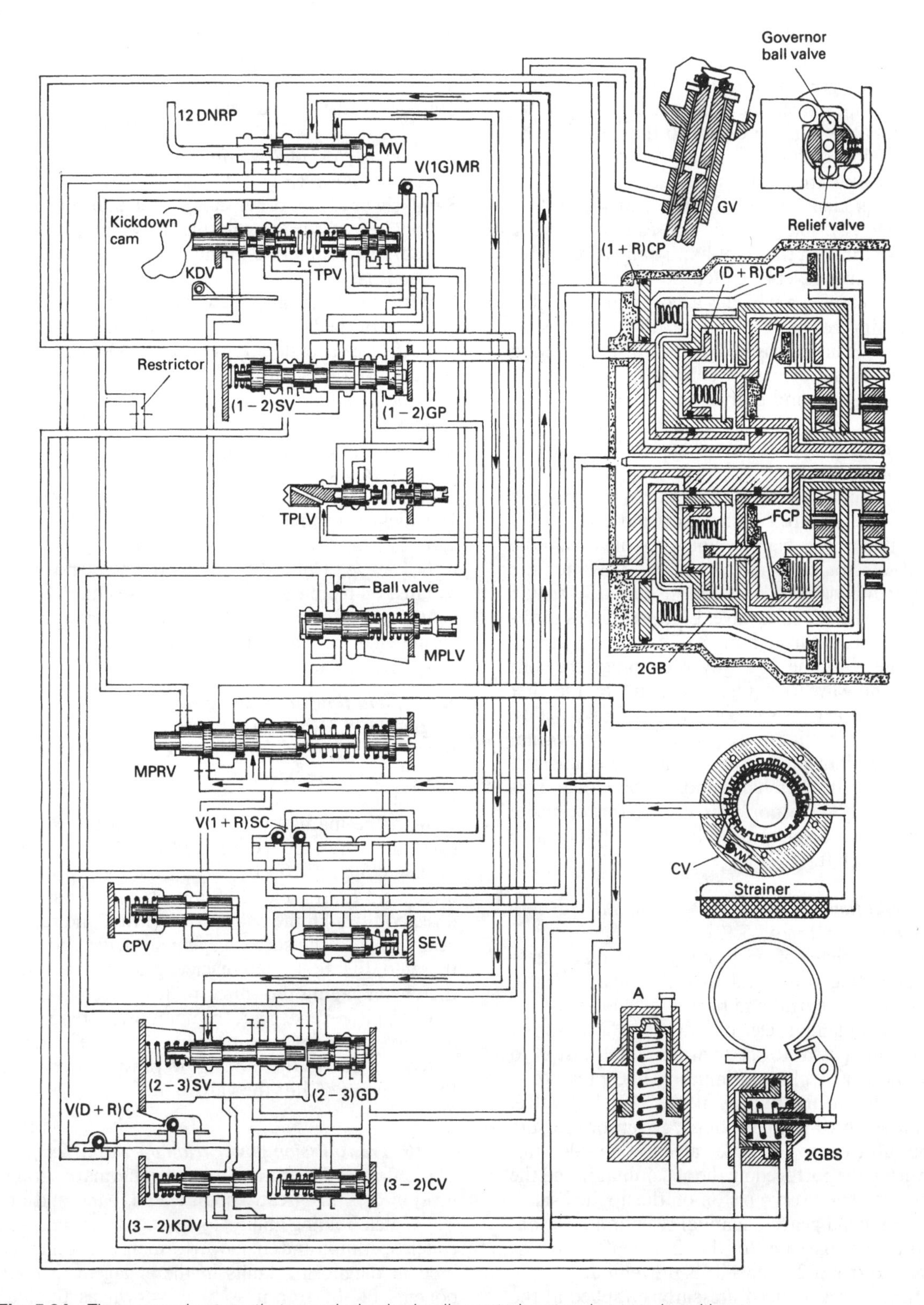

Fig. 5.24 Three speed automatic transmission hydraulic control system in neutral position

2–3 shift valves and to the forward clutch piston which clamps the friction clutch plates.

Increased governor pressure acting on the 2–3 governor plug moves the adjacent 2–3 shift valve over. This allows line pressure to flow around the wasted region of the shift valve to the 3–2 kick-down valve and from there to the second gear band servo spring chamber side (release) of the piston. Simultaneously, line pressure passes from the 2–3 shift valve through the right hand ball valve orifice for direct and reverse clutch to the direct and reverse clutch piston which is accordingly engaged.

The main pressure regulating valve will be subjected to full line pressure acting on its left hand end so that line pressure reduction will be as in drive range — second gear.

5.8.6 *Fluid flow in first gear — manual selection* (Figs 5.24 and 5.15)

With the manual selector in '1' position, line pressure passes to the forward clutch piston and accordingly applies the clutch plates. Line pressure from the manual selector valve moves the ball valve for the first gear manual range so that it cuts off throttle pressure to the 1–2 shift valve. Line pressure is therefore able to pass through the ball valve for first and reverse gear to the 1–2 governor plug, the soft engagement valve and finally passing to the first and reverse brake piston to engage the brake plates.

Consequently, line pressure will fill the normal throttle pressure lines of 1–2 shift valve and will react against the left hand end of the valve. This then prevents governor pressure at the opposite end acting on the governor plug moving the valve for a 1–2 upshift.

5.8.7 *Fluid flow in second gear — manual selection* (Figs 5.24 and 5.19)

During this phase of gear change, the main pressure regulating valve and throttle pressure valve operate as for neutral and park.

When the manual selector valve is in '2' position, line pressure passes to the forward clutch piston to apply the clutch plates. Similarly, line pressure is also directed to the middle of the 1–2 shift valve. When road speed is high enough, governor pressure will be sufficient to push the valve to one side, thus uncovering the port feeding the accumulator and the second gear band servo piston on the applied side.

In the second gear — manual selection position, line pressure passing to the 2–3 shift valve is blocked so that a 2–3 upshift is prevented.

In first gear a reduced pressure is applied at the end of the main pressure regulating valve, causing a reduced line pressure to be created. Likewise in second gear full line pressure will act behind the main pressure regulating valve so that the line pressure is further reduced.

5.8.8 *Fluid flow in drive range — second gear — forced kickdown* (Figs 5.24 and 5.15(b))

Similarly as for all other manual selector positions, the main pressure regulating valve and throttle pressure valve operate in the same way as for neutral and park.

With the manual selector valve in D and the accelerator pedal fully depressed, the kickdown valve reduced waste aligns with the kickdown line outlet port thus causing throttle pressure to flow into the kickdown lines.

Kickdown pressure now flows to 1–2 and 2–3 governor plugs assisting the throttle pressure applied on the 1–2 and 2–3 shift valves. When the governor pressure is low enough, throttle pressure and kickdown pressure overcomes governor pressure, causing the 2–3 governor plug to move to the right. As a result, the 2–3 shift valve moves to exhaust oil from the drive and reverse clutch piston chamber and the second gear band servo piston spring chamber causing a 3–2 downshift to occur.

5.8.9 *Fluid flow in reverse gear* (Figs 5.24, 5.21(a) and 5.14(a))

In reverse gear the pressure regulator valve and throttle pressure valve operate in a similar way to neutral.

With the manual selection valve in the R position, line pressure is directed to the drive and reverse clutch piston by way of the ball valves for direct and reverse clutch. Similarly, line pressure is directed through the ball valves for first and reverse gear brake to the soft engagement valve and from there to the first and reverse gear brake piston thereby engaging the clutch plates.

Whilst in reverse gear, the main pressure regulating valve provides maximum line pressure increase, this being due to there being no pressure acting on the left hand end of the valve.

5.8.10 *Transmission power train operating faults*

The effective operation of an automatic transmission depends greatly upon clutch, band and one way clutch holding ability, torque converter one way clutch operation and engine performance. The method used in diagnosing faults in the engagement components of the transmission is known as the *stall test*. This test entails accelerating the engine with the throttle wide open to maximum speed while the torque converter turbine is held stationary.

Stall test procedure

1 Drive car or run engine until engine and transmission has attained normal working temperatures.
2 Check the level of fluid in the transmission box and correct if necessary.
3 Apply the hand brake and chock the wheels.
4 Connect a tachometer via leads to the coil ignition terminals.
5 Apply the foot brake, select D range and fully press the accelerator pedal down for a period not exceeding 10 seconds to avoid overheating the transmission fluid (this is very important).
6 Quickly observe the highest engine speed reached on the tachometer and immediately release the throttle pedal.
7 Shift selector lever to N and allow transmission fluid to cool at least two minutes or more before commencing next test.
8 Repeat tests 5 and 6 in '1' and R range.

Interpreting stall test results A typical stall test maximum engine speed could be 2300 rev/min ±100. If the actual stall speed differs from the recommended value (i.e. 2300 rev/min), Table 5.3 should be used as a guide to trace the fault. The stall test therefore helps to determine if the fault is due to the engine, the converter or the transmission assembly.

Note The reason why a slipping torque converter stator drags down the engine's maximum speed is because the spinning stator makes the converter behave as a fluid coupling (no torque multiplication), causing the fluid to have a retarding effect on the impeller.

By performing the stall test in D, '1' and R range, observing in which range or ranges the slippage occurs and comparing which clutch or band operates in the slipping range enables the effective components to be eliminated and the defective components to be identified (see Table 5.4).

Table 5.3 Table of stall tests

Test results	Possible causes
Below 1600 rev/min	Stator slip
Approximately 2100 rev/min	Poor engine performance
Above 2500 rev/min	Transmission slip

Examples

a) Slip in R can be the drive and reverse clutch or first and reverse brake. Engage '1' range. If slip still occurs, first and reverse brake must be slipping.
b) Slip in D can be forward clutch or one way clutch. Engage '1'. If slip still occurs, forward clutch must be slipping.
c) Slip in R can be the drive and reverse clutch or first and reverse brake. Engage '1' range. If there is no slip the drive and reverse clutch could be slipping.

Road test for defective torque converter A road test enables a seized or slipping stator to be engaged, whereas the stall test can only indicate a possible slipping stator. The symptoms for a faulty stator one way clutch are shown in Table 5.5.

If the converter one way clutch has seized, the vehicle will have poor high speed performance because the stator reaction above the coupling point speed hinders the circulation of fluid if it is not able to freewheel. Conversely, if the converter one way clutch is slipping there will be no stator reaction for the fluid and therefore no torque multiplication so that the acceleration will be sluggish up to about 50 km/h.

5.9 The continuously variable belt and pulley transmission

The continuously variable transmission CVT, as used by Ford and Fiat, is based simply on the

Table 5.4 Table of possible faults

Selector position	Possible fault
R	(D + R)C or (1 + R)B
D	FC or OWC
1	FC, OWC or (1 + R)C

Table 5.5 Table of symptoms for a faulty one way clutch

	Vehicle response	
Fault	0–50 km/h	above 50 km/h
Slipping stator	Very sluggish No hill start possible	Drives normally
Seized stator	Drives normally	Loss of power Severe overheating

principle of a belt running between two V-shaped pulleys which is designed so that the effective belt contact diameter settings can be altered to produce a stepless change in the input to output pulley shaft speed.

Van Doorne Transmissie in Holland has been mainly responsible for the development of the steel belt which is the key component in the transmission. At present the steel belt power output capacity is suitable for engine sizes up to 1.6 litres but there does not appear to be any reason why uprated steel belts cannot be developed.

This type of transmission does not suffer from the limitations of the inefficient torque converter which is almost universely used by automatic transmissions incorporating epicyclic gear trains operated by multiplate clutches and band brakes.

5.9.1 Stepless speed ratios (Figs 5.25 and 5.26)

The transmission consists basically of a pair of variable width vee-shaped pulleys which are interconnected by a composite steel belt. Each pulley consists of two shallow half cones facing each other and mounted on a shaft, one being rigidly attached to it whereas the other half is free to slide axially on linear ball splines (Fig. 5.25). The variable speed ratios are obtained by increasing or decreasing the effective wrap contact diameter of the belt with the primary input pulley producing a corresponding reduction or enlargement of the secondary output pulley working diameter. The belt variable wrap contact diameter for both primary and secondary pulleys is obtained by the wedge shaped belt being supported between the inclined adjacent walls of the two half pulleys.

When the primary input half pulleys are brought axially closer, the wedge or vee-shaped belt running between them is squeezed and is forced to ride up the tapered walls to a larger diameter. Conversely, since the belt is endless and inextensible, the secondary output half pulleys are compelled to separate, thus permitting the belt wrap to move inwards to a smaller diameter.

Alternatively, drawing the secondary output half pulleys closer to each other enlarges the belt's running diameter at that end. Accordingly it must reduce the primary input pulley wrap diameter at the opposite end. A one to one speed ratio is obtained when both primary and secondary pulleys are working at the same belt diameter (Fig. 5.26). A speed ratio reduction (underdrive) occurs when the primary input pulley operates at a larger belt contact diameter than the secondary output pulley (Fig. 5.26). Conversely, a speed ratio increase (over drive) is achieved if the belt contact with the primary pulley is at a smaller diameter relative to the secondary pulley wrap diameter (Fig. 5.26).

In the case of the Ford Fiesta, the pulleys provide a continuously variable range of ratios from bottom 2.6:1 to a super overdrive top of 0.445:1.

An intermediate gear reduction of about 1.4:1 between the belt output pulley shaft and the final drive crownwheel is also provided so that the transmission can be made to match the engine's power output and the car's design expectation.

5.9.2 Belt design (Figs 5.26 and 5.27)

Power is transmitted from the input to the output pulley through a steel belt which resembles a steel necklace of thin trapezoidal plates strung together between two multistrip bands made from flexible high strength steel (Fig. 5.27). There are 300 plates, each plate being roughly 2 mm thick, 25 mm wide and 12 mm deep, so that the total length of the endless belt is approximately 600 mm. Each band is composed of 10 continuous strips 0.18 mm thick. Also made of high strength steel, they fit into location slots on either side of the plates, their purpose being to guide the plates, whereas it is the plates' function to transmit the drive by pushing. Another feature of the plates is that they are embossed in a dimple form to assist in the automatic alignment of the plates as they flex around the pulley.

Contact between the belt plates and pulley is provided by the tapered edges of the plates which match the inclination of the pulley walls. When in drive, both primary and secondary sliding half pulleys are forced against the belt so that different plates are in contact and are wedged between the vee profile of the pulley at any one time.

Consequently, the grip produced between the plates and pulley walls also forces the plates together so that in effect they become a continuous strut which transmits drive in compression (unlike the conventional belt which transfers power under tension (Fig. 5.26)). The function of the non-drive side of the belt, usually referred to as the slack side, is only to return the plate elements back to the beginning of the drive (compressive) side of the pulleys.

The relative movement between the band strips and the plates by this design is very small, therefore frictional losses are low. Nevertheless the transmission efficiency is only 92% with a one to one speed ratio dropping to something like 86% at pull-away, when the speed ratio reduction is 2.6:1.

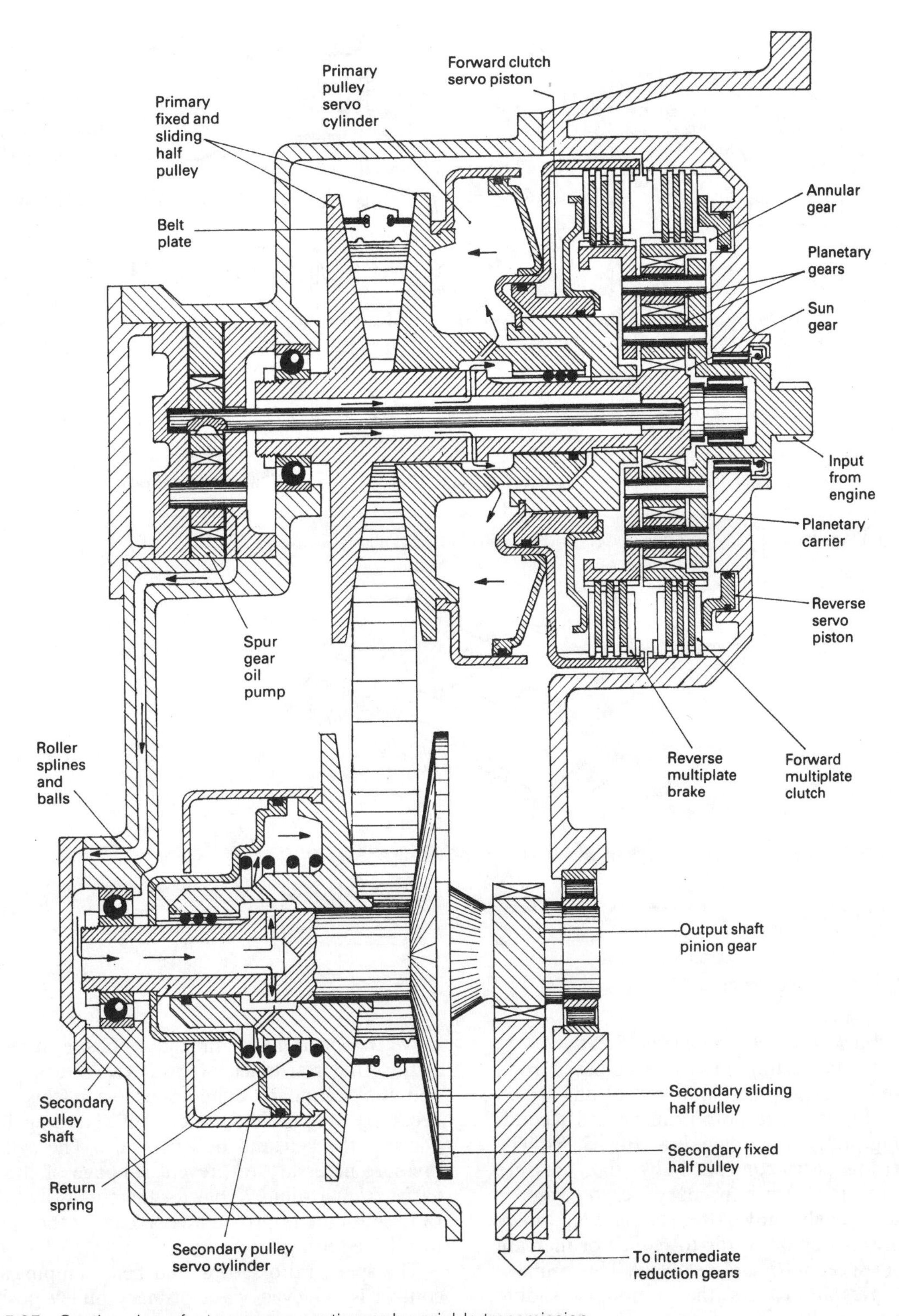

Fig. 5.25 Section view of a transverse continuously variable transmission

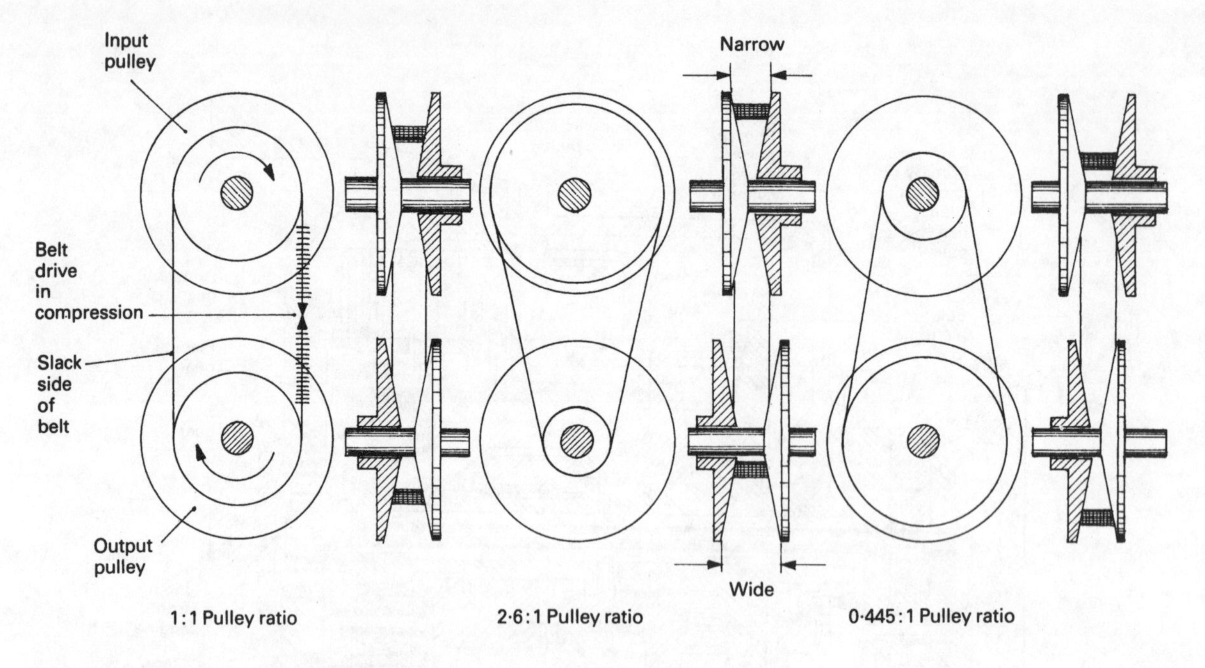

Fig. 5.26 Illustration of pulley and belt under- and overdrive speed ratios

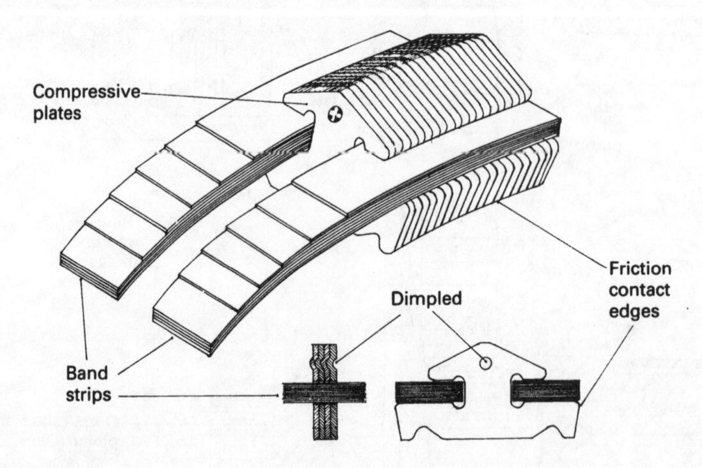

Fig. 5.27 Steel belt construction

5.9.3 Hydraulic control system (Fig. 5.28)

The speed ratio setting control is achieved by a spur type hydraulic pump and control unit which supplies oil pressure to both primary and secondary sliding pulley servo cylinders (Fig. 5.28). The ratio settings are controlled by the pressure exerted by the larger primary servo cylinder which accordingly moves the sliding half pulley axially inwards or outwards to reduce or increase the output speed setting respectively. This primary cylinder pressure causes the secondary sliding pulley and smaller secondary servo cylinder to move proportionally in the opposite direction against the resistance of both the return spring and the secondary cylinder pressure, this being necessary to provide the correct clamping loads between the belt and pulleys' walls. The cylinder pressure necessary to prevent slippage of the belt varies from around 22 bar for the pull away lowest ratio setting to approximately 8 bar for the highest overdrive setting.

The speed ratio setting and belt clamping load control is achieved via a primary pulley position senser road assembly.

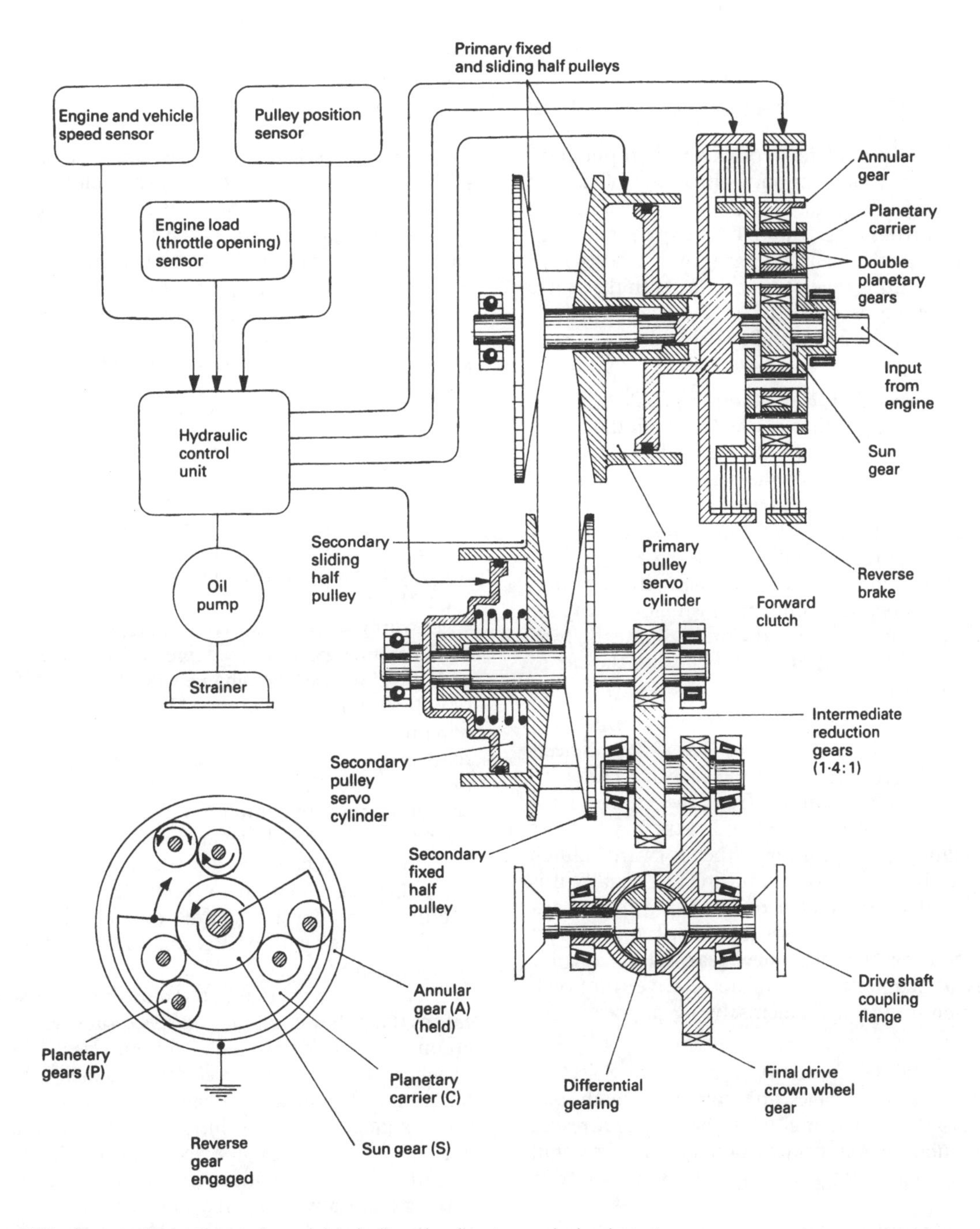

Fig. 5.28 Transaxle continuously variable belt and pulley transmission layout

However, engine and road speed signals are provided by a pair of pitot tubes which sense the rate of fluid movement, this being a measure of speed, be it either under the influence of fluid flow caused by the engine's input or by the output drive relating to vehicle speed.

5.9.4 Epicyclic gear train construction and description (Figs 5.25 and 5.28)

Drive in both forward and reverse direction is obtained by a single epicyclic gear train controlled by a forward multiplate clutch and a reverse multiplate brake, both of which are of the wet type (immersed in oil) (Fig. 5.25). The forward clutch is not only used for engagement of the drive but also to provide an initial power take-up when driving away from rest.

The epicyclic gear train consists of an input planetary carrier, which supports three sets of double planetary gears, and the input forward clutch plates. Surrounding the planetary gears is an internally toothed annulus gear which also supports the rotating reverse brake plates. In the centre of the planetary gears is a sun gear which is attached to the primary pulley drive shaft.

Neutral or park (N or P position) (Fig. 5.28) When neutral or park position is selected, both the multiple clutch and brake are disengaged. This means that the annulus gear and the planetary gears driven by the input planetary carrier are free to revolve around the sun gear without transmitting any power to the primary pulley shaft.

The only additional feature when park position is selected is that a locking pawl is made to engage a ring gear on the secondary pulley shaft, thereby preventing it from rotating and causing the car to creep forward.

Forward drive (D or L position) (Fig. 5.28) Selecting D or L drive energizes the forward clutch so that torque is transmitted from the input engine drive to the right and left hand planetary carriers and planet pins, through the forward clutch clamped drive and driven multiplates. Finally it is transferred by the clutch outer casing to the primary pulley shaft. The forward gear drive is a direct drive causing the planetary gear set to revolve bodily at engine speed with no relative rotational movement of the gears themselves.

Reverse drive (R position) (Fig. 5.28) Selecting reverse gear disengages the forward clutch and energizes the reverse multiplate brake. As a result, the annular gear is held stationary and the input from the engine rotates the planetary carrier (Fig. 5.28).

The forward clockwise rotation of the carrier causes the outer planet gears to rotate on their own axes as they are compelled to roll round the inside of internally toothed annular gear in an anticlockwise direction.

Motion is then transferred from the outer planet gears to the sun gear via the inner planet gears. Because they are forced to rotate clockwise, the meshing sun gear is directionally moved in the opposite sense anticlockwise, that is in the reverse direction to the input drive from the engine.

5.9.5 Performance characteristics (Fig. 5.29)

With D drive selected and the car at a standstill with the engine idling, the forward clutch is just sufficiently engaged to produce a small amount of transmission drag (point 1). This tends to make the car creep forwards which can be beneficial when on a slight incline (Fig. 5.29). Opening the throttle slightly fully engages the clutch, causing the car to move positively forwards (point 2). Depressing the accelerator pedal further sets the speed ratio according to the engine speed, road speed and the driver's requirements. The wider the throttle is opened the lower the speed ratio setting will be and the higher the engine speed and vice versa. With a light constant throttle opening at a minimum of about 1700 rev/min (point 3) the speed ratio moves up to the greatest possible ratio for a road speed of roughly 65 km/h which can be achieved on a level road. If the throttle is opened still wider (point 4) the speed ratio setting will again change up, but at a higher engine speed. Fully depressing the accelerator pedal will cause the engine speed to rise fairly rapidly (point 5) to about 4500 rev/min and will remain at this engine speed until a much higher road speed is attained. If the engine speed still continues to rise the pulley system will continue to change up until maximum road speed (point 6) has been reached somewhere near 5000 rev/min.

Partially reducing the throttle open then causes the pulley combination to move up well into the overdrive speed ratio setting, so that the engine speed decreases with only a small reduction in the car's cruising speed (point 7).

Even more throttle reduction at this road speed causes the pulley combination speed ratio setting to go into what is known as a backout upshift (point 8), where the overdrive speed ratio reaches its maximum limit. Opening the throttle wide again brings about a kickdown downshift (point 9) so that there is a surplus of power for acceleration. A further feature which provides engine braking when driving fast on winding and hilly slopes is through the selection of L range; this changes the form of driving by preventing an upshift when the throttle is eased and in fact causes the pulley combination to move the speed ratio towards an underdrive situation (point 10), where the engine operates between 3000 and 4000 rev/min over an extensive road speed range.

The output torque developed by this continuously variable transmission approaches the ideal

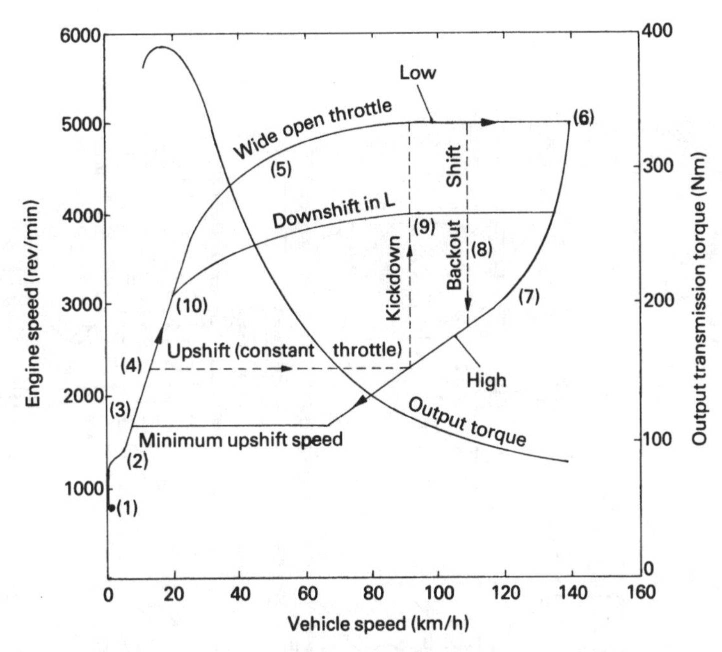

Fig. 5.29 CVT speed and torque performance characteristics

constant power curve (Fig. 5.29) in which the torque produced is inversely proportional to the car's road speed.

5.10 Five speed automatic transmission with electronic-hydraulic control

5.10.1 Automatic transmission gear train system (Fig. 5.30)

This five speed automatic transmission system is broadly based on a ZF design. Power is supplied though a hydrodynamic three element torque converter incorporating an integral disc type lock-up clutch. The power drive is then directed though a Ravigneaux type dual planetary gear train which provides five forward gears and one reverse gear; it then passes to the output side via a second stage single planetary gear train. The Ravigneaux planetary gear train has both large and small input sun gears, the large sun gears mesh with three long planet gears whereas the small sun gears mesh with three short planet gears; both the long and the short planet gears are supported on a single gear carrier. A single ring-gear meshing with the short planet gear forms the output side of the planetary gear train. Individual gear ratios are selected by applying the input torque to either the pinion carrier or one of the sun gears and holding various other members stationary.

5.10.2 Gear train power flow for individual gear ratios

D drive range — first gear (Fig. 5.31) With the position selector lever in D drive range, the one way clutch (OWC) holds the front planet carrier while multiplate clutch (B) and the multiplate brake (G) are applied. Power flows from the engine to the torque converter pump wheel, via the fluid media to the output turbine wheel. It is then directed by way of the input shaft and the applied multiplate clutch (B) to the front planetary large sun gear (S_L). With the front planet carrier (C_F) held stationary by the locked one way clutch (OWC), power passes from the large sun gear (S_L) to the long planet gears (P_L) in an anticlockwise direction. The long planet gear (P_L) therefore drives the short planet gears (P_S) in a clockwise direction thus compelling the front annular ring gear (A_F) to move in a clockwise direction. Power thus flows from the front annular ring gear (A_F) though the rear intermediate shaft to the rear planetary gear annular ring gear (A_R) in a clockwise direction. With the rear sun gear (S_R) held stationary by the applied multiplate brake (G), the rear planet gears (P_R) are forced to roll around the fixed sun gear in a clockwise direction, this in turn compels the rear planet carrier (C_R) and the output shaft also to rotate in a clockwise direction at a much reduced speed. Thus a two stage speed reduction produces an overall underdrive

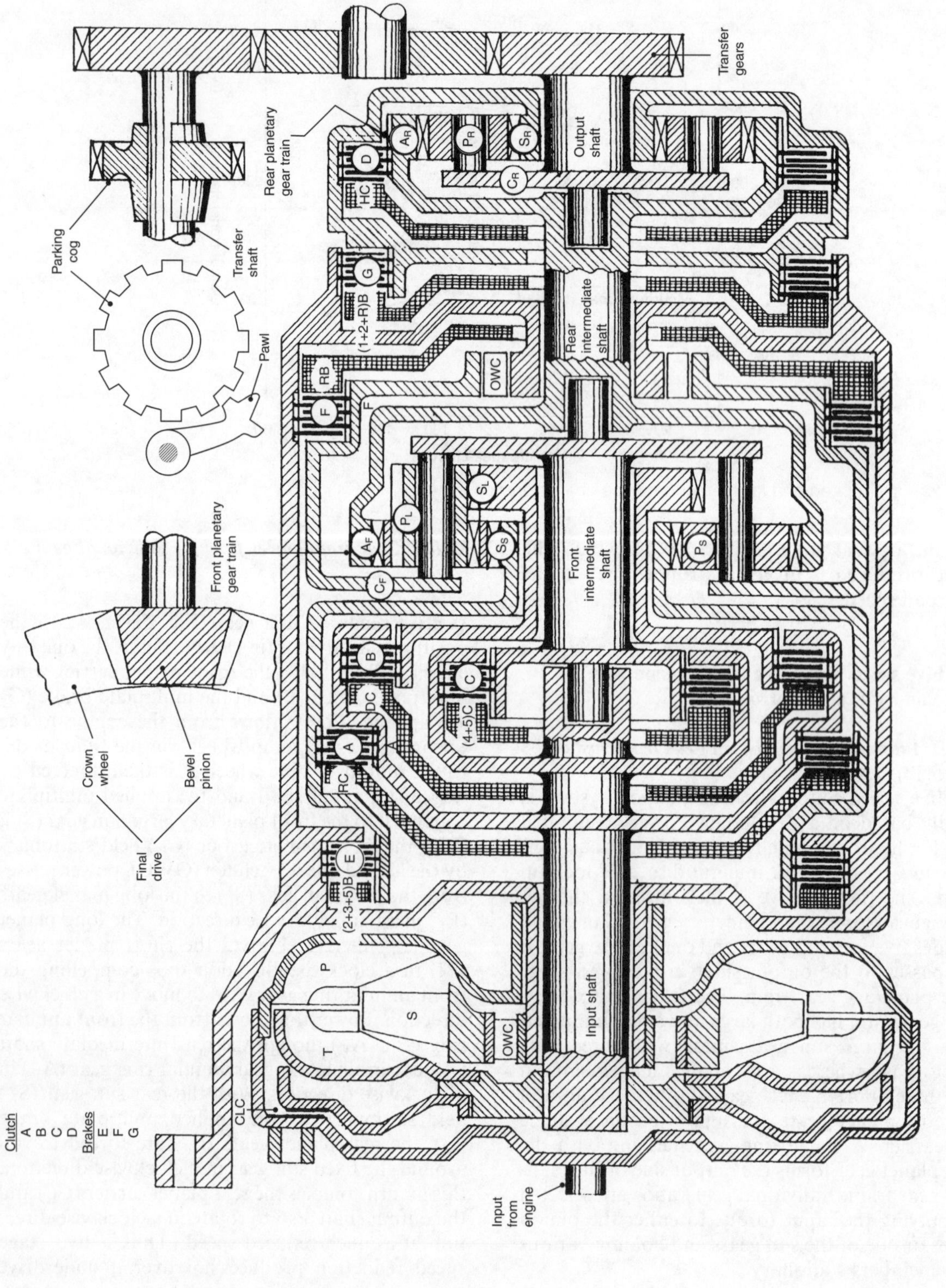

Fig. 5.30 Five speed and reverse automatic transmission (transaxle/longitudinal) layout

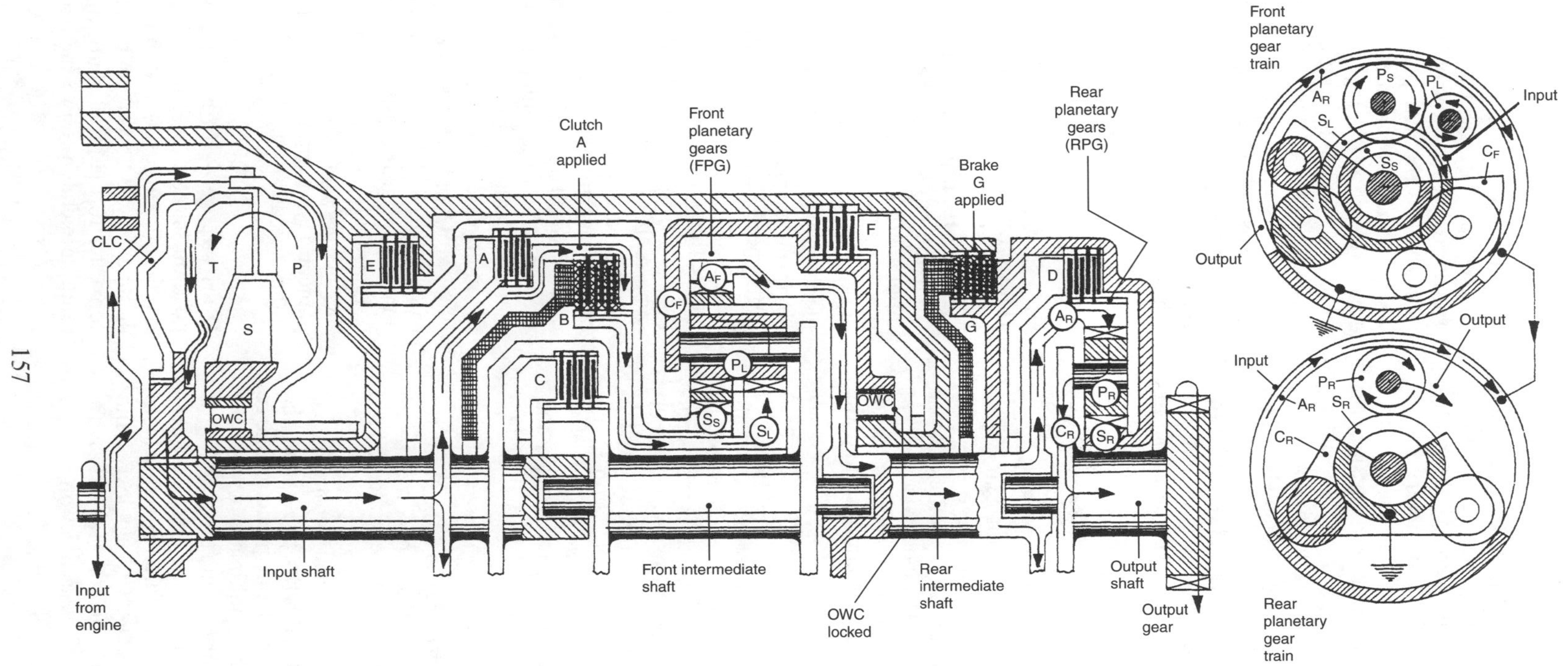

Fig. 5.31 Five speed and reverse automatic transmission power flow first gear

first gear. If the '2' first gear is selected multiplate brake (F) is applied in addition to the multiplate clutch (B) and multiplate brake (G). As a result instead of the one way clutch (OWC) allowing the vehicle to freewheel on overrun, the multiplate brake (F) locks the front planetary carrier (C_F) to the casing. Consequently a positive drive exists between the engine and transmission on both drive and overrun: it thus enables engine braking to be applied to the transmission when the transmission is overrunning the engine.

D drive range — second gear (Fig. 5.32) With the position selector lever in D drive range, multiplate clutch (C) and multiplate brakes (B) and (G) are applied.

Power flows from the engine via the torque converter to the input shaft, it then passes via the multiplate clutch (B) to the first planetary large sun gear (S_L). With the multiplate brake (E) applied, the front planetary small sun gear (S_S) is held stationary. Consequently the large sun gear (S_L) drives the long plant gears (P_L) anticlockwise and the short planet gears (P_S) clockwise, and at the same time, the short planet gears (P_S) are compelled to roll in a clockwise direction around the stationary small sun gear (S_S).

The drive then passes from the front planetary annular ring gear (A_F) to the rear planetary annular ring gear (A_R) via the rear intermediate shaft. With the rear sun gear (S_R) held stationary by the applied multiplate brake (G) the clockwise rotation of the rear annular ring gear (A_R) compels the rear planet gears (P_R) to roll around the held rear sun gear (S_R) in a clockwise direction taking with it the rear carrier (C_R) and the output shaft at a reduced speed. Thus the overall gear reduction takes place in both front and rear planetary gear trains.

D drive range — third gear (Fig. 5.33) With the position selector lever in D drive range, multiplate clutches (B) and (D), and multiplate brake (E) are applied.

Power flows from the engine via the torque converter to the input shaft, it then passes via the multiplate clutch (B) to the front planetary large sun gear (S_L). With the multiplate brake (E) applied, the front planetary small sun gear (S_S) is held stationary. This results in the large sun gear (S_L) driving the long planet gears (P_L) anticlockwise and the short planet gears (P_S) clockwise, and simultaneously, the short planet gears (P_S) are compelled to roll in a clockwise direction around the stationary small sun gear (S_S). Consequently, the annular ring gear (A_F) is also forced to rotate in a clockwise direction but at a reduced speed to that of the input large sun gear (S_L). The drive is then transferred from the front planetary annular ring gear (A_F) to the rear planetary annular ring gear (A_R) via the rear intermediate shaft. With the multiplate clutch (D) applied the rear planetary sun gear (S_R) and rear annular ring gear (A_R) are locked together, thus preventing the rear planet gears from rotating independently on their axes. The drive therefore passes directly from the rear annular ring gear (A_R) to the rear carrier (C_R) and output shaft via the jammed rear planet gears. Thus it can be seen that the overall gear reduction is obtained in the front planetary gear train, whereas the rear planetary gear train only provides a one-to-one through drive.

D drive range — fourth gear (Fig. 5.34) With the positive selector lever in D drive range, multiplate clutches (B), (C) and (D) are applied. Power flows from the engine via the torque converter to the input shaft, it then passes via the multiplate clutch (B) to the front planetary large sun gear (S_L) and via the multiplate clutch (C) to the front planetary planet-gear carrier (C_F). Consequently both the large sun gear and the planet carrier rotate at the same speed thereby preventing any relative planetary gear motion, that is, the gears are jammed. Hence the output drive speed via the annular ring gear (A_F) and the rear intermediate shaft is the same as that of the input shaft speed. Power is then transferred to the rear planetary gear train by way of the front annular ring gear (A_F) and rear intermediate shaft to the rear planetary annular ring gear (A_R) and rear intermediate shaft to the rear planetary annular ring gear (A_R). However, with the multiplate clutch (D) applied, the rear annular ring gear (A_R) becomes locked to the rear sun gear (S_R); the drive therefore flows directly from the rear annular ring gear to the rear planet carrier (C_R) and output shaft via the jammed planet gears. Thus there is no gear reduction in both front and rear planetary gear trains, hence the input and output rotary speeds are similar.

D drive range — fifth gear (Fig. 5.35) With the position selector lever in D drive range, multiplate clutches (C) and (D) and multiplate brake (E) are applied. Power flows from the engine via the torque converter to the input shaft, it then passes via the multiplate clutch (C) to the front planetary planet

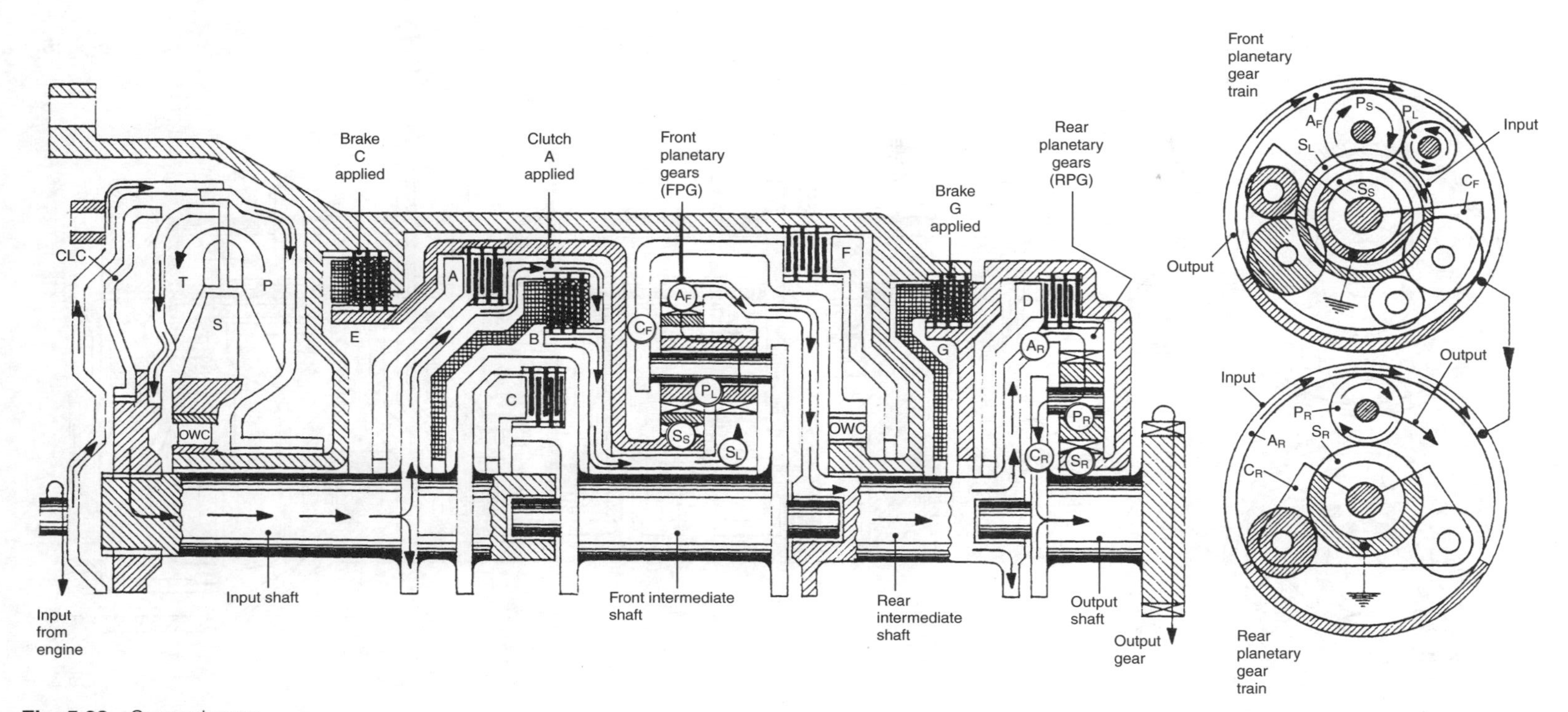

Fig. 5.32 Second gear

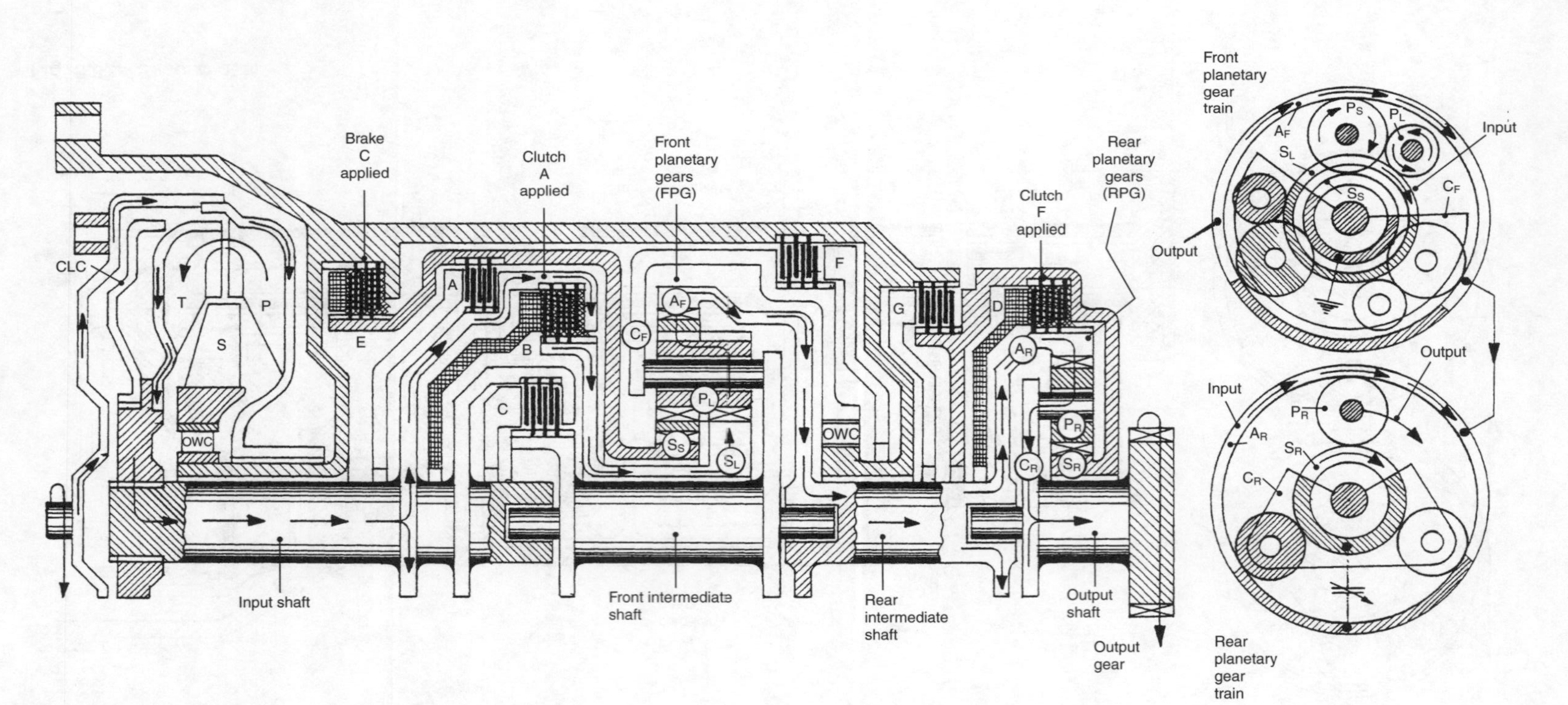

Fig. 5.33 Third gear

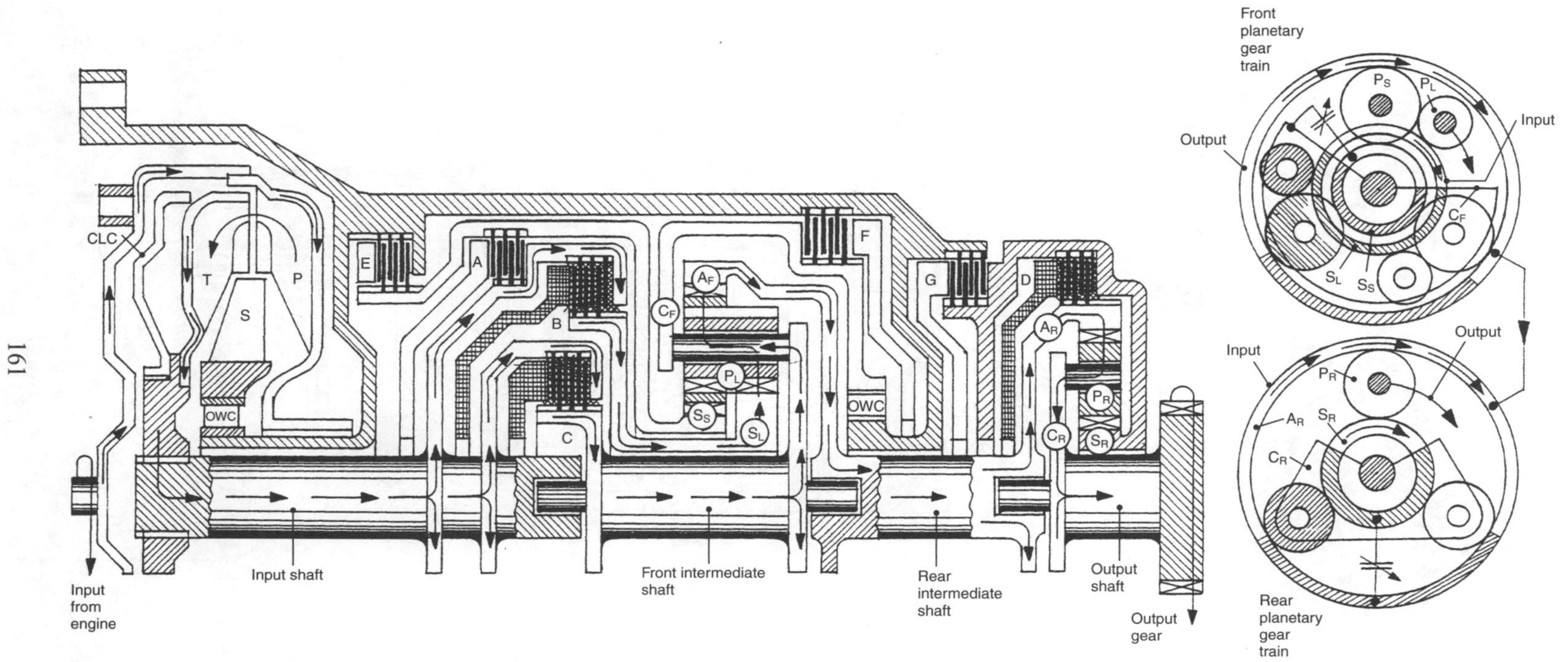

Fig. 5.34 Fourth gear

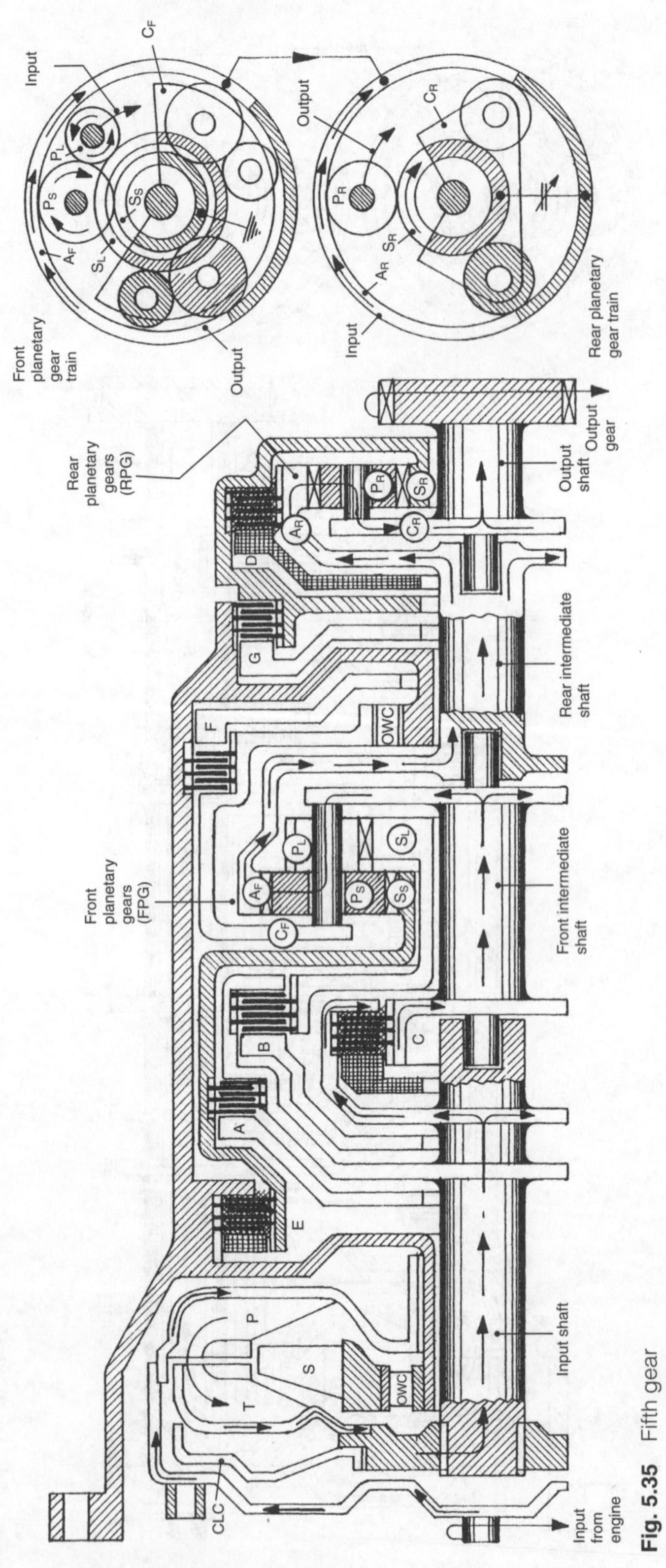

Fig. 5.35 Fifth gear

carrier (C_F). With the multiplate brake (E) applied the front planetary short sun gear (S_S) remains stationary. As a result the planet gear carrier (C_F) and both long and short planet gear pins are driven around in a clockwise direction; it thus compels the short planet gears (P_S) to roll clockwise around the fixed small sun gear (S_S). It also causes the annulus ring gear (A_F) to revolve around its axis; however, this will be at a speed greater than the input planet carrier (C_F). Note that the long planet gears (P_L) and large sun gear (S_L) revolve but are both inactive. The drive then passes from the front planetary annular ring gear (A_F) to the rear planetary annular ring gear (A_R) via the rear intermediate shaft. With the multiplate clutch (D) applied both rear annular ring gear (A_R) and rear sun gear (S_R) are locked together. Hence the rear planet gears sandwiched between both the sun and the annular gears also jam; the drive therefore is passed directly though the jammed rear planetary gear train cluster to the output shaft without a change in speed. An overall speed step-up is thus obtained, that is, an overdrive fifth gear is achieved, the step-up taking place only in the first stage planetary gear train, the second planetary gear train providing only a through one-to-one drive.

R reverse gear (Fig. 5.36) With the position selector in reverse R position, the multiplate brakes (F) and (G) and the multiplate clutch (A) are applied. Power flows from the engine to the torque converter to the input shaft, it then passes via the multiplate clutch (A) to the front planetary small sun gear (S_S). With the multiplate clutch (F) applied, the front planet gear carrier (C_F) is held stationary, and the drive passes from the clockwise rotating small sun gear (S_S) to the short planet gears (P_S) making the latter rotate anticlockwise. As a result the internal toothed front annular ring gear (A_F) will also be compelled to rotate anticlockwise. The drive then passes from the front planetary annular ring gear (A_F) to the rear planetary annular ring gear (A_R) via the rear intermediate shaft. With the rear sun gear (S_R) held stationary by the applied multiplate brake (G) the anticlockwise rotation of the rear annular ring gear (A_R) compels the rear planet gear (P_R) to roll around the held rear sun gear (S_R) in an anticlockwise direction taking with it the rear carrier (C_R) and the output shaft at a reduced speed.

Thus the direction of drive is reversed in the first planetary gear train, and there is an under-drive gear reduction in both planetary gear trains.

5.10.3 Gear change–hydraulic control (Fig. 5.30)

The shifting from one gear ratio to another is achieved by a sprag type one way clutch (when shifting from first to second gear and vice versa), four rotating multiplate clutches 'A, B, C and D' and three held multiplate brakes 'E, F and G'. The multiplate clutches and brakes are engaged by electro-hydraulic control, hydraulic pressure being supplied by the engine driven fluid pump. To apply a clutch or brake, pressurized fluid from the hydraulic control unit is directed to an annular shaped piston chamber causing the piston to clamp together the drive and driven friction disc members of the multiplate clutch. Power therefore is able to be transferred from the input to the output clutch members while these members rotate at different speeds. Shifting from one ratio to another takes place by applying and releasing various multiplate clutches/brakes. During an up or down gear shift such as 2–3, 3–4, 4–5 or 5–4, 4–3, 3–2 one clutch engages while another clutch disengages. To achieve an uninterrupted power flow, the disengaging clutch remains partially engaged but at a much reduced clamping pressure, whereas the engaging clutch clamping pressure rise takes place in a phased pattern.

5.10.4 Upshift clutch overlap control characteristics (Fig. 5.37 (a–c))

The characteristics of a gear ratio upshift is shown in Fig. 5.37(a), it can be seen with the vehicle accelerating, and without a gear change the engine speed steadily rises; however, during a gear ratio upshift transition phase, there is a small rise in engine speed above that of the speed curve when there is no gear ratio change taking place. This slight speed upsurge is caused by a small amount of slip overlap between applying and releasing the clutches. Immediately after the load transference phase there is a speed decrease and then a steady speed rise, this being caused by the full transmitted driving load now pulling down the engine speed, followed by an engine power recovery which again allows the engine speed to rise.

When a gear upshift is about to commence the engaging clutch pressure Fig. 5.37(b) rises sharply from residual to main system pressure for a short period of time, it then drops rapidly to just under half the main system pressure and remains at this value up to the load transfer phase. Over the load transfer phase the engaging clutch pressure rises fairly quickly; however, after this phase the pressure rise is at a much lower rate. Finally a small pressure

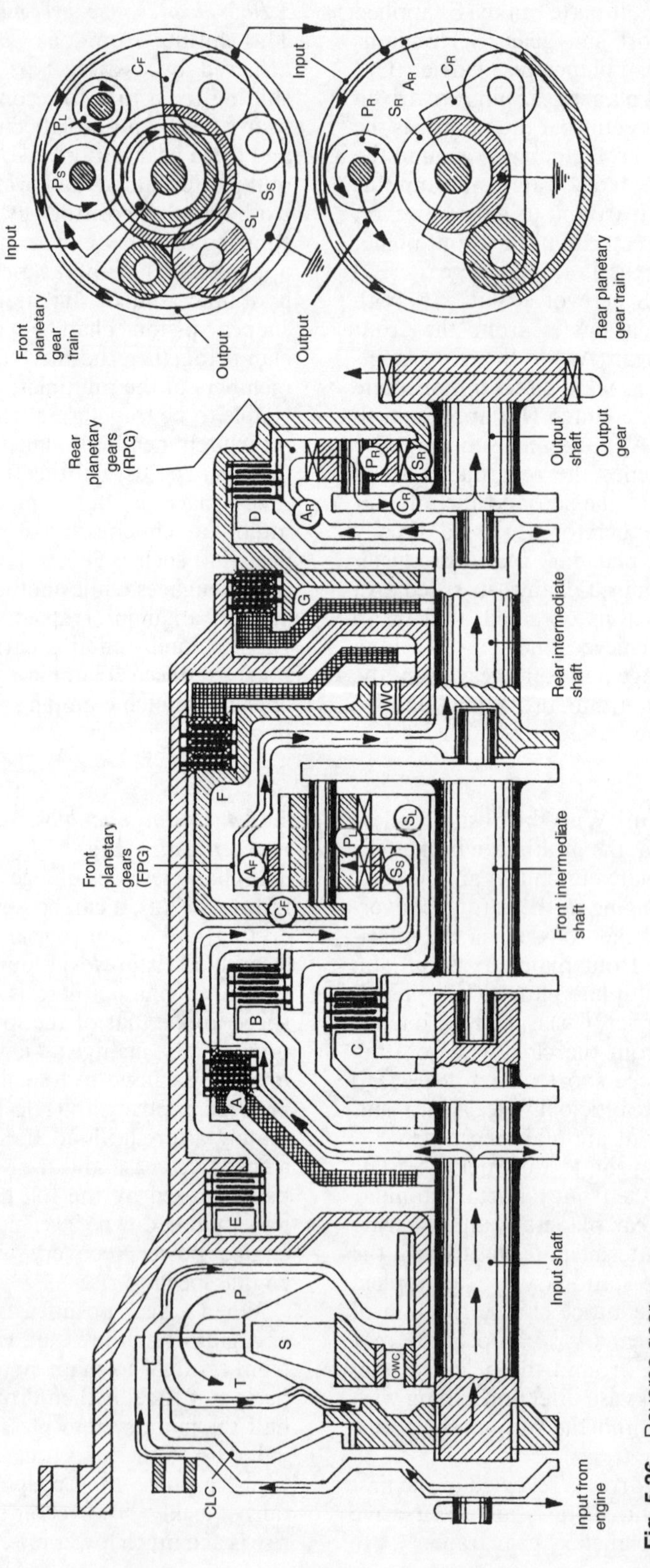

Fig. 5.36 Reverse gear

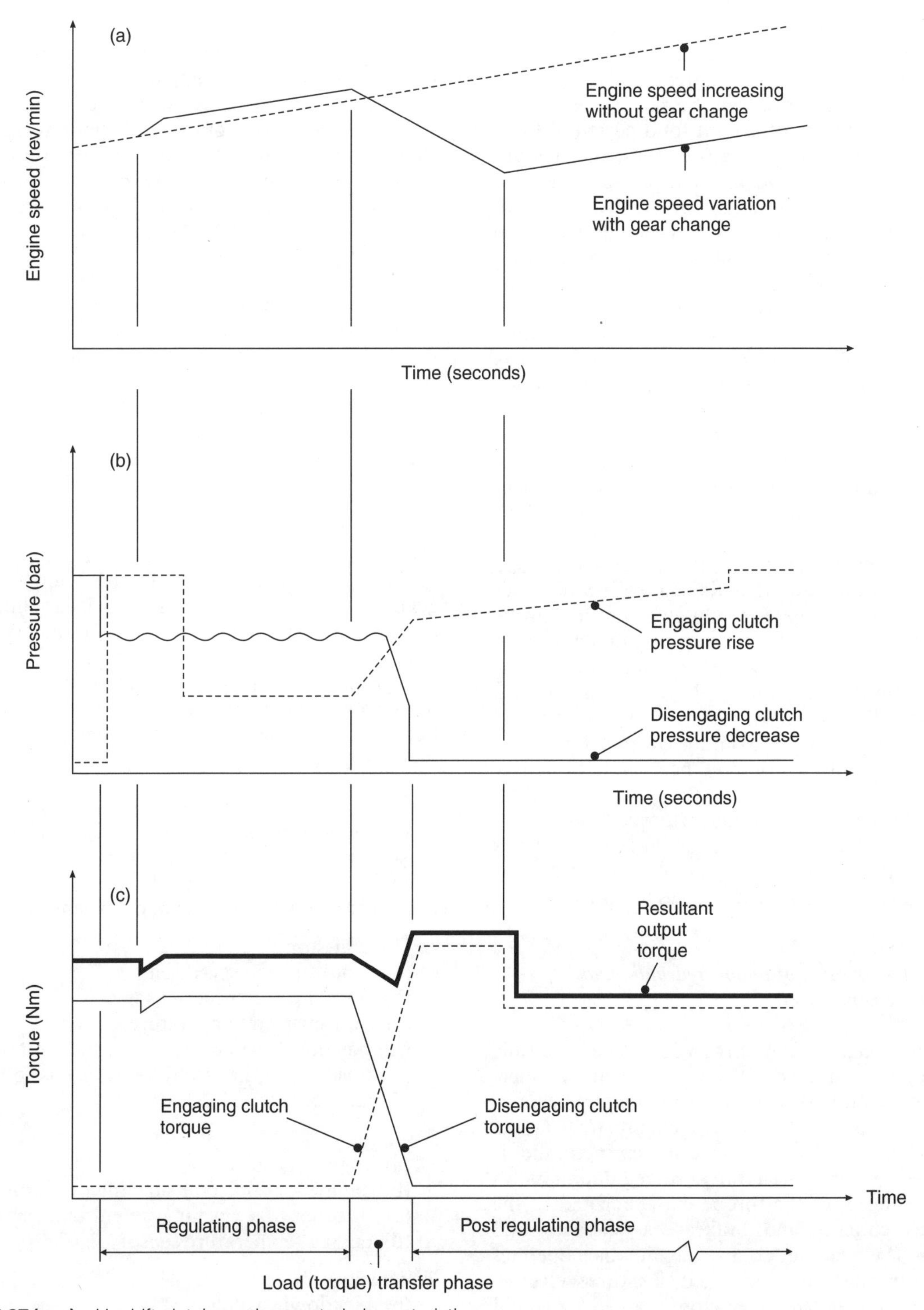

Fig. 5.37 (a–c) Upshift clutch overlap control characteristics

jump brings it back to the main system pressure. Between the rise and fall of the engaging clutch pressure, the disengaging clutch pressure falls to something like two thirds of the main systems pressure, it then remains constant for a period of time. Near the end of the load transfer phase the pressure collapses to a very low residual pressure where it remains during the time the clutch is disengaged. Fig. 5.37(b) therefore shows a pressure overlap between the disengaging clutch pressure decrease and the engaging clutch pressure increase over the load transfer period. The consequence of too much pressure overlap would be to cause heavy binding of the clutch and brake multiclutch plate members and high internal stresses in the transmission power line, whereas insufficient pressure overlap causes the engine speed to rise when driving though the load transfer period. Fig. 5.37(c) shows how the torque load transmitted by the engaging and disengaging clutches changes during a gear ratio upshift. It shows a very small torque dip and recovery for the disengaging clutch after the initial disengaging clutch pressure drop, then during the load transfer phase the disengaging clutch output torque declines steeply while the engaging clutch output torque increases rapidly. The resultant transmitted output torque over the load transfer phase also shows a dip but recovers and rises very slightly above the previous maximum torque, this being due to the transmission now being able to deliver the full engine torque.

Finally the transmitted engine torque drops a small amount at the point where the engine speed has declined to its minimum, it then remains constant as the engine speed again commences to rise.

5.10.5 Description of major hydraulic and electronic components

Hydraulic control unit (Fig. 5.38) The hydraulic control unit is housed in the oil-pan position underneath the transmission gears. A fluid pump operates the hydraulic circuitry; it is driven directly from the engine via the torque converter casing, fluid is directed by way of a pressure regulation valve to the interior of the torque converter and to the various clutches and brakes via passages and valves. The hydraulic control circuit which operates the gear shifts are activated by three electro-magnetic operated open/close valves (solenoid valves) and four electro-magnetic progressive opening and closing regulation valves (electronic pressure regulation valve 'EPRV'). Both types of valves are energized by the electronic transmission control unit 'ETCU' which in turn receives input signals from various speed, load, temperature and accelerator pedal sensors; all of these sensors are simultaneously and continuously monitoring the changing parameters. In addition a position selector lever or button operated by the driver relays the different driving programs to the electronic transmission control unit 'ETCU'.

Electronic transmission control unit The function of the electronic transmission control unit 'ETCU' is to collect, analyse and process all the input signals and to store program data so that the appropriate hydraulic circuit pressures will bring about transmission gear changes to match the engine speed and torque, vehicle's weight and load, driver's requirements and road conditions.

Control program The stored program provides data to give favourable shift characteristics for gears and the torque converter lock-up clutch, it co-ordinates parameters for pressure calculations, engine manipulation and synchronizing gear change phases, it provides regulation parameters for smooth gear shifts and the converter lock-up and finally it has built-in parameters for fault diagnoses.

Transmission input signal sensors The various signals which activate the electronic transmission control unit 'ETCU' can be divided into three groups: (1) transmission, (2) engine and (3) vehicle:

1 Transmission
  (a) input turbine speed sensor
  (b) output drive speed sensor
  (c) transmission temperature sensor
  (d) position switch signalling-selector lever position to the electronic transmission control unit

2 Engine
  (a) engine speed sensor
  (b) engine load-injector opening duration
  (c) throttle valve opening-potentiometer
  (d) engine temperature sensor

3 Vehicle
  (a) kickdown switch
  (b) position PRND432 indicator
  (c) manual gear selection program
  (d) brake light switch

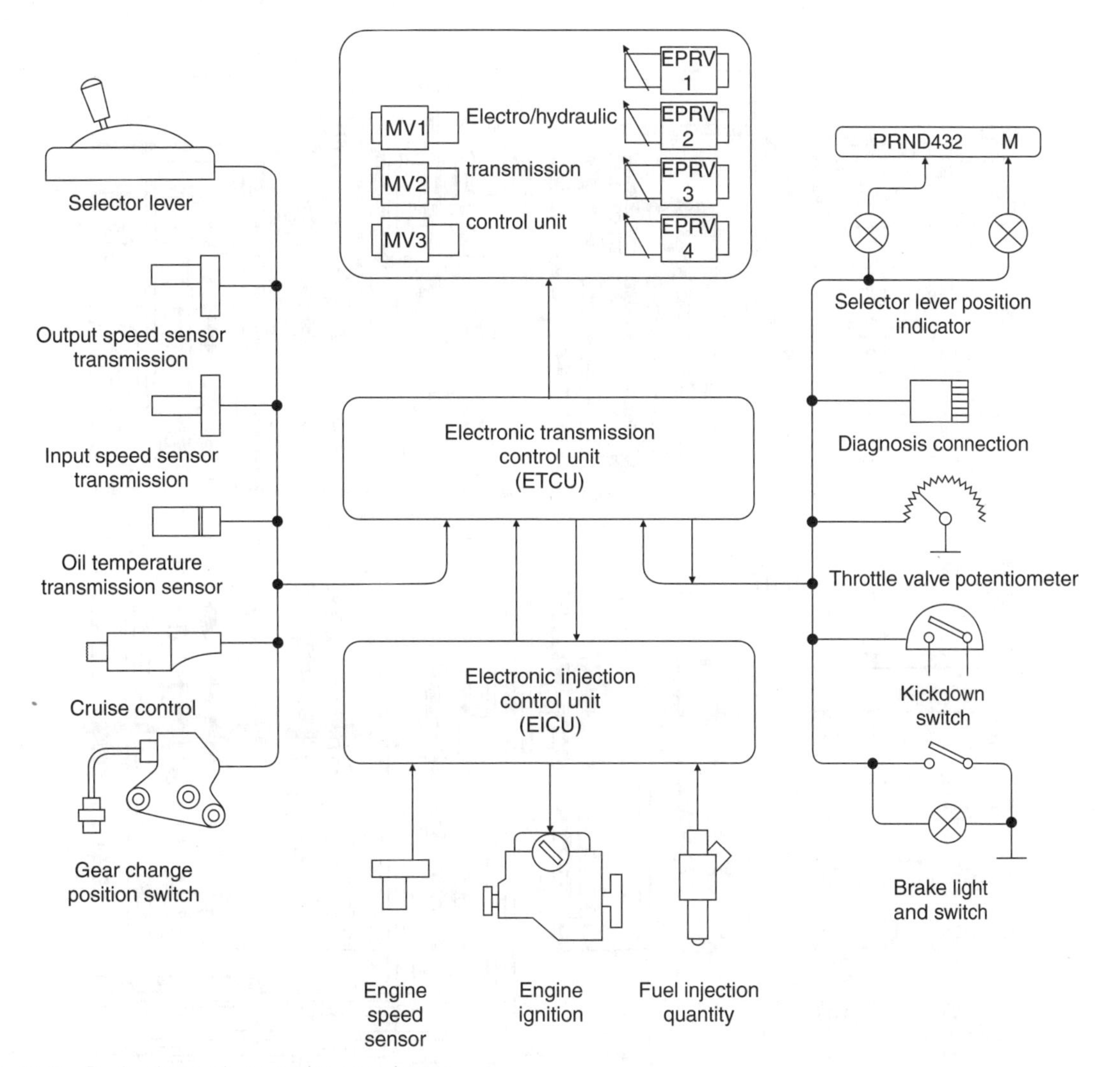

Fig. 5.38 Basic electronic control system layout

5.10.6 Description and function of the electro/hydraulic valves

Solenoid (electro-magnetic) valves (MV1, MV2 and MV3) (Fig. 5.39) The solenoid valves MV1, MV2 and MV3 are electro-magnetic disc armature operated ball-type valves which are energized by current supplied by the electronic transmission control unit. The ball-valve is either in the open or closed position, when the valve is de-energized the ball-valve closes the inlet port and vents the outlet port whereas when energized the ball-valve blocks the vent port and opens the inlet port. Solenoid valve MV1 when energized activates shift valve SV-1 and SV-3.

Solenoid valve MV2 activates shift valve SV-2 and shuts down the switch over function of shift valve SV-3. Solenoid valve MV3 activates the traction/coasting valve (T/C)V.

Electronic pressure regulating valves (EPRV-1, EPRV-2, EPRV-3 and EPRV-4) (Fig. 5.39) The electronic pressure regulating valves EPRV-(1–4) are variable pressure electro-magnetic cylindrical armature operated needle-type valves, the output pressure delivered being determined by the magnitude of the current supplied at any one time by the electronic transmission control unit to the electronic pressure regulating valves. With increasing current the taper-needle valve orifice is enlarged, this increases the fluid spill and correspondingly reduces the actuating control pressure delivered to the various valves responsible for gear shifts.

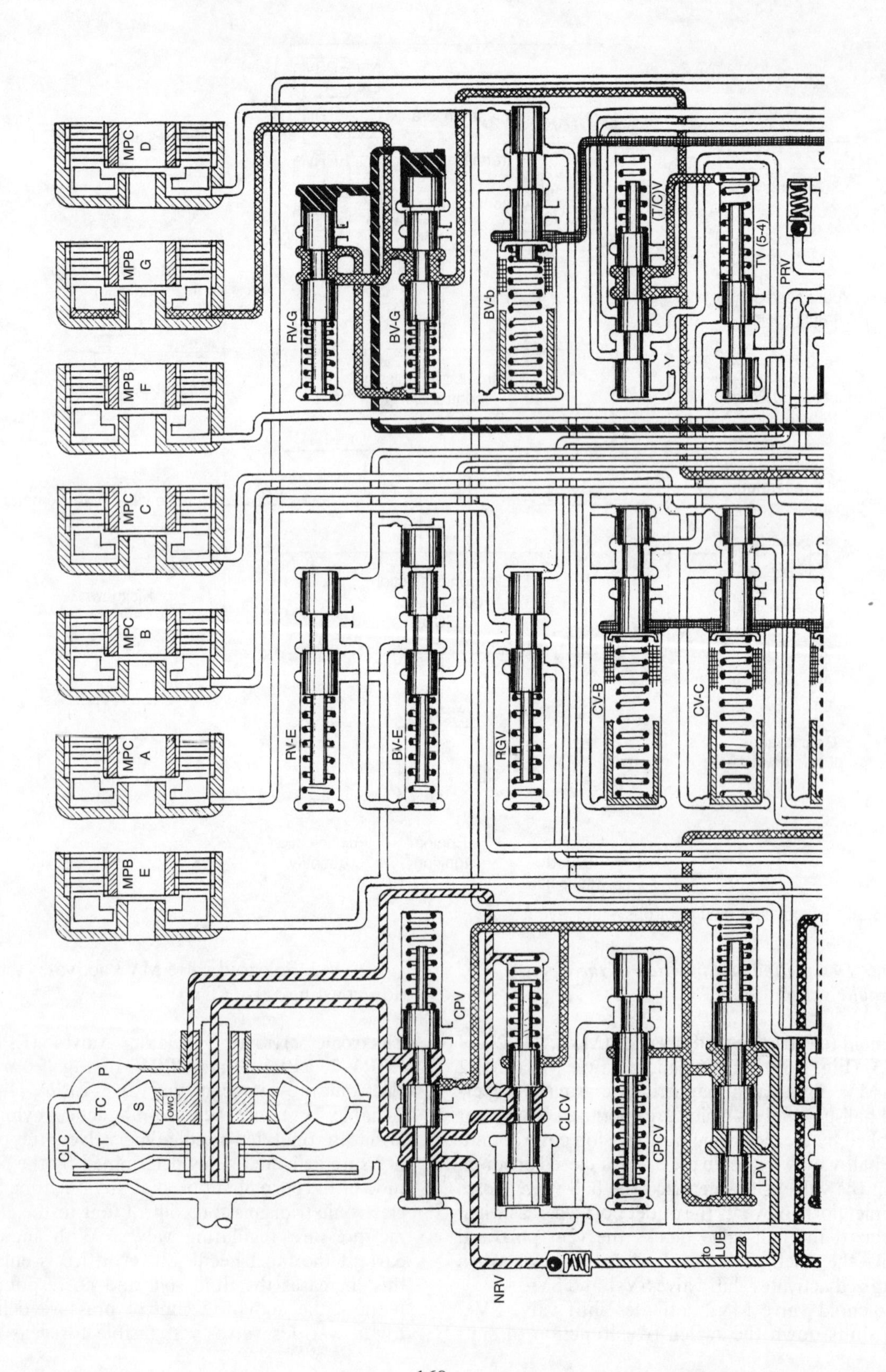
MPC D
MPB G
MPB F
MPC C
MPC B
MPC A
MPB E
RV-G
BV-G
BV-b
(T/C)V
TV (5-4)
PRV
Y
RV-E
BV-E
RGV
CV-B
CV-C
CPV
CLCV
CPCV
LPV
to LUB
NRV
CLC
T
TC
P
S
OWC

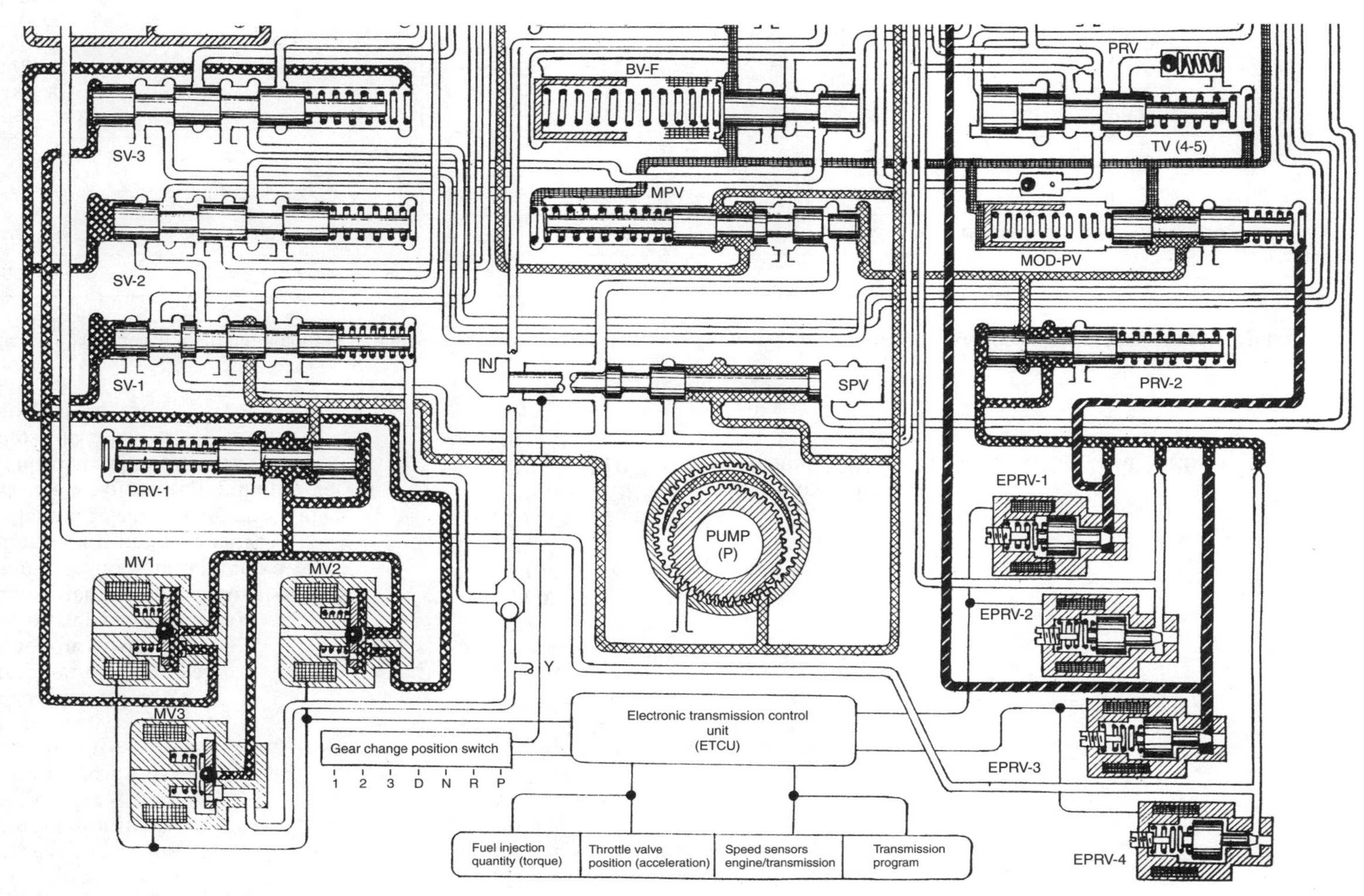

Fig. 5.39 Hydraulic/electronic transmission control system – Neutral position

Hydraulic/electronic automatic transmission control system abbreviation key for Figs 5.39–5.46.

A list of key components and abbreviations used in the description of the electro/hydraulic control system is as follows:

1 fluid pressure pump	P
2 selector position valve	SPV
3 main pressure valve	MPV
4 pressure reducing valves	PRV-1 and PRV-2
5 modulation pressure valve	MOD-PV
6 shift valve	SV-1, SV-2 and SV-3
7 reverse gear valve	RGV
8 clutch valves	CV-A, CV-B, CV-C and CV-D
9 brake valves	BV-E, BV-F and BV-G
10 retaining valves	RV-E and RV-G
11 traction/coasting valve	(T/C)V
12 traction valves	TV (4–5) and TV (5–4)
13 converter pressure valve	CPV
14 converter pressure control valve	CPCV
15 converter lock-up clutch valve	CLCV
16 lubrication pressure valve	LPV
17 solenoid (electro-magnetic) valves	MV1, MV2 and MV3
18 electronic pressure regulating valves	EPRV-1, EPRV-2, EPRV-3 and EPRV-4
19 multiplate clutch/brake	MPC-A, MPC-B, MPC-C and MPC-D/MPB-E, MPB-F and MPB-G
20 pressure relief valve	PRV
21 non-return valve	NRV

Table 5.6 Hydraulic/electronic automatic transmission control system solenoid valve, clutch and brake engagement sequence for different gear ratios for Figs 5.39–5.46

Solenoid valve – clutch – brake engagement sequence

Gear range	Solenoid valve logic							Clutch and brake logic							
	Solenoid valves			Pressure regulating valves				Clutch				Brake			
	1	2	3	1	2	3	4	A	B	C	D	E	F	G	OWC
N/P neutral/park	A	A		A		A								A	
D 1st gear	A	A		A		A			A					A	A
D 2nd gear	A	A		A	A	A			A			A		A	
D 3rd gear		A		A	A				A		A	A			
D 4th gear				A			A		A	A	A				
D 5th gear	A		A	A	A		A			A	A	A			
2 1st gear	A	A		A		A			A				A	A	(A)
R = reverse	A			A		A		A					A	A	

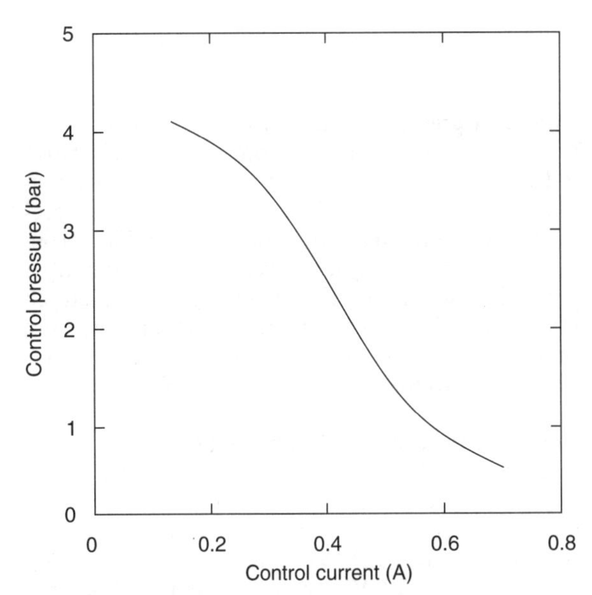

Fig. 5.40 Electronic pressure regulating valve current-pressure characteristics

Thus control pressure delivered is inversely proportional to the amount of current supplied, that is, as the current rises the pressure decreases and vice versa. The characteristics of control pressure versus control current is shown in Fig. 5.40.

5.10.7 Description and function of pump and hydraulic valves

Pump (P) (Fig. 5.39) This internal gear crescent-type pump consists of an internal toothed-spur ring gear which runs outside but in mesh with a driving external toothed-spur gear, so that its axis of rotation is eccentric to that of the driving gear. Due to their eccentricity, there is a space between the external and internal gears which is occupied by a fixed spacer block known as the crescent whose function is to separate the inlet-output port areas. The rotation of the gears creates a low pressure area at the inlet suction end of the crescent which draws in fluid. As the gear wheels rotate, oil will be trapped between the teeth of the inner driver gear and the inside crescent side walls, and between teeth of the outer gear and the outside crescent side wall. These teeth will then carry this fluid around to the other end of the crescent where it will then be discharged at pressure by both set of teeth into the output port.

Selector position valve (SPV) (Fig. 5.39) This valve is indirectly operated by the driver to select the forward and reverse direction of drive and the neutral or park positions.

Main pressure valve (MPV) (Fig. 5.39) The main pressure valve 'MPV' regulates the fluid pressure supply produced by the internal gear crescent pump; it is a variable pressure limiting valve which relates to driving conditions and the driver's demands.

Pressure reduction valve (PRV-1) (Fig. 5.39) The pressure reduction valve 'PRV-1' reduces the main fluid pressure supply to an approximate constant 5 bar output pressure which is the necessary fluid pressure supply to operate the solenoid valves MV1, MV2 and MV3.

Pressure reduction valve (PRV-2) (Fig. 5.39) The pressure reduction valve 'PRV' reduces the main fluid pressure supply to an approximate constant 5 bar output pressure which is the necessary fluid pressure supply to operate the electronic pressure regulation valves EPRV-1, EPRV-2, EPRV-3 and EPRV-4.

Modulation pressure value (MOD-PV) (Fig. 5.39) The modulation pressure valve is actuated by the electronic pressure regulator valve EPRV-1, it produces an output pressure which rises proportional to engine torque. The modulation pressure is conveyed to the main pressure valve and to each of the clutch valves, its purpose being to raise the system's pressure and to maximize the opening of the clutch valves with increased engine load so that a higher supply pressure reaches the appropriate multiplate clutch or/and brake.

Shift valves (SV-1, SV-2 and SV-3) (Fig. 5.39) The shift valves are actuated by the various solenoid valves MV1, MV2 and MV3: the function of a shift valve is to convey system pressure to the relevant operating circuit controlling the application or release of the various multiplate clutches/brakes.

Reverse gear valve (RGV) (Fig. 5.39) The reverse gear valve functions as a shift valve for selecting reverse gear; it also acts as safety valve for the forward gears by interrupting system pressure reaching clutch 'A', thus preventing the reverse gear being accidentally engaged whenever the vehicle is moving in a forward direction.

Clutch/brakes (CV-A, CV-B, CV-C, CV-D/BV-E, BV-F and BV-G) (Fig. 5.39) The clutch valves control the engagement and disengagement of the multiplate clutches and brakes. These valves are variable pressure reduction valves which are actuated by the appropriate solenoid valves, electronic pressure regulator valves, traction valves and shift valves and are responsible for producing the desired clutch pressure variations during each gear shift phase. Clutch valves CV-B, CV-C and CV-F are influenced by modulation pressure which resists the partial closure of the clutch valves, hence it permits relatively high fluid pressure to reach these multiplate clutches and brake when large transmission torque is being transmitted.

Retaining valves (RV-E and RV-G) (Fig. 5.39) In addition to the electronic pressure regulator valve which actuates the clutch valves, the retaining valves RV-E and RV-G modify the opening and closing phases of the clutch valves in such a way as to cause a progressive build-up or a rapid collapse of operating multiplate clutch/brake fluid pressure during engagement or disengagement respectively.

Traction/coasting valve (T/C-V) (Fig. 5.39) The traction coasting valve T/C-V cuts out the regulating action of the traction valve TV (5–4) and shifts the traction valve TV (4–5) into the shut-off position when required.

Traction valve (TV) (4–5) (Fig. 5.39) The traction valve TV (4–5) controls the main system fluid pressure to the multiplate-clutch MPC-B via the traction valve TV (5–4) and clutch valve CV-B and hence blocks the fluid pressure reaching the multiplate clutch CV-B when there is a upshift from fourth to fifth gear.

Traction valve (TV) (5–4) (Fig. 5.39) The traction valve TV (5–4) is another form of clutch valve, its function being to supply system pressure to the multiplate clutch MPC-B via clutch valve CV-B when there is a downshift from fifth to fourth gear.

Converter pressure valve (CPV) (Fig. 5.39) The converter pressure valve 'CPV' supplies the torque converter with a reduced system pressure to match the driving demands, that is, driving torque under varying driving conditions, it also serves as a pressure limiting valve to prevent excessive pressure build-up in the torque converter if the system pressure should become unduly high. The valve in addition vents the chamber formed on the drive-plate side of the lock-up clutch when the torque converter pressure control valve is actuated.

Converter pressure control valve (CPCV)
(Fig. 5.39) The converter pressure control valve 'CPCV' is actuated by the electronic pressure regulation valve 'EPRV-4', its purpose being to prevent the converter pressure valve 'CPV' from supplying reduced system pressure to the chamber formed between the drive-plate and lock-up clutch and to vent this space. As a result the fluid pressure on the torque converter side of the lock-up clutch is able to clamp the latter to the drive-plate.

Converter lock-up clutch valve (CLCV)
(Fig. 5.39) The converter lock-up clutch valve 'CLCV' is actuated with the converter pressure control valve 'CPCV' by the electronic pressure regulation valve 'EPRV-4'. The converter lock-up clutch valve 'CLCV' when actuated changes the direction of input flow at reduced system pressure from the drive-plate to the turbine wheel side of the lock-up clutch. Simultaneously the converter pressure valve 'CPV' is actuated, this shifts the valve so that the space between the drive-plate and lock-up clutch face is vented. As a result the lock-up clutch is forced hard against the drive-plate thus locking out the torque converter function and replacing it with direct mechanical drive via the lock-up clutch.

Lubrication pressure valve (LPV) (Fig. 5.39) The lubrication pressure valve 'LPV' supplies fluid lubricant at a suitable reduced system pressure to the internal rubbing parts of the transmission gear train.

5.10.8 Operating description of the electro/hydraulic control unit

To simplify the various solenoid valve, clutch and brake engagement sequences for each gear ratio Table 5.6 has been included.

Neutral and park position (Fig. 5.39) With the selector lever in neutral or park position, fluid is delivered from the oil-pump to the selector position valve 'SPV', modulation pressure valve 'MOD-V', pressure reduction valves 'PRV-1' and 'PRV-2', shift valve 'SV-1', traction/coasting valve '(T/C)V' and clutch valve 'CV-G'. Regulating fluid pressure is supplied to the torque converter 'TC' via the converter pressure valve 'CPV' and to the lubrication system by way of the lubrication pressure valve 'LPV'. At the same time regulated constant fluid pressure (5 bar) is supplied to the solenoid valves 'MV1, MV2 and MV3' via the pressure reduction valve 'PRV-1', and the electronic pressure regulating valves 'EPRV-(1–4)' via the pressure reduction valve 'PRV-2'. In addition controlling modulation pressure is relayed to the spring chamber of clutch valves 'CV-B, CV-C and CV-D' and brake valve 'CV-F' via the modulation pressure valve 'MOD-PV'. Neutral and parking position has the following multiplate clutch solenoid valves and electronic pressure regulator valves activated:

1 multiplate brake 'MPB-G'.
2 solenoid valves 'MV1 and MV3'.
3 electronic pressure regulating valves 'EPRV-1 and EPRV-2'.

First gear (Fig. 5.41) Engagement of first gear is obtained by applying the one way clutch 'OWC' and multiplate clutch and multiplate brake 'MPC-B and MPB-G' respectively. This is achieved in the following manner:

1 *Moving selector position valve 'SPV' into D drive range.* Fluid pressure from the selector position valve 'SPV' then passes via the traction valves 'TV (4–5) and TV (5–4)' respectively to clutch valve 'CV-B', it therefore permits fluid pressure to apply the multiplate clutch 'MPC-B'.
2 *Energizing solenoid valves 'MV1 and MV2' opens both valves.* Solenoid valve 'MV1' applies a reduced constant fluid pressure to the left-hand side of shift valves 'SV-1 and SV-3'. Shift valve 'SV-1' shifts over to the right-hand side against the tension of the return spring blocking the fluid pressure passage leading to clutch valve 'CV-D', however shift valve 'SV-3' cannot move over since a similar reduced constant pressure is introduced to the spring end of the valve via solenoid valve 'MV2'. Solenoid valve 'MV2' applies reduced constant pressure to the left-hand side of shift valve 'SV-2' and the right-hand side of shift valve 'SV-3'; this pushes the shift valve 'SV-2' to the right and so prevents shift valve 'SV-3' also being pushed to the right by fluid pressure from solenoid valve 'MV1' as previously mentioned.
3 Electronic pressure regulator valve 'EPRV-1' supplies a variable regulated fluid pressure to the modulation pressure valve 'MOD-PV', this pressure being continuously adjusted by the electronic transmission control unit 'ETCU' to suit the operating conditions. Electronic pressure regulating valve 'EPRV-3 supplies a variable controlling fluid pressure to brake and retaining valves 'BV-G and RV-G' respectively, enabling fluid pressure to apply the multiplate brake 'MPB-G'.

Second gear (Fig. 5.42) Engagement of second gear is obtained by applying multiplate clutch 'MPC-B' and the multiplate brakes 'MPB-E and MPB-G'. This is achieved in the following manner with the selector position valve in the D drive range:

1 Multiplate clutch and brake 'MPC-B and MPB-G' respectively applied as for first gear.
2 Solenoid valves 'MV1 and MV2' are energized, thus opening both valves. Fluid pressure from 'MV1' is applied to the left-hand side of both 'SV-1 and SV-3'; however, only valve SV-1 shifts over to the right-hand side. At the same time fluid pressure from solenoid valve 'MV2' shifts valve 'SV-2' over against the return-spring tension and also pressurizes the spring end of shift valve 'SV-3'. This prevents shift valve 'SV-3' moving over to the right-hand side when fluid pressure from solenoid valve 'MV-1' is simultaneously applied at the opposite end.
3 The electronic pressure regulating valves 'EPRV-1 and EPRV-3' have their controlling current reduced, thereby causing an increase in line pressure to the modulation valve MOD-PV and to both brake and retaining valves 'BV-G and RV-B' respectively. Consequently line pressure continues to apply the multiplate brake 'MPB-G'.
4 The electronic pressure regulating valve 'EPRV-2' has its controlling current reduced, thus progressively closing the valve, consequently there will be an increase in fluid pressure acting on the right-hand side of both brake and retaining valves 'BV-E and RV-E' respectively. As a result the brake valve 'BV-E' opens to permit line pressure to actuate and apply the multiplate brake 'MPB-E'.

Third gear (Fig. 5.43) Engagement of third gear is obtained by applying the multiplate clutches 'MPC-B and MPC-D' and the multiplate brake MPB-E.

The shift from second to third gear is achieved in the following manner with the selector position valve in the D drive range:

1 Multiplate clutch 'MPC-B' and multiplate brake 'MPB-E' are applied as for second gear.
2 Solenoid valve 'MV2' remains energized thus keeping the valve open as for first and second gear.

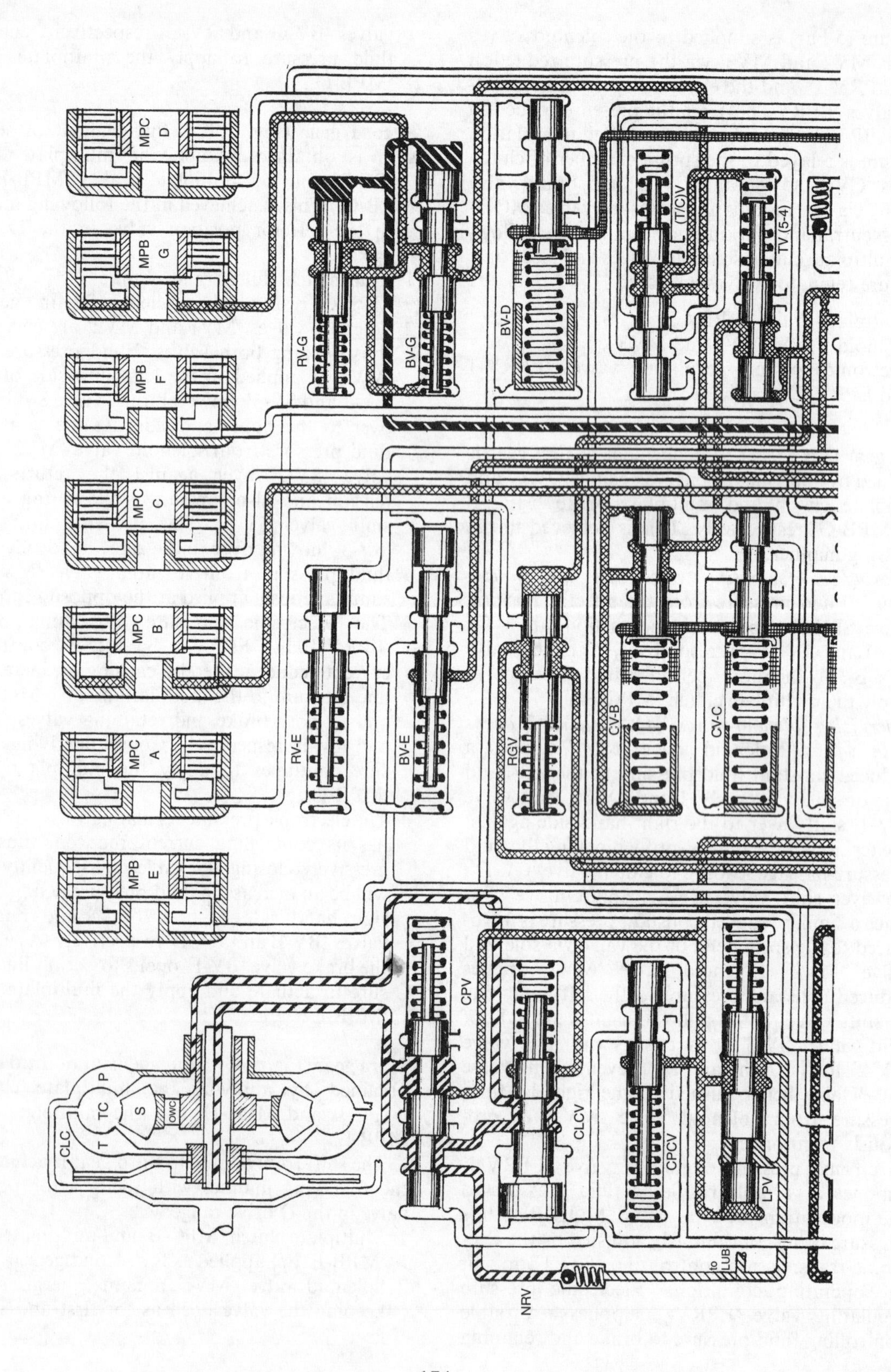
MPC
D
MPB
G
MPB
F
MPC
C
MPC
B
MPC
A
MPB
E
RV-G
BV-G
BV-D
(T/C)V
Y
TV (5-4)
RV-E
BV-E
RGV
CV-B
CV-C
CPV
CLCV
CPCV
LPV
NRV
to LUB
CLC
T
TC
P
S
OWC

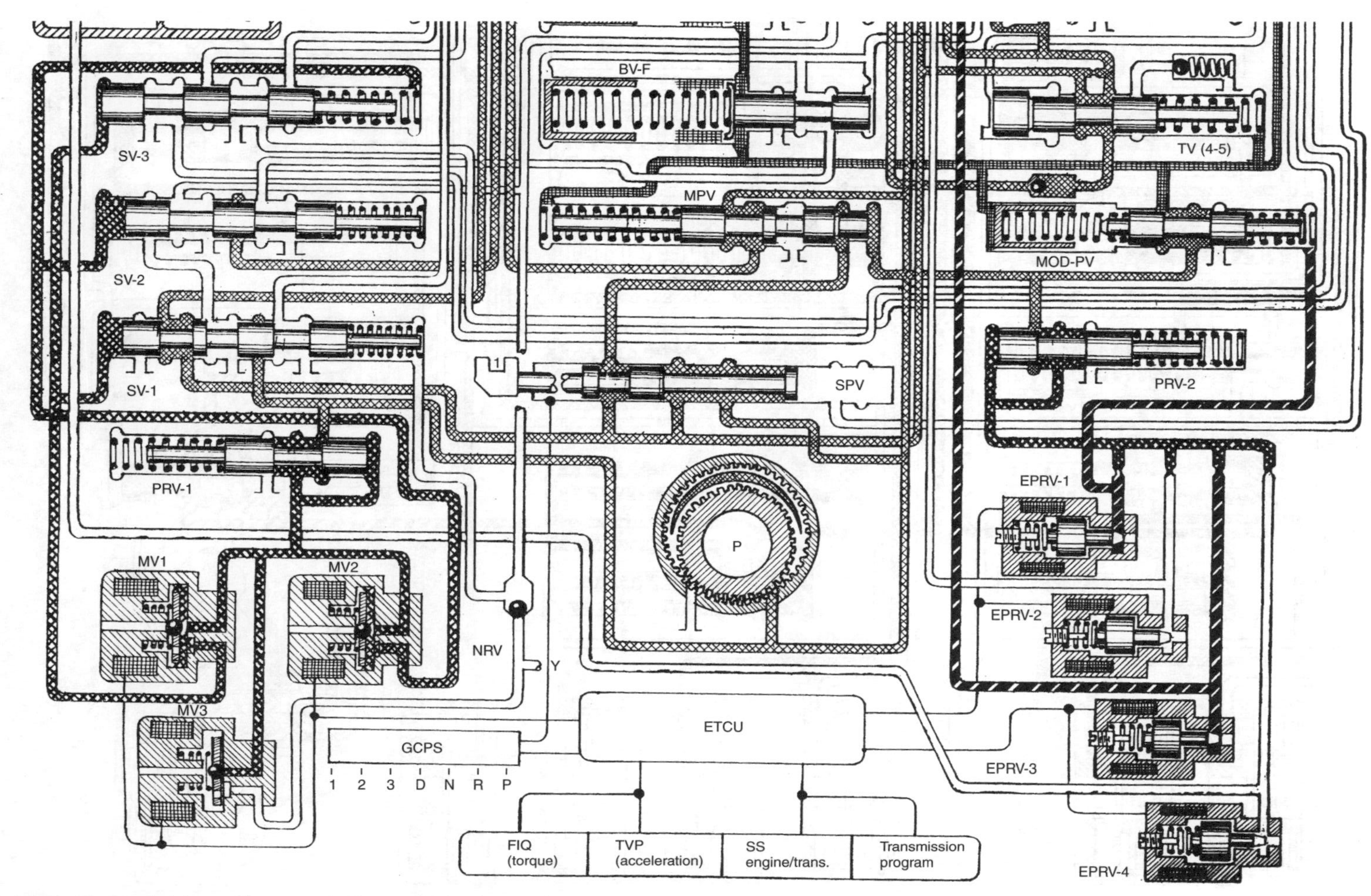

Fig. 5.41 Hydraulic/electronic transmission control system – first gear

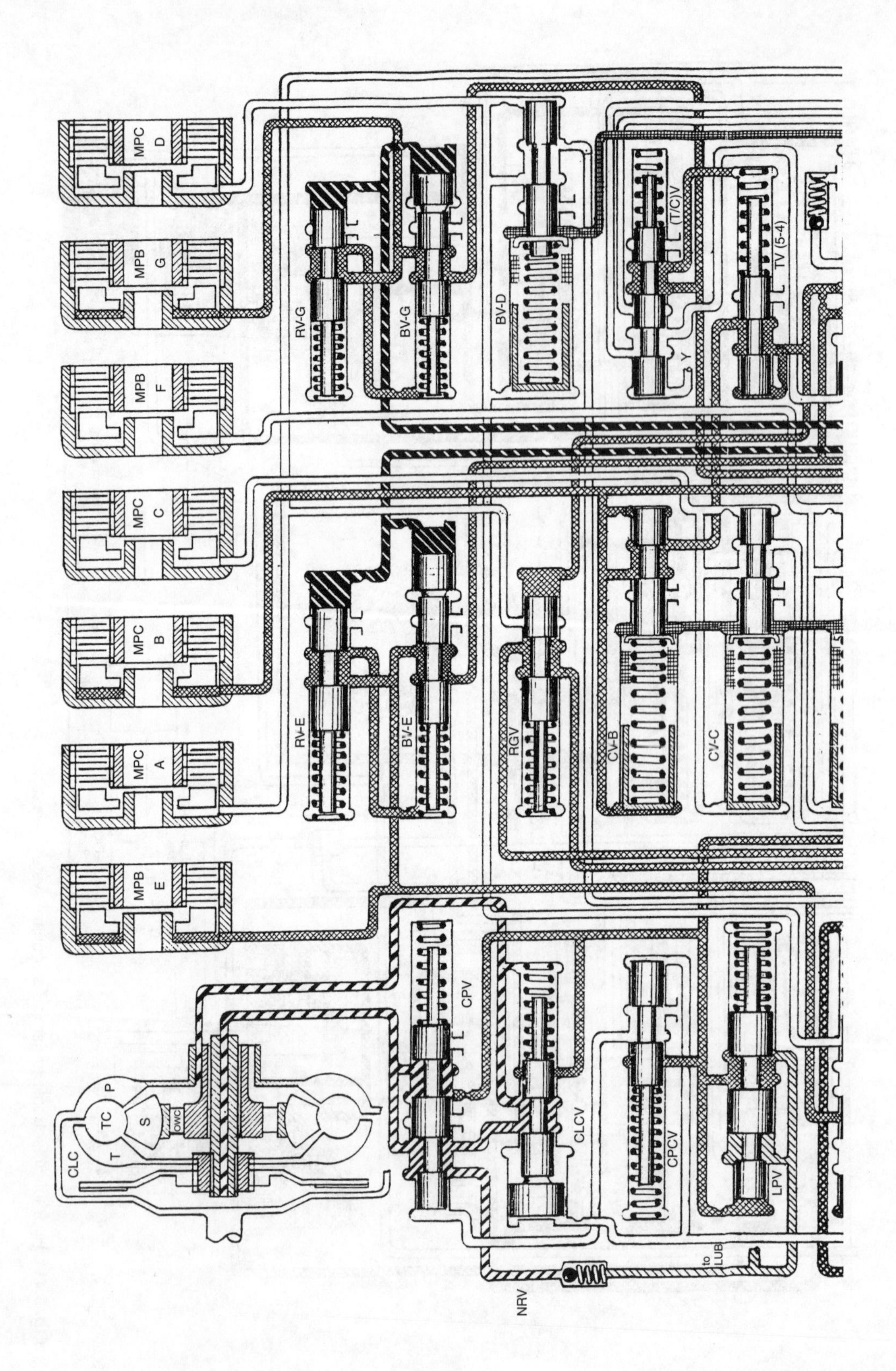
MPC
D
MPB
G
MPB
F
MPC
C
MPC
B
MPC
A
MPB
E
RV-G
BV-G
BV-D
(T/C)V
Y
TV (5-4)
RV-E
BV-E
RGV
CV-B
CV-C
CPV
CLCV
CPCV
LPV
CLC
TC
P
T
S
OWC
NRV
to
LUB

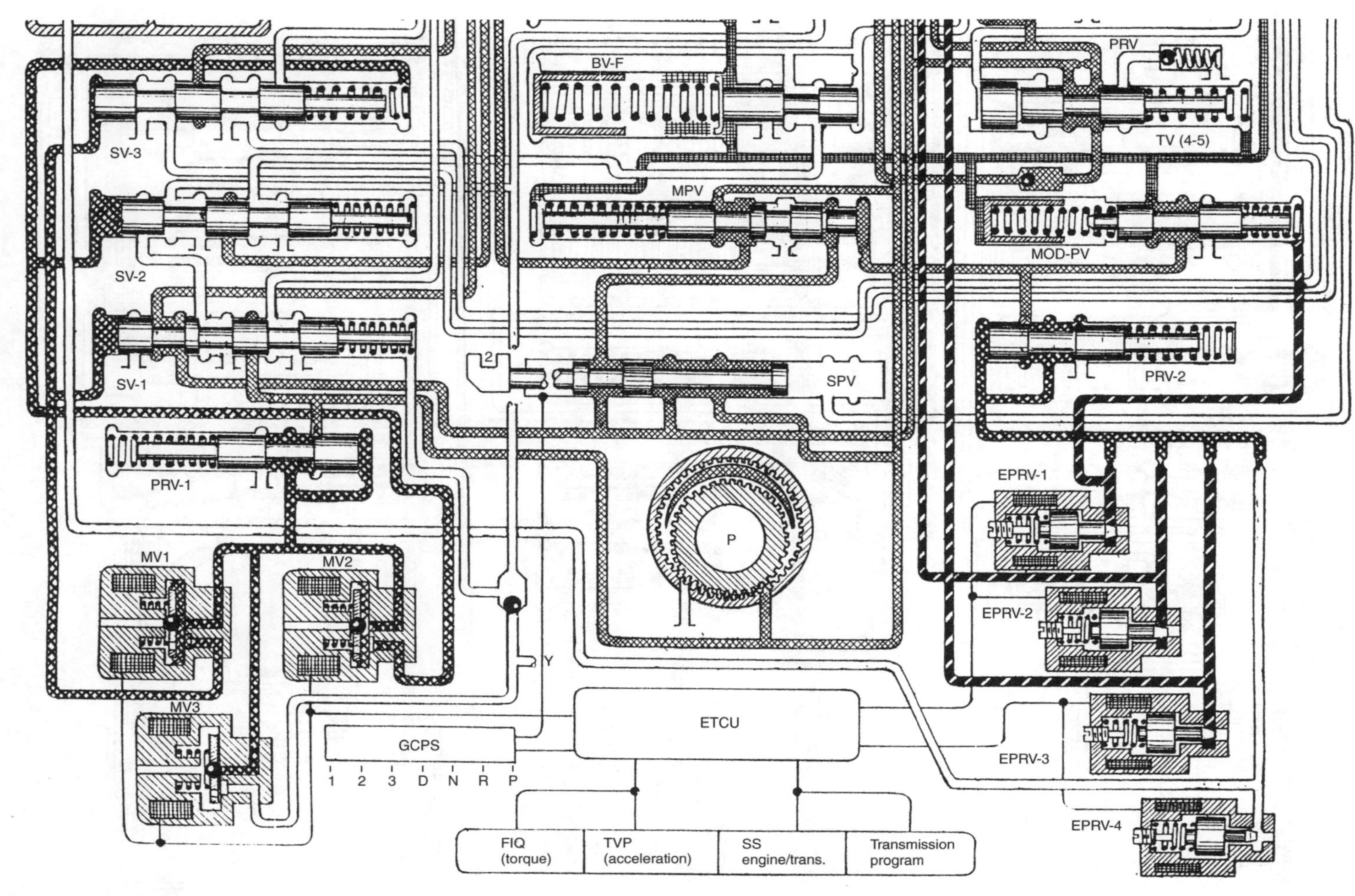

Fig. 5.42 Hydraulic/electronic transmission control system – second gear

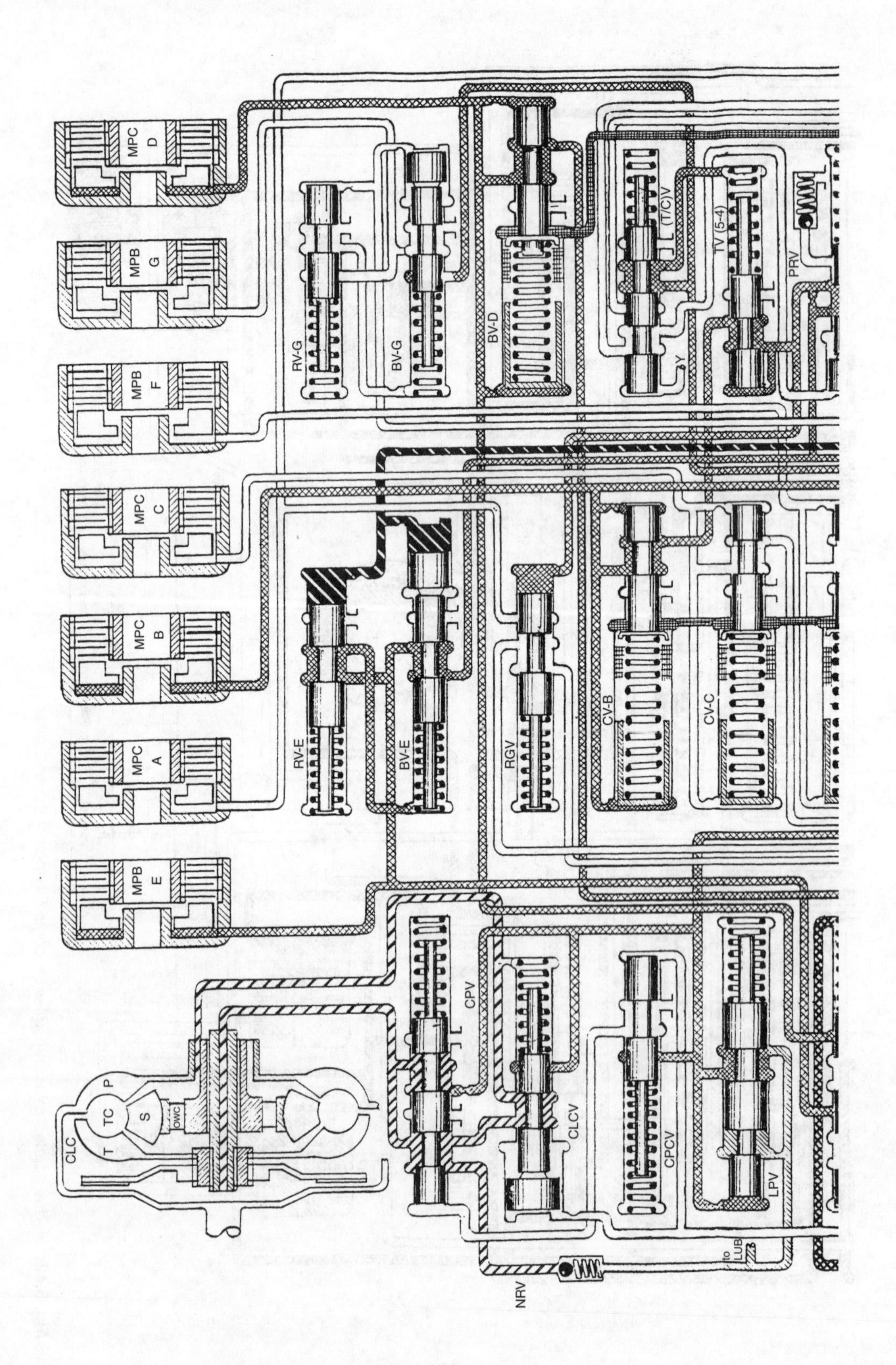
MPC
D
MPB
G
MPB
F
MPC
C
MPC
B
MPC
A
MPB
E
RV-G
BV-G
BV-D
(T/C)V
TV (5-4)
PRV
Y
RV-E
BV-E
RGV
CV-B
CV-C
CLC
T
TC
P
S
OWC
CPV
CLCV
CPCV
LPV
to
LUB
NRV

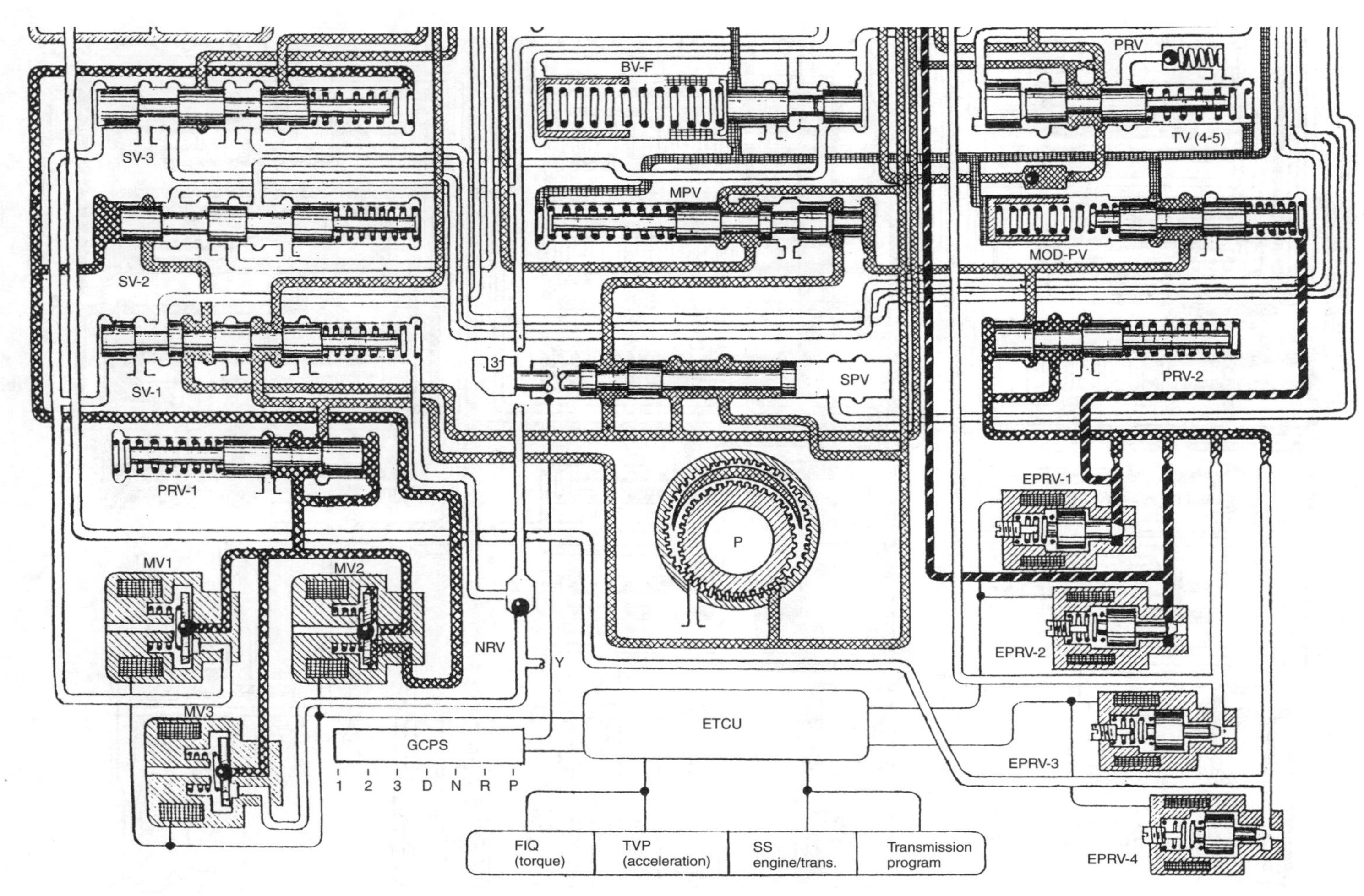

Fig. 5.43 Hydraulic/electronic transmission control system – third gear

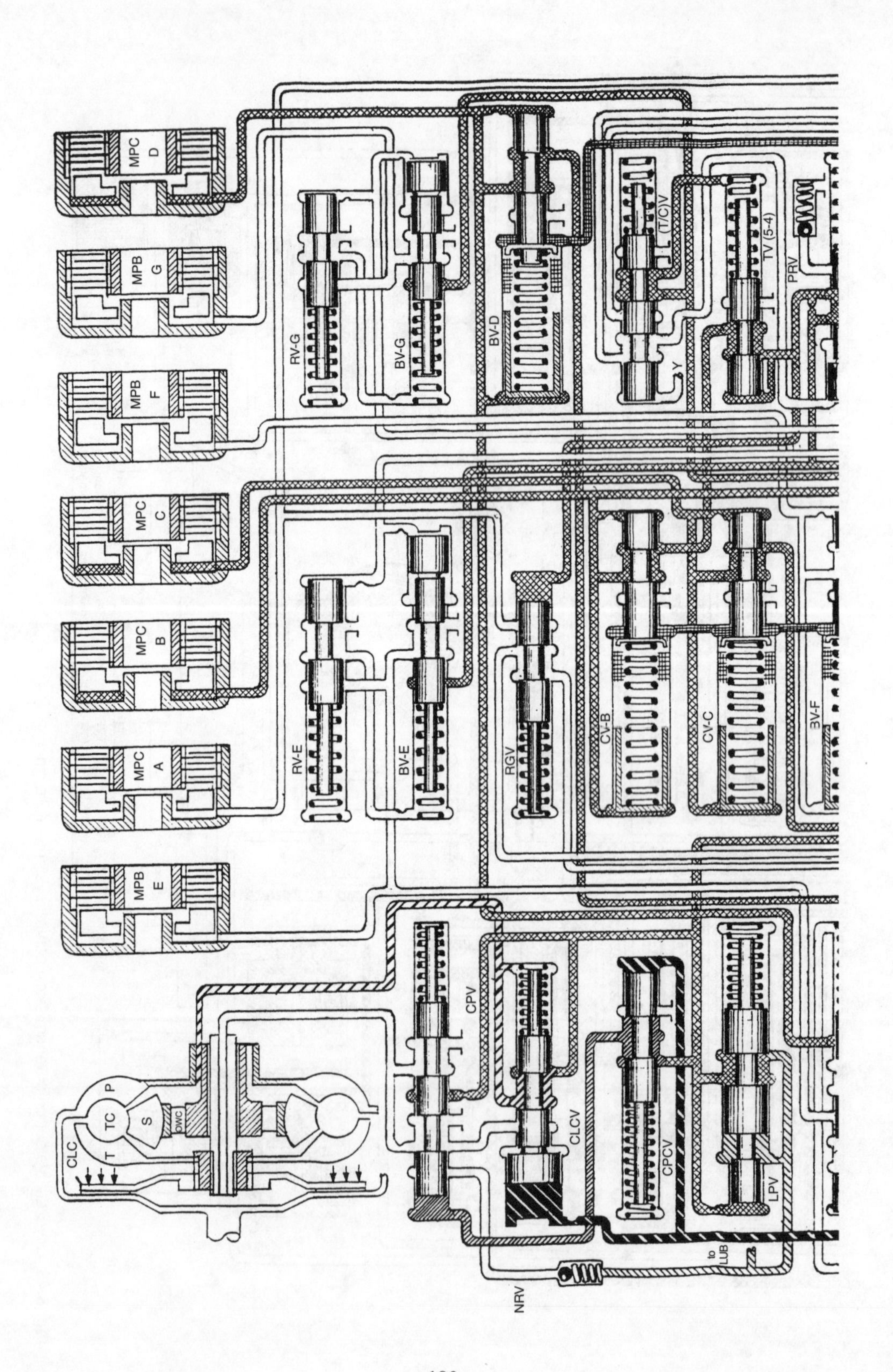
MPC D
MPB G
MPB F
MPC C
MPC B
MPC A
MPB E
RV-G
BV-G
BV-D
(T/C)V
Y
TV (5-4)
PRV
RV-E
BV-E
RGV
CV-B
CV-C
BV-F
CPV
CLCV
CPCV
LPV
to LUB
NRV
CLC
T
TC
P
S
OWC

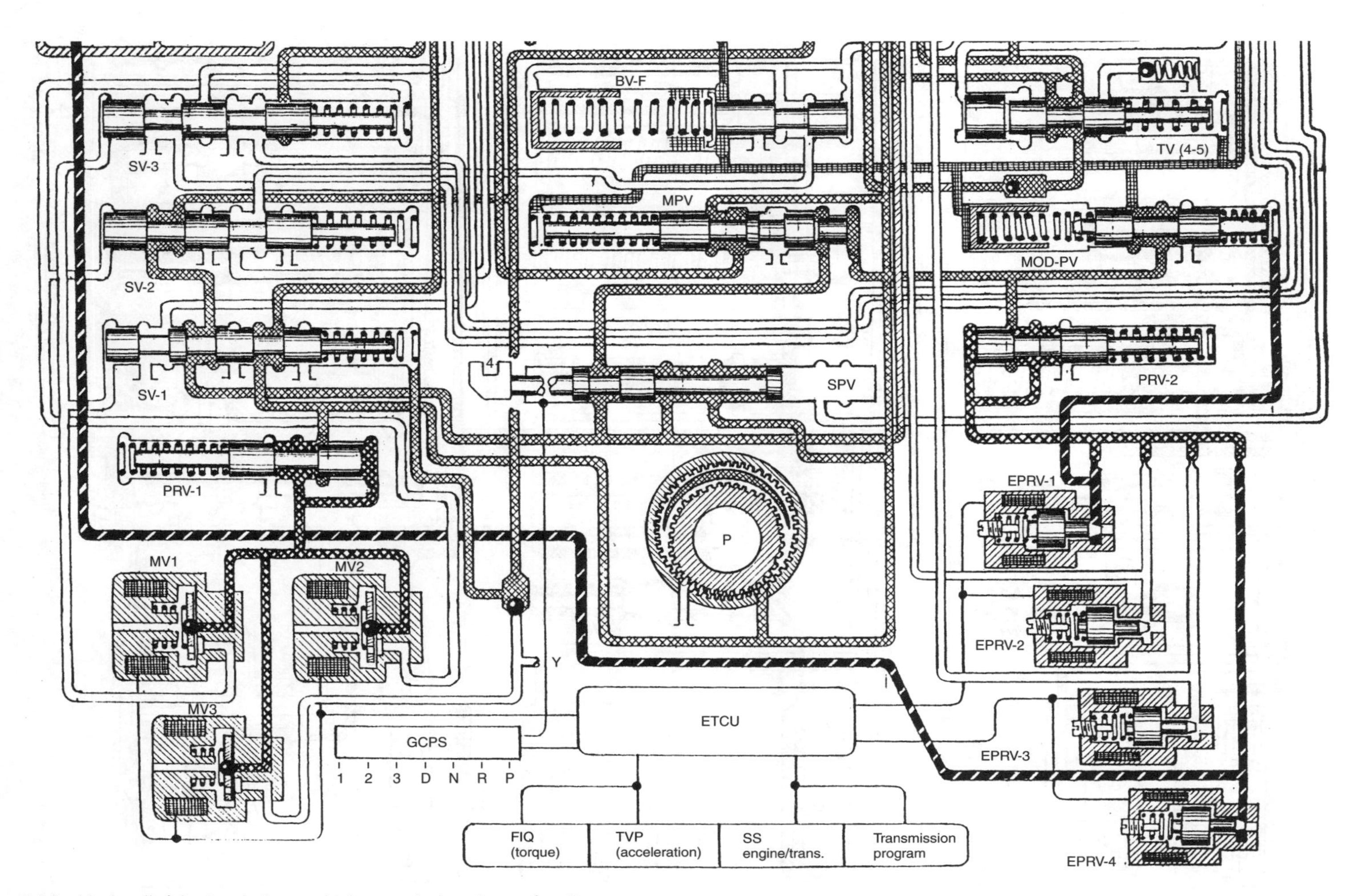

Fig. 5.44 Hydraulic/electronic transmission control system – fourth gear

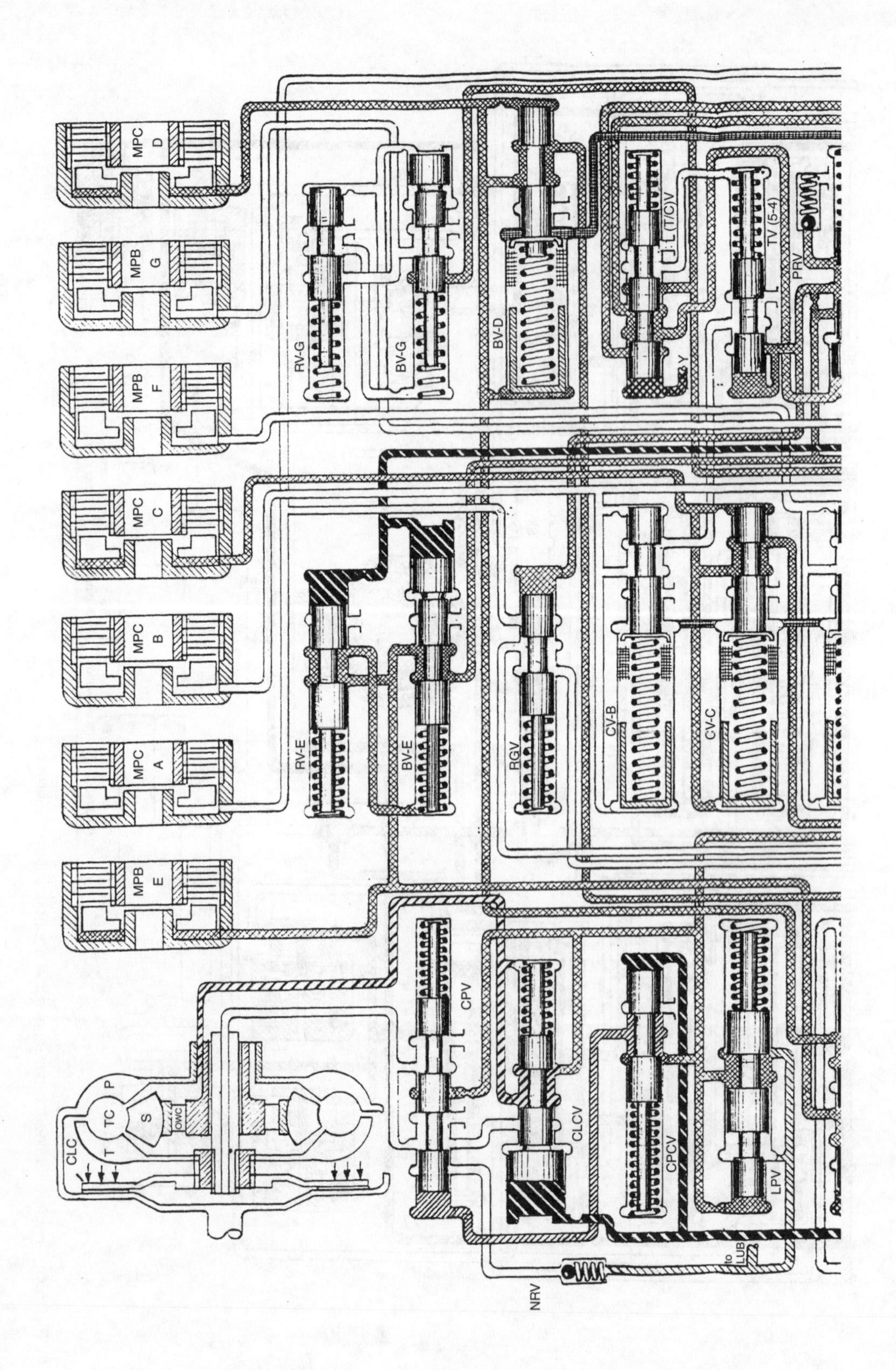
MPC D
MPB G
MPB F
MPC C
MPC B
MPC A
MPB E
RV-G
BV-G
BV-D
(T/C)V
TV (5-4)
PRV
Y
RV-E
BV-E
RGV
CV-B
CV-C
CPV
CLCV
CPCV
LPV
NRV
to LUB
CLC
T
TC
P
S
OWC

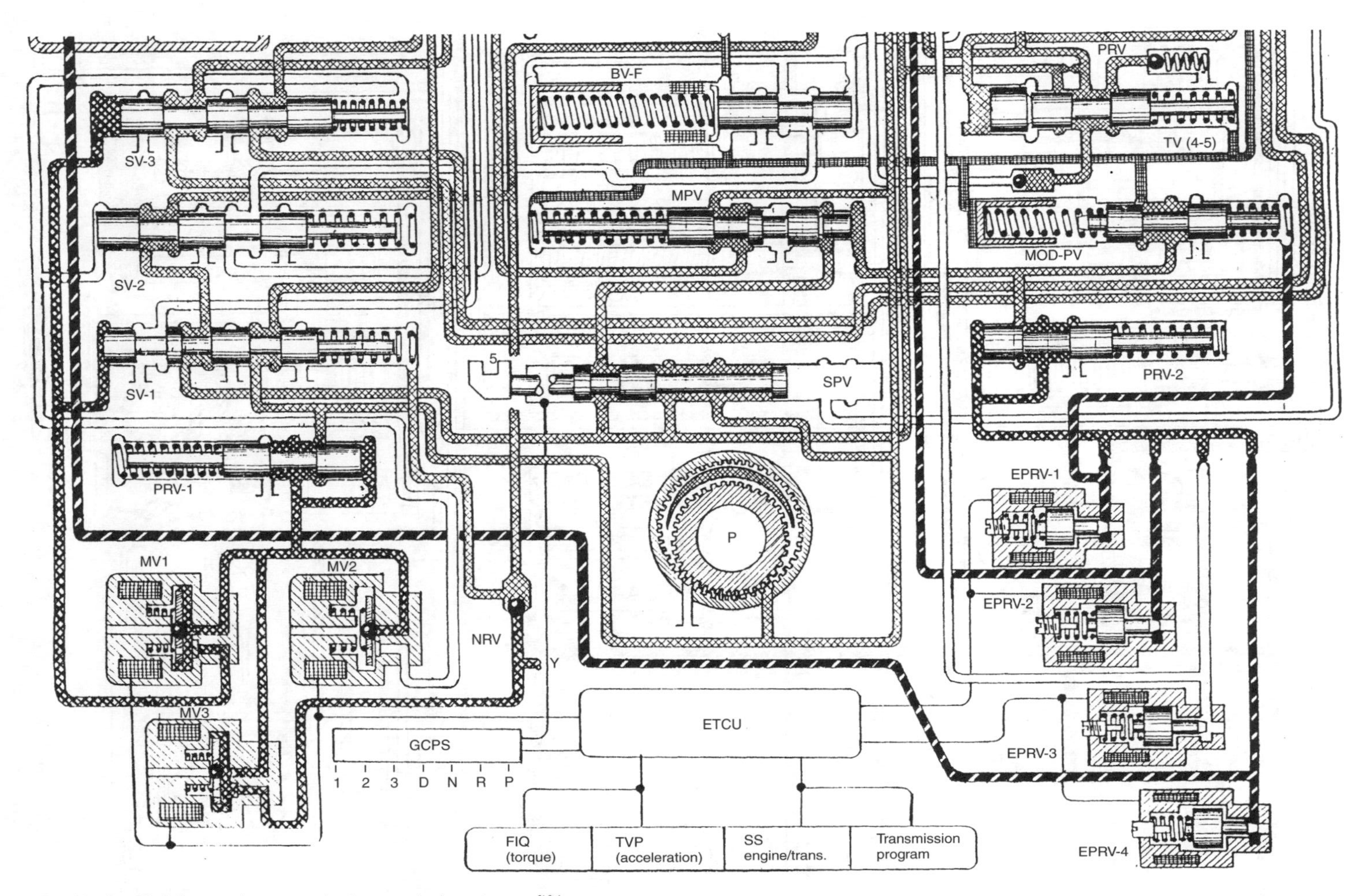

Fig. 5.45 Hydraulic/electronic transmission control system – fifth gear

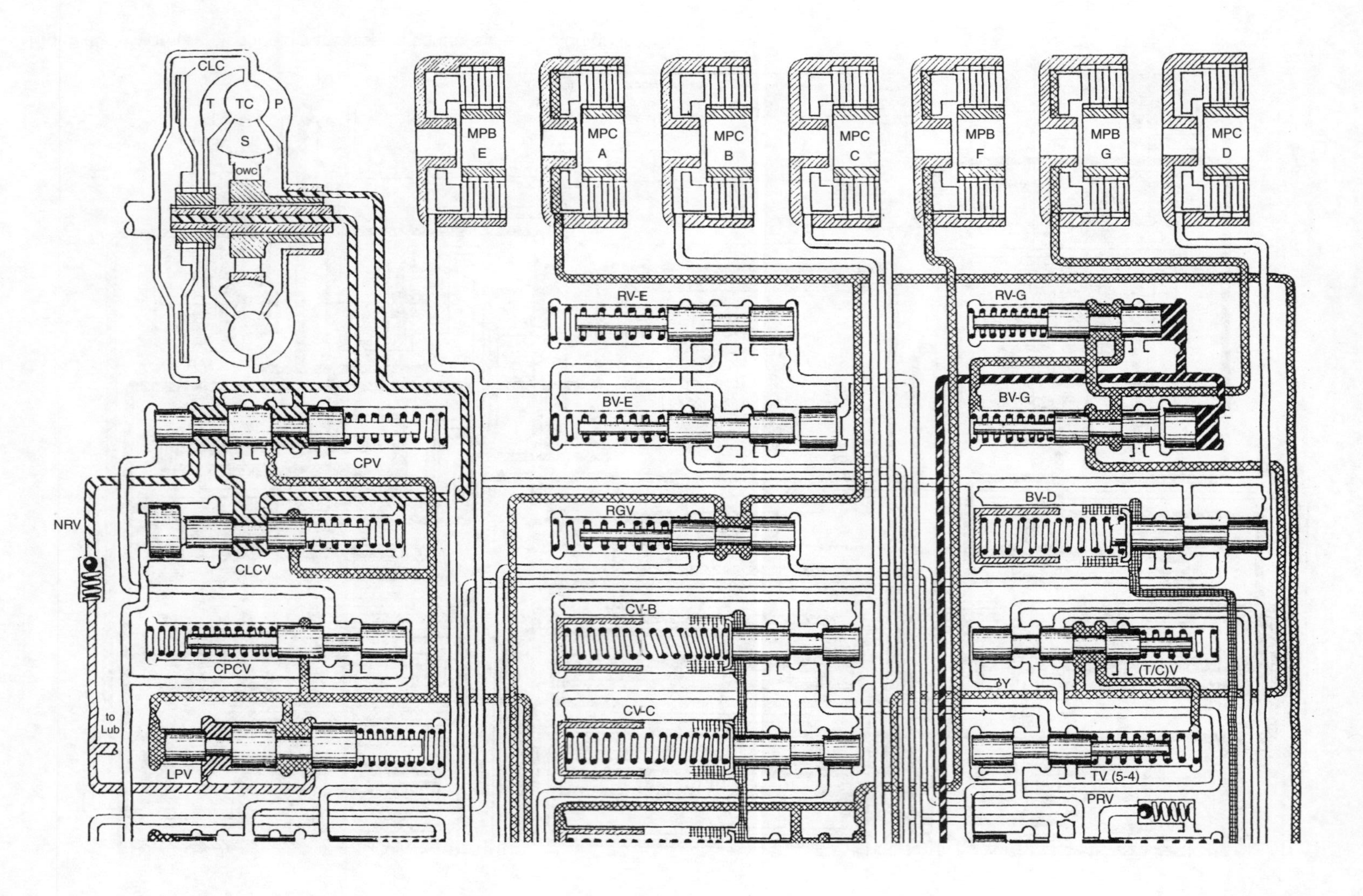
CLC
T
TC
P
S
OWC
MPB E
MPC A
MPC B
MPC C
MPB F
MPB G
MPC D
RV-E
BV-E
RGV
RV-G
BV-G
BV-D
CPV
NRV
CLCV
CPCV
to Lub
LPV
CV-B
CV-C
Y
(T/C)V
TV (5-4)
PRV

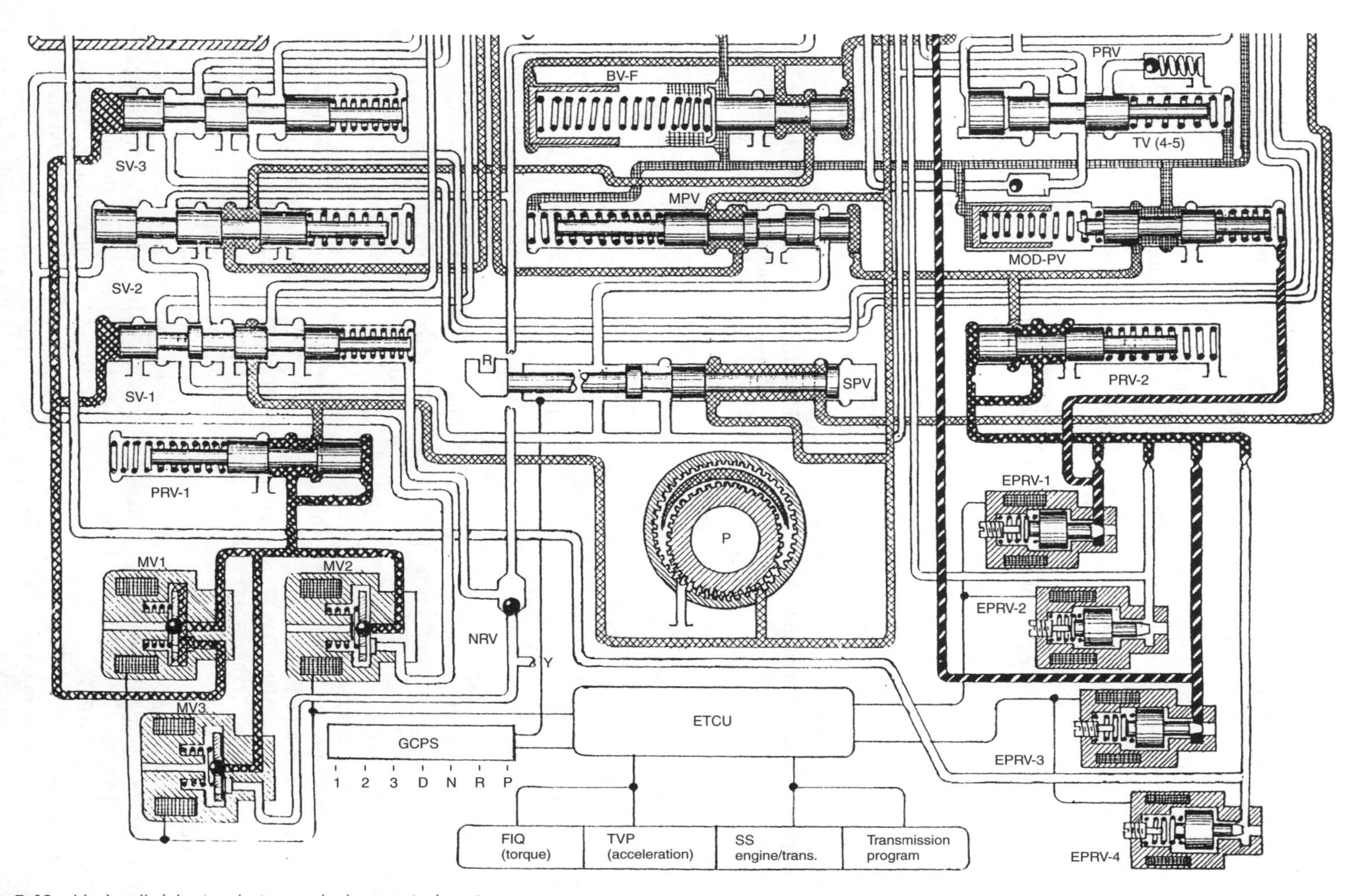

Fig. 5.46 Hydraulic/electronic transmission control system – reverse gear

3 Solenoid valve 'MV3' is in the de-energized state, it therefore blocks line pressure reaching traction/coasting valve '(T/C)V' via passage 'Y-Y'.
4 Electronic pressure regulating valves 'EPRV-1 and EPRV-2' de-energized, this closes the valves and increases their respective regulating fluid pressure as for second gear.
5 Electronic pressure regulating valve 'EPRV-3' control current is increased, this causes the valve to open and the regulating fluid pressure to collapse. The returning spring now moves the clutch and retaining valves 'CV-G and RV-G' respectively over to the right-hand side. Brake valve 'BV-G' now blocks the line pressure reaching the multiplate clutch MPB-G and releases (exhausts) the line pressure imposed on the annular shaped brake piston; the multiplate brake 'MPB-G' is therefore disengaged.
6 Solenoid valve 'MV1' is de-energized, this permits the shift valve 'SV-1' to return to the left-hand side. Subsequently line pressure now passes via the shift valve 'SV-1' to the clutch valve 'CV-D' and hence applies the multiplate clutch 'MPC-D'.

Fourth gear (Fig. 5.44) Engagement of fourth gear is obtained by applying the multiplate clutches 'MPC-B, MPC-C and MPC-D'.

The shift from third to fourth gear is achieved in the following manner with the selector position valve in the D drive range:
1 Multiplate clutches 'MPC-B and MPC-D' applied as for third gear.
2 Solenoid valves 'MV1 and MV3' de-energized and closed as for third gear.
3 Electronic pressure regulating valve 'EPRV-1' de-energized and partially closed, whereas 'EPRV-3' remains energized and open, both valves operating as for third gear.
4 Electronic pressure regulating valve 'EPRV-2' now progressively energizes and opens, this removes the control pressure from brake and retaining valves 'BV-E and RV-E' respectively. Line pressure to brake valve 'BV-E' is now blocked causing the release (exhausting) of fluid pressure via the brake valve 'BV-E' and the disengagement of the multiplate brake 'MPB-E'.
5 Fluid pressure now passes though to the multiplate clutch 'MPC-C' via shift valves 'SV-1 and SV-2', and clutch-valve 'CV-C'. Subsequently, the multiplate clutch 'MPC-C' is applied to complete the gear shift from third to fourth gear.
6 Electronic pressure regulating valve 'EPRV-4' de-energizes and progressively closes. Control pressure now shifts converter pressure control valve 'CPCV' to the left-hand side and converter lock-up clutch 'CLCV' to the right-hand side. Fluid pressure is thus supplied via the converter lock-up clutch valve 'CLCV' to the torque converter 'TC', whereas fluid pressure reaching the left-hand side of the torque converter lock-up clutch chamber is now blocked by the converter lock-up clutch valve 'CLCV' and exhausted by the converter pressure valve 'CPV'. As a result fluid pressure within the torque converter pushes the lock-up clutch hard against the impeller rotor casing. Subsequently the transmission drive, instead of passing via fluid media from the impeller-rotor casing to the turbine-rotor output shaft, is now diverted directly via the lock-up clutch from the impeller-rotor casing to the turbine-rotor output shaft.

Fifth gear (Fig. 5.45) Engagement of fifth gear is obtained by applying the multiplate clutches 'MPC-C and MPC-D' and the multiplate brake 'MPB-E'.

The shift from fourth to fifth gear is achieved in the following manner with the selector position valve 'SPV' in the D drive range:
1 Multiplate clutches 'MPC-C and MPC-D' applied as for fourth gear.
2 Solenoid valve 'MV2' de-energized as for fourth gear.
3 Solenoid valve 'MV3' is energized, this allows fluid pressure via passage 'Y-Y' to shift traction/coasting valve '(T/C)V' over to the right-hand side. As a result fluid pressure is released (exhausts) from the spring side of the traction valve 'TV (5–4)', hence fluid pressure acting on the left-hand end of the valve now enables it to shift to the right-hand side.
4 Solenoid valve 'MV1' is energized, this pressurizes the left-hand side of the shift valves 'SV-1 and SV-3'. However, 'SV-1' cannot move over due to the existing fluid pressure acting on the spring end of the valve, whereas 'SV-3' is free to shift to the right-hand end. Fluid pressure from the clutch valve 'CV-E' now passes via shift valve 'SV-3' and traction/coasting valve '(T/C)V' to the traction valve 'TV (4–5)' causing the latter to shift to the right-hand side. Consequently traction valve 'TV (4–5)' now blocks the main fluid pressure passing through the clutch valve 'CV-B' and simultaneously releases the multiplate clutch 'MPC-B' by exhausting the fluid pressure being applied to it.
5 Electronic pressure regulating valve 'EPRV-2' de-energized and partially closed. Controlled

fluid pressure now passes to the right-hand end of the clutch valve 'CV-E' and retaining valve 'RV-E', thus causing both valves to shift to the left-hand end. Fluid pressure is now permitted to apply the multiplate brake 'MPB-E' to complete the engagement of fifth gear.

6 Electronic pressure regulating valve 'EPRV-4' de-energized as for fourth gear. This causes the converter lock-up clutch 'CLC' to engage thereby by-passing the torque converter 'TC' fluid drive.

Reverse gear (Fig. 5.46) Engagement of reverse gear is obtained by applying the multiplate clutch 'MPC-A' and the multiplate brakes 'MPB-F and MPB-G'.

The shift from neutral to reverse gear is achieved in the following manner with the selector position valve 'SPV' moved to reverse drive position 'R'.

1 Multiplate brake 'MPB-G' applied as for neutral and park position.
2 Solenoid valve 'MV1' energized thus opening the valve. Constant fluid pressure now moves shift valves 'SV-1 and SV-3' over to the right-hand side.
3 Electronic pressure regulating valve 'EPRV-1' de-energized as for neutral position.
4 Electronic pressure regulating valves 'EPRV-3' de-energized and closed. Controlling fluid pressure is relayed to the brake valve 'BV-G' and retaining valve 'RV-G'. Both valves shift to the left-hand side thus permitting fluid pressure to reach and apply the multiplate brake 'MPB-G'.
5 Selector position valve 'SPV' in reverse position diverts fluid pressure from the fluid pump, directly to multiplate clutch 'MPC-A' and indirectly to multiplate brake 'MPB-F' via the selector position valve 'SPV', reverse gear valve 'RGV', shift valve 'SV-2' and the clutch valve 'CV-F'. Both multiplate clutch 'MPC-A' and multiplate brake 'MPB-F' are therefore applied.

5.11 Semi-automatic (manual gear change two pedal control) transmission system

5.11.1 Description of transmission system (Fig. 5.48)

The system being described is broadly based on the ZF Man Tip Matic/ZF AS Tronic 12 speed twin countershaft three speed constant mesh gearbox with a front mounted two speed 'splitter' gear change and a rear positioned single stage two speed epicyclic gear 'range' change; however, the basic concept has been modified and considerably simplified in this text.

Gear changes are achieved by four pneumatically operated power cylinders and pistons which are attached to the ends of the three selector rods, there being one power cylinder and piston for each of the splitter and range selector rods and two for the three speed and reverse constant mesh two piece selector rod. Gear shifts are actuated by inlet and exhaust solenoid control valves which supply and release air to the various shift power cylinders as required (see Fig. 5.48).

A multiplate transmission brake with its inlet and exhaust solenoid control valves are provided to shorten the slow down period of the clutch, input shaft and twin countershaft assembly during the gear change process.

A single plate dry friction clutch is employed but instead of having a conventional clutch pedal to control the engagement and disengagement of the power flow, a pneumatic operated clutch actuator with inlet and exhaust solenoid control valves are used. Thus the manual foot control needed for driving away from rest and changing gear is eliminated.

Gradual engagement of the power flow via the clutch when pulling away from a standstill and smooth gear shift changes are achieved via the wheel speed and engine speed sensors, air pressure sensors and the electronic diesel control unit (EDCU): this being part of the diesel engine management equipment, they all feed signals to the electronic transmission control unit (ETCU). This information is then processed so that commands to the various solenoid control valves can be made to produce the appropriate air pressure delivery and release to meet the changing starting and driving conditions likely to be experienced by a transmission system. A gear selector switch control stick provides the driver with a hand control which instructs the electronic transmission control unit (ETCU) to make an up and down gear shift when prevailing engine torque and road resistance conditions are matched.

5.11.2 Splitter gear change stage (Fig. 5.47)

Power flows via the clutch and input shaft to the splitter synchromesh dog clutch. The splitter synchromesh dog clutch can engage either the left or right hand matching dog clutch teeth on the central splitter gear mounted on the input shaft to obtain a low splitter gear ratio, or to the central third gear

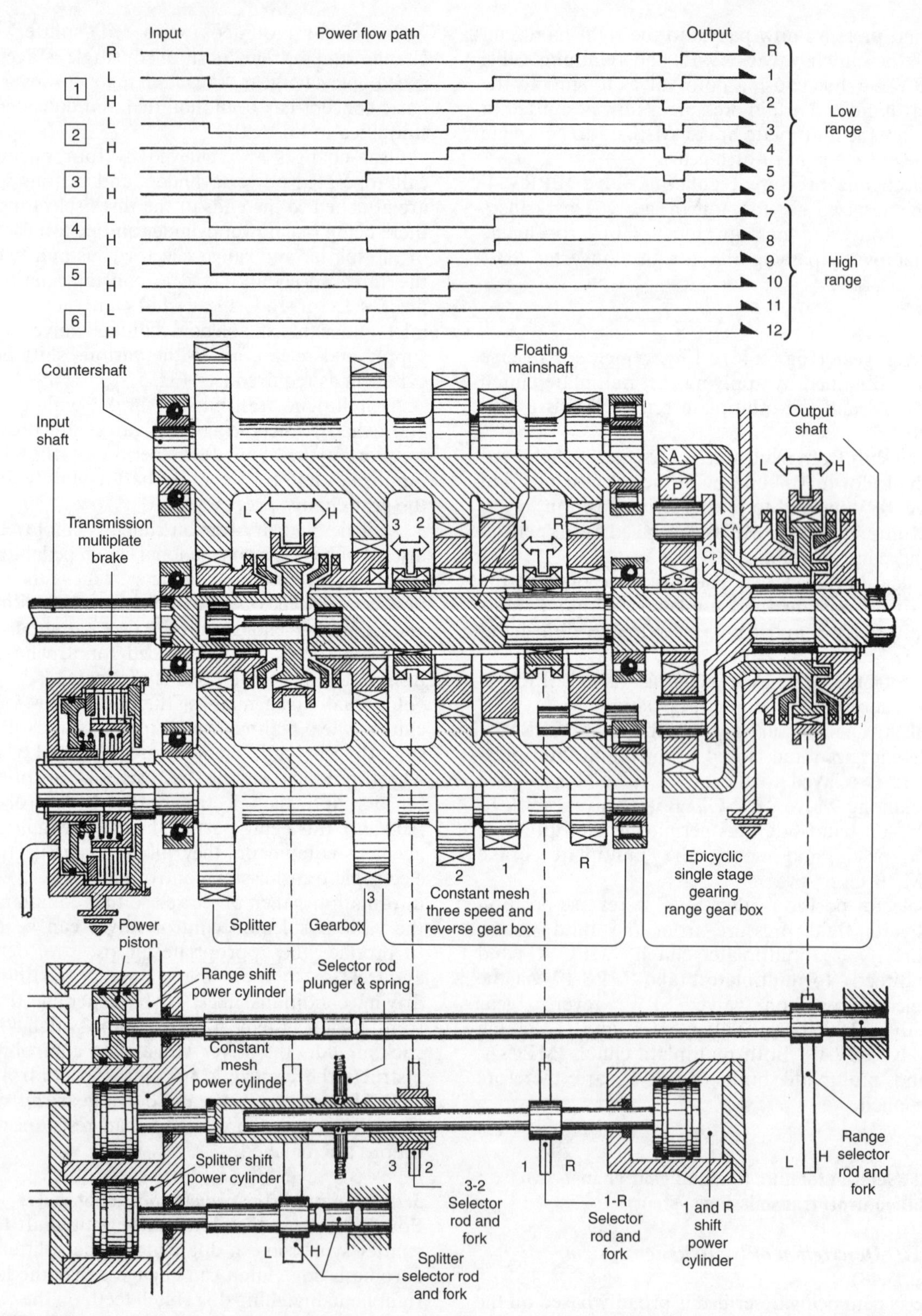

Fig. 5.47 Twin countershaft 12 speed constant mesh gearbox with synchromesh two speed splitter and range changes

mounted on the mainshaft to obtain the high splitter gear ratio. Power is now able to pass via the twin countershafts to each of the mainshaft constant mesh central gears by way of the constant mesh gears 1, 2, 3 and R.

5.11.3 Constant mesh 1-2-3 and R gear stage (Fig. 5.47)

The selection and engagement of one of the sliding dog clutch set of teeth either with R, 1, 2 or 3 floating mainshaft central constant mesh gears permits the drive path to flow from the twin countershaft gears via the mainshaft to the epicylic range change single stage gear train.

5.11.4 Range change gear stage (Fig. 5.47)

Low range gear selection With the synchromesh dog clutch hub moved to the left-hand side, the internal toothed annular gear (A) will be held stationary; the drive from floating mainshaft is therefore compelled to pass from the central sun gear (S) to the output shaft via the planet gear carrier (C_P) (see Fig. 5.47). Now since the annular gear is held stationary, the planet gears (P) are forced to rotate on their axes and also to roll around the internal teeth of the annular gear (A), consequently the planet carrier (C_P) and output shaft will now rotate at a lower speed than that of the sun gear (S) input.

High range gear selection With the syncromesh dog clutch hub moved to the right-hand side, the annular gear (A) becomes fixed to the output shaft, therefore the drive to the planet gears (P) via the floating mainshaft and sun gear (S) now divides between the planet gear carrier (C_P) and the annular gear carrier (C_A) which are both fixed to the output shaft (see Fig. 5.47). As a result the planet gears (P) are prevented from rotating on their axes so that while the epicyclic gear train is compelled to revolve as one rigid mass, it therefore provides a one-to-one gear ratio stage.

5.11.5 Clutch engagement and disengagement (Fig. 5.48)

With the ignition switched on and the first gear selected the clutch will automatically and progressively take up the drive as the driver depresses the accelerator pedal. The three basic factors which determine the smooth engagement of the transmission drive are vehicle load, which includes pulling away from a standstill and any road gradient, vehicle speed and engine speed. Thus the vehicle's resistance to move is monitored in terms of engine load by the electronic diesel control unit 'EDCU' which is part of the diesel engine's fuel injection equipment, and engine speed is also monitored by the EDCU, whereas vehicle speed or wheel speed is monitored by the wheel brake speed sensors. These three factors are continuously being monitored, the information is then passed on to the electronic transmission control unit 'ETCU' which processes it so that commands can be transferred in the form of electric current to the inlet and exhaust clutch actuator solenoid control valves.

Engagement and disengagement of clutch when pulling away from a standstill (Fig. 5.48) With the vehicle stationary, the ignition switched on and first gear selected, the informed ETCU energizes and opens the clutch solenoid inlet control valve whereas the exhaust control valve remains closed (see Fig. 5.48). Compressed air now enters the clutch cylinder actuator, this pushes the piston and rod outwards causing the clutch lever to pivot and to pull back the clutch release bearing and sleeve. As a result the clutch drive disc plate and input shaft to the gearbox will be disengaged from the engine. As the driver depresses the accelerator pedal the engine speed commences to increase (monitored by the engine speed sensor), the ETCU now progressively de-energizes the solenoid controlled clutch inlet valve and conversely energizes the solenoid controlled exhaust valve. The steady release of air from the clutch actuator cylinder now permits the clutch lever, release bearing and sleeve to move towards the engagement position where the friction drive plate is progressively squeezed between the flywheel and the clutch pressure plate. At this stage the transmission drive can be partially or fully taken up depending upon the combination of engine speed, load and wheel speed.

As soon as the engine speed drops below some predetermined value the ETCU reacts by de-energizing and closing the clutch exhaust valve and energizing and opening the clutch inlet valve, thus compressed air will again enter the clutch actuator cylinder thereby causing the friction clutch drive plate to once more disengage.

Note a built-in automatic clutch re-adjustment device and wear travel sensor is normally incorporated within the clutch unit.

Engagement and disengagement of the clutch during a gear change (Fig. 5.48) When the driver moves the gear selector stick into another gear position

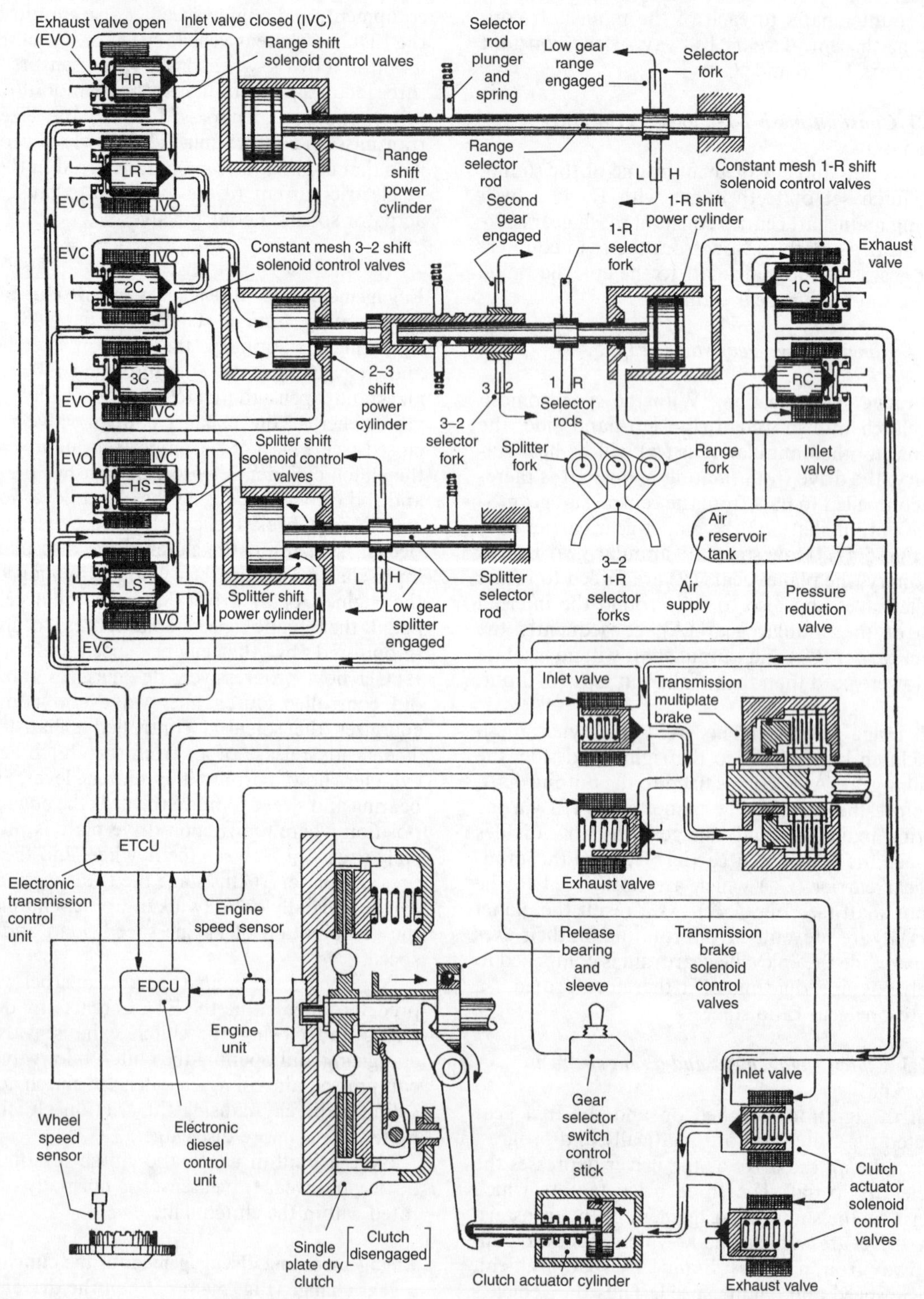

Fig. 5.48 A simplified electro/pneumatic gear shift and clutch control

with the vehicle moving forwards, the ETCU immediately signals the clutch solenoid control valves to operate so that the compressed air can bring about the disengagement and then engagement of the clutch drive plate for sufficient time (programmed time setting) for the gear shift to take place (see Fig. 5.48). This is achieved in the first phase by de-energizing and closing the clutch solenoid exhaust valve and correspondingly energizing and opening the inlet valve, thus permitting the compressed air to enter the clutch actuator cylinder and to release the clutch. The second phase de-energizes and closes the inlet valve and then energizes and opens the exhaust valve so that the clutch release mechanism allows the clutch to engage the transmission drive.

5.11.6 Transmission brake (Figs 5.47 and 5.48)
This is a compressed air operated multiplate brake. Its purpose is to rapidly reduce the free spin speed of the driven disc plate, input shaft and twin countershaft masses when the clutch is disengaged thus enabling fast and smooth gear shifts to be made.

When a gear shift change is about to be made the driver moves the gear selector stick to a new position. This is signalled to the ETCU, and one outcome is that the transmission brake solenoid control inlet valve is energized to open (see Fig. 5.48). It thus permits compressed air to enter the piston chamber and thereby to squeeze together the friction disc plate so that the freely spinning countershafts are quickly dragged down to the main shaft's speed, see Fig. 5.47. Once the central gears wedged in between the twin countershafts have unified their speed with that of the mainshaft, then at this point the appropriate constant mesh dog clutch can easily slide into mesh with it adjacent central gear dog teeth. Immediately after the gear shift the transmission brake inlet valve closes and the exhaust valve opens to release the compressed air from the multiplate clutch cylinder thereby preventing excessive binding and strain imposed to the friction plates and assembly.

5.11.7 Splitter gear shifts (Fig. 5.48)
The splitter gear shift between low and high gear ratio takes place though a synchromesh type dog clutch device. Note for all the gear changes taking place in the gearbox, the splitter gears are constantly shifted from low to high going up the gear ratios or from high to low going down the gear ratios. With ignition switched on and the gear selector stick positioned say in low gear, the ETCU signals the splitter solenoid control to close and open the inlet and exhaust valves respectively for the high splitter gear solenoid control, and at the same time to close and open the exhaust and inlet valves respectively for the low splitter gear solenoid control (see Fig. 5.48). The splitter shift power cylinder will now operate, compressed air will be released from the left-hand side and simultaneously compressed air will be introduced to the right-hand side of the splitter shift power cylinder; the piston and selector rod now smoothly shift to the low splitter gear position. Conversely if high splitter gear was to be selected, the reverse would happen to the solenoid control valves with regards to their opening and closing so that the piston and selector rod would in this case move to the right.

5.11.8 Range gear shifts (Figs 5.47 and 5.48)
The range gear shift takes place though a single stage epicyclic gear train and operates also via a synchromesh type dog clutch mechanism.

Going though the normal gear change sequence from 1 to 12 the first six gear ratios one to six are obtained with the range gear shift in the low position and from seventh to twelfth gear in high range shift position, see Fig. 5.47.

With the ignition switched on and the gear selector stick moved to gear ratios between one and six the low range gear shift will be selected, the ETCU activates the range shift solenoid control valves such that the high range inlet and exhaust valves are closed, and opened respectively, whereas the low range inlet valve is opened and exhaust valve is closed, see Fig. 5.48. Hence compressed air is exhausted from the left hand side of the range shift power cylinder and exposed to fresh compressed air on the right-hand side. Subsequently the piston and selector rod is able to quickly shift to the low range position.

A similar sequence of events takes place if the high range gear shift is required except the opening and closing of the valves will be opposite to that needed for the low range shift.

5.11.9 Constant mesh three speed and reverse gear shift (Figs 5.47 and 5.48)
These gear shifts cover the middle section of the gearbox which involves the engagement and disengagement of the various central mainshaft constant mesh gears via a pair of sliding dog clutches. There is a dog clutch for engagement and disengagement for gears 1-R and similarly a second dog clutch for gears 2–3.

To go though the complete gear ratio steps, the range shift is put initially into 'low', then the splitter gear shifts are moved alternatively into low and high as the constant mesh dog clutch gears are shifted progressively up; this is again repeated but the second time with the range shift in high (see Fig. 5.47). This can be presented as range gear shifted into 'low', 1 gear constant mesh low and high splitter, 2 gear constant mesh low and high splitter, and 3 gear constant mesh low and high splitter gear; at this point the range gear is shifted into 'high' and the whole sequence is repeated, 1 constant mesh gear low and high splitter, 2 constant mesh gear low and high splitter and finally third constant mesh gear low and high splitter; thus twelve gear ratios are produced thus:

First six overall gear ratios = splitter gear (L and H)S × constant mesh gears (1, 2 and 3) × range gear low (LR)

Second six overall gear ratios = splitter gear (L and H) × constant mesh gears (1, 2 and 3) × range gear high (HR).

1 $OGR = LS \times CM\ (1) \times LR$
2 $OGR = HS \times CM\ (1) \times LR$
3 $OGR = LS \times CM\ (2) \times LR$
4 $OGR = LS \times CM\ (2) \times LR$
5 $OGR = LS \times CM\ (3) \times LR$
6 $OGR = HS \times CM\ (3) \times LR$
Low range

7 $OGR = LS \times CM\ (1) \times HR$
8 $OGR = HS \times CM\ (1) \times HR$
9 $OGR = LS \times CM\ (2) \times HR$
10 $OGR = HS \times CM\ (2) \times HR$
11 $OGR = LS \times CM\ (3) \times HR$
12 $OGR = HS \times CM\ (3) \times HR$
High range

where OGR = overall gear ratio
CM = constant mesh gear ratio
LS/HS = low or high splitter gear ratio
LS/HR = low or high range gear ratio

Assume that the ignition is switched on and the vehicle is being driven forwards in low splitter and low range shift gear positions (see Fig. 5.48). To engage one of the three forward constant mesh gears, for example, the second gear, then the gear selector stick is moved into 3 gear position (low splitter, low range 2 gear). Immediately the ETCU signals the constant mesh 3–2 shift solenoid control valves by energizing the 2 constant mesh solenoid control so that its inlet valve opens and its exhaust valve closes; at the same time, the 3 constant mesh solenoid control is de-energized so that its inlet valve closes and the exhaust valve opens (see Fig. 5.48). Accordingly, the 2–3 shift power cylinder will be exhausted of compressed air on the right-hand side, while compressed air is delivered to the left-hand side of the cylinder, the difference in force between the two sides of the piston will therefore shift the 3–2 piston and selector rod into the second gear position. It should be remembered that during this time period, the clutch will have separated the engine drive from the transmission and that the transmission brake will have slowed the twin countershafts sufficiently for the constant mesh central gear being selected to equalize its speed with the mainshaft speed so that a clean engagement takes place. If first gear was then to be selected, the constant mesh 3–2 shift solenoid control valves would both close their exhaust valves so that compressed air enters from both ends of the 2–3 shift power cylinder, it therefore moves the piston and selector rod into the neutral position before the 1-R shift solenoid control valves are allowed to operate.

6 Transmission bearings and constant velocity joints

6.1 Rolling contact bearings

Bearings which are designed to support rotating shafts can be divided broadly into two groups; the plain lining bearing, known as the *journal bearing*, and the *rolling contact bearing*. The fundamental difference between these bearings is how they provide support to the shaft. With plain sleeve or lining bearings, metal to metal contact is prevented by the generation of a hydrodynamic film of lubricant (oil wedge), which supports the shaft once operating conditions have been established. However, with the rolling contact bearing the load is carried by balls or rollers with actual metal to metal contact over a relatively small area.

With the conventional journal bearing, starting friction is relatively high and with heavy loads the coefficient of friction may be in the order of 0.15. However, with the rolling contact bearing the starting friction is only slightly higher than the operating friction. In both groups of bearings the operating coefficients will be very similar and may range between 0.001 and 0.002. Hydrodynamic journal bearings are subjected to a cyclic projected pressure loading over a relatively large surface area and therefore enjoy very long life spans. For example, engine big-ends and main journal bearings may have a service life of about 160 000 kilometres (100 000 miles). Unfortunately, the inherent nature of rolling contact bearing raceway loading is of a number of stress cycles of large magnitude for each revolution of the shaft so that the life of these bearings is limited by the fatigue strength of the bearing material.

Lubrication of plain journal bearings is very important. They require a pressurized supply of consistent viscosity lubricant, whereas rolling contact bearings need only a relatively small amount of lubricant and their carrying capacity is not sensitive to changes in lubricant viscosity. Rolling contact bearings have a larger outside diameter and are shorter in axial length than plain journal bearings.

Noise levels of rolling contact bearings at high speed are generally much higher than for plain journal bearings due mainly to the lack of a hydrodynamic oil film between the rolling elements and their tracks and the windage effects of the ball or roller cage.

6.1.1 *Linear motion of a ball between two flat tracks* (Fig. 6.1)

Consider a ball of radius r_b placed between an upper and lower track plate (Fig. 6.1). If the upper track plate is moved towards the right so that the ball completes one revolution, then the ball has rolled along the lower track a distance $2\pi r_b$ and the upper track has moved ahead of the ball a further distance $2\pi r_b$. Thus the relative movement, L, between both track plates will be $2\pi r_b + 2\pi r_b$, which is twice the distance, l, travelled forward by the centre of the ball. In other words, the ball centre will move forward only half that of the upper to lower relative track movement.

i.e. $$\frac{l}{L} = \frac{2\pi r_b}{4\pi r_b} = \frac{1}{2}$$

$$\therefore l = \frac{L}{2}$$

6.1.2 *Ball rolling contact bearing friction* (Fig. 6.2(a and b))

When the surfaces of a ball and track contact under load, the profile a–b–c of the ball tends to flatten out and the profile a–e–c of the track becomes concave (Fig. 6.2(a)). Subsequently the pressure between the contact surfaces deforms them into a common elliptical shape a–d–c. At the same time, a bulge will be established around the contact edge of the ball due to the displacement of material.

If the ball is made to roll forward, the material in the front half of the ball will be subjected to increased compressive loading and distortion whilst that on the rear half experiences pressure release (Fig. 6.2(b)). As a result, the stress distribution over the contact area will be constantly varying.

The energy used to compress a perfect elastic material is equal to that released when the load is removed, but for an imperfect elastic material (most materials), some of the energy used in straining the material is absorbed as internal friction (known as *elastic hysteresis*) and is not released when the load is removed. Therefore, the energy absorbed by the ball and track when subjected to a compressive load, causing the steel to distort, is greater than that released as the ball moves forward. It is this missing

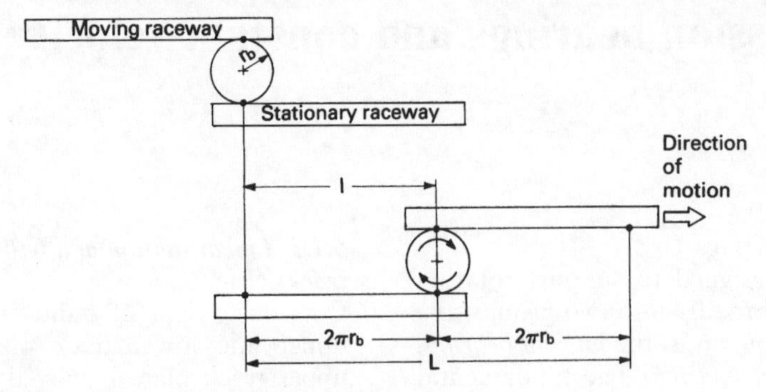

Fig. 6.1 Relationship of rolling element and raceway movement

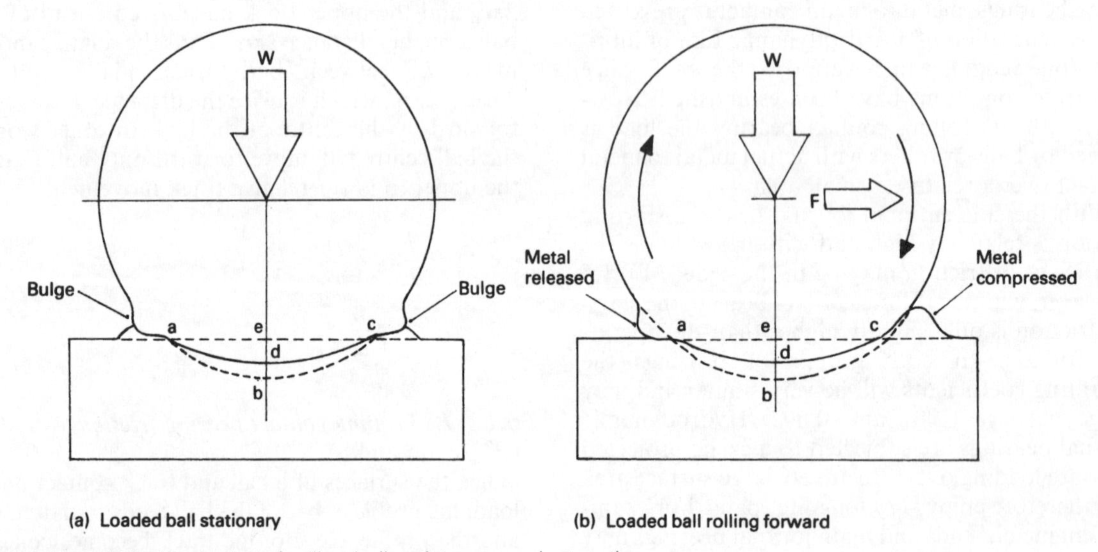

Fig. 6.2 (a and b) Illustration of rolling ball resistance against motion

energy which creates a friction force opposing the forward motion of the ball.

Owing to the elastic deformation of the contact surfaces of the ball and track, the contact area will no longer be spherical and the contact profile arc will therefore have a different radius to that of the ball (Fig. 6.2(b)). As a result, the line a–e–c of the undistorted track surface is shorter in length than the rolling arc profile a–d–c. In one revolution the ball will move forward a shorter distance than if the ball contact contour was part of a true sphere. Hence the discrepancy of the theoretical and actual forward movement of the ball is accommodated by slippage between the ball and track interfaces.

6.1.3 Radial ball bearings (Fig. 6.3)

The essential elements of the multiball bearing is the inner externally grooved and the outer internally grooved ring races (tracks). Lodged between these inner and outer members are a number of balls which roll between the grooved tracks when relative angular motion of the rings takes place (Fig. 6.3(a)). A fourth important component which is not subjected to radial load is the ball cage or retainer whose function is to space the balls apart so that each ball takes its share of load as it passes from the loaded to the unloaded side of the bearing. The cage prevents the balls piling up and rubbing together on the unloaded bearing side.

Contact area The area of ball to track groove contact will, to some extent, determine the load carrying capacity of the bearing. Therefore, if both ball and track groove profiles more or less conform, the bearing load capacity increases. Most radial ball bearings have circular grooves ground in the inner and outer ring members, their radii being 2–4% greater than the ball radius so that ball to track contact, friction, lubrication and cooling can be controlled (Fig. 6.3(a)). An unloaded bearing produces a ball to track point contact, but as the load is increased, it changes to an elliptical contact area (Fig. 6.3(a)). The outer ring contact area will be larger than that of the inner ring since the track curvature of the outer ring is in effect concave and that of the inner ring is convex.

Bearing failure The inner ring face is subjected to a lesser number of effective stress cycles per revolution of the shaft than the corresponding outer ring race, but the maximum stress developed at the inner race because of the smaller ball contact area as opposed to the outer race tends to be more critical in producing earlier fatigue in the inner race than that at the outer race.

Lubrication Single and double row ball bearings can be externally lubricated or they may be prepacked with grease and enclosed with side covers to prevent the grease escaping from within and at the same time stopping dust entering the bearing between the track ways and balls.

6.1.4 Relative movement of radial ball bearing elements (Fig. 6.3(b))

The relative movements of the races, ball and cage may be analysed as follows:

Consider a ball of radius r_b revolving N_b rev/min without slip between an inner rotating race of radius r_i and outer stationary race of radius r_o (Fig. 6.3(b)). Let the cage attached to the balls be at a pitch circle radius r_p and revolving at N_c rev/min.

Linear speed of ball $= 2\pi r_b N_b (m/s)$ (1)

Linear speed of inner race $= 2\pi r_i N_i (m/s)$ (2)

Linear speed of cage $= 2\pi r_p N_c (m/s)$ (3)

Pitch circle radius $r_p = \frac{r_i + r_o}{2} (m)$ (4)

But the linear speed of the cage is also half the speed of the inner race

i.e. $\frac{2\pi r_i N_i}{2}$

Hence Linear speed of cage $= \pi r_i N_i (m/s)$ (5)

If no slip takes place,

Linear speed of ball = Linear speed of inner race

$$2\pi r_b N_b = 2\pi r_i N_i$$

$$\therefore N_b = \frac{r_i}{r_b} N_i (\text{rev/min}) \quad (6)$$

Linear speed of cage = Half inner speed of inner race

$$2\pi r_p N_c = \pi r_i N_i$$

$$\text{Hence} \quad N_c = \frac{\pi}{2\pi} \frac{r_i}{r_p} N_i$$

$$\therefore N_c = \frac{r_i}{r_p} \frac{N_i}{2} (\text{rev/min}) \quad (7)$$

Example A single row radial ball bearing has an inner and outer race diameter of 50 and 70 mm respectively.

If the outer race is held stationary and the inner race rotates at 1200 rev/min, determine the following information:

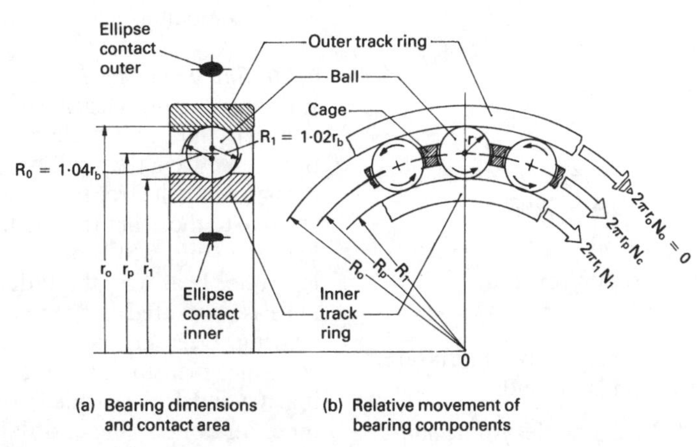

Fig. 6.3 (a and b) Deep groove radial ball bearing terminology

(a) The number of times the balls rotate for one revolution of the inner race.
(b) The number of times the balls rotate for them to roll round the outer race once.
(c) The angular speed of balls.
(d) The angular speed of cage.

(a) Diameter of balls $= r_o - r_i$

$$= 35 - 25 = 10\,\text{mm}$$

Assuming no slip,

$$\text{Number of ball rotations} \times \text{Ball circumference} = \text{Inner race circumference}$$

$$\text{Number of ball rotations},\ 2\pi r_b = 2\pi r_i$$

$$\therefore \text{Number of ball revolutions} = \frac{2\pi r_i}{2\pi r_b} = \frac{r_i}{r_b}$$

$$= \frac{25}{5} = 5 \text{ revolutions}$$

(b)
$$\text{Number of ball rotations} \times \text{Ball circumference} = \text{Outer race circumference}$$

$$\text{Number of ball rotations},\ 2\pi r_b = 2\pi r_o$$

$$\therefore \text{Number of rotations} = \frac{2\pi r_o}{2\pi r_b} = \frac{r_o}{r_b}$$

$$= \frac{35}{5} = 7 \text{ revolutions}$$

(c) Ball angular speed
$$N_b = \frac{r_i}{r_b} N_i$$

$$= \frac{25}{5} \times 1200$$

$$= 6000 \text{ rev/min}$$

(d)
$$r_p = \frac{r_i + r_o}{2} = \frac{25 + 35}{2}$$

$$= 30 \text{ mm}$$

Cage angular speed
$$N_c = \frac{r_i}{r_p}\frac{N_1}{2}$$

$$= \frac{25}{30} \times \frac{1200}{2}$$

$$= 500 \text{ rev/min}$$

6.1.5 Bearing loading

Bearings used to support transmission shafts are generally subjected to two kinds of loads:

1 A load (force) applied at right angles to the shaft and bearing axis. This produces an outward force which is known as a radial force. This kind of loading could be caused by pairs of meshing spur gears radially separating from each other when transmitting torque (Fig. 6.4).
2 A load (force) applied parallel to the shaft and bearing axis. This produces an end thrust which is known as an axial force. This kind of loading could be caused by pairs of meshing helical gears trying to move apart axially when transmitting torque (Fig. 6.4).

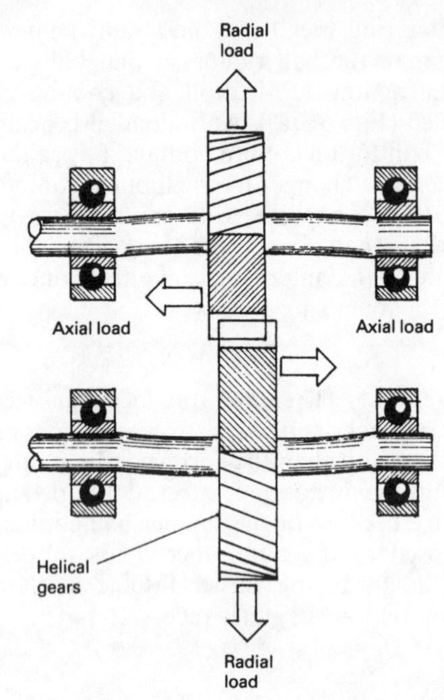

Fig. 6.4 Illustration of radial and axial bearing loads

When both radial and axial loads are imposed on a ball bearing simultaneously they result in a single combination load within the bearing which acts across the ball as shown (Fig. 6.6).

6.1.6 Ball and roller bearing internal clearance

Internal bearing clearance refers to the slackness between the rolling elements and the inner and outer raceways they roll between. This clearance is measured by the free movement of one raceway ring relative to the other ring with the rolling elements in between (Fig. 6.5). For ball and cylindrical (parallel) roller bearings, the radial or diametrical clearance is measured perpendicular to the axis of the bearing. Deep groove ball bearings also have axial clearance measured parallel to the axis of the bearing. Cylindrical (parallel) roller bearings without inner and outer ring end flanges do not have axial clearance. Single row angular contact bearings and

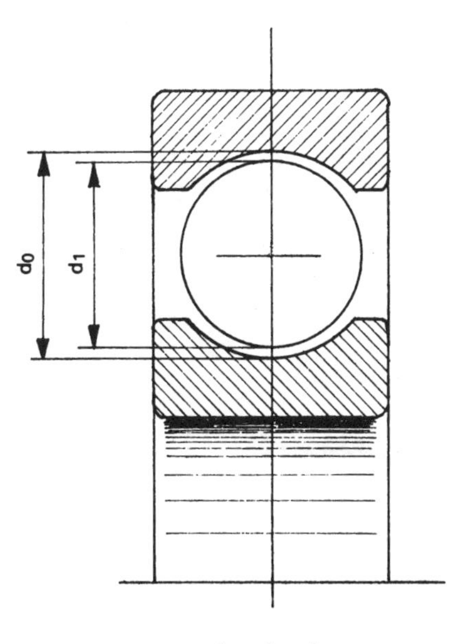

$d_c = d_0 - d_1$

where d_0 = Distance between tracks
d_1 = Ball diameter
d_c = Diametric clearance

Fig. 6.5 Internal bearing diametric clearance

taper roller bearings do have clearance slackness or tightness under operating conditions but this cannot be measured until the whole bearing assembly has been installed in its housing.

A radial ball bearing working at operating temperature should have little or no diametric clearance, whereas roller radial bearings generally operate more efficiently with a small diametric clearance.

Radial ball and roller bearings have a much larger initial diametric clearance before being fitted than their actual operating clearances.

The difference in the initial and working diametric clearances of a bearing, that is, before and after being fitted, is due to a number of reasons:

1 The compressive interference fit of the outer raceway member when fitted in its housing slightly reduces diameter.
2 The expansion of the inner raceway member when forced over its shift minutely increases its diameter.

The magnitude of the initial contraction or expansion of the outer and inner raceway members will depend upon the following:

a) The rigidity of the housing or shaft; is it a low strength aluminium housing, moderate strength cast iron housing or a high strength steel housing? Is it a solid or hollow shaft; are the inner and outer ring member sections thin, medium or thick?
b) The type of housing or shaft fit; is it a light, medium or heavy interference fit?

The diametric clearance reduction when an inner ring is forced over a solid shaft will be a proportion of the measured ring to shaft interference.

The reductions in diametric clearance for a heavy and a thin sectioned inner raceway ring are roughly 50% and 80% respectively. Diametric clearance reductions for hollow shafts will of course be less.

Working bearing clearances are affected by the difference in temperature between the outer and inner raceway rings which arise during operation. Because the inner ring attached to its shaft is not cooled so effectively as the outer ring which is supported in a housing, the inner member expands more than the outer one so that there is a tendency for the diametric clearance to be reduced due to the differential expansion of the two rings.

Another reason for having an initial diametric clearance is it helps to accommodate any inaccuracies in the machining and grinding of the bearing components.

The diametric clearance affects the axial clearance of ball bearings and in so doing influences their capacity for carrying axial loads. The greater the diametric clearance, the greater the angle of ball contact and therefore the greater the capacity for supporting axial thrust (Fig. 6.6).

Bearing internal clearances have been so derived that under operating conditions the existing clearances provide the optimum radial and axial load carrying capacity, speed range, quietness of running and life expectancy. As mentioned previously, the diametric clearance is greatly influenced by the type of fit between the outer ring and its housing and the inner ring and its shaft, be they a slip, push, light press or heavy press interference.

The tightness of the bearing fit will be determined by the extremes of working conditions to which the bearing is subjected. For example, a light duty application will permit the bearing to be held with a relatively loose fit, whereas for heavy conditions an interference fit becomes essential.

To compensate for the various external fits and applications, bearings are manufactured with different diametric clearances which have been standardized by BSI and ISO. Journal bearings are made with a range of diametrical clearances, these clearances being designated by a series of codes shown below in Table 6.1.

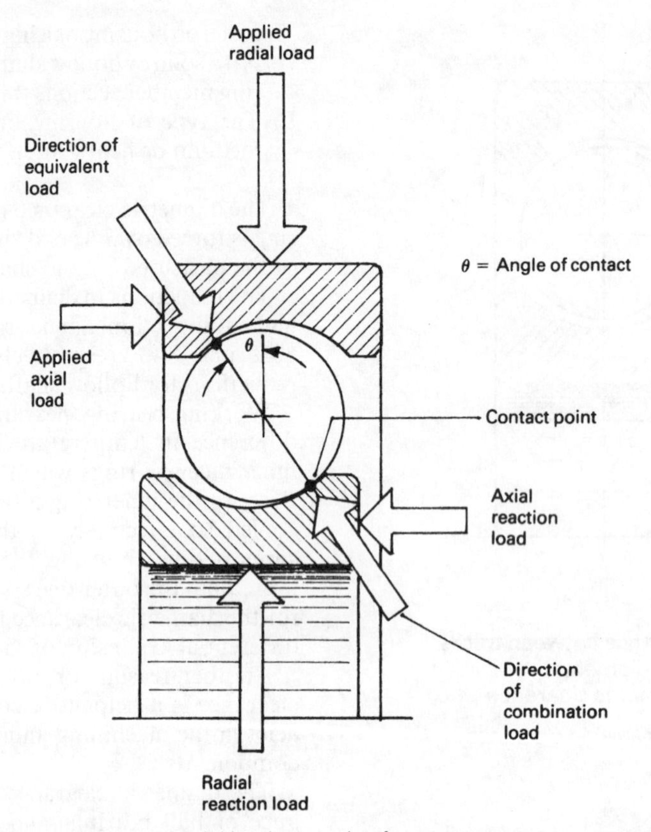

Fig. 6.6 Effects of diametric clearance and axial load on angle of contact

Table 6.1 Journal bearing diametrical clearances

BSI Designation	ISO Group	SKF Designation	Hoffmann Designation
—	Group 2	C2	0
DC2	Normal group	Normal	00
DC3	Group 3	C3	000
—	Group 4	C4	0000

Note The lower the number the smaller is the bearing's diametric clearance. In the new edition of BS 292 these designations are replaced by the ISO groups. For special purposes, bearings with a smaller diametric clearance such as Group 1 and larger Group 5 are available.

The diametrical clearances 0, 00, 000 and 0000 are usually known as *one dot, two dot, three dot* or *four dot* fits. These clearances are identified by the appropriate code or number of polished circles on the stamped side of the outer ring.

The applications of the various diametric clearance groups are compared as follows:

Group 2 These bearings have the least diametric clearance. Bearings of this group are suitable when freedom from shake is essential in the assembled bearing. The fitting interference tolerance prevents the initial diametric clearance being eliminated. Very careful attention must be given to the bearing housing and shaft dimensions to prevent the expansion of the inner ring or the contraction of the outer ring causing bearing tightness.

Normal group Bearings in this group are suitable when only one raceway ring has made an interference fit and there is no appreciable loss of clearance due to temperature differences. These diametric clearances are normally adopted with radial ball bearings for general engineering applications.

Group 3 Bearings in this group are suitable when both outer and inner raceway rings have made an interference fit or when only one ring has an interference fit but there is likely to be some loss of clearance due to temperature differences. Roller bearings and ball bearings which are subjected to axial thrust tend to use this diametric clearance grade.

Group 4 Bearings in this group are suitable when both outer and inner bearing rings are an interference fit and there is some loss of diametric clearance due to temperature differences.

6.1.7 Taper roller bearings

Description of bearing construction (Fig. 6.7) The taper roller bearing is made up of four parts; the inner raceway and the outer raceway, known respectively as the *cone* and *cup,* the taper rollers shaped as frustrums of cones and the cage or roller retainer (Fig. 6.8). The taper rollers and both inner and outer races carry load whereas the cage carries no load but performs the task of spacing out the rollers around the cone and retaining them as an assembly.

Taper roller bearing true rolling principle (Fig. 6.8(a and b)) If the axis of a cylindrical (parallel) roller is inclined to the inner raceway axis, then the relative rolling velocity at the periphery of both the outer and inner ends of the roller will tend to be different due to the variation of track diameter (and therefore circumference) between the two sides of the bearing. If the mid position of the roller produced true rolling without slippage, the portion of the roller on the large diameter side of the tracks would try to slow down whilst the other half of the roller on the smaller diameter side of the tracks would tend to speed up. Consequently both ends would slip continuously as the central raceway member rotated relative to the stationary outer race members (Fig. 6.8(a)).

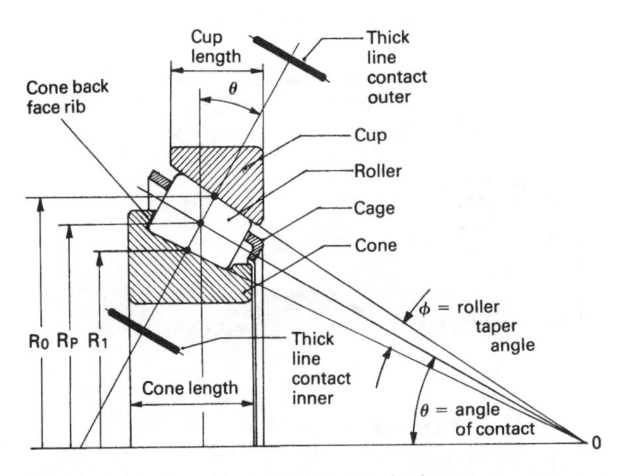

Fig. 6.7 Taper rolling bearing terminology

The design geometry of the taper roller bearing is therefore based on the cone principle (Fig. 6.8(b)) where all projection lines, lines extending from the cone and cup working surfaces (tracks), converge at one common point on the axis of the bearing.

With the converging inner and outer raceway (track) approach, the track circumferences at the large and small ends of each roller will be greater and smaller respectively. The different surface velocities on both large and small roller ends will be accommodated by the corresponding change in track circumferences. Hence no slippage takes place, only pure rolling over the full length of each roller as they revolve between their inner and outer tracks.

Angle of contact (Fig. 6.7) Taper roller bearings are designed to support not only radial bearing loads but axial (thrust) bearing loads.

The angle of bearing contact Θ, which determines the maximum thrust (axial) load, the bearing can accommodate is the angle made between the perpendiculars to both the roller axis and the inner cone axis (Fig. 6.7). The angle of contact Θ is also half the pitch cone angle. These angles can range from as little as 7½ to as much as 30°. The standard or normal taper roller bearing has a contact angle of 12–16° which will accommodate moderate thrust (axial) loads. For large and very large thrust loads, medium and steep contact angle bearings are available, having contact angles in the region of 20 and 28° respectively.

Area of contact (Fig. 6.7) Contact between roller and both inner cone and outer cup is of the line form without load, but as the rollers become loaded the elastic material distorts, producing a thick line contact area (Fig. 6.7) which can support very large combinations of both radial and axial loads.

Cage (Fig. 6.7) The purpose of the cage container is to equally space the rollers about the periphery of the cone and to hold them in position when the bearing is operating. The prevention of rolling elements touching each other is important since adjacent rollers move in opposite directions at their points of closest approach. If they were allowed to touch they would rub at twice the normal roller speed.

The cage resembles a tapered perforated sleeve (Fig. 6.7) made from a sheet metal stamping which

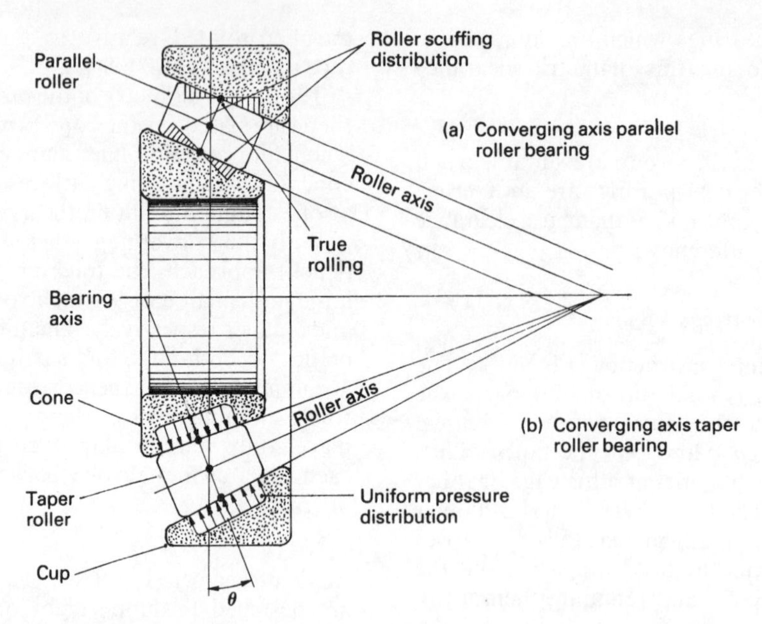

Fig. 6.8 Principle of taper rolling bearing

has a series of roller pockets punched out by a single impact of a multiple die punch.

Although the back cone rib contributes most to the alignment of the rollers, the bearing cup and cone sides furthest from the point of bearing loading may be slack and therefore may not be able to keep the rollers on the unloaded side in their true plane. Therefore, the cage (container) pockets are precisely chamfered to conform to the curvature of the rollers so that any additional corrective alignment which may become necessary is provided by the individual roller pockets.

Positive roller alignment (Fig. 6.9) Both cylindrical parallel and taper roller elements, when rolling between inner and outer tracks, have the tendency to skew (tilt) so that extended lines drawn through their axes do not intersect the bearing axis at the same cone and cup projection apex. This problem has been overcome by grinding the large end of each roller flat and perpendicular to its axis so that all the rollers square themselves exactly with a shoulder or rib machined on the inner cone (Fig. 6.9). When there is any relative movement between the cup and cone, the large flat ends of the rollers make contact with the adjacent shoulder (rib) of the cone, compelling the rollers to positively align themselves between the tapered faces of the cup and cone without the guidance of the cage. The magnitude of the roller-to-rib end thrust, known as the *seating force,* will depend upon the taper roller contact angle.

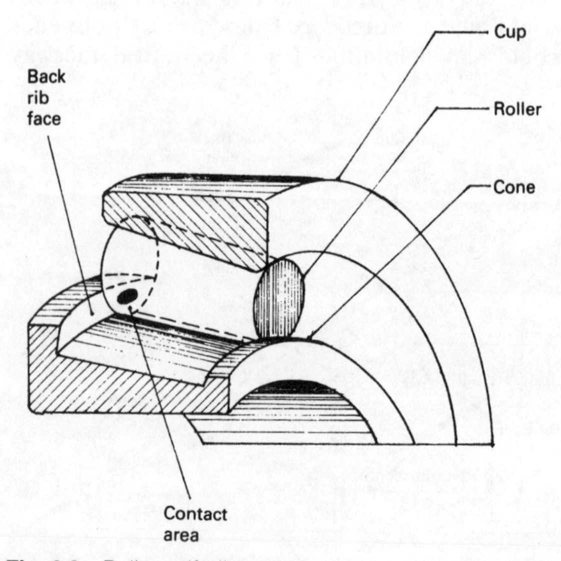

Fig. 6.9 Roller self-alignment

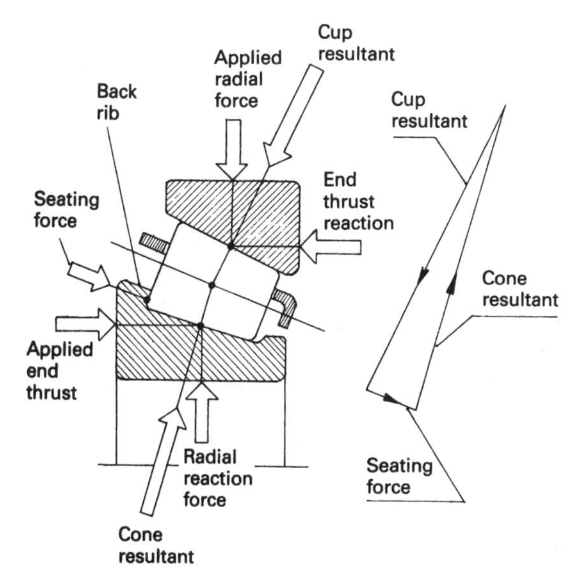

Fig. 6.10 Force diagram illustrating positive roller alignment seating force

Self-alignment roller to rib seating force (Fig. 6.10) To make each roller do its full share of useful work, positive roller alignment is achieved by the large end of each roller being ground perpendicular to its axis so that when assembled it squares itself exactly with the cone back face rib (Fig. 6.10).

When the taper roller bearing is running under operating conditions it will generally be subjected to a combination of both radial and axial loads. The resultant applied load and resultant reaction load will be in apposition to each other, acting perpendicular to both the cup and cone track faces. Since the rollers are tapered, the direction of the perpendicular resultant loads will be slightly inclined to each other, they thereby produce a third force parallel to the rolling element axis. This third force is known as the *roller-to-rib seating force* and it is this force which provides the rollers with their continuous alignment to the bearing axis. The magnitude of this *roller-to-rib seating force* is a function of the included taper roller angle which can be obtained from a triangular force diagram (Fig. 6.10). The diagram assumes that both the resultant applied and reaction loads are equal and that their direction lies perpendicular to both the cup and cone track surface. A small roller included angle will produce a small rib seating force and vice versa.

6.1.8 Bearing materials

Bearing inner and outer raceway members and their rolling elements, be they balls or rollers, can be made from either a case hardening alloy steel or a through hardened alloy steel.

a) The case hardened steel is usually a low alloy nickel chromium or nickel-chromium molybdenum steel, in which the surface only is hardened to provide a wear resistance outer layer while the soft, more ductile core enables the bearing to withstand extreme shock and overloading.
b) The through hardened steel is generally a high carbon chromium steel, usually about 1.0% carbon for adequate strength, together with 1.5% chromium to increase hardenability. (This is the ability of the steel to be hardened all the way through to a 60–66 Rockwell C scale.)

The summary of the effects of the alloying elements is as follows:

Nickel increases the tensile strength and toughness and also acts as a grain refiner. Chromium considerably hardens and raises the strength with some loss in ductibility, whilst molybdenum reduces the tendency to temper-brittleness in low nickel low chromium steel.

Bearing inner and outer raceways are machined from a rod or seamless tube. The balls are produced by closed die forging of blanks cut from bar stock, are rough machined, then hardened and tempered until they are finally ground and lapped to size.

Some bearing manufacturers use case-hardened steel in preference to through-hardened steel because it is claimed that these steels have hard fatigue resistant surfaces and a tough crack-resistant core. Therefore these steels are able to withstand impact loading and prevent fatigue cracks spreading through the core.

6.1.9 Bearing friction

The friction resistance offered by the different kinds of rolling element bearings is usually quoted in terms of the coefficient of friction so that a relative comparison can be made. Bearing friction will vary to some extent due to speed, load and lubrication. Other factors will be the operating conditions which are listed as follows:

1 Starting friction will be higher than the dynamic normal running friction.
2 The quantity and viscosity of the oil or grease; a large amount of oil or a high viscosity will increase the frictional resistance.
3 New unplanished bearings will have higher coefficient of friction values than worn bearings which have bedded down.

4 Preloading the bearing will initially raise the coefficient of friction but under working conditions it may reduce the overall coefficient value.
5 Pre-lubricated bearings may have slightly higher coefficients than externally lubricated bearings due to the rubbing effect of the seals.

Coefficient of friction — average values for various bearing arrangements

Self-alignment ball bearings	=	0.001
Cylindrical roller bearing	=	0.0011
Thrustball bearings	=	0.0013
Single row deep grooveball bearings	=	0.0015
Taper and spherical roller bearings	=	0.0018

6.1.10 Ball and roller bearing load distribution (Fig. 6.11)

When either a ball or roller bearing is subjected to a radial load, the individual rolling elements will not be loaded equally but will be loaded according to their disposition to the direction of the applied load. Applying a radial load to a bearing shaft pushes the inner race towards the outer race in the direction of the load so that the balls or rollers in one half of the bearing do not support any load whereas the other half of the bearing reacts to the load (Fig. 6.11(a)). The distribution of load on the reaction side of the bearing will vary considerably with the diametrical rolling element clearance and the mounting rigidity preventing deformation of the bearing assembly.

If the internal radial clearances of the rolling elements are zero and the inner and outer bearing races remain true circles when loaded, the load distribution will span the full 180° so that approximately half the balls or rollers will, to some extent, share the radial load (Fig. 6.11(b)). Conversely slackness or race circular distortion under load will reduce the projected load zone so that the rolling elements which provide support will be very much more loaded resulting in considerably more shaft deflection under load. Lightly preloaded bearings may extend the radial load zone to something greater than 180° but less than 360° (Fig. 6.11(c)). This form of initial bearing loading will eliminate gear mesh teeth misalignment due to shaft deflection under operating conditions. Heavy bearing preloading may extend the load zone to 360° (Fig. 6.11(d)); this degree of preloading should only be used for severe working conditions or where large end thrust is likely to be encountered and must be absorbed without too much axial movement.

End thrusts (axial loads), unlike radial loads, produce a uniform load distribution pattern around the bearing (Fig. 6.11(e)). Deep groove radial ball bearings can tolerate light end thrust. Angular contact ball bearings are capable of supporting medium axial loads. Taper roller bearings, be they normal or steep angled, can operate continuously under heavy and very heavy end loads respectively. Only if the shaft being supported deflects will the end load distribution become uneven.

6.1.11 Bearing fatigue

Fatigue in ball or roller bearings is caused by repeated stress reversals as the rolling elements move around the raceways under load. The periodic elastic compression and release as the rolling elements make their way around the tracks will ultimately overwork and rupture the metal just below the surface. As a result, tiny cracks propagate almost parallel to the surface but just deep enough to be invisible. With continuous usage the alternating stress cycles will cause the cracks to extend, followed by new cracks sprouting out from the original ones. Eventually there will be a network of minute interlinking cracks rising and merging together on the track surface. Subsequently, under further repeating stress cycles, particles will break away from the surface, the size of material leaving the surface becoming larger and larger. This process is known as *spalling* of the bearing and eventually the area of metal which has come away will end the effective life of the bearing. If bearing accuracy and low noise level is essential the bearing will need to be replaced, but if bearing slackness and noise can be accepted, the bearing can continue to operate until the rolling elements and their tracks find it impossible to support the load.

6.1.12 Rolling contact bearing types

Single row deep grooved radial ball bearing (Fig. 6.12) These bearings are basically designed for light to medium radial load operating conditions. An additional feature is the depth of the grooves combined with the relatively large size of the balls and the high degree of conformity between balls and grooves which gives the bearing considerable thrust load carrying capacity so that the bearing will operate effectively under both radial and axial loads.

These bearings are suitable for supporting gearbox primary and secondary shafts etc..

Single row angular contact ball bearing (Fig. 6.13) Bearings of this type have ball tracks which are so

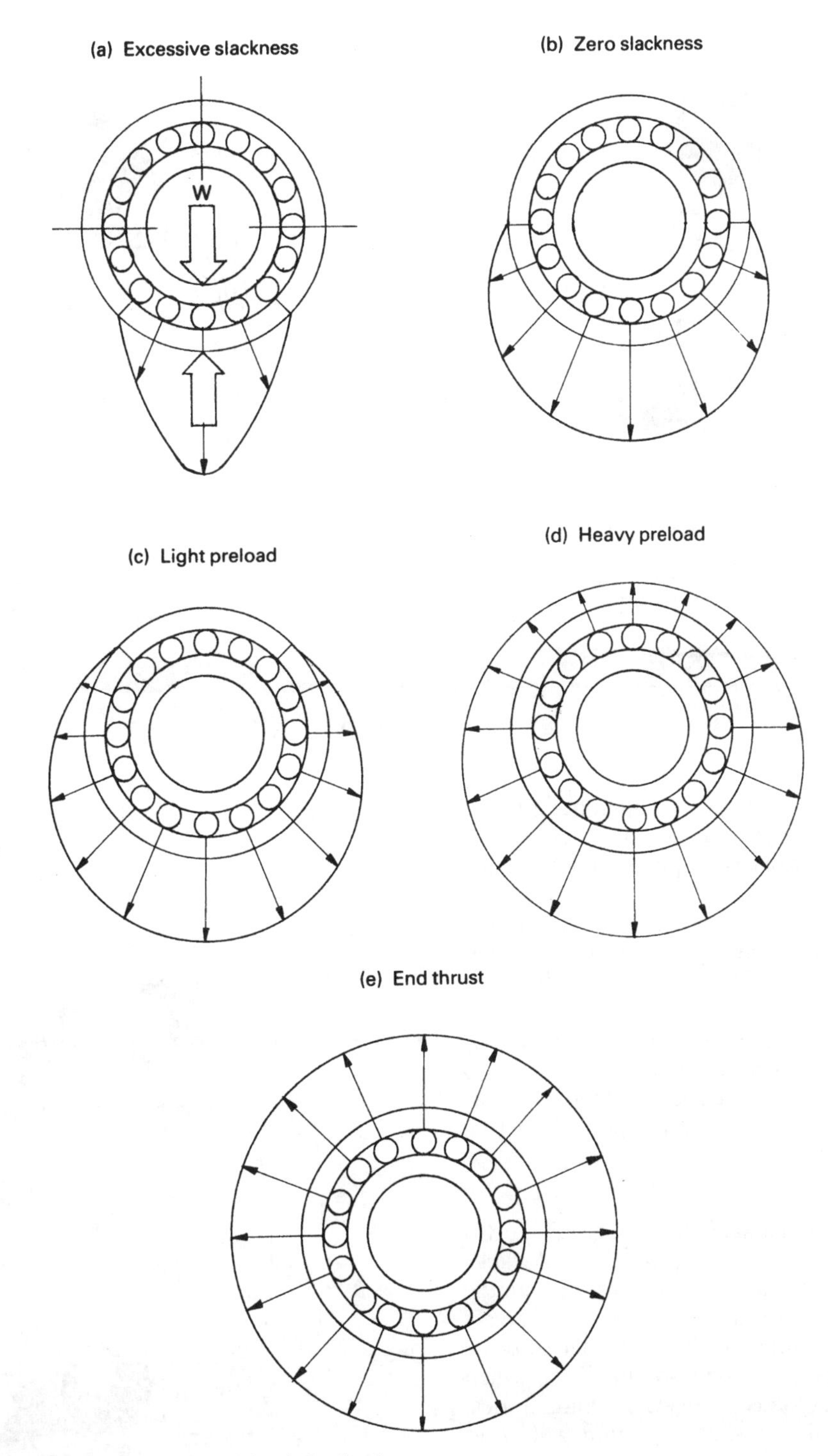

Fig. 6.11 (a–e) Bearing radial and axial load distribution

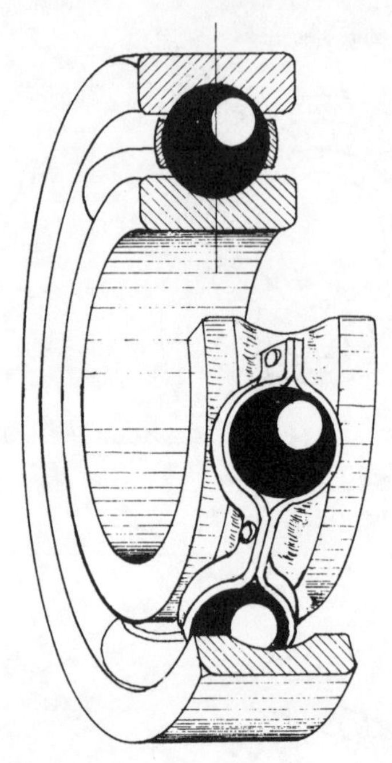

Fig. 6.12 Single row deep groove radial ball bearing

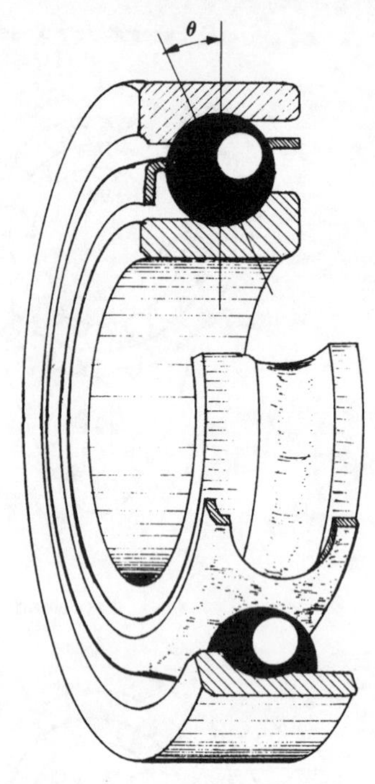

Fig. 6.13 Single row deep angular contact ball bearing

disposed that a line through the ball contact forms an acute angle with the bearing shaft axes. Ball to track ring contact area is elliptical and therefore with the inclined contact angle this bearing is particularly suitable for heavy axial loads. Adjustment of these bearings must always be towards another bearing capable of dealing with axial loads in the opposite direction. The standard contact angle is 20°, but for special applications 12, 15, 25 and 30° contact angle bearings are available. These bearings are particularly suited for supporting front and rear wheel hubs, differential cage housings and steering box gearing such as the rack and pinion.

Double row angular contact ball bearings (Fig. 6.14) With this double row arrangement, the ball tracks are ground so that the lines of pressure through the balls are directed towards two comparatively widely separated points on the shaft. These bearings are normally preloaded so that even when subjected to axial loads of different magnitudes, axial deflection of the shaft is minimized. End thrust in both axial directions can be applied and at the same time very large radial loads can be carried for a relatively compact bearing assembly.

A typical application for this type of bearing would be a semi- or three-quarter floating outer

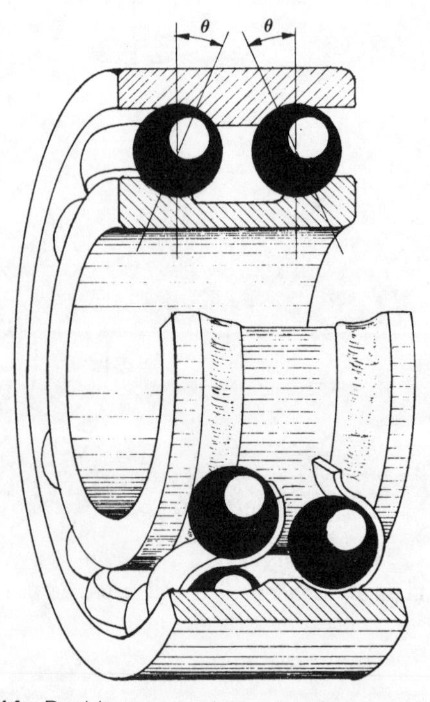

Fig. 6.14 Double row angular contact ball bearing

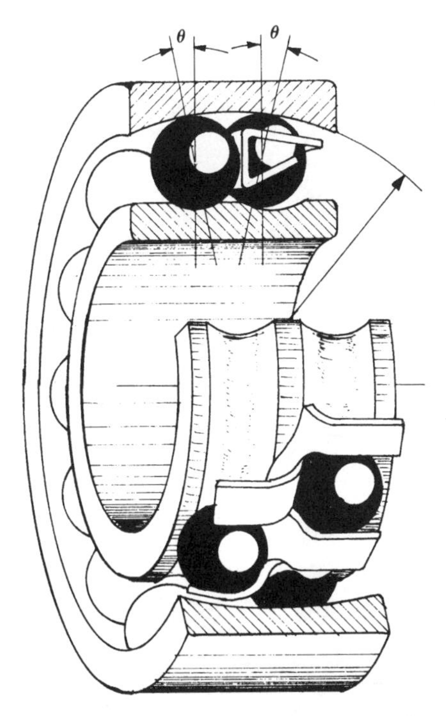

Fig. 6.15 Double row self-alignment ball bearing

half shaft bearing, gearbox secondary output shaft bearing etc.

Double row self-aligning ball bearing (Fig. 6.15) This double row bearing has two rows of balls which operate in individual inner raceway grooves in conjunction with a common spherical outer raceway ring. The spherical outer track enables the inner ring and shaft to deflect relative to the outer raceway member, caused by the balls not only rolling between and around their tracks but also across the common outer circular track. Thus the self-aligning property of the bearing automatically adjusts any angular deflection of the shaft due to mounting errors, whip or settlement of the mounting. It also prevents the bearing from exerting a bending influence on the shaft. The radial load capacity for a single row self-aligning bearing is considerably less than that for the deep groove bearing due to the large radius of the outer spherical race providing very little ball to groove contact. This limitation was solved by having two staggered rows of balls to make up for the reduced ball contact area.

Note that double row deep groove bearings are not used because radial loads would be distributed unevenly between each row of balls with a periodic shaft deflection. They are used for intermediate propellor shaft support, half shafts and wherever excessive shaft deflection is likely to occur.

Single row cylindrical roller bearing (Fig. 6.16) In this design of roller bearing, the rollers are guided by flanges, one on either the inner or outer track ring. The other ring does not normally have a flange. Consequently, these bearings do not take axial loads and in fact permit relative axial deflection of shafts and bearing housing within certain limits. These bearings can carry greater radial loads than the equivalent size groove bearing and in some applications both inner and outer ring tracks are flanged to accommodate very light axial loads.

Bearings of this type are used in gearbox and final drive transmissions where some axial alignment may be necessary.

Single row taper roller bearing (Fig. 6.17) The geometry of this class of bearing is such that the axes of its rollers and conical tracks form an angle with the shaft axis. The taper roller bearing is therefore particularly adaptable for applications where large radial and axial loads are transmitted simultaneously. For very severe axial loads, steep taper angle bearings are available but to some extent this is at the expense of the bearing's radial load carrying capacity. With taper bearings, adjustment must always be towards another bearing capable of dealing with axial forces acting in the opposite direction. This is a popular bearing for medium and heavy duty wheel hubs, final drive pinion shafts, the differential cage and crownwheel bearings, for heavy duty gearbox shaft support and in-line injection pump camshafts.

Double row taper roller bearing (Fig. 6.18) These bearings have a double cone and two outer single cups with a row of taper rollers filling the gap between inner and outer tracks on either side. The compactness of these bearings makes them particularly suitable when there is very little space and where large end thrusts must be supported in both axial directions. Thus in the case of a straddled final drive pinion bearing, these double row taper bearings are more convenient than two single row bearings back to back. Another application for these double tow taper roller bearings is for transversely mounted gearbox output shaft support.

Double row spherical roller bearing (Fig. 6.19) Two rows of rollers operate between a double

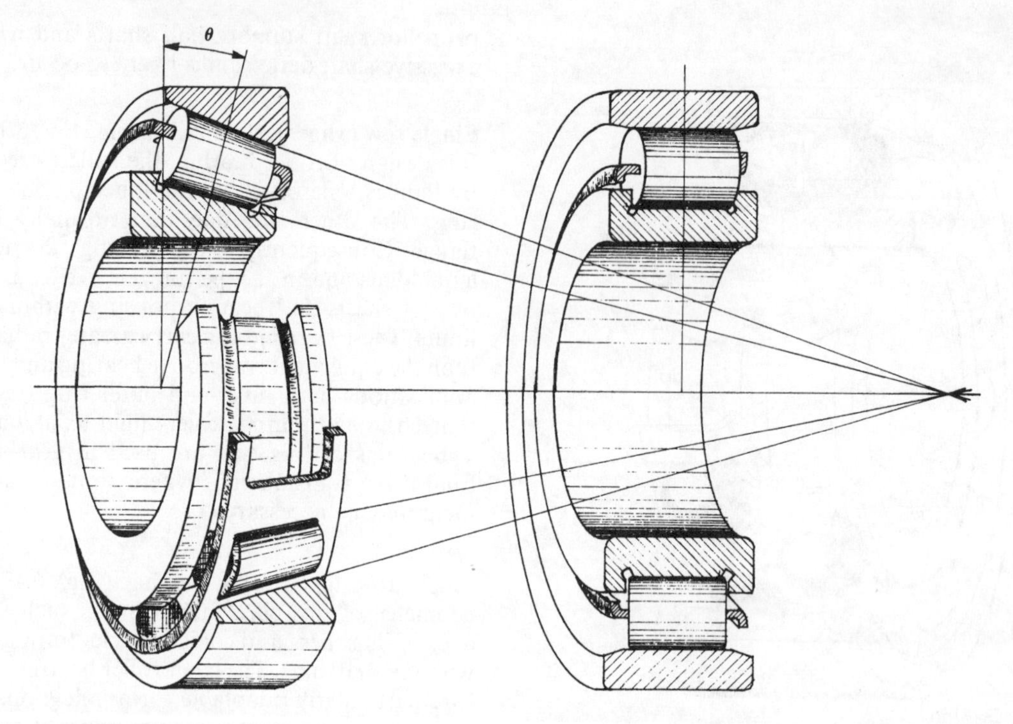

Fig. 6.17 Single row taper roller bearing

Fig. 6.16 Single row cylindrical roller bearing

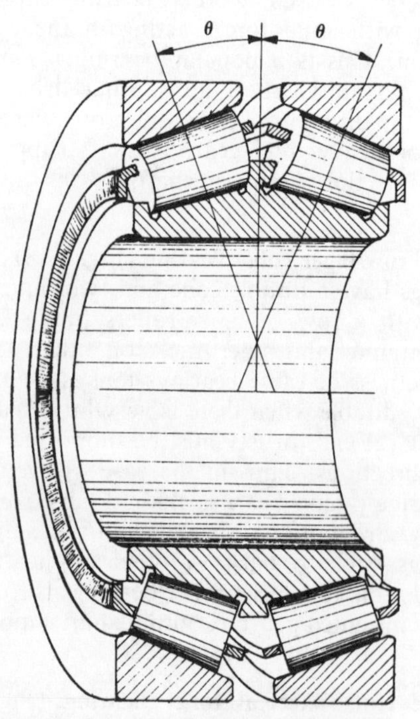

Fig. 6.18 Double row taper roller bearing

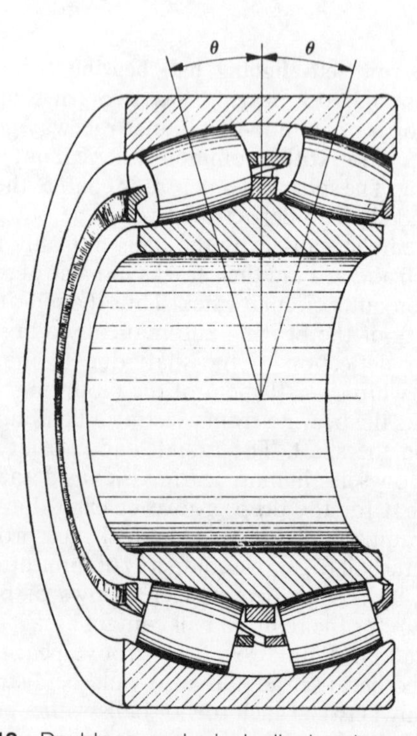

Fig. 6.19 Double row spherical roller bearing

grooved inner raceway and a common spherically shaped outer raceway ring. With both spherical rollers having the same radii as the outer spherical raceway, line contact area is achieved for both inner and outer tracks. The inner double inclined raceway ring retains the two rows of rollers within their tracks, whereas the outer spherical track will accommodate the rollers even with the inner track ring axis tilted relative to the outer track ring axis. This feature provides the bearing with its self-alignment property so that a large amount of shaft deflection can be tolerated together with its capacity, due to roller to track conformity, to operate with heavy loads in both radial and axial directions. This type of bearing finds favour where both high radial and axial loads are to be supported within the constraints of a degree of shaft misalignment.

Single row thrust ball bearing (Fig. 6.20) These bearings have three load bearing members, two grooved annular disc plates and a ring of balls lodged between them. A no-load-carrying cage fourth member of the bearing has two functions; firstly to ease assembly of the bearing when being installed and secondly to evenly space the balls around their grooved tracks. Bearings of this type operate with one raceway plate held stationary while the other one is attached to the rotating shaft.

In comparison to radial ball bearings, thrust ball bearings suffer in operation from an inherent increase in friction due to the balls sliding between the grooved tracks. To minimize the friction, the groove radii are made 6–8% larger than the radii of the balls so that there is a reduction in ball contact area. Another limitation of these bearings is that they do not work very satisfactorily at high rotative speeds since with increased speed the centrifugal force pushes the balls radially outwards, so causing the line of contact, which was originally in the middle of the grooves, to shift further out. This in effect increases ball to track sliding and subsequently the rise in friction generates heat. These bearings must deal purely with thrust loads acting in one direction and they can only tolerate very small shaft misalignment. This type of bearing is used for injection pump governor linkage axial thrust loads, steering boxes and auxiliary vehicle equipment.

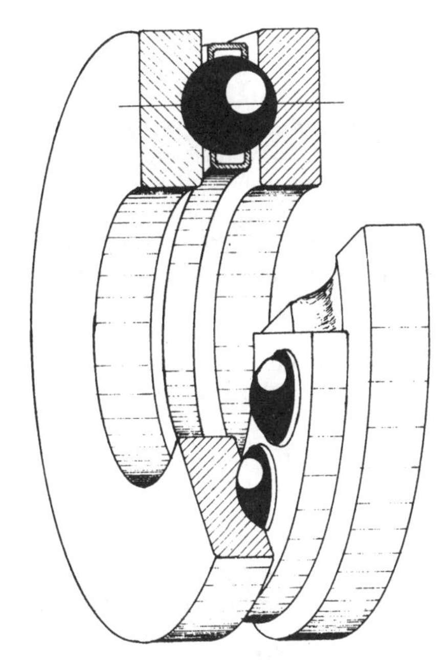

Fig. 6.20 Single row thrust bearing

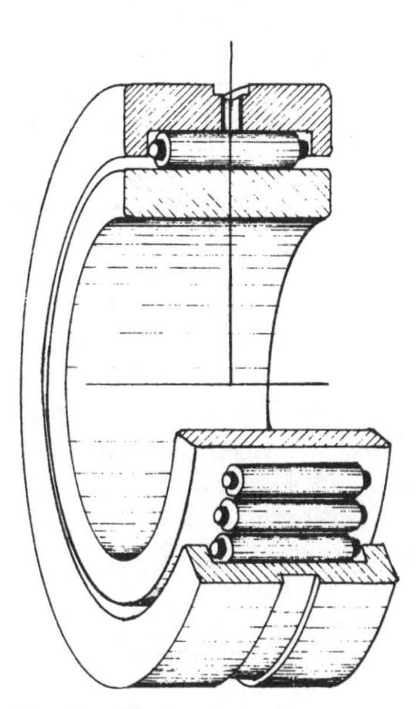

Fig. 6.21 Needle roller bearing

Needle roller bearings (Fig. 6.21) Needle roller bearings are similar to the cylindrical roller bearing but the needle rollers are slender and long and there is no cage (container) to space out the needles around the tracks. The bearing has an inner plain raceway ring. The outer raceway is shouldered either side to retain the needles and has a circular groove machined on the outside with two or four radial holes to provide a passageway for needle lubrication. The length to diameter ratio for the needles usually lies between 3 to 8 and the needle roller diameter normally ranges from 1.5 to 4.5 mm. Sometimes there is no inner raceway ring and the needles operate directly upon the shaft. To increase the line contact area, and therefore the load carrying capacity, needles are made relatively long, but this makes the needle sensitive to shaft misalignment which may lead to unequal load distribution along the length of each needle. Another inherent limitation with long needles is that they tend to skew and slide causing friction losses and considerable wear. The space occupied by a complete needle roller bearing is generally no more than that of a hydrodynamic plain journal bearing. Bearings of this type are well suited for an oscillating or fluctuating load where the needles operate for very short periods before the load or motion reverses, thereby permitting the needles to move back into their original position, parallel to the shaft axis.

Because the needles are not separated from each other, there is a tendency for them to rub together so that friction can be relatively high. To improve lubrication, the needle ends are sometimes tapered or stepped so that oil or grease may be packed between the ends of the needles and the adjacent raceway shoulders.

Needle roller bearings are used for universal joints, gearbox layshafts, first motion shafts, mainshaft constant mesh gear wheel bearings, two stroke heavy duty connecting rod small bearings etc.

Water pump spindle double row ball bearing (Fig. 6.22) In cases where it is necessary to have two bearings spaced apart to support a lightly loaded shaft, such as that used for a water pump and fan, it is sometimes convenient to dispense with the inner raceway ring and mount the balls directly onto the raceway grooves formed on the shaft itself. The space between the shaft and housing is then charged with grease and is fully sealed so that no further attention is required during the working life of the bearings. Such arrangements not only reduce the size of the bearings and housing but also reduce the cost of the assembly.

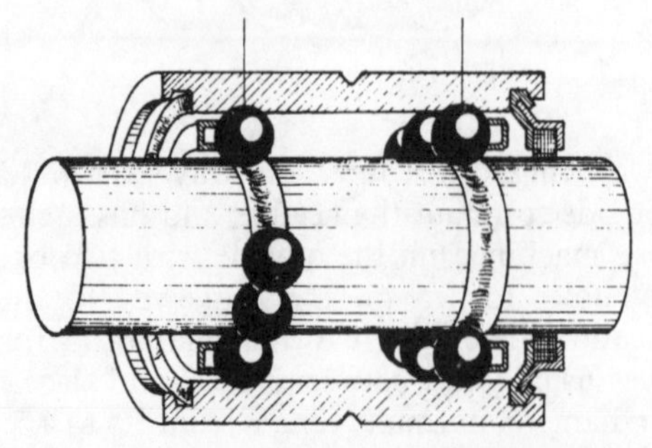

Fig. 6.22 Water pump spindle double row ball bearing

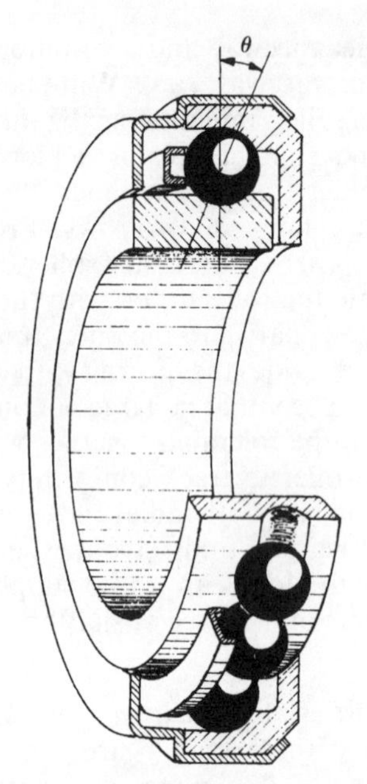

Fig. 6.23 Clutch release thrust ball bearing

Clutch release thrust ball bearing (Fig. 6.23) Where only pure thrust loads are to be coped with for relatively short periods at high speeds, the use of angular contact ball bearings is generally preferred to the single row parallel disc type ball thrust bearing. This configuration has a deep grooved inner race ring and an angular contact outer race ring with a thrust flange on one side. These inner and outer ball track bearing arrangements do not suffer from high rotative speed effects and will operate over a long period of time under moderate random thrust loading, such as when a clutch release mechanism is engaged and released. Sealing the assembled bearing is a steel cover pressing which is designed to retain the pre-packed grease and to exclude dirt getting to the balls and grooved tracks.

6.1.13 Structural rigidity and bearing preloading (Fig. 6.24)

The universal practice of preloading bearing assembly supporting a shaft or hub (Fig. 6.24) raises the rigidity of the bearing assembly so that its deflection under operating conditions is minimized. Insufficient structural rigidity of a shaft or hub assembly may be due to a number of factors which can largely be overcome by preloading the bearings. Bearing

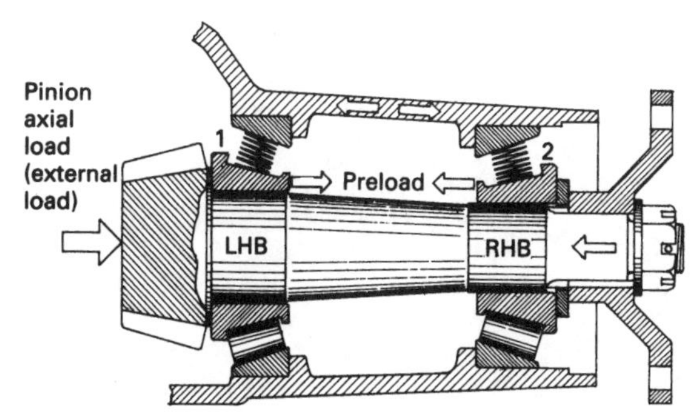

Fig. 6.24 Final drive pinion bearing spring preload analogy

preload goes a long way towards compensating for the following inherent side effects which occur during service.

1 The actual elasticity of the roller elements and their respective tracks cause the bearings to deflect both radially and axially in proportion to the applied load and could amount to a considerable increase in shaft movement under working conditions.
2 As the rolling elements become compressed between the inner and outer tracks, the minute surface irregularities tend to deform under the loaded half of the bearing so that the inner raceway ring or cone member centre axis becomes eccentric to that of the outer ring or cup member.
3 If the structure of the housing which contains the bearing is not sufficiently substantial or is made from soft or low strength aluminium alloy, it may yield under heavy loads so that the bearing roller elements become loose in their tracks.
4 Temperature changes may cause the inner or outer track members to become slack in their housings once they have reached operating temperature conditions even though they may have had an interference fit when originally assembled.
5 Over the working life of a bearing the metal to metal contact of the rolling elements and their raceways will planish the rolling surfaces so that bearing slackness may develop.

6.1.14 Bearing selection (Fig. 6.25)

The rigidity of a rolling contact bearing to withstand both radial and axial loads simultaneously is a major factor in the type of bearing chosen for a particular application. With straight cut gear teeth, pairs of meshing gears are forced apart due to the leverage action when torque is applied so that radial loads alone are imposed onto the bearings. However the

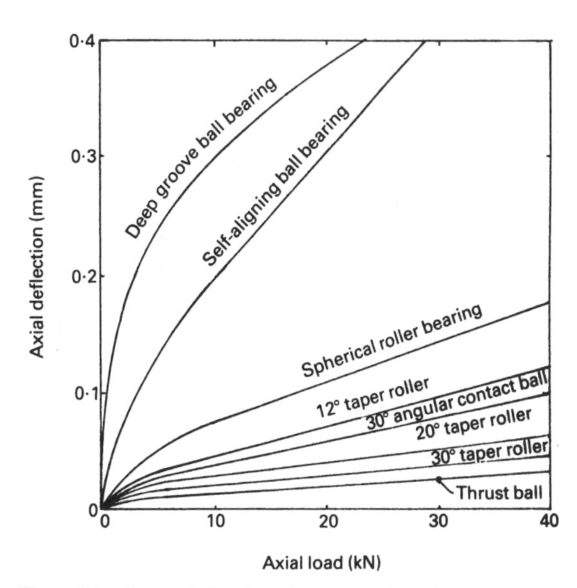

Fig. 6.25 Load-deflection characteristics of several rolling contact bearings (70 mm inner ring shaft diameter)

majority of transmission gear trains have either helical cut teeth or are bevel gears. In either case, end thrust is generated which must be absorbed by the bearings to prevent the gears separating in an axial direction. Bearings are therefore designed not only to carry radial loads but also to support various amounts of axial thrust. As can be seen in Fig. 6.25 the various types of rolling contact bearings offer a range of axial load–deflection characteristics. The least rigid bearing constructions are the deep groove ball and the self-alignment ball bearings, whereas the roller type bearings, with the exception of the angular contact ball and pure thrust ball bearings, provide considerably more axial stiffness. Furthermore, the ability for taper roller bearings to increase their axial load capacity depends to some extent on the angle of bearing contact. The larger the angle, the greater the axial load carrying capacity for a given axial deflection will be. The radial load–deflection characteristics follow a very similar relationship as the previous axial ones with the exception of the pure thrust ball bearing which cannot support radial loads.

6.1.15 Preloading ball and taper roller bearings

An understanding of the significance of bearing preloading may be best visualized by considering a final drive bevel pinion supported between a pair of taper roller bearings (Fig. 6.24). Since the steel of which the rollers, cone and cup are made to obey Hooke's

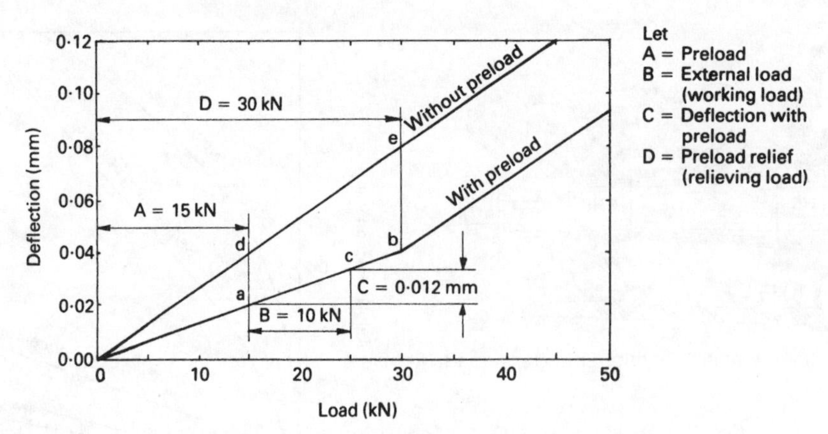

Fig. 6.26 Comparison of bearing load–deflection graph with and without preload

law, whereby strain is directly proportional to the stress producing it within the elastic limit of the material, the whole bearing assembly can be given the spring analogy treatment (Fig. 6.25). The major controlling factor for shaft rigidity is then the stiffness (elastic rate) of the bearing which may be defined as the magnitude of the force exerted per unit of distortion,

i.e. $S = W/x$ (N/m)
where S = (N/m)
W = applied force (N)
x = deflection (mm)

Let the bevel pinion nut be tightened so that each roller bearing is squeezed together axially 0.04 mm when subjected to a preload of 15 kN. Each bearing will have a stiffness of

$$S = W/x$$
$$= 15/0.04 = 375 \text{ kN/mm}$$

Now if an external force (crownwheel tooth load) exerts an outward axial thrust to the right of magnitude 15 kN, the left hand bearing (1), will be compressed a further 0.04 mm whereas the preloaded right hand bearing (2) will be released 0.04 mm, just to its unloaded free position. Thus the preloaded assembly has increased its bearing stiffness to $\frac{30}{0.04} = 750$ kN/mm, which is twice the stiffness of the individual bearings before they were preloaded.

Fig. 6.26 shows the relationship between axial load and deflection for bearings with and without preload. With the preloaded bearing assembly, the steepness of the straight line (O–a–c–b) for the left hand bearing and shaft is only half of the unpreloaded assembly (O–d–e) and its stiffness is 750 kN/mm (double that of the unpreloaded case). Once the outer right hand bearing has been relieved of all load, the stiffness of the whole bearing assembly reverts back to the nil preload assembly stiffness of 375 kN/mm. The slope now becomes parallel to the without preload deflection load line. The deflection of the bearing assembly for an external axial force of 10 kN when imposed on the pinion shaft in the direction towards the right hand side can be read off the graph vertically between the preload and working load points (a and c) giving a resultant deflection of 0.012 mm.

For the designer to make full use of bearing preload to raise the rigidity of the pinion shaft bearing assembly, the relieving load (outer bearing just unloaded) should exceed the working load (external force) applied to the shaft and bearing assembly.

The technique of reducing bearing axial deflection against an applied end thrust so that the bearing assembly in effect becomes stiffer can be appreciated another way by studying Fig. 6.27.

Suppose a pair of taper bearings (Fig. 6.24) are subjected to a preload of 15 kN. The corresponding axial deflection will be 0.04 mm according to the linear deflection–load relationship shown. If an external axial load of 10 kN is now applied to the pinion shaft so that it pushes the shaft towards the bearings, the load on the left hand bearing (1) will instantaneously increase to 25 kN. The increased deflection of bearing (1) accompanying the increase in load will cause the right hand bearing (2) to lose some of its preload and hence some of its deflection. Simultaneously the change of preload on bearing (2) will influence the load acting on bearing (1) and hence the deflection of this bearing. Once

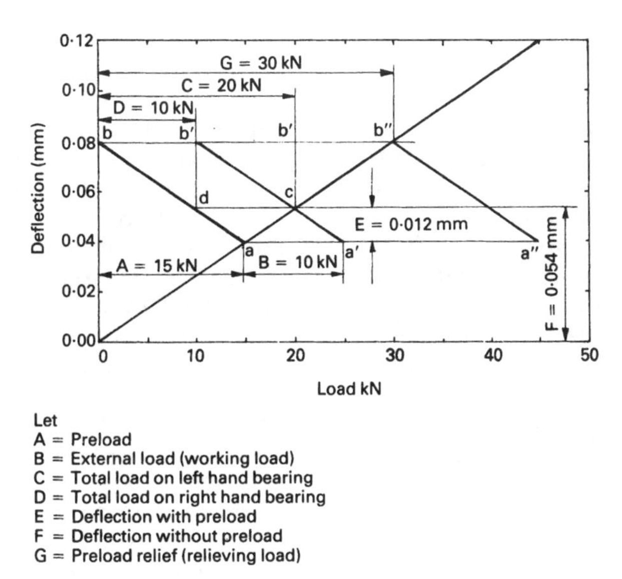

Fig. 6.27 Bearing load–deflection graph using inverted preload curve to obtain working load deflection

equilibrium between the two bearings has been established, the rollers of bearing (1) will support a load of less than 25 kN and those of bearing (2) will carry a load of less than 15 kN.

The distribution of the applied load between the two bearings at equilibrium may be determined by inverting the deflection–load curve from zero to preload deflection, a–b. This represents the reduction in preload of the right hand bearing (2) as the external axial load forces the pinion shaft towards the left hand bearing. The inverted right hand bearing preload curve can be shifted horizontally from its original position a–b where it intersects the full curve at point a to a new position a′–b′, the distance a to a′ being equal to the external load applied to the pinion shaft (working load). Note that point a′ is the instantaneous load on bearing (1) before equilibrium is established. The intersection of the shifted inverted curve with the full curve point c represents the point of equilibrium for bearing (1), the total left hand bearing load. The axial deflection of the pinion shaft under the applied load of 10 kN is thus equal to the vertical reading between point a and c, that is 0.012 mm. The equilibrium point for bearing (2) can be found by drawing a horizontal line from point c to intersect at point d on the original inverted curve, so that point (d) becomes the total right hand bearing load.

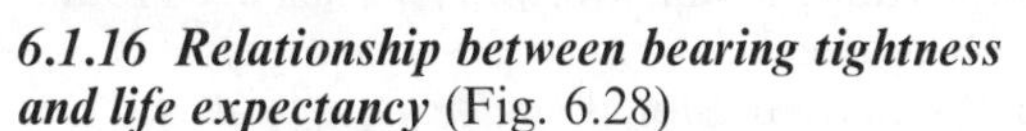

6.1.16 Relationship between bearing tightness and life expectancy (Fig. 6.28)

Taper roller and angular contact ball bearing life is considerably influenced by the slackness or tightness

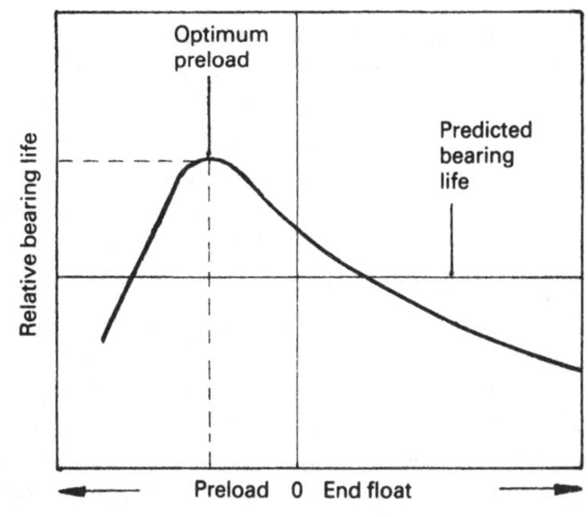

Fig. 6.28 Effect of bearing tightness or slackness on life expectancy

to which the bearings are originally set (Fig. 6.28). The graph shows that if the bearings are heavily preloaded the excessive elastic distortion, and possibly the breakdown in lubrication, will cause the bearings to wear rapidly. Likewise excessive end float causes roller to track misalignment and end to end shock loading with much reduced service life. However it has been found that a small degree of bearing preload which has taken up all the free play when stationary loosens off under working conditions so that the rollers will have light positive contact with their tracks. This results in pure rolling and hence optimum bearing life.

6.2 The need for constant velocity joints

Universal joints are necessary to transmit torque and rotational motion from one shaft to another when their axes do not align but intersect at some point. This means that both shafts are inclined to each other by some angle which under working conditions may be constantly varying.

Universal joints are incorporated as part of a vehicle's transmission drive to enable power to be transferred from a sprung gearbox or final drive to the unsprung axle or road wheel stub shaft.

There are three basic drive applications for the universal joint:

1 propellor shaft end joints between longitudinally front mounted gearbox and rear final drive axle,
2 rear axle drive shaft end joints between the sprung final drive and the unsprung rear wheel stub axle,
3 front axle drive shaft end joints between the sprung front mounted final drive and the unsprung front wheel steered stub axle.

Universal joints used for longitudinally mounted propellor shafts and transverse rear mounted drive shafts have movement only in the vertical plane. The front outer drive shaft universal joint has to cope with movement in both the vertical and horizontal plane; it must accommodate both vertical suspension deflection and the swivel pin angular movement to steer the front road wheels.

The compounding of angular working movement of the outer drive shaft steering joint in two planes imposes abnormally large and varying working angles at the same time as torque is being transmitted to the stub axle. Because of the severe working conditions these joints are subjected to special universal joints known as *constant velocity joints*. These have been designed and developed to eliminate torque and speed fluctuations and to operate reliability with very little noise and wear and to have a long life expectancy.

6.2.1 Hooke's universal joint (Figs 6.29 and 6.30)

The Hooke's universal joint comprises two yoke arm members, each pair of arms being positioned at right angles to the other and linked together by an intermediate cross-pin member known as the spider. When assembled, pairs of cross-pin legs are supported in needle roller caps mounted in each yoke arm, this then permits each yoke member to swing at right angles to the other.

Because pairs of yoke arms from one member are situated in between arms of the other member, there will be four extreme positions for every revolution when the angular movement is taken entirely by only half of the joint. As a result, the spider cross-pins tilt back and forth between these extremes so that if the drive shaft speed is steady throughout every complete turn, the drive shaft will accelerate and decelerate twice during one revolution, the magnitude of speed variation becoming larger as the drive to driven shaft angularity is increased.

Hooke's joint speed fluctuation may be better understood by considering Fig. 6.29. This shows the drive shaft horizontal and the driven shaft inclined downward. At zero degree movement the input yoke cross-pin axis is horizontal when the drive shaft and the output yoke cross-pin axis are vertical. In this position the output shaft is at a minimum. Conversely, when the input shaft has rotated a further 90°, the input and output yokes and cross-pins will be in the vertical and horizontal position respectively. This produces a maximum output shaft speed. A further quarter of a turn will move the joint to an identical position as the initial position so that the output speed will be again at a minimum. Thus it can be seen that the cycle of events repeat themselves every half revolution.

Table 6.2 shows how the magnitude of the speed fluctuation varies with the angularity of the drive to driven shafts.

Table 6.2 Variation of shaft angle with speed fluctuation

Shaft angle (deg)	5	10	15	20	25	30	35	40
% speed fluctuation	0.8	3.0	6.9	12.4	19.7	28.9	40.16	54

The consequences of only having a single Hooke's universal joint in the transmission line can be appreciated if the universal joint is considered as the link between the rotating engine and the vehicle in motion, moving steadily on the road. Imagine the engine's revolving inertia masses rotating at some constant speed and the vehicle itself travelling along uniformly. Any cyclic speed variation caused by the angularity of the input and output shafts will produce a correspondingly periodic driving torque fluctuation. As a result of this torque variation, there will be a tendency to wind and unwind the drive in proportion to the working angle of the joint, thereby imposing severe stresses upon the transmission system. This has been found to produce uneven wear on the driving tyres.

To eliminate torsional shaft cyclic peak stresses and wind-up, universal joints which rotate uniformly during each revolution become a necessity.

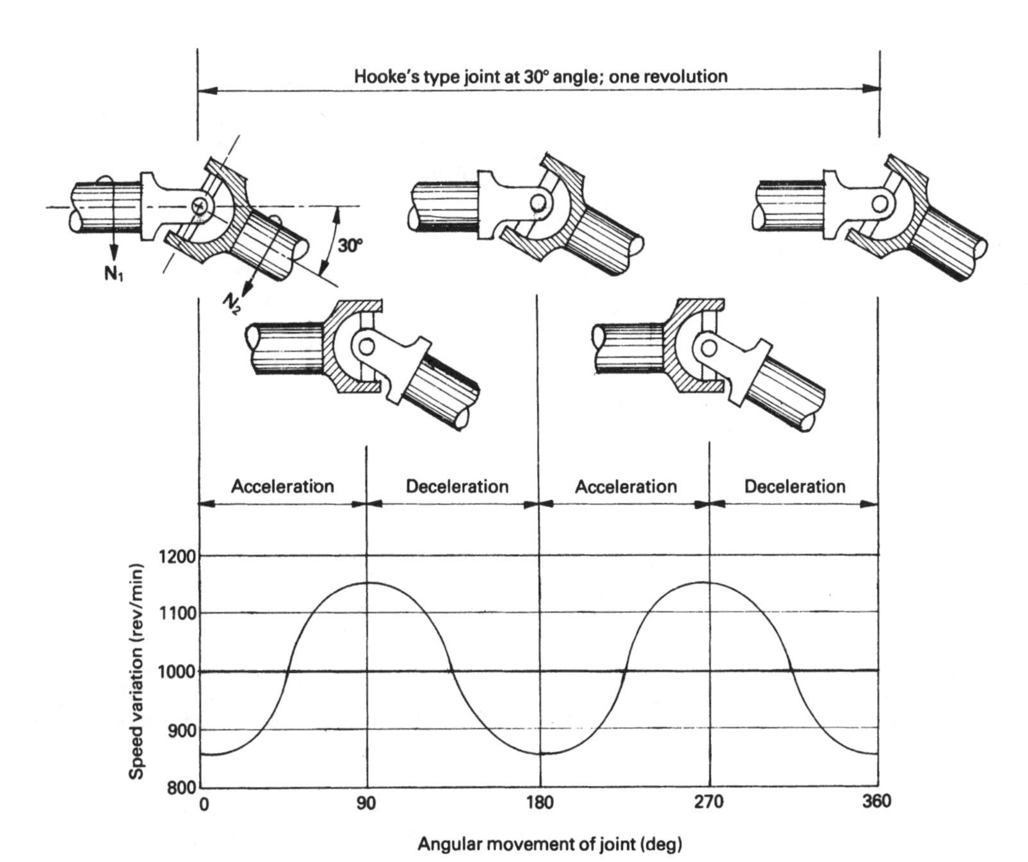

Fig. 6.29 Hooke's joint cycle of speed fluctuation for 30° shaft angularity

6.2.2 Hooke's joint cyclic speed variation due to drive to driven shaft inclination (Fig. 6.30)

Consider the Hooke's joint shown in Fig. 6.30(a) with the input and output yokes in the horizontal and vertical position respectively and the output shaft inclined Θ degrees to the input shaft.

Let ω_i = input shaft angular velocity (rad/sec)
ω_o = output shaft angular velocity (rad/sec)
Θ = shaft inclination (deg)
R = pitch circle joint radius (mm)

Then
Linear velocity of point $(p) = \omega_i y$
and
Linear velocity of point $(p) = \omega_o R$.

Since these velocities are equal,

$$\omega_o R = \omega_i y$$

$$\therefore \omega_o = \omega_i \frac{y}{R}$$

but $\frac{y}{R} = \cos\Theta$.

Thus $\omega_o = \omega_i \cos\Theta$

but $\omega_i = \frac{2\pi}{60} N_i$.

So $\frac{2\pi}{60} N_o = \frac{2\pi}{60} N_i \cos\Theta$

Hence $N_o = N_i \cos\Theta$ (this being a minimum) (1).

If now the joint is rotated a quarter of a revolution (Fig. 6.30(b)) the input and output yoke positions will be vertical and horizontal respectively.

Then
Linear velocity of point $(p) = \omega_o y$
also
Linear velocity of point $(p) = \omega_i R$.

Since these velocities are equal,

$$\omega_o y = \omega_i R$$

$$\omega_o = \omega_i \frac{R}{y}$$

but $\frac{R}{y} = \frac{1}{\cos\Theta}$.

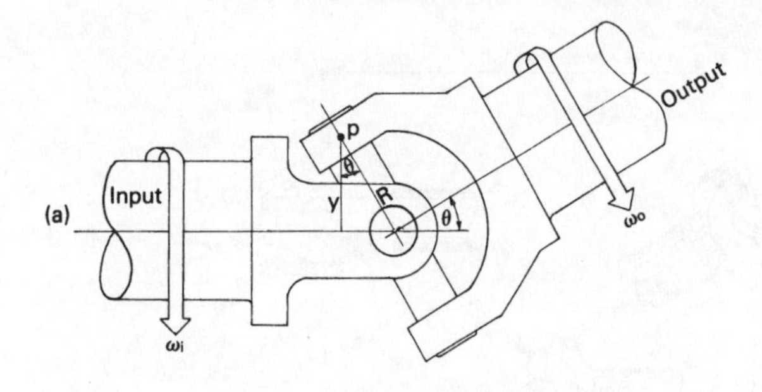

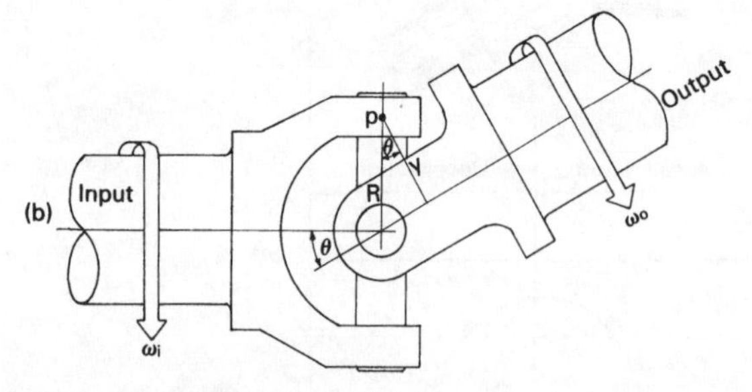

Fig. 6.30 (a and b) Hooke's joint geometry

Thus $\omega_o = \dfrac{\omega_i}{\cos\Theta}$

$$\frac{2\pi}{60}N_o = \frac{2\pi}{60}\frac{N_i}{\cos\Theta}$$

$$N_o = \frac{N_i}{\cos\Theta}\text{(this being a maximum)} \quad (2)$$

Note

1 When $y = R$ the angular instantaneous velocities will be equal.
2 When y is smaller than R, the output instantaneous velocity will be less than the output.
3 When y is larger than R, the output instantaneous velocity will be greater than the input.

Example 1 A Hooke's universal joint connects two shafts which are inclined at 30° to each other. If the driving shaft speed is 500 rev/min, determine the maximum and minimum speeds of the driven shaft.

$$\text{Minimum speed} \quad N_o = N_i \cos 30$$
$$= 500 \times 0.866$$
$$= 433 \text{ (rev/min)}$$

$$\text{Maximum speed} \quad N_o = \frac{N_i}{\cos 30}$$
$$= \frac{500}{0.866}$$
$$= 577 \text{ (rev/min)}$$

Example 2 A Hooke's universal joint connects two shafts which are inclined at some angle. If the input and output joint speeds are 500 and 450 rev/min respectively, find the angle of inclination of the output shaft.

$$N_o = N_i \cos\Theta$$

$$\cos\Theta = \frac{N_o}{N_i}$$

$$\text{Hence} \quad \cos\Theta = \frac{450}{500} = 0.9$$

Therefore $\Theta = 25°50'$

6.2.3 Constant velocity joints

Constant velocity joints imply that when two shafts are inclined at some angle to one another and they are coupled together by some sort of joint, then a uniform input speed transmitted to the output shaft produces the same angular output speed throughout one revolution. There will be no angular acceleration and deceleration as the shafts rotate.

6.2.4 Double Hooke's type constant velocity joint (Figs 6.31 and 6.32)

One approach to achieve very near constant velocity characteristics is obtained by placing two Hooke's type joint yoke members back to back with their yoke arms in line with one another (Fig. 6.31). When assembled, both pairs of outer yoke arms will be at right angles to the arms of the central double yoke member. Treating this double joint combination in two stages, the first stage hinges the drive yoke and driven central double yoke together, whereas the second stage links the central double yoke (now drive member) to the driven final output yoke. Therefore the second stage drive half of the central double yoke is positioned a quarter of a revolution out of phase with the first stage drive yoke (Fig. 6.32).

Consequently when the input and output shafts are inclined to each other and the first stage driven central double yoke is speeding up, the second stage driven output yoke will be slowing down. Conversely when the first stage driven member is reducing speed the second stage driven member increases its speed; the speed lost or gained by one half of the joint will equal that gained or lost by the second half of the joint respectively. There will therefore be no cyclic speed fluctuation between input and output shafts during rotation.

An additional essential feature of this double joint is a centring device (Fig. 6.31) normally of the ball and socket spring loaded type. Its function is to maintain equal angularity of both the input and

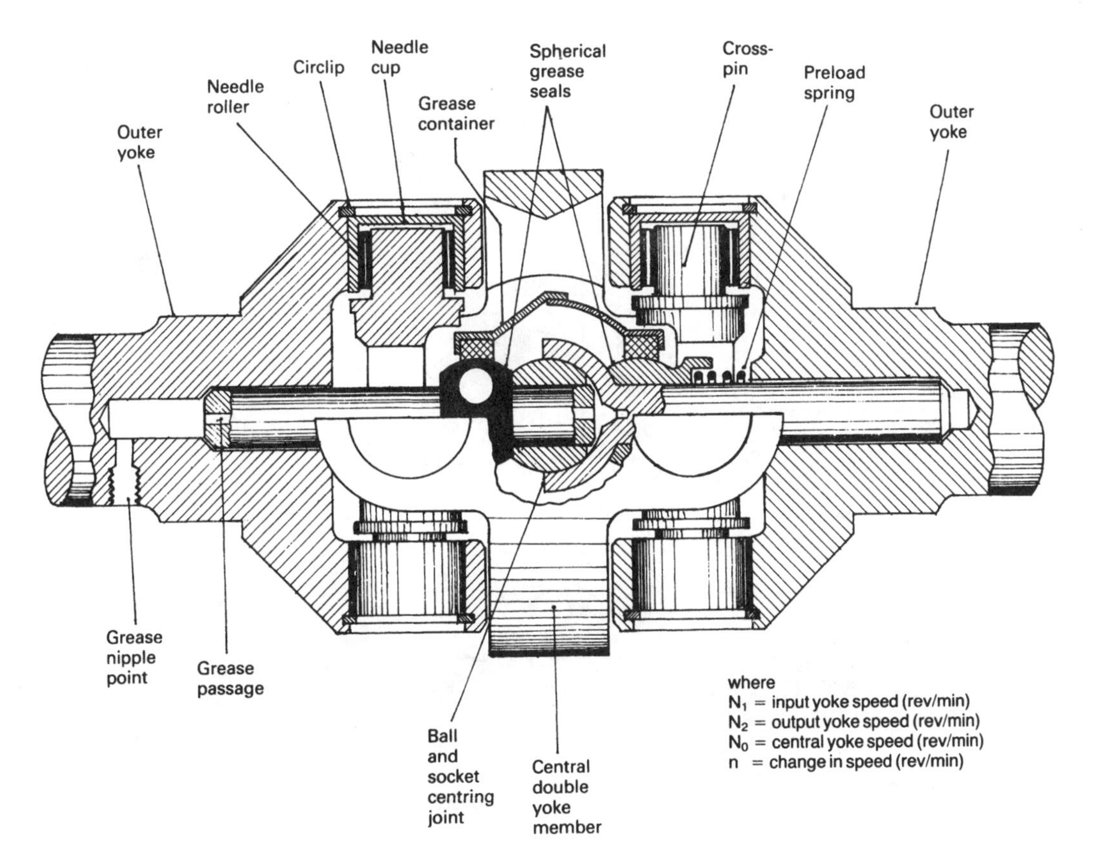

Fig. 6.31 Double Hooke's type constant velocity joint

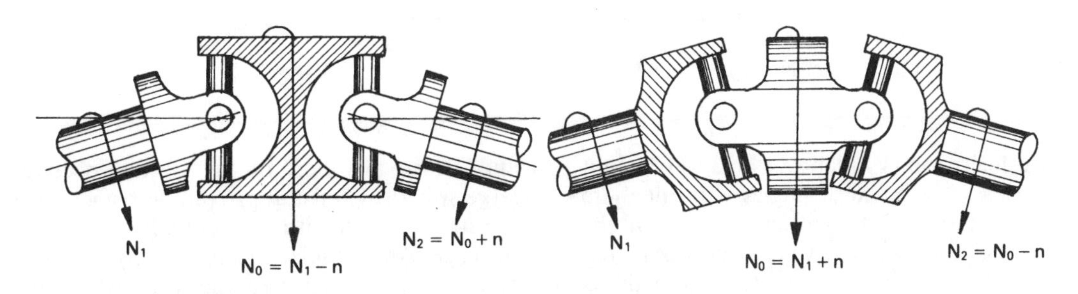

Fig. 6.32 Double Hooke's type joint shown in two positions 90° out of phase

output shafts relative to the central double yoke member. This is a difficult task due to the high end loads imposed on the sliding splined joint of the drive shaft when repeated suspension deflection and large drive torques are being transmitted simultaneously. However, the accuracy of centralizing the double yokes is not critical at the normal relatively low drive shaft speeds.

This double Hooke's joint is particularly suitable for heavy duty rigid front wheel drive live axle vehicles where large lock-to-lock wheel swivel is necessary. A major limitation with this type of joint is its relatively large size for its torque transmitting capacity.

6.2.5 Birfield joint based on the Rzeppa Principle (Fig. 6.33)

Alfred Hans Rzeppa (pronounced *sheppa*), a Ford engineer in 1926, invented one of the first practical

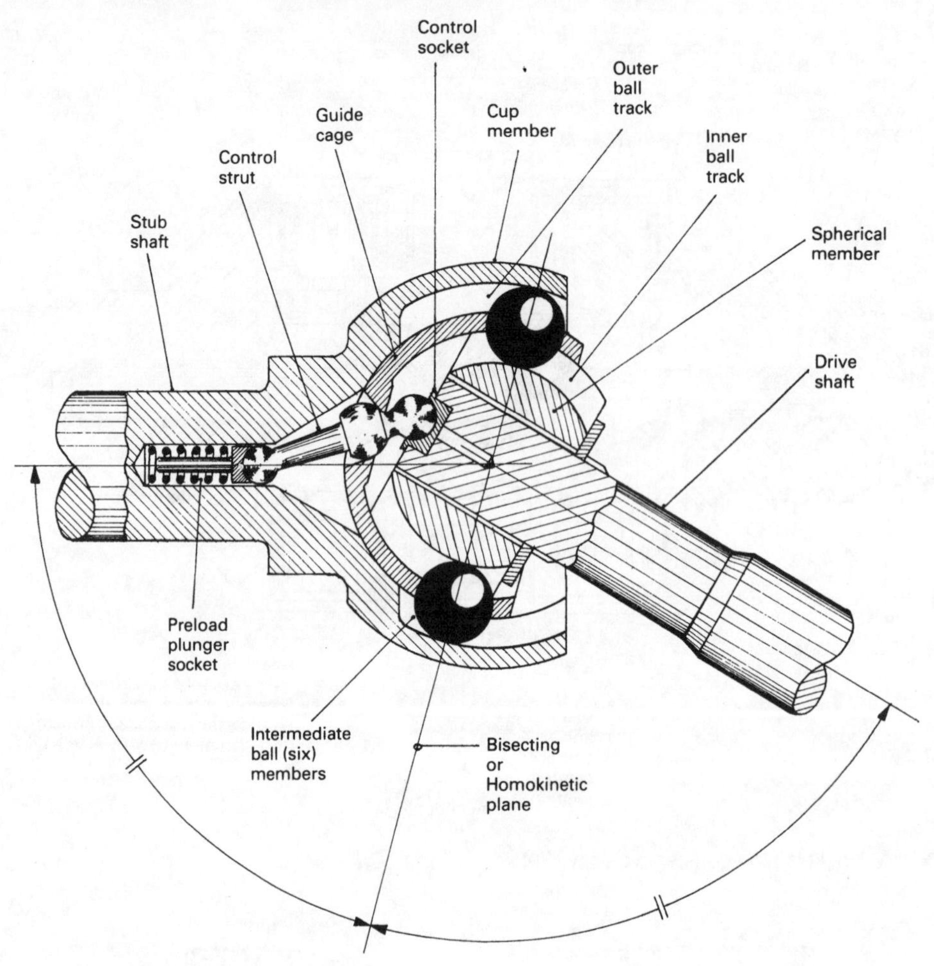

Fig. 6.33 Early Rzeppa constant velocity joint

constant velocity joints which was able to transmit torque over a wide range of angles without there being any variation in the rotary motion of the output shafts. An improved version was patented by Rzeppa in 1935. This joint used six balls as intermediate members which where kept at all times in a plane which bisects the angle between the input and output shafts (Fig. 6.33). This early design of a constant velocity joint incorporated a controlled guide ball cage which maintained the balls in the bisecting plane (referred to as the median plane) by means of a pivoting control strut which swivelled the cage at an angle of exactly half that made between the driving and driven shafts. This control strut was located in the centre of the enclosed end of the outer cup member, both ball ends of the strut being located in a recess and socket formed in the adjacent ends of the driving and driven members of the joint respectively. A large spherical waist approximately midway along the strut aligned with a hole made in the centre of the cage. Any angular inclination of the two shafts at any instant deflected the strut which in turn proportionally swivelled the control ball cage at half the relative angular movement of both shafts. This method of cage control tended to jam and suffered from mechanical wear.

Joint construction (Fig. 6.34) The Birfield joint, based on the Rzeppa principle and manufactured by Hardy Spicer Limited, has further developed and improved the joint's performance by generating converging ball tracks which do not rely on a controlled ball cage to maintain the intermediate ball members on the median plane (Fig. 6.34(b)). This

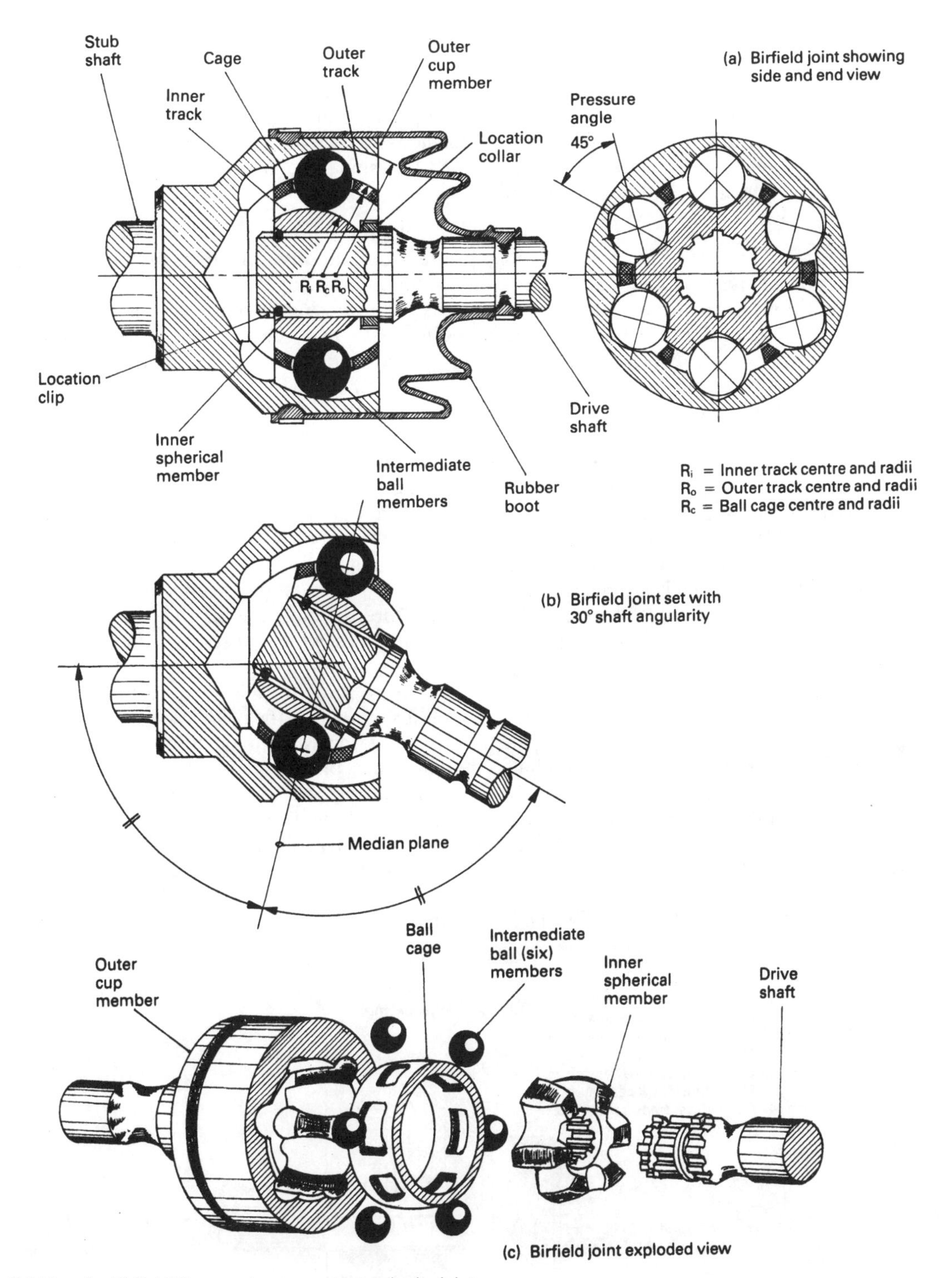

Fig. 6.34 (a–c) Birfield Rzeppa type constant velocity joint

joint has an inner (ball) input member driving an outer (cup) member. Torque is transmitted from the input to the output member again by six intermediate ball members which fit into curved track grooves formed in both the cup and spherical members. Articulation of the joint is made possible by the balls rolling inbetween the inner and outer pairs of curved grooves.

Ball track convergence (Figs 6.34 and 6.35) Constant velocity conditions are achieved if the points of contact of both halves of the joint lie in a plane which bisects the driving and driven shaft angle, this being known as the median plane (Fig. 6.34(b)). These conditions are fulfilled by having an intermediate member formed by a ring of six balls which are kept in the median plane by the shape of the curved ball tracks generated in both the input and output joint members.

To obtain a suitable track curvature in both half, inner and outer members so that a controlled movement of the intermediate balls is achieved, the tracks (grooves) are generated on semicircles. The centres are on either side of the joint's geometric centre by an equal amount (Figs 6.34(a) and 6.35). The outer half cup member of the joint has the centre of the semicircle tracks offset from the centre of the joint along the centre axis towards the open mouth of the cup member, whilst the inner half spherical member has the centre of the semicircle track offset an equal amount in the opposite direction towards the closed end of the joint (Fig. 6.35).

When the inner member is aligned inside the outer one, the six matching pairs of tracks form grooved tunnels in which the balls are sandwiched.

The inner and outer track arc offset centre from the geometric joint centre are so chosen to give an angle of convergence (Fig. 6.35) marginally larger than 11°, which is the minimum amount necessary to positively guide and keep the balls on the median plane over the entire angular inclination movement of the joint.

Track groove profile (Fig. 6.36) The ball tracks in the inner and outer members are not a single semicircle arc having one centre of curvature but instead are slightly elliptical in section, having effectively two centres of curvature (Fig. 6.36). The radius of curvature of the tracks on each side of the ball at the four pressure angle contact points is larger than the ball radius and is so chosen so that track contact occurs well within the arc grooves, so that groove edge overloading is eliminated. At the same time the ball contact load is taken about one third below and above the top and bottom ball tips so that compressive loading of the balls is considerably reduced. The pressure angle will be equal in the inner and outer tracks and therefore the balls are all under pure compression at all times which raises the limiting stress and therefore loading capacity of the balls.

The ratio of track curvature radius to the ball radius, known as the *conformity ratio,* is selected so that a 45° pressure angle point contact is achieved, which has proven to be effective and durable in transmitting the torque from the driving to the driven half members of the joint (Fig. 6.36).

As with any ball drive, there is a certain amount of roll and slide as the balls move under load to and fro along their respective tracks. By having a pressure angle of 45°, the roll to sliding ratio is roughly 4:1.

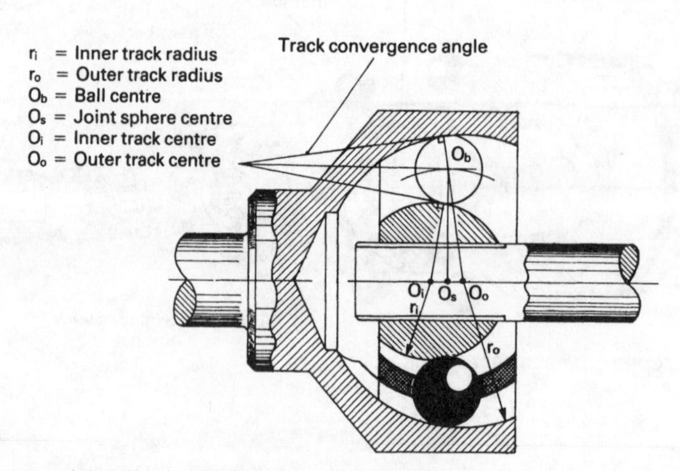

Fig. 6.35 Birfield Rzeppa type joint showing ball track convergence

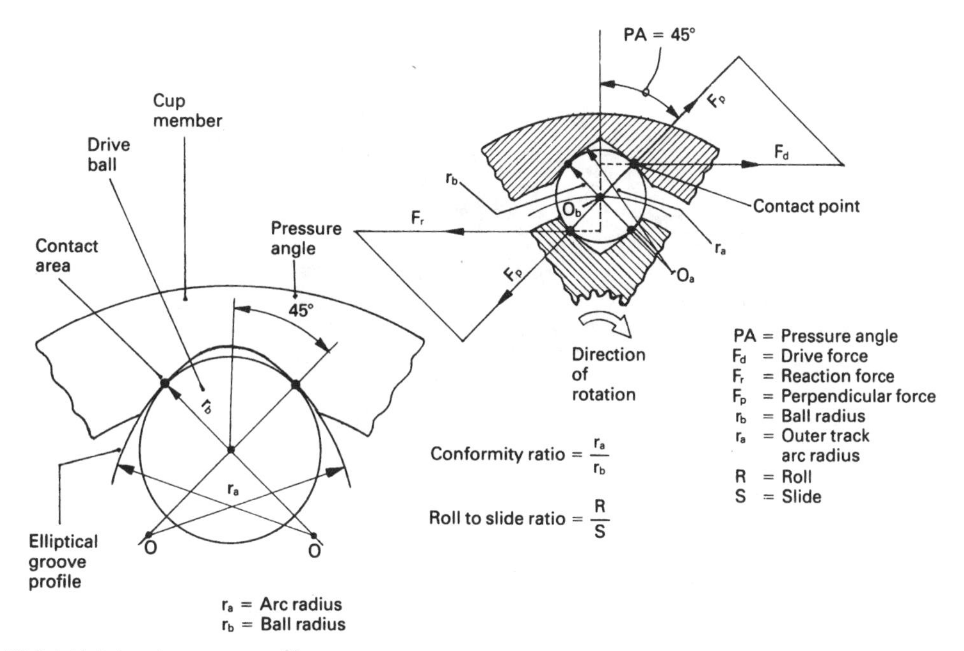

Fig. 6.36 Birfield joint rack groove profile

This is sufficient to minimize the contact friction during any angular movement of the joint.

Ball cage (Fig. 6.34(b and c)) Both the inner drive and outer driven members of the joint have spherical external and internal surfaces respectively. Likewise, the six ball intermediate members of this joint are positioned in their respective tracks by a cage which has the same centre of arc curvature as the input and output members (Fig. 6.34(c)). The cage takes up the space between the spherical surfaces of both male inner and female outer members. It provides the central pivot alignment for the two halves of the joint when the input and output shafts are inclined to each other (Fig. 6.34(b)).

Although the individual balls are theoretically guided by the grooved tracks formed on the surfaces of the inner and outer members, the overall alignment of all the balls on the median plane is provided by the cage. Thus if one ball or more tends not to position itself or themselves on the bisecting plane between the two sets of grooves, the cage will automatically nudge the balls into alignment.

Mechanical efficiency The efficiency of these joints is high, ranging from 100% when the joint working angle is zero to about 95% with a 45° joint working angle. Losses are caused mainly by internal friction between the balls and their respective tracks, which is affected by ball load, speed and working angle and by the viscous drag of the lubricant, the latter being dependent to some extent by the properties of the lubricant chosen.

Fault diagnoses Symptoms of front wheel drive constant velocity joint wear or damage can be narrowed down by turning the steering to full lock and driving round in a circle. If the steering or transmission now shows signs of excessive vibration or clunking and ticking noises can be heard coming from the drive wheels, further investigation of the front wheel joints should be made. Split rubber gaiters protecting the constant velocity joints can considerably shorten the life of a joint due to exposure to the weather and abrasive grit finding its way into the joint mechanism.

6.2.6 Pot type constant velocity joint (Fig. 6.37)
This joint manufactured by both the Birfield and Bendix companies has been designed to provide a solution to the problem of transmitting torque with varying angularity of the shafts at the same time as accommodating axial movement.

There are four basic parts to this joint which make it possible to have both constant velocity characteristics and to provide axial plunge so that the effective drive shaft length is able to vary as the angularity alters (Fig. 6.37):

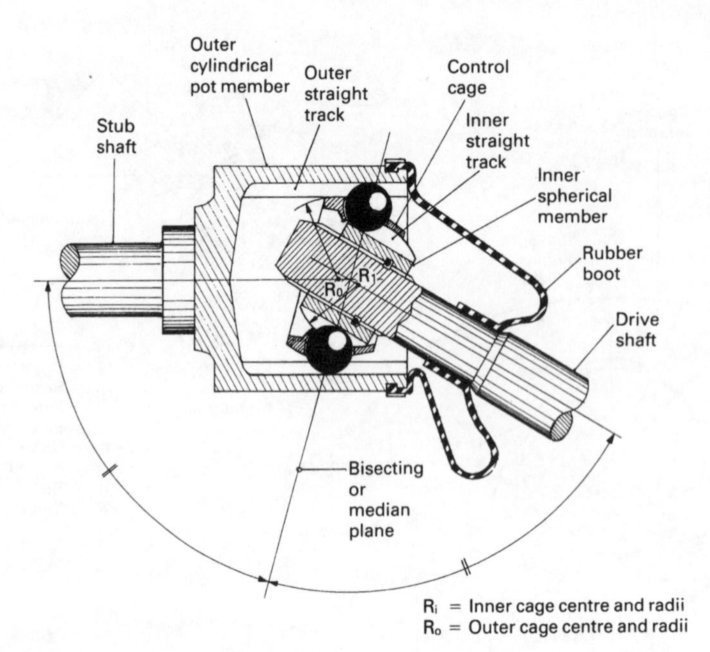

Fig. 6.37 Birfield Rzeppa pot type joint

1 A pot input member which is of cylindrical shape forms an integral part of the final drive stub shaft and inside this pot are ground six parallel ball grooves.
2 A spherical (ball) output member is attached by splines to the drive shaft and ground on the external surface of this sphere are six matching straight tracked ball grooves.
3 Transmitting the drive from the input to the output members are six intermediate balls which are lodged between the internal and external grooves of both pot and sphere.
4 A semispherical steel cage positions the balls on a common plane and acts as the mechanism for automatically bisecting the angle between the drive and driven shafts (Fig. 6.38).

It is claimed that with straight cut internal and external ball grooves and a spherical ball cage which is positioned over the spherical (ball) output member that a truly homokinetic (bisecting) plane is achieved at all times. The joint is designed to have a maximum operating angularity of 22°, 44° including the angle, which makes it suitable for independent suspension inner drive shaft joints.

6.2.7 Carl Weiss constant velocity joint

(Figs 6.38 and 6.40)

A successful constant velocity joint was initially invented by Carl W. Weiss of New York, USA, and was patented in 1925. The Bendix Products Corporation then adopted the Weiss constant velocity principle, developed it and now manufacture this design of joint (Fig. 6.38).

Joint construction and description With this type of time constant velocity joint, double prong (arm) yokes are mounted on the ends of the two shafts transmitting the drive (Fig. 6.37). Ground inside each prong member are four either curved or straight ball track grooves (Fig. 6.39). Each yoke arm of one member is assembled inbetween the prong of the other member and four balls located in adjacent grooved tracks transmit the drive from one yoke member to the other. The intersection of each matching pair of grooves maintains the balls in a bisecting plane created between the two shafts, even when one shaft is inclined to the other (Fig. 6.40). Depending upon application, some joint models have a fifth centralizing ball inbetween the two yokes while the other versions, usually with straight ball tracks, do not have the central ball so that the joint can accommodate a degree of axial plunge, especially if, as is claimed, the balls roll rather than slide.

Carl Weiss constant velocity principle (Fig. 6.41) Consider the geometric construction of the upper half of the joint (Fig. 6.41) with ball track

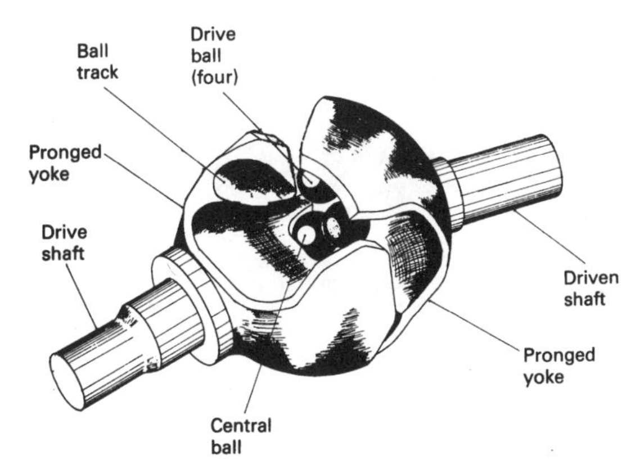

Fig. 6.38 Pictorial view of Bendix Weiss constant velocity type joint

curvatures on the left and right hand yokes to be represented by circular arcs with radii (r) and centres of curvature L and R on their respective shaft axes when both shafts are in line. The centre of the joint is marked by point O and the intersection of both the ball track arc centres occurs at point P. Triangle L O P equals triangle R O P with sides L P and R P being equal to the radius of curvature. The offset of the centres of track curvature from the joint centre are L O and R O, therefore sides L P and R P are also equal. Now, angles L O P and R O P are two right angles and their sum of $90° + 90°$ is equal to the angle L O R, that is 180°, so that point P lies on a perpendicular plane which intersects the centre of the joint. This plane is known as the *median* or *homokinetic plane*.

If the right hand shaft is now swivelled to a working angle its new centre of track curvature will be R′ and the intersection point of both yoke ball track curvatures is now P′ (Fig. 6.41). Therefore triangle L O P′ and R O P′ are equal because both share the same bisecting plane of the left and right hand shafts. Thus it can be seen that sides L P′ and R P′ are also equal to the track radius of curvature r and that the offset of the centres of O R′ and O R are equal to L O. Consequently, angle L O P′ equals angle R′ O P′ and the sum of the angles L O P′ and R′ O P′ equals angle L O R′ of $180 - \Theta$. It therefore follows that angle L O P′ equals angle R′ O P′ which is $(180 - \Theta)/2$. Since P′ bisects the angle made between the left and right hand shaft axes it must lie on the median (homokinetic) plane.

The ball track curvature intersecting point line projected to the centre of the joint will always be half the working angle Θ made between the two shaft axes and fixes the position of the driving balls. The geometry of the intersecting circular arcs therefore constrains the balls at any instant to be in the median (homokinetic) plane.

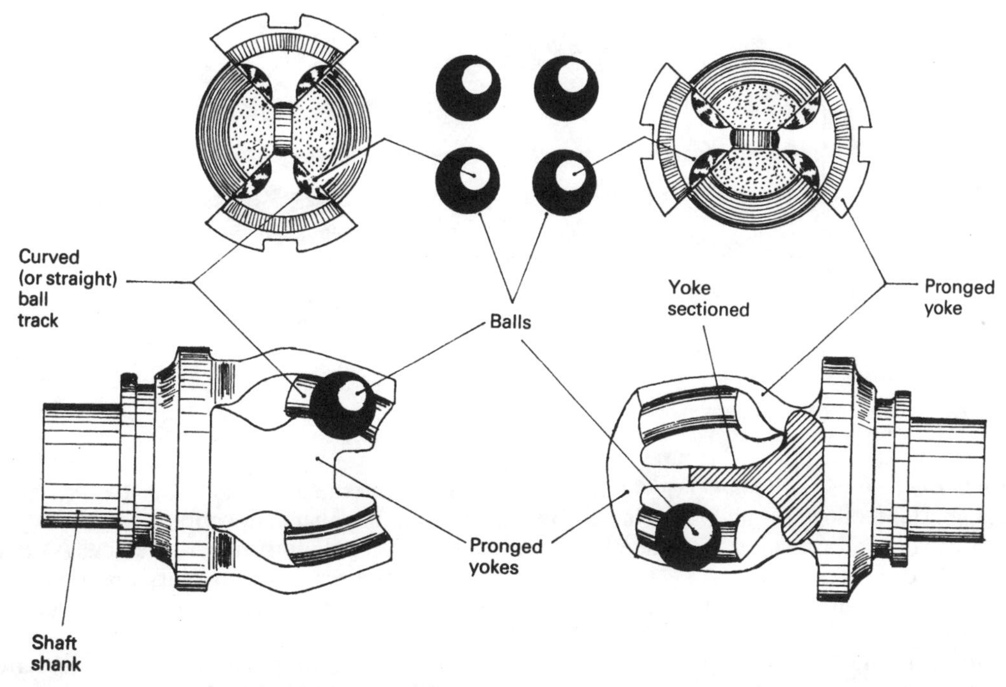

Fig. 6.39 Side and end views of Carl Weiss type joint

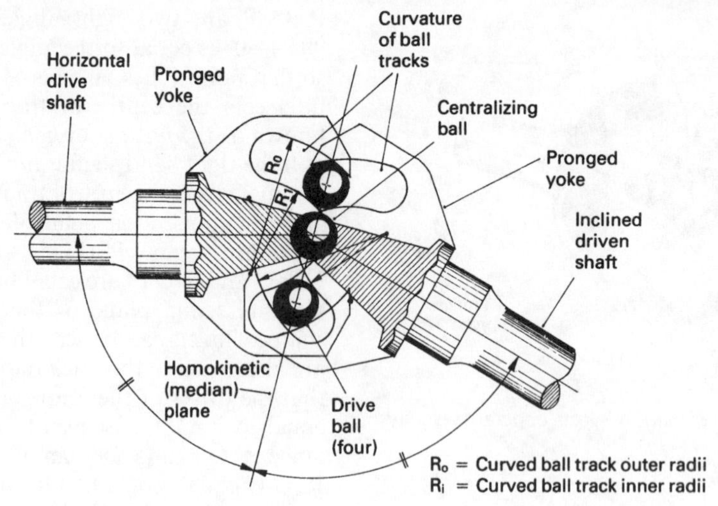

Fig. 6.40 Principle of Bendix Weiss constant velocity type joint

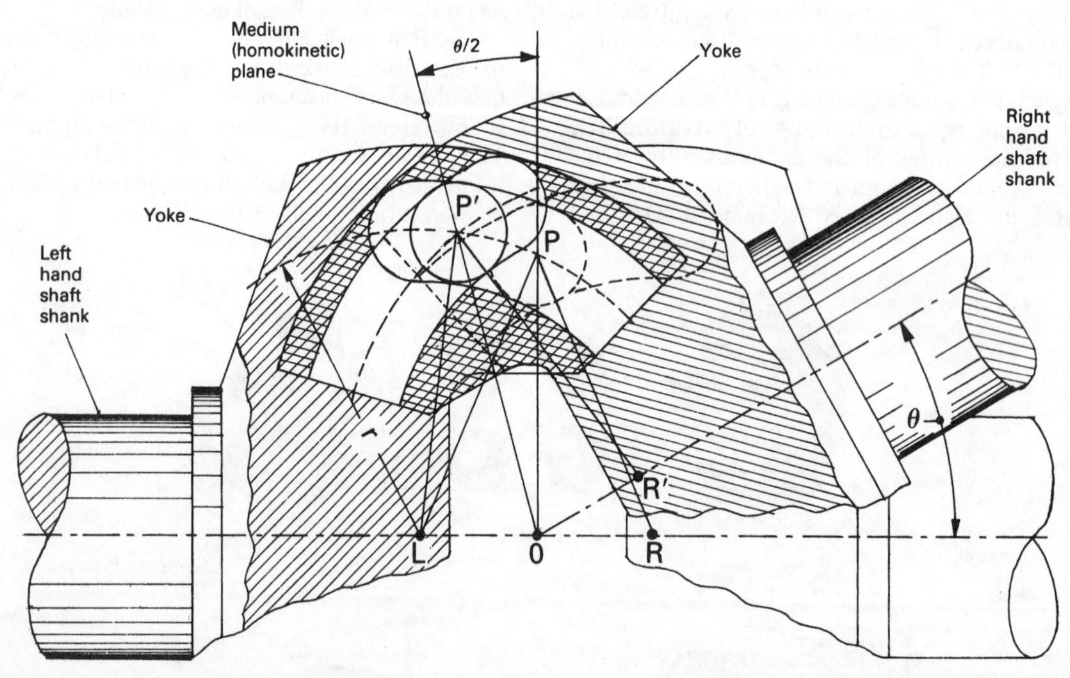

Fig. 6.41 Geometry of Carl Weiss type joint

6.2.8 Tracta constant velocity joint (Fig. 6.42)
The tracta constant velocity joint was invented by Fennille in France and was later manufactured in England by Bendix Ltd.

With this type of joint there are four main components: two outer yoke jaw members and two intermediate semispherical members (Fig. 6.41(a)). Each yoke jaw engages a circular groove machined on the intermediate members. In turn both intermediate members are coupled together by a swivel tongue (spigot joint) and grooved ball (slotted joint).

In some ways these joints are very similar in action to a double Hooke's type constant velocity joint.

Relative motion between the outer jaw yokes and the intermediate spherical members is via the yoke jaw fitting into circular grooves formed in each intermediate member. Relative movement between adjacent intermediate members is provided by a double tongue formed on one member slotting into a second circular groove and cut at right angles to the jaw grooves (Fig. 6.42(b)).

When assembled, both the outer yoke jaws are in alignment, but the central tongue and groove part of the joint will be at right angles to them (Fig. 6.43 (a and b)). If the input and output shafts are inclined at some working angle to each other, the driving intermediate member will accelerate and decelerate during each revolution. Owing to the central tongue and groove joint being a quarter of a revolution out of phase with the yoke jaws, the corresponding speed fluctuation of the driven intermediate and output jaw members exactly counteract and neutralize the input half member's speed variation. Thus the output speed changes will be identical to that of the input drive.

Relative motion between members of this type of joint is not of a rolling nature but one of sliding. Therefore friction losses will be slightly higher than for couplings which incorporate intermediate ball members, but the large flat rubbing surfaces in contact enables large torque loads to be transmitted. The size of these joints are fairly large compared to other types of constant velocity joint arrangements but it is claimed that these joints provide constant velocity rotation at angles up to 50°. A tracta joint incorporated in a rigid front wheel drive axle is shown in Fig. 6.42(c and d).

6.2.9 Tripot universal joint (Fig. 6.43)

Instead of having six or four ball constant velocity joints, a low cost semi-constant velocity joint providing axial movement and having only three bearing contact points has been developed. This joint is used at the inner final drive end of a driving shaft of independent suspension as it not only accommodates continuous variations in shaft working angles, but also longitudinal length changes both caused by road wheel suspension vertical flexing.

One version of the tripot joint incorporates a three legged spider (tripole) mounted on a splined hub which sits on one end of the drive shaft (Fig. 6.43(a and b)). Each of the spider legs supports a semispherical roller mounted on needle bearings. The final drive stub shaft is integral with the pot housing and inside of this pot are ground roller track grooves into which the tripole rollers are lodged.

In operation, the stub shaft and pot transfers the drive via the grooves, rollers and spider to the output drive shaft.

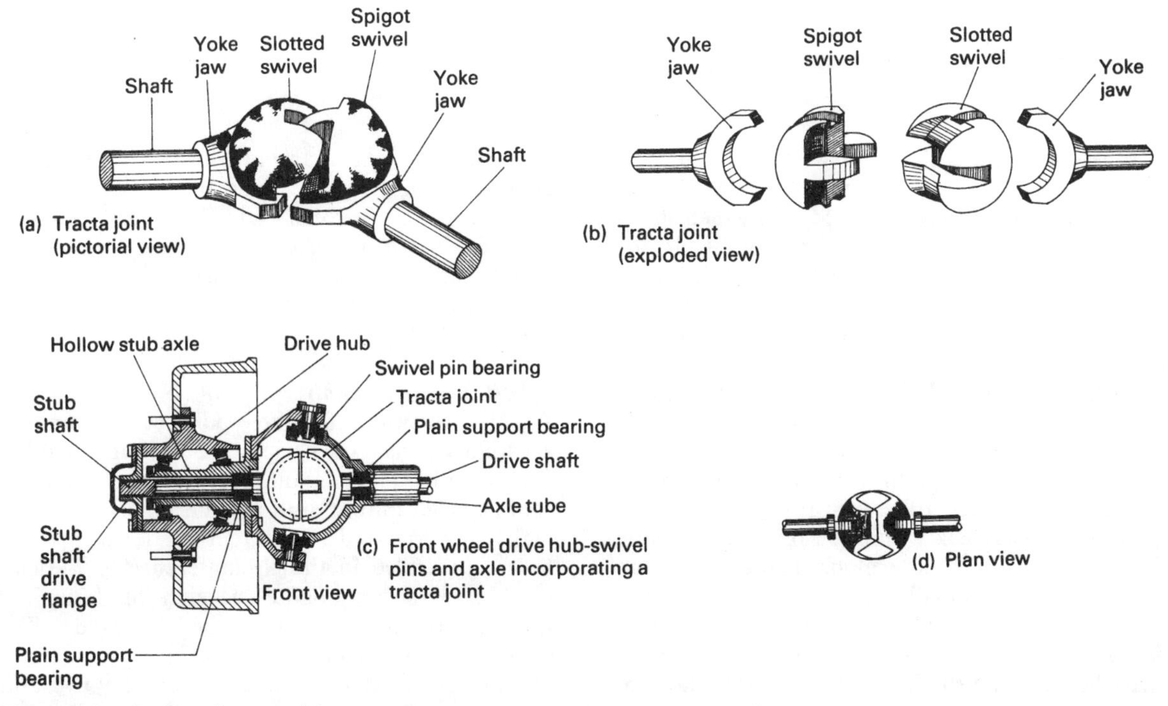

Fig. 6.42 (a–d) Bendix tracta joint

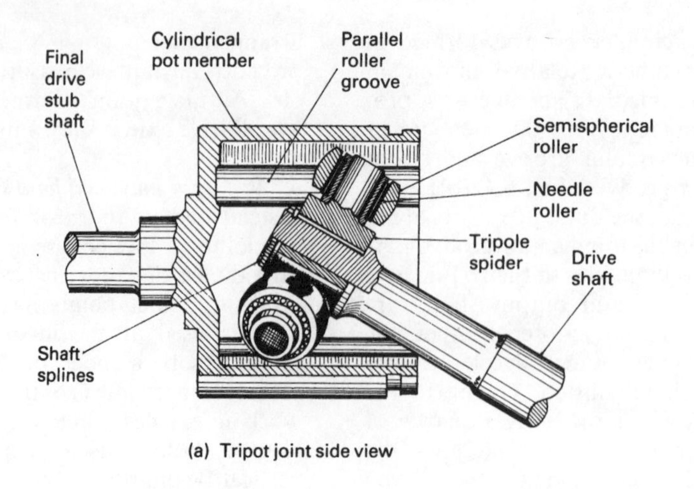

(a) Tripot joint side view

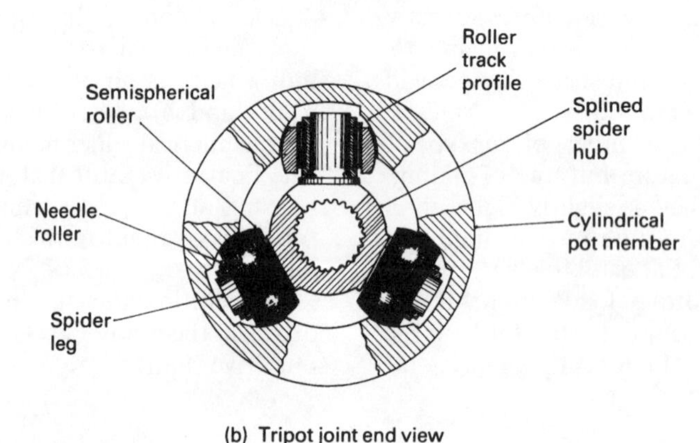

(b) Tripot joint end view

Fig. 6.43 (a and b) Tripot type universal joint

When there is angularity between the final drive stub shaft and drive shaft, the driven shaft and spider will rotate on an inclined axis which intersects the stub shaft axis at some point. If the motion of one roller is followed (Fig. 6.43(a)), it will be seen that when the driven shaft is inclined downwards, when one spider leg is in its lowest position, its rollers will have moved inwards towards the blind end of the pot, but as the spider leg rotates a further 180° and approaches its highest position the roller will have now moved outwards towards the mouth of the pot. Thus as the spider revolves each roller will roll to and fro in its deep groove track within the pot. At the same time that the rollers move along their grooves, the rollers also slide radially back and forth over the needle bearings to take up the extended roller distance from the centre of rotation as the angularity between the shafts becomes greater and vice versa as the angle between the shafts decreases. Because the rollers are attached to the driven shaft through the rigid spider, the point of contact between the three rollers and their corresponding grooves do not produce a plane which bisects the angle between the driving and driven shafts. Therefore this coupling is not a true constant velocity joint.

6.2.10 Tripronged universal joint (Fig. 6.44)
Another version of the three point contact universal joint consists of a triple prong input member (Fig. 6.45(b)) forming an integral part of the drive shaft and an output stub axle cup member inside which a tripole spider is located Fig. 6.44(a). Three holes are drilled in the circumference of the cup member to accommodate the ends of the spider legs, these being rigidly attached by welds (Fig. 6.44(a and c)). Mounted over each leg is a roller spherical ring which is free to both revolve and slide.

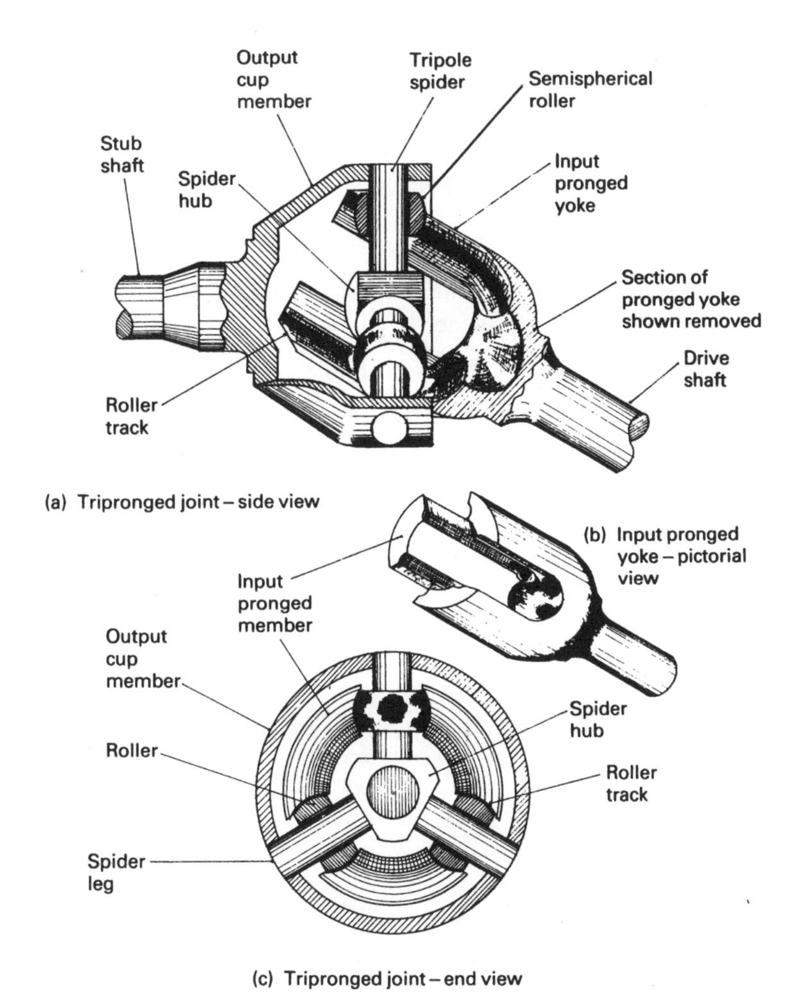

Fig. 6.44 (a–c) Tripronged type universal joint

When assembled, the input member prongs are located in between adjacent spider legs and the roller aligns the drive and driven joint members by lodging them in the grooved tracks machined on each side of the three projecting prongs (Fig. 6.44(c)).

The input driveshaft and pronged member imparts driving torque through the rollers and spider to the output cup and stub axle member. If there is an angle between the drive and driven shafts, then the input drive shaft will swivel according to the angularity of the shafts. Assuming that the drive shaft is inclined downwards (Fig. 6.44(a)), then the prongs in their highest position will have moved furthest out from the engaging roller, but the rollers in their lowest position will be in their deepest position along the supporting tracks of the input member.

As the shaft rotates, each roller supported and restrained by adjacent prong tracks will move radially back and forth along their respective legs to accommodate the orbiting path made by the rollers about the output stub axle axis. Because the distance of each roller from the centre of rotation varies from a maximum to a minimum during one revolution, each spider leg will produce an acceleration and deceleration over the same period.

This type of joint does not provide true constant velocity characteristics with shaft angularity since the roller plane does not exactly bisect the angle made between the drive and driven shaft, but the joint is tolerant to longitudinal plunge of the drive shaft.

7 Final drive transmission

7.1 Crownwheel and pinion axle adjustments

The setting up procedure for the final drive crownwheel and pinion is explained in the following sequence:

1 Remove differential assembly with shim preloaded bearings.
2 Set pinion depth.
3 Adjust pinion bearing preload.
   a) Set pinion bearing preloading using spacer shims.
   b) Set pinion bearing preloading using collapsible spacer.
4 Adjust crownwheel and pinion backlash and differential bearing preload.
   a) Set differential cage bearing preload using shims.
   b) Set crownwheel and pinion backlash using shims.
   c) Set crownwheel and pinion and preloading differential bearing using adjusting nuts.
5 Check crownwheel and pinion tooth contact.

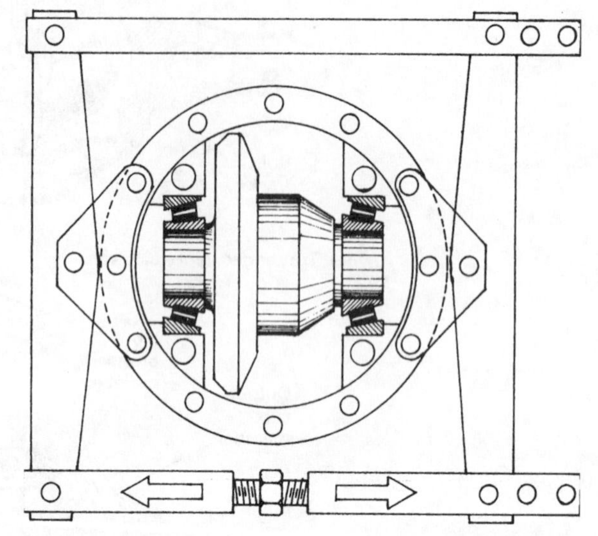

Fig. 7.1 Stretching axle housing to remove differential cage assembly

7.1.1 Removing differential assembly with shim preloaded bearings (Fig. 7.1)

Before removing the differential assembly from the final drive housing, the housing must be expanded to relieve the differential cage bearing preload. Spreading the housing is achieved by assembling the housing stretcher plates (Fig. 7.1) to the housing, taking up the turnbuckle slack until it is hand tight and tightening the turnbuckle with a spanner by three to four flats of the hexagonal until the differential cage bearing end thrust is removed. Never stretch the housing more than 0.2 mm, otherwise the distortion may become permanent. The differential cage assembly can then be withdrawn by levering out the unit.

7.1.2 Setting pinion depth (Fig. 7.2)

Press the inner and outer pinion bearing cups into the differential housing and then lubricate both bearings. Slip the standard pinion head spacer (thick shim washer) and the larger inner bearing over the dummy pinion and align assembly into the pinion housing (Fig. 7.2). Slide the other bearing and centralizing cone handle over the pinion shank, then screw on the preloading sleeve. Hold the handle of the dummy pinion while winding round the preload sleeve nut until the sleeve is screwed down to the first mark for re-used bearings or second for new bearings. Rotate dummy pinion several times to ensure bearings seat properly. Check the bearing preload by placing a preload gauge over the preload sleeve nut and read off the torque required to rotate the dummy pinion. (A typical preload torque would be 2.0–2.4 Nm.)

Place the stepped gauge block and dial indicator magnetic stand onto the surface plate then swing the indicator spindle over selected gauge step and zero indicator gauge.

Clean the driving pinion head and place the magnetic dial gauge stand on top of the pinion head. Move the indicator arm until the spindle of the gauge rests on the centre of one of the differential bearing housing bores. Slightly swing the gauge across the bearing housing bore until the minimum reading at the bottom of the bore is obtained. Repeat the check for the opposite bearing bore. Add the two readings together and divide by two to obtain a mean reading. This is the *pinion cone distance correction factor*.

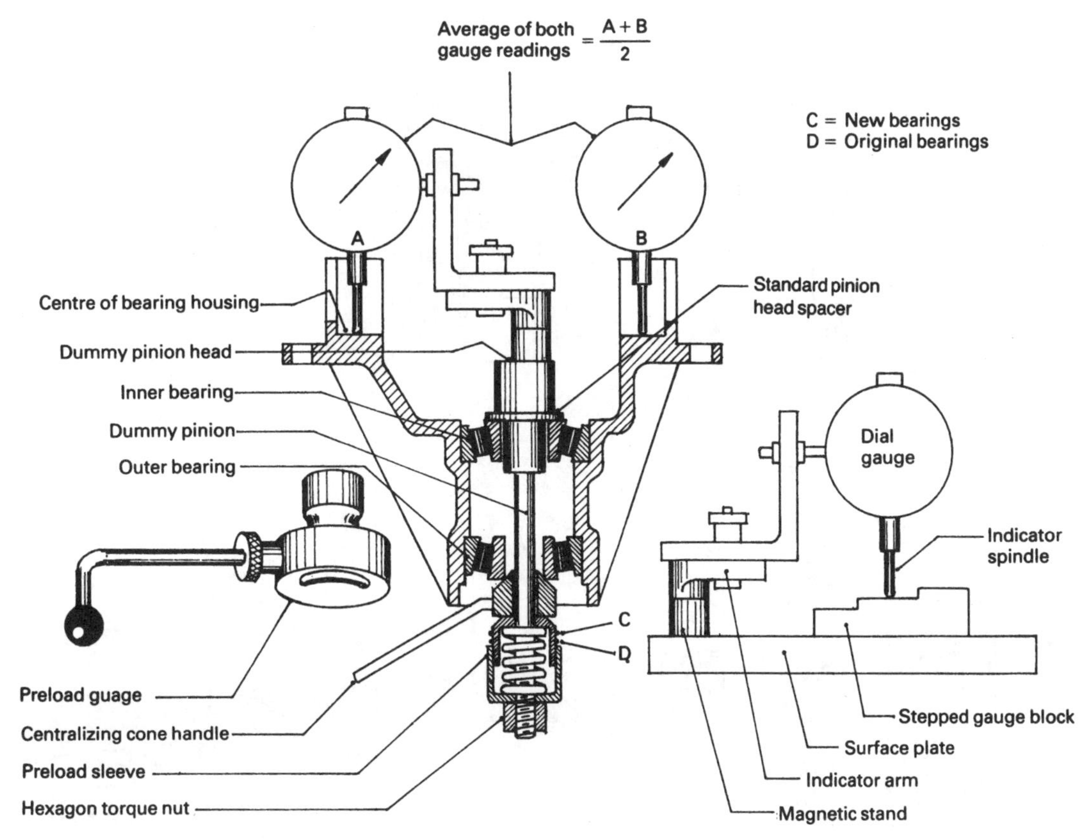

Fig. 7.2 Setting pinion depth dummy pinion

Etched on the pinion head is either the letter N (normal) or a number with either a positive or negative sign in front which provides a correction factor for deviations from the normal size within the production tolerance for the pinion cone distance.

If the etched marking on the pinion face is N (normal), there should be no change in pinion head washer thickness.

If the etched marking on the pinion face is positve (+) (pinion head height oversize), reduce the size of the required pinion head washer by the amount marked.

If the etched marking on the pinion face is negative (−) (pinion head height undersize), increase size of the required pinion head washer by the amount marked.

The numbers range between 5 and 30 (units are hundredths of millimetres). So, +20 means subtracting 20/100 mm, i.e. 0.2 mm subtracted from pinion head washer thickness, or −5 means adding 5/100 mm, i.e. 0.05 mm added to pinion head washer thickness.

Calculating pinion head washer thickness For example,

Average clock bearing bore reading	=	0.05 mm
Pinion head standard washer thickness	=	1.99 mm
Pinion cone distance correction factor	=	0.12 mm
Required pinion head washer thickness	=	2.12 mm

7.1.3 Adjusting pinion bearing preload

Setting pinion bearing preload using spacer shims (Fig. 7.3(a)) Slip the correct pinion head washer over pinion shank and then press on the inner bearing cone. Oil the bearing and fit the pinion assembly into the housing. Slide on the bearing spacer with the small end towards the drive flange. Fit the old preload shim next to the spacer, oil and fit outer

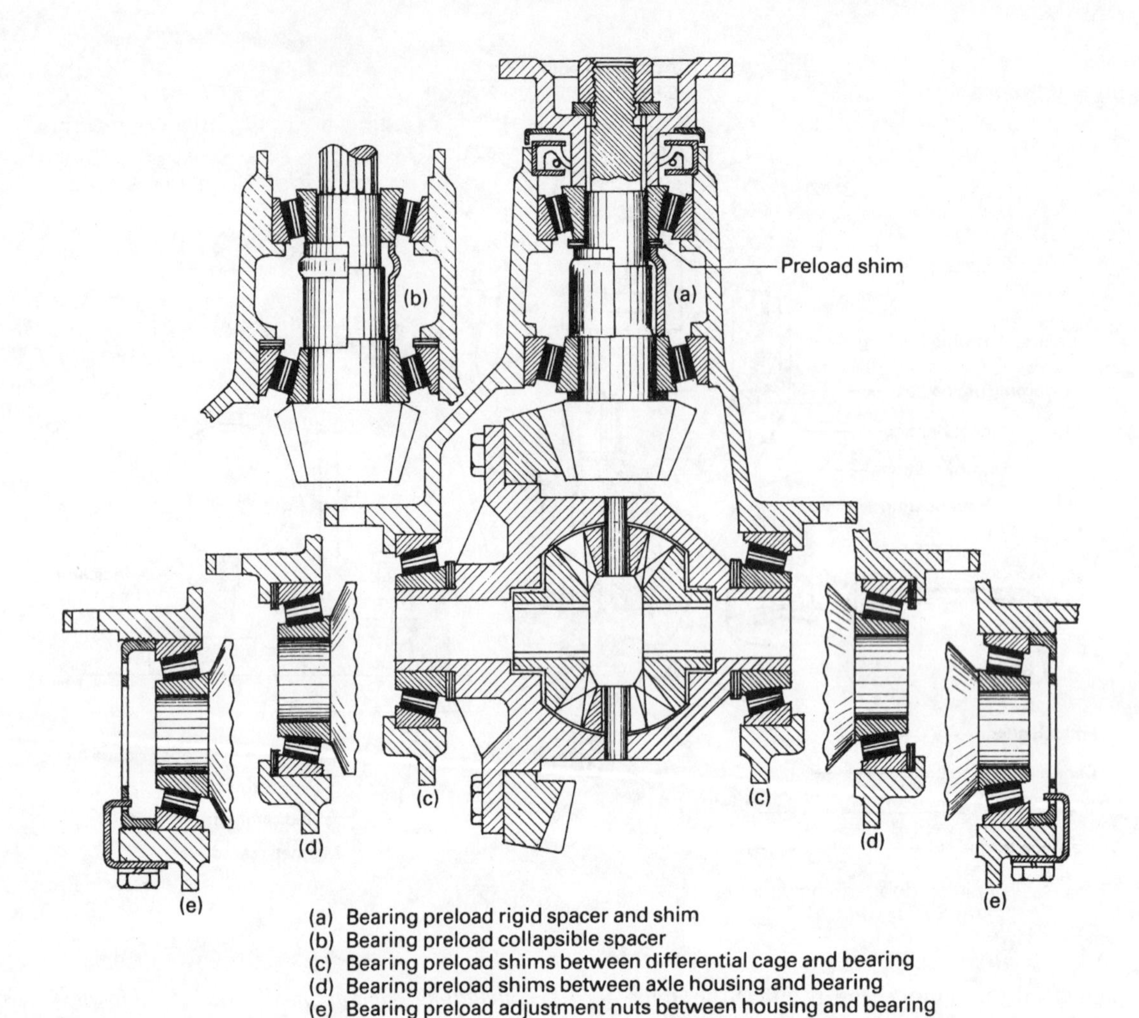

Fig. 7.3 (a–e) Crownwheel and pinion adjustment methods

bearing to pinion shank. Assemble the pinion drive flange, washer and nut (Fig. 7.3(a)).

Using a torque wrench gradually tighten the nut to the correct torque (about 100–130 Nm). Rotate the pinion several times so that the bearings settle to their running conditions and then check the preload resistance using the preload gauge attached to the pinion nut or drive flange. Typical bearing preload torque ranges 15–25 Nm. If necessary, increase or decrease the spacer shim thickness to keep within the specified preload.

If the preload is high, increase spacer shim thickness. Alternatively if the preload is low, decrease the shim thickness. Note that 0.05 mm shim thickness is approximately equivalent to 0.9 Nm pinion preload torque.

To alter pinion preload, remove pinion nut, flange, washer and pinion outer bearing. If preloading is high, add to the original spacer shim thickness, but if preload is too low, remove original shim and fit a thinner one.

Once the correct pinion preload has been obtained, remove pinion nut, washer and drive flange. Fit a new oil seal and finally reassemble. Retighten drive flange nut to the fully tight setting (i.e. 120 Nm) if a castlenut is used instead of a self-locking nut fit split pin.

Setting pinion bearing preload using collapsible spacer (Fig. 7.3(b)) Fit the selected pinion head spacer washer to the pinion and press on inner pinion bearing cone. Press both pinion bearing cones into housing. Fit the outer bearing cone to its cup in the pinion housing and locate a new oil seal in the housing throat with the lip towards the bearing and press it in until it contacts the inner shoulder. Lightly oil the seal.

Install the pinion into the final drive housing with a new collapsible spacer (Fig. 7.3(b)). Fit the drive flange and a new retainer nut. Tighten the nut until a slight end float can be felt on the pinion.

Attach the pinion preload gauge to the drive flange and measure the oil seal drag (usually around 0.6 Nm). To this oil sealed preload drag add the bearing preload torque of 2.2.–3.0 Nm.

i.e. Total preload = Oil seal drag + Bearing drag

$$= 0.6 + 2.5 = 3.1\ \text{Nm}$$

Gradually and carefully tighten the drive flange nut, twisting the pinion to seat the bearings, until the required preload is obtained. Frequent checks must be taken with the preload gauge and if the maximum preload is exceeded the collapsible spacer must be renewed. Note that slackening off the drive flange nut will only remove the established excessive preload and will not reset the required preload.

7.1.4 Adjust crownwheel and pinion backlash and differential bearing preload

Setting differential cage bearing preload using shims (Figs 7.3(c and d) and 7.4) Differential bearing preload shims may be situated between the differential cage and bearings (Fig. 7.3(c)) or between axle housing and bearings (Fig. 7.3(d)). The method of setting the differential bearing preload is similar in both arrangements, but only the case of shims between the axle housing and bearing will be described.

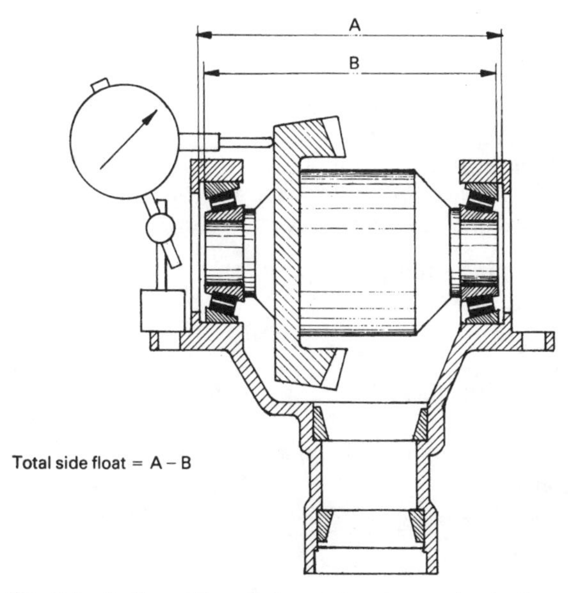

Fig. 7.4 Setting differential cage bearing preload using shims

With the pinion removed, press both differential bearing cones onto the differential cage and slip the bearing cups over rollers and cones. Lower the differential and crownwheel assembly with bearing cups but without shims into the final drive axle housing. Install the dial indicator on the final drive housing with the spindle resting against the back face of the crownwheel. Insert two levers between the final drive housing and the differential cage assembly, fully moving it to one side of the housing. Set the indicator to zero and then move the assembly to the other side and record the reading, which will give the total side float between the bearings as now assembled and the abutment faces of the final drive housing. A preload shim thickness is then added to the side float between the differential bearings and final drive housing. This normally amounts to a shim thickness of 0.06 mm added to both sides of the differential. The total shim thickness required between the differential bearings and final drive housing can then be divided according to the crownwheel and pinion backlash requirements as under setting backlash with shims.

Calculating total shim thickness for differential bearings For example,

Differential side float	=	1.64 mm
Differential bearing preload allowance (2 × 0.06)	=	0.12 mm
Total differential bearing pack thickness	=	1.76 mm

Setting crownwheel and pinion backlash using shims (Figs 7.5 and 7.6) After the pinion depth has been set, place the differential assembly with the bearing cups but without shims into the final drive housing, being sure that all surfaces are absolutely clean. Install a dial indicator on the housing with spindle resting against the back of the crownwheel (Fig. 7.5). Insert two levers between the housing and the crownwheel side of the differential assembly. Move the differential away from the pinion until the opposite bearing cup is seated against the housing. Set the dial indicator to zero. The levers are now transferred to the opposite side of the differential cage so that the whole unit can now be pushed

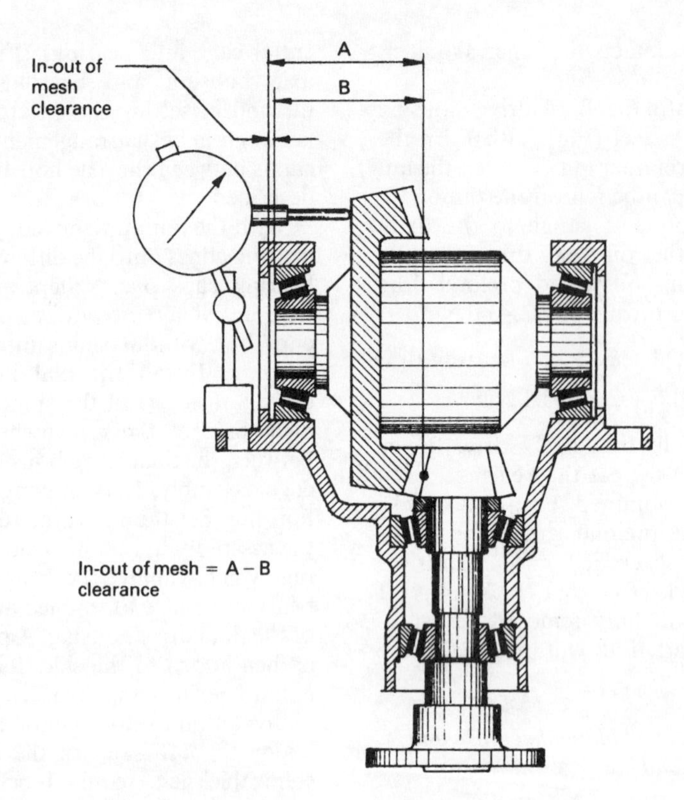

Fig. 7.5 Setting crownwheel and pinion backlash using shims

towards the pinion until both crownwheel and pinion teeth fully mesh. Observe the dial indicator reading, which is the *in-out* of mesh clearance between the crownwheel and pinion teeth (shims removed). This denotes the thickness of shims minus the backlash allowance to be placed between the final drive housing and the bearing cone on the crownwheel side of the differential cage to obtain the correct backlash.

Backlash allowance is either etched on the crownwheel or it may be assumed a movement of 0.05 mm shim thickness from one differential bearing to the other will vary the backlash by approximately 0.05 mm.

Example From the following data determine the shim pack thickness to be placed on both sides of the crownwheel and differential assembly between the bearing and axle housing.

Differential side float with shims removed	= 1.64 mm
Differential bearing preload allowance each side	= 0.06 mm
In-out of mesh clearance	= 0.62 mm
Backlash allowance	= 0.12 mm

In-out of mesh clearance	= 0.62 mm
Differential bearing preload allowance (add)	= 0.06 mm
Backlash allowance (subtract)	= 0.12 mm
Required shim pack crownwheel side	= 0.56 mm

Total differential side float	= 1.64 mm
Differential bearing preload allowance (add)	= 0.06 mm
Crownwheel side shim pack without preload and allowance (subtract)	= 0.50 mm
Required shim pack opposite crownwheel side	= 1.20 mm

Alternatively,

Required shim pack opposite crownwheel side = Total differential bearing pack thickness – Shim pack crownwheel side

= (1.64 + 0.12) – 0.56
= 1.76 – 0.56
= 1.20 mm

To check the crownwheel and pinion backlash, attach the dial gauge magnetic stand on the axle housing flange with the dial gauge spindle resting against one of the crownwheel teeth so that some sort of gauge reading is obtained (Fig. 7.6). Hold the pinion stationary and rock the crownwheel backwards and forwards observing the variation in gauge reading, this valve being the backlash between the crownwheel and pinion teeth. A typical backlash will range between 0.10 and 0.125 mm for original bearings or 0.20 and 0.25 mm for new bearings.

Setting crownwheel and pinion backlash and preloading differential bearings using adjusting nuts (Figs 7.6 and 7.7) Locate the differential bearing caps on their cones and position the differential assembly in the final drive housing. Refit the bearing caps with the mating marks aligned and replace the bolts so that they just nip the caps in position. Screw in the adjusting nuts whilst rotating the crownwheel until there is just a slight backlash. Bolt the spread gauge to the centre bolt hole of the bearing cap and fit an inverted bearing cap lock tab to the other cap (Fig. 7.7). Ensure that the dial gauge spindle rests against the lock tab and set the gauge to zero. Mount the backlash gauge magnetic base stand on the final drive housing flange so that the dial spindle rests against a tooth at right angles to it and zero the gauge (Fig. 7.6). Screw in the adjusting nuts until a backlash of 0.025 to 0.05 mm is indicated when rocking the crownwheel. Swing the backlash gauge out of position.

Screw in the adjusting nut on the differential side whilst rotating the crownwheel until a constant cap spread (preload) of 0.20–0.25 mm is indicated for new bearings, or 0.10–0.125 mm when re-using the original bearings.

Swing the backlash gauge back into position and zero the gauge. Hold the pinion and rock the crownwheel. The backlash should now be 0.20–0.25 mm for new bearings or 0.10–0.125 mm with the original bearings.

If the backlash is outside these limits, adjust the position of the crownwheel relative to the pinion by slackening the adjusting nut on one side and tightening the nut on the other side so that the cap spread remains unaltered. The final tightening must always be made to the nut on the crownwheel side.

Refit the lock tabs left and right hand and torque down the cap bolts; a typical value for a car axle is 60–70 Nm.

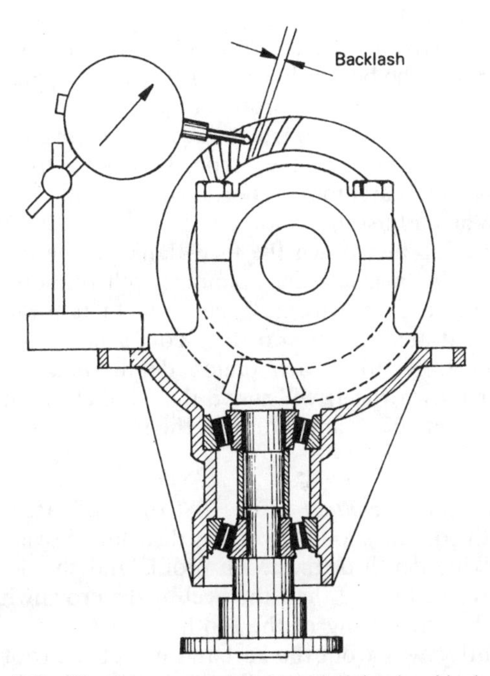

Fig. 7.6 Checking crownwheel and pinion backlash

7.1.5 Gear tooth terminology (Fig. 7.8(a))

Pitch line The halfway point on the tooth profile between the face and the flank is called the *pitch line*.

Tooth root The bottom of the tooth profile is known as the *tooth root*.

Tooth face The upper half position of the tooth profile between the pitch line and the tooth tip is called the *tooth face*.

Tooth flank The lower half position of the tooth profile between the pitch line and the tooth root is called the *tooth flank*.

Tooth heel The outer large half portion of the crownwheel tooth length is known as the *heel*.

Tooth toe The inner half portion of the crownwheel tooth length is known as the *toe*.

Drive side of crownwheel tooth This is the convex side of each crownwheel tooth wheel which receives the contact pressure from the pinion teeth when the engine drives the vehicle forward.

Coast side of crownwheel tooth This is the concave side of each crownwheel tooth which contacts the

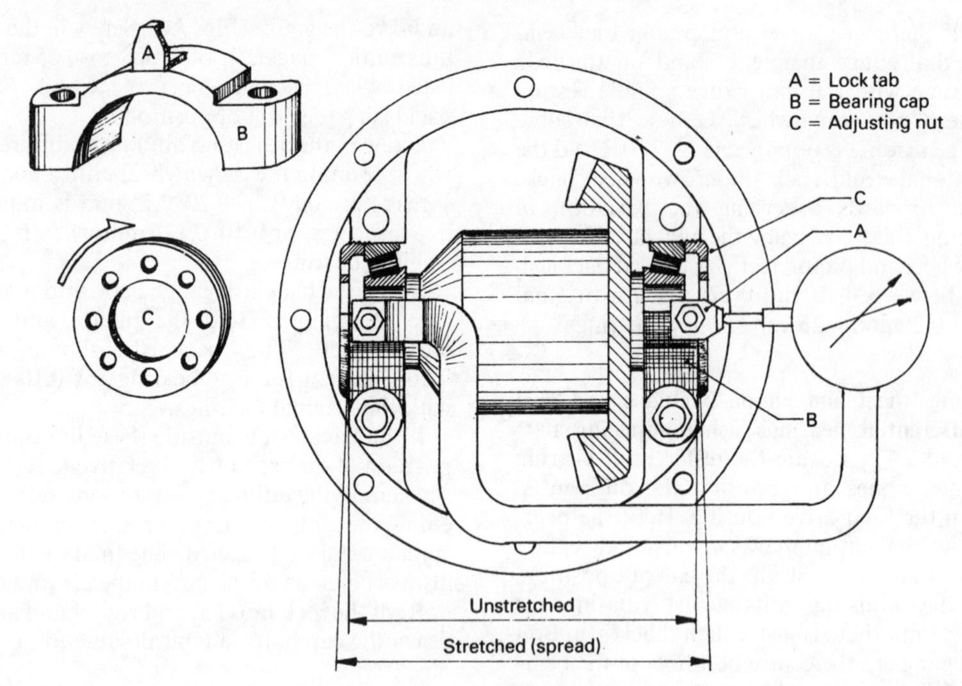

Fig. 7.7 Setting differential cage bearing preload using adjusting nuts

pinion teeth when the transmission overruns the engine or the vehicle is being reversed.

Crownwheel and pinion backlash The free clearance between meshing teeth is known as *backlash*.

7.1.6 Checking crownwheel and pinion tooth contact

Prepare crownwheel for examining tooth contact marks (Fig. 7.8) After setting the correct backlash, the crownwheel and pinion tooth alignment should be checked for optimum contact. This may be achieved by applying a marking cream such as Prussian blue, red lead, chrome yellow, red or yellow ochre etc. to three evenly spaced groups of about six teeth round the crownwheel on both drive coast sides of the teeth profiles. Apply a load to the meshing gears by holding the crownwheel and allowing it to slip round while the pinion is turned a few revolutions in both directions to secure a good impression around the crownwheel. Examine the tooth contact pattern and compare it to the recommended impression.

Understanding tooth contact marks (Fig. 7.8(a–f)) If the crownwheel to pinion tooth contact pattern is incorrect, there are two adjustments that can be made to change the position of tooth contact. These adjustments are of backlash and pinion depth.

The adjustment of backlash moves the contact patch lengthwise back and forth between the toe heel of the tooth. Moving the crownwheel nearer the pinion decreases the backlash, causing the contact patch to shift towards the toe portion of the tooth. Increasing backlash requires the crownwheel to be moved sideways and away from the pinion. This moves the contact patch nearer the heel portion of the tooth.

When adjusting pinion depth, the contact patch moves up and down the face–flank profile of the tooth. With insufficient pinion depth (pinion too far out from crownwheel) the contact patch will be concentrated at the top (face zone) of the tooth. Conversely, too much pinion depth (pinion too near crownwheel) will move the contact patch to the lower root (flank zone) of the tooth.

Ideal tooth contact (Fig. 7.8(b)) The area of tooth contact should be evenly distributed over the working depth of the tooth profile and should be nearer to the toe than the heel of the crownwheel tooth. The setting of the tooth contact is initially slightly away from the heel and nearer the root to compensate for any deflection of the bearings,

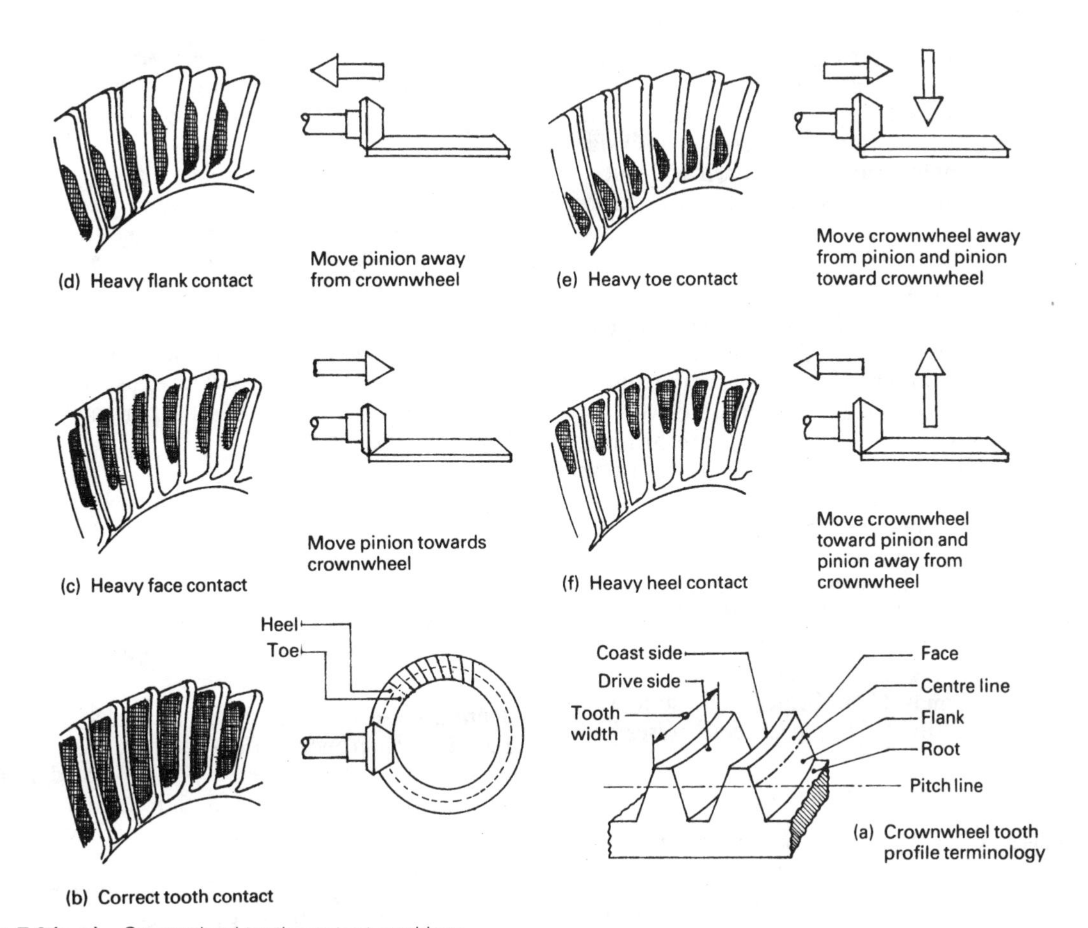

Fig. 7.8 (a–e) Crownwheel tooth contact markings

crownwheel, pinion and final drive housing under operating load conditions, so that the pressure contact area will tend to spread towards the heel towards a more central position.

Heavy face (high) tooth contact (Fig. 7.8(c)) Tooth contact area is above the centre line and on the face of the tooth profile due to the pinion being too far away from the crownwheel (insufficient pinion depth). To rectify this condition, move the pinion deeper into mesh by using a thicker pinion head washer to lower the contact area and reset the backlash.

Heavy flank (low) tooth contact (Fig. 7.8(d)) Tooth contact area is below the centre line and on the flank of the tooth profile due to the pinion being too far in mesh with the crownwheel (too much pinion depth). To rectify this condition, move the pinion away from the crownwheel using a thinner washer between the pinion head and inner bearing cone to raise the contact area and then reset the backlash.

Heavy toe contact (Fig. 7.8(e)) Tooth contact area is concentrated at the small end of the tooth (near the toe). To rectify this misalignment, increase backlash by moving the crownwheel and differential assembly away from the pinion, by transferring shims from the crownwheel side of the differential assembly to the opposite side, or slacken the adjusting nut on the crownwheel side of the differential and screw in the nut on the opposite side an equal amount. If the backlash is increased above the maximum specified, use a thicker washer (shim) behind the pinion head in order to keep the backlash within the correct limits.

Heavy heel contact (Fig. 7.8(f)) Tooth contact area is concentrated at the large end of the tooth which is near the heel. To rectify this misalignment, decrease backlash by moving the crownwheel nearer the pinion (add shims to the crownwheel side of the differential and remove an equal thickness of shims from the opposite side) or slacken the differential side adjusting nut and tighten the crownwheel side nut an equal amount. If the backlash is reduced below the minimum specified, use a thinner washer (shim) behind the pinion head.

7.1.7 Final drive axle noise and defects

Noise is produced with all types of meshing gear teeth such as from spur, straight or helical gears and even more so with bevel gears where the output is redirected at right angles to the input drive.

Vehicle noises coming from tyres, transmission, propellor shafts, universal joints and front or rear wheel bearings are often mistaken for axle noise, especially tyre to road surface rumbles which can sound very similar to abnormal axle noise. Listening for the noise at varying speeds and road surfaces, on drive and overrun conditions will assist in locating the source of any abnormal sound.

Once all other causes of noise have been eliminated, axle noise may be suspected. The source of axle noise can be divided into gear teeth noises and bearing noise.

Gear noise Gear noise may be divided into two kinds:

1 Broken, bent or forcibly damaged gear teeth which produce an abnormal audible sound which is easily recognised over the whole speed range.
   a) Broken or damaged teeth may be due to abnormally high shock loading causing sudden tooth failure.
   b) Extended overloading of both crownwheel and pinion teeth can be responsible for eventual fatigue failure.
   c) Gear teeth scoring may eventually lead to tooth profile damage. The causes of surface scoring can be due to the following:
      i) Insufficient lubrication or incorrect grade of oil
      ii) Insufficient care whilst running in a new final drive
      iii) Insufficient crownwheel and pinion backlash
      iv) Distorted differential housing
      v) Crownwheel and pinion misalignment
      vi) Loose pinion nut removing the pinion bearing preload.
2 Incorrect meshing of crownwheel and pinion teeth. Abnormal noises produced by poorly meshed teeth generate a very pronounced cyclic pitch whine in the speed range at which it occurs whilst the vehicle is operating on either drive or overrun conditions.

Noise on drive If a harsh cyclic pitch noise is heard when the engine is driving the transmission it indicates that the pinion needs to be moved slightly out of mesh.

Noise on overrun If a pronounced humming noise is heard when the vehicle's transmission overruns the engine, this indicates that the pinion needs to be moved further into mesh.

Slackness in the drive A pronounced time lag in taking the drive up accompanied by a knock when either accelerating or decelerating may be traced to end play in the pinion assembly due possibly to defective bearings or incorrectly set up bearing spacer and shim pack.

Bearing noise Bearings which are defective produce a rough growling sound that is approximately constant in volume over a narrow speed range. Driving the vehicle on a smooth road and listening for rough transmission sounds is the best method of identifying bearing failure.

A distinction between defective pinion bearings or differential cage bearings can be made by listening for any constant rough sound. A fast frequency growl indicates a failed pinion bearing, while a much slower repetition growl points to a defective differential bearing. The difference in sound is because the pinion revolves at about four times the speed of the differential assembly.

To distinguish between differential bearing and half shaft bearing defects, drive the vehicle on a smooth road and turn the steering sharply right and left. If the half shaft bearings are at fault, the increased axle load imposed on the bearing will cause a rise in the noise level, conversely if there is no change in the abnormal rough sound the differential bearings should be suspect.

Defective differential planet and sun gears The sun and planet gears of the differential unit very rarely develop faults. When differential failure does occur, it is usually caused by shock loading, extended overloading and seizure of the differential planet gears to the cross-shaft resulting from excessive wheel spin and consequently lubrication breakdown.

A roughness in the final drive transmission when the vehicle is cornering may indicate defective planet/sun gears.

7.2 Differential locks

A differential lock is desirable, and in some cases essential, if the vehicle is going to operate on low traction surfaces such as sand, mud, wet or waterlogged ground, worn slippery roads, ice bound roads etc. at relatively low speeds.

Drive axle differential locks are incorporated on heavy duty on/off highway and cross-country vehicles to provide a positive drive between axle half shafts when poor tyre to ground traction on one wheel would produce wheel spin through differential bevel gear action.

The differential lock has to be engaged manually by cable or compressed air, whereas the limited slip or viscous coupling differential automatically operates as conditions demand.

All differential locks are designed to lock together two or more parts of the differential gear cluster by engaging adjacent sets of dog clutch teeth. By this method, all available power transmitted to the final drive will be supplied to the wheels. Even if one wheel loses grip, the opposite wheel will still receive power enabling it to produce torque and therefore tractive effect up to the limit of the tyres' ability to grip the road. Axle wind-up will be dissipated by wheel bounce, slippage or scuffing. These unwanted reactions will occur when travelling over slippery soft or rough ground where true rolling will be difficult. Since the tyre tread cannot exactly follow the contour of the surface it is rolling over, for very brief periodic intervals there will be very little tyre to ground adhesion. As a result, any build up of torsional strain between the half shafts will be continuously released.

7.2.1 Differential lock mechanism

(Figs 7.9 and 7.10)

One example of a differential lock is shown in Fig. 7.9. In this layout a hardened and toughened flanged side toothed dog clutch member is clamped and secured by dowls between the crownwheel and differential cage flanges. The other dog clutch member is comprised of a sleeve internally splined to slot over the extended splines on one half shaft. This sleeve has dog teeth cut at one end and the double flange formed at the end to provide a guide groove for the actuating fork arm.

Engagement of the differential lock is obtained when the sleeve sliding on the extended external splines of the half shaft is pushed in to mesh with corresponding dog teeth formed on the flanged member mounted on the crownwheel and cage. Locking one half shaft to the differential cage prevents the bevel gears from revolving independently within the cage. Therefore, the half shafts and cage

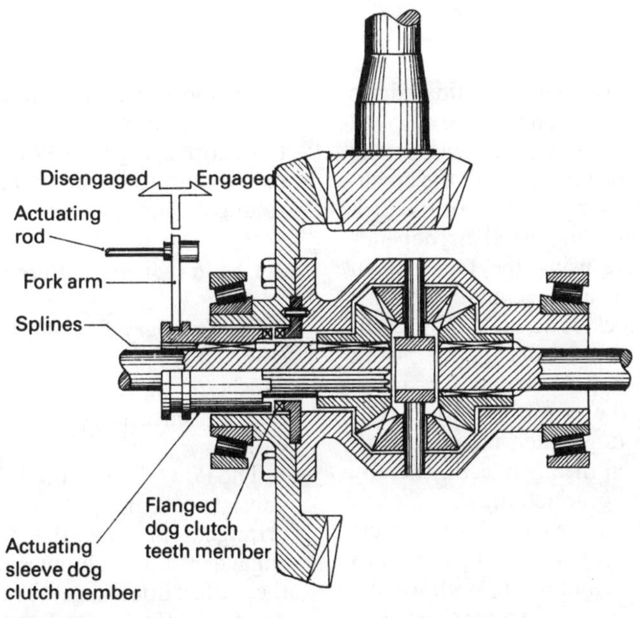

Fig. 7.9 Differential lock mechanism

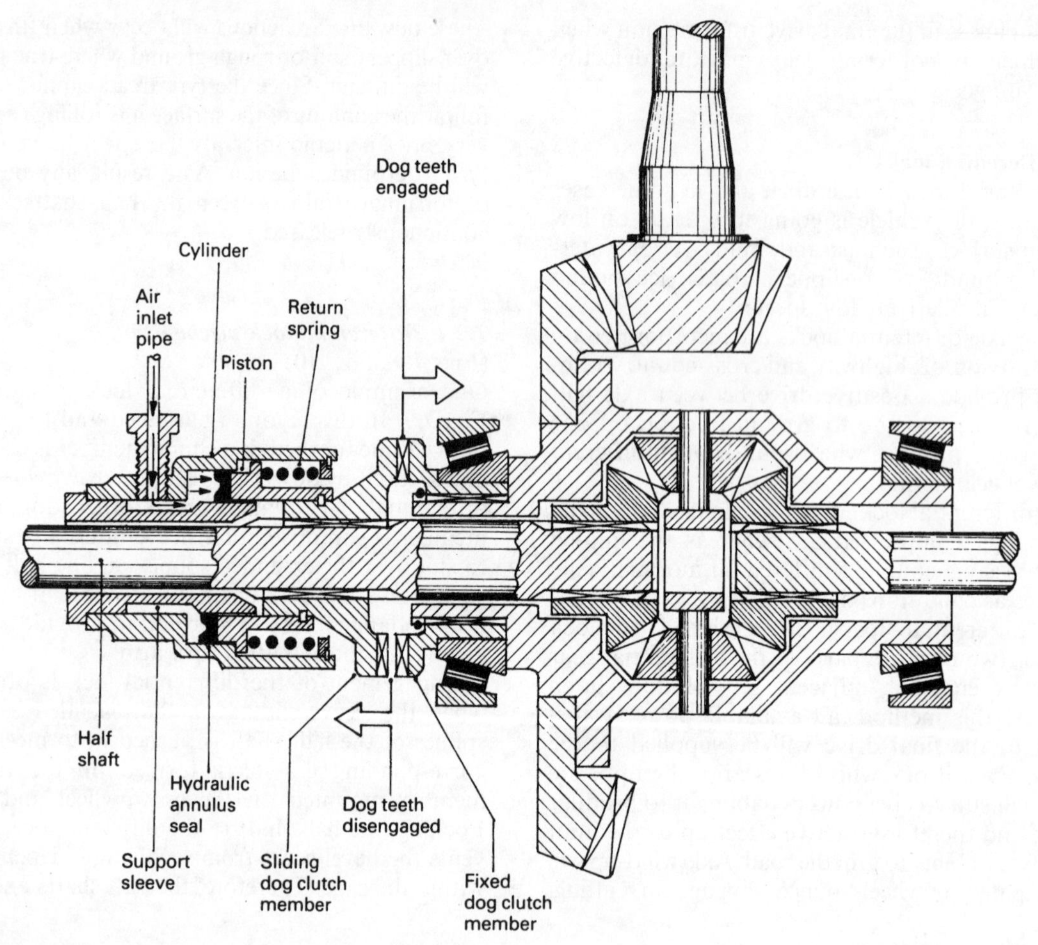

Fig. 7.10 Differential lock mechanism with air control

will be compelled to revolve with the final drive crownwheel as one. The lock should be applied when the vehicle is just in motion to enable the tooth to align, but not so fast as to cause the crashing of misaligned teeth. The engagement of the lock can be by cable, vacuum or compressed air, depending on the type of vehicle using the facility. An alternative differential lock arrangement is shown in Fig. 7.10 where the lock is actuated by compressed air operating on an annulus shaped piston positioned over one half shaft. When air pressure is supplied to the cylinder, the piston is pushed outwards so that the sliding dog clutch member teeth engage the fixed dog clutch member teeth, thereby locking out the differential gear action.

When the differential lock is engaged, the vehicle should not be driven fast on good road surfaces to prevent excessive tyre scrub and wear. With no differential action, relative speed differences between inner and outer drive wheels can only partially be compensated by the tyre tread having sufficient time to distort and give way in the form of minute hops or by permitting the tread to skid or bounce while rolling in slippery or rough ground conditions.

7.3 Skid reducing differentials

7.3.1 Salisbury Powr-Lok limited slip differential (Fig. 7.11)

This type of limited slip differential is produced under licence from the American Thornton Axle Co.

The Powr-Lok limited slip differential essentially consists of an ordinary bevel gear differential arranged so that the torque from the engine engages friction clutches locking the half shafts to the differential cage. The larger the torque, the greater the locking effect (Fig. 7.11).

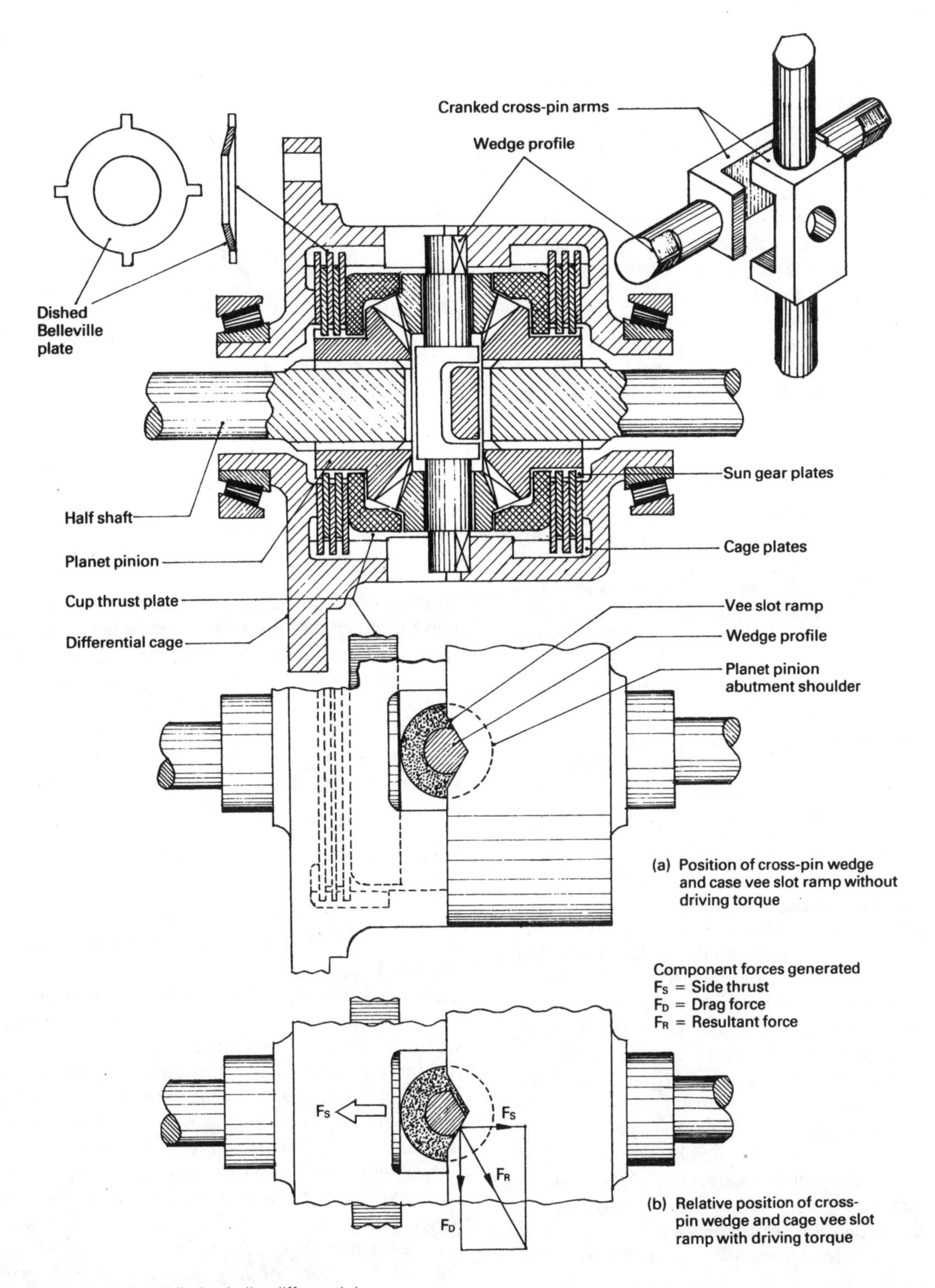

Fig. 7.11 Multiclutch limited slip differential

There are three stages of friction clutch loading:

1 Belleville spring action,
2 Bevel gear separating force action,
3 Vee slot wedging action.

Belleville spring action (Fig. 7.11) This is achieved by having one of the clutch plates dished to form a Belleville spring so that there is always some spring axial loading in the clutch plates. This then produces a small amount of friction which tends to lock the half shaft to the differential cage when the torque transmitted is very low. The spring thus ensures that when adhesion is so low that hardly any torque can be transmitted, some drive will still be applied to the wheel which is not spinning.

Bevel gear separating force action (Fig. 7.11) This arises from the tendency of the bevel planet pinions in the differential cage to force the bevel sun gears outwards. Each bevel sun gear forms part of a hub which is internally splined to the half shaft so that it is free to move outwards. The sun gear hub is also splined externally to align with one set of clutch plates, the other set being attached by splines to the differential cage. Thus the extra outward force exerted by the bevel pinions when one wheel tends to spin is transmitted via cup thrust plates to the clutches, causing both sets of plates to be camped together and thereby preventing relative movement between the half shaft and cage.

Vee slot wedging action (Fig. 7.11(a and b)) When the torque is increased still further, a third stage of friction clutch loading comes into being. The bevel pinions are not mounted directly in the differential cage but rotate on two separate arms which cross at right angles and are cranked to avoid each other. The ends of these arms are machined to the shape of a vee wedge and are located in vee-shaped slots in the differential cage. With engine torque applied, the drag reaction of the bevel planet pinion cross-pin arms relative to the cage will force them to slide inwards along the ramps framed by the vee-shaped slots in the direction of the wedge (Fig. 7.11(a and b)). The abutment shoulder of the bevel planet pinions press against the cup thrust plates and each set of clutch plates are therefore squeezed further together, increasing the multiclutch locking effect.

Speed differential and traction control (Fig. 7.12) Normal differential speed adjustment takes place continuously, provided the friction of the multi-plate clutches can be overcome. When one wheel spins the traction of the other wheel is increased by an amount equal to the friction torque generated by the clutch plates until wheel traction is restored. A comparison of a conventional differential and a limited slip differential tractive effort response against varying tyre to road adhesion is shown in Fig. 7.12.

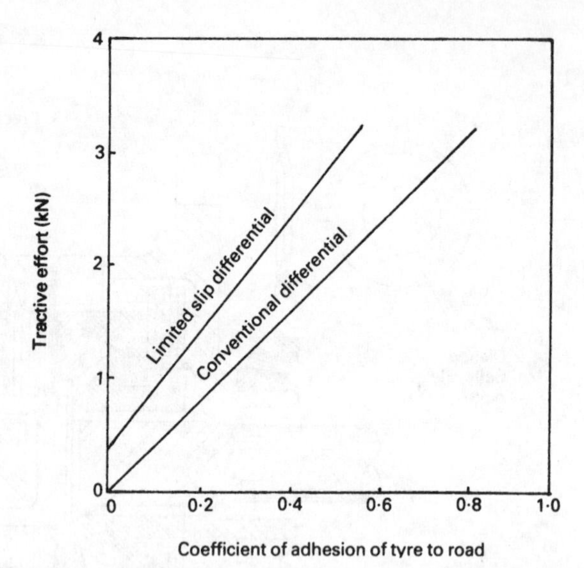

Fig. 7.12 Comparison of tractive effort and tyre to road adhesion for both conventional and limited slip differential

7.3.2 Torsen worm and wheel differential

Differential construction (Figs 7.13 and 7.14) The Torsen differential has a pair of worm gears, the left hand half shaft is splined to one of these worm gears while the right hand half shaft is splined to the other hand (Fig. 7.13). Meshing with each worm gear on each side is a pair of worm wheels (for large units triple worm wheels on each side). At both ends of each worm wheel are spur gears which mesh with adjacent spur gears so that both worm gear and half shafts are indirectly coupled together.

Normally with a worm gear and worm wheel combination the worm wheel is larger than the worm gear, but with the Torsen system the worm gear is made larger than the worm wheel. The important feature of any worm gear and worm wheel is that the teeth are cut at a helix angle such that the worm gear can turn the worm wheel but the worm wheel cannot rotate the worm gear. This is achieved with the Torsen differential by giving the

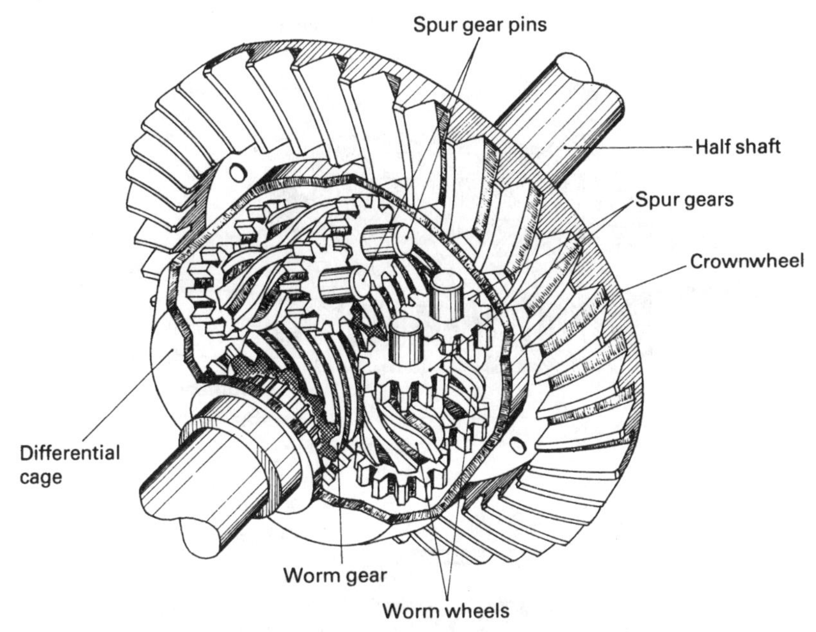

Fig. 7.13 Pictorial view of Torsen worm and spur gear differential

worm gear teeth a fine pitch while the worm wheel has a coarse pitch.

Note that with the conventional meshing spur gear, be it straight or helical teeth, the input and output drivers can be applied to either gear. The reversibility and irreversibility of the conventional bevel gear differential and the worm and worm wheel differential is illustrated in Fig. 7.14 by the high and low mechanical efficiencies of the two types of differential.

Differential action when moving straight ahead (Fig. 7.15) When the vehicle is moving straight ahead power is transferred from the propellor shaft to the bevel pinion and crownwheel. The crownwheel and differential cage therefore revolve as one unit (Fig. 7.15). Power is divided between the left and right hand worm wheel by way of the spur gear pins which are attached to the differential cage. It then flows to the pair of meshing worm gears, where it finally passes to each splined half shaft. Under these conditions, the drive in terms of speed and torque is proportioned equally to both half shafts and road wheels. Note that there is no relative rotary motion between the half shafts and the differential cage so that they all revolve as a single unit.

Differential action when cornering (Fig. 7.15) When cornering, the outside wheel of the driven axle will

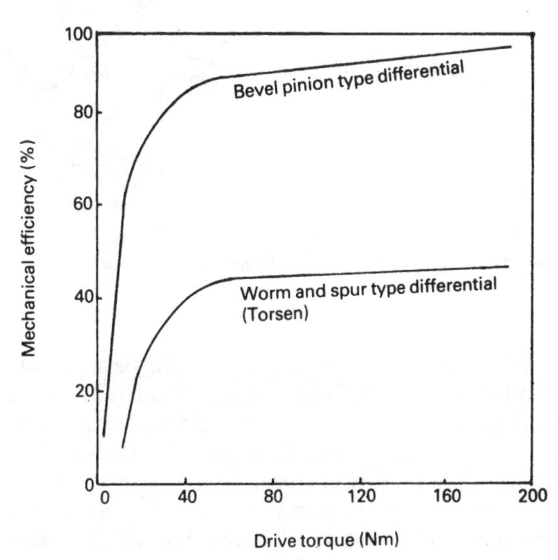

Fig. 7.14 Comparison of internal friction expressed in terms of mechanical efficiency of both bevel pinion type and worm and spur type differentials

tend to rotate faster than the inside wheel due to its turning circle being larger than that of the inside wheel. It follows that the outside wheel will have to rotate relatively faster than the differential cage, say by +20 rev/min, and conversely the inside wheel has to reduce its speed in the same proportion, of say −20 rev/min.

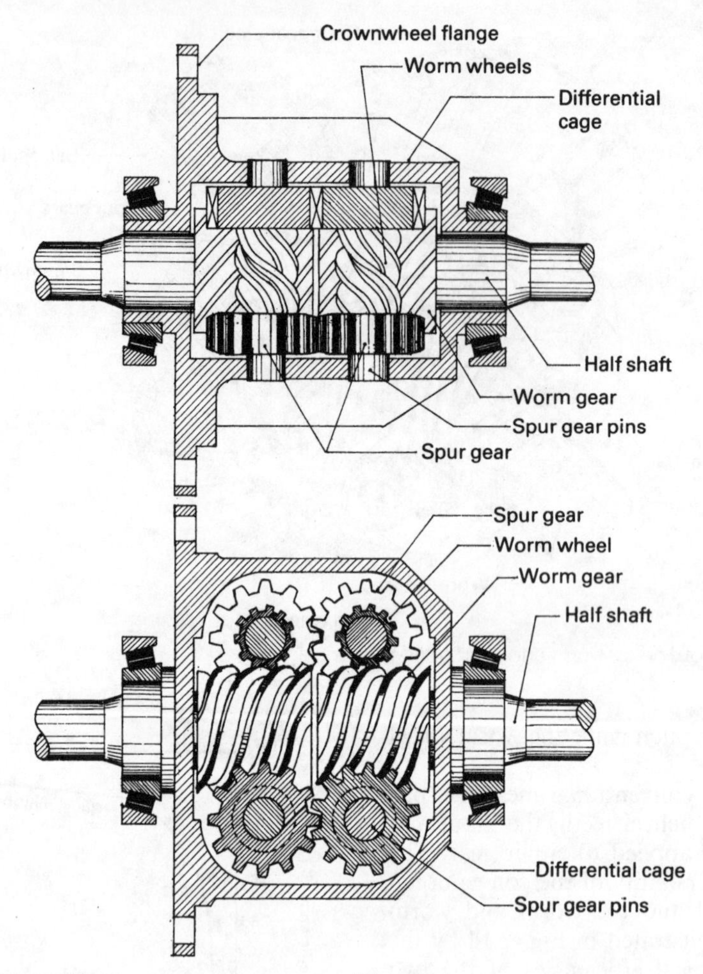

Fig. 7.15 Sectioned views of Torsen worm and spur gear differential

When there is a difference in speed between the two half shafts, the faster turning half shaft via the splined worm gears drives its worm wheels about their axes (pins) in one direction of rotation. The corresponding slower turning half shaft on the other side drives its worm wheels about their axes (pins) in the opposite direction but at the same speed (Fig. 7.15).

Since the worm wheels on opposite sides will be revolving at the same speed but in the opposite sense while the vehicle is cornering they can be simply interlinked by pairs of meshing spur gears without interfering with the independent road speed requirements for both inner and outer driving road wheels.

Differential torque distribution (Fig. 7.15) When one wheel loses traction and attempts to spin, it transmits drive from its set of worm gears to the worm wheels. The drive is then transferred from the worm wheels on the spinning side to the opposite (good traction wheel) side worm wheels by way of the bridging spur gears (Fig. 7.15). At this point the engaging teeth of the worm wheel with the corresponding worm gear teeth jam. Thus the wheel which has lost its traction locks up the gear mechanism on the other side every time there is a tendency for it to spin. As a result of the low traction wheel being prevented from spinning, the transmission of torque from the engine will be concentrated on the wheel which has traction.

Another feature of this mechanism is that speed differentiation between both road wheels is maintained even when the wheel traction differs considerably between wheels.

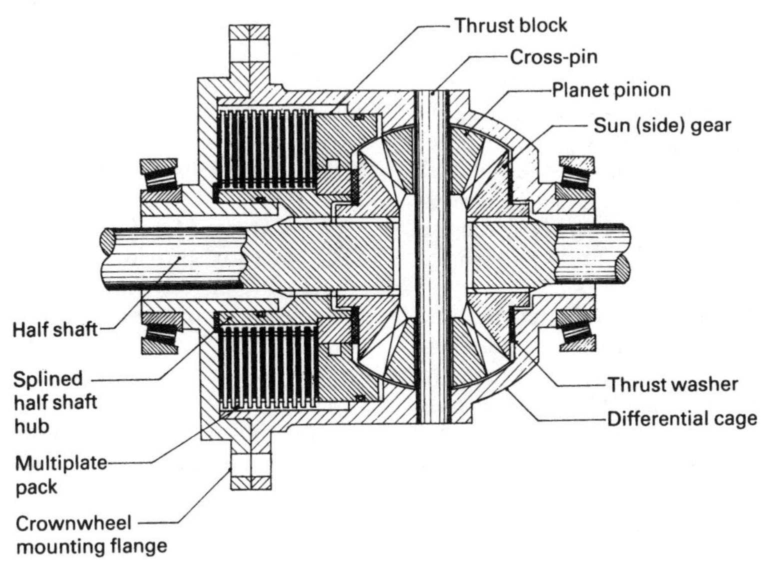

Fig. 7.16 Viscous coupling differential

7.3.3 Viscous coupling differential

Description of differential and viscous coupling (Figs 7.16 and 7.17) The crownwheel is bolted to the differential bevel gearing and multiplate housing. Speed differentiation is achieved in the normal manner by a pair of bevel sun (side) gears, each splined to a half shaft. Bridging these two bevel sun gears are a pair of bevel planet pinions supported on a cross-pin mounted on the housing cage. A multiplate back assembly is situated around the left hand half shaft slightly outboard from the corresponding sun gear (Fig. 7.16).

The viscous coupling consists of a series of spaced interleaved multiplates which are alternatively splined to a half shaft hub and the outer differential cage. The cage plates have pierced holes but the hub plates have radial slots. Both sets of plates are separated from each other by a 0.25 mm gap. Thus the free gap between adjacent plates and the interruption of their surface areas with slots and holes ensures there is an adequate storage of fluid between plates after the sealed plate unit has been filled and that the necessary progressive viscous fluid torque characteristics will be obtained when relative movement of the plates takes place.

When one set of plates rotate relative to the other, the fluid will be sheared between each pair of adjacent plate faces and in so doing will generate an opposing torque. The magnitude of this resisting torque will be proportional to the fluid viscosity and the relative speed difference between the sets of plates. The dilatent silicon compound fluid which has been developed for this type of application has the ability to maintain a constant level of viscosity throughout the operating temperature range and life expectancy of the coupling (Fig. 7.17).

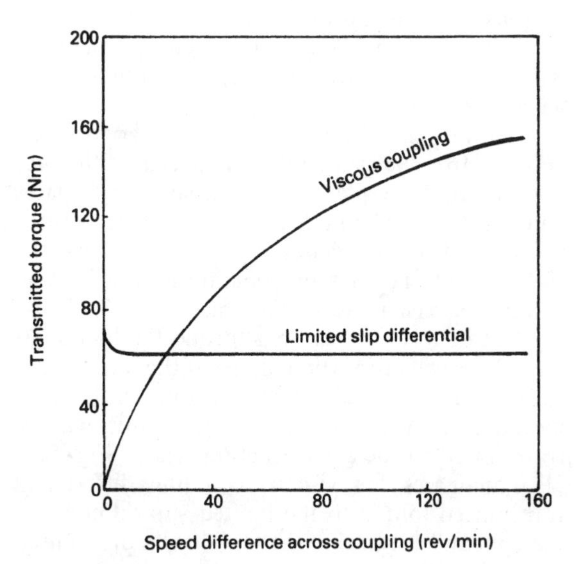

Fig. 7.17 Comparison of torque transmitted to wheel having the greater adhesion with respect to speed difference between half shafts for both limited slip and viscous coupling

Speed differential action (Fig. 7.16) In the straight ahead driving mode the crownwheel and differential cage driven by the bevel pinion act as the input to the differential gearing and in so doing the power path transfers to the cross-pin and bevel planet gears. One of the functions of these planet gears is to link (bridge) the two sun (side) gears so that the power flow is divided equally between the sun gears and correspondently both half shafts (Fig. 7.16).

When rounding a bend or turning a corner, the outer wheel will have a greater turning circle than the inner one. Therefore the outer wheel tends to increase its speed and the inner wheel decrease its speed relative to the differential cage rotational speed. This speed differential is made possible by the different torque reactions each sun gear conveys back from the road wheel to the bevel planet pinions. The planet gears 'float' between the sun gears by rotating on their cross-pin, thus the speed lost relative to the cage speed by the inner road wheel and sun gear due to the speed retarding ground reaction will be that gained by the outer road wheel and sun gear.

Viscous coupling action (Figs 7.16 and 7.17) In the situation when one wheel loses traction caused by possibly loose soil, mud, ice or snow, the tyre–road tractive effort reaction is lost. Because of this lost traction there is nothing to prevent the planet pinions revolving on their axes, rolling around the opposite sun gear, which is connected to the road wheel sustaining its traction, with the result that the wheel which has lost its grip will just spin (race) with no power being able to drive the good wheel (Fig. 7.16). Subsequently, a speed difference between the cage plates and half shaft hub plates will be established and in proportion to this relative speed, the two sets of coupling plates will shear the silicon fluid and thereby generate a viscous drag torque between adjacent plate faces (Fig. 7.17). As a result of this viscous drag torque the half shaft hub plates will proportionally resist the rate of fluid shear and so partially lock the differential gear mechanism. A degree of driving torque will be transmitted to the good traction wheel. Fig. 7.17 also compares the viscous coupling differential transmitted torque to the limited slip differential. Here it can be seen that the limited slip differential approximately provides a constant torque to the good traction wheel at all relative speeds, whereas the viscous coupling differential is dependent on speed differences between both half shafts so that the torque transmitted to the wheel supplying tractive effort rises with increased relative speed between the half shaft and differential cage.

7.4 Double reduction axles

7.4.1 The need for double reduction final drives

The gearbox provides the means to adjust and match the engine's speed and torque so that the vehicle's performance responds to the driver's expectations under the varying operating conditions. The gearbox gear reduction ratios are inadequate to supply the drive axle with sufficient torque multiplication and therefore a further permanent gear reduction stage is required at the drive axle to produce the necessary road wheel tractive effect. For light vehicles of 0.5–2.0 tonne, a final drive gear reduction between 3.5:1 and 4.5:1 is generally sufficient to meet all normal driving conditions, but with commercial vehicles carrying considerably heavier payloads a demand for a much larger final drive gear reduction of 4.5–9.0:1 is essential. This cannot be provided by a single stage final drive crownwheel and pinion without the crownwheel being abnormally large. Double reduction axles partially fulfil the needs for heavy goods vehicles operating under normal conditions by providing two stages of gear reduction at the axle.

In all double reduction final drive arrangements the crownwheel and pinion are used to provide one stage of speed step down. At the same time the bevel gearing redirects the drive perpendicular to the input propellor shaft so that the drive then aligns with the axle half shafts.

7.4.2 Double reduction axles with first stage reduction before the crownwheel and pinion

Double reduction with spur gears ahead of bevel gears (Fig. 7.18) With a pair of helical gears providing the first gear reduction before the crownwheel and pinion, a high mounted and compact final drive arrangement is obtained. This layout has the disadvantage of the final gear reduction and thus torque multiplication is transmitted through the crownwheel and pinion bevel gears which therefore absorbs more end thrust and is generally considered to be less efficient in operation compared to helical spur type gears. The first stage of a double reduction axle is normally no more than 2:1 leaving the much larger reduction for the output stage.

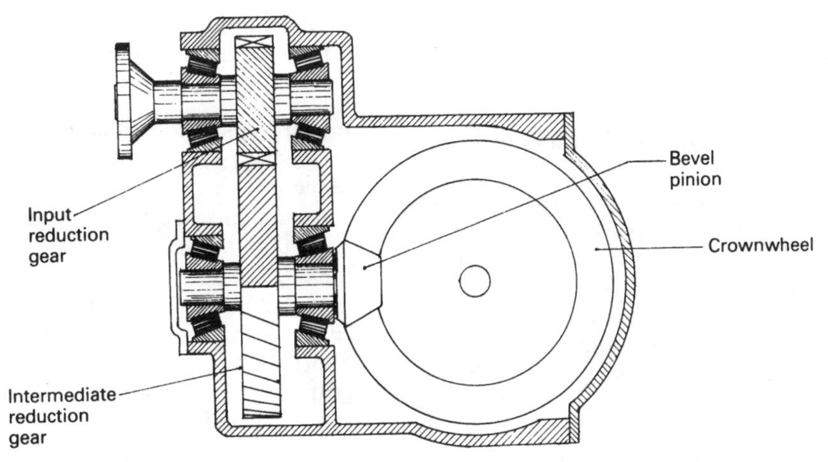

Fig. 7.18 Final drive spur double reduction ahead of bevel pinion

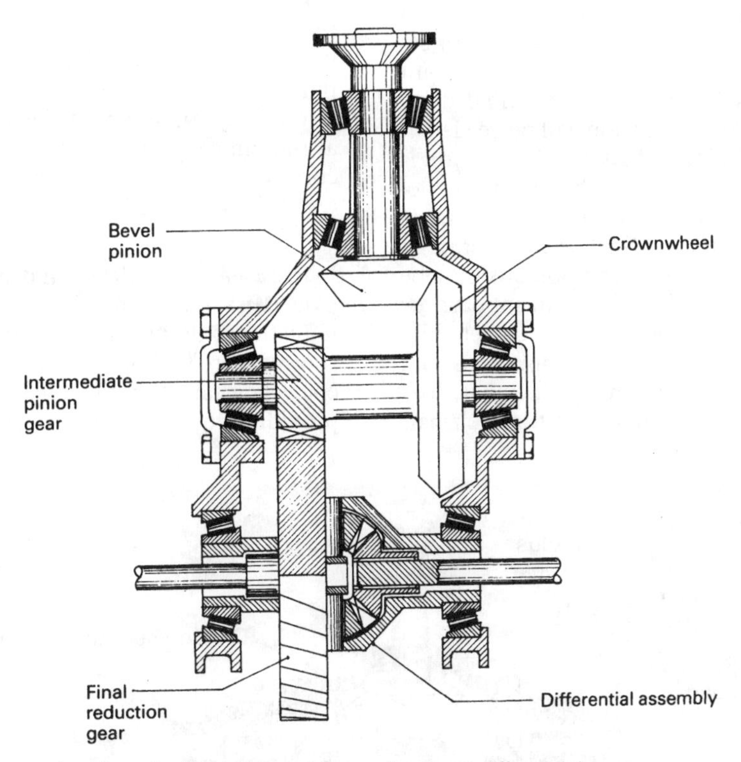

Fig. 7.19 Final drive spur double reduction between crownwheel and differential

Double reduction with bevel gears ahead of spur gears (Fig. 7.19) A popular double reduction arrangement has the input from the propellor shaft going directly to the bevel pinion and crownwheel. The drive is redirected at right angles to that of the input so making it flow parallel to the half shafts, the first stage gear reduction being determined by the relative sizes (number of teeth) of the bevel gears. A helical pinion gear mounted on the same shaft as the crownwheel meshes with a helical gear wheel bolted to the differential cage. The combination of these two gear sizes provides the second stage gear reduction. Having the bevel gears ahead of the helical gears ensures that only a proportion of torque multiplication will be constrained by them, while the helical gears will absorb the full torque reaction of the final gear reduction.

7.4.3 Inboard and outboard double reduction axles

Where very heavy loads are to be carried by on-off highway vehicles, the load imposed on the crownwheel and pinion and differential unit can be reduced by locating a further gear reduction on either side of the differential exit. If the second gear reduction is arranged on both sides close to the differential cage, it is referred to as an inboard reduction. They can be situated at the wheel ends of the half shafts, where they are known as outboard second stage gear reduction. By having the reduction directly after the differential, the increased torque multiplication will only be transmitted to the half shafts leaving the crownwheel, pinion and differential with a torque load capacity proportional to their gear ratio. The torque at this point may be smaller than with the normal final drive gear ratio since less gear reduction will be needed at the crownwheel and pinion if a second reduction is to be provided. Alternatively, if the second reduction is in the axle hub, less torque will be transmitted by the half shafts and final drive differential and the dimensions of these components can be kept to a minimum. Having either an inboard or outboard second stage gear reduction enables lighter crownwheel and pinion combinations and differential assembly to be employed, but it does mean there are two gear reductions for each half shaft, as opposed to a single double reduction drive if the reduction takes place before the differential.

Inboard epicyclic double reduction final drive axle (Scammell) (Fig. 7.20) With this type of double reduction axle, the first stage conforms to the conventional crownwheel and pinion whereas the second stage reduction occurs after passing through the differential. The divided drive has a step down gear reduction via twin epicyclic gear trains on either side of the differential cage (Fig. 7.20). Short shafts connect the differential bevel sun gears to the pinion sun gear of the epicyclic gear train. When drive is being transmitted, the rotation of the sun gears rotates the planet pinions so that they are forced to roll 'walk' around the inside of the reaction annulus gear attached firmly to the axle casing. Support to the planet pinions and their pins is given by the planet carrier which is itself mounted on a ball race. Thus when the planet pinions are made to rotate on their own axes they also bodily rotate about the same axis of rotation as the sun gear, but at a reduced speed, and in turn convey power to the half shafts splined to the central hub portion of the planet carriers.

Inboard epicyclic differential and double reduction axle (Kirkstall) (Fig. 7.21) This unique double reduction axle has a worm and worm wheel first stage gear reduction. The drive is transferred to an epicyclic gear train which has the dual function of providing the second stage gear reduction while at

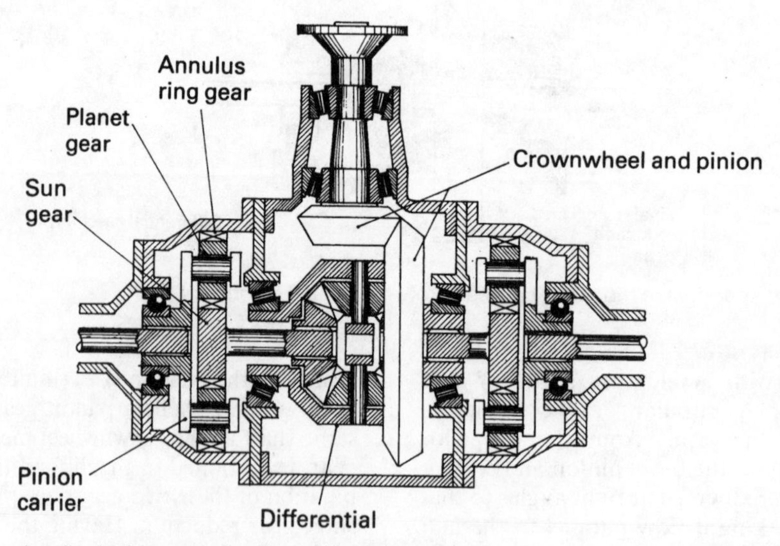

Fig. 7.20 Inboard epicyclic double reduction final drive axle

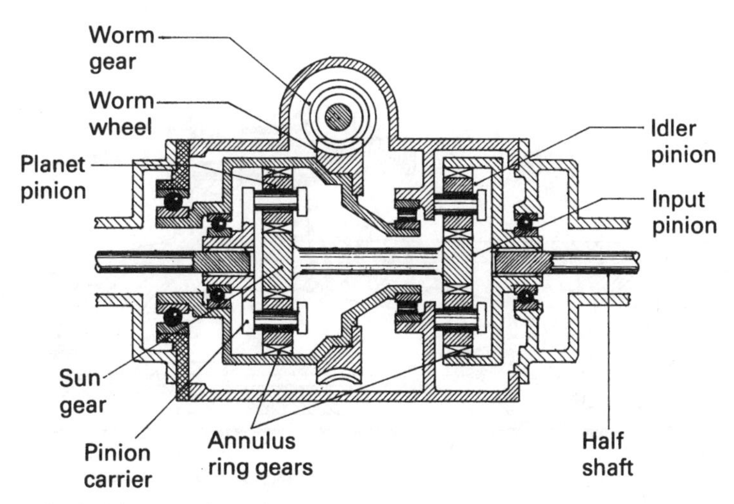

Fig. 7.21 Inboard epicyclic double reduction axle

the same time performing as the final drive differential (Fig. 7.21).

Principle of operation Power is transmitted from the propellor shaft to the worm and worm wheel which produces a gear reduction and redirects the drive at right angles and below the worm axis of rotation (Fig. 7.21). The worm wheel is mounted on the annulus carrier so that they both rotate as one. Therefore the three evenly spaced planet pinions meshing with both the annulus and the sun gear are forced to revolve and move bodily on their pins in a forward direction. Since the sun gear is free to rotate (not held stationary) it will revolve in a backward direction so that the planet carrier and the attached left hand half shaft will turn at a reduced speed relative to the annulus gear.

Simultaneously, as the sun gear and shaft transfers motion to the right hand concentric gear train central pinion, it passes to the three idler pinions, compelling them to rotate on their fixed axes, and in so doing drives round the annulus ring gear and with it the right hand half shaft which is splined to it.

The right hand gear train with an outer internal ring gear (annulus) does not form an epicyclic gear train since the planet pins are fixed to the casing and do not bodily revolve with their pins (attached to a carrier) about some common centre of rotation. It is the purpose of the right hand gear train to produce an additional gear reduction to equalize the gear reduction caused by the planet carrier output on the left hand epicyclic gearing with the sun gear output on the right hand side.

Differential action of the epicyclic gears The operation of the differential is quite straight forward if one imagines either the left or right hand half shaft to slow down as in the case when they are attached to the inner wheel of a cornering vehicle.

If when cornering the left hand half shaft slows down, the planet carrier will correspondingly reduce speed and force the planet pinions revolving on their pins to spin at an increased speed. This raises the speed of the sun gear which indirectly drives, in this case, the outer right hand half shaft at a slightly higher speed. Conversely, when cornering if the right hand half shaft should slow down, it indirectly reduces the speed of the central pinion and sun gear. Hence the planet pinions will not revolve on their pins, but will increase their speed at which they also roll round the outside of the sun gear. Subsequently the planet pins will drive the planet carrier and the left hand half shaft at an increased speed.

7.4.4 Outboard double reduction axles

Outboard epicyclic spur gear double reduction axle (Fig. 7.22)

Description of construction (Fig. 7.22) A gear reduction between the half shaft and road wheel hub may be obtained through an epicyclic gear train. A typical step down gear ratio would be 4:1. The sun gear may be formed integrally with or it may be splined to the half shaft (Fig. 7.22). It is made to engage with three planet gears carried on pins fixed to and rotating with the hub, thus driving

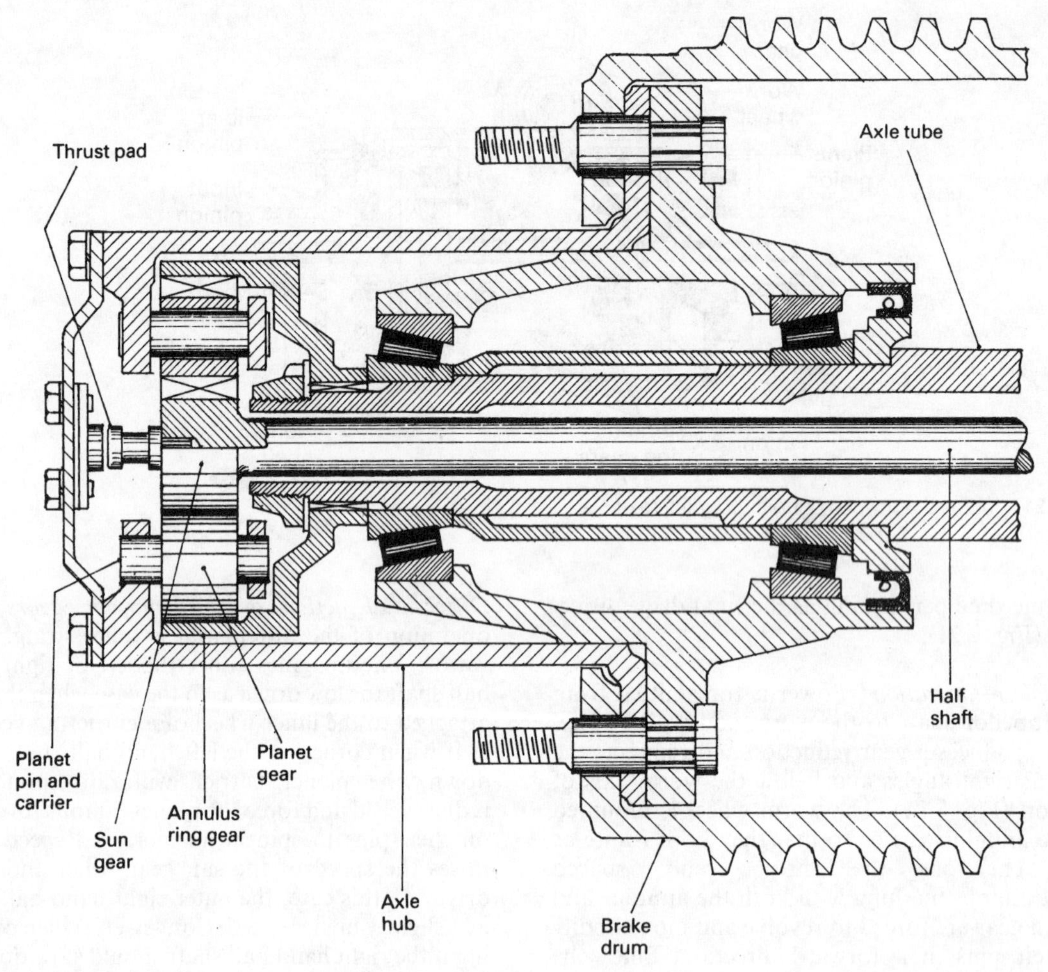

Fig. 7.22 Outboard epicyclic spur double reduction axle

the latter against the reaction of an outer annulus splined to the stationary axle tube. The sun wheel floats freely in a radial direction in mesh with the planet pinions so that driving forces are distributed equally on the three planet pinions and on their axes of rotation. A half shaft and sun gear end float is controlled and absorbed by a thrust pad mounted on the outside end cover which can be initially adjusted by altering the thickness of a shim pack.

Description of operation (Fig. 7.22) In operation, power flows from the differential and half shaft to the sun gear where its rotary motion is distributed between the three planet pinions. The forced rotation of these planet pinions compels them to roll around the inside of the reaction annulus ring gear (held stationary) so that their axes of rotation, and the planet pins, are forced to revolve about the sun gear axis. Since the planet pins are mounted on the axle hub, which is itself mounted via a fully floating taper bearing arrangement on the axle tube, the whole hub assembly will rotate at a much reduced speed relative to the half shaft's input speed.

Outboard epicyclic bevel gear single and two speed double reduction axle (Fig. 7.23) This type of outboard double reduction road wheel hub employs bevel epicyclic gearing to provide an axle hub reduction. To achieve this gear reduction there are two bevel sun (side) gears. One is splined to and mounted on the axle tube and is therefore fixed. The other one is splined via the sliding sleeve dog clutch to the half shaft and so is permitted to revolve (Fig. 7.23). Bridging both of these bevel sun gears are two planetary bevel gears which are supported on a cross-pin mounted on the axle hub.

The planetary bevel gear double reduction axle hub may be either two speed, as explained in the following text, or a single speed arrangement in which the half shaft is splined permanently and directly to the outer sun gear.

High ratio (Fig. 7.23) High ratio is selected and engaged by twisting the speed selector eccentric so that its offset peg pushes the sliding sleeve outwards (to the left) until the external teeth of the dog clutch move out of engagement from the sun gear and into engagement with the internal teeth formed inside the axle hub end plate. Power is transferred from the differential and half shaft via the sleeve dog clutch directly to the axle hub without producing any gear reduction.

Low ratio (Fig. 7.23) When low ratio is engaged, the sleeve dog clutch is pushed inwards (to the right) until the external teeth of the dog clutch are moved out of engagement from the internal teeth of the hub plate and into engagement with the internal teeth of the outer bevel sun gear. The input drive is now transmitted to the half shaft where it rotates the outer bevel sun gear so that the bevel planet gears are compelled to revolve on the cross-pin. In doing so they are forced to roll around the fixed inner bevel sun gear. Consequently, the cross-pin which is attached to the axle hub is made to revolve about the half shaft but at half its speed.

7.5 Two speed axles

The demands for a truck to operate under a varying range of operating conditions means that the overall transmission ratio spread needs to be extensive, which is not possible with a single or double reduction final

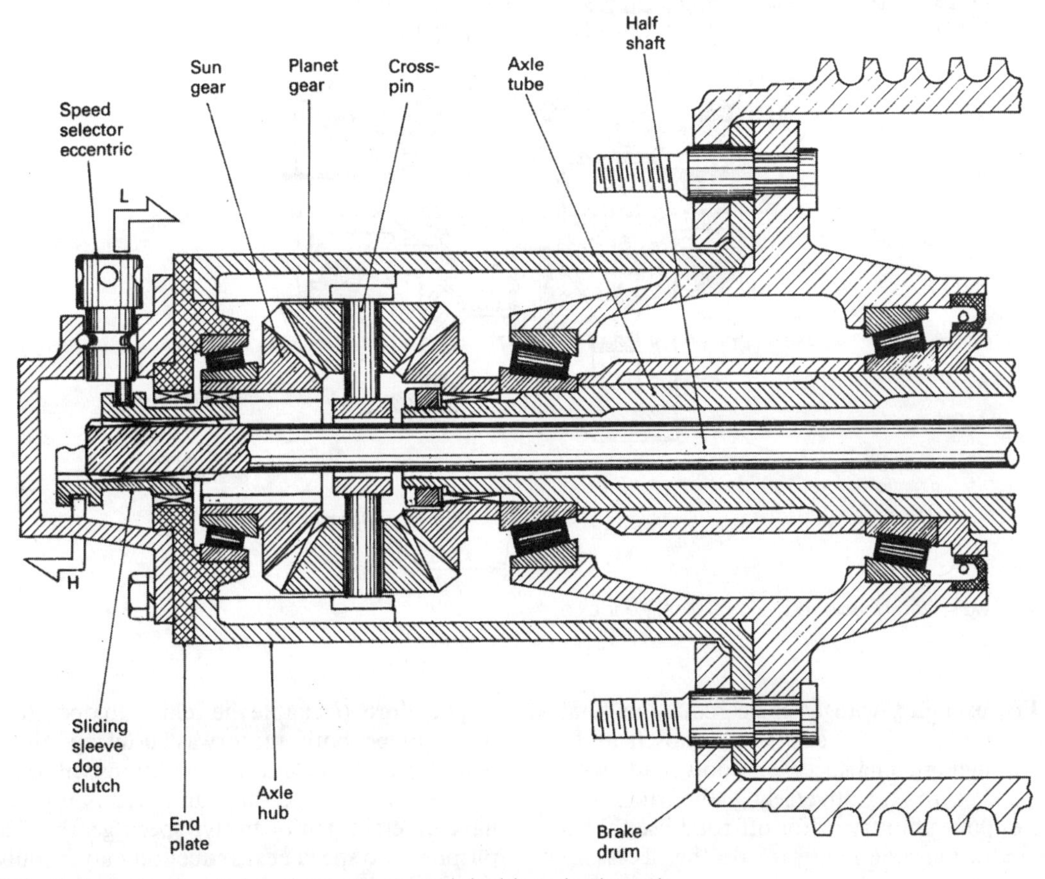

Fig. 7.23 Outboard epicyclic bevel gear two speed double reduction axle

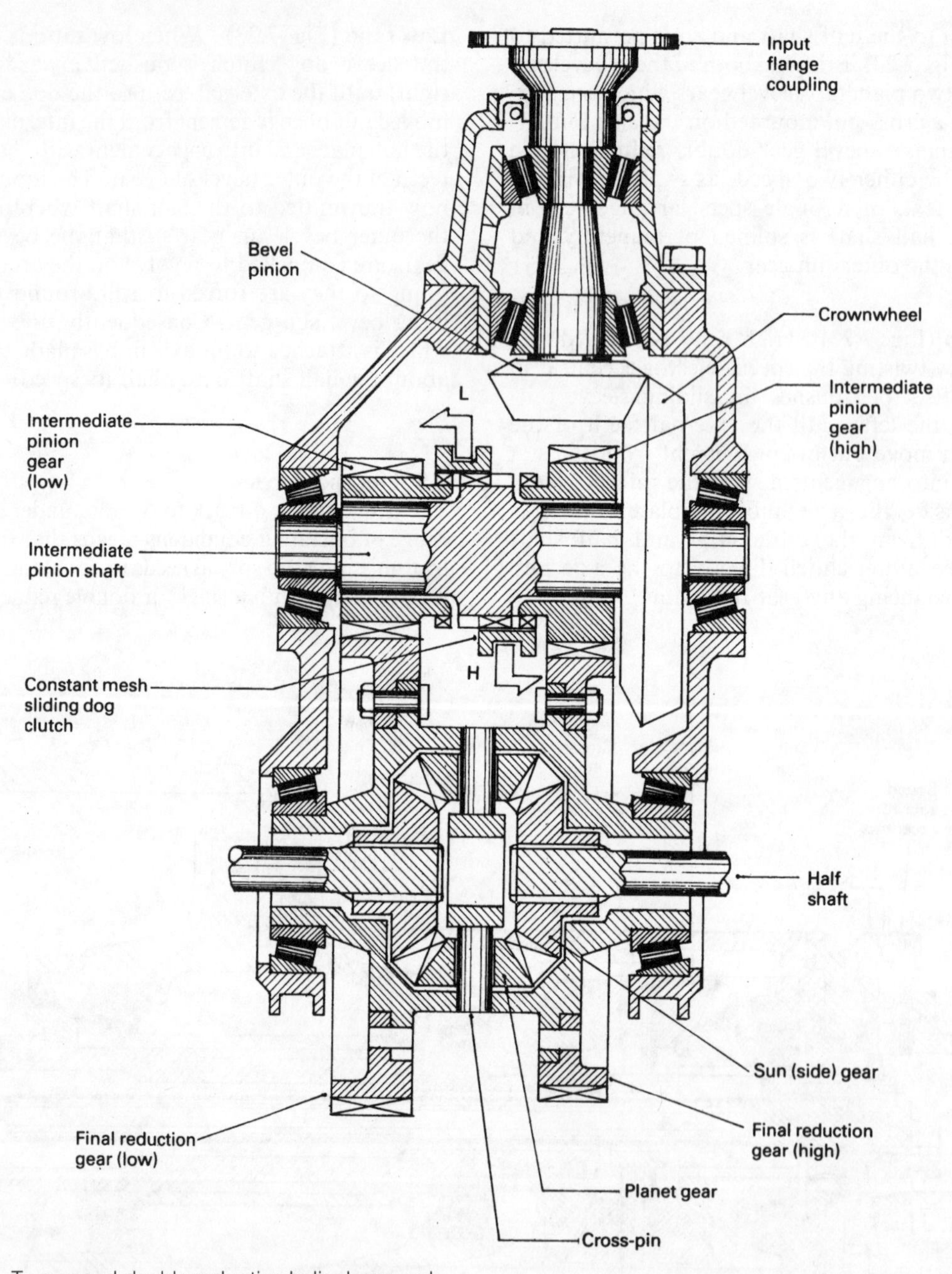

Fig. 7.24 Two speed double reduction helical gear axle

drive. For example, with a single reduction final drive the gear reduction can be so chosen as to provide a high cruising speed on good roads with a five speed gearbox. Conversely, if the truck is to be used on hilly country or for off-road use then a double reduction axle may provide the necessary gear reduction.

Therefore, to enable the vehicle to operate effectively under both motorway cruising and town stopping and accelerating conditions without overloading or overspeeding and without having to have an eight, ten or twelve speed gearbox, a dual purpose two speed gear reduction may be built into the final drive axle.

Combining a high and low ratio in the same axle doubles the number of gears available from the standard gearbox. The low range of gears will then provide the maximum pulling power for heavy duty operations on rough roads, whereas the high range of gears allows maximum speed when conditions are favourable. From the wide range of gear ratios the driver can choose the exact combination to suit any conditions of load and road so that the engine will always operate at peak efficiency and near to its maximum torque speed band.

7.5.1 Two speed double reduction helical gear axle (Rockwell-Standard) (Fig. 7.24)

This two speed double reduction helical gear axle has a conventional crownwheel and bevel pinion single speed first reduction with a second stage speed reduction consisting of two pairs of adjacent pinion and wheel helical cut gears. These pinions mounted on the crownwheel support shaft act as intermediate gears linking the crownwheel to the differential cage final reduction wheel gears (Fig. 7.24).

Low ratio (Fig. 7.24) Low ratio is engaged when the central sliding dog clutch splined to the crownwheel shaft slides over the selected (left hand) low speed smaller pinion dog teeth. Power from the propellor shaft now flows to the bevel pinion where it is redirected at right angles to the crownwheel and shaft. From here it passes from the locked pinion gear and crownwheel to the final reduction wheel gear bolted to the differential cage. The drive is then divided via the differential cross-pin and planet pinions between both sun gears where it is transmitted finally to the half shafts and road wheels.

High ratio (Fig. 7.24) High ratio is engaged in a similar way as for low ratio but the central sliding dog clutch slides in the opposite direction (right hand) over the larger pinion dog teeth. The slightly larger pinion meshing with a correspondently smaller differential wheel gear produces a more direct second stage reduction and hence a higher overall final drive axle gear ratio.

7.5.2 Two speed epicyclic gear train axle (Eaton) (Fig. 7.25)

With this arrangement an epicyclic gear train is incorporated between the crownwheel and differential cage (Fig. 7.25).

High ratio (Fig. 7.25) When a high ratio is required, the engagement sleeve is moved outwards from the differential cage so that the dog teeth of both the sleeve and the fixed ring teeth disengage. At the same time the sun gear partially slides out of mesh with the planet pinions and into engagement with the outside pinion carrier internal dog teeth. Subsequently, the sun gear is free to rotate. In addition the planet pinions and carrier are locked to the sun gear, so that there can be no further relative motion within the epicyclic gear train (i.e. annulus, planet pinions, carrier, differential cage and sun gear). In other words, the crownwheel and differential cage are compelled to revolve as one so that the final drive second stage gear reduction is removed.

Low ratio (Fig. 7.25) When the engagement sleeve is moved inwards its dog clutch teeth engage with the stationary ring teeth and the sun gear is pushed fully in to mesh with the planet pinion low ratio that has been selected. Under these conditions, the input drive from the propellor shaft to the bevel pinion still rotates the crownwheel but now the sun gear is prevented from turning. Therefore the rotating crownwheel with its internal annulus ring gear revolving about the fixed sun gear makes the planet pinions rotate on their own axes (pins) and roll around the outside of the held sun gear. As a result of the planet pinions meshing with both the annulus and sun gear, and the crownwheel and annulus rotating while the sun gear is held stationary, the planetary pinions are forced to revolve on their pins which are mounted on one side of the differential cage. Thus the cage acts as a planet pinion carrier and in so doing is compelled to rotate at a slower rate relative to the annulus gear speed. Subsequently, the slower rotation of the differential cage relative to that of the crownwheel produces the second stage gear reduction of the final drive.

7.6 The third (central) differential

7.6.1 The necessity for a third differential

When four wheel drive cars or tandem drive axle bogie trucks are to be utilized, provision must be provided between drive axles to compensate for any difference in the mean speeds of each drive axle as opposed to speed differentiation between pairs of axle road wheels.

Speed difference between driving axles are influenced by the following factors:

1 Speed variation between axles when a vehicle moves on a curved track due to the slight differ-

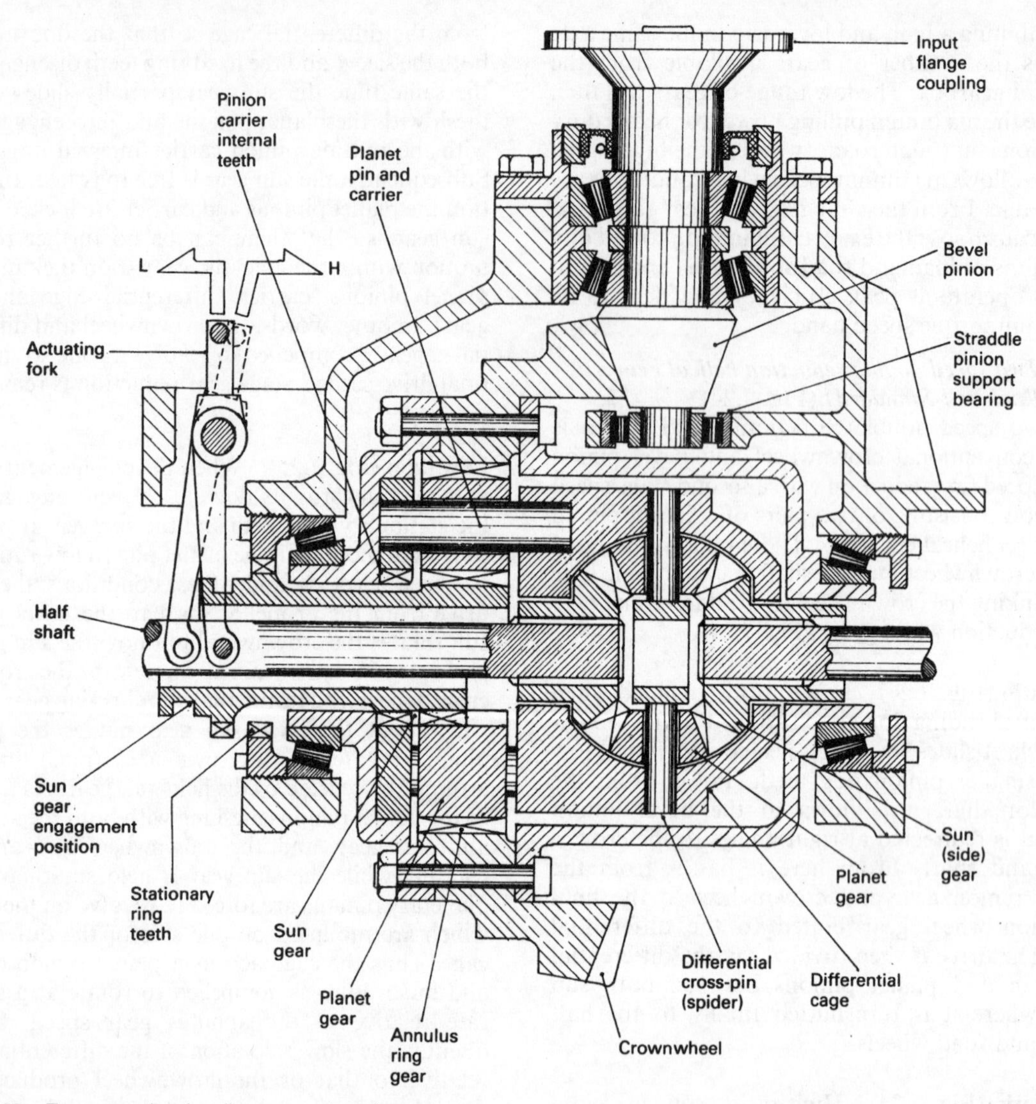

Fig. 7.25 Two speed epicyclic gear train axle

ence in rolling radius of both axles about some instantaneous centre of rotation.

2 Small road surface irregularities, causing pairs of driving wheels to locally roll into and over small dips and humps so that each pair of wheels are actually travelling at different speeds at any one moment.
3 Tyres which have different amounts of wear or different tread patterns and construction such as cross-ply and radials, high and low profile etc. and are mixed between axles so that their effective rolling radius of the wheel and tyre combination varies.
4 Uneven payload distribution will alter the effective rolling radius of a wheel and tyre so that heavily laden axles will have smaller rolling radii wheels and therefore complete more revolutions over a given distance than lightly laden axles.
5 Unequal load distribution between axles when accelerating and braking will produce a variation of wheel effective rolling radius.
6 Loss of grip between pairs of road wheels produces momentary wheel spin and hence speed differences between axles.

7.6.2 Benefits of a third differential (Fig. 7.26)

Operating a third differential between front and rear wheel drive axles or rear tandem axles has certain advantages:

1 The third differential equally divides driving torque and provides speed differentiation between both final drive axles so that the relative torque and speed per axle are better able to meet the individual road wheel requirements, thereby minimizing tyre distortion and scrub.
2 Transmission torsional wind-up between axles is minimized (Fig. 7.26) since driving and reaction torques within each axle are not opposing but are permitted to equalize themselves through the third differential.
3 Odd tyres with different diameters are interchangeable without transmission wind-up.
4 Tractive effect and tyre grip is shared between four wheels so that wheel traction will be more evenly distributed. Therefore the amount of tractive effect per wheel necessary to propel a vehicle can be reduced.
5 Under slippery, snow or ice conditions, the third differential can generally be locked-out so that if one pair of wheels should lose traction, the other pair of wheels are still able to transmit traction.

7.6.3 Inter axle with third differential

Description of forward rear drive axle (Fig. 7.27) A third differential is generally incorporated in the forward rear axle of a tandem bogie axle drive layout because in this position it can be conveniently arranged to extend the drive to the rear axle (Fig. 7.27).

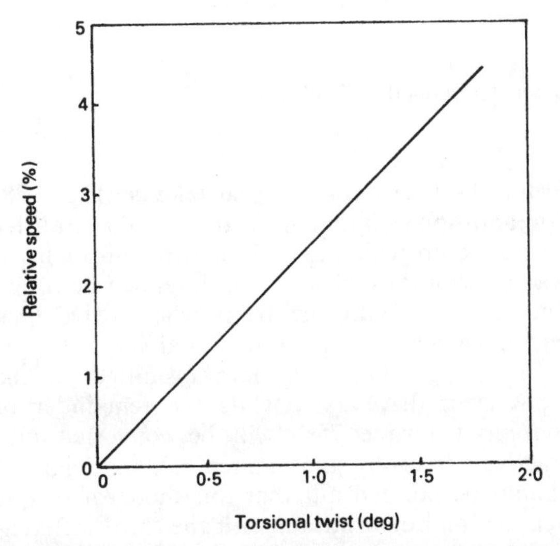

Fig. 7.26 Relationship of relative speeds of double drive axles and the amount of transmission twist

Power from the gearbox propellor shaft drives the axle input shaft. Support for this shaft is provided by a ball race mounted in the casing at the flanged end and by a spigot bearing built into the integral sun gear and output shaft at the other end. Bevel planet pinions supported on the cross-pin spider splined to the input shaft divide the drive between both of the bevel sun gears. The left hand sun gear is integral with the input helical gear and is free to rotate relative to the input shaft which it is mounted on, whereas the right hand bevel sun gear is integral with the output shaft. This output shaft is supported at the differential end by a large taper roller bearing and by a much smaller parallel roller bearing at the opposite flanged output end.

A tandem axle transmission arrangement is shown in Fig. 7.28(a) where D_1, D_2 and D_3 represent the first axle, second axle and inter axle differential respectively.

When power is supplied to the inter axle (forward rear axle) through the input shaft and to the bevel planet pinion via the cross-pin spider, the power flow is then divided between both sun gears. The drive from the left hand sun gear then passes to the input helical gear to the final drive bevel pinion helical gear where it is redirected at right angles by the crownwheel and pinion to the axle differential and half shafts.

At the same time the power flowing to the right hand sun gear goes directly to the output shaft flange where it is then transmitted to the rear axle via a pair of universal joints and a short propellor shaft.

Third differential action (Fig. 7.27) When both drive axles rotate at the same speed, the bevel planet pinions bridging the opposing sun gears bodily move around with the spider but do not revolve on their own axes. If one axle should reduce its speed relative to the other one, the planet pinions will start to revolve on their cross-pins so that the speed lost by one sun gear relative to the spider's input speed will be gained by the other sun gear.

Therefore the third differential connecting the two axles permits each axle mean speed to automatically adjust itself to suit the road operating conditions without causing any torsional wind-up between axle drives.

Third differential lock-out (Fig. 7.27) For providing maximum traction when road conditions are unfavourable such as driving over soft, slippery or steep ground, a differential lock-out clutch is incorporated. When engaged this device couples

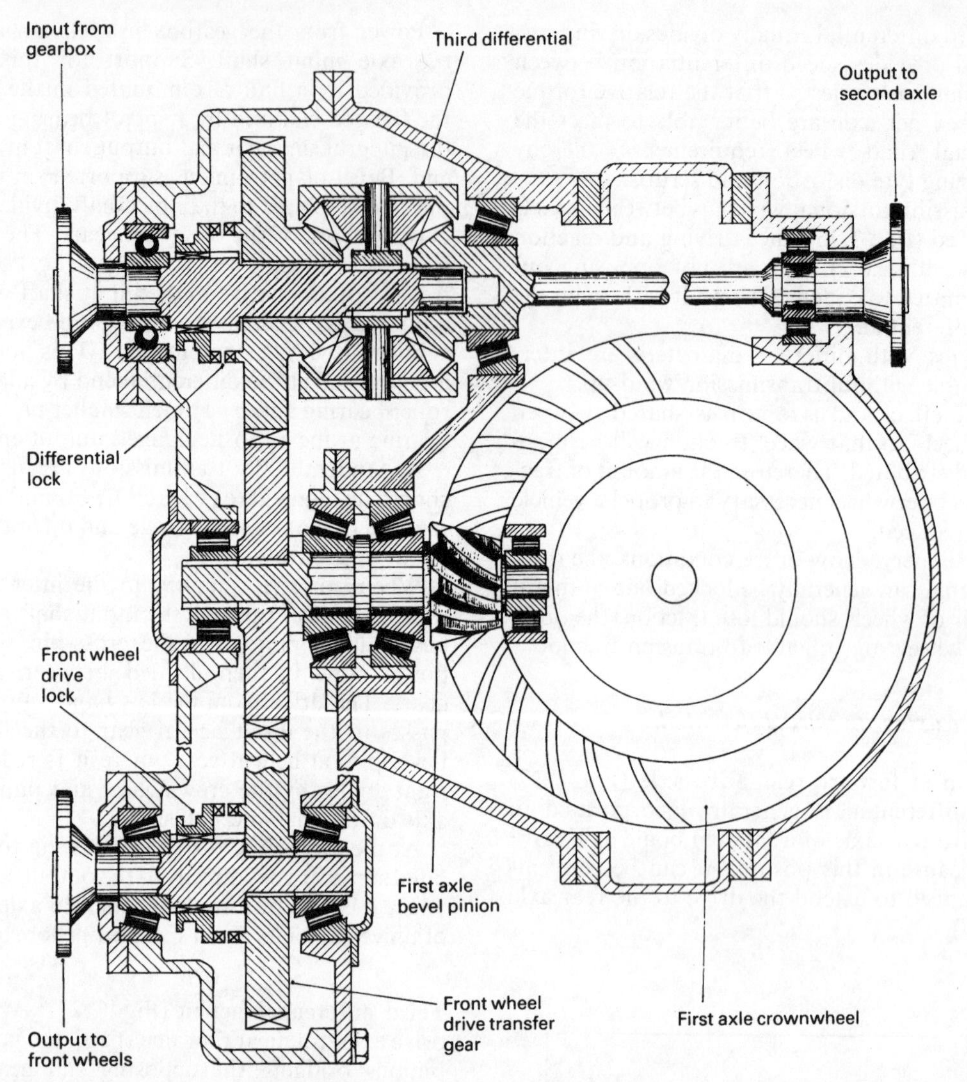

Fig. 7.27 Final drive with third differential and lock and optional transfer gearing for front

the input shaft directly with the input helical gear and left hand bevel sun gear so that the differential planet pinions are prevented from equally dividing the input torque between the two axles at the expense of axle speed differentiation. Consequently, when the third differential is locked out each axle is able to deliver independently to the other axle tractive effect which is only limited by the grip between the road wheels and the quality of surface it is being driven over. It should be observed that when the third differential lock-out is engaged the vehicle should only be operated at slow road speeds, otherwise excessive transmission wind-up and tyre wear will result.

Front wheel drive transfer gear take-up (Fig. 7.28) An additional optional feature is the transfer gear take-up which is desirable for on-off highway applications where the ground can be rough and uneven. With the front wheel drive lock clutch engaged, 25% of the total input torque from the gearbox will be transmitted to the front steer drive axle, while the remainder of the input torque 75% will be converted into tractive effect by the tandem axles. Again it should be pointed out that this mode of torque delivery and distribution with the third differential locked-out must only be used at relatively low speeds.

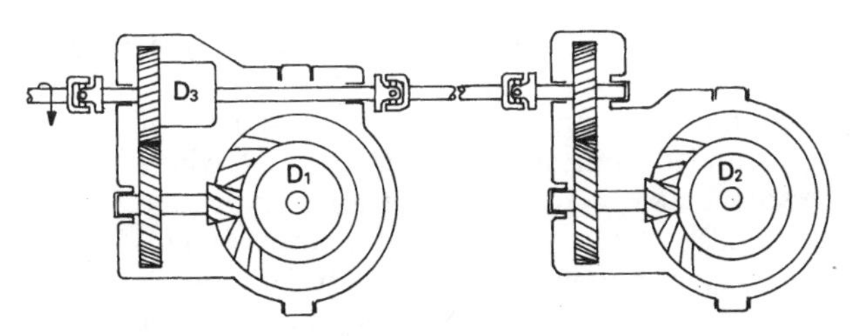

(a) Tandem spiral gear final-drive axle layout

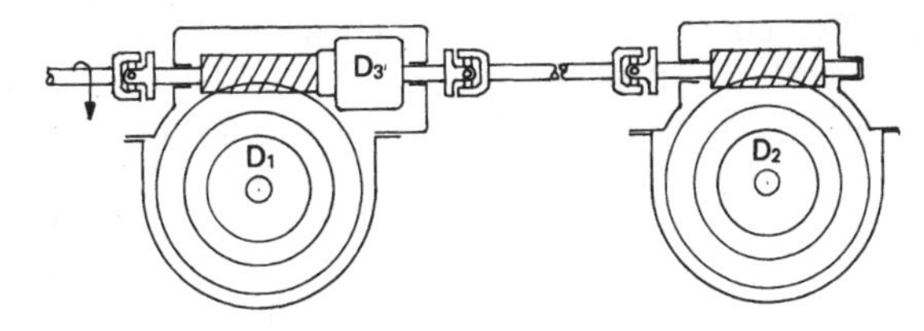

(b) Tandem worm final-drive axle layout

Fig. 7.28 (a and b) Tandem drive axle layout

*7.6.4 **Worm and worm wheel inter axle with third differential*** (Fig. 7.29)

Where large final drive gear reductions are required which may range from 5:1 to 9:1, either a double reduction axle must be used or alternatively a worm and worm wheel can provide a similar step down reduction. When compared with the conventional crownwheel and pinion final drive gear reduction the worm and worm wheel mechanical efficiency is lower but with the double reduction axle the worm and worm wheel efficiency is very similar to the latter.

Worm and worm wheel axles usually have the worm underslung when used on cars so that a very low floor pan can be used. For heavy trucks the worm is arranged to be overslung, enabling a large ground to axle clearance to be achieved.

When tandem axles are used, an inter axle third differential is necessary to prevent transmission wind-up. This unit is normally built onto the axle casing as an extension of the forward axle's worm (Fig. 7.29).

The worm is manufactured with a hollow axis and is mounted between a double taper bearing to absorb end thrust in both directions at one end and a parallel roller bearing at the other end which just sustains radial loads. The left hand sun gear is attached on splines to the worm but the right hand sun gear and output shaft are mounted on a pair of roller and ball bearings.

Power flow from the gearbox and propellor shaft is provided by the input spigot shaft passing through the hollow worm and coming out in the centre of the bevel gear cluster where it supports the internally

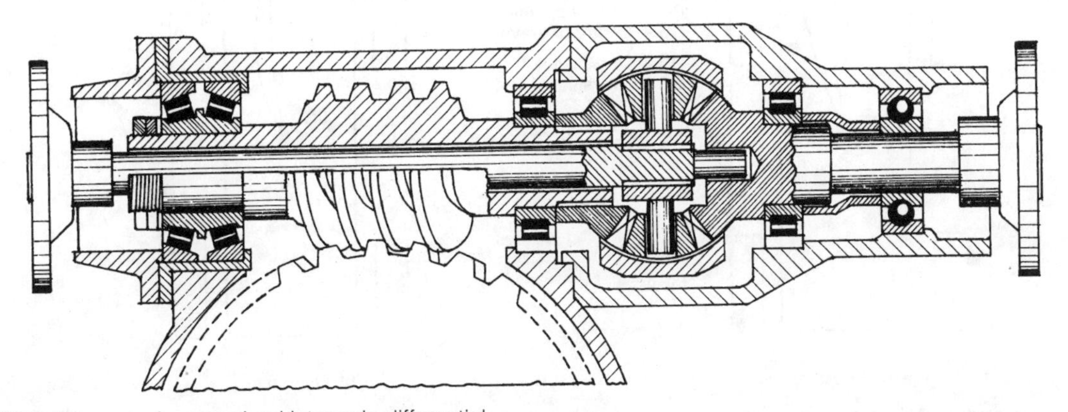

Fig. 7.29 Worm and worm wheel inter axle differential

splined cross-pin spider and their corresponding planet pinions. Power is then split between the front axle (left hand) sun gear and worm and the rear axle (right hand) sun gear and output shaft, thus transmitting drive to the second axle.

Consequently if the two axle speeds should vary, as for example when cornering, the planet pinions will revolve on their axes so that the sun gears are able to rotate at speeds slightly above and below that of the input shaft and spider, but at the same time still equally divide the torque between both axles.

Fig. 7.28(b) shows the general layout of a tandem axle worm and worm wheel drive where D_1, D_2 and D_3 represent the first axle, second axle and inter axle differentials respectively.

7.7 Four wheel drive arrangements

7.7.1 Comparison of two and four wheel drives

The total force that a tyre can transmit to the road surface resulting from tractive force and cornering for straight and curved track driving is limited by the adhesive grip available per wheel.

When employing two wheel drive, the power thrust at the wheels will be shared between two wheels only and so may exceed the limiting traction for the tyre and condition of the road surface. With four wheel drive, the engine's power will be divided by four so that each wheel will only have to cope with a quarter of the power available, so that each individual wheel will be far below the point of transmitting its limiting traction force before breakaway (skid) is likely to occur.

During cornering, body roll will cause a certain amount of weight transfer from the inner wheels to the outer ones. Instead of most of the tractive effort being concentrated on just one driving wheel, both front and rear outer wheels will share the vertical load and driving thrust in proportion to the weight distribution between front and rear axles. Thus a four wheel drive (4WD) when compared to a two wheel drive (2WD) vehicle has a much greater margin of safety before tyre to ground traction is lost.

Transmission losses overall for front wheel drive (FWD) are in the order of 10%, whereas rear wheel drive (RWD) will vary from 10% in direct fourth gear to 13% in 1st, 2nd, 3rd, and 5th indirect gears. In general, overall transmission losses with four wheel drive (4WD) will depend upon the transmission configuration and may range from 13% to 15%.

7.7.2 Understeer and oversteer characteristics (Figs 7.30 and 7.31)

In general, tractive or braking effort will reduce the cornering force (lateral force) that can be generated

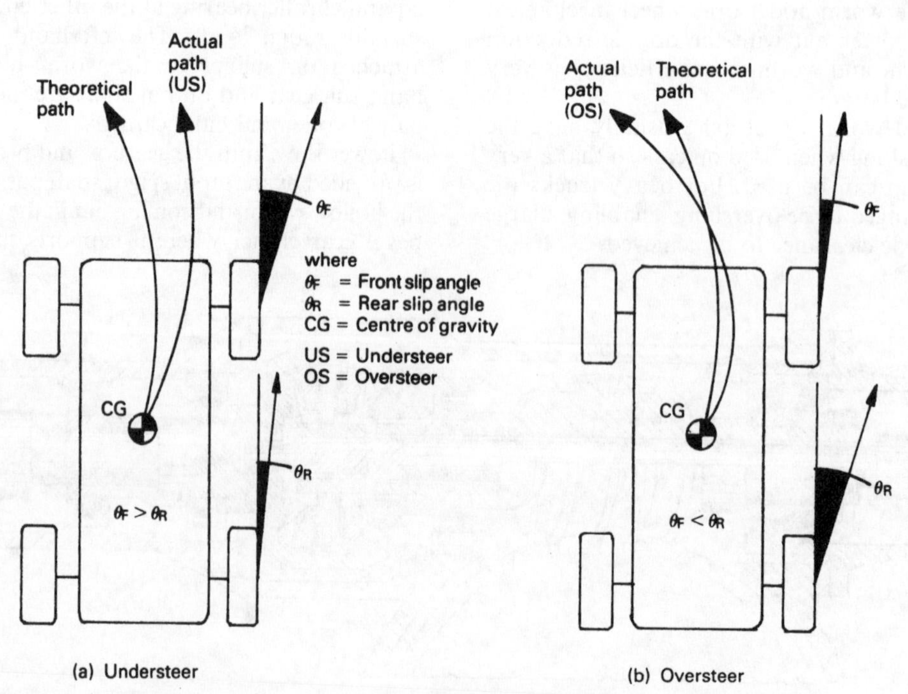

Fig. 7.30 (a and b) The influence of front and rear tyre slip angles on steering characteristics

for a given slip angle by the tyre. In other words the presence of tractive or braking effort requires larger slip angles to be produced for the same cornering force; it reduces the cornering stiffness of the tyres. The ratio of the slip angle generated at the front and rear wheels largely determines the vehicle's tendency to oversteer or understeer (Fig. 7.30).

The ratio of the front to rear slip angles when greater than unity produces understeer,

i.e. Ratio $\frac{\Theta_F}{\Theta_R} < 1.$

When the ratios of the front to rear slip angles are less than unity oversteer is produced,

i.e. Ratio $\frac{\Theta_F}{\Theta_R} > 1.$

If the slip angle of the rear tyres is greater than the front tyres the vehicle will tend to oversteer, but if the front tyres generate a greater slip angle than the rear tyres the vehicle will have a bias to understeer.

Armed with the previous knowledge of tyre behaviour when tractive effort is present during cornering, it can readily be seen that with a rear wheel drive (RWD) vehicle the tractive effort applied to propel the vehicle round a bend increases the slip angle of the rear tyres, thus introducing an oversteer effect. Conversely with a front wheel drive (FWD) vehicle, the tractive effort input during a turn increases the slip angle of the front tyres so producing an understeering effect.

Experimental results (Fig. 7.31) have shown that rear wheel drive (RWD) inherently tends to give oversteering by a small slightly increasing amount, but front and four wheel drives tend to understeer by amounts which increase progressively with speed, this tendency being slightly greater for the front wheel drive (FWD) than for the four wheel drive (4WD).

7.7.3 ***Power loss*** (Figs 7.32 and 7.33)

Tyre losses become greater with increasing tractive force caused partially by tyre to surface slippage. This means that if the total propulsion power is shared out with more driving wheels less tractive force will be generated per wheel and therefore less overall power will be consumed. The tractive force per wheel generated for a four wheel drive compared to a two wheel drive vehicle will only be half as great for each wheel, so that the overall tyre to road slippage will be far less. It has been found that the power consumed (Fig. 7.32) is least for the front wheel drive and greatest for the rear wheel drive, while the four wheel drive loss is somewhere in between the other two extremes.

The general relationship between the limiting tractive power delivered per wheel with either propulsion or retardation and the power loss at the wheels is shown to be a rapidly increasing loss as the power delivered to each wheel approaches the limiting adhesion condition of the road surface. Thus with a dry road the power loss is relatively small with

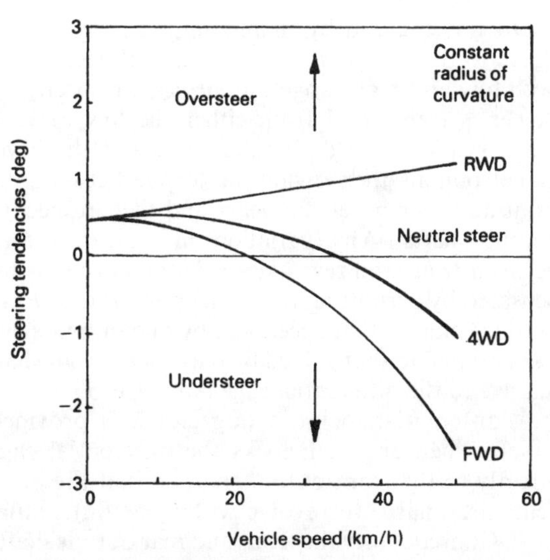

Fig. 7.31 Comparison of the over- and understeer tendency of RWD, FWD and 4WD cars on a curved track

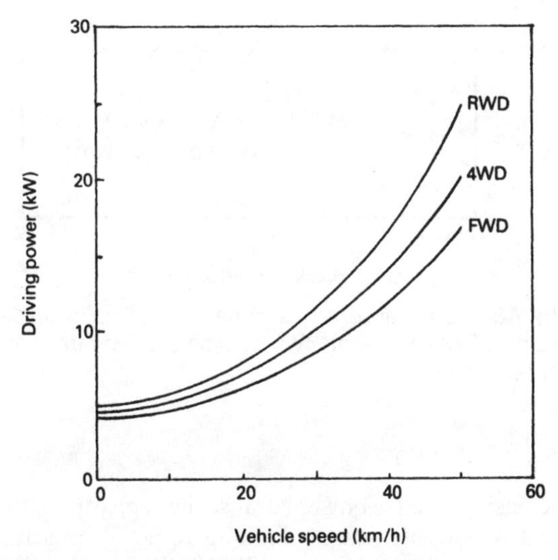

Fig. 7.32 Comparison of the power required to drive RWD, FWD and 4WD cars on a curved track at various speeds

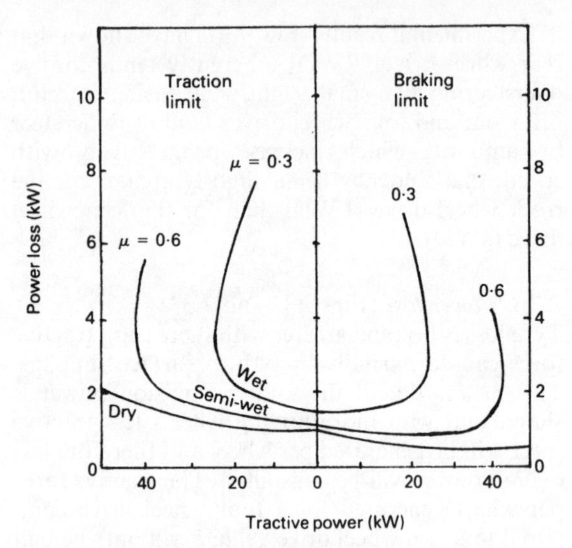

Fig. 7.33 Relationship of tractive power and power loss for different road conditions

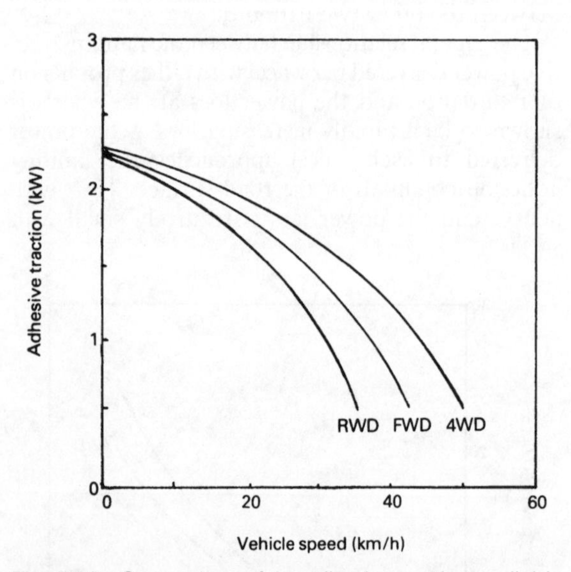

Fig. 7.34 Comparison of the adhesive traction available to Drive, RWD, FWD and 4WD cars on a curved track at various speeds

increasing tractive power because the tyre grip on the road is nowhere near its limiting value. With semi-wet or wet road surface conditions the tyre's ability to maintain full grip deteriorates and therefore the power loss increases at a very fast rate (Fig. 7.33).

7.7.4 Maximum speed (Fig. 7.34)

If friction between the tyre and road sets the limit to the maximum stable speed of a car on a bend, then the increasing centrifugal force will raise the cornering force (lateral force) and reduce the effective tractive effort which can be applied with rising speed (Fig. 7.34). The maximum stable speed a vehicle is capable of on a curved track is highest with four wheel drive followed in order by the front wheel drive and rear wheel drive.

7.7.5 Permanent four wheel drive transfer box (Land and Range Rover) (Fig. 7.35)

Transfer gearboxes are used to transmit power from the gearbox via a step down gear train to a central differential, where it is equally divided between the front and rear output shafts (Fig. 7.35). Power then passes through the front and rear propellor shafts to their respective axles and road wheels. Both front and rear coaxial output shafts are offset from the gearbox input to output shafts centres by 230 mm.

The transfer box has a low ratio of 3.32:1 which has been found to suit all vehicle applications. The high ratio uses alternative 1.003:1 and 1.667:1 ratios to match the Range Rover and Land Rover requirements respectively. This two stage reduction unit incorporates a three shaft six gear layout inside an aluminium housing. The first stage reduction from the input shaft to the central intermediate gear provides a 1.577:1 step down. The two outer intermediate cluster gears mesh with low and high range output gears mounted on an extension of the differential cage.

Drive is engaged by sliding an internally splined sleeve to the left or right over dog teeth formed on both low and high range output gears respectively. Power is transferred from either the low or high range gears to the differential cage and the bevel planet pinions then divide the torque between the front and rear bevel sun gears and their respective output shafts. Any variation in relative speeds between front and rear axles is automatically compensated by permitting the planet pinions to revolve on their pins so that speed lost by one output shaft will be equal to that gained by the other output shaft relative to the differential cage input speed.

A differential lock-out dog clutch is provided which, when engaged, locks the differential cage directly to the front output shaft so that the bevel gears are unable to revolve within the differential cage. Consequently the front and rear output shafts are compelled to revolve under these conditions at the same speed.

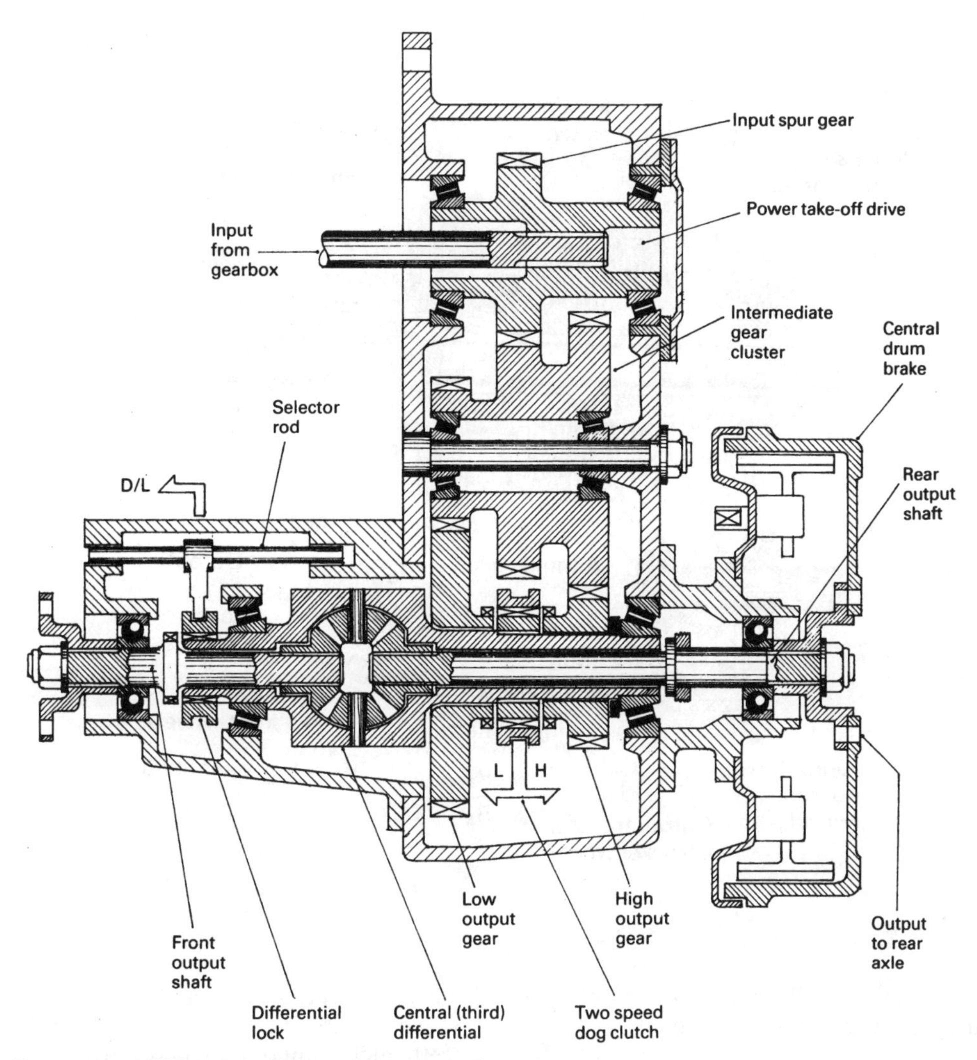

Fig. 7.35 Permanent 4WD Land and Range Rover type of transfer box

A power take-off coupling point can be taken from the rear of the integral input gear and shaft. There is also a central drum parking brake which locks both front and rear axles when applied.

It is interesting that the low range provides an overall ratio down to 40:1, which means that the gearbox, transfer box and crownwheel and pinion combined produce a gear reduction for gradient ability up to 45°.

7.7.6 Third (central) differential with viscous coupling

Description of third differential and viscous coupling (Fig. 7.36) The gearbox mainshaft provides the input of power to the third differential (sometimes referred to as the central differential). This shaft is splined to the planet pinion carrier (Fig. 7.36). The four planet pinions are supported on the carrier mesh on the outside with the internal teeth of the annulus ring gear, while on the inside the teeth of the planet pinions mesh with the sun gear teeth. A hollow shaft supports the sun gear. This gear transfers power to the front wheels via the offset input and output sprocket wheel chain drive. The power path is then completed by way of a propellor shaft and two universal joints to the front crownwheel and pinion. Mounted on a partially tubular shaped carrier is the annulus ring gear which transfers power from the planet pinions directly to the output shaft of the transfer box unit. Here the power is conveyed to the rear axle

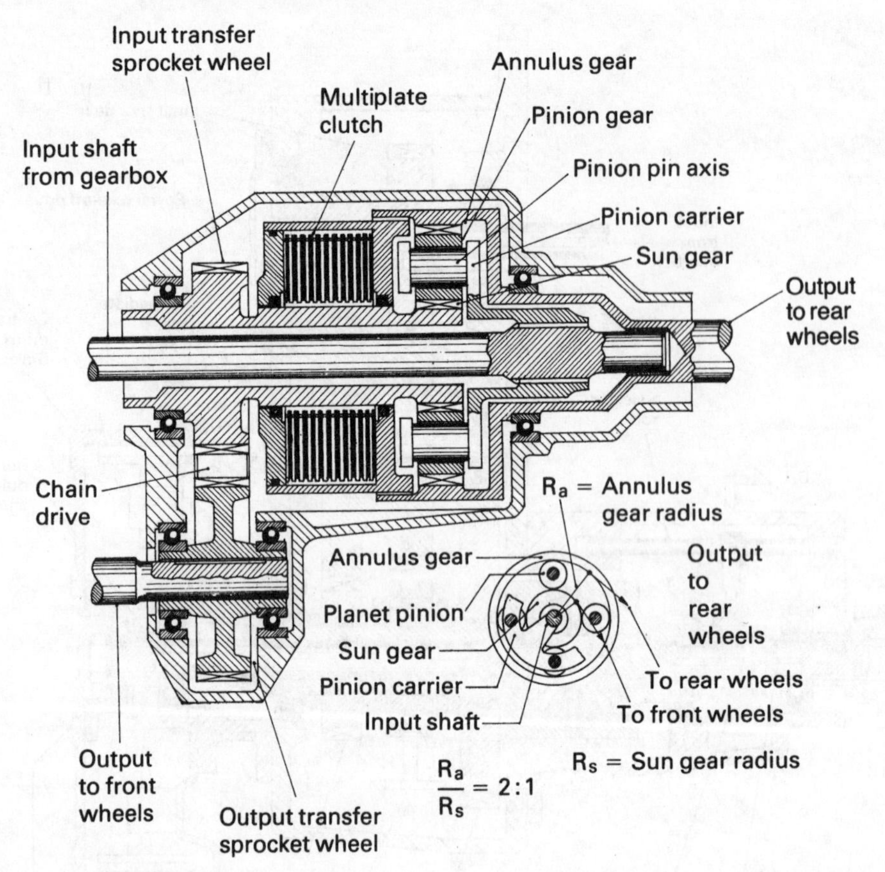

Fig. 7.36 Third differential with viscous coupling

by a conventional propellor shaft and coupled at either end by a pair of universal joints.

Speed balance of third differential assembly with common front and rear wheel speed (Fig. 7.36) Power from the gearbox is split between the sun gear, taking the drive to the front final drive. The annulus gear conveys power to the rear axle. When the vehicle is moving in the straight ahead direction and all wheels are rotating at the same speed, the whole third differential assembly (the gearbox mainshaft attached to the planet carrier), planet pinions, sun gear and annulus ring gear will all revolve at the same speed.

Torque distribution with common front and rear wheel speed (Fig. 7.36) While rear and front propellor shafts turn at the same speed, the torque split will be 66% to the rear and 34% to the front, determined by the 2:1 leverage ratio of this particular epicyclic gear train. This torque distribution is achieved by the ratio of the radii of the meshing teeth pitch point of both planet to annulus gear and planet to sun gear from the centre of shaft rotation. Since the distance from the planet to annulus teeth pitch point is twice that of the planet to sun teeth pitch point, the leverage applied to the rear wheel drive will be double that going to the front wheel drive.

Viscous coupling action (Fig. 7.36) Built in with the epicyclic differential is a viscous coupling resembling a multiplate clutch. It comprises two sets of mild steel disc plates; one set of plates are splined to the hollow sun gear shafts while the other plates are splined to a drum which forms an extension to the annulus ring gear. The sun gear plates are disfigured by circular holes and the annulus drum plates have radial slots. The space between adjacent plates is filled with a silicon fluid. When the front and rear road wheels are moving at slightly different speeds, the sun and annulus gears are permitted to revolve at speeds relative to the input planet carrier speed and yet still transmit power without causing any transmission wind-up.

Conversely, if the front or rear road wheels should lose traction and spin, a relatively large speed difference will be established between the sets of plates attached to the front drive (sun gear) and those fixed to the rear drive (annulus gear). Immediately the fluid film between pairs of adjacent plate faces shears, a viscous resisting torque is generated which increases with the relative plate speed. This opposing torque between plates produces a semi-lock-up reaction effect so that tractive effort will still be maintained by the good traction road wheel tyres. A speed difference will always exist between both sets of plates when slip occurs between the road wheels either at the front or rear. It is this speed variation that is essential to establish the fluid reaction torque between plates, and thus prevent the two sets of plates and gears (sun and annulus) from racing around relative to each other. Therefore power will be delivered to the axle and road wheels retaining traction even when the other axle wheels lose their road adhesion.

7.7.7 ***Longitudinal mounted engine with integral front final drive four wheel drive layout*** (Fig. 7.37)

The power flow is transmitted via the engine to the five speed gearbox input primary shaft. It then transfers to the output secondary hollow shaft by way of pairs of gears, each pair combination having different number of teeth to provide the necessary range of gear ratios (Fig. 7.37). The hollow secondary shaft extends rearwards to the central differential cage. Power is then divided by the planet pinions between the left and right hand bevel sun gears. Half the power flows to the front crownwheel via the long pinion shaft passing through the centre of the secondary hollow output shaft while the other half flows from the right hand sun gear to the rear axle via the universal joints and propellor shaft.

When the vehicle is moving forward in a straight line, both the front and rear axles rotate at one common speed so that the axle pinions will revolve at the same speed as the central differential cage. Therefore the bevel gears will rotate bodily with the cage but cannot revolve relative to each other.

Steering the vehicle or moving onto a bend or curved track will immediately produce unequal turning radii for both front and rear axles which meet at some common centre (instantaneous centre). Both axles will be compelled to rotate at slightly different speeds. Due to this speed variation between front and rear axles, one of the central differential sun gears will tend to rotate faster than its cage while the other one will move correspondently slower than its cage. As a result, the sun gears will force the planet pinions to revolve on their pins and at the same time revolve bodily with the cage. This speed difference on both sides of the differential is automatically absorbed by the revolving planet pinions now being permitted to move relative to the sun gears by rolling on their toothed faces. By these means, the bevel gears enable both axles to rotate at speeds demanded by their instantaneous rolling radii at any one moment without causing torsional wind-up. If travelling over very rough, soft, wet or steep terrain, better traction may be achieved with the central differential locked-out.

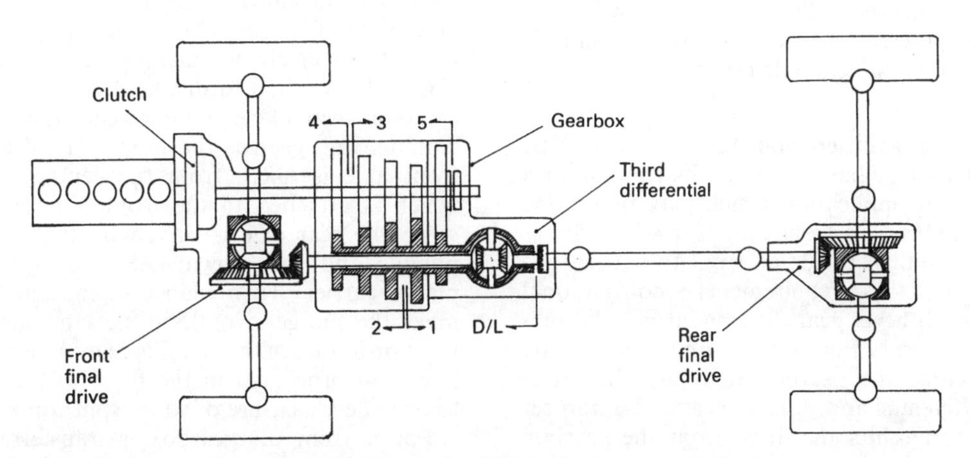

Fig. 7.37 Longitudinally mounted engine with integral front final drive four wheel drive system

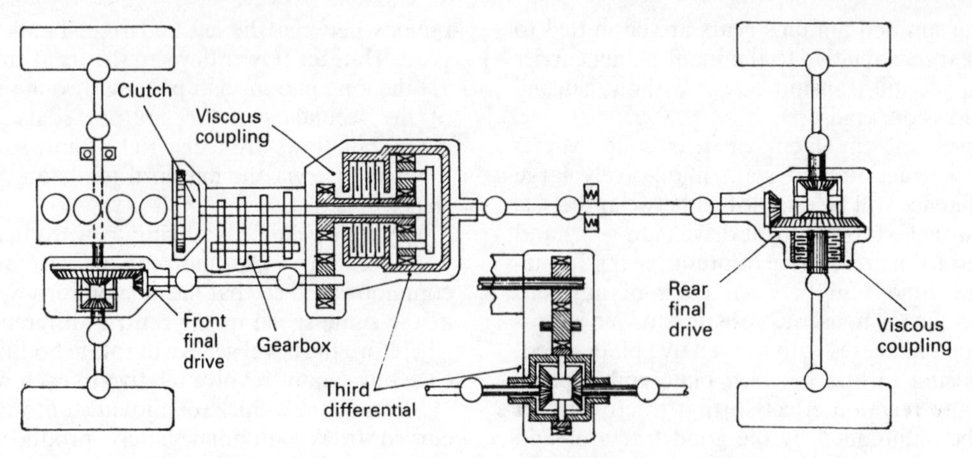

Fig. 7.38 Longitudinally mounted engine with independent front final drive four wheel drive system

7.7.8 Longitudinal mounted engine with independent front axle four wheel drive layout (Fig. 7.38)

Epicyclic gear central differential (Fig. 7.38) A popular four wheel drive arrangement for a front longitudinally mounted engine has a transfer box behind its five speed gearbox. This incorporates a viscous coupling and an epicyclic gear train to split the drive torque, 34% to the front and 66% to the rear (Fig. 7.38). A chain drives a forward facing drive shaft which provides power to the front differential mounted beside the engine sump. The input drive from the gearbox mainshaft directly drives the planet carrier and pinions. Power is diverted to the front axle through the sun gear and then flows to the hollow output shaft to the chain sprockets. Output to the rear wheels is taken from the annulus ring gear and carrier which transmits power directly to the rear axle. To minimize wheel spin between the rear road wheels a combined differential and viscous coupling is incorporated in the rear axle housing.

Bevel gear central differential (Fig. 7.38) In some cases vehicles may have a weight distribution or a cross-counting application which may find 50/50 torque split between front and rear wheel drives more suitable than the 34/66 front to rear torque split. To meet these requirements a conventional central (third) bevel gear differential may be preferred, see insert in Fig. 7.38. Again a transfer box is used behind the gearbox to house the offset central differential and transfer gears. The transfer gear train transmits the drive from the gearbox mainshaft to the central differential cage. Power then passes to the spider cross-pins which support the bevel planet pinions. Here the torque is distributed equally between the front and rear bevel sun gears, these being connected indirectly through universal joints and propellor shafts to their respective axles. When the vehicle is moving along a straight path, the planet pinions do not rotate but just revolve bodily with the cage assembly.

Immediately the vehicle is manoeuvred or is negotiating a bend, the planet pinions commence rotating on their own pins and thereby absorb speed differences between the two axles by permitting them not only to turn with the cage but also to roll round the bevel sun gear teeth at the same differential. However, they are linked together by bevel gearing which permits them independently to vary their speeds without torsional wind-up and tyre scuffing.

7.7.9 Transversely mounted engine with four wheel drive layout (Fig. 7.39)

One method of providing four wheel drive to a front transversely mounted engine is shown in Fig. 7.39. A 50/50 torque split is provided by an epicyclic twin planet pinion gear train using the annulus ring gear as the input. The drive to the front axle is taken from the central sun gears which is attached to the front differential cage, while the rear axle is driven by the twin planet pinions and the crownwheel, which forms the planet carrier. Twin planet pinions are used to make the sun gear rotate in the same direction of rotation as that of the annulus gear. A viscous coupling is incorporated in the front axle differential to provide a measure of wheel spin control.

Power from the gearbox is transferred to the annulus ring gear by a pinion and wheel, the ring

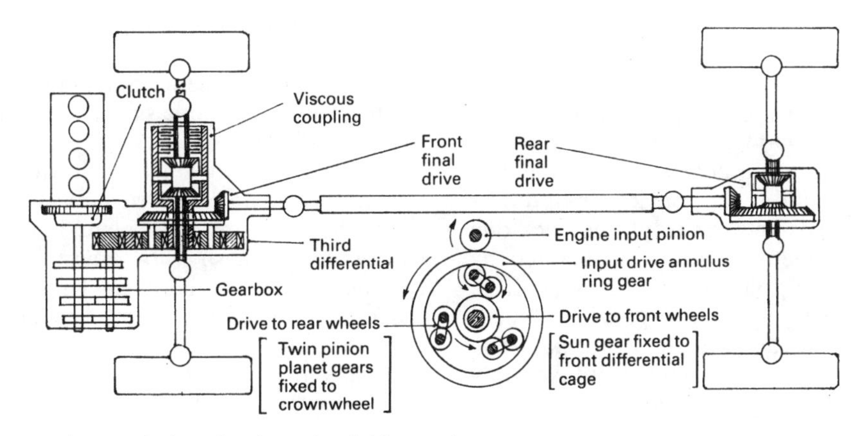

Fig. 7.39 Transversely mounted engine four wheel drive system

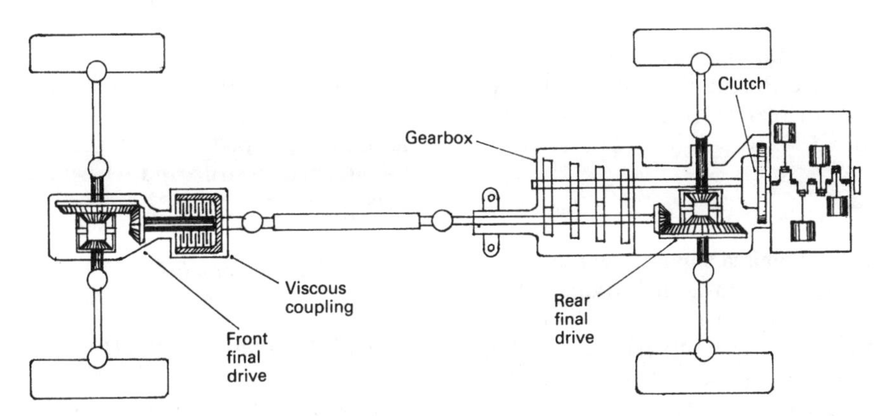

Fig. 7.40 Rear mounted engine four wheel drive system

gear having external teeth to mesh with the input pinion from the gearbox and internal teeth to drive the twin planet gears. Rotation of the annulus ring gear drives the outer and inner planet pinions and subsequently rotates the planet carrier (crownwheel in this case). The front crownwheel and pinion redirect the drive at right angles to impart motion to the propellor shaft. Simultaneously the inner planet pinion meshes with the central sun gear so that it also relays motion to the front differential cage.

7.7.10 Rear mounted engine four wheel drive layout (Fig. 7.40)

This arrangement has an integral rear engine and axle with the horizontal opposed four cylinder engine mounted longitudinally to the rear of the drive shafts and with the gearbox forward of the drive shafts (Fig. 7.40). Power to the rear axle is taken directly from the gearbox secondary output shaft to the crownwheel and pinion through 90° to the wheel hubs. Similarly power to the front axle is taken from the front end of the gearbox secondary output shaft to the front axle assembly comprised of the crownwheel and pinion differential and viscous coupling.

The viscous relative speed-sensitive fluid coupling has two independent perforated and slotted sets of steel discs. One set is attached via a splined shaft to a stub shaft driven by the propellor shaft from the gearbox, the other to the bevel pinion shank of the front final drive. The construction of the multi-interleaf discs is similar to a multiplate clutch but there is no engagement or release mechanism. Discs always remain equidistant from each other and power transmission is only by the silicon fluid which stiffens and produces a very positive fluid drag between plates. The sensitivity and effectiveness of the transference of torque is dependent upon the diameter and number of plates (in this case 59 plates), size of perforated holes and slots, surface roughness of the plates as well as temperature and generated pressure of fluid.

The drive to the front axle passes through the viscous coupling so that when both front and rear axle speeds are similar no power is transmitted to the front axle. Inevitably, in practice small differences in wheel speeds between front and back due to variations of effective wheel radii (caused by uneven load distribution, different tyre profiles, wear and cornering speeds) will provide a small degree of continuous drive to the front axle. The degree of speed sensitivity is such that it takes only one eighth of a turn in speed rotational difference between each end of the coupling for the fluid to commence to stiffen. Only when there is a loss of grip through the rear wheels so that they begin to slip does the mid-viscous coupling tend to lock-up to provide positive additional drive to the front wheels. A mechanical differential lock can be incorporated in the front or rear axles for travelling over really rough ground.

7.8 Electro-hydraulic limited slip differential

A final drive differential allows the driving wheels on each side of a vehicle to revolve at their true rolling speed without wheel slip when travelling along a straight uneven surface, a winding road or negotiating a sharp corner. If the surface should be soft, wet, muddy, or slippery for any other reason, then one or the other or even both drive wheels may lose their tyre to ground traction, the vehicle will then rely on its momentum to ride over these patchy slippery low traction surfaces. However if the vehicle is travelling very slowly and the ground surface is particularly uneven, soft or slippery, then loss of traction of one of the wheels could easily be sufficient to cause the wheel to spin and therefore to transmit no drive. Unfortunately, due to the inherent design of the bevelled gear differential the traction delivered at the good gripping wheel will be no more than that of the tyre that has lost its grip. A conventional bevelled gear differential requires that each sun (side) gear provides equal driving torque to each wheel and at the same time opposite sun (side) gears provide reaction torque equal to the driving torque of the opposite wheel. Therefore as soon as one wheel' loses ground traction its opposite wheel, even though it may have a firm tyre to ground contact, is only able to produce the same amount of effective traction as the wheel with limited ground grip.

7.8.1 Description of multiplate clutch mechanism (Fig. 7.41)

To overcome this deficiency a multiplate wet clutch is incorporated to one side of the differential cage, see Fig. 7.41. One set of the clutch plates have internal spin teeth which mesh with splines formed externally on an extended sun (side) gear, whereas the other set of inter disc plates have external spline teeth which mesh with internal splines formed inside the differential cage. Thus, when there are signs of any of the wheels losing their grip the clutch plates are automatically clamped partially or fully together. The consequence of this is to partially or fully lock both left and right hand side output drive shafts together so that the loss of drive of one drive wheel will not affect the effectiveness of the other wheel. To activate the engagement and release of the multiplate clutch, a servo-piston mounted on the right hand side bearing support flange is used: the piston is stepped and has internal seals for each step so that hydraulic fluid is trapped between the internal and external stepped piston and bearing support flange respectively.

7.8.2 Operating conditions

Normal differential action (Fig. 7.41) With good road wheel grip the multiplate clutch is disengaged by closing the delivery solenoid valve and releasing fluid to the reservoir tank via the open return solenoid valve. Under these conditions when there is a difference in speed between the inner and outer road wheels, the bevel-planet pinions are free to revolve on their axes and hence permit each sun (side) gear to rotate at the same speed as its adjacent road wheel thereby eliminating any final drive transmission windup and tyre scrub.

One wheel on the threshold of spinning (Fig. 7.41) If one wheel commences to spin due to loss of traction the wheel speed sensor instantly detects the wheel's acceleration and signals the ECU; the computer then processes this information and taking into account that a small amount of slip improves the tyre to ground traction will then energize and de-energize the delivery and return solenoid valves respectively. Fluid will now be pumped from the power assistant steering systems pump to the servo-piston, the pressure build up against the piston will engage and clamp the multiplate clutch via the thrust-pins and plate so that the differential

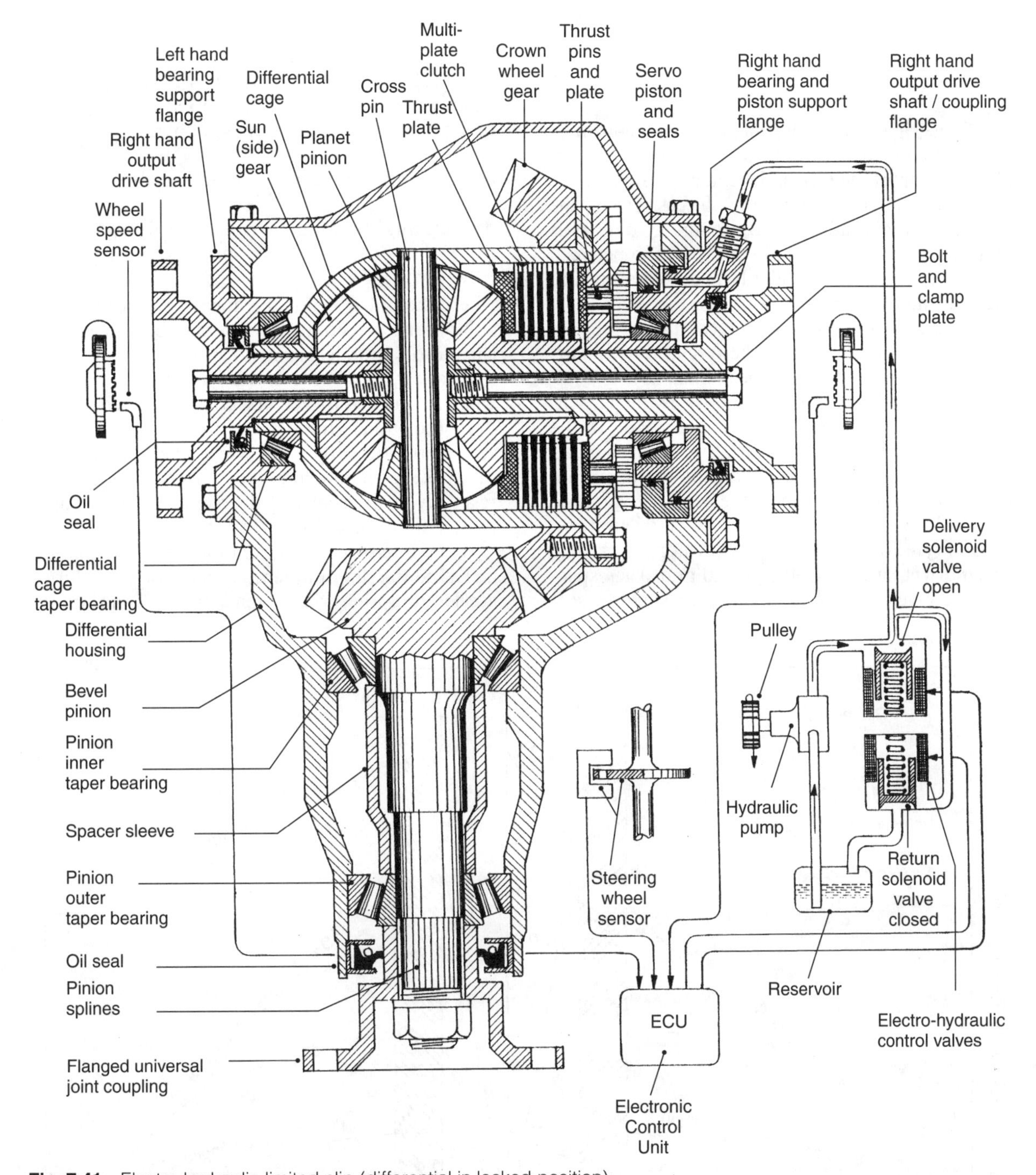

Fig. 7.41 Electro-hydraulic limited-slip (differential in locked position)

cage now is able to provide the reaction torque for the other wheel still delivering traction to the ground.

The ECU is able to take into account the speed of the vehicle and if the vehicle is turning gently or sharply which is monitored by the individual brake speed sensors and the steering wheel accelerator sensor. These two sensors therefore indirectly control the degree of lock-up which would be severe when pulling away from a standstill but would ease up with increased vehicle speed and when negotiating a bend.

7.9 Tyre grip when braking and accelerating with good and poor road surfaces (Fig. 7.42)

The function of the tyre and tread is to transfer the accelerating and decelerating forces from the wheels to the road. The optimum tyre grip is achieved when there is about 15–25% slip between the tyre tread and road under both accelerating and decelerating driving conditions, see Fig. 7.42.

Tyre grip is a measure of the coefficient of friction (μ) generated between the tyre and road surface at any instant, this may be defined as $\mu = F/W$ where F is the frictional force and W is the perpendicular force between the tyre and road. If the frictional and perpendicular forces are equal ($\mu = 1.0$) the tyre tread is producing its maximum grip, whereas if $\mu = 0$ then the grip between the tyre and road is zero, that is, it is frictionless. Typical tyre to road coefficient for a good tarmac dry and wet surface would be 1.0 and 0.7 respectively, conversely for poor surfaces such as soft snow covered roads the coefficient of friction would be as low as 0.2.

Wheel slip for accelerating and decelerating is usually measured as the slip ratio or the percentage of slip and may be defined as follows:

accelerating slip ratio = road speed/tyre speed

decelerating (braking) slip ratio = tyre speed/road speed

where the tyre speed is the linear periphery speed.

Note the percentage of slip may be taken as the slip ratio $\times$ 100. There is no slippage or very little that takes place between the tyre and road surface when a vehicle is driven at a constant speed along a dry road, under these conditions the slip ratio is zero (slip ratio = 0). Conversely heavy acceleration or braking may make the wheels spin or lock respectively thus causing the slip ratio to approach unity (slip ratio = 1.0).

If the intensity of acceleration or deceleration is increased the slip ratio tends to increase since during acceleration the wheels tend to slip and in the

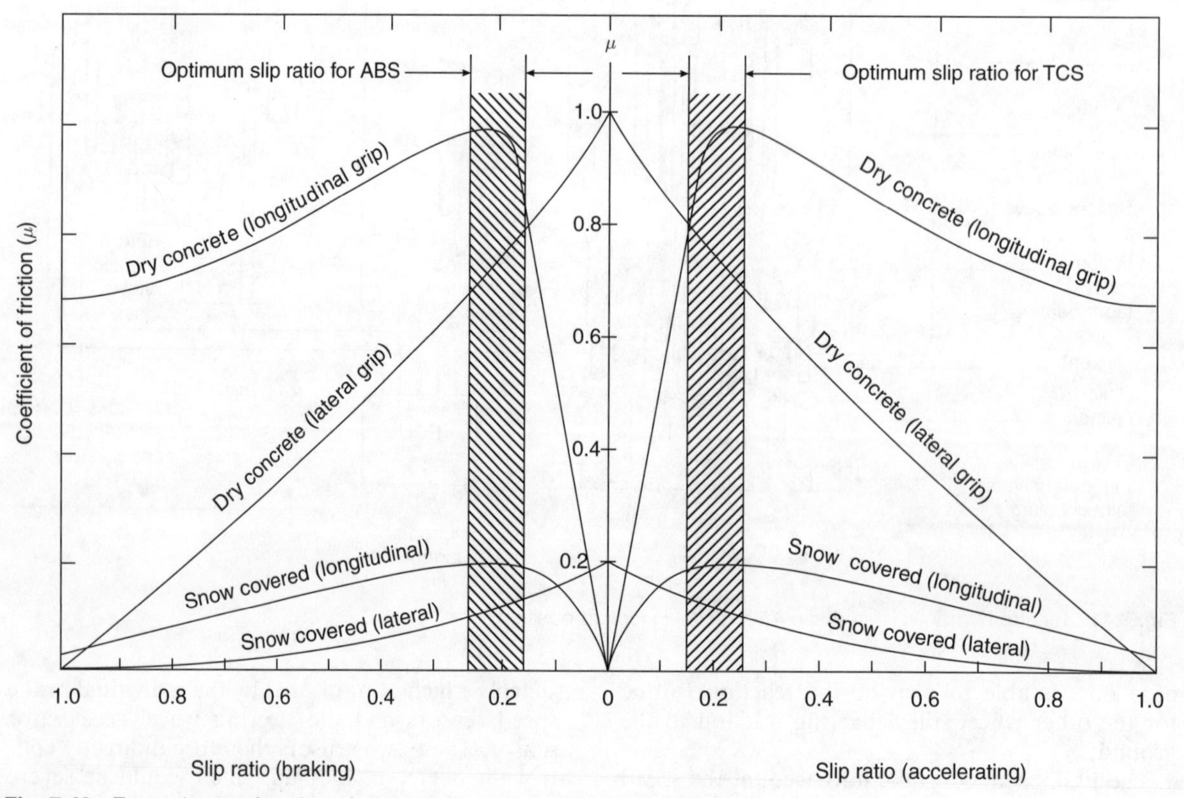

Fig. 7.42 Tyre grip as a function of slip ratio for various driving conditions

extreme spin, whereas during deceleration (braking) the wheels tend to move slower than the vehicle speed and under very heavy braking will lock, that is stop rotating and just slide along the surface. When considering the relationship between tyre slip and grip it should be observed that the tyre grip measurements are in two forms, longitudinal (lengthways) forces and lateral (sideways) forces, see Fig. 7.42. In both acceleration and deceleration in the longitudinal direction mode a tyre tends to produce its maximum grip (high μ) with a slip ratio of about 0.2 and as the slip ratio decreases towards zero, the tyre grip falls sharply; however, if the slip ratio increases beyond the optimum slip ratio of 0.2 the tyre grip will tend to decrease but at a much slower rate. With lateral direction grip in terms of sideways force coefficient of friction, the value of μ (grip) is much lower than for the forward rolling frictional grip and the maximum grip (high μ) is now produced with zero tyre slip. Traction control systems (TCS) respond to wheel acceleration caused by a wheel spinning as its tyre loses its grip with the road surface, as opposed to antilock braking systems (ABS) which respond to wheel deceleration caused by a wheel braking and preventing the wheel from turning and in the limit making it completely lock.

7.10 Traction control system

With a conventional final drive differential the torque output from each driving wheel is always equal. Thus if one wheel is driven over a slippery patch, that wheel will tend to spin and its adjacent sun (side) gear will not now be able to provide the reaction torque for the other (opposite) sun (side) gear and driving road wheel. Accordingly, the output torque on the other wheel which still has a good tyre to surface grip will be no more than that of the slipping wheel and it doesn't matter how much the driver accelerates to attempt to regain traction, there still will be insufficient reaction torque on the spinning side of the differential for the good wheel to propel the vehicle forward.

One method which may be used to overcome this loss of traction when one wheel loses its road grip, is to simply apply the wheel brake of the wheel showing signs of spinning so that a positive reaction torque is provided in the differential; this counteracts the delivery to the good wheel of the half share of the driving torque being supplied by the combined engine and transmission system in terms of tractive effort between the tyre and ground.

To achieve this traction control an electronic control unit (ECU) is used which receives signals from individual wheel speed sensors, and as soon as one of the driving wheels tends to accelerate (spin due to loss of tyre to ground traction), the sensor's generated voltage change is processed by the ECU computer, and subsequently current is directed to the relevant traction solenoid valve unit so that hydraulic brake pressure is transmitted to the brake of the wheel about to lose its traction. As soon as the braked wheel's speed has been reduced to a desirable level, then the ECU signals the traction solenoid valve unit to release the relevant wheel brake.

7.10.1 Description of system (Fig. 7.43)

This traction control system consists of: an electric motor driven hydraulic pump which is able to generate brake pressure independently to the foot brake master cylinder and a pressure storage accumulator; a traction boost unit which comprises a cylinder housing, piston and poppet valve, the purpose of which is to relay hydraulic pressure to the appropriate wheel brake caliper and at the same time maintain the traction boost unit circuit fluid separate from the foot brake master cylinder fluid system; a pair of traction solenoid valve units each having an outlet and return valve regulates the cut-in and -out of the traction control; an electronic control unit (ECU) is provided and individual wheel speed sensors which monitor the acceleration of both driven and non-driving wheels. Should the speed of either of the driven wheels exceed the mean wheel speed of the non-driven wheels by more than about 20%, then the ECU will automatically apply the appropriate wheel brake via the traction solenoid valves and traction boost device.

7.10.2 Operating conditions

Foot brake applied (Fig. 7.43(a)) With the foot pedal released brake fluid is able to flow freely between the master cylinder and both brake calipers via the open traction boost unit. The boost piston will be in its outer-most position thereby holding the poppet valve in its fully open position. When the foot pedal is pushed down, brake fluid pressure will be transmitted though the fluid via the traction boost poppet valve to both of the wheel calipers thus causing the brakes to be applied.

Traction control system applied (off-side slipping wheel braked) (Fig. 7.43(b)) When one of the driving wheels begins to spin (off-side wheel in this example) the adjacent speed sensor voltage change signals the ECU, immediately it computes

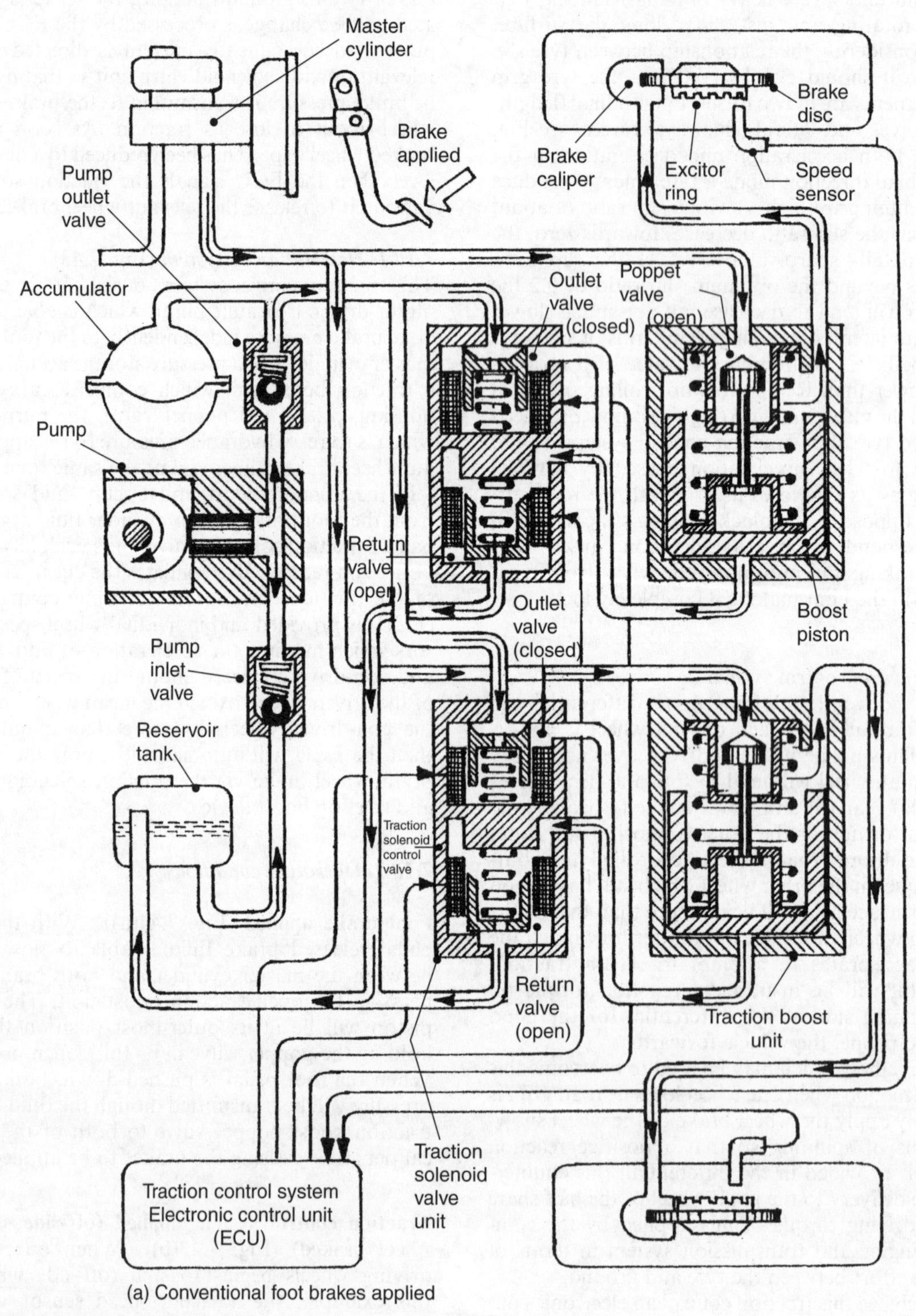

Fig. 7.43 (a and b) Traction control system

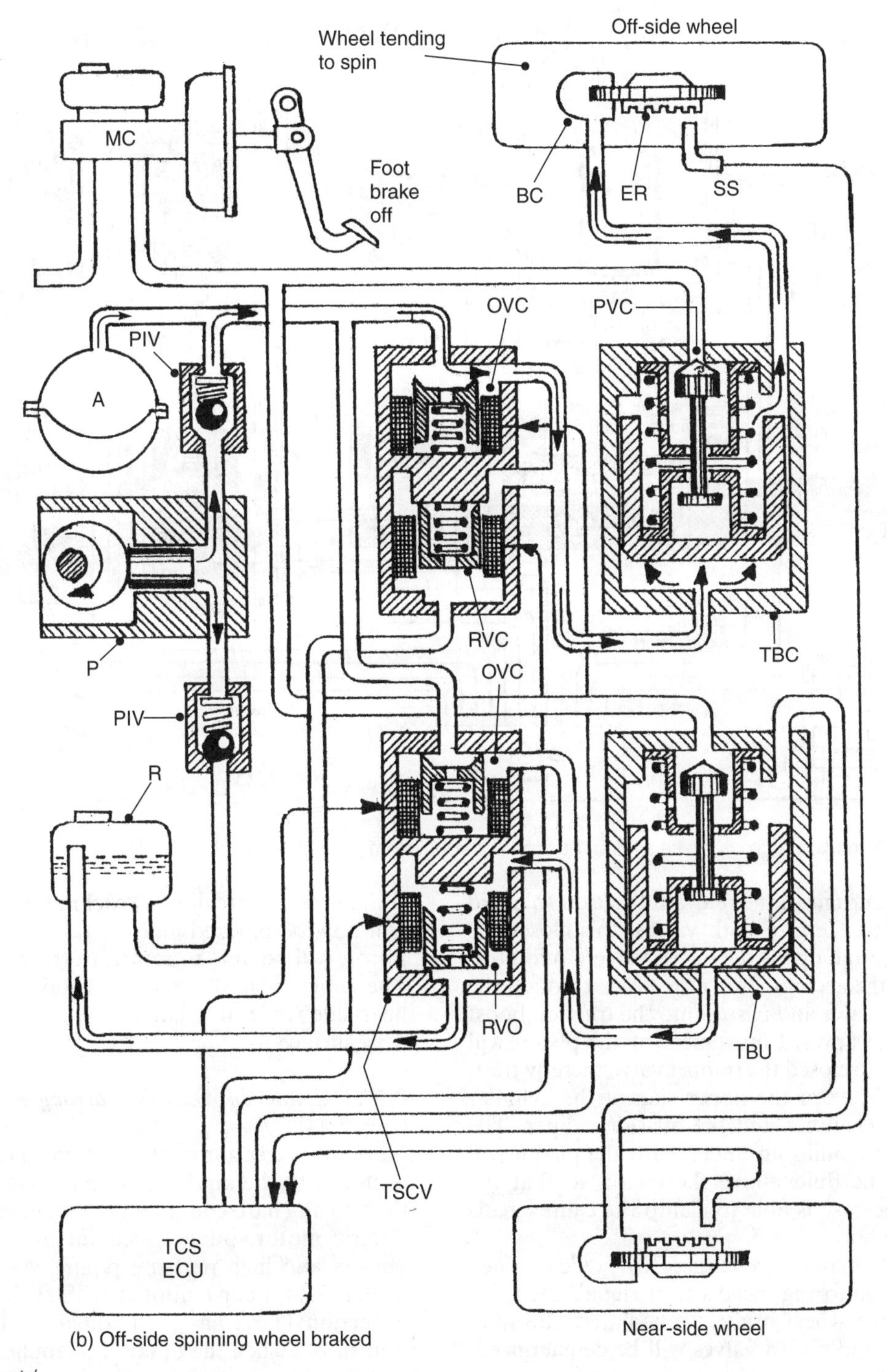

(b) Off-side spinning wheel braked

Fig. 7.43 *contd*

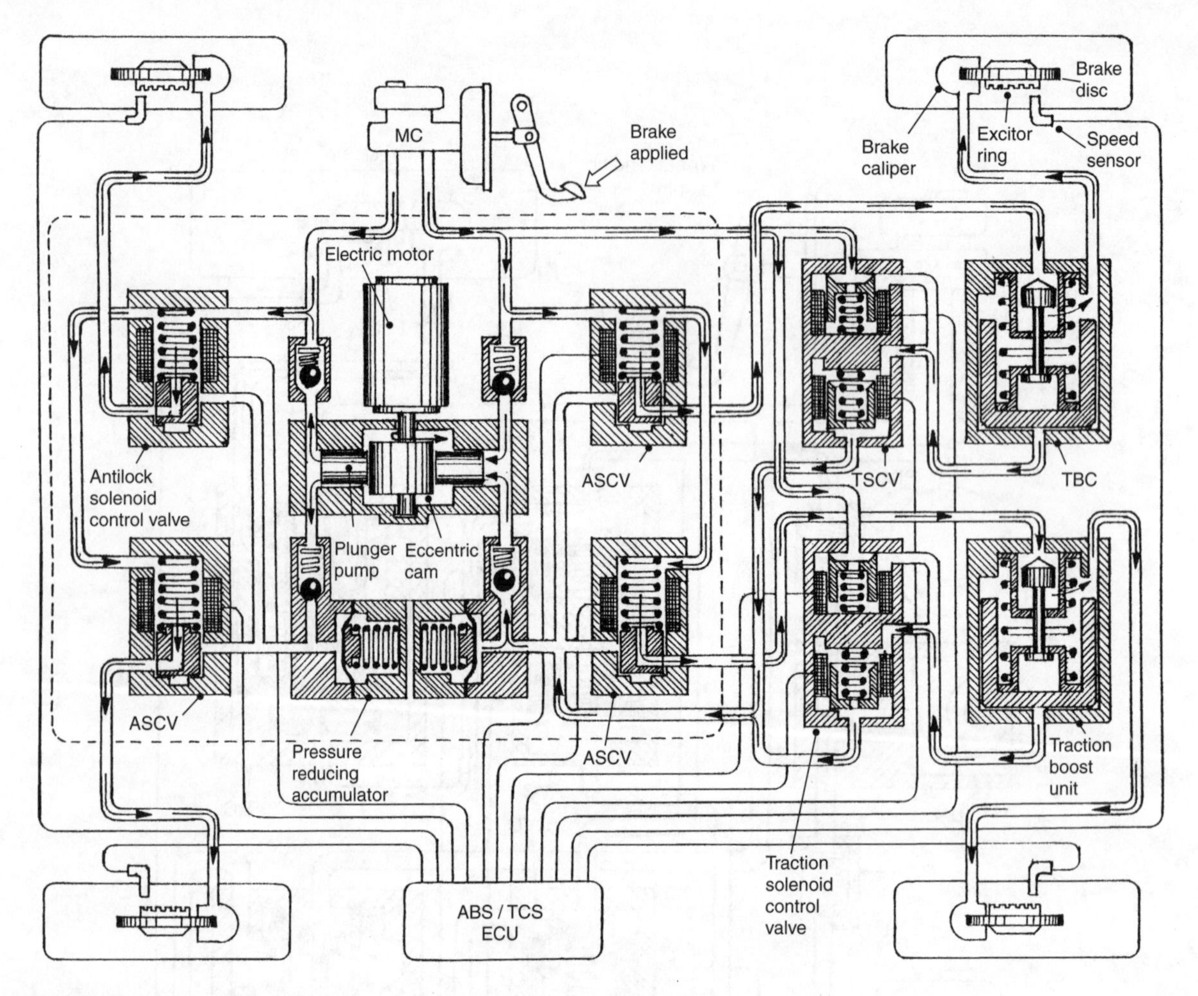

Fig. 7.44 Combined antilock brake system/traction control system (brakes applied)

and relays current to the relevant traction solenoid valve unit to energize both valves, this closes the return valve and opens the outlet valve. Fluid pressure from the accumulator now flows through the open outlet valve and passes into the traction boost cylinder, the upward movement of the piston will instantly snap closed the poppet valve thereby trapping fluid between the upper side of the cylinder and piston chamber and the off-side caliper. The fluid pressure build up underneath the piston will pressurize the fluid above the piston so that the pressure increase is able to clamp the caliper pads against the brake disc.

As the wheel spin speed reduces to a predetermined value the monitoring speed sensor signals the ECU to release the wheel brake, immediately the solenoid outlet and return valves will be de-energized, thus causing them to close and open respectively. Fluid pressure previously reaching the boost piston will now be blocked and the fluid underneath the piston will be able to return to the reservoir tank. The same cycle of events will take place for the near-side wheel if it happens to move over a slippery surface.

7.10.3 Combined ABS/TCS arrangement

(Fig. 7.44)

Normally a traction control system (TCS) is incorporated with the antilock braking system (ABS) so that it can share common components such as the electric motor, pump, accumulator, wheel brake sensors and high pressure piping. As can be seen in Fig. 7.44 a conventional ABS system described in section 11.7.2 has been added to. This illustration shows when the brakes are applied fluid pressure is transmitted indirectly through the antilock solenoid control valve to the front brake calipers; however, assuming a rear wheel drive, fluid pressure also is transmitted to the rear brakes via the antilock solenoid control valve and then through the traction boost unit to the wheel brake calipers thus applying the brakes.

ABS operating (Fig. 7.44) When the wheel brake speed sensor signals that a particular wheel is tending towards wheel lock, the appropriate antilock solenoid control valve will be energized so that fluid pressure to that individual wheel brake is blocked and the entrapped fluid pressure is released to the pressure reducing accumulator (note Fig. 7.44 only shows the system in the foot brake applied position).

TCS operating (Fig. 7.44) If one of the wheel speed sensors signals that a wheel is moving towards slip and spin the respective traction solenoid control valve closes its return valve and opens its outlet valve; fluid pressure from the pump now provides the corresponding boost piston with an outward thrust thereby causing the poppet valve to close (note Fig. 7.44 only shows the system in the foot brake applied position). Further fluid pressure acting on the head of the piston now raises the pressure of the trapped fluid in the pipe line between the boost piston and the wheel caliper. Accordingly the relevant drive wheel is braked to a level that transmits a reaction torque to the opposite driving wheel which still retains traction.

One limitation of a brake type traction control is that a continuous application of the TCS when driving over a prolonged slippery terrain will cause the brake pads and disc to become excessively hot; it thus may lead to brake fade and a very high wear rate of the pads and disc.

8 Tyres

8.1 Tractive and braking properties of tyres

8.1.1 Tyre grip

Tyres are made to grip the road surface when the vehicle is being steered, accelerated, braked and/or negotiating a corner and so the ability to control the tyre to ground interaction is of fundamental importance. Road grip or friction is a property which resists the sliding of the tyre over the road surface due to a retardant force generated at the tyre to ground contact area. The grip of different tyres sliding over various road surface finishes may be compared by determining the coefficient of friction for each pair of rubbing surfaces.

The coefficient of friction may be defined as the ratio of the sliding force necessary to steadily move a solid body over a horizontal surface to the normal reaction supporting the weight of the body on the surface (Fig. 8.1).

$$\text{i.e. Coefficient of friction } (\mu) = \frac{\text{Frictional force}}{\text{Normal reaction}} = \frac{F}{W}$$

where μ = coefficient of friction
F = frictional force (N)
W = normal reaction (weight of body) (N)

Strictly speaking, the coefficient of friction does not take into account the surface area tread pattern which maximizes the interlocking mechanism

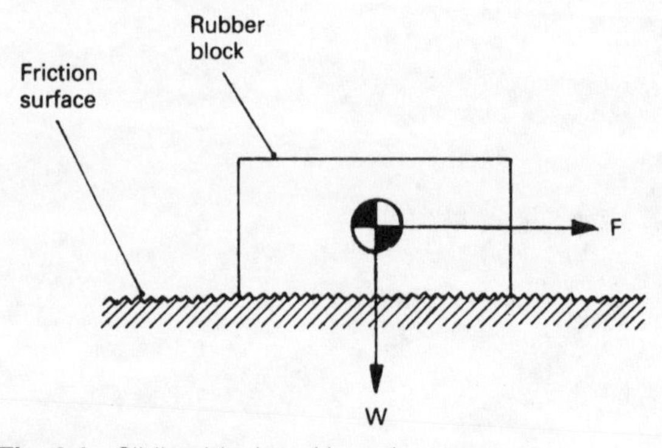

Fig. 8.1 Sliding block and board

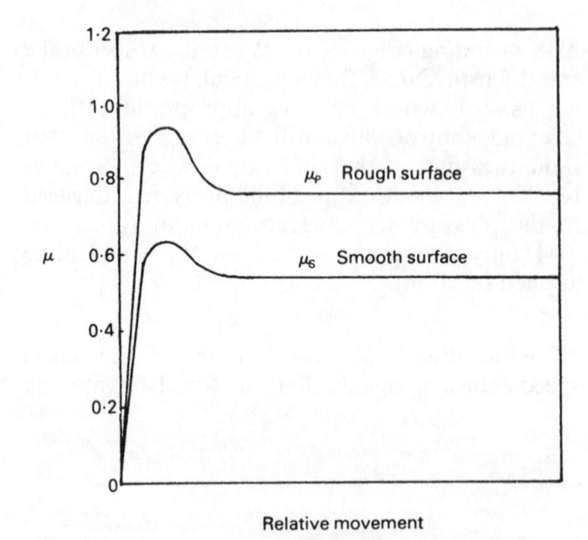

Fig. 8.2 Variation of friction with relative movement

between the flexible tread elements and road. Therefore when dealing with tyres it is usual to refer to the *coefficient of adhesive friction*. The maximum coefficient of adhesive friction created between a sliding tread block and a solid surface occurs under conditions of slow movement or creep (Fig. 8.2). This critical stage is known as the *peak coefficient*, μ_p, and if the relative movement of the rubber on the surface is increased beyond this point the friction coefficient falls. It continues to fall until bodily sliding occurs, this stage being known as the *sliding coefficient* μ_s. Sliding friction characteristics are consistent with the behaviour of rolling tyres.

A modern compound rubber tyre will develop a higher coefficient of friction than natural rubber. In both cases their values decrease as the road surface changes from dry to a wet condition. The rate of fall in the coefficient of friction is far greater with a worn tyre tread as opposed to a new tyre as the degree of road surface wetness increases (Fig. 8.3).

It has been found that the frictional grip of a bald tyre tread on a rough dry road surface is as good or even better than that achieved with a new tread (Fig. 8.3). The reason for this unexpected result is due to the greater amount of rubber interaction with the ground surface for a given size of contact patch. It therefore develops a larger reaction force which

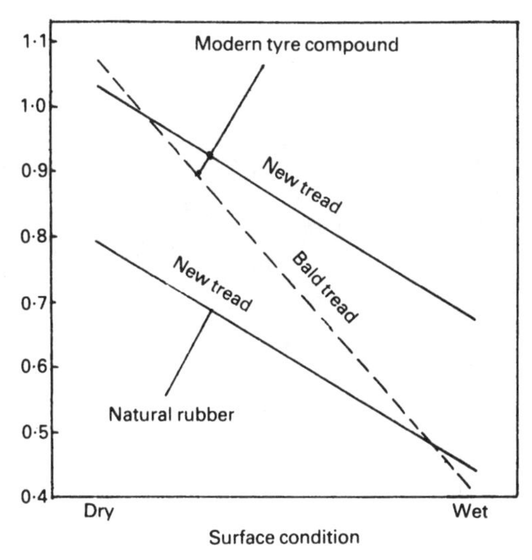

Fig. 8.3 Effect of surface condition on the coefficient of adhesive friction with natural and synthetic rubber using new and bald tyre treads

opposes the movement of the tyre. Under ideal road conditions and the amount of deformable rubber actually in contact with the road maximized for a given contact path area, the retarding force which can be generated between the tyre and ground can equal the vertical load the wheel supports. In other words, the coefficient of adhesive friction can reach a value of 1.0. However, any deterioration in surface roughness due to surface ridges being worn, or chippings becoming submerged in asphalt, or the slightest amount of wetness completely changes the situation. A smooth bald tyre will not be able to grip the contour of the road, whereas the tyre with a good tread pattern will easily cope and maintain a relatively high value of retardant force.

When transmitting tractive or braking forces, the tyre is operating with slip or creep. It is believed that the maximum friction is developed when a maximum number of individual tread elements are creeping at or near an optimum speed relative to the ground. The distribution by each element of the tread is not equal nor is it uniform throughout the contact patch. The frictional forces developed depend upon the pressure distribution within the contact patch area and the creep effects. Once bodily slippage begins to occur in one region of the contact area, the progression to the fully sliding condition of the contact area as a whole is extremely rapid.

Under locked wheel conditions, the relative sliding speed between a tyre tread and the road is the speed of the vehicle. If, however, the braking is such that the wheels are still rotating, the actual speed between the tyre tread and the road must be less than that of the vehicle. Even on surfaces giving good braking when wet, maximum coefficients occur at around 10–20% slip. This means that the actual speed between the tyre tread and the road is around one eighth of the vehicle speed or less. Under these conditions, it is possible to visualize that the high initial peak value occurs because the actual tyre ground relative speed relates to a locked wheel condition at a very low vehicle speed (Fig. 8.3).

The ability to utilize initial peak retardation under controlled conditions is a real practical asset to vehicle retardation and, because the tyre is still rolling, to vehicle directional control.

Braking effectiveness can therefore be controlled and improved if the wheels are prevented from completely locking in contrast to the wheels actually being locked when the brakes are applied. Thus when braking from different speeds (Fig. 8.4) it can be seen that the unlocked wheels produce a higher peak coefficient of adhesive friction as opposed to the locked condition which generates only a sliding coefficient of adhesive friction. In both situations the coefficient of adhesive friction decreases as the speed from which braking first commences increases.

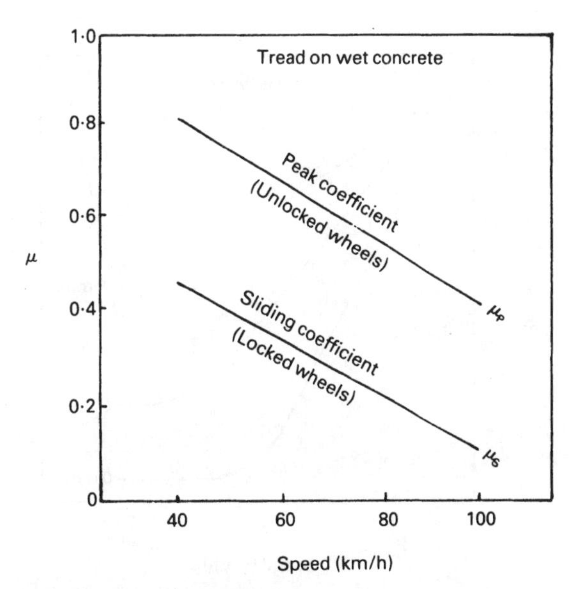

Fig. 8.4 Effect of speed on both peak and sliding coefficient of adhesive friction

8.1.2 Grip control

Factors influencing the ability of a tyre to grip the road when being braked are:

a) the vehicle speed,
b) the amount of tyre wear,
c) the nature of the road surface,
d) the degree of surface wetness.

Vehicle speed (Fig. 8.4) Generally as the speed of the vehicle rises, the time permitted for tread to ground retardation is reduced so that the grip or coefficient of adhesive friction declines (Fig. 8.4).

Tyre wear (Fig. 8.5) As the tyre depth is reduced, the ability for the tread to drain off water being swept in front of the tread is reduced. Therefore with increased vehicle speed inadequate drainage will reduce the tyre grip when braking (Fig. 8.5).

Road surface wetness (Fig. 8.6) The reduction in tyre grip when braking from increased vehicle speed drops off at a much greater rate as the rainfall changes from light rain, producing a surface water depth of 1 mm, to a heavy rainstorm flooding the road to a water depth of about 2.5 mm (Fig. 8.6).

Road surface texture (Fig. 8.7) A new tyre braked from various speeds will generate a higher peak coefficient of adhesive friction with a smaller fall off at the higher speeds on wet rough surfaces compared to braking on wet smooth surfaces (Fig. 8.7).

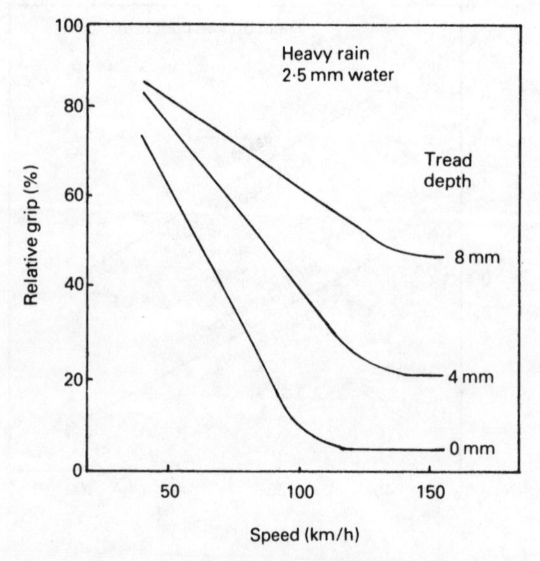

Fig. 8.5 Effect of speed on relative tyre grip with various tread depth when braking on a wet road

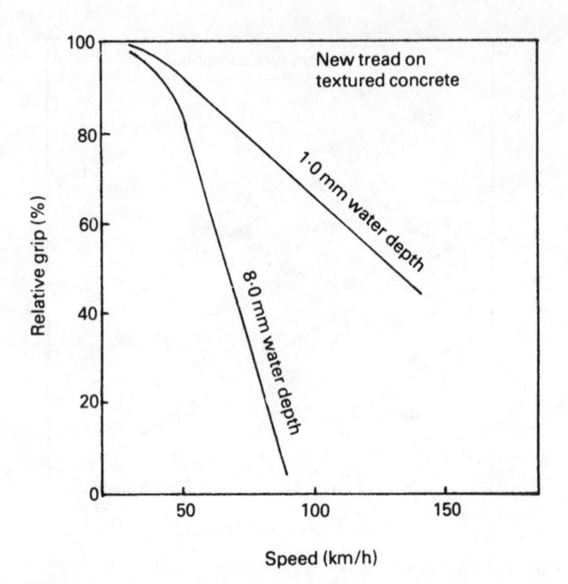

Fig. 8.6 Effect of speed on relative tyre grip with various road surface water depths

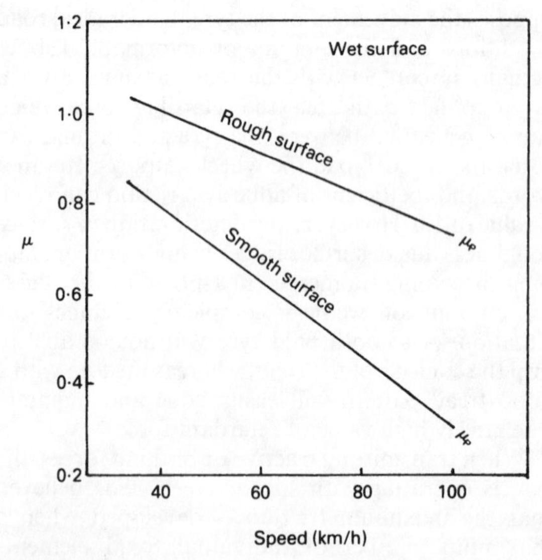

Fig. 8.7 Effect of speed on the coefficient of adhesive friction with both wet rough and smooth surfaces

The reduction in the coefficient of adhesive friction when braking with worn tyres on both rough and particularly smooth wet surfaces will be considerably greater.

8.1.3 Road surface texture (Fig. 8.8)

A road surface finish may be classified by its texture which may be broadly divided in *macrotexture,*

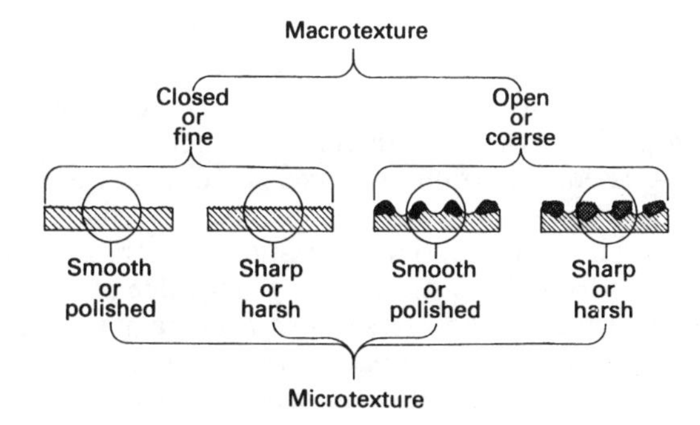

Fig. 8.8 Terminology and road surface texture

which represents the surface section peak to valley ripple or roughness, and *microtexture* which is a measure of the smoothness of the ripple contour (Fig. 8.8). Further subdivisions may be made; macrotexture may range from closed or fine going onto open or coarse whereas microtexture may range from smooth or polished extending to sharp or harsh.

For good tyre grip under dry and wet conditions the road must fulfil two requirements. Firstly, it must have an open macrotexture to permit water drainage. Secondly, it should have a microtexture which is harsh; the asperities of the texture ripples should consist of many sharp points that can penetrate any remaining film of water and so interact with the tread elements. If these conditions are fulfilled, a well designed tyre tread will provide grip not only under dry conditions but also in wet weather. A worn road surface may be caused by the hard chippings becoming embedded below the soft asphalt matrix or the microtexture of these chippings may become polished. In the case of concrete roads, the roughness of the brushed or mechanically ridged surface may become blunted and over smooth. To obtain high frictional grip over a wide speed range and during dry and wet conditions, it is essential that the microtexture is harsh so that pure rubber to road interaction takes place.

8.1.4 Braking characteristics on wet roads (Fig. 8.9)

Maximum friction is developed between a rubber tyre tread and the road surface under conditions of slow movement or creep. A tyre's braking response on a smooth wet road with the vehicle travelling at a speed, say 100 km/h, will show the following characteristics (Fig. 8.9).

When the brakes are in the first instance steadily applied, the retardation rate measured as a fraction of the gravitational acceleration (g m/s^2) will rise rapidly in a short time interval up to about 0.5 g. This phase of braking is the normal mode of braking when driving on motorways. In traffic, it enables the

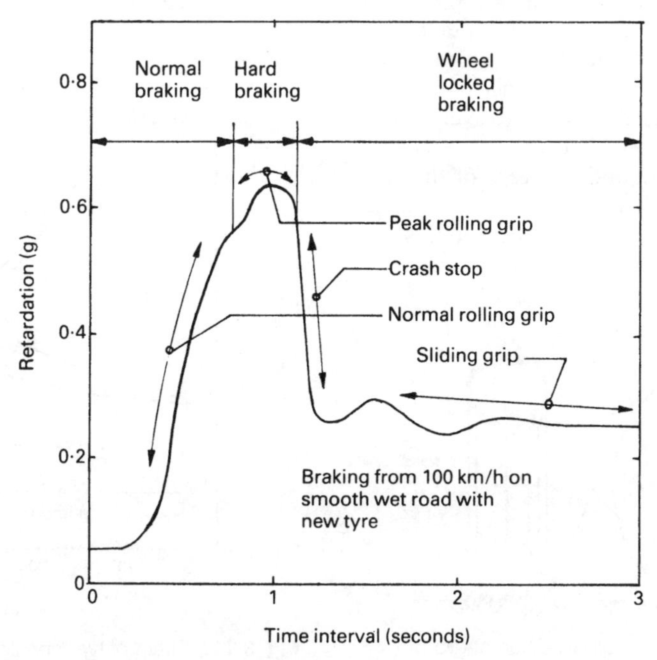

Fig. 8.9 Possible retardation braking cycle on a wet road

driver to reduce the vehicle speed fairly rapidly with good directional stability and no wheel lock taking place. If an emergency braking application becomes necessary, the driver can raise the foot brake effort slightly to bring the vehicle retardation to its peak value of just over 0.6 g, but then should immediately release the brake, pause and repeat this on-off sequence until the road situation is under control. Failing to release the brake will lock the wheels so that the tyre road grip changes from one of rolling to sliding. As the wheels are prevented from rotating, the braking grip generated between the contact patches of the tyres drops drastically as shown in the crash stop phase. If the wheels then remain locked, the retardation rate will steady at a much lower value of just over 0.2 g. The tyres will now be in an entirely sliding mode, with no directional stability and with a retardation at about one third of the attainable peak value. With worn tyre treads the braking characteristics of the tyres will be similar but the braking retardation capacity is considerably reduced.

8.1.5 Rolling resistance (Figs 8.10 and 8.11)

When a loaded wheel and tyre is compelled to roll in a given direction, the tyre carcass at the ground interface will be deflected due to a combination of the vertical load and the forward rolling effect on the tyre carcass (Fig. 8.10). The vertical load tends to flatten the tyre's circular profile at ground level, whereas the forward rolling movement of the wheel will compress and spread the leading contact edge and wall in the region of the tread. At the same time, the trailing edge will tend to reduce its contact pressure and expand as it is progressively freed from the ground reaction. The consequences of the continuous distortion and recovery of the tyre carcass at ground level means that energy is being used in rolling the tyre over the ground and it is not all returned as strain energy as the tyre takes up its original shape. (Note that this has nothing to do with a tractive force being applied to the wheel to propel it forward.) Unfortunately when the carcass is stressed, the strain produced is a function of the stress. On releasing the stress, because the tyre material is not perfectly elastic, the strain lags behind so that the strain for a given value of stress is greater when the stress is decreasing than when it is increasing. Therefore, on removing the stress completely, a residual strain remains. This is known as hysteresis and it is the primary cause of the rolling resistance of the tyre.

The secondary causes of rolling resistance are air circulation inside the tyre, fan effect of the rotating tyre by the air on the outside and the friction between the tyre and road caused by tread slippage. A typical analysis of tyre rolling resistance losses at high speed can be taken as 90–95% due to internal hysteresis, 2–10% due to friction between the tread and ground, and 1.5–3.5% due to air resistance.

Rolling resistance is influenced by a number of factors as follows:

a) cross-ply tyres have higher rolling resistance than radial ply (Fig. 8.11),
b) the number of carcass plies and tread thickness increase the rolling resistance due to increased hysteresis,
c) natural rubber tyres tend to have lower rolling resistance than those made from synthetic rubber,

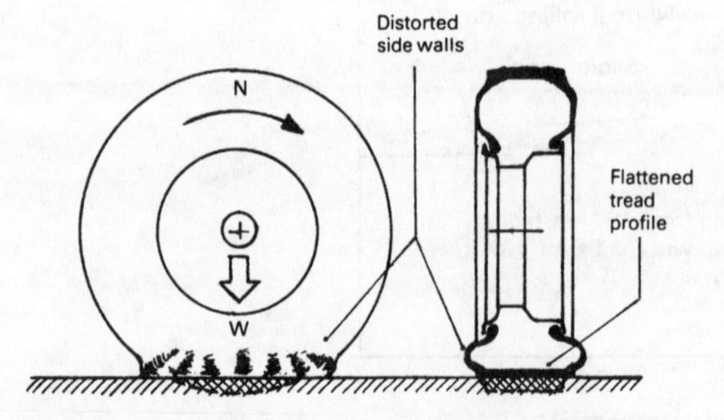

Fig. 8.10 Illustration of side wall distortion at ground level

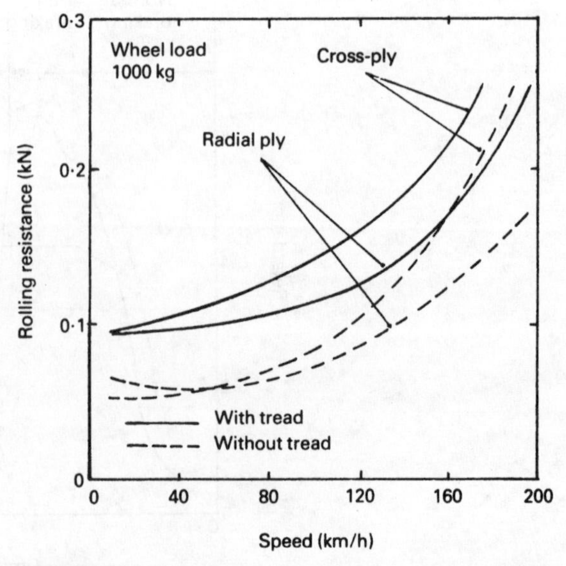

Fig. 8.11 Effect of tyre construction on rolling resistance

d) hard smooth dry surfaces have lower rolling resistances than rough or worn out surfaces,
e) the inflation pressure decreases the rolling resistance on hard surfaces,
f) higher driving speed increases the rolling resistance due to the increase in work being done in deforming the tyre over a given time (Fig. 8.11),
g) increasing the wheel and tyre diameter reduces the rolling resistance only slightly on hard surfaces but it has a pronounced effect on soft ground,
h) increasing the tractive effort also raises the rolling resistance due to the increased deformation of the tyre carcass and the extra work needed to be done.

8.1.6 Tractive and braking effort (Figs 8.12, 8.13, 8.14, 8.15, 8.16 and 8.17)

A tractive effort at the tyre to ground interface is produced when a driving torque is transmitted to the wheel and tyre. The twisting of the tyre carcass in the direction of the leading edge of the tread contact patch is continuously opposed by the tyre contact patch reaction on the ground. Before it enters the contact patch region a portion of the tread and casing will be deformed and compressed. Hence the distance that the tyre tread travels when subjected to a driving torque will be less than that in free rolling (Fig. 8.12).

If a braking torque is now applied to the wheel and tyre, the inertia on the vehicle will tend to pull the wheel forward while the interaction between the tyre contact patch and ground will oppose this motion. Because of this action, the casing and tread elements on the leading side of the tyre become stretched just before they enter the contact patch region in contrast with the compressive effect for driving tyres (Fig. 8.13). As a result, when braking torque is applied the distance the tyre moves will be greater than when the tyre is subjected to free rolling only. The loss or gain in the distance the tread

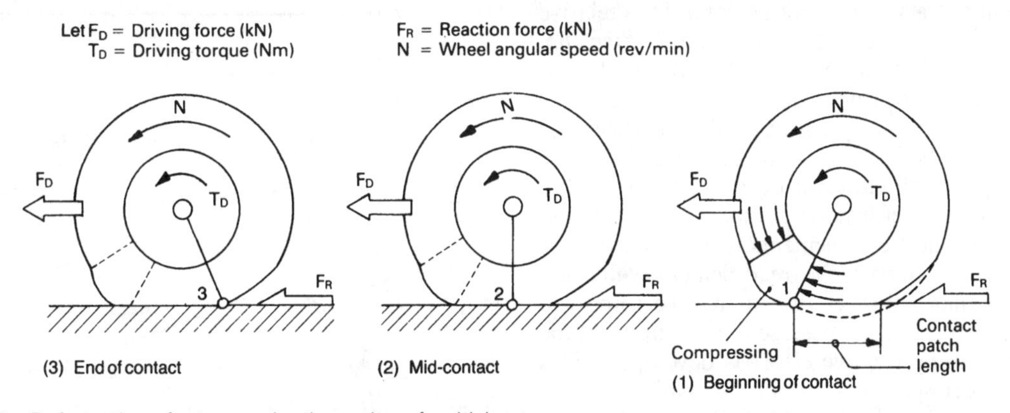

Fig. 8.12 Deformation of a tyre under the action of a driving torque

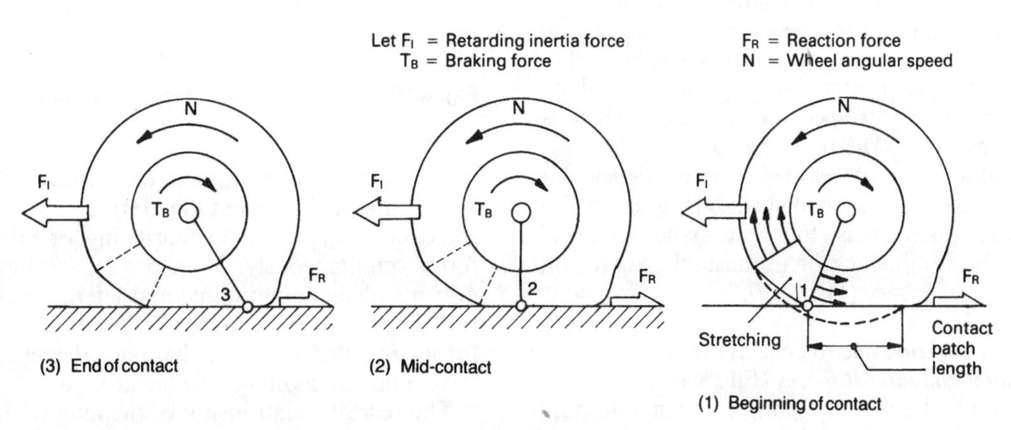

Fig. 8.13 Deformation of a tyre under the action of a braking torque

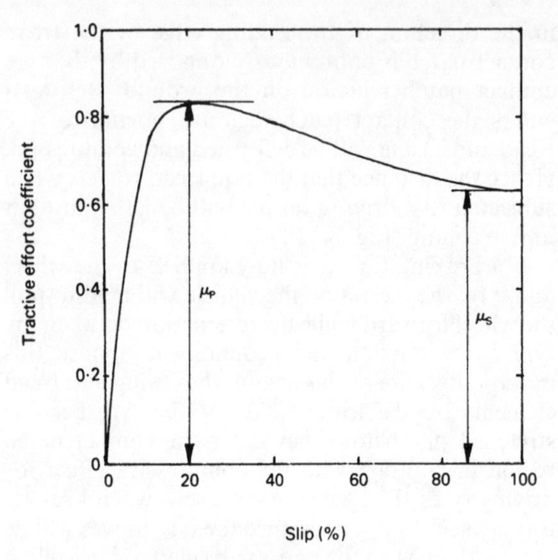

Fig. 8.14 Effect of tyre slip on tractive effort

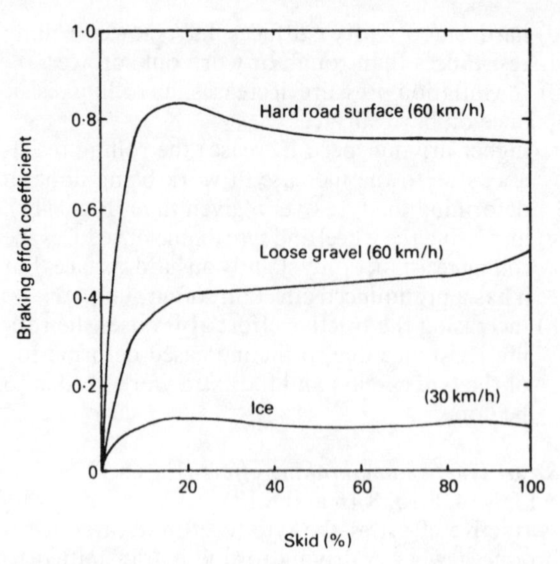

Fig. 8.15 Effect of ground surface on braking effort

travels under tractive or braking conditions relative to that in free rolling is known as *deformation slip,* and it can be said that under steady state conditions slip is a function of tractive or braking effort.

When a driving torque is applied to a wheel and tyre there will be a steep initial rise in tractive force matched proportionally with a degree of tyre slip, due to the elastic deformation of the tyre tread. Eventually, when the tread elements have reached their distortion limit, parts of the tread elements will begin to slip so that a further rise in tractive force will produce a much larger increase in tyre slip until the peak or limiting tractive effort is developed. This normally corresponds to on a hard road surface to roughly 15–20% slip (Fig. 8.14). Beyond the peak tractive effort a further increase in slip produces an unstable condition with a considerable reduction in tractive effort until pure wheel spin results (the tyre just slides over the road surface). A tyre subjected to a braking torque produces a very similar braking effort response with respect to wheel slip, which is now referred to as *skid.* It will be seen that the maximum braking effort developed is largely dependent upon the nature of the road surface (Fig. 8.15) and the normal wheel loads (Fig. 8.16), whereas wheel speed has more influences on the unstable skid region of a braking sequence (Fig. 8.17).

8.1.7 Tyre reaction due to concurrent longitudinal and lateral forces (Fig. 8.18)

A loaded wheel and tyre rolling can generate only a limited amount of tread to ground reaction to

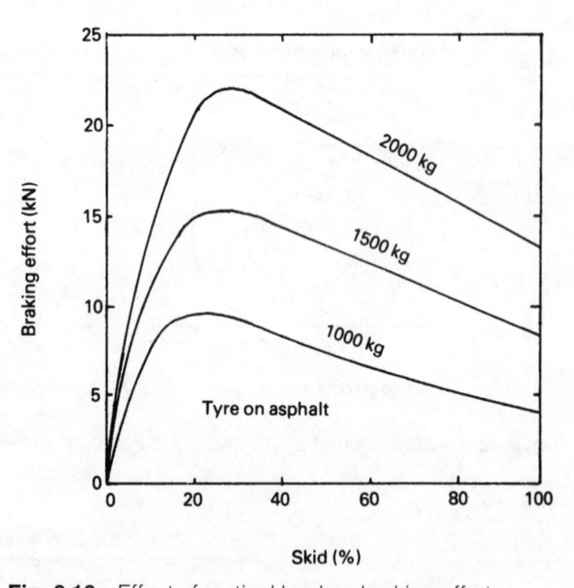

Fig. 8.16 Effect of vertical load on braking effort

resist the tyre slipping over the surface when the tyre is subjected to longitudinal (tractive or braking) forces and lateral (side) (cornering or crosswind) forces simultaneously. Therefore the resultant components of the longitudinal and lateral forces must not exceed the tread to ground resultant reaction force generated by all of the tread elements within the contact area biting into the ground.

The relative relationship of the longitudinal and lateral forces acting on the tyre can be shown by

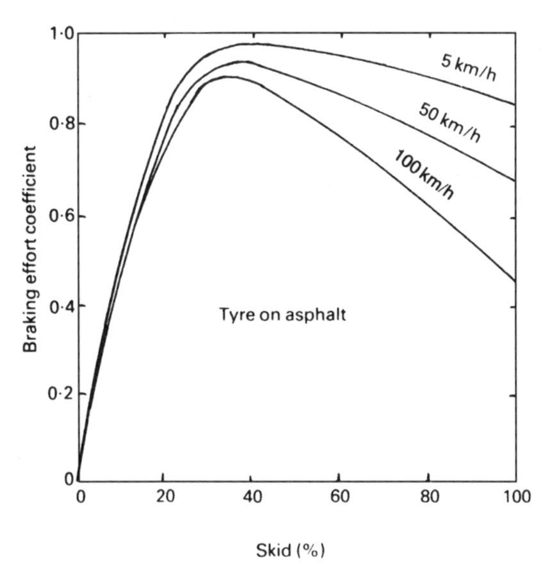

Fig. 8.17 Effect of vehicle speed on braking effort

resolving both forces perpendicularly to each other within the boundary of limiting reaction force circle (Fig. 8.18(a and b)). This circle with its vector forces shows that when longitudinal forces due to traction or braking forces is large (Fig. 8.18(c and d)), the tyre can only sustain a much smaller side force. If the side force caused either by cornering or a crosswind is large, the traction or braking effort must be much reduced.

8.2 Tyre materials

8.2.1 The structure and properties of rubber (Figs 8.19, 8.20 and 8.21)

The outside carcass and tread of a tyre is made from a rubber compound that is a mix of several substances to produce a combination of properties necessary for the tyre to function effectively. Most metallic materials are derived from simple molecules held together by electrostatic bonds which sustain only a limited amount of stretch when subjected to tension (Fig. 8.19). Because of this, the material's elasticity may be restricted to something like 2% of its original length. Rubber itself may be either natural or synthetic in origin. In both cases the material consists of many thousands of long chain molecules all entangled together. When stretched, the giant rubber molecules begin to untangle themselves from their normal coiled state and in the process of straightening out, provide a considerable amount of extension which may be of the order of 300% of the material's original length. Thus it is not the electrostatic bonds being stretched

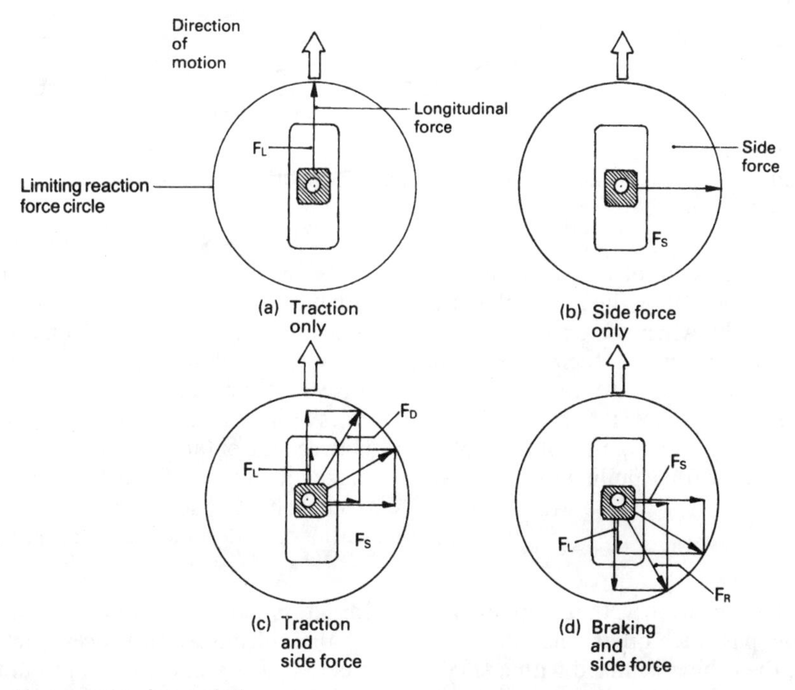

Fig. 8.18 (a–d) Limiting reaction force circle

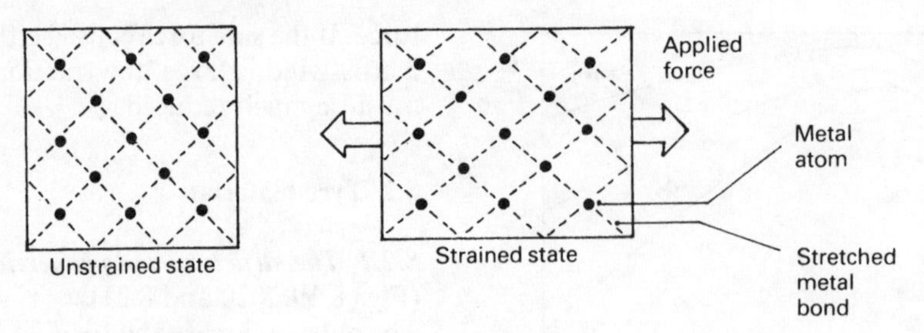

Fig. 8.19 Metal atomic lattice network

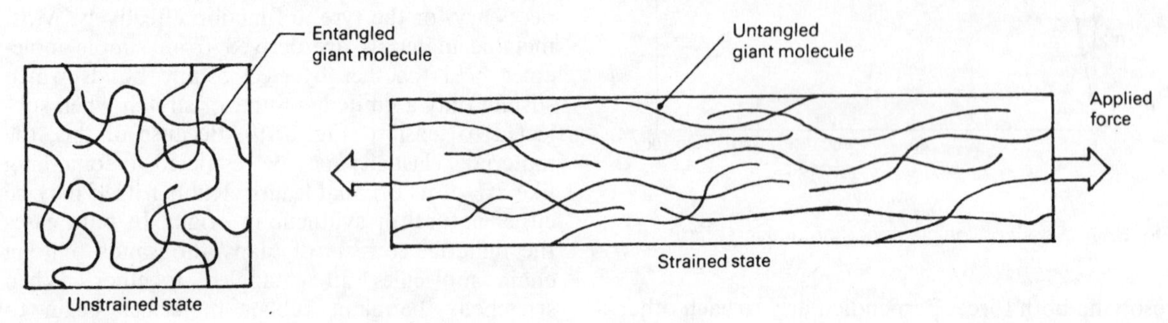

Fig. 8.20 Raw rubber network of long chain molecules

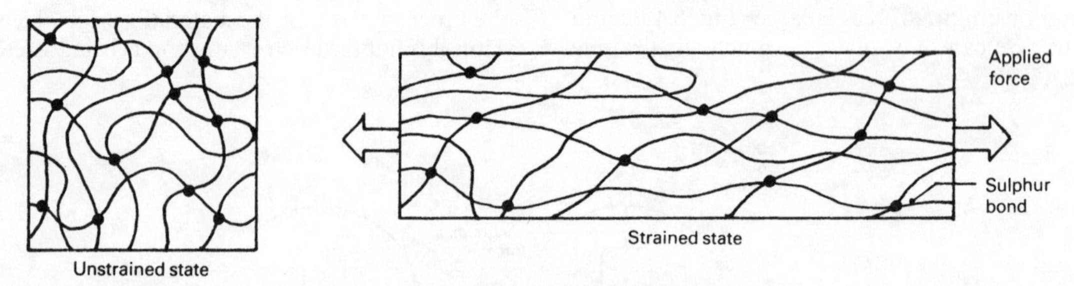

Fig. 8.21 Vulcanized rubber cross-linked network of long chain molecules

but the uncoiling and aligning of the molecules in the direction of the forces pulling the material apart (Fig. 8.20). Consequently, when the tensile force is removed the molecules revert to their free state and thereby draw themselves into an entangled network again. Hence it is not the bonds being stretched but the uncoiling and aligning of the molecules in the direction of the force pulling the material apart.

Vulcanization To reduce the elasticity and to increase the strength of the rubber, that is to restrict the molecules sliding past each other when the substance is stretched, the rubber is mixed with a small amount of sulphur and then heated, usually under pressure. The chemical reaction produced is known either as curing or more commonly as vulcanization (named after Vulcan, the Roman god of fire). As a result, the sulphur molecules form a network of cross-links between some of the giant rubber molecules (Fig. 8.21). The outcome of the cross-linking between the entangled long chain molecules is that it makes it more difficult for these molecules to slip over each other so that the rubber becomes stronger with a considerable reduction in flexibility.

Initiators and accelerator To start off and speed up the vulcanization process, activators such as a metallic zinc oxide are used to initiate the reaction and an organic accelerator reduces the reaction time and temperature needed for the sulphur to produce a cross-link network.

Carbon black Vulcanized rubber does not have sufficient abrasive resistance and therefore its rate of wear as a tyre tread material would be very high. To improve the rubber's resistance against wear and tear about a quarter of a rubber compound content is made up of a very fine carbon powder known as carbon black. When it is heated to a molten state the carbon combines chemically with the rubber to produce a much harder and tougher wear resistant material.

Oil extension To assist in producing an even dispersion of the rubber compound ingredients and to make processing of the tyre shape easier, an emulsion of hydrocarbon oil is added (up to 8%) to the rubber latex to dilute or extend the rubber. This makes the rubber more plastic as opposed to elastic with the result that it becomes tougher, offers greater wear resistance and increases the rubber's hysteresis characteristics thereby improving its wet grip properties.

Anti-oxidants and -ozonates Other ingredients such as an anti-oxidant and anti-ozonate are added to preserve the desirable properties of the rubber compound over its service life. The addition of anti-oxidants and -ozonates (1 or 2 parts per 100 parts of rubber) prevents heat, light and particularly oxygen ageing the rubber and making it hard and brittle.

8.2.2 Mechanical properties

To help the reader understand some of the terms used to define the mechanical properties of rubber the following brief definitions are given:

Material resilience This is the ability for a solid substance to rebound or spring back to its original dimensions after being distorted by a force. A material which has a high resilience generally has poor road grip as it tends to spring away from the ground contact area as the wheel rolls forward.

Material plasticity This is the ability for a solid material to deform without returning to its original shape when the applied force is removed. A material which has a large amount of plasticity promotes good road grip as each layer of material tends to cling to the road surface as the wheel rolls.

Material hysteresis This is the sluggish response of a distorted material taking up its original form so that some of the energy put into deforming the carcass, side walls and tread of a tyre at the contact patch region will still not be released when the tyre has completed one revolution and the next distortion period commences. As the cycle of events continues, more and more energy will be absorbed by the tyre, causing its temperature to rise. If this heat is not dissipated by the surrounding air, the inner tyre fabric will eventually become fatigued and therefore break away from the rubber encasing it, thus destroying the tyre. For effective tyre grip a high hysteresis material is necessary so that the distorted rubber in contact with the ground does not immediately spring away from the surface but is inclined to mould and cling to the contour of the road surface.

Material fatigue This is the ability of the tyre structure to resist the effects of repeated flexing without fracture, particularly with operating temperatures which may reach something of the order of 100°C for a heavy duty tyre although temperatures of 80–85°C are more common.

8.2.3 Natural and synthetic rubbers

Synthetic materials which have been developed as substitutes for natural rubber and have been utilized for tyre construction are listed with natural rubber as follows:

a) Natural rubber (NR)
b) Chloroprene (Neoprene) rubber (CR)
c) Styrene–butadiene rubber (SBR)
d) Polyisoprene rubber (IR)
e) Ethylene propylene rubber (EPR)
f) Polybutadiene rubber (BR)
g) Isobutene–isoprene (Butyl) rubber (IIR)

Natural rubber (NR) Natural rubber has good wear resistance and excellent tear resistance. It offers good road holding on dry roads but retains only a moderately good grip on wet surfaces. One further merit is its low heat build-up, but this is contrasted by high gas permeability and its resistance to ageing and ozone deterioration is only fair. The side walls and treads have been made from natural rubber but nowadays it is usually blended with other synthetic rubbers to exploit their desirable properties and to minimize their shortcomings.

Chloroprene (Neoprene) rubber (CR) This synthetic rubber is made from acetylene and hydrochloric acid. Wear and tear resistance for this rubber compound, which was one of the earliest to compete with natural rubber, is good with a reasonable road surface grip. A major limitation is its inability to bond with the carcass fabric so a natural rubber film has to be interposed between the cords and the Neoprene covering. Neoprene rubber has a moderately low gas permeability and does not show signs of weathering or ageing throughout a tyre's working life. When blended with natural rubber it is particularly suitable for side wall covering.

Styrene–butadiene rubber (SBR) Compounds of this material are made from styrene (a liquid) and butadiene (a gas). It is probably the most widely used synthetic rubber within the tyre industry. Styrene–butadiene rubber (SBR) forms a very strong bond to fabrics and it has a very good resistance to wear, but suffers from poor tear resistance compared to natural rubber. One outstanding feature of this rubber is its high degree of energy absorption or high hysteresis and low resilience. It is these properties which give it exceptional grip, especially on wet surfaces. Due to the high heat build up, SBR is restricted to the tyre tread while the side walls are normally made from low hysteresis compounds which provide greater rebound response and run cooler. Blending SBR with NR enables the best properties of both synthetic and natural rubber to be utilized so that only one rubber compound is necessary for some types of car tyres. The high hysteresis obtained with SBR is partially achieved by using an extra high styrene content and by adding a large proportion of oil to extend the compound, the effects being to increase the rubber plastic properties and to lower its resilience (i.e. reduce its rebound response).

Polyisoprene rubber (IR) This compound has very similar characteristics to natural rubber but has improved wear and particularly tear resistance with a further advantage of an extremely low heat build up with normal tyre flexing. These properties make this material attractive when blended with natural rubber and styrene–butadiene rubber to produce tyre treads with very high abrasion resistance. For heavy duty application such as track tyres where high temperatures and driving on rough terrains are a problem, this material has proved to be successful.

Ethylene propylene rubber (EPR) The major advantage of this rubber compound is its ability to be mixed with large amounts of cheap carbon black and oil without destroying its rubbery properties. It has excellent abrasive ageing and ozone resistance with varying road holding qualities in wet weather depending upon the compound composition. Skid resistance on ice has also been varied from good to poor. A great disadvantage, however, is that the rubber compound bonds poorly to cord fabric. Generally, the higher the ethylene content the higher the abrasive resistance, but at the expense of a reduction in skid resistance on ice. Rubber compounds containing EPR have not proved to be successful up to the present time.

Polybutadiene rubber (BR) This rubbery material has outstanding wear resistance properties and is exceptionally stable with temperature changes. It has a high resilience that is a low hysteresis level.

When blended with SBR in the correct proportions, it reduces the wet road holding slightly and considerably improves its ability to resist wear. Because of its high resilience (large rebound response), if mixed in large proportions, the road holding in wet weather can be relatively poor. It is expensive to produce. When it is used for tyres it is normally mixed with SBR in the proportion of 15 to 50%.

Isobutene–isoprene (Butyl) rubber (IIR) Rubber of this kind has exceptionally low permeability to gas. In fact it retains air ten times longer than tubes made from natural rubber, with the result that it has been used extensively for tyre inner tubes and for linings of tubeless tyres. Unfortunately it will not blend with SBR and NR unless it is chlorinated, but in this way it can be utilized as an inner tube lining material for tubeless tyres. The resistance to wear is good and it has a high hysteresis so that it responds more like plastic than rubber to distortion at ground level. Road grip is good for both dry and wet conditions. When mixed with carbon black its desirable properties are generally improved. Due to its high hysteresis tyre treads made from this material do not generate noise in the form of squeal since it does not readily give out energy to the surroundings.

8.2.4 Summary of the merits and limitations of natural and synthetic rubber compounds

Some cross-ply tyres are made from one compound from bead to bead, but the severity of the carcass flexure with radial ply tyres encourages the manufacturers of tyres to use different rubber composition for various parts of the tyre structure so that their properties match the duty requirements of each functional part of the tyre (i.e. tread, side wall, inner lining, bead etc.).

Side walls are usually made from natural rubber blended with polybutadiene rubber (BR) or styrene–butadiene rubber (SBR) or to a lesser extent Neoprene or Butyl rubber or even natural rubber alone. The properties needed for side wall material are a resistance against ozone and oxygen attack, a high fatigue resistance to prevent flex cracking and good compatibility with fabrics and other rubber compounds when moulded together.

Tread wear fatigue life and road grip depends to a great extent upon the surrounding temperatures, weather conditions, be they dry, wet, snow or ice bound, and the type of rubber compound being used. A comparison will now be made with natural rubber and possibly the most important synthetic rubber, styrene–butadiene (SBR). At low temperatures styrene–butadiene (SBR) tends to wear more than natural rubber but at higher temperatures the situation reverses and styrene–butadiene rubber (SBR) shows less wear than natural rubber. As the severity of the operating condition of the tyre increases SBR tends to wear less relative to NR. The fatigue life of all rubber compounds is reduced as the degree of cyclic distortion increases. For small tyre deflection SBR has a better fatigue life but when deflections are large NR provides a longer service life. Experience on ice and snow shows that NR offers better skid resistance, but as temperatures rise above freezing, SBR provides an improved resistance to skidding. This cannot be clearly defined since it depends to some extent on the amount of oil extension (plasticizer) provided in the blending in both NR and SBR compounds. Oil extension when included in SBR and NR provides similarly improved skid resistance and in both cases becomes inferior to compounds which do not have oil extension.

Two examples of typical rubber compositions suitable for tyre treads are:

a)	High styrene butadiene rubber	31%
	Oil extended butadiene rubber	31%
	Carbon black	30%
	Oil	6%
	Sulphur	2%
b)	Styrene butadiene rubber	45%
	Natural rubber	15%
	Carbon black	30%
	Oil	8%
	Sulphur	2%

8.3 Tyre tread design

8.3.1 Tyre construction

The construction of the tyre consists basically of a carcass, inner beads, side walls, crown belt (radials) and tread.

Carcass The carcass is made from layers of textile core plies. Cross-ply tyres tend to still use nylon whereas radial-ply tyres use either raylon or polyester.

Beads The inside diameter of both tyre walls support the carcass and seat on the wheel rim. The edges of the tyre contacting the wheel are known as beads and moulded inside each bead is a strengthening endless steel wire cord.

Side walls The outside of the tyre carcass, known as the side walls, is covered with rubber compound. Side walls need to be very flexible and capable of protecting the carcass from external damage such as cuts which can occur when the tyre is made to climb up a kerb.

Bracing belt Between the carcass and tyre tread is a crown reinforcement belt made from either synthetic fabric cord such as raylon or for greater strength steel cores. This circumferential endless cord belt provides the rigidity to the tread rubber.

Tread The outside circumferential crown portion of the tyre is known as the tread. It is made from a hard wearing rubber compound whose function is to grip the contour of the road.

8.3.2 Tyre tread considerations

The purpose of a pneumatic tyre is to support the wheel load by a cushion of air trapped between the well of the wheel rim and the toroid-shaped casing known as the carcass. Wrapped around the outside of the tyre carcass is a thick layer of rubber compound known as the tread whose purpose is to protect the carcass from road damage due to tyre impact with the irregular contour of the ground and the abrasive wear which occurs as the tyre rolls along the road. While the wheel is rotating the tread provides driving, braking, cornering and steering grip between the tyre and ground. Tyre grip must be available under a variety of road conditions such as smooth or rough hard roads, dry or wet surfaces, muddly tracks, fresh snow or hard packed snow and ice and sandy or soft soil terrain. Tread grip may be defined as the ability of a rolling tyre to continuously develop an interaction between the individual tread elements and the ground so that any longitudinal (driving) or lateral (side) forces imposed on the wheel will not be able to make the tread in contact with the ground slide.

A tyre tread pattern has two main functions:

1 to provide a path for drainage of water which might become trapped between the tyre contact patch and the road,
2 to provide tread to ground bite when the wheel is subjected to both longitudinal and lateral forces under driving conditions.

8.3.3 Tread bite

Bite is obtained by selecting a pattern which divides the tread into many separate elements and providing each element with a reasonably sharp well defined edge. Thus as the wheel rotates these tread edges engage with the ground to provide a degree of mechanical tyre to ground interlock in addition to the frictional forces generated when transmitting tractive or braking forces.

The major features controlling the effectiveness of the tread pattern in wet weather are:

1 drainage grooves or channels,
2 load carrying ribs,
3 load bearing blocks,
4 multiple microslits or sipes.

Tread drainage grooves (Fig. 8.22(a, b, c and d)) The removal of water films from the tyre to ground interface is greatly facilitated by having a number of circumferential grooves spaced out across the tread width (Fig. 8.22(a)). These grooves enable the leading elements of the tread to push water through the enclosed channels made by the road sealing the underside of the grooves. Water therefore emerges from the trailing side of the contact patch in the form of jets. If these grooves are to be effective, their total cross-sectional area should be adequate to channel all the water immediately ahead of the leading edge of the contact patch away. If it cannot cope the water will become trapped between the tread ribs or blocks so that these elements lift and become separated from the ground, thus reducing the effective area of the contact patch and the tyre's ability to grip the ground.

To speed up the water removal process under the contact patch, lateral grooves may be used to join together the individual circumferential grooves and to provide a direct side exit for the outer circumferential grooves. Normally many grooves are preferred to a few large ones as this provides a better drainage distribution across the tread.

Tread ribs (Fig. 8.22(a and b)) Circumferential ribs not only provide a supportive wearing surface for the tyre but also become the walls for the drainage grooves (Fig. 8.22(a and b)). Lateral (transverse) ribs or bars provide the optimum bite for tractive and braking forces but circumferential ribs are most effective in controlling cornering and steering stability. To satisfy both longitudinal and lateral directional requirements which may be acting concurrently on the tyre, ribs may be arranged diagonally or in the form of zig-zag circumferential ribs to improve the wiping effect across the tread surface under wet conditions. It is generally better to break the tread pattern into many narrow ribs than a few wide ones, as this prevents the formation of hydrodynamic water wedges which may otherwise tend to develop

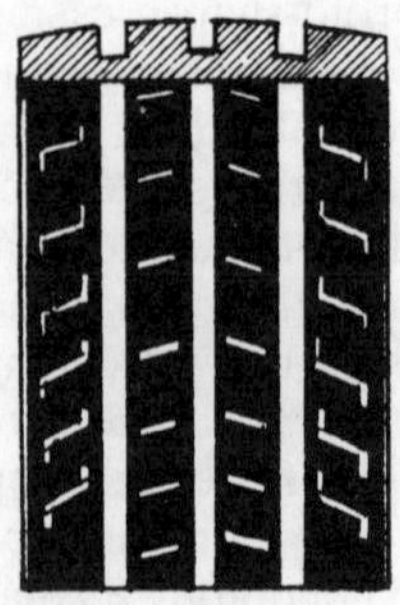

(a) Circumferential straight grooves and ribs with multiple sipes

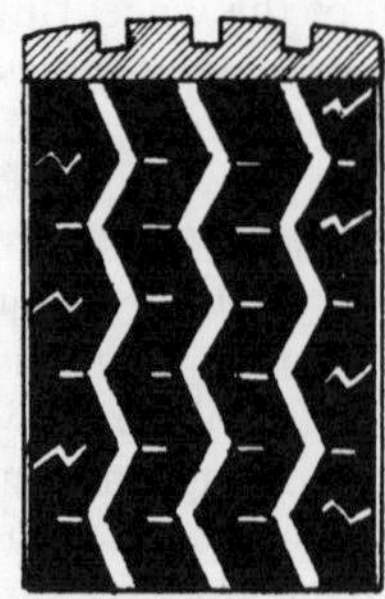

(b) Circumferential zig-zag grooves and ribs with multiple sipes

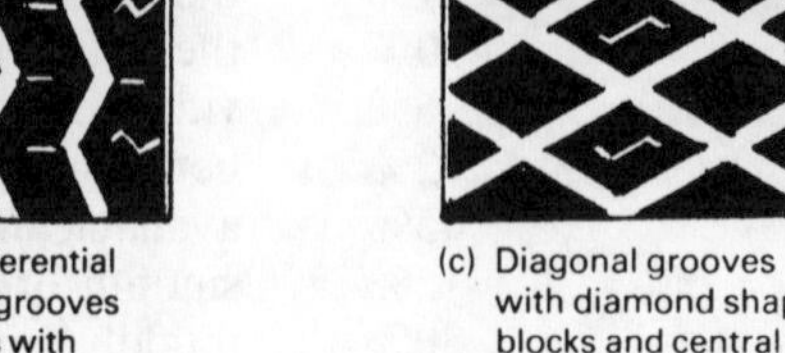

(c) Diagonal grooves with diamond shaped blocks and central sipes

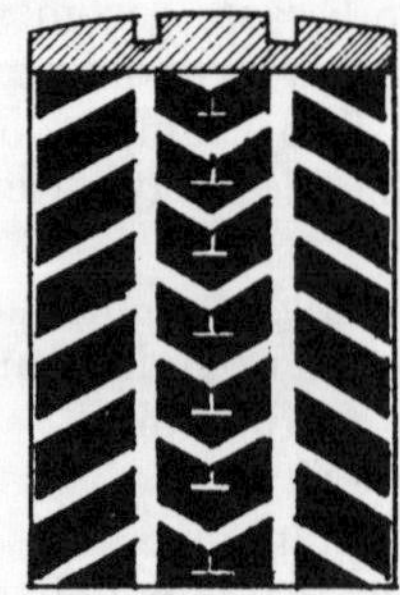

(d) Diagonal bars with central vee blocks and sipes

Fig. 8.22 (a–d) Basic tyre tread patterns

with the consequent separation of the tread elements from the road.

Tread blocks (Figs 8.22(c and d) and 8.23(a and b)) If longitudinal circumferential grooves in the middle of the tread are complemented by lateral (transverse) grooves channelled to the tread shoulders, then with some tread designs the drainage of water can be more effective at speed. The consequences of both longitudinal and lateral drainage channels is that the grooves encircle portions of the tread so that they become isolated island blocks (Fig. 8.22 (c and d)). These blocks can be put to good use as they provide a sharp wiping and biting edge where the interface of the tread and ground meet. To improve their biting effectiveness for tractive and braking forces as well as steering and cornering forces, these forces may be resolved into diagonal resultants so that the blocks are sometimes arranged in an oblique formation. A limitation to the block pattern concept is caused by inadequate support around the blocks so that under severe operating conditions, the bulky rubber blocks tend to bend and distort. This can be partially overcome by incorporating miniature buttresses between the drainage grooves which lean between blocks so that adjacent blocks support each other. At the same time, drainage channels which burrow below the high mounted buttresses are prevented from closing. Tread blocks in the form of bars, if arranged in a herringbone fashion, have proved to be effective on rugged ground. Square or rhombus-shaped blocks provide a tank track unrolling action greatly reducing movement in the tread contact area. This pattern helps to avoid the break-up on the top layer of sand or soil and thus prevents the tyre from digging into the ground. Because of the inherent tendency of the individual blocks to bend somewhat when they are subjected to ground reaction forces, they suffer from toe to heel rolling action which causes blunting of the leading edge and trailing edge feathering. Generally tyres which develop this type of wear provide a very good water sweeping action when new, which permits the tread elements to bite effectively into the ground, but after the tyre has been on the road for a while, the blunted leading edge allows water to enter underneath the tread elements. Consequently the slightest amount of water interaction between the block elements and ground reduces the ability for the tread to bite and in the extreme cases under locked wheel braking conditions a hydrodynamic water wedge action may result, causing a mild form of aquaplaning to take place.

For tread block elements to maintain their wiping action on wet surfaces, wear should be from toe to heel (Fig. 8.23(a)). If, however, wear occurs in the reverse order, that is from heel to toe (Fig. 8.23(b)), the effectiveness of the tread pattern will be severely reduced since the tread blocks then become the platform for a hydrodynamic water wedge which at speed tries to lift the tread blocks off the ground.

Tread slits or sipes (Figs 8.22(a, b, c and d) and 8.24(a, b and c)) Microslits, or *sipes* as they are commonly called, are incisions made at the surface of the tyre tread, going down to the full depth of the tread grooves. They resemble a knife cut, except that instead of being straight they are mostly of a zig-zag nature (Fig. 8.22(a, b, c and d)). Normally these sipes terminate within the tread elements, but sometimes one end is permitted to intersect the side wall of a drainage groove. In some tread patterns the sipes are all set at a similar angle to each other, the zig-zag shape providing a large number of edges which point in various directions. Other designs have sets of sipes formed at different angles to each other so that these sipes are effective whichever way the wheel points and whatever the direction the ground reaction forces operate.

Sipes or slits in their free state are almost closed, but as they move into the contact patch zone the ribs or blocks distort and open up (Fig. 8.24(a)). Because of this, the sipe lips scoop up small quantities of water which still exist underneath the tread. This wiping action enables some biting edge reaction with the ground. Generally, the smaller the sipes are and more numerous they are the greater will be their effective contribution to road grip. The

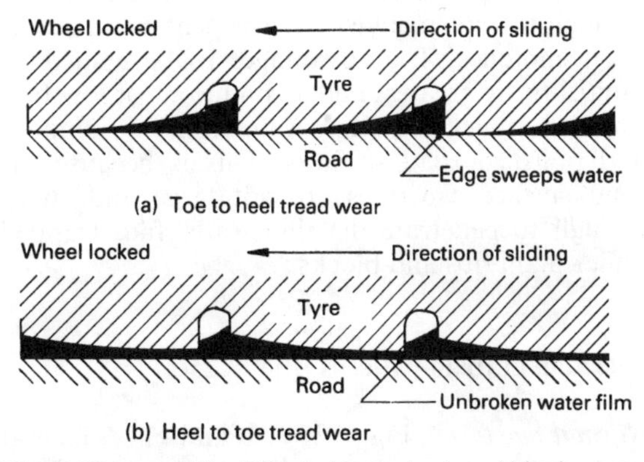

Fig. 8.23 (a and b) Effect of irregular tread block wear

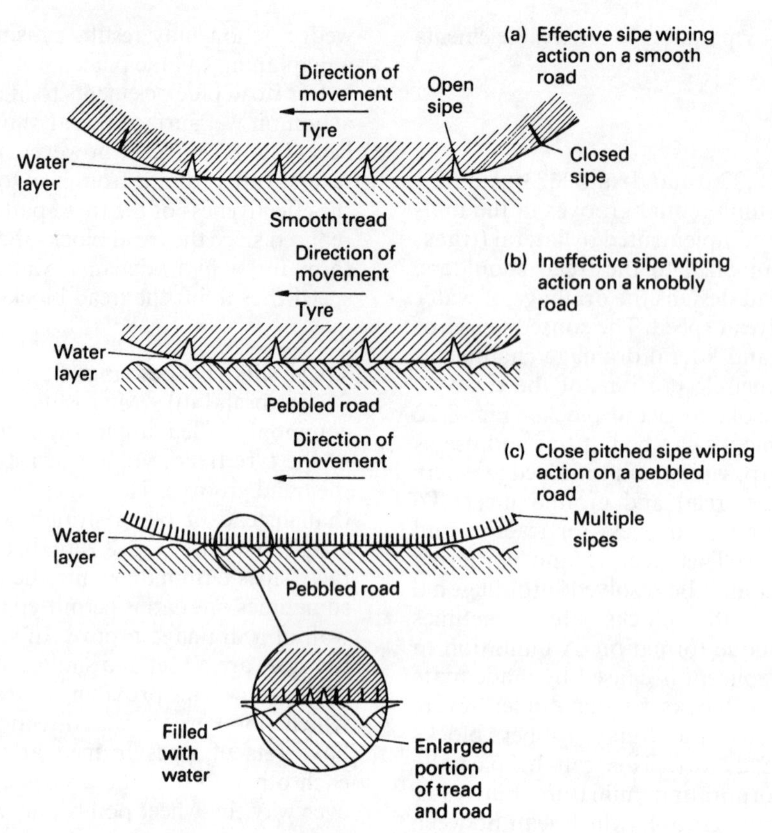

Fig. 8.24 (a–c) Effectiveness of microslits on wet road surfaces

normal spacing of sipes (microslits) on a tyre tread makes them ineffective on a pebbled road surface because there will be several pebbles between the pitch of the sipes (Fig. 8.24(b)), and water will lie between these rounded stones, therefore only a few of the stones will be subjected to the wiping edge action of the opened lips. An alternative method to improve the wiping process would be to have many more wiping slits (Fig. 8.24(c)), but this is very difficult to implement with the present manufacturing techniques. The advantages to be gained by multislits are greatest under conditions of low friction associated with thin water films on smooth and polished road surfaces. This is because the road surface asperities are not large and sharp enough to penetrate the thin water film trapped under plain ribs and blocks.

Selection of tread patterns (Fig. 8.25(a–1))

Normal car tyres (Fig. 8.25(a, b and c)) General duty car tyres which are capable of operating effectively at all speeds tend to have tread blocks situated in an oblique fashion with a network of surrounding drainage grooves which provide both circumferential and lateral water release.

Winter car tyres (Fig. 8.25(d, e and f)) Winter car tyres are normally very similar to the general duty car tyre but the tread grooves are usually wider to permit easier water dispersion and to provide better exposure of the tread blocks to snow and soft ice without sacrificing too much tread as this would severely reduce the tyre's life.

Truck tyres (Fig. 8.25(g and h)) Truck tyres designed for steered axles usually have circumferential zig-zag ribs and grooves since they provide very good lateral reaction when being steered on curved tracks. Drive axle tyres, on the other hand, are designed with tread blocks with adequate grooving so that optimum traction grip is obtained under both dry and wet conditions. Some of these tyres also have provision for metal studs to be inserted for severe winter hard packed snow and ice conditions.

Fig. 8.25 (a–l) Survey of tyre tread patterns

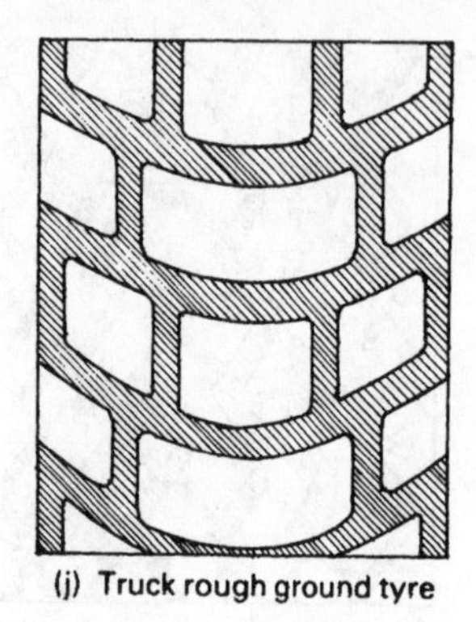

(j) Truck rough ground tyre

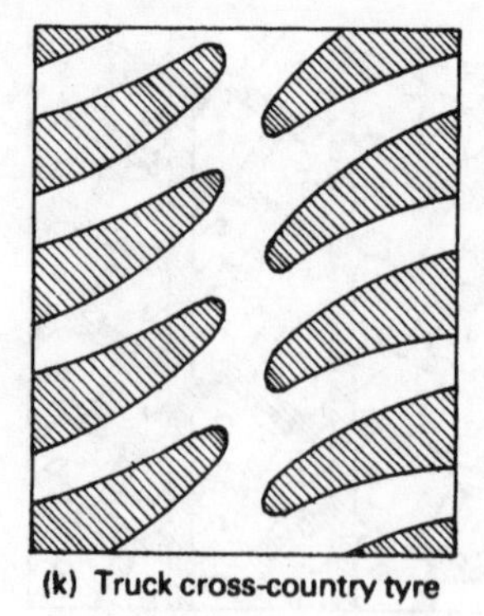

(k) Truck cross-country tyre

(l) Tractor cross-country tyre

Fig. 8.25 *contd*

Off/on road vehicles (Fig. 8.25(i)) Off/on road vehicle tyres usually have a much simpler bold block tread with a relatively large surrounding groove. This enables each individual block to react independently with the ground and in this manner bite and exert traction on soil which may be hard on the surface but soft underneath without break-up of the top layer, thus preventing the tyre digging in. The tread pattern blocks are also designed to be small enough to operate on hard surfaced roads at moderate speeds without excessive ride harshness.

Truck and tractor off road and cross-country tyres (Fig. 8.25(j, k and l)) Truck or tractor tyres designed for building sites or quarries generally have slightly curved rectangular blocks separated with wide grooves to provide a strong flexible casing and at the same time present a deliberately penetrating grip. Cross-country tyres which tend to operate on soft soil tend to prefer diagonal bars either merging into a common central rib or arranged with separate overlapping diagonal bars, as this configuration tends to provide exceptionally good traction on muddy soil, snow and soft ice.

8.3.4 The three zone concept of tyre to ground contact on a wet surface (Fig. 8.26)

The interaction of a tyre with the ground when rolling on a wet surface may be considered in three phases (Fig. 8.26):

Leading zone of unbroken water film (1) The leading zone of the tread contacts the stagnant water film covering the road surface and displaces the majority of the water into the grooves between the ribs and blocks of the tread pattern.

Intermediate region of partial breakdown of water film (2) The middle zone of the tread traps and reduces the thickness of the remaining water

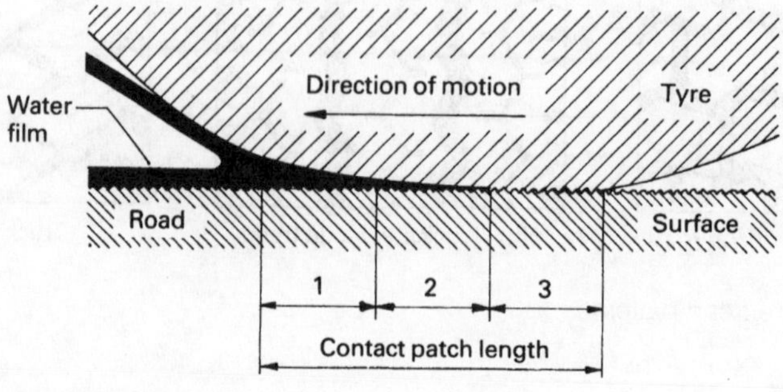

Fig. 8.26 Tyre to ground zones of interaction

between the faces of the ribs or blocks and ground so that some of the road surface asperities now penetrate through the film of water and may actually touch the tread. It is this region which is responsible for the final removal of water and is greatly assisted by multiple sipes and grooved drainage channels. If the ribs and blocks are insufficiently relieved with sipes and grooves it is possible that under certain conditions aquaplaning may occur in this region.

The effectiveness of this phase is determined to some extent by the texture of the road surface, as this considerably influences the dryness and potency of the third road grip phase.

Trailing zone of dry tyre to road contact (*3*) The water film has more or less been completely squeezed out at the beginning of this region so that the faces of the ribs and blocks bearing down on the ground are able to generate the bite which produces the tractive, braking and cornering reaction forces.

8.3.5 Aquaplaning (hydroplaning) (Fig. 8.27)
The performance of a tyre rolling on wet or semi-flooded surface will depend to some degree upon the tyre profile tread pattern and wear. If a smooth tread is braked over a very wet surface, the forward rotation of the tyre will drag in the water immediately in front between the tread face and ground and squeeze it so that a hydrodynamic pressure is created. This hydrodynamic pressure acts between the tyre and ground, its magnitude being proportional to the square of the wheel speed. With the wheel in motion, the water will form a converging wedge between the tread face and ground and so exert an upthrust on the underside of the tread. As a result of the pressure generated, the tyre tread will tend to separate itself from the ground. This condition is known as aquaplaning or hydroplaning. If the wheel speed is low only the front region of the tread rides on the wedge of water, but if the speed is rising the water wedge will progressively extend backward well into the contact patch area (Fig. 8.27). Eventually the upthrust created by the product of the hydrodynamic pressure and contact area equals the vertical wheel load. At this point the tyre is completely supported by a cushion of water and therefore provides no traction or directional control. If the tread has circumferential (longitudinal) and transverse (lateral) grooves of adequate depth then the water will drain through these passages at ground level so that aquaplaning is minimized even at high speeds. As the tyre tread wears the critical speed at which aquaplaning occurs becomes much lower. On very wet roads a bald tyre is certain to be subjected to aquaplaning at speeds above 60 km/h and therefore the vehicle when driven has no directional stability. Low aspect ratio tyres may find it difficult to channel the water away from the centre of the tread at a sufficiently high

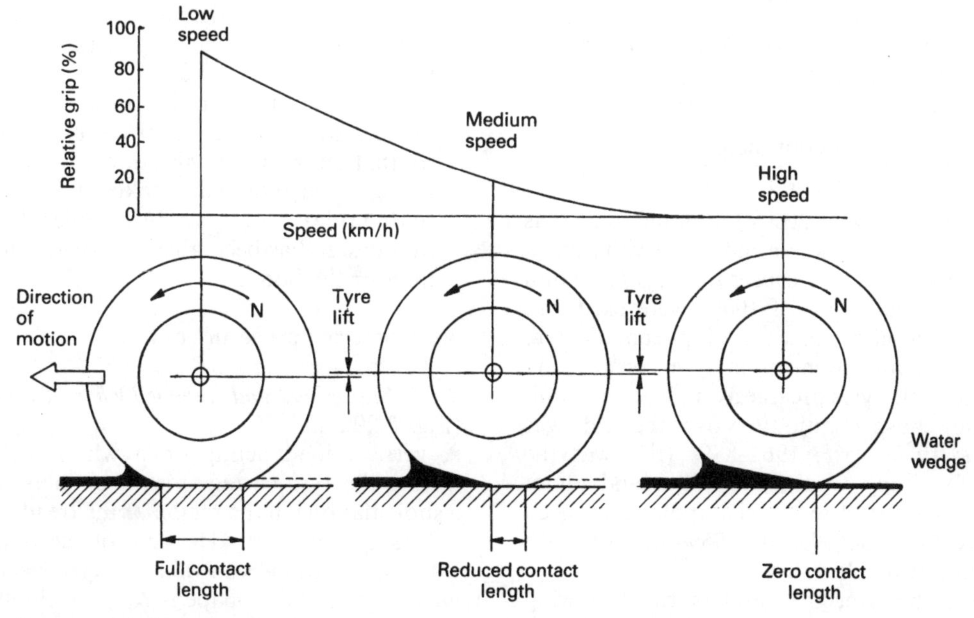

Fig. 8.27 Tyre aquaplaning

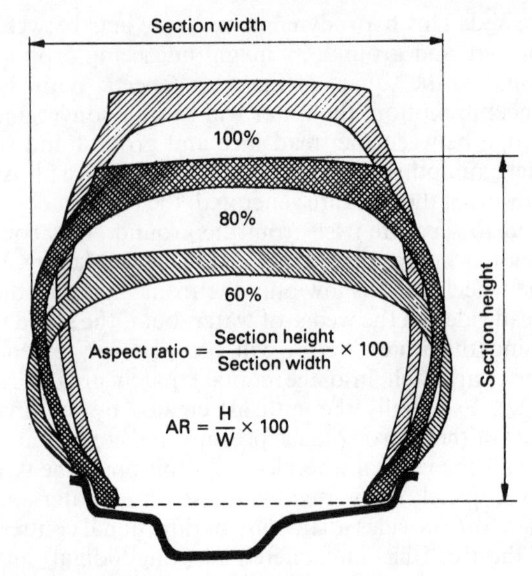

Fig. 8.28 Tyre profiles with different aspect ratios

rate and therefore must rely more on the circumferential grooves than on transverse grooving.

8.3.6 Tyre profile and aspect ratio (Fig. 8.28)
The profile of a tyre carcass considerably influences its rolling and handling behaviour. Because of the importance of the tyre's cross-sectional configuration in predicting its suitability and performance for various applications, the aspect ratio was introduced. This constant for a particular tyre may be defined as the ratio of the tyre cross-sectional height (the distance between the tip of the tread to the bead seat) to that of the section width (the outermost distance between the tyre walls) (Fig. 8.28).

i.e. $\text{Aspect ratio} = \frac{\text{Section height}}{\text{Section width}} \times 100$

A tyre with a large aspect ratio is referred to as a *high aspect ratio profile tyre* and a tyre with a small aspect is known as a *low aspect ratio profile*. Until about 1934 aspect ratios of 100% were used, but with the better understanding of pneumatic tyre properties and improvement in tyre construction lower aspect ratio tyres became available. The availability of lower aspect ratio tyres over the years was as follows; 1950s–95%, 1962–88% (this was the standard for many years), 1965–80% and about 1968–70%. Since then for special applications even lower aspect ratios of 65%, 60%, 55% and even 50% have become available.

Lowering the aspect ratio has the following effects:

1 The tyre side wall height is reduced which increases the vertical and lateral stiffness of the tyre.
2 A shorter and wider contact patch is established. The overall effect is to raise the load carrying capacity of the tyre.
3 The wider contact patch enables larger cornering forces to be generated so that vehicles are able to travel faster on bends.
4 The shorter and wider contact patch decreases the pneumatic trail which correspondingly reduces and makes more consistent the self-aligning torque.
5 The shorter and broader contact patch will, under certain driving conditions, reduce the slip angles generated by the tyre when subjected to side forces. Accordingly this reduces the tread distortion and as a result scuffing and wear will decrease.
6 With an increase in vertical stiffness and a reduction in tyre deflection with lower aspect ratio tyres, less energy will be dissipated by the tyre casing so that rolling resistance will be reduced. This also results in the tyre being able to run continuously at high speeds at lower temperatures which tends to prolong the tyre's life.
7 The increased lateral stiffness of a low profile tyre will increase the sensitivity to camber variations and quicken the response to steering changes.
8 Wider tyre contact patches make it more difficult for water drainage at speed particularly in the mid tread region. Hence the tread pattern design with low profile tyres becomes more critical on wet roads, if their holding is to match that of higher aspect ratio tyres.
9 The increased vertical stiffness of the tyre reduces static deflection of the tyre under load, so that more road vibrations are transmitted through the tyre. This makes it a harsher ride so that ride comfort is reduced unless the suspension design has been able to provide more isolation for the body.

8.4 Cornering properties of tyres

8.4.1 Static load and standard wheel height (Figs 8.29 and 8.30)
A vertical load acting on a wheel will radially distort the tyre casing from a circular profile to a short flat one in the region of the tread to ground interface (Fig. 8.29). The area of the tyre contact with the ground is known as the *tyre contact patch area;* its plan view shape is roughly elliptical. The consequence of this tyre deflection is to reduce the

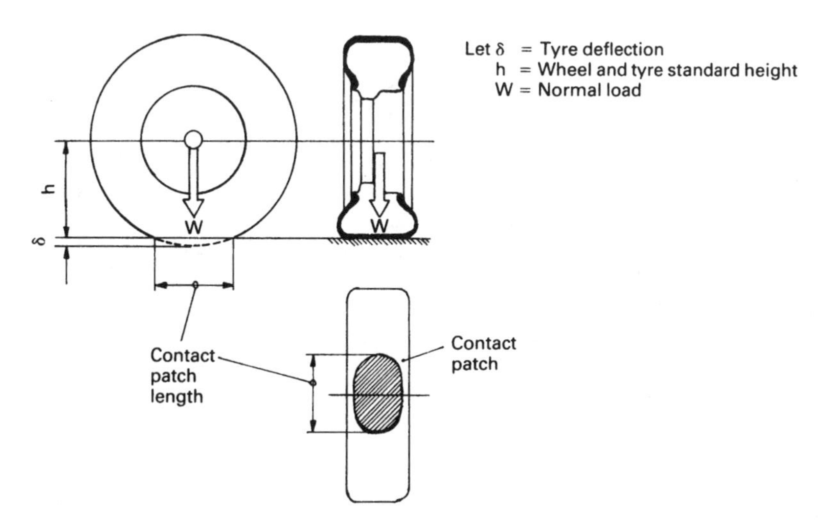

Fig. 8.29 Illustration of static tyre deflection

standard height of the wheel, that is the distance between the wheel axis and the ground. Generally tyre deflection will be proportional to the radial load imposed on the wheel; increasing the tyre inflation pressure reduces the tyre deflection for a given vertical load (Fig. 8.30). Note that there is an initial deflection (Fig. 8.30) due to the weight of the wheel and tyre alone. The steepness of the load deflection curve is useful in estimating the static stiffness of the tyres which can be interpreted as a measure of its vibration and ride qualities.

8.4.2 Tyre contact patch (Figs 8.29 and 8.31)
The downward radial load imposed on a road wheel causes the circular profile of the tyre in contact with the ground to flatten and spread towards the front and rear of its natural rolling plane. When the wheel is stationary, the interface area between the tyre and ground known as the contact patch will take up an elliptical shape (Fig. 8.29), but if the wheel is now subjected to a side thrust the grip between the tread and ground will distort the patch into a semibanana configuration (Fig. 8.31 (a)). It is the ability of the tyre contact patch casing, and elements of the tread to comply and change shape due to the imposed reaction forces, which gives tyres their steering properties. Generally, radial ply tyres form longer and broader contact patches than their counterpart cross-ply tyres, hence their superior road holding.

8.4.3 Cornering force (Fig. 8.31(a and b))
Tyres are subjected not only to vertical forces but also to side (lateral) forces when the wheels are in motion due to road camber, side winds, weight transfer and centrifugal force caused by travelling round bends and steering the vehicle on turns. When a side force, sometimes referred to as lateral force, is imposed on a road wheel and tyre, a reaction between the tyre tread contact patch and road surface will oppose any sideway motion. This resisting force generated at the tyre to road interface is known as the cornering force (Fig. 8.31(a and b)), its magnitude being equal to the lateral force but it acts in the opposite sense. The amount of

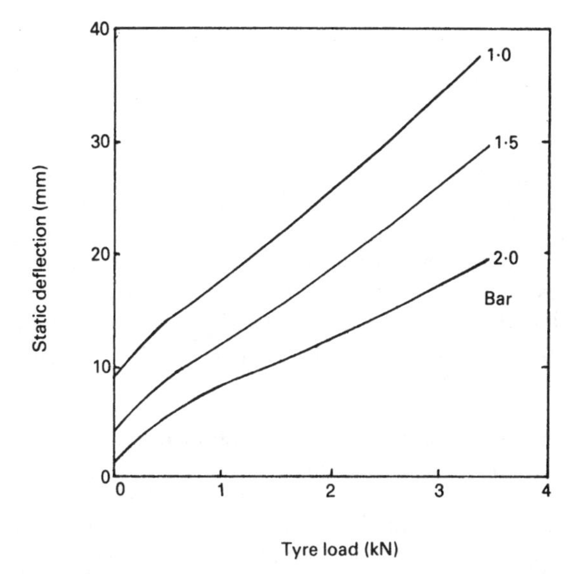

Fig. 8.30 Effect of tyre vertical load on static deflection

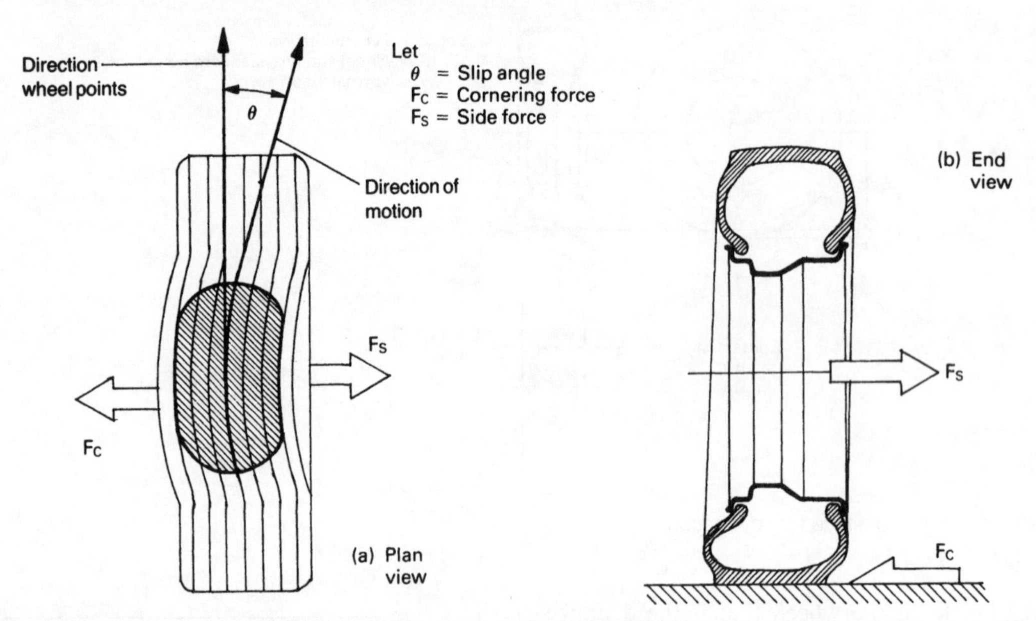

Fig. 8.31 Tyre tread contact patch distortion when subjected to a side force

cornering force developed increases roughly in proportion with the rise in lateral force until the grip between the tyre tread and ground diminishes. Beyond this point the cornering force cannot match further increases in lateral force with the result that tyre breakaway is likely to occur. Note that the greater the cornering force generated between tyre and ground, the greater the tyre's grip on the road.

The influencing factors which determine the amount of cornering force developed between the tyre and road are as follows:

Slip angle Initially the cornering force increases linearly with increased slip angle, but beyond about four degrees slip angle the rise in cornering force is non-linear and increases at a much reduced rate (Fig. 8.32), depending to a greater extent on tyre design.

Vertical tyre load As the vertical or radial load on the tyre is increased for a given slip angle, the cornering force rises very modestly for small slip angles but at a far greater rate with larger slip angles (Fig. 8.33).

Tyre inflation pressure Raising the tyre inflation pressure linearly increases the cornering force for a given slip angle (Fig. 8.34). These graphs also show that increasing the tyre slip angle considerably raises the cornering forces generated.

8.4.4 Slip angle (Fig. 8.31(a))

Any lateral force applied to a road wheel will tend to push the supple tyre walls sideways, but the opposing tyre to ground reaction causes the tyre contact patch to take up a curved distorted shape. As a result, the rigid wheel will point and roll in the direction it is steered, whereas the tyre region in contact with the ground will follow the track continuously laid down by the deformed tread path of the contact patch (Fig. 8.31(a)). The angle made between the direction of the wheel plane and that which it travels is known as the slip angle. Provided the slip angle is small, the compliance of the tyre construction will allow each element of tread to remain in contact with the ground without slippage.

8.4.5 Cornering stiffness (cornering power) (Fig. 8.32)

When a vehicle travels on a curved path, the centrifugal force (lateral force) tends to push sideways each wheel against the opposing tyre contact patch to ground reaction. As a result, the tyre casing and tread in the region of the contact patch very slightly deform into a semicircle so that the path followed by the tyre at ground level will not be quite the same as the direction of the wheel points. The resistance offered by the tyre crown or belted tread region by the casing preventing it from deforming and generating a slip angle is a measure of the tyre's cornering power. The cornering power, nowadays more usually termed *cornering stiffness*, may be defined as the cornering force required to be developed for every degree of slip angle generated.

i.e.

$$\text{Cornering stiffness} = \frac{\text{Cornering force}}{\text{Slip angle}} (\text{kN/deg})$$

In other words, the cornering stiffness of a tyre is the steepness of the cornering force to slip angle curve normally along its linear region (Fig. 8.32). The larger the cornering force needed to be generated for one degree of slip angle the greater the cornering stiffness of the tyre will be and the smaller the steering angle correction will be to sustain the intended path the vehicle is to follow. Note that the supple flexing of a radial ply side wall should not be confused with the actual stiffness of the tread portion of the tyre casing.

8.4.6 Centre of pressure (Fig. 8.35)

When a wheel standing stationary is loaded, the contact patch will be distributed about the geometric centre of the tyre at ground level, but as the wheel rolls forward the casing supporting the tread is deformed and pushed slightly to the rear (Fig. 8.35). Thus in effect the majority of the cornering force generated between the ground and each element of tread moves from the static centre of pressure to some dynamic centre of pressure behind the vertical centre of the tyre, the amount of displacement corresponding to the wheel construction, load, speed and traction. The larger area of tread to ground reaction will be concentrated behind the static centre of the wheel and the actual distribution of cornering force from front to rear of the contact patch is shown by the shaded area between the centre line of the tyre and the cornering force plotted line. The total cornering force is therefore roughly proportional to this shaded area and its resultant dynamic position is known as the centre of pressure (Fig. 8.35).

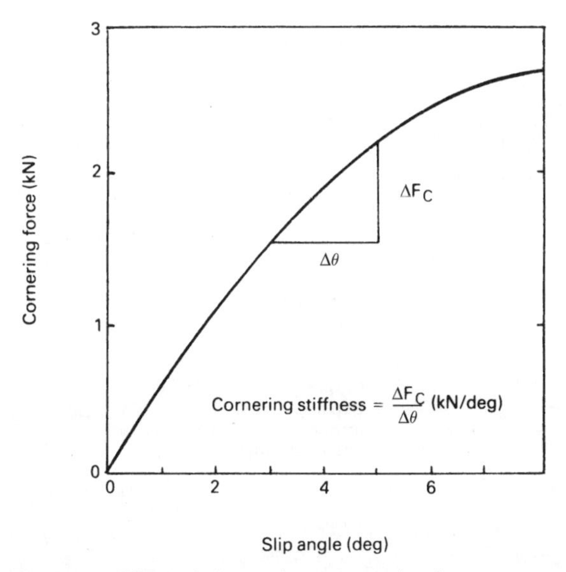

Fig. 8.32 Effect of slip angle on cornering force

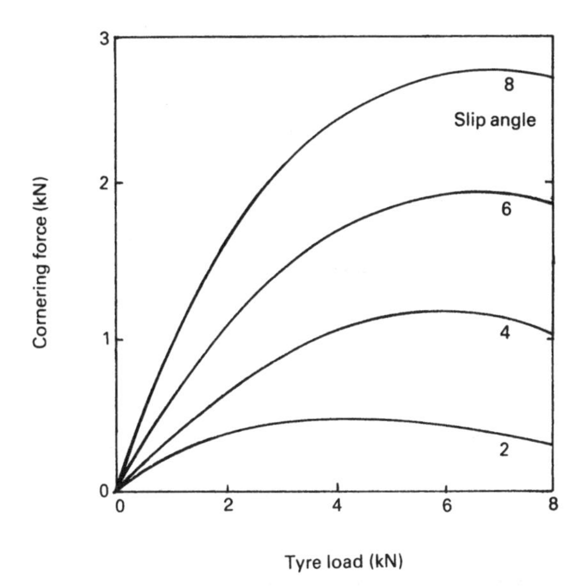

Fig. 8.33 Effect of tyre vertical load on cornering force

8.4.7 Pneumatic trail (Fig. 8.35)

The cornering force generated at any one time will be approximately proportional to the shaded area between the tyre centre line and the cornering force plotted line so that the resultant cornering force (centre of pressure) will act behind the static centre of contact. The distance between the static and dynamic centres of pressure is known as the *pneumatic trail* (Fig. 8.35), its magnitude being dependent upon the degree of creep between tyre and ground, the vertical wheel load, inflation pressure, speed and tyre constriction. Generally with the longer contact patch, radial ply tyres have a greater pneumatic trail than those of the cross-ply construction.

8.4.8 Self-aligning torque (Fig. 8.35)

When a moving vehicle has its steering wheels turned to negotiate a bend in the road, the lateral (side) force generates an equal and opposite

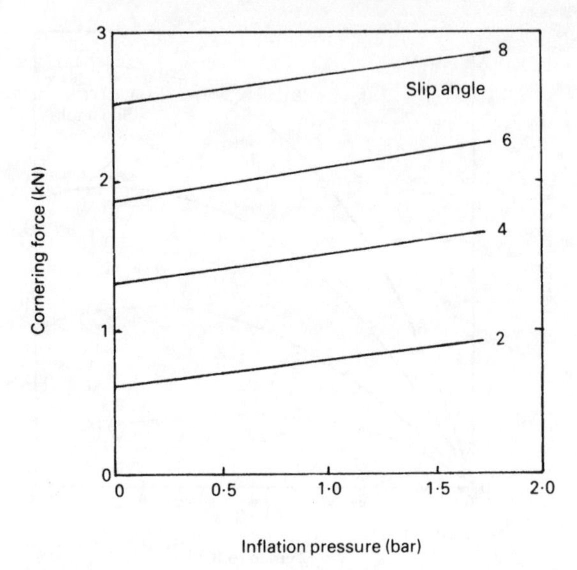

Fig. 8.34 Effect of tyre inflation pressure on cornering force

reaction force at ground level known as the *cornering force*. As the cornering force centre of pressure is to the rear of the geometric centre of the wheel and the side force acts perpendicularly through the centre of the wheel hub, the offset between the these two forces, known as the pneumatic trail, causes a moment (couple) about the geometric wheel centre which endeavours to turn both steering wheels towards the straight ahead position. This self-generating torque attempts to restore the plane of the wheels with the direction of motion and it is known as the *self-aligning torque* (Fig. 8.35). It is this inherent tyre property which helps steered tyres to return to the original position after negotiating a turn in the road. The self-aligning torque (SAT) may be defined as the product of the cornering force and the pneumatic trail.

i.e. $T_{SAT} = F_c \times t_p$ (Nm)

Higher tyre loads increase deflection and accordingly enlarge the contact patch so that the pneumatic trail is extended. Correspondingly this causes a rise in self-aligning torque. On the other hand increasing the inflation pressure for a given tyre load will shorten the pneumatic trail and reduce the self-aligning torque. Other factors which influence self-aligning torque are load transfer during braking, accelerating and cornering which alter the contact patch area. As a general rule, anything which increases or decreases the contact patch length raises or reduces the self-aligning torque respectively. The self-aligning torque is little affected with small slip angles when braking or accelerating, but with larger slip angles braking decreases the aligning torque and acceleration increases it (Fig. 8.36).

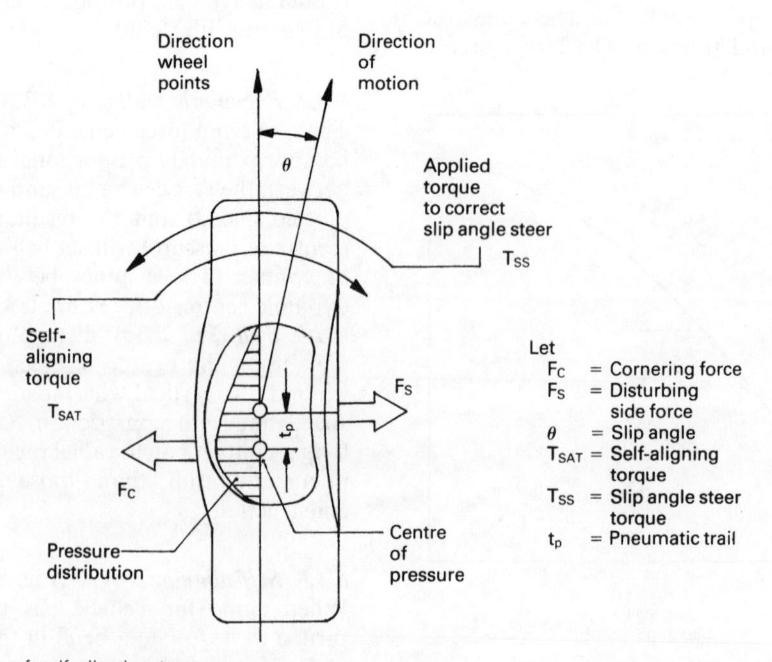

Fig. 8.35 Illustration of self-aligning torque

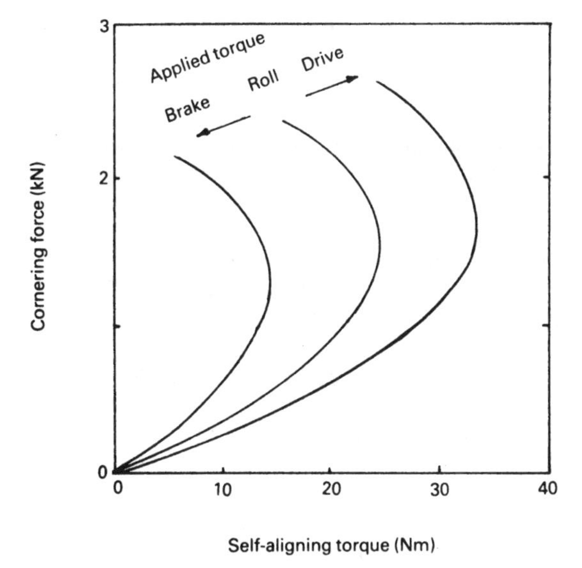

Fig. 8.36 Variation of self-aligning torque with cornering force

Static steering torque, that is the torque needed to rotate the steering when the wheels are not rolling, has nothing to do with the generated self-aligning torque when the vehicle is moving. The heavy static steering torque experienced when the vehicle is stationary is due to the distortion of the tyre casing and the friction created between the tyre tread elements being dragged around the wheels' point of pivot at ground level. With radial ply tyres the more evenly distributed tyre to ground pressure over the contact patch makes manoeuvring the steering harder than with cross-ply tyres when the wheels are virtually stationary.

8.4.9 Camber thrust (Figs 8.37 and 8.38)

The tilt of the wheel from the vertical is known as the camber. When it leans inwards towards the turning centre it is considered to be negative and when the top of the wheel leans away from the turning centre it is positive (Fig. 8.37). A positive camber reduces the cornering force for a given slip angle relative to that achieved with zero camber but negative camber raises it.

Constructing a vector triangle of forces with the known vertical reaction force and the camber inclination angle, and projecting a horizontal component perpendicular to the reaction vector so that it intersects the camber inclination vector, enables the magnitude of the horizontal component, known as camber thrust, to be determined (Fig. 8.37). The camber thrust can also be calculated as the product of the reaction force and the tangent of the camber angle.

i.e. Camber thrust = Wheel reaction × $\tan\beta$

The total lateral force reaction acting on the tyre is equal to the sum of the cornering force and camber thrust.

i.e. $F = F_c \pm F_t$

Where F = total lateral force

F_c = cornering force

F_t = camber thrust

When both forces are acting in the same direction, that is with the wheel tilting towards the centre of the turn, the positive sign should be used, if the wheel tilts outwards the negative sign applies (Fig. 8.38).

Thus negative camber increases the lateral reaction to side forces and positive camber reduces it.

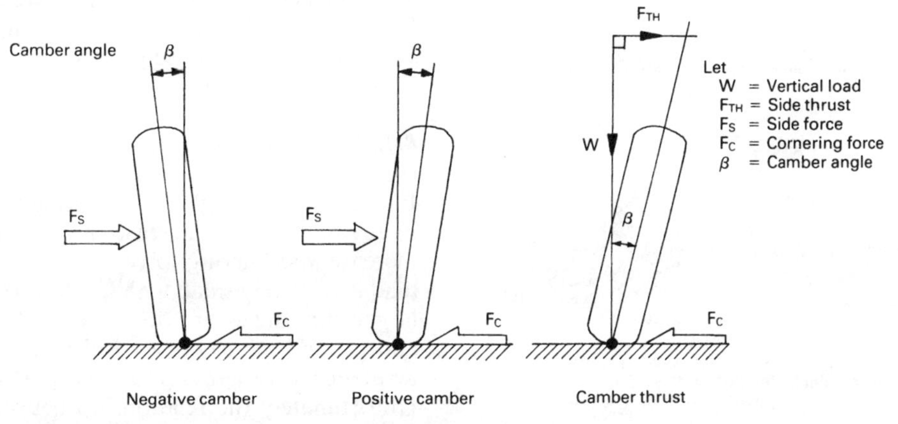

Fig. 8.37 Illustrating positive and negative camber and camber thrust

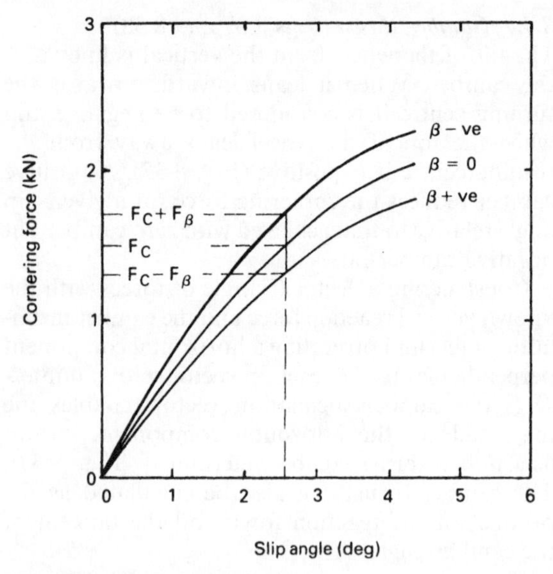

Fig. 8.38 Effect of slip angle on cornering force with various camber angles

8.4.10 Camber scrub (Fig. 8.39)

When a wheel is inclined to the vertical it becomes cambered and a projection line drawn through the wheel axis will intersect the ground at some point. Thus if the wheel completes one revolution a cone will be generated about its axis with the wheel and tyre forming its base.

If a vehicle with cambered wheels is held on a straight course each wheel tread will advance along a straight path. The distance moved along the road will correspond to the effective rolling radius at the mid-point of tyre contact with the road (Fig. 8.39). The outer edge of the tread (near the apex) will have a smaller turning circumference than the inner edge (away from apex). Accordingly, the smaller outer edge will try to speed up while the larger inner edge will tend to slow down relative to the speed in the middle of the tread. As a result, the tread portion in the outer tread region will slip forward, the portion of tread near the inner edge will slip backwards and only in the centre of tread will true rolling be achieved.

To minimize tyre wear due to camber scrub modern suspensions usually keep the wheel camber below 1½ degrees. Running wheels with a slight negative camber on bends reduces scrub and improves tyre grip whereas positive camber increases tread scrub and reduces tyre to road grip.

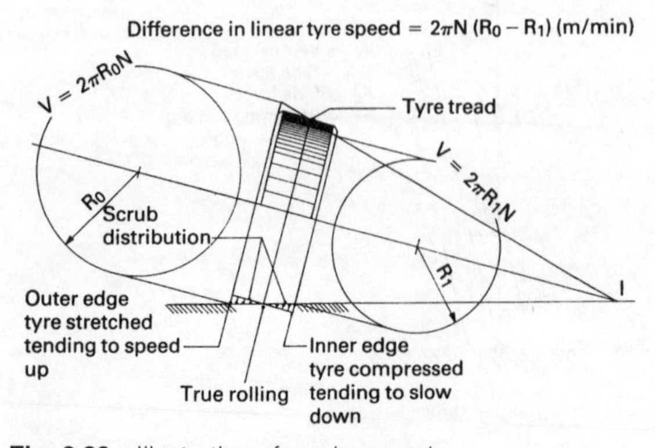

Fig. 8.39 Illustration of camber scrub

8.4.11 Camber steer (Fig. 8.40)

When a vehicle's wheels are inclined (cambered) to the vertical, the rolling radius is shorter on one side of the tread than on the other. The tyre then forms part of a cone and tries to rotate about its apex (Fig. 8.40(a and b)). Over a certain angular motion of the wheel, a point on the larger side of the tyre will move further than a point on the smaller side of the tyre and this causes the wheel to deviate from the straight ahead course to produce camber steer. Positive camber will make the wheels turn away from each other (Fig. 8.40(b)), i.e. *toe-out*, whereas negative camber on each side will make the wheels turn towards each other, i.e. *toe-in*. This is one of the reasons why the wheel track has to be set to match the design of suspension to counteract the inherent tendency of the wheels to either move away or towards each other.

Slightly inclining both wheels so that they lean towards the centre of turn reduces the angle of turn needed by the steered wheels to negotiate a curved path since the tyres want to follow the natural directional path of the generated cone (Fig. 8.41(a)). Conversely, if the wheels lean outwards from the centre of turn the tyres are compelled to follow a forced path which will result in a greater steering angle and consequently a degree of camber scrub (Fig. 8.41(b)).

8.4.12 Lateral weight transfer (Figs 8.42 and 8.43)

For a given slip angle the cornering force generally increases with the increase in vertical load. This increase in cornering force with respect to vertical load is relatively small with small slip angles, but as larger slip angles are developed between the tyre and ground increased vertical load enables much greater cornering forces to be generated (Fig. 8.42). Unfortunately the relationship between cornering force and vertical load is non-linear. This is because

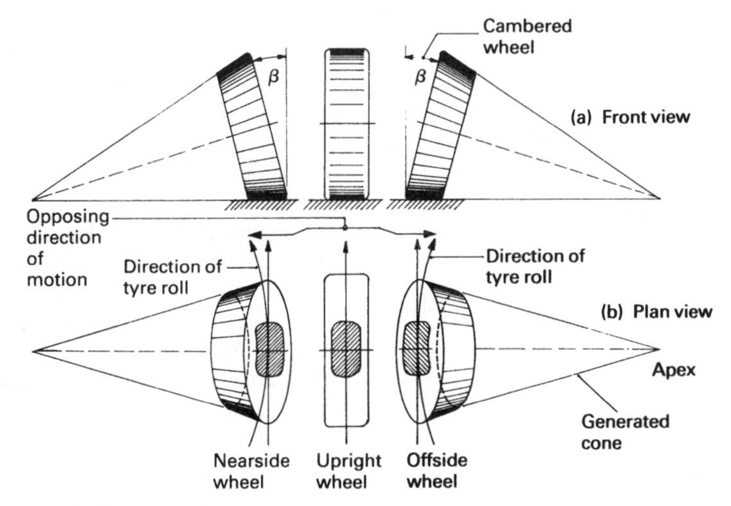

Fig. 8.40 Camber steer producing toe-out

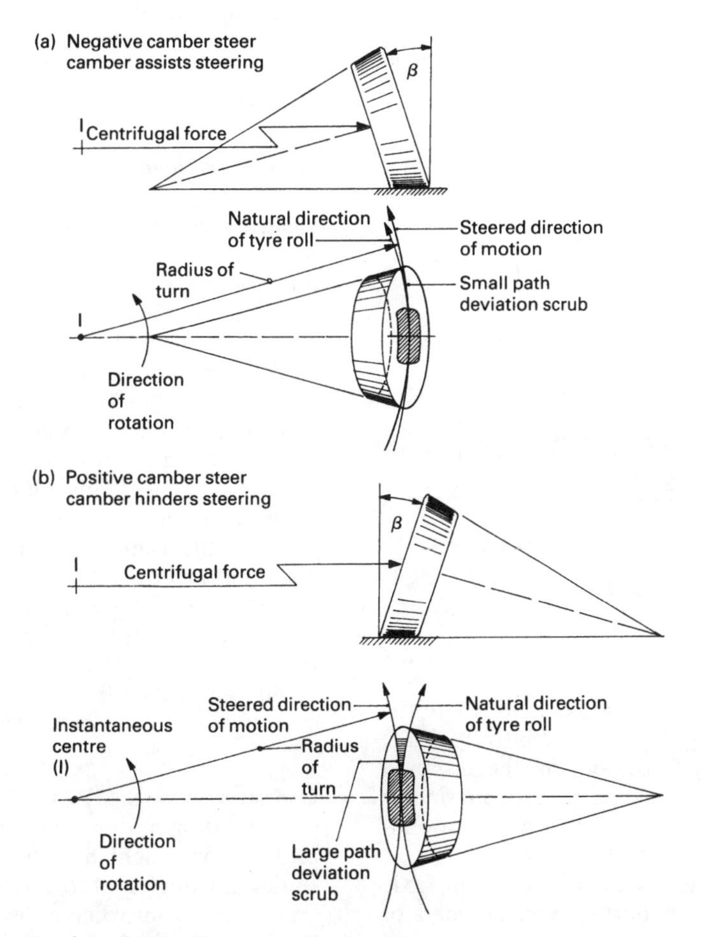

Fig. 8.41 (a and b) Principle of camber steer

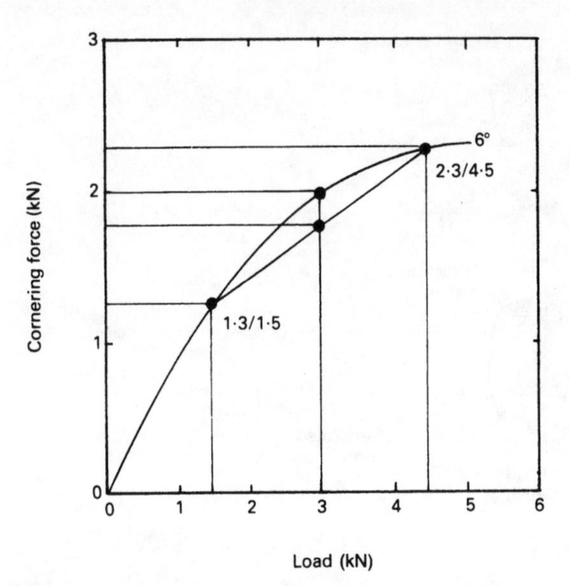

Fig. 8.42 Effect of transverse load transfer on the cornering force developed by a pair of tyres attached to axle

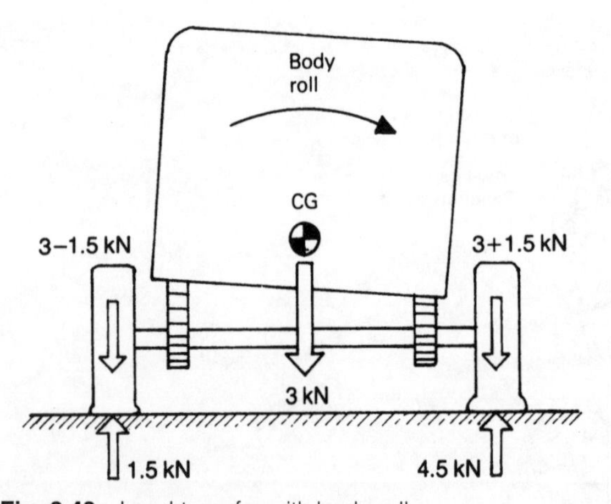

Fig. 8.43 Load transfer with body roll

an initial increase in vertical wheel load where the curve rise is steep produces a relatively large increase in cornering force, but as the imposed loading on the wheel becomes much larger a similar rise in vertical load does not produce a corresponding proportional increase in cornering force.

Consider a pair of tyres on a beam axle (Fig. 8.43), each with a normal vertical load of 3 kN. The cornering force per tyre with this load will be 2 kN for a given slip angle of 6°. If the vehicle is subjected to body roll under steady state movement on a curved track, then there will be certain amount of lateral weight transfer. Thus if the normal load on the inside wheel is reduced by 1.5 kN, the load on the outer wheel will be increased by the same amount.

As a result the total cornering force of the two tyres subjected to body roll will be $1.3 + 2.3 = 3.6$ kN (Fig. 8.42) which is less than the sum of both tyre cornering forces when they support their normal vertical load of $2 \times 2 = 4$ kN. The difference between the normal and body roll tyre loading thus reduces the cornering force capability for a given slip angle by 0.4 kN. This demonstrates that a pair of tyres on the front or rear axle to develop the required amount of cornering force to oppose a given centrifugal force and compensate for lateral weight transfer must increase the slip angles of both tyres. Thus minimizing body roll will reduce the slip angles necessary to sustain a vehicle at a given speed on a circular track.

8.5 Vehicle steady state directional stability

8.5.1 Directional stability along a straight track

Neutral steer (Fig. 8.44) Consider a vehicle moving forward along a straight path and a side force due possibly to a gust of wind which acts through the vehicle's centre of gravity which for simplicity is assumed to be mid-way between the front and rear axles. If the side force produces equal steady state slip angles on the front and rear tyres, the vehicle will move on a new straight line path at an angle to the original in proportion to the slip angles generated (Fig. 8.44). This motion is without a yaw velocity; a rotation about a vertical axis passing through the centre of gravity, and therefore is known as *neutral steer*.

Note that if projection lines are drawn perpendicular to the tyre tread direction of motion when the front and rear tyres are generating equal amounts of slip angle, then these lines never meet and there cannot be any rotational turn of the vehicle.

Oversteer (Fig. 8.45) If, due possibly to the suspension design, tyre construction and inflation pressure or weight distribution, the mean steady static slip angles of the rear wheels are greater than at the front when a disturbing side force acts through the vehicle centre of gravity, then the path

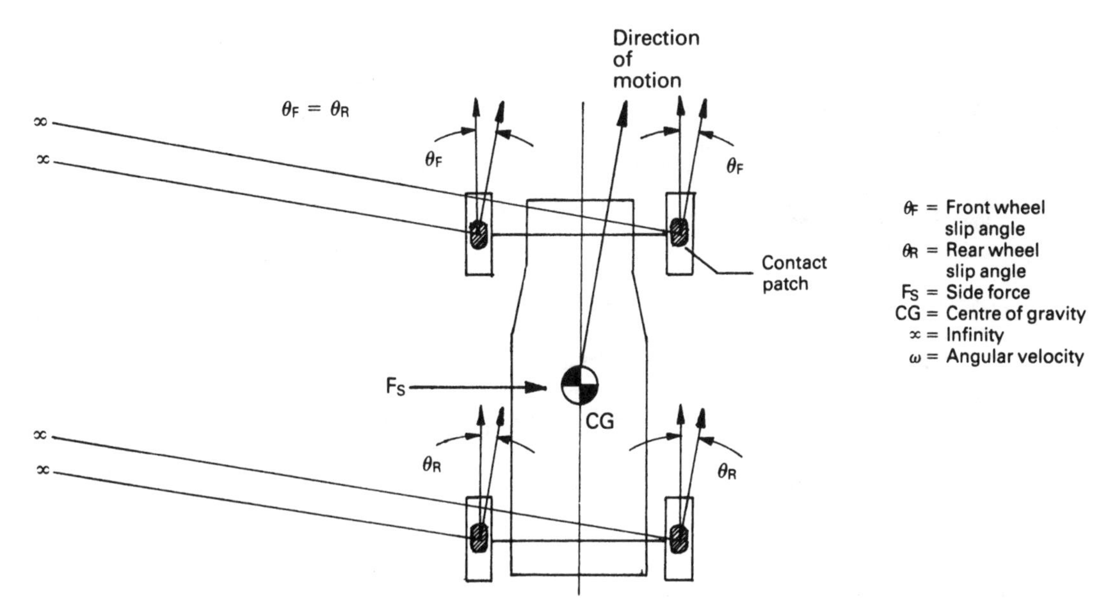

Fig. 8.44 Neutral steer on straight track

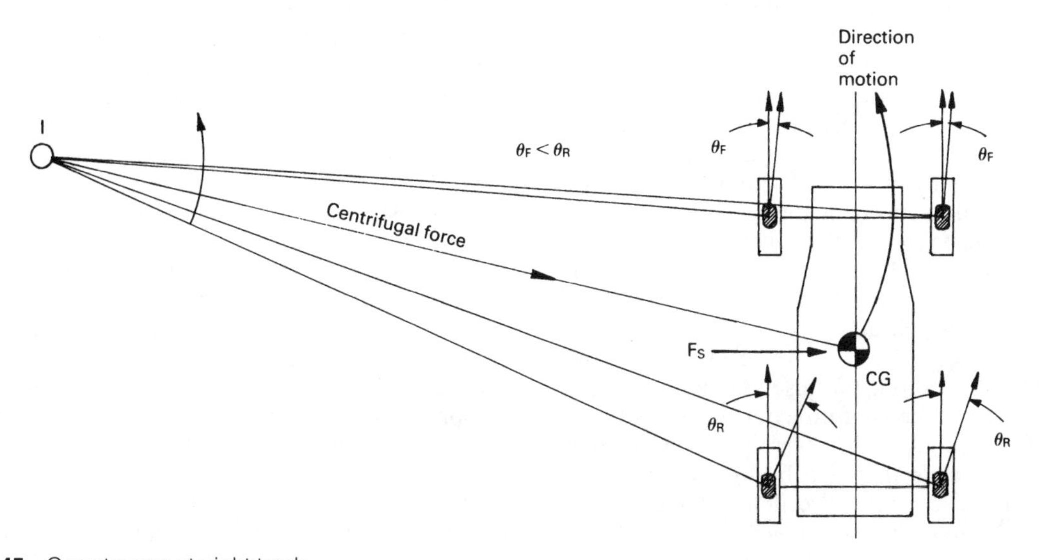

Fig. 8.45 Oversteer on straight track

of the vehicle is in a curve towards the direction of the applied side force (Fig. 8.45). The reason for this directional instability can be better understood if projection lines are drawn perpendicular to the direction the tyres roll with the generated slip angles. It can be seen that these projection lines roughly intersect each other at some common point known as the instantaneous centre, and therefore a centrifugal force will be produced which acts in the same direction as the imposed side force. Thus the whole vehicle will tend to rotate about this centre so that it tends to swing towards the disturbing force. To correct this condition known as *oversteer,* the driver therefore has to turn the steering in the same direction as the side force away from the centre of rotation.

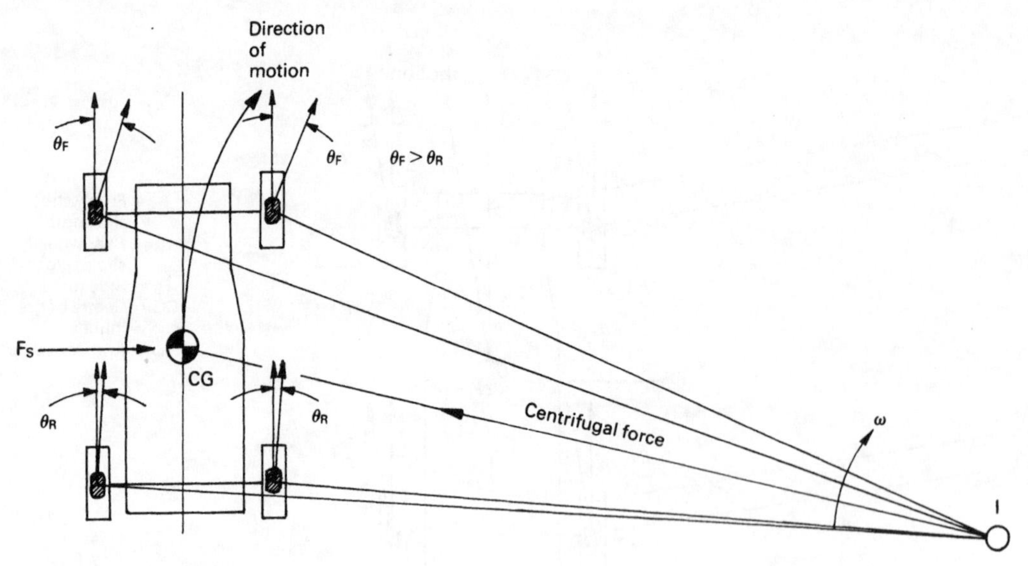

Fig. 8.46 Understeer on straight track

Understeer (Fig. 8.46) Now consider the situation of a vehicle initially moving along a straight path when a disturbing side force is imposed through the vehicle's centre of gravity. This time there is a larger slip angle on the front tyres than at the rear (Fig. 8.46). Again project lines perpendicularly to the tyre tread direction of motion when they are generating their slip angles but observe that these projections meet approximately at a common point on the opposite side to that of the side force. The vehicle's directional path is now a curve away from the applied side force so that a centrifugal force will be produced which acts in opposition to the disturbing side force. Thus the vehicle will be encouraged to rotate about the instantaneous centre so that it moves in the same direction as the disturbing force. Correction for this steering condition which is known as *understeer* is achieved by turning the steering in the opposite direction to the disturbing force away from the instantaneous centre of rotation. It is generally agreed that an oversteer condition is dangerous and undesirable, and that the slip angles generated on the front wheels should be slightly larger than at the rear to produce a small understeer tendency.

8.5.2 Directional stability on a curved track

True rolling of all four wheels can take place when projection lines drawn through the rear axle and each of the front wheel stub axles all meet at a common point somewhere along the rear axle projected line. This steering layout with the front wheels pivoted at the ends of an axle beam is known as the *Ackermann principle,* but strictly it can only be applied when solid tyres are used and when the vehicle travels at relatively slow speeds. With the advent of pneumatic tyres, the instantaneous centre somewhere along the extended projection from the rear axle now moves forwards relative to the rear axle. The reason for the positional change of the instantaneous centre is due to the centrifugal force produced by the vehicle negotiating a corner generating an opposing cornering force and slip angle under each tyre. Therefore projection lines drawn perpendicular to the direction each wheel tyre is moving due to the slip angles now converge somewhere ahead of the rear axle. This is essential if approximate true rolling conditions are to prevail with the vehicle travelling at speed.

Oversteer (Fig. 8.47) If the slip angles of the rear wheel tyres are made greater than on the front tyres when the vehicle is turning a corner (Fig. 8.47), the projected lines drawn perpendicular to the direction of motion of each tyre corresponding to its slip angle will all merge together at some common point (dynamic instantaneous centre) forward of the rear axle, further in and therefore at a shorter radius of turn than that produced for the Ackermann instantaneous centre for a given steering wheel angle of turn.

Under these driving conditions the vehicle will tend to steer towards the bend. Because the radius of the turn is reduced, the magnitude of the

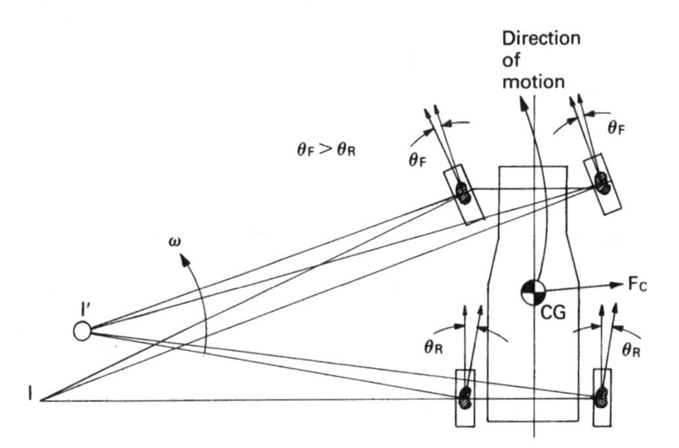

Fig. 8.47 Oversteer on turns

centrifugal force acting through the vehicle centre of gravity will be larger; it therefore raises the oversteer tendency of the vehicle. At higher vehicle speeds on a given circular path, the oversteer response will become more pronounced because the rise in centrifugal force will develop more tyre to ground reaction and correspondently increase the slip angles at each wheel. This is an unstable driving condition since the vehicle tends to turn more sharply into the bend as the speed rises unless the lock is reduced by the driver. For a rear wheel drive vehicle the application of tractive effort during a turn reduces the cornering stiffness and increases the slip angles of the rear wheels so that an oversteering effect is produced.

Understeer (Figs 8.48 and 8.49) If the slip angles generated on the front wheel tyres are larger than those on the rear tyre when the vehicle is turning a corner (Fig. 8.48) then projection lines drawn perpendicular to the direction of motion of each tyre, allowing for its slip angle, will now all intersect approximately at one point also forward of the rear axle, but further out at a greater radius of turn than that achieved with the Ackermann instantaneous centre.

With the larger slip angles generated on the front wheels the vehicle will tend to steer away from the bend. Because the radius of turn is larger, the magnitude of the centrifugal force produced at the centre of gravity of the vehicle will be less than for the oversteer situation. Thus the understeer tendency generally is less severe and can be corrected by turning the steering wheels more towards the bend. If tractive effort is applied when negotiating a circular path with a front wheel drive vehicle, the cornering stiffness of the front tyres is reduced. As a result, the slip angles are increased at the front, thereby introducing an understeer effect.

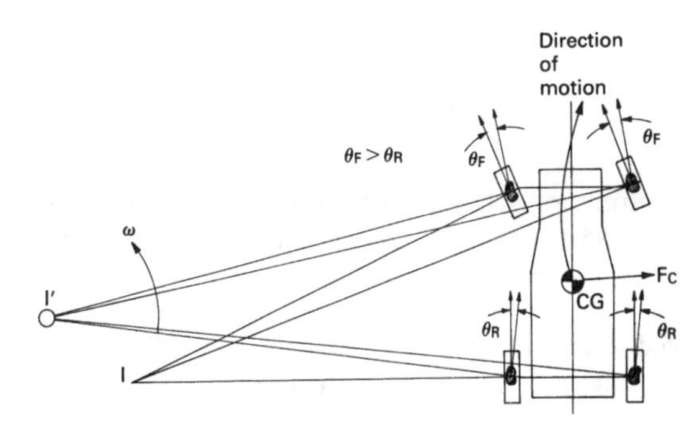

Fig. 8.48 Understeer on turns

A comparison between the steered angle of the front wheels or driver's steering wheel angle and vehicle speed for various steering tendencies is shown in Fig. 8.49. It can be seen that neutral steer maintains a constant steering angle throughout the vehicle's speed range, whereas both under- and oversteer tendencies increase with speed. An important difference between over- and understeer is that understeer is relatively progressive as the speed rises but oversteer increases rapidly with speed and can become dangerous.

8.6 Tyre marking identification (Tables 8.1 and 8.2)

To enable a manufacturer or customer to select the recommended original tyre or to match an equivalent tyre based on the vehicle's application

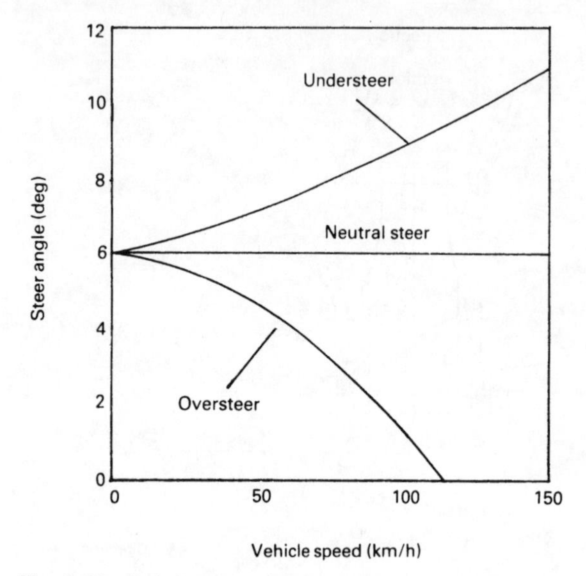

Fig. 8.49 Relationship of steer angle speed and vehicle speed of neutral steer, understeer and oversteer

requirement, wheel and tyre dimensions, tyre profile, maximum speed and load carrying capacity, a standard marking code has been devised.

8.6.1 Car tyres

Current tyres are marked in accordance with the standards agreed by the European Tyre and Rim Technical Organisation. Tyres with cross-ply construction and normal 82% aspect ratio do not indicate these features but radial construction and lower aspect ratios are indicated. Tyre section width, speed capacity, wheel rim diameter and tread pattern are always indicated.

Example 1
a) 165 SR 13 Mx
b) 185/70 VR 15 XWX

165 or 185 = nominal section width of tyre in millimetres
70 = 70% aspect ratio (Note no figures following 165 indicates 82% aspect ratio)
S or V = letter indicates speed capability (S = 180, V = 210 km/h)
R = radial construction
13 or 15 = nominal wheel rim diameter in inches
MX, XWX = manufacturer's tread pattern

In some instances section width is indicated in inches.

Example 2 6.45 Q 14

6.45 = nominal section width of tyre in inches
Q = letter indicates speed capability (speed symbol Q = 160 km/h)
14 = nominal wheel rim diameter in inches

Note No aspect ratio or construction indicated. Therefore assume 82% aspect ratio and cross-ply construction.

A revised form of marking has been introduced to include the maximum speed and load carrying capacity of the tyre under specified operating conditions.

A letter symbol indicates the maximum speed (Table 8.1) and a numerical code will identify the load carrying capacity (Table 8.2).

Example of new form of marking 205/70 R 13 80 S MXV

205 = normal section width in millimetres
70 = 70% aspect ratio
R = radial construction
13 = nominal wheel rim diameter in inches
80 = load index (from Table 8.2: 80 = 450 kg)
S = speed symbol (from Table 8.1: S = 180 km/h)
MXV = manufacturer's tread pattern code

Table 8.1 Speed symbols (SS)

Speed symbol (SS)	Speed (km/h)	SS	Speed (km/h)	SS	Speed (km/h)	SS	Speed (km/h)
A4	20	E	70	L	120	R	170
A6	30	F	80	M	130	S	180
A8	40	G	90	N	140	T	190
B	50	J	100	P	150	U	200
C	60	K	110	Q	160	H	210

(V = over 210)

Table 8.2 Load index (LI)

LI	kg	LI	kg	LI	kg	LI	kg
10	60	80	450	150	3350	220	25000
20	80	90	600	160	4500	230	33500
30	106	100	800	170	6000	240	45000
40	140	110	1060	180	8000	250	60000
50	190	120	1400	190	10600	260	80000
60	250	130	1900	200	14000	270	106000
70	335	140	2500	210	19000	280	140000

8.6.2 Light, medium and heavy truck tyres

Truck tyres sometimes include ply rating which indicates the load carrying capacity.

Example 10 R 20.0 PR12 XZA

10 = nominal section width of tyre in inches
R = radial construction
20.0 = nominal wheel rim diameter in inches
PR12 = ply rating
XZA = manufacturer's tread pattern

The revised form of marking indicates the load carrying capacity and speed capability for both single and twin wheel operation. The ply rating has been superseded by a load index because with improved fabric materials such as rayon, nylon and polyester as opposed to the original cotton cord ply, fewer ply are required to obtain the same strength using cotton as the standard, and therefore the ply rating does not give an accurate indication of tyre load bearing capacity.

Example 295/70 R 22.5 Tubeless 150/140L XZT

295 = nominal section width of tyre in millimetres
70 = 70% aspect ratio
R = radial construction
22.5 = nominal rim diameter in inches
150 = load index for singles (from Table 8.2: 150 = 3350 kg per tyre)
140 = load index for twins (from Table 8.2: 140 = 2500 kg per tyre)
L = speed symbol (from Table 8.1: L = 120 km/h)
XZT = manufacturer's tread pattern

8.7 Wheel balancing

The wheel and tyre functions are the means to support, propel and steer the vehicle forward and backward when rolling over the road surface. In addition the tyre cushions the wheel and axle from all the shock impacts caused by the roughness of the road contour. For the wheel and tyre assembly to rotate smoothly and not to generate its own vibrations, the wheel assembly must be in a state of rotatory balance.

An imbalance of the mass distribution around the wheel may be caused by a number of factors as follows:

a) tyre moulding may not be fitted concentric on the wheel rim,
b) wheel lateral run out or buckled wheel rim,
c) tyre walls, crown tread thickness may not be uniform all the way round the carcass when manufactured,
d) wheel lock when braking may cause excessive tread wear over a relatively small region of the tyre,
e) side wall may scrape the curb causing excessive wear on one side of the tyre,
f) tyre over or under inflation may cause uneven wear across the tread,
g) tyre incorrectly assembled on wheel relative to valve.

Whichever reason or combination of reasons has caused the uneven mass concentration (or lack of mass) about the wheel, one segment of the wheel and tyre will become lighter and therefore the tyre portion diametrically opposite will be heavier. Hence the heavy region of the tyre can be considered as a separate mass which has no diametrically opposing mass to counteract this inbalance.

Consequently the heavier regions of the wheel and tyre assembly when revolving about its axis (the axle or stub axle) will experience a centrifugal force. This force will exert an outward rotating pull on the support axis and bearings. The magnitude of this outward pull will be directly proportional to the out of balance mass, the square of the wheel rotational speed, and inversely proportional to the radius at which the mass is concentrated from its axis of rotation.

i.e. Centrifugal force $(F) = \dfrac{m\,V^2}{R}$ (N)

where F = centrifugal force (N)
m = out of balance mass (kg)
V = linear wheel speed (m/s)
R = radius at which mass is concentrated from the axis of rotation (m)

Example If, due to excessive braking, 100 g of rubber tread has been removed from a portion of the tyre tread 250 mm from the centre of rotation, determine when the wheel has reached a speed of 160 km/h the following:

a) angular speed of wheel in revolutions per minute,
b) centrifugal force.

Linear speed of wheel $V = \dfrac{160 \times 10^3}{60}$

$= 2666.666$ m/min

or $V = \dfrac{2666.666}{60}$

$= 44.444$ m/s

a) Angular speed of wheel $N = \frac{V}{\pi D}$

$= \frac{2666.666}{\pi 0.5}$

$= 1697.65 \text{ rev/min}$

b) Centrifugal force $F = \frac{m V^2}{R}$

$= \frac{0.1 (44.444)^2}{0.25}$

$= 790.1 \text{ N}$

From this calculation based on a vehicle travelling at a speed of one hundred miles per hour (160 km/h) and a typical wheel size for a car, the hundred gramme imbalance of the tyre produces a radial outward pull on the wheel axis of 790 Newtons. The magnitude of this 790 Newton force can be best appreciated by converting it to weight (mass) (79 kg) and then imagine lifting and carrying 79 one kilogramme bags of sugar for some distance.

8.7.1 Cyclic movement of a heavy spot on a wheel relative to the ground (Fig. 8.50)

When a road wheel rolls over a flat surface for one complete revolution, a point P on its circumference starting and finishing at ground level plots a curve known as a cycloid which represents the changing linear speed of the point P during each cycle of rotation (Fig. 8.50). For the short time point P is at ground level, its velocity remains at zero and at its highest position from the ground its forward velocity will be at a maximum. The average forward velocity of point P is at mid-height axle level, this also being the vehicle's forward speed. Thus the top of the tyre moves at twice the speed of the vehicle and in the same direction.

If point P is a heavy spot on the tyre, it will accelerate from zero to a maximum velocity for half a revolution and then decelerate to zero velocity again to complete the second half revolution. Since this spot has mass and changes its velocity, it will be subjected to a varying acceleration force which acts in a direction tangential to this curve. Consequently the direction of the inertia pull caused by this heavy spot constantly changes as the wheel moves forwards. The greatest reaction experienced on the wheel occurs within the short time the heavy spot decelerates downwards to ground level, momentarily stops, changes its direction and accelerates upwards. Hence at the end of each cycle and the beginning of the next there will be a tendency to push down and then lift up the tyre from the ground. At very low speeds this effect may be insignificant but as the vehicle speed increases, the magnitude of the accelerating force acting on this out of balance mass rises and thereby produces the periodic bump and bounce or jerking response of the tyre.

The balancing of rotating masses can be considered in two stages: firstly the static balance in one plane of revolution, this form of balance is known as *static balance,* and secondly the balance in more than one plane of revolution, commonly referred to as *dynamic balance*.

8.7.2 Static balance (Fig. 8.51)

This form of imbalance is best illustrated when a wheel and tyre assembly has been mounted on the hub of a wheel balancing machine which is then spun around by hand and released. The momentum put

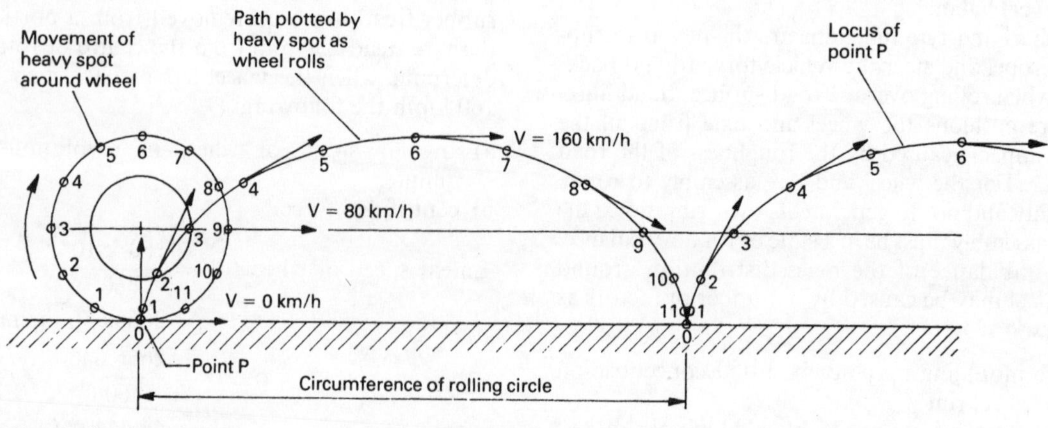

Fig. 8.50 Cyclic movement of a heavy spot on wheel relative to the road

into rotating the wheel tends to spin it a few times. It then stops momentarily and starts to oscillate to and fro with decreasing amplitude until eventually coming to rest. If a chalk mark is made on the tyre at its lowest point and the wheel is now turned say 90° and then released again, it will immediately commence to rotate on its own, one way and then the other way, until coming to rest with the chalk mark at the lowest point as before. This demonstrates that the heaviest part of the wheel assembly will always gravitate to the lowest position. If a small magnetic weight is placed on the wheel rim diametrically opposite the heavy side of the wheel and it has been chosen to be equivalent to the out of balance mass, then when the wheel is rotated to any other position, it remains in that position without any tendency to revolve on its own. If, however, there is still a slight movement of the wheel, or if the wheel wants to oscillate faster than before the magnetic weight was attached, then in the first case the balancing weight is too small and in the second case too large. This process of elimination by either adding or reducing the amount of weight placed opposite the heavy side of the wheel and then moving round the wheel about a quarter of a turn to observe if the wheel tries to rotate on its own is a common technique used to check and correct any wheel imbalance on one plane. When the correct balancing weight has been determined replace the magnetic weight with a clip-in weight of similar mass. With a little experience this trial and error method of statically balancing the wheel can be quick, simple and effective.

The consequences of a statically unbalanced wheel and tyre is that the heavy side of the wheel will pull radially outwards as it orbits on a fixed circular path around its axis of rotation, due to the centrifugal force created by the heavier side of the wheel (Fig. 8.51). If the swivel pins and the centre of the unbalanced mass are offset to each other,

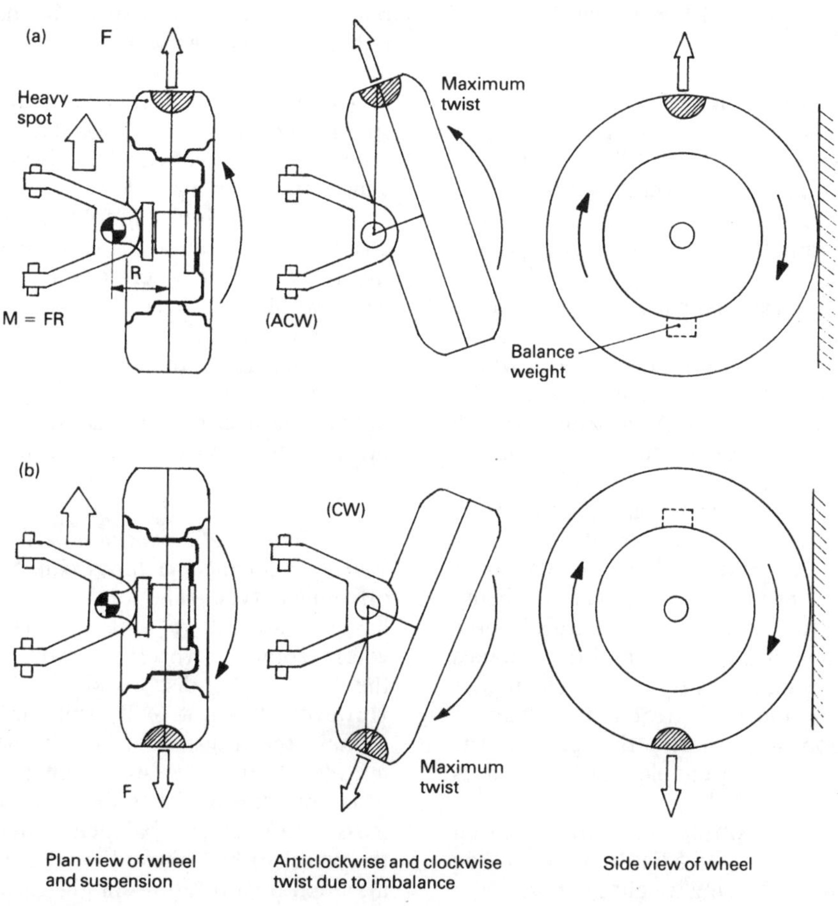

Fig. 8.51 (a and b) Illustration of static wheel imbalance

then when the heavy spot is in the horizontal plane pointing towards the front of the vehicle a moment of force is produced ($M = FR$) which will endeavour to twist the stub axle and wheel assembly anticlockwise about the swivel pins (Fig. 8.51(a)). As the wheel rolls forward a further half turn, the heavy spot will now face towards the rear so that the stub axle and wheel assembly will try to swivel in the opposite direction (clockwise) (Fig. 8.5(b)). Hence with a statically unbalanced tyre the stub axle will twist about its pivot every time the heavier side of the wheel completes half a revolution between the extremities in the horizontal plane. The oscillations generated will thus be transmitted to the driver's steering wheel in the form of tremors which increase in frequency and magnitude as the vehicle's speed rises. If there is a substantial amount of swivel pin or kingpin wear, the stub axle will be encouraged to move vertically up or down on its supporting joints. This might convey vibrations to the body via the suspension which could become critical if permitted to resonate with possibly the unsprung or sprung parts of the vehicle.

8.7.3 Dynamic balance (Fig. 8.52)

If a driven drum is made to engage the tread of the tyre so that the wheel is spun through a speed range there is a likelihood that the wheel will develop a violent wobble which will peak at some point as the wheel speed rises and then decreases as the speed is further increased.

This generated vibration is caused by the balance weight having been placed correctly opposite the heavy spot of the tyre but on the wheel rim which may be in a different rotational plane to that of the original out of balance mass. As a result the tyre heavy spots pull outwards in one plane while the balance weight of the wheel rim, which is being used to neutralize the heavy region of the tyre, pulls radially outwards in a second plane. Consequently, due to the offset of the two masses, a rocking couple is produced, its magnitude being proportional to the product of centrifugal force acting through one of the masses and the distance between the opposing forces ($C = FX$). The higher the wheel speed and the greater the distance between the opposing forces, the greater the magnitude of the rocking couple will be which is causing the wheel to wobble.

The effects of the offset statically balanced masses can be seen in Fig. 8.52(a, b and c). When the heavy spot and balancing weight are horizontal (Fig. 8.52(a)), the mass on the outside of the wheel points in the forward direction of the vehicle and the mass on the inside of the wheel points towards the rear so that the wheel will tend to twist in an anticlockwise direction about the swivel pins. With a further 180° rotation of the wheel, the weights will again be horizontal but this time the outer weight has moved to the rear and the inner weight towards the front of the vehicle. Thus the sense of the unbalanced rocking couple will have changed to a clockwise one. For every revolution of the wheel, the wheel will rock in both a clockwise and anticlockwise direction causing the driver's steering wheel to jerk from side to side (Fig. 8.52(c)). Note that when the masses have moved into a vertical position relative to the ground, the swivel pins constrain the rocking couple so that no movement occurs unless the swivel ball joints or kingpins are excessively worn.

The characteristics of the resulting wheel wobble caused by both static and dynamic imbalance can be distinguished by the steady increase in the amplitude of wheel twitching about the swivel pins with rising wheel speed in the case of static unbalanced wheels, whereas with dynamic imbalance the magnitude of wheel twitching rises to a maximum and then declines with further wheel speed increase (Fig. 8.53). Thus with dynamic imbalance, a wheel can be driven at road speeds which are on either side of the critical period of oscillation (maximum amplitude) without sensing any undue instability. If the wheel is driven within the relative narrow critical speed range violent wheel wobble results.

Slackness in the swivel pins or steering linkage ball joints with unbalanced tyres will promote excessive wheel twitch or wobble, resulting not only in the steering wheel sensing these vibrations, but causing heavy tyre tread scrub and wear.

8.7.4 Methods of balancing wheels

Wheel balancing machines can be of the on- or off-vehicle type. The on-vehicle wheel balancer has the advantage that it balances the wheel whole rotating wheel assembly which includes the hub, brake disc or drum, wheel and tyre. However, it is not really suitable for drive axles because the transmission drive line does not permit the wheel hub to spin freely (which is essential when measuring the imbalance of any rotating mass). Off-vehicle balancing machines require the wheel to be removed from the hub and to be mounted on a rotating spindle forming part of the balancing equipment.

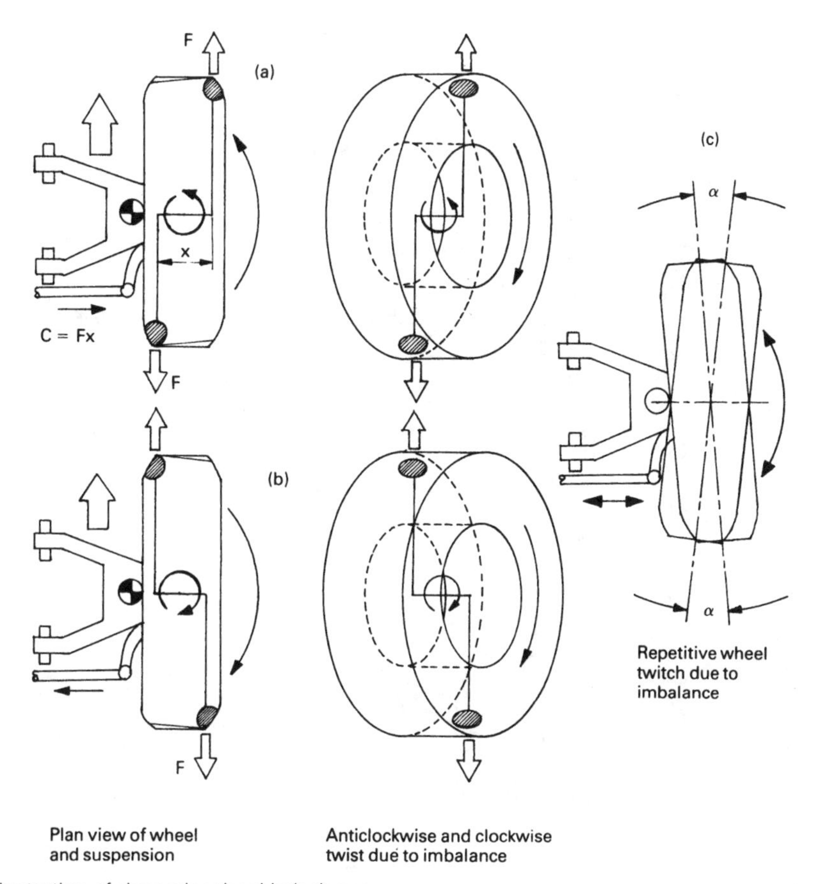

Fig. 8.52 (a–c) Illustration of dynamic wheel imbalance

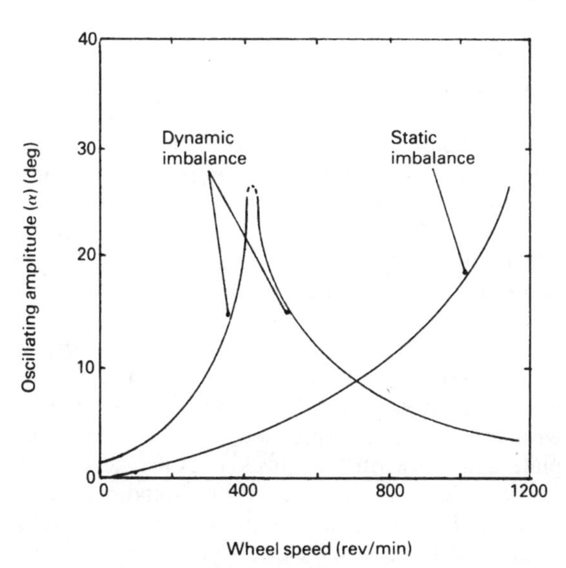

Fig. 8.53 Relationship of wheel speed and oscillating amplitude for both static and dynamic imbalance

Balancing machine which balances statically and dynamically in two separate planes (Fig. 8.54) The wheel being balanced is mounted on the spindle of the mainshaft which is supported by a pair of spaced out ball bearings. This machine incorporates a self-aligning ball bearing at the wheel end mounted rigidly to the balancing machine frame, whereas the rear bearing is supported between a pair of stiff opposing springs which are themselves attached to the balancing machine frame. An electric motor supplies the drive to the spindle by way of the engagement drum rubbing hard against the tyre tread of the wheel assembly being balanced.

When the wheel and tyre is spun and the assembly commences to wobble about the self-aligning bearing, the restraining springs attached to the second bearing absorb the out of balance forces and the deflection of the mainshaft and spindle.

An electro-magnetic moving coil vibration detector (*transducer*) is installed vertically between the second bearing and the machine frame. When

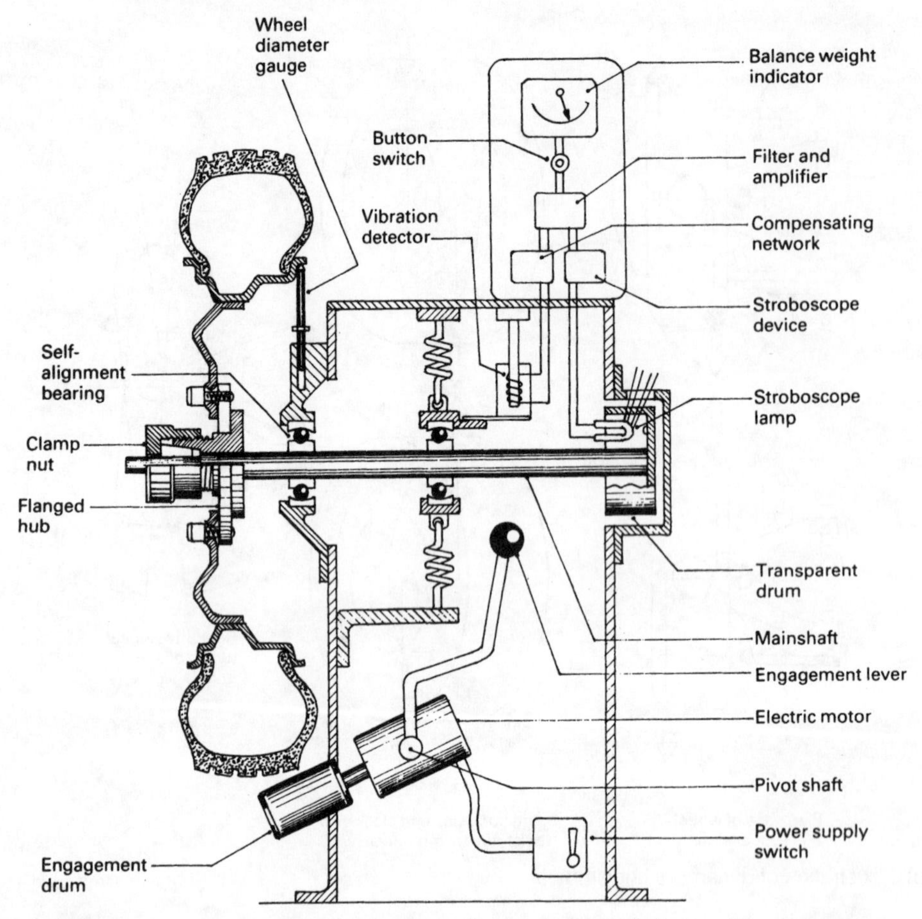

Fig. 8.54 Wheel balancing machine which balances statically and dynamically in two different planes

the wheel assembly wobbles, the armature (rod) in the centre of the transducer coil moves in and out of a strong magnetic field provided by the permanent magnet. This causes the armature coil to generate a voltage proportional to the relative movement of the rod. The output signal from the detector is a direct measure of the imbalance of the wheel assembly. It is therefore fed into a compensating network which converts the signals into the required balance weight to be attached to the outside of the wheel rim in the left hand plane. These modified, but still very weak, electrical signals are then passed through a filter which eliminates unwanted side interference. They are then amplified so that they can activate the stroboscope device and the weight indicator meter.

The weight indicator meter computes the voltage amplitude signal coming from the detector and, when calibrated, indicates the size of the weight to be added to the plane of balance, in this case the outside of the wheel. Conversely, the stroboscope determines the angular phase of the balance weight on the wheel. This is achieved by the sinusoidal voltage being converted into a sharply defined bright flash of light in the stroboscope lamp. A rotating numbered transparent drum is illuminated by the stroboscope flash and the number which appears on the top of the drum relates to the phase position of the required balance weight.

Mounting of wheel on balancing machine spindle (Fig. 8.54) Mount the wheel onto the flanged multi-hole steel plate. Align the wheel stud holes with corresponding threaded holes in the flange plate and screw on the wheel studs provided. Slide the wheel hub assembly along the spindle until the inside of the wheel rim just touches the adjustable distance rod and then lock the hub to the spindle via the sleeve nut. The positioning of the wheel assembly relative to the supporting self-aligning bearing is important since the inside wheel rim now is in the same rotating plane as the centre of the bearing. Any couple which might have been formed when the balance weights were attached to the inside of the wheel rim are eliminated as there is now no offset.

Dynamic balance setting (Fig. 8.54) To achieve dynamic balance, switch on the power, pull the drive roller lever until the roller is in contact with the tyre and allow the wheel to attain full speed. Once maximum speed has been reached, push the lever so that the roller is freed from the tyre. If the wheel assembly is unbalanced the wheel will pass through a violent period of wobble and then it will steady again as the speed falls. While the wheel is vibrating, the magnitude and position of the imbalance can be read from the meter and from the stroboscope disc aperture respectively. A correction factor is normally given for the different wheel diameters which must be multiplied by the meter reading to give the actual balance weight. Select the nearest size of balance weight to that calculated, then rotate the wheel by hand to the number constantly shown on the stroboscope disc when the wheel was spinning and finally attach the appropriate balance weight to the top of the wheel on the outside (away from the machine). Thus the outer half of the wheel is balanced.

Static balance setting (Fig. 8.54) Static balance is obtained by allowing the wheel to settle in its own position when it will naturally come to rest with the heaviest point at bottom dead centre. Select a magnetic weight of say 50 grammes and place this on the inside rim at top dead centre and with this in position turn the wheel a quarter of a revolution. If the magnetic weight is excessive, the weighted side will naturally gravitate towards the bottom but if it is insufficient, the weight will rise as the wheel slowly revolves. Alter the size of the magnetic weight and repeat the procedure until there is no tendency for the wheel to rotate on its own whatever its position. The wheel is now statically and dynamically balanced and a quick check can be made by repeating the spin test for dynamic balance. Once the correct static balance weight has been found, replace the magnetic weight by a clip-on type.

Balancing machine which dynamically balances in two planes (Fig. 8.55) The machine is so constructed that the wheel being balanced is mounted on the spindle of the mainshaft which is supported by a pair of spaced out ball bearings housed in a cylindrical cradle, which itself is supported on four strain rods which are reduced in diameter in their mid region (Fig. 8.55). An electric motor supplies the drive to the mainshaft via a rubberized flat belt and pulleys.

Vibration detectors are used to sense the out of balance forces caused by the imbalance of the wheel assembly. They are normally of electromechanical moving coil type transducers. The tranducer consists of a small armature in the form of a stiff rod which contains a light weight coil. The rod is free to move in a strong magnetic field supplied by a permanent magnet. The armature rods are rigidly attached to the mainshaft and bearing cradle and the axes of the rods are so positioned as to coincide with the direction of vibration. The housing and permanent magnets of the detectors are mounted onto the supporting frame of the cradle. The relative vibratory motion of the armature rod to the casing causes the armature coil to generate a voltage proportional to the relative vibrational velocity.

The output signals from the two detectors are fed into a compensating network and then into the selector switch. The compensating network is so arranged that the output signals are proportional to the required balance weights in the left and right hand balancing planes respectively. The output voltages from the selector switch are very small signals and will include unwanted frequency components. These are eliminated by the filter. At the same time the signals are amplified so that they can operate the stroboscope device and both weight indicating meters. These weight indicating meters measure the amplitude of the voltage from the detectors and when calibrated indicate the actual weights to be added in each plane. The stroboscope device changes the sinusoidal voltage into a sharply defined pulse which occurs at the same predetermined point in every cycle. This pulse is converted into a very bright flash of light in the stroboscope lamp when focused on the rotating numbered transparent drum; one number will appear on the top of the apparently stationary surface. The number is a measure of the relative phase position of the voltage and is arranged to indicate the position of the required balance weight.

Dynamic balance setting in two planes (Fig. 8.55) Mount the wheel over the spindle and slide the conical adaptor towards the wheel so that its taper enters the central hole of the wheel. Screw the sleeve nut so that the wheel is centralized and wedged against the flanged hub (Fig. 8.55). Any

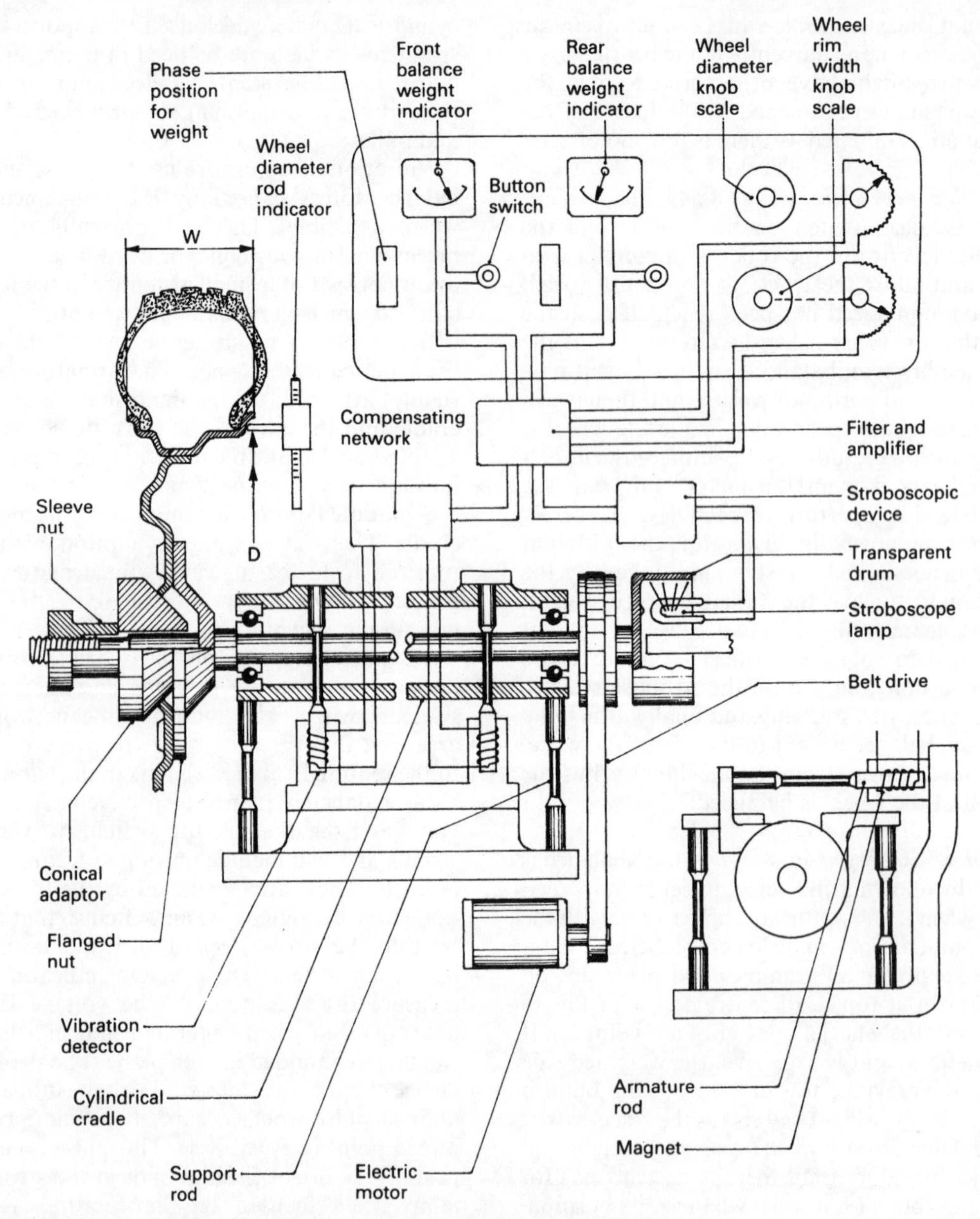

Fig. 8.55 Wheel balancing machine which dynamically balances in two planes

existing balance weights should now be removed and the wheel should also be brushed clean.

Before the wheel assembly is actually balanced on the machine, the two basic wheel dimensions must be programmed into the electronic network circuit. This is carried out by simply moving the wheel diameter indicator probe against the inside of the wheel rim and reading off the wheel diameter and also measuring with a caliper gauge normally provided the wheel width. The wheel diameter and wheel rim width measurements are then set by rotating the respective potentiometer knob on the display console to these dimensions so that the electronic network is altered to correct for changes in the centrifugal force and rocking couples which will vary with different wheel sizes.

Balancing machines of this type usually measure and provide correction for wheel imbalance for both static and dynamic balance with respect to both the outer and inner wheel rim rotating planes.

The start button is now pressed to energize the electric motor. As a result the drive belt will rotate the mainshaft and spin the wheel assembly under test.

First the state of wheel balance in the outer wheel rim plane of rotation is measured by pressing the outer rim weight indicator meter switch. The meter pointer will align on the scale with size of balance weight required in grammes; the stroboscope indicator window will also show a number which corresponds to the wheel position when a balance weight is to be attached.

Once the balance weight size and angular position has been recorded, the wheel assembly is brought to a standstill by pressing the stop button and then releasing it when the wheel just stops rotating.

The wheel should now be rotated by hand until the number previously observed through the stroboscope window again appears, then attach the selected size of balance weight to the top of the outer wheel rim.

The whole procedure of spinning the wheel assembly is then repeated but on the second time the inner rim weight indicator button is pressed so that the balance weight reading and the phase position of the wheel refer to the inner rim rotating plane. Again the wheel is braked, then turned by hand to the correct phase position given by the stroboscope number. Finally the required balance weight is attached, this time to the inner wheel rim at the top of the wheel.

8.7.5 Wheel and tyre run-out

Before proceeding to balance the wheel asseembly the wheel should be checked for run-out in both lateral and radial directions relative to the axis of rotation. If the wheel or tyre run-out is excessive it should be corrected before the wheel assembly is balanced.

Lateral run-out (Fig. 8.56) If the wheel being examined has been jacked clear of the ground and when spun appears to wobble so that the wheel rim or tyre wall moves axially inward or outward in a wavy fashion lateral (sideway), run-out is taking place. This may be caused by either the wheel rim being buckled or the tyre being fitted unevenly around the rim of the wheel so that the wheel assembly will produce dynamic imbalance. Deflating the tyre and repositioning the bead against the inside of the wheel rim will usually correct any tyre lateral run-out. Lateral run-out of the wheel itself should be no greater than 2.0 mm.

Radial run-out (Fig. 8.57) If with the wheel jacked clear of the ground, the wheel and tyre assembly appears to lift and fall every time the wheel completes one revolution, then the distance from the axis of rotation to the tyre tread instead of being constant around the periphery of the tyre is varying.

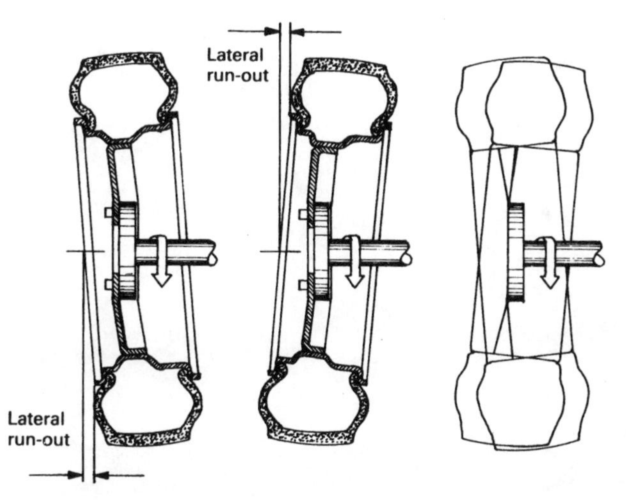

Fig. 8.56 Illustration of lateral tyre run-out

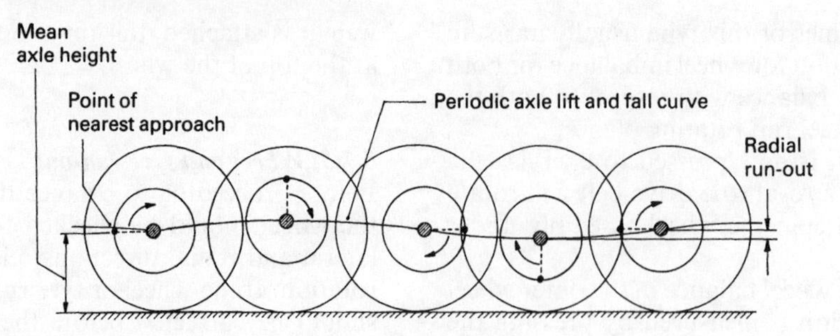

Fig. 8.57 Illustration of radial tyre run-out

This error is usually caused by the tyre having been fitted eccentrically about the wheel rim so that when the wheel assembly is spun, radial run-out will be observed, and as a result, the wheel assembly will be in a state of static imbalance. Tyre eccentricity can usually be cured by repositioning the tyre on the wheel rim. The maximum wheel eccentricity should not exceed 2.0 mm.

9 Steering

9.1 Steering gearbox fundamental design

9.1.1 Steering gearbox angular ratios

The steering gearbox has two main functions: it produces a gear reduction between the input steering wheel and the output drop arm (Pitman arm) and it redirects the input to output axis of rotation through a right angle.

Generally, the steering road wheel stub axles must be capable of twisting through a maximum steering angle of 40° either side of the straight ahead position. The overall angular gear ratio of a steering gearbox may be as direct as 12:1 for light small vehicles or as indirect as 28:1 for heavy vehicles. Therefore, lock to lock drop arm angular displacement amounts to 80° and with a 12:1 and 28:1 gear reduction the number of turns of the steering wheel would be derived as follows:

Lock to lock steering wheel turns for 12:1 $= \frac{80 \times 12}{360}$

$= 2.66$ revolutions

Lock to lock steering wheel turns for 28:1 reduction $= \frac{80 \times 28}{360}$

$= 6.22$ revolutions

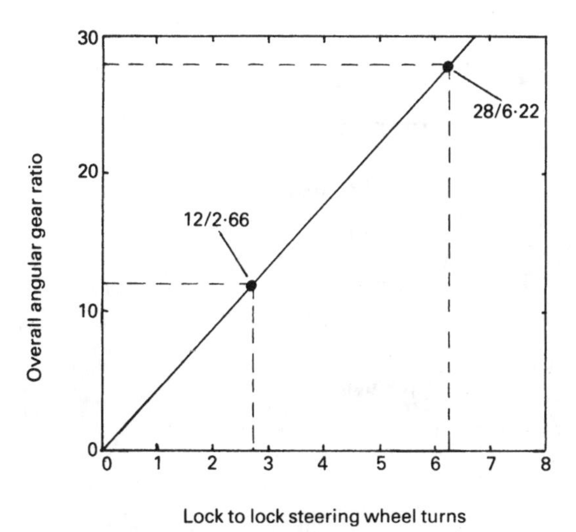

Fig. 9.1 Relationship of overall angular gear ratio and steering wheel lock to lock revolutions

From these results plotted in Fig. 9.1 it can be seen that the 12:1 reduction needs the steering wheel to be rotated 1.33 turns from the straight ahead position. The 28:1 reduction will require more than twice this angular displacement, namely 3.11 turns. Thus with the 12:1 gear reduction, the steering may be heavy but can be turned from the straight ahead position to full lock and back again relatively quickly. However the 28:1 reduction will provide a light steering wheel but the vehicle will be compelled to corner much slower if the driver is to be able to complete the manoeuvre safely.

9.1.2 Screw and nut steering gear mechanism (Fig. 9.2)

To introduce the principles of the steering gearbox, the screw and nut type mechanism will be examined as this is the foundation for all the other types of steering box gear reduction mechanisms.

A screw is made by cutting an external spiral groove around and along a cylindrical shaft, whereas a nut is produced by cutting a similar spiral groove on the internal surface of a hole made in a solid block.

The thread profile produced by the external and internal spiral grooves may take the form of a vee, trapezoidal, square or semicircle, depending upon the actual application.

A nut and screw combination (Fig. 9.2) is a mechanism which increases both the force and

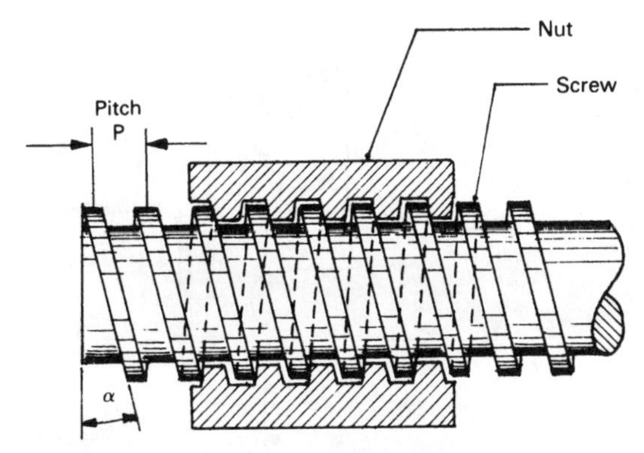

Fig. 9.2 Screw and nut friction steering gear mechanism

movement ratios. A small input effort applied to the end of a perpendicular lever fixed to the screw is capable of moving a much larger load axially along the screw provided that the nut is prevented from rotating.

If the screw is prevented from moving longitudinally and it revolves once within its nut, the nut advances or retracts a distance equal to the axial length of one complete spiral groove loop. This distance is known as the thread pitch or lead (p).

The inclination of the spiral thread to the perpendicular of the screw axis is known as the helix angle (α). The smaller the helix angle the greater the load the nut is able to displace in an axial direction. This is contrasted by the reduced distance the nut moves forwards or backwards for one complete revolution of the screw.

The engaged or meshing external and internal spiral threads may be considered as a pair of infinitely long inclined planes (Fig. 9.3(a and b)). When the nut is prevented from turning and the screw is rotated, the inclined plane of the screw slides relative to that of the nut. Consequently, a continuous wedge action takes place between the two members in contact which compels the nut to move along the screw.

Because of the comparatively large surface areas in contact between the male and female threads and the difficulty of maintaining an adequate supply of lubricant between the rubbing faces, friction in this mechanism is relatively high with the result that mechanical efficiency is low and the rate of wear is very high.

A major improvement in reducing the friction force generated between the rubbing faces of the threads has been to introduce a series of balls (Fig. 9.4) which roll between the inclined planes as the screw is rotated relatively to the nut.

The overall gear ratio is achieved in a screw and nut steering gearbox in two stages. The first stage occurs by the nut moving a pitch length for every one complete revolution of the steering wheel. The second stage takes place by converting the linear movement of the nut back to an angular one via an integral rocker lever and shaft. Motion is imparted to the rocker lever and shaft by a stud attached to the end of the rocker lever. This stud acts as a pivot and engages the nut by means of a slot formed at right angles to the nut axis.

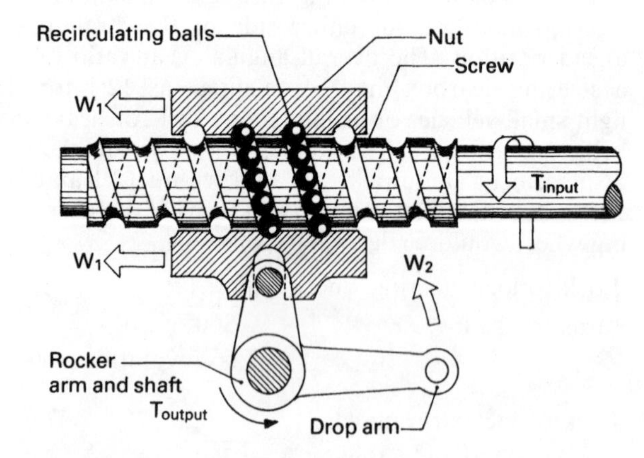

Fig. 9.4 Screw and nut recirculating ball low friction gear mechanism

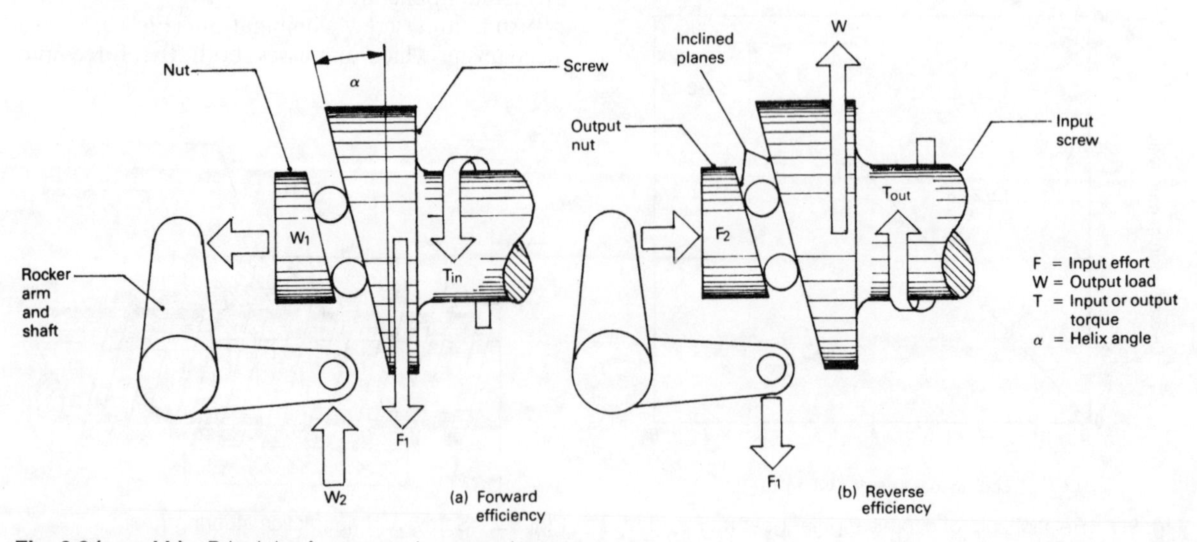

Fig. 9.3 (a and b) Principle of screw and nut steering gear

Forward and reverse efficiency The forward efficiency of a steering gearbox may be defined as the ratio of the output work produced at the drop arm to move a given load to that of the input work done at the steering wheel to achieve this movement.

i.e. Forward efficiency =

$$\frac{\text{Output work at drop arm}}{\text{Input work at steering wheel}} \times 100$$

Conversely the reverse efficiency of a steering gearbox is defined as the ratio of the output work produced at the steering wheel rim causing it to rotate against a resisting force to that of the input work done on the drop arm to produce this movement.

i.e. Reverse efficiency =

$$\frac{\text{Output work at steering wheel}}{\text{Input work at drop arm}} \times 100$$

A high forward efficiency means that very little energy is wasted within the steering gearbox in overcoming friction so that for a minimum input effort at the steering wheel rim a maximum output torque at the drop arm shaft will be obtained.

A small amount of irreversibility is advantageous in that it reduces the magnitude of any road wheel oscillations which are transmitted back to the steering mechanism. Therefore the vibrations which do get through to the steering wheel are severely damped.

However, a very low reverse efficiency is undesirable because it will prevent the self-righting action of the kingpin inclination and castor angle straightening out the front wheels after steering the vehicle round a bend.

Relationship between the forward and reverse efficiency and the helix angle (Figs 9.3, 9.4 and 9.5) The forward efficiency of a screw and nut mechanism may be best illustrated by considering the inclined plane (Fig. 9.3(a)). Here the inclined plane forms part of the thread spiral of the screw and the block represents the small portion of the nut. When the inclined plane (wedge) is rotated anticlockwise (moves downwards) the block (nut) is easily pushed against whatever load is imposed on it. When the screw moves the nut the condition is known as the *forward efficiency*.

In the second diagram (Fig. 9.3(b)) the block (nut) is being pressed towards the right which in turn forces the inclined plane to rotate clockwise (move upward), but this is difficult because the helix angle (wedge angle) is much too small when the nut is made to move the screw. Thus when the mechanism is operated in the reverse direction the efficiency (reverse) is considerably lower than when the screw is moving the nut. Only if the inclined plane angle was to be increased beyond 40° would the nut be easily able to rotate the screw.

The efficiency of a screw and nut mechanism will vary with the helix angle (Fig. 9.5). It will be at a maximum in the region of 40–50° for both forward and reverse directions and fall to zero at the two extremes of 0 and 90° (helix angle). If both forward and reverse efficiency curves for a screw and nut device were plotted together they would both look similar but would appear to be out of phase by an amount known as the friction factor.

Selecting a helix angle that gives the maximum forward efficiency position (A) produces a very high reverse efficiency (A′) and therefore would feed back to the driver every twitch of the road wheels caused by any irregularities on the road surface. Consequently it is better to choose a smaller helix angle which produces only a slight reduction in the forward efficiency (B) but a relatively much larger reduced reverse efficiency (B–B′). As a result this will absorb and damp the majority of very small vibrations generated by the tyres rolling over the road contour as they are transmitted through the steering linkage to the steering gearbox.

A typical value for the helix angle is about 30° which produces forward and reverse efficiencies of about 55% and 30% without balls respectively. By incorporating recirculating balls between the screw and nut (Fig. 9.4) the forward and reverse efficiencies will rise to approximately 80% and 60% respectively.

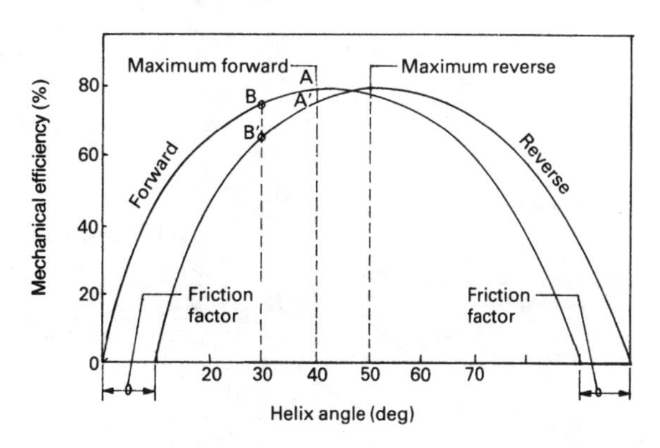

Fig. 9.5 Efficiency curves for a screw and nut recirculating ball steering gear

Summary and forward and reverse efficiency The efficiency of a screw and nut mechanism is relatively high in the forward direction since the input shaft screw thread inclined plane angle is small. Therefore a very large wedge action takes place in the forward direction. In the reverse direction, taking the input to be at the steering box drop arm end, the nut threads are made to push against the steering shaft screw threads, which in this sense makes the inclined plane angle very large, thus reducing the wedge advantage. Considerable axial force on the nut is necessary to rotate the steering shaft screw in the reverse direction, hence the reverse efficiency of the screw and nut is much lower than the forward efficiency.

9.1.3 Cam and peg steering gearbox (Fig. 9.6)

With this type of steering box mechanism the conventional screw is replaced by a cylindrical shaft supported between two angular contact ball bearings (Fig. 9.6). Generated onto its surface between the bearings is a deep spiral groove, usually with a variable pitch. The groove has a tapered side wall profile which narrows towards the bottom.

Positioned half-way along the cam is an integral rocker arm and shaft. Mounted at the free end of the rocker arm is a conical peg which engages the tapered sides of the groove. When the camshaft is rotated by the steering wheel and shaft, one side of the spiral groove will screw the peg axially forward or backward, this depending upon the direction the cam turns. As a result the rocker arm is forced to pivot about its shaft axis and transfers a similar angular motion to the drop arm which is attached to the shaft's outer end.

To increase the mechanical advantage of the cam and peg device when the steering is in the straight ahead position, the spiral pitch is generated with the minimum pitch in the mid-position. The pitch progressively increases towards either end of the cam to give more direct steering response at the expense of increased steering effort as the steering approaches full lock.

Preload adjustment of the ball races supporting the cam is provided by changing the thickness of shim between the end plate and housing. Spring loaded oil seals are situated at both the drop arm end of the rocker shaft and at the input end of the camshaft.

Early low efficiency cam and peg steering boxes had the peg pressed directly into a hole drilled in the rocker arm, but to improve efficiency it is usual

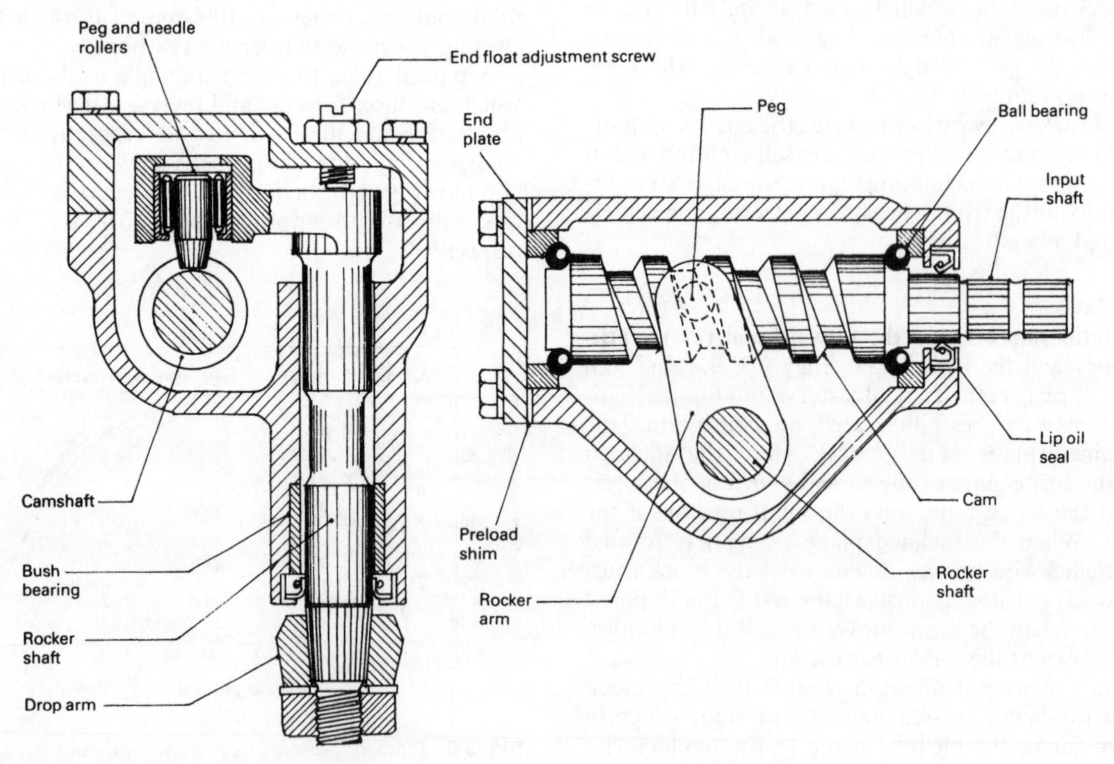

Fig. 9.6 Cam and peg steering type gearbox

to support the peg with needle rollers assembled inside an enlarged bore machined through the rocker arm. For heavy duty applications, and where size permits, the peg can be mounted in a parallel roller race with a combined radial and thrust ball race positioned at the opposite end to the peg's tapered profile. An alternative high efficiency heavy duty arrangement for supporting the peg uses opposing taper roller bearings mounted directly onto the rocker arm, which is shaped to form the inner tracks of the bearings.

Cam and peg mechanisms have average forward and reverse efficiencies for pegs that are fixed in the rocker arm of 50% and 30% respectively, but needle mounted pegs raise the forward efficiency to 75% and the reverse to 50%.

To obtain the correct depth of peg to cam groove engagement, a rocker shaft end play adjustment screw is made to contact a ground portion of the rocker shaft upper face.

The rocker shaft rotates in a bronze plain bearing at the drop arm end and directly against the bearing bore at the cam end. If higher efficiency is required, the plain bush rocker shaft bearing can be replaced by needle bearings which can raise the efficiency roughly 3–5%.

9.1.4 Worm and roller type steering gearbox (Fig. 9.7)

This steering gear consists of an hourglass-shaped worm (sometimes known as the cam) mounted between opposing taper roller bearings, the outer race of which is located in the end plate flange and in a supporting sleeve at the input end of the worm shaft (Fig. 9.7). Shims are provided between the end plates and housing for adjusting the taper roller bearing preload and for centralizing the worm relative to the rocker shaft.

Engaging with the worm teeth is a roller follower which may have two or three teeth. The roller follower is carried on two sets of needle rollers supported on a short steel pin which is located between the fork arm forged integrally with the rocker shaft.

In some designs the needle rollers are replaced by ball races as these not only support radial loads but also end thrust, thereby substantially reducing frictional losses.

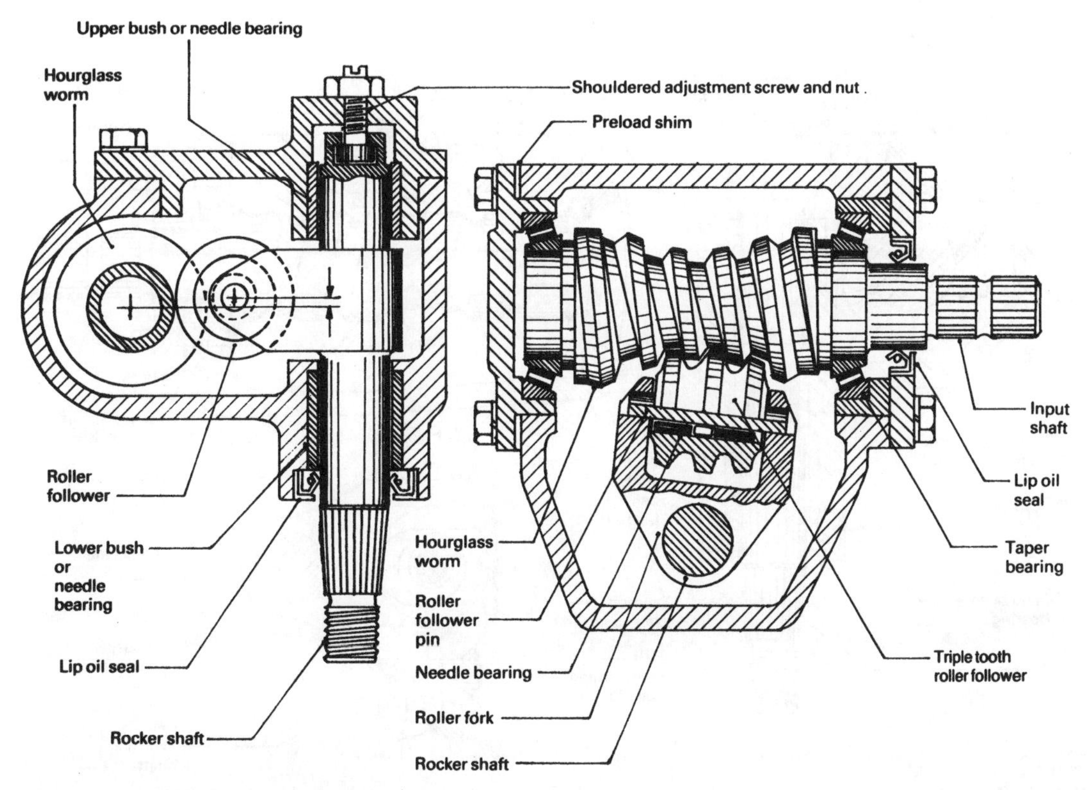

Fig. 9.7 Worm and roller type steering gearbox

The rocker shaft is supported on two plain bushes; one located in the steering box and the other in the top cover plate. End thrust in both directions on the rocker shaft is taken by a shouldered screw located in a machined mortise or 'T' slot at one end of the rocker shaft.

To adjust the depth of mesh of the worm and roller (Fig. 9.7), move the steering wheel to the mid-position (half the complete number of turns of the steering wheel from lock to lock), screw in the end thrust shouldered screw until all free movement is taken up and finally tighten the lock nut (offset distance being reduced).

Centralization of the cam in relation to the rocker shaft roller is obtained when there is an equal amount of backlash between the roller and worm at a point half a turn of the steering wheel at either side of the mid-position. Any adjustment necessary is effected by the transference from one end plate to the other of the same shims as those used for the taper bearing preload (i.e. the thickness of shim removed from one end is added to the existing shims at the other).

The forward and reverse efficiencies of the worm roller gear tend to be slightly lower than the cam and peg type of gear (forward 73% and reverse 48%) but these efficiencies depend upon the design to some extent. Higher efficiencies can be obtained by incorporating a needle or taper roller bearing between the rocker shaft and housing instead of the usual plain bush type of bearing.

9.1.5 Recirculating ball nut and rocker lever steering gearbox (Fig. 9.8)

Improvement in efficiency of the simple screw and nut gear reduction is achieved with this design by replacing the male and female screw thread by semicircular grooves machined spirally onto the input shaft and inside the bore of the half nut and then lodging a ring of steel balls between the internal and external grooves within the nut assembly (Fig. 9.8).

The portion of the shaft with the spiral groove is known as the *worm*. It has a single start left hand spiral for right hand drive steering and a right hand spiral for left hand drive vehicles.

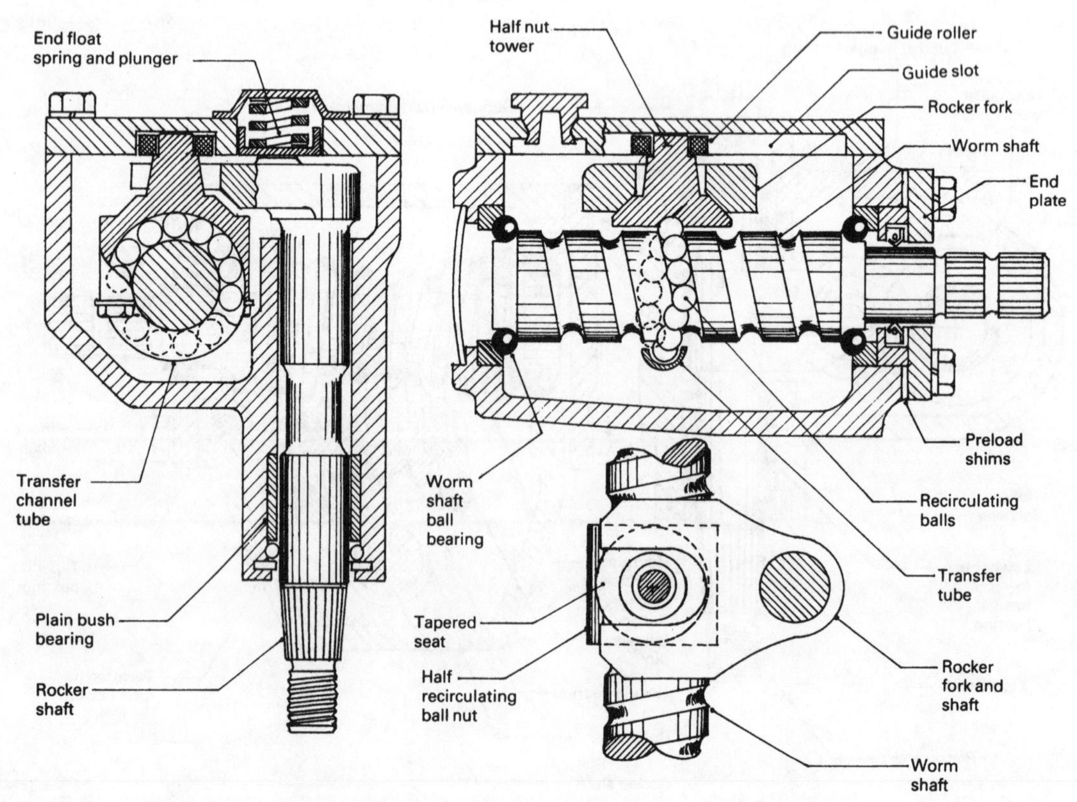

Fig. 9.8 Recirculating ball nut and rocker lever steering type gearbox

The worm shaft is supported between two sets of ball races assembled at either end normally in an aluminium housing. Steel shims sandwiched between the detachable plate at the input end of the shaft provide adjustment of the bearing preload. Situated on the inside of the end plate is a spring loaded lip seal which contacts the smooth surface portion of the worm shaft.

Assembled to the worm is a half nut with a detachable semicircular transfer tube secured to the nut by a retainer and two bolts. The passage formed by the grooves and transfer tube is fitted with steel balls which are free to circulate when the worm shaft is rotated.

The half nut has an extended tower made up of a conical seat and a spigot pin. When assembled, the conical seat engages with the bevel forks of the rocker lever, whereas a roller on the nut spigot engages a guide slot machined parallel to the worm axis in the top cover plate. When the worm shaft is rotated, the spigot roller engaged in its elongated slot prevents the nut turning. Movement of the nut along the worm will result in a similar axial displacement for the spigot roller within its slot.

End float of the rocker lever shaft is controlled by a spring loaded plunger which presses the rocker lever bevel forks against the conical seat of the half nut.

The rocker lever shaft is supported directly in the bore of the housing material at the worm end but a bronze bush is incorporated in the housing at the drop arm end of the shaft to provide adequate support and to minimize wear. An oil seal is fitted just inside the bore entrance of the rocker shaft to retain the lubricant within the steering box housing.

The worm shaft has parallel serrations for the attachment of the steering shaft, whereas the rocker shaft to drop arm joint is attached by a serrated taper shank as this provides a more secure attachment.

Forward and reverse efficiencies for this type of recirculating ball and rocker lever gear is approximately 80% and 60% respectively.

9.1.6 Recirculating ball rack and sector steering gearbox (Fig. 9.9)

To reduce friction the conventional screw and nut threads are replaced by semicircular spiral grooves (Fig. 9.9). These grooves are machined externally around and along the cylindrically shaped shaft which is known as the worm and a similar groove is machined internally through the bore of the nut.

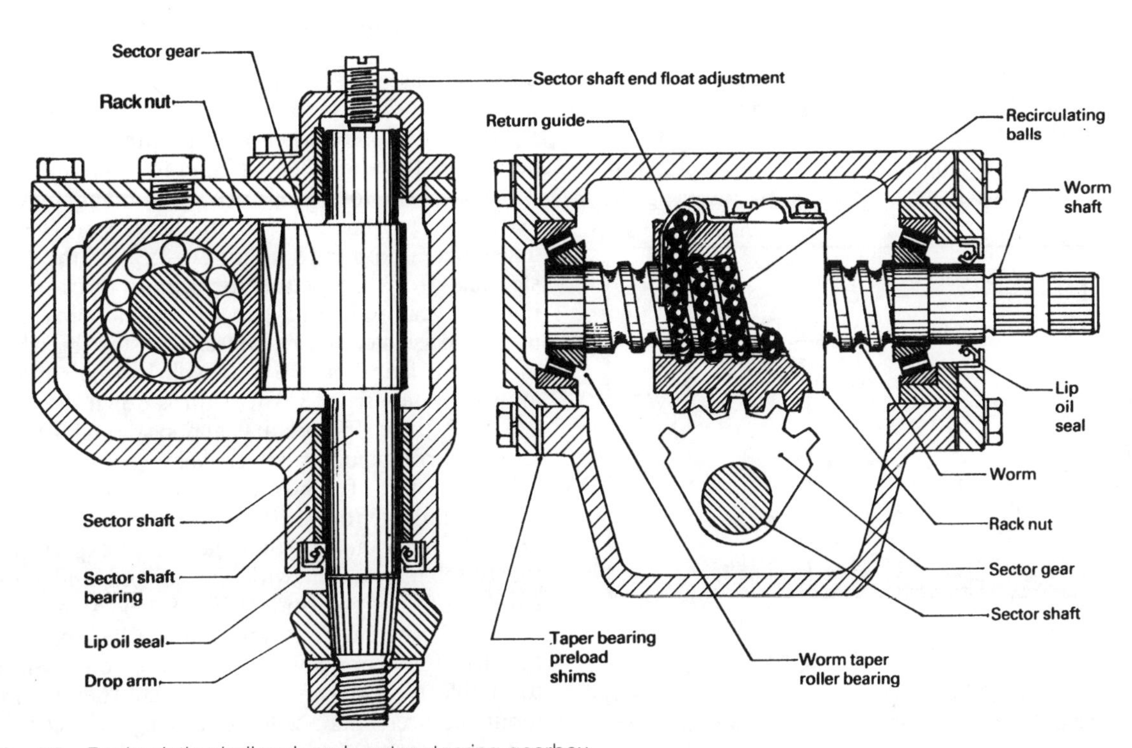

Fig. 9.9 Recirculating ball rack and sector steering gearbox

Engagement of the worm and nut is obtained by lodging a series of steel balls between the two sets of matching semicircular spiral grooves.

There are two separate ball circuits within the ball nut, and when the steering wheel and worm rotates, the balls roll in the grooves against the nut. This causes the nut to move along the worm. Each ball rotates one complete loop around the worm after which it enters a ball return guide. The guide deflects the balls away from the grooved passages so that they move diagonally across the back of the nut. They are then redirected again into the grooved passages on the other side of the nut.

One outer face of the rectangular nut is machined in the shape of teeth forming a gear rack. Motion from the nut is transferred to the drop arm via a toothed sector shaft which meshes with the rack teeth, so that the linear movement of the nut is converted back to a rotary motion by the sector and shaft.

An advantage of this type of steering gear is that the rack and sector provides the drop arm with a larger angular movement than most other types of mechanisms which may be an essential feature for some vehicle applications. Because of the additional rack and sector second stage gear reduction, the overall forward and reverse efficiencies are slightly lower than other recirculating ball mechanisms. Typical values for forward and reverse efficiencies would be 70% and 45% respectively.

9.2 The need for power assisted steering

(Figs 9.10 and 9.11)

With manual steering a reduction in input effort on the steering wheel rim is achieved by lowering the steering box gear ratio, but this has the side effect of increasing the number of steering wheel turns from lock so that manoeuvring of the steering will take longer, and accordingly the vehicle's safe cornering speed has to be reduced.

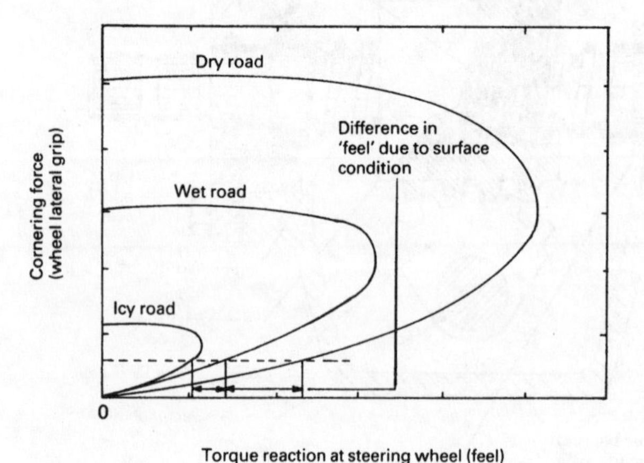

Fig. 9.10 Typical relationship of tyre grip on various road surfaces and the torque reaction on the driver's steering wheel

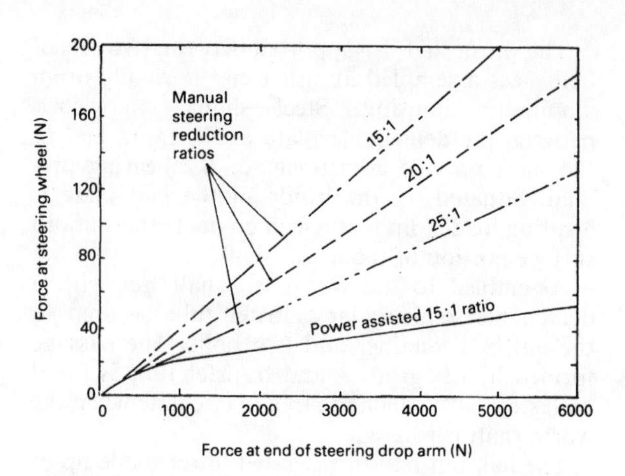

Fig. 9.11 Comparison of manual steering with different reduction gear ratio and power assisted steering

With the tendency for more weight to be put on the front steering wheels of front wheel drive cars and the utilization of radial ply tyres with greater tyre width, larger static turning torques are required. The driver's expectancy for faster driving and cornering makes power assisted steering desirable and in some cases essential if the driver's ability to handle the vehicle is to match its performance.

Power assistance when incorporated on passenger cars reduces the driver's input to something like 25–30% of the total work needed to manoeuvre it. With heavy trucks the hydraulic power (servo) assistance amounts to about 80–85% of the total steering effort. Consequently, a more direct steering box gear reduction can be used to provide a more precise steering response. The steering wheel movement from lock to lock will then be reduced approximately from 3½ to 4 turns down to about 2½ to 3 turns for manual and power assistance steering arrangements respectively.

The amount of power assistance supplied to the steering linkage to the effort put in by the driver is normally restricted so that the driver experiences the tyres' interaction with the ground under the varying driving conditions (Fig. 9.10). As a result there is sufficient resistance transmitted back to the driver's steering wheel from the road wheels to enable the driver to sense or feel the steering input requirements needed effectively to steer the vehicle.

The effects of reducing the driver's input effort at the steering wheel with different steering gear overall gear ratios to overcome an output opposing resistance at the steering box drop arm is shown in Fig. 9.11. Also plotted with these manual steering gear ratios is a typical power assisted steering input effort curve operating over a similar working load output range. This power assisted effort curve shows that for very low road wheel resistance roughly up to 1000 N at the drop arm, the input effort of 10 to 20 N is practically all manual. It is this initial manual effort at the steering wheel which gives the driver his sense of feel or awareness of changes in resistance to steering under different road surface conditions, such as whether the ground is slippery or not.

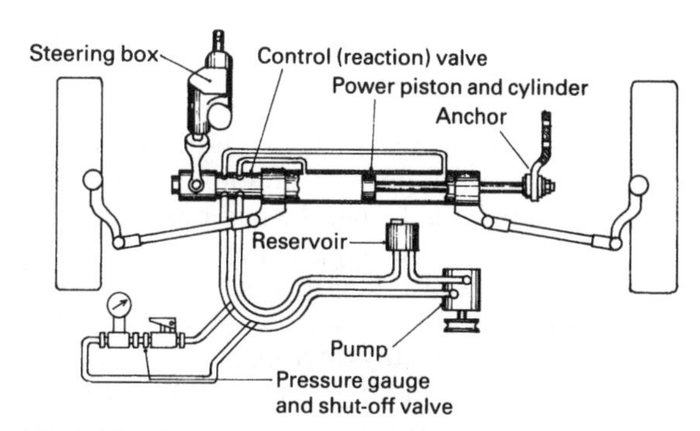

Fig. 9.13 Steering box with external directly coupled power assisted steering utilized with independent front suspension

9.2.1 External direct coupled power assisted steering power cylinder and control valve

Description of power assisted steering system (Figs 9.12, 9.13 and 9.14) This directly coupled power assisted system is hydraulic in operation. The power assisted steering layout (Fig. 9.14) consists of a moving power cylinder. Inside this cylinder is a double acting piston which is attached to a ramrod anchored to the chassis by either rubber bushes or a ball joint. One end of the power cylinder is joined to a spool control valve which is supported by the steering box drop arm and the other end of the power cylinder slides over the stationary ramrod. When the system is used on a commercial vehicle with a rigid front axle beam (Fig. 9.12), the steering drag link is coupled to the power cylinder and control valve by a ball joint. If a car or van independent front suspension layout is used (Fig. 9.13), the power cylinder forms a middle moveable steering member with each end of the split track rods attached by ball joints at either end. The power source comes from a hydraulic pump mounted on the engine, and driven by it a pair of flexible hydraulic pipes connect the pump and a fluid reservoir to the spool control valve which is mounted at one end of the power cylinder housing. A conventional steering box is used in the system so that if the hydraulic power should fail the steering can be manually operated.

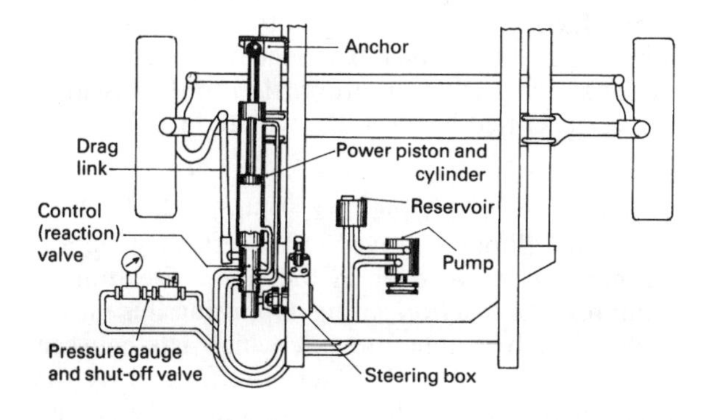

Fig. 9.12 Steering box with external directly coupled power assisted steering utilized with rigid axle front suspension

With the removal of any steering wheel effort a pre-compressed reaction spring built into the control valve (Fig. 9.14) holds the spool in the neutral position in addition to a hydraulic pressure which is directed onto reaction areas within the control valve unit. Provided the steering effort is less than that required to overcome the preload of the reaction spring, the spool remains central and the fluid is permitted to circulate from the pump through the valve and back to the reservoir. Under these conditions there will be no rise in hydraulic pressure and the steering will be manually operated. Consequently, the pump will be running light and therefore will consume very little power.

When the steering effort at the driver's wheel is greater than the preload stiffness of the reaction spring, the spool valve will move slightly to one side. This action partially traps fluid and prevents it returning to the reservoir so that it now pressurizes one side or the other of the double acting piston, thereby providing the power assistance necessary to move the steering linkage. The more the spool valve misaligns itself from the central position the greater the restriction will be for the fluid to return to the reservoir and the larger the pressure build up will be on one side or the other of the double acting piston to apply the extra steering thrust to turn the steering road wheels.

Operation of control valve and power piston (Fig. 9.14)

Neutral position (Fig. 9.14(a)) With the valve spool in the neutral position and no power assistance being used, fluid from the pump passes freely from the right hand supply port and annular groove in the valve housing, across the spool valve middle land to the return groove and port in the valve housing, finally returning to the reservoir. At the same time fluid passes from both the spool grooves to passages leading to the left and right hand power cylinder chambers which are sealed off from each other by the double acting piston. Thus whatever the position of the piston in the power cylinder when the spool is in the central or neutral position, there will be equal pressure on either side of the double acting piston. Therefore the piston will remain in the same relative position in the cylinder until steering corrections alter the position of the spool valve.

Right hand steering movement (Fig. 9.14(b)) If the drop arm pushes the ball pin to the right, the spool control edges 1 and 3 now overlap with the valve housing lands formed by the annular grooves. The fluid flows from the supply annular groove into the right hand spool groove where it then passes along passages to the right hand cylinder chamber where the pressure is built up to expand the chamber.

The tendency for the right hand cylinder chamber to expand forces fluid in the left hand contracting cylinder chamber to transfer through passages to the left hand spool groove. It then passes to the valve housing return annular groove and port back to the reservoir. Note that the ramrod itself remains stationary, whereas the power cylinder is the moving member which provides the steering correction.

Left hand steering movement (Fig. 9.14(c)) Movement of the drop arm to the left moves the spool with it so that control edges 2 and 4 now overlap with the adjacent valve housing lands formed by the annular grooves machined in the bore. Fluid flows from the supply annular groove in the valve housing to the axial passage in the spool and is then diverted radially to the valve body feed annular groove and the spool left hand groove. Fluid continues to flow along the passage leading to the left hand power cylinder chamber where it builds up pressure. As a result the left hand chamber expands, the right hand chamber contracts, fluid is thus displaced from the reducing space back to the right hand spool groove, it then flows out to the valve housing return groove and port where finally it is returned to the reservoir.

Progressive power assistance (Fig. 9.14(a)) While the engine is running and therefore driving the hydraulic power pump, fluid enters the reaction chamber via the axial spool passage.

Before any spool movement can take place relative to the valve housing to activate the power assistance, an input effort of sufficient magnitude must be applied to the drop arm ball pin to compress the reaction spring and at the same time overcome the opposing hydraulic pressure built up in the reaction chamber. Both the reaction spring and the fluid pressure are utilized to introduce a measure of resistance at the steering wheel in proportion to the tyre to ground reaction resistance when the steered road wheels are turned and power assistance is used.

Progressive resistance at the steering wheel due to the hydraulic pressure in the reaction chamber can be explained in the following ways:

Right hand spool reaction (Fig. 9.14(b)) Consider the drop arm ball pin initially moved to the right. The reaction ring will also move over and slightly compress the reaction spring. At the same time the hydraulic pressure in the reaction chamber will oppose this movement. This is because the pressure acts between the area formed by the annular shoulder in the valve chamber housing taking the reaction spring thrust, and an equal projected area acting on the reaction ring at the opposite end of the chamber. The greater the hydraulic pressure the larger the input effort must be to turn the steering wheel so that the driver experiences a degree of feel at the steering wheel in proportion to the resisting forces generated between the tyre and road.

Left hand spool reaction (Fig. 9.14(c)) If the drop arm and ball pin is moved to the left, the reaction washer will move over in the same direction to compress the reaction spring. Opposing this movement is the hydraulic pressure which acts between the reaction washer shoulder area formed by the reduced diameter of the spool spindle and an equal projected area of the reaction ring situated at the opposite end. If the steering wheel effort is removed, the hydraulic pressure in the reaction

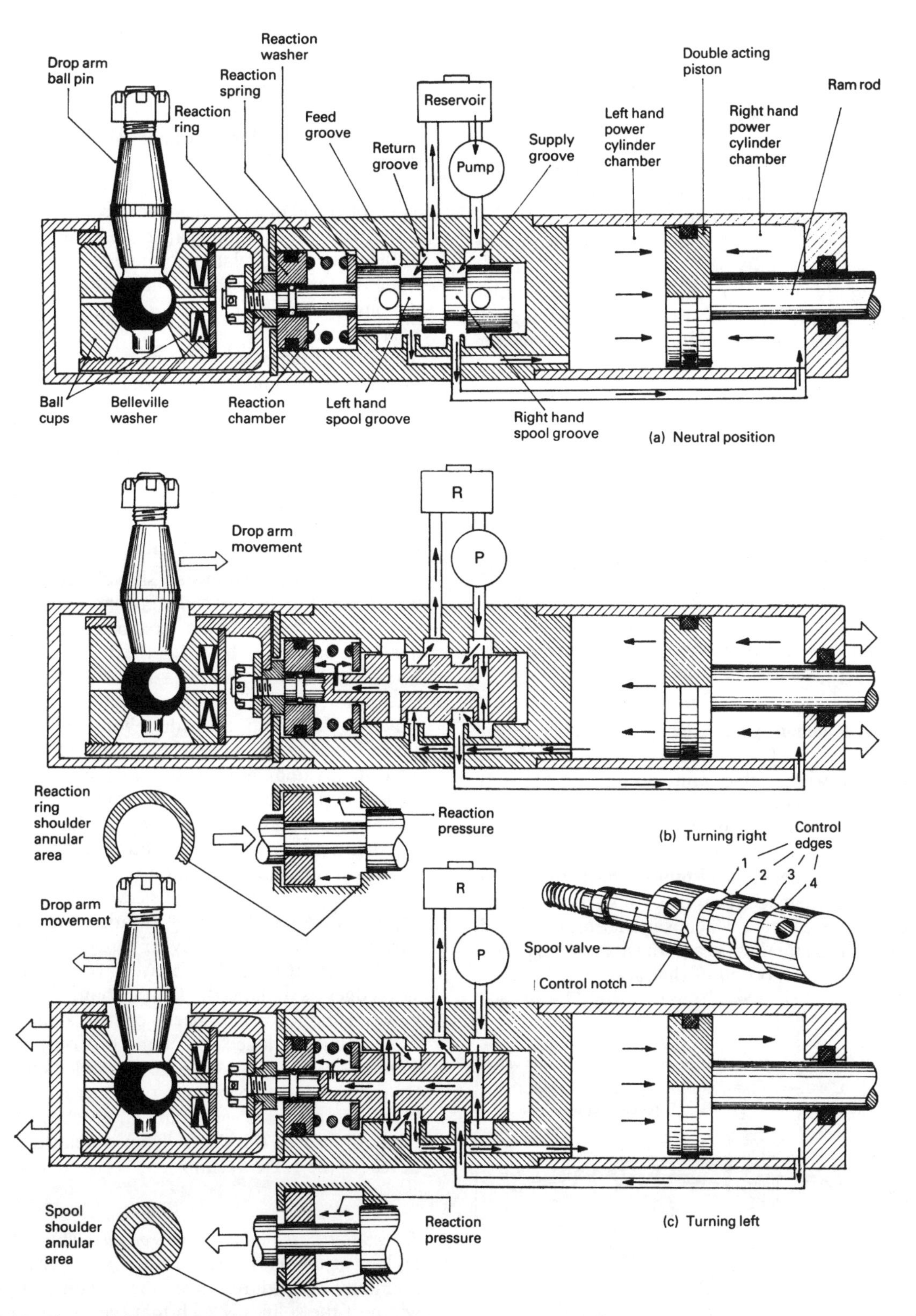

Fig. 9.14 External directly coupled power assisted steering

chamber will react between the reaction ring, the reaction washer housing and spool shoulders, and thereby attempt to move the spool back to its original central or neutral position.

Correction for the variation in cross-section areas on opposite side of the power piston (Fig. 9.14(b and c)) To counteract the reduction in effective area on the ramrod side of the double acting piston, the annular shoulder area of the spool (Fig. 9.14(b)) is made slightly larger than the reaction ring annular shoulder area (Fig. 9.14(c)) machined in the reaction chamber. Consequently, a greater opposing hydraulic reaction will be created when turning the steering to the left to oppose the full cross-sectional area of the power piston compared to the situation when the steering is turned to the right and a reduced power piston cross-sectional area due to the ram is exposed to the hydraulic pressure in the power cylinder. In this way a balanced self-centralizing response is obtained in whatever position the road steering wheels may be positioned.

9.2.2 Rack and pinion power assisted steering gear power cylinder and control valve

Rack and pinion gear (Fig. 9.15) This power assisted steering system is comprised of a rack and pinion gear with double acting power (servo) piston mounted on the rack and a rotary valve coaxial with the extended pinion shaft (Fig. 9.15).

Helical teeth are cut on the 3% nickel, 1% chromium case hardened steel pinion shaft, they mesh with straight cut teeth on a 0.4% carbon manganese silicon steel rack which is induction hardened. To accommodate these gears, the axis of the pinion is inclined at 76° to that of the rack.

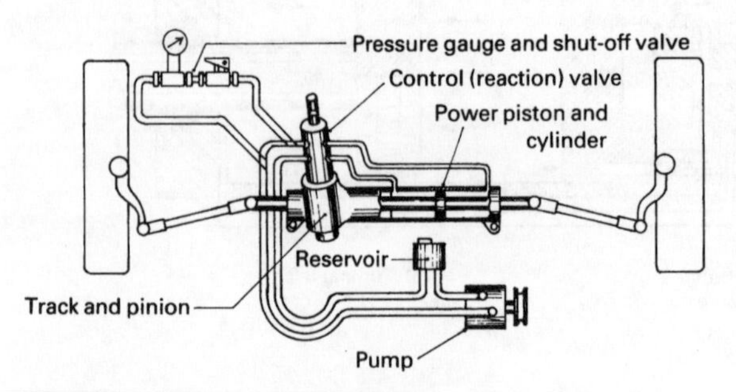

Fig. 9.15 Integral rack and pinion power assisted steering utilized with independent front suspension

Description of rotary control valve (Fig. 9.16(a, b and c)) The three major components of the rotary type control valve are the rotor shaft, the torsion bar and the valve sleeve (Fig. 9.16(a)).

Slots are milled longitudinally on the periphery of the rotor shaft (Fig. 9.16(b)) and on the inner surface of the valve sleeve of which the rotor is assembled and in which it is free to rotate. The sleeve is rotated by the pinion gear shaft. Limited rotation of the rotor shaft, relative to the pinion gear, occurs when the torsion bar that connects the rotor shaft to the pinion shaft is angularly deflected. Hence when steering effort is applied, the torsion bar twists and the slots on the rotor move relative to those in the sleeve to allow fluid to pass to one side of the double acting piston which operates inside the power cylinder. The direction of rotation of the rotor relative to the sleeve determines which side of the double acting piston the fluid will act.

The rotor shaft of the valve forms part of the 0.5% carbon steel rotor shaft and is connected by a torsion bar to the pinion shaft. The outer case-hardened 0.15% carbon steel sleeve which is coaxial with the rotor floats on the ground surface of the rotor shaft, there being a diametrical clearance of 0.004 to 0.012 mm between the rotor and sleeve. The sleeve is connected by a steel trim pin screw to the pinion shaft. This pin, the threaded end of which is in a tapped radial hole in the pinion shaft, has a spherical head that fits with no clearance in a radial hole drilled in the sleeve at the pinion end. The axis of the head of this pin is eccentric to that of the threaded shank. Hence, when the pinion shaft and rotor shaft have been locked in the central position of rotation, the correct angular position of the sleeve, relative to that of the rotor, can be set by rotating the pin. This permits the valve assembly to be trimmed, so that the division of fluid flow between rotor and sleeve slot edges is balanced. The amount of opening between the rotor and sleeve control slots is equal to the angular deflection of the torsion bar. Four square section Teflon seals are assembled into annular grooves in the periphery of the sleeve. Between these seals are three wider annular grooves, again on the periphery of the sleeve, from which fluid enters or leaves the valve assembly.

Rotor shaft to pinion shaft coupling (Fig. 9.16(a)) Splines at one end of the rotor shaft register in an internally splined recess in the pinion shaft. The width of the splines is such that the torsion bar can twist a total of seven degrees before the splines contact one another. Manual steering effort is transferred from the rotor shaft through the 1% chromium-molybdenum-steel torsion bar, which has an approximate waist diameter of 5 mm, to the pinion shaft. The splined coupling between these two members ensures that, in the event of failure of the power assistance, the steering gear can be operated manually without overstressing the torsion bar.

Rotor and sleeve longitudinal and annular grooves (Fig. 9.16(b)) Six equally spaced longitudinal slots are milled on the circumference of the rotor. Three of the rotor slots are longer than the other three and these two lengths are disposed alternately around the rotor periphery. There are also six matching equally spaced groove slots with closed-off ends in the bore of the valve sleeve. The angular relationship of these two sets of slots controls the flow of fluid from the pump to the power cylinder and from the cylinder to the reservoir.

The positioning of the ports and annular grooves around the valve sleeve are now considered from the pinion end of the rotor shaft.

The first valve port feeds or returns fluid from the left hand side of the power cylinder, the second port delivers fluid from the pump, the third port feeds or returns fluid from the right hand side of the power cylinder and the fourth port acts as the return passage to the reservoir.

Each of the first three ports (counting from the left) communicates with corresponding annular grooves in the outer periphery of the sleeve, while the fourth port, through which fluid is returned to the reservoir, communicates with an annular space between the right hand end of the valve housing and the end of the sleeve.

Radial holes in the central sleeve groove connect the pump pressurized fluid to the short supply slots on the periphery of the rotor shaft. The sleeve grooves on either side are connected through small radial holes to the three longer slots on the rotor surface. These longer slots provide a return passage to the annular space on the right of the valve which then leads back to the reservoir.

Whenever the pump is operating, fluid passes through the delivery port in the sleeve and is then transferred to the three short rotor slots. Thereafter the fluid either passes directly to the longer slots, and hence to the return port, or flows through one feed and return port going to the power cylinder, while fluid from the other end of the cylinder is returned through the longer slots to the reservoir.

Torsion bar stiffness (Fig. 9.16(a)) When torque is applied at the steering wheel, the grip of the tyres on the road causes the torsion bar to twist. Relative movement between the rotor shaft and sleeve upsets the balanced flow of fluid and pressure at the rotary control valve going to both sides of the power piston. Therefore the deflection rate of the torsion bar is important in determining the torque required to steer the vehicle. To transmit the desired input torque from the steering wheel and rotor shaft to the pinion shaft, a standard 108 mm long torsion bar is used. Its diameter and stiffness are 5.2 mm and 1.273 N/deg and the actual diameter of the torsion bar can be changed to suit the handling requirements of the vehicle.

Rotor slot edge control (Fig. 9.16(b)) If the valve rotor and sleeve slot control edges were sharp as shown in the end view (Fig. 9.16(b)), the area through which the fluid flowed would vary directly with the deflection of the torsion bar, and hence the applied torque. Unfortunately, the relationship between pressure build-up and effective valve flow area is one in which the pressure varies inversely as the square of the valve opening area. This does not provide the driver with a sensitive feel at the steering wheel rim. The matching of the pressure build-up, and hence valve opening area, relative to the driver's input effort at the steering wheel rim has been modified from the simple sharp edge slots on the rotor, to contoured stepped edges (Fig. 9.16(b)) which provide a logarithmic change of area so that the increase of pressure with rim effort is linear. This linear relationship is retained up to 28 bar which is the limit for driving a large car under all road conditions. Above 28 bar the rise of pressure with applied steering wheel torque is considerably steepened to ensure that parking efforts are kept to a minimum. The relationship between the angular displacement of the rotary valve and the hydraulic pressure applied to one side of the power cylinder piston is shown in Fig. 9.17.

An effort of less than 2.5 N at the rim of the steering wheel is sufficient to initiate a hydraulic pressure differential across the double acting power piston. During manoeuvres at very low speeds when parking, a manual effort of about 16 N is required. When the car is stationary on a dry road, an effort of 22 N is sufficient to move the steering from the straight ahead position to full lock position.

Operation of the rotary control valve and power piston

Neutral position (Fig. 9.16(a)) With no steering effort being applied when driving along a straight

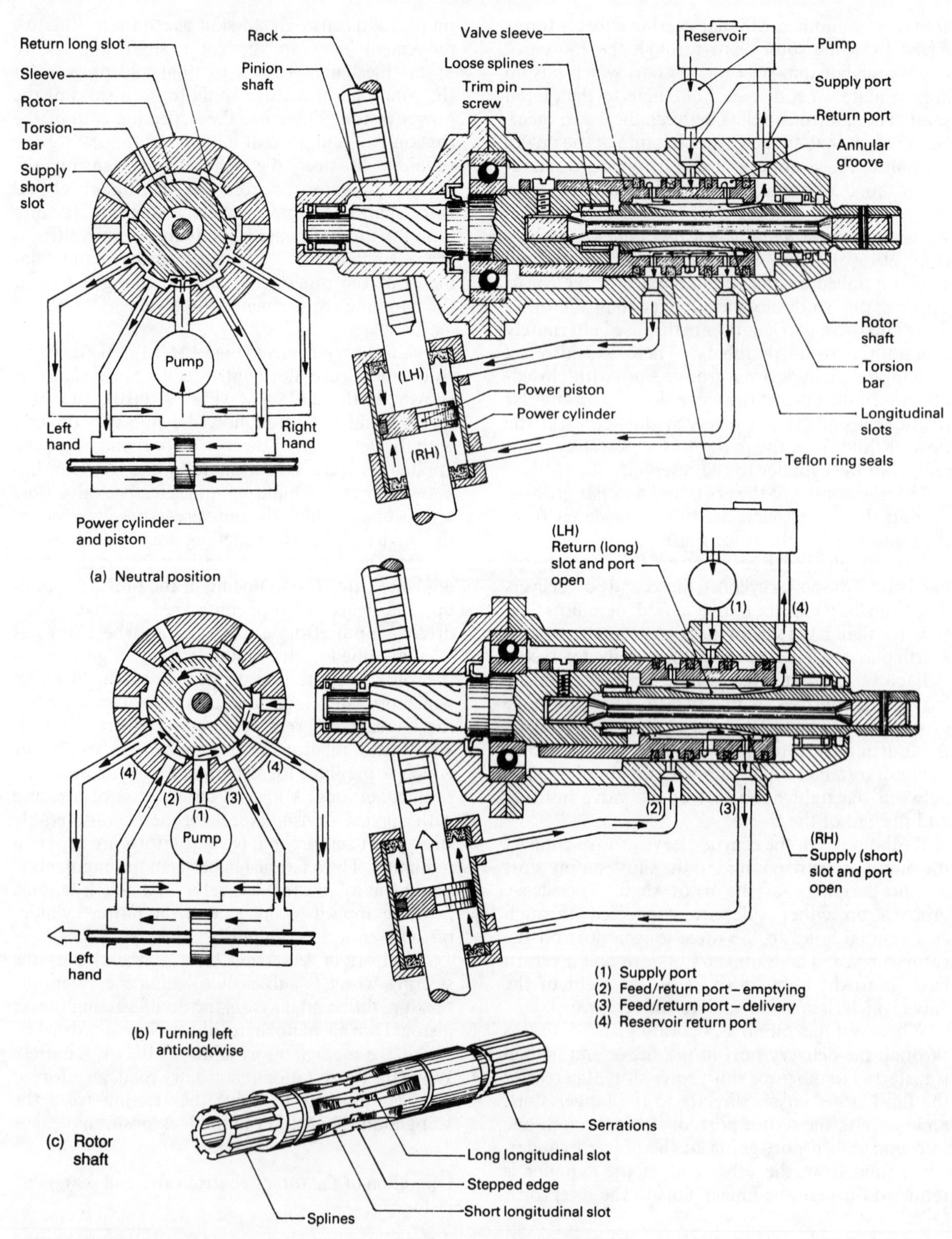

Fig. 9.16 (a–d) Rack and pinion power assisted steering with rotary control valve

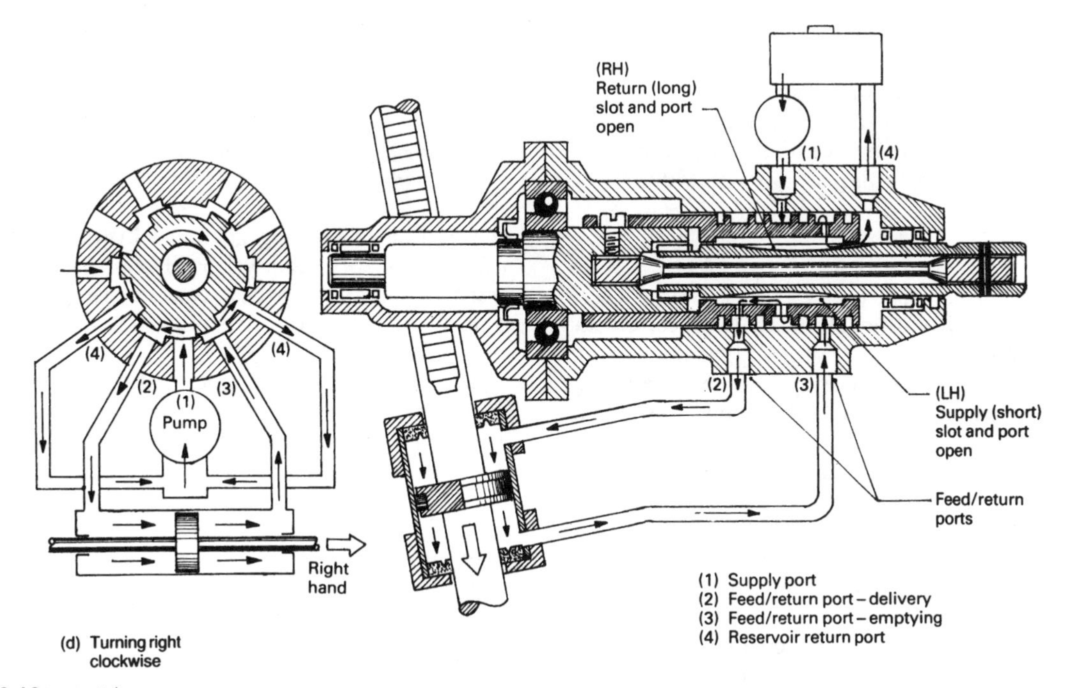

Fig. 9.16 *contd*

track, the longitudinal lands formed by the slots milled on the rotor periphery angularly align with the internal sleeve slots so that equal space exists between the edges of adjacent rotor and sleeve slots.

Fluid therefore flows from the pump to the delivery port into the short slots in the rotor. Some fluid then passes to the feed/return slots in the sleeve and out to the ports communicating with either side of the power cylinder. The majority of the fluid will pass between the edges of both the rotor and sleeve slots to the rotor long slots where it then flows out through the return port back to the reservoir. Because fluid is permitted to circulate from the pump to the reservoir via the control valve, there is no pressure build-up across the power piston, hence the steering remains in the neutral position.

Anticlockwise rotation of the steering wheel (Fig. 9.16(b)) Rotating the steering wheel anticlockwise twists the rotor shaft and torsion bar so that the leading edges of the rotor lands align with corresponding lands on the sleeve, thereby blocking off the original fluid exit passage. Fluid now flows from the pump delivery port to the short slots on the rotor. It then passes between the trailing rotor and sleeve edges, through to the sleeve slots and from there to the left hand side of the power cylinder via the feed/return pipe so that the pressure on this side of the piston rises. At the same time fluid will be displaced from the right hand cylinder end passing through to the sleeve slots

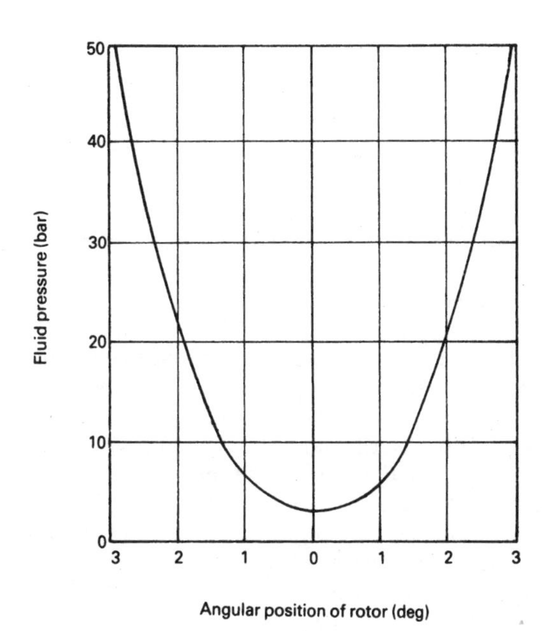

Fig. 9.17 Relationship of fluid pressure delivered to the power cylinder from the control valve and the angular deflection of the control valve and torsion bar

via the right hand feed/return pipe and port. The flow of fluid continues, passing between the trailing land edges of the rotor and sleeve to the long rotor slots and out through the return pipe back to the reservoir. Thus a pressure difference is established across the double acting piston, to provide power assistance.

Clockwise rotation of the steering wheel (Fig. 9.16(c)) Rotating the steering wheel clockwise angularly deflects the rotor so that the leading edges of the rotor lands overlap with corresponding internal lands on the sleeve. Fluid now flows from the pump delivery port into the short rotor slots and out to the right hand feed/return port to the power cylinder via the gap created between the trailing edges of the rotor and sleeve lands. Pressurizing the right hand side of the power cylinder pushes fluid out from the left hand side of the cylinder, through the feed/return pipe and port into the sleeve slots, through the enlarged gap created between the trailing rotor and sleeve edges and into the long rotor slots. It is then discharged through the return port and pipe back to the reservoir.

Progressive power assistance (Fig. 9.16(b and c)) When the steering wheel is turned left or right, that is, anticlockwise or clockwise, the rotor shaft which is rigidly attached to the steering column shaft rotates a similar amount. A rotary movement is also imparted through the torsion bar to the pinion shaft and the valve sleeve as these members are locked together. However, due to the tyre to ground resistance, the torsion bar will twist slightly so that the rotation of the pinion and sleeve will be less than that of the rotor input shaft. The greater the road wheel resistance opposing the turning of the front wheel, the more the torsion bar will twist, and therefore the greater the misalignment of the rotor and sleeve slots will be. As a result, the gap between the leading edges of both sets of slots will become larger, with a corresponding increase in fluid pressure entering the active side of the power cylinder.

As the steering manoeuvres are completed, the initially smaller sleeve angular movement catches up with the rotor movement because either the road wheel resistance has been overcome or steering wheel turning effort has been reduced. Consequently, the reduced torque now acting on the steering column shaft enables the torsion bar to unwind (i.e. straighten out). This causes the power assistance to be reduced in accordance with the realignment or centralization of the rotor slots relative to the sleeve lands.

9.2.3 Integral power assisted steering gear power cylinder and control valve

Description of steering gear and hydraulic control valve (Figs 9.18 and 9.19) The integral power assisted steering gearbox can be used for both rigid front axle and independent front suspension (Fig. 9.18) layouts.

The rack and sector recirculating ball steering gear, power cylinder and hydraulic control valves are combined and share a common housing (Fig. 9.19(a)). The power piston in this arrangement not only transforms hydraulic pressure into force to assist the manual input effort but it has two other functions:

1 it has a rack machined on one side which meshes with the sector,
2 it has a threaded axial bore which meshes via a series of recirculating balls with the input worm shaft.

The input end of the worm shaft, known as the worm head, houses two shuttle valve pistons which have their axes at right angles to the worm shaft. Since they are assembled within the worm head they rotate with it.

Drive is transferred from the hollow input shaft to the worm shaft through a torsion bar. Movement of the shuttle valves relative to the worm shaft which houses the valves is obtained by the hollow double pronged input shaft. Each prong engages with a transverse hole situated mid-way between the shuttle valve ends.

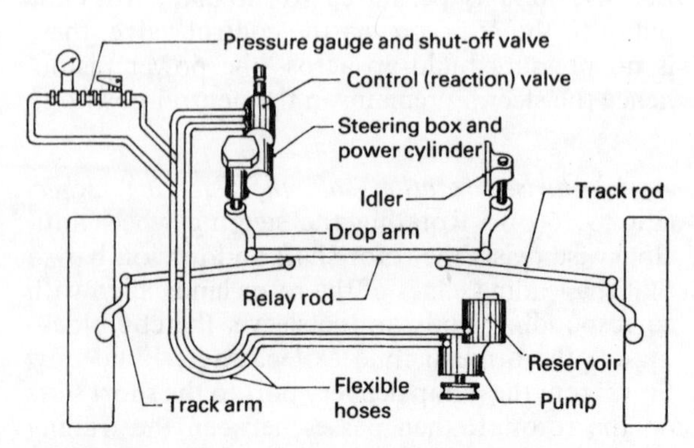

Fig. 9.18 Integral steering gearbox and power assisted steering utilized with independent front suspension

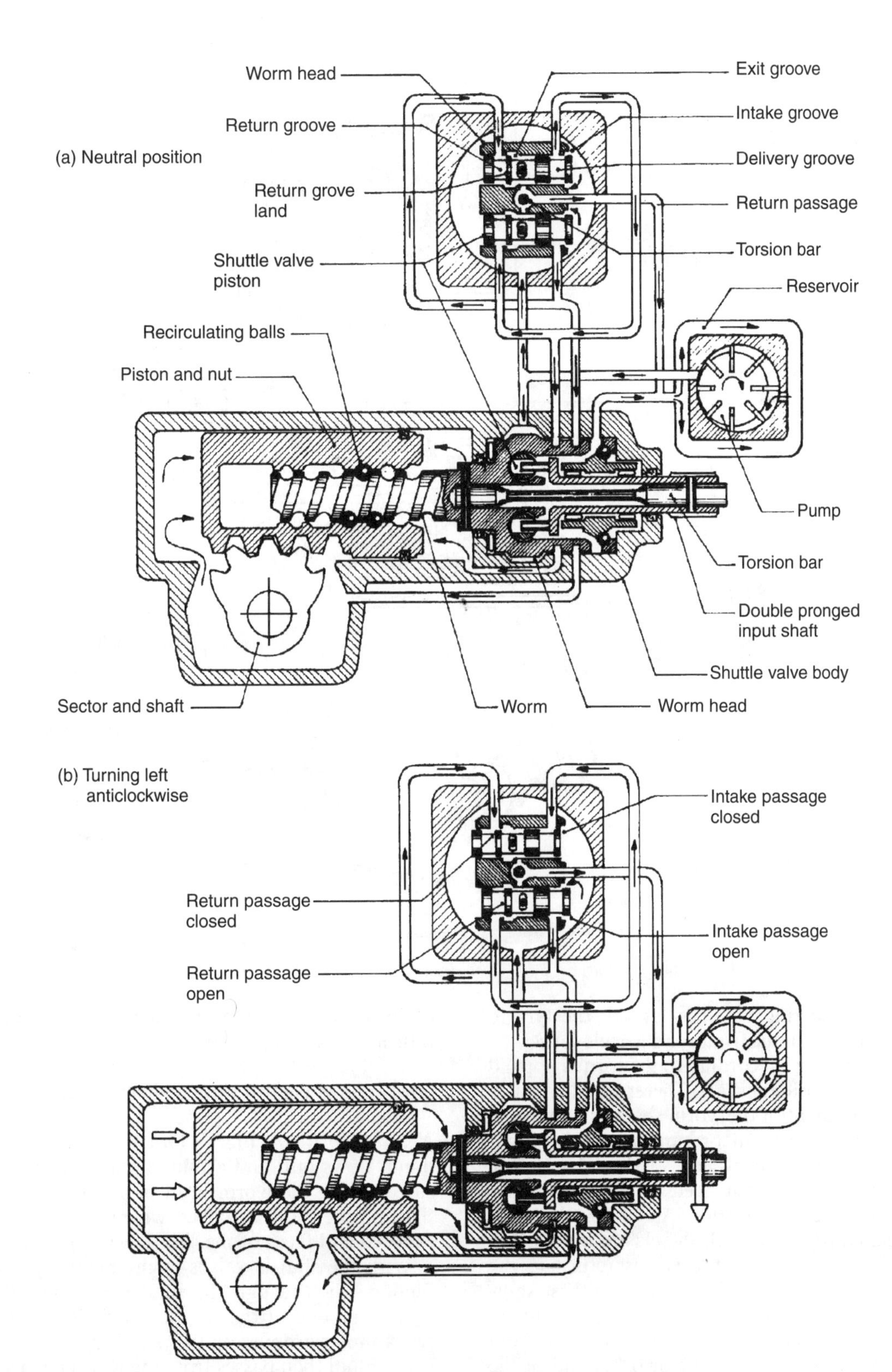

Fig. 9.19 (a–c) Integral power assisted steering gear power cylinder and control valve

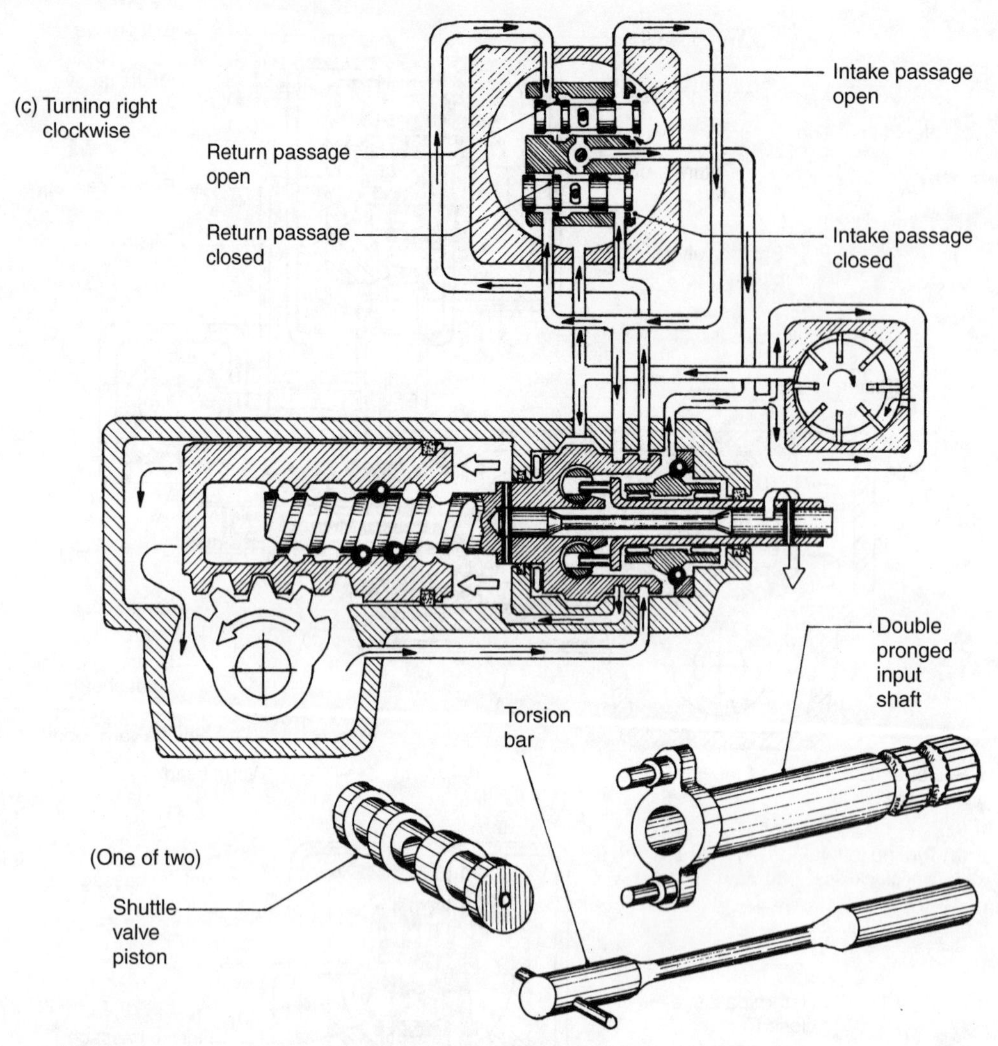

Fig. 9.19 *contd*

When the steering wheel is turned, the tyre to ground reaction on the front road wheels causes the torsion bar to twist according to the torque applied on the steering column shaft. Therefore the relative angular movement of the worm shaft to that of the input shaft increases in proportion to the input torque at the steering wheel, so that the shuttle valves will both be displaced an equal amount from the mid-neutral position. As soon as the steering wheel effort is released, the elastic torsion bar ensures that the two shuttle valves return to the neutral or mid position. The function of these shuttle valves is to transfer fluid under pressure, in accordance to the steering input torque, from the pump delivery port to one or other end of the integral power cylinder whilst fluid from the opposite end of the cylinder is released and returned to the reservoir.

Operation of control valve and power piston

Neutral position (Fig. 9.19(a)) Fluid from the pump flows into and around an annular chamber surrounding the worm head in a plane similar to that of the shuttle valves where it acts on the exposed end faces of the shuttle valve pistons.

With the shuttle valves in the neutral position, fluid moves through the intake passages on the right hand end of the shuttle valve pistons, to the two annular grooves on the periphery of the worm head. Fluid then passes from the worm head annular grooves to the left hand side of the power piston via the horizontal long passage and sector chamber, and to the right hand piston face directly by way of the short passage. From the worm head grooves fluid will also flow into the shuttle valve return grooves, over each return groove land which is aligned with the exit groove, to the middle waisted region of the shuttle valve and into the torsion bar and input shaft chamber. Finally fluid moves out from the return pipe back to the pump reservoir.

Turning left (anticlockwise rotation) (Fig. 9.19(b)) An anticlockwise rotation of the steering wheel against the front wheel to ground opposing resistance distorts the torsion bar as input torque is transferred to the worm shaft via the torsion bar. The twisting of the torsion bar means that the worm shaft also rotates anticlockwise, but its angular movement will be less than the input shaft displacement. As a result, the prongs of the input shaft shift the upper and lower shuttle valves to the left and right respectively. Accordingly this movement closes both the intake and return passages of the upper shuttle valve and at the same time opens both the intake and return passages of the lower shuttle valve.

Fluid can now flow from the pump into the worm head annular space made in the outer housing. It then passes from the lower shuttle valve intake to the right hand worm head annular groove. The transfer of fluid is complete when it enters the left hand power cylinder via the sector shaft. The amount of power assistance is a function of the pressure build-up against the left side of the piston, which corresponds to the extent of the shuttle valve intake passage opening caused by the relative angular movements of both the input shaft and worm shaft.

Movement of the power piston to the right displaces fluid from the right hand side of the power cylinder, where it flows via the worm head annular groove to the lower shuttle valve return passage to the central torsion bar and input shaft chamber. It then flows back to the reservoir via the flexible return pipe.

Turning right (clockwise rotation) (Fig. 9.19(c)) Rotating the steering wheel in a clockwise direction applies a torque via the torsion bar to the worm in proportion to the tyre to ground reaction and the input effort. Due to the applied torque, the torsion bar twists so that the angular movement of the worm shaft lags behind the input shaft displacement. Therefore the pronged input will rotate clockwise to the worm head.

With a clockwise movement of the input shaft relative to the worm head, the upper shuttle valve piston moves to the right and the lower shuttle valve piston moves to the left. Consequently, the upper shuttle valve opens both the intake and return passages but the lower shuttle valve closes both the intake and return passages.

Under these conditions fluid flows from the pump to the annular space around the worm head in the plane of the shuttle valves. It then enters the upper valve intake, fills the annular valve space and passes around the left hand worm head groove. Finally, fluid flows through the short horizontal passage into the right hand side of the power cylinder where, in proportion to the pressure build-up, it forces the piston to the left. Accordingly the meshing rack and sector teeth compel the sector shaft to rotate anticlockwise.

At the same time as the fluid expands the right hand side of the power cylinder, the left hand side of the power cylinder will contract so that fluid will be displaced through the long horizontal passage to the worm head right hand annular groove. Fluid then flows back to the reservoir via the upper shuttle valve return groove and land, through to the torsion bar and input shaft chamber and finally back to the reservoir.

9.2.4 Power assisted steering lock limiters (Fig. 9.20(a and b))

Steering lock limiters are provided on power assisted steering employed on heavy duty vehicles to prevent excessive strain being imposed on the steering linkage, the front axle beam and stub axles and the supporting springs when steering full lock is approached. It also protects the hydraulic components such as the pump and the power cylinder assembly from very high peak pressures which could cause damage to piston and valve seals.

Power assisted steering long stem conical valve lock limiter The lock limiters consist of a pair of conical valves with extended probe stems located in the sector shaft end cover (Fig. 9.20(a and b)). Each valve is made to operate when the angular movement of the sector shaft approaches either steering lock, at which point a cam profile machined on the end of the sector shaft pushes open one or other of the limiting valves. Opening one of the limiter valves releases the hydraulic pressure in the power cylinder end which is supplying the assistance; the

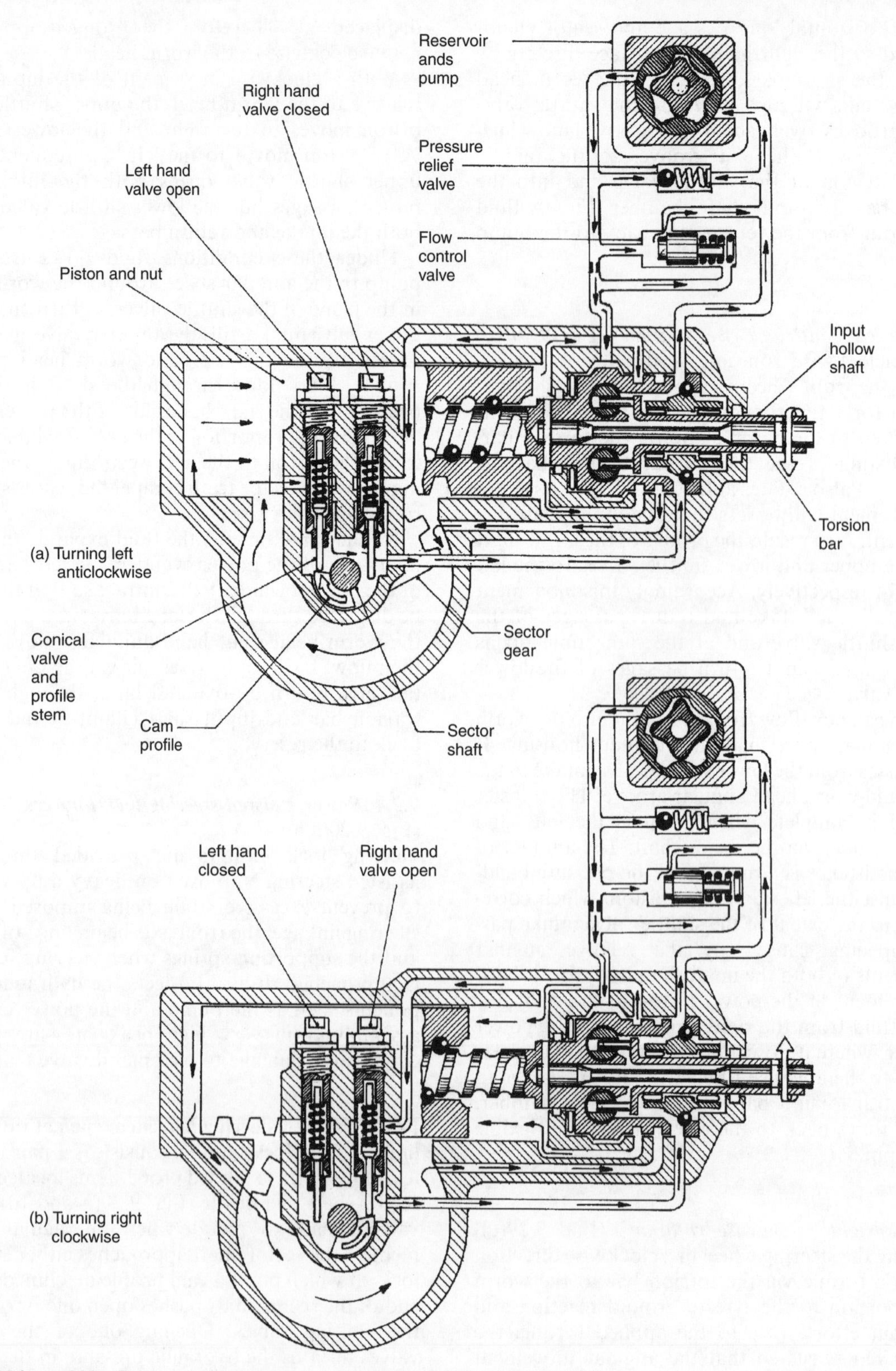

Fig. 9.20 (a and b) Power assisted steering long stem conical valve lock limiter

excess fluid is then permitted to flow back to the reservoir via the control housing.

Turning left (anticlockwise steering rotation) (Fig. 9.20(a)) Rotation of the input shaft anticlockwise applies both manual and hydraulic effort onto the combined power piston and nut of the steering box so that it moves to the right within the cylinder. Just before the steering reaches full lock, one of the sector cam faces contacts the corresponding valve stem and pushes the conical valve off its seat. Pressurized fluid will immediately escape past the open valve through to the return chamber in the control valve housing, where it flows back to the reservoir. Therefore, any further rotation of the sector shaft will be entirely achieved by a considerable rise in manual effort at the steering wheel, this being a warning to the driver that the steering has reached maximum lock.

Turning right (clockwise steering rotation) (Fig. 9.20(b)) Rotation of the steering box input shaft clockwise screws the worm out from the piston and nut and actuates the control valve so that hydraulic pressure builds up on the right hand end of the piston. As the sector shaft rotation approaches maximum lock, the sector cam meets the valve stem, presses open the valve against the valve return spring tension and causes the hydraulic pressure in the right hand cylinder chamber to drop. The excess fluid will now flow back to the reservoir via the right hand end annular chamber in the control valve housing. The driver will immediately experience a considerable increase in manual effort at the steering wheel, indicating that the road wheels have been rotated to near enough maximum lock.

Power assisted steering double ball valve lock limiter This lock limiter consists of a simple double ball valve located in the blank end of the integral piston and nut. To control the stroke of the piston an adjustable stop pin is mounted in the enclosed end of the power cylinder housing, while the right hand piston movement is limited by the stop pin mounted in the end of the worm shaft.

Turning left (anticlockwise steering rotation) (Fig. 9.21(a)) Rotation of the steering input shaft anticlockwise causes both manual and hydraulic effort to act on the combined power piston and nut, moving it towards the right. As the steering lock movement is increased, the piston approaches the end of its stroke until the right hand ball valve contacts the worm shaft stop pin, thereby forcing the ball off its seat. The hydraulic pressure existing on the left side of the piston, which has already opened the left hand side ball valve, is immediately permitted to escape through the clearance formed between the internal bore of the nut and the worm shaft. Fluid will now flow along the return passage leading to the control reaction valve and from there it will be returned to the reservoir. The release of the fluid pressure on the right side of the piston therefore prevents any further hydraulic power assistance and any further steering wheel rotation will be entirely manual.

Turning right (clockwise steering rotation) (Fig. 9.21(b)) Rotation of the steering box input shaft clockwise screws the worm out from the piston and nut. This shifts the shuttle valve pistons so that the hydraulic pressure rises on the right hand end of the piston. Towards the end of the left hand stroke of the piston, the ball valve facing the blind end of the cylinder contacts the adjustable stop pin. Hydraulic pressure will now force the fluid from the high pressure end chamber to pass between the worm and the bore of the nut to open the right hand ball valve and to escape through the left hand ball valve into the sector gear chamber. The fluid then continues to flow along the return passage going to the control reaction valve and from there it is returned to the reservoir. The circulation of fluid from the pump through the piston and back to the reservoir prevents further pressure build-up so that the steering gearbox will only operate in the manual mode. Hence the driver is made aware that the road wheels have been turned to their safe full lock limit.

9.2.5 Roller type hydraulic pump

(Fig. 9.22(a and b))

The components of this pump (Fig. 9.22 (a and b)) consist of the stationary casing, cam ring and the flow and pressure control valve. The moving parts comprise of a rotor carrier mounted on the drive shaft and six rollers which lodge between taper slots machined around the rotor blank. The drive shaft itself is supported in two lead-bronze bushes, one of which is held in the body and the other in the end cover. A ball bearing at the drive end of the shaft takes the load if it is belt and pulley driven.

The rotor carrier is made from silicon manganese steel which is heat treated to a moderate hardness.

The rotor slots which guide the rollers taper in width towards their base, but their axes instead of being radial have an appreciable trailing angle so as to provide better control over the radial movement of the rollers. The hollow rollers made of case- hardened steel are roughly 10 mm in diameter and there are three standard roller lengths of 13, 18 and 23 mm to accommodate three different capacity pumps.

The cam ring is subjected to a combined rolling and sliding action of the rollers under the generated pressure. To minimize wear it is made from heat treated nickel-chromium cast iron. The internal profile of the cam ring is not truly cylindrical, but is made up from a number of arcs which are shaped to maximize the induction of delivery of the fluid as it circulates through the pump.

To improve the fluid intake and discharge flow there are two elongated intake ports and two similar discharge ports at different radii from the shaft axes. The inner ports fill or discharge the space between the rollers and the bottoms of their slots and the outer ports feed or deliver fluid in the space formed between the internal cam ring face and the lobes of the rotor carrier. The inner elongated intake port has a narrow parallel trailing (transition) groove at one end and a tapered leading (timing) groove at the other end. The inner discharge port has only a tapered trailing (timing) groove at one end. These secondary circumferential groove extensions to the main inner ports provide a progressive fluid intake and discharge action as they are either sealed or exposed by the rotor carrier lobes and thereby reduce shock and noise which would result if these ports were suddenly opened or closed, particularly if air has become trapped in the rotor carrier slots.

Operating cycle of roller pump (Fig. 9.22(a and b)) Rotation of the drive shaft immediately causes the centrifugal force acting on the rollers to move them outwards into contact with the internal face of the cam ring. The functioning of the pump can be considered by the various phases of operation as

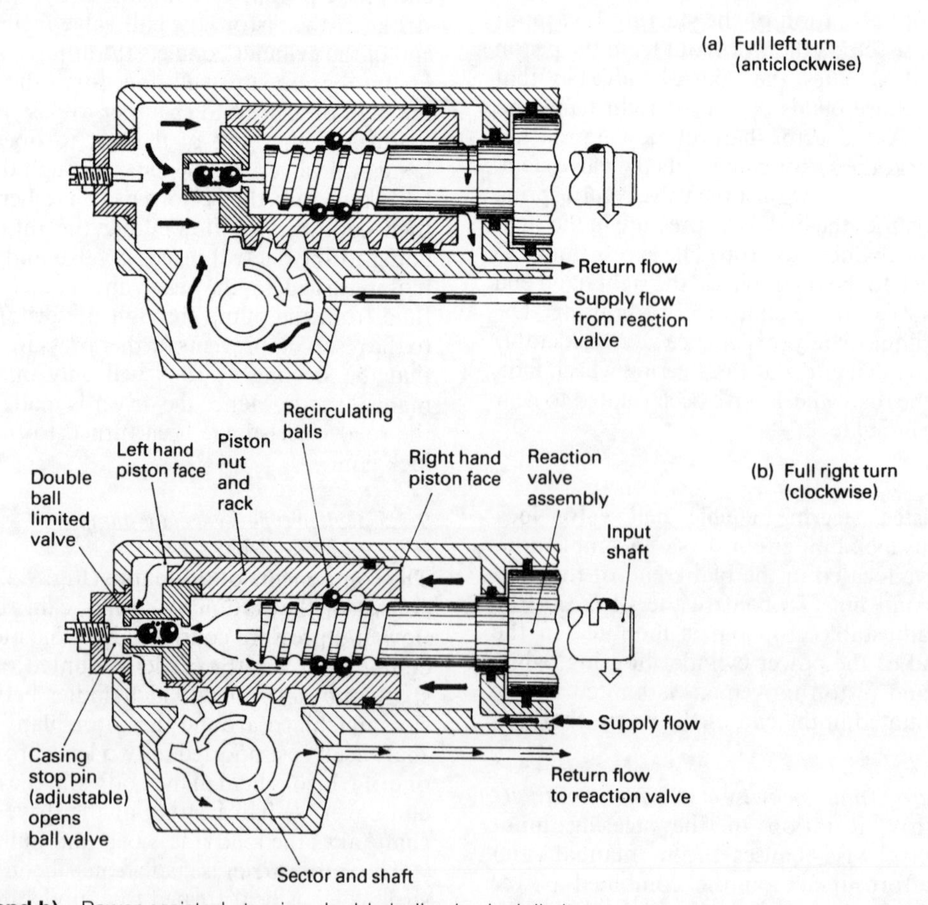

Fig. 9.21 (a and b) Power assisted steering double ball valve lock limit

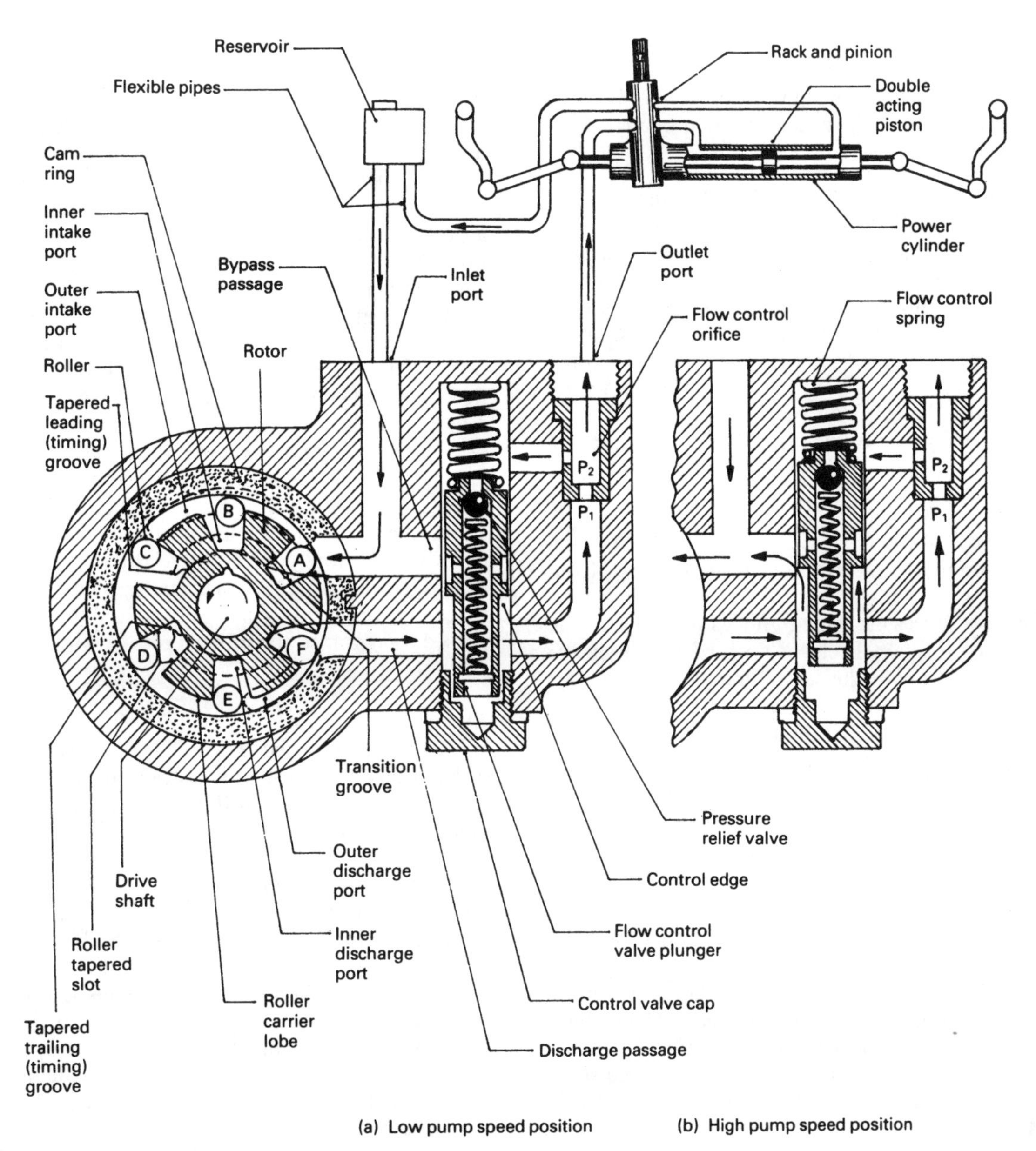

Fig. 9.22 (a and b) Power assisted roller type pump and control valve unit

an individual roller moves around the internal cam face through positions A, B, C, D, E and F.

Filling phase (Fig. 9.22(a)) As the roller in position A moves to position B and then to position C, the space between the eccentric mounted rotor carrier lobe and cam face increases. Therefore the volume created between adjacent rollers will also become greater. The maximum chamber volume occurs between positions C and D. As a result, the pressure in these chambers will drop and thus induce fluid from the intake passages to enter by way of the outer chamber formed by the rotor lobe and the cam face and by the inner port into the tapered roller slot region. Filling the two regions of the chamber separately considerably speeds up the fluid intake process.

Pressurization phase (Fig. 9.22(a)) With further rotation of the rotor carrier, the leading edge of the rotor slot just beyond position C is just on the point of closing the intake ports, and the space formed between adjacent rollers at positions C and D starts to decrease. The squeezing action pressurizes the fluid.

Discharge phase (Fig. 9.22(a)) Just beyond roller position D the inner discharge port is uncovered by the trailing edge of the rotor carrier slot. This immediately enables fluid to be pushed out through the inner discharge port. As the rotor continues to rotate, the roller moves from position D to E with a further decrease in radial chamber space so that there is a further rise in fluid pressure. Eventually the roller moves from position E to F. This uncovers the outer discharge port so that an increased amount of fluid is discharged into the outlet passage.

Transition phase (Fig. 9.22(a)) The roller will have completed one revolution as it moves from position F to the starting position at A. During the early part of this movement the leading edge of the rotor slot position F closes both of the discharge ports and at about the same time the trailing edge of the rotor slot position A uncovers the transition groove in readiness for the next filling phase. The radial space between the rotor lobe and internal cam face in this phase will be at a minimum.

Flow and pressure control valves

Description of the flow and pressure control valve unit (Fig. 9.22(a and b)) The quantity of fluid discharged from the roller type pump and the build-up in fluid pressure both increase almost directly with rising pump rotor speed. These characteristics do not meet the power assisted steering requirements when manoeuvring at low speed since under these conditions the fluid circulation is restricted and a rise in fluid pressure is demanded to operate the power cylinders double acting piston. At high engine and vehicle speed when driving straight ahead, very little power assistance is needed and it would be wasteful for the pump to generate high fluid pressures and to circulate large amounts of fluid throughout the hydraulic system. To overcome the power assisted steering mismatch of fluid flow rate and pressure build-up, a combined flow control and pressure relief valve unit is incorporated within the cast iron pump housing. The flow control valve consists of a spring loaded plunger type valve and within the plunger body is a ball and spring pressure relief valve. Both ends of the plunger valve are supplied with pressurized fluid from the pump. Situated in the passage which joins the two end chambers of the plunger is a calibrated flow orifice. The end chamber which houses the plunger return spring is downstream of the flow orifice.

Fluid from the pump discharge ports moves along a passage leading into the reduced diameter portion of the flow control plunger (Fig. 9.22(a)). This fluid circulates the annular space surrounding the lower part of the plunger and then passes along a right angled passage through a calibrated flow orifice. Here some of the fluid is diverted to the flow control plunger spring chamber, but the majority of the fluid continues to flow to the outlet port of the pump unit, where it then goes through a flexible pipe to the control valve built into the steering box (pinion) assembly. When the engine is running, fluid will be pumped from the discharge ports to the flow control valve through the calibrated flow orifice to the steering box control valve. It is returned to the reservoir and then finally passed on again to the pump's intake ports.

Principle of the flow orifice (Fig. 9.22(a and b)) With low engine speed (Fig. 9.22(a)), the calibrated orifice does not cause any restriction or apparent resistance to the flow of fluid. Therefore the fluid pressure on both sides of the orifice will be similar, that is P_1.

As the pump speed is raised (Fig. 9.22(b)), the quantity of fluid discharged from the pump in a given time also rises, this being sensed by the flow orifice which cannot now cope with the increased amount of fluid passing through. Thus the orifice becomes a restriction to fluid flow, with the result that a slight rise in pressure occurs on the intake side of the orifice and a corresponding reduction in pressure takes place on the outlet side. The net outcome will be a pressure drop of P_1–P_2, which will now exist across the orifice. This pressure differential will become greater as the rate of fluid circulation increases and is therefore a measure of the quantity of fluid moving through the system in unit time.

Operation of the flow control valve (Fig. 9.22 (a and b)) When the pump is running slowly the pressure drop across the flow orifice is very small so that the plunger control spring stiffness is sufficient to fully push the plunger down onto the valve cap stop (Fig. 9.22(a)). However, with rising pump speed the flow rate (velocity) of the fluid increases and so does the pressure difference between both sides of the orifice. The lower pressure P_2 on the output side of the orifice will be applied against the plunger crown in the control spring chamber, whereas the higher fluid pressure P_1 will act underneath the plunger against the annular shoulder area and on the blanked off stem area of the plunger. Eventually, as the flow rate rises and the pressure difference becomes more pronounced, the hydraulic pressure acting on the lower part of the plunger P_1 will produce an upthrust which equals the downthrust of the control spring and the fluid pressure P_2. Consequently any further increase in both fluid velocity and pressure difference will cause the flow control plunger to move back progressively against the control spring until the shouldered edge of the plunger uncovers the bypass port (Fig. 9.22(b)). Fluid will now easily return to the intake side of the pump instead of having to work its tortuous way around the complete hydraulic system. Thus the greater the potential output of the pump due to its speed of operation the further back the plunger will move and more fluid will be bypassed and returned to the intake side of the pump. This means in effect that the flow output of the pump will be controlled and limited irrespective of the pump speed (Fig. 9.23). The maximum output characteristics of the pump are therefore controlled by two factors; the control spring stiffness and the flow orifice size.

Operation of the pressure relief valve (Fig. 9.22 (a and b)) The pressure relief valve is a small ball and spring valve housed at one end and inside the plunger type flow control valve at the control spring chamber end (Fig. 9.22(a)). An annular groove is machined on the large diameter portion of the plunger just above the shoulder. A radial relief hole connects this groove to the central spring housing.

With this arrangement the ball relief valve is subjected to the pump output pressure on the downstream (output) side of the flow orifice.

If the fluid output pressure exceeds some predetermined maximum, the ball will be dislodged from its seat, permitting fluid to escape from the control spring chamber, through the centre of the plunger and then out by way of the radial hole and annular groove in the plunger body. This fluid is then returned to the intake side of the pump via the bypass port.

Immediately this happens, the pressure P_2 in the control spring chamber drops, so that the increased pressure difference between both ends of the flow control plunger pushes back the plunger. As a result the bypass port will be uncovered, irrespective of the existing flow control conditions, so that a rapid pressure relief by way of the flow control plunger shoulder edge is obtained. It is the ball valve which senses any peak pressure fluctuation but it is the flow control valve which actually provides the relief passage for the excess of fluid. Once the ball valve closes, the pressure difference across the flow orifice for a given flow rate is again established so that the flow control valve will revert back to its normal flow limiting function.

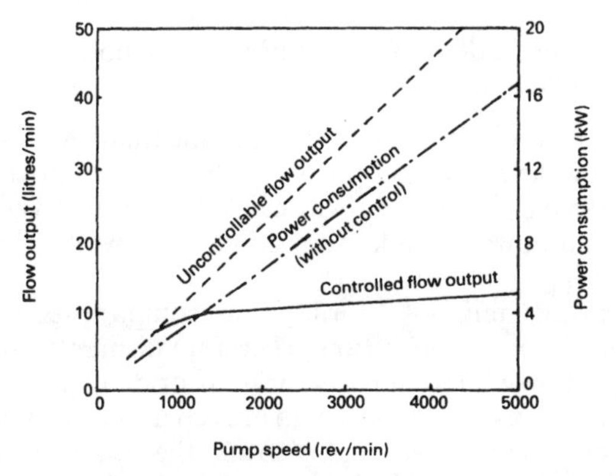

Fig. 9.23 Typical roller pump flow output and power consumption characteristics

9.2.6 Fault diagnosis procedure

Pump output check (Figs 9.12, 9.13, 9.15 and 9.18)

1 Disconnect the inlet hose which supplies fluid pressure from the pump to the control (reaction) valve, preferably at the control valve end.
2 Connect the inlet hose to the pressure gauge end of the combined pressure gauge and shut-off valve tester and then complete the hydraulic circuit by joining the shut-off valve hose to the control valve.
3 Top up the reservoir if necessary.
4 Read the maximum pressure indicated on type rating plate of pump or manufacturer's data.
5 Start the engine and allow it to idle with the shut-off valve in the open position.
6 Close the shut-off valve and observe the maximum pressure reached within a maximum time span of 10 seconds. Do not exceed 10 seconds, otherwise the internal components of the pump will be overworked and will heat up excessively with the result that the pump will be damaged.

7 The permissible deviation from the rated pressure may be ±10%. If the pump output is low, the pump is at fault whereas if the difference is higher, check the functioning of the flow and pressure control valves.

An average maximum pressure figure cannot be given as this will depend upon the type and application of the power assistant steering. A typical value for maximum pressure may range from 45 bar for a ram type power unit to anything up to 120 bar or even more with an integral power unit and steering box used on a heavy commercial vehicle.

Power cylinder performance check (Figs 9.12, 9.13, 9.15 and 9.18)

1 Connect the combined pressure gauge and shut-off valve tester between the pump and control valve as under pump output check.
2 Open shut-off valve, start and idle the engine and turn the steering from lock to lock to bleed out any trapped air.
3 Turn the steering onto left hand full lock. Hold the steering on full lock and check pressure reading which should be within 10% of the pump output pressure.
4 Turn the steering onto the opposite lock and again check the pump output pressure.
5 If the pressure difference between the pump output and the power cylinder on both locks is greater than 10% then the power cylinder is at fault and should be removed for inspection.
6 If the pressure is low on one lock only, this indicates that the reaction control valve is not fully closing in one direction.

A possible cause of uneven pressure is that the control valve is not centralizing or that there is an internal fault in the valve assembly.

Binding check A sticking or binding steering action when the steering is moved through a portion of a lock could be due to the following:

a) Binding of steering joint ball joints or control valve ball joint due to lack of lubrication. Inspect all steering joints for seizure and replace where necessary.
b) Binding of spool or rotary type control valve. Remove and inspect for burrs wear and damage.

Excessive free-play in the steering If when turning the driving steering wheel, the play before the steering road wheels taking up the response is excessive check the following;

1 worn steering track rod and drag link ball joints if fitted,
2 worn reaction control valve ball pin and cups,
3 loose reaction control valve location sleeve.

Heavy steering Heavy steering is experienced over the whole steering from lock to lock, whereas binding is normally only experienced over a portion of the front wheel steering movement. If the steering is heavy, inspect the following items:

1 External inspection — Check reservoir level and hose connections for leakage. Check for fan belt slippage or sheared pulley woodruff key and adjust or renew if necessary.
2 Pump output — Check pump output for low pressure. If pressure is below recommended maximum inspect pressure and flow control valves and their respective springs. If valve's assembly appears to be in good condition dismantle pump, examine and renew parts as necessary.
3 Control valve — If pump output is up to the manufacturer's specification dismantle the control valve. Examine the control valve spool or rotor and their respective bore. Deep scoring or scratches will allow internal leaks and cause heavy steering. Worn or damaged seals will also cause internal leakage.
4 Power cylinder — If the control valve assembly appears to be in good condition, the trouble is possibly due to excessive leakage in the power cylinder. If there is excessive internal power cylinder leakage, the inner tube and power piston ring may have to be renewed.

Noisy operation To identify source of noise, check the following:

1 Reservoir fluid level — Check the fluid level as a low level will permit air to be drawn into the system which then will cause the control valve and power cylinder to become noisy while operating.
2 Power unit — Worn pump components will cause noisy operation. Therefore dismantle and examine internal parts for wear or damage.
3 If the reservoir and pump are separately located, check the hose supply from the reservoir to pump for a blockage as this condition will cause air to be drawn into the system.

Steering chatter If the steering vibrates or chatters check the following:

1 power piston rod anchorage may be worn or requires adjustment,
2 power cylinder mounting may be loose or incorrectly attached.

9.3 Steering linkage ball and socket joints

All steering linkage layouts are comprised of rods and arms joined together by ball joints. The ball joints enable track rods, drag-link rods and relay rods to swivel in both the horizontal and vertical planes relative to the steering arms to which they are attached. Most ball joints are designed to tilt from the perpendicular through an inclined angle of up to 20° for the axle beam type front suspension, and as much as 30° in certain independent front suspension steering systems.

9.3.1 Description of ball joint (Fig. 9.24(a–f))

The basic ball joint is comprised of a ball mounted in a socket housing. The ball pin profile can be divided into three sections; at one end the pin is parallel and threaded, the middle section is tapered and the opposite end section is spherically shaped. The tapered middle section of the pin fits into a similarly shaped hole made at one end of the steering arm so that when the pin is drawn into the hole by the threaded nut the pin becomes wedged.

The spherical end of the ball is sandwiched between two half hemispherical socket sets which may be positioned at right angles to the pin's axis (Fig. 9.24(a and b)). Alternatively, a more popular arrangement is to have the two half sockets located axially to the ball pin's axis, that is, one above the other (Fig. 9.24(c–f)).

The ball pins are made from steel which when heat treated provide an exceptionally strong tough core with a glass hard surface finish. These properties are achieved for normal manual steering applications from forged case-hardened carbon (0.15%) manganese (0.8%) steel, or for heavy duty power steering durability from forged induction hardened 3% nickel 1% chromium steel. For the socket housing which might also form one of the half socket seats, forged induction hardened steels such as a 0.35% carbon manganese 1.5% steel can be used. A 1.2% nickel 0.5% chromium steel can be used for medium and heavy heavy duty applications.

9.3.2 Ball joint sockets (Fig. 9.24(c–f))

Modern medium and heavy duty ball and socket joints may use the ball housing itself as the half socket formed around the neck of the ball pin. The other half socket which bears against the ball end of the ball pin is generally made from oil impregnated sintered iron (Fig. 9.24(c)); another type designed for automatic chassis lubrication, an induction hardened pressed steel half socket, is employed (Fig. 9.24(d)). Both cases are spring loaded to ensure positive contact with the ball at all times. A helical (slot) groove machined across the shoulder of the ball ensures that the housing half socket and ball top face is always adequately lubricated and at the same time provides a bypass passage to prevent pressurization within the joint.

Ball and socket joints for light and medium duty To reduce the risk of binding or seizure and to improve the smooth movement of the ball when it swivels, particularly if the dust cover is damaged and the joint becomes dry, non-metallic sockets are preferable. These may be made from moulded nylon and for some applications the nylon may be impregnated with molybdenum disulphide. Polyurethane and Teflon have also been utilized as a socket material to some extent. With the nylon sockets (Fig. 9.24(e)) the ball pin throat half socket and the retainer cap is a press fit in the bore of the housing end float. The coil spring accommodates initial settling of the nylon and subsequent wear and the retainer cap is held in position by spinning over a lip on the housing. To prevent the spring loaded half socket from rotating with the ball, two shallow tongues on the insert half socket engage with slots in the floating half socket. These ball joints are suitable for light and medium duty and for normal road working conditions have an exceptionally longer service life.

For a more precise adjustment of the ball and socket joint, the end half socket may be positioned by a threaded retainer cap (Fig. 9.24(f)) which is screwed against the ball until all the play has been taken up. The cap is then locked in position by crimping the entrance of the ball bore. A Belleville spring is positioned between the half socket and the screw retainer cap to preload the joint and compress the nylon.

9.3.3 Ball joint dust cover (Fig. 9.24(c–f))

An important feature for a ball type joint is its dust cover, often referred to as the *boot* or *rubber gaiter*, but usually made from either polyurethane or nitrile rubber mouldings, since both these materials have a high resistance to attack by ozone and do not tend to crack or to become hard and brittle at low temperature. The purpose of the dust cover is

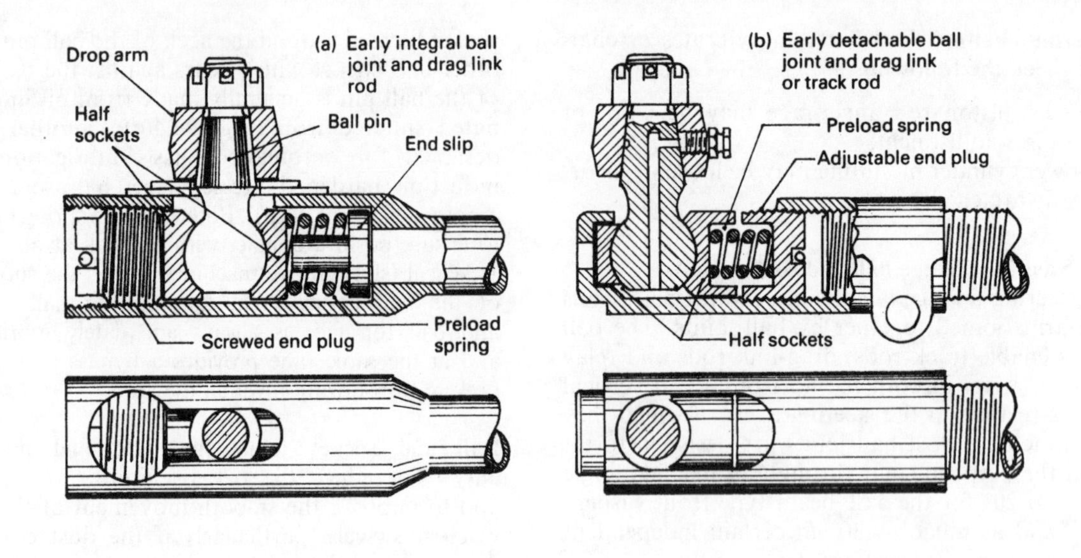

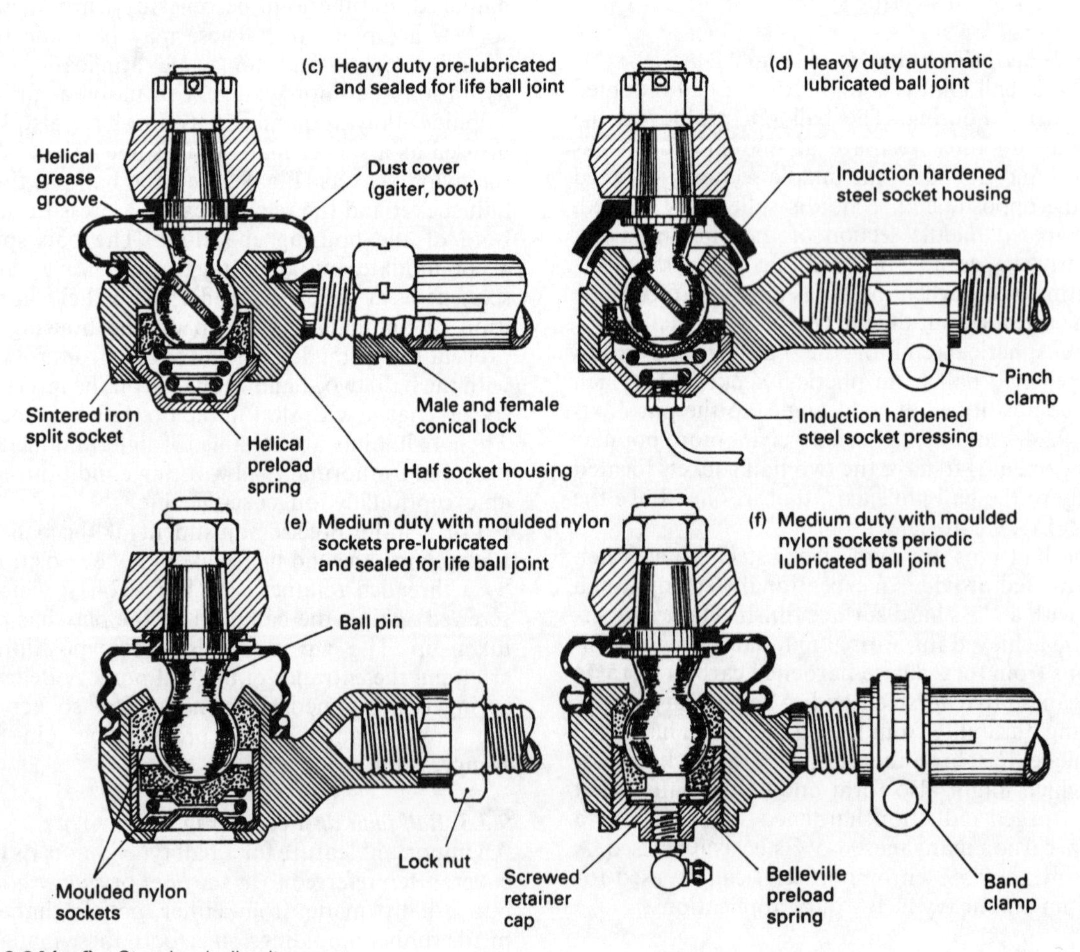

Fig. 9.24 (a–f) Steering ball unit

to exclude road dirt moisture and water, which if permitted to enter the joint would embed itself between the ball and socket rubbing surfaces. The consequence of moisture entering the working section of the joint is that when the air temperature drops the moisture condenses and floods the upper part of the joint. If salt products and grit are sprayed up from the road, corrosion and a mild grinding action might result which could quickly erode the glass finish of the ball and socket surfaces. This is then followed by the pitting of the spherical surfaces and a wear rate which will rapidly increase as the clearance between the rubbing faces becomes larger.

Slackness within the ball joint will cause wheel oscillation (shimmy), lack of steering response, excessive tyre wear and harsh or notchy steering feel.

Alternatively, the combination of grease, grit, water and salts may produce a solid compound which is liable to seize or at least stiffen the relative angular movement of the ball and socket joint, resulting in steering wander.

The dust boot must give complete protection against exposure from the road but not so good that air and the old grease cannot be expelled when the joint is recharged, particularly if the grease is pumped into the joint at high pressure, otherwise the boot will burst or it may be forced off its seat so that the ball and socket will become exposed to the surroundings.

The angular rotation of the ball joint, which might amount to 40° or even more, must be accommodated. Therefore, to permit relative rotation to take place between the ball pin and the dust cover, the boot makes a loose fit over the ball pin and is restrained from moving axially by the steering arm and ball pin shoulder while a steel ring is moulded into the dust cover to prevent the mouth of the boot around the pin spreading out (Fig. 9.24(c–f)). In contrast, the dust cover makes a tight fit over the large diameter socket housing by a steel band which tightly grips the boot.

9.3.4 Ball joint lubrication

Before dust covers were fitted, ball joints needed to be greased at least every 1600 kilometres (1000 miles). The advent of dust covers to protect the joint against dirt and water enabled the grease recharging intervals to be extended to 160 000 kilometres (10 000 miles). With further improvements in socket materials, ball joint design and the choice of lubricant the intervals between greasing can be extended up to 50 000 kilometres (30 000 miles) under normal road working conditions. With the demand for more positive and reliable steering, joint lubrication and the inconvenience of periodic off the road time, automatic chassis lubrication systems via plastic pipes have become very popular for heavy commercial vehicles so that a slow but steady displacement of grease through the ball joint system takes place. The introduction to split socket mouldings made from non-metallic materials has enabled a range of light and medium duty ball and socket joints to be developed so that they are grease packed for life. They therefore require no further lubrication provided that the boot cover is a good fit over the socket housing and it does not become damaged in any way.

9.4 Steering geometry and wheel alignment

9.4.1 Wheel track alignment using Dunlop optical measurement equipment — calibration of alignment gauges

1 Fit contact prods onto vertical arms at approximately centre hub height.
2 Place each gauge against the wheel and adjust prods to contact the wheel rim on either side of the centre hub.
3 Place both mirror and view box gauges on a level floor (Fig. 9.25(b)) opposite each other so that corresponding contact prods align and touch each other. If necessary adjust the horizontal distance between prods so that opposing prods are in alignment.
4 Adjust both the mirror and target plate on the viewbox to the vertical position until the reflection of the target plate in the mirror is visible through the periscope tube.
5 Look into the periscope and swing the indicator pointer until the view box hairline is positioned in the centre of the triangle between the two thick vertical lines on the target plate.
6 If the toe-in or -out scale hairline does not align with the zero reading on the scale, slacken off the two holding down screws and adjust indicator pointer until the hairline has been centred. Finally retighten screws.

Toe-in or -out check (Fig. 9.25(a, b and c))

1 Ensure that tyre pressures are correct and that wheel bearings and track rod ends are in good condition.
2 Drive or push the vehicle in the forward direction on a level surface and stop. Only take

readings with the vehicle rolled forward and never backwards as the latter will give a false toe angle reading.

3 With a piece of chalk mark one tyre at ground level.
4 Place the mirror gauge against the left hand wheel and the view box gauge against the right hand wheel (Fig. 9.25(b)).
5 Push each gauge firmly against the wheels so that the prods contact the wheel on the smooth surface of the rim behind the flanged turnover since the edge of the latter may be slightly distorted due to the wheel scraping the kerb when the vehicle has been parked. Sometimes gauges may be held against the wheel rim with the aid of rubber bands which are hooked over the tyres.
6 Observe through the periscope tube the target image. Swing the indicator pointer to and fro over the scale until the hairline in the view box coincides with the centre triangle located between the thick vertical lines on the target plate which is reflected in the mirror.
7 Read off the toe-in or -out angle scale in degrees and minutes where the hairline aligns with the scale.
8 Check the toe-in or -out in two more positions by pushing the vehicle forward in stages of a third of a wheel revolution observed by the chalk mark on the wheel. Repeat steps 4 to 7 in each case and record the average of the three toe angle readings.
9 Set the pointer on the dial calculator to the wheel rim diameter and read off the toe-in

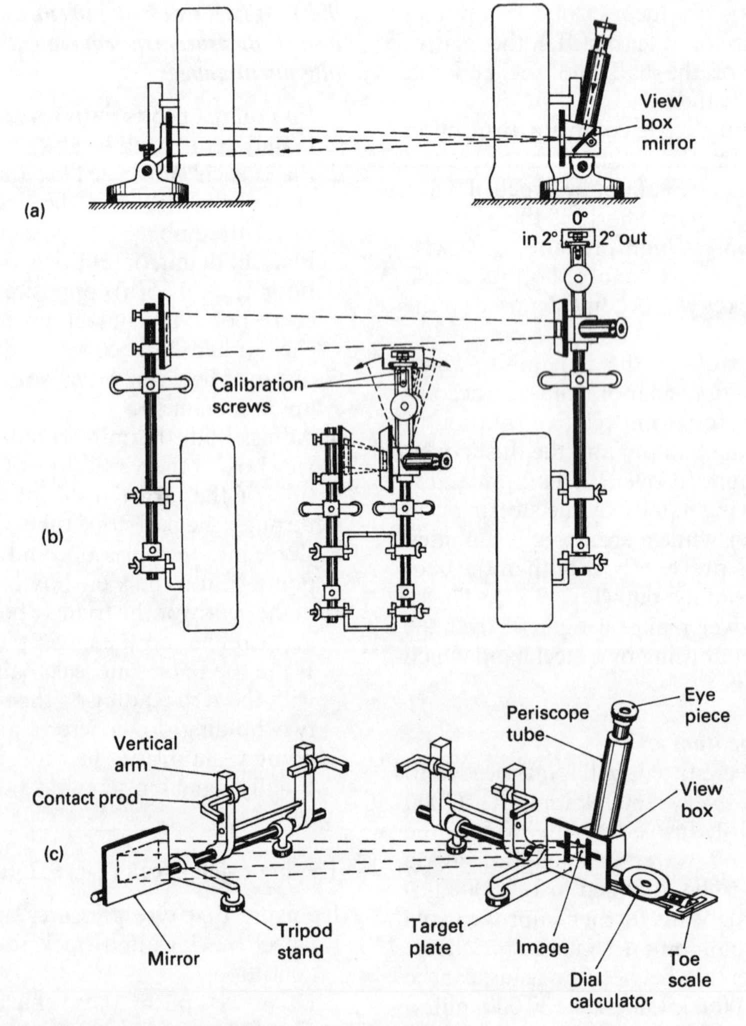

Fig. 9.25 (a–c) Wheel track alignment using the Dunlop equipment

or -out in millimetres opposite the toe angle reading obtained on the toe-in or -out scale. Alternatively, use Table 9.1 to convert the toe-in or -out angle to millimetres.

10 If the track alignment is outside the manufacturer's recommendation, slacken the track rod locking bolts or nuts and screw the track rods in or out until the correct wheel alignment is achieved. Recheck the track toe angle when the track rod locking devices have been tightened.

9.4.2 Wheel track alignment using Churchill line cord measurement equipment

Calibration of alignment gauges
(Fig. 9.26(a))

1 Clamp the centre of the calibration bar in a vice.
2 Attach an alignment gauge onto each end of the calibration bar.
3 Using the spirit bubble gauge, level both of the measuring gauges and tighten the clamping thumbscrews.
4 Attach the elastic (rubber) calibration cord between adjacent uncoloured holes formed in each rotor.
5 Adjust measuring scale by slackening the two wing nuts positioned beneath each measuring scale, then move the scale until the zero line aligns exactly with the red hairline on the pointer lens. Carefully retighten the wing nuts so as not to move the scale.
6 Detach the calibration cord from the rotors and remove the measuring gauges from calibration bar.

Toe-in or -out check (front or rear wheels)
(Fig. 9.26(a))

1 Position a wheel clamp against one of the front wheels so that two of the threaded contact studs mounted on the lower clamp arm rest inside the rim flange in the lower half of the wheel. For aluminium wheels change screw studs for claw studs provided in the kit.
2 Rotate the tee handle on the centre adjustment screw until the top screw studs mounted on the upper clamp arm contact the inside rim flange in the upper half of the wheel. Fully tighten centre adjustment screw tee handle to secure clamp to wheel.
3 Repeat steps 1 and 2 for opposite side front wheel.
4 Push a measuring gauge over each wheel clamp stub shaft and tighten thumbscrews. This should not prevent the measuring gauge rotating independently to the wheel clamp.
5 Attach the elastic cord between the uncoloured hole in the rotor of each measuring gauge.
6 Wheel lateral run-out is compensated by the following procedure of steps 7–10.
7 Lift the front of the vehicle until the wheels clear the ground and place a block underneath one of the wheels (in the case of front wheel drive vehicles) to prevent it from rotating.
8 Position both measuring gauges horizontally and hold the measuring gauge opposite the blocked wheel. Slowly rotate the wheel one complete revolution and observe the measuring gauge reading which will move to and fro and record the extreme of the pointer movement on the scale. Make sure that the elastic cord does not touch any part of the vehicle or jack.
9 Further rotate wheel in the same direction until the mid-position of the wheel rim lateral run-out is obtained, then chalk the tyre at the 12 o'clock position.

Table 9.1 Conversion of degrees to millimetres

Degree	Rim size					
	10″ mm	12″ mm	13″ mm	14″ mm	15″ mm	16″ mm
5	0.40	0.48	0.53	0.57	0.60	0.64
10	0.80	0.96	1.06	1.13	1.21	1.28
15	1.20	1.44	1.59	1.70	1.81	1.92
20	1.60	1.92	2.12	2.27	2.42	2.56
25	2.00	2.40	2.65	2.84	3.02	3.20
30	2.40	2.88	3.19	3.40	3.62	3.84
35	2.80	3.36	3.72	3.97	4.22	4.48
40	3.20	3.84	4.25	4.54	4.83	5.12
45	3.60	4.32	4.78	5.11	5.43	5.76
50	4.00	4.80	5.31	5.67	6.03	6.40
55	4.40	5.28	5.84	6.24	6.64	7.04
1.00	4.80	5.76	6.37	6.81	7.24	7.68
1.05	5.20	6.24	6.90	7.38	7.85	8.32
1.10	5.60	6.72	7.43	7.95	8.45	8.96
1.15	6.00	7.20	7.96	8.51	9.06	9.60
1.20	6.40	7.68	8.49	9.07	9.66	10.24
1.25	6.80	8.16	9.03	9.64	10.25	10.88
1.30	7.20	8.64	9.56	10.21	10.86	11.52
1.35	7.60	9.12	10.09	10.78	11.47	12.16
1.40	8.00	9.60	10.62	11.35	12.08	12.80
1.45	8.40	10.08	11.15	11.91	12.68	13.44
1.50	8.80	10.56	11.68	12.48	13.28	14.08
1.55	9.20	11.04	12.21	13.05	13.89	14.72
2.00	9.60	11.52	12.75	13.62	14.49	15.36

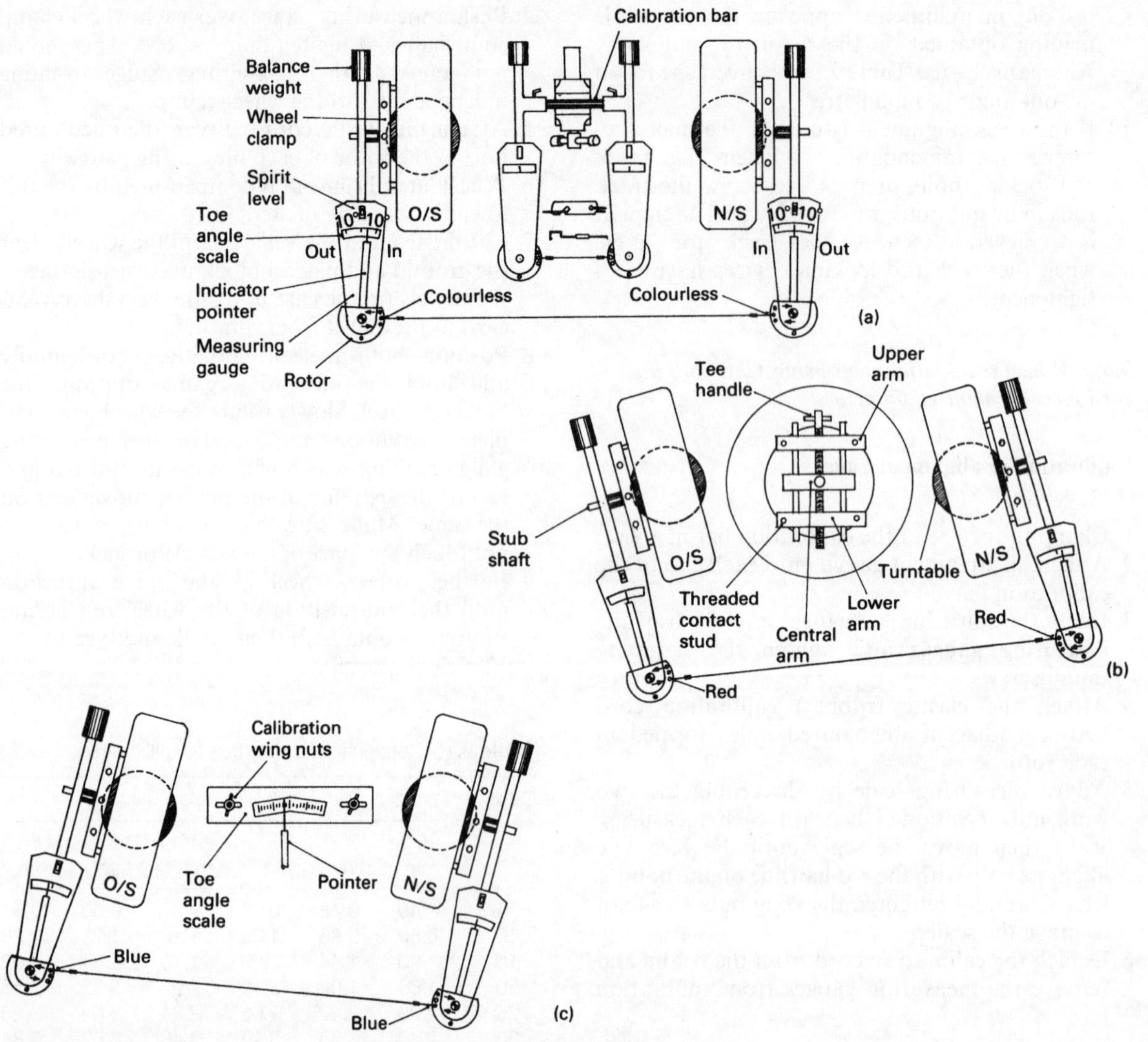

Fig. 9.26 (a–c) Wheel track alignment using the Churchill equipment

10 Repeat steps 7 to 9 for the opposite side front wheel.
11 Position each front wheel with the chalk mark at 12 o'clock.
12 Utilize the brake pedal depressor to prevent the wheels from rotating.
13 Slide a turntable underneath each front wheel, remove the locking pins and then lower the front wheels onto both turntables.
14 Bounce the front of the vehicle so that the suspension quickly settles down to its normal height.
15 Tilt each measuring gauge to the horizontal position by observing when the spirit level bubble is in the mid-position. Lock the measuring gauge in the horizontal position with the second thumbscrew.
16(a) Observe the left and right toe angle reading and add them together to give the combined toe angle of the front wheels.
(b) Alternatively, turn one road wheel until its measuring gauge pointer reads zero, then read the combined toe angle on the opposite side measuring gauge (front wheels only).

17 Using Table 9.1 provided, convert the toe angle into track toe-in or -out in millimetres and compare with the manufacturer's recommendations.

Toe-out on turns check (Fig. 9.26(b and c))

1 After completing the toe-in or -out check, keep the wheel clamp and measuring gauge assembly attached to each front wheel. Maintain the mid-wheel lateral run-out position with the tyre at the 12 o'clock position and ensure the brake pedal depressor is still applied.
2 Transfer the position of the elastic cord hook ends attached to the measuring gauge rotors from the uncoloured holes to the red holes which are pitched 15° relative to the uncoloured holes.
3 Rotate the right hand (offside) wheel in the direction the arrow points on the measuring gauge facing the red hole in which the cord is hooked until the scale reads zero. At this point the right hand wheel (which becomes the outer wheel on the vehicle turning circle) will have been pivoted 15°. Make sure that the cord does not touch any obstruction.
4 Observe the reading on the opposite left hand (near side) measuring gauge scale, which is the toe-out turns angle for the left hand (near side) wheel (the inner wheel on the vehicle's turning circle).
5 Change the cord to the blue holes in each measuring gauge rotor.
6 Rotate the left hand (near side) wheel in the direction the arrow points on the measuring gauge facing the blue hole until the hairline pointer on the left hand measuring gauge reads zero.
7 Read the opposite right hand (offside) wheel measuring gauge scale which gives the toe-out on turns for the right hand (offside) wheel (the inner wheel on the vehicle's turning circle).
8 Compare the left and right hand toe-out turns readings which should be within one degree of one another.

9.4.3 Front to rear wheel misalignment
(Fig. 9.27(a))

An imaginary line projected longitudinally between the centre of the front and rear wheel tracks is known as the vehicle's *centre line* or the *axis of symmetry* (Fig. 9.27(a)). If the vehicle's body and suspension alignment is correct, the vehicle will travel in the same direction as the axis of symmetry. When the wheel axles at the front and rear are misaligned, the vehicle will move forward in a skewed line relative to the axis of symmetry. This second directional line is known as the *thrust axis* or *driving axis*. The angle between the axis of symmetry and the thrust axis is referred to as the *thrust axis deviation* which will cause the front and rear wheels to be laterally offset to each other when the vehicle moves in the straight ahead direction.

If the vehicle has been constructed and assembled correctly the thrust axis will coincide with the axis of symmetry, but variation in the rear wheel toe angles relative to the axis of symmetry will cause the vehicle to be steered by the rear wheels. As a result, the vehicle will tend to move in a forward direction and partially in a sideway direction. The vehicle will therefore tend to pull or steer to one side and when driving round a bend the steering will oversteer on one lock and understeer on the opposite lock. In the case of Fig. 9.27(a), with a right hand lateral offset the vehicle will understeer on left hand bends and oversteer on right hand turns.

The self-steer effect of the rear wheels due to track or axle misalignment will conflict with the suspension geometry such as the kingpin inclination and castor which will therefore attempt to direct the vehicle along the axis of symmetry. Consequently, the tyres will be subjected to excessive scrub.

Thrust axis deviation may be produced by body damage displacing the rear suspension mounting points, rear suspension worn bushes, poorly located leaf springs and distorted or incorrectly assembled suspension members.

Front to rear alignment check using Churchill line cord measurement equipment (Fig. 9.28(a and b))

1 Check rear wheel toe angle by using the procedure adopted for front wheel toe angle measurement (Fig. 9.26(a)). Use the convention that toe-in is positive and toe-out is negative.
2 Keep the wheel clamp and measuring gauge assembly on both rear wheels.
3 Attach a second pair of wheel clamps to both front wheels.
4 Remove the rear wheel toe elastic cord from the two measuring gauges.
5 Hook a front to rear alignment elastic cord between the stub shaft deep outer groove of the front wheel clamps and the single hole in the measuring gauge rotors set at 90° from the middle hole of the three closely spaced holes (Fig. 9.28(a and b)).

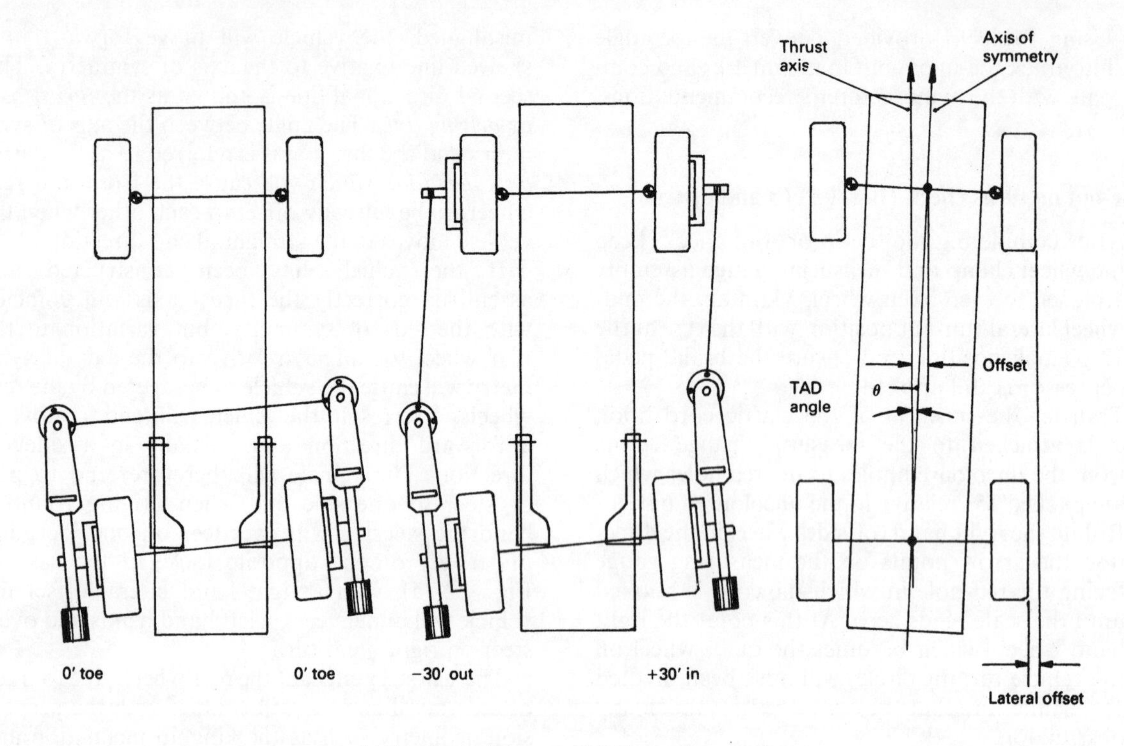

(a) Front to rear alignment with rigid rear axle suspension

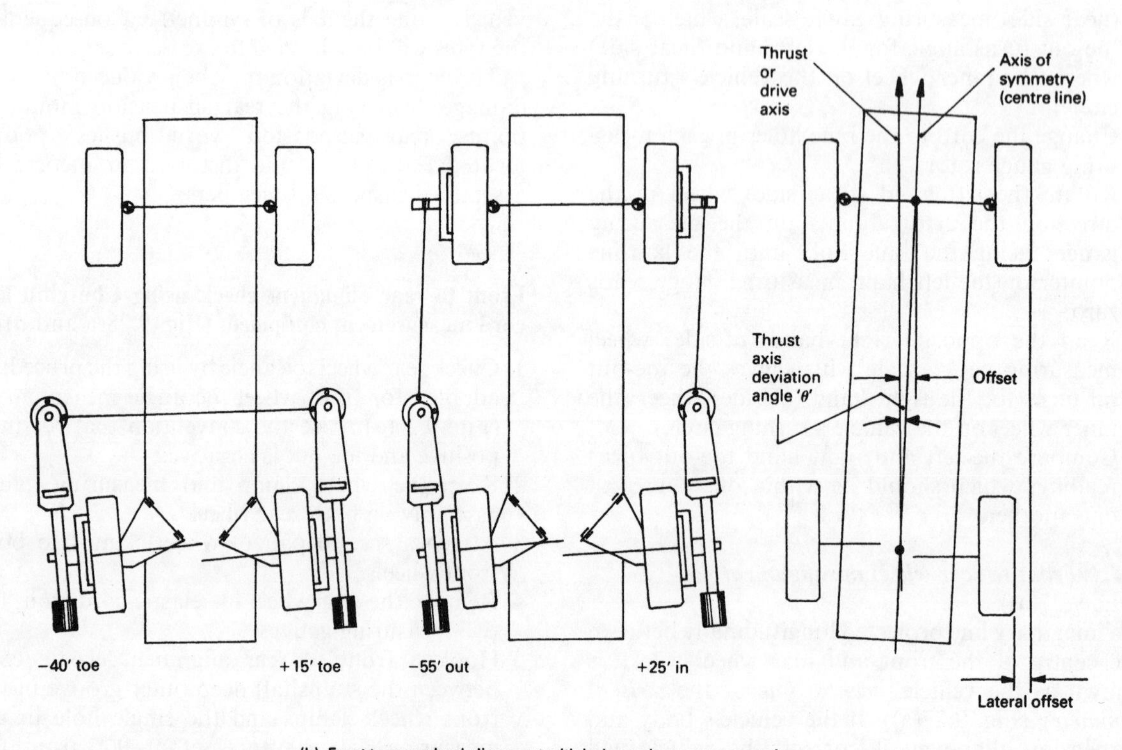

(b) Front to rear wheel alignment with independent rear suspension

Fig. 9.27 (a and b) Front to rear wheel alignment procedure

6 Apply a slight tension to the front to rear alignment cord using the metal plate adjusters.
7 With all four wheels pointing in the straight ahead direction, read and record the left and right hand measuring gauge scales (Fig. 9.27 (a and b)). To determine the thrust axis deviation (TAD) angle subtract the left reading from the right reading and divide the difference of the reading by two.

i.e.

$$\text{Thrust axis deviation (TAD) angle} = \frac{R - L}{2}$$

where R = Right hand measuring gauge reading
L = Left hand measuring gauge reading

8 The lateral offset can be approximately determined from the formula

$$\text{Lateral offset} = \text{Wheel base} \times \tan\left(\frac{R - L}{2}\right)$$

however, the makers of the equipment supply Table 9.2 to simplify the conversion from thrust axis deviation angle to lateral offset.

Example 1 (Fig. 9.27(b)) Determine the rear wheel toe-in or -out and the front to rear lateral offset for a 3000 mm wheelbase vehicle having a rigid rear axle and 13 inch diameter wheels from the following information:

Left rear wheel toe angle reading = 0′
Right rear wheel toe angle reading = 0′
Left side front to rear measurement reading — out (out = negative) = −30′
Right side front to rear measurement reading — in (in = positive) = +30′

a) Toe-in or -out:
Rear wheel combined toe angle = 0′ + 0′ = 0′
Thus wheels are *parallel*.

b) Lateral offset:

$$\text{Thrust axis deviation} = \frac{R - L}{2} = \frac{+30' - (-30')}{2} = \frac{+30' + 30'}{2} = \frac{60'}{2} = 30'$$

Note A minus minus makes a plus; −(−) = +

From lateral offset Table 9.2, a thrust axis deviation of 30′ for a wheel base of 3000 mm is equivalent to a lateral offset to the right of 22 mm when the vehicle moves in a forward direction.

Example 2 (Fig. 9.27(b)) Determine the rear wheel toe-in or -out and the front to rear lateral offset of a vehicle having independent rear suspension from the following data:

Wheelbase = 3400 mm
Wheel diameter = 13 inches
Left rear wheel toe angle reading — out (out = negative) = −40′
Right rear wheel toe angle reading — in (in = positive) = +15′
Left side front to rear measuring gauge reading — out = −55′
Right side front to rear measuring gauge reading — in = +25′

a) Toe-in or -out:
Rear wheel combined toe angle = (−40′) + (+15′) = 25′
From toe conversion table a toe angle of −25′ for a 13 inch diameter wheel is equivalent to a toe-out of 2.65 mm.

b) Lateral offset:

$$\text{Thrust axis deviation (TAD) angle} = \frac{R - L}{2} = \frac{+25' - (-55')}{2} = \frac{+25' + 55'}{2} = \frac{80'}{2} = 40'$$

From lateral offset Table 9.2, a thrust axis deviation of 40′ for a wheelbase of 3400 mm is equivalent to a lateral offset to the right of 33.5 mm when the vehicle is moving in the forward direction.

9.4.4 Six wheel vehicle with tandem rear axle steering geometry (Fig. 9.28)

For any number of road wheels on a vehicle to achieve true rolling when cornering, all projected lines drawn through each wheel axis must intersect at one common point on the inside track, this being the instantaneous centre about which the vehicle travels. In the case of a tandem rear axle arrangement in which the axles are situated parallel to each

Table 9.2 Lateral offset tables

Lateral offset of front wheels in relation to rear wheels (Measurements in millimeters)						
Wheelbase mm	10°	20°	30°	40°	50°	60°
1800	4.0	7.5	11.5	15.0	19.0	23.0
2000	4.5	8.5	13.0	17.5	22.0	26.0
2200	5.0	10.0	15.0	20.0	24.5	30.0
2400	5.5	11.0	16.5	22.0	27.5	33.5
2600	6.0	12.0	18.5	24.5	30.5	37.0
2800	6.5	13.5	20.0	26.5	33.5	40.5
3000	7.0	14.5	22.0	29.0	36.5	44.0
3200	8.0	15.5	23.5	31.5	39.0	47.5
3400	8.5	17.0	25.0	33.5	42.0	51.0
3600	9.0	18.0	27.0	36.0	45.0	54.5
3800	9.5	19.0	28.5	38.5	48.0	58.0

(Measurements in inches)						
Wheelbase ft in	10°	20°	30°	40°	50°	60°
6 0	0.2	0.3	0.5	0.6	0.8	0.9
6 6	0.2	0.3	0.5	0.7	0.8	1.0
7 0	0.2	0.4	0.6	0.7	0.9	1.1
7 6	0.2	0.4	0.6	0.8	1.0	1.2
8 0	0.2	0.4	0.7	0.9	1.1	1.3
8 6	0.2	0.5	0.7	1.0	1.2	1.4
9 0	0.3	0.5	0.8	1.0	1.3	1.5
9 6	0.3	0.5	0.8	1.1	1.4	1.6
10 0	0.3	0.6	0.9	1.2	1.5	1.7
10 6	0.3	0.6	0.9	1.2	1.5	1.9
11 0	0.3	0.7	1.0	1.3	1.6	2.0
11 6	0.3	0.7	1.0	1.4	1.7	2.1

A negative (−) value indicates front wheels offset to left
A positive (+) value indicates front wheels offset to right

other, lines projected through the axles would never meet and in theory true rolling cannot exist. However, an approximate instantaneous centre for the steered vehicle can be found by projecting a line mid-way and parallel between both rear axles, this being assumed to be the common axis of rotation (Fig. 9.30). Extended lines passing through both front wheel stub axles, if made to intersect at one point somewhere along the common projected single rear axle line, will then produce very near true rolling condition as predicted by the Ackermann principle.

Improvements in rear axle suspension design have introduced some degree of roll steer which minimizes tyre scrub on the tandem axle wheels.

This is achieved by the camber of the leaf springs supporting the rear axles changing as the body rolls so that both rear axles tend to skew in the plan view

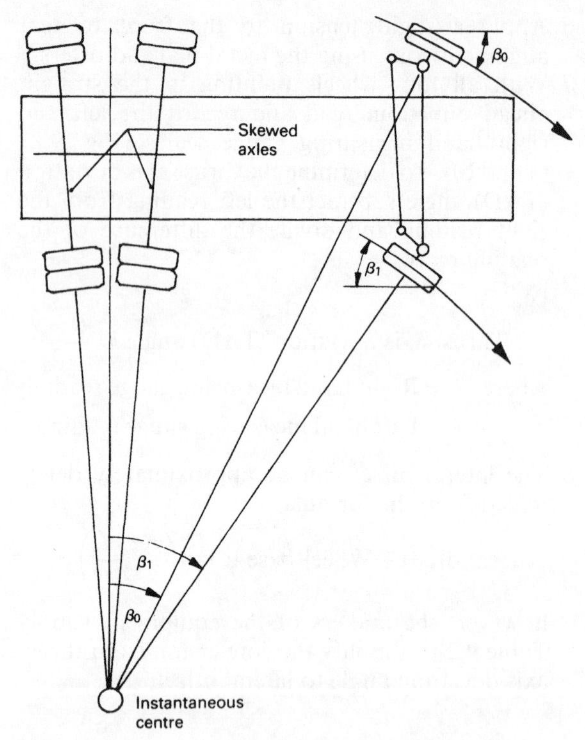

Fig. 9.28 Six wheel tandem rear axle vehicle steering geometry

so that the imaginary extended lines drawn through both rear axles would eventually meet. Unfortunately lines drawn through the front steered stub axles and the rear skewed axles may not all meet at one point. Nevertheless, they may almost merge so that very near true rolling can occur for a large proportion of the steering angle when the vehicle is in motion. The remainder of the rear axle wheel misalignment is absorbed by suspension spring distortion, shackle joints or torque arm rubber joints, and tyre compliance or as undesirable tyre scrub.

9.4.5 Dual front axle steering

Operating large rigid trucks with heavy payloads makes it necessary in addition to utilizing tandem axles at the rear to have two axles in the front of the vehicle which share out and support the load.

Both of the front axles are compelled to be steer axles and therefore need to incorporate steering linkage such as will produce true or near true rolling when the vehicle is driven on a curved track.

The advantages gained by using dual front steering axles as opposed to a single steer axle are as follows:

1 The static payload is reduced per axle so that static and dynamic stresses imposed on each axle assembly are considerably lessened.
2 Road wheel holding is improved with four steered wheels as opposed to two, particularly over rough ground.
3 Road wheel impact shocks and the subsequent vibrations produced will be considerably reduced as the suspension for both sets of wheels share out the vertical movement of each axle.
4 Damage to one axle assembly or a puncture to one of the tyres will not prevent the vehicle being safely steered to a standstill.
5 Tyre wear rate is considerably reduced for dual axle wheels compared to single axle arrangements for similar payloads. Because the second axle wheels have a smaller turning angle relative to the foremost axle wheels, the tyre wear is normally less with the second axle road wheels.

A major disadvantage with dual front axles is that it is unlikely in practice that both instantaneous centres of the first and second stub axle turning circles will actually intersect at one point for all angles of turn. Therefore tyre scrub may be excessive for certain angles of steering wheel rotation.

Dual front axle steering geometry (Fig. 9.29) When a pair of axles are used to support the front half of a vehicle each of these axles must be steered if the vehicle is to be able to negotiate a turning circle.

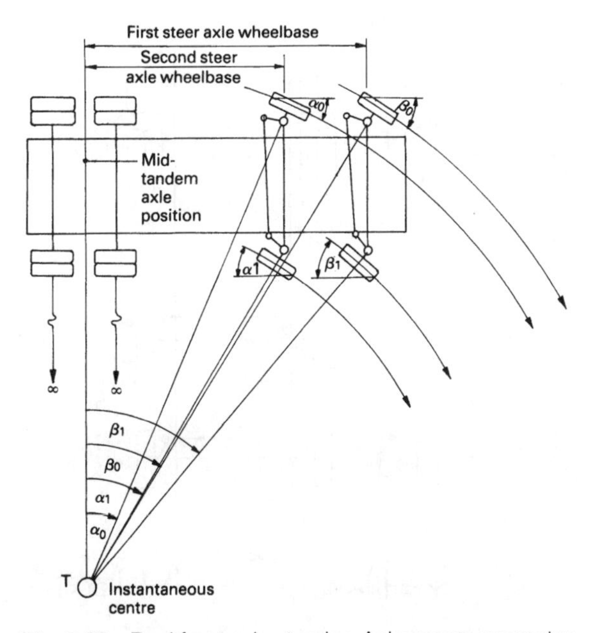

Fig. 9.29 Dual front axle steering Ackermann geometry

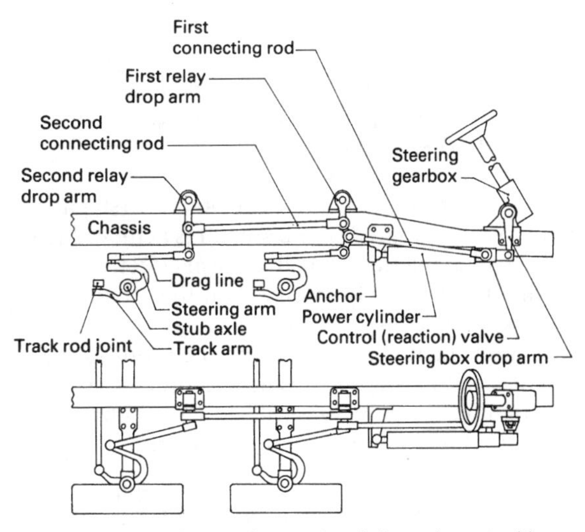

Fig. 9.30 Dual front axle steering linkage layout with power assistance

For a dual front axle vehicle to be steered, the Ackermann principle must apply to each of the front axles. This means that each axle has two wheels pivoted at each end of its beam. To enable true rolling of the wheels to take place when the vehicle is travelling along a curved track, lines drawn through each of these four stub axles must intersect at a common centre of rotation, somewhere along the extended line drawn between the tandem rear axles (Fig. 9.29).

Because the wheelbase between the first front axle is longer than the second front axle, relative to the mid-tandem axle position, the turning angles of both first front wheels will be greater than those of the second front axle wheels. The correct angle difference between the inner and outer wheels of each axle is obtained with identical Ackermann linkage settings, whereas the angle differential between the first and second axles is formed by the connecting rod ball joint coupling location on both relay drop arms being at different distances from their respective pivot point.

The dual steering linkage with power assistance ram usually utilizes a pair of swing relay drop arms bolted onto the chassis side member with their free ends attached to each axle drag link (Fig. 9.30).

The input work done to operate the steering is mainly supplied by the power cylinder which is coupled by a ball joint to the steering gearbox drop arm at the front and the power piston rod is anchored through a ball joint and bracket to the chassis at the rear end. To transfer the driver's input effort and power assistance effort to both steer axles, a forward connecting rod links the front end of the power cylinder to the first relay drop arm. A second relay connecting rod then joins both relay arms together.

A greater swivel movement of the first pair of stub axles compared to the second is achieved (Fig. 9.31) by having the swing drop arm effective length of the first relay AB shorter than the second relay arm A′B′. Therefore the second relay arm push or pull movement will be less than the input swing of the first relay arm. As a result, the angular swing of the first relay, $\Theta = 20°$, will be less than for the second relay arm angular displacement, $\Theta' = 14°$.

Dual front axle alignment checks using Dunlop optical measurement equipment (Fig. 9.32(a–d))

1 Roll or drive forward. Check the toe-in or -out of both pairs of front steering wheels and adjust track rods if necessary (Fig. 9.32(a)).
2 Assemble mirror gauge stand with the mirror positioned at right angles to the tubular stand. Position the mirror gauge against a rear axle wheel (preferably the nearest axle to the front) with the mirror facing towards the front of the vehicle (Fig. 9.32(b)).
3 Place the view box gauge stand on the floor in a transverse position at least one metre in front of the vehicle so that the view box faces the mirror (Fig. 9.32(c)). Move the view box stand across until the reflected image is centred in the view box with a zero reading on the scale. Chalk mark the position of the view box tripod legs on the ground.
4 Bring the mirror gauge stand forward to the first steer axle wheel and place gauge prods against wheel rim (Fig. 9.34(c)).
5 If both pairs of steer axle wheels are set parallel (without toe-in or -out), set the pointer on the toe angle scale to zero. Conversely, if both steer axles have toe-in or -out settings, move the pointer on the toe angle scale to read half the toe-in or -out

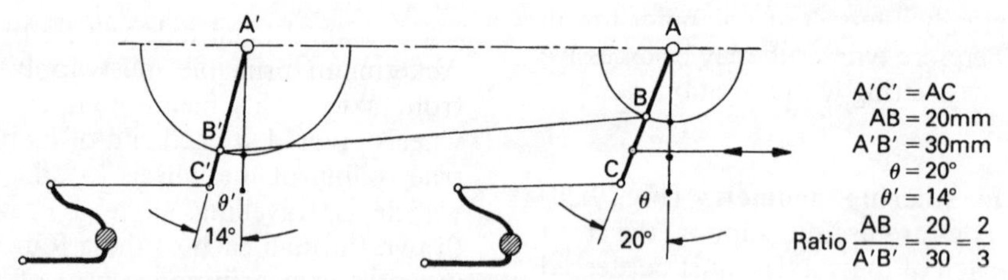

Fig. 9.31 Dual front axle steering interconnecting relay linkage principle

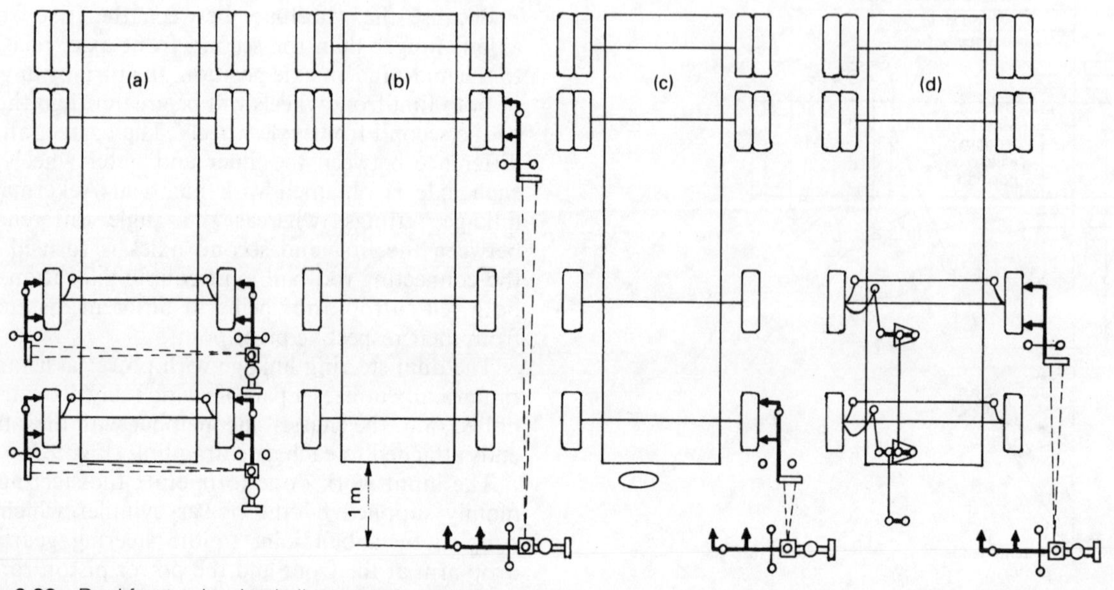

Fig. 9.32 Dual front axle wheel alignment procedure

figure, i.e. with a track toe angle of 30′ set the pointer to read 15′.

6 Look through the periscope tube and with an assistant move the driver's steering wheel in the cab until the hairline is central in the view box (Fig. 9.32(c)). At the same time make sure that the mirror gauge prods are still in contact with the front wheel rim. The first front steer axle is now aligned to the first rigid rear axle.

7 Move the mirror gauge from the first steer axle wheel back to the second steer axle wheel and position the prods firmly against the wheel rim (Fig. 9.32(d)).

8 Look into the periscope. The hairline in the view box should be centrally positioned with the toe angle pointer still in the same position as used when checking the first steer axle (Fig. 9.32(d)).

If the hairline is off-centre, the relay connecting rod between the two relay idler arms should be adjusted until the second steer axle alignment relative to the rear rigid axle and the first steer axle has been corrected. Whilst carrying out any adjustment to the track rods or relay connecting rod, the overall wheel alignment may have been disturbed. Therefore a final check should be made by repeating all steps from 2 to 8.

9.4.6 Steer angle dependent four wheel steering system (Honda)

This steer angle dependent four wheel steering system provides dual steering characteristics enabling *same direction steer* to take place for small steering wheel angles. This then changes to *opposite direction steer* with increased steering wheel deviation from the straight ahead position. Both of these steer characteristics are explained as follows:

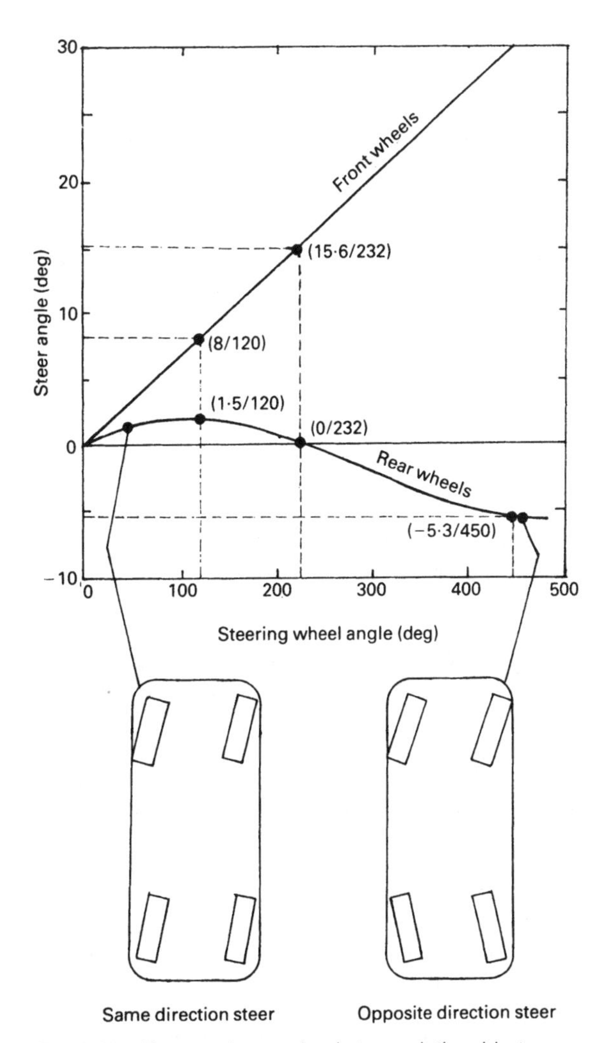

Fig. 9.33 Front and rear wheel steer relationship to driver's steering wheel angular movement

Opposite direction steer (Fig. 9.33) At low speed and large steering wheel angles the rear wheels are turned by a small amount in the opposite direction to the front wheels to improve manoeuvrability when parking (Fig. 9.32). In effect opposite direction steer reduces the car's turning circle but it does have one drawback; the rear wheels tend to bear against the side of the kerb. Generally there is sufficient tyre sidewall distortion and suspension compliance to accommodate the wheel movement as it comes into contact with the kerb so that only at very large steering wheel angles can opposite direction steer becomes a serious problem.

Same direction steer (Fig. 9.33) At high speed and small steering wheel angles the rear wheels are turned a small amount in the same direction as the front wheels to improve both steering response and stability at speed (Fig. 9.33). This feature is particularly effective when changing lanes on motorways. Incorporating a same direction steer to the rear wheels introduces an understeer characteristic to the car because it counteracts the angular steering movement of the front wheels and consequently produces a stabilizing influence in the high speed handling of the car.

Front and rear road wheel response relative to the steering wheel angular movement (Fig. 9.33) Moving the steering wheel approximately 120° from its central position twists the front wheels 8° from the straight ahead position. Correspondingly, it moves the eccentric shaft peg to its maximum horizontal annular gear offset, this being equivalent to a maximum 1.5° same direction steer for the rear road wheels (Fig. 9.33).

Increasing the steering wheel rotation to 232° turns the front wheels 15.6° from the straight ahead position which brings the planetary peg towards the top of the annular gear and in vertical alignment with the gear's centre. This then corresponds to moving the rear wheels back to the straight ahead position (Fig. 9.33).

Further rotation of the steering wheel from the straight ahead position orbits the planetary gear over the right hand side of the annular gear. Accordingly the rear wheels steadily move to the opposite direction steer condition up to a maximum of 5.3° when the driver's steering wheel has been turned roughly 450° (Fig. 9.33).

Four wheel steer (FWS) layout (Fig. 9.34) The steering system is comprised of a rack and pinion front steering box and a rear epicyclic steering box coupled together by a central drive shaft and a pair of Hooke's universal end joints (Fig. 9.35). Both front and rear wheels swivel on ball suspension joints which are steered by split track rods actuating steering arms. The front road wheels are interlinked by a rack and transverse input movement to the track rods is provided by the input pinion shaft which is connected to the driver's steering wheel via a split steering shaft and two universal joints. Steering wheel movement is relayed to the rear steering box by way of the front steering rack which meshes with an output pinion shaft. This movement of the front rack causes the output pinion and centre drive shaft to transmit motion to the rear steering box. The rear steering box mechanism then converts the angular input shaft motion to a transverse linear movement. This is then conveyed to the rear wheel swivels by the stroke rod and split track rods.

Rear steering box construction (Fig. 9.35) The rear steering box is basically formed from an epicyclic gear set consisting of a fixed internally toothed annular ring gear in which a planetary gear driven by an eccentric shaft revolves (Fig. 9.35). Motion is transferred from the input eccentric shaft to the planetary gear through an offset peg attached to a disc which is mounted centrally on the eccentric shaft. Rotation of the input eccentric shaft imparts movement to the planetary gear which is forced to orbit around the inside of the annular gear. At the same time, motion is conveyed to the guide fork via a second peg mounted eccentrically on the face of the planetary gear and a slider plate which fits over the peg (Fig. 9.35). Since the slider plate is located between the fork fingers, the rotation of the planetary gear and peg causes the slider plate to move in both a vertical and horizontal direction. Due to the construction of the guide fork, the slider plate is free to move vertically up and down but is constrained in the horizontal direction so that the stroke rod is compelled to move transversely to and fro according to the angular position of the planetary gear and peg.

Adopting this combined epicyclic gear set with a slider fork mechanism enables a small same direction steer movement of the rear wheels to take place for small deviation of the steering wheel from the straight ahead position. The rear wheels then progressively change from a same direction steer movement into an opposite steer displacement with larger steering angles.

The actual steering wheel movement at which the rear wheels change over from the same direc-

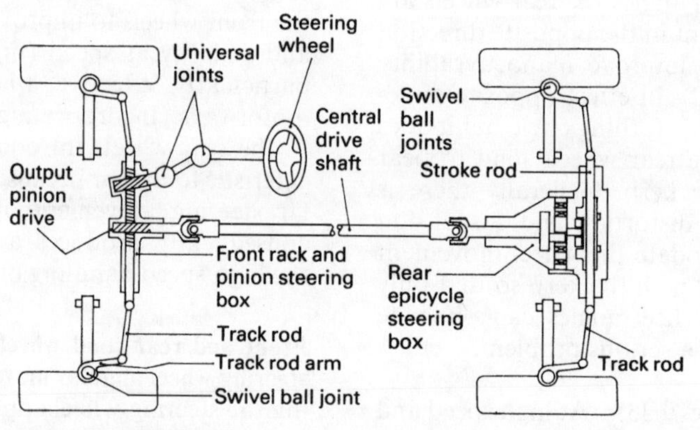

Fig. 9.34 Four wheel steering (4WS) system

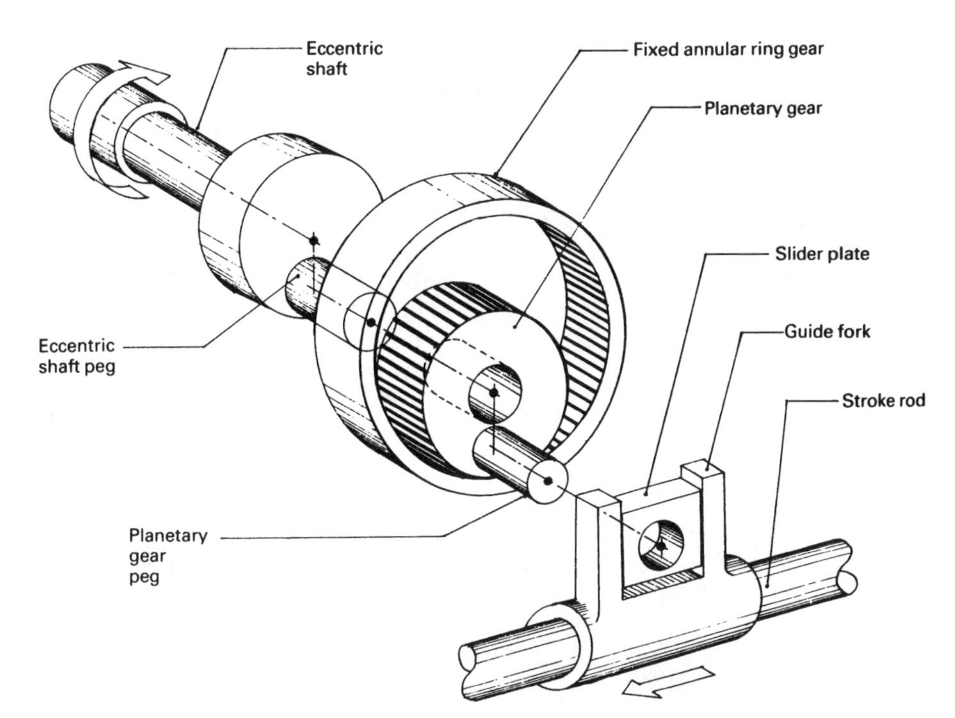

Fig. 9.35 Epicyclic rear steering box

tion steer to the opposite direction steer and the magnitude of the rear wheel turning angles relative to both conditions are dependent upon the epicyclic gear set gear ratio chosen.

Rear steering box operation (Fig. 9.36(a–e)) The automatic correction of the rear road wheels from a same direction steer to opposite direction steer with increasing front road wheel turning angle and vice versa is explained by Fig. 9.36(a–e).

Central position With the steering wheel in the straight ahead position, the planetary gear sits at the bottom of the annular gear with both eccentric shaft and planetary pegs located at bottom dead centre in the mid-position (Fig. 9.36(a)).

90° eccentric shaft peg rotation Rotating the eccentric shaft through its first quadrant (0–90°) in a chosen direction from the bottom dead centre position compels the planetary gear to roll in an anticlockwise direction up the left hand side of the annular ring gear. This causes the planetary peg and the stroke rod to be displaced slightly to the left (Fig. 9.36(b)) and accordingly makes the rear wheels move to a same direction steer condition.

180° eccentric shaft peg rotation Rotating the eccentric shaft through its second quadrant (90–180°) causes the planetary gear to roll anticlockwise inside the annular gear so that it moves with the eccentric peg to the highest position. At the same time, the planetary peg orbits to the right hand side of the annular gear centre line (Fig. 9.36(c)) so that the rear road wheels turn to the opposite direction steer condition.

270° eccentric shaft peg rotation Rotating the eccentric shaft through a third quadrant (180–270°) moves the planetary gears and the eccentric shaft peg to the 270° position, causing the planetary peg to orbit even more to the right hand side (Fig. 9.36(d)). Consequently further opposite direction steer will be provided.

360° eccentric shaft peg rotation Rotating the eccentric shaft through a fourth quadrant (270–360°) completes one revolution of the eccentric shaft. It therefore brings the planetary gear back to the base of the annular ring gear with the eccentric shaft peg in its lowest position (Fig. 9.36(e)). The planetary peg will have moved back to the central position, but this time the peg is in its highest position. The front to rear wheel steering drive gearing is normally so arranged that

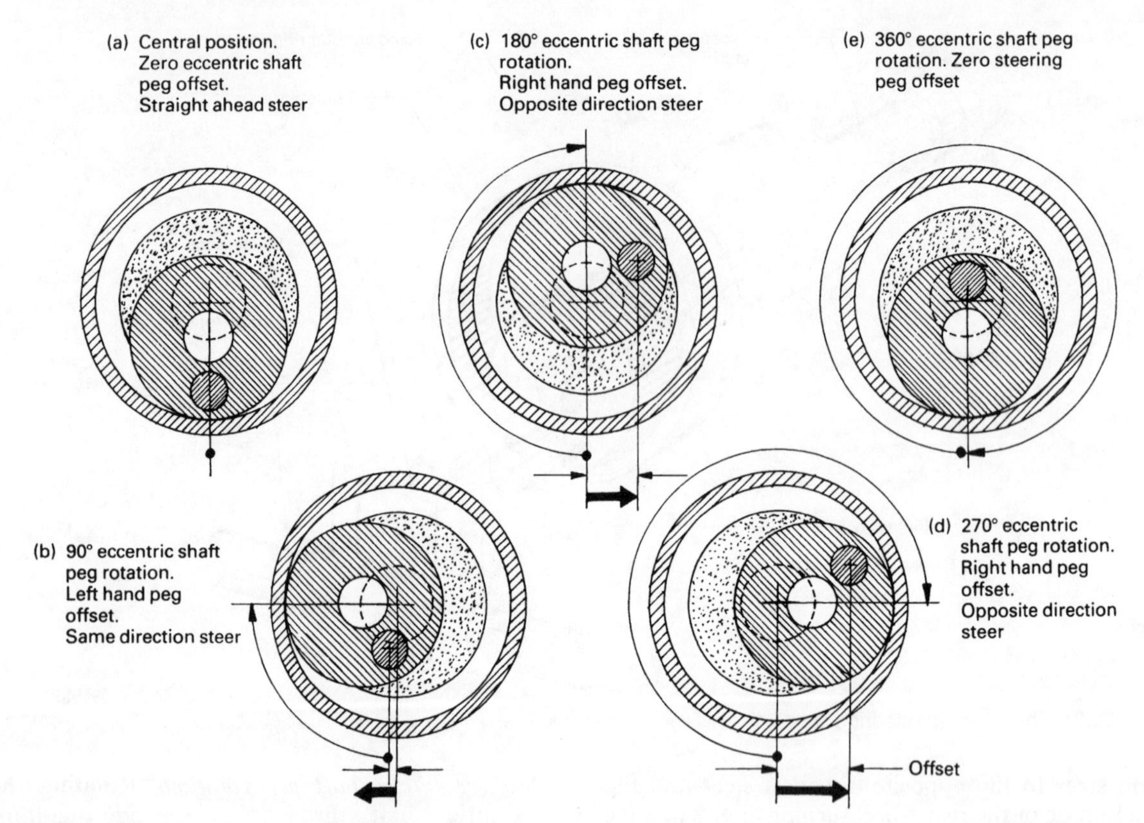

Fig. 9.36 (a–d) Principle of rear steering box mechanism

the epicyclic gear set does not operate in the fourth quadrant even under full steering lock conditions.

9.5 Variable-ratio rack and pinion

(Fig. 9.37(a–d))

Variable-ratio rack and pinion can be made to improve both manual and power assisted steering operating characteristics. For a manual rack and pinion steering system it is desirable to have a moderately high steering ratio to provide an almost direct steering response while the steering wheel is in the normally 'central position' for straight ahead driving and for very small steering wheel angular correction movement. Conversely for parking manoeuvres requiring a greater force to turn the steering wheel on either lock, a more indirect lower steering ratio is called for to reduce the steering wheel turning effort. However, with power assisted steering the situation is different; the steering wheel response in the straight ahead driving position still needs to be very slightly indirect with a relatively high steering ratio, but with the power assistance provided the off-centre steering response for manoeuvring the vehicle can be made more direct compared with a manual steering with a slightly higher steering ratio. The use of a more direct low steering ratio when the road wheels are being turned on either lock is made possible by the servo action of the hydraulic operated power cylinder and piston which can easily overcome the extra tyre scrub and swivel-pin inclination resisting force. The variable-ratio rack is achieved by having tooth profiles of different inclination along the length of the rack, accordingly the pitch of the teeth will also vary over the tooth span.

With racks designed for manual steering the centre region of the rack has wide pitched teeth with a 40° flank inclination, whereas the teeth on either side of the centre region of the rack have a closer pitch with a 20° flank inclination. Conversely, power assisted steering with variable-ratio rack and pinion (see Fig. 9.37(c)) has narrow pitch teeth with 20° flank inclination in the central region; the tooth profile then changes to a wider pitch with 40° flank inclination away from the central region of the rack for both steering locks.

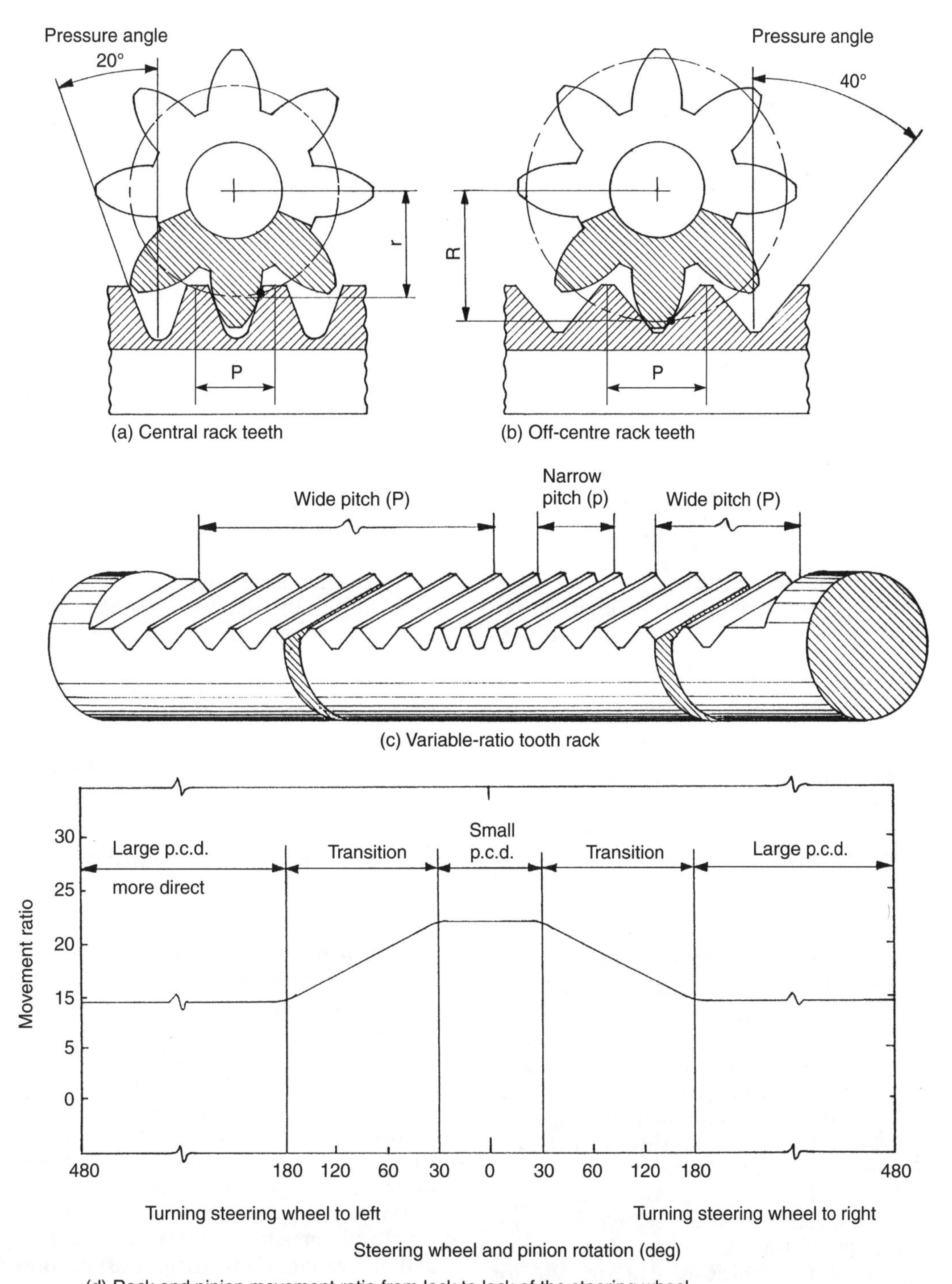

Fig. 9.37 (a–d) Variable ratio rack and pinion steering suitable for power assisted steering

With variable-ratio rack and pinion involute teeth the rack has straight sided teeth. The sides of the teeth are normal to the line of action, therefore, they are inclined to the vertical at the pressure angle. If the rack has narrow pitch 'p' 20° pressure-angle teeth, the pitch circle diameter (2R) of the pinion will be small, that is, the point of contact of the meshing teeth will be close to the tip of the rack teeth (Fig. 9.37(a)), whereas with wide pitched 'P' 40° pressure-angle tooth contact between teeth will be near the root of the rack teeth (Fig. 9.37(b)) so its pitch circle diameter (2R) will be larger.

The ratio of steering wheel radius to pinion pitch circle radius (tooth contact radius) determines the movement ratio. Thus the smaller the pitch circle radius of the pinion for a given steering wheel size, the greater will be the movement ratio (see Fig. 9.37(d)), that is, a smaller input effort will be needed to steer the vehicle, but inversely, greater will be the steering wheel movement relative to the vehicle road wheel steer angle.

This design of rack and pinion tooth profile can provide a movement-ratio variation of up to 35% with the number of steering wheel turns limited to 2.8 from lock to lock.

9.6 Speed sensitive rack and pinion power assisted steering

9.6.1 Steering desirability

To meet all the steering requirements the rack and pinion steering must be precise and direct under normal driving conditions, to provide a sense of feel at the steering wheel and for the steering wheel to freely return to the straight ahead position after the steering has been turned to one lock or the other. The conventional power assisted steering does not take into account the effort needed to perform a steering function relative to the vehicle speed, particularly it does not allow for the extra effort needed to turn the road wheels when manoeuvring the vehicle for parking.

The 'ZF Servotronic' power assisted steering is designed to respond to vehicle speed requirements, 'not engine speed', thus it provides more steering assistance when the vehicle is at a standstill or moving very slowly than when travelling at speed; at high speed the amount of steering assistance may be tuned to be minimal, so that the steering becomes almost direct as with a conventional manual steering system.

9.6.2 Design and construction (Fig. 9.38(a–d))

The 'ZF Servotronic' speed-sensitive power assisted steering uses a conventional rotary control valve, with the addition of a reaction-piston device which modifies the servo assistance to match the driving mode.

The piston and rotary control valve assembly comprises a pinion shaft, valve rotor shaft with six external longitudinal groove slots, valve sleeve with six matching internal longitudinal groove slots, torsion bar, reaction-piston device and an electro-hydraulic transducer. The reaction-piston device is supported between the rotary valve rotor and valve sleeve, and guided internally by the valve rotor via three axially arranged ball grooves and externally guided by the valve sleeve through a multi-ball helix thread.

The function of the reaction-piston device is to modify the fluid flow gap formed between the valve rotor and sleeve longitudinal groove control edges for different vehicle driving conditions.

An electronic control unit microprocessor takes in speed frequency signals from the electronic speedometer, this information is then continuously evaluated, computed and converted to an output signal which is then transmitted to the hydraulic transducer mounted on the rotary control valve casing. The purpose of this transducer is to control the amount of hydraulic pressure reaching the reaction-piston device based on the information supplied to the electronic control unit.

9.6.3 Operation of the rotary control valve and power cylinder

Neutral position (Figs 9.38(a) and 9.39(a)) With the steering wheel in its central free position, pressurized fluid from the pump enters the valve sleeve, passes though the gaps formed between the longitudinal groove control edges of both sleeve and rotor, then passes to both sides of the power cylinder. At the same time fluid will be expelled via corresponding exit 'sleeve/rotor groove' control-edge gaps to return to the reservoir. The circulation of the majority of fluid from the pump to the reservoir via the control valve prevents any build-up of fluid pressure in the divided power cylinder, and the equalization of the existing pressure on both sides of the power piston neutralizes any 'servo' action.

Anticlockwise rotation of the steering wheel (turning left — low speed) (Figs 9.38(b) and 9.39(b)) Rotating

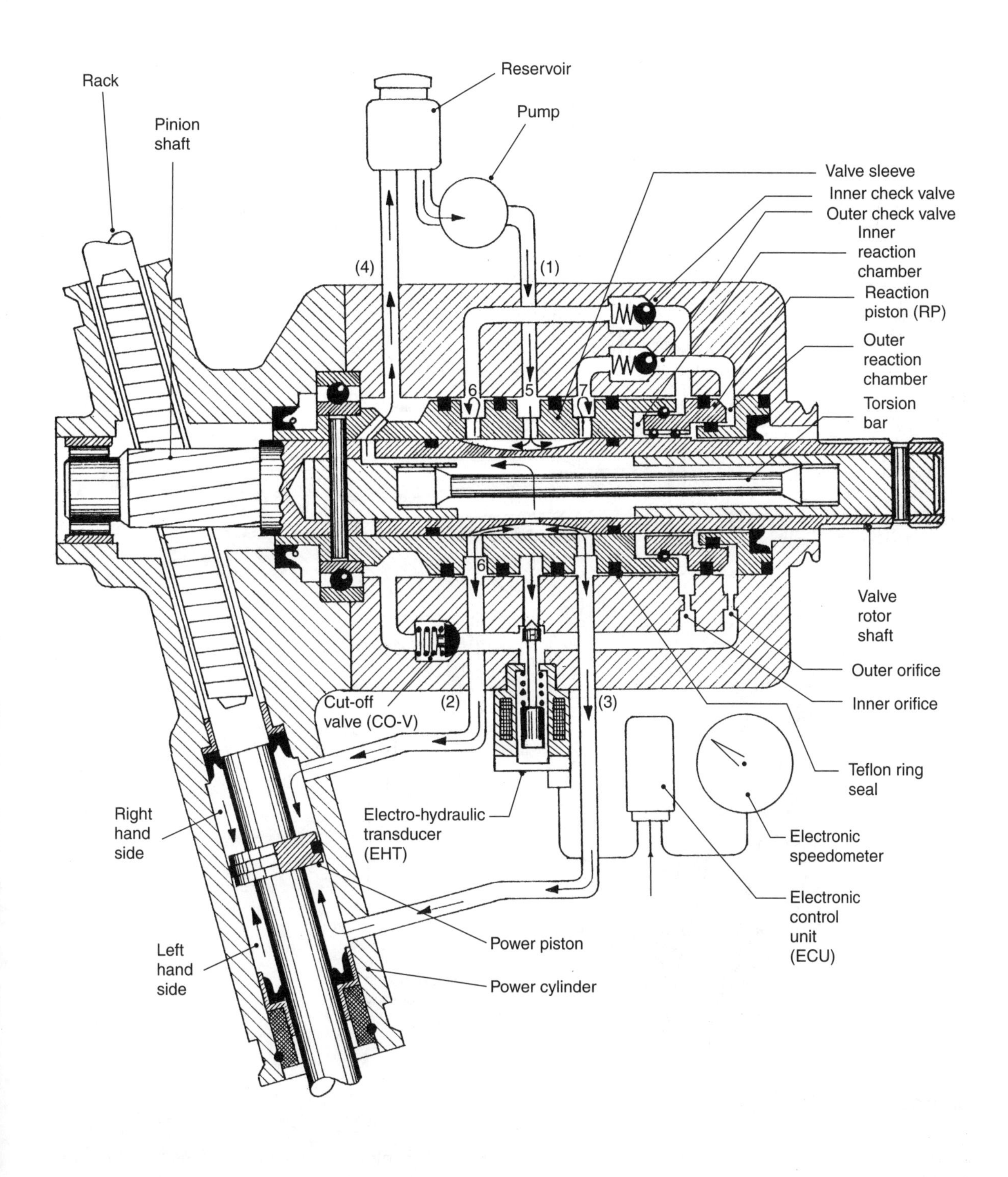

(a) Neutral position

Fig. 9.38 (a–d) Speed sensitive rack and pinion power assisted steering with rotary reaction control valve

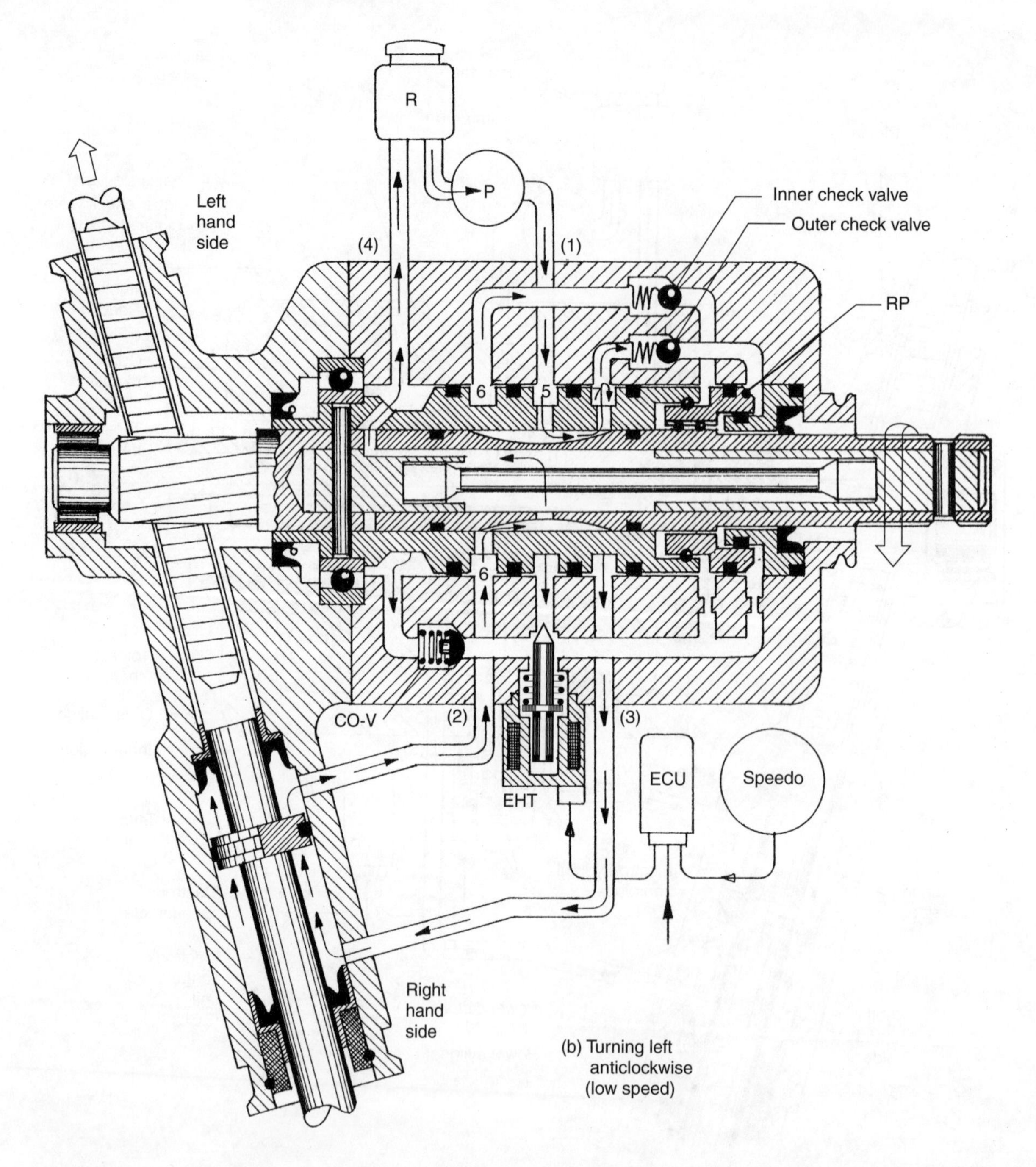

Fig. 9.38 *contd*

the steering wheel in an anticlockwise direction twists the control valve rotor against the resistance of the torsion bar until the corresponding leading edges of the elongated groove in the valve rotor and sleeve align. At this point the return path to the exit port '4' is blocked by control edges '2' while fluid from the pump enters port '1'; it then passes in between the enlarged control-edge gaps to come out of port '3', and finally it flows into the right-hand power cylinder chamber.

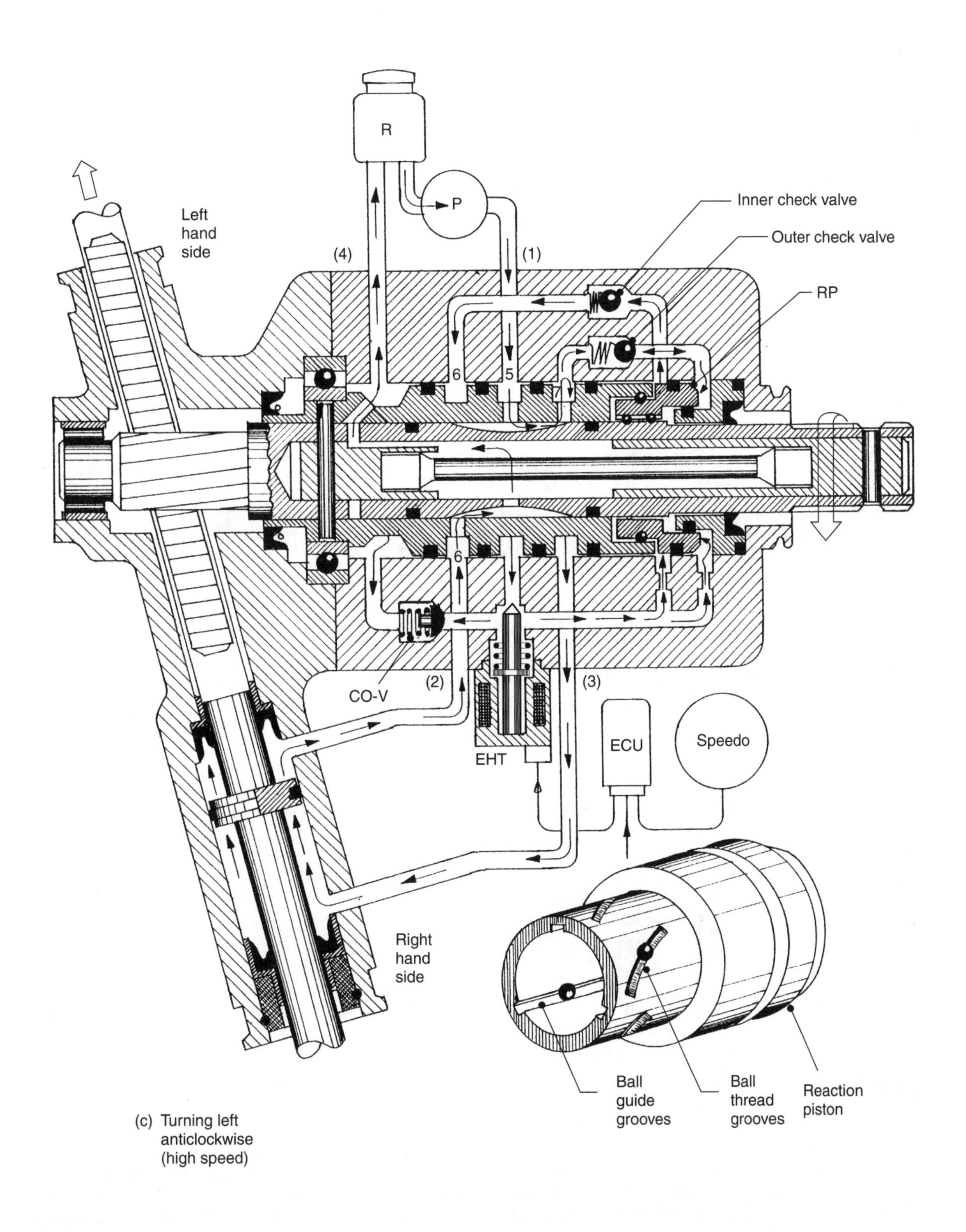

(c) Turning left anticlockwise (high speed)

Fig. 9.38 *contd*

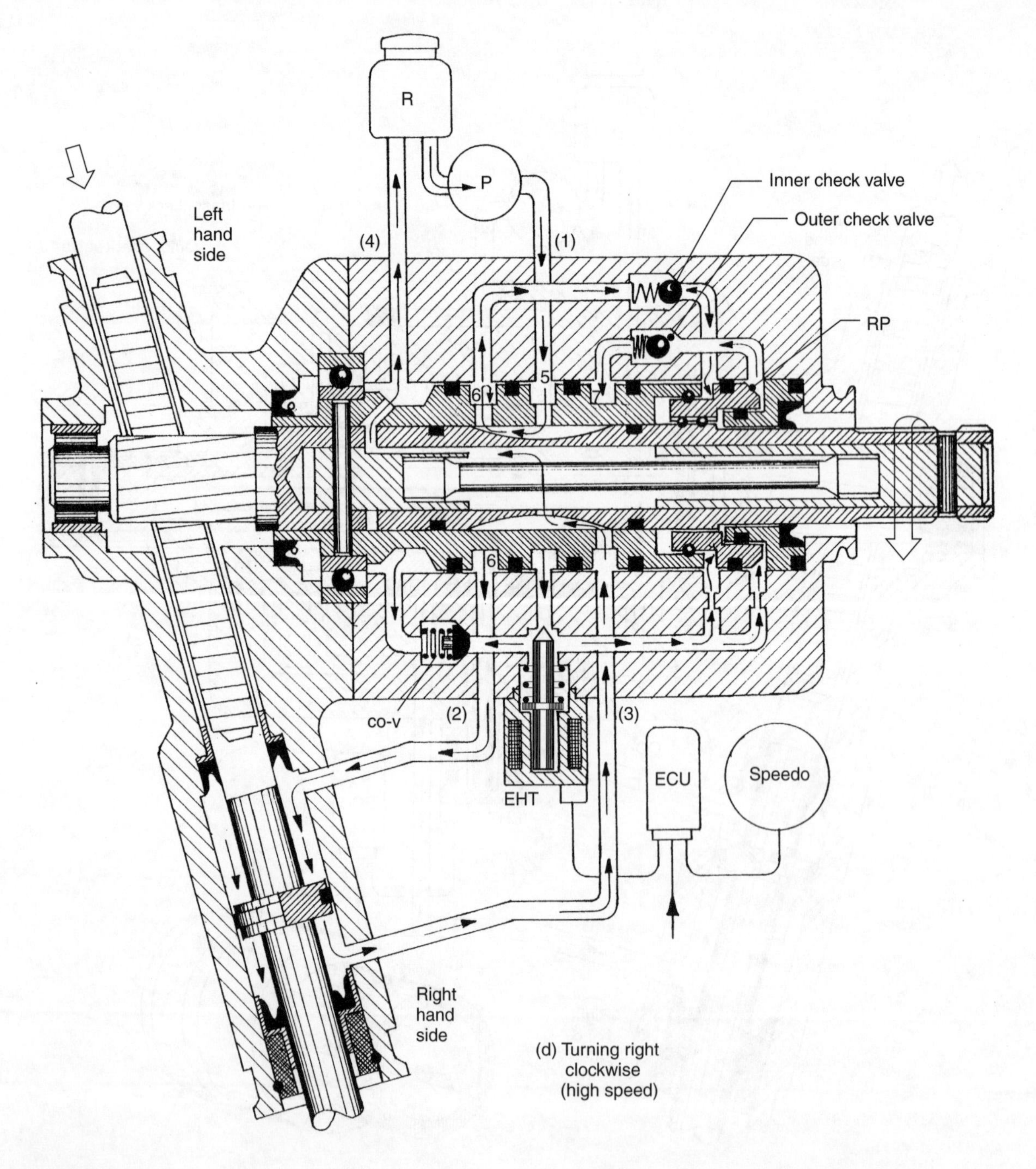

Fig. 9.38 *contd*

Conversely fluid from the left hand side power cylinder chamber is pushed towards port '2' where it is expelled via the enlarged trailing control-edge gap to the exit port '4', then is returned to the reservoir. The greater the effort by the driver to turn the steering wheel, the larger will be the control-edge gap made between the valve sleeve and rotor and greater will be the pressure imposed on the right hand side of the power piston.

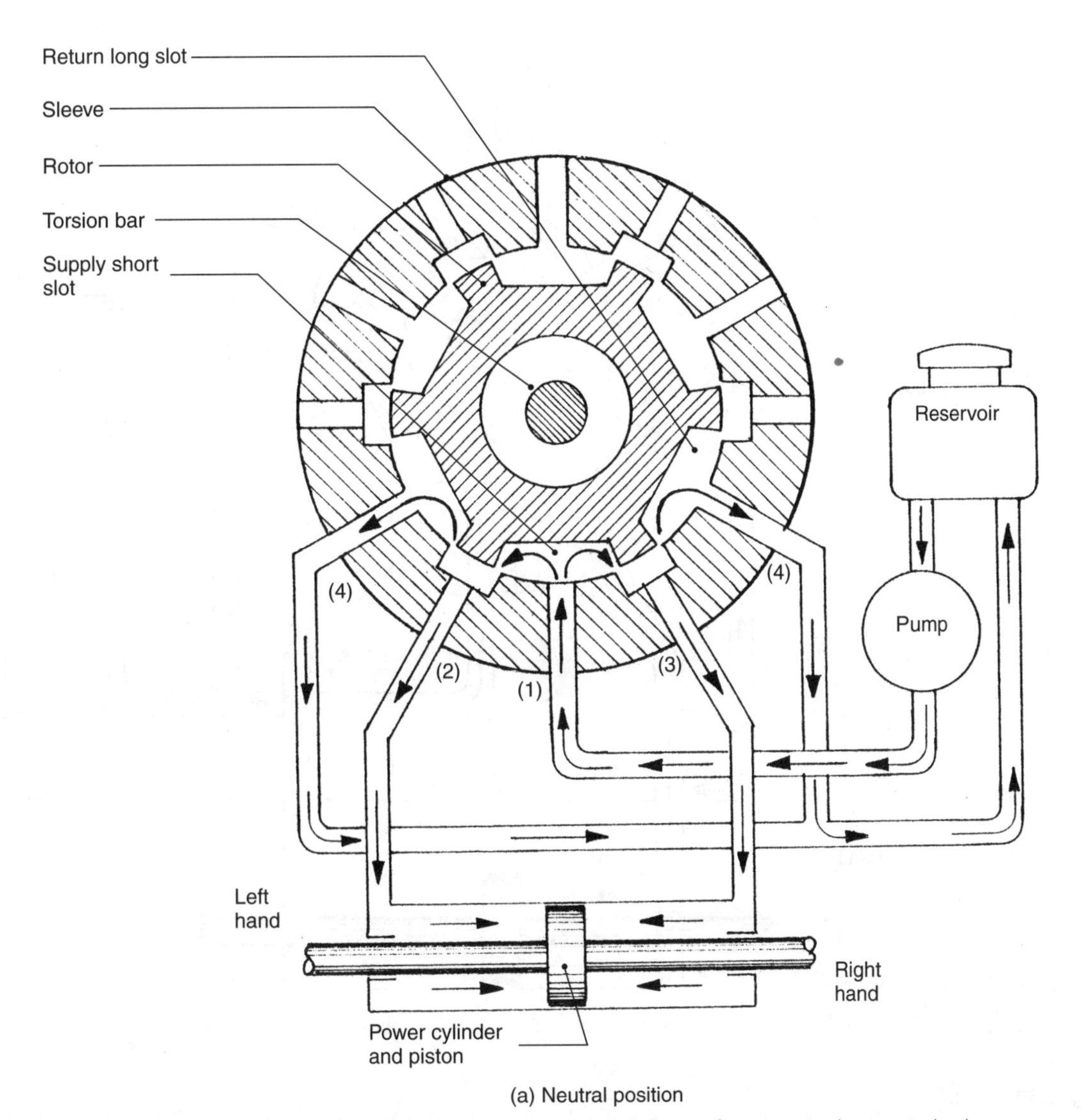

Fig. 9.39 (a–c) Rack and pinion power assisted steering sectional end views of rotary reaction control valve

When the vehicle is stationary or moving very slowly and the steering wheel is turned to manoeuvre it into a parking space or to pull out from a kerb, the electronic speedometer sends out its minimal frequency signal to the electronic control unit. This signal is processed and a corresponding control current is transmitted to the electro-hydraulic transducer. With very little vehicle movement, the control current will be at its maximum; this closes the transducer valve thus preventing fluid pressure from the pump reaching the reaction valve piston device and for fluid flowing to and through the cut-off valve. In effect, the speed sensitive rotary control valve under these conditions now acts similarly to the conventional power assisted steering; using only the basic rotary control valve, it therefore is able to exert relatively more servo assistance.

Anticlockwise rotation of the steering wheel (turning left — high speed) (Figs 9.38(c) and 9.39(b)) With increasing vehicle speed the frequency of the electronic speedometer signal is received by the electronic control unit; it is then processed and converted to a control current and relayed to the electro-hydraulic transducer. The magnitude of this control current decreases with rising vehicle speed,

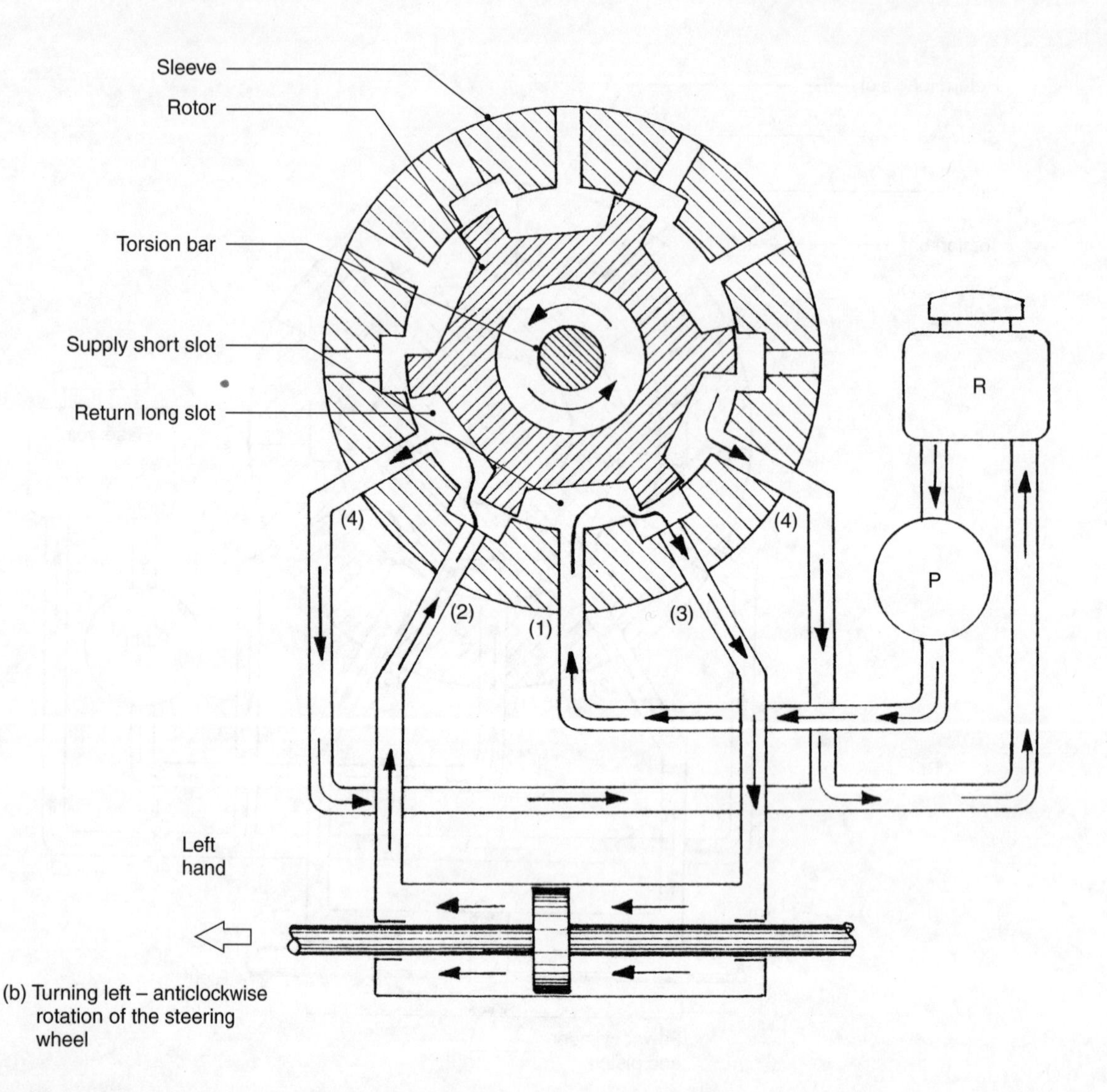

(b) Turning left – anticlockwise rotation of the steering wheel

Fig. 9.39 *contd*

correspondingly the electro-hydraulic transducer valve progressively opens thus permitting fluid to reach the reaction piston at a pressure determined by the transducer-valve orifice opening. If the steering wheel is turned anticlockwise to the left (Fig 3.38(c)), the fluid from the pump enters radial groove '5', passes along the upper longitudinal groove to radial groove '7', where it circulates and comes out at port '3' to supply the right hand side of the power cylinder chamber with fluid.

Conversely, to allow the right hand side cylinder chamber to expand, fluid will be pushed out from the left hand side cylinder chamber; it then enters port '2' and radial groove '6', passing through the lower longitudinal groove and hollow core of the rotor valve, finally returning to the reservoir via port '4'. Fluid under pressure also flows from radial groove '7' to the outer chamber check valve to hold the ball valve firmly on its seat. With the electro-hydraulic transducer open fluid under pump pressure will now flow from radial grooves '5' to the inner and outer reaction-piston device orifices. Fluid passing though the inner orifice circulates around the reaction piston and then passes to the inner reaction chamber check valve where it pushes the ball off its seat. Fluid then escapes through this open check valve back to the reservoir by way of the radial groove '6' through the centre of the valve rotor and out via port '4'. At the same time fluid flows to the outer piston

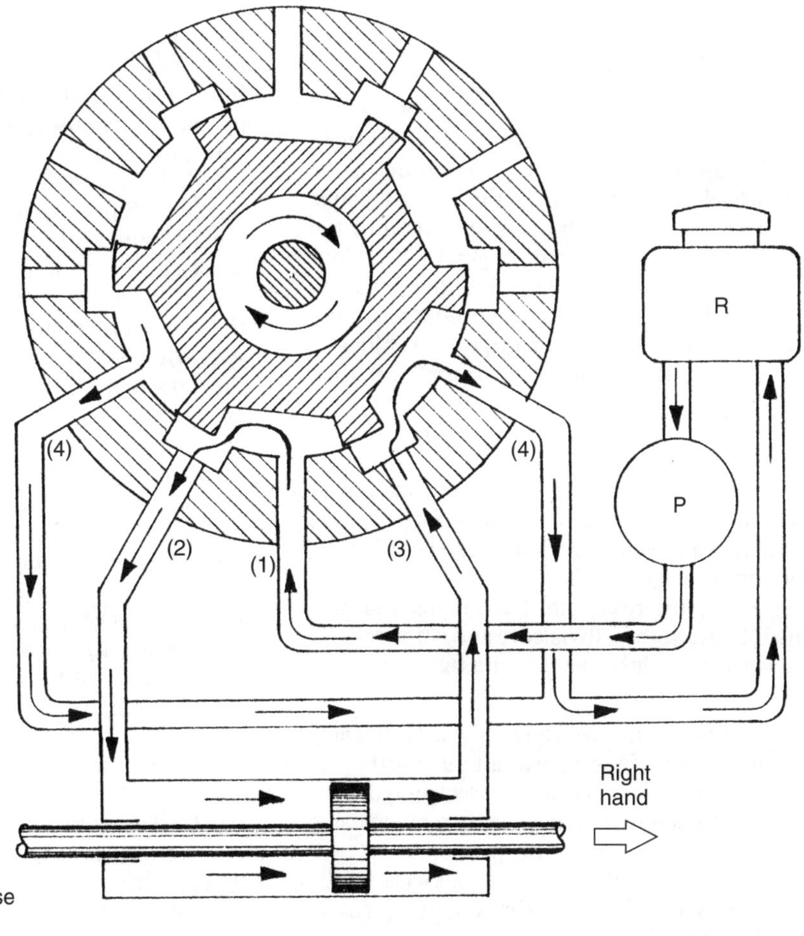

(c) Turning right – clockwise rotation of the steering wheel

Fig. 9.39 *contd*

reaction chamber and to the right hand side of the outer check valve via the outer orifice, but slightly higher fluid pressure from port '7' acting on the opposite side of the outer check valve prevents the valve opening. However, the fluid pressure build-up in the outer piston reaction chamber will tend to push the reaction piston to the left hand side, consequently due to the pitch of the ball-groove helix, there will be a clockwise opposing twist of the reaction piston which will be transmitted to the valve rotor shaft. Accordingly this reaction counter twist will tend to reduce the fluid gap made between the valve sleeve and rotor longitudinal control edges; it therefore brings about a corresponding reaction in terms of fluid pressure reaching the left hand side of the power piston and likewise the amount of servo assistance.

In the high speed driving range the electro-hydraulic transducer control current will be very small or even nil; it therefore causes the transducer valve to be fully open so that maximum fluid pressure will be applied to the outer reaction piston. The resulting axial movement of the reaction piston will cause fluid to be displaced from the inner reaction chamber through the open inner reaction chamber check valve, to the reservoir via the radial groove '6', lower longitudinal groove, hollow rotor and finally the exit port '4'.

As a precaution to overloading the power steering, when the reaction piston fluid pressure reaches its pre-determined upper limit, the cut-off valve opens to relieve the pressure and to return surplus fluid to the reservoir.

Clockwise rotation of the steering wheel (turning right — low speed) (Fig. 9.39(c)) Rotation of the steering wheel clockwise twists the control valve against the resistance of the torsion bar until the corresponding leading control edges of the elongated grooves in the valve rotor and sleeve are aligned. When the leading groove control edges align, the return path to the exit port '3' is blocked while fluid from the pump enters port '1'; it then passes inbetween the enlarged control-edge gap to come out of port '2' and finally flows into the left hand power cylinder chamber.

Conversely, fluid from the right hand side power cylinder chamber is displaced towards port '3'where it is expelled via the enlarged gap made between the trailing control edges to the exit port '4'; the fluid then returns to the reservoir. The greater the misalignment between the valve sleeve and rotor control edges the greater will be the power assistance.

Clockwise rotation of the steering wheel (turning right — high speed) (Figs 9.38(d) and 9.39(c)) With increased vehicle speed the electro-hydraulic transducer valve commences to open thereby exposing the reaction piston to fluid supply pressure.

If the steering wheel is turned clockwise to the right (Fig. 9.38 (d)), the fluid from the pump enters the radial groove '5', passes along the upper longitudinal grooves to radial groove '6' where it circulates and comes out at port '2' to supply the power cylinder's left hand side chamber with fluid.

Correspondingly fluid will be displaced from the power cylinder's right hand chamber back to the reservoir via port '3' and groove '7', passing through to the lower longitudinal groove and hollow core of the rotor valve to come out at port '4'; from here it is returned to the reservoir.

Fluid under pressure will also flow from radial groove '6' to the reaction piston's outer chamber check valve thereby keeping the ball valve in the closed position. Simultaneously, with the electro-hydraulic transducer open, fluid will flow from radial groove '5' to the inner and outer reaction-piston orifices. Fluid under pressure will also pass though the outer orifice, and circulates around the reaction piston before passing to the reaction piston's outer chamber check valve; since the fluid pressure on the spring side of the check valve ball is much lower, the ball valve is forced to open thus causing fluid to be returned to the reservoir via the radial groove '7', lower elongated rotor groove, hollow rotor core and out via port '4'. At the same time fluid flows to the inner chamber of the reaction piston via its entrance orifice. Therefore, the pressure on the spring side of its respective ball check valve remains higher thus preventing the ball valve opening. Subsequently pressure builds up in the inner chamber of the reaction piston, and therefore causes the reaction piston to shift to the right hand side; this results in an anticlockwise opposing twist to the reaction piston due to the ball-groove helices. Accordingly the reaction counter twist will reduce the flow gap between corresponding longitudinal grooves' control edges so that a reduced flow will be imposed on the left hand side of the power cylinder. Correspondingly an equal quantity of fluid will be displaced from the reaction piston outer chamber which is then returned to the reservoir via the now open outer check valve. Thus as the electro-hydraulic transducer valve progressively opens with respect to vehicle speed, greater will be the fluid pressure transmitted to the reaction piston inner chamber and greater will be the tendency to reduce the flow gap between the aligned sleeve and rotor valve control edges, hence the corresponding reduction in hydro-servo assistance to the steering.

9.6.4 Characteristics of a speed sensitive power steering system (Fig. 9.40)
Steering input effort characteristics relative to vehicle speed and servo pressure assistance are shown in Fig. 9.40. These characteristics are derived from the microprocessor electronic control unit which receives signals from the electronic speedometer and transmits a corresponding converted electric current to the electro-hydraulic transducer valve attached to the rotary control valve casing. Accordingly, the amount the electro-hydraulic transducer valve opens controls the degree of fluid pressure reaction on the modified rotary control valve (Fig. 9.38(c)). As a result the amount of power assistance given to the steering system at different vehicle speeds can be made to match more closely the driver's input to the vehicle's resistance to steer under varying driving conditions.

Referring to Fig. 9.40 at zero vehicle speed when turning the steering, for as little an input steering wheel torque of 2 Nm, the servo fluid pressure rises to 40 bar and for only a further 1 Nm input rise (3 Nm in total) the actuating pressure can reach 94 bar. For a vehicle speed of 20 km/h the rise in servo pressure is less steep, thus for an input effort torque of 2 Nm the actuating pressure has only risen to

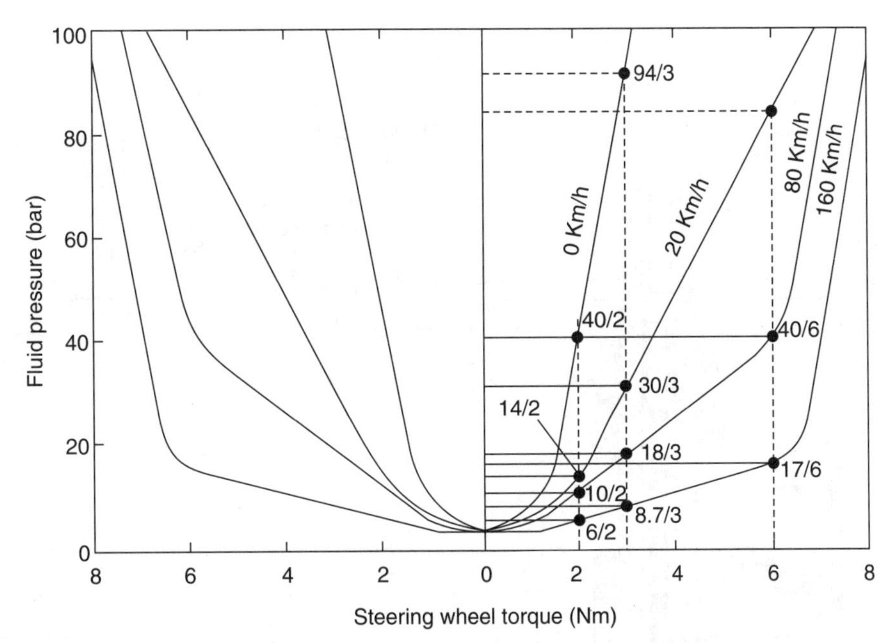

Fig. 9.40 Speed sensitive power steering steering wheel torque to servo fluid pressure characteristics for various road speeds

about 14 bar and for an input of 3 Nm the pressure just reaches 30 bar. With a higher vehicle speed of 80 km/h the servo pressure assistance is even less, only reaching 10, 18 and 40 bar for an input torque of 2, 3 and 6 Nm respectively; however, beyond an input torque of 6 Nm the servo pressure rises very steeply. Similarly for a vehicle speed of 160 km/h the rise in servo pressure assistance for an input torque rise ranging from 2 to 6 Nm only increases from 6 to 17 bar respectively, again beyond this input torque the servo pressure rises extremely rapidly. These characteristics demonstrate that there is considerable servo pressure assistance when manoeuvring the vehicle at a standstill or only moving slowly; conversely there is very little assistance in the medium to upper speed range of a vehicle, in fact the steering is almost operated without assistance unless a very high input torque is applied to the steering wheel in an emergency.

9.7 Rack and pinion electric power assisted steering

The traditional hydraulic actuated power assisted steering requires weighty high pressure equipment, which incorporates an engine driven high pressure pump, fluid reservoir and filter, reaction valve, high pressure hoses, servo cylinder, piston, ram and a suitable fluid. There is a tendency for fluid to leak due to severe overloading of the steering linkage when driving against and over stone kerbs and when manoeuvring the car during parking in confined spaces. The electric power assisted steering unit is relatively light, compact, reliable and requires a maximum current supply of between 40 and 80 amperes when parking (depending on the weight imposed on the front road wheels) and does not consume engine power as is the case of a hydraulic power assisted steering system which does apply a relatively heavy load on the engine.

9.7.1 Description and construction (Fig. 9.41)

The essentials of a rack and pinion electric power assisted steering comprises an input shaft attached to the steering wheel via an intermediate shaft and universal joint and a integral output shaft and pinion which meshes directly with the steering rack, see Fig. 9.41. A torsion bar mounted in the centre of the hollow input shaft joins the input and output shafts together and transfers the driver's manual effort at the steering wheel to the pinion output shaft. Electrical servo assistance is provided by an electric motor which supplies the majority of the steering torque to the output pinion shaft when the car's steering is being manoeuvred. Torque is transferred from the electric motor to the output pinion shaft through a ball bearing supported worm gear and a worm wheel mounted and attached to the output pinion shaft.

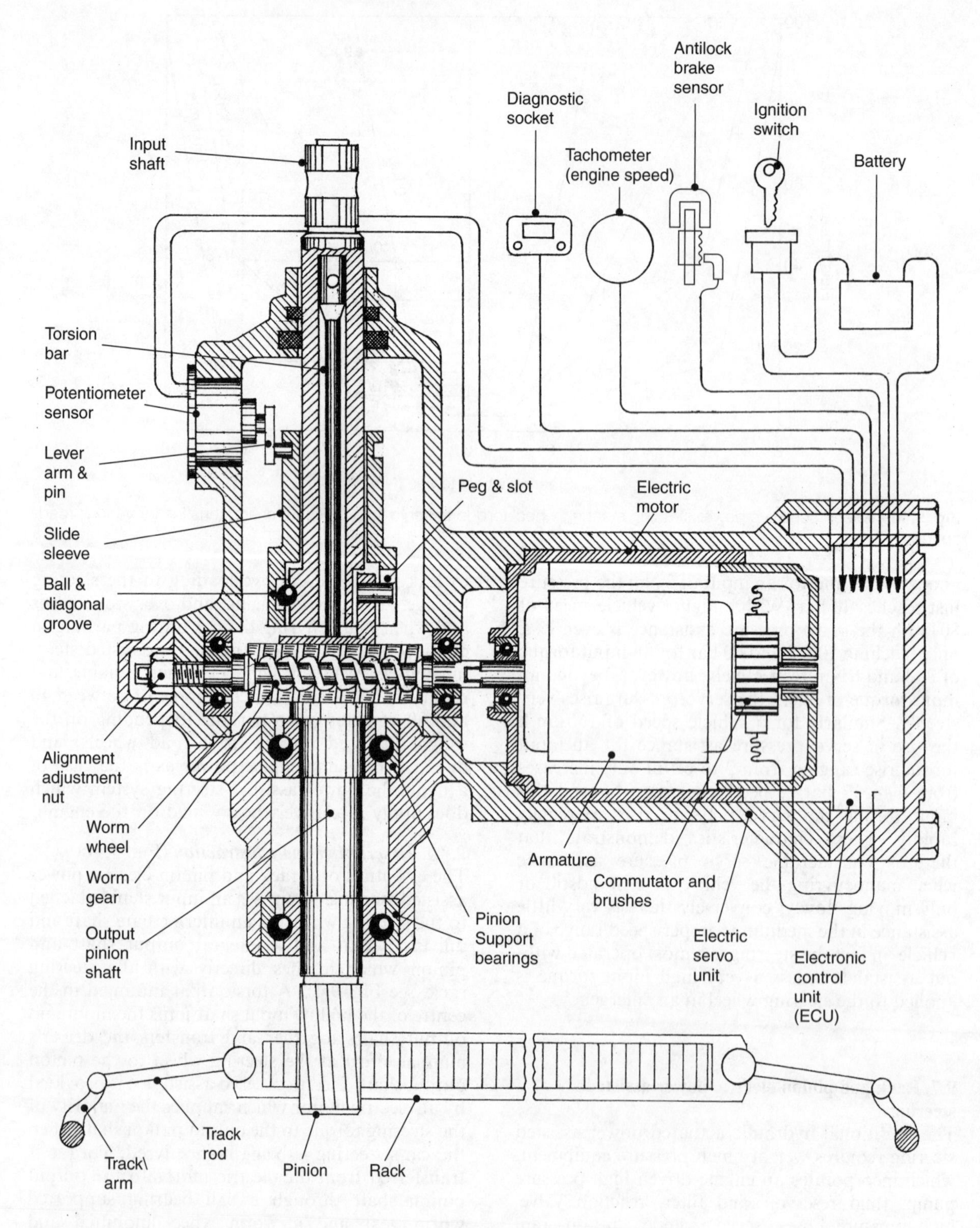

Fig. 9.41 Rack and pinion electric power assisted steering system

Relative angular misalignment between the input and output shafts is measured by transforming this angular movement into an axial linear movement along the input shaft by means of a slide sleeve, control ball, internal diagonal groove and a peg and slot. The slide sleeve which fits over the input shaft can move axially relative to the input shaft and rotates with the output shaft due to the peg and slot. Proportionate axial movement of the slide sleeve to the misalignment of the input to the output shafts is achieved by the internal diagonally formed groove in the slide sleeve and the control ball held in the shoulder part of the input shaft. Any axial slide-sleeve movement is registered by the rotary potentiometer (variable resistor) through the potentiometer arm and pin which is located in the slide sleeve's external groove.

When the steering is initially turned against the tyre to road surface grip resistance, the input torque applied to the steering is transferred to the pinion output shaft through the central torsion bar. The torsional twist of the torsion bar, that is, the angular misalignment of the input and output shafts, is proportional to the input effort at the steering wheel before the servo electric motor responds and supplies the extra input torque to the pinion output shaft to produce the desired amount of steering turn by the front road wheels. Should the electric servo assistance fail for any reason, then the steering input effort will be entirely provided by the driver though the torsion bar; under these conditions however the driver will experience a much heavier steering. A limit to the maximum torsion bar twist is provided when protruding ridges formed on the input and output shafts butt with each other.

An electronic control unit which is a microprocessor takes in information from various electrical sensors and then translates this from a programmed map into the required steering assistance to be delivered by the servo electric motor. Mechanical power is supplied by a servo electric motor which is able to change its polarity so that it can rotate either in a clockwise or anticlockwise direction as commanded by the direction of steering turn, the drive being transferred from the output pinion shaft via a warm gear and warm wheel. The large gear reduction ratio provided with this type of drive gearing enables the warm wheel to rotate at a much reduced speed to that of the warm gear and enables a relatively large torque to be applied to the output pinion shaft with a moderately small electric motor.

Steering wheel torque is monitored in terms of relative angular misalignment of the input and output shafts by the slide-sleeve movement, this is then converted into an electrical signal via the interlinked rotary potentiometer sensor. Engine and road speed sensors enable the electronic control unit to provide speed-sensitive assistance by providing more assistance at low vehicle speed when manoeuvring in a restricted space and to reduce this assistance progressively with rising speed so that the driver experiences a positive feel to the steering wheel. Note the engine and vehicle speeds are monitored by the tachometer and anti-lock brake sensors respectively.

9.7.2 Operating principle (Figs 9.42(a–c))

Neutral position (Fig. 9.42(b)) When the input and output shafts are aligned as when the steering wheel is in a neutral no turning effort position, the control ball will be in the central position of the diagonal control groove. Correspondingly the potentiometer lever arm will be in the horizontal position, with zero signal feed current to the electronic control unit and the power supply from the electronic control unit to the servo electric motor switched off. Note the potentiometer is calibrated with the wiper arm in its mid-track position to signal a zero feed current.

Clockwise right hand turn (Fig. 9.42(a)) When the steering wheel is turned clockwise to give a right hand turn, the input torque applied by the steering wheel causes a relative angular misalignment between the input and output shaft, this being proportional to the degree of effort the driver applies. As a result the control ball rotates clockwise with the input shaft relative to the output shaft, and since the slide sleeve cannot rotate independently to the output pinion shaft due to the peg and slot, the flanks of the diagonal groove are compelled to slide past the stationary control ball, thus constraining the slide sleeve to an axial upward movement only.

Accordingly the rotary potentiometer lever arm will twist anticlockwise thereby causing the wiper arm to brush over the wire or ceramic resistive track. The change in resistance and current flow signals to the electronic control unit that servo assistance is required, being in proportion to the amount the slide sleeve and rotary potentiometer moves. Once the initial effort at the steering wheel has been applied the torsional twist of the torsion bar relaxes; this reduces the relative misalignment of the input and output shafts so that the rotary potentiometer lever arm moves to a reduced feed

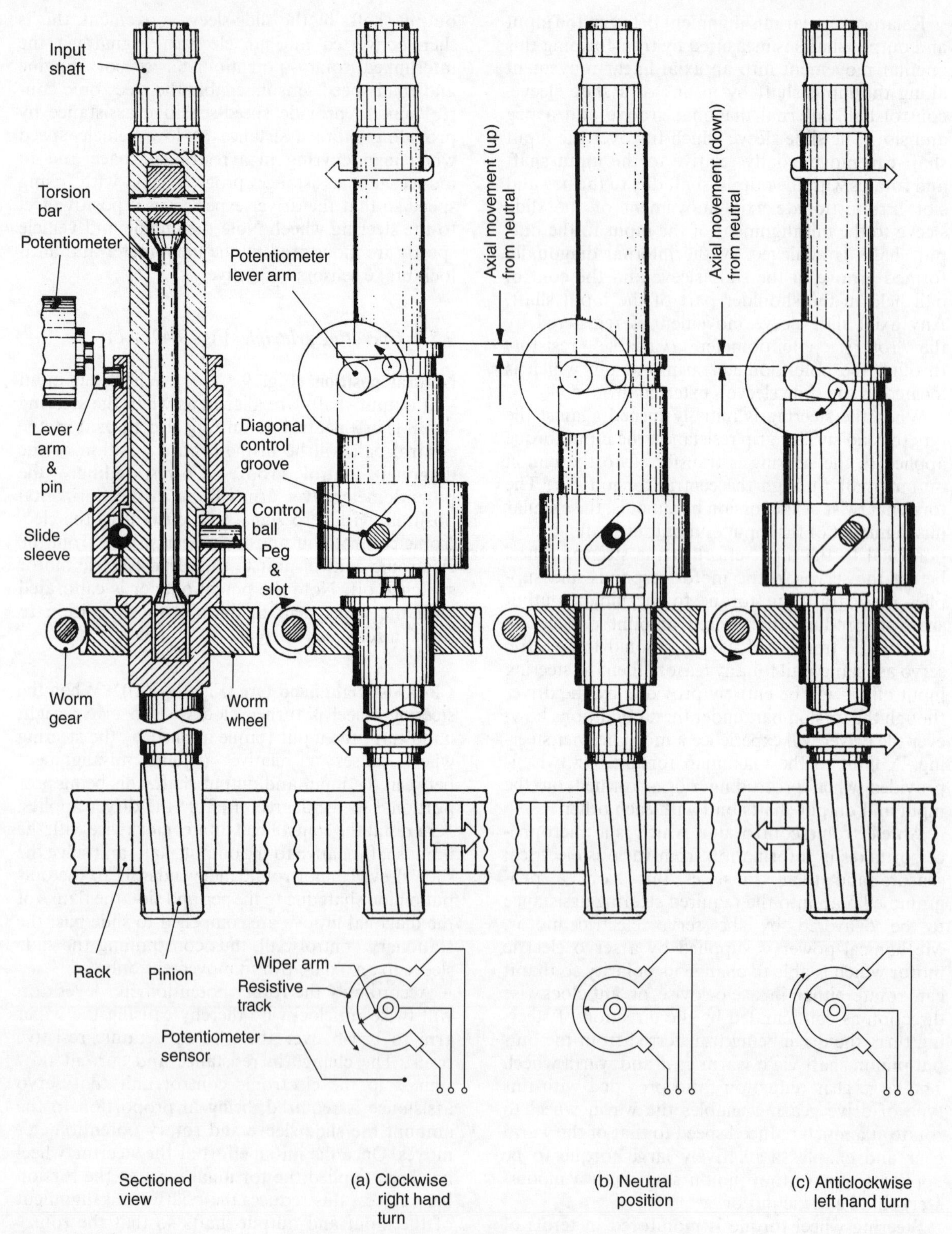

Fig. 9.42 (a–c) Operating principles for a rack and pinion electric power assisted steering

current position or even to zero feed current position. At this point the electronic control unit switched 'off' the electrical supply to the servo electric motor so that servo assistance via the warm gear and warm wheel to output pinion shaft comes to an abrupt end.

Anticlockwise left hand turn (Fig. 9.42(c)) When the steering wheel is turned anticlockwise to negotiate a left hand turn, the input effect applied by the driver to the steering wheel causes a relative angular misalignment between the input and output shafts, the relative twist of the torsion bar being proportional to the driver's input effort on the steering wheel. Due to the rotary movement of the input shaft, and control ball relative to the pinion output shaft, the diagonal groove in the sleeve will be forced to move over the stationary control ball in a downward axial direction since the peg and slot only permits the slide sleeve to move axially. The vertical downward displacement of the sleeve is relayed to the rotary potentiometer lever arm which will now partially rotate in a clockwise direction; its wiper arm will therefore brush over the resistive track, and an appropriate signal current will then be fed to the electronic control unit. The servo electric motor is then switched on, and thereby rotates the worm gear and in turn the worm wheel but at much reduced speed (due to the very large gear reduction ratio provided by a worm gear and worm wheel) in an anticlockwise direction. As the input torque effort by the driver on the steering wheel is reduced almost to nil, the relative misalignment of the input and output shaft will likewise be reduced; correspondingly the rotary potentiometer wiper arm will move to its mid-resistance position signalling zero current feed to the electronic control unit; it therefore switches off and stops the servo electric motor.

10 Suspension

10.1 Suspension geometry

The stability and effective handling of a vehicle depends upon the designers' selection of the optimum steering and suspension geometry which particularly includes the wheel camber, castor and kingpin inclination. It is essential for the suspension members to maintain these settings throughout their service life.

Unfortunately, the pivoting and swivelling joints of the suspension system are subject to both wear and damage and therefore must be checked periodically. With the understanding of the principles of the suspension geometry and their measurements it is possible to diagnose and rectify steering and suspension faults. Consideration will be given to the terminology and fundamentals of suspension construction and design.

10.1.1 Suspension terminology

Swivel joints or king pins These are the points about which the steering wheel stub axles pivot.

Pivot centre The point where the swivel ball joint axis or kingpin axis projects and intersects the ground.

Contact patch This is the flattened crown area of a tyre which contacts the ground.

Contact centres This is the tyre contact patch central point which is in contact with the ground.

Track This is the transverse distance between both steering wheel contact centres.

10.1.2 Wheel camber angle (Figs 10.1 and 10.2)

Wheel camber is the lateral tilt or sideway inclination of the wheel relative to the vertical (Fig. 10.1). When the top of the wheel leans inwards towards the body the camber is said to be negative, conversely an outward leaning wheel has positive camber.

Road wheels were originally positively cambered to maintain the wheel perpendicular to the early highly cambered roads (Fig. 10.2) and so shaped as to facilitate the drainage of rain water. With modern underground drainage, road camber has been greatly reduced or even eliminated and therefore wheel camber has been reduced to something like ½ to 1½ degrees.

The axis of rotation of a cambered wheel if projected outwards will intersect the ground at the apex of a cone generated if the wheel was permitted to roll freely for one revolution. The wheel itself then resembles the frustrum of a cone (Fig. 10.1). The path taken by the cambered wheel (frustrum of a cone) if free to roll would be a circle about the apex. Consequently both front wheels will tend to steer outwards in opposite directions as the vehicle moves forwards. In practice, the track rods and ball joints are therefore preloaded as they restrain the wheels from swivelling away from each other when the vehicle is in motion. If both wheels have similar camber angles, their outward pull on the track rods will be equal and therefore balance out. If one wheel is slightly more cambered than the other, due maybe to body roll with independent suspension or because of misalignment, the steering wheels will tend to wander or pull to one side as the vehicle is steered in the straight ahead position.

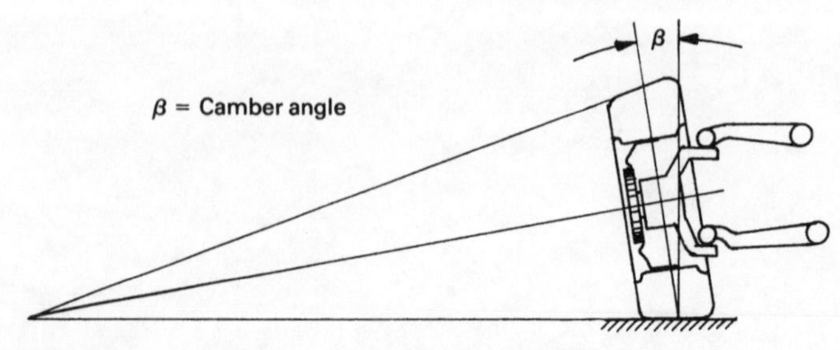

Fig. 10.1 Wheel camber geometry

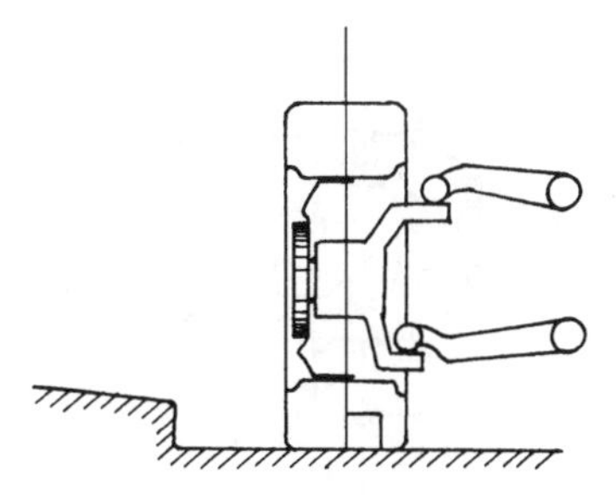

Fig. 10.2 Road camber

A negatively cambered wheel leaning towards the radius of a curved track or bend increases its cornering power and reduces the tyre contact patch slip angle for a given cornering force compared to a wheel rolling in an upright position. Conversely, a positively cambered wheel leaning away from the centre of rotation reduces its cornering power and increases the tyre slip angle for a similar cornering force compared to a wheel rolling perpendicular to the ground.

To provide a small amount of understeer, the front wheels are normally made to generate a greater slip angle than the rear wheels by introducing positive wheel camber on the front wheels and maintaining the rear wheels virtually perpendicular to the ground.

When cornering with positive camber angles on both front wheels, the inner and outer wheels will lean inwards and outwards respectively relative to the centre of rotation of the turn. At the same time, body roll transfers weight from the inner wheel to the outer one. As a result the inner wheel will generate less slip angle than the outer wheel because it provides an inward leaning, more effective tyre grip with less vertical load than that of the less effective outward leaning tyre, which supports a greater proportion of the vehicle's weight. The front cambered tyres will generate on average more slip angle than the upright rear wheels and this causes the vehicle to have an understeer cornering tendency.

Steered positive cambered wheels develop slightly more slip angle than uncambered wheels. When they are subjected to sudden crosswinds or irregular road ridging, the tyres do not instantly deviate from their steered path, with the result that a more stable steering is achieved.

With the adoption of wider tyres as standard on some cars, wheel camber has to be kept to a minimum to avoid excessive edge wear on the tyres unless the suspension has been designed to cope with the new generation of low profile wide tread width tyre.

10.1.3 Swivel or kingpin inclination

(Figs 10.3–10.7)

Swivel pin or kingpin inclination is the lateral inward tilt (inclination) from the top between the upper and lower swivel ball joints or the kingpin to the vertical (Fig. 10.3). If the swivel ball or pin axis is vertical (perpendicular) to the ground, its contact centre on the ground would be offset to the centre of the tyre contact patch (Fig. 10.4). The offset between the pivot centre and contact patch centre is equal to the radius (known as the scrub radius) of a semicircular path followed by the rolling wheels when being turned about their pivots. When turning the steering the offset scrub produces a torque T created by the product of the offset radius r and the opposing horizontal ground reaction force F (i.e. $T = Fr$ (Nm)). A large pivot to wheel contact centre offset requires a big input torque to overcome the opposing ground reaction, therefore the steering will tend to be heavy. No offset (zero offset radius) (Fig. 10.5) prevents the tread rolling and instead causes it to scrub as the wheel is steered so that at low speed the steering also has a heavy response. A compromise is usually made by offsetting the pivot and contact wheel centres to roughly 10–25% of the tread width for a standard sized tyre. This small offset permits the pivot axis to remain within the contact patch, thereby enabling a rolling movement to still take place when the wheels are pivoted so that tyre scruff and creep (slippage) are minimized. One other

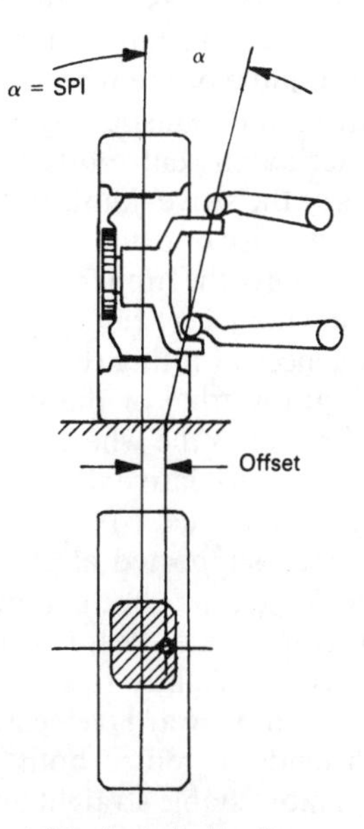

Fig. 10.3 Swivel (king) pin inclination

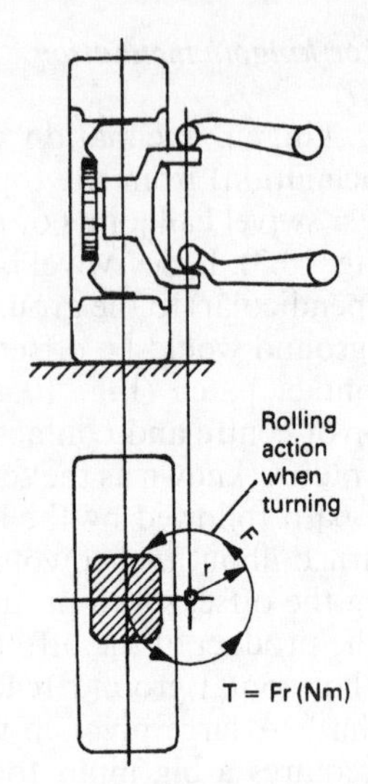

Fig. 10.4 Swivel (king) pin vertical axis offset

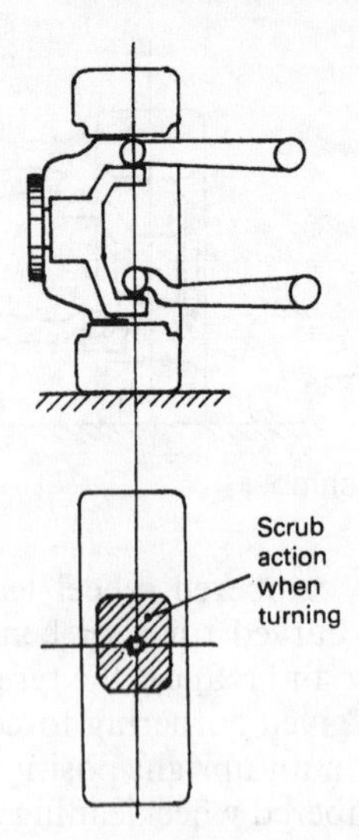

Fig. 10.5 Dished wheel centre point steering

effect of a large pivot to contact centre offset is when one of the wheels hits an obstacle like a bump or pothole in the road; a large opposing twisting force would be created momentarily which would be relayed back to the driver's steering wheel in a twitching fashion.

To reduce or even eliminate pivot to wheel centre offset, the whole stub axle, hub bearing assembly and disc or drum would have to be positioned within the centre region of the wheel rim and also extend, and therefore protrude, beyond the wheel rim flange (Fig. 10.5). A dished wheel arrangement of this type is known as centre point steering because both pivot centre and contact patch centres coincide in the middle of the wheel.

The alternative and realistic way of reducing the pivot to contact patch centre offset is to laterally incline the axis of the swivel joints so that the whole hub assembly and disc or drum is positioned inside the wheel and only the upper swivel joint may protrude outside the wheel rim.

The consequences of tilting the swivel pin axis is the proportional lowering of the stub axle axis in the horizontal plane as the wheel assembly swivels about its pivot points relative to the straight ahead position (Fig. 10.6(a and b)). Because the road wheels are already supported at ground level, the reverse happens, that is, both upper and lower wishbone arms or axle beam which supports the vehicle body are slightly raised. This unstable state produces a downward vehicle weight component which tends to return both steered wheel assemblies to a more stable straight ahead position. In other words, the pivot inclination produces a self-centring action which is independent of vehicle speed or traction but is dependent upon the weight concentration on the swivel joints and their inclination. A very large swivel ball or pin inclination produces an excessively strong self-centring effect which tends to kick back on turns so that the swivel ball or pin inclination angle is usually set between 5 and 15°. A typical and popular value would be something like 8 or 12°.

The combination of both camber and swivel joint inclination is known as the *included angle* and the intersection of both of these axes at one point at ground level classifies this geometry as *centre point steering* (Fig. 10.7). In practice, these centre lines projected through the ball joints or pins and through the centre of the wheel are made to meet at some point below ground level. Thus an offset exists between the projected lines at ground level, which produces a small twisting movement when the wheels are steered. As a result, the wheels tend to roll about a circular path with the offset as its radius, rather than twist about its swivel centre with a continuous slip-grip action which occurs when there is no offset as with the centre point steering geometry.

10.1.4 Castor angle (Figs 10.8 and 10.9)

The inclination of the swivel ball joint axis or kingpin axis in the fore and aft direction, so that the tyre contact centre is either behind or in front of the imaginary pivot centre produced to the ground, is known as the *castor angle* (Fig. 10.8(b and c)). Positive castor angle is established when the wheel contact centre trails behind the pivot point at

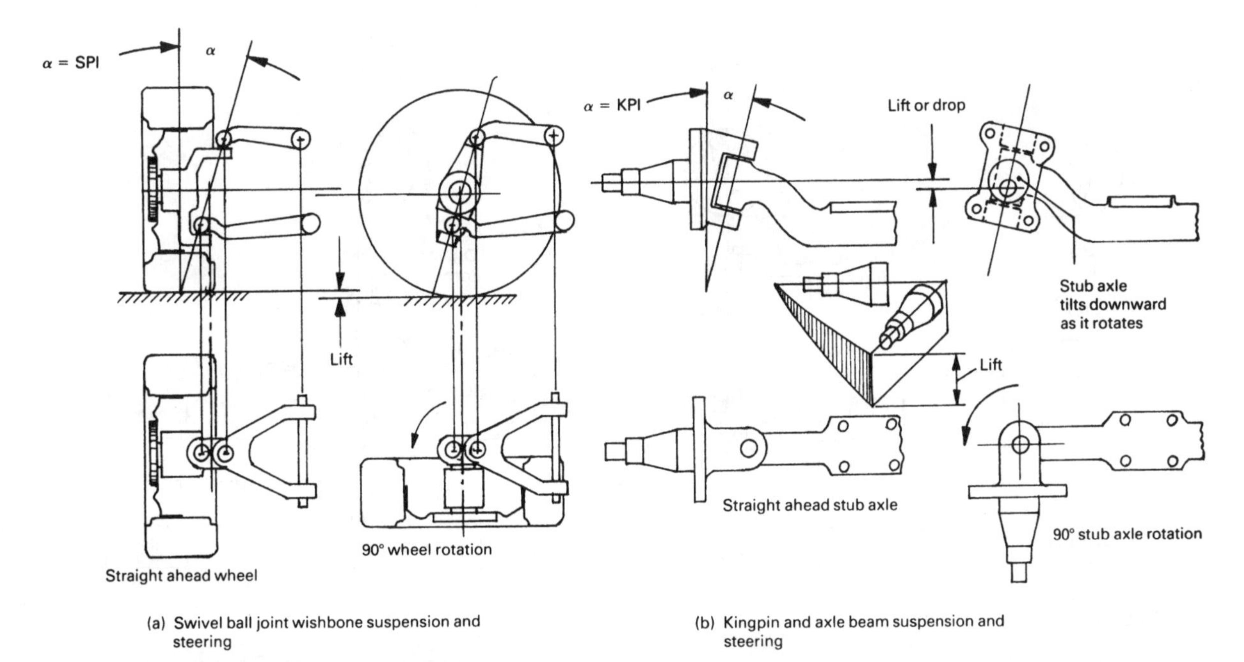

Fig. 10.6 (a and b) Swivel and kingpin inclination self-straightening tendency

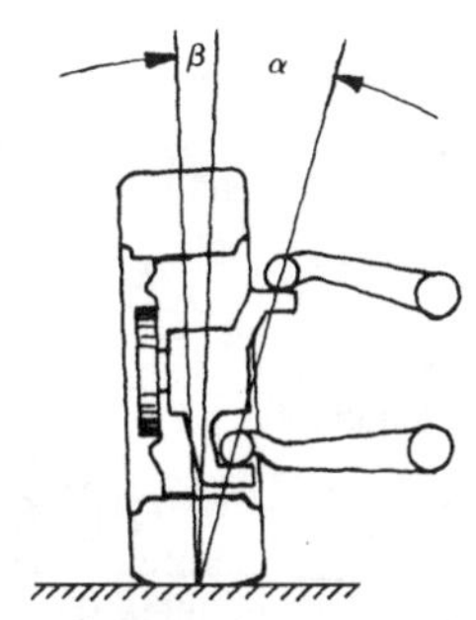

Fig. 10.7 Camber and swivel pin inclination centre point steering

ground level (Fig. 10.8(b)). Negative castor angle exists if the wheel contact centre leads the pivot axis intersection at ground level (Fig. 10.8(c)).

If the pivot centre and wheel contact patch centre coincide the castor is nil (Fig. 10.8(a)). Under these conditions the steered wheels become unstable as they tend to twitch from side to side when the vehicle travels along a straight path.

A rear wheel drive vehicle has the front wheel steer pivot axis inclined backward to produce positive castor (Fig. 10.9(a)). As the vehicle is propelled from the rear (the front wheels are pushed by the driving thrust transmitted by the rear drive wheels), it causes the front wheels to swing around their pivot axis until the tyre contact centre trails directly behind. This action takes place because the drag force of the front tyres on the road causes both tyres to move until they are in a position where no out of balance force exists, that is, positioned directly to the rear of the pivot swivel balls or pin axis.

With front wheel drive vehicles the situation is different because the driving torque is transmitted through the steered front wheels (Fig. 10.9(b)). By inclining the pivot axis forwards, a negative castor is produced and instead of the pivot axis being pushed by the rear wheel drive thrust, traction is now transmitted through the front wheels so that the pivot axis is pulled forwards. The swivel balls or pin mounting swing to the rear of the contact patch centre, due to the vehicle rolling resistance acting through the rear wheels, opposing any forward motion.

The effects of castor angle can be seen in Fig. 10.9(a and b), when the steering is partially turned on one lock. The trail or lead distance between the pivot centre and contact patch centre rotates as the steered wheels are turned so that the forward driving force F_D and the equal but opposite ground reaction F_R are still parallel but are now offset by a distance x. Therefore a couple (twisting movement) M is generated of magnitude $M = Fx$, where $F = F_D = F_R$. With the vehicle in motion, the couple M will continuously try to reduce itself to zero by eliminating the offset x. In other words, the driving and reaction forces F_D and F_R are at all

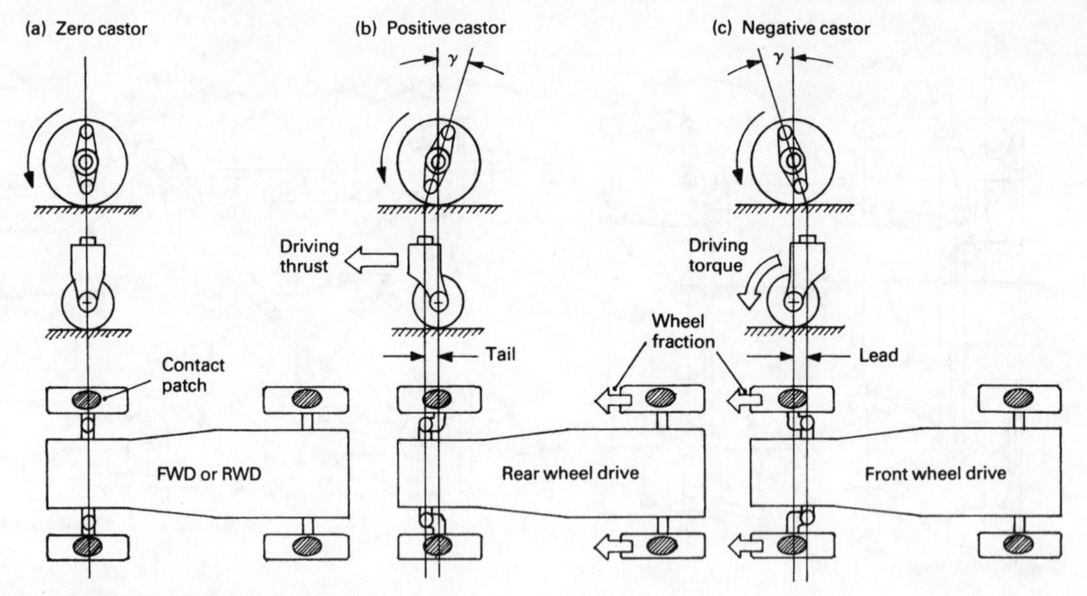

Fig. 10.8 Castor angle steering geometry

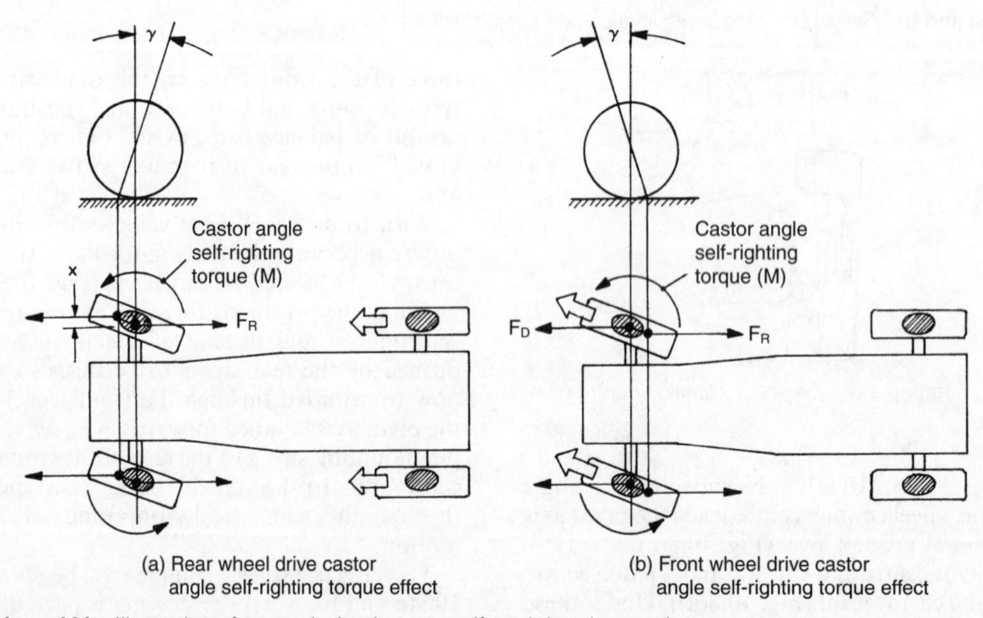

Fig. 10.9 (a and b) Illustration of steered wheel castor self-straightening tendency

times tending to align themselves with the wheels rolling when the steering has been turned to one lock. As a result the trailing or leading offset x produces a self-righting effect to the steered wheels. The greater the angle the wheels have been steered, the larger the pivot centre to contact patch centre offset x and the greater the castor self-centring action will be. The self-righting action which tends to straighten out the steering after it has been turned from the straight position, increases with both wheel traction and vehicle speed.

10.1.5 Swivel joint positive and negative offset (Figs 10.10–10.15)

When one of the front wheels slips during a brake application, the inertia of the moving mass will tend to swing the vehicle about the effective wheel which is bringing about the retardation because

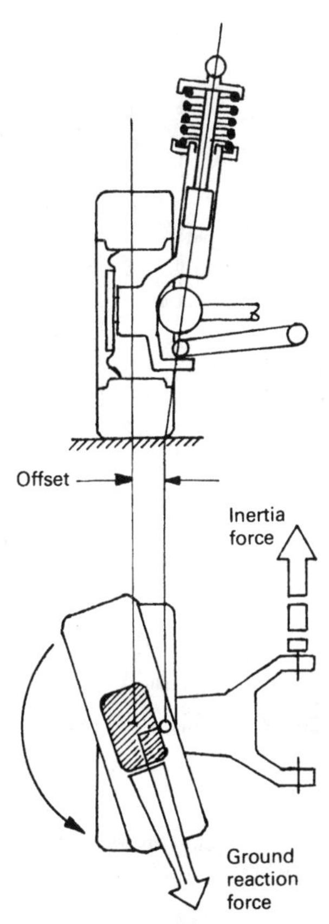

Fig. 10.10 Swivel pin inclination positive offset

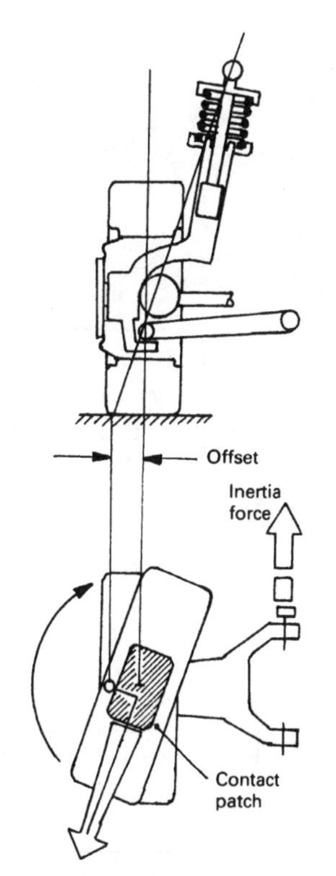

Fig. 10.11 Swivel pin inclination negative offset

there is very little opposing resistance from the wheel on the opposite side (Fig. 10.12).

If the offset of the swivel ball joints is on the inside of the tyre contact patch the swivel inclination is known as *positive offset* (Fig. 10.10). When the wheels are braked the positive offset distance and the inertia force of the vehicle produce a turning movement which makes the wheels pivot about the contact patch centre in an outward direction at the front (Fig. 10.10). If the off side (right) wheel moves onto a slippery patch, the vehicle will not only veer to the left, due to the retarding effect of the good braked wheel preventing the vehicle moving forward, but the near side (left) wheel will also turn and steer to the left (Fig. 10.13). Therefore the positive offset compounds the natural tendency for the vehicle to swerve towards the left if the right hand wheel skids instead of continuing on a stable straight ahead path.

Arranging for the swivel ball joint inclination centre line to intersect the ground on the outside of the contact patch centre produces what is known as *negative offset* (Fig. 10.11). With negative offset the

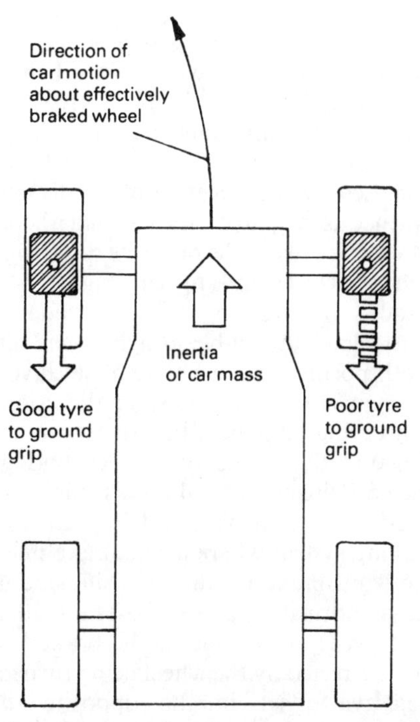

Fig. 10.12 Directional stability when one wheel skids whilst being braked

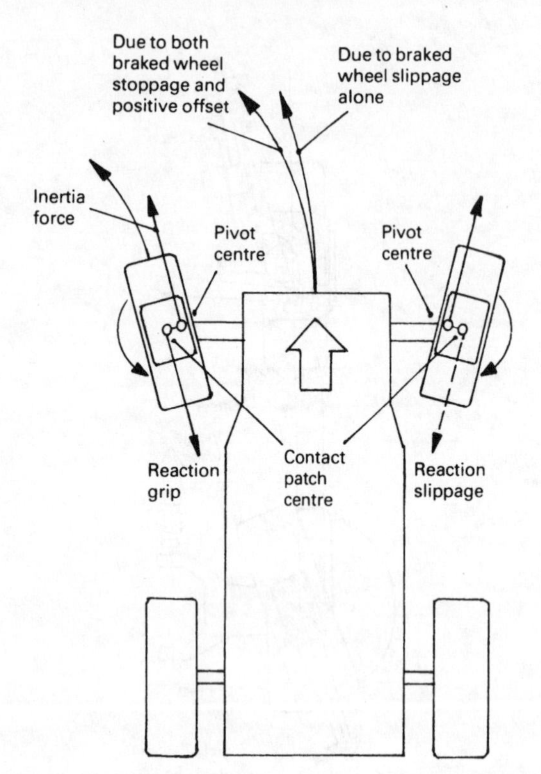

Fig. 10.13 Directional stability with positive offset when one wheel skids whilst being braked

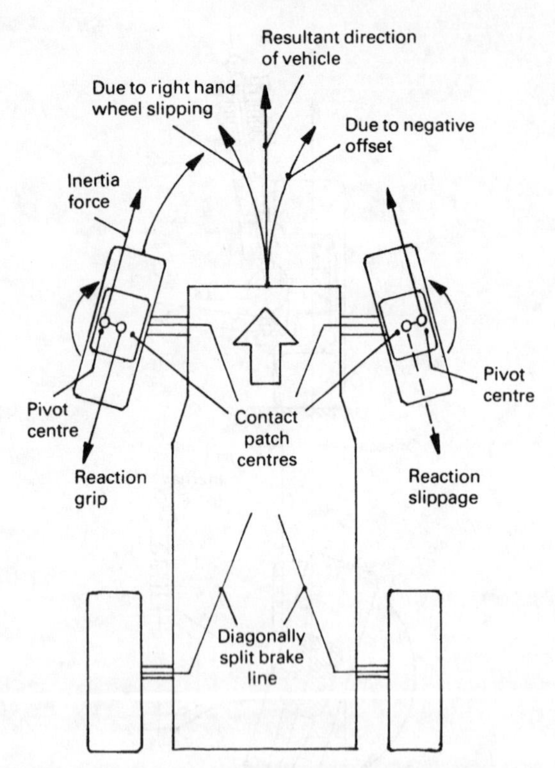

Fig. 10.14 Directional stability with negative offset when one wheel skids whilst being braked

momentum of the vehicle will produce a turning moment that makes the wheels swivel inwards at the front about the contact patch centre (Fig. 10.11) because the swivel ball joints and stub axle assembly are being pulled forwards and around the patch centre caused by the negative offset distance. The consequence of negative offset is that the effective braked wheel twists in the opposite direction to that to which the vehicle tends to veer (Fig. 10.14) and so counteracts the swerving tendency, enabling the vehicle to remain in a stable straight ahead direction.

In both positive and negative offset layouts, the skidding wheel turns in the same direction as the initial swerving tendency, but since it is not contributing greatly to the tyre to ground grip, its influence on directional stability is small.

The effect of negative offset is ideal for a split line braking system where if one brake line should fail, the front brake on the opposite side will still operate as normal (Fig. 10.14). The tendency for the car to veer to the side of the braked wheel is partially corrected by the wheel being turned due to the negative offset in the opposite direction (inwards), away from the direction in which the car wants to swerve.

When cornering, the sideways distortion of the tyre walls will misalign the wheel centre to that of the tread centre so that the swivel ball joint inclination offset will alter. The outer front wheel which supports the increase in weight due to body roll reduces positive offset (Fig. 10.15(a)), while negative offset becomes larger (Fig. 10.15(b)) and therefore makes it easier for the car to be steered when negotiating a bend in the road.

10.1.6 MacPherson strut friction and spring offset (Figs 10.16 and 10.17)

The MacPherson strut suffers from stickiness in the sliding motion of the strut, particularly under light load with an extended strut since the cylinder rod bearing and the damper piston will be closer together. Because the alignment of the strut depends upon these two sliding members, extending and reducing their distance will increase the side loading under these conditions.

The problem of reducing friction between the inner and outer sliding members is largely overcome in two ways:

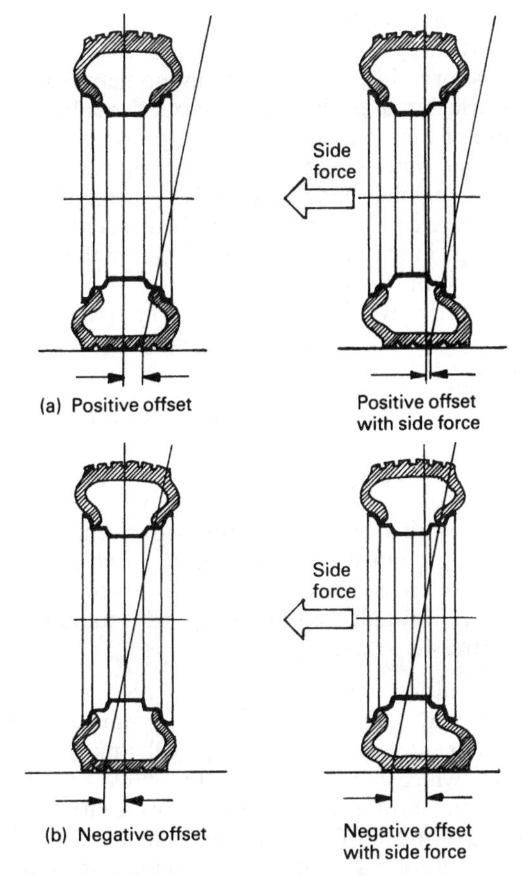

Fig. 10.15 (a and b) Swivel pin inclination offset change when cornering

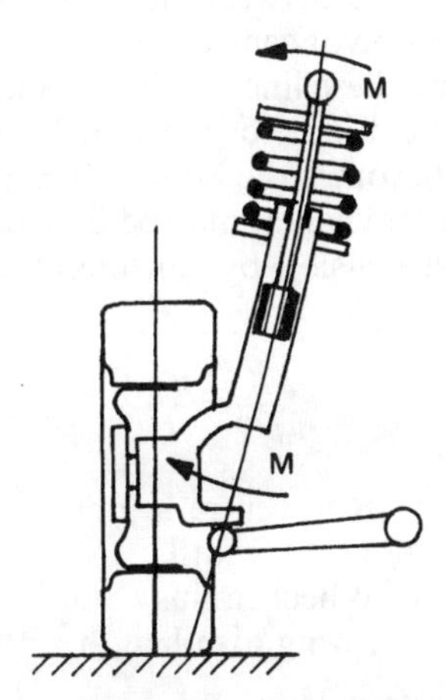

Fig. 10.16 Concentric coil spring and swivel pin axes permit bending moment reaction

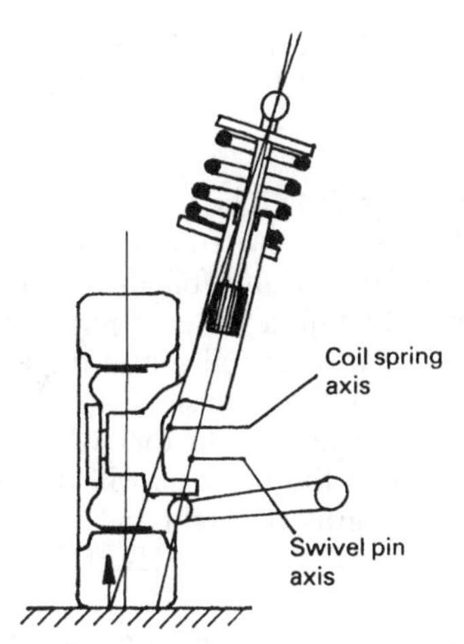

Fig. 10.17 Coil spring to swivel pin axis offset counteracts bending moment

(a) By reducing the friction, particularly with any initial movement, using a condition which is known as *stiction*. This is achieved by facing the bearing surfaces with impregnated poly-tetra-fluorethytene (PTFE) which gives the rubbing pairs an exceptionally low coefficient of friction.
(b) By eliminating the bending moment on the strut under normal straight ahead driving although there will be a bending moment under cornering conditions.

The tendency for the strut to bend arises because the wheel is offset sideways from the strut, causing the stub axle to act as a cantilever from the base of the strut to the wheel it supports, with the result the strut bends in a curve when extended or under heavy loads (Fig. 10.16).

A simple solution which is commonly applied to reduce the bending moment on the strut is to angle the axis of the coil spring relative to the swivel joint axis causing the spring to apply a bending moment in the opposite sense to the vehicle load bending moment (Fig. 10.17). Under normal conditions this coil spring axis tilt is sufficient to neutralize the bending moment caused by the inclined strut and the stub axle offset, but the forces involved while cornering produce much larger bending moments which are absorbed by the rigidity of the strut alone.

10.2 Suspension roll centres

Roll centres (Fig. 10.29) The roll centre of a suspension system refers to that centre relative to the ground about which the body will instantaneously rotate. The actual position of the roll centre varies with the geometry of the suspension and the angle of roll.

Roll axis (Fig. 10.29) The roll axis is the line joining the roll centres of the front and the rear suspension. Roll centre height for the front and rear suspension will be quite different; usually the front suspension has a lower roll centre than that at the rear, causing the roll axis to slope down towards the front of the vehicle. The factors which determine the inclination of the roll axis will depend mainly on the centre of gravity height and weight distribution between front and rear axles of the vehicle.

10.2.1 Determination of roll centre height
(Fig. 10.18)

The determination of the roll centre height can be best explained using the three instantaneous centre method applied to the swing axle suspension, which is the basic design used for the development of almost any suspension geometry (Fig. 10.18).

A vehicle's suspension system involves three principal items; the suspended body B, the supporting wheels W and the ground G which provides the reaction to the downward load of the vehicle.

If a body which is suspended between two pairs of wheels is to be capable of rolling relative to the ground, then there must be three instantaneous centres as follows:

1 I_{BG} the instantaneous centre of the body relative to the ground which is more commonly known as the body roll centre,
2 I_{WB} the instantaneous centre of the wheel relative to the body which is the swing arm point of pivot,
3 I_{WG} the instantaneous centre of the wheel relative to the ground which is the contact centre between the tyre and ground. It therefore forms a pivot permitting the top of the wheel to tilt laterally inwards or outwards.

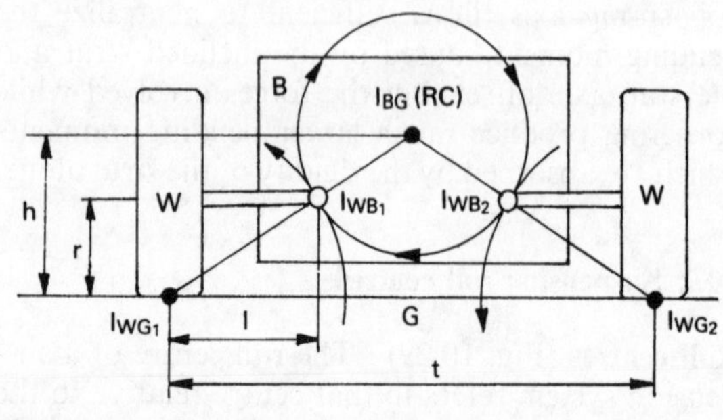

Fig. 10.18 Short swing axle

10.2.2 Short swing arm suspension
(Fig. 10.18)

When cornering, an overturning moment is generated which makes the body roll outwards from the centre of turn. The immediate response is that the inner and outer swing arm rise and dip respectively at their pivoted ends so that the inner and outer wheels are compelled to tilt on their instantaneous tyre to ground centres, I_{WG_1} and I_{WG_2}, in the opposite direction to the body roll.

For effective body roll to take place there must be two movements within the suspension geometry:

1 The swing arm pivot instantaneous centres I_{WB_1} and I_{WB_2} rotate about their instantaneous centres I_{WG_1} and I_{WG_2} in proportion to the amount of body roll.
2 The swing arm pivot instantaneous centres I_{WB_1} and I_{WB_2} move on a circular path which has a centre derived by the intersecting projection lines drawn through the tyre to ground instantaneous centres I_{WG_1} and I_{WG_2}.

The tilting, and therefore rotation, of both swing arms about the tyre to ground instantaneous centres I_{WG_1} and I_{WG_2} will thus produce an arc which is tangential to the circle on which the swing arm pivot instantaneous centres I_{WB_1} and I_{WB_2} touch. Therefore, the intersecting point I_{BG}, where the projection lines which are drawn through the wheel to ground contact points and the swing arm pivots meet, is the instantaneous centre of rotation for the body relative to the ground. This point is usually referred to as the *body roll centre*.

Thus the body roll centre may be found by drawing a straight line between the tyre contact centre and swing arm pivot centre of each half suspension and projecting these lines until they intersect somewhere near the middle of the vehicle. The point of intersection becomes the body roll centre.

The roll centre height may be derived for a short swing arm suspension by consideration of similar triangles:

$$\frac{h}{t/2} = \frac{r}{l}$$

where h = Roll centre height
t = Track width
r = Wheel radius
l = Swing arm length

Hence $h = \dfrac{tr}{2l}$

10.2.3 Long swing arm suspension (Fig. 10.19)

The long swing arm suspension is very similar to the short swing arm arrangement previously described, but the arms extend to the opposite side of the body relative to its wheel it supports and therefore both arms overlap with each other (Fig. 10.19).

The roll centre is determined by joining the tyre contact centre and the swing arm pivot centre by a straight line for each half suspension. The point where these lines meet is the body roll centre and its distance above or below the ground is known as the roll centre height. Because the long swing arm suspension has a much longer arm than used on the short swing arm layout, the slope of the lines joining the tyre contact centre and swing arm pivot is not so steep. Therefore the crossover point which determines the body roll centre height is lower for the long swing arm than for the short swing arm suspension.

The inherent disadvantage of the short swing arm suspension is that there is too much camber change with body roll and there is a tendency for the axle arms to jack the body up when cornering. Whereas the long swing arm suspension would meet most of the requirements for a good quality ride, it is impractical for a front suspension layout as it would not permit the engine to be situated relatively low between the two front wheels.

10.2.4 Transverse double wishbone suspension (Figs 10.20, 10.21 and 10.22)

If lines are drawn through the upper and lower wishbone arms and extended until they meet either inwards (Fig. 10.20) or outwards (Fig. 10.21), their intersection point becomes a virtual instantaneous centre for an imaginary (virtual) triangular swing arm suspension. The arc scribed by the wishbone arms pivoting relative to the body is almost identical to that of the imaginary or virtual arm which swings about the instantaneous virtual centres I_{BW_1} and I_{BW_2} for small movements of the suspension. Therefore, the body roll centre for a transverse double wishbone suspension can be derived similarly to a long swing arm suspension.

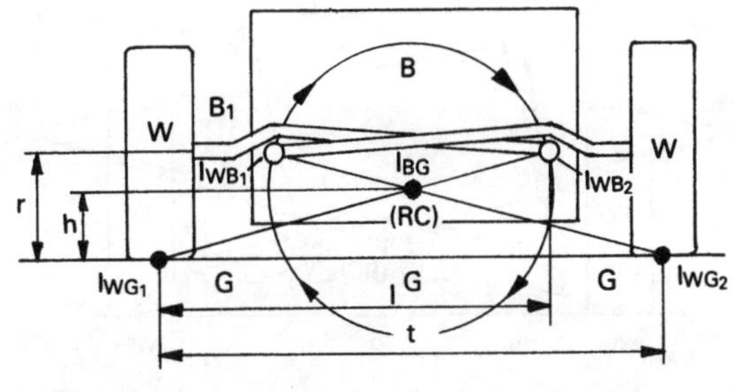

Fig. 10.19 Long swing axle

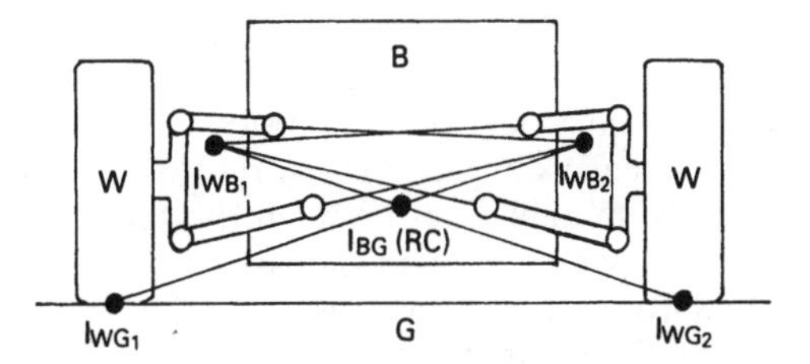

Fig. 10.20 Inward converging transverse double wishbone

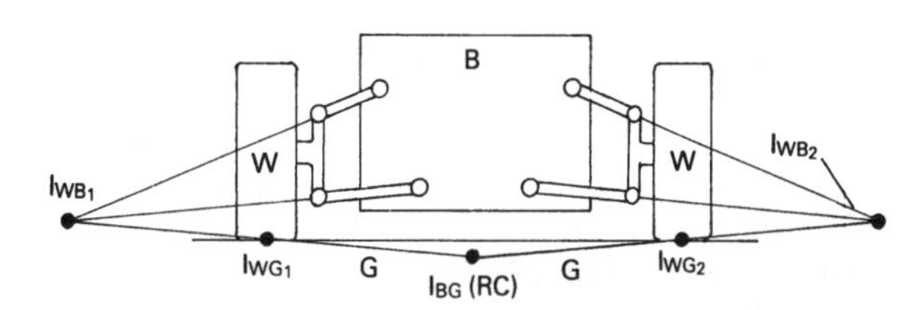

Fig. 10.21 Outward converging transverse double wishbone

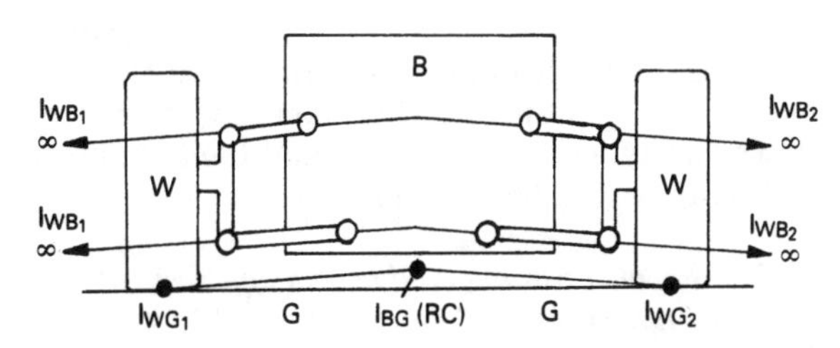

Fig. 10.22 Parallel transverse double wishbone

For inwardly converging transverse upper and lower wishbone arm suspension (Fig. 10.20) the body roll centre can be derived in two stages. Firstly, extend straight lines through the wishbone arms until they meet somewhere on the opposite side of the body at their virtual instantaneous centres I_{WB_1} and I_{WB_2}. Secondly, draw straight lines between the tyre contact centres I_{WG_1} and I_{WG_2} and the virtual centres I_{BW_1} and I_{BW_2} for each half suspension. The point where these inclined lines intersect is therefore the body roll centre I_{BG}.

For outward converging transverse upper and lower wishbone arm suspension (Fig. 10.21) the body roll centre is found again by drawing two sets of lines. Firstly project straight lines through the wishbone arms for each side of the vehicle until they meet somewhere on the outside of each wheel at their virtual instantaneous centres I_{WB_1} and I_{WB_2}. Next draw straight lines between the tyre contact centres I_{WG_1} and I_{WG_2} and the virtual centres I_{WB_1} and I_{WB_2} for each half suspension, and at the same time extend these lines until they intersect near the middle of the vehicle. This point therefore becomes the body roll centre I_{BG}. It can be seen that inclining the wishbone arms so that they either converge inward or outward produces a corresponding high and low roll centre height.

With parallel transverse upper and lower wishbone arms suspension (Fig. 10.22) lines drawn through the double wishbone arms would be parallel. They would never meet and so the virtual instantaneous centres I_{WB_1} and I_{WB_2} would tend to infinity ∞. Under these circumstances, lines normally drawn between the tyre contact centres I_{WG_1} and I_{WG_2} and the virtual instantaneous centres I_{WB_1} and I_{WB_2} would slope similarly to the wishbone extended lines. Consequently, the downwardly inclined parallel wishbone suspension predicts the tyre contact centre to virtual centre extended lines which meet at the roll centre would meet just above ground level. Therefore if the parallel wishbone arms were horizontally instead of downwardly inclined to the ground then the body roll centre would be at ground level.

10.2.5 Parallel trailing double arm and vertical pillar strut suspension (Figs 10.23 and 10.24)

In both examples of parallel double trailing arm (Fig. 10.23) and vertical pillar strut (Fig. 10.24) suspensions their construction geometry becomes similar to the parallel transverse double wishbone layout, due to both vertical stub axle members moving parallel to the body as they deflect up and down. Hence looking at the suspension from the front, neither the double trailing arms (Fig. 10.23) nor the sliding pillar (Fig. 10.24) layout has any transverse swing tendency about some imaginary pivot. Lines drawn through the two trailing arm pivot axes or sliding pillar stub axle, which represent the principle construction points for determining the virtual swing arm centres, project to infinity. The tyre contact centre to virtual instantaneous centre joining lines projected towards the middle of the vehicle will therefore meet at ground level, thus setting the body roll centre position. Inclining the trailing arm pivot axes or the vertical sliding pillar axis enables the roll centre height to be varied proportionally.

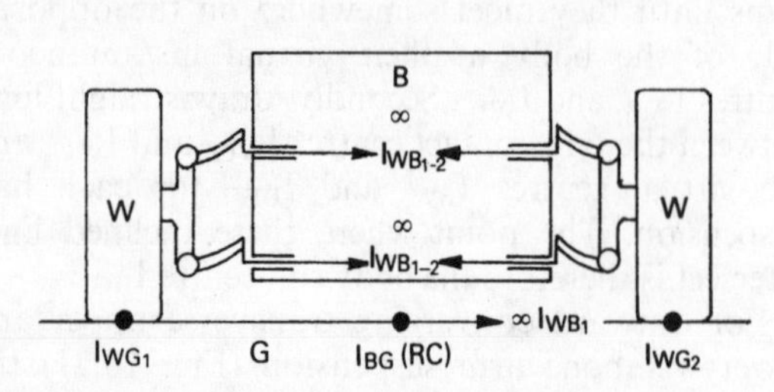

Fig. 10.23 Parallel trailing double arm

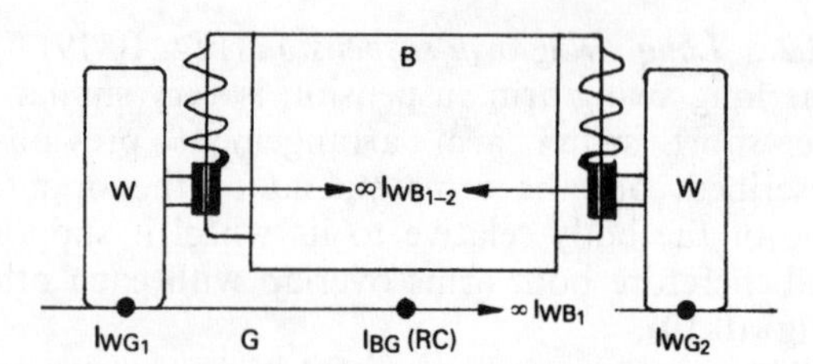

Fig. 10.24 Vertical pillar strut

10.2.6 MacPherson strut suspension (Fig. 10.25)

To establish the body roll centre height of any suspension, two of the three instantaneous centres, the tyre contact centre and the swing arm virtual centre must first be found. If straight lines are drawn between, and in some cases projected beyond, these instantaneous centres the third instantaneous centre which is the body roll centre becomes the point where both lines intersect.

The tyre contact centres (instantaneous centres I_{WG_1} and I_{WG_2}) where the wheels pivot relative to the ground are easily identified as the centres of the tyre where they touch the ground, but the second instantaneous virtual centre can only be found once the virtual or imaginary equivalent swing arm geometry has been identified.

For the MacPherson strut suspension (Fig. 10.25) the vertical swing arm and pivot centres I_{BW_1} and I_{BW_2} are obtained for each half suspension by projecting a line perpendicular to the direction

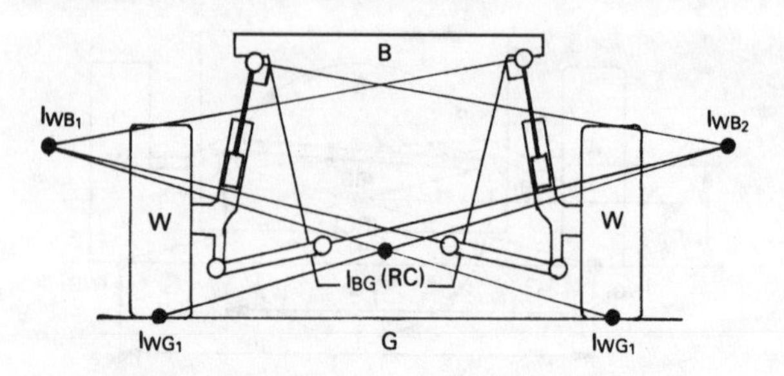

Fig. 10.25 MacPherson strut

of strut slide at the upper pivot. A second line is then drawn through and beyond the lower control arm until it intersects the first line. This point is the instantaneous virtual centre about which the virtual swing arm pivots.

Straight lines are then drawn for each half suspension between the tyre contact centre and the virtual swing arm centre. The point of intersection of these two lines will then be the third instantaneous centre I_{BG}, commonly referred to as the body roll centre.

10.2.7 Semi-trailing arm rear suspension (Fig. 10.26)

A semi-trailing arm suspension has the rear wheel hubs supported by a wishbone arm pivoted on an inclined axis across the body (Fig. 10.26(a)).

If lines are projected through the wishbone arm pivot axis and the wheel hub axis they will intersect at the virtual instantaneous centres I_{BW_1} and I_{BW_2} (Fig. 10.26(a and b)). The distance between these centres and the wheel hub is the transverse equivalent (virtual) swing arm length a. Projecting a third line perpendicular to the wheel hub axis so that it intersects the skewered wishbone arm axis produces the equivalent fore and aft (trailing) swing arm length b for the equivalent (virtual) semi-trailing triangular arm (Fig. 10.26(c)). The movement of this virtual swing arm changes the wheel camber and moves the wheel hub axis forward as the wheel deflects in bump or bounce from the horizontal position.

The body roll centre can now be determined by drawing a rear view of both virtual swing arms (Fig. 10.26(b)) and then drawing lines between each half swing arm instantaneous pivot centres I_{WB_1} and I_{WB_2} and the tyre contact centres I_{WG_1} and I_{WG_2}. The point where these two sloping lines cross over can then be defined as the body roll centre I_{BG}.

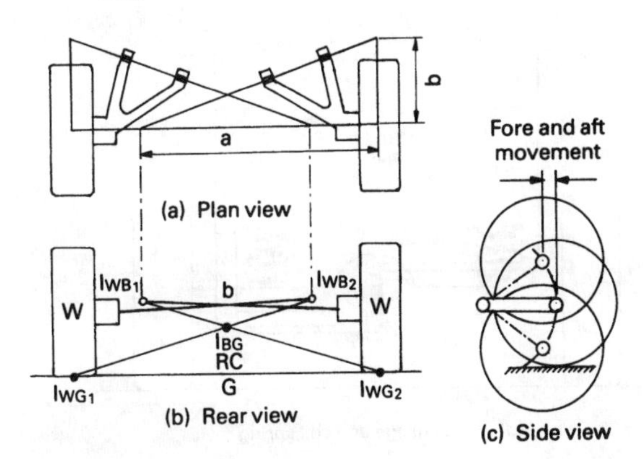

Fig. 10.26 Semi-trailing arm

10.2.8 High load beam axle leaf spring sprung body roll stability (Fig. 10.27)

The factors which influence the resistance to body roll (Fig. 10.27) are as follows:

a) The centrifugal force acting through the centre of gravity of the body load.
b) The arm length from the centre of load to the effective roll centre h_1 or h_2.
c) The spring stiffness in Newtons/metre of vertical spring deflection.
d) The distance between the centres of both springs known as the spring stability base t_s.
e) The distance between road wheel centres known as the tyre stability base t_w.

Considering the same side force acting through the centre of gravity of the body load and similar spring stiffness for both under- and over-slung springs (Fig. 10.27), two fundamental observations can be made.

Firstly it can be seen (Fig. 10.27) that with overslung springs the body roll centre RC_1 is much higher than that for underslung springs RC_2 and therefore the overslung springs provide a smaller overturning arm length h_1 as opposed to h_2 for the underslung springs. As a result, the high roll centre with the small overturning arm length offers a greater resistance to body roll than a low roll centre with a long overturning arm.

Secondly it can be seen (Fig. 10.27) that the triangular projection lines produced from the centre of gravity through the centres of the springs to

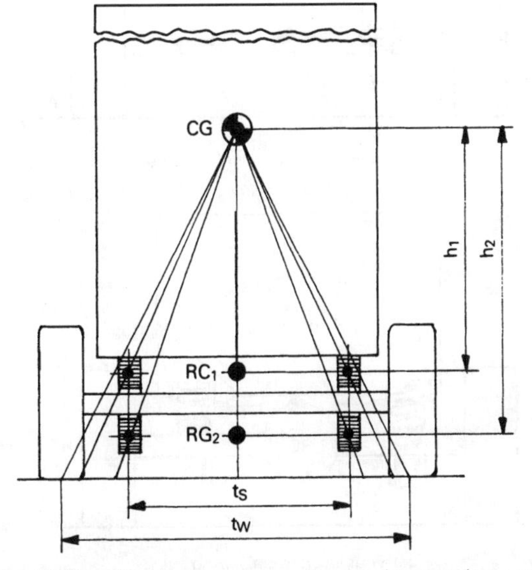

Fig. 10.27 Effects of under- and over-slung springs on the roll centre height

the ground provide a much wider spring stability base for the high mounted springs compared to the low mounted underslung springs. In fact the overslung spring centre projection lines nearly approach the tyre stability base width t_w which is the widest possible for such an arrangement without resorting to outboard spring seats.

10.2.9 Rigid axle beam suspension

(Fig. 10.28(a–d))
An axle beam suspension is so arranged that both wheel stub axles are rigidly supported by a common transverse axle beam member which may be a steered front solid axle beam, a live rear axle hollow circular sectioned casing or a DeDion tubular axle beam.

With a rigid axle beam suspension there cannot be any independent movement of the two stub axles as is the case with a split swing axle layout. Therefore any body roll relative to the ground must take place between the axle beam and the body itself. Body roll can only take place about a mechanical pivot axis or about some imaginary axis somewhere near mid-spring height level.

Methods used to locate and control the axle movement are considered as follows:

Longitudinally located semi-elliptic springs (Fig. 10.28(a)) When semi-elliptic leaf springs support the body, the pivoting point or body roll centre will be roughly at spring-eye level but this will become lower as the spring camber (leaves bow) changes from positive upward bowed leaves when unloaded to negative downward bowed leaves with increased payload.

Transverse located Panhard rod (Fig. 10.28(b)) The use of coil springs to support the body requires some form of lateral body to axle restraint if a torque tube type axle is to be utilized. This may be provided by a diagonally positioned Panhard rod attached at its ends to both the axle and body. When the body tilts it tends to move sideways and either lifts or dips depending which way the side force is applied. Simultaneously the body will roll about the mid-position of the Panhard rod.

Diagonally located tie rods (Fig. 10.28(c)) To provide both driving thrust and lateral support for

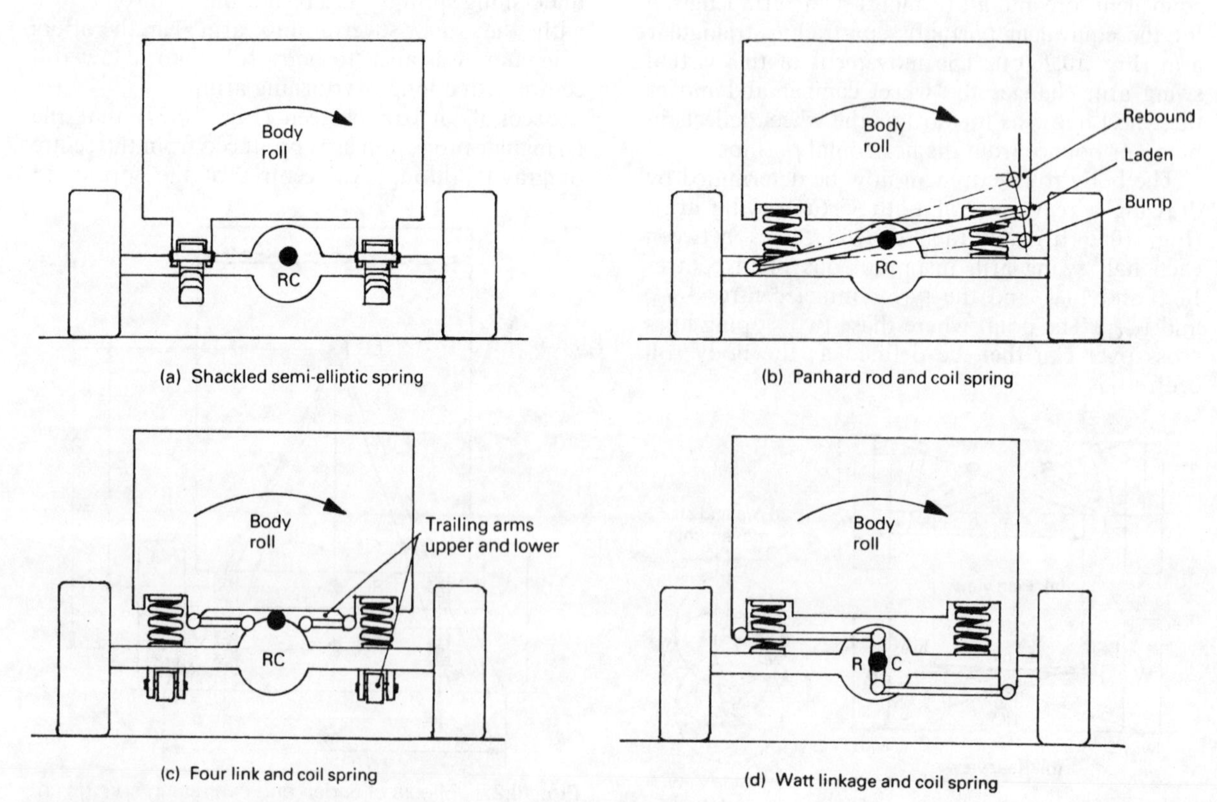

Fig. 10.28 (a–d) Body roll centres for rigid beam axle suspensions

a helical coil spring live axle layout, a trailing four link suspension may be adopted which has a pair of long lower trailing arms which absorb both the driving and braking torque reactions and a pair of short upper diagonally located tie rods to control any lateral movement. Any disturbing side forces which attempt to make the body tilt sideways will cause it to roll about a centre roughly in line with the upper tie rod height.

Transverse Watt linkage (Fig. 10.28(d)) An alternative arrangement for controlling the sideways movement for a coil spring suspension when used in conjunction with either a live axle or a DeDion tube is the Watt linkage. Suspension linkages of this type consist of a pair of horizontal tie rods which have their outer ends anchored to the body and their inner ends coupled to a central balance lever which has its pivot attachment to the axle beam. If the body is subjected to an overturning moment it will result in a body roll about the Watt linkage balance lever pivot point. This instantaneous centre is therefore the body roll centre.

10.3 Body roll stability analysis

When a vehicle turns a corner the centrifugal force produced acts outwards through the centre of gravity of the sprung mass, but it is opposed by the tyre to ground reaction so that the vehicle will tend to overturn. An overturning moment is therefore generated which tends to transfer weight from the inner wheels to the outside wheels. At the same time due to the flexibility and softness of the suspension, the body rolls so that in effect it overhangs and imposes an additional load to the outer wheels.

The opposition to any body roll will be shared out between the front and rear suspension according to their roll resistance. Thus if the front suspension roll stiffness with an anti-roll bar is twice that of the rear, then the front wheels will sustain two thirds of the roll couple while the rear ones only carry one third.

10.3.1 Body roll couple (Fig. 10.29)

The body roll couple (moment) M consists of two components:

Centrifugal moment about the roll centre = Fa (Nm)

Transverse displacement moment $= w\ a\ \tan\Theta \simeq Wa\Theta$ (Nm)

where F = centrifugal side force
a = distance between the centre of gravity and roll centre
w = unsprung weight
Θ = angle of body roll

Hence

Total roll movement or couple $M = Fa + Wa\Theta = (F + W\Theta)a$ (Nm)

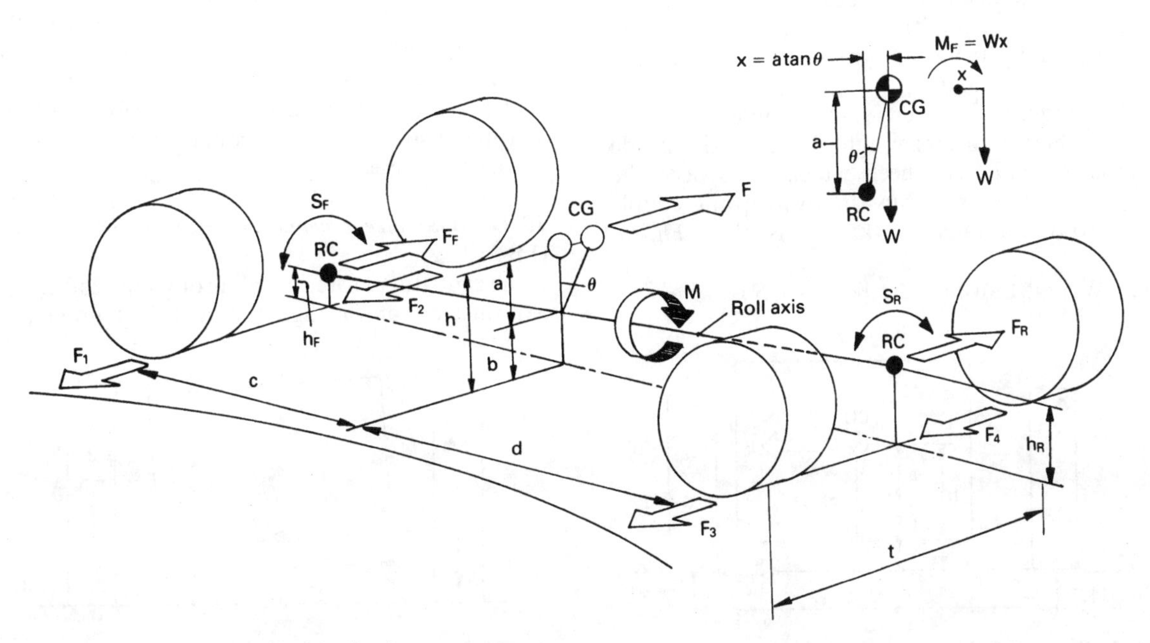

Fig. 10.29 Body roll centres and roll axis

The sum of these couples are resisted by the springs in proportion to their torsional stiffness at the front and rear.

Body roll stiffness (Fig. 10.29) The body roll stiffness is defined as the roll couple produced per degree of body roll.

i.e. $\text{Roll stiffness} = \frac{\text{Roll couple}}{\text{Roll angle}} \text{(Nm/deg)}$

hence $S = \frac{M}{\Theta} \text{(Nm/deg)}$

where S = roll stiffness (Nm/deg)
M = roll couple (Nm)
Θ = angle of roll (deg)

The fraction of torsional stiffness for the front and rear suspensions will therefore be:

$$S_F = \frac{S_F}{S_F + S_R} \left(\frac{\text{Nm/deg}}{\text{Nm/deg}}\right)$$

$$S_R = \frac{S_R}{S_R + S_F}$$

where S_F = fraction of front torsional stiffness
S_R = fraction of rear torsional stiffness

10.3.2 Body overturning couple (Fig. 10.30)

The centrifugal force F created when a vehicle is travelling on a circular track acts through the body's centre of gravity CG at some height h and is opposed by the four tyre to ground reaction forces F_1, F_2, F_3 and F_4.

Consequently an overturning couple Fh is produced which transfers weight W from the inside wheels to the outer wheels which are spaced the track width t apart. Thus the overturning couple will also be equivalent to Wt, that is, $Wt = Fh$.

i.e. $\text{Weight transferred } W = \frac{Fh}{t} \text{(N)}$

It should be noted that the centre of gravity height h is made up from two measurements; the distance between the ground and the body roll centre b and the distance between the roll centre and the centre gravity a.

Therefore
Total body roll couple $= Fh = F(a + b)$ (N)
$M = Fa + Fb$ (N)

10.3.3 Body roll weight transfer (Fig. 10.31)

The product Fa is the overturning couple rotating about the roll centre which causes the body to roll. This couple is opposed by a reaction couple Rt where R is the vertical reaction force due to the weight transfer and t is the wheel track width.

Therefore $Rt = Fa$

$$R = \frac{Fa}{t} \text{(N)}$$

This shows that as the distance between the ground and the body roll centre known as the couple arm becomes smaller, the overturning couple and therefore the body roll will also be reduced in the same proportion. Thus if the couple arm a is reduced to zero the reaction force R will likewise approach zero. A small couple is desirable so that the driver experiences a sense of body roll as a warning for cornering stability. If both roll centre and centre of gravity height coincided there would be no indication to the driver that the lateral forces acting on the body were reaching the limit of the tyre to ground sideway grip. Consequently suspensions in which the centre of gravity and the roll centre are at the same height can cause without warning a sudden tyre to ground breakaway when cornering at speed.

10.3.4 Body direct weight transfer couple (Fig. 10.32)

If the centrifugal force acted through the roll centre axis instead of through the centre of gravity, a

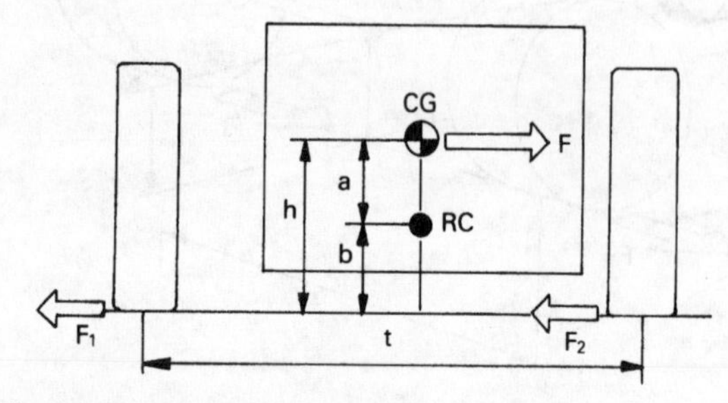

Fig. 10.30 Overturning couple

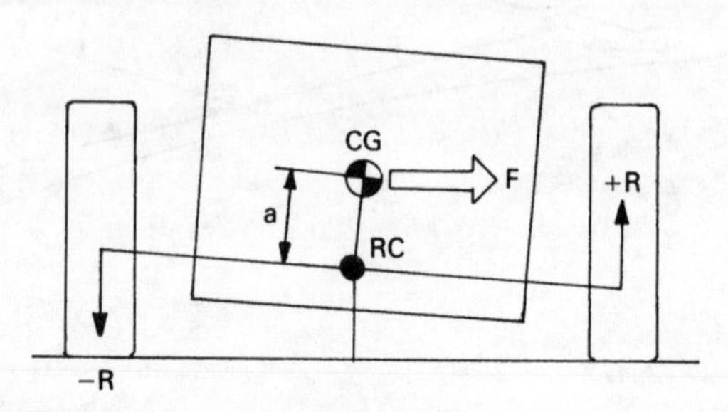

Fig. 10.31 Body roll weight transfer

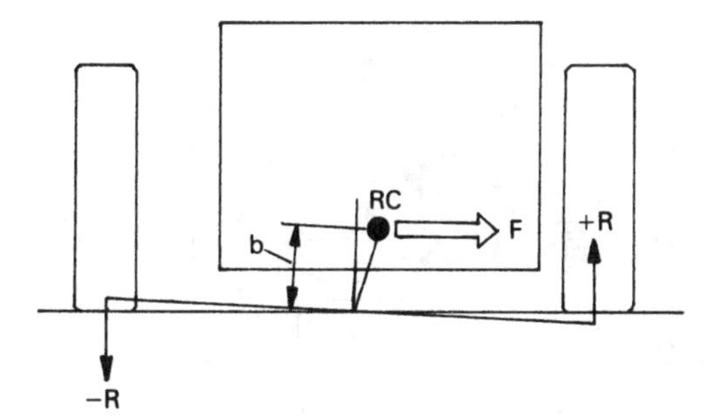

Fig. 10.32 Direct weight transfer

moment F_b about the ground would be produced so that a direct transference of weight from the inner to the outer wheels occurs. The reaction to this weight transfer for a track width t is a resisting moment R_t which is equal but opposite in sense to the moment F_b.

Hence $Rt = Fb$,

therefore $R = \frac{Fb}{t}$ (N)

If the fore and aft weight distribution is proportional between the front and rear axle roll centres, the centrifugal force F acting through the roll centre axis would be split into two forces F_F and F_R which act outwards from the front and rear roll centres.

Thus $$R_F = \frac{F_F\,b_F}{t}\text{(N)}$$

$$R_R = \frac{F_R\,b_R}{t}\text{(N)}$$

where R, R_F and R_R = Total, front and rear vertical reaction forces respectively

Thus lowering the body roll centre correspondingly reduces the vertical reaction force R and by having the roll centre at ground level the direct weight transfer couple will be eliminated.

Therefore if the roll axis slopes from the ground upwards from front to rear, all the direct weight transfer couple will be concentrated on the rear wheels.

10.3.5 Body roll couple distribution (Fig. 10.29)

The body roll couple on the front and rear tyres is proportional to the front and rear suspension stiffness fraction.

i.e. Roll couple on front tyres

$$M_F = \frac{S_F}{S_F + S_R}(F + W\Theta)a + F_F h_F\text{(Nm)}$$

Roll couple on rear tyres

$$M_R = \frac{S_R}{S_R + S_F}(F + W\Theta)a + F_R h_R\text{(Nm)}$$

Body roll angle The body roll angle may be defined as the roll couple per unit of roll stiffness

i.e. $$\text{Total roll angle} = \frac{\text{Roll couple}}{\text{Roll stiffness}}\ \frac{\text{Nm}}{\text{Nm/deg}}$$

$$= \frac{M}{S_F + S_R}\text{(deg)}$$

10.3.6 Body roll weight transfer (Fig. 10.29)

The body roll weight transferred may be defined as the roll couple per unit width of track

i.e. Total roll weight transfer

$$= \frac{\text{Roll couple}}{\text{Track width}}\left(\frac{\text{Nm}}{\text{m}}\right)$$

hence $W = \frac{M}{t}$ (N)

Front suspension weight transfer

$$W_F = \frac{S_F}{S_F + S_R} \times \frac{M}{t}\text{(N)}$$

Rear suspension weight transfer

$$W_R = \frac{S_R}{S_R + S_F} \times \frac{M}{t}\text{(N)}$$

where W, W_F and W_R = Total, front and rear weight transfer respectively (N)
t = Wheel tract (m)

10.3.7 Lateral (side) force distribution (Fig. 10.33)

The total lateral resisting forces generated at all tyre to ground interfaces must equal the centrifugal

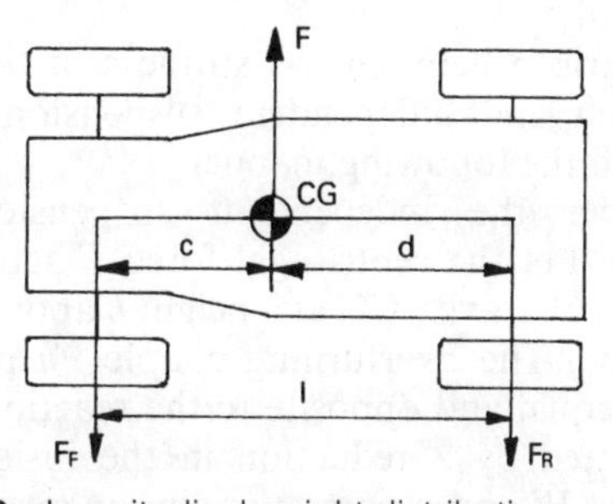

Fig. 10.33 Longitudinal weight distributions

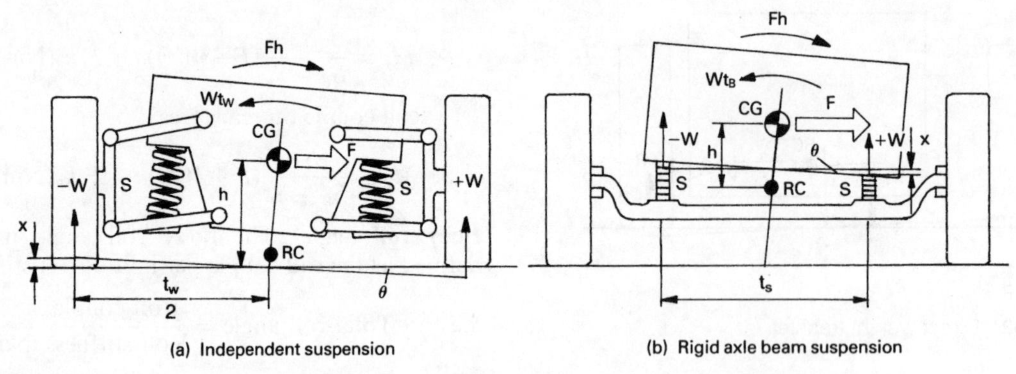

Fig. 10.34 (a and b) Comparison of rigid and independent suspension body roll stiffness

force acting through the body's centre of gravity. Thus the fore and aft position of the centre of gravity determines the weight distribution between the front and rear wheels and therefore the proportion of cornering force necessary to be generated by their respective tyres.

If F_F and F_R are the front and rear tyre to ground cornering forces, then taking moments about F_R

$$F_F l = F_b$$

Therefore $$F_F = \frac{F_b}{l} \text{(N)}$$

$$F_R l = F_a$$

Therefore $$F_R = \frac{F_a}{l} \text{(N)}$$

Thus the amount of load and cornering force carried by either the front or rear tyres is proportional to the distance the centre of gravity is from the one or the other axle. Normally there is slightly more weight concentrated at the front half of the vehicle so that greater cornering forces and slip angles are generated at the front wheels compared to the rear.

10.3.8 Comparison of rigid axle beam and independent suspension body roll stiffness

(Fig. 10.24)

A comparison between roll stiffness of both rigid axle beam and independent suspension can be derived in the following manner:

Consider the independent suspension (Fig. 10.34(a)). Let the centrifugal force F act through the centre of gravity CG at a height h above the roll centre RC. The overturning couple Fh produced must be equal and opposite to the reaction couple Wt_w created by a reduction in the inside wheel reaction $-W$ and a corresponding increase in the outside wheel reaction $+W$ between the effective spring span t_w.

If the vertical spring stiffness is S N/m and the vertical deflection at the extremes of the spring span is x m then the angle of body roll Θ degrees can be derived as follows:

$$\tan\Theta = \frac{x}{t_w/2} = \frac{2x}{t_w} \quad (1)$$

Weight transfer $$W = x\,S$$

Therefore Overturning couple $= Fh$

and Reaction couple $= Wt = Sxt$

(since $W = Sx$)

$$\therefore Fh = Sxt_w$$

$$\text{or } x = \frac{Fh}{t_w S} \quad (2)$$

From (1) $$\tan\Theta = \frac{2x}{t_w}$$

but $$x = \frac{Fh}{St_w}$$

so $$\tan\Theta = \frac{2\,Fh}{t\,St_w}$$

When Θ is small, $\tan\Theta \simeq \Theta$

$$\therefore \Theta = \frac{2Fh}{St_w{}^2} \quad (3)$$

This formula shows that the body roll angle is proportional to both centrifugal force F and the couple arm height h but it is inversely proportional to both the spring stiffness k and the square of the spring span $t_w{}^2$, which in this case is the wheel track.

i.e. $\Theta \propto F$, $\Theta \propto h$, $\Theta \propto \frac{1}{S}$ and $\Theta \propto \frac{1}{t_{w^2}}$

A similar analysis can be made for the rigid axle beam suspension (Fig. 10.34(b)), except the spring span then becomes the spring base t_s instead of t_w. Because the spring span for a rigid axle beam suspension is much smaller than for an independent suspension ($t_w^2 \gg t_s^2$), the independent wide spring span suspension offers considerably more roll resistance than the narrow spring span rigid axle beam suspension and is therefore preferred for cars.

10.4 Anti-roll bars and roll stiffness (Fig. 10.35)

10.4.1 Anti-roll bar function

A torsion anti-roll bar is incorporated into the suspension of a vehicle to enable low rate soft springs to be used which provides a more comfortable ride under normal driving conditions. The torsion bar does not contribute to the suspension spring stiffness (the suspension's resistance to vertical deflection) as its unsprung weight is increased or when the driven vehicle is subjected to dynamic shock loads caused possibly by gaps or ridges where concrete sections of the road are joined together. However, the anti-roll bar does become effective if one wheel is raised higher than the other (Fig. 10.35) as the vehicle passes over a hump in the road or the body commences to roll while cornering. Under these conditions, the suspension spring stiffness (total spring rate) increases in direct proportion to the relative difference in deflection of each pair of wheels when subjected to the bump and rebound of individual wheels or body roll when the vehicle is moving on a circular path.

10.4.2 Anti-roll bar construction (Fig. 10.36)

Generally the anti-roll bar is formed from a medium carbon steel solid circular sectioned rod which is positioned transversely and parallel to the track (Fig. 10.36) so that it nearly spans the distance between the road wheels (Fig. 10.35). The bar is bent at both ends in right angles to form cracked arms. These arms can then be actuated by short link rods attached to the unsprung portion of the suspension such as the axle beam or transverse wishbone arms for independent suspension. The main transverse span of the rod is supported by rubber bearings positioned on the inside of the cranked arms at each end. These bush bearings are either mounted directly onto the body structure when incorporated

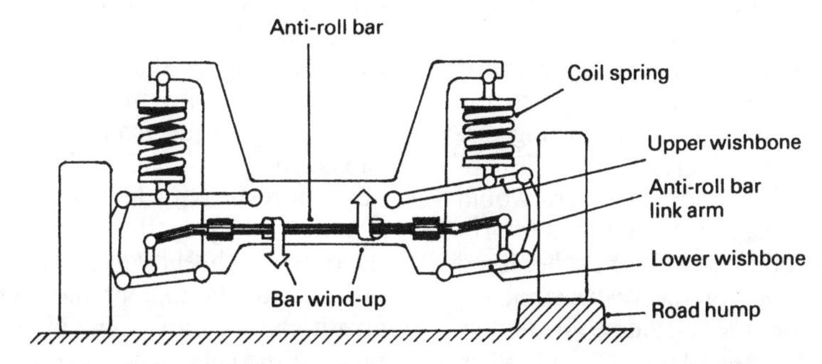

Fig. 10.35 Transverse double wishbone coil spring independent suspension with anti-roll bar

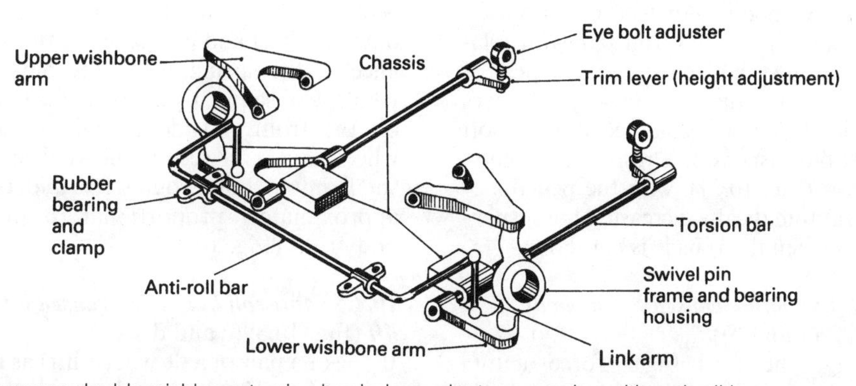

Fig. 10.36 Transverse double wishbone torsion bar independent suspension with anti-roll bar

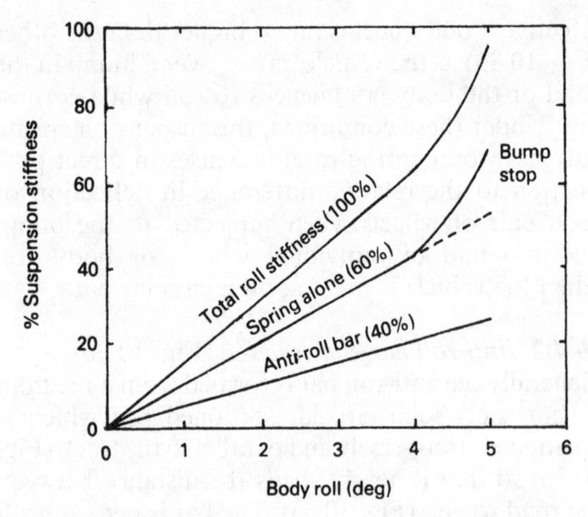

Fig. 10.37 Relationship of body roll and suspension spring and anti-roll bar stiffness

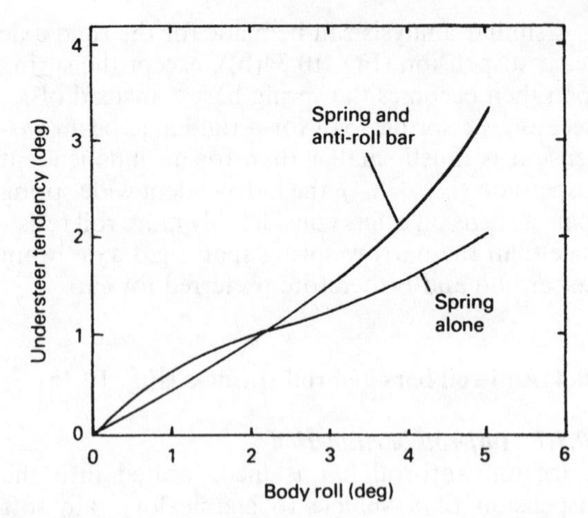

Fig. 10.38 Relationship of body roll and the understeer tendency with and without an anti-roll bar

on cars (Fig. 10.35) or indirectly for commercial vehicles (Fig. 10.39) on short vertical arms which provide a swing attachment to the chassis.

10.4.3 Anti-roll bar operating principle

When a pair of road wheels supported on an axle travel over a bumpy road one or other wheel will lift and fall as they follow the contour of the road surface. If the springs were relatively hard, that is they have a high spring rate, then the upthrust caused by the bumps would be transmitted to the body which would then lift on the side being disturbed. Thus the continuous vertical deflection of either wheel when the vehicle moves forward would tend to make the body sway from side to side producing a very uncomfortable ride. On the other hand if softer springs were used for the suspension, the small road surface irregularities would be adequately absorbed by the springs and dampers, but when cornering there would be insufficient spring stiffness to resist the overturning moment; this would therefore permit excessive body roll which could not be tolerated. Incorporating an anti-roll bar with relatively soft suspension springs mostly overcomes the difficulties discussed and therefore greatly improves the vehicle's ride. This is possible because the soft springs improve the suspension's response on good straight roadways (Fig. 10.37), with the benefits of the anti-roll bar automatically increasing the suspension roll stiffness when the vehicle is cornering.

10.4.4 Anti-roll bar action caused by the body rolling (Fig. 10.39(a and b))

When cornering, the centrifugal force acting through the centre of gravity of the sprung body produces an overturning moment created by its offset to the body's roll centre which will therefore tend to make the body roll (Fig. 10.39(a and b)). The rolling body will tilt the transverse span of the roll bar with it so that the cranked arms on the outside wheel to the turn will be depressed downward, whereas the cranked arm on the opposite end near the inside wheel to the turn will tend to rise. The consequence of this misalignment of the anti-roll bar arms is that the two cranked arms will rotate in opposite directions to each other and so transmit a torque from the inside wheel which is subjected to less load to the outside wheel which is now more heavily loaded. The effect of the torsional wind-up in the bar is that it tries to rotate the outside wheel cranked arm and since the arm is attached to the axle or indirectly to the wishbone arm it cannot move. The alternative is for the roll bar and the rubber bearing mount near the outside wheel to lift in proportion to the degree of twisting torque. It therefore counteracts some of the downward push due to the increase in weight to the outside wheel and as a result stiffens the roll resistance of the springing on the outside wheel as a whole. Consequently a larger slip angle is generated on the front outside wheel relative to the rear wheel, and as a result, the vehicle will develop a small initial but progressive understeer tendency approximately proportional to the amount the body rolls (Fig. 10.38).

10.4.5 Anti-roll bar action caused by single wheel lift (Fig. 10.39(c and d))

If one of a pair of axle wheels lifts as it climbs over a bump (Fig. 10.39(c)) in the road, then the vertical

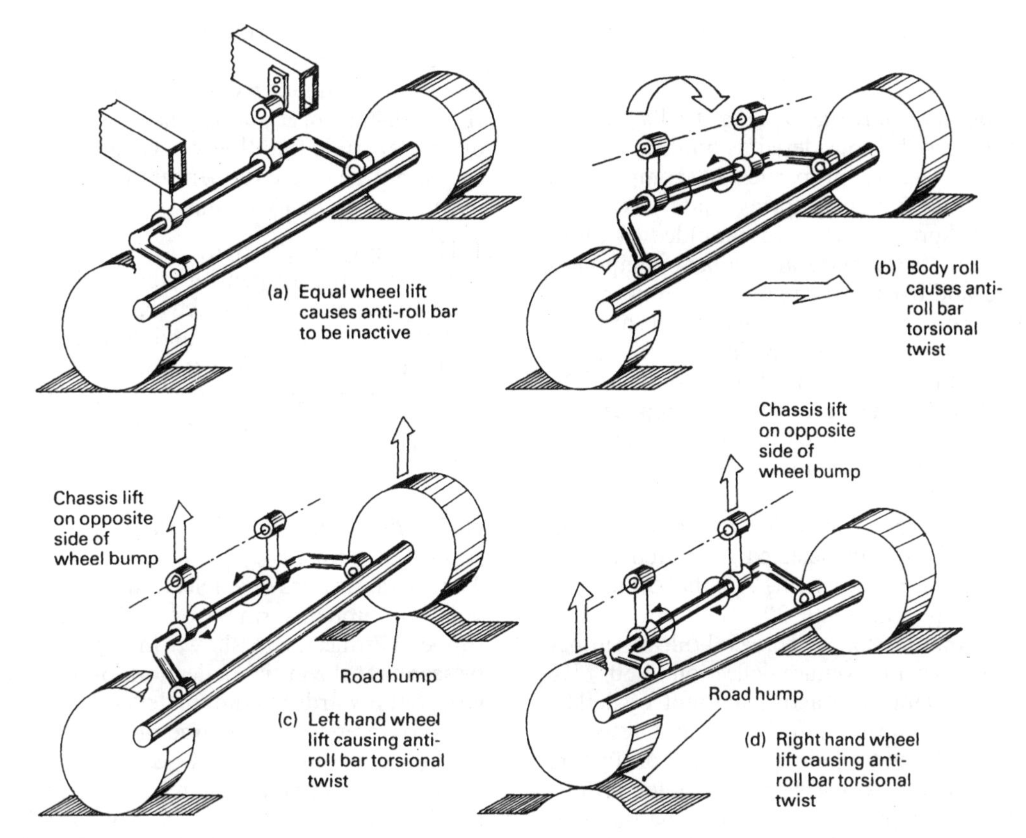

Fig. 10.39 (a–d) Anti-roll bar action

deflection of the wheel and spring raises and rotates the anti-roll bar's cranked arm on that side so that the transverse span of the bar is twisted. The bar is therefore subjected to a torque which is proportional to its angle of rotation.

This twisting torque is transferred to the opposite cranked arm which then applies a downward force onto the axle and wheel. However, because the wheel cannot sink into the ground, the reaction occurs on the rubber bearing mount arm which therefore tends to lift up the side of the chassis on the opposite side to the vertically deflected wheel. As a result, both sides of the chassis (body) will have been raised, thereby enabling the vehicle's body to remain upright instead of tilting to one side. Similarly, if the opposite wheel hits an obstacle in the road (Fig. 10.39(d)), the torsional wind-up of the bar transfers an upward thrust to the other side, which again tends to lift the chassis on the undisturbed wheel side and so maintains the sprung chassis and body on an even keel (Fig. 10.39(c)).

10.5 Rubber spring bump or limiting stops

10.5.1 Bump stop function

(Figs 10.40 and 10.42)

Suspension bump and body roll control depends upon the stiffness of both the springs and anti-roll bar over the normal operating conditions, but if the suspension deflection approaches maximum bump or roll the bump stop (Fig. 10.40(a, b, c and d)) becomes active and either suddenly or progressively provides additional resistance to the full deflection of the wheel and axle relative to the body (Fig. 10.42). The bump stop considerably stiffens the resisting spring rate near the limit of its vertical movement to prevent shock impact and damage to the suspension components. The stop also isolates the sprung and unsprung members of the suspension under full deflection conditions so that none of the noise or vibrations are transmitted through to the body structure. In essence the bump stop enables an anti-roll bar to be used which has a slightly lower spring rate, therefore permitting a more cushioned ride for a moderate degree of body roll.

10.5.2 Bump stop construction (Fig. 10.40(a–d))

Bump stops may be considered as limiting springs as they have elastic properties in compression similar to other kinds of spring materials. Solid and hollow spring stops are moulded without reinforcement from natural rubber compound containing additives to increase the ozone resistance. The deflection characteristics for a given size of rubber stop spring are influenced by the hardness of the rubber, this being controlled to a large extent by the proportion of sulphur and carbon black which is mixed into the rubber compound. The most common rubber compound hardness used for a rubber spring stop is quoted as a shore hardness of 65; other hardness ranging from 45 to 75 may be selected to match a particular operating requirement. A solid cylindrical rubber block permits only 20% deflection when loaded in compression, whereas hollow rubber spring stops have a maximum deflection of 50–75% of their free height. The actual amount of deflection for a given spring stop height and response to load will depend upon a number of factors such as the rubber spring stop size, outer profile, wall thickness, shape of inner cavities, hardness of rubber compound and the number of convolution folds.

Bump rubber spring stops may be solid and conical in shape or they may be hollow and cylindrical or rectangular shaped with a bellow profile (Fig. 10.40(a, b, c and d)) having either a single, double or triple fold (known as convolutions). The actual profile of the rubber bump stop selected will depend upon the following:

1 How early in the deflection or load operating range of the suspension the rubber begins to compress and become active.
2 Over what movement and weight change the bump stop is expected to contribute to the sudden or progressive stiffening of the suspension so that it responds to any excessive payload, impact load and body roll.

10.5.3 Bump stop characteristics
(Figs 10.41 and 10.42)

The characteristics of single, double and triple convolution rubber spring stops, all using a similar rubber hardness, are shown in Fig. 10.41. It can be seen that the initial deflection for a given load is large but towards maximum deflection there is very little compression for a large increase in load. The relation between load and deflection for bump is not quite the same on the release rebound so that the two curves form what is known as a hysteresis loop. The area of this loop is a measure of the energy absorbed and the internal damping within

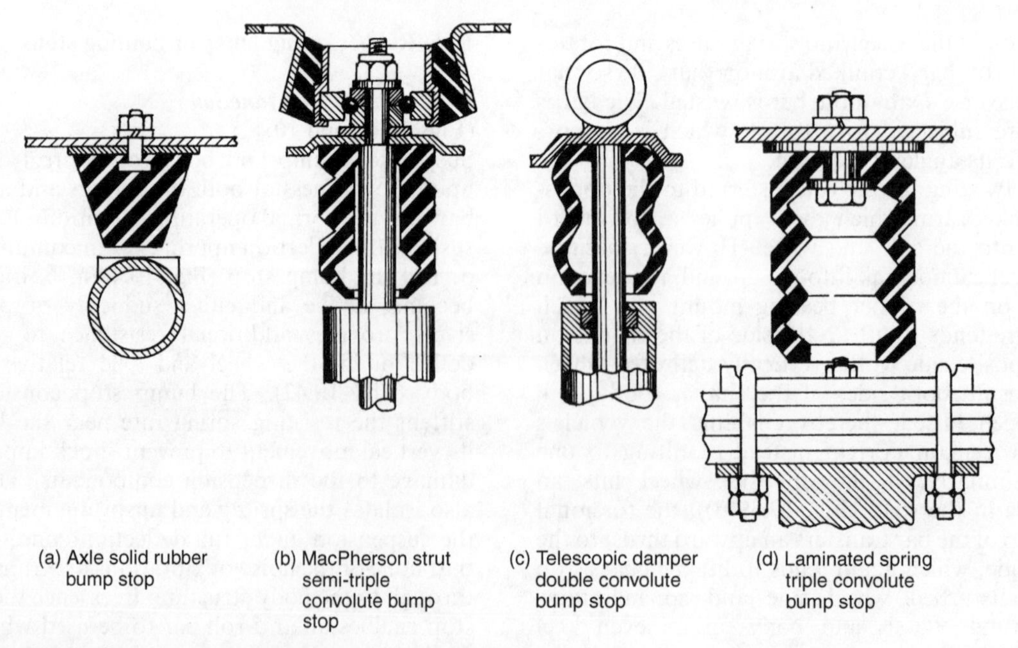

Fig. 10.40 (a–d) Suspension bump stop limiter arrangements

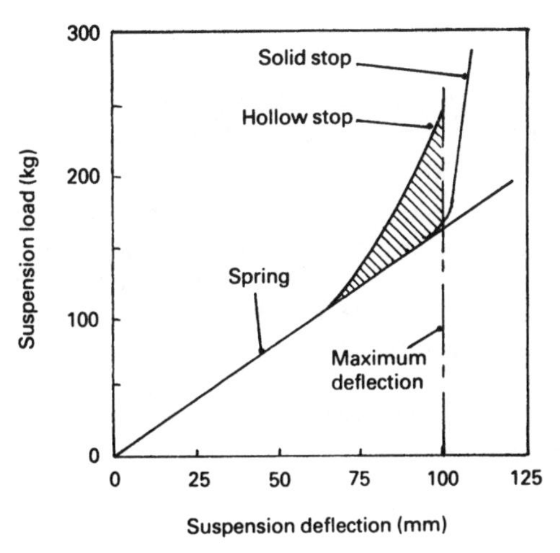

Fig. 10.41 Characteristics of hollow rubber single, double and triple convolute progressive bump stops

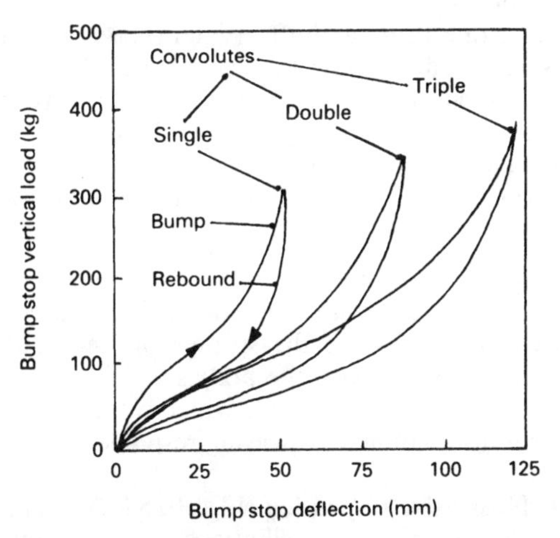

Fig. 10.42 Combined characteristics of suspension spring and rubber bump stops

the rubber in one cycle of compression and expansion of the rubber spring stop. For hollow rubber spring stops they always end in a point; this means for any load change there will be some spring deflection.

Fig. 10.42 shows how the bump spring stop deviates from the main spring load-deflection curve at about two-thirds maximum deflection and that the resultant stiffness (steepness of curve) of the steel spring, be it leaf, coil or torsion bar, and that of the bump spring stops considerably increases towards full load.

10.6 Axle location

10.6.1 Torque arms (Figs 10.28(c) and 10.44)
Torque arms, sometimes known as radius arms or rods, are mounted longitudinally on a vehicle between the chassis/body structure and axle or unsprung suspension member. Its purpose is to permit the axle to move up and down relative to the sprung chassis/body and to maintain axle alignment as the torque arm pivots about its pin, ball or conical rubber joint. Sometimes the upper torque rods are inclined diagonally to the vehicle's lengthwise axis to provide lateral axle stability (Figs 10.28(c) and 10.44). These arms form the link between the unsprung suspension members and the sprung chassis/body frame and are therefore able to transmit both driving and braking forces and to absorb the resulting torque reactions.

10.6.2 Panhard rod (Fig. 10.28(b))
Panhard rods, also known as transverse control rods or arms, are positioned across and between both rear wheels approximately parallel to the axle (Fig. 10.28(b)). One end of the rod is anchored to one side of the axle span while the other end is anchored to the body structure; both attachments use either pin or ball type rubber joints. A Panhard rod restrains the body from moving sideways as the vehicle is subjected to lateral forces caused by sidewinds, inclined roads and centrifugal forces when cornering. When the body is lowered, raised or tilted relative to the axle, the Panhard rod is able to maintain an approximate transverse axle alignment (Fig. 10.28(b)) relative to the chassis/body thus relieving the suspension springs from side loads.

10.6.3 Transverse located Watt linkage
(Fig. 10.43(a, b and c))
A Watt linkage (Fig. 10.43) was the original mechanism adopted by James Watt to drive his beam steam engine. This linkage is comprised of two link rods pivoting on the body structure at their outer ends and joined together at their inner ends by a coupler or equalizing arm which is pivoted at its centre to the middle of the rear axle. When in mid-position the link rods are parallel whereas the equalizing arm is perpendicular to both (Fig. 10.43(b)).

If vertical movement of the body occurs either towards bump (Fig. 10.43(c)) or rebound (Fig. 10.43(a)) the end of the link rods will deviate an equal amount away from the central pivot point of the coupling arm. Thus the left hand upper link rod

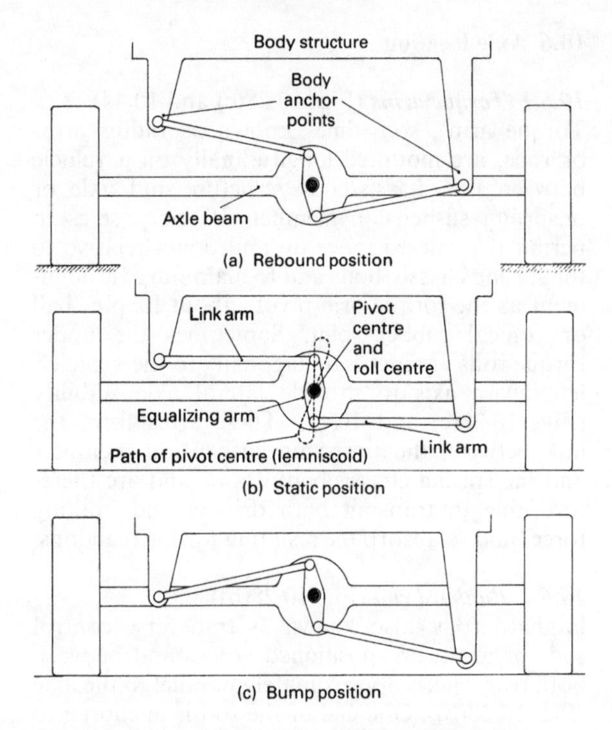

Fig. 10.43 (a–c) Transversely located Watt linkage

will tend to pull towards the left and the right hand lower link rod will apply an equal pull towards the right. The net result will be to force the equalizing arm to rotate anticlockwise to accommodate the inclination to the horizontal of both link rods. If the left hand link rod were made the lower link and the right hand rod the upper link, then the direction of tilt for the equalizing arm would now become clockwise.

For moderate changes in the inclination of the link rods, the body will move in a vertical straight line, thus maintaining a relatively accurate body to axle lateral alignment. Excessive up and down movement of the body will cause the pivot centre to describe a curve resembling a rough figure eight, a configuration of this description being known as a *lemniscoid* (Fig. 10.43(b)).

Under body roll conditions when cornering, the whole body relative to the axle and wheels will be restrained to rotate about the equalizing arm pivot centre at mid-axle height; this point therefore becomes the roll centre for the rear end of the body.

A similar Watt linkage arrangement can be employed longitudinally on either side of the wheels to locate the axle in the fore and aft direction.

10.7 Rear suspension arrangements

10.7.1 Live rigid axle rear suspension

Suspension geometry characteristics of a live axle are as follows:

1 Wheel camber is zero irrespective if the vehicle is stationary or moving round a bend in the road.
2 If one wheel moves over a hump or dip in the road then the axle will tilt causing both wheels to become cambered.
3 Because both wheels are rigidly joined together the wheel track remains constant under all driving conditions.
4 Because the axle casing, half shafts and final drive are directly supported by the wheels, the unsprung weight of a live axle is very high.
5 With a live rigid axle, which is attached to the body by either leaf or coil springs, the body will tilt about some imaginary roll centre roughly mid-way between the upper and lower spring anchorage points.
6 Horizontal fore and aft or lateral body location is achieved by using the leaf springs themselves as restraining members or, in the case of coil springs which can only support the vehicle's vertical load and therefore cannot cope with driving thrust and side loads, horizontally positioned control rods.

Without accurate control of horizontal body movement relative to the axle casing caused by vertical deflection of the springs or longitudinal and transverse forces, the body's weight distribution would be unpredictable which would result in poor road holding and steering response.

Hotchkiss drive suspension (Fig. 10.84(a)) This is the conventional semi-elliptic spring suspension which has each spring positioned longitudinally on each side of the axle and anchored at the front end directly to a spring hanger attached to the body structure and at the rear end indirectly via swing shackle plates to the rear spring hangers, the axle being clamped to the springs somewhere near their mid-span position. Thus fore and aft driving and braking forces are transmitted through the front half of the springs and lateral forces are accommodated by the rigidity of the spring leaves and spring anchorage.

Four link coil spring live axle rear suspension (Fig. 10.44) Substituting coil springs for semi-elliptic springs requires a separate means of locating and maintaining body and axle alignment when

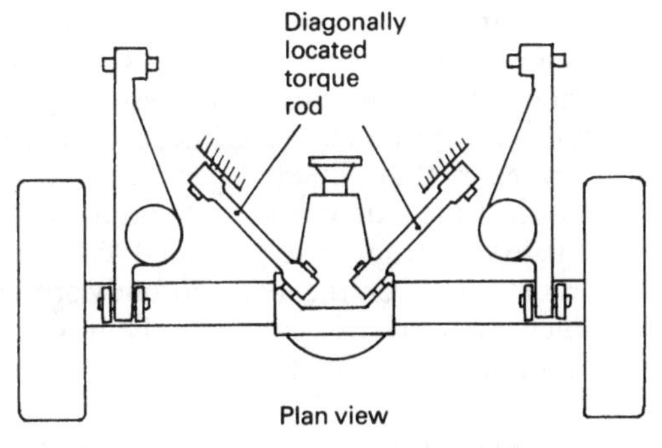

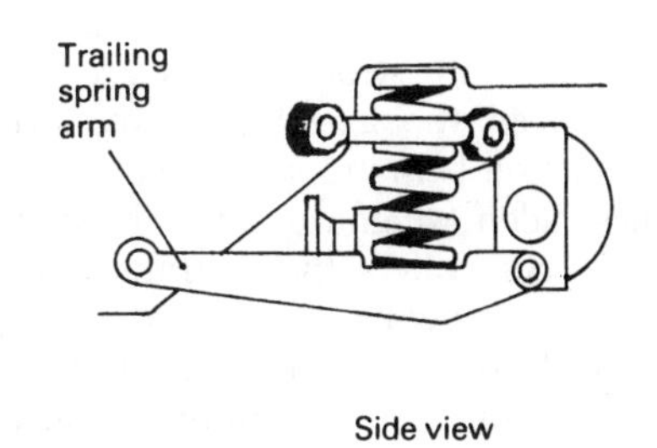

Fig. 10.44 Four link coil spring live axle rear suspension

subjected to longitudinal and transverse forces caused by spring deflection, body roll or driving and braking thrust loads.

The locating links are comprised of a pair of long trailing lower arms and a pair of short diagonally positioned upper torque arms (Fig. 10.44). Rubber pin joints secure the forward ends of the arms to the body structure but the lower rear ends are attached underneath the axle tubes as far apart as possible and the upper short torque arms attached much closer together onto the final drive housing. The coil springs are mounted between the upper body structure and the lower pressed steel trailing arms. These springs only provide vertical support and cannot restrain any horizontal movement on their own. Spring deflection due to a change in laden weight causes both sets of arms to swivel together, thereby preventing the axle assembly rotating and possibly making the universal joints operate with very large angles. Both driving and braking thrust are transmitted through the lower trailing arms which usually are of a length equal to roughly half the wheel track so that when the arms swing the change in wheelbase is small. The upper arms are normally inclined at 45° to the car's centre line axis so that they can absorb any axle reaction torque tending to rotate the axle, and at the same time prevent relative lateral movement between the body and axle. Body roll or axle tilt are permitted due to the compliance of the rubber pin joints.

A relatively high roll centre is obtained which will be roughly at the upper torque arm height.

Torque tube rear wheel drive suspension (Fig. 10.45)
One of the major problems with the Hotchkiss drive layout is that the axle torque reaction tends to spin the axle casing when transmitting drive torque in the opposite direction to the rotating wheels and when braking to twist the axle casing in the same direction as the revolving wheels. The result is a considerably distorted semi-elliptic spring and body to axle misalignment. To overcome this difficulty, a rigid tube may be bolted to the front of the final drive pinion housing which extends to the universal joint at the rear of the gearbox or a much shorter tube can be used which is supported at its front end by a rubber pin or ball joint attached to a reinforcing cross-member

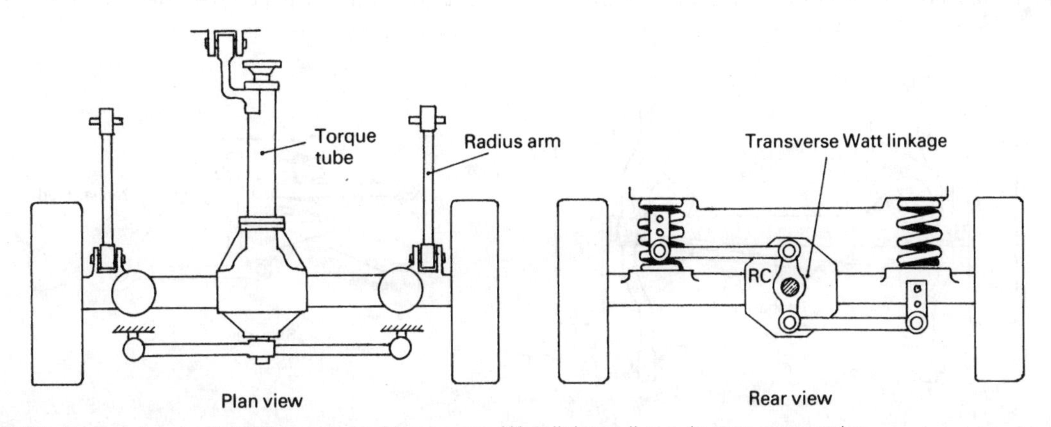

Fig. 10.45 Torque tube with trailing arm and transverse Watt linkage live axle rear suspension

(Fig. 10.45). On either side of the torque tube is a trailing arm which locates the axle and also transmits the driving and braking thrust between the wheels and body. Coil springs are mounted vertically between the axle and body structure, their only function being to give elastic support to the vehicle's laden weight. Lateral body to axle alignment is controlled by a transverse Watt linkage. The linkage consists of an equalizing arm pivoting centrally on the axle casing with upper and lower horizontal link arms anchored at their outer ends by rubber pin joints to the body structure. Thus when the springs deflect or the body rolls, the link arms will swing about their outer body location centres causing the equalizing arm to tilt and so restrain any relative lateral body to axle movement without hindering body vertical displacement.

With the transversely located Watt linkage, the body roll centre will be in the same position as the equalizing arm pivot centre. The inherent disadvantages of this layout are still the high amount of unsprung weight and the additional linkage required for axle location.

10.7.2 Non-drive rear suspension

The non-drive (dead) rear axle does not have the drawback of a large unsprung weight and it has the merit of maintaining both wheels parallel at all times. There is still the unwanted interconnection between the wheels so that when one wheel is raised off the ground the axle tilts and both wheels become cambered.

The basic function of a rear non-drive rear suspension linkage is to provide a vertical up and down motion of the axle relative to the body as the springs deflect and at the same time prevent longitudinal and lateral axle misalignment due to braking thrust, crosswinds or centrifugal side force.

Five link coil spring leading and trailing arm Watt linkage and Panhard rod non-drive axle rear suspension (Fig. 10.46) One successful rigid axle beam and coil spring rear suspension linkage has incorporated a Watt linkage parallel to each wheel to control the axle in the fore and aft direction (Fig. 10.46). A transversely located Panhard rod connected between the axle and body structure is also included to restrict lateral body movement when it is subjected to side thrust.

Trailing arms with central longitudinal wishbone and anti-roll tube non-drive axle rear suspension (Fig. 10.47) A rectangular hollow sectioned axle beam spans the two wheels and on either side are mounted a pair of coil springs. A left and right hand trailing arm links the axle beam to the body structure via rubber bushed pivot pins located at both ends of the arms at axle level (Fig. 10.47). To locate the axle beam laterally and to prevent it rotating when braking, an upper longitudinal wishbone arm ('A' arm) is mounted centrally between the axle and body structure. The 'A' arm maintains the axle beam spring mounting upright as the spring deflects in either bump or rebound, thus preventing the helical coil springs bowing. It also keeps the axle beam aligned laterally when the body is subjected to any side forces caused by sloping roads, crosswinds and centrifugal force.

Situated just forward of the axle beam is a transverse anti-roll tube welded to the inside of each trailing arm. When body roll occurs while the car is cornering, the inner and outer trailing arms will tend to lift and dip respectively. This results in both trailing arms twisting along their length. Therefore the anti-roll tube, which is at right angles to the arms, will be subjected to a torque which will be resisted by the tube's torsional stiffness. This torsional resistance thus contributes to the coil spring

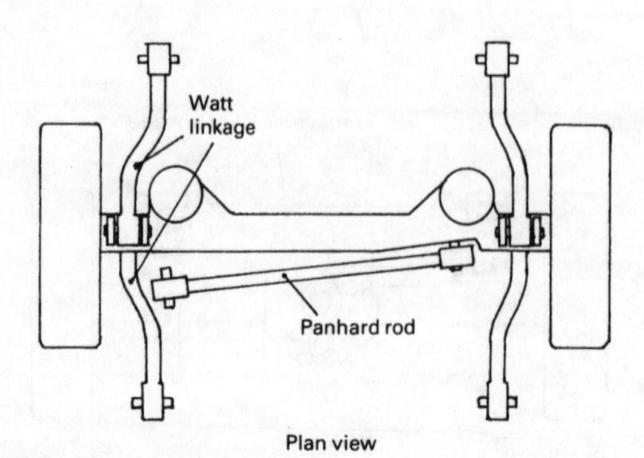

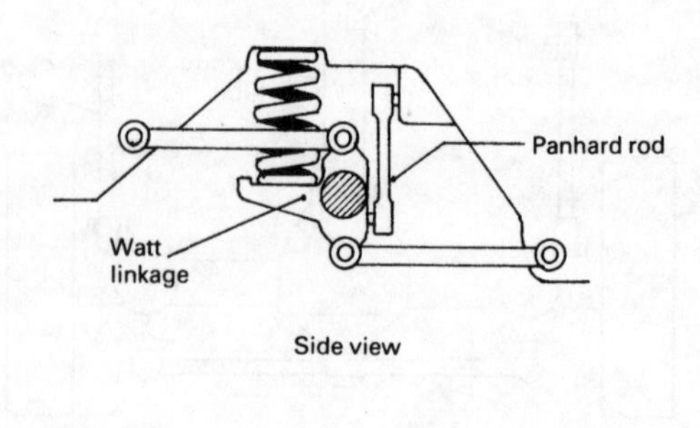

Fig. 10.46 Five link coil spring leading and trailing arm Watt linkage and Panhard rod dead axle rear suspension

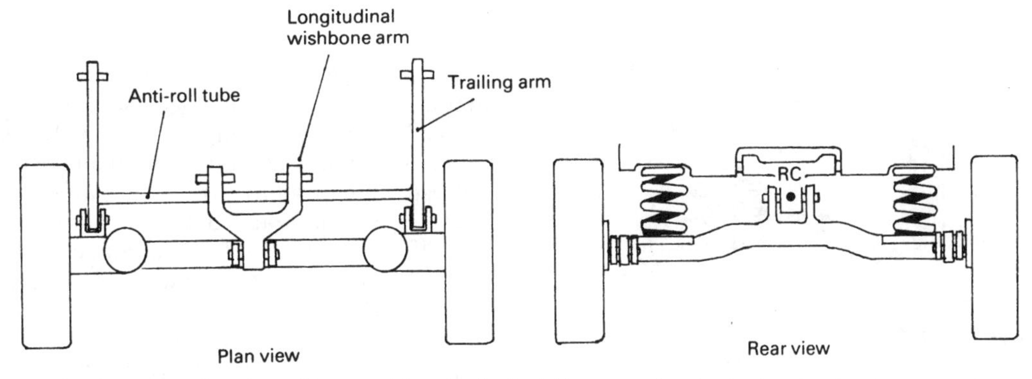

Fig. 10.47 Trailing arm coil spring with central longitudinal wishbone and anti-roll tube dead axle suspension

roll stiffness and increases in proportion to the angle of roll. With this type of suspension the unsprung weight is minimized and the wheels remain perpendicular to the ground under both laden weight and body roll changes.

Trailing arm and torsion bar spring with non-drive axle rear suspension (Fig. 10.48) The coil springs normally intrude into the space which would be available for passengers or luggage, therefore torsion bar springs transversely installed in line with the pivots of the two trailing arms provide a much more compact form of suspension springing (Fig. 10.48). During roll of the body, and also when the wheels on each side are deflected unequally, the axle beam is designed to be loaded torsionally, to increase the torsional flexibility and to reduce the stress in the material. The axle tube which forms the beam is split underneath along its full length. This acts as an anti-roll bar or stabilizer when the springs are unevenly deflected. The pivot for each trailing arm is comprised of a pair of rubber bushes pressed into each end of a transverse tube which forms a cross-member between the two longitudinal members of the floor structure of the body. The inner surface of the rubber bush is bonded to a hexagonal steel sleeve which is mounted on a boss welded to the outside of the trailing arm. In the centre of the trailing arm boss is a hexagonal hole which receives the similar shaped end of the torsion bar. To prevent relative movement between the male and female joint made between the boss and torsion bar, a bolt locked by a nut in a tapped radial hole in the boss presses against one of the flats on the torsion bar.

One torsion bar spring serves both suspension arms so that a hexagon is forged mid-way between the ends of the bar. It registers in a hexagonal hole formed in the steel collar inserted in and spot

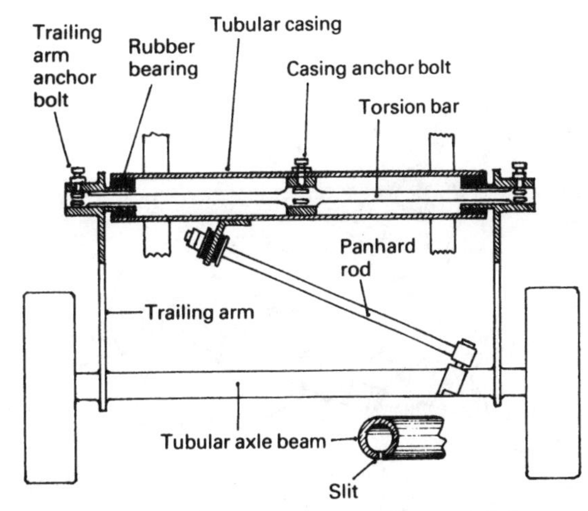

Fig. 10.48 Trailing arm and torsion bar spring with dead axle rear suspension

welded to the transverse tube that houses the torsion bar spring. Again the torsion bar and collar are secured by a radial bolt locked by a nut.

In the static laden position a typical total angular deflection of the spring would be 20° and at full bump about 35°. To give lateral support for the very flexible trailing arms a Panhard rod is diagonally positioned between the trailing arms so that it is anchored at one end to the axle beam and at the other end to the torsion bar tubular casing. All braking torque reaction is absorbed by both trailing arms.

Trailing arm and coil spring twist axle beam non-drive axle rear suspension (Fig. 10.49(a, b and c)) The pivoting trailing arms are joined together at their free ends by an axle beam comprised of a tubular torsion bar enclosed by a inverted 'U' channel steel section, the ends of the beam being

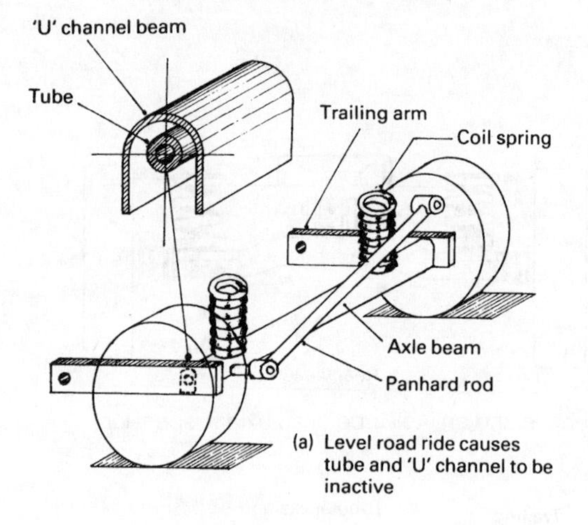

(a) Level road ride causes tube and 'U' channel to be inactive

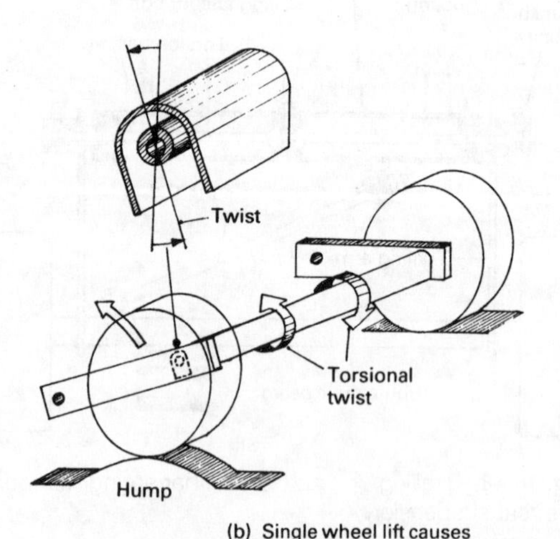

(b) Single wheel lift causes tube and 'U' channel to twist

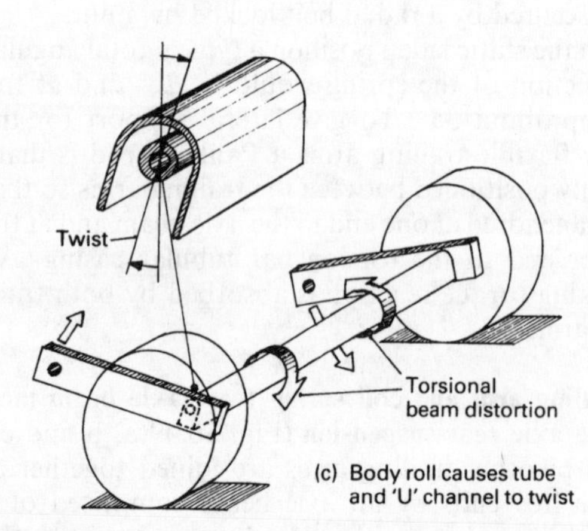

(c) Body roll causes tube and 'U' channel to twist

Fig. 10.49 (a–c) Trailing arm twist axle beam rear suspension

butt welded to the insides of the both trailing arms (Fig. 10.49(a, b and c)).

When both wheels are deflected an equal amount, caused by increased laden weight only, the coil springs are compressed (Fig. 10.49(a)). If one wheel should be raised more than the other, its corresponding trailing arm rotates about its pivot causing the axle beam to distort to accommodate the difference in angular rotation of both arms (Fig. 10.49(b)). Consequently the twisted axle beam tube and outer case section will transfer the torsional load from the deflected trailing arm to the opposite arm. This will also cause the undeflected arm to rotate to some degree, with the result that the total body sway is reduced.

During cornering when the body rolls, the side of the body nearest the turn will lift and the opposite side will dip nearer to the ground (Fig. 10.49(c)). Thus the inner trailing arm will be compelled to rotate clockwise, whereas the outer trailing arm rotates in the opposite direction anticlockwise. As a result of this torsional wind-up of the axle beam, the outer wheel and trailing arm will tend to prevent the inner trailing arm from rotating and lifting the body nearest the turn. Hence the body roll tendency will be stabilized to some extent when cornering.

With this axle arrangement much softer coil springs can be used to oppose equal spring deflection when driving in the straight ahead direction than could otherwise be employed if there were no transverse interconnecting beam.

Strut and link non-drive rear independent suspension (Fig. 10.50) With this suspension the wheel hub carrier's up and down motion is guided by the strut's sliding action which takes place between its piston and cylinder. The piston rod is anchored by a rubber pivot to the body structure and the cylinder member of the strut is rigidly attached to the wheel hub carrier (Fig. 10.50). A transverse link (wishbone arm) connects the lower part of the hub carrier to the body, thereby constraining all lateral movement between the wheels and body. The swing link arm and sliding strut member's individual movements combine in such a way that the hub carrier's vertical motion between bump and rebound produces very little change to the static wheel camber, either when the laden weight alters or when cornering forces cause the body to roll.

Braking fore and aft inertia forces are transmitted from the body to the hub carrier and wheel by trailing radius arms which are anchored at their

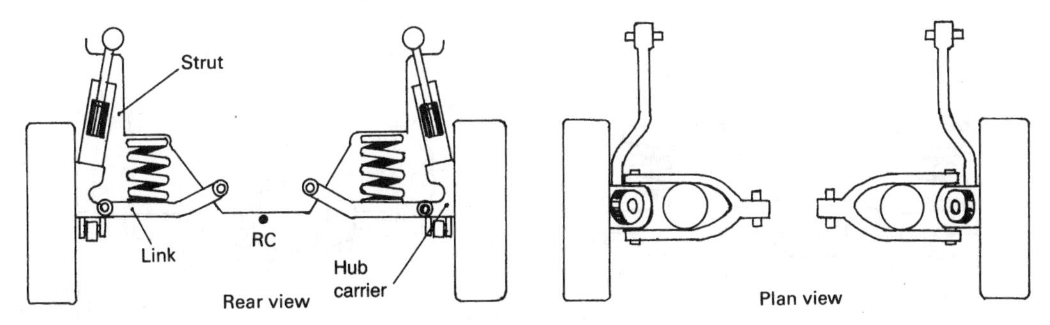

Fig. 10.50 Strut and link non-drive independent rear suspension

forward ends by rubber pin joints to the body under-structure. Owing to the trailing radius arms being linked between the body and the underside of each wheel hub carrier, deflection of the coil springs will cause a small variation in wheel toe-in to occur between the extremes in vertical movement.

The positioning of the body roll centre height will be largely influenced by the inclination of the swing arm relative to the horizontal; the slope of these transverse arms are usually therefore chosen so that the roll centre height is just above ground level.

10.7.3 Rear wheel drive suspension

Swing arm rear wheel drive independent suspension (Fig. 10.51) This suspension normally takes the form of a pair of triangular transverse ('A' arm) swing arm members hinging on inboard pivot joints situated on either side of the final drive with their axes parallel to the car's centre line (Fig. 10.51). Coil springs are mounted vertically on top of the swing arm members near the outer ends. The wheels are supported on drive hubs mounted on ball or tapered roller bearings located within the swing arm frame.

Each drive shaft has only one universal joint mounted inboard with its centre aligned with that of the swing arm pivot axes. If the universal joints and swing arm pivot axes are slightly offset (above and below in diagram), the universal joints must permit a certain amount of sliding action to take place to compensate for any changes in drive shaft length as the spring deflects. Usually the outer end of the drive shaft forms part of the stub axle wheel hub.

Any increase in static vehicle weight causes the swing arms to dip so that the wheels which were initially perpendicular to the road now become negatively cambered, that is, both wheels lean towards the body at the top. Consequently, when the body rolls during cornering conditions, the inner and outer wheels relative to the turn become cambered negatively and positively respectively; they both lean towards the centre of rotation. With a change in static vehicle weight both swing arms pivot and dip an equal amount which reduces the wheel track width. Similarly, if the body rolls the inner swing arm pivot centre rises and the outer swing arm pivot drops, so in fact both the swing arm pivots tend to rotate about their roll centres thus reducing the width of the wheel track again. Both wheels at all times will remain parallel as there is no change in wheel toe-in or -out.

Low pivot split axle coil spring rear wheel drive independent suspension (Fig. 10.52) The conventional transverse swing arm suspension suffered from three major limitations:

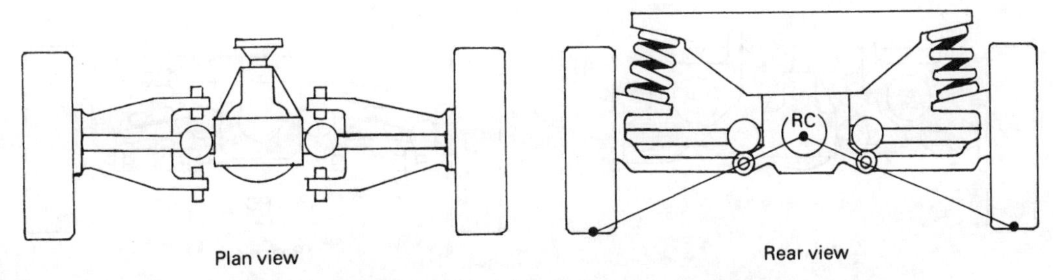

Fig. 10.51 Transverse swing arm coil spring rear wheel drive independent suspension

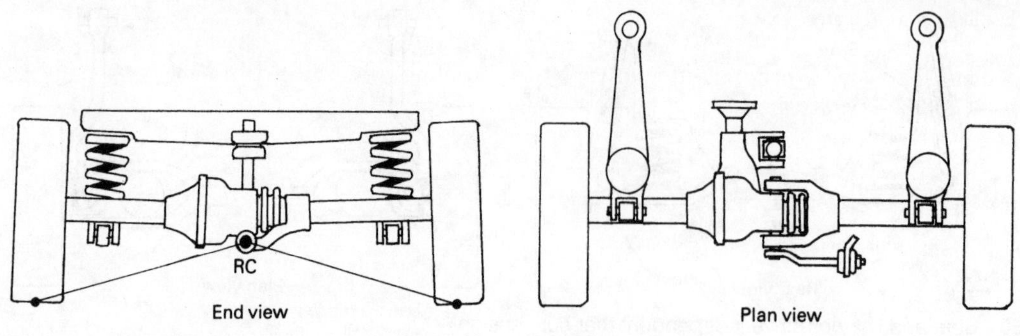

Fig. 10.52 Low pivot split axle coil spring rear wheel drive suspension

1 The swing arms were comparatively short because the pivot had to be mounted on either side of the final drive housing; it therefore caused a relatively large change in wheel camber as the car's laden weight increased or when wheel bounce occurred.
2 Due to the projection lines extending from the tyre to ground centre contact to and beyond the swing arm pivot centres, the body roll centre with this type of suspension was high.
3 There was a tendency when cornering for the short swing arms to become jacked up and with the load concentrated on the outside, the highly positively cambered wheel reduced its ability to hold the road so that the rear end of the car was subjected to lateral breakaway.

To overcome the shortcomings of the relatively large change in wheel camber and the very high roll centre height, the *low pivot split axle suspension* was developed.

With this modified swing axle arrangement the axle is split into two, with the adjacent half-axles hinged on a common pivot axis below the final drive housing (Fig. 10.52). A vertical strut supports the final drive assembly; at its upper end it is mounted on rubber discs which bear against the rear cross-member and at its lower end it is anchored to a pin joint situated on the hinged side of the final drive pinion housing. The left hand half-axle casing houses a drive shaft, crownwheel and differential unit. A single universal joint is positioned inside the casing so that it aligns with the pivot axis of the axles. The right hand half-axle houses its own drive shaft and a rubber boot protects the final drive assembly from outside contamination, such as dirt and water. A horizontal arm forms a link between the pivot axis and body structure and controls any lateral movement of the body relative to the axles. Fore and aft support for each half-axle is given by trailing radius arms which also carry the vertically positioned coil springs. The body roll centre thus becomes the pivot axis for the two half-axles which is considerably lower than for the conventional double pivot short swing arm suspension.

Trailing arm rear wheel drive independent suspension (Fig. 10.53) The independent trailing arm suspension has both left and right hand arms hinged on an axis at right angles to the vehicle centre line (Fig. 10.53). Each arm, which is generally semi-triangular shaped, is attached to two widely spaced pivot points mounted on the car's rear subframe. Thus the trailing arms are able to transfer the drive thrust from the wheel and axle to the body structure, absorb both drive and braking torque reactions and to restrain transverse body movement when the vehicle is subjected to lateral forces. The

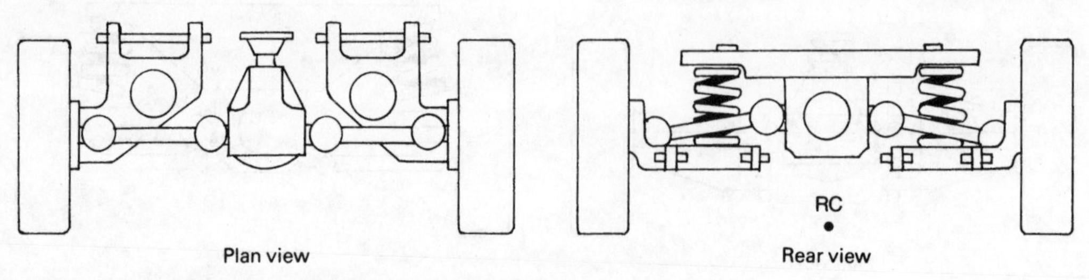

Fig. 10.53 Trailing arm coil spring rear wheel drive independent suspension

rear ends of each arm support a live wheel hub, the drive being transmitted from the final drive to each wheel via drive shafts and inner and outer universal joints to accommodate the angular deflection of the trailing arms. The inner joints also incorporate a sliding joint to permit the effective length of the drive shafts to vary as the trailing arms articulate between bump and rebound.

When the springs deflect due to a change in laden weight, both wheels remain perpendicular to the ground. When the body rolls on a bend, the inner wheel becomes negatively cambered and the outside wheel positively cambered; both wheels lean away from the turn. Spring deflection, caused by either an increase in laden weight or wheel impact, does not alter the wheel track toe-in or -out or the wheel track width, but body roll will cause the wheel track to widen slightly.

Semi-trailing arm rear wheel drive independent suspension (Fig. 10.54) With the semi-trailing arm suspension each arm pivots on an axis which is inclined (skewed) to something like 50 to 70 degrees to the car's centre line axis (Fig. 10.54). The pivot axes of these arms are neither transverse nor longitudinally located but they do lie on an axis which is nearer the trailing arm pivot axis (which is at right angles to the car's centre line axis). Consequently the arms are classified as semi-trailing.

Swivelling of these semi-trailing arms is therefore neither true transverse or true trailing but is a combination of both. The proportion of each movement of the semi-trailing arm will therefore depend upon its pivot axis inclination relative to the car's centre line. With body roll the transverse swing arm produces positive camber on the inside wheel and negative camber on the outer one (both wheels lean inwards when the body rolls), whereas with a trailing arm negative camber is produced on the inside wheel and positive camber on the outer one (both wheels lean outwards with body roll).

Skewing the pivot axis of the semi-trailing arm suspension partially neutralizes the inherent tendencies when cornering for the transverse swing arm wheels to lean towards the turn and for the trailing arm wheels to lean away from the turn. Therefore the wheels remain approximately perpendicular to the ground when the car is subjected to body roll.

Because of the relatively long effective swing arm length of the semi-trailing arm, only a negligible change to negative camber on bump and positive camber on rebound occurs when both arms deflect together. However, there is a small amount of wheel toe-in produced on both inner and outer wheels for both bump and rebound arm movement, due to the trailing arm swing action pulling the wheel forward as it deflects and at the same time the transverse arm swing action tilting the wheel laterally.

By selecting an appropriate semi-trailing arm pivot axis inclination, an effective swing arm length can be produced to give a roll centre height somewhere between the ground and the pivot axis of the arms. By this method the slip angles generated by the rear tyres can be adjusted to match the understeer cornering characteristics required.

Transverse double link arm rear wheel drive independent suspension (Figs 10.55 and 10.56) This class of suspension may take the form of an upper and lower wishbone arm linking the wheel hub carrier to the body structure via pivot joints provided at either end of the arms. Drive shafts transfer torque from the sprung final drive unit to the wheel hub through universal joints located at the inner and outer ends of the shafts. Driving and braking thrust and torque reaction is transferred through the wide set wishbone pivot joints. One form of transverse double link rear wheel drive independent suspension uses an inverted semi-elliptic spring for its upper arm (Fig. 10.55).

A double wishbone layout has an important advantage over the swing axle and trailing arm arrangements in that the desired changes of wheel camber, relative to motions of the suspension, can

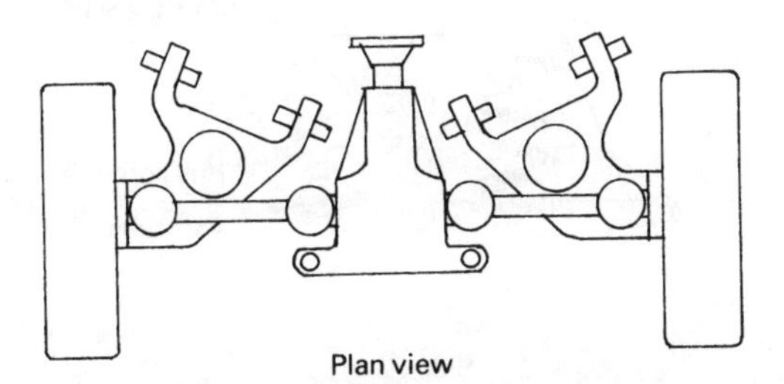

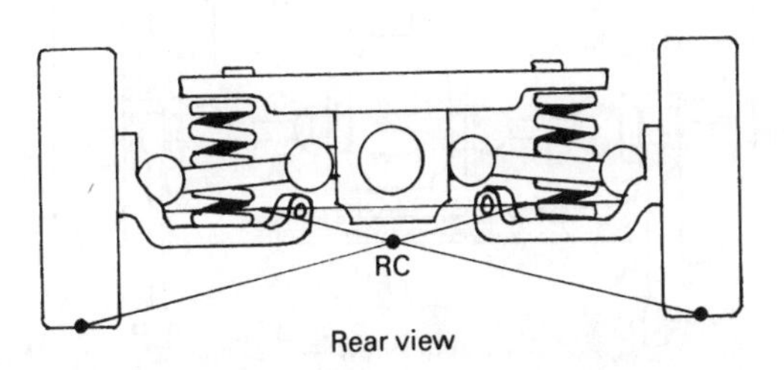

Fig. 10.54 Semi-trailing arm coil spring rear wheel drive independent suspension

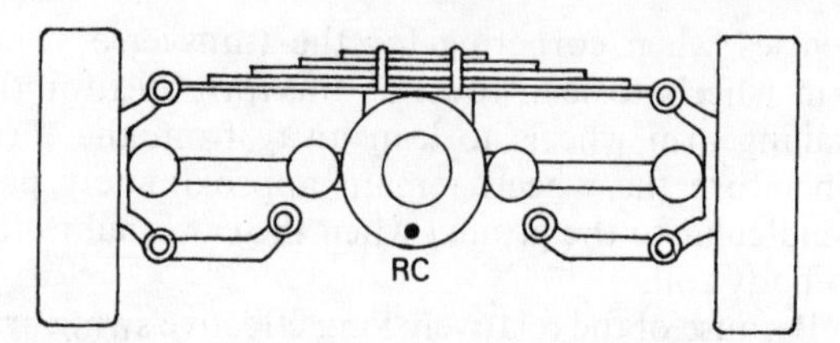

Fig. 10.55 Transverse swing arm and inverted semi-elliptic spring rear wheel drive independent suspension

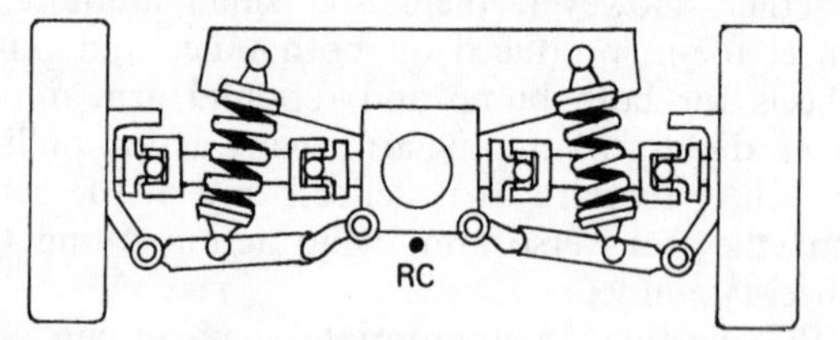

Fig. 10.56 Transverse swing arm and double universal joint load bearing drive shaft rear independent suspension

be obtained more readily. With swing axles, camber changes tend to be too great, and the roll centre too high. Wheels located by trailing arms assume the inclination of the body when it rolls, thereby reducing the cornering forces that the tyres produce. Generally, transverse double link arm suspensions are designed to ensure that, when cornering, the outer wheel should remain as close to the vertical as possible.

A modified version (Fig. 10.56) of the transverse double link suspension comprises a lower transverse forked tubular arm which serves mainly to locate the wheel transversely; longitudinal location is provided by a trailing radius arm which is a steel pressing connecting the outer end of the tubular arm to the body structure. With this design the upper transverse link arm has been dispensed with, and a fixed length drive shaft with Hooke's universal joints at each end now performs the task of controlling the wheel hub carrier alignment as the spring deflects. Compact twin helical coil springs are anchored on both sides of the lower tubular forked arms with telescopic dampers positioned in the middle of each spring.

DeDion axle rear wheel drive suspension (Figs 10.57 and 10.58) The DeDion axle is a tube (sometimes rectangular) sectioned axle beam with cranked (bent) ends which are rigidly attached on either side to each wheel hub. This permits the beam to clear the final drive assembly which does not form part of the axle beam but is mounted independently on the underside of the body structure (Figs 10.57 and 10.58).

To attain good ride characteristics the usual sliding couplings at the drive shaft to the wheels are dispensed with in this design since when transmitting drive or braking torque, such couplings generate considerable frictional resistance which opposes the sliding action. A sliding joint is provided in the axle tube to permit wheel track variation during suspension movement (Fig. 10.57). Axle lateral location is therefore controlled by the drive shafts which are permitted to swing about the universal joint centres but are prevented from extending or contracting in length. Fore and aft axle location is effected by two Watt linkages. These comprise two lower trailing fabricated pressed steel arms, which also serve as the lower seats for the coil springs. Their rear ends are carried on pivots below the hub carriers. The other parts of the Watt linkage consist of two rearward extending tubular arms, each attached to a pivot above the hub carrier. The upper and lower unequal length link arm pivot centres on the body structure are arranged in such a way that the axle has a true vertical movement as the spring deflects so that there are no roll steer effects. When the body rolls

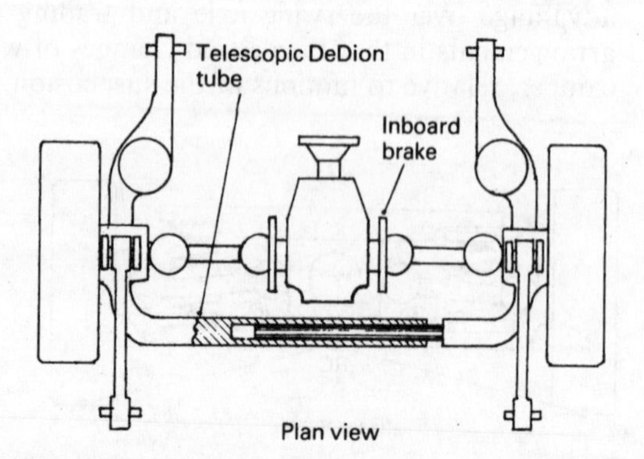

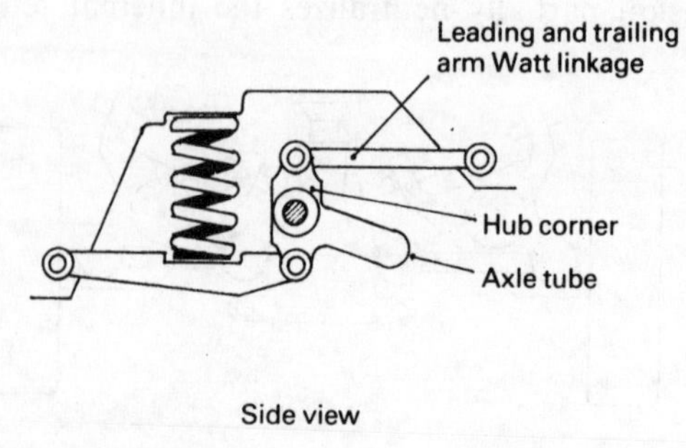

Fig. 10.57 DeDion axle with leading and trailing arm Watt linkage rear suspension

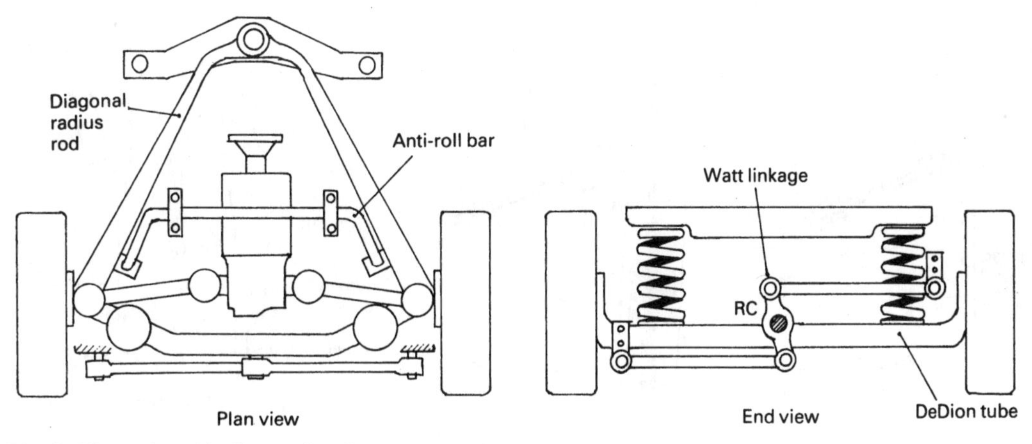

Fig. 10.58 DeDion tube with diagonal radius arms and Watt transverse linkage rear suspension

one hub carrier tends to rotate relative to the other, which is permitted by the sliding joint in the axle tube. The inner and outer sliding joints of the axle tube are supported on two widely spaced bronze bushes. The internal space between the inner and outer axle tube is filled about two thirds full of oil and lip seals placed on the outboard end of each bearing bush prevents seepage of oil. A rubber boot positioned over the axle sliding joint prevents dirt and water entering between the inner and outer tube members.

A DeDion axle layout reduces the unsprung suspension weight for a rear wheel drive car, particularly if the brakes are situated inboard. It keeps both road wheels parallel to each other under all driving conditions and transfers the driving and braking torque reactions directly to the body structure instead of by the conventional live axle route by way of the axle casing and semi-elliptic springs or torque rods to the body. The wheels do not remain perpendicular to the ground when only one wheel lifts as it passes over a hump or dip in the road. The body roll centre is somewhere near the mid-height position of the wheel hub carrier upper and lower link arm pivot points; a typical roll centre height from the ground would be 316 mm.

An alternative DeDion axle layout forms a triangle with the two diagonal radius arms which are rigidly attached to it (Fig. 10.58). The apex where the two radius arms meet is ahead of the axle and is pivoted by a ball joint to the body cross-member so that the driving and braking thrust is transferred from the axle to the body structure via the diagonal arms and single pivot. A transverse Watt linkage mounted parallel and to the rear of the axle beam controls lateral body movement relative to the axle. Therefore the body is constrained to roll on an axis which passes between the front pivot supporting the radius arms and the central Watt linkage pivot to the rear of the axle.

The sprung final drive which is mounted on the underside of the rear axle arch transmits torque to the unsprung wheels by way of the drive shaft and their inner and outer universal joints. The effective length of the drive shaft is permitted to vary as the suspension deflects by adopting splined couplings or pot type joints for both inner universal joints.

10.8 Suspension design consideration

10.8.1 Suspension compliance steer

(Fig. 10.59(a and b))

Rubber bush type joints act as the intermediates between pivoting suspension members and the body to reduce the transmission of road noise from the tyres to the body. The size, shape and rubber hardness are selected to minimize noise vibration and ride hardness by operating in a state of compressive or torsional distortion.

If the rubber joints are subjected to any abnormal loads, particularly when the suspension pivots are being articulated, the theoretical geometry of the swing members may be altered so that wheel track misalignment may occur.

The centrifugal force when cornering can produce lateral accelerations of 0.7 to 8.0 g which is sufficient to compress and distort the rubber and move the central pin off-centre to the outer hole which supports the rubber bush.

With transverse or semi-trailing arms suspension (Fig. 10.59(a)) the application of the brakes retards the rotation of the wheels so that they lag behind the inertia of the body mass which is still trying to

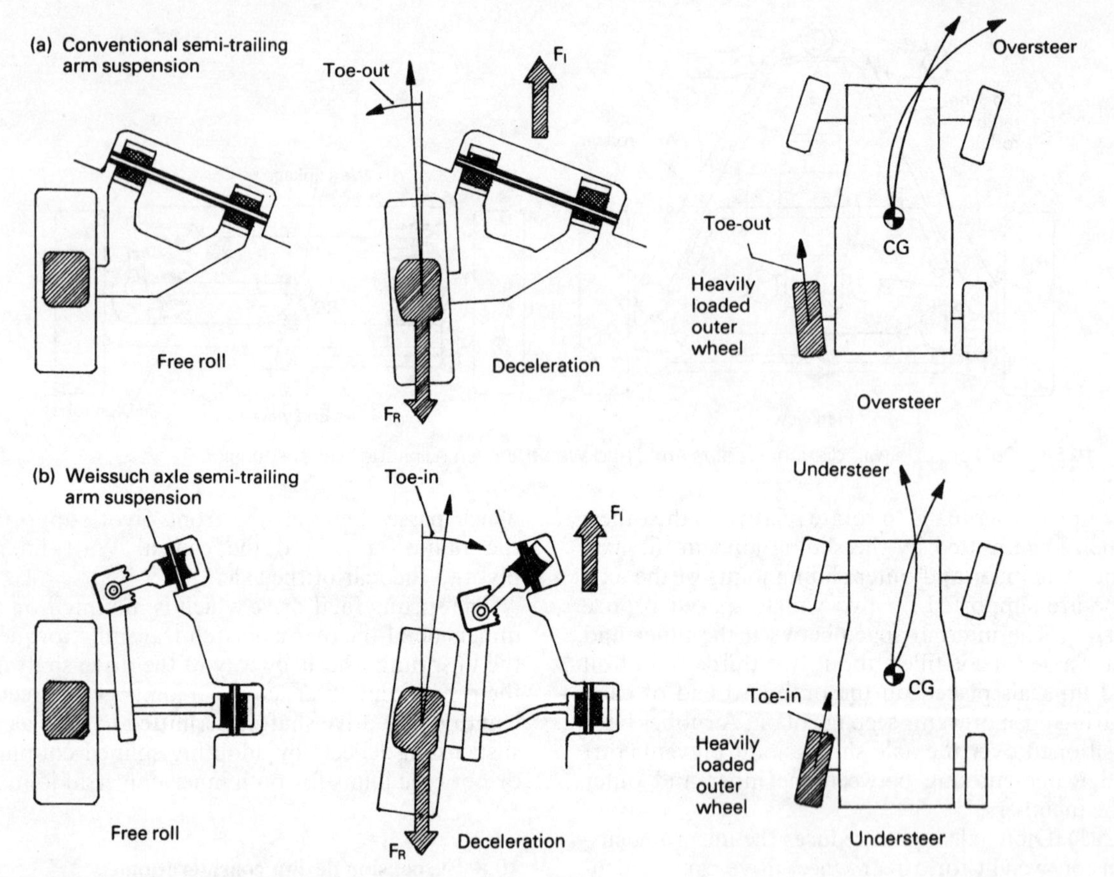

Fig. 10.59 (a and b) Semi-trailing suspension compliance steer

thrust itself forward. Consequently the opposing forces between the body and suspension arms will distort the rubber joint, causing the suspension arms to swing backwards and therefore make the wheel track toe outwards.

The change in the wheel track alignment caused by the elastic deflection of the suspension rubber pivot joints is known as *suspension compliance steer* since it introduces an element of self-steer to vehicle.

Compliance steer is particularly noticeable on cornering if the brakes are being applied since the heavily loaded outside rear wheel and suspension is then subjected to both lateral forces and fore and aft force which cause an abnormally large amount of rubber joint distortion and wheel toe-out (Fig. 10.59(a)), with the result that the steering will develop an unstable oversteer tendency.

A unique approach to compliance steer is obtained with the Weissuch axle used on some Porsche cars (Fig. 10.59(b)). This rear transverse upper and lower double arm suspension has an additional lower two piece link arm which takes the reaction for both the accelerating and decelerating forces of the car. The lower suspension links consist of a trailing tubular steel member which carries the wheel stub axle and the transverse steel plate arm. The trailing member has its front end pivoted to a short torque arm which is anchored to the body by a rubber bush and pin joint pivoted at about 30° to the longitudinal car axis. When the car decelerates the drag force pulls on the rubber bush pin joint (Fig. 10.59(b)) so that the short torque arm is deflected backward. As a result, the transverse steel plate arm distorts towards the rear and the front end of the trailing tubular member supporting the wheel is drawn towards the body, thus causing the wheel to toe-in. Conversely, when the car is accelerated the wheel tends to toe-out, but this is compensated by the static (initial) toe-in which is enough to prevent them toeing-out under driven conditions. The general outcome of the lower transverse and trailing link arm deflection is that when cornering the more heavily loaded outside wheel will toe-in and therefore counteract some of the front wheel steer, thus producing a degree of understeer.

10.8.2 Suspension roll steer

(Fig. 10.60(a, b and c))

When a vehicle is cornering the body tilts and therefore produces a change in its ground height between the inside and outside wheels. By careful design, the suspension geometry can be made to alter the tracking direction of the vehicle. This self-steer effect is not usually adopted on the front suspension as this may interfere with steering geometry but it is commonly used for the rear suspension to increase or decrease the vehicle's turning ability in proportion to the centrifugal side force caused by cornering. Because it affects the steering handling characteristics when cornering it is known as *roll oversteer* and *roll understeer* respectively.

Roll steer can be designed to cancel out large changes in tyre slip angles when cornering, particularly for the more heavily loaded outer rear wheel since the slip angle also increases roughly in proportion to the magnitude of the side force.

The amount of side force created on the front or rear wheels is in proportion to the load distribution on the front and rear wheels. If the car is lightly laden at the front the rear wheels generate a greater slip angle than at the front, thus producing an oversteer tendency. When the front is heavily loaded, the front end has a greater slip angle and so promotes an understeer response.

The object of roll steer on the rear wheels is for the suspension geometry to alter in such a way that

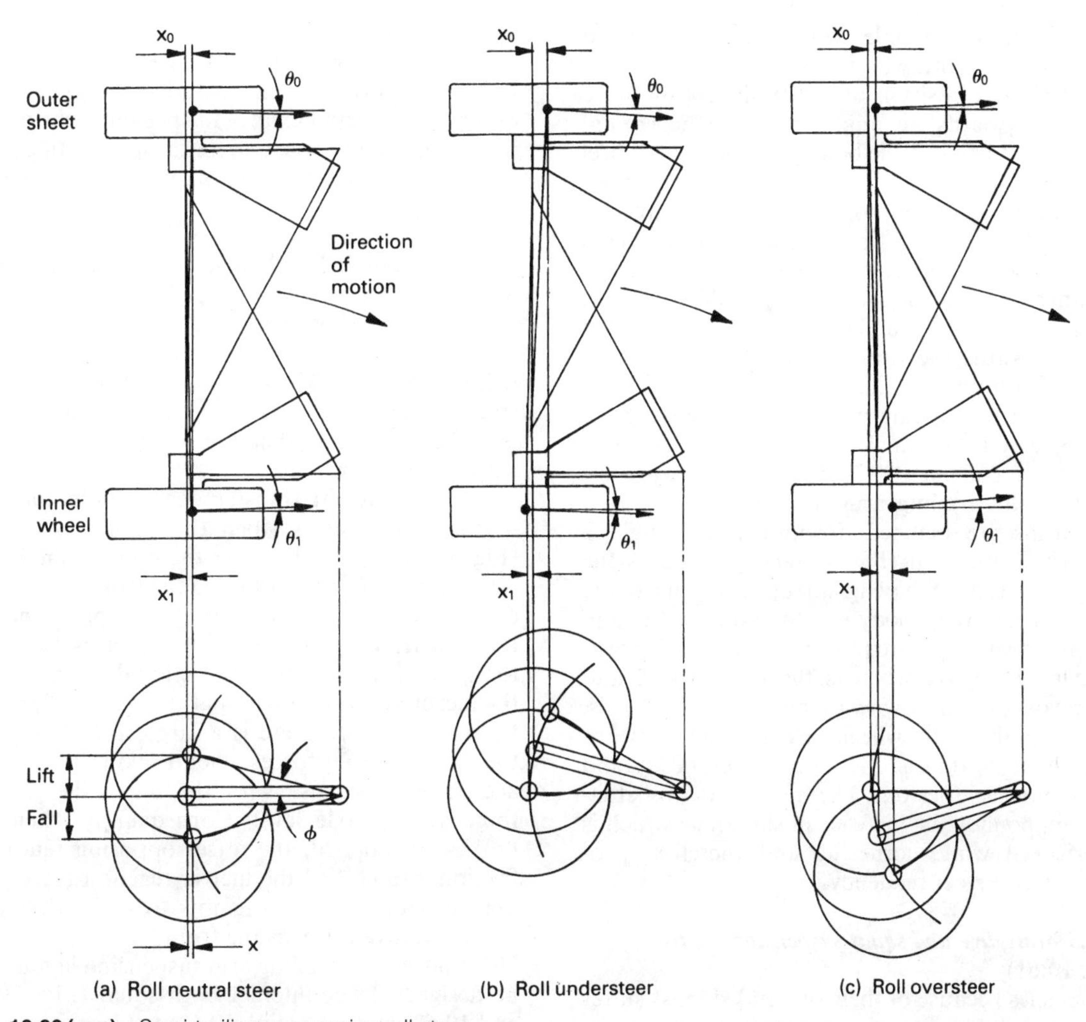

Fig. 10.60 (a–c) Semi-trailing suspension roll steer

the rear wheels steer the back end of the vehicle either outwards or inwards to compensate for the deviation in directional steer caused by changes in tyre slip angle.

A good example which illustrates suspension roll steer is with the semi-trailing arm steer rear suspension (Fig. 10.60(a, b and c)). If the body tilts when the vehicle corners the arms swing about their pivots so that the wheel axle attached to their free ends scribes circular arcs as they deflect up or down.

When the body rolls with the trailing arms set horizontally in their static position (Fig. 10.60(a)), the outer wheel and arm swings upwards towards the body whereas the inner wheel and arm rotates downwards and away from the body.

The consequence of the movement of the arms is that both axles move forward a distance x but because the axles of both trailing arms pivot at an inclined angle to the central axis of the vehicle the axis of end wheel axle will be slightly skewed inward so that both wheels now toe-in.

If the static position of the trailing arms were now set upwards an angle from the horizontal (Fig. 10.60(b)), when the body rolls the outer wheel and arm swing further upwards, whereas the inner wheel swings in the opposite direction (downwards towards the horizontal position). The outcome is that the outer wheel axle moves forwards whereas the inner wheel axle moves slightly to the rear. As a result, both the outer and inner stub axles skew the wheels towards the turn so that the outer wheel track toes-in and the inner wheel toes-out. Thus the change in tracking would tend to counteract any increase in slip angle due to cornering and so cause more understeer.

Setting the trailing arm static position so that both arms are inclined downwards an angle ϕ from the horizontal (Fig. 10.60(c)) produces the opposite effect to having an upward tilt to the trailing arms. With body roll the outer wheel and arm now swings towards the horizontal and moves backwards slightly whereas the inner wheel and arm pivots further downward and moves forwards. Consequently both wheels are skewed outward from the turn, that is, the inner wheel toes-in and the outer wheel toes-out. The tracking in this situation compounds the increase in slip angle which is experienced while cornering and therefore produces an oversteer tendency.

10.8.3 Anti-dive and squat suspension control (Fig. 10.61)

All vehicles because of their suspended mass suffer from weight transfer when they are either accelerated, as when pulling away from a standstill, or when retarding while being braked.

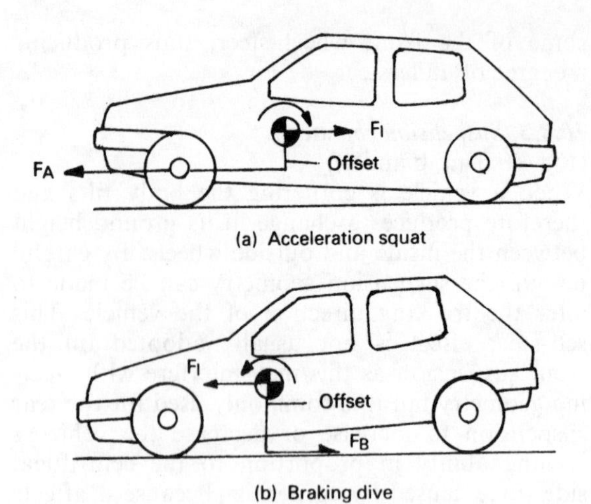

Fig. 10.61 (a and b) Vehicle squat and dive

A vehicle driven from a standstill (Fig. 10.61(a)) experiences a rapid change in speed in a short time interval so that a large horizontal accelerating force F_A is delivered at axle level to overcome the opposing body's inertia force F_I which acts in the opposite sense through the centre of gravity but which is generally situated well above axle height somewhere between the two axles. Due to the vertical offset distance between the accelerating force F_A and the inertia reaction force F_I, a pitch moment will be produced which transfers weight from the front to the rear wheels as the front of the car lifts and the rear sinks, thereby making the car body squat at the rear.

Likewise weight transfer occurs from the rear to the front wheels when the vehicle is braked (Fig. 10.61(b)) which causes the body to pitch forward so that the rear rises and the front suspension dips, which gives a front nose dive appearance to the vehicle. The forces involved when braking are the ground level retarding frictional force F_B and the inertia reaction force F_I at the centre of gravity height. Therefore there is a larger offset between the two opposing forces when braking than when accelerating because with the latter the driving force acts at axle level. Consequently when the brakes are applied, the offset opposing retarding frictional force and the inertia reaction force produce a couple which attempts to make the body pitch and dive towards the front.

A leading and trailing arm suspension layout can be designed to counteract both squat (Figs 10.62 and 10.63) and dive (Fig. 10.64) tendencies.

When the vehicle accelerates forwards, the reaction to the driving torque pivots the suspension arm about the axle in the opposite direction to the input driving torque. Thus in the case of a front wheel drive vehicle (Fig. 10.62) the arm swings downwards and opposes the front upward lift caused by the reluctant inertia couple. Likewise with a rear wheel drive vehicle (Fig. 10.63) the reaction to the driving torque swivels the suspension arm upward and so resists the rearward pitch caused by the weight transference from the front to the rear axle.

For both drive acceleration and braking the amount of squat and dive is controlled by the length of the leading and trailing arms. The shorter they are, the greater their resistance to weight transference will be, and from that point of view alone, the better the quality of ride will be.

A large number of modern suspensions are based on trailing or semi-trailing arm designs which can build in anti-squat and -dive control but leading arm front suspension has inherent undesirable features and therefore is rarely used. However, anti-squat and -dive control can be achieved by producing a virtual lead arm front suspension, that is, by arranging the swing axis of a double wishbone arm suspension to converge longitudinally along the wheelbase at some point.

The double transverse wishbone arm suspension geometry (Fig. 10.65) is laid out so that the top wishbone arm axis tilts downwards and the lower slightly upward towards the rear so that lines drawn through these pivot axes intersect somewhere towards the rear.

When the brakes are applied, the body will tend to pitch downward at the front but the clamped disc caliper or back plate will attempt to rotate with the road wheel. The result is that the upper and lower wishbone pivot axis converging projections form in effect an imaginary leading arm of length R which tends to swing upwards to the rear about the wheel axle. It therefore imparts an upthrust which opposes and cancels the downward pitch of the body.

Similar results can be obtained with the MacPherson strut suspension (Fig. 10.66) where the strut is made to tilt backward from the top and the lower transverse wishbone arm pivot axis tilts upwards to the rear. A line drawn perpendicular to the strut through the top pivot will then intersect a line projecting from the wishbone pivot axis. The distance between the strut to wishbone ball pivot and the meeting point of the two rearward projected lines therefore provides the effective trailing arm length or swing radius R. Thus an anti-dive torque T is produced of magnitude FR which opposes the forward transfer of weight when braking.

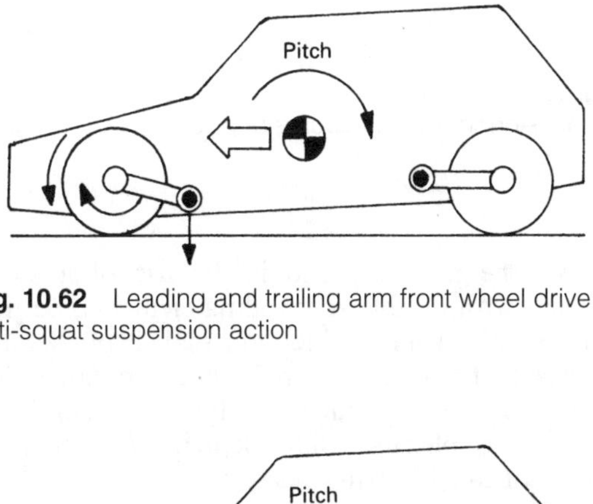

Fig. 10.62 Leading and trailing arm front wheel drive anti-squat suspension action

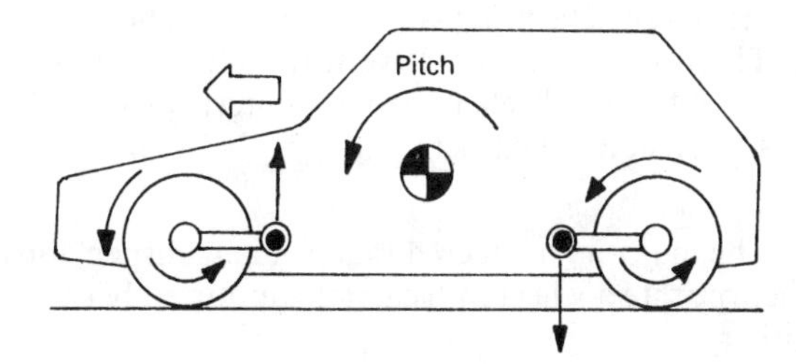

Fig. 10.64 Leading and trailing arm brake anti-drive suspension

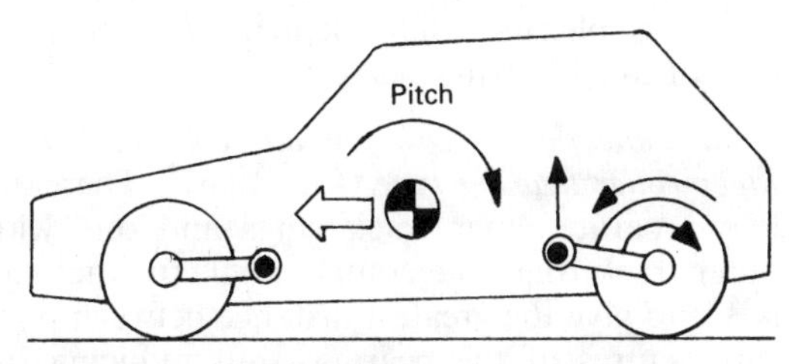

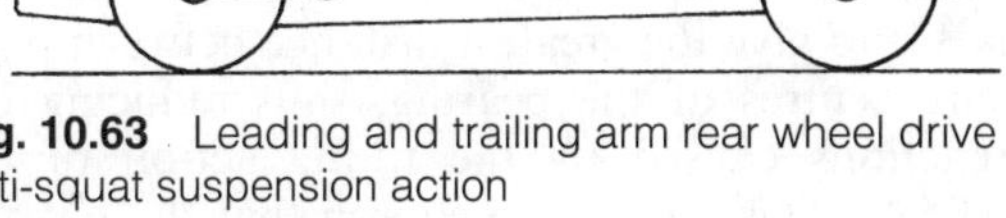

Fig. 10.63 Leading and trailing arm rear wheel drive anti-squat suspension action

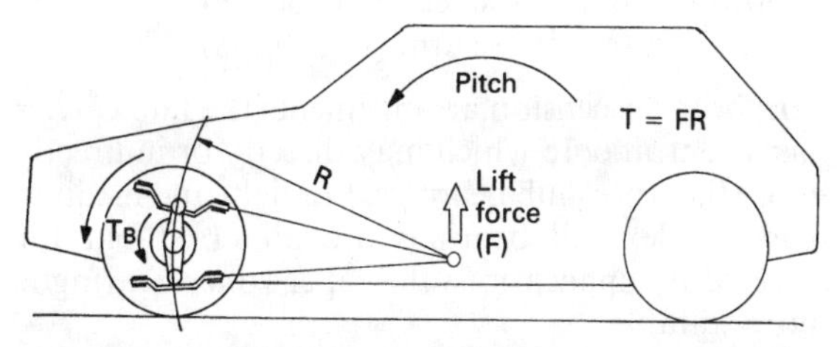

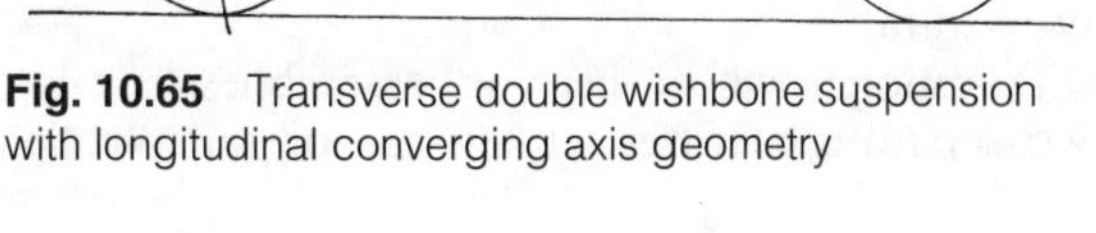

Fig. 10.65 Transverse double wishbone suspension with longitudinal converging axis geometry

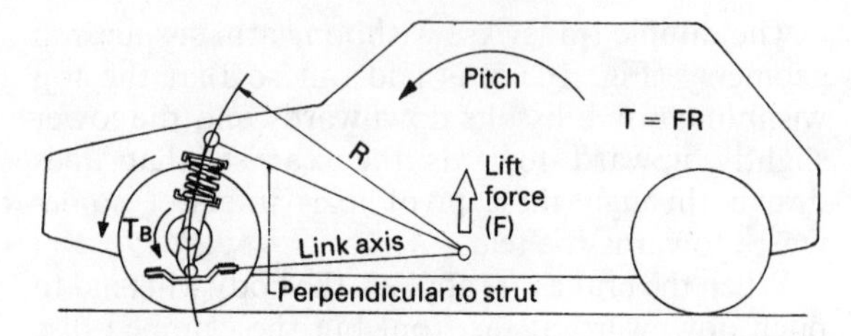

Fig. 10.66 MacPherson strut suspension with longitudinal converging axis geometry

Unfortunately the amount of anti-dive control must be limited since the upward swing of the imaginary trailing arm rotates the steering swivel joints so that the castor angle changes from positive to negative, thus destabilizing the steering firmness and so producing steering reaction and wander. Normally front suspension design restricts the anti-dive control to within 50 to 70% and the rear suspension may provide a 100% cancellation of brake dive.

10.8.4 Front wheel drive independent suspension wheel bearing arrangements (Figs 10.67 and 10.68)

With a front wheel drive independent suspension two major functions must be fulfilled:

1 The wheels must be able to turn about their swivel pins simultaneously as the suspension members deflect between bump and rebound.
2 The transmission of drive torque from the final drive to the wheels must be uninterrupted as the suspension members move between their extremes.

The majority of steered independent suspensions incorporate a wheel hub carrier supported between either;

a) an upper and lower ball and socket joint mounted between a pair of transverse arms (Fig. 10.67),
b) a leg strut mounted on a swivel bearing and a lower ball and socket joint located at the free end of a transverse arm (Fig. 10.68).

In both suspension arrangements the hub carrier has a central bore which may directly or indirectly house the wheel hub bearings. For light and medium loads, roller ball bearings are preferred but for heavy duty applications the taper roller bearing is more suitable.

Traditional wheel bearing assemblies employ two separate bearings; either ball or taper roller

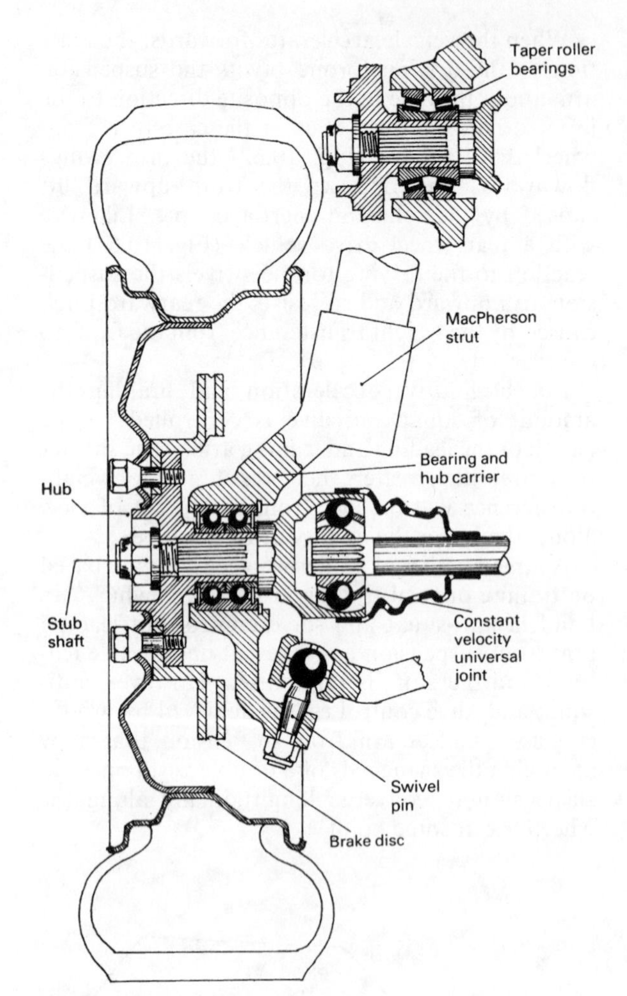

Fig. 10.67 Front wheel drive MacPherson strut suspension with single double row ball or roller wheel bearings

types. The present trend is the use of a single bearing with double row rolling elements, be they ball or taper rollers which are sealed, pre-set and lubricated for life. The preference is because they provide a more compact and cheaper assembly.

These double row rolling element single bearings can be of the following classes:

1 *Detachable double row angular contact ball or taper roller bearing type* (Fig. 10.67). There are two separate inner track rings and one wide outer track ring. The contact angle for the balls is 32° to give the greatest distance between pressure centres of the bearing, thus reducing the reactions caused by the tilting action of the wheels. This angle is so chosen that the bearing

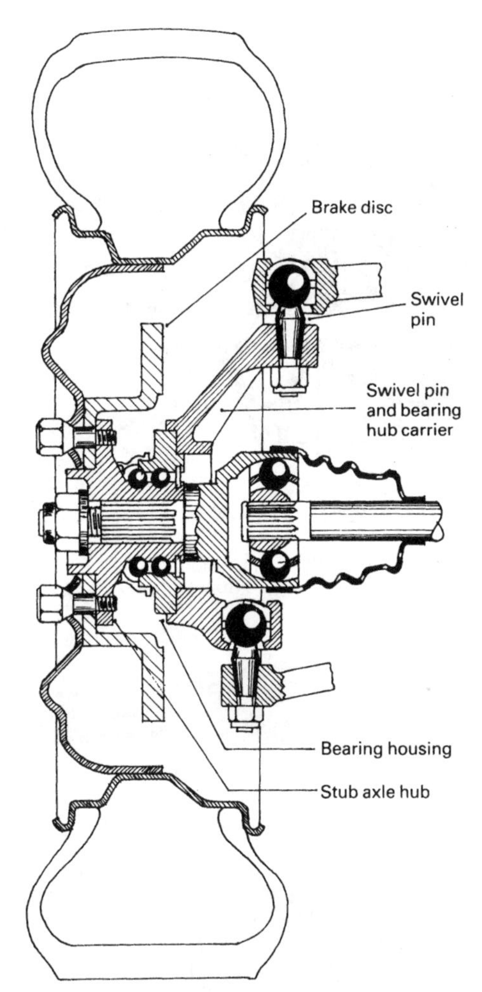

Fig. 10.68 Front wheel drive transverse wishbone suspension with fully integrated double low wheel bearings

has sufficient radial load capacity to withstand the weight imposed on the wheel and also to provide adequate axial load carrying capacity under cornering conditions. The cage that separates the balls is made from Nylon and does not, if damaged, affect the bearing performance. Preloading of the ball or taper roller bearings is set at the factory, therefore no adjustment is required after the bearing is assembled to its hub.

When assembled, the inner track rings are a force fit over the hub sleeve which is internally splined to the constant velocity joint's output stub shaft and the outer track ring is a press fit inside the hub carrier bore.

2 *Fully integrated double row angular contact ball bearing type* (Fig. 10.68). With this arrangement the inner track ring is extended on the outside with a flange to locate and support the wheel while its middle is bored and splined to accommodate the constant velocity joint splined output shaft. Thus the inner bearing member (track ring) takes over the whole function of the normal drive wheel hub. The outer track ring also supports both rows of balls and it is enlarged in the centre to provide a flange which aligns accurately within the wheel hub carrier's bore. Thus the inner and outer bearing members are integral parts of the wheel hub and bearing housing attached to the hub carrier respectively.

In both bearing arrangements the stub shaft nut is fully tightened to prevent axial movement between the hub and stub shaft and also, in the case of the detachable double row bearing, to secure its position.

10.9 Hydrogen suspension

10.9.1 Hydrogen interconnected suspension (Moulton–Dunlop) (Fig. 10.69(a, b and c))

The spring unit is comprised of a nitrogen filled spherical spring chamber welded to a double conical shaped displacement chamber (Fig. 10.69(a, b and c)). A hydraulic damper in the form of a pair of rubber compression blocks separates both spherical spring and displacer chambers, its function being to control the flow of fluid as it passes to and fro between the two chambers. The displacer chamber is sealed at its lower end by a load absorbing nylon reinforced rubber diaphragm which rolls between the conical piston and the tapered displacer chamber skirt as the suspension deflects up and down when the wheels pass over any irregularities on the road surface.

Within the spherical spring chamber is a butyl-rubber diaphragm which separates the sphere into a nitrogen charged (17.5 bars) upper region (the spring media) which is sealed for life, and the lower region which is filled with fluid. Initially fluid is pumped into the displacer chamber until it reaches the nitrogen charging pressure. Then it will compress and lift the separator diaphragm off the bottom of the sphere. Since the gas and fluid pressures on both sides of the diaphragm are equal, the separator diaphragm is not subjected to heavy loads, in fact it only functions as a flexible wall to keep the gas and fluid apart. A water based fluid containing 50% industrial alcohol and a small percentage of anti-corrosion additive is pumped into

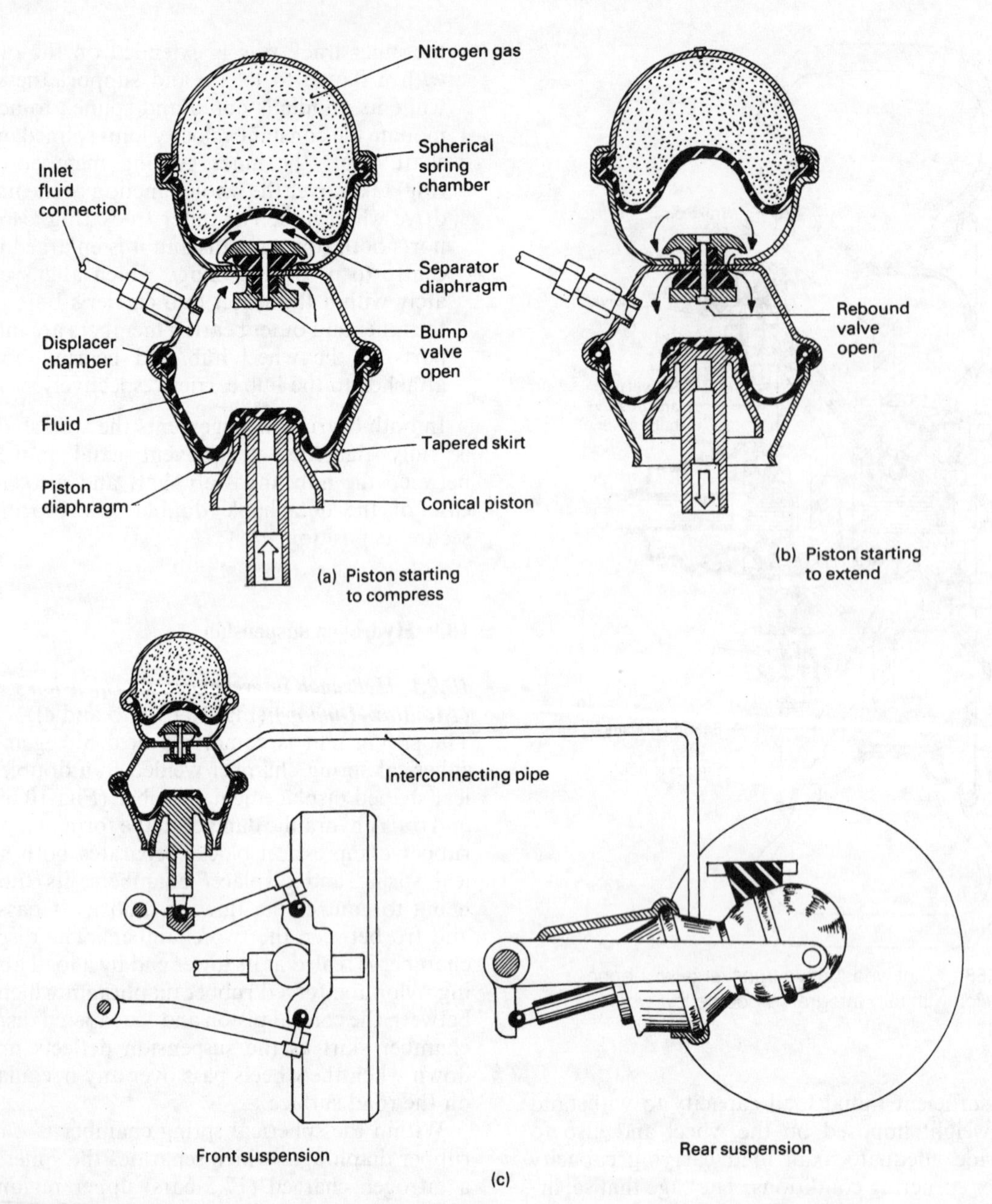

Fig. 10.69 (a–c) Hydrogas interconnected suspension system

the system to a pressure of 23 bars with the car in the unladen state, this being the condition in which the car's body to ground height is checked.

One advantage in using a rolling diaphragm type displacer instead of a piston and cylinder is that a water based fluid can be utilized as opposed to an oil which would not have such stable viscosity characteristics.

Taper rate (Fig. 10.69(a and b)) The effective area of piston compressing the fluid is that projected area of the displacer diaphragm which is not supported by the internal tapered skirt of the displacer chamber. Therefore, as the load on the displacer piston increases and the piston is pushed further into the chamber, less of the displacer diaphragm will be supported by the chamber's skirt and more will form part of the projected effective piston area. The consequence of the diaphragm piston contracting within the displacer chamber is that the load-bearing area of the piston is increased due to the diaphragm rolling away from its supporting tapered chamber skirt. As a result the resistance offered by the fluid against the inward movement of the piston rises. In other words, due to the tapered chamber's skirt, the spring rate (stiffness) increases in proportion to the spring's deflection.

Spring compressing due to bump response (Fig. 10.69(a)) When the tyre of the wheel hits a hump in the road, the whole wheel assembly attached to the suspension rises rapidly. This causes the displacer piston to move further into the displacer chamber. Consequently fluid in the displacer chamber will be displaced and pushed into the spherical spring chamber via the transfer port and bump valve. The rapid transfer of fluid into the spring chamber compresses the separator diaphragm against the nitrogen gas and the resilience of the gas therefore absorbs the impact shock. If there was no elastic media between the body structure and the deflecting suspension, any sudden upward movement would be transmitted directly to the body structure and passengers thus producing a very uncomfortable ride.

In actual fact movement of the fluid from the displacer chamber into the spring chamber takes place in three stages:

1 If the road bumps are very small and the vehicle is moving slowly, sufficient fluid flows through the permanently open transfer hole to equalize the pressure on both sides of this restriction.
2 If the road bumps are more severe the increased pressure build-up in the displacer chamber will be sufficient to lift the flaps on the rubber bump valve off a second pair of bleed holes. Additional fluid can now flow into the spring chamber in a shorter time span.
3 If the roughness of the road surface worsens or the speed at which the vehicle travels increases even more, then there will be a continuous rise in pressure of the fluid trapped in the displacer chamber. As a result of the extreme pressure build-up, the rubber bump valve itself will be progressively lifted from its seat to permit more fluid to enter the spring chamber. Thus in total more fluid is transferred from the displacer chamber to the spring chamber in a given time, but the built-in opposing resistance to the flow of fluid produces a measure of damping which slows down the violent uplifts caused by the impact of the tyre with obstacles in the road.

Spring extending due to rebound response (Fig. 10.69(b)) After the wheel has passed over a hump in the road the bounce action of the nitrogen gas pushes some of the fluid from the spring chamber back to the displacer chamber causing the displacer piston to extend from the displacer chamber.

The return of fluid from the spring chamber to the displacer chamber takes place in two stages:

1 If the bumps in the road are small or the vehicle is moving very slowly, then only a small amount of fluid needs to be transferred back to the displacer chamber in a given time. The movement of this fluid out of the spring chamber can be coped with adequately by the permanently open transfer hole. This means the damping action takes place as fluid is bypassed through the permanent bleed hole for low speed conditions.
2 If the bumps in the road are larger and the speed of the vehicle is higher, then the highly pressurized fluid in the spring chamber will lift the rubber rebound valve progressively from its seat, thus permitting a greater rate of flow of fluid back into the displacer chamber.

Because the progressive opening of the rubber valve is pressure sensitive, the flow of fluid is restricted and it is this tendency to slow down the fluid movement that produces the retarding effect on the rebound expanding gas.

Comparison of bump and rebound fluid damping control The extension (rebound) of the displacer piston is slightly slower than on contraction (bump) because there is not an intermediate flap valve second stage opening as there is on bump. Thus for small deflections of the displacer piston the permanent bleed transfer hole controls the movement of fluid in both bump and rebound directions. For more rapid displacement of fluid on rebound there is only the rebound compressive rubber block valves which regulate the flow of fluid in the extending direction, this being equivalent to both the flap valve and compressive rubber block valve opening on the contracting (bump) stroke.

Bump and pitch mode (Fig. 10.69(c)) When the front or rear wheel passes over a bump, the contraction of the displacer piston inside the displacer chamber at that wheel causes fluid displacement through the interconnecting pipe to the other wheel spring unit on the same side of the vehicle.

This movement of fluid into the other spring unit's displacer chamber extends its displacer piston within the chamber and thereby lifts the suspension and body up to the same level as that at the car's opposite end. Fluid movement from one suspension spring unit to the other therefore prevents pitch and enables the car to ride at a level attitude.

At moderate speeds the fluid is simply displaced from front to rear spring unit and vice versa, the fluid pressure remaining constant so that the coupled nitrogen gas springs are not further deflected.

Roll or bounce mode (Fig. 10.69(c)) If the body rolls due to cornering or the car bounces as a whole, then both front and rear suspensions are deflected together. The simultaneous fluid displacements increase the fluid pressure and dynamically compress and contract both of the nitrogen springs. Thus with the inward movement of the pistons the projected effective piston areas increase so that a larger fluid area has to be lifted. Consequently both the front and rear spring stiffness on the side of the body furthest away from the turn considerably increase the suspension's resistance to roll.

Similarly, if the body bounces at both ends together, then the spring stiffness rates increase as the displacer pistons approach their inner dead centres so that a much greater resistance against the downward movement of the body occurs if the bounce becomes violent.

10.10 Hydropneumatic automatic height correction suspension (Citroen) (Figs 10.70, 10.71 and 10.72)

The front suspension may be either a MacPherson strut (Fig. 10.70) or a transverse double wishbone arm arrangement (Fig. 10.71(a)), whereas the rear suspension is of the trailing arm type. Front and rear anti-roll bars are incorporated to increase the body roll stiffness and to actuate both front and rear height correction valves.

Spring unit (Fig. 10.71(a)) The suspension spring units (Fig. 10.71(a)) comprise two main parts;

1 a steel spherical canister containing a rubber diaphragm which separates the nitrogen spring media from the displacement fluid;
2 a steel cylinder and piston which relays the suspension's vertical deflection movement to the rubber diaphragm by displacing the fluid.

When the wheel meets a hump in the road, the piston is pushed inwards so that it displaces fluid from the cylinder into the sphere. Consequently the flexible rubber diaphragm squeezes the nitrogen gas into a small space (Fig. 10.71(b)). If the wheel hits a pot hole, the pressurized gas expands and forces fluid from the sphere into the cylinder, thereby making the piston move outward. By this method of changing the volume of fluid entering the sphere, the gas either is compressed or expanded relative to the initial charge pressure so that the resilience of the gas prevents the force of the road shocks from transferring to the body structure.

Pump accumulator and pressure regulator (Fig. 10.70) The initial fluid pressure source comes from a seven piston swashplate engine-driven hydraulic pump which is able to provide a continuous flow of fluid at a predetermined pressure. The pump feeds the spherically shaped accumulator which uses nitrogen as the spring media and a rubber diaphragm to accommodate the volume of stored fluid. The accumulator stores the highly pressurized fluid and can immediately deliver fluid to the system in the event of a sudden demand. It also permits the pump to idle and therefore eliminates repeated cutting in and out.

When the pump is idling the pressure generated is only enough to return the fluid to the reservoir through the pressure regulator. The pressure regulator and accumulator unit control the minimum pressure necessary for the operation of the system and the maximum pressure needed to charge the accumulator and to limit the maximum pressure delivered by the pump (the pump cut-in pressure of 140–150 bar and the cut-out pressure of 165–175 bar).

Height correction valve (Fig. 10.72(a, b, c and d)) Automatic height correction is achieved by varying the volume of incompressible fluid between the sphered diaphragm and the piston. Increased vehicle weight lowers the body, thus causing the suspension arms to deflect and at the same time rotate the anti-roll bar. The angular rotation of the anti-roll bar is a measure of the suspension's vertical deflection relative to the vehicle's normal static height. This movement is relayed to the height correction valve via a torsional control rod clamped to the anti-roll bar at one end and to a control rod lever which is attached to the height correction valve at the other end.

To avoid continuous height correction every time a pair of wheels roll over a hump or dip in the road, a delayed response is introduced to the

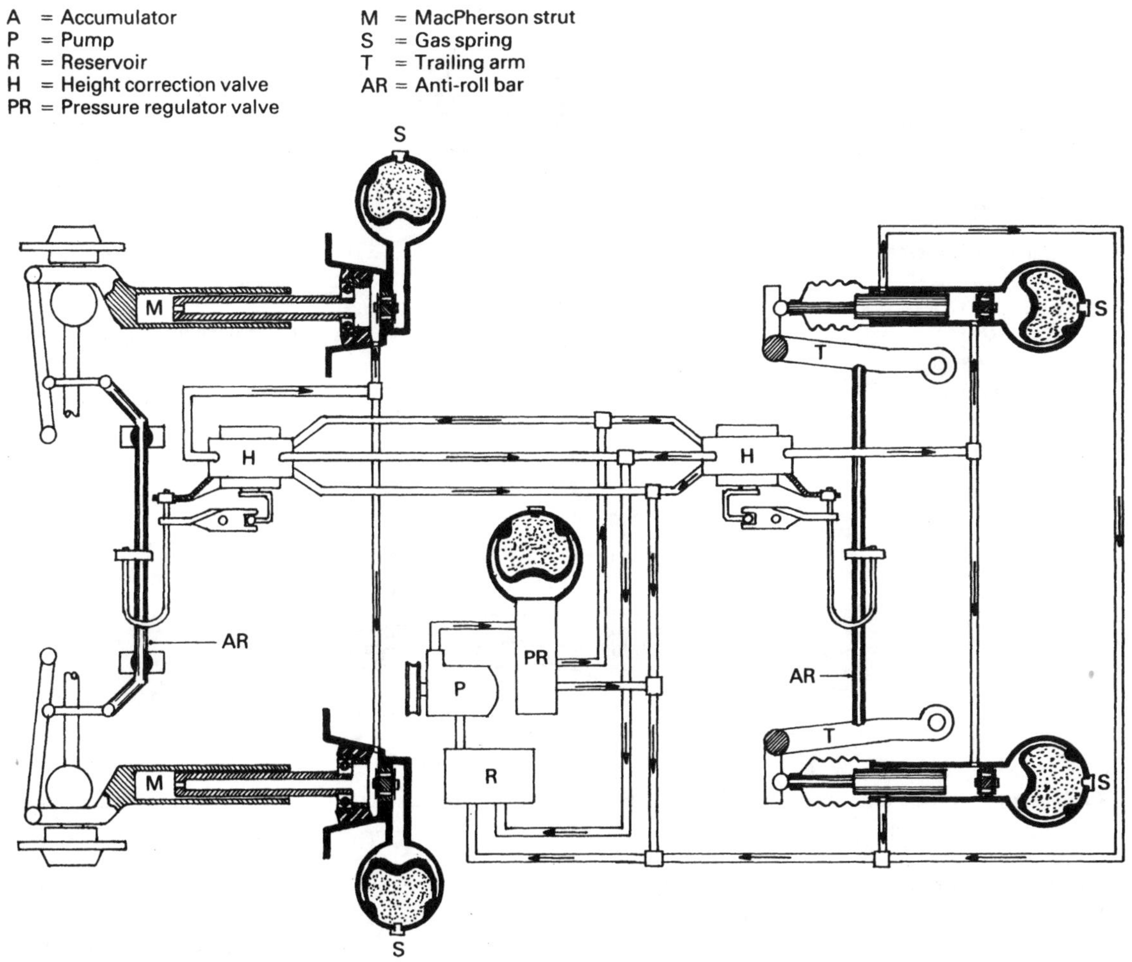

A = Accumulator
P = Pump
R = Reservoir
H = Height correction valve
PR = Pressure regulator valve
M = MacPherson strut
S = Gas spring
T = Trailing arm
AR = Anti-roll bar

Fig. 10.70 General layout of the hydropneumatic automatic height correction suspension

height correction valves so that the spring unit cylinder is not being charged on bump or discharged on rebound. Height correction will therefore be achieved only after a small time pause during which time the suspension will have had time to adjust to a change in the loads imposed on the spring units. Once the spring unit cylinder has been fully recharged, or discharged to bring the suspension height back to the standard setting, the height correction valve is made to respond immediately by either moving from inlet charging to neutral cut-off or from exhaust discharge to neutral cut-off position.

Charging the spring unit (spool valve movement from neutral cut-off to inlet open) (Fig. 10.72(a)) An increase in car load causes the lower transverse arm to pivot and the anti-roll bar to rotate. At the same time the control rod twists and tries to tilt the control rod lever, thereby transmitting an axial load to the height correction spool valve. The effect of shifting the spool valve to the left hand side is to move it from the cut-off position to the inlet open position. An increased amount of fluid is now forced between the piston and diaphragm causing the vehicle to rise until the anti-roll bar, which is rotating in the opposite direction, pulls the spool valve back to the neutral cut-off position. The return to the cut-off position is rapid because the spool valve does not offer any resistance in this direction, and the vehicle height will have been brought back to its normal position. To slow down the movement from cut-off to inlet charge positions, the disc valve in the right hand diaphragm chamber is closed. Therefore, the only way the fluid can be transferred from the right to

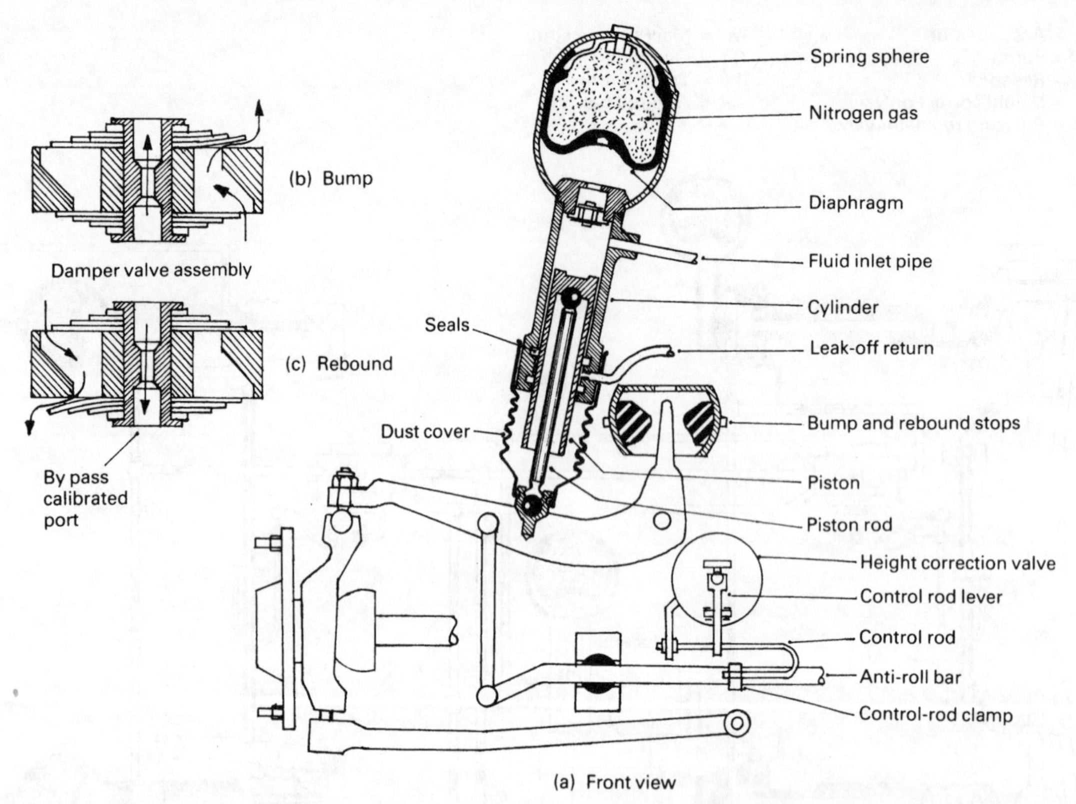

Fig. 10.71 (a–c) Detailed arrangement of hydropneumatic spring unit using a transverse double arm linkage suspension

the left hand chamber is through the restricted passage with the result that the spool valve shift movement is very sluggish.

Discharge the spring unit (spool valve movement from neutral cut-off to exhaust open) (Fig. 10.72(b)) Decreasing the car load has the reverse effect to increasing the load. This time the spool valve moves from the neutral cut-off position to the exhaust open position. The excess fluid between the piston and diaphragm is now expelled to the reservoir tank and the suspension spring unit contracts until the body to ground height has been corrected, at which point the spool valve again will be in the neutral cut-off position. Similarly the discharge process is also slowed down so that the valve does not respond to small changes in dynamic loads caused by suspension vibration as the wheels travel over the road surface irregularities.

Spool valve movement from inlet charge to neutral cut-off (Fig. 10.72(c)) Once the spring unit cylinder has been fully recharged with fluid, the anti-roll bar will have rotated sufficiently to make the spool valve alter its direction of slide towards the neutral cut-off position. This return movement of the spool valve to the cut-off position is rapidly speeded up because the left hand disc valve is in the open position so that when the spool valve first starts to change its direction of slide, fluid in the unrestricted passage will force the right hand valve off its seat. As a result, fluid movement from the left hand to the right hand diaphragm chamber takes place through both the restricted and unrestricted passages, speeding up the fluid transfer and accordingly the spool valve movement to the neutral cut-off position. As soon as the spool valve reaches its cut-off position the disc valve in the left hand diaphragm chamber re-seats. This action stops the spool valve overshooting its cut-off position and therefore avoids the valve going through a second recharge and discharge cycle of correction.

Spool valve movement from exhaust discharge to neutral cut-off (Fig. 10.72(d)) A rapid closing of the exhaust valve takes place once the fluid in an over-charged cylinder has been permitted to escape back to the reservoir thus restoring the suspension

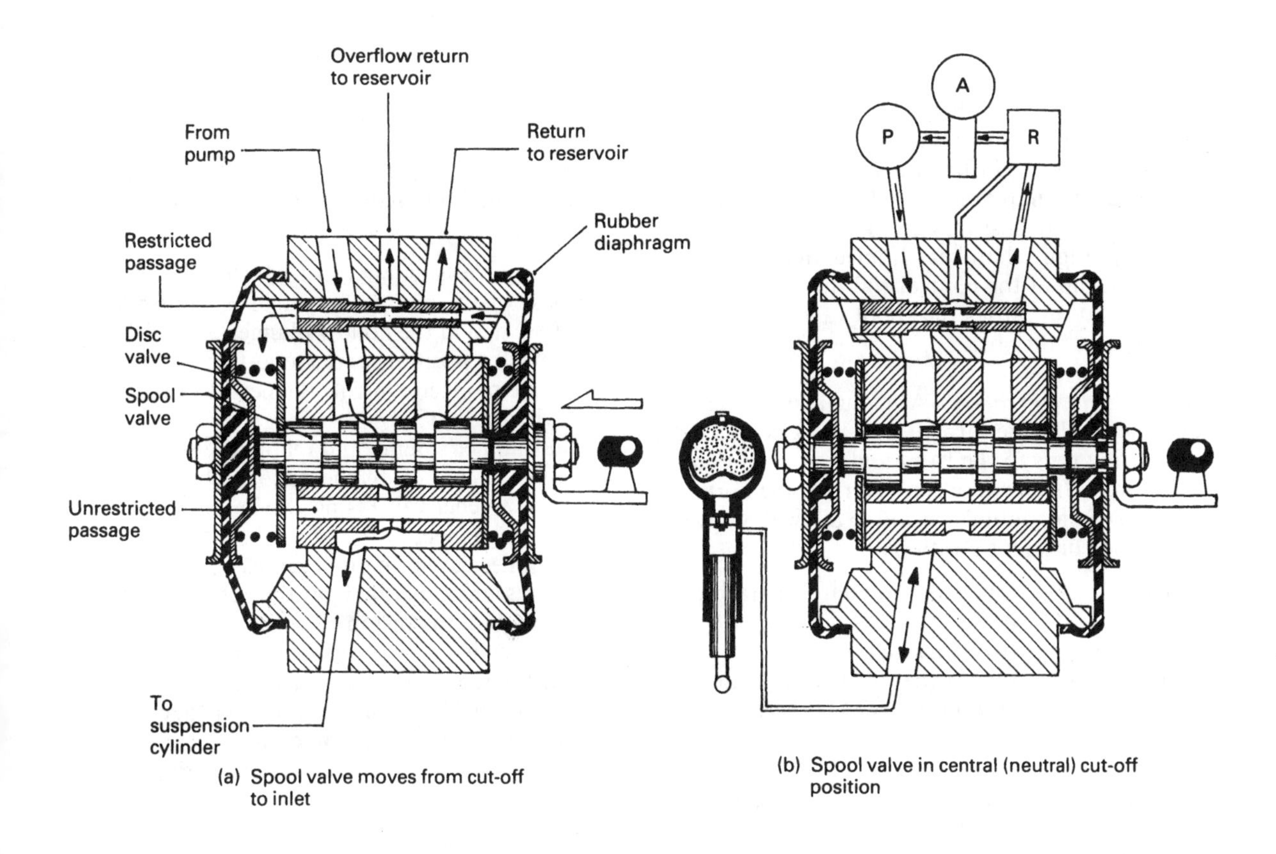

(a) Spool valve moves from cut-off to inlet

(b) Spool valve in central (neutral) cut-off position

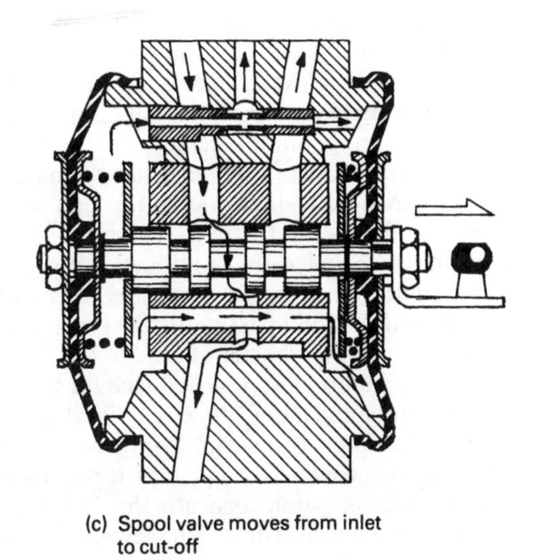

(c) Spool valve moves from inlet to cut-off

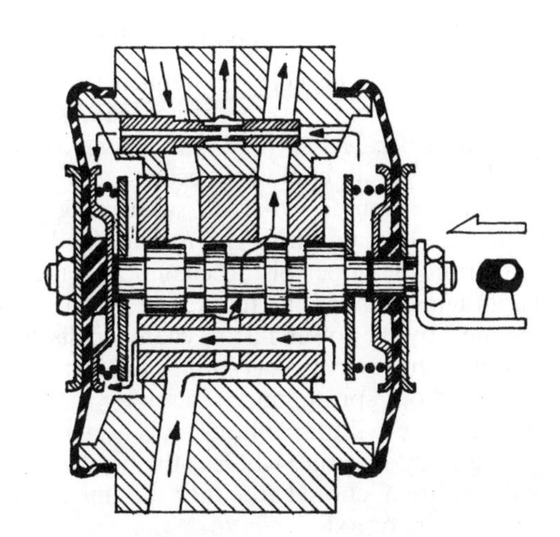

(d) Spool valve moves from exhaust to cut-off

Fig. 10.72 (a–d) Height correction valve action

to its standard height. The ability for the spool valve to respond quickly and close off the exhaust valve is due to the right hand disc valve being open. Thus fluid in the unrestricted passage is permitted to push open the right hand disc valve, this allows fluid to readily move through both the restricted and unrestricted passages from the right to left hand diaphragm chamber. Immediately the torsional wind-up of the control rod due to the anti-roll bar rotation causes the spool valve to shift to the neutral cut-off position.

Manual height correction A manual control lever is provided inside the car, the lever being connected by actuating rods to the front and rear height correction units. Its purpose is to override the normal operation of the spool valve and to allow the driver to select five different positions:

Normal — this is the standard operating position
High or low — two extreme positions
Two positions — intermediate between normal and high

10.10.1 Hydropneumatic self-levelling spring unit

(Figs 10.73(a and b) and 10.74(a, b and c))
This constant height spring unit consists of two sections;

1 a pneumatic spring and hydraulic damper system,
2 a hydraulic constant level pump system.

An approximately constant frequency of vibration for the sprung mass, irrespective of load, is obtained by having two gas springs, a main gas spring, in which the gas is contained behind a diaphragm, and a correction gas reservoir spring (Fig. 10.73(a, b and c)). The main spring is controlled by displacing fluid from the upper piston chamber to the spring diaphragm chamber and the correction gas spring is operated by the lower piston chamber discharging fluid into the reservoir gas spring chamber.

The whole spring unit resembles a telescopic damper. The cylindrical housing is attached to the sprung body structure whereas the piston and integral rod are anchored to either the unsprung suspension arm or axle.

The housing unit comprises four coaxial cylinders;

1 the central pump plunger cylinder with the lower conical suction valve and an upper one way pump outlet disc valve mounted on the piston,
2 the piston cylinder which controls the gas springs and damper valves,
3 the inner gas spring and reservoir chamber cylinder,
4 the outer gas spring chamber cylinder which is separated from hydraulic fluid by a flexible diaphragm.

The conical suction valve which is mounted in the base of the plunger's cylinder is controlled by a rod located in the hollow plunger. A radial bleed port or slot position about one third of the way down the plunger controls the height of the spring unit when in service.

The damper's bump and rebound disc valves are mounted in the top of the piston cylinder and an emergency relief valve is positioned inside the hollow pump plunger at the top.

The inner gas spring is compressed by hydraulic fluid pressure generated by the retraction of the space beneath the piston.

The effective spring stiffness (rate) is the sum of the stiffnesses of the two gas springs which are interconnected by communication passages. Therefore the stiffness increase of load against deflection follows a steeper curve than for one spring alone.

Gas spring and damper valve action (Fig. 10.73 (a and b)) There are two inter-related cycles; one is effected by the pressure generated above the piston and the other relates to the pressure developed below the piston.

When, during bump travel (Fig. 10.73(a)), the piston and its rod move upwards, hydraulic fluid passes through the damper bump valve to the outer annular main gas spring chamber and compresses the gas spring. Simultaneously as the load beneath the piston reduces, the inner gas spring and reservoir expand and fluid passes through the transfer port in the wall to fill up the enlarging lower piston chamber cylinder. Thus the deflection of the diaphragm against the gas produces the elastic resilience and the fluid passing through the bump valve slows down the transfer of fluid to the gas spring so that the bump vibration frequency is reduced.

On rebound (Fig. 10.73(b)) fluid is displaced from the outer spring chamber through the damper rebound valve into the upper piston cylinder and at the same time fluid beneath the piston is pushed out of the lower piston chamber into the inner gas spring chamber where it now compresses the inner gas spring.

Likewise fluid which is being displaced from the main gas spring to the upper piston chamber

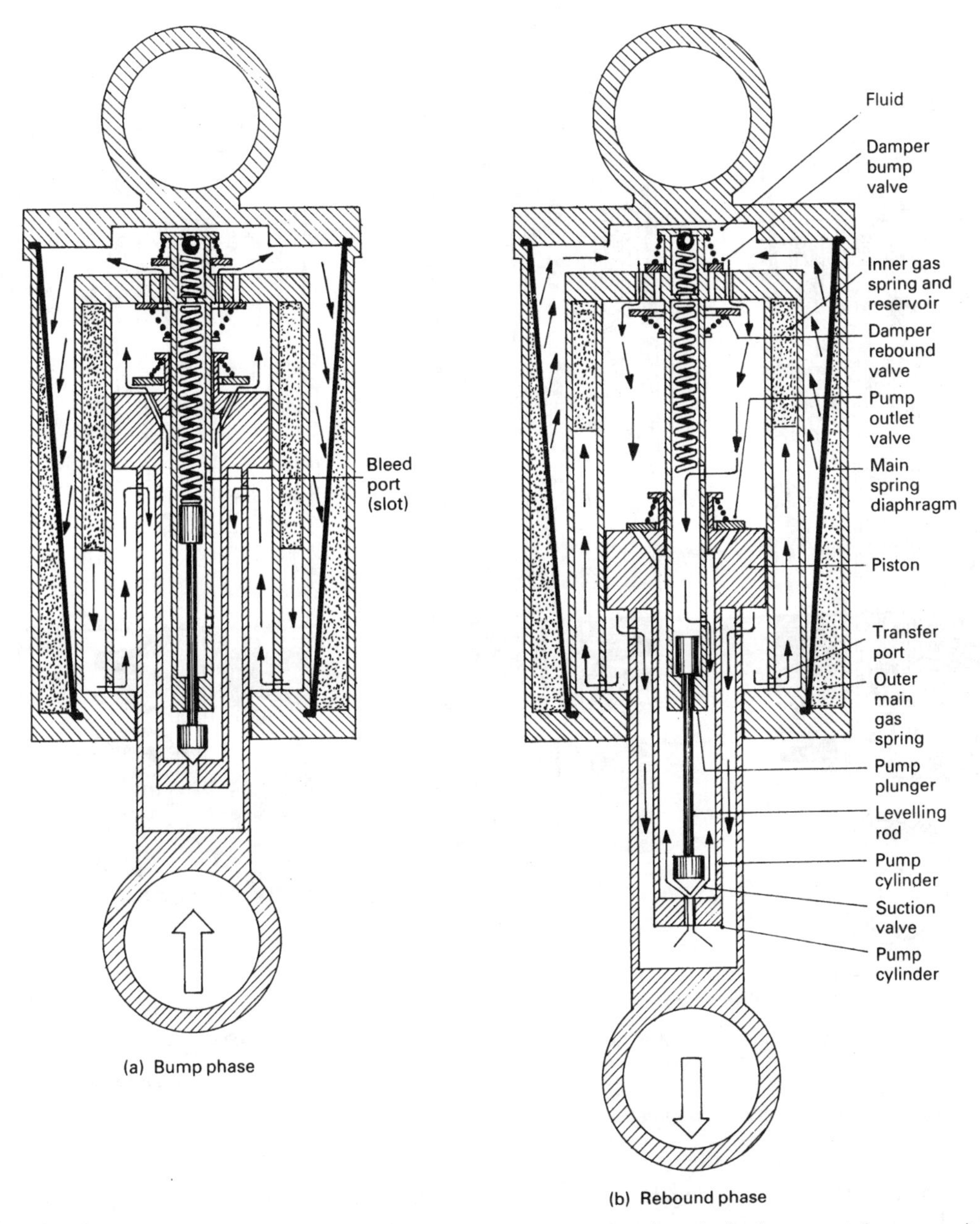

Fig. 10.73 (a and b) Exaggerated diagrams illustrating the self-levelling action of a hydropneumatic suspension unit

experiences an increased resistance due to the rebound valve passage restriction so that the fluid transfer is achieved over a longer period of time.

Pump self-levelling action (Figs 10.74(a, b and c) and 10.73(a and b)) The movement of the piston within its cylinder also causes the pump plunger to be actuated. During bump travel (Figs 10.73(a) and 10.74(a)) the plunger chamber space is reduced, causing fluid to be compressed and pushed out from below to above the piston via the pump outlet valve. On rebound (Fig. 10.74(c)), the volume beneath the piston is replenished. However, this action only takes place when the piston and rod have moved up in the cylinder beyond the designed operating height.

The conical suction valve, which is mounted in the base of the plunger's cylinder and is controlled

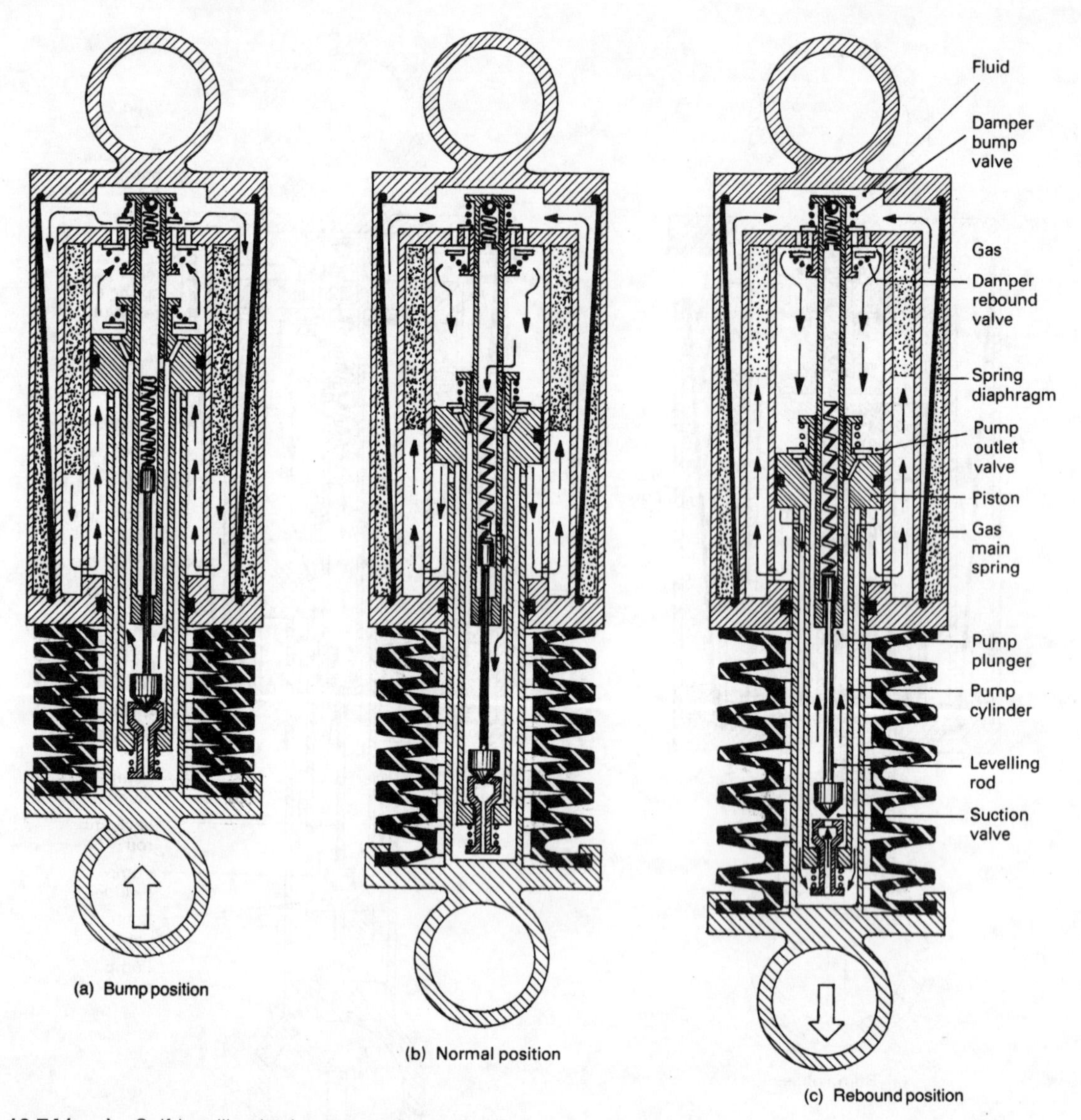

Fig. 10.74 (a–c) Self-levelling hydropneumatic suspension

by a rod located in the hollow plunger, and also a radial bleed port or slot, positioned about one third of the way down the plunger, control the height of spring unit when in service.

The damper's bump and rebound disc valves are mounted in the top of the piston cylinder and an emergency relief valve is positioned inside the hollow pump plunger at the top.

The inner gas spring is compressed by hydraulic fluid pressure generated by the retraction of the space beneath the piston.

The pumping action is provided by the head of the plunger's small cross-sectional area pushing down onto the fluid in the pump chamber during the bump travel (Fig. 10.74(a)). This compels the fluid to transfer through the pump outlet valve into the large chamber above the piston. The pressure of the fluid above the piston and that acting against the outer gas spring diaphragm is the pressure necessary to support the vehicle's unsprung mass which bears down on the spring unit. During rebound travel (Fig. 10.74(c)), the fluid volume in the pump chamber increases while the volume beneath the piston decreases. Therefore some of the fluid in the chamber underneath the piston will be forced into the inner gas spring chamber against the trapped gas, whilst the remainder of the excess fluid will be transferred from the lower piston chamber through a passage that leads into an annular chamber that surrounds the pump chamber. The pressurized fluid surrounding the pump chamber will then force open the conical suction valve permitting fluid to enter and fill up the pump chamber as it is expanded during rebound (Fig. 10.74(c)). This sequence of events continues until the piston has moved far enough down the fixed pump plunger to expose the bleed port (or slot) in the side above the top of the piston (Figs 10.74(c) and 10.73(b)).

At this point the hollow plunger provides a connecting passage for the fluid so that it can flow freely between the upper piston chamber and the lower plunger chamber. Therefore, as the piston rod contracts on bump, the high pressure fluid in the plunger chamber will be discharged into the upper piston chamber by not only the pump outlet valve but also by the plunger bleed port (slot) (Fig. 10.74(a)). However, on the expansion stroke some of the pressurized fluid in the upper piston chamber can now return to the plunger chamber and thereby prevents the conical suction valve opening against the pressure generated in the lower piston chamber as its volume decreases. The plunger pumping action still continues while the spring unit height contracts, but on extension of the spring unit (Fig. 10.74(c)) the fluid is replenished not from the lower piston chamber as before but from the upper piston chamber so that the height of the spring unit cannot increase the design spring unit length.

When the spring unit is extended past the design height the underside of the piston increases the pressure on the fluid in the reservoir chamber and at the same time permits fluid to bleed past the conical suction valve into the plunger chamber. If the spring unit becomes fully extended, the suction valve is lifted off its seat, enabling the inner spring chamber to be filled with fluid supplied from the lower piston chamber and the plunger chamber.

10.11 Commercial vehicle axle beam location

An axle beam suspension must provide two degrees of freedom relative to the chassis which are as follows:

1 Vertical deflection of axle due to static load or dynamic bump and rebound so that both wheels can rise and fall together.
2 Transverse axle twist to permit one wheel to rise while the other one falls at the same time as the vehicle travels over uneven ground.

In addition, the suspension must be able to restrain all other axle movements relative to the chassis and the construction should be such that it is capable of supporting the forces and moments that are imposed between the axle and chassis.

Both vertical axle deflection and transverse axle tilt involve some sort of rotational movement of the restraining and supporting suspension members, be they the springs themselves or separate arm members they must be able to swing about some pivot point.

The two basic methods of providing articulation of suspension members is the pivot pin joint and the ball and socket joint. These joints may either be rigid metal, semi-rigid plastic or flexible rubber, their selection and adoption being determined by the vehicle's operating requirements.

To harness the axle so that it is able to transfer accelerating effort from the wheels to the chassis and vice versa, the suspension must have built-in members which can absorb the following forces and moments;

1 vertical forces caused by vehicle laden weight,
2 longitudinal forces caused by tractive and braking effort,
3 transverse forces caused by centrifugal force, side slopes and lateral winds,
4 rotational torque reactions caused by driving and braking efforts.

10.11.1 Multi-leaf spring eye support
(Fig. 10.75(a, b and c))

Axle location by multi-leaf springs relies on the spring eyes having sufficient strength and support to cope with the vehicle's laden weight driving and braking thrust and lateral forces. Springs designed

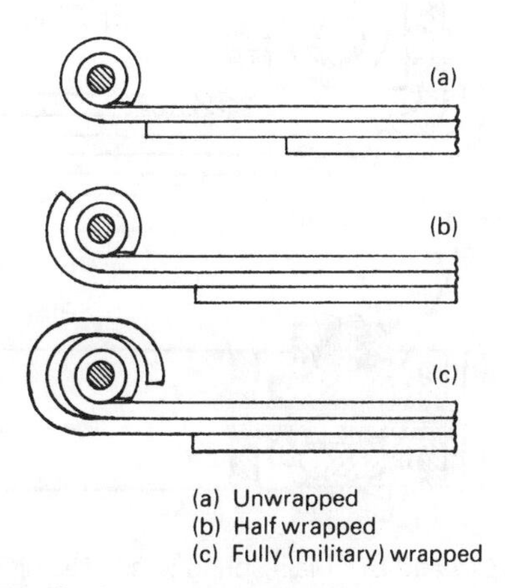

Fig. 10.75 Spring eye protection

for cars and light vans generally need only a single main leaf (Fig. 10.75(a)) wrapped around the bush and shackle pin alone, but for heavy duty conditions it is desirable to have the second leaf wrapped around the main leaf to give it additional support.

If a second leaf were to be wrapped tightly around the main leaf eye, then there could not be any interleaf sliding which is essential for multi-leaf spring flexing to take place. As a compromise for medium duty applications, a partial or half-wrapped second leaf may be used (Fig. 10.75(b)) to support the main leaf of the spring. This arrangement permits a small amount of relative lengthwise movement to occur when the spring deflects between bump and rebound. For heavy duty working conditions, the second leaf may be wrapped loosely in an elongated form around the main lead eye (Fig. 10.75(c)). This allows a degree of relative movement to occur, but at the same time it provides backup for the main leaf eye. If the main leaf should fracture at some point, the second leaf is able to substitute and provide adequate support; it therefore prevents the axle becoming out of line and possibly causing the vehicle to steer out of control.

10.11.2 Transverse and longitudinal spring, axle and chassis attachments (Figs 10.76–10.83)

For small amounts of transverse axle twist, rubber bushes supporting the spring eye-pins and shackle plates are adequate to absorb linkage misalignment, and in extreme situations the spring leaves themselves can be made to distort and accommodate axle transverse swivel relative to the chassis frame. In certain situations where the vehicle is expected to operate over rough ground additional measures may have to be taken to cope with very large degrees of axle vertical deflection and transverse axle tilt.

The semi-elliptic spring may be attached to the chassis and to the axle casing in a number of ways to accommodate both longitudinal spring leaf camber (bow) change due to the vehicle's laden weight and transverse axle tilt caused by one or other wheel rising or falling as they follow the contour of the ground.

Spring leaf end joint attachments may be of the following kinds;

a) cross-pin anchorage (Fig. 10.76),
b) pin and fork swivel anchorage (Fig. 10.77),
c) bolt and fork swivel anchorage (Fig. 10.78),
d) pin and ball swivel anchorage (Fig. 10.79),
e) ball and cap swivel anchorage (Fig. 10.80).

Alternatively, the spring leaf attachment to the axle casing in the mid-span region may not be a direct clamping arrangement, but instead may be through some sort of pivoting device to enable a relatively large amount of transverse axle tilt to be

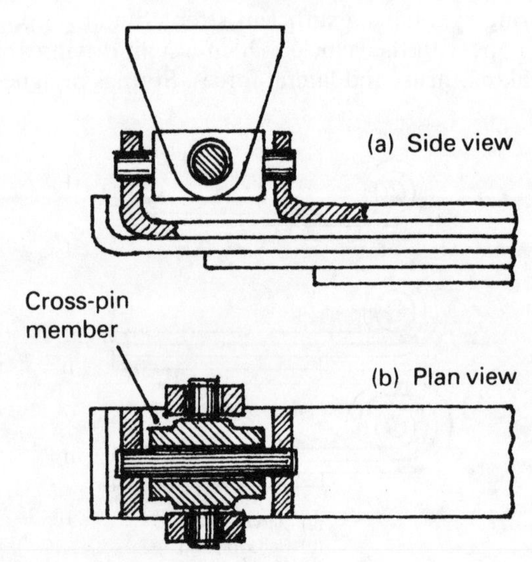

Fig. 10.76 (a and b) Main spring to chassis hinged cross-pin anchorage

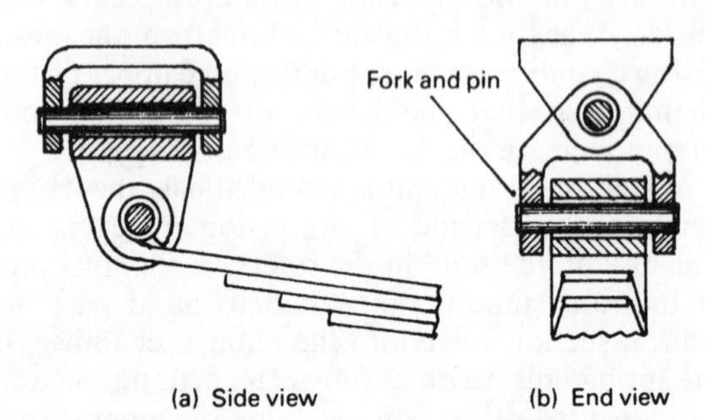

Fig. 10.77 Main spring to chassis pin and fork swivel anchorage

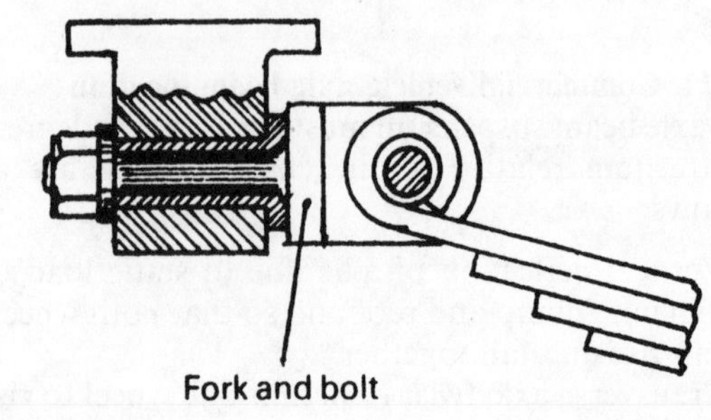

Fig. 10.78 Main spring to chassis bolt and fork swivel anchorage

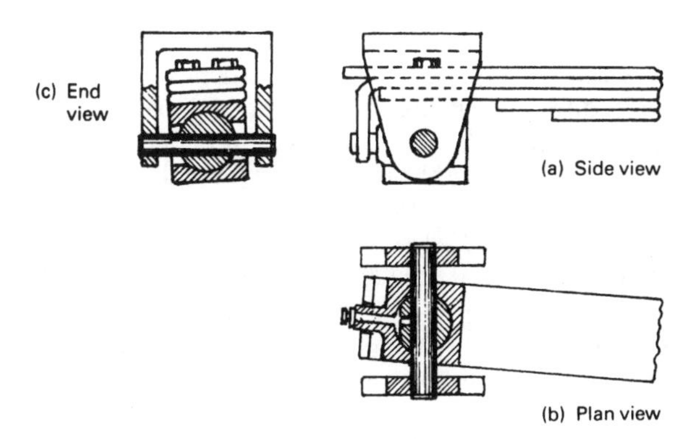

Fig. 10.79 (a–c) Main spring to chassis pin and spherical swivel anchorage

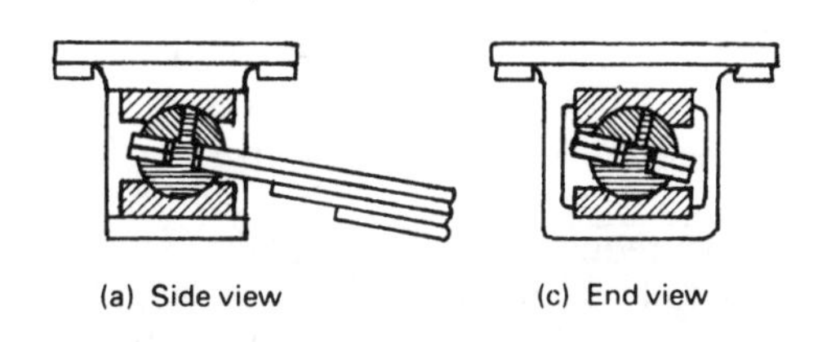

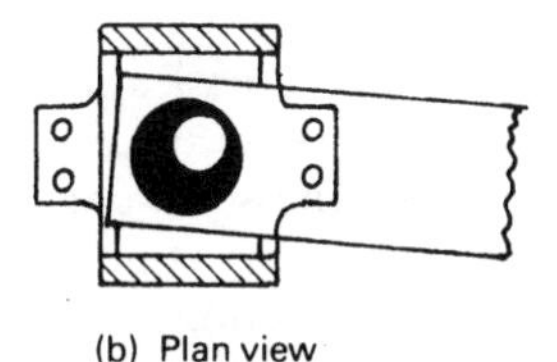

Fig. 10.80 (a–c) Main spring to chassis spherical swivel anchorage

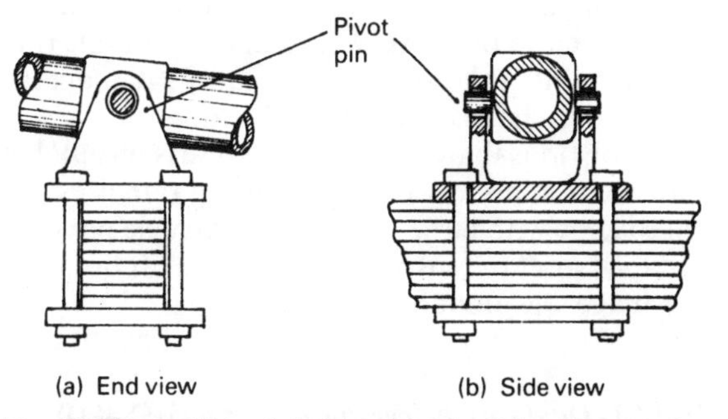

Fig. 10.81(a and b) Axle to spring pivot pin seat mounting

accommodated. Thus transverse axle casing to spring relative movement can be achieved by either a pivot pin (Fig. 10.81) or a spherical axle saddle joint (Fig. 10.82) arrangement. Likewise for reactive balance beam shackle plate attachments the

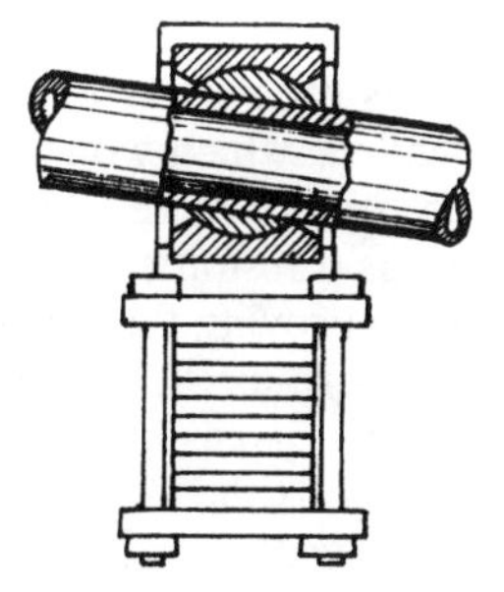

Fig. 10.82 Axle to spring spherical seat mounting

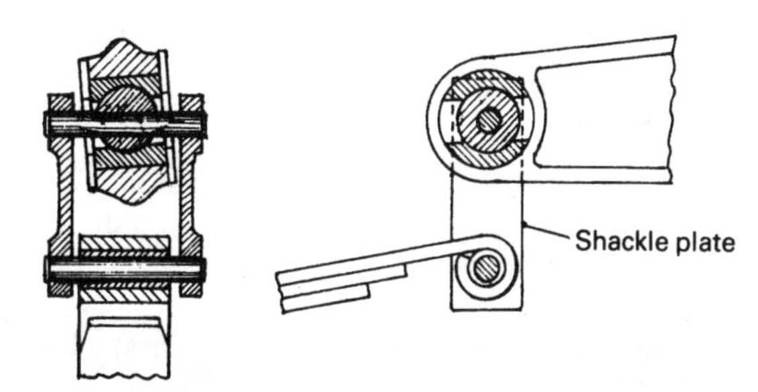

Fig. 10.83 Tandem axle balance beam to shackle plate spherical joint

joints may also be of the spherical ball and cap type joint (Fig. 10.83).

10.12 Variable rate leaf suspension springs

The purpose of the suspension is to protect the body from the shocks caused by the vehicle moving over an uneven road surface. If the axle were bolted directly to the chassis instead of through the media of the springs, the vehicle chassis and body would try to follow a similar road roughness contour and would therefore lift and fall accordingly. With increased speed the wheel passing over a bump would bounce up and leave the road so that the grip between the tyre and ground would be lost. Effectively no tractive effort, braking retardation or steering control could take place under these conditions.

A suspension system is necessary to separate the axle and wheels from the chassis so that when the wheels contact bumps in the road the vertical deflection is absorbed by the elasticity of the spring material, the strain energy absorbed by the springs on impact being given out on rebound but under damped and controlled conditions. The deflection of the springs enables the tyres to remain in contact with the contour of the road under most operating conditions. Consequently the spring insulates the body from shocks, protects the goods being transported and prevents excessively high stresses being imposed on the chassis which would lead to fatigue failure. It also ensures that the driver is cushioned from road vibrations transmitted through the wheels and axle, thereby improving the quality of the ride. The use of springs permits the wheels to follow the road contour and the chassis and body to maintain a steady mean height as the vehicle is driven along the road. This is achieved by the springs continuously extending and contracting between the axle and chassis, thereby dissipating the energy imparted to the wheels and suspension assembly.

A vehicle suspension is designed to permit the springs to deflect from an unladen to laden condition and also to allow further deflection caused by a wheel rapidly rolling over some obstacle or pot hole in the road so that the impact of the unsprung axle and wheel responds to bump and rebound movement. How easily the suspension deflects when loaded statically or dynamically will depend upon the stiffness of the springs (spring rate) which is defined as the load per unit deflection.

i.e. $$\text{Spring stiffness or rate } S = \frac{\text{Applied load}}{\text{Deflection}} = \frac{W}{x}\ (\text{N/m})$$

A low spring stiffness (low spring rate) implies that the spring will gently bounce up and down in its free state which has a low natural frequency of vibration and therefore provides a soft ride. Conversely a high spring stiffness (high spring rate) refers to a spring which has a high natural frequency of vibration which produces a hard uncomfortable ride if it supports only a relatively light load. Front and rear suspensions have natural frequencies of vibration roughly between 60 and 90 cycles per minute. The front suspension usually has a slightly lower frequency than the rear. Typical suspension natural frequencies would be 75/85 cycles per minute for the front and rear respectively. Spring frequencies below 60 cycles per minute promote car sickness whereas frequencies above 90 cycles per minute tend to produce harsh bumpy rides. Increasing the vehicle load or static deflection for a given set of front and rear spring stiffness reduces the ride frequency and softens the ride. Reducing the laden vehicle weight raises the frequency of vibration and the ride hardness.

Vehicle laden weight, static suspension deflection, spring stiffness and ride comfort are all inter-related and produce conflicting characteristics.

For a car there is not a great deal of difference between its unladen and fully laden weight; the main difference being the driver, three passengers, luggage and full fuel tank as opposed to maybe a half full fuel tank and the driver only. Thus if the car weighs 1000 kg and the three passengers, luggage and full fuel tank weigh a further 300 kg, the ratio of laden and unladen weight will be 1300/1000 = 1.3:1. Under these varying conditions, the static suspension deflection can be easily accommodated by soft low spring rates which can limit the static suspension deflection to a maximum of about 50 mm with very little variation in the natural frequency of vibration of the suspension system. For a heavy goods vehicle, if the unladen weight on one of the rear axles is 2000 kg and its fully laden capacity is 10 000 kg, then the ratio of laden to unladen weight would be 10 000/2000 = 5:1. It therefore follows that if the spring stiffness for the axle suspension is designed to give the best ride with the unladen axle, a soft low spring rate would be required. Unfortunately, as the axle becomes fully laden, the suspension would deflect maybe five times the unladen static deflection of, say, 50 mm which would amount to 250 mm. This large change from unladen to fully laden chassis height would cause considerable practical complications and therefore could not be acceptable.

If the suspension spring stiffnesses were to be designed to give the best ride when fully laden, the change in suspension deflection could be reduced to something between 50 and 75 mm when fully laden. The major disadvantage of utilizing high spring rates which give near optimum ride conditions when fully laden would be that when the axle is unladen, the stiffness of the springs would be far too high so that a very hard uncomfortable ride would result, followed by mechanical damage to the various chassis and body structures.

It is obvious that a single spring rate is unsuitable and that a dual or progressive spring rate is essential to cope with large variations in vehicle payload and to restrict the suspension's vertical lift or fall to a manageable amount.

10.12.1 Dual rate helper springs (Fig. 10.84(a))

This arrangement is basically a main semi-elliptic leaf spring with a similar but smaller auxiliary spring located above the main spring. This spring is anchored to the chassis at the front via a shackle pin to the spring hanger so that the driving thrust can be transmitted from the axle and wheel to the chassis. The rear end of the spring only supports

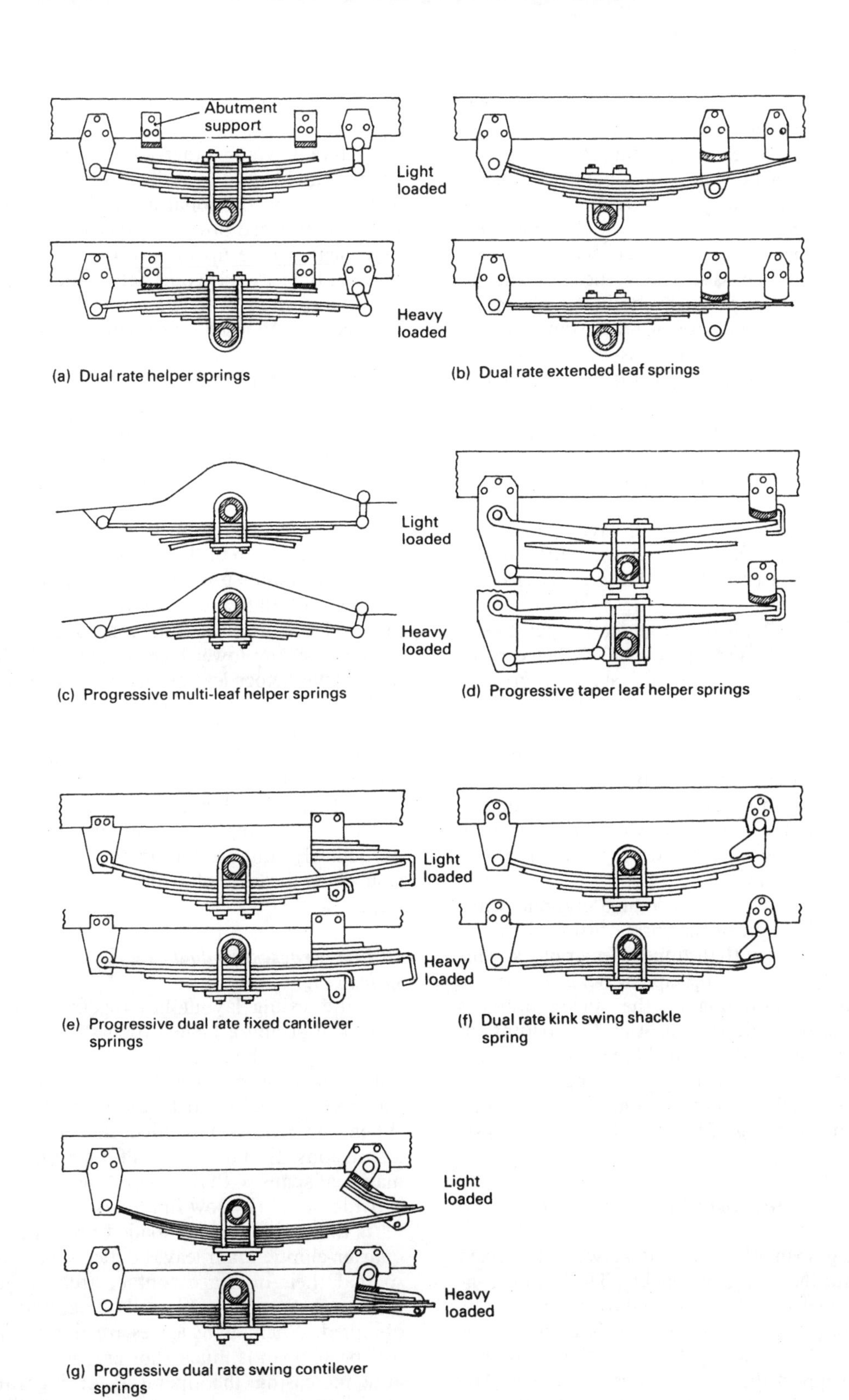

Fig. 10.84 (a–g) Variable rate leaf spring suspension

the downward load and does not constrain the fore and after movement of the spring.

In the unladen state only the main spring supports the vehicle weight and any payload carried (Fig. 10.84(a)) is subjected to a relatively soft ride. Above approximately one third load, the ends of the auxiliary helper spring contact the abutments mounted on the chassis. The vertical downward deflection is now opposed by both sets of springs which considerably increase the total spring rate and also restrict the axle to chassis movement. The method of providing two spring rates, one for lightly laden and a second for near fully laden condition, is adopted by many heavy goods vehicles.

10.12.2 Dual rate extended leaf springs (Fig. 10.84(b))

With this semi-elliptic leaf spring layout the axle is clamped slightly offset to the mid-position of the spring. The front end of the spring is shackled to the fixed hanger, whereas the rear end when unloaded bears against the outer slipper block. The full span of the spring is effective when operating the vehicle partially loaded. A slight progressive stiffening of the spring occurs with small increases in load, due to the main spring blade rolling on the curved slipper pad from the outermost position towards its innermost position because of the effective spring span shortening. Hence the first deflection stage of the spring provides a very small increase in spring stiffness which is desirable to maintain a soft ride.

Once the vehicle is approximately one third laden, the deflection of the spring brings the main blade into contact with the inner slipper block. This considerably shortens the spring length and the corresponding stiffening of the spring prevents excessive vertical deflection. Further loading of the axle will make the main blade roll round the second slipper block, thereby providing the second stage with a small amount of progressive stiffening. Suspension springing of this type has been successful on heavy on/off road vehicles.

10.12.3 Progressive multi-leaf helper springs (Fig. 10.84(c))

The spring span is suspended between the fixed hanger and the swinging shackle. The spring consists of a stack of leaves clamped together near the mid-position, with about two thirds of the leaves bowed (cambered) upward so that their tips contact and support the immediate leaf above it. The remainder of the leaves bow downward and so do not assist in supporting the body weight when the car or van is only partially laden. As the vehicle becomes loaded, the upper spring leaves will deflect and curve down on either side of the axle until their shape matches the first downward set lower leaf. This provides additional upward resistance to the normally upward bowed (curved) leaves so that as more leaves take up the downward bowed shape more of the leaves become active and contribute to the total spring stiffness. This progressive springing has been widely used on cars and vans.

10.12.4 Progressive taper leaf helper springs (Fig. 10.84(d))

Under light loads a small amount of progressive spring stiffening occurs as the rear end of the main taper leaf rolls from the rearmost to the frontmost position on the curved face of the slipper block, thereby reducing the effective spring length. The progressive action of the lower helper leaf is caused by the normally upward curved main taper leaf flexing and flattening out as heavier loads are imposed on the axle. The consequences are that the main spring lower face contact with the upper face of the helper leaf gradually spreads outwards and therefore provides additional and progressive support to the main taper leaf.

The torque rod is provided to transmit the driving force to the chassis and also forms the cranked arms of an anti-roll bar in some designs.

This progressive spring stiffening arrangement is particularly suitable for tractor unit rear suspension where the rates of loaded to unloaded weight is large.

10.12.5 Progressive dual rate fixed cantilever spring (Fig. 10.84(e))

This interesting layout has the front end of the main leaf spring attached by a shackle pin to the fixed hanger. The main blade rear tip contacts the out end of a quarter-elliptic spring, which is clamped and mounted to the rear spring hanger. When the axle is unloaded the effective spring length consists of both the half- and quarter-elliptic main leaf spans so that the combined spring lengths provides a relative low first phase spring rate.

As the axle is steadily loaded both the half- and quarter-elliptic main leaves deflect and flatten out so that their interface contact area progressively moves forwards until full length contact is obtained. When all the leaves are aligned the effective spring span is much shorter, thereby considerably increasing the operating spring rate. This spring suspension concept has been adopted for the rear spring on some tractor units.

10.12.6 Dual rate kink swing shackle spring (Fig. 10.84(f))

Support for the semi-elliptic spring is initially achieved in the conventional manner; the front end of the spring is pinned directly to the front spring hanger and indirectly via the swinging shackle plates to the rear spring hanger. The spring shackle plates have a right angled abutment kink formed on the spring side of the plates.

In the unladen state the cambered (bowed) spring leaves flex as the wheel rolls over humps and dips, causing the span of the spring to continuously extend and contract. Thus the swinging shackle plates will accommodate this movement. As the axle becomes laden, the cambered spring leaves straighten out until eventually the kink abutment on the shackle plates contact the upper face of the main blade slightly in from the spring eye. Any further load increase will kink the main leaf, thereby shortening the effective spring span and resulting in the stiffening of the spring to restrict excessive vertical deflection. A kink swing shackle which provides two stages of spring stiffness is suitable for vans and light commercial vehicles.

10.12.7 Progressive dual rate swing contilever springs (Fig. 10.84(g))

This dual rate spring has a quarter-elliptic spring pack clamped to the spring shackle plates. In the unloaded condition the half-elliptic main leaf and the auxiliary main leaf tips contact each other. With a rise in axle load, the main half-elliptic leaf loses its positive camber and flattens out. At the same time the spring shackle plates swing outward. This results in both main spring leaves tending to roll together thereby progressively shortening the effective spring leaf span. Instead of providing a sudden reduction in spring span, a progressive shortening and stiffening of the spring occurs. Vans and light commercial vehicles have incorporated this unusual design of dual rate springing in the past, but the complicated combined swing shackle plate and spring makes this a rather expensive way of extending the spring rate from unladen to fully laden conditions.

10.13 Tandem and tri-axle bogies

A heavy goods vehicle is normally laden so that about two thirds or more of the total load is carried by the rear axle. Therefore the concentration of weight over a narrow portion of the chassis and on one axle, even between twin wheels, can be excessive.

In addition to the mechanical stresses imposed on the vehicle's suspension system, the subsoil stress distribution on the road for a single axle (Fig. 10.85(a)) is considerably greater than that for a tandem axle bogie (Fig. 10.85(b)) for similar payloads. Legislation in this country does not normally permit axle loads greater than ten tonne per axle. This weight limit prevents rapid deterioration of the road surface and at the same time spreads the majority of load widely along the chassis between two or even three rear axles.

The introduction of more than two axles per vehicle poses a major difficulty in keeping all the wheels in touch with the ground at the same time, particularly when driving over rough terrains (Fig. 10.86). This problem has been solved largely by having the suspensions of both rear axles interconnected so that if one axle rises relative to the chassis the other axle will automatically be lowered and wheel to road contact between axles will be fully maintained.

If twin rear axles are used it is with conventional half-elliptic springs supported by fixed front spring hangers and swinging rear spring shackle plates. If they are all mounted separately onto the chassis, when moving over a hump or dip in the road the front or rear axle will be lifted clear of the ground (Fig. 10.87) so that traction is lost for that particular axle and its wheels. The consequences of one or the other pairs of wheels losing contact with the road surface are that road-holding ability will be greatly reduced, large loads will suddenly be imposed on a single axle and an abnormally high amount of tyre scruffing will take place.

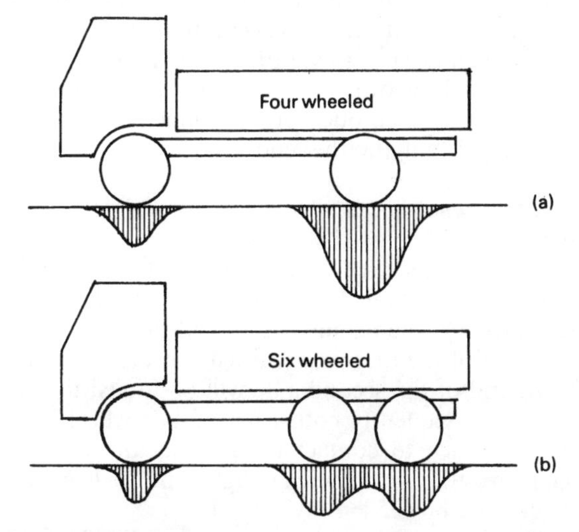

Fig. 10.85 (a and b) Road stress distribution in subsoil underneath road wheels

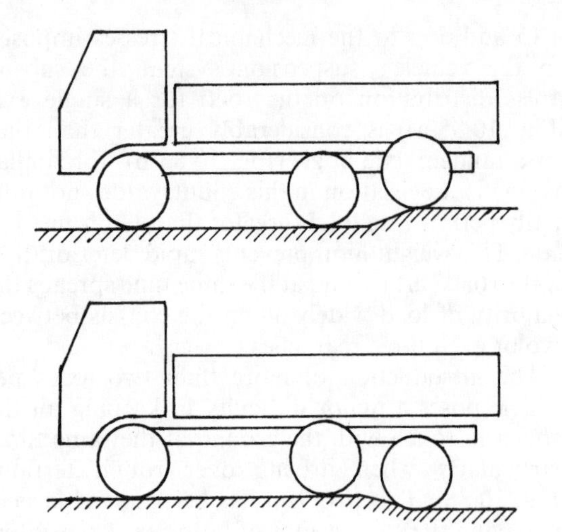

Fig. 10.86 Illustrating the need for tandem axle articulation

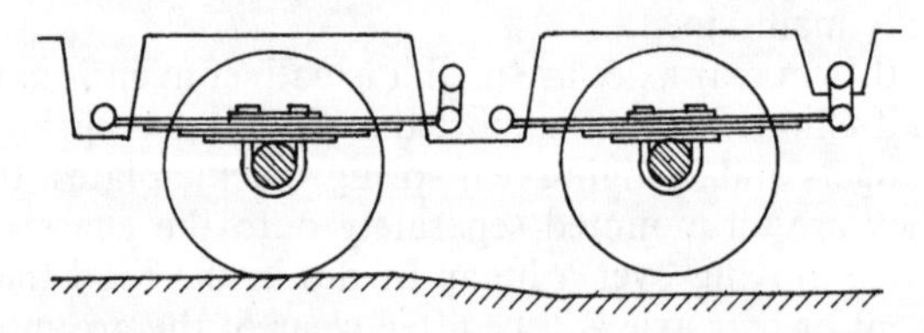

Fig. 10.87 Uncompensated twin axle suspension

To share out the vehicle's laden weight between the rear tandem axles when travelling over irregular road surfaces, two basic suspension arrangements have been developed:

1 pivoting reactive or non-reactive balance beam which interconnects adjacent first and second semi-elliptic springs via their shackle plates,
2 a central pivoting single (sometimes double) vertical semi-elliptic spring which has an axle clamped to it at either end.

10.13.1 Equalization of vehicle laden weight between axles (Figs 10.88 and 10.89)

Consider a reactive balance beam tandem axle bogie rolling over a hump or dip in the road (Fig. 10.88). The balance beam will tilt so that the rear end of the first axle is lifted upwards and the front end of the second axle will be forced downward. Consequently both pairs of axle wheels will be compelled to contact the ground and equally share out the static laden weight imposed on the whole axle bogie.

The tilting of the balance beam will lift the first axle a vertical distance $h/2$, which is half the hump

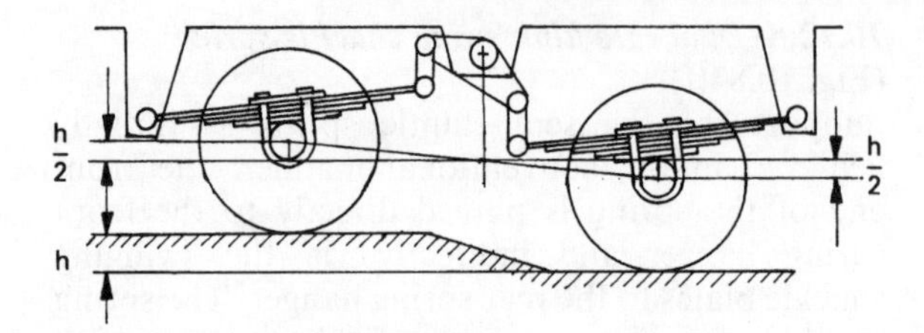

Fig. 10.88 Payload distribution with reactive balance beam and swing shackles

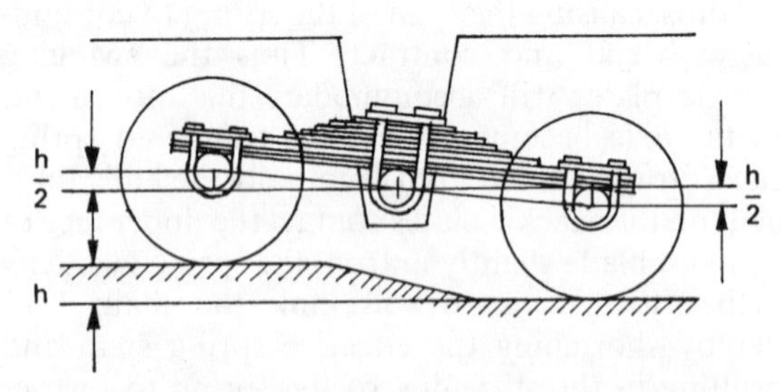

Fig. 10.89 Payload distribution with single inverted semi-elliptic spring

or dip's vertical height. The second axle will fall a similar distance $h/2$. The net result is that the chassis with the tandem axle bogie will only alter its height relative to ground by half the amount of a single axle suspension layout (Fig. 10.88). Thus the single axle suspension will lift or lower the chassis the same amount as the axle is raised or lowered from some level datum, whereas the tandem axle bogie only changes the chassis height relative to the ground by half the hump lift or dip drop.

In contrast to the halving of the vertical lift or fall movement of the chassis with tandem axles, there are two vertical movements with a tandem axle as opposed to one for a single axle each time the vehicle travels over a bump. Thus the frequency of the chassis vertical lift or fall with tandem axles will be twice that for a single axle arrangement.

Similar results will be achieved if a central pivoting inverted transverse spring tandem axle bogie rides over a hump or dip in the road (Fig. 10.89). Initially the first axle will be raised the same distances as the hump height h, but the central pivot will only lift half the amount $h/2$. Conversely if the first axle goes into a dip, the second axle will be above the first axle by the height of the dip, but the chassis will only be lowered by half this vertical movement $h/2$. Again the frequency of lift and fall of the chassis as the tandem axles move over the irregularities in the road will be double the frequency compared to a single axle suspension.

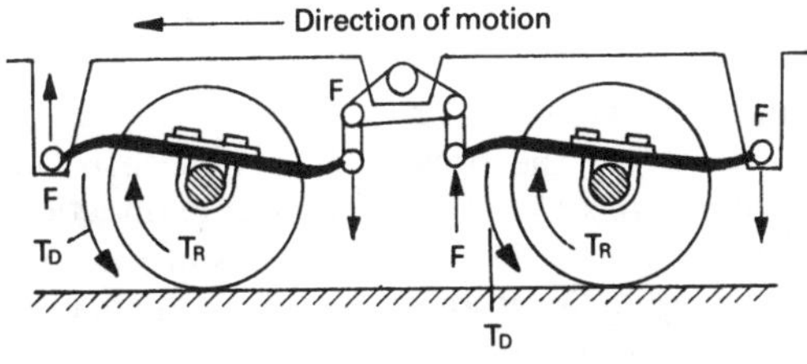

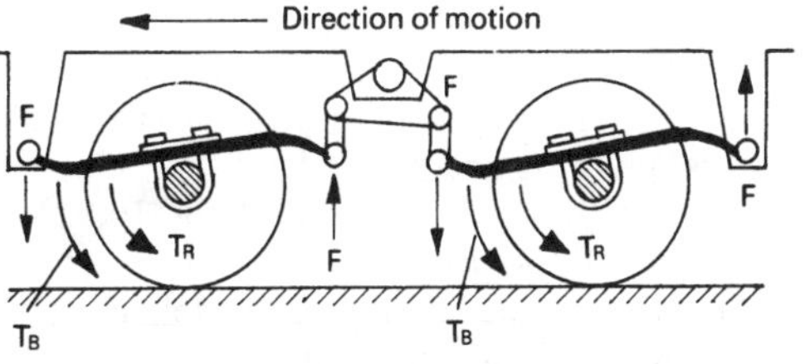

Fig. 10.90 (a and b) Reactive balance beam tandem axle suspension

10.13.2 Reactive balance beam tandem axle bogie suspension (Fig. 10.90(a and b))

Suspension arrangements of this type distribute the laden weight equally between the two axles due to the swing action of the balance beam (Fig. 10.90 (a and b)). The balance beam tilts according to the reaction load under each axle so that, within the chassis to ground height variation limitations, it constantly adjusts the relative lift or fall of each axle to suit the contour of the road.

Unfortunately the driving and braking torques produce unequal reaction through the spring linkage. Therefore under these conditions the vehicle's load will not be evenly distributed between axles.

Consider the situation when tractive effort is applied at the wheels when driving away from a standstill (Fig. 10.90(a)). Under these conditions the driving axle torque T_D produces an equal but opposite torque reaction T_R which tends to make the axle casing rotate in the opposite direction to that of the axle shaft and wheel. Subsequently the front spring ends of both axles tend to be lifted by force *F*, and the rear spring ends are pulled downwards by force *F*. Hence the overall reaction at each spring to chassis anchor point causes the balance beam to tilt anticlockwise and so lift the chassis away from the first axle, whereas the second axle is drawn towards the chassis. This results in the contact reaction between wheel and ground for the first axle to be far greater than for the second axle. In fact the second axle may even lose complete contact with the road.

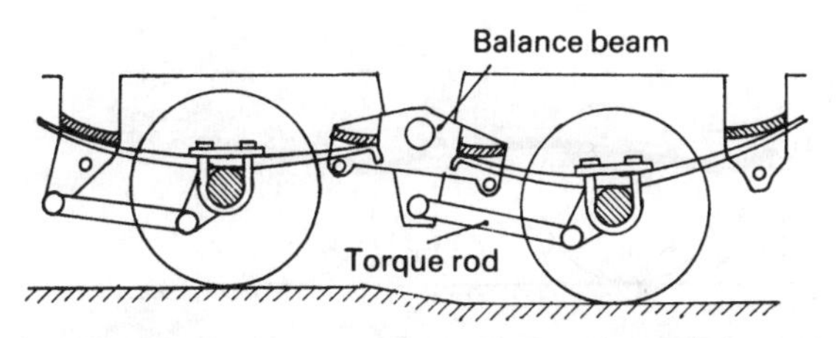

Fig. 10.91 Reactive balance beam with slipper contact blocks and torque arms tandem axle suspension

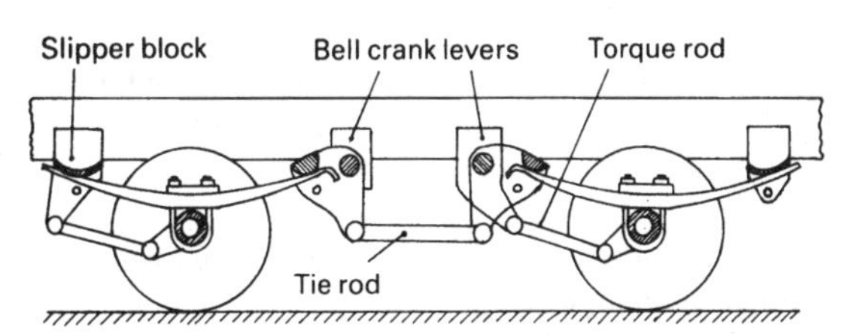

Fig. 10.92 Tandem wide spread reactive bell crank lever taper leaf spring

Conversely if the brakes are applied (Fig. 10.90(b)), the retarding but still rotating wheels will tend to drag the drum or disc brake assembly round with the axle casing T_R. The rotation of the axle casing in the same direction of rotation as the wheels means that the front spring ends of both axles will be pulled downward by force *F*. The corresponding rear spring ends will be lifted upward by the reaction force *F*. Thus in contrast to the driving torque directional reaction, the braking torque T_B will tilt the balance beam clockwise so that the second axle and wheel will tend to move away from the chassis, thereby coming firmly into contact with the road surface. The first axle and wheel will move further towards the chassis so that very little grip between the tyre and road occurs. In practice the upward lift of the first wheel and axle will cause the tyres to move in a series of hops and rebounds which will result in heavily loading the second axle, reducing the overall braking effectiveness and causing the first axle tyres to be subjected to excessive scuffing.

A reactive balance beam tandem axle bogie suspension using tapered leaf springs and torque arms to transmit the driving and braking forces and torques is shown in Fig. 10.91. With this layout driving and braking torque reactions will cause similar unequal load distribution.

To enable a wide spread axle to be used on trailers, the conventional reactive balance beam interconnecting spring linkage has been modified so that laden vehicle weight can still be shared equally between axles. Thus instead of the central balance beam (Fig. 10.90) there are now two bell crank levers pivoting back to back on chassis spring hangers with a central tie rod (Fig. 10.92).

In operation, if the front wheel rolls over an obstacle its supporting spring will deflect and apply an upward thrust against the bell crank lever slipper. Accordingly, a clockwise turning moment will be applied to the pivoting lever. This movement is then conveyed to the rear bell crank lever via the tie rod, also making it rotate clockwise. Consequently the rear front end of the spring will be lowered, thus permitting the rear wheels to keep firmly in contact with the road while the chassis remains approximately horizontal.

10.13.3 Non-reactive bell crank lever and rod tandem axle bogie suspension

(Fig. 10.93(a and b))

To overcome the unequal load distribution which occurs with the reactive balance beam suspension when either driving or braking, a non-reactive bell crank lever and rod linkage has been developed which automatically feeds similar directional reaction forces to both axle rear spring end supports (Fig. 10.93(a and b)).

Both axle spring end reactions are made to balance each other by a pair of bell crank levers mounted back to back on the side of the chassis via pivot pins. Each axle rear spring end is attached by a shackle plate to the horizontal bell crank lever ends while the vertical bell crank lever ends are interconnected by a horizontally positioned rod.

When the vehicle is being driven (Fig. 10.93(a)) both axle casings react by trying to rotate in the opposite direction to that of the wheels so that the axle springs at their rear ends are pulled downward. The immediate response is that both bell crank levers will tend to twist in the opposite direction to each other, but this is resisted by the connecting rod which is put into compression. Thus the rear end of each axle spring remains at the same height relative to the chassis and both axles will equally share the vehicle's laden weight.

Applying the brakes (Fig. 10.93(b)) causes the axle casings to rotate in the same direction as the wheels so that both axle springs at their rear ends will tend to lift. Both rear spring ends are attached to the horizontal ends of the bell crank levers. Therefore they will attempt to rotate in the opposite direction to each other, but any actual movement is prevented by the interconnected rod which will be subjected to a tensile force. Therefore equal braking torques are applied to each axle and equal turning moments are imposed on each bell crank lever which neutralizes any brake reaction in the suspension linkage. Since there is no interference with the suspension height adjustment during braking, the load distribution will be equalized between axles, which will greatly improve brake performance.

10.13.4 Inverted semi-elliptic spring centrally pivoted tandem axle bogie suspension

(Figs 10.94, 10.95 and 10.96)

This type of tandem axle suspension has either one or two semi-elliptic springs mounted on central pivots which form part of the chassis side members. The single springs may be low (Fig. 10.94) or high (Fig. 10.95) mounted. To absorb driving and braking torque reaction, horizontally positioned torque arms are linked between the extended chassis side members and the axle casing. If progressive slipper spring ends are used (Fig. 10.95), double torque arms are inclined so that all driving and braking torque reactions are transmitted through these arms and only the vehicle's laden vertical load is carried by the springs themselves.

Articulation of the axles is achieved by the inverted springs tilting on their pivots so that one axle will be raised while the other one is lowered when negotiating a hump or dip in the road. As the axles move up and down relative to the central pivots, the torque arms will also pivot on their

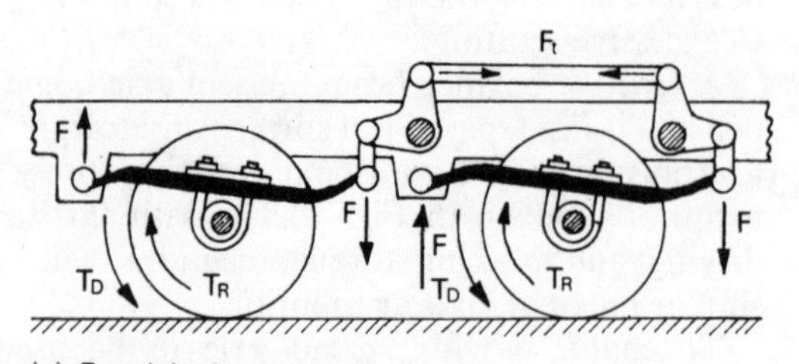

(a) Equal driving torque distribution

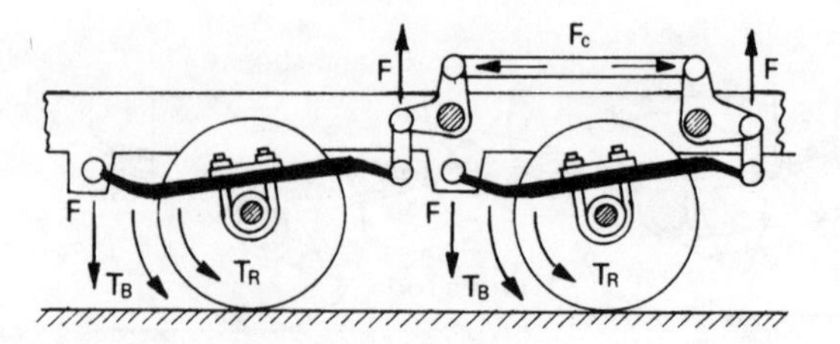

(b) Equal braking torque distribution

Fig. 10.93 (a and b) Non-reaction bell crank lever and rod

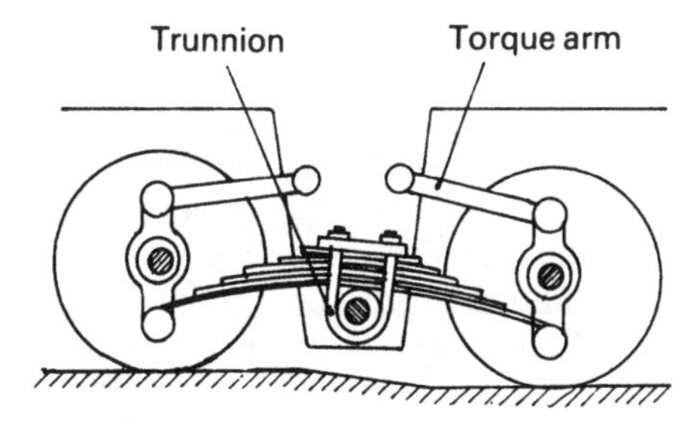

Fig. 10.94 Low mounted single inverted semi-elliptic spring with upper torque rods

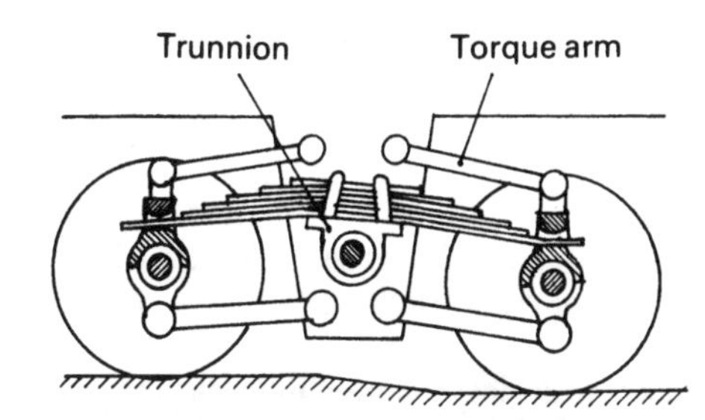

Fig. 10.95 High mounted single inverted semi-elliptic spring with lower torque rods

rubber end joints. Therefore the axle casing vertical arms will remain approximately upright at all times.

Any driving or braking reaction torque is transmitted through both the springs and torque arms to the central spring pivot and torque rod joint pins mounted on the reinforced and extended chassis side members. Very little interference is experienced with the load distribution between the two axles when the vehicle is being accelerated or retarded.

For heavy duty cross-country applications the double inverted semi-elliptic spring suspension is particularly suitable (Fig. 10.96). The double inverted spring suspension and the central spring pivots, enable the springs to swivel a large amount (up to a 500 mm height difference between opposite axles) about their pivots when both pairs of axle wheels roll continuously over very uneven ground. This arrangement tolerates a great deal more longitudinal axle articulation than the single inverted spring and torque arm suspension.

Large amounts of transverse (cross) articulation are made possible by attaching the upper and lower spring ends to a common gimbal bracket which is loosely mounted over the axle casing (Fig. 10.96). The gimbal brackets themselves are supported on horizontal pivot pins anchored rigidly to the casing. This allows the axle to tilt transversely relative to the bracket's springs and chassis without causing any spring twist or excessive stress concentrations between flexing components.

10.13.5 Alternative tandem axle bogie arrangements

Leading and trailing arms with inverted semi-elliptic spring suspension (Fig. 10.97) An interesting tandem axle arrangement which has been used for recovery vehicles and tractor units where the laden to unladen ratio is high is the inverted semi-elliptic spring with leading and trailing arm (Fig. 10.97). The spring and arms pivot on a central chassis member; the arm forms a right angle with its horizontal portion providing the swing arm, while the vertical upper portion is shaped to form a curved slipper block bearing against the end of the horizontal semi-elliptic leaf spring.

The upper faces of the horizontal swing arm are also curved and are in contact with a centrally mounted 'V'-shaped member which becomes effective only when the tandem axle bogie is about half laden. Initially in the unladen state, both swing arms are supported only by the full spring length; this therefore provides a relatively low spring stiffness. As the axles become loaded, the leading and

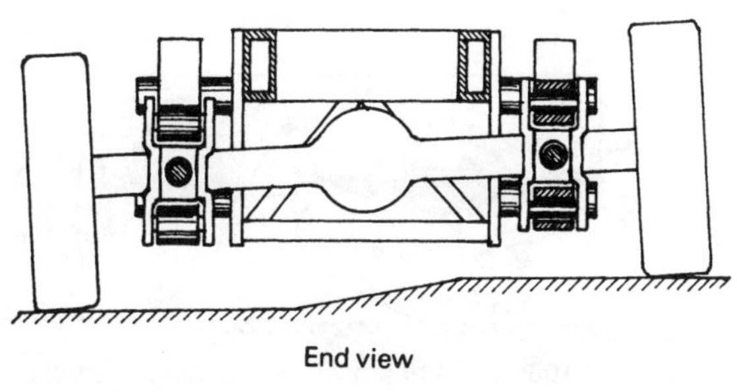

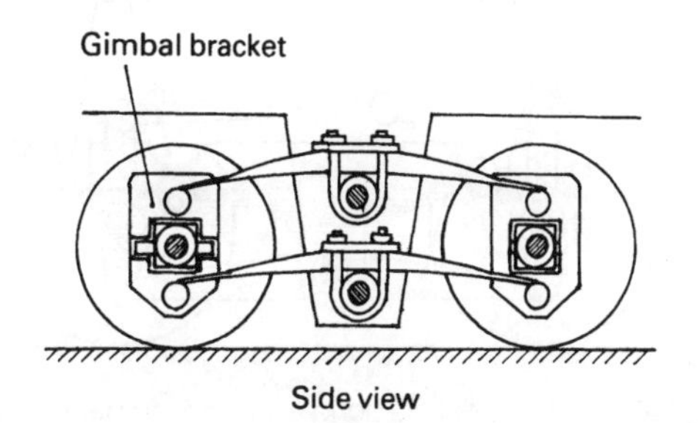

Fig. 10.96 Double inverted semi-elliptic spring

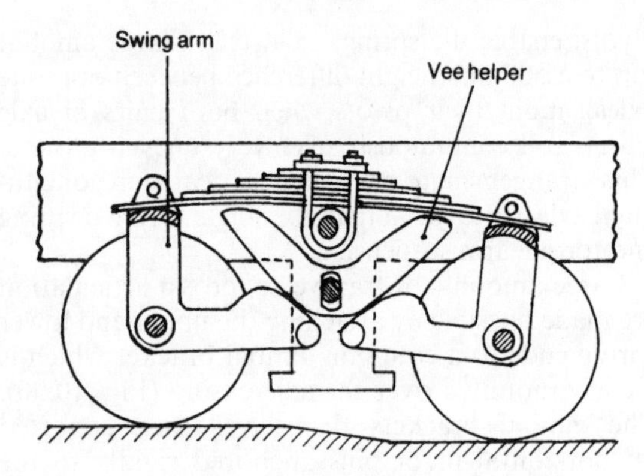

Fig. 10.97 Leading and trailing arms with inverted semi-elliptic spring

trailing arms pivot and swing upward, thereby steadily pushing the central 'V' helper member into contact with the main spring leaf over a much shorter blade span. The rolling contact movement between the upper and lower faces of the swing arms and the central 'V' helper member produce a progressive stiffening of the main spring under laden conditions.

Hendrickson long equalization balance beam with single semi-elliptic springs (Fig. 10.98) This tandem suspension arrangement uses a low mounted, centrally pivoted long balance beam spanning the distance between axles and high mounted leading and trailing torque rods (Fig. 10.98). A semi-elliptic spring supports the vehicle's payload. It is anchored at the front end to a spring hanger and at the rear bears against either the outer or both inner and outer curved slipper hangers. The balance beam is attached to the spring by the 'U' bolts via its pivot mount.

The spring provides support for the vehicle's weight and transmits the accelerating or decelerating thrusts between the axles and chassis. The balance beam divides the vehicle's laden weight between the axles and in conjunction with the torque rods absorbs the driving and braking torque reaction. The two stage spring stiffness is controlled by the effective spring span, which in the unladen condition spans the full spring length to the outer slipper block and in the laden state is shortened as the spring deflects, so that it now touches the inner slipper block spring hanger. For some cross-country applications the outer slipper block hanger is not incorporated so that there is only a slight progressive stiffening due to the spring blade to curved slipper block rolling action as the spring deflects with increasing load. With this four point chassis frame mounting and rigid balance beam, both the springs and the chassis are protected against concentrated stress which therefore makes this layout suitable for on/off rigid six or eight wheel rigid tracks.

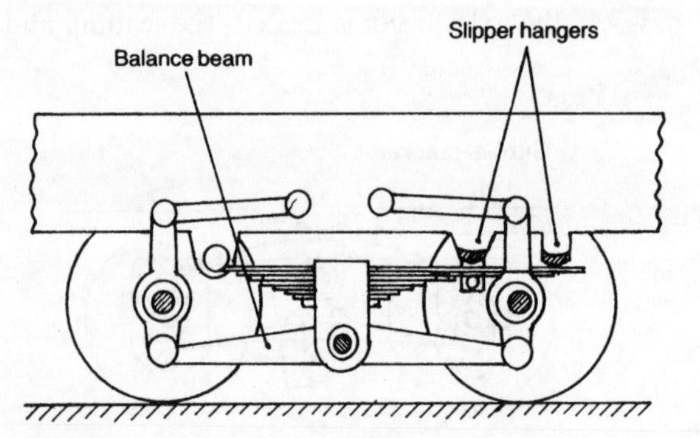

Fig. 10.98 Hendrickson long equalization balance beam with single semi-elliptic spring

Pivot beam with single semi-elliptic spring (Figs 10.99 and 10.100) This kind of suspension has a single semi-elliptic spring attached at the front end directly to a spring hanger and at the rear to a pivoting beam which carries the trailing axle (Fig. 10.99).

With a conventional semi-elliptic spring suspension, the fixed and swing shackles both share half ($\frac{1}{2}W$) of the reaction force imposed on the chassis caused by an axle load W.

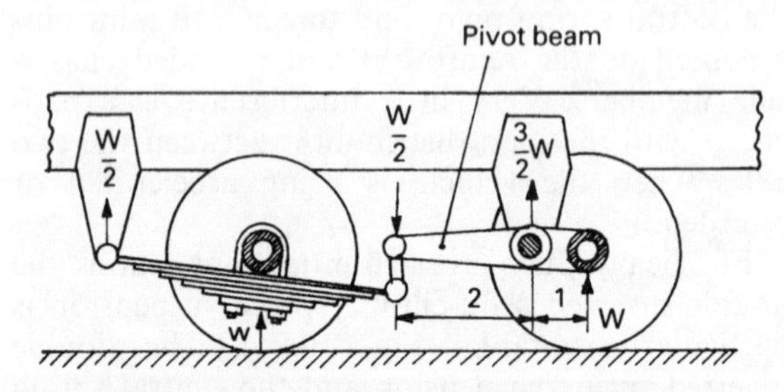

Fig. 10.99 Pivot beam with single semi-elliptic spring

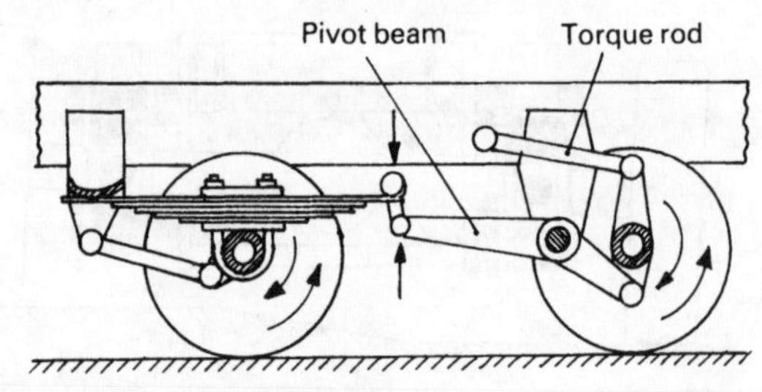

Fig. 10.100 Pivot beam with semi-elliptic spring and torque rod

With the pivoting balance beam coupled to the tail-end of the spring, half the leading axle load ($\frac{1}{2}W$) reacting at the swing shackle is used to balance the load supported by the trailing axle. For the chassis laden weight to be shared equally between axles, the length of beam from the pivot to the shackle plate must be twice the trailing distance from the pivot to the axle. This means that if the load reaction at each axle is W, then with the leading axle clamped to the centre of the spring span and with a pivot beam length ratio of 2:1 the upward reaction force on the front spring hanger will be $\frac{1}{2}W$ and that acting through the pivot $1\frac{1}{2}W$, giving a total upward reaction force of $\frac{1}{2}W + 1\frac{1}{2}W = 2W$. In other words, the downward force at the front of the pivot beam caused by the trailing axle supported by the pivot is balanced by the upward force at the rear end of the spring caused by the load on the leading axle. Thus if the front wheel lifts as it rolls over a bump, the trailing end of the spring rises twice as much as the axle. It attempts to push the trailing axle down so that its wheels are in hard contact with the ground.

With the second axle mounted between the lower trailing arm and the upper torque rod (Fig. 10.100), most of the driving and braking torque reaction is neutralized. Only when accelerating with a single drive axle is there some weight transfer from the non-drive axle (second) to the drive axle (first).

By arranging the first axle to be underslung (Fig. 10.100) instead of overslung (Fig. 10.99), a wider spring base projected to the ground will result in greater roll resistance.

Trailing arm with progressive quarter-elliptic spring (Fig. 10.101) Each axle is carried on a trailing arm; the arms on one side are interconnected by a spring in such a way that the upward reaction at one wheel increases the downward load on the other (Fig. 10.101). The inverted quarter-elliptic spring is clamped to the rear trailing arm.

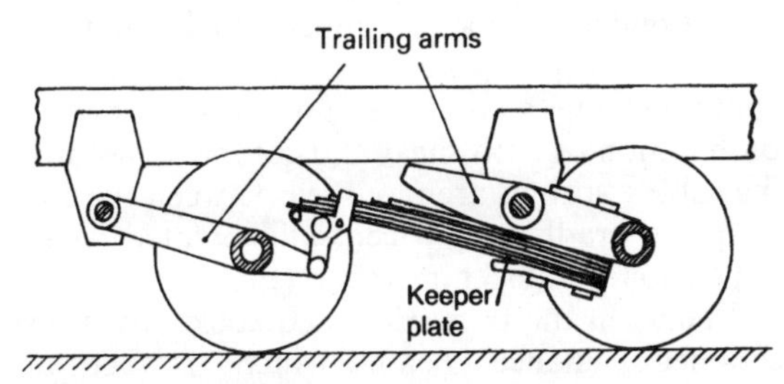

Fig. 10.101 Trailing arm with progressive quarter-elliptic spring

Its leading end is shackled to a bracket on the front trailing arm. Both trailing arms are welded fabricated steel members of box-section. The attachment of the quarter-elliptic springs to the rear trailing arms is so arranged that as the spring deflects on bump a greater length of spring comes into contact with the curved surface of the arm, thereby reducing the effective spring length with a corresponding increase in stiffness. On rebound, the keeper plate beneath the spring is extended forward and curved downward so that there is some progressive stiffening of the spring also on rebound. With this effective spring length control, the trailer will ride softly and easily when unladen and yet the suspension will be able to give adequate upward support when the trailer is fully laden.

Tri-axle semi-trailer suspension (Fig. 10.102(a and b)) Tri-axle bogies are used exclusively on trailers. Therefore all these axles are dead and only laden weight distribution and braking torque reaction need to be considered.

The reactive balance beam interlinking between springs is arranged in such a way that an upward reaction at one wheel increases the downward load on the other, so that each of the three axles supports one third of the laden load (Fig. 10.102(a)).

The load distribution between axles is not quite so simple when the vehicle is being braked, owing to torque reaction making the axle casings rotate in the opposite sense to that of the road wheels. Consequently the foremost end of each spring tends to pull downwards while the rearmost spring ends push upwards. Accordingly the balance beams will react and therefore tilt clockwise. The net change in axle height relative to the chassis is that as the first axle is raised slightly so that tyre to road contact is reduced, the second axle experiences very little height change since the spring front end is made to dip while the rear end is lifted, and the third axle is forced downwards which increases the axle load and the tyre to road contact grip. This uneven axle load distribution under braking conditions is however acceptable since it does not appear to greatly affect the braking efficiency or to cause excessive tyre wear.

One problem with tri-axle trailers is that it is difficult and even impossible to achieve true rolling for all wheels when moving on a curved track due to the large wheel span of the three parallel axles, thus these layouts can suffer from excessive tyre scrub. This difficulty can be partially remedied by using only single wheels on the foremost and rearmost axles with the conventional twin wheels on

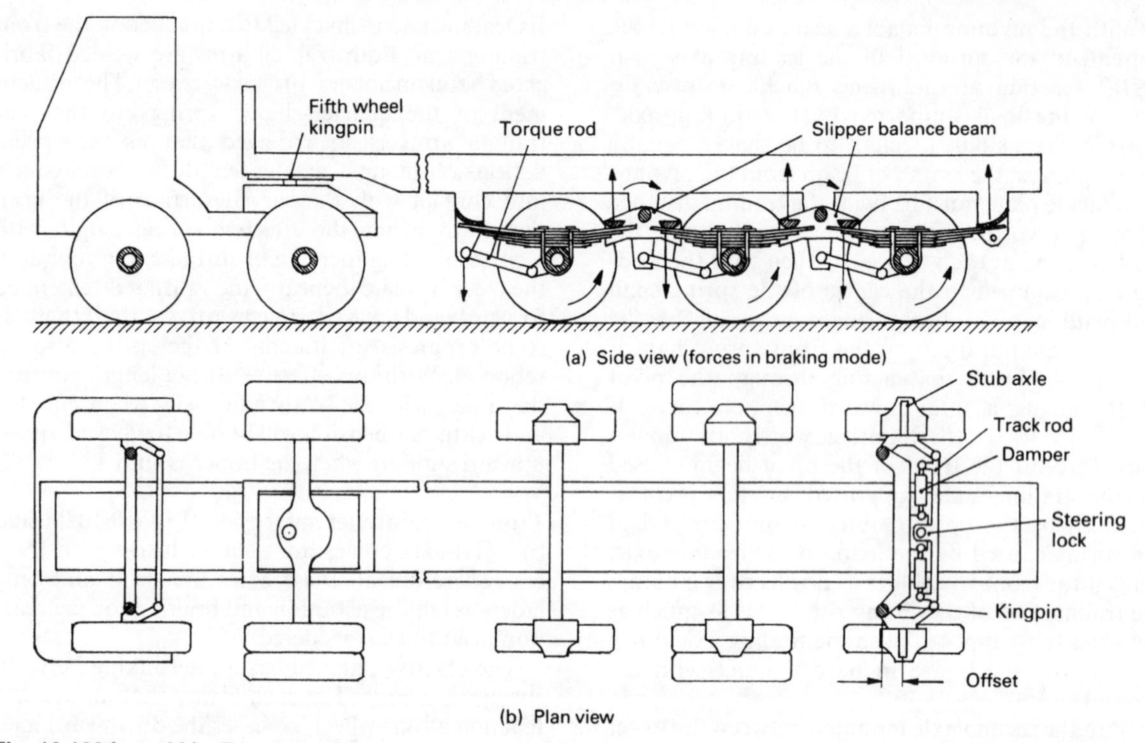

Fig. 10.102 (a and b) Tri-axle semi-trailer with self-steer axle

the middle axle (Fig. 10.102(b)). An alternative and more effective method is to convert the third axle into a self-steer one. Self-steer axles, when incorporated as part of the rearmost axle, not only considerably reduce tyre scrub but also minimize trailer cut-in because of the extent that the rear end is kicked out when cornering. Not only do self-steer axles improve tri-axle wheel tracking but they are also justified for tandem axle use.

Self-steer axle (Fig. 10.102(b)) The self-steer axle has a conventional axle beam with kingpin bosses swept forward to that of the stub axle centre line to provide the offset positive castor trail (Fig. 10.102(b)). Consequently the cornering side thrust on the tyre walls causes the wheels to turn the offset kingpins into line with the vehicle's directional steered path being followed. Excessive movement of either wheel about its kingpin is counteracted by the opposite wheel through the interconnecting track rod, while the trail distance between the kingpin and stub axle provides an automatic self-righting action when the vehicle comes out of a turn.

Possible oscillation on the stub axles is absorbed by a pair of heavy-duty dampers which become very effective at speed, particularly if the wheels are out of balance or misaligned.

Since the positive castor trail is only suitable for moving in the forward direction, when the vehicle reverses the wheels would tend to twitch and swing in all directions. Therefore, when the vehicle is being reversed, the stub axles are locked by a pin in the straight ahead position, this operation being controlled by the driver in the cab. The vehicle therefore behaves as if all the rear wheels are attached to rigid axles.

10.14 Rubber spring suspension

10.14.1 Rubber springs mounted on balance beam with stabilizing torque rods (Fig. 10.103)

Suspension rubber springs are made from alternatively bonded layers of rubber blocks and steel reinforcement plates sandwiched between inclined mounting plates so that the rubber is subjected to a combination of both shear and compressive forces. The rubber springs are mounted between the chassis spring cradle and a centrally pivoted wedge-shaped load transfer member (Fig. 10.103). The load between the two axles is equalized by a box-sectioned balance beam which is centrally mounted by a pivot to the load transfer member. To eliminate brake torque reaction, upper 'A'

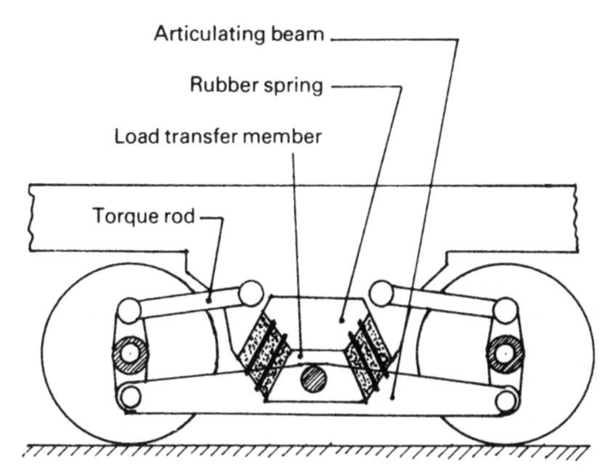

Fig. 10.103 Rubber spring mounted on balance beam with leading and trailing torque arms

brackets or torque arms are linked between the axles and chassis. With a pair of inclined rubber springs positioned on both sides of the chassis, loading of the axles produces a progressive rising spring rate due to the stress imposed into the rubber, changing from shear to compression as the laden weight rises.

The axles are permitted to articulate to take up any variation in road surface unevenness independently of the amount the laden weight of the vehicle has caused the rubber springs to deflect.

All pivot joints are rubber bushed to eliminate lubrication.

These rubber spring suspensions can operate with a large amount of axle articulation and are suitable for non-drive tandem trailers, rigid trucks with tandem drive axles and bulk carrier tankers.

10.14.2 Rubber spring mounted on leading and trailing arms interlinked by balance beam

(Fig. 10.104(a, b and c))

This tandem axle suspension is comprised of leading and trailing swing arms pivoting at their inner ends on the downward extending chassis frame with their outer ends clamped to the axle casings (Fig. 10.104(a)). The front and rear rubber springs are sandwiched between swing arm rigid mounting plates and a centrally pivoting balance beam. When in position these springs are at an inclined angle and are therefore subjected to a combination of compression and shear force.

When the swing arms articulate the spring mounting plate faces swivel and move in arcs. Thus the nature of the spring loading changes from a mainly shear action with very little compressive loading when the axles are unladen (Fig. 10.104(b)) to much greater compressive loading and very little shear as the axles become fully laden (Fig. 10.104(c)). Since the rubber springs are about 14 times stiffer in compression than shear, the springs become progressively harder as the swing arms deflect with increasing laden weight. If the first axle is deflected upward as it moves over a bump, the increased compressive load acting on the spring will tilt the balance beam so that an equal increase in load will be transferred to the second axle.

Because the axles are mounted at the ends of the swing arms and the springs are positioned nearer to the pivot centres, axle movement will be greater than spring deflection. Therefore the overall suspension spring stiffness is considerably reduced for the ratio of axle and spring plate distance from the swing arm pivot centre which accordingly lowers the bounce frequency by 30%. Both leading and trailing swing arms absorb the braking torque reaction so that load distribution between axles will be approximately equal.

10.14.3 Willetts (velvet ride) leading and trailing arm torsional rubber spring suspension

(Fig. 10.105(a and b))

The tandem suspension consists of leading and trailing swing arms. These arms are mounted back to back with their outer ends attached to the first and second drive axles whereas the swivel ends are supported on central trunnion pivot tubes which are mounted on a frame cross-member on either side of the chassis (Fig. 10.105(a)).

Torque arms attached to the suspension cross-member and to brackets in the centres of each axle casing assist the swing arms to transfer driving and braking torque reaction back to the chassis. These stabilizing torque arms also maintain the axles at the correct angular position. Good drive shaft geometry during articulation is obtained by the torque arms maintaining the axles at their correct angular position. Panhard rods (transverse tracking arms) between the frame side-members and the axle casings provide positive axle control and wheel tracking alignment laterally.

The spring consists of inner and outer annular shaped rubber members which are subjected to both torsional and vertical static deflection (Fig. 10.105(b)). The inner rubber member is bonded on the inside to the pivot tube which is supported by the suspension cross-member and on the outside to a steel half shell.

The outer rubber member is bonded on the inside to a median ring and on the outside to two half shells. The inside of the median ring is profiled

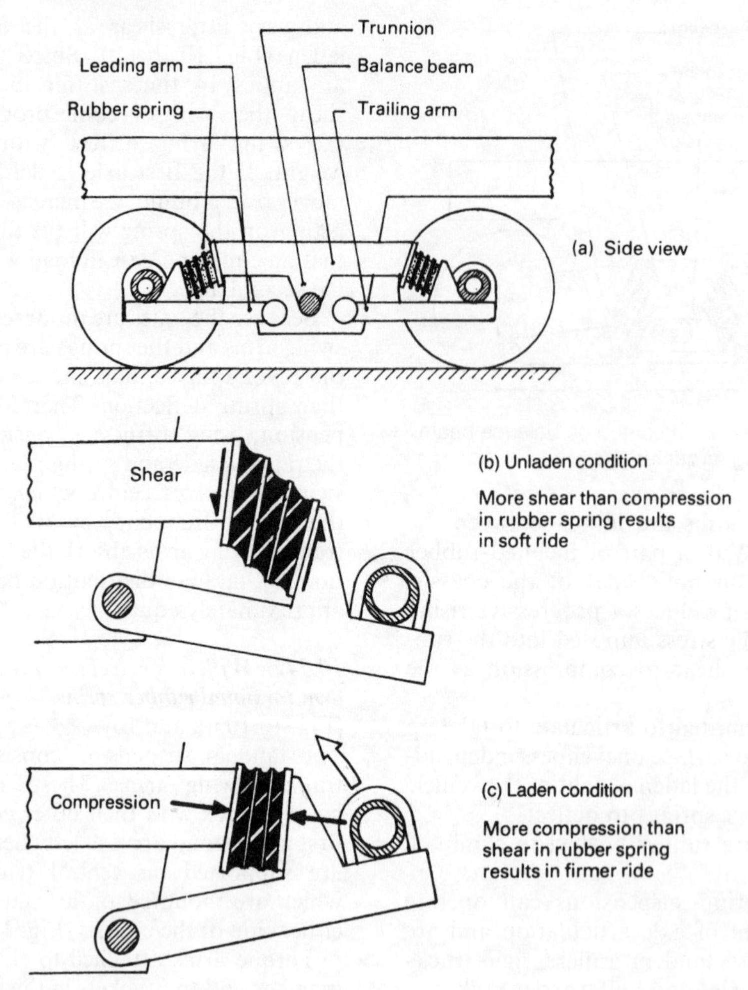

Fig. 10.104 (a–c) Rubber springs mounted leading and trailing arms interlinked by rocking beam

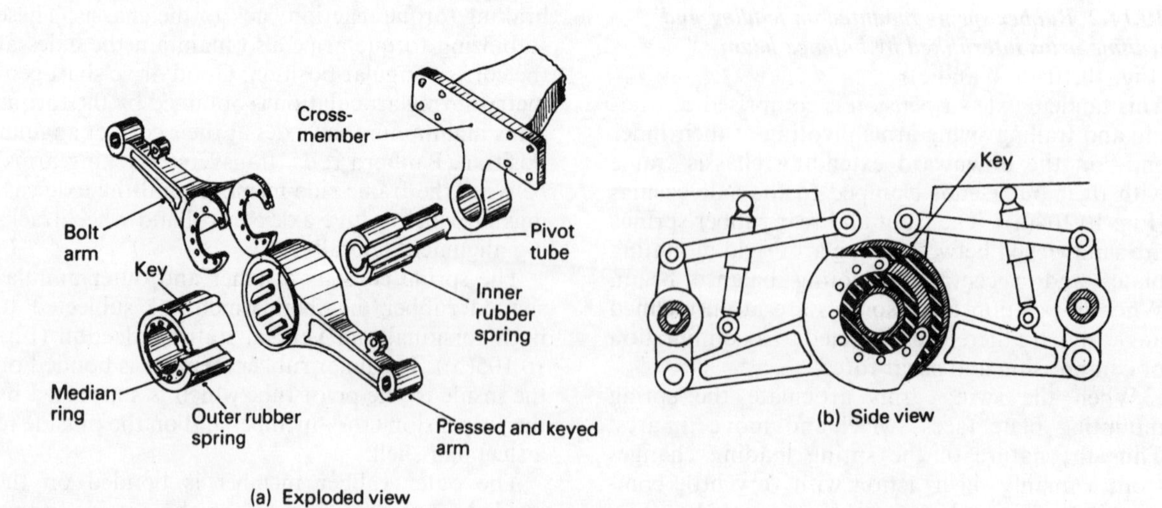

Fig. 10.105 (a and b) Willetts (velvet) leading and trailing arm torsional rubber spring suspension

to the same shape as the inner rubber member and half shell thus preventing inter rotation between the inner and outer rubber members. Key abutments are formed on the circumference of each outer half shell. These keys are used to locate (index) the outer rubber spring members relative to the trailing pressed (keyed) swing arm (Fig. 10.105(b)).

When assembled, the outer rubber member fits over the inner rubber member and half shell, whereas the trailing swing arm spring aperture is a press fit over the outer pair of half shells which are bonded to the outer rubber member. The leading swing arm side plates fit on either side of the median ring and aligned bolt holes enable the two members to be bolted together (Fig. 10.105(a and b)).

Load equalization between axles is achieved by torsional wind-up of the rubber spring members. Thus any vertical deflection of one or other swing arm as the wheels roll over any bumps on the road causes a torque to be applied to the rubber members. Accordingly an equal torque reaction will be transferred through the media of the rubber to the other swing arm and axle. As a result, each axle will support an equal share of the laden weight. Therefore contact and grip between wheels of both axles will be maintained at all times.

The characteristic of this springing is a very low stiffness in the unladen state which therefore provides a soft ride. A progressive spring stiffening and hardness of ride occurs as the swing arms are made to deflect against an increase in laden weight. An overall cushioned and smoothness of ride results.

An additional feature of this suspension geometry is that when weight is transferred during cornering from the inside to the outside of the vehicle, the deflection of the swing arms spreads the outer pair of wheels and draws the inner pair of wheels closer together. As a result smaller turning circles can be achieved without excessive tyre scrub.

10.15 Air suspensions for commercial vehicles

A rigid six wheel truck equipped with pairs of air springs per axle is shown in Fig. 10.106. The front suspension has an air spring mounted between the underside of each chassis side-member and the transverse axle beam, and the rear tandem suspension has the air springs mounted between each trailing arm and the underside of the chassis (Figs 10.107 and 10.108).

Air from the engine compressor passes through both the unloader valve and the pressure regulator valve to the reservoir tank. Air is also delivered to the brake system reservoir (not shown). Once the compressed air has reached some pre-determined upper pressure limit, usually between 8 and 8.25 bar, the unloader valve exhausts any further air delivery from the pump directly to the atmosphere, thereby permitting the compressor to 'run light'. Immediately the air supply to the reservoir has dropped to a lower limit of 7.25 bar, the unloader valve will automatically close its exhaust valve so that air is now transferred straight to the reservoir to replenish the air consumed. Because the level of air pressure demanded by the brakes is greater than that for the suspension system, a pressure regulator valve is incorporated between the unloader valve and suspension reservoir valve, its function being to reduce the delivery pressure for the suspension to approximately 5.5 bar.

Air now flows from the suspension reservoir through a filter and junction towards both the front and rear suspensions by way of a single

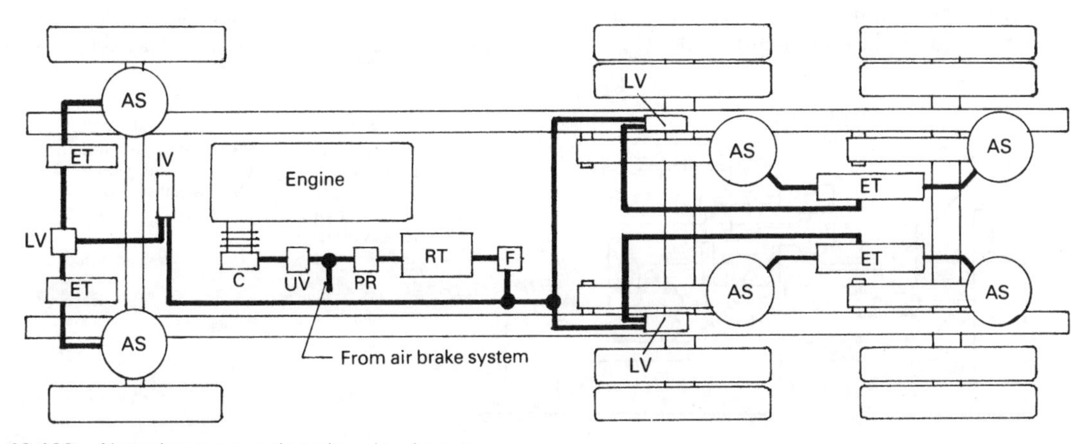

Fig. 10.106 Air spring suspension plan view layout

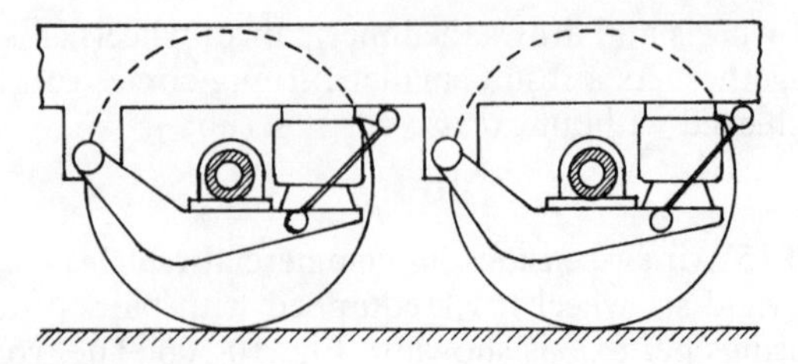

Fig. 10.107 Tandem trailing arm rolling diaphragm air sprung suspension

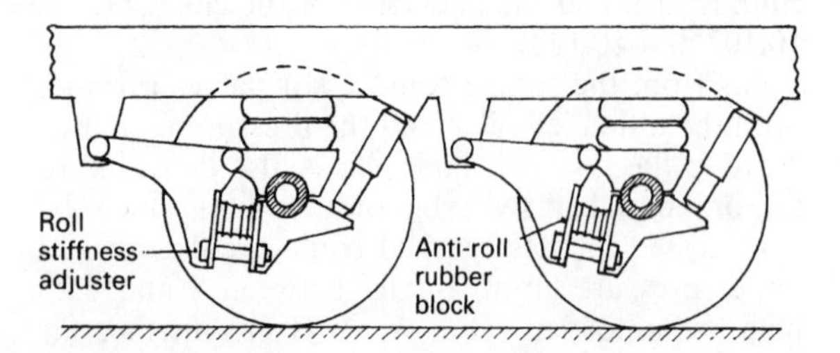

Fig. 10.108 Tandem trailing arm bellows spring suspension with rubber anti-roll blocks

central levelling valve at the front (Fig. 10.109) and a pair of levelling valves on each side of the first tandem axle. These levelling valves are bolted to the chassis, but they are actuated by an arm and link rod attached to the axles. It is the levelling valves' function to sense any change in the chassis to axle height and to increase or decrease the air pressure supply passing to the air springs, thereby raising or reducing the chassis height respectively. The air pressure actually reaching the springs may vary from 5.5 bar fully laden down to 2.5 bar when the vehicle is empty.

To improve the quality of ride, extra volume tanks can be installed in conjunction with the air springs to increase the volume of air in the system. This minimizes changes in overall pressure and reduces the spring rate (spring stiffness), thus enabling the air springs to provide their optimum frequency of spring bounce.

An additional feature at the front end of the suspension is an isolating valve which acts both as a junction to split the air delivery to the left and right hand air springs and to permit air to pass immediately to both air springs if there is a demand for more compressed air. This valve also slows down the transfer of air from the outer spring to the inner spring when the body rolls while the vehicle is cornering.

10.15.1 Levelling valve (Figs 10.109 and 10.110(a and b))

A pre-determined time delay before air is allowed to flow to or from the air spring is built into the valve unit. This ensures that the valves are not operated by axle bump or rebound movement as the vehicle rides over rough road surfaces, or by increased loads caused by the roll of the body on prolonged bends or on highly cambered roads.

The valve unit consists of two parts; a hydraulic damper and the air control valve (Fig. 10.110(a and b)). Both the damper and the valves are actuated by the horizontal operating lever attached to the axle via a vertical link rod. The operating lever pivots on a cam spindle mounted in the top of the valve assembly housing. The swing movement of the operating lever is relayed to the actuating arm through a pair of parallel positioned leaf springs fixed rigidly against the top and bottom faces of the flat cam, which forms an integral part of the spindle.

When the operating lever is raised or lowered, the parallel leaf springs attached to the lever casing pivot about the cam spindle. This causes both leaf springs to deflect outwards and at the same time

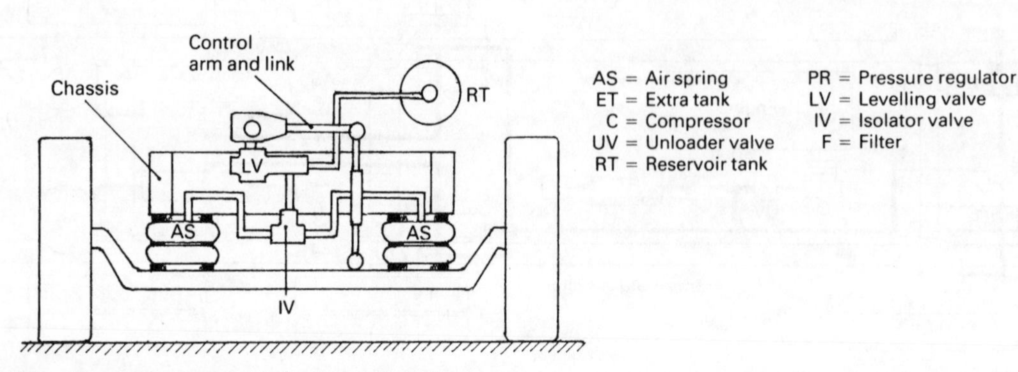

Fig. 10.109 Air spring suspension front view layout

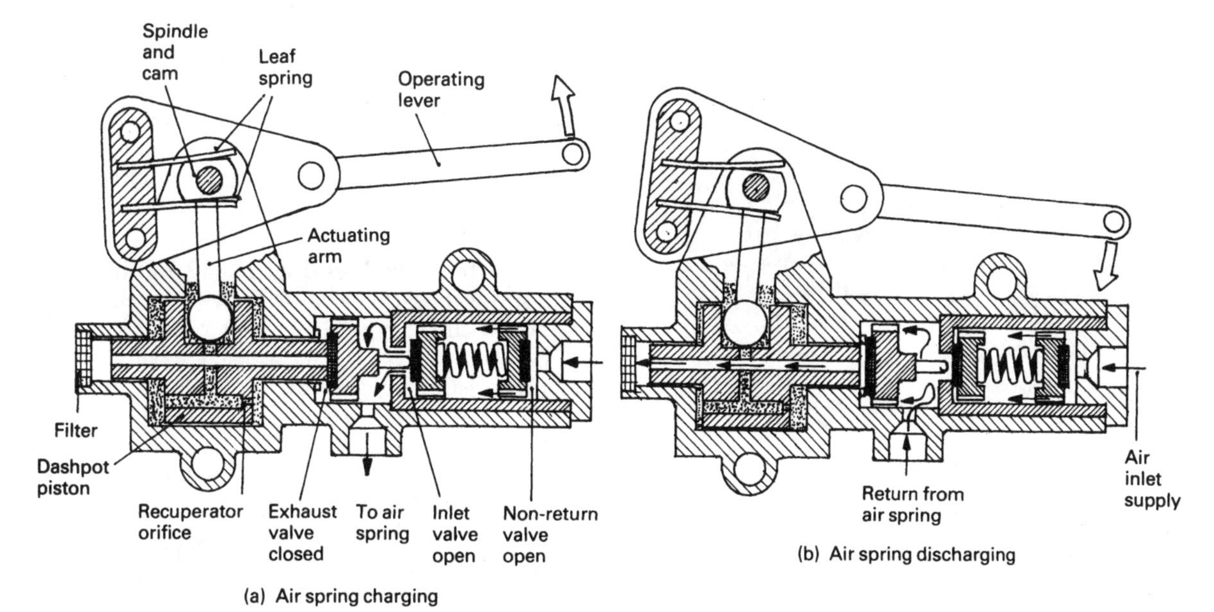

Fig. 10.110 (a and b) Levelling air control valve

applies a twisting movement to the cam spindle. It therefore tends to tilt the attached actuating arm and accordingly the dashpot piston will move either to the right or left against the fluid resistance. There will be a small time delay before the fluid has had time to escape from the compressed fluid side of the piston to the opposite side via the clearance between the piston and cylinder wall, after which the piston will move over progressively. A delay of 8 to 12 seconds on the adjustment of air pressure has been found suitable, making the levelling valve inoperative under normal road surface driving conditions.

Vehicle being loaded (Fig. 10.110(a)) If the operating lever is swung upward, due to an increase in laden weight, the piston will move to the right, causing the tubular extension of the piston to close the exhaust valve and the exhaust valve stem to push open the inlet valve. Air will then flow past the non-return valve through the centre of the inlet valve to the respective air springs. Delivery of air will continue until the predetermined chassis-to-axle height is reached, at which point the lever arm will have swung down to move the piston to the left sufficiently to close the inlet valve. In this phase, the springs neither receive nor lose air. It is therefore the normal operating position for the levelling valve and springs.

Vehicle being unloaded (Fig.10.110(b)) If the vehicle is partially unloaded, the chassis will rise relative to the axle, causing the operating arm to swing downward. Consequently, the piston will move to the left so that the exhaust valve will now reach the end of the cylinder. Further piston movement to the left will pull the tubular extension of the piston away from its rubber seat thus opening the exhaust valve. Excessive air will now escape through the centre of the piston to the atmosphere until the correct vehicle height has been established. At this point the operating lever will begin to move the piston in the opposite direction, closing the exhaust valve. This cycle of events will be repeated as the vehicle's laden weight changes. A non-return valve is incorporated on the inlet side to prevent air loss from the spring until under maximum loading or if the air supply from the reservoir should fail.

10.15.2 Isolating valve (Fig. 10.111(a and b))
An isolating valve is necessary when cornering to prevent air being pumped from the spring under compression to that under expansion, which could considerably reduce body roll resistance.

The valve consists of a T-piece pipe air supply junction with a central cylinder and plunger valve (Fig. 10.111(a and b)).

When the air springs are being charged, compressed air enters the inlet part of the valve from

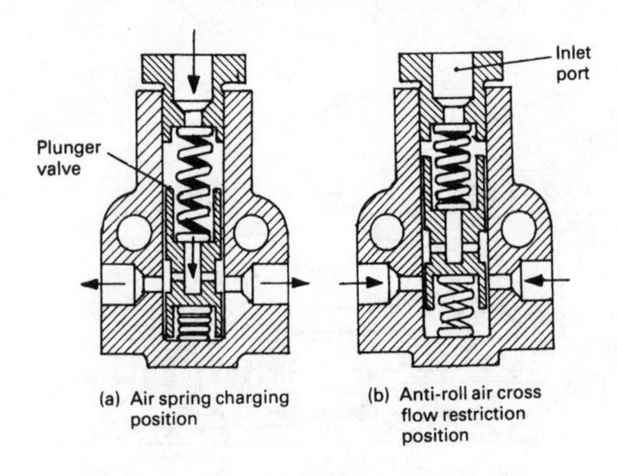

Fig. 10.111 (a and b) Isolator valve

the levelling valve and pushes the shuttle valve towards the end of its stroke against the spring situated between the plunger and cylinder blank end (Fig. 10.111(a)). Air will pass through the centre of the valve and come out radially where the annular groove around the valve aligns with the left and right hand output ports which are connected by pipe to the air springs.

Once the levelling valve has shut off the air supply to the air springs, the shuttle valve springs are free to force the shuttle valve some way back towards the inlet port. In this position the shuttle skirt seals both left and right hand outlet ports (Fig. 10.111(b)) preventing the highly pressurized outer spring from transferring its air charge to the expanded inner spring (which is subjected to much lower pressure under body roll conditions).

The shuttle valve is a loose fit in its cylinder to permit a slow leakage of air from one spring to the other should one spring be inflated more rapidly than the other, due possibly to uneven loading of the vehicle.

10.15.3 Air spring bags (Figs 10.112 and 10.113) Air spring bags may be of the two or three convoluted bellows (Fig. 10.112) or rolling lobe (diaphragm) type (Fig. 10.113), each having distinct characteristics. In general, the bellows air spring

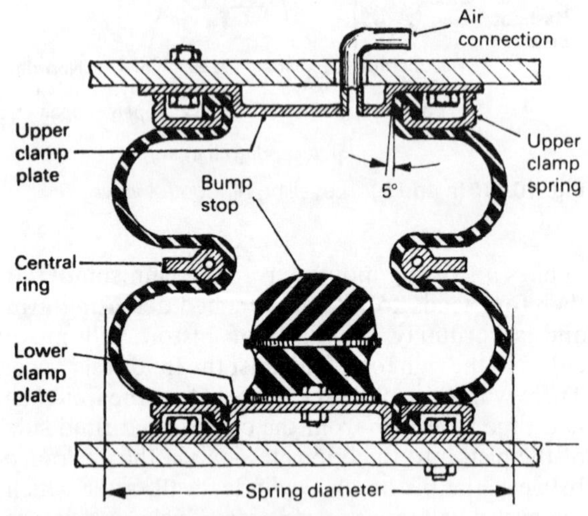

Fig. 10.112 Involute bellow spring

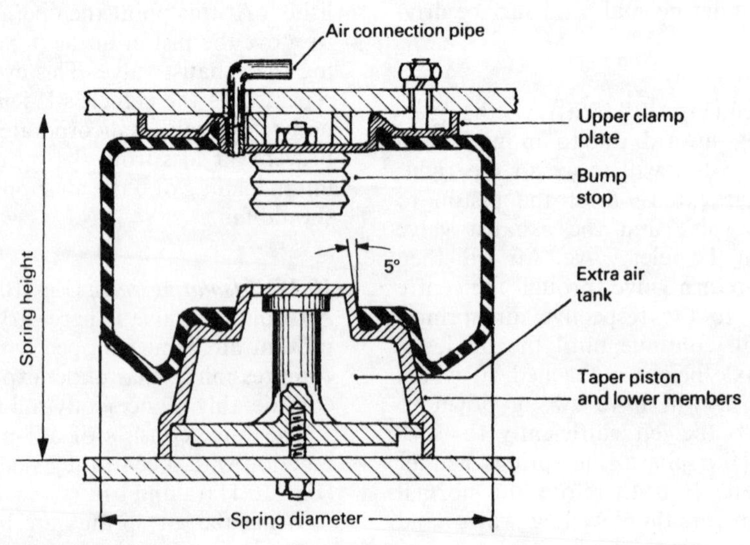

Fig. 10.113 Rolling diaphragm spring

(Fig. 10.112) is a compact flexible air container which may be loaded to relatively high load pressures. Its effective cross-sectional area changes with spring height — reducing with increase in static height and increasing with a reduction in static height. This is due to the squeezing together of the convolutes so that they spread further out. For large changes in static spring height, the three convolute bellows type is necessary, but for moderate suspension deflection the twin convolute bellow is capable of coping with the degree of expansion and contraction demanded.

With the rolling diaphragm or lobe spring (Fig. 10.113) a relatively higher installation space must be allowed at lower static pressures. Progressive spring stiffening can be achieved by tapering the skirt of the base member so that the effective working cross-sectional area of the rolling lobe increases as the spring approaches its maximum bump position.

The normal range of natural spring frequency for a simply supported mass when fully laden and acting in the direct mode is 90–150 cycles per minute (cpm) for the bellows spring and for the rolling lobe type 60–90 cpm. The higher natural frequency for the bellow spring compared to the rolling lobe type is due mainly to the more rigid construction of the convolute spring walls, as opposed to the easily collapsible rolling lobe.

As a precaution against the failure of the supply of air pressure for the springs, a rubber limit stop of the progressive type is assembled inside each air spring, and compression of the rubber begins when about 50 mm bump travel of the suspension occurs.

The springs are made from tough, nylon-reinforced Neoprene rubber for low and normal operating temperature conditions but Butyl rubber is sometimes preferred for high operating temperature environments.

An air spring bag is composed of a flexible cylindrical wall made from reinforced rubber enclosed by rigid metal end-members. The external wall profile of the air spring bag may be plain or bellow shaped. These flexible spring bags normally consist of two or more layers of rubber coated rayon or nylon cord laid in a cross-ply fashion with an outside layer of abrasion-resistant rubber and sometimes an additional internal layer of impermeable rubber to minimize the loss of air.

In the case of the bellow type springs, the air bags (Fig. 10.112) are located by an upper and lower clamp ring which wedges their rubber moulded edges against the clamp plate tapered spigots. The rolling lobe bag (Fig. 10.113) relies only upon the necks of the spring fitting tightly over the tapered and recessed rigid end-members. Both types of spring bags have flat annular upper and lower regions which, when exposed to the compressed air, force the pliable rubber against the end-members, thereby producing a self-sealing action.

10.15.4 Anti-roll rubber blocks (Fig. 10.108)

A conventional anti-roll bar can be incorporated between the trailing arms to increase the body roll stiffness of the suspension or alternatively built-in anti-roll rubber blocks can be adopted (Fig. 10.108). During equal bump or rebound travel of each wheel the trailing arms swing about their front pivots. However, when the vehicle is cornering, roll causes one arm to rise and the other to fall relative to the chassis frame. Articulation will occur at the rear end of the trailing arm where it is pivoted to the lower spring base and axle member. Under these conditions, the trailing arm assembly adjacent to the outer wheel puts the rubber blocks into compression, whereas in the other trailing arm, a tensile load is applied to the bolt beneath the rubber block. As a result, the total roll stiffness will be increased. The stiffness of these rubber blocks can be varied by adjusting the initial rubber compressive preload.

10.15.5 Air spring characteristics (Figs 10.114, 10.115, 10.116 and 10.117)

The bounce frequency of a spring decreases as the sprung weight increases and increases as this weight is reduced. This factor plays an important part in the quality of ride which can be obtained on a heavy goods or passenger vehicle where there could be a fully laden to unladen weight ratio of up to 5:1.

An inherent disadvantage of leaf, coil and solid rubber springs is that the bounce frequency of vibration increases considerably as the sprung spring mass is reduced (Fig. 10.114). Therefore, if a heavy goods vehicle is designed to give the best ride frequency, say 60 cycles per minute fully laden, then as this load is removed, the suspension's bounce frequency could rise to something like 300 cycles per minute when steel or solid rubber springs are used, which would produce a very harsh, uncomfortable ride. Air springs, on the other hand, can operate over a very narrow bounce frequency range with considerable changes in vehicle laden weight, say 60–110 cycles per minute for a rolling lobe air spring (Fig. 10.114). Consequently the quality of ride with air springs is maintained over a wide range of operating conditions.

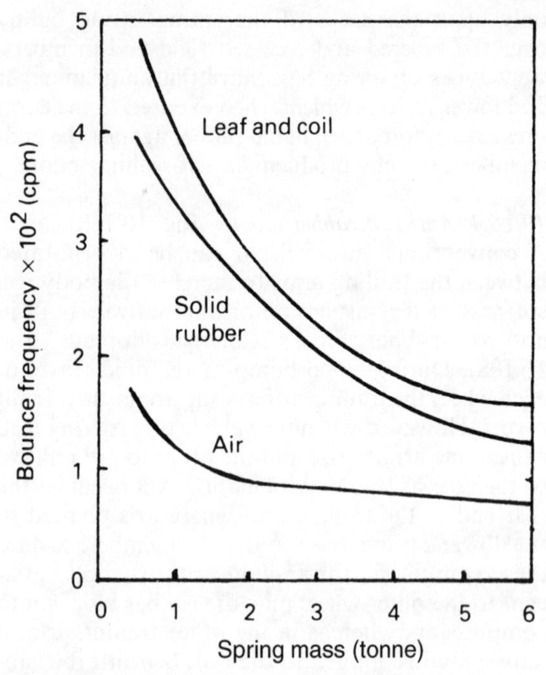

Fig. 10.114 Effects and comparison of payload on spring frequency for various types of spring media

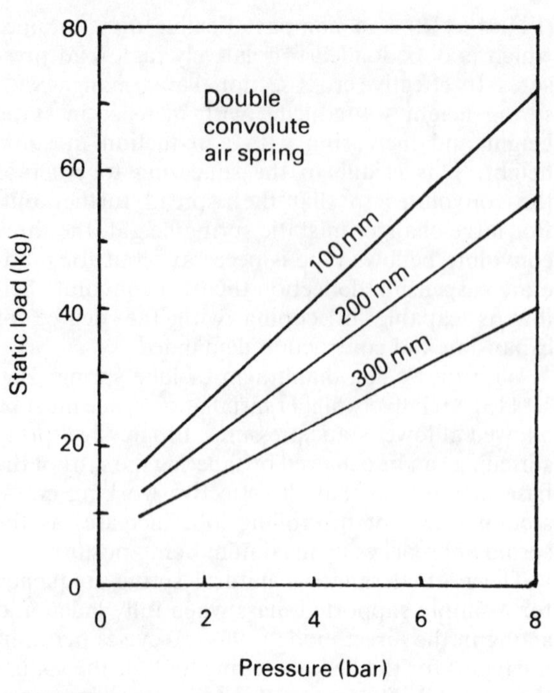

Fig. 10.116 Effects of static payload on spring air pressure for various spring static heights

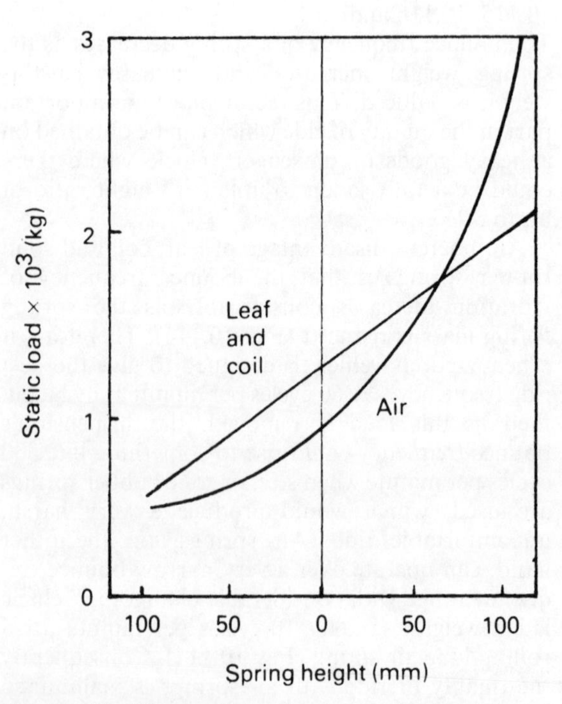

Fig. 10.115 Effects of static load on spring height

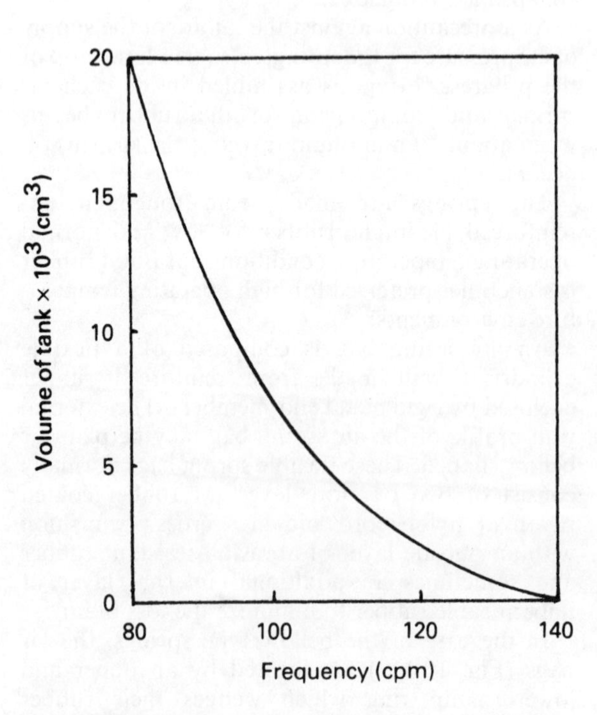

Fig. 10.117 Relationship of extra air tank volume and spring frequency

Steel springs provide a direct rise in vertical deflection as the spring mass increases, that is, they have a constant spring rate (stiffness) whereas air springs have a rising spring stiffness with increasing load due to their effective working area enlarging as the spring deflects (Fig. 10.115). This stiffening characteristic matches far better the increased resistance necessary to oppose the spring deflection as it approaches the maximum bump position.

To support and maintain the spring mass at constant spring height, the internal spring air pressure must be increased directly with any rise in laden weight. These characteristics are shown in Fig. 10.116 for three different set optimum spring heights.

The spring vibrating frequency will be changed by varying the total volume of air in both extra tank and spring bag (Fig. 10.117). The extra air tank capacity, if installed, is chosen to provide the optimum ride frequency for the vehicle when operating between the unladen and fully laden conditions.

10.16 Lift axle tandem or tri-axle suspension

(Figs 10.118, 10.119 and 10.120)

Vehicles with tandem or tri-axles which carry a variety of loads ranging from compact and heavy to bulky but light may under-utilize the load carrying capacity of each axle, particularly an empty return journey over a relatively large proportion of the vehicle's operating time.

When a vehicle carries a full load, a multi-axle suspension is essential to meet the safety regulations, but the other aspects are improved road vibration isolation from the chassis, better road holding and adequate ride comfort.

If a conventional multi-axle suspension is operated below half its maximum load carrying capacity, the quality of road holding and ride deteriorates, suspension parts wear rapidly, and increased wheel bounce causes a rise in tyre scrub and subsequent tyre tread wear.

Conversely, reducing the number of axles and wheels in contact with the road when the payload is decreased extends tyre life, reduces rolling forward resistance of the vehicle and therefore improves fuel consumption.

10.16.1 Balance beam lift axle suspension arrangement (Figs 10.118 and 10.119)

A convenient type of tandem suspension which can be adapted so that one of the axles can be simply and rapidly raised or lowered to the ground

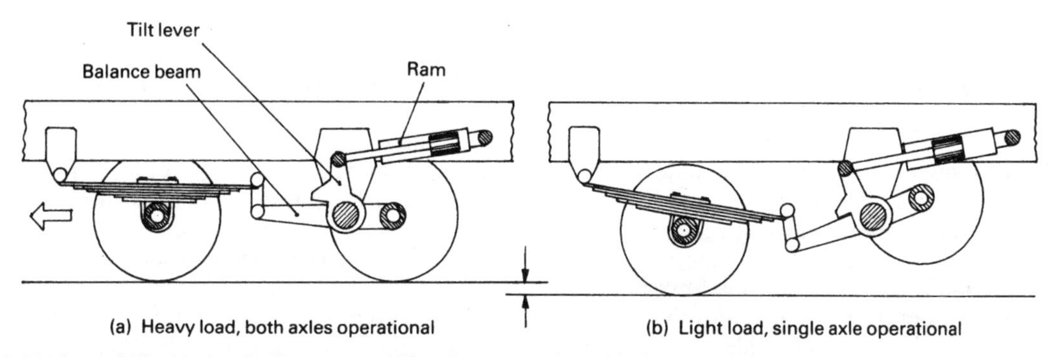

Fig. 10.118 (a and b) Hydraulically operated lift axle suspension with direct acting ram

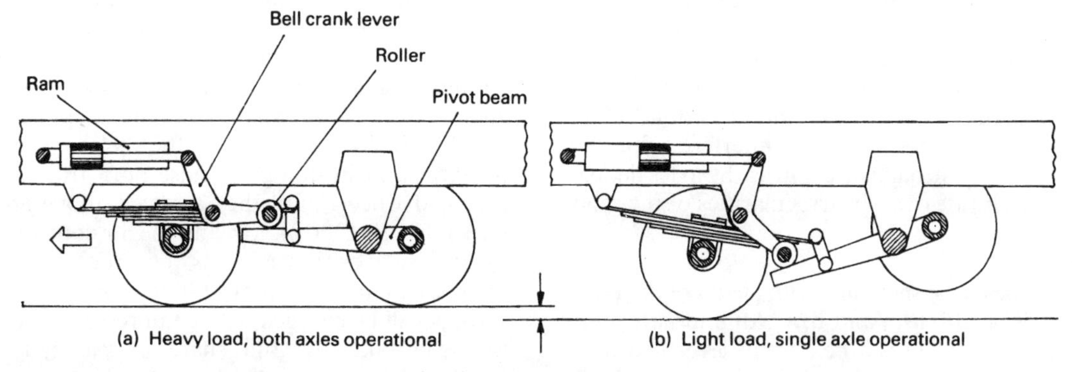

Fig. 10.119 (a and b) Hydraulically operated lift axle suspension with bell-crank lever and ram

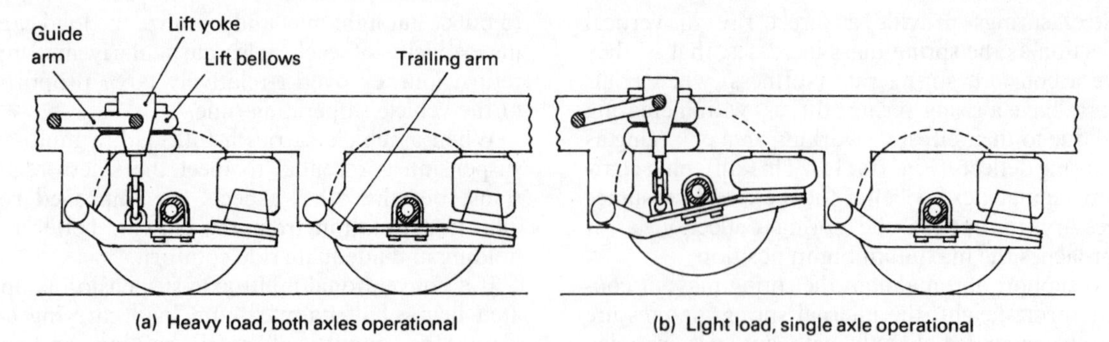

Fig. 10.120 (a and b) Pneumatically operated lift axle suspension

without having to make major structural changes is the *semi-elliptic spring and balance beam combination* (Figs 10.118 and 10.119). Raising the rearmost of the two axles from the ground is achieved by tilting the balance beam anticlockwise so that the forward part of the balance beam appears to push down the rear end of the semi-elliptic spring. In effect, what really happens is the balance beam pivot mounting and chassis are lifted relative to the forward axle and wheels. Actuation of the balance beam tilt is obtained by a power cylinder and ram, anchored to the chassis at the cylinder end, whilst the ram-rod is connected either to a tilt lever, which is attached indirectly to the balance beam pivot, or to a bell crank lever, which relays motion to the extended forward half of the balance beam.

Balance beam suspension with tilt lever axle lift (Fig. 10.118(a and b)) With the tilt lever axle lift arrangement, applying the lift control lever introduces fluid under pressure to the power cylinder, causing the ram-rod to extend. This forces the tilt lever to pivot about its centre of rotation so that it bears down on the left hand side of the beam. Consequently the balance beam is made to take up an inclined position (Fig. 10.118(b)) which is sufficient to clear the rear road wheels off the ground. When the axle is lowered by releasing the hydraulic pressure in the power cylinder, the tilt lever returns to its upright position (Fig. 10.118(a)) and does not then interfere with the articulation of the balance beam as the axles deflect as the wheels ride over the irregularities of the road surface.

Balance beam suspension with bell-crank lever axle lift (Fig. 10.119(a and b)) An alternative lift axle arrangement uses a bell-crank lever to transmit the ram-rod force and movement to the extended front end of the balance beam. When hydraulic pressure is directed to the power cylinder, the bell-crank lever is compelled to twist about its pivot, causing the roller to push down and so roll along the face of the extended balance beam until the rear axle is fully raised (Fig. 10.119(b)). Removing the fluid pressure permits the weight of the chassis to equalize the height of both axles again and to return the ram-rod to its innermost position (Fig. 10.119(a)). Under these conditions the bell-crank lever roller is lifted clear of the face of the balance beam. This prevents the oscillating motion of the balance beam being relayed back to the ram in its cylinder.

10.16.2 Pneumatically operated lift axle suspension (Fig. 10.120(a and b))

A popular lift axle arrangement which is used in conjunction with a trailing arm air spring suspension utilizes a separate single air bellow situated at chassis level in between the chassis side-members. A yoke beam supported by the lift air bellows spans the left and right hand suspension trailing arms, and to prevent the bellows tilting as they lift, a pair of pivoting guide arms are attached to the lift yoke on either side. To raise the axle wheels above ground level, the manual air control valve is moved to the raised position; this causes compressed air to exhaust from the suspension air springs and at the same time allows pressurized air to enter the lift bellows. As the air pressure in the lift bellows increases, the bellows expand upward, and in doing so, raise both trailing arm axle and wheels until they are well above ground level (Fig. 10.120(b)). Moving the air control valve to 'release' position reverses the process. Air will then be exhausted from the lift bellows while the air springs will be charged with compressed air so that the axle takes its full share of payload (Fig. 10.120(a)).

An additional feature of this type of suspension is an overload protection where, if the tandem suspension is operating with one axle lifted and receives loads in excess of the designed capacity, the second axle will automatically lower to compensate.

10.17 Active suspension

An ideal suspension system should be able to perform numerous functions that are listed below:

1. To absorb the bumps and rebounds imposed on the suspension from the road.
2. To control the degree of body roll when cornering.
3. To maintain the body height and to keep it on an even keel between light and full load conditions.
4. To prevent body dive and squat when the car is rapidly accelerated or is braked.
5. To provide a comfortable ride over rough roads yet maintain suspension firmness for good steering response.
6. To isolate small and large round irregularities from the body at both low and high vehicle speeds.

These demands on a conventional suspension are only partially achieved as to satisfy one or more of the listed requirements may be contrary to the fulfilment of some of the other desired suspension properties. For example, providing a soft springing for light loads will excessively reduce the body height when the vehicle is fully laden, or conversely, stiffening the springing to cope with heavy loads will produce a harsh suspension under light load conditions. Accordingly, most conventional suspensions may only satisfy the essential requirements and will compromise on some of the possibly less important considerations. An active suspension will have built into its design means to satisfy all of the listed demands; however, even then it may not be possible due to the limitations of a design and cost to meet and overcome all of the inherent problems experienced with vehicle suspension. Thus it would be justified to classify most suspensions which have some form of height levelling and anti-body roll features as only semi-active suspensions.

For an active suspension to operate effectively various sensors are installed around the vehicle to monitor changing driving conditions; the electrical signals provided by these sensors are continuously fed to the input of an electronic control unit microprocessor. The microprocessor evaluates and processes the data supplied by the sensors on the changing speed, loads, and driving conditions imposed on the suspension system. On the basis of these data and with the aid of a programmed map memory, calculations are made as to what adjustments should be made to the suspension variables. These instructions are then converted into electrical output signals and are then directed to the various levelling and stiffening solenoid control valves. The purpose of these control valves is to deliver or exit fluid to or from the various parts of a hydraulic controlled self-levelling suspension system.

10.17.1 Description and application of sensors

A list of sensors which can be used are given below; however, a limited combination of these sensors may only be installed depending on the sophistication of the suspension system adopted:

1. body height sensor
2. steering wheel sensor
3. longitudinal acceleration sensor
4. lateral acceleration sensor
5. brake pressure sensor
6. brake pedal sensor
7. acceleration pedal sensor
8. load sensor
9. vehicle speed sensor
10. mode selector

Height sensor (Fig. 10.121) The linear variable differential sensor is often used to monitor vertical height movement as there is no contact between moving parts; it therefore eliminates any problems likely to occur due to wear. It is basically a transformer having a central primary winding and two

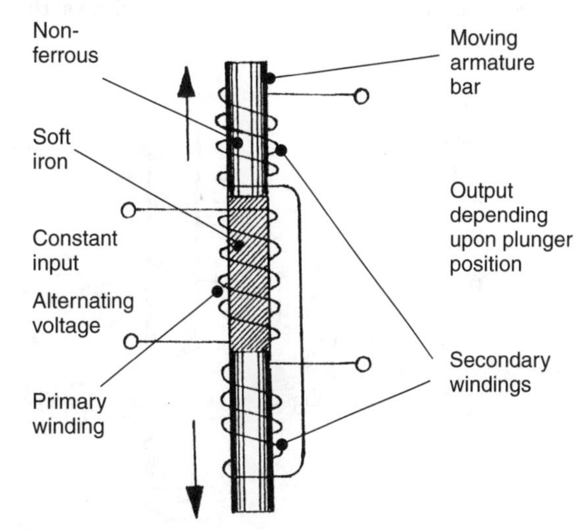

Fig. 10.121 Height sensor (linear variable differential type)

secondary windings connected in series in opposition to each other. An alternative input supply voltage is applied to the primary winding; this produces a magnetic flux which cuts through the secondary winding thereby inducing an alternative voltage into the secondary winding. The difference between the voltage generated in each secondary winding therefore becomes the output signal voltage. With the non-ferromagnetic/soft iron armature bar in the central position each secondary winding will generate an identical output voltage so that the resultant output voltage becomes zero. However when the armature (attached to the lower suspension arm) moves up or down as the body height changes the misalignment of the soft iron/non-ferromagnetic armature causes the output voltage to increase in one winding and decrease in the other, the difference in voltage increasing in direct proportion to the armature displacement. This alternative voltage is then converted to a direct voltage before entering the electronic-control unit.

Steering sensor (Fig. 10.122) This sensor monitors the angular position of the steering wheel and the rate of change of the steering angle. The sensor comprises a slit disc attached to the steering column and rotates with the steering wheel and a fixed 'U' shaped detector block containing on one side three phototransistors and on the other side three corresponding light-emitting diodes. The disc rotates with the steering column and wheel and at the same time the disc moves between the light-emitting diode and the phototransistor block overhang. When the column is turned the rotating slotted disc alternatively exposes and blocks the light-emitting beams directed towards the phototransistors; this interruption of the light beams generates a train of logic pulses which are then processed by the microprocessor to detect the steering angle and the rate of turn. To distinguish which way the steering wheel is turned a left and right hand phototransistor is included, and a third phototransistor is located between the other two to establish the neutral straight ahead position. The difference in time between light beam interruptions enables the microprocessor to calculate the angular velocity of the driving wheel at any one instance in time. In some active suspension systems, when the angular velocity exceeds a pre-fixed threshold the electronic-control unit switches the suspension to a firm ride mode.

Acceleration sensor (Fig. 10.123) A pendulum strain gauge type acceleration sensor is commonly used for monitoring body acceleration in both longitudinal and lateral directions. It is comprised of a leaf spring rigidly supported at one end with a mass attached at its free end. A thin film strain gauge wired in the form of a wheatstone bridge circuit is bonded to the leaf spring on one side, two of the four resistors are passive whereas the other two are active. As the vehicle is accelerated the pendulum due to the inertia of the mass will reluctantly hold back thus causing the spring to deflect. The pair of active resistor arms therefore become strained (stretch) and hence alter their resistance, thus producing an imbalance to the wheatstone bridge circuit resulting in an output voltage proportional to the magnitude of the acceleration. When using this type of sensor for monitoring

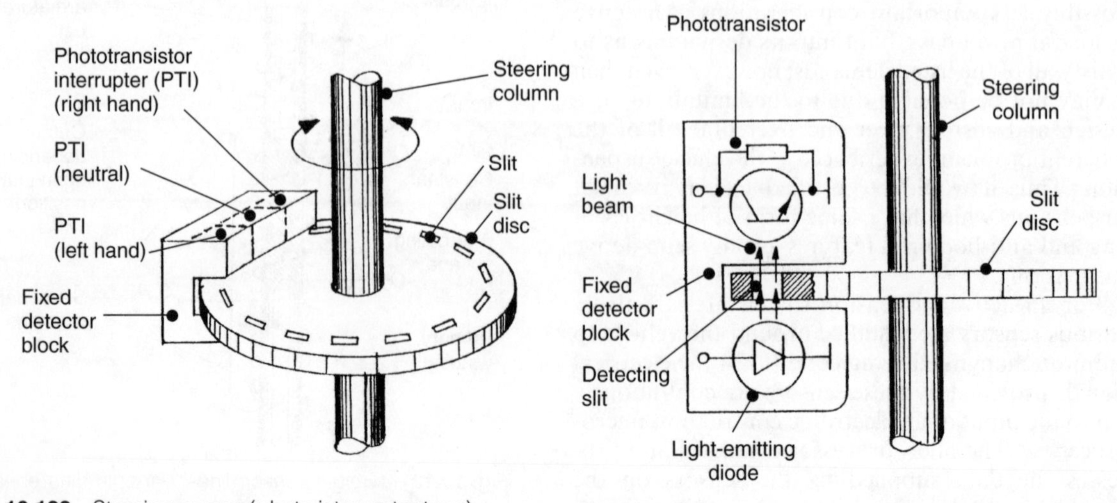

Fig. 10.122 Steering sensor (photo interrupter type)

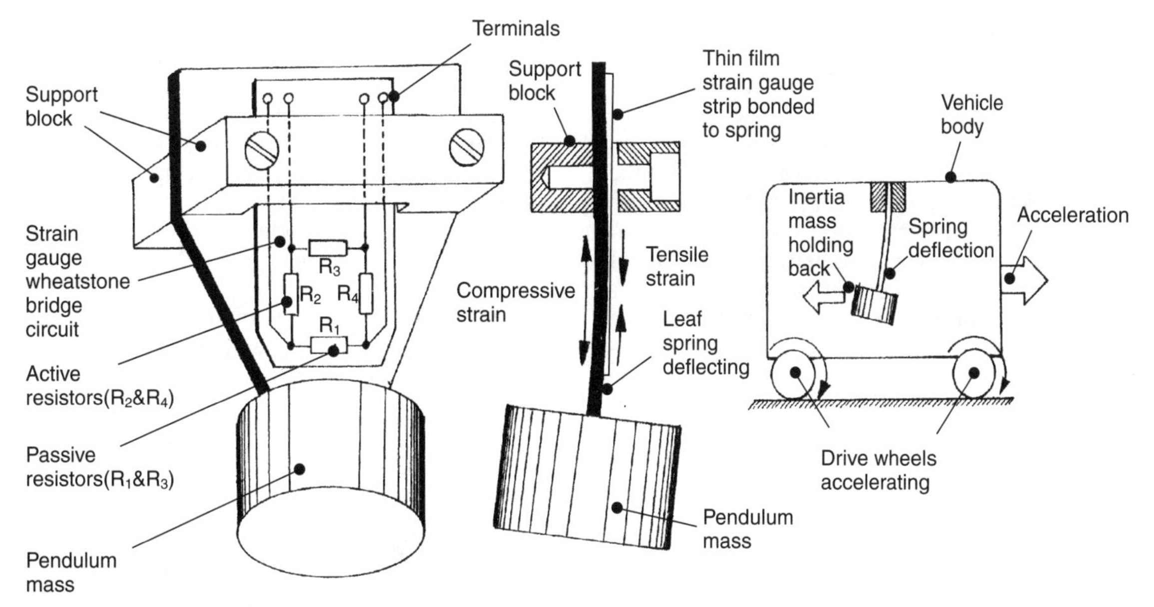

Fig. 10.123 Acceleration sensor (pendulum strain gauge type)

lateral acceleration, it should be installed either near the front or rear to enable it to sense the swing of the body when the car is cornering it, there is also a measure in the degree of body yaw.

Brake pedal/pressure sensor These sensors are used to indicate the driver's intentions to brake heavily by either monitoring the brake pedal movement or in the form of a pressure switch tapped in to the hydraulic brake circuit. With the pressure switch method the switch is set to open at some predetermined brake-line pressure (typically about 35 bar); this causes the input voltage to the electronic-control unit to rise. Once 5 volts is reached (usual setting) the electronic-control unit switches the suspension to 'firm' ride mode. When the braking pressure drops below 35 bar the pressure switch closes again; this grounds the input to the electronic-control unit and causes its output voltage to the solenoid control valves also to collapse, and at this point the suspension reverts to 'soft' ride mode.

Acceleration pedal sensor These sensors can be of the simple rotary potentiometer attached to the throttle linkage indicating the throttle opening position. A large downward movement or a sudden release of the accelerator pedal signals to the electronic-control unit that the driver intends to rapidly accelerate or decelerate, respectively. When accelerating hard the rapid change in the potentiometer resistance and hence input voltage signals the electronic-control unit to switch the suspension to firm ride mode.

Load sensor Load sensors are positioned on top of the strut actuator cylinder; its purpose is to monitor the body load acting down on each strut actuator.

Vehicle speed sensor Vehicle speed can be monitored by the speedometer or at the transmission end by an inductive pick-up or Hall effect detector which produces a series of pulses whose frequency is proportional to vehicle speed. Once the vehicle speed exceeds some predetermined value the electronic-control unit automatically switches the suspension to 'firm' ride mode. As vehicle speed decreases, a point will be reached when the input to the electronic-control unit switches the suspension back to 'soft' ride mode.

Mode selector This dashboard mounted control switches the suspension system via the electronic-control unit to either a comfort (soft) ride mode for normal driving conditions or to a sports (firm) ride mode. However, if the vehicle experiences severe driving conditions while in the comfort ride mode, the electronic-control unit overrides the mode selector and automatically switches the suspension to sports (firm) ride mode.

10.17.2 Active hydro/coil spring suspension system (Fig. 10.124(a–h))

A typical fully active hydraulic self-levelling suspension system utilizing strut actuators consisting of a cylinder, piston and ram-shaft installed between each spring and its body support. A hydraulic pump driven from the engine supplies high pressure fluid to the accumulator and to the individual strut actuators via a pressure regulating valve and a levelling valve. The purpose of the accumulator is to store fluid at maximum pressure so that it can instantly be discharged to the various strut actuator cylinders when commanded; this would not be possible without an excessive time delay since the pump could not generate and discharge sufficient quantity of high pressure fluid in the time-span required to maintain the car on a level keel. When the engine is running, high pressure fluid is supplied to each strut actuator to bring the body level up to its design height via the levelling control valve. Note some systems may have more than one body height setting.

Unequal weight distribution levelling control (Fig 10.124(a and b)) Uneven weight distribution is automatically compensated for by the levelling control

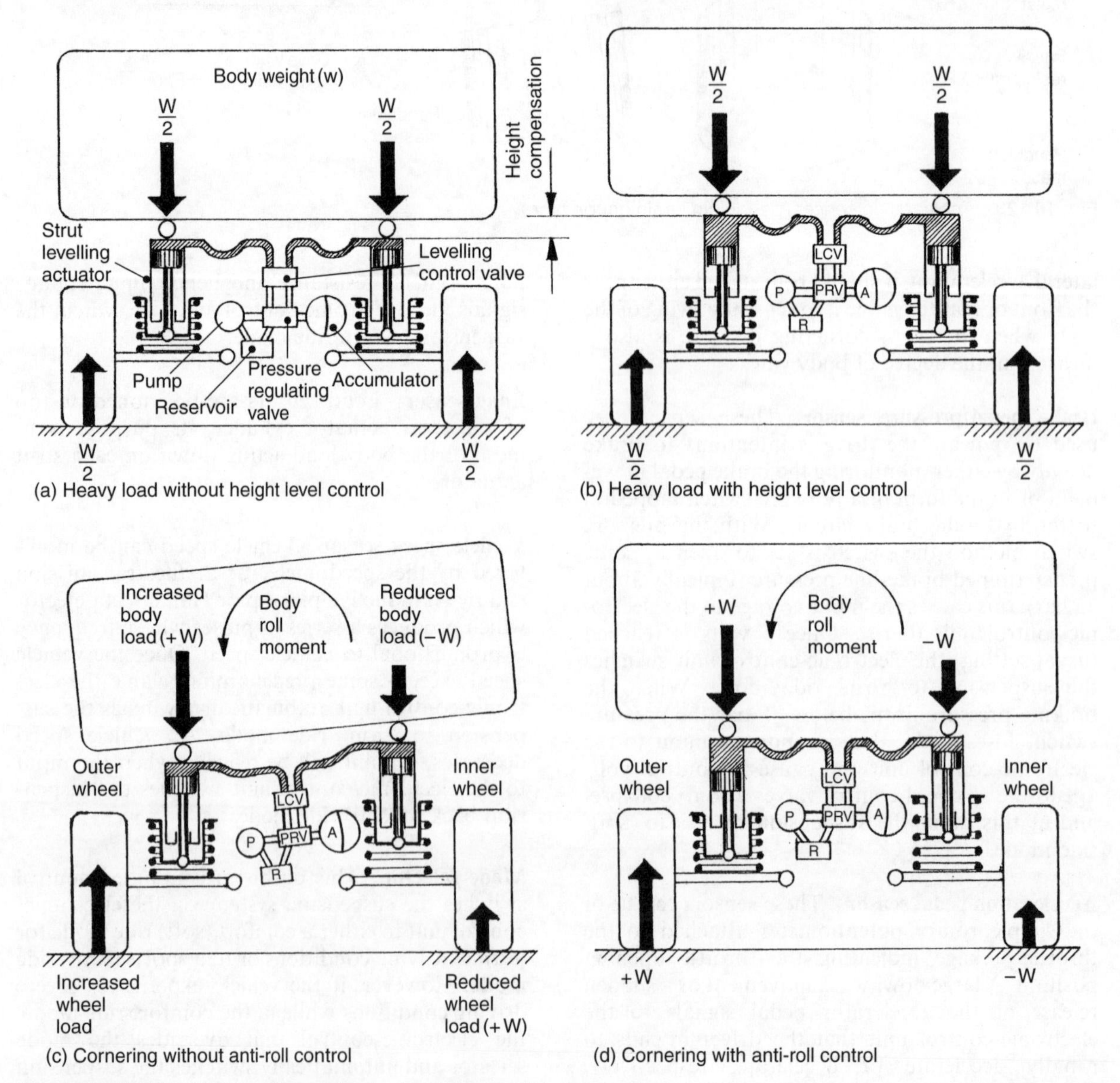

Fig. 10.124 (a–g) Active self–levelling hydraulic/coil spring suspension

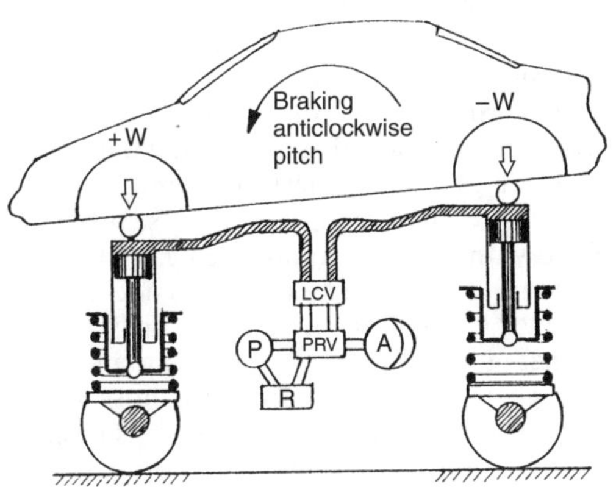

(e) Heavy braking without anti-dive control

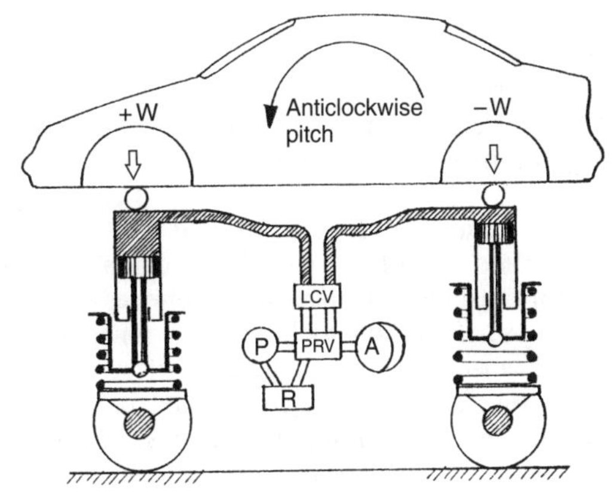

(f) Heavy braking with anti-dive control

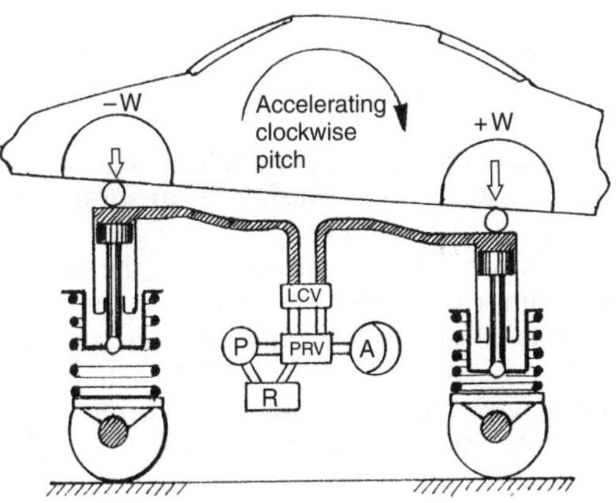

(g) Heavy acceleration without anti-squat control

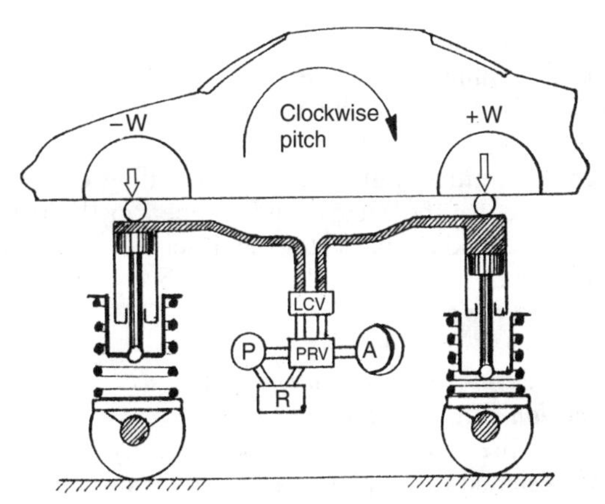

(h) Heavy acceleration with anti-squat control

Fig. 10.124 *contd*

valve which allows fluid to enter or exit from the individual strut actuators. Figure 10.124(a) shows the body height if there was no level control or if the self-levelling height is set to a low level, whereas Fig. 10.124(b) shows the suspension level raised against the compressed road springs to compensate for the heavy load. Longitudinal levelling also takes place between the front and rear suspension strut actuators when, for instance, all the passenger seats are occupied and the boot is full of luggage.

Body roll control (Fig. 10.124(c and d)) When the car is negotiating a corner the body tends to tilt so that the inner and outer wheel loads are reduced and increased, respectively. This lateral load transfer shown in Fig. 10.124(c) compresses the outer springs and expands the inner springs thus causing the body to roll and to become uncomfortable for the driver and passengers. To compensate for the weight transfer, fluid is pumped or released into the outer strut actuators via the levelling control valve until it has lifted the body on the outside to the same height as the inside (see Fig. 10.124(d)). Usually a small angular roll is deliberately allowed to provide the driver with a sense of caution.

Anti-dive control (Fig. 10.124(e and f)) If the car is braked rapidly there is a tendency for the body to pitch forwards, that is, the front of the body temperately dives downwards and the rear lifts (see Fig. 10.124(e)); the dive experienced is due to the longitudinal weight transfer since the body mass wants to continue moving forwards but the road wheels and the unsprung suspension mass are being retarded by the action of the braking force. To overcome this inherent deficiency in the suspension design which occurs usually when soft springs are used, fluid is rapidly transferred into both of the front strut actuator cylinders (see Fig. 10.124(f)), thereby correcting the front to rear tilt of the body over the braking period and then releasing the excess fluid from the front actuators when normal driving resumes.

Anti-squat control (Fig. 10.124(g and h)) If a car is accelerated rapidly, particularly when pulling away from a standstill, there is a proneness for the body due to its inertia to hold back whereas the propelled wheels and unsprung part of suspension tend to move ahead of the interlinked body. This results in the body tilting backwards so that it squats heavily on the rear axles and wheels (see Fig. 10.124(g)). To correct this ungainly stance when the car is being accelerated, fluid is quickly displaced from the accumulator and pump through the open levelling control valve into the rear strut actuator cylinders; this levels the body longitudinally (see Fig. 10.124(h)). Once the acceleration sensor detects a reduction in acceleration, the electronic-control unit signals the levelling control valve to return the excess fluid trapped in the rear actuator cylinders back to the reservoir so that under steady driving conditions the body remains parallel to the road.

10.17.3 Semi-active controlled hydro/gas suspension (Fig. 10.125(a and b))

Most of the self-levelling suspension layouts are only designed to achieve semi-active suspension control as there is a cost factor and it might not be justified to produce a near perfect suspension since there is always some inherent deficiencies due to other factors built into the body/suspension/transmission design. One version of a semi-active suspension is shown in Fig. 10.125(a, b) here, instead of using steel coil or leaf springs a hydro/gas spring is employed. These spring units basically consist of two hemispherical chambers separated by a flexible diaphragm; the outer sealed chamber is filled with pressurized nitrogen gas which acts as a spring media whereas the underside chamber is connected through the fluid to the height adjusting strut actuators and the supply pump and accumulator. There are three springs for both front and rear pair of suspensions. The system is designed to give two spring rates, these are 'sports' (stiff springing) and 'comfort' (light springing) and there are two damping modes, firm and soft.

Wheel deflection is absorbed through the height actuator cylinder and piston ram so that as a wheel goes over a bump or pot hole, the suspension swing-arm will tilt correspondingly up or down. An upward movement of a swing-arm and ram will displace fluid into the chamber underneath the diaphragm thereby causing it to compress the nitrogen gas. If however the wheel goes into a pot hole the downward movement of the swing-arm and ram causes the effective cylinder space to increase, this reduces the fluid pressure and permits the hydrogen gas to expand by pushing down the diaphragm, and fluid will therefore be displaced back into the height actuator cylinder.

Height and levelling control (Fig. 10.125(a and b)) With the engine running fluid is pumped into the accumulator and into all four height actuators and

Soft ride third spring
Restriction damper
Left hand hydro/gas spring
Right hand hydro/gas spring
Height actuator
Stiffening plunger valve
Levelling control valve
Swing arm
Pressure regulating valve
Hydraulic pump
Reservoir
Accumulator

(a) Soft ride interlinked three spring suspension mode

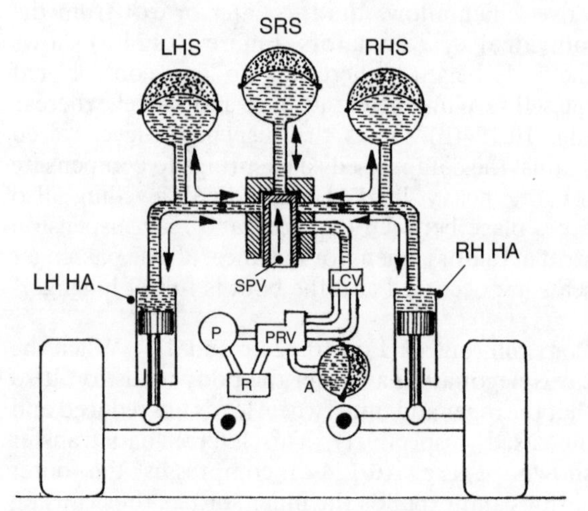

(b) Firm ride independent two spring suspension mode

Fig. 10.125 (a and b) Semi-active hydro/gas suspension

into each lower spring unit chamber until the predetermined body height setting is reached, at which point the levelling control valve blocks any further supply of fluid to the system. If the load changes due to more or less passengers and luggage, the levelling control valve will automatically permit fluid to enter or leave the spring/actuator system to maintain the optimum programmed body height setting.

Comfort (light springing) ride (Fig. 10.125(a)) For a comfortable ride under normal driving road conditions the spring loaded plunger is pulled outwards (downward). Both height actuators are interconnected with the fluid supply and the soft-ride third spring; the latter increases the volume of compressible gas by 50%, thus reducing the spring stiffness, thereby providing greater ride comfort. When the body rolls during cornering fluid is transferred from the outer to the inner height actuator cylinders but this is slowed down to some extent by the restrictor dampers.

Sports (stiff springing) ride (Fig. 10.125(b)) If the car is to be driven fast or is moving sharply around a bend the suspension system can be switched to sports (firm) ride by way of pushing the stiffening plunger inwards (upwards), see Fig. 10.125(b). This has the effect of blocking the third spring fluid movement from both adjacent right and left hand wheel springs. All three springs are now isolated from each other so that the vertical deflection of each wheel strut actuator is confined entirely to its own gas filled spring. Accordingly the spring rate for vertical piston ram movement is stiffened for each wheel suspension, and this also provides a degree of body roll control.

10.17.4 Semi-active hydro/gas electronic controlled suspension system (Fig. 10.126)

The system shown in Fig. 10.126 illustrates a front double wishbone suspension and a rear semi-trailing swing-arm suspension, each wheel having its own levelling strut actuator and hydro/gas spring unit. A third soft ride hydro/gas unit is shared between the front pair of suspensions and similarly a third unit is installed for the rear suspension. An electronic-control unit microprocessor is incorporated in the system which takes in signals from the various sensors; this information is then processed and converted to electrical instructions to the various solenoid control valves. Note that some of the sensors shown in Fig. 10.126 may not always be included in a semi-active suspension system, the actual sensors chosen will depend upon the degree of control sophistication to be built into the suspension design. The sensors shown are listed as follows: height sensors, load sensors, steering sensor, longitudinal and lateral acceleration sensors, acceleration pedal sensor, brake pressure and pedal sensors and vehicle speed sensor. For instance a simple semi-active suspension can operate effectively with just four sensors such as height sensors, steering sensor, brake pressure sensor and vehicle speed sensor. The electronic-control unit output directs the energizing and de-energizing of the front and rear solenoid controlled levelling valves and the front and rear solenoid stiffening valves. There is a mode selector which enables the driver to put the suspension in either sports (firm) ride or comfort (soft) ride; however, when switched to comfort ride, if driving conditions become harsh the suspension automatically reverts to sports (stiff springing) ride, but it will eventually change back to comfort ride when normal driving conditions prevail.

Comfort (soft) ride mode (Fig. 10.126) When switched to comfort ride mode the electronic-control unit will supply current to energize both the front and rear solenoid controlled delivery valves incorporated in both of the levelling control valve units and also to supply current to energize the front and rear stiffening solenoid valve units (note Fig. 10.126 shows the front levelling control valve and the front stiffening valves open to fluid delivery). Fluid is now permitted to flow to each wheel strut actuator cylinder and its corresponding hydro/gas spring unit and in addition to the both front and rear soft ride third spring units until the preset body height level is reached. The appropriate sensors now signal the electronic-control unit to switch off the power supply to all the solenoid valves; the correct quantity of fluid is therefore contained in the four actuator cylinders for the conditions prevailing at the time. Should the body load be reduced such as when a passenger gets out, the electronic-control unit signals both the solenoid controlled return valves installed in the front and rear levelling control valve units to open (note Fig. 10.126 shows the rear levelling control return valve open for the exit of fluid), fluid will now return to the reservoir until the body height sensors signal that the correct body height has settled again to the manufacturer's setting, and at this point the solenoid return valves close. This cycle of events for delivery and releasing fluid to the spring and levelling system is continuous. When driving conditions change

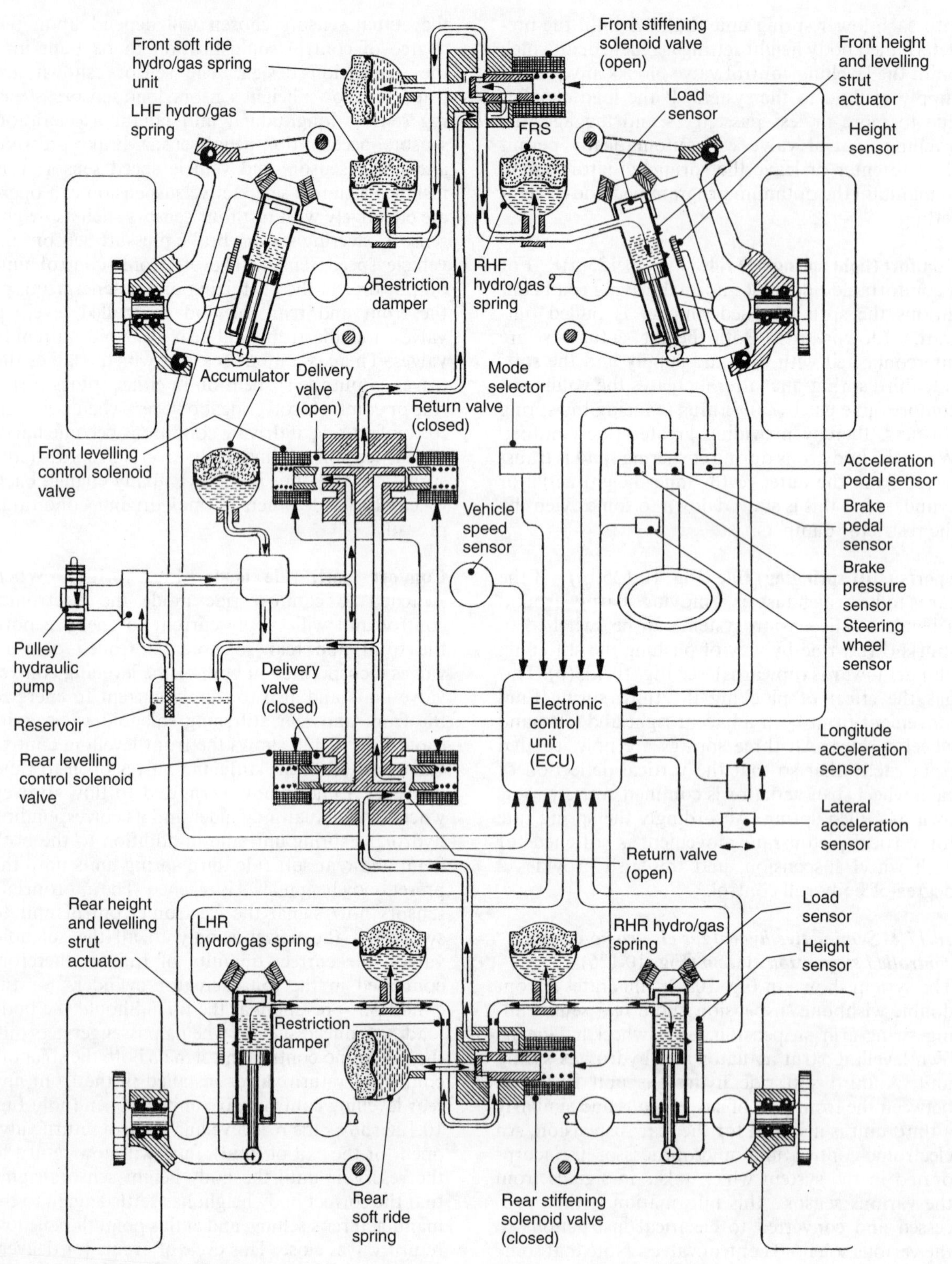

Fig. 10.126 Semi-active hydro/gas electronic controlled suspension

such as during rapid acceleration, hard braking, fast cornering or high speed driving, the front and rear solenoid controlled stiffening valves are signalled to de-energize; this permits the stiffening plunger valves to close, thereby isolating the front and rear interconnected three spring units from each other, thus automatically changing the suspension mode to sports (firm) ride (note Fig. 10.126 shows the rear stiffening solenoid valve closed as for sports (stiff springing) ride mode).

Sports (firm) ride mode (Fig. 10.126) If the driver switches the suspension to sports (stiff springing) ride mode the electronic-control unit will instruct the front and rear stiffening valve unit to de-energize and push in the plunger valve. As a result each wheel spring and strut actuator becomes isolated from the spring opposite it and also from the third stiffening spring unit. As a result the reduction in diaphragm area exposed to the nitrogen gas (now 50% less) increases the stiffness rate of each spring/actuator unit; it therefore provides a firm ride for good steering response, and the prevention of fluid moving between the right and left hand strut actuators increases the suspension's resistance to body roll.

10.18 Electronic controlled pneumatic (air) suspension for on and off road use

A pneumatic (air) controlled suspension system provides a variable spring rate so that a constant suspension frequency is obtained between light and heavy load conditions. Additional telescopic dampers are also installed to improve the ride quality and comfort. A driver's height control is also provided to enable the body height to be adjusted for specific purposes such as loading the boot or cargo space, towing a trailer, driving over rough terrain, or muddy ground, or travelling though flooded areas.

10.18.1 Description of system (Fig. 10.127)

The basic air suspension system consists of four rolling lobe air springs mounted between the chassis and each of the suspension lower swing arms. These springs work within an air pressure ranging between 6 and 10 bar, thus for low height settings and light loads the pressure needed may be down to 6 bar but for higher height settings and heavy loads the pressure could rise to just under 10 bar. Compressed air is supplied via a single cylinder compressor driven by an electric motor and a reservoir tank is provided to store compressed air for instant use. A unloader valve is provided to safeguard the compressor and system from overload. There are four ride height solenoid valves, one for each air spring unit and an inlet and outlet solenoid valve which controls the air supply to and from the system. In addition there are inlet and outlet exhaust solenoid valves which control the release of excess air to the atmosphere. An air silica gel drier dries the wet freshly compressed air before it passes though the various valves and enters the reservoir tank and air spring units.

There are various sensors and switches which provide essential information for the ECU to process and to make corrective decisions to the levelling of the vehicle's body, these are as follows:

Compressor pressure switch A compressor pressure switch monitors the reservoir tank air pressure and provides a signal to the electronic-control unit (ECU) to switch the compressor's electric motor on when the pressure drops below 7.5 bar and to switch it off as the pressure reaches 10 bar.

Height sensors The air spring unit height is monitored by individual height sensors which provide voltage signals to the ECU, the ECU then computes this informs to determine when the pre-determined spring unit height has been reached, and when to switch 'on' or 'off' the ride height solenoid valves. Each wheel suspension is constantly moving up and down so that the different heights measured over a period of 12 sec are recorded to enable the ECU to calculate if more or less air is needed to fill any of the rolling lobe springs.

Engine speed sensor The engine speed sensor takes its readings from the alternator; it is used to indicate if the engine is running as the ECU will switch 'off' the compressor's electric motor if the engine is not running or if the engine speed drops below 500 rev/min to prevent the electric motor draining the battery.

Vehicle speed sensor The vehicle speed sensor takes its readings from the speedometer, to monitor road speed. When the vehicle's speed exceeds 80 km/h the ECU lowers the ride height by about 20 mm to reduce aerodynamic drag and to improve the road holding stability.

Hand brake switch The hand brake switch is used to inform the ECU that the vehicle has stopped and is stationary and only then can the lower access height setting mode be actuated for safety reasons if required.

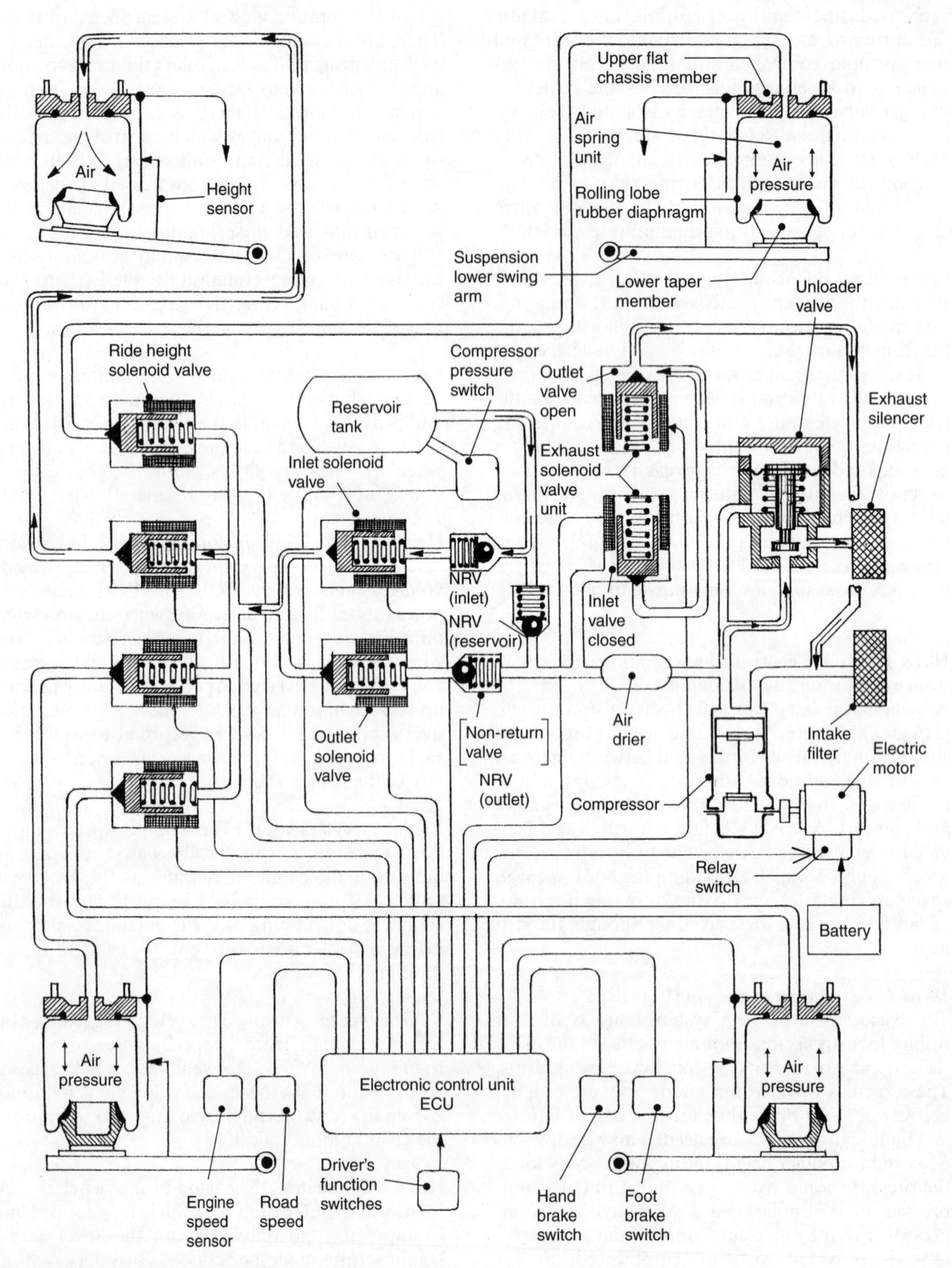

Fig. 10.127 Electronic controlled pneumatic (air) suspension

Driver's function switches The driver's function switches are dash board mounted and are used to select the different height programmes in the ECU when in manual mode. These are standard ride height used for trailer towing, high profile mode for off-road driving and a low profile mode for load accessing.

10.18.2 Operating conditions (Fig. 10.127)

Compressor charging When the ignition switch is on and the engine is running the ECU signals the electric-motor relay switch to close and to energize the electric motor; the compressor will then rotate and begin to charge the reservoir tank via the reservoir non-return valve (NRV). Under these conditions both the inlet and the outlet solenoid valves are de-energized, thereby closing off the ride height solenoid valves and air spring units from the rest of the system. The exhaust outlet solenoid valve is de-energized and is closed, and the inlet solenoid valve is energized and therefore opens, thus permitting air pressure to force down the unloader valve diaphragm against the spring tension, hence closing the unloader valve and thus preventing air discharging into the atmosphere (opposite to that shown in Fig. 10.127). As the air pressure in the reservoir approaches 10 bar the compressor pressure switch signals the ECU to switch 'off' the compressor's electric motor.

Raising and lowering height levels If the rolling lobe air springs are in the low profile access mode or if there has been a loss or expulsion of air previously from the springs, then each height sensor will signal to the ECU to energize and open the inlet solenoid valve and each of the ride height solenoid valves, hence permitting the reservoir tank to discharge additional air into the four air spring bags. Once the air spring units have risen to their setting mode the ECU de-energizes and closes the inlet and ride height solenoid valves thereby retaining the air mass within the spring units. Should the air spring height be too high for a particular ride height setting due to a reduction in load or that a different height profile mode has been selected, then the ECU will energize and open the four ride height valves, the outlet solenoid valve, the inlet and outlet exhaust solenoid valves. Air will now be released from the air spring bags, where it then escapes to the compressors intake and intake filters and via the open unloader valve into the atmosphere; this will continue until the air springs are restored to their pre-programmed height at which point the ECU will cut 'off' the current supply to close all of these valves.

Individual spring levelling If one corner of the vehicle dips more than the others (see left hand top corner Fig. 10.127), possibly due to uneven passenger or load distribution, the relevant height sensors detect this in the from of a voltage change. This is therefore passed on to the ECU and accordingly energizes and opens the inlet solenoid valve and the appropriate ride height inlet solenoid valve; once again pressurized air will be delivered from the reservoir tank via the inlet NRV to the particular spring to compensate for the body sag until the body is level again.

Unloader valve action When the compressor is charging the system the exhaust inlet and outlet valves are open and closed respectively, if the discharge pressure from the compressor should exceed its maximum safe limit of around 10 bar, then air pressure acting with the spring underneath the diaphragm lifts it up against the resisting downward acting air pressure thus enabling the valve to open and to unload the system.

11 Brake system

11.1 Braking fundamentals

11.1.1 The energy of motion and work done in braking (Fig. 11.1)

A moving vehicle possesses kinetic energy whose value depends on the weight and speed of the vehicle. The engine provides this energy in order to accelerate the vehicle from a standstill to given speed, but this energy must be partially or totally dissipated when the vehicle is slowed down or brought to a standstill. Therefore it is the function of the brake to convert the kinetic energy possessed by the vehicle at any one time into heat energy by means of friction (Fig. 11.1).

The equation for kinetic energy, that is the energy of motion, may be given by

$$U_k = \tfrac{1}{2} mV^2$$

where U_k = kinetic energy of vehicle (J)
m = mass of vehicle (Kg)
V = speed of vehicle (m/s)

The work done in bringing the vehicle to rest is given by

$$U_w = Fs$$

where U_w = work done (J)
F = average braking force (N)
s = distance travelled (m)

When braking a moving vehicle to a standstill, the work done by the brake drums must equal the initial kinetic energy possessed by the vehicle so that

$$U_w = U_k,$$

$$Fs = \tfrac{1}{2} mV^2$$

$$\therefore \text{Average brake force} \quad F = \frac{mV^2}{2s} \text{(N)}$$

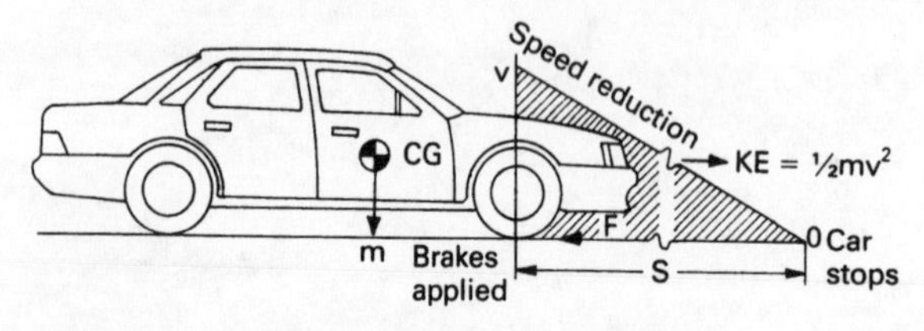

Fig. 11.1 Illustration of braking conditions

Example (Fig. 11.1) A car of mass 800 kg is travelling at 36 km/h. Determine the following:

a) the kinetic energy it possesses,
b) the average braking force to bring it to rest in 20 metres.

a) $V = \dfrac{36 \times 1000}{60 \times 60} = 10\,\text{m/s}$

$$\text{Kinetic energy} = \tfrac{1}{2} mV^2 = \tfrac{1}{2} \times 800 \times 10^2 = 40\,\text{kJ}$$

b) Work done to stop car = change in vehicle's kinetic energy

$$Fs = \tfrac{1}{2} mV^2$$

$$20F = 40\,000$$

$$\therefore F = \frac{40\,000}{20} = 2000\,\text{N} = 2\,\text{kN}$$

11.1.2 Brake stopping distance and efficiency

Braking implies producing a force which opposes the motion of the vehicle's wheels, thereby reducing the vehicle speed or bringing it to a halt. The force or resistance applied to stop a vehicle or reduce its speed is known as the braking force.

The braking efficiency of a vehicle is defined as the braking force produced as a percentage of the total weight of the vehicle, that is

$$\text{Braking efficiency} = \frac{\text{Braking force}}{\text{Weight of vehicle}} \times 100.$$

When the braking force is equal to the whole weight of the vehicle being braked, the braking efficiency is denoted as 100%. The braking efficiency is generally less than 100% because of insufficient road adhesion, the vehicle is on a down gradient or the brake system is ineffective.

The brake efficiency is similar to the coefficient of friction which is the ratio of the frictional force to the normal load between the rubbing surfaces.

$$\text{i.e.} \quad \text{Coefficient of friction} = \frac{\text{Friction force}}{\text{Normal load}}$$

$$\text{that is} \quad \mu = \frac{F}{N}$$

$$\text{i.e.} \quad \eta = \frac{F}{N} = \mu$$

where μ = coefficient of friction
η = brake efficiency
F = friction force
N = normal load.

Thus a braking efficiency of 100% is equal to a coefficient of friction of one.

i.e. $\eta(100\%) = \frac{F}{N} = \mu = 1$

11.1.3 Determination of brake stopping distance (Fig. 11.1)

A rough estimate of the performance of a vehicle's brakes can be made by applying one of the equations of motion assuming the brakes are 100%.

i.e. $V^2 = U^2 + 2gs$

where U = initial braking speed (m/s)
V = final speed (m/s)
g = deceleration due to gravity ($9.81\,m/s^2$)
s = stopping distance (m)

If the final speed of the vehicle is zero (i.e. V = 0)

then $0 = U + 2gs$

and $s = \frac{U^2}{2g} = \frac{U^2}{2 \times 9.81} \simeq \frac{U^2}{20}$

To convert km/h to m/s;

$$\text{U(m/s)} = \frac{1000}{60 \times 60} U = 0.28\ U\ \text{(km/h)}$$

$$\therefore s = \frac{(0.28\ U)^2}{2g} = \frac{(0.28)^2 U^2}{2 \times 9.81} = 0.004\ U^2 \text{(m)}$$

Example Calculate the minimum stopping distance for a vehicle travelling at 60 km/h.

Stopping distance $s = 0.004\ U^2 \text{(m)}$
$= 0.004 \times 60^2 = 14.4\,\text{m}$

11.1.4 Determination of brake efficiency (Fig. 11.1)

The brake efficiency can be derived from the kinetic energy equation and the work done in bringing the vehicle to a standstill.

Let F = braking force (N),
μ = coefficient of friction,
W = vehicle weight (N),
U = initial braking speed (m/s),
m = vehicle mass (kg),
s = stopping distance (m),
η = brake efficiency.

Then equating work and kinetic energy,

$$Fs = \tfrac{1}{2} mU^2$$

but $m = \frac{W}{g}$

$$\therefore Fs = \frac{WU^2}{2g}$$

$$\therefore s = \frac{WU^2}{2Fg}$$

but $\frac{F}{W} = \mu = \eta$

Thus $s = \frac{U^2}{2g\eta}$

$$\therefore \eta = \frac{U^2}{2gs} \times 100$$

but $U\text{(m/s)} = 0.28\ U\text{(km/h)}$

Hence $\eta = \frac{(0.28\ U)^2}{2 \times 9.81s} \times 100 = 0.4\frac{U^2}{s}\%$

Example Determine the braking efficiency of a vehicle if the brakes bring the vehicle to rest from 60 km/h in a distance of 20 metres.

$$\eta = \frac{0.4\ U^2}{s} = \frac{0.4 \times 60^2}{20}$$

$\therefore \eta = 72\%$

A table of vehicle stopping distances for various vehicle speeds and brake efficiencies is shown in Table 11.1.

11.1.5 Adhesion factor

The stopping distance of a wheel is greatly influenced by the interaction of the rotating tyre tread

Table 11.1 Stopping distances for various vehicle speeds and brake efficiencies

Vehicle speed	Stopping distance for various braking efficiencies (m)					
km/h	100%	90%	80%	70%	60%	50%
10	0.4	0.4	0.5	0.6	0.7	0.8
20	1.6	1.8	2.0	2.3	2.7	3.2
30	3.6	4.0	4.5	5.1	6.0	7.2
40	6.4	7.1	8.0	9.1	10.7	12.8
50	10.0	11.1	12.5	14.3	16.7	20.0
60	14.4	16.0	18.0	80.6	24.0	28.8
70	19.6	21.8	24.5	28.0	32.7	39.2
80	25.6	28.4	32.0	36.6	42.7	51.2
90	32.4	36.0	40.5	46.3	54.0	64.8
100	40.0	44.4	50.0	57.1	66.7	80.0

Table 11.2 Adhesion factors for various road surfaces

No.	Material	Condition	Adhesion factor
1	Concrete, coarse asphalt	dry	0.8
2	Tarmac, gritted bitumen	dry	0.6
3	Concrete, coarse asphalt	wet	0.5
4	Tarmac	wet	0.4
5	Gritted bitumen tarmac	wet	0.3
6	Gritted bitumen tarmac	greasy	0.25
7	Gritted bitumen, snow compressed	greasy	0.2
8	Gritted bitumen, snow compressed	dry	0.15
9	Ice	wet	0.1

and the road surface. The relationship between the decelerating force and the vertical load on a wheel is known as the *adhesion factor* (μ_a). This is very similar to the coefficient of friction (μ) which occurs when one surface slides over the other, but in the case of a correctly braked wheel, it should always rotate right up to the point of stopping to obtain the greatest retarding resistance.

Typical adhesion factors for various road surfaces are given in Table 11.2.

11.2 Brake shoe and pad fundamentals

11.2.1 Brake shoe self-energization (Fig. 11.2)

The drum type brake has two internal semicircular shoes lined with friction material which matches up to the internal rubbing face of the drum. The shoes are mounted on a back plate, sometimes known as a torque plate, between a pivot anchor or wedge type abutment at the lower shoe ends, and at the upper shoe top end by either a cam or hydraulic piston type expander. For simplicity the expander in Fig. 11.2 is represented by two opposing arrows and the shoe linings by two small segmental blocks in the mid-region of the shoes.

When the drum is rotating clockwise, and the upper tips of the shoes are pushed apart by the expander force F_e, a normal inward reaction force N will be provided by the drum which resists any shoe expansion.

As a result of the drum sliding over the shoe linings, a tangential frictional force $F_t = \mu N$ will be generated between each pair of rubbing surfaces.

The friction force or drag on the right hand shoe (Fig. 11.2) tends to move in the same direction as its shoe tip force F_e producing it and accordingly helps to drag the shoe onto the drum, thereby effectively raising the shoe tip force above that of the original expander force. The increase in shoe tip force above that of the input expander force is described as *positive servo,* and shoes which provide this self-energizing or servo action are known as *leading shoes.*

i.e. $F_L = F_e + F_t$

where F_L = leading shoe tip resultant force

Likewise considering the left hand shoe (Fig. 11.2) the frictional force or drag F_t tends to oppose and cancel out some of the shoe tip force F_e producing it. This causes the effective shoe tip force to be less than the expander input force. The resultant reduction in shoe tip force below that of the initial

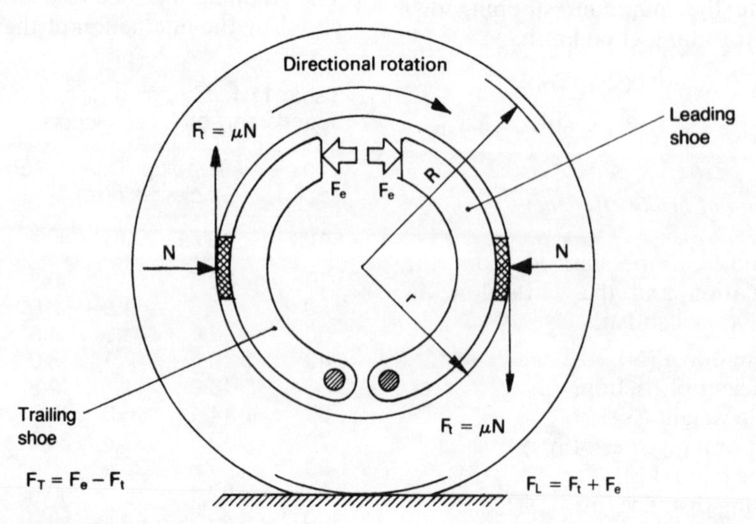

Fig. 11.2 Drum and shoe layout

input tip force is described as *negative servo,* and shoe arrangements which have this de-energizing property are known as *trailing shoes.*

i.e. $F_T = F_e - F_t$

where F_T = trailing shoe tip resultant force

The magnitude of the self-energizing action is greatly influenced by the rubbing surface temperature, dampness, wetness, coefficient of friction and speed of drum rotation.

Changing the direction of rotation of the drum causes the original leading and trailing shoes to reverse their energizing properties, so that the leading and trailing shoes now become trailing and leading shoes respectively.

The shoe arrangement shown in Fig. 11.2 is described as a *leading-trailing shoe drum brake.*

Slightly more self-energizing is obtained if the shoe lining is heavily loaded at the outer ends as opposed to heavy mid-shoe loading.

11.2.2 Retarding wheel and brake drum torques (Fig. 11.2)

The maximum retarding wheel torque is limited by wheel slip and is given by

$$T_w = \mu_a WR \text{ (Nm)}$$

where T_w = wheel retarding torque (Nm)
μ_a = adhesion factor
W = vertical load on wheel (N)
R = wheel rolling radius (m)

Likewise the torque produced at this brake drum caused by the frictional force between the lining and drum necessary to bring the wheel to a standstill is given by

$$T_B = \mu N r \text{ (Nm)}$$

where T_B = brake drum torque (Nm)
μ = coefficient of friction between lining and drum
N = radial force between lining and drum (N)
r = drum radius (m)

Both wheel and drum torques must be equal up to the point of wheel slip but they act in the opposite direction to each other. Therefore they may be equated.

i.e. $T_B = T_w$

$\mu N r = \mu_a WR$

$\therefore$ Force between lining and drum $N = \dfrac{\mu_a WR}{\mu r}$ (N)

Example A road wheel has a rolling radius of 0.2 m and supports a load of 5000 N and has an adhesion factor of 0.8 on a particular road surface. If the drum radius is 0.1 m and the coefficient of friction between the lining and drum is 0.4, determine the radial force between the lining and drum.

$$N = \frac{\mu_a WR}{\mu r} = \frac{0.8 \times 5000 \times 0.2}{0.4 \times 0.1} = 20\,000 \text{ N or } 20 \text{ kN}$$

11.2.3 Shoe and brake factors (Fig. 11.2)

If the brake is designed so that a low operating force generates a high braking effort, it is said to have a high self-energizing or servo action. This desirable property is obtained at the expense of stability because any frictional changes disproportionately affect torque output. A brake with little self-energization, while requiring a higher operating force in relation to brake effort, is more stable in operation and is less affected by frictional changes.

The multiplication of effort or self-energizing property for each shoe is known as the *shoe factor.*

The shoe factor S is defined as the ratio of the tangential drum drag at the shoe periphery F_t to the force applied by the expander at the shoe tip F_e.

i.e. $\text{Shoe factor} = \dfrac{\text{Tangential drum force}}{\text{Shoe tip force}}$

$$S = \frac{F_t}{F_e}$$

The combination of different shoe arrangements such as leading and trailing shoes, two leading shoes, two trailing shoes etc. produces a brake factor B which is the sum of the individual shoe factors.

Brake factor = Sum of shoe factors

i.e. $B = (S_L + S_T)$, $2S_L$, $2S_T$ and $(S_p + S_s)$

11.2.4 Drum shoe arrangements (Fig. 11.3(a–c))

Leading and trailing shoe brakes (Fig. 11.3(a)) If a single cylinder twin piston expander (double acting) is mounted between two shoe tips and the opposite shoe tips react against a fixed abutment, then the leading shoe is forced against the drum in the forward rotation direction, whilst the trailing shoe works against the rotation direction producing

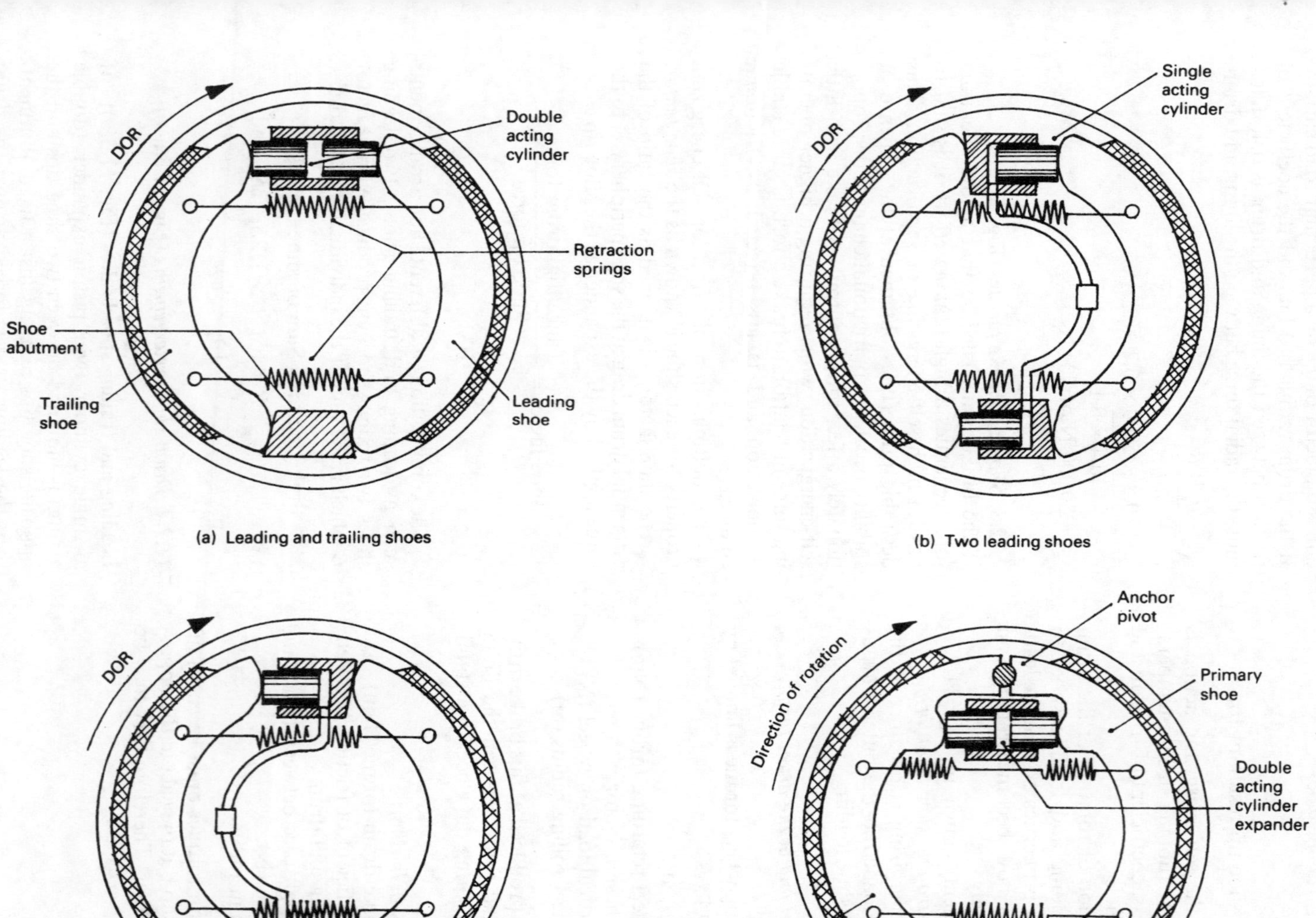

Fig. 11.3 (a–d) Various brake shoe arrangements

much less frictional drag. Such an arrangement provides a braking effect which is equal in both forward and reverse motion. Rear wheel brakes incorporating some sort of hand brake mechanism are generally of the lead and trailing shoe type.

Two leading shoe brakes (Fig. 11.3(b)) By arranging a pair of single piston cylinders (single acting) diametrically opposite each other with their pistons pointing in the direction of drum rotation, then when hydraulic pressure is applied, the drum to lining frictional drag force pulls the shoes in the same direction as the shoe tip piston forces, thus causing both shoes to become self-energizing. Such a layout is known as a *two leading shoe drum type brake*. In reverse, the braking force is reduced due to the drag force opposing the piston tip forces; both shoes in effect then have a trailing action. Two leading shoe brakes are possibly still the most popular light commercial type front wheel brake.

Two trailing shoe brakes (Fig. 11.3(c)) If now two separate single acting cylinders are mounted between the upper and lower shoe tips so that both pistons counteract the rotational forward direction of the drum, then the resultant lining drag force will be far less for each shoe, that is, there is a negative servo condition.

Brakes with this layout are therefore referred to as *two trailing shoe brakes*. This arrangement is suitable for application where lining stability is important and a servo assisted booster is able to compensate for the low resultant drag force relative to a given input shoe tip force. A disadvantage of a two trailing shoe brake is for the same brake effect as a two leading shoe brake; much higher hydraulic line pressures have to be applied.

Duo-servo shoe brakes (Fig. 11.3(d)) A double acting cylinder expander is bolted to the back plate and the pistons transmit thrust to each adjacent shoe, whereas the opposite shoe tip ends are joined together by a floating adjustment link. On application of the brake pedal with the vehicle being driven forward, the pistons move both shoes into contact with the revolving drum. The shoe subjected to the piston thrust which acts in the same direction as the drum rotation is called the *primary shoe* and this shoe, when pulled around with the drum, transfers a considerable force to the adjacent shoe tip via the floating adjustment link. This second shoe is known as the *secondary shoe* and its initial movement with the drum pushes it hard against the anchor pin, this being permitted by the pistons themselves floating within the cylinder to accommodate any centralization which might become necessary. Under these conditions a compounding of both the primary circumferential drag force and that produced by the secondary shoe itself takes place so that a tremendous wedge or self-wrapping effect takes place far in excess of that produced by the two expander pistons alone. These brakes operate equally in the forward or reverse direction. Duo-servo shoe brakes give exceptionally good performance but are very sensitive to changes in shoe lining properties caused by heat and wetness.

Because the secondary shoe performs more work and therefore wears quicker than the primary shoe, lining life is equalized as far as possible by fitting a thick secondary shoe and a relatively thin primary shoe.

11.2.5 The principle of the disc brake (Fig. 11.4(a, b and c))

The disc brake basically consists of a rotating circular plate disc attached to and rotated by the wheel hub and a bridge member, known as the caliper, which straddles the disc and is mounted on the suspension carrier, stub axle or axle casing (Fig. 11.4(b)). The caliper contains a pair of pistons and friction pads which, when the brakes are applied, clamp the rotating disc, causing it to reduce speed in accordance to the hydraulic pressure behind each piston generated by the pedal effort.

The normal clamping thrust N on each side of the disc (Fig. 11.4(b and c)) acting through the pistons multiplied by the coefficient of friction μ generated between the disc and pad interfaces produces a frictional force $F = \mu N$ on both sides of the disc. If the resultant frictional force acts through the centre of the friction pad then the mean distance between the centre of pad pressure and the centre of the disc will be

$$\frac{R_2 - R_1}{2} = R.$$

Accordingly, the frictional braking torque (Fig. 11.4(a)) will be dependent upon twice the frictional force (both sides) and the distance the pad is located from the disc centre of rotation. That is,

$$\text{Braking torque} = 2\mu N\left(\frac{R_2 - R_1}{2}\right)\text{(Nm)}$$

$$\text{i.e.} \quad T_B = 2\mu NR \text{ (Nm)}$$

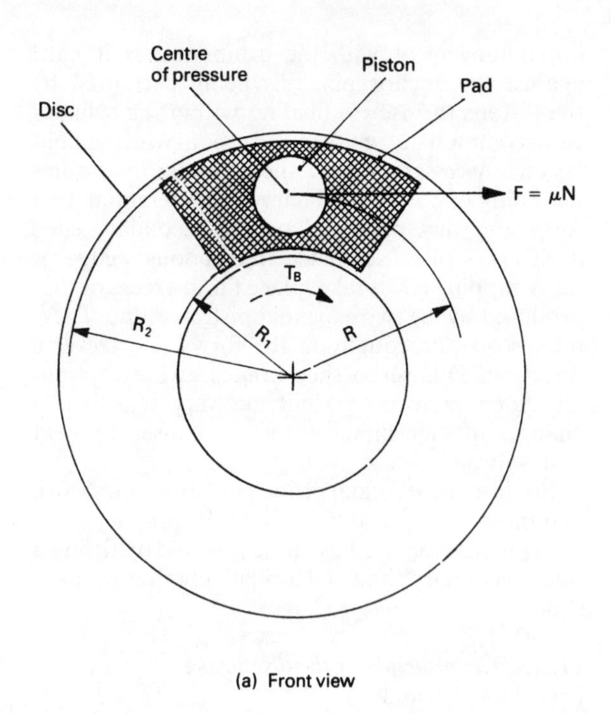

(a) Front view

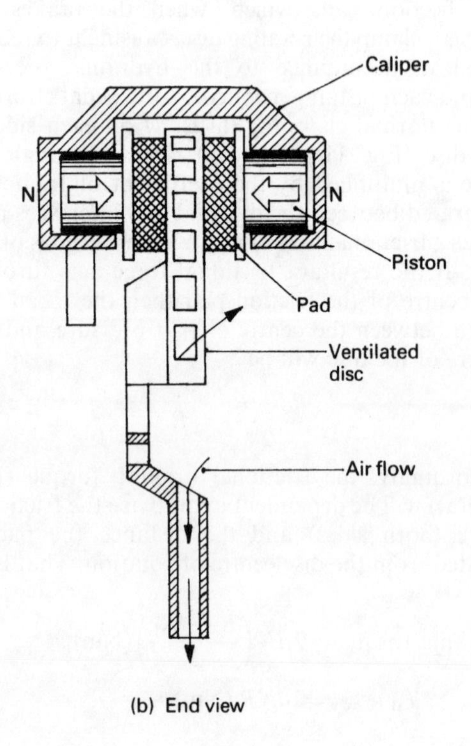

(b) End view

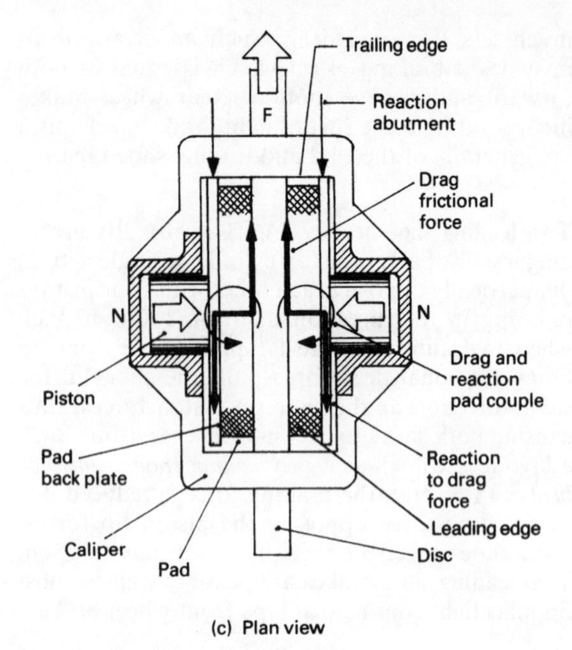

(c) Plan view

Fig. 11.4 (a–c) Disc and pad layout

Example If the distance between the pad's centre of pressure and the centre of disc rotation is 0.12 m and the coefficient of friction between the rubbing faces is 0.35, determine the clamping force required to produce a braking torque of 84 Nm.

$$T_B = 2\mu NR$$

$$\therefore \text{ Clamping force } N = \frac{T_B}{2\mu R}$$

$$= \frac{84}{2 \times 0.35 \times 0.12}$$

$$= 1000\,\text{N}$$

11.2.6 Disc brake pad alignment (Fig. 11.4)

When the pads are initially applied they are loaded against the disc with uniform pressure, but a small tilt tendency between the leading and trailing pad edges caused by frictional pad drag occurs. In addition the rate of wear from the inner to the outer pad edges is not uniform. The bedding-in conditions of the pads will therefore be examined in the two parts as follows:

1 Due to the thickness of the pad there is a small offset between the pad/disc interface and the pad's back plate reaction abutment within the caliper (Fig. 11.4(c)). Consequently, a couple is

produced which tends to tilt the pad into contact with the disc at its leading edge harder compared to the trailing edge. This in effect provides a very small self-energizing servo action, with the result that the wear rate at the leading edge is higher than that at the trailing edge.

2 The circular distance covered by the disc in one revolution as it sweeps across the pad face increases proportionately from the inner to the outer pad's edges (Fig. 11.4(a)). Accordingly the rubbing speed, and therefore the work done, increases from the inner to the outer pad edges, with the result that the pad temperature and wear per unit area rises as the radial distance from the disc centre becomes greater.

11.2.7 Disc brake cooling (Fig. 11.4)

The cooling of the brake disc and its pads is achieved mostly by air convection, although some of the heat is conducted away by the wheel hub. The rubbing surface between the rotating disc and the stationary pads is exposed to the vehicle's frontal airstream and directed air circulation in excess of that obtained between the drum and shoe linings. Therefore the disc brake is considerably more stable than the drum brake under continued brake application. The high conformity of the pad and disc and the uniform pressure enable the disc to withstand higher temperatures compared to the drum brake before thermal stress and distortion become pronounced. Because there is far less distortion with discs compared to drums, the disc can operate at higher temperatures. A further feature of the disc is it expands towards the pads, unlike the drum which expands away from the shoe linings. Therefore, when hot, the disc brake reduces its pedal movement whereas the drum brake increases its pedal movement.

Cast ventilated discs considerably improve the cooling capacity of the rotating disc (Fig. 11.4(b)). These cast iron discs are in the form of two annular plates ribbed together by radial vanes which also act as heat sinks. Cooling is effected by centrifugal force pushing air through the radial passages formed by the vanes from the inner entrance to the outer exit. The ventilated disc provides considerably more exposed surface area, producing something like a 70% increase in convection heat dissipation compared to a solid disc of similar weight. Ventilated discs reduce the friction pad temperature to about two-thirds that of a solid disc under normal operating conditions. Pad life is considerably increased with lower operating temperatures, but there is very little effect on the frictional properties of the pad material. Ventilated wheels have very little influence on the disc cooling rate at low speeds. At very high speeds a pressure difference is set up between the inside and outside of the wheel which forces air to flow through the vents towards the disc and pads which can amount to a 10% improvement in the disc's cooling rate. The exposure of the disc and pads to water and dirt considerably increases pad wear.

The removal of dust shields will increase the cooling rate of the disc and pad assembly but it also exposes the disc and pads to particles of mud, dust and grit which adhere to the disc. This will cause a reduction in the frictional properties of the rubbing pairs. If there has to be a choice of a lower working temperature at the expense of contaminating the disc and pads or a higher working temperature, the priority would normally be in favour of protecting the rubbing surfaces from the atmospheric dust and from the road surface spray.

11.2.8 A comparison of shoe factors and shoe stability (Fig. 11.5)

A comparison of different brake shoe arrangements and the disc brake can be made on a basis of shoe factor, *S*, or output torque compared against the variation of rubbing coefficient of frictions (Fig. 11.5). The coefficient of friction for most linings and pads ranges between 0.35 and 0.45, and it can be seen that within the normal coefficient of friction working range the order of smallest to greatest shoe factor is roughly as follows in Table 11.3.

This comparison shows that the torque output (shoe factor) for a single or two trailing shoes is only approximately one-third of the single or two leading shoe brake, and that the combination of a leading and trailing shoe is about twice that of the two trailing shoe, or roughly two-thirds of the two leading shoe arrangement (Fig. 11.5). The disc and pad's performance is very similar to the two trailing shoe layout, but with higher coefficients of friction the disc brake shoe factor rises at a faster rate than that of the two trailing shoe brake. Overall, the duo-servo shoe layout has a superior shoe factor relative to all other arrangements, amounting to roughly five times that of the two trailing shoes and just under twice that of the two leading shoe brake.

Conversely, the lining or pad stability, that is, the ability of the shoes or pads to maintain approximately the same shoe factor if there is a small change in the coefficient, due possibly to wetness or an increase in the friction material temperature, alters in the reverse order as shown in Table 11.3.

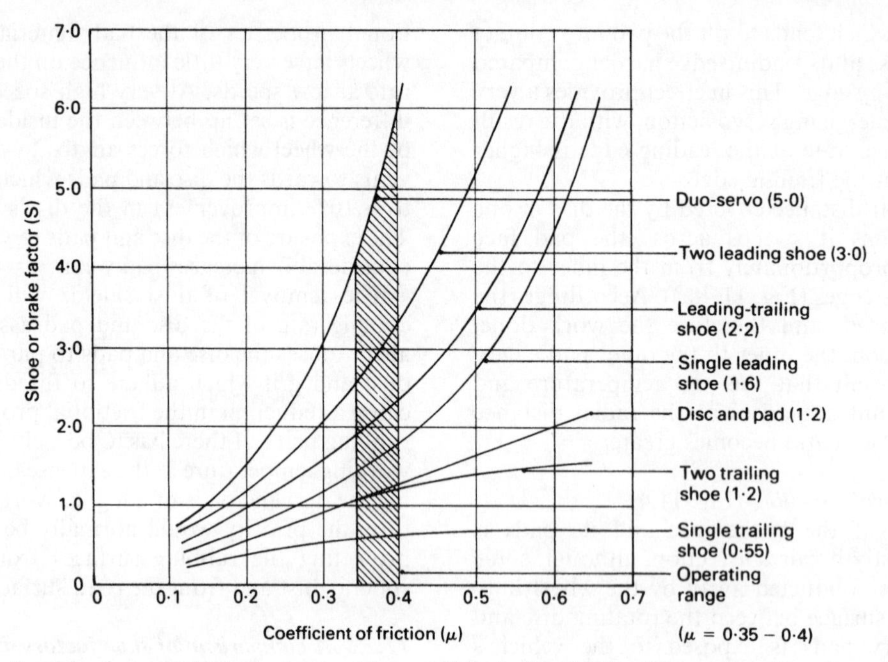

Fig. 11.5 Relationship of shoe or brake factors and the coefficient of friction for different shoe layouts and the disc brake

Generally, brakes with very high shoe factors are unstable and produce a relatively large change in shoe factor (output torque) for a small increase or decrease in the coefficient of friction between the rubbing surfaces. Layouts which have low shoe factors tend to produce a consistent output torque for a considerable shift in the coefficient of friction. Because of the instability of shoe layouts with high shoe factors, most vehicle designers opt for the front brakes to be either two leading shoes or disc and pads, and at the rear a leading and trailing shoe system. They then rely on vacuum or hydraulic servo assistance or full power air operation. Thus having, for example, a combined leading and trailing shoe brake provides a relatively high leading shoe factor but with only a moderate degree of stability, as opposed to a very stable trailing shoe which produces a very low shoe factor. The properties of each shoe arrangement complement the other to produce an effective and a reliable foundation brake. Leadings and trailing shoe brakes are still favoured on the rear wheels since they easily accommodate the hand brake mechanism and produce an extra self-energizing effect when the hand brake is applied, which in the case of the disc and pad brake is not obtainable and therefore requires a considerable greater clamping force for wheel lock condition.

11.2.9 Properties of friction lining and pad materials

Friction level (Fig. 11.6) The average coefficient of friction with modern friction materials is between 0.3 and 0.5. The coefficient of friction should be sufficiently high to limit brake pedal effort and to reduce the expander leverage on commercial vehicles, but not so high as to produce *grab,* and in the extreme case cause lock or *sprag* so that rotation of the drum becomes impossible. The most suitable grade of friction material must be used to match the degree of self-energization created by the shoe and pad configuration and applications.

Resistance to heat fade (Fig. 11.6) This is the ability of a lining or pad material to retain its coefficient of friction with an increase in rubbing temperature. The maximum brake torque the lining or pad is to absorb depends on the size and type of brake, gross vehicle weight, axle loading, the front to rear braking ratio and the maximum attainable speed. A good quality material should retain its friction level throughout the working temperature range of the drum and shoes or disc and pads. A reduction in the frictional level in the

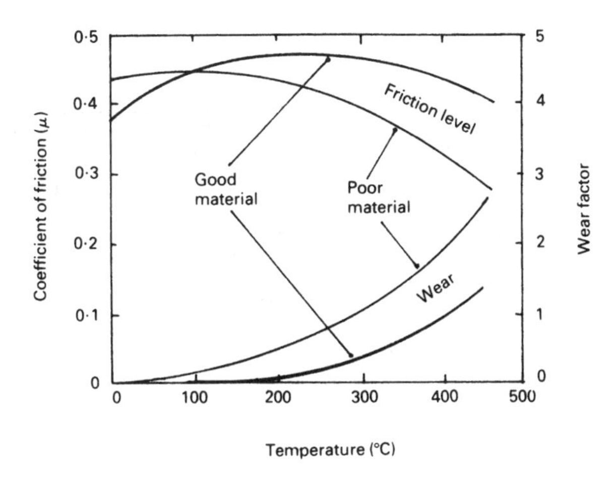

Fig. 11.6 Effects of temperature on the coefficient of friction

higher temperature range may be tolerated, provided that it progressively decreases, because a rapid decline in the coefficient of friction could severely reduce the braking power capability when the vehicle is being driven on long descents or subjected to continuous stop-start journey work. The consequences of a fall in the friction level will be greater brake pedal effort with a very poor retardation response. It has been established that changes in the frictional level which occur with rising working temperatures are caused partly by the additional curing of the pad material when it heats up in service and partly because chemical changes take place in the binder resin.

Recovery from fade (Fig. 11.6) This is a measure of the ability of a friction material to revert to its original friction level upon cooling after brake lining or pad temperature fade has taken place. The frictional characteristics of a good quality material will return on cooling, even after being subjected to repeatedly severe heating, but an inferior material may have poor recovery and the friction level may be permanently altered. Poor recovery is caused principally by a chemical breakdown in the ingredients. This may cause hardening, cracking, flaking, charring or even burning of the linings or pads. If the linings or pads are using thermoplastic binder resins a deposit may form on the rubbing surfaces which may distort the friction properties of the material.

Resistance to wear (Fig. 11.6) The life of a friction material, be it a lining or pad, will depend to a great extent upon the rubbing speed and pressure. The wear is greatly influenced by the working temperature. At the upper limits of the temperature range, the lining or pad material structure is weakened, so that there is an increase in the shear and tear action at the friction interface resulting in a higher wear rate.

Resistance to rubbing speed (Fig. 11.7) The coefficient of friction between two rubbing surfaces should in theory be independent of speed, but it has been found that the intensity of speed does tend to slightly reduce the friction level, particularly at the higher operating temperature range. Poor friction material may show a high friction level at low rubbing speeds, which may cause judder and grab when the vehicle is about to stop, but suffers from a relatively rapid decline in the friction lever as the rubbing speed increases.

Resistance to the intensity of pressure (Fig. 11.8) By the laws of friction, the coefficient of friction should not be influenced by the pressure holding the rubbing surfaces together, but with developed friction materials which are generally compounds held together with resin binders, pressure between the rubbing surfaces does reduce the level of friction. It has been found that small pressure increases at relative low pressures produce a marked reduction in the friction level, but as the intensity of

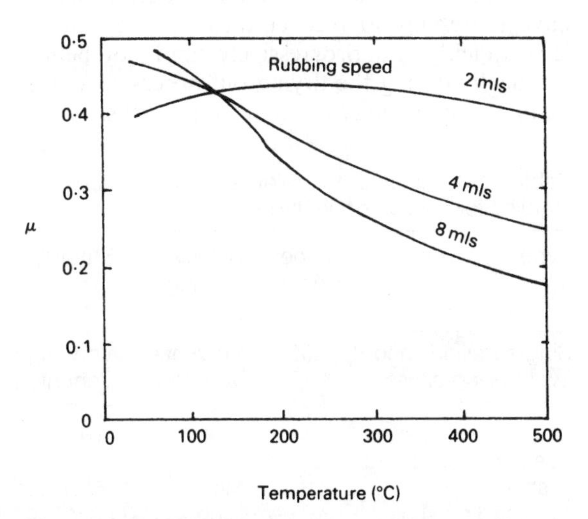

Fig. 11.7 Effects of rubbing speed on the level of friction over the temperature range

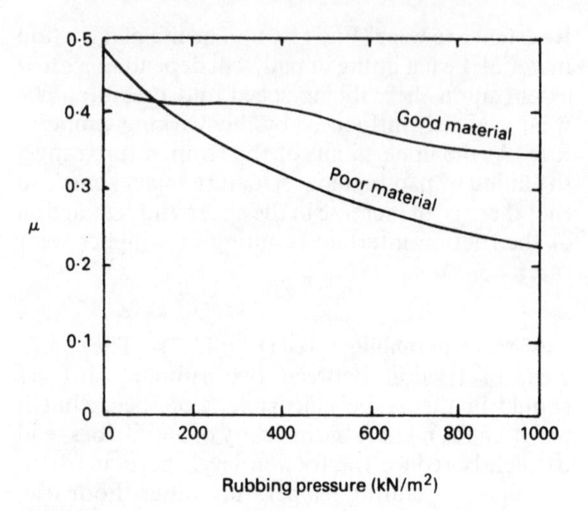

Fig. 11.8 Effects of rubbing pressure on the coefficient of friction

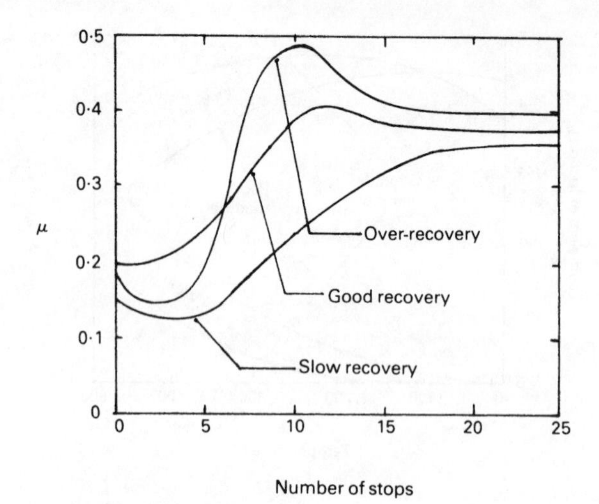

Fig. 11.9 Effects of water contamination on the material's friction recovery over a period of vehicle stops

pressure becomes high the decrease in friction level is much smaller. A pressure-stable lining will produce deceleration proportional to the pedal effort, but pressure-sensitive materials will require a relatively greater pedal force for a given braking performance. Disc brakes tend to operate better when subjected to high rubbing pressures, whereas shoe linings show a deterioration in performance when operating with similar pressures.

Resistance to water contamination (Fig. 11.9) All friction materials are affected by water contamination to some extent. Therefore, a safe margin of friction level should be available for wet conditions, and good quality friction materials should have the ability to recover their original friction level quickly and progressively (and not behave erratically during the drying out process). A poor quality material may either recover very slowly or may develop over-recovery tendency (the friction level which is initially low due to the wetness rises excessively during the drying out period, falling again as the lining or pad dries out completely). Over-recovery could cause brake-grab and even wheel-lock, under certain driving conditions.

Table 11.3 Shoe factor, relative braking power and stability for various brake types

Type of brake	Shoe factor	Relative braking power	Stability
Single trailing shoe	0.55	Very low	Very high
Two trailing shoes	1.15	Very low	Very high
Disc and pad	1.2	Low	High
Single leading shoe	1.6	High	Low
Leading and trailing shoes	2.2	Moderate	Moderate
Two leading shoes	3.0	High	Low
Duo-servo shoes	5.0	Very high	Very low

Resistance to moisture sensitivity The effects of atmospheric dampness, humidity or dew may increase the friction level for the first few applications, with the result that the brakes may become noisy and develop a tendency to grab for a short time. Moisture-sensitive friction materials should not be used on brakes which have high self-energizing characteristics.

Friction materials Materials which may be used for linings or pads generally have their merits and limitations. Sintered metals tend to have a long life but have a relatively low coefficient of friction. Ceramics mixed with metals have much higher coefficient of friction but are very rigid and therefore must be made in sections. They tend to be very harsh on the drums and disc, causing them to suffer from much higher wear rates than the asbestos-based materials. There has been a tendency to produce friction materials which contain much less asbestos and much more soft metal, such as brass zinc inserts or aluminium granules. Non-asbestos materials are now available which contain DuPont's *Kevlar*, a high strength aramid fibre. One manufacturer uses this high strength fibre in pulp form as the main body for the friction material, whereas another manufacturer uses a synthetically created body fibre derived from molten blast-furnace slag reinforced with Kevlar for the main body. Some non-asbestos materials do suffer from a drastic reduction in the coefficient of friction when operating in winter temperatures which, if not catered for in the brake design, may not be adequate for overnight parking brake hold.

11.3 Brake shoe expanders and adjusters

11.3.1 Self-adjusting sector and pawl brake shoe mechanism (Fig. 11.10(a, b and c))

With this leading and trailing shoe rear wheel brake layout the two shoes are actuated by opposing twin hydraulic plungers.

A downward hanging hand brake lever pivots from the top of the trailing shoe. A toothed sector lever pivots similarly from the top of the leading shoe, but its lower toothed sector end is supported and held in position with a spring loaded toothed pawl. Both shoes are interlinked with a strut bar.

Hand brake operation When the hand brake lever is applied the cable pulls the hand lever inwards, causing it to react against the strut. As it tilts it forces the trailing shoe outwards to the drum. At the same time the strut is forced in the opposite direction against the sector lever. This also pushes the leading shoe via the upper pivot and the lower toothed pawl towards the drum. The hand brake shoe expander linkage between the two shoes

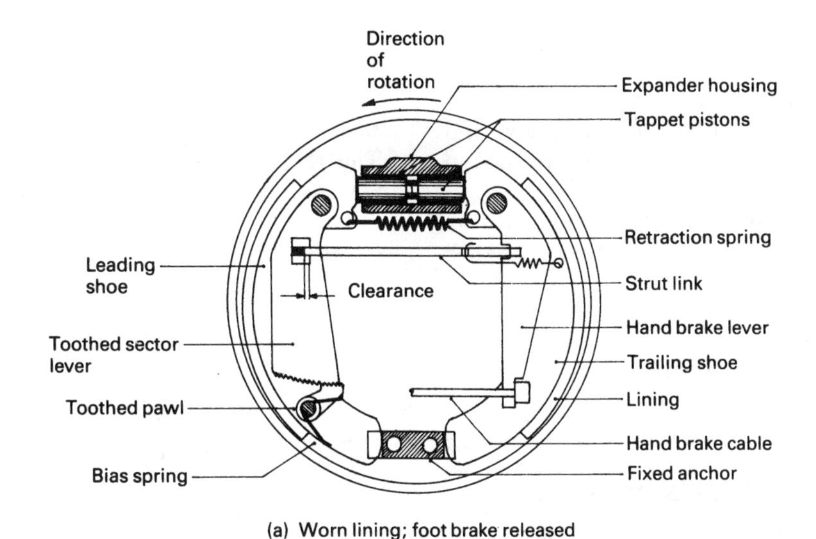

(a) Worn lining; foot brake released

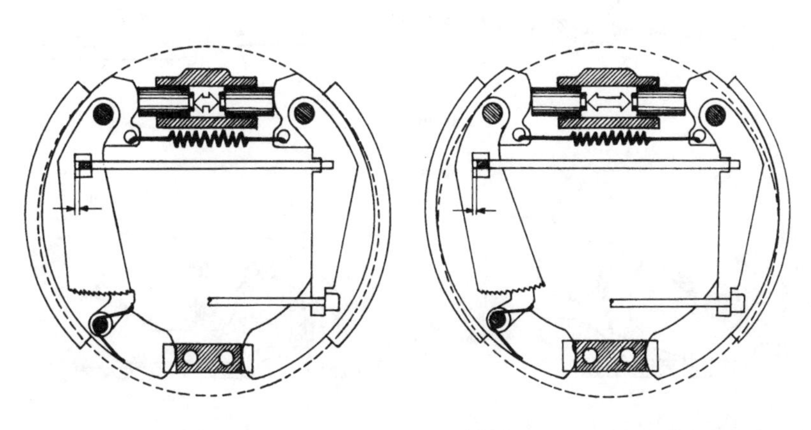

(b) New lining; foot brake applied

(c) Worn lining; foot brake applied

Fig. 11.10 (a–c) Self-adjusting sector and pawl shoes with forward full hand brake

therefore floats and equalizes the load applied by each shoe to the drum.

Automatic adjuster operation

Brake application with new linings (Fig. 11.10(b)) When the foot brake is applied, hydraulic pressure forces the twin plungers apart so that the shoes are expanded against the drum. If the linings are new and there is very little lining to shoe clearance, then the outward movement of the leading and trailing shoes will not be sufficient for the clearance between the rectangular slot in the sector lever and the strut inner edge to be taken up. Therefore the shoes will return to their original position when the brakes are released.

Brake application with worn lining (Fig. 11.10(c)) Applying the foot brake with worn linings makes the brake shoes move further apart. The first part of the outward movement of the leading shoe takes up the clearance between the strut's inner edge and the adjacent side of the rectangular slot formed in the sector lever. As the shoe moves further outwards, the strut restrains the sector lever moving with the leading shoe, so that it is forced to swivel clockwise about its pivot. This permits the spring-loaded pawl to ride over the sector teeth until the shoe contacts the drum. At this point the pawl teeth drop into corresponding teeth on the sector, locking the sector lever to the leading shoe.

When the brakes are released, the shoes are pulled inwards by the retraction spring, but only back to where the rectangular slot contacts the outer edge of the strut (Fig. 11.10(a)). The next time the brakes are applied the shoes will not move far enough out for the slot to strut clearance to be taken up. When the brakes are released the shoe returns to the previous position and the sector to pawl ratchet action does not occur.

As the lining wears, eventually there will be sufficient slot to strut end clearance for the ratchet action to take place and for the pawl to slide over an extra sector tooth before re-engaging the sector in a more advanced position.

11.3.2 Self-adjusting quadrant and pinion brake shoe mechanism (Fig. 11.11(a–d))

This rear wheel brake layout incorporates leading and trailing brake shoes which have a hydraulically operated foot brake system and a mechanically actuated hand brake. The brake shoes are mounted

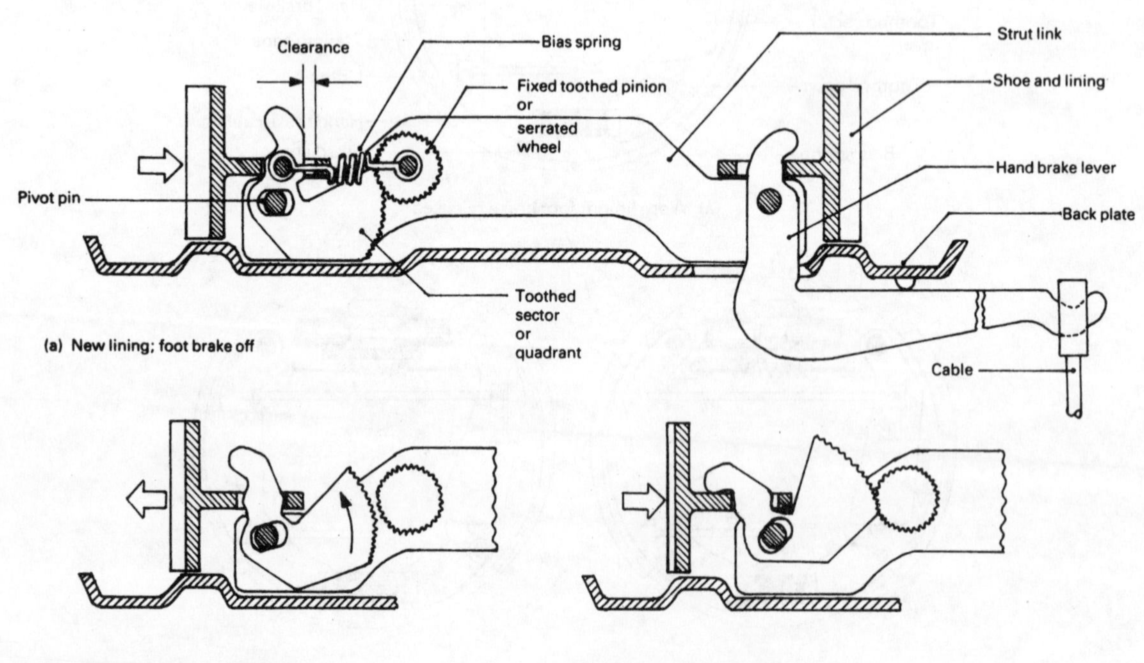

Fig. 11.11 (a–d) Self-adjusting sector and pinion brake shoes with cross-pull hand brake

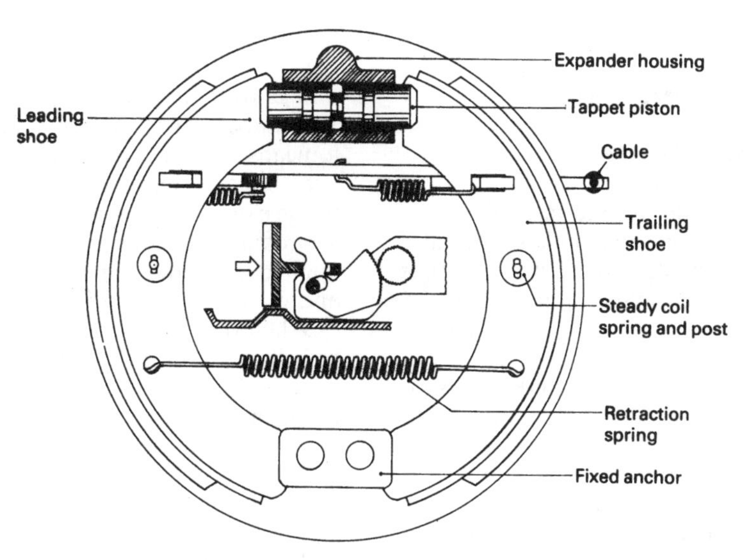

(c) Half worn lining; foot brake off

Fig. 11.11 *contd*

on a back plate and the lower shoe tips are prevented from rotating by a fixed anchor abutment plate riveted to the back plate. The upper shoe tips are actuated by twin hydraulic plungers and a hand brake strut and lever mechanism which has a built-in automatic shoe clearance adjustment device.

The hand brake mechanism consists of a strut linking the two shoes together indirectly via a hand brake lever on the trailing shoe and a quadrant lever on the leading shoe.

Brake application with new linings (Fig. 11.11(a)) When the foot pedal is depressed, the hydraulic plungers are pushed apart, forcing the shoes into contact with the drum. When the brakes are released the retraction spring pulls the shoe inwards until the rectangular slot in the shoe web contacts the outer edge of the quadrant lever. The lever is then pushed back until the teeth on its quadrant near the end of the quadrant mesh with the fixed pinion teeth (or serrations). The position on the quadrant teeth where it meshes with the pinion determines the amount the shoe is permitted to move away from the drum and the gap between the quadrant's inner edge and the slot contact (the lining to shoe clearance).

Brake application with half worn linings (Fig. 11.11 (b and c)) When the foot brake applied, hydraulic pressure forces the shoe plungers outwards. The leading shoe moves out until the clearance between the inner edge of the quadrant lever and slot touch. Further outward shoe movement disengages the quadrant lever teeth from the adjacent pinion teeth and at the same time twists the lever. When the brakes are released the retraction spring pulls the shoes together. Initially the leading shoe web slot contacts the outer edge of the quadrant lever, and then further shoe retraction draws the quadrant lever teeth into engagement with the fixed pinion teeth, but with half worn linings the quadrant will mesh with the pinion somewhere mid-way between the outer edges of the quadrant. Consequently the shoes will only be allowed to move part of the way back to maintain a constant predetermined lining to drum clearance.

Brake application with fully worn linings (Fig. 11.11(d)) When the brakes are operated with fully worn linings the shoes move outwards before they contact the drum. During this outward movement the rear end of the slot contacts the inner edge of the quadrant lever, disengaging it from the pinion. At the same time the quadrant lever rotates until the fingered end of the lever touches the side of the shoe web. Releasing the brakes permits the shoes to retract until the quadrant lever contacts the pinion at its least return position near the quadrant's edge, furthest away from the new lining retraction position.

If any more lining wear occurs, the quadrant is not able to compensate by moving into a more raised position and therefore the master cylinder pedal movement will become excessive, providing a warning that the linings need replacing.

11.3.3 Strut and cam brake shoe expander (Fig. 11.12(a and b))

This type of shoe expander is used in conjunction with leading and trailing shoe brakes normally operated by air pressure-controlled brake actuators connected to a lever spline mounted on the camshaft, which is itself supported on a pair of plain bronze bushes.

The camshaft mounted in the expander housing is splined at its exposed end to support and secure the actuating lever (Fig. 11.12(a)). The other end, which is enclosed, supports an expander cam which has two spherical recesses to accommodate a pair of ball-ended struts. The opposite strut ends are located inside a hollow tappet plunger (follower). Mounted on the end of each tappet plunger is a tappet head abutment which guides and supports the twin web shoes. This construction enables the linings to follow the drum shape more accurately. The tappet head abutments are inclined to provide a means for self-centralizing the brake shoes after each brake application. The cam strut lift relative to the camshaft angular movement tends to give an approximately constant lift rate for the normal angular operating range of the cam between new and worn linings conditions.

When the brakes are applied (Fig. 11.12(b)) the camshaft is rotated, causing the struts to move outwards against the hollow tapper plungers. The tappet head abutments force the shoes into contact with the drum, thereby applying the brakes.

11.3.4 Wedge shoe adjuster unit (Fig. 11.13)

The adjuster housing is made from malleable iron and is spigoted and bolted firmly to the back plate (Fig. 11.13). A hardened steel wedge is employed

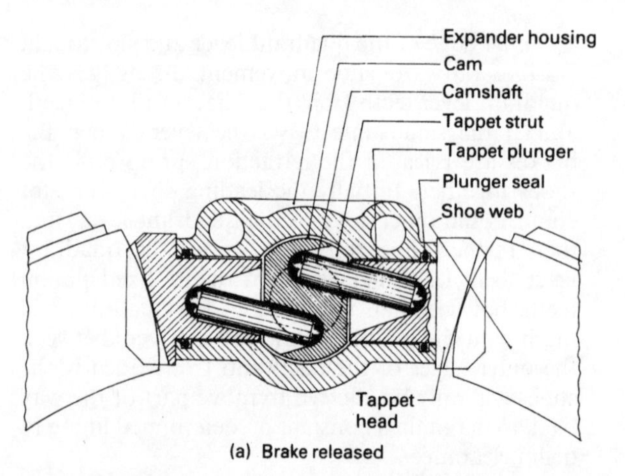

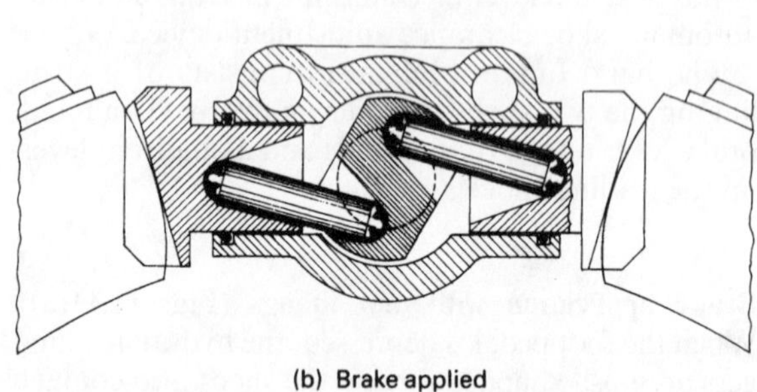

Fig. 11.12 (a and b) Strut and cam brake shoe expander

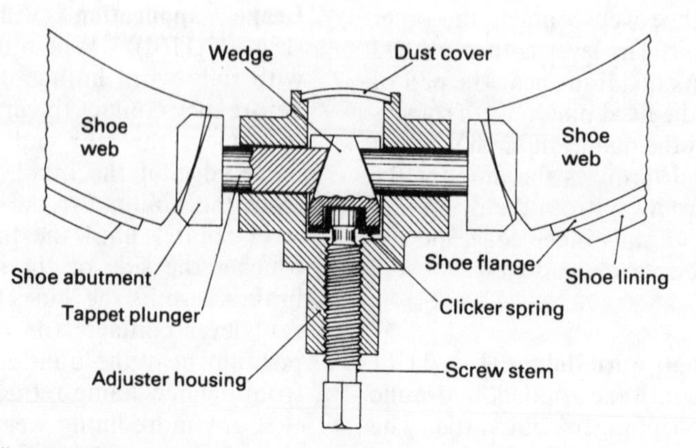

Fig. 11.13 Wedge shoe adjuster unit

with a screw adjuster stem rotating within the wedge which does not rotate, but moves at right angles to the inclined faced tappet plungers. So that accurate adjustment for each brake assembly is possible, a clicker spring is located between the screwed stem and the wedge. This spring fits onto two flats provided on the screw stem (not shown). The clicker spring has two embossed dimples which align and clip into shallow holes formed in the back of the wedge when the shoes are being adjusted; they therefore enable a fine adjustment to be made while also preventing the adjuster screw stem unwinding on its own while in service.

11.3.5 S cam shoe expander (Fig. 11.14)

Cam expander requirements The object of a cam brake shoe expander (Fig. 11.14(a and b)) is to convert an input camshaft leverage torque into a shoe tip force. The shape of the cam profile plays a large part in the effective expansion thrust imposed on the shoe tips as the shoe linings wear and the clearance between the drum and linings increase.

Early S-shaped cams were derived from an Archimedean spiral form of locus which gives a constant rate of lift per degree of cam rotation, but varying cam radius. The present tendency is for the S cam to be generated from an involute spiral (Fig. 11.14(a)) which gives a slight reduction in lift per degree of cam rotation, but maintains a constant cam effective radius so that the shoe tip force always acts in the same direction relative to the cam shoe roller, no matter which part of the cam profile is in contact with the roller (Fig. 11.14(b)). By these means the shoe tip force will remain approximately constant for a given input torque for the whole angular movement of the cam between new and worn lining conditions. Note this does not mean that the effective input torque will be constant. This depends upon the push or pull rod and the slack adjuster lever remaining perpendicular to each other which is unlikely.

Cam profile geometry The involute to a base circle is generated when a straight line is rolled round a circle without slipping; points on the line will trace out an involute. The involute profile may be produced by drawing a base circle and a straight line equal to its circumference and dividing both into the same number of equal parts (Fig. 11.14(a)). From the marked points on the circle draw tangents to represent successive positions of the generated line. Step off the unwrapped portion of the circumference along each tangent and then plot a smooth curve passing through the extending tangential lengths. The locus generated is the involute to base circle, this shape being the basic shape of the so-called S cam.

Cam and shoe working conditions With new shoe linings the leading shoe works harder than the trailing shoe so that initially the leading shoe wear will be higher than that of the trailing shoe. If there is adequate camshaft to bush bearing clearance, shoe wear will eventually be sufficient to permit the camshaft to float between the shoe tips, allowing the trailing shoe to produce the same friction drag as the leading shoe, thus producing the *equal work* condition.

If the shoe tip force applied by the cam is equal, then the camshaft floats on its bearing. In practice, because the shoe tip force is not always equal, a resultant reaction force input will be provided by the camshaft to maintain equilibrium. Therefore the frictional force between the shaft and bearing can be significant in the mechanical losses between the input camshaft torque and the shoe tip force. The input camshaft torque may be derived from both the shaft frictional torque and the cam to roller contact torque (Fig. 11.14(c)).

Let μ_C = coefficient of camshaft to bearing friction
R = resultant camshaft radial load (N)
r_c = camshaft radius (m)
r_b = base circle radius (m)
F_1 and F_2 = roller contact forces (N)

Then

Camshaft frictional torque $= \mu_c Rr_c$ (Nm)

and

Cam to roller torque $= (F_1 + F_2)r_b$ (Nm)

Therefore

$$\text{Input camshaft torque } T_c = \mu_c Rr_c + (F_1 + F_2)r_b \text{ (Nm)}$$

Cam design considerations To give the highest shoe factor, that is the maximum shoe frictional drag to input torque, a low rate of cam lift is desirable. This conflicts with the large total lift needed to utilize the full lining thickness which tends to be limited to 19 mm.

Typical rates of cam lift vary from 0.2 to 0.4 mm/deg which correspond to brake factors of about 12 to 16 with the involute cam profile.

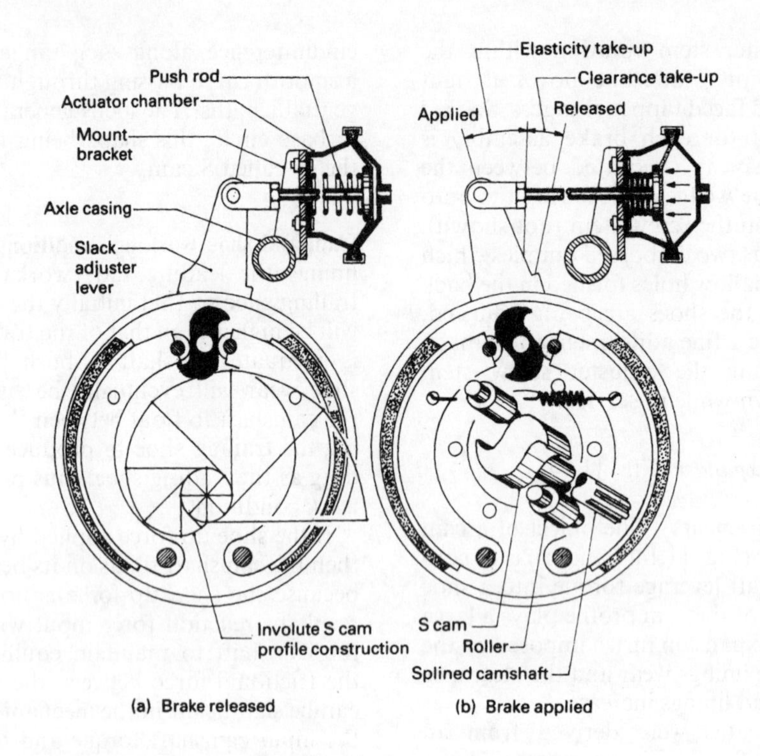

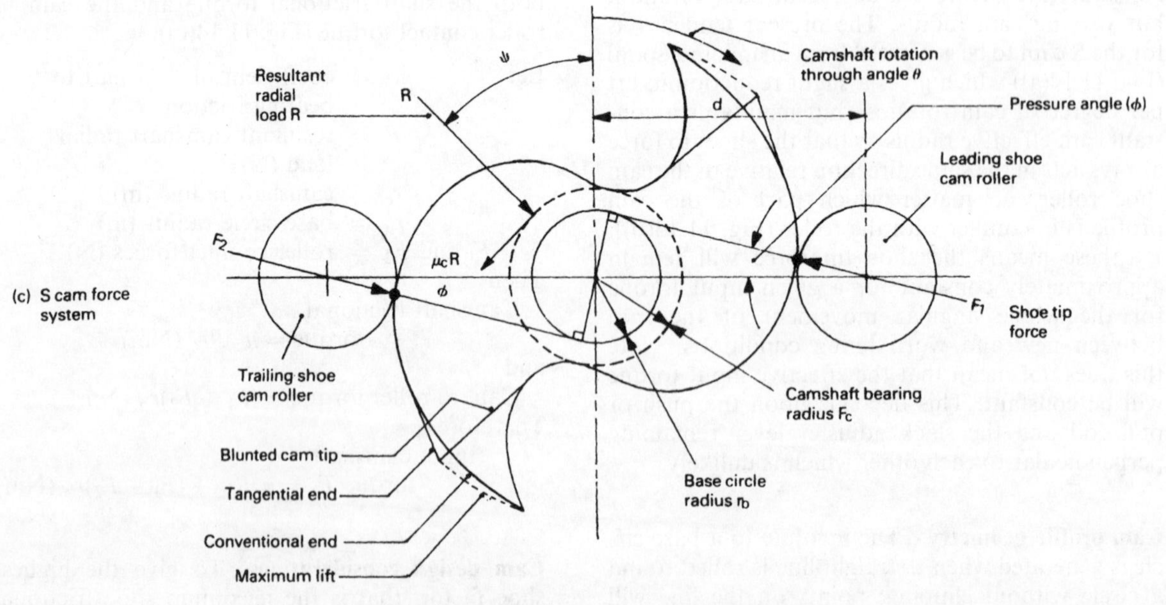

Fig. 11.14 (a–c) Air operated foundation brake assembly

As the cam lifts (Fig. 11.14(c)) the pressure angle ψ which is made between the cam and roller centre lines and the base circle tangential line decreases. For the cam to be self-returning the pressure angle should not be permitted to be reduced below 10°.

One approach to maximize cam lift without the rollers falling off the end of the cam in the extreme wear condition is to use the involute cam up to the point where the lining rivets would contact the drum and relining would be required. Beyond this point the cam is continued in a straight line, tangential to the cam profile (Fig. 11.14(c)). By this method, total cam lift is achieved for the normal thickness of lining within the designed angular movement of the cam, which is not possible with the conventional involute cam. Shoe tip force efficiency does drop off in the final tangential lift cam range but this is not a serious problem as it is very near the end of the linings' useful life. One important outcome of altering the final involute profile is that the blunting of the cam tips considerably strengthens the cam.

11.3.6 Wedge type brake shoe expander with automatic adjustment (Fig. 11.15)

The automatic brake shoe adjustment provides a self-adjusting mechanism actuated by the expander movement during the on/off brake application cycle, enabling a predetermined lining to drum clearance to be maintained. When a brake application takes place, an adjusting pawl mechanism senses the movement of an adjusting sleeve located in one of the wedge expander plungers. If the sleeve travel exceeds 1.52 mm, the spring-loaded pawl acting on the sleeve teeth drops into the next tooth and automatically makes the adjusting screw wind out a predetermined amount. An approximate 1.14 mm lining to drum clearance will be maintained when the brakes are released, but if the adjusting sleeve and plunger outward travel is less than 1.52 mm, then the whole plunger assembly will move back to its original position without any adjustment being made.

Description (Fig. 11.15) The automatic adjustment is built into one of the expander plungers. With this construction the adjusting screw is threaded into an adjusting sleeve, the sleeve being a free fit inside the hollow plunger. A hollow cap screw, spring, and an adjusting pawl are preassembled and act as a plunger guide. The end of the adjusting pawl has sawtooth type teeth which engage corresponding helical teeth on the outside of the adjusting sleeve.

Operation (Fig. 11.16) As the brakes are applied, the plunger sleeve and screw move outwards and the sloping face of the teeth on the adjusting sleeve lifts the adjusting pawl against the spring. When the brake is released, the rollers move down both the central wedge and the two outer plungers' inclined planes to their fully released position. As the linings wear, both the plunger strokes and resulting pawl lift gradually increase until the pawl climbs over and drops into the next tooth space. The next time, when the brake is released and the plunger is pushed back into its bore, the upright face of the pawl teeth prevents the sleeve moving directly back. It permits the sleeve to twist as its outside helical teeth slide through the corresponding guide pawl teeth in its endeavour for the whole plunger assembly to contract inwards to the off position, caused by the inward pull of the shoe retraction spring. The partial rotation of the adjusting sleeve unscrews and advances the adjusting screw to a new position. This reduces the lining clearance. This cycle of events is repeated as the lining wears. The self-adjustment action only operates in the forward vehicle direction. Once the brake shoes have been installed and manually adjusted no further attention is necessary until the worn linings are replaced.

11.3.7 Manual slack adjuster (Fig. 11.17)

Purpose A slack adjuster is the operating lever between the brake actuator chamber push rod and the camshaft. It is used with 'S' type cam shoe expanders and features a built-in worm and worm wheel adjustment mechanism enabling adjustment to be made without involving the removal and alteration of the push rod length.

Operation (Fig. 11.17) The slack adjuster lever incorporates in its body a worm and worm wheel type adjuster (Fig. 11.17). The slack adjuster lever is attached indirectly to the splined camshaft via the internal splines of the worm wheel located inside the slack adjuster body. For optimum input leverage the slack adjuster lever and the push rod should be set to maintain an inside angle just greater than 90° with the brakes fully applied. Once the push rod length has been set, further angular adjustment of the 'S' cam is made by rotating the worm shaft so that the large gear reduction between the meshing worm and worm wheel will slowly turn the worm wheel and camshaft until the cam flanks take up the excess shoe lining to drum clearance. Owing to the low reverse efficiency of the worm and worm wheel gearing, the worm and shaft will not normally rotate on its own. To prevent the possibility of the worm and shaft unwinding, caused perhaps by transmitted oscillatory movement of the slack adjuster during periods of applying and releasing the brakes, a lock sleeve is utilized.

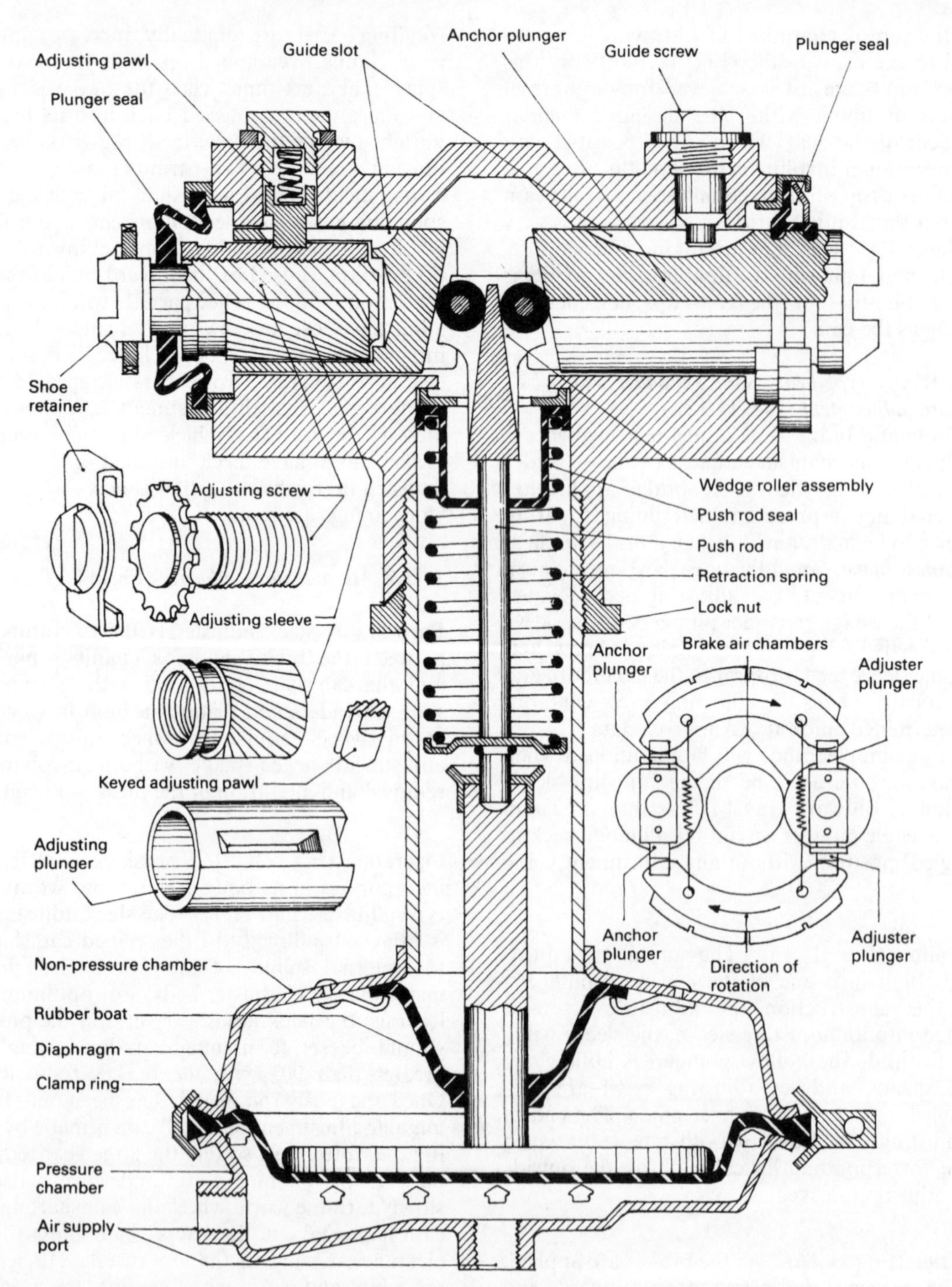

Fig. 11.15 Twin wedge foundation brake expander and automatic adjuster

Adjustment (Fig. 11.17) Cam adjustment is provided by the hexagon head of the worm shaft situated on the side of the slack adjuster body (Fig. 11.17). To adjust the cam relative to the slack adjuster lever, the lock sleeve is depressed against the worm lock spring by a suitable spanner until the worm shaft is free to turn. The worm shaft is then rotated with the spanner until all the excess

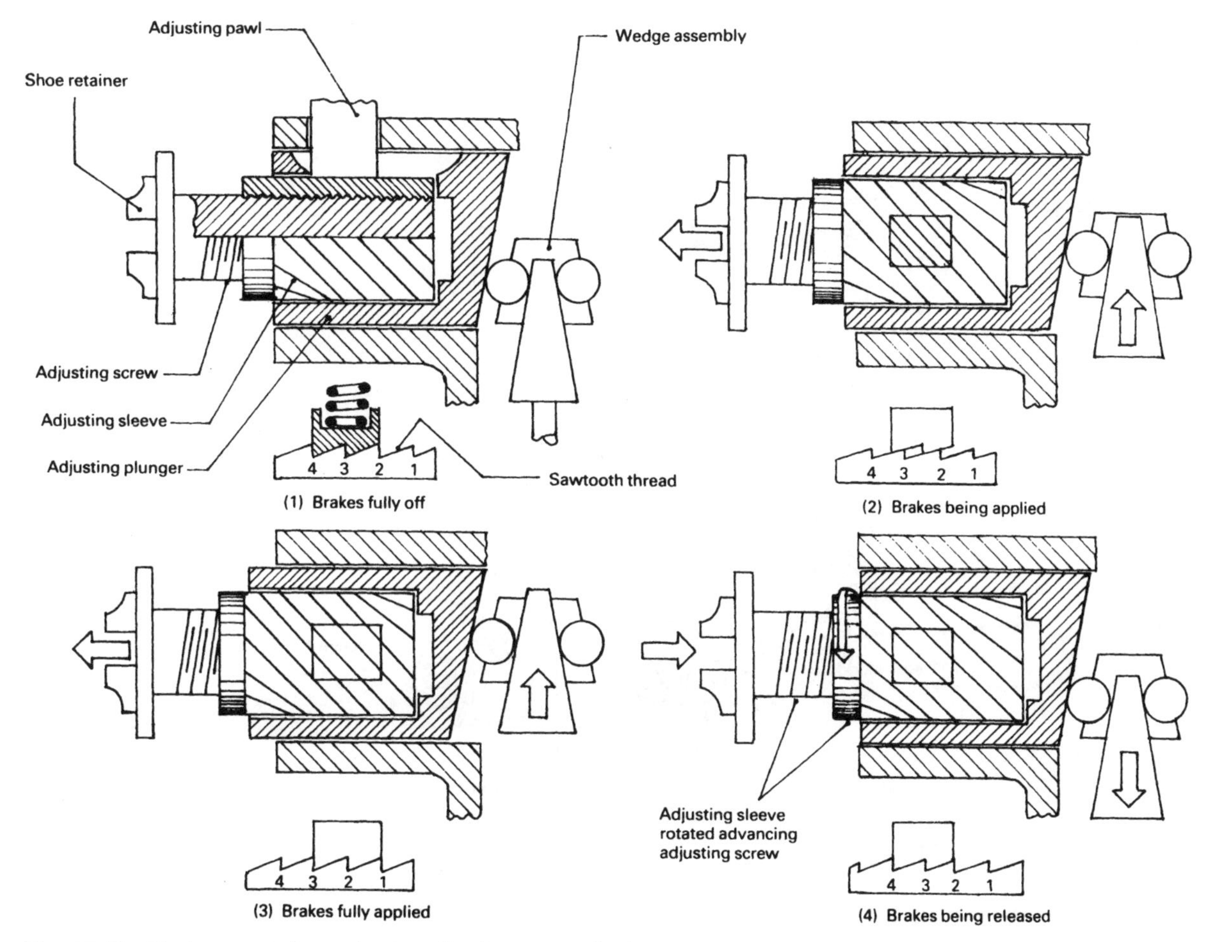

Fig. 11.16 Wedge expander and automatic clearance adjuster

shoe lining to drum clearance is eliminated. The worm shaft is then prevented from unwinding by the worm lock spring forcing the lock sleeve against the hexagonal head of the worm shaft. The removal of the spanner permits the worm lock spring to push the internally serrated lock sleeve up to and over the hexagonal bolt head. To prevent the lock sleeve rotating, a guide pin fixed in the slack adjuster body aligns with a slot machined on the sleeve.

11.3.8 Automatic slack adjuster
(Fig. 11.18(a, b, c and d))

Purpose Once set up automatic slack adjusters need no manual adjustment during the life of the brake linings. Self-adjustment takes place whilst the brakes are released (when the clearance between lining and drum exceeds 1.14 mm). This designed clearance ensures that there is no brake drag and adequate cooling exists for both shoe and drum.

Operation (Fig. 11.18)

Automatic adjustment When the foot or hand brake valve is operated, the brake actuator chamber push rod connected to the slack adjuster will rotate the slack adjuster body and camshaft in an anticlockwise direction (Fig. 11.18(a)). The restraining link rod of the ratchet lever which is connected to some fixed point on the axle will cause the ratchet lever to rotate clockwise relative to the slack adjuster body, and thus the ratchet pawl will ride up one of the teeth of the ratchet wheel.

When adjustment is required, the relative movement between the ratchet lever and the main slack adjuster body is sufficient for the pawl to ride over one complete tooth space and engage the next tooth, following which, on release of the brake (Fig. 11.18(b)), as the slack adjuster moves to its 'off' position, the ratchet wheel will be rotated by the amount of the one tooth space. As the ratchet

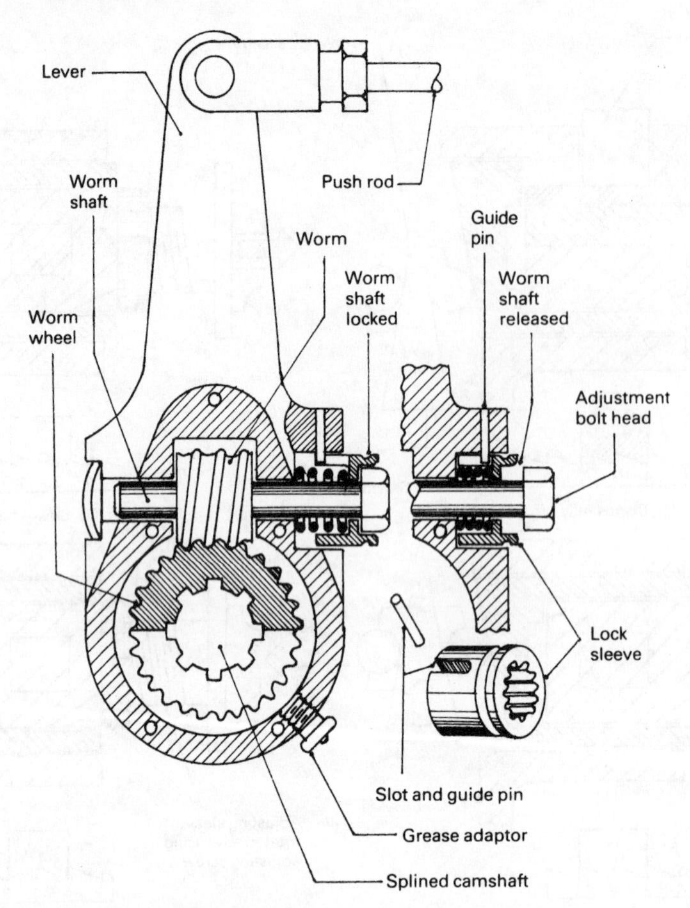

Fig. 11.17 Manual slack adjuster (Clayton Dewandre)

wheel rotates, so too will the ratchet worm, which in turn transmits motion to the ratchet worm wheel and splined main worm shaft. Thus the main worm, and subsequently the main worm wheel, are rotated, causing the slack adjuster lever to take up a new position relative to the camshaft splined into the main worm wheel.

Adjustment will only take place if the pawl moves sufficiently around the ratchet wheel to collect the next tooth. This can only occur if the relative movement between the ratchet lever and the main slack adjuster body exceeds an angle of 8.18°.

Initial manual adjustment (Fig. 11.18(c)) Manual adjustment after fitting new shoe linings can be made by unscrewing the hexagonal cap from the extended slack adjuster body. This releases the ratchet worm wheel from the ratchet worm by the action of the spring behind the worm wheel, which causes the worm wheel to slide outward along the splines to the stop washer. A spanner is then fitted over the square on the end of the ratchet worm wheel and rotated until the shoe to drum excess clearance has been taken up. After the shoes have been manually adjusted, the ratchet worm wheel should be pushed back until it meshes with the ratchet worm. The end cap should then be replaced.

11.4 Disc brake pad support arrangements

11.4.1 Swing yoke type brake caliper (Fig. 11.19)
This is a lightweight, single cylinder, disc brake caliper. The caliper unit consists of a yoke made from a rigid steel pressing, a cylinder assembly, two pads and a carrier bracket which is bolted to the suspension hub carrier. The cylinder is attached by a tongue and groove joint to one side of the yoke frame whilst the yoke itself is allowed to pivot at

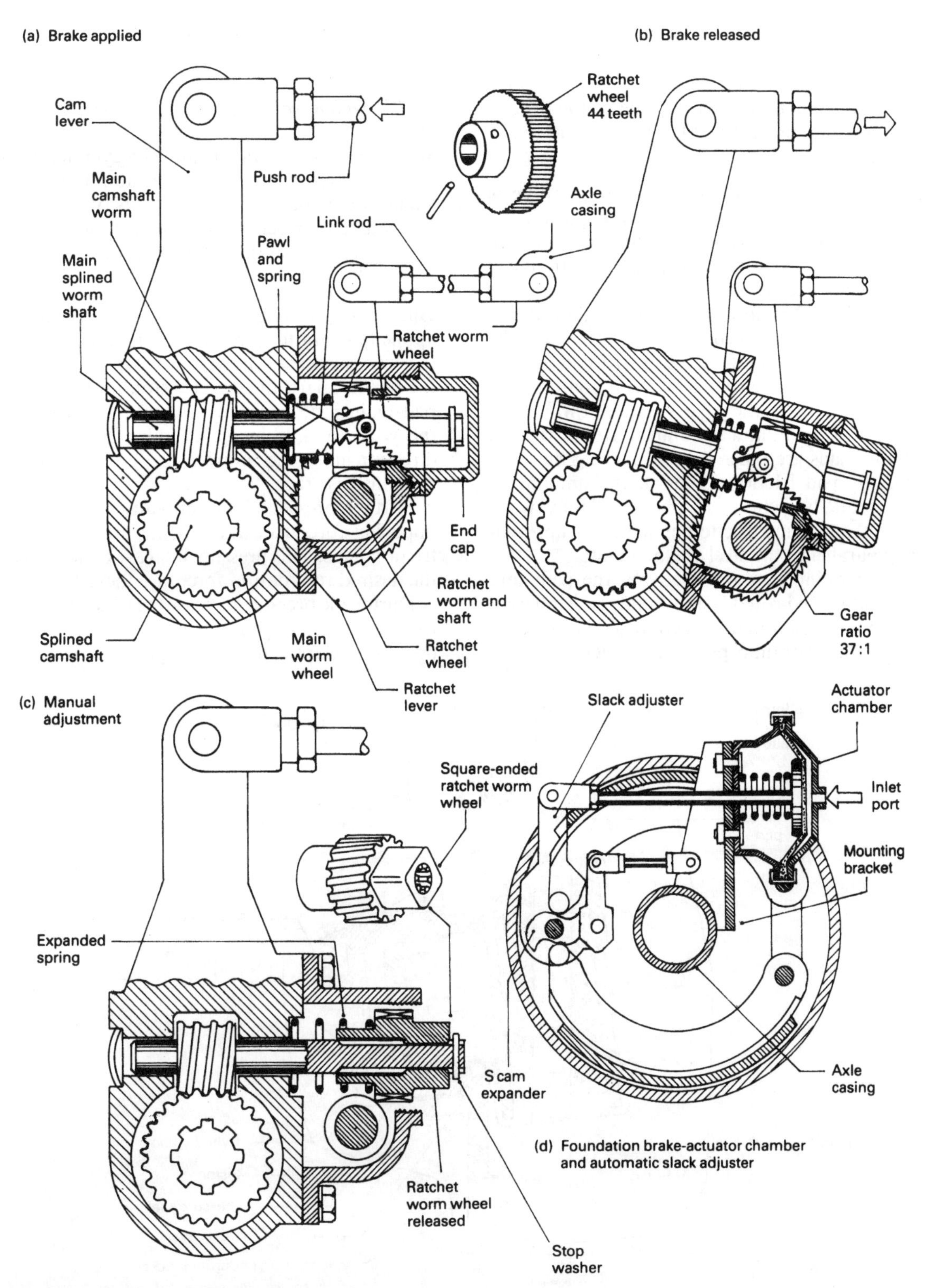

Fig. 11.18 (a–d) Automatic slack adjuster

one end on it supporting carrier bracket. The disc is driven by the transmission drive shaft hub on which it is mounted and the lining pads are positioned and supported on either side of the disc by the rectangular aperature in the yoke frame.

Operation (Fig. 11.19) When the foot brake is applied generated hydraulic pressure pushes the piston and inboard pad against their adjacent disc face. Simultaneously, the hydraulic reaction will move the cylinder in the opposite direction away from the disc. Consequently, as the outboard pad and cylinder body are bridged by the yoke, the latter will pivot, forcing the outboard pad against the opposite disc face to that of the inboard pad.

As the pads wear, the yoke will move through an arc about its pivot, and to compensate for this tilt the lining pads are taper shaped. During the wear life of the pad friction material, the amount of taper gradually reduces so that in a fully worn state the remaining friction material is approximately parallel to the steel backing plate.

The operating clearance between the pads and disc is maintained roughly constant by the inherent distortional stretch and retraction of the pressure seals as the hydraulic pressure is increased and reduced respectively, which accordingly moves the piston forwards and back.

11.4.2 Sliding yoke type brake caliper
(Fig. 11.20)

With this type of caliper unit, the cylinder body is rigidly attached to the suspension hub carrier, whereas the yoke steel pressing fits over the cylinder body and is permitted to slide between parallel grooves formed in the cylinder casting.

Operation (Fig. 11.20) When the foot brake is applied, hydraulic pressure is generated between the two pistons. The hydraulic pressure pushes the piston apart, the direct piston forces the direct pad against the disc whilst the indirect piston forces the yoke to slide in the cylinder in the opposite direction until the indirect pad contacts the outstanding disc face.

Further pressure build-up causes an equal but opposing force to sandwich the disc between the friction pads. This pressure increase continues until the desired retardation force is achieved.

During the pressure increase the pressure seals distort as the pistons move apart. When the hydraulic pressure collapses the rubber pressure seals retract

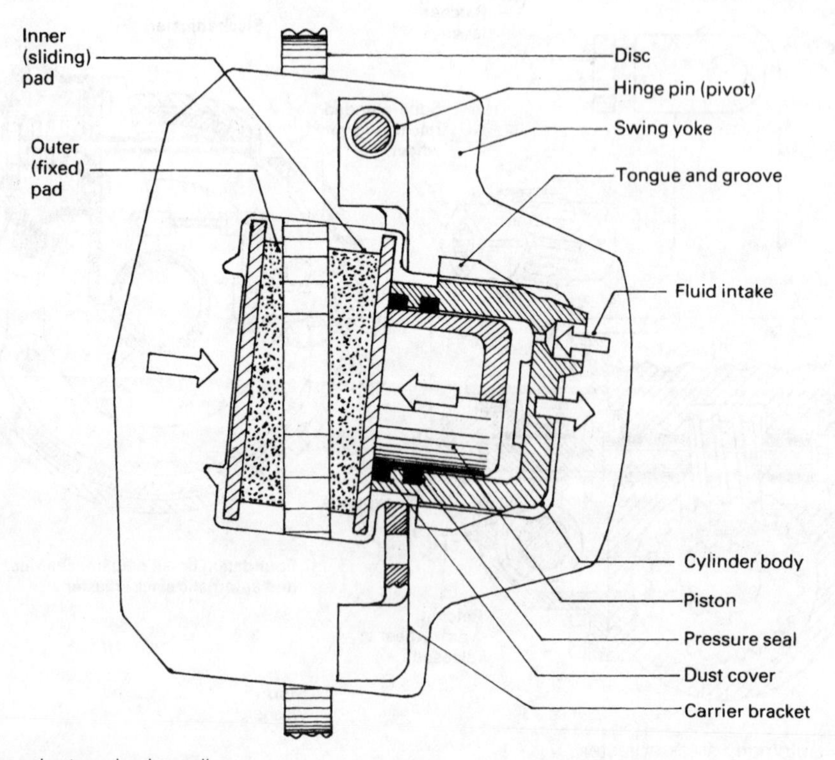

Fig. 11.19 Swing yoke type brake caliper

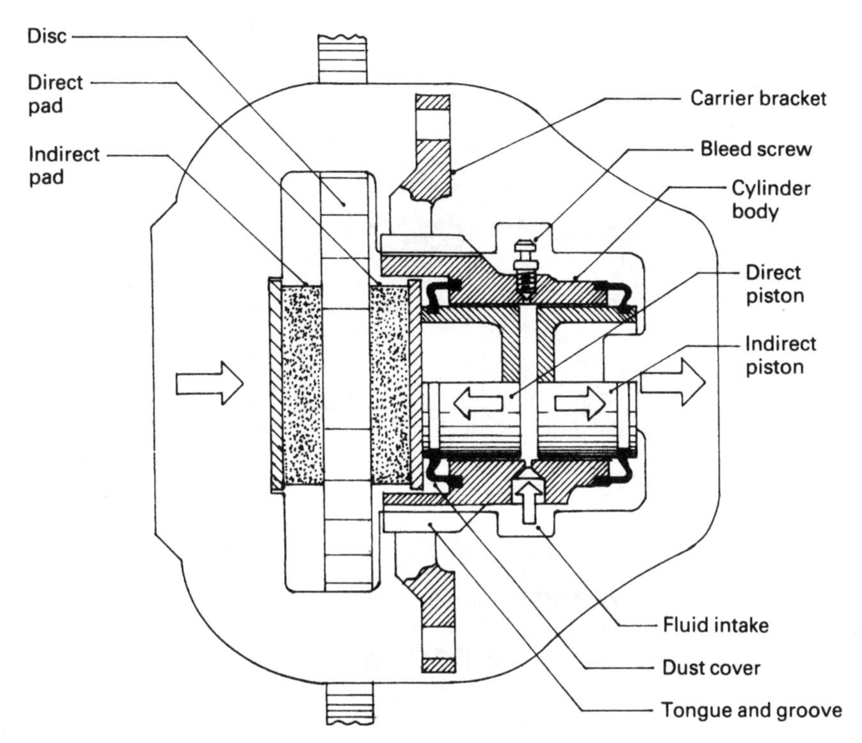

Fig. 11.20 Sliding yoke type brake caliper

and withdraw the pistons and pads from the disc surface so that friction pad drag is eliminated.

Yoke rattle between the cylinder and yoke frame is reduced to a minimum by inserting either a wire or leaf type spring between the sliding joints.

11.4.3 Sliding pin type brake caliper (Fig. 11.21)
The assembled disc brake caliper unit comprises the following; a disc, a carrier bracket, a cylinder caliper bridge, piston and seals, friction pads and a pair of support guide pins.

The carrier bracket is bolted onto the suspension hub carrier, its function being to support the cylinder caliper bridge and to absorb the brake torque reaction.

The cylinder caliper bridge is mounted on a pair of guide pins sliding in matched holes machined in the carrier bracket. The guide pins are sealed against dirt and moisture by dust covers so that equal frictional sliding loads will be maintained at all times. On some models a rubber bush sleeve is fitted to one of the guide pins to prevent noise and to take up brake deflection.

Frictional drag of the pads is not taken by the guide pins, but is absorbed by the carrier bracket. Therefore the pins only support and guide the caliper cylinder bridge.

As with all other types of caliper units, pad to disc free clearance is obtained by the pressure seals which are fitted inside recesses in the cylinder wall and grip the piston when hydraulic pressure forces the piston outwards, causing the seal to distort. When the brakes are released and the pressure is removed from the piston crown, the strain energy of the elastic rubber pulls back the piston until the pressure seal has been restored to its original shape.

Operation (Fig. 11.21) When the foot brake is applied, the hydraulic pressure generated pushes the piston and cylinder apart. Accordingly the inboard pad moves up to the inner disc face, whereas the cylinder and bridge react in the opposite sense by sliding the guide pins out from their supporting holes until the outboard pad touches the outside disc face. Further generated hydraulic pressure will impose equal but opposing forces against the disc faces via the pads.

11.4.4 Sliding cylinder body type brake caliper (Fig. 11.22)
This type of caliper unit consists of a carrier bracket bolted to the suspension hub carrier and a single piston cylinder bridge caliper which straddles

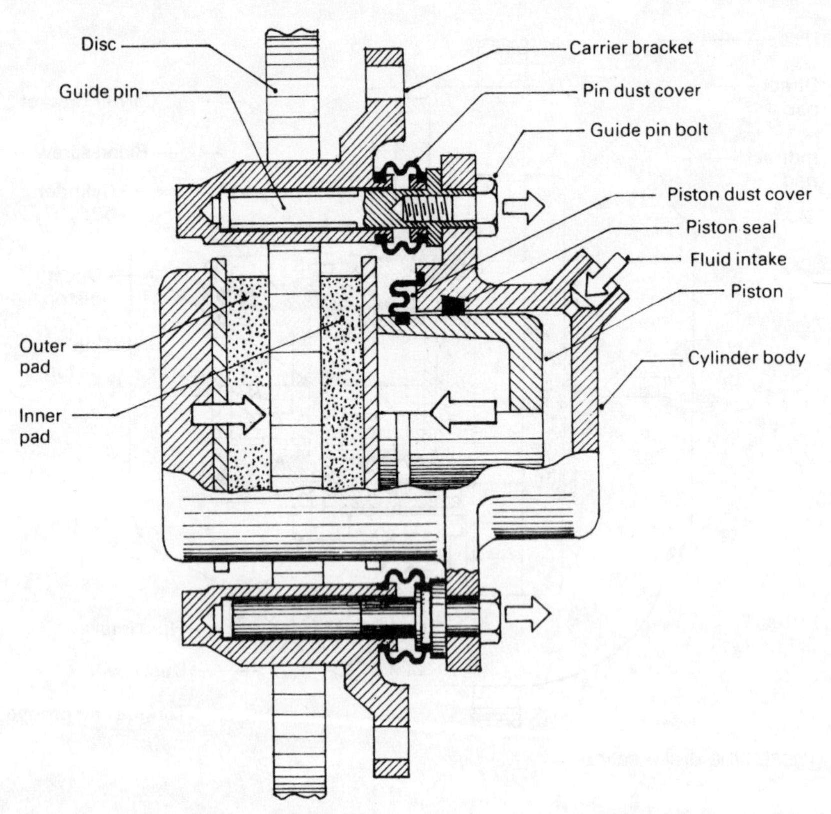

Fig. 11.21 Slide pin type brake caliper

the disc and is allowed to slide laterally on guide keys positioned in wedge-shaped grooves machined in the carrier bracket.

Operation (Fig. 11.22) When the foot brake is applied, the generated hydraulic pressure enters the cylinder, pushing the piston with the direct acting pad onto the inside disc face. The cylinder body caliper bridge is pushed in the opposite direction. As a result, the caliper bridge reacts and slides in its guide groove at right angles to the disc until the indirect pad contacts the outside disc face, thereby equalling the forces acting on both sides of the disc.

A pad to disc face working clearance is provided when the brakes are released by the retraction of the pressure seal, drawing the piston a small amount back into the cylinder after the hydraulic pressure collapes.

To avoid vibration and noise caused by relative movement between the bridge caliper and carrier bracket sliding joint, anti-rattle springs are normally incorporated alongside each of the two-edge-shaped grooves.

11.4.5 Twin floating piston caliper disc brake with hand brake mechanism (Fig. 11.23)

This disc brake unit has a pair of opposing pistons housed in each split half-caliper. The inboard half-caliper is mounted on a flanged suspension hub carrier, whereas the other half straddles the disc and is secured to the rotating wheel hub. Lining pads bonded to steel plates are inserted on each side of the disc between the pistons and disc rubbing face and are held in position by a pair of steel pins and clips which span the two half-calipers. Brake fluid is prevented from escaping between the pistons and cylinder walls by rubber pressure seals which also serve as piston retraction springs, while dirt and moisture are kept out by flexible rubber dust covers.

Foot brake application (Fig. 11.23) Hydraulic pressure, generated when the foot brake is applied, is transferred from the inlet port to the central half-caliper joint, where it is then transmitted along passages to the rear of each piston.

As each piston moves forward to take the clearance between the lining pads and disc, the piston

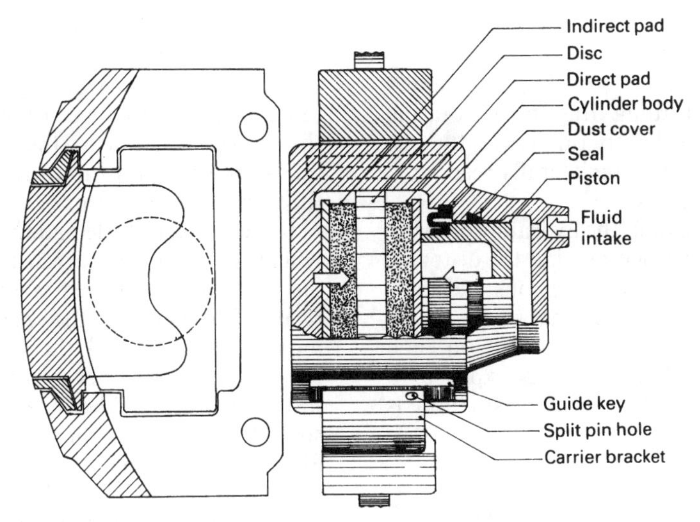

Fig. 11.22 Slide cylinder body brake caliper

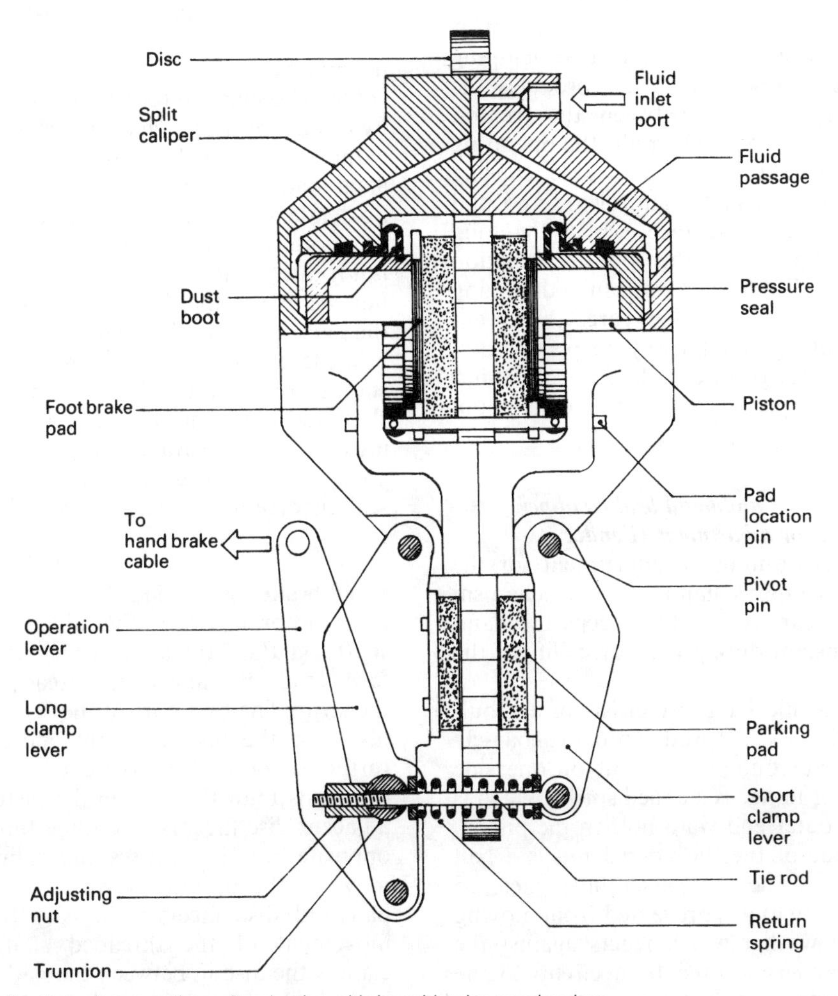

Fig. 11.23 Twin floating piston caliper disc brake with hand brake mechanism

pressure seals are distorted. Further pressure build-up then applies an equal but opposite force by way of the lining pads to both faces of the disc, thereby creating a frictional retarding drag to the rotating disc. Should the disc be slightly off-centre, the pistons will compensate by moving laterally relative to the rubbing faces of the disc.

Releasing the brakes causes the hydraulic pressure to collapse so that the elasticity within the distorted rubber pressure seals retracts the pistons and pads until the seals convert to their original shape.

The large surface area which is swept on each side of the disc by the lining pads is exposed to the cooling airstream so that heat dissipation is maximized.

Hand brake application (Fig. 11.23) The hand brake mechanism has a long and short clamping lever fitted with friction pads on either side of the disc and pivots from the lower part of the caliper. A tie rod with an adjusting nut links the two clamping levers and, via an operating lever, provides the means to clamp the disc between the friction pads. Applying the hand brake pulls the operating lever outwards via the hand brake cable, causing the tie rod to pull the short clamp lever and pad towards the adjacent disc face, whilst the long clamp and pad is pushed in the opposite direction against the other disc face. As a result, the lining pads grip the disc with sufficient force to prevent the car wheels rolling on relatively steep slopes.

To compensate for pad wear, the adjustment nut should be tightened periodically to give a maximum pad to disc clearance of 0.1 mm.

11.4.6 Combined foot and hand brake caliper with automatic screw adjustment (Bendix)

This unit provides automatic adjustment for the freeplay in the caliper's hand brake mechanism caused by pad wear. It therefore keeps the hand brake travel constant during the service life of the pads.

The adjustment mechanism consists of a shouldered nut which is screwed onto a coarsely threaded shaft. Surrounding the nut on one side of the shoulder or flange is a coiled spring which is anchored at its outer end via a hole in the piston. On the other side of the shouldered nut is a ball bearing thrust race. The whole assembly is enclosed in the hollow piston and is prevented from moving out by a thrust washer which reacts against the thrust bearing and is secured by a circlip to the interior of the piston.

Foot brake application (Fig. 11.24(a)) When the hydraulic brakes are applied, the piston outward movement is approximately equal to the predetermined clearance between the piston and nut with the brakes off, but as the pads wear, the piston takes up a new position further outwards, so that the normal piston to nut clearance is exceeded.

If there is very little pad wear, hydraulic pressure will move the piston forward until the pads grip the disc without the thrust washer touching the ball race. However, as the pads wear, the piston moves forward until the thrust washer contacts the ball race. Further outward movement of the piston then forces the thrust washer ball race and shouldered nut together in an outward direction. Since the threaded shaft is prevented from rotating by the strut and cam, the only way the nut can move forward is by unwinding on the screw shaft. Immediately the nut attempts to turn, the coil spring uncoils and loses its grip on the nut, permitting the nut to screw out in proportion to the piston movement.

On releasing the foot brake, the collapse of the hydraulic pressure enables the pressure seals to withdraw the pads from the disc. Because the axial load has been removed from the nut, there is no tendency for it to rotate and the coil spring therefore contracts, gripping the nut so that it cannot rotate. Note that the outward movement of the nut relative to the threaded shaft takes up part of the slack in the mechanical linkage so that the hand brake lever movement remains approximately constant throughout the life of the pads. The threaded shaft and nut device does not influence the operating pad to disc clearance when the hydraulic brakes are applied as this is controlled only by the pressure seal distortion and elasticity.

Hand brake application (Fig. 11.24(b)) Applying the hand brake causes the cable to rotate the camshaft via the cam lever, which in turn transfers force from the cam to the threaded shaft through the strut. The first part of the screwed shaft travel takes up the piston to nut end-clearance. With further screw shaft movement the piston is pushed outwards until the pad on the piston contacts the adjacent disc face. At the same time an equal and opposite reaction causes the caliper cylinder to move in the opposite direction until the outside pad and disc face touch. Any further outward movement of the threaded shaft subsequently clamps the disc in between the pads. Releasing the hand brake lever relaxes the pad grip on the disc

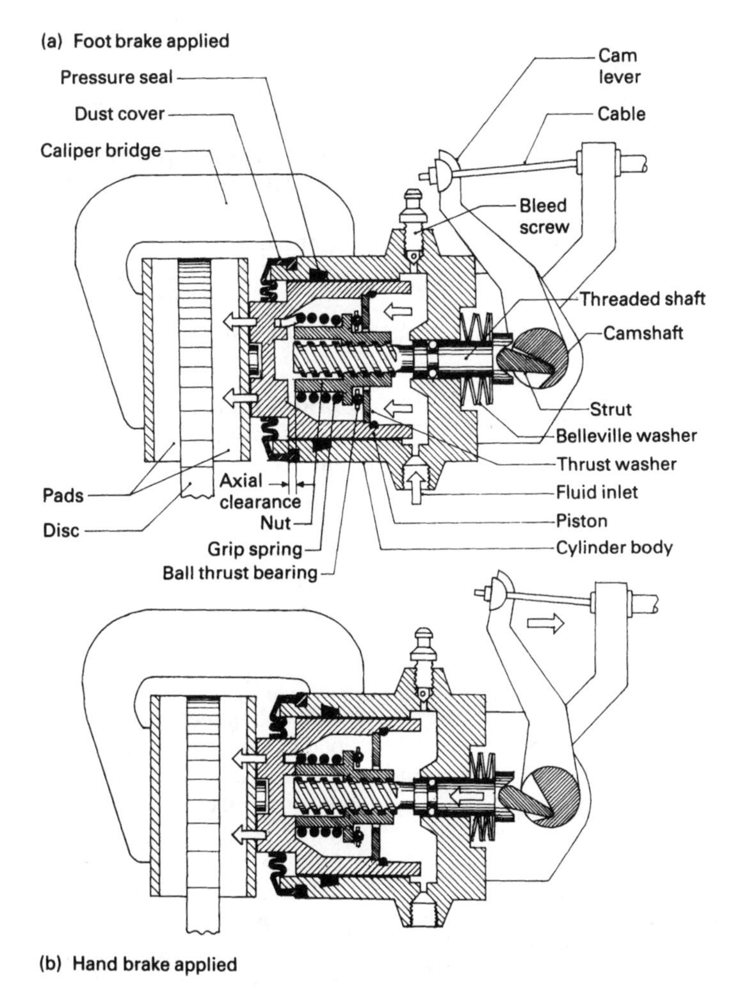

Fig. 11.24 (a and b) Combined foot and hand brake caliper with automatic screw adjustment

with the assistance of the Belleville washers which draws back the threaded shaft to the 'off' position to avoid the pads binding on the disc.

11.5 Dual- or split-line braking systems

Dual- or split-line braking systems are used on all cars and vans to continue to provide some degree of braking if one of the two hydraulic circuits should fail. A tandem master cylinder is incorporated in the dual-line braking system, which is in effect two separate master cylinder circuits placed together end on so that it can be operated by a common push rod and foot pedal. Thus, if there is a fault in one of the hydraulic circuits, the other pipe line will be unaffected and therefore will still actuate the caliper or drum brake cylinders it supplies.

11.5.1 Front to rear brake line split (Fig. 11.25(a))

With this arrangement, the two separate hydraulic pipe lines of the tandem master cylinder are in circuit with either both the front or rear caliper or shoe expander cylinders. The weakness with this pipe line split is that roughly two-thirds of the braking power is designed to be absorbed by the front calipers, and only one-third by the rear brakes. Therefore if the front brakes malfunction, the rear brake can provide only one-third of the original braking capacity.

11.5.2 Diagonally front to rear brake split (Fig. 11.25(b))

To enable the braking effort to be more equally shared between each hydraulic circuit (if a fault should occur in one of these lines), the one front

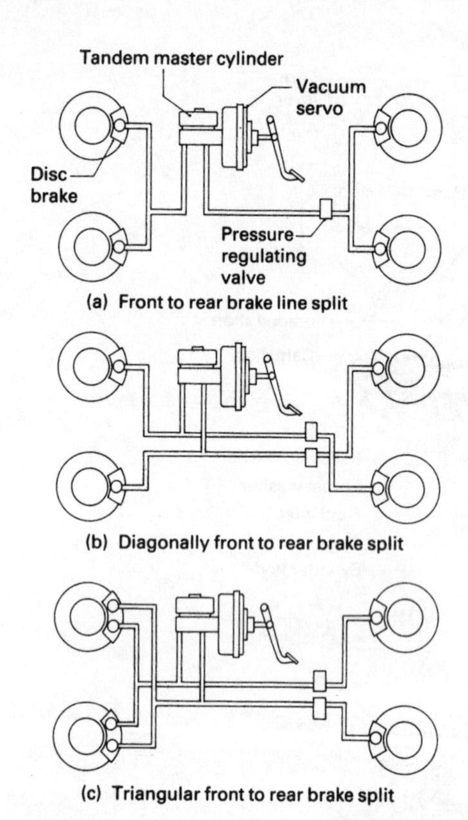

Fig. 11.25 (a–c) Dual- or split-line braking systems

and one diagonally opposite rear wheel are connected together. Each hydraulic circuit therefore has the same amount of braking capacity and the ratio of front to rear braking proportions do not influence the ability to stop. A diagonal split also tends to retard a vehicle on a relatively straight line on a dry road.

11.5.3 Triangular front to rear brake split (Fig. 11.25(b))

This hydraulic pipe line system uses front calipers which have two independent pairs of cylinders, and at the rear conventional calipers or drum brakes. Each fluid pipe line circuit supplies half of each front caliper and one rear caliper or drum brake cylinder. Thus a leakage in one or the other hydraulic circuits will cause the other three pairs of calipers or cylinders or two pairs of caliper cylinders and one rear drum brake cylinder to provide braking equal to about 80% of that which is possible when both circuits are operating. When one circuit is defective, braking is provided on three wheels; it is then known as a *triangular split*.

11.5.4 Compensating port type tandem master cylinder (Fig. 11.26(a–d))

Tandem master cylinders are employed to operate dual-line hydraulic braking systems. The master cylinder is composed of a pair of pistons functioning within a single cylinder. This enables two independent hydraulic cylinder chambers to operate. Consequently, if one of these cylinder chambers or part of its hydraulic circuit develops a fault, the other cylinder chamber and circuit will still continue to effectively operate.

Brakes off (Fig. 11.26(a)) With brakes in the 'off' position, both primary and secondary pistons are pushed outwards by the return springs to their respective stops. Under these conditions fluid is permitted to circulate between the pressure chambers and the respective piston recesses via the small compensating port, reservoir supply outlet and the large feed ports for both primary and secondary brake circuits.

Brakes applied (Fig. 11.26(b)) When the foot pedal is depressed, the primary piston moves inwards and, at the same time, compresses both the intermediate and secondary return springs so that the secondary piston is pushed towards the cylinder's blanked end.

Initial movement of both pistons causes their respective recuperating seals to sweep past each compensating port. Fluid is trapped and, with increased piston travel, is pressurized in both the primary and secondary chambers and their pipe line circuits, supplying the front and rear brake cylinders. During the braking phase, fluid from the reservoir gravitates and fills both of the annular piston recesses.

Brakes released (Fig. 11.26(a)) When the foot pedal effort is removed, the return springs rapidly expand, pushing both pistons outwards. The speed at which the swept volume of the pressure chambers increases will be greater than the rate at which the fluid returns from the brake cylinders and pipe lines. Therefore a vacuum is created within both primary and secondary pressure chambers.

As a result of the vacuum created, each recuperating seal momentarily collapses. Fluid from the annular piston recess is then able to flow through the horizontal holes in the piston head, around the inwardly distorted recuperating seals and into their respective pressure chambers. This extra fluid

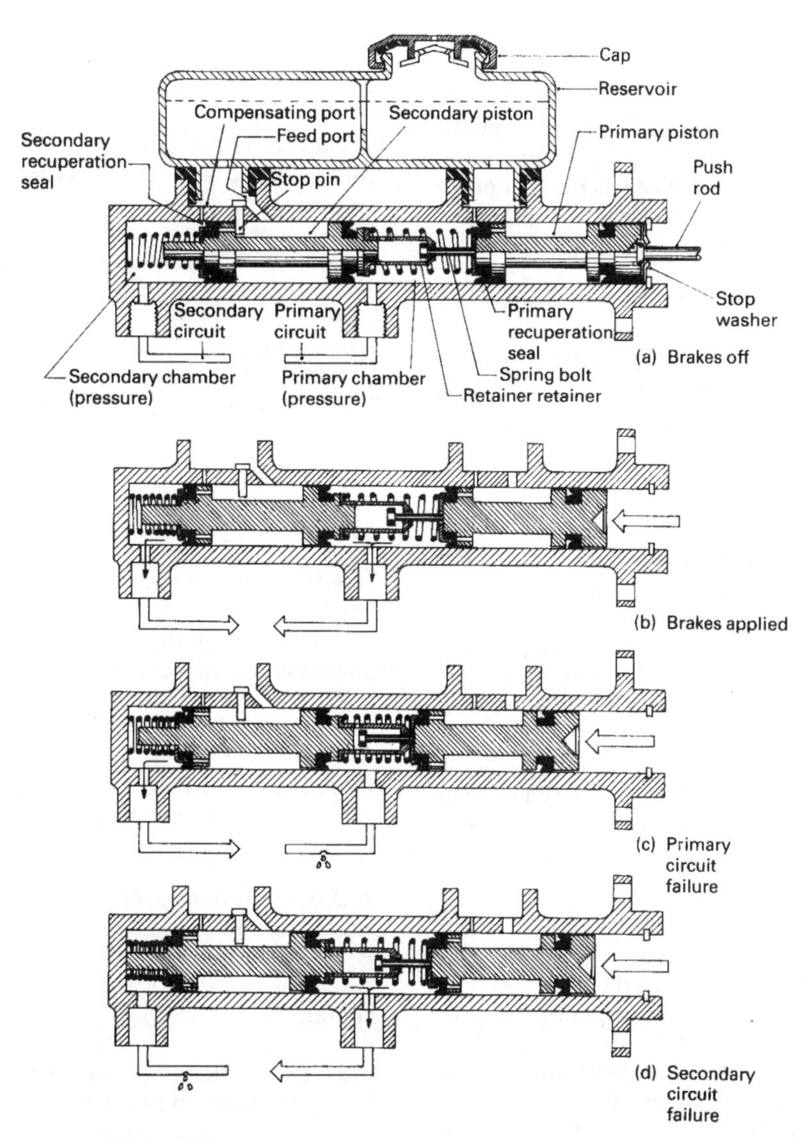

Fig. 11.26 (a–d) Tandem master cylinder

entering both pressure chambers compensates for any fluid loss within the brake pipe line circuits or for excessive shoe to drum clearance. But, if too much fluid is induced in the chambers, some of this fluid will pass back to the reservoir via the compensating ports after the return springs have fully retracted both pistons.

Failure in the primary circuit (Fig. 11.26(c)) Should a failure (leakage) occur in the primary circuit, there will be no hydraulic pressure generated in the primary chamber. When the brake pedal is depressed, the push rod and primary piston will move inwards until the primary piston abuts the secondary spring retainer. Further pedal effort will move the secondary piston recuperating seal beyond the compensating port, thereby pressurizing the fluid in the secondary chamber and subsequently transmitting this pressure to the secondary circuit pipe line and the respective brake cylinders.

Failure in the secondary circuit (Fig. 11.26(d)) If there is a failure (leakage) in the secondary circuit, the push rod will move the primary piston inwards until its recuperating seal sweeps past the compensating port, thus trapping the existing fluid in the primary chamber. Further pedal effort increases the pressure in the primary chamber and at the same time both pistons, separated by the primary chamber fluid, move inwards unopposed until the secondary piston end stop contacts the cylinder's blanked end. Any more increase in braking effort raises the primary chamber pressure, which accordingly pressurizes the primary circuit brake cylinders.

The consequence of a failure in the primary or secondary brake circuit is that the effective push rod travel increases and a greater pedal effort will need to be applied for a given vehicle retardation compared to a braking system which has both primary and secondary circuits operating.

11.5.5 Mecanindus (roll) pin type tandem master cylinder incorporating a pressure differential warning actuator (Fig. 11.27(a–d))

The tandem or split master cylinder is designed to provide two separate hydraulic cylinder pressure chambers operated by a single input push rod. Each cylinder chamber is able to generate its own fluid pressure which is delivered to two independent brake pipe line circuits. Thus if one hydraulic circuit malfunctions, the other one is unaffected and will provide braking to the wheel cylinders forming part of its system.

Operation of tandem master cylinder

Brakes off (Fig. 11.27(a)) With the push rod fully withdrawn, both primary and secondary pistons are forced outwards by the return springs. This outward movement continues until the central poppet valve stems contact their respective Mecanindus (roll) pins. With further withdrawal the poppet valves start opening until the front end of each elongated slot also contacts their respective roll pins, at which point the valves are fully open. With both valves open, fluid is free to flow between the primary and secondary chambers and their respective reservoirs via the elongated slot and vertical passage in the roll pins.

Brakes applied (Fig. 11.27(b)) When the brake pedal is applied, the push rod and the primary return spring pushes both pistons towards the cylinder's blank end. Immediately both recuperating poppet valves are able to snap closed. The fluid trapped in both primary and secondary chambers is then squeezed, causing the pressure in the primary and secondary pipe line circuits to rise and operate the brake cylinders.

Brakes released (Fig. 11.27(a)) Removing the foot from the brake pedal permits the return spring to push both pistons to their outermost position. The poppet valve stem instantly contacts their respective roll pins, causing both valves to open. Since the return springs rapidly push back their pistons, the volume increase in both the primary and secondary chambers exceeds the speed of the returning fluid from the much smaller pipe line bore, with the result that a depression is created in both chambers. Fluid from the reservoir flows via the elongated slot and open poppet valve into the primary and secondary chambers to compensate for any loss of fluid or excessive shoe to drum or pad to disc clearance. This method of transferring fluid from the reservoir to the pressure chamber is more dynamic than relying on the collapse and distortion of the rubber pressure seals as in the conventional master cylinder.

Within a very short time the depression disappears and fluid is allowed to flow freely to and fro from the pressure chambers to compensate for fluid losses or fluid expansion and contraction caused by large temperature changes.

11.5.6 Operation of the pressure differential warning actuator

As a warning to the driver that there is a fault in either the primary or secondary hydraulic braking circuits of a dual-line braking system, a pressure differential warning actuator is usually incorporated as an integral part of the master cylinder or it may be installed as a separate unit (Fig. 11.27).

The switch unit consists of a pair of opposing balance pistons spring loaded at either end so that they are normally centrally positioned. Mounted centrally and protruding at right angles into the cylinder is an electrical conducting prod, insulated from the housing with a terminal formed at its outer end. The terminal is connected to a dashboard warning light and the electrical circuit is completed by the earth return made by the master cylinder.

Operation (Fig. 11.27(b)) If, when braking, both hydraulic circuits operate correctly, the opposing fluid pressure imposed on the outer ends of the balance piston will maintain the pistons in their equilibrium central position.

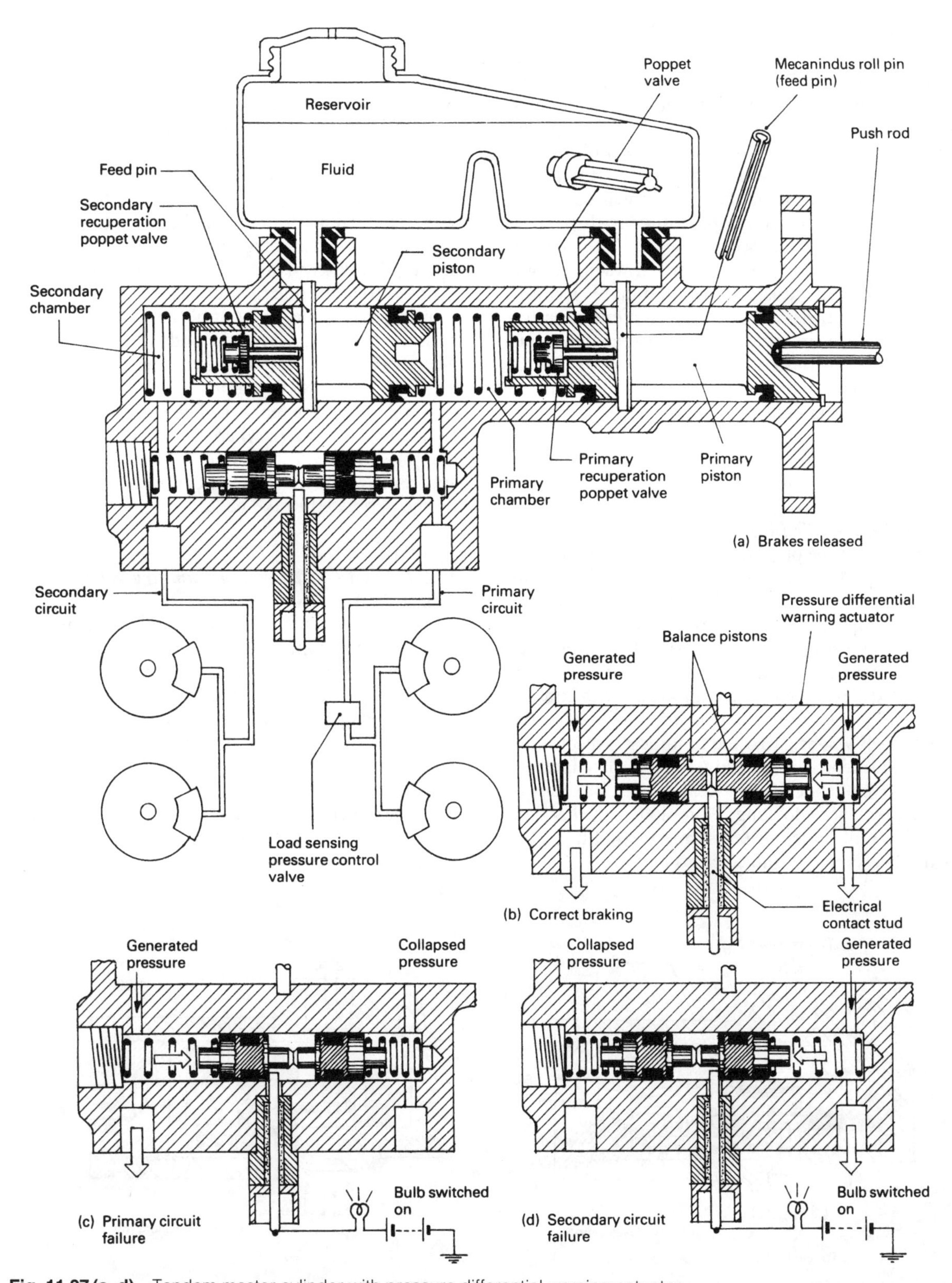

Fig. 11.27 (a–d) Tandem master cylinder with pressure differential warning actuator

Should one or the other of the dual circuits develop a pressure drop fault due to fluid leakage (Fig. 11.27(c and d)), then if the pressure difference of 10 bar or more exists between the two circuits, an imbalance of the fluid pressure applied against the outer ends of the pistons will force both pistons to move in the direction of the faulty circuit. The sideways movement of the pistons will cause the shoulder of the correctly operating circuit balance piston to contact the protruding prod, thus automatically completing the dashboard warning light electrical circuit, causing it to illuminate. Removing the brake pedal effort causes the fluid pressure in the effective circuit to collapse, thereby enabling the balance pistons to move back to their central positions. This interrupts the electrical circuit so that the warning light switches off.

11.6 Apportioned braking

11.6.1 Pressure limiting valve (Fig. 11.28)

The object of the pressure limiting valve is to interrupt the pressure rise of fluid transmitted to the rear wheel brakes above some predetermined value, so that the rear brakes will be contributing a decreasing proportion of the total braking with further increased pedal effort and master cylinder generated line pressure. By imposing a maximum brake line pressure to the rear brake cylinders, the rear wheels will be subjected to far less overbraking when the vehicle is heavily braked. It therefore reduces the tendency for rear wheel breakaway caused by wheels locking. Note that with this type of valve unit under severe slippery conditions the rear wheels are still subjected to lock-up.

Operation (Figs 11.28 and 11.29) Under light brake pedal application, fluid pressure from the master cylinder enters the valve inlet port and passes through the centre and around the outside of the plunger on its way to the outlet ports via the wasted region of the plunger (Fig. 11.28(a)).

When heavy brake applications are made (Fig. 11.28(b)), the rising fluid pressure acting on the large passage at the rear of the plunger displaces the plunger assembly. Instantly the full cross-sectional area equivalent to the reaction piston is exposed to hydraulic pressure, causing the plungers to move forward rapidly until the plunger end seal contacts the valve seat in the body of the valve unit. The valve closing pressure is known as the *cut-off pressure*. Under these conditions the predetermined line pressure in the rear pipe line will be maintained constant (Fig. 11.29), whereas the front brake pipe line pressure will continue to rise unrestricted, according to the master cylinder pressure generated by the depressed brake pedal.

11.6.2 Load sensing pressure limiting valve (Fig. 11.20)

To take into account the weight distribution between the front and rear wheels between an unladen and fully laden vehicle, a load sensing valve may be incorporated in the pipe line connecting the master cylinder to the rear wheel brakes. The function of the valve is to automatically separate the master cylinder to rear brake pipe line by closing a cut-off valve when the master cylinder's generated pressure reaches some predetermined maximum. This cut-off pressure will vary according to the weight imposed on the rear axle.

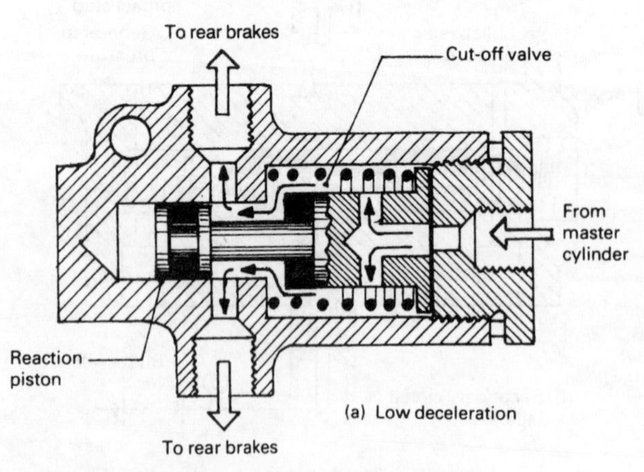

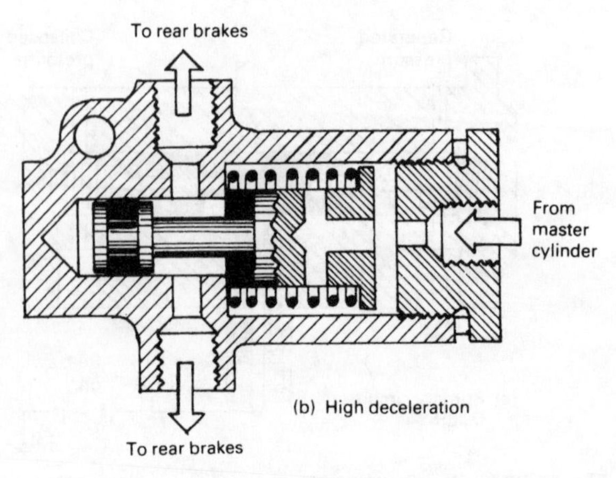

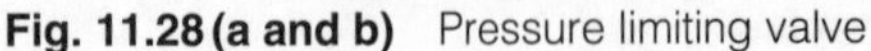
Fig. 11.28 (a and b) Pressure limiting valve

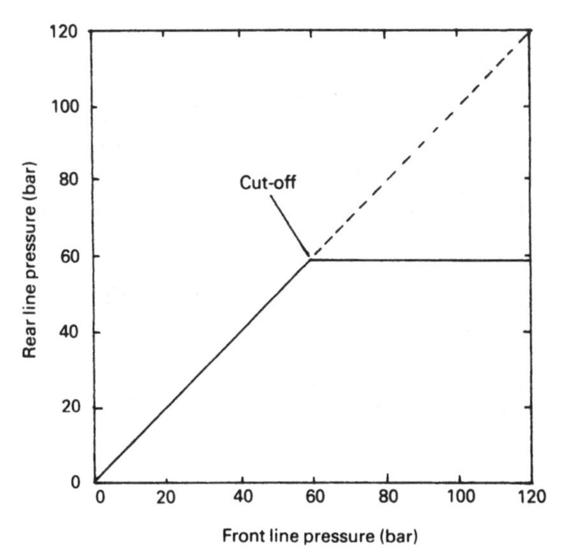

Fig. 11.29 Pressure limiting valve front to rear brake line characteristics

Operation (Figs 11.30 and 11.31) This valve device consists of a plunger supporting a rubber face valve which is kept open by the tension of a variable rate leaf spring. The inward thrust on the plunger keeping the end face valve open is determined by the leaf spring pre-tension controlled by the rear suspension's vertical deflection via the interconnecting spring and rod link. When the vehicle's rear suspension is unloaded, the leaf spring will be partially relaxed, but as the load on the rear axle increases, the link spring and rod pulls the leaf spring towards the valve causing it to stiffen and increase the inward end thrust imposed on the plunger.

With a light brake pedal application, fluid pressure generated by the master cylinder enters the inlet port and passes around the wasted plunger on its way out to the rear brake pipe line (Fig. 11.30(a)).

If a heavy brake application is made (Fig. 11.30(b)), the rising fluid pressure from the master cylinder passes through the valve from the inlet to the outlet ports until the pressure creates a force at the end of the plunger (Force = Pressure × Area) which opposes the spring thrust, pushing back the plunger until the face valve closes. Any further fluid pressure rise will only be transmitted to the front brake pipe lines, whereas the sealed-off fluid pressure in the rear brake pipe lines remains approximately constant.

If the load on the rear axle alters, the vertical deflection height of the suspension will cause the leaf spring to stiffen or relax according to any axle load increase or decrease.

A change in leaf spring tension therefore alters the established pressure (Fig. 11.31) (at which point the cut-off valve closes) and the maximum attainable pressure trapped in the rear brake pipe lines.

11.6.3 Load sensing progressive pressure limiting valve (Fig. 11.32)

The load sensing progressive pressure limiting valve regulates the fluid pressure transmitted to the rear brake cylinders once the master cylinder's generated pressure has risen above some predetermined value corresponding to the weight carried on the rear axle.

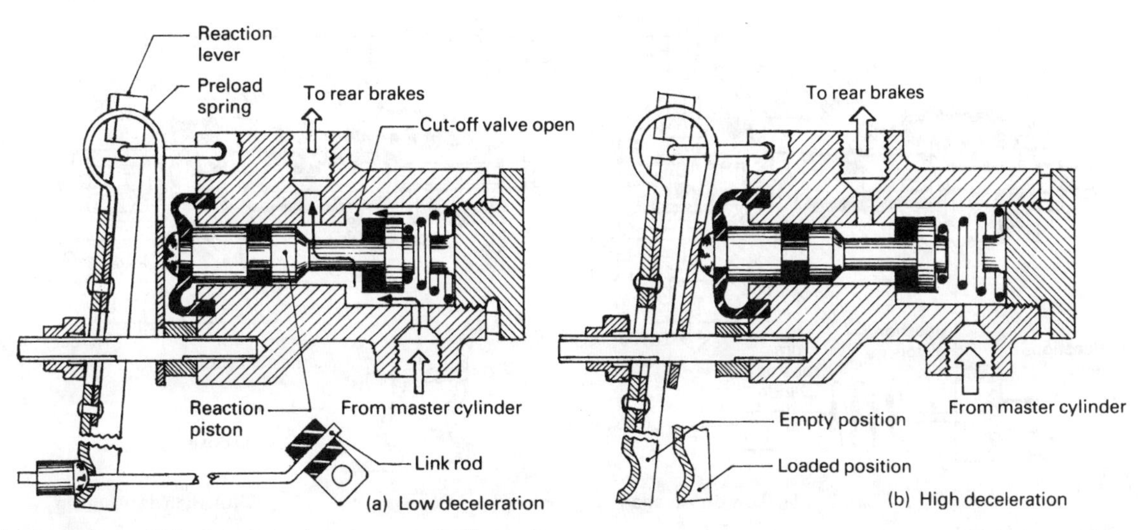

Fig. 11.30 (a and b) Load sensing pressure limiting valve

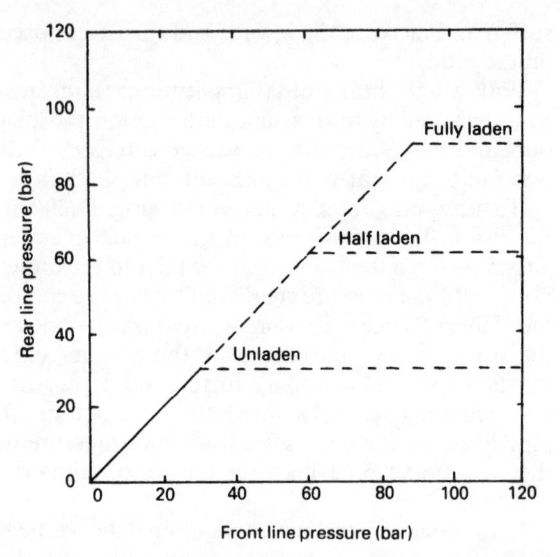

Fig. 11.31 Load sensing pressure limiting valve front to rear brake line characteristics

The reduced rate of pressure increase, in proportion to the pedal effort in the rear brake pipe line, provides a braking ability for both the front and rear brakes which approximately matches the load distribution imposed on the front and rear wheels, so that the tendency for the rear wheels to be either under or over braked is considerably reduced.

Operation (Fig. 11.32) When the foot pedal is applied lightly (Fig. 11.32(a)), pressure generated by the master cylinder will be transferred through the centre of the stepped reaction piston, between the cone and seat and to the rear brake pipe lines.

If the brake pedal is further depressed (Fig. 11.32(b)), increased fluid pressure acting on the large piston area produces an end force, which, when it exceeds the opposing link spring tension and fluid pressure acting on the annular piston face, causes the stepped piston to move outwards. This outward movement of the piston continues until the valve stem clears the cylinder's blanked end, thereby closing the valve. The valve closure is known as the *cut-off point* since it isolates the rear brake pipe lines from the master cylinder delivery.

Further generation of master cylinder pressure exerted against the annular piston face produces an increase in force which moves the piston inwards, once again opening the valve. The hydraulic connection is re-established, allowing the rear brake pipe line fluid pressure to increase. However, the pressure exerted against the end face of the piston immediately becomes greater than the spring force and hydraulic force pushing on the annular piston face, and so the piston moves outwards, again closing the valve.

Every time the valve is opened with rising master cylinder pressure, the rear brake pipe line pressure increases in relation to the previous closing of the valve. Over a heavy braking pressure rise phase the piston oscillates around a position of balance, causing a succession of valve openings and closings. It subsequently produces a smaller pressure rise in the rear brake pipe line than with the directly connected front brake pipe lines.

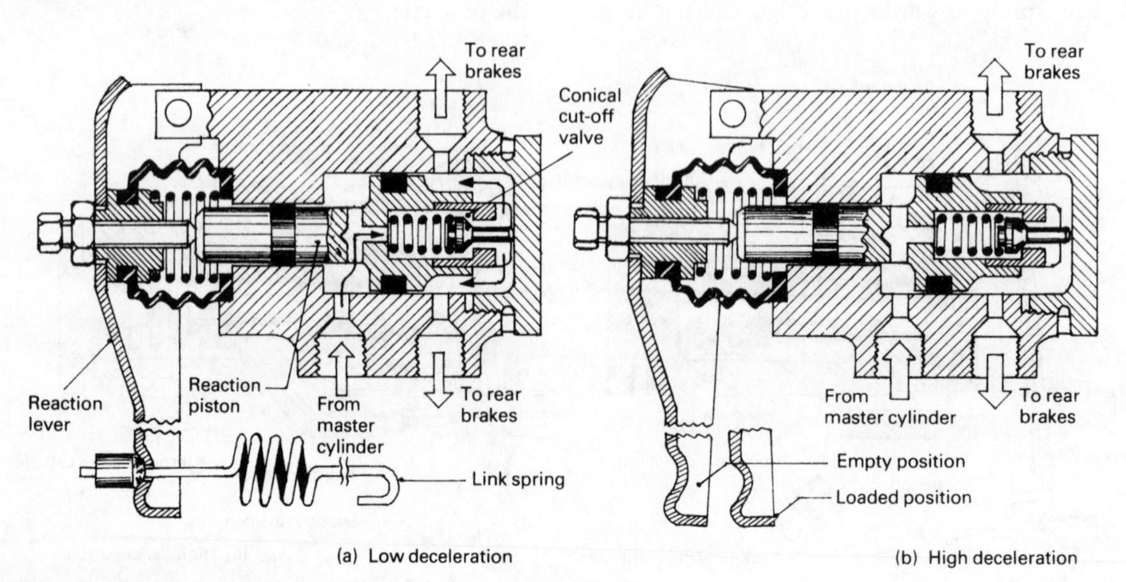

Fig. 11.32 (a and b) Load sensing progressive pressure limiting valve

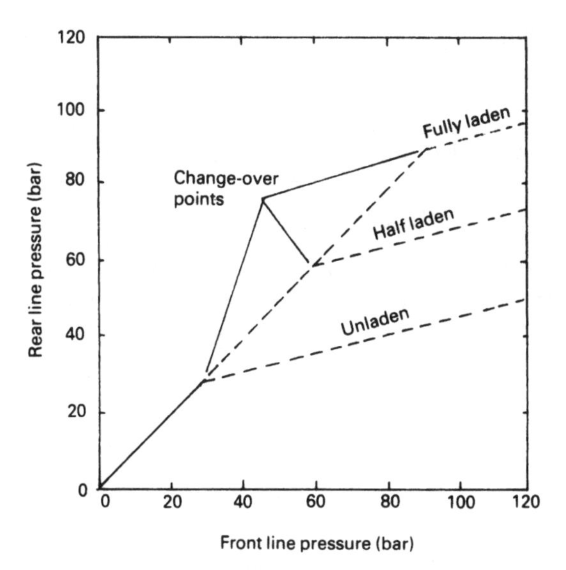

Fig. 11.33 Load sensing and progressive pressure limiting valve front to rear brake line characteristics

The ratio of the stepped piston face areas determines the degree of rear brake pipe line increase with respect to the front brake pipe lines (Fig. 11.33).

The cut-off or change point depends on the tensioning of the pre-setting spring which varies with the rear suspension deflection. The brake force distribution between the front and rear brakes is not only affected by the static laden condition, but even more so by the dynamic weight transference from the rear to the front axle.

11.6.4 Inertia pressure limiting valve (Fig. 11.34)

The inertia pressure limiting valve is designed to restrict the hydraulic line pressure operating the rear wheel brakes when the deceleration of the vehicle exceeds about 0.3 g. In preventing a further rise in the rear brake line pressure, the unrestricted front brake lines will, according to the hydraulic pressure generated, increase their proportion of braking relative to the rear brakes.

Operation (Figs 11.34 and 11.35) The operating principle of the inertia valve unit relies upon the inherent inertia of the heavy steel ball rolling up an inclined ramp when the retardation of the vehicle exceeds some predetermined amount (Fig. 11.34(b)). When this happens, the weight of the ball is removed from the stem of the disc valve, enabling the return spring fitted between the inlet port and valve shoulder to move the valve into the cut-off position.

At this point, the fluid trapped in the rear brake pipe line will remain constant (Fig. 11.35), but fluid flow between the master cylinder and front brakes is unrestricted and therefore will continue to rise with increased pedal force. As a result, the front brakes will contribute a much larger proportion of the total braking effort than the rear brakes.

When the vehicle has slowed down sufficiently or even stopped, the steel ball will gravitate to its lowest point, thereby pushing open the cut-off valve. Fluid is now free again to move from the master cylinder to the rear wheel brakes (Fig. 11.34(a)).

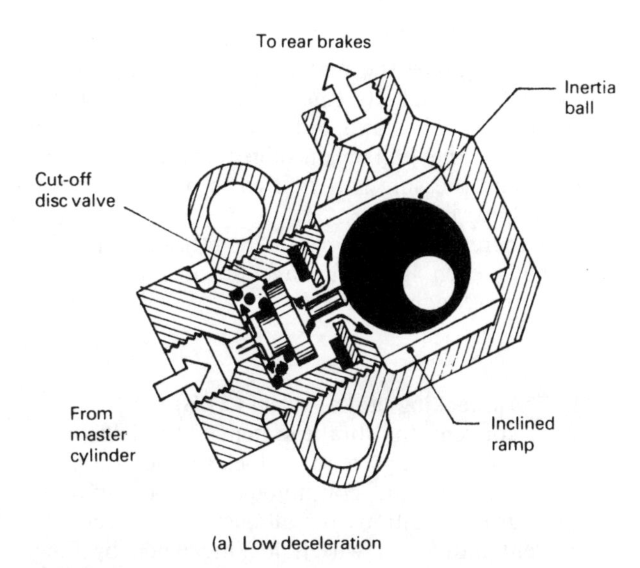

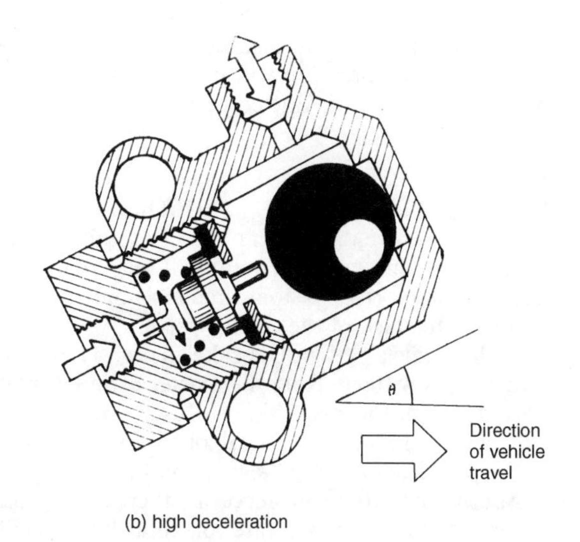

Fig. 11.34 (a and b) Inertia pressure limiting valve

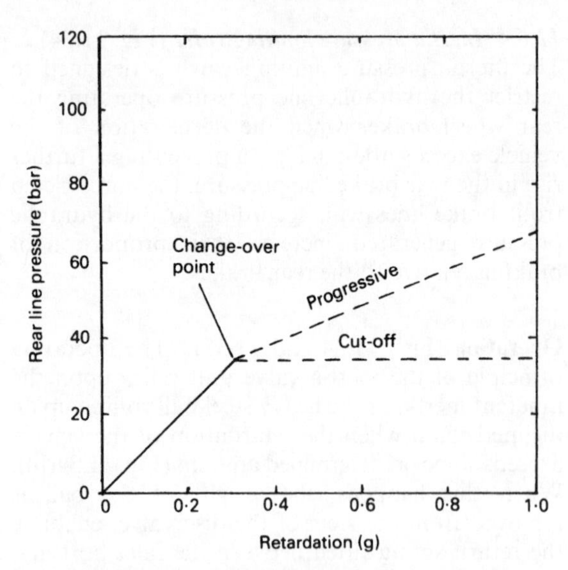

Fig. 11.35 Inertia pressure limiting valve and inertia progressive pressure limiting front to rear brake line

11.6.5 Inertia and progressive pressure limiting valve (Fig. 11.36)

The inertia and progressive pressure limiting valve unit enables the braking power between the front and rear wheels to be adjusted to match the weight transference from the rear wheels to the front wheels in proportion to the vehicle's deceleration rate. This two stage valve unit allows equal fluid pressure to flow between the front and rear brakes for light braking, but above some predetermined deceleration of the vehicle, the direct pressure increase to the rear brakes stops. With moderate to heavy braking, the front brake line pressure will equal the master cylinder generated pressure. The rear brake line pressure will continue to increase but at a much slower rate compared to that of the front brakes.

Operation (Figs 11.35 and 11.36) This inertia and progressive valve unit differs from the simple inertia pressure limiting valve because it incorporates a stepped piston (two piston dimeters) and the ball performs the task of the cut-off valve.

If the vehicle is lightly braked (Fig. 11.36(a)), fluid will flow freely from the master cylinder inlet port, through the dispersing diffuser, around the ball, along the piston central pin passage to the outlet port leading to the rear wheel brakes.

As the brake pedal force is increased (Fig. 11.36(b)), the vehicle's rate of retardation will cause the ball to continue to move forward by rolling up the inclined ramp until it seals off the central piston passage. This state is known as the *point of cut-off*.

Further foot pedal effort directly increases the front brake pipe line pressure and the pressure in the ball chamber. It does not immediately increase the rear brake pipe line pressure on the output of the valve.

Under these conditions, the trapped cut-off pressure in the rear brake lines reacts against the large piston cross-sectional area, whereas the small piston cross-sectional area on the ball chamber side of the piston is subjected to the master cylinder hydraulic pressure.

As the master cylinder's generated pressure rises with greater foot pedal effort, the input force produced on the small piston side (Input force = Master cylinder pressure × Small piston area) will increase until it exceeds the opposing output force produced on the large piston area (Output force = Rear brake line pressure × Large piston area).

A further rise in master cylinder pressure which will also be experienced in the ball chamber pushes the stepped piston backwards. Again, the rear brake line pressure will start to rise (Fig. 11.35), but at a reduced rate determined by the ratio of the small piston area to large piston area, i.e. A_S/A_L. For example, if the piston area ratio is 2:1, then the rear brake line pressure increase will be half the input master cylinder pressure rise.

i.e. $P_o = \frac{P_i}{2}$

where P_i = input pressure
P_o = output pressure

To safeguard the rear brake pipe lines, should the piston reach its full extent of its travel, the centre pin will stand out from the piston. Consequently the ball is dislodged from its seat so that fluid pressure is permitted to pass to the rear brake pipe lines.

If there are two separate rear brake pipe line circuits, each line will have its own rear brake pressure reducing valve.

11.7 Antilocking brake system (ABS)

With conventional brake systems one of the road wheels will always tend to lock sooner than the other, due to the continuously varying tyre to road grip conditions for all the road wheels. To prevent individual wheels locking when braking, the pedal should not be steadily applied but it

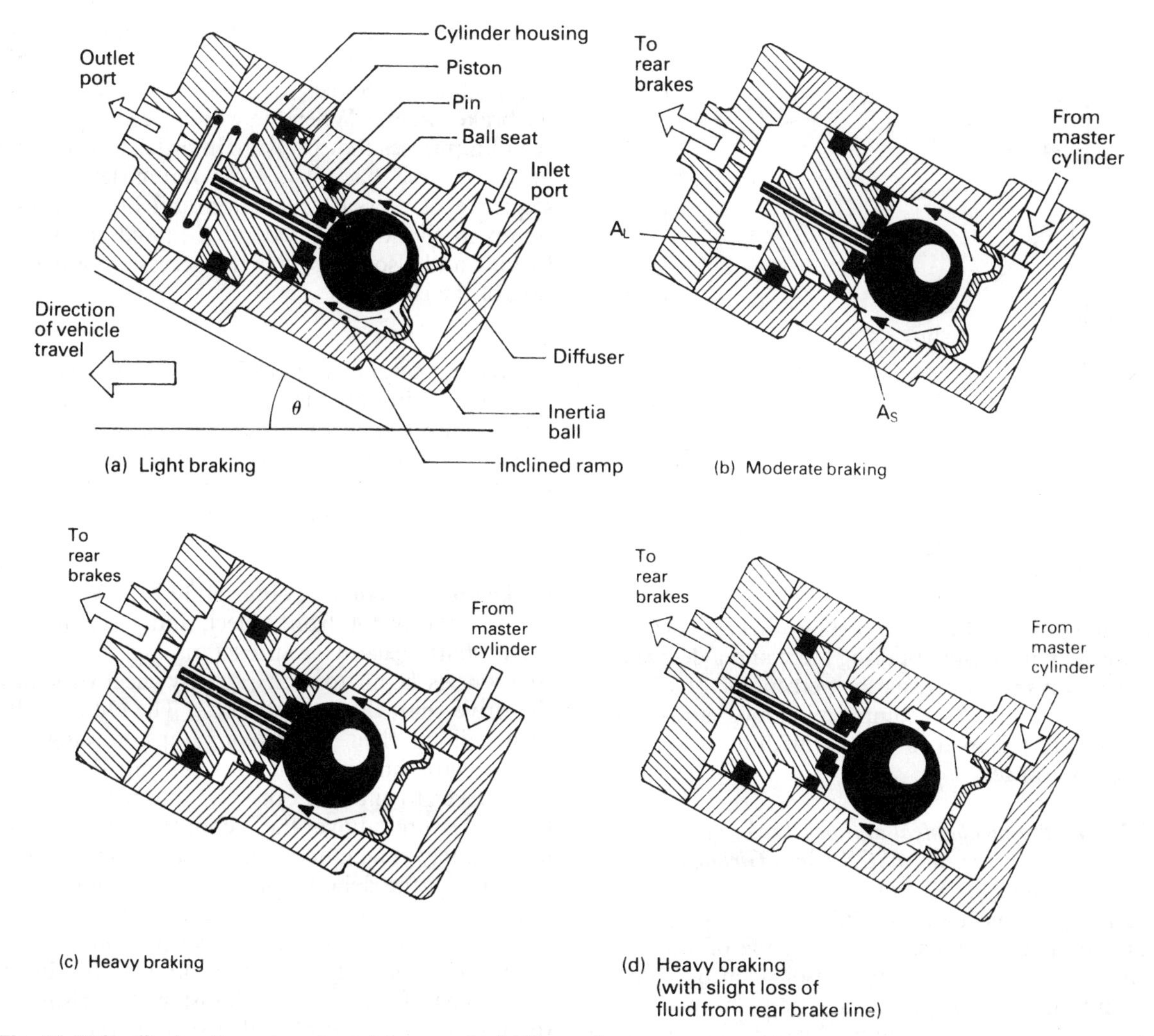

Fig. 11.36 (a–d) Inertia and progressive pressure limiting valve

should take the form of a series of impulses caused by rapidly depressing and releasing the pedal. This technique of pumping and releasing the brake pedal on slippery roads is not acquired by every driver, and in any case is subjected to human error in anticipating the pattern of brake pedal application to suit the road conditions. An antilock brake system does not rely on the skill of the driver to control wheel lock, instead it senses individual wheel slippage and automatically superimposes a brake pipe line pressure rise and fall which counteracts any wheel skid tendency and at the same time provides the necessary line pressure to retard the vehicle effectively.

When no slip takes place between the wheel and road surface, the wheel's circumference (periphery) speed and the vehicle's speed are equal. If, when the brakes are applied, the wheel circumference speed is less than the vehicle speed, the speed difference is the slip between the tyre and road surface. When the relative speeds are the same the wheels are in a state of pure rolling. When the wheels stop rotating with the vehicle continuing to move forward the slip is 100%, that is, the wheel has locked.

To attain optimum brake retardation of the vehicle, a small amount of tyre to ground slip is necessary to provide the greatest tyre tread to road surface interaction. For peak longitudinal braking force an approximately 15% wheel slip is necessary (Fig. 11.37), whereas steerability when braking depends upon a maximum sideways tyre to ground resistance which is achieved only with the

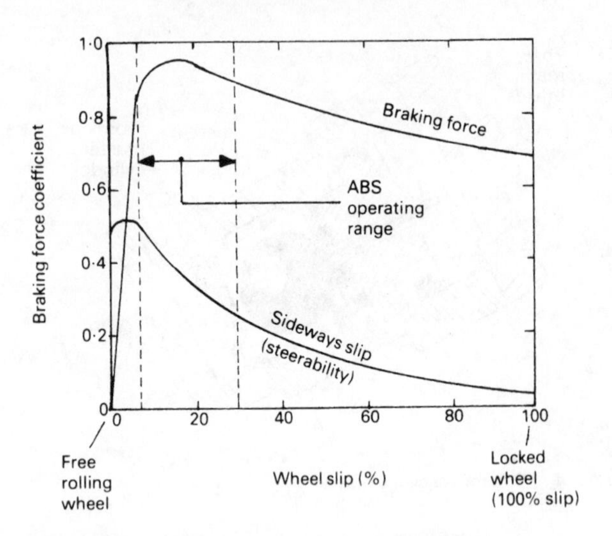

Fig. 11.37 Relationship of braking force coefficient and wheel slip

minimum of slip (Fig. 11.37). Thus there is conflict between an increasing braking force and a decreasing sideways resistance as the percentage of wheel slip rises initially. As a compromise, most anti-skid systems are designed to operate within an 8–30% wheel slip range.

11.7.1 Hydro-mechanical antilock brake system (ABS) suitable for cars (SCS Lucas Girling) (Figs 11.38 and 11.39)

This hydro-mechanical antilock braking system has two modular units, each consisting of an integrated flywheel decelerating sensor, cam operated piston type pump and the brake pressure modulator itself (Fig. 11.38). Each modulator controls the adjacent wheel brake and the diagonally opposite rear wheel via an apportioning valve. The modular flywheel sensor is driven by a toothed belt at 2.8 times the wheel speed. The flywheel sensor determines when the front wheel is approaching a predetermined deceleration. In response to this the modulator reduces the pressure in the respective brake circuits. When the wheel speeds up again, the pump raises that pressure in order to bring the braking force back to a maximum level. This sequence of pressure reduction and build-up can be up to five times a second to avoid the wheel locking and also to provide the necessary deceleration of the car.

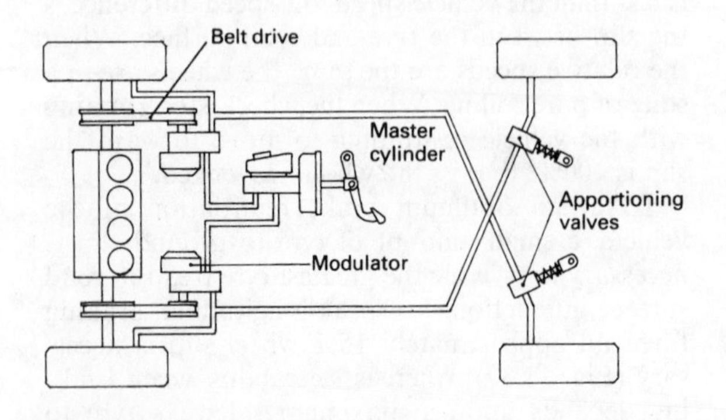

Fig. 11.38 Stop control braking system (SCS) layout

Braking as normal (Fig. 11.39(a)) Under normal braking conditions, the master cylinder fluid output is conveyed to the wheel brakes through the open cut-off valve. The dump valve is closed and the pump piston is held out of engagement from the rotating eccentric cam by the return spring.

Brake pressure reducing (Fig. 11.39(b)) When the deceleration of the front wheel, and therefore the drive shaft, exceeds a predetermined maximum (the wheels begin to lock), the flywheel overruns the drive shaft due to its inertia. The clutch balls then roll up their respective ramps, forcing the flywheel to slide inwards and causing the dump valve lever to tilt and open the dump valve. The fluid pressure above the deboost piston drops immediately. The much higher brake line pressure underneath the deboost piston and the pump piston forces the pump piston against its cam and raises the deboost piston. Fluid above both pistons is displaced back to the reservoir via the dump valve. The effect of the deboost piston rising is to close the ball cut-off valve so that the master cylinder pipe line fluid output and the wheel cylinder pipe line input become isolated from each other. As a result, the sealed chamber space below the deboost piston is enlarged, causing a rapid reduction in the fluid pressure delivered to the wheel cylinders and preventing the wheels connected to this brake circuit locking.

Brake pressure increasing (Fig. 11.39(c)) The pressure reduction resulting from the previous phase releases the brakes and allows the wheel to accelerate to the speed of the still decelerating flywheel. When the drive shaft and the flywheel are at roughly equal speeds, the clutch balls roll down their respective ramps, enabling the dump valve lever return spring to slide the flywheel over. The dump valve lever then pivots and closes the needle-type dump valve. The flywheel is again coupled to

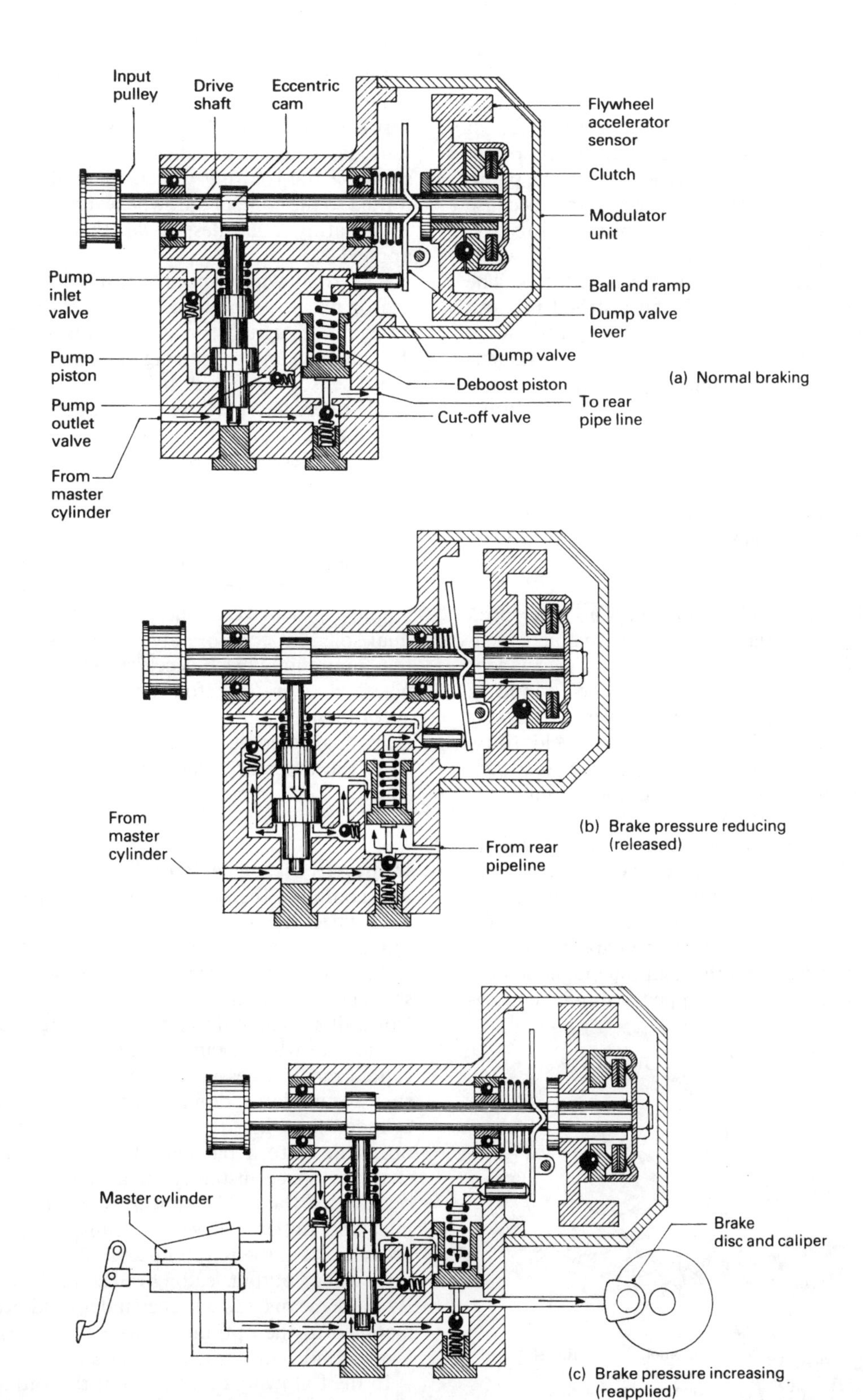

Fig. 11.39 (a–c) Antilock braking system (ABS) for front wheel drive

the drive shaft so that its speed rises with the drive shaft. At the same time the pump piston commences to build up pressure above the deboost piston by the action of the pump inlet and outlet valves. The output pressure generated by the pump pushes the deboost piston downward and, because the space underneath the deboost piston forms part of the brake pipe line circuit leading to the wheel cylinders, the total fluid volume is reduced. The brake pipe line pressure will be restored in steps due to the pump action until the downward movement of the deboost piston stem once again opens the cut-off valve. The pump piston then disengages and thereafter further pressure rise in the brake pipe line will be provided by the master cylinder in the normal way.

11.7.2 Hydraulic-electric antilock brake system (ABS) suitable for cars (Bosch) (Figs 11.40 and 11.41)

Speed sensor and excitor (Fig. 11.40) The speed sensor uses the variable reluctance magnetic sensing principle, whereby a cylindrical permanent magnetic core with a coil wire wound around it, mounted on the stationary hub carrier, axle casing or back plate, produces a magnetic field (flux) which overlaps the rotating excitor ring. The excitor may be of the tooth ring or rib-slot ring type attached to the rotating wheel hub or drive shaft. A number of teeth or slots are arranged radially which, with the speed of rotation of the road wheel, determine the frequency of the signal transmitted to the electronic-control unit. As the wheel and excitor revolve, the teeth and gaps or ribs and slots of the excitor pass through the magnetic field of the sensor. The coil wrapped around the magnetic cone senses the changing intensity of the magnetic field as the teeth or ribs pass through the flux lines and so an alternating voltage is induced in the coil, whose frequency is proportional to the speed of the rotating wheel. The voltage is transmitted to the control unit whenever the road wheels are rotating, regardless of whether the brakes are applied.

The road wheel speed measured by the speed sensor provides the wheel deceleration and wheel acceleration signals for the electronic-control unit. The merging and processing of the individual wheel speed sensor signals by the control unit provide a single reference speed which is roughly the vehicle speed. A comparison of any individual wheel speed with the reference speed supplies the wheel to road slip (wheel tending to lock) signal.

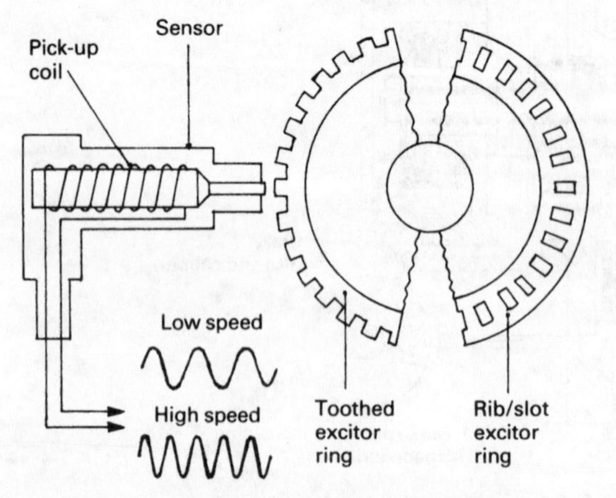

Fig. 11.40 Magnetic speed sensor and excitor

Electronic-control unit (Fig. 11.41(a)) The function of the electronic-control unit is to receive, amplify, process, compute and energize the individual solenoid control valves. That is, to evaluate the minimum wheel deceleration and maximum wheel acceleration for optimum braking and accordingly supply the energizing current to the individual solenoid control valves so they can regulate the necessary wheel cylinder pipe line pressures.

Hydro/electric modulator (Fig. 11.41(a)) This unit combines the solenoid control valves; one for each wheel, an accumulator for each of the dual-brake circuits and a twin cylinder return flow pump driven from an electric motor. The solenoid valve switches half or fully on and off through the control unit's solid-state circuits, causing the master cylinder to wheel cylinder fluid supply to be interrupted many times per second. The reduced pressure accumulator rapidly depressurizes the wheel cylinder pipe line fluid when the solenoid valve opens the return passage, due to the diaphragm chamber space instantly enlarging to absorb the outflow of fluid. The return flow pump, with its inlet and outlet ball valves, transfers fluid under pressure from the reducer accumulator to the master cylinder output leading to the brake cylinders. By these means, the wheel cylinder fluid pressure is matched to the optimum braking severity relative to the condition of the road surface.

In the following description of the anti-skid system operating, only one wheel is considered for simplicity.

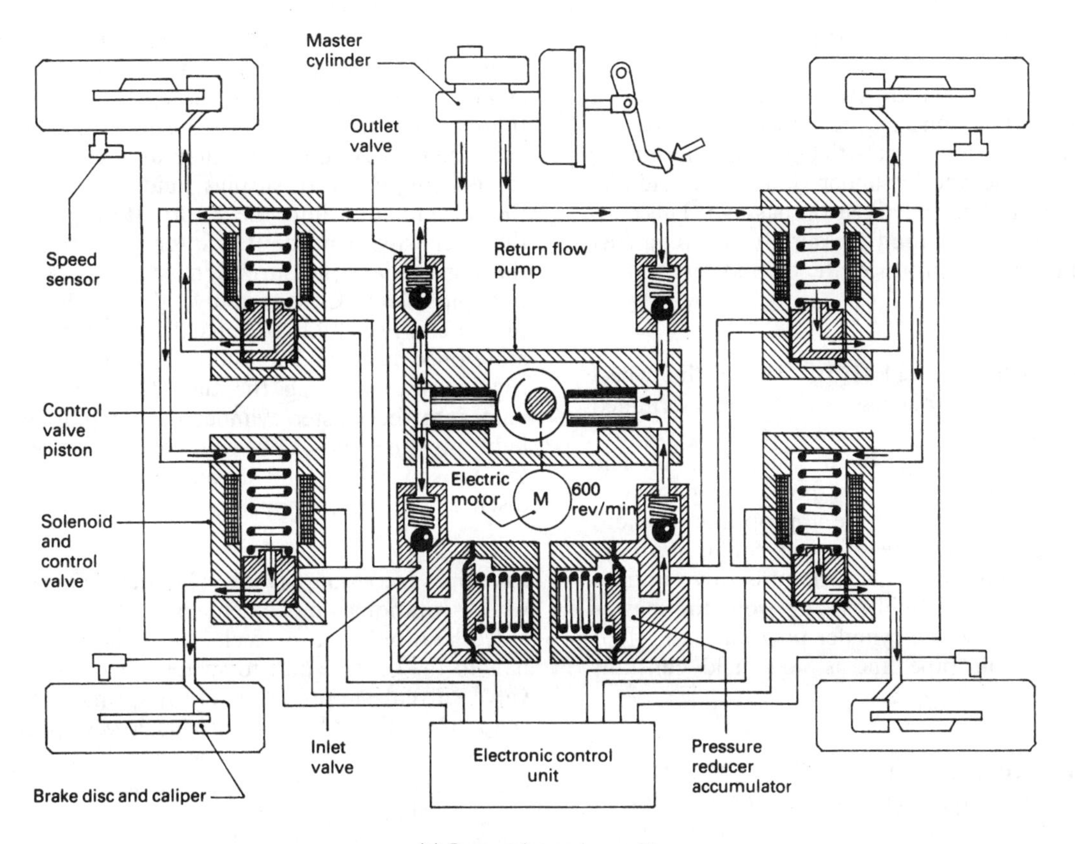

(a) Pressure increasing position

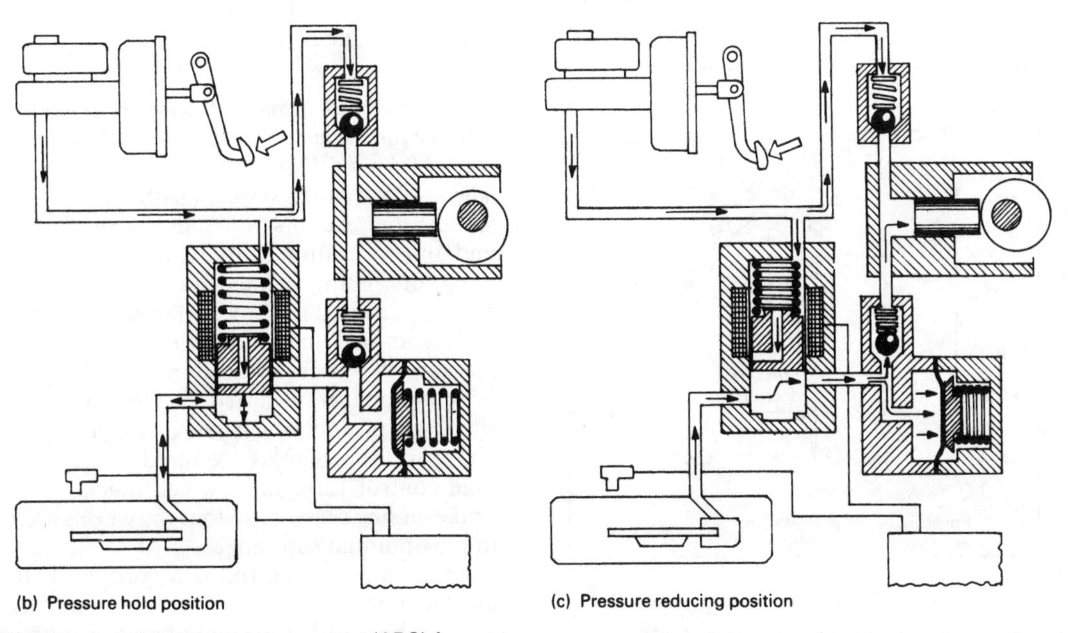

(b) Pressure hold position

(c) Pressure reducing position

Fig. 11.41 (a–c) Antilocking brake system (ABS) for cars

Normal braking conditions (Fig. 11.41(a)) Under normal braking conditions, the solenoid is disengaged and the armature valve is held in its lowest position by the return spring. When the brakes are applied, fluid flows unrestricted from the master cylinder to the wheel cylinder via the solenoid piston armature type valve central passage. This continues until the required pressure build-up against the caliper piston produces the desired retardation to the vehicle.

Pressure hold (Fig. 11.41(b)) When the wheel deceleration approaches some predetermined value, the speed sensor signals to the computer control unit the danger of the wheel locking. The control unit immediately responds by passing a small electric current to the appropriate solenoid valve. Accordingly, the solenoid coil is partially energized. This raises the armature valve until it blocks the flow of fluid passing from the master cylinder to the wheel cylinder pipe line. The fluid pressure in the pipe line is now held constant (Fig. 11.42).

Pressure reducing (Fig. 11.41(c)) Should the wheel sensor still signal an abnormally rapid speed reduction likely to cause the wheel to lock, the control unit increases the supply of current to the solenoid coil, causing the armature valve to lift still further to a position where it uncovers the return flow passage. The 'hold' line pressure collapses instantly because the highly pressurized fluid is able to escape into the pressure reducer accumulator. At the same time as the accumulator is being charged, surplus fluid is drawn from the accumulator into the return flow pump via the inlet valve whence it is discharged back into the appropriate pressurized master cylinder output pipe line. Consequently, the reduction in pressure (Fig. 11.42) permits the wheel to accelerate once again and re-establish its grip with the road surface. During the time fluid is pumped back into the master cylinder output pipe line, a light pressure pulsation will be experienced on the foot pedal by the driver due to the cyclic discharge of the pump.

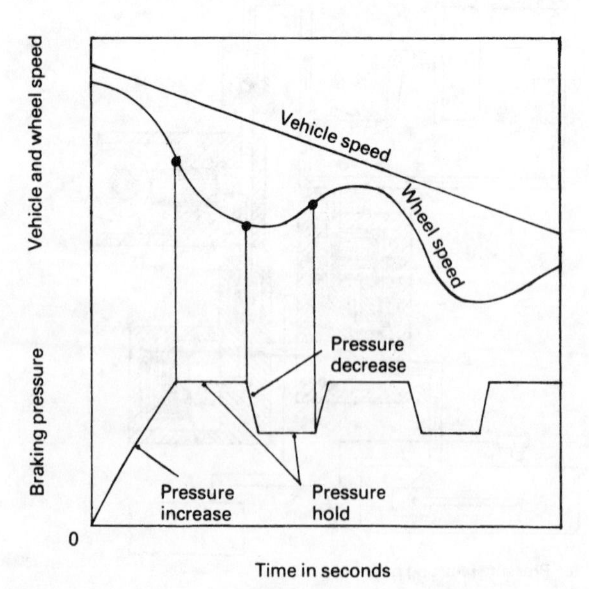

Fig. 11.42 Typical antilock brake system (ABS) pressure, wheel and vehicle speed characteristics with respect to time

Pressure increasing (Fig. 11.41(a)) Once the wheel rotational movement has changed from a deceleration back to acceleration, the sensor signals to the control unit to switch off the solenoid valve current supply. The return spring instantly snaps the solenoid valve into its lowest position and once again the fluid passage between the master cylinder output pipe line and the wheel caliper cylinder pipe line is re-established, causing the brake to be re-applied (Fig. 11.42). The sensitivity and response time of the solenoid valve is such that the pulsating regulation takes place four to ten times per second.

11.7.3 Air/electric antilock brake system (ABS) suitable for commercial vehicles (WABCO)

(Figs 11.43 and 11.44)

The antilock brake system (ABS) consists of wheel sensors and excitors which detect the deceleration and an acceleration of individual wheels by generating alternating voltages the frequency of which are proportional to the wheel speed (Fig. 11.43(a)).

Sensors on each wheel (Fig. 11.40) continually measure the wheel speed during braking and this information is transmitted to an electronic (processor) control unit which senses when any wheel is about to lock. Signals are rapidly relayed to solenoid control valve units which quickly adjust the brake air line pressure so that the wheels are braked in the optimum slip range.

Each wheel is controlled according to the grip available between its tyre and the road. By these means, the vehicle is brought to a halt in the shortest time without losing vehicle stability and steerability.

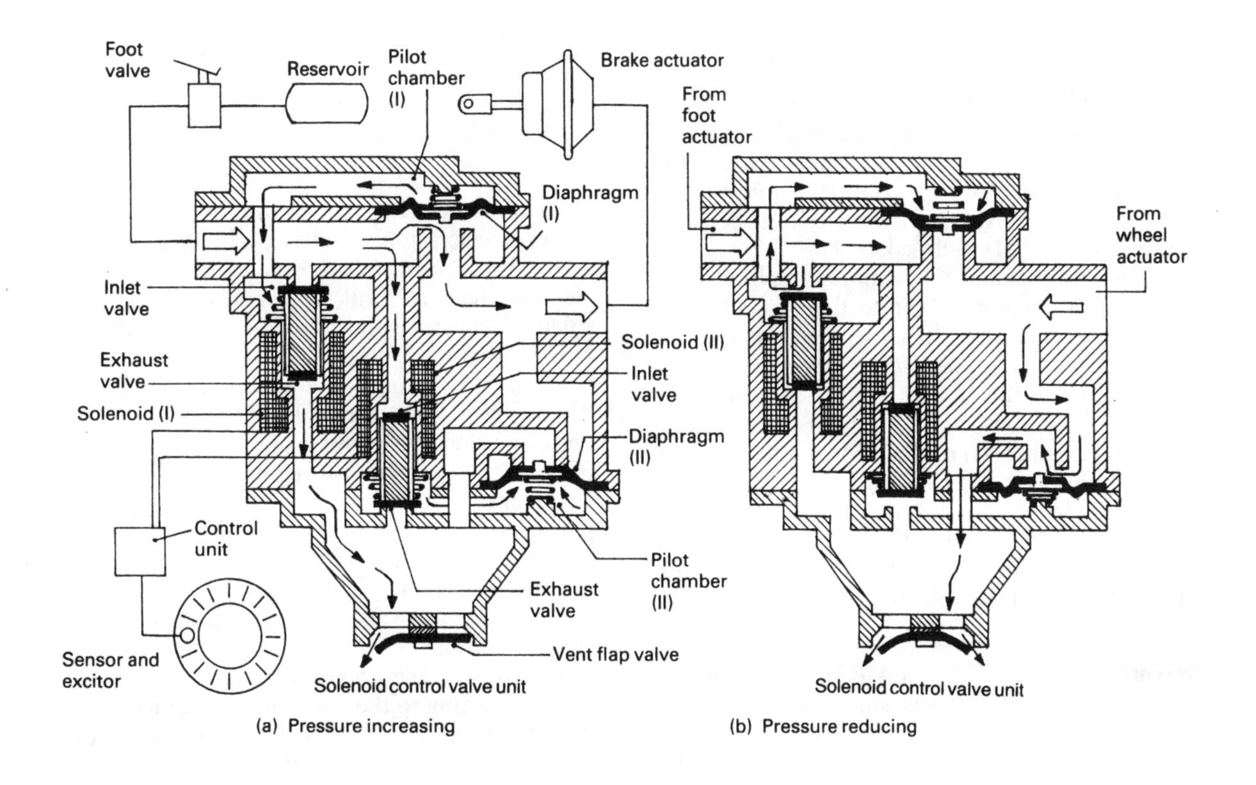

(a) Pressure increasing

(b) Pressure reducing

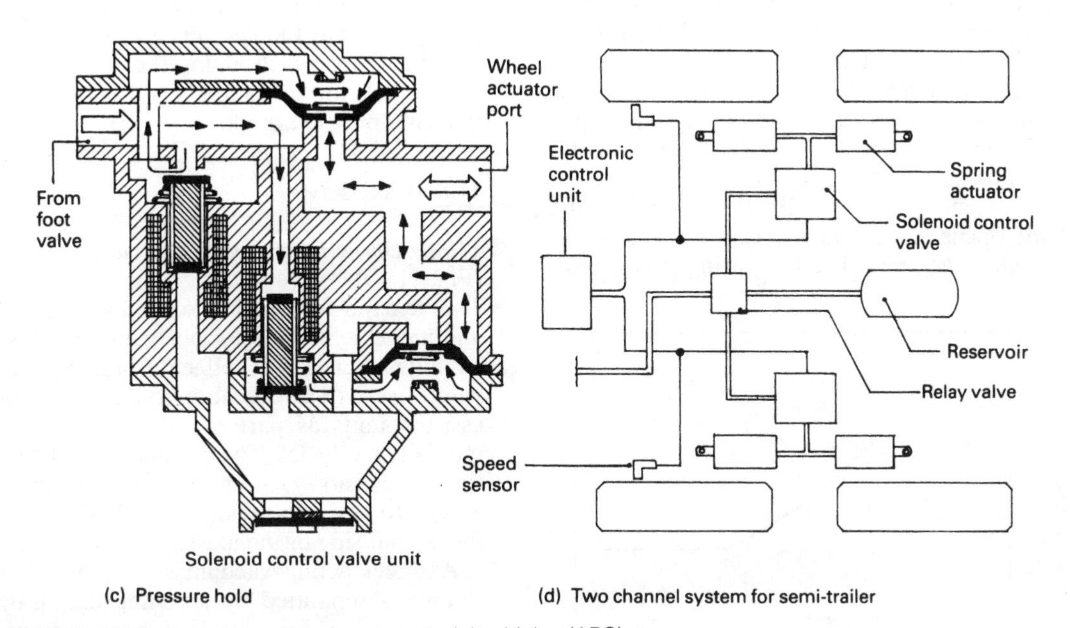

(c) Pressure hold

(d) Two channel system for semi-trailer

Fig. 11.43 (a–d) Antilock brake system for commercial vehicles (ABS)

Pressure increasing (Fig. 11.43(a)) When the foot pedal is depressed, initially both solenoids are switched off so that their armatures are moved to their outermost position by the return springs. Consequently the first solenoid's inlet valve (I) is closed and its exhaust valve (I) is open whereas the second solenoid valve's inlet valve (II) is open and its exhaust valve (II) is closed.

Under these conditions, pilot chamber (I) is exhausted of compressed air so that air delivered from the foot valve enters the solenoid control valve unit inlet port and pushes open diaphragm (I) outlet passage, enabling compressed air to be supplied to the wheel brake actuator. At the same time pilot chamber (II) is filled with compressed air so that diaphragm (II) closes off the exhaust passage leading to the atmosphere. As a result, the foot pedal depression controls the rising air pressure (Fig. 11.44) delivered from the foot valve to the wheel actuator via the solenoid control valve unit.

Pressure reducing (Fig. 11.43(b)) As soon as wheel deceleration or wheel slip threshold values are exceeded, the sensor transmits this information to the electronic-control unit which signals to the solenoid valve unit to reduce the wheel actuator pipe line air pressure.

Both solenoids are energized. This opens inlet valve (I) whilst inlet valve (II) is closed and exhaust valve (II) is opened. The open inlet valve (I) allows air to enter and pressurize pilot chamber (I) so that diaphragm (I) closes the outlet passage, thus preventing any more air from the foot valve passing through to the outlet passage port.

At the same time, solenoid (II) closes inlet valve (II) and opens exhaust valve (II). This exhausts air from pilot chamber (II), permitting compressed air from the wheel actuator to push open diaphragm (II) outlet exhaust passage, causing the air pressure in the actuator pipe line to reduce quickly (Fig. 11.44).

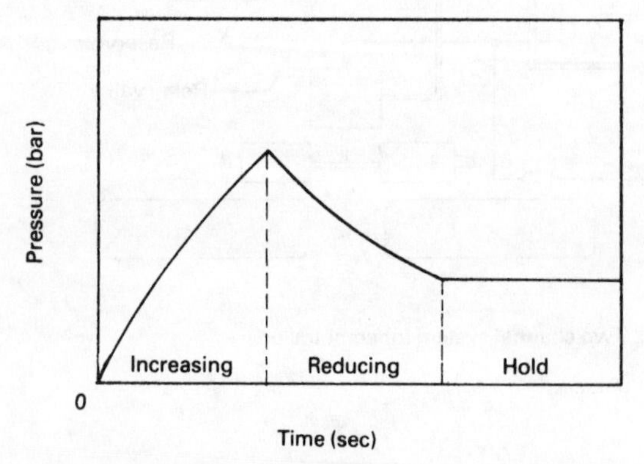

Fig. 11.44 Air/electric antilock brake system (ABS) pressure/time characteristics

Pressure hold (Fig. 11.43(c)) When the road wheel acceleration reaches some predetermined value, the sensor relays this information to the electronic-control unit, which in turn signals the solenoid control valve unit to hold the remaining pipe line actuator pressure.

Solenoid (I) remains energized but solenoid (II) is de-energized. Therefore solenoid (I) inlet valve (I) and exhaust valve (I) remain open and closed respectively. Inlet valve (II) allows compressed air into pilot chamber (I) so that diaphragm (I) closes the outlet passage leading to the wheel actuator pipe line.

Conversely, solenoid (II) is now de-energized causing its return spring to move the armature so that the inlet valve (II) opens and exhaust valve (II) closes. Compressed air from the foot valve now flows through the open inlet valve (II) along the passage leading to the underside of diaphragm (II), thus keeping the outlet exhaust passage closed. Compressed air at constant pressure (Fig. 11.44) is now trapped between both closed diaphragm outlet passages and the wheel actuator pipe line. This pipe line pressure is maintained until the sensor signals that the wheel is accelerating above its threshold, at which point the electronic-control unit signals the solenoid control valve to switch to its rising pressure mode.

11.8 Brake servos

11.8.1 Operating principle of a vacuum servo (Fig. 11.45)

The demand for a reduction in brake pedal effort and movement, without losing any of the sensitivity and response to the effective braking of cars and vans, has led to the adoption of vacuum servo assisted units as part of the braking system for most light vehicles. These units convert the induction manifold vacuum energy into mechanical energy to assist in pressurizing the brake fluid on the output side of the master cylinder.

A direct acting vacuum servo consists of two chambers separated by a rolling diaphragm and power piston (Fig. 11.45). The power piston is coupled to the master cylinder outer primary piston by a power push rod. The foot pedal is linked through a pedal push rod indirectly to the power piston via a vacuum-air reaction control valve.

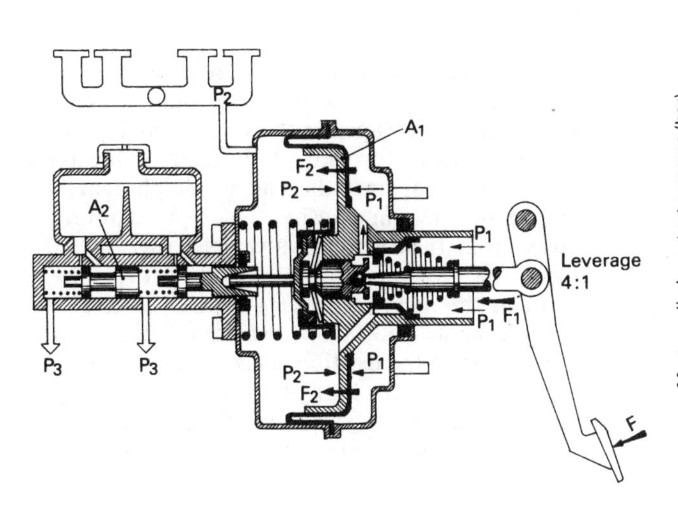

(a) Operating principle of a vacuum servo

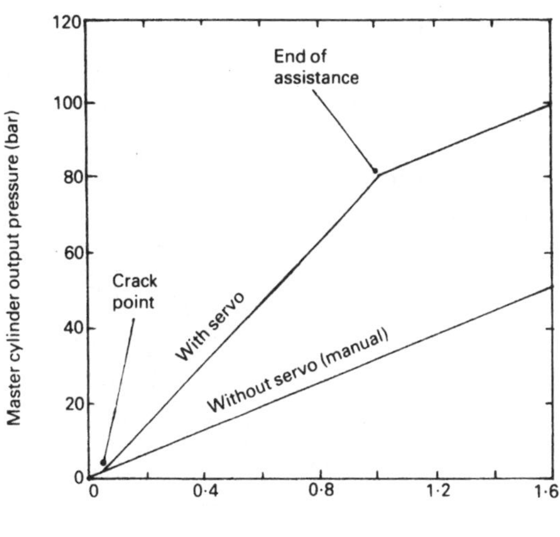

(b) Vacuum servo input to output characteristics

Fig. 11.45 (a and b) Operating principle and characteristics of a vacuum servo

When the brakes are in the 'off' position, both sides of the power piston assembly are subjected to induction manifold pressure. When the brakes are applied, the vacuum in the front chamber remains undisturbed, whilst the vacuum in the rear chamber is replaced by atmospheric air closing the vacuum supply passage, followed by the opening of the air inlet passage to the rear chamber. The resulting difference of pressure across the power piston causes it to move towards the master cylinder, so that the thrust imposed on both the primary and secondary pistons in the master cylinder generates fluid pressure for both brake lines.

The operating principle of the vacuum servo is best illustrated by the following calculation:

Example (Fig. 11.45(a)) A direct acting vacuum servo booster has a 200 mm diameter power piston suspended on both sides by the induction manifold vacuum (depression), amounting to a gauge reading of 456 mm Hg, that is 0.6 bar below atmospheric pressure.
(*Note* 1 bar = 760 mm Hg = 100 KN/m^2).

The foot pedal leverage ratio is 4:1 and the master cylinder has 18 mm diameter.

Determine the following when a pedal effort of 300 N is applied and the rear power piston chamber which was occupied with manifold vacuum is now replaced by atmospheric air (Fig. 11.45(a)).

a) The push rod thrust and generated primary and secondary hydraulic brake line pressures due only to the foot pedal effort.
b) The power push rod thrust and the generated fluid pressures in the pipe lines due only to the vacuum servo action.
c) The total pedal push rod and power piston thrust and the corresponding generated fluid pressure in the pipe lines when both foot pedal and servo action are simultaneously applied to the master cylinder.

Let F = foot pedal effort (N)
F_1 = pedal push rod thrust (N)
F_2 = power piston thrust (N)
P_1 = pressure in the rear chamber (kN/m^2)
P_2 = manifold pressure (kN/m^2)
P_3 = fluid generated pressure (kN/m^2)
A_1 = cross-sectional area of power piston (m^2)
A_2 = cross-sectional area of master cylinder bore (m^2)

a) Pedal push rod thrust $F_1 = F \times 4$
$= 300 \times 4$
$= 1200$ N or 1.2 kN

Master cylinder fluid pressure P_3 $= \dfrac{F_1}{A_2}$

$$= \frac{1.2}{\frac{\pi}{4}(0.018)^2}$$

$= 4715.7$ kN/m^2 or 47.2 bar

b) Power piston thrust $F_2 = A_1(P_1 - P_2)$

$$= \frac{\pi}{4}(0.2)^2(100 - 40)$$

$$= 1.88\,\text{kN}$$

Master cylinder fluid pressure P_3

$$= \frac{F_2}{A_2}$$

$$= \frac{1.88}{\frac{\pi}{4}(0.018)^2}$$

$$= 7387.93\,\text{kN/m}^2 \text{ or } 73.9\,\text{bar}$$

c) Total power piston and pedal push rod thrust

$$= F_1 + F_2$$

$$= 1.2 + 1.88$$

$$= 3.08\,\text{kN}$$

Total master cylinder fluid pressure P_3

$$= \frac{F_3}{A_2}$$

$$= \frac{3.08}{\frac{\pi}{4}(0.018)^2}$$

$$= 12103.635\,\text{kN/m}^2 \text{ or } 121.04\,\text{bar}$$

11.8.2 Direct acting suspended vacuum-assisted brake servo unit (Fig. 11.46(a, b and c))

Brake pedal effort can be reduced by increasing the leverage ratio of the pedal and master cylinder to wheel cylinder piston sizes, but this is at the expense of lengthening the brake pedal travel, which unfortunately extends the brake application time. The vacuum servo booster provides assistance to the brake pedal effort, enabling the ratio of master cylinder to wheel cylinder piston areas to be reduced. Consequently, the brake pedal push rod effective stroke can be reduced in conjunction with a reduction in input foot effort for a given rate of vehicle deceleration.

Operation

Brakes off (Fig. 11.46(a)) With the foot pedal fully released, the large return spring in the vacuum chamber forces the rolling diaphragm and power piston towards and against the air/vac chamber stepped steel pressing.

When the engine is running, the vacuum or negative pressure (below atmospheric pressure) from the induction manifold draws the non-return valve away from its seat, thereby subjecting the whole vacuum chamber to a similar negative pressure to that existing in the manifold.

When the brake pedal is fully released, the outer spring surrounding the push rod pulls it and the relay piston back against the valve retaining plate. The inlet valve formed on the end of the relay piston closes against the vac/air diaphragm face and at the same time pushes the vac/air diaphragm away from the vacuum valve. Negative pressure from the vacuum chamber therefore passes through the inclined passage in the power piston around the seat of the open vacuum valve where it then occupies the existing space formed in the air/vac chamber to the rear of the rolling diaphragm. Hence with the air valve closed and the vacuum valve open, both sides of the power piston are suspended in vacuum.

Brakes applied (Fig. 11.46(b)) When the foot pedal is depressed the pedal push rod moves towards the diaphragm power piston, pushing the relay piston hard against the valve retaining plate. Initially the vac/air diaphragm closes against the vacuum valve's seat and with further inward push rod movement the relay piston inlet seat separates from the vac/air diaphragm face. The air/vac chamber is now cut off from the vacuum supply and atmospheric air is now free to pass through the air filter, situated between the relay piston inlet valve seat and diaphragm face, to replace the vacuum in the air/vac chamber. The difference in pressure between the low primary vacuum chamber and the high pressure air/vac chamber causes the power piston and power push rod to move forward against the master cylinder piston so the fluid pressure is generated in both brake circuits to actuate the front and rear brakes.

Brake held on (Fig. 11.46(c)) Holding the brake pedal depressed momentarily continues to move the power piston with the valve body forward under the influence of the greater air pressure in the air/vac chamber, until the rubber reaction pad is compressed by the shoulder of the power piston against the opposing reaction of the power push rod. As a result of squeezing the outer rubber rim of the reaction pad, the rubber distorts and extrudes towards the centre and backwards in the relay piston's bore. Subsequently, only the power piston and valve body move forward whilst the relay piston and pedal push rod remain approximately in the same position until the air valve seat closes against the vac/air diaphragm face. More

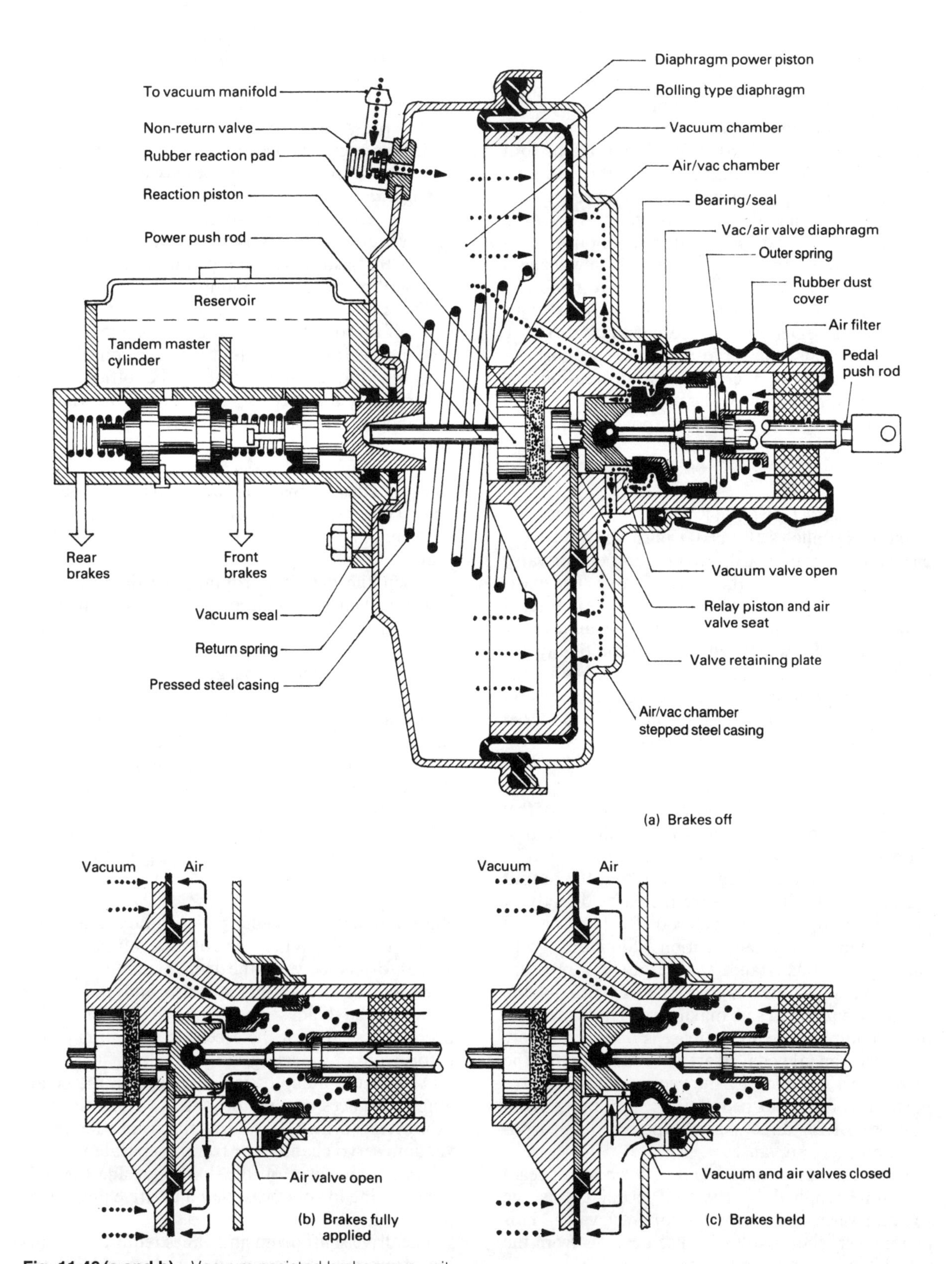

Fig. 11.46 (a and b) Vacuum-assisted brake servo unit

atmospheric air cannot now enter the air chamber so that there is no further increase in servo power assistance. In other words, the brakes are on hold. The reaction pad action therefore provides a progressive servo assistance in relation to the foot pedal effort which would not be possible if only a simple reaction spring were positioned between the reaction piston and the relay piston.

If a greater brake pedal effort is applied for a given hold position, then the relay piston will again move forward and compress the centre region of the reaction pad to open the air valve. The extra air permitted to enter the air/vac chamber therefore will further raise the servo assistance proportionally. The cycle of increasing or decreasing the degree of braking provides new states of hold which are progressive and correspond to the manual input effort.

Brakes released (Fig. 11.46(a)) Releasing the brake pedal allows the pedal push rod and relay piston to move outwards; first closing the air valve and secondly opening the vacuum valve. The existing air in the air/vac chamber will then be extracted to the vacuum chamber via the open vacuum valve, the power piston's inclined passage, and finally it is withdrawn to the induction manifold. As in the brakes 'off' position, both sides of the power piston are suspended in vacuum, thus preparing the servo unit for the next brake application.

Vacuum servo operating characteristics (Fig. 11.45(b)) The benefits of vacuum servo assistance are best shown in the input to output characteristic graphs (Fig. 11.45(b)). Here it can be seen that the output master cylinder line pressure increases directly in proportion to the pedal push rod effort for manual (unassisted) brake application. Similarly, with vacuum servo assistance the output line pressure rises, but at a much higher rate. Eventually the servo output reaches its maximum. Thereafter any further output pressure increase is obtained purely by direct manual pedal effort at a reduced rate. The extra boost provided by the vacuum servo in proportion to the input pedal effort may range from 1½:1 to 3:1 for direct acting type servos incorporated on cars and vans.

Servo assistance only begins after a small reaction force applied by the foot pedal closes the vacuum valve and opens the air inlet valve. This phase where the servo assistance deviates from the manual output is known as the *crack point*.

11.8.3 Types of vacuum pumps

(Fig. 11.47(a, b and c))
For diesel engines which develop very little manifold depression, a separate vacuum pump driven from the engine is necessary to operate the brake servo. Vacuum pumps may be classified as *reciprocating diaphragm* or *piston* or *rotary vane* types.

In general, for high speed operation the vane type vacuum pump is preferred and for medium speeds the piston type pump is more durable than the diaphragm vacuum pump.

These pumps are capable of operating at depressions of up to 0.9 bar below atmospheric pressure. One major drawback is that they are continuously working and cannot normally be offloaded by interrupting the drive or by opening the vacuum chamber to the atmosphere.

Reciprocating diaphragm or piston type vacuum pump (Fig. 11.47(a and b)) These pumps operate very similarly to petrol and diesel engine fuel lift pumps.

When the camshaft rotates, the diaphragm or piston is displaced up and down, causing air to be drawn through the inlet valve on the downstroke and the same air to be pushed out on the upward stroke through the discharge valve.

Consequently, a depression is created within the enlarging diaphragm or piston chamber causing the brake servo chamber to become exhausted (drawn out) of air, thereby providing a pressure difference between the two sides of the brake servo which produces the servo power.

Lubrication is essential for plungers and pistons but the diaphragm is designed to operate dry.

Rotary vane type vacuum pump (rotary exhauster) (Fig. 11.47(c)) When the rotor revolves, the cell spaces formed between the drum blades on the inlet port side of the casing increase and the spaces between the blades on the discharge port side decrease, because of the eccentric mounting of the rotor drum in its casing.

As a result, a depression is created in the enlarging cell spaces on the inlet side, causing air to be exhausted (drawn out) directly from the brake vacuum servo chamber or from a separate vacuum reservoir. However on the discharge side the cells are reducing in volume so that a positive pressure is produced.

The drive shaft drum and vanes require lubricating at pressure or by gravity or suction from the

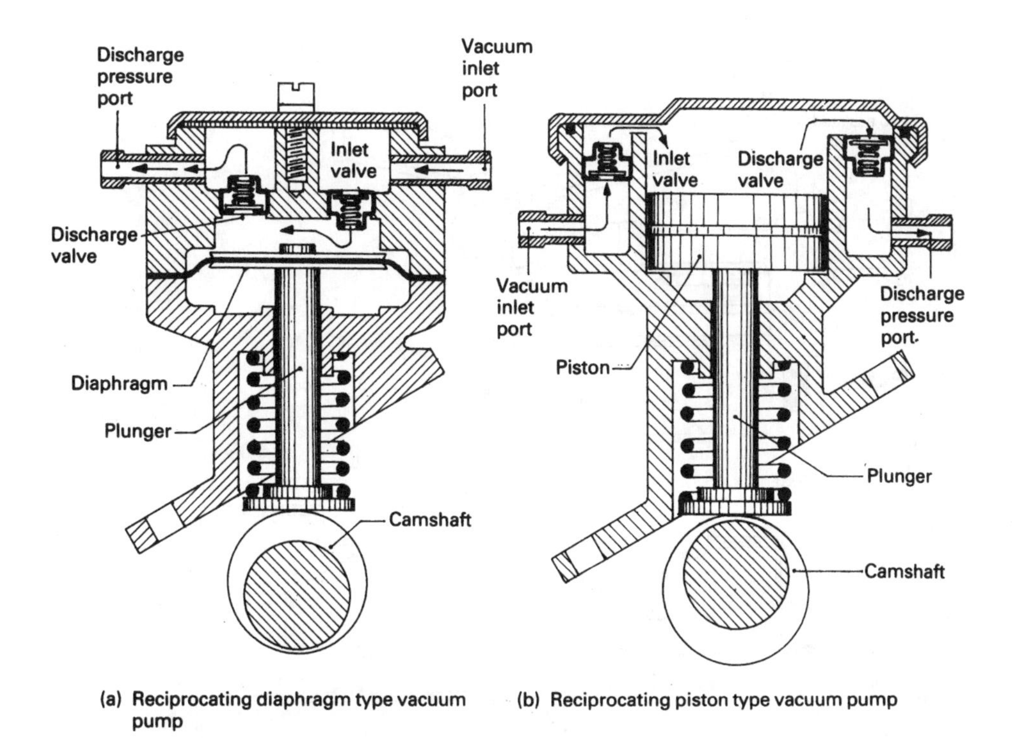

(a) Reciprocating diaphragm type vacuum pump

(b) Reciprocating piston type vacuum pump

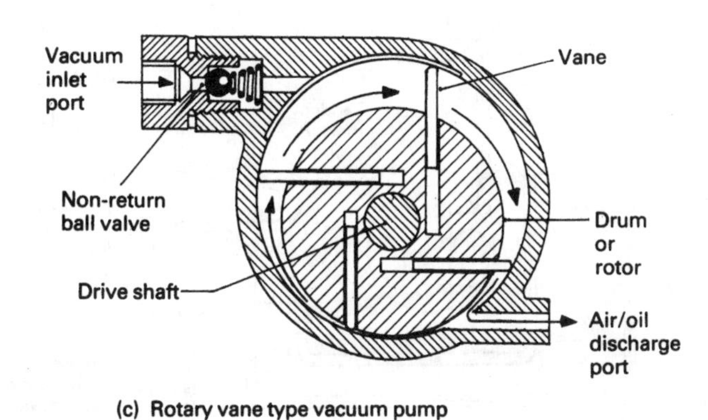

(c) Rotary vane type vacuum pump

Fig. 11.47 (a–c) Types of vacuum pumps

engine oil supply. Therefore, the discharge port returns the oil-contaminated air discharge back to the engine crank case.

11.8.4 Hydraulic servo assisted steering and brake system

Introduction to hydraulic servo assistance (Fig. 11.48) The alternative use of hydraulic servo assistance is particularly suited where emission control devices to the engine and certain types of petrol injection system reduce the available intake manifold vacuum, which is essential for the effective operation of vacuum servo assisted brakes. Likewise, diesel engines, which produce very little intake manifold vacuum, require a separate vacuum source such as a vacuum pump (exhauster) to operate a vacuum servo unit; therefore, if power assistant steering is to be incorporated it becomes economical to utilize the same hydraulic pump (instead of a vacuum pump) to energize both the steering and brake servo units.

The hydraulic servo unit converts supplied fluid energy into mechanical work by imposing force

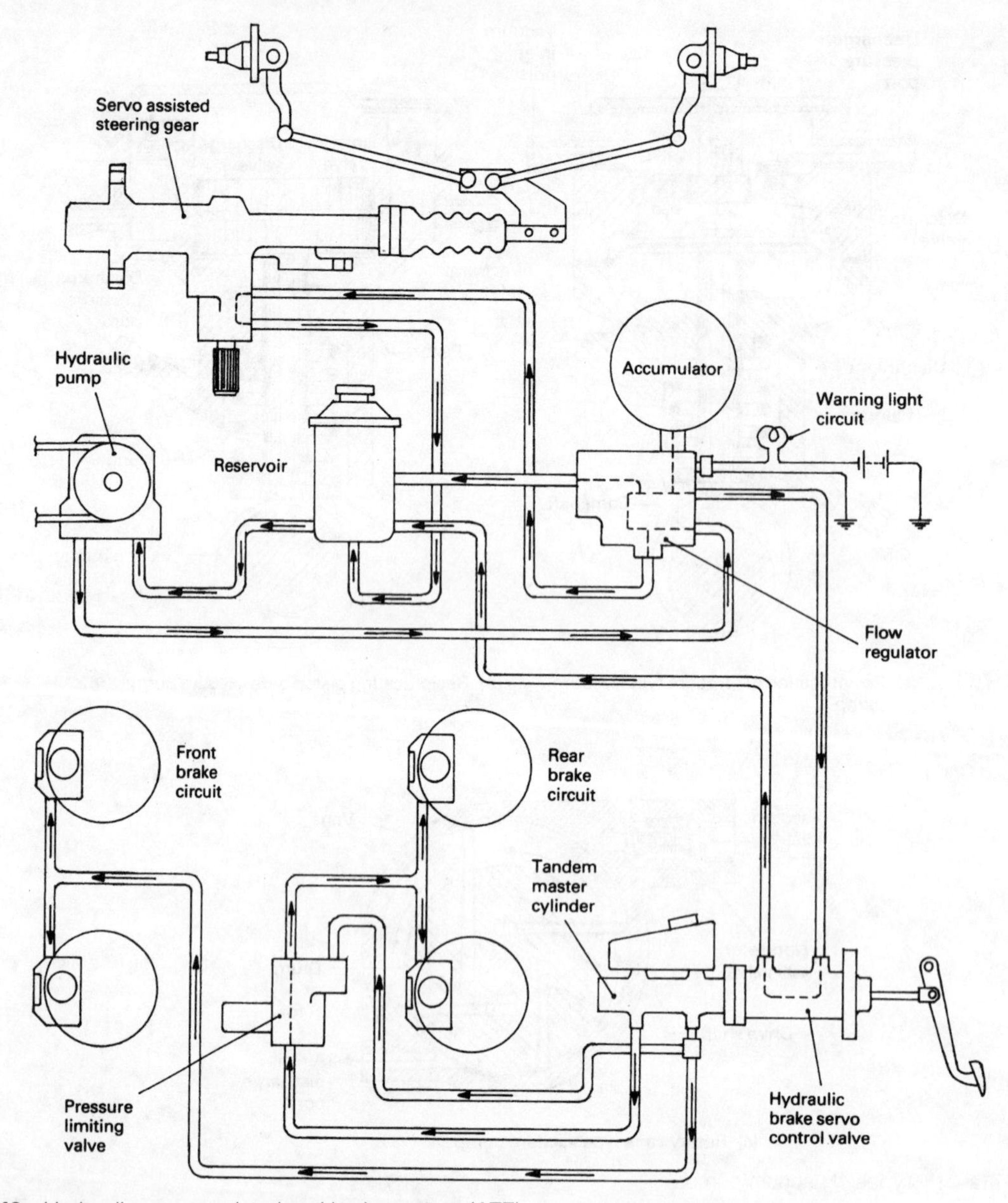

Fig. 11.48 Hydraulic servo-assisted and brake system (ATE)

and movement to a power piston. A vane type pump provides the pressure energy source for both the power assisted steering and for the brake servo. When the brake accumulator is being changed approximately 10% of the total pump output is used, the remaining 90% of the output returns to the power steering system. When the accumulator is fully charged, 100% of the pump output returns via the power steering control unit to the reservoir. Much higher operating pressures are used in a hydraulic servo compared to the vacuum type servo. Therefore the time needed to actuate the brakes is shorter.

The proportion of assistance provided to the pedal effort is determined by the cross-sectional area ratio of both the power piston and reaction piston. The larger the power piston is relative to the reaction piston, the greater the assistance will be and vice versa.

In the event of pump failure the hydraulic accumulator reserves will still provide a substantial number of power assisted braking operations.

Pressure accumulator with flow regulator and cut-out valve unit (Fig. 11.49(a and b)) The accumulator provides a reserve of fluid under pressure if the engine should stall or in the event of a failure of the source of pressure. This enables several brake applications to be made to bring the vehicle safely to a standstill.

The pressure accumulator consists of a spherical container divided in two halves by a rubber diaphragm. The upper half, representing the spring media, is pressurized to 36 bar with nitrogen gas and the lower half is filled with the operating fluid under a pressure of between 36 and 57 bar. When the accumulator is charged with fluid, the diaphragm is pushed back, causing the volume of the nitrogen gas to be reduced and its pressure to rise. When fluid is discharged, the compressed nitrogen gas expands to compensate for these changes and the flexible diaphragm takes up a different position of equilibrium. At all times both gas and fluid pressures are similar and therefore the diaphragm is in a state of equilibrium.

Accumulator being charged (Fig. 11.49(a)) When the accumulator pressure drops to 36 bar, the cut-out spring tension lifts the cut-out plunger against the reduced fluid pressure. Immediately the cut-out ball valve opens and moves from its lower seat to its uppermost position. Fluid from the vane type pump now flows through the cut-out valve, opens the non-return conical valve and permits fluid to pass through to the brake servo unit and to the under side of the accumulator where it starts to compress the nitrogen gas. The store of fluid energy will therefore increase. At the same time, the majority of fluid from the vane type pump flows to the power assisted steering control valve by way of the flutes machined in the flow regulator piston.

Accumulator fully charged (Fig. 11.49(b)) When the accumulator pressure reaches its maximum 57 bar, the cut-out valve ball closes due to the fluid pressure pushing down the cut-out plunger. At the same time, pressurized fluid in the passage between the non-return valve and the rear of the flow regulating piston is able to return to the reservoir via the clearance between the cut-out plunger and guide bore. The non-return valve closes and the fluid pressure behind the flow regulating piston drops. Consequently the fluid supplied from the pump can now force the flow regulator piston further back against the spring so that the total fluid flow passes unrestricted to the power assisted steering control valve.

Hydraulic servo unit (Fig. 11.50(a, b and c)) The hydraulic servo unit consists of a power piston which provides the hydraulic thrust to the master cylinder. A reaction piston interprets the response from the brake pedal input effort and a control tube valve, which actuates the pressurized fluid delivery and release for the servo action.

Brakes released (Fig. 11.50(a)) When the brake pedal is released, the push rod reaction piston and control tube are drawn towards the rear, firstly causing the radial supply holes in the control tube to close and secondly opening the return flow hole situated at the end of the control tube. The pressurized fluid in the operating chamber escapes along the centre of the control tube out to the low pressure chamber via the return flow hole, where it then returns to the fluid reservoir (container). The power piston return spring pushes the power piston back until it reaches the shouldered end stop in the cylinder.

Brakes normally applied (Fig. 11.50(b)) When the brake pedal is depressed, the reaction piston and control tube move inwards, causing the return flow hole to close and partially opening the control tube supply holes. Pressurized fluid from either the accumulator or, when its pressure is low, from the pump, enters the control tube central passage and passes out into the operating chamber. The pressure build-up in the operating chamber forces the power piston to move away from the back end of the cylinder. This movement continues as long as the control tube is being pushed forwards (Fig. 11.50(b)).

Holding the brake pedal in one position prevents the control tube moving further forwards. Consequently the pressure build-up in the operating chamber pushes the power piston out until the radial supply holes in both the power piston and control tube are completely misaligned. Closing the radial supply holes therefore produces a state of balance between the operating chamber fluid thrust and the pressure generated in the tandem master cylinder.

The pressure in the operating chamber is applied against both the power piston and the reaction piston so that a reaction is created opposing the pedal effort in proportion to the amount of power assistance needed at one instance.

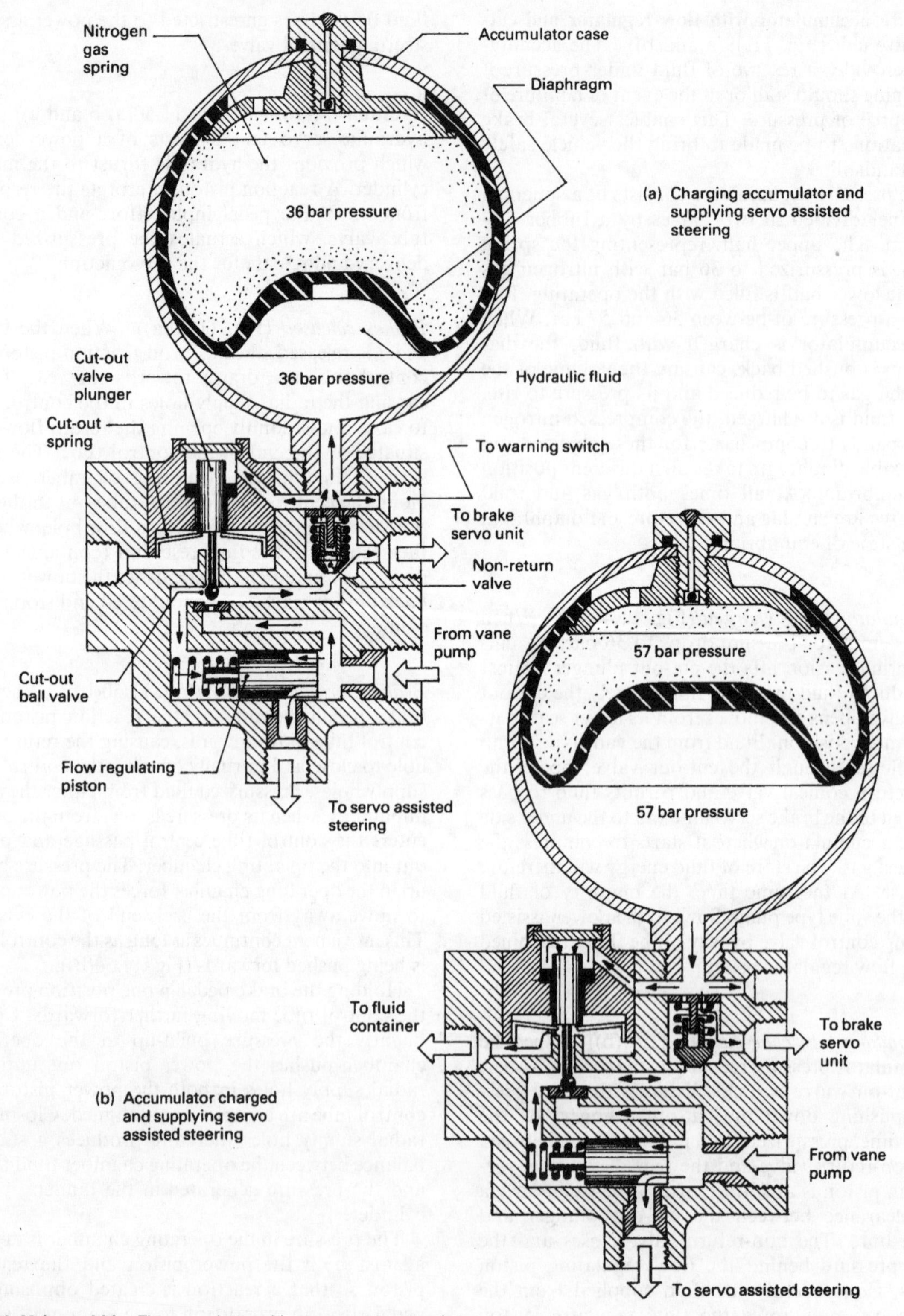

Fig. 11.49 (a and b) Flow regulator with pressure accumulator

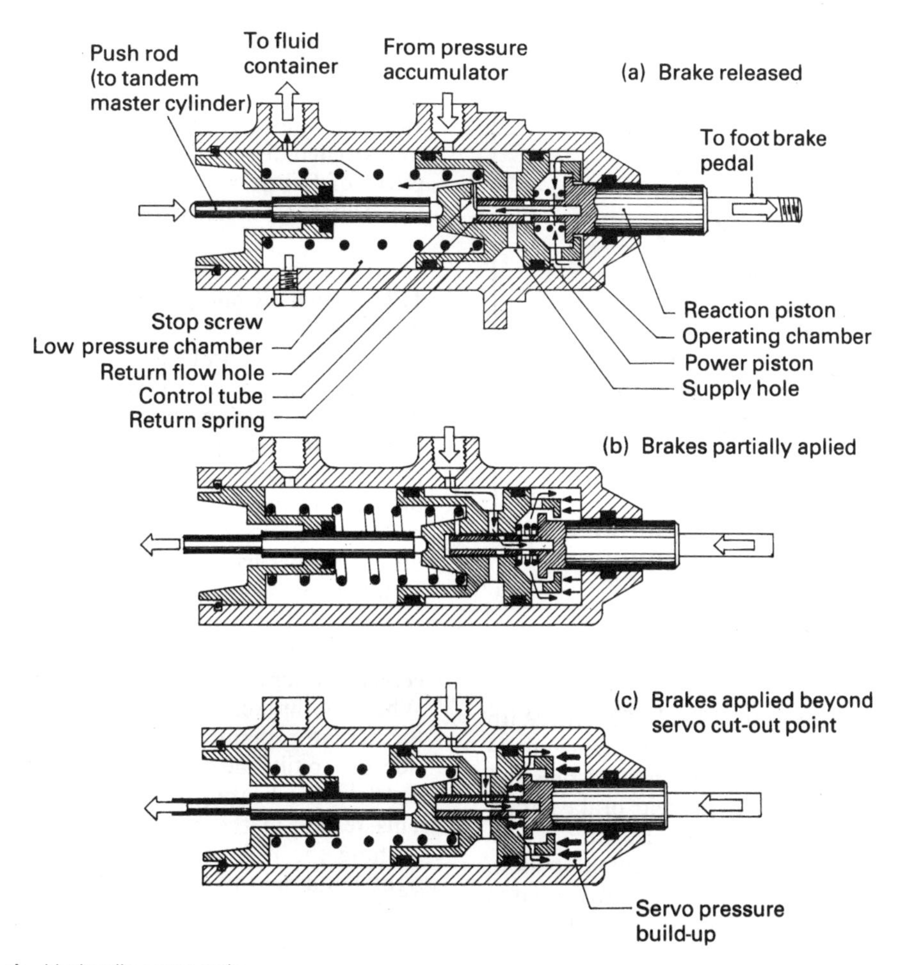

Fig. 11.50 (a–c) Hydraulic servo unit

Braking beyond the cut-out point (Fig. 11.50(c)) When the accumulator cut-out pressure is reached, the control tube touches the power piston, causing the radial supply holes in the control tube to fully align with the power piston. Under these conditions, the accumulator is able to transfer its maximum pressure to the operating chamber. The power piston is therefore delivering its greatest assistance. Any further increase in master cylinder output line pressure is provided by the brake pedal effort alone, as shown in Fig. 11.51, at the minimum and maximum cut-out pressures of 36 and 57 bar respectively.

Rear brake circuit pressure regulator and cut-out device (Fig. 11.52(a, b and c)) The rear brake pressure regulator and cut-off device provide an increasing front to rear line pressure ratio, once the line pressing in the rear pipe line has reached some predetermined minimum value. In other words, the pressure rise in both front and rear pipe lines increases equally up to some pre-set value, but beyond this point, the rear brake pipe line pressure increases at a much reduced rate relative to the front brakes. An additional feature is that if the front brake circuits should develop some fault, then automatically the pressure regulator is bypassed to ensure that full master cylinder fluid pressure is able to operate the rear brakes.

Low brake fluid pressure (Fig. 11.52(a)) When the brakes are lightly applied, the pressure in the front pipe line circuit pushes the cut-off piston over against the opposing spring force. Simultaneously, fluid from the master cylinder enters the inlet port, passes through the open pressure reducing valve, then flows around the wasted cut-off piston on its way out to the rear brake pipe line circuit.

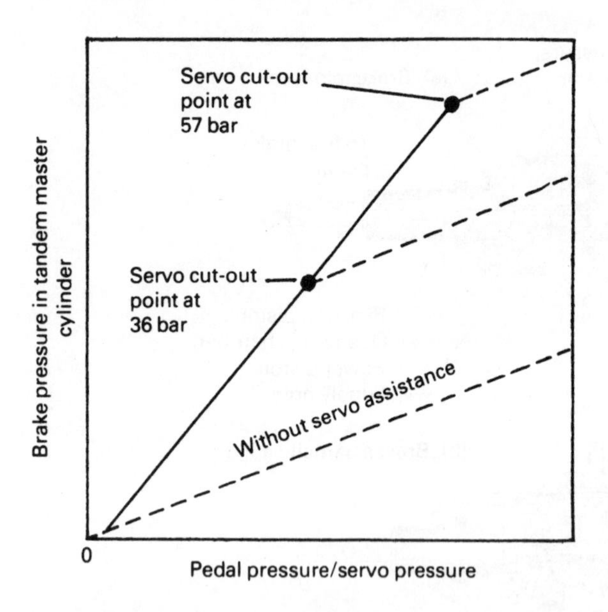

Fig. 11.51 Hydraulic servo action pressure characteristics

High braking fluid pressure (Fig. 11.52(b)) With increased foot pedal effort, the fluid pressure entering the inlet port and passing through the pressure reducing valve, on the way to the rear brake circuit outlet port, rises proportionally. Eventually the resultant force imposed on the stepped piston, caused by the fluid pressure acting on the large and small surface areas of the piston, pushes it outwards against the resistance of the preload spring until the pressure reducing valve closes. Further master cylinder generated pressure acting on the annular face of the stepped piston forces the piston to move in the opposite direction, thereby increasing the rear brake pipe line fluid pressure on the large surface area side of the piston, but at a reduced rate to that of the master cylinder output fluid pressure. The pressure reducing valve is immediately dislodged from its seat. The pressure reducing valve opens and closes repeatedly with rising master cylinder output fluid pressure until the reduced pressure on the large surface area output side of the piston has adjusted itself. These pressure characteristics are shown in Fig. 11.52(d).

Front brake circuit fail condition (Fig. 11.52(c)) If the front brake circuit should fail, the pressure imposed on the cut-off piston collapses, enabling the spring at the opposite end to push over the cut-off piston so that the left hand side of the shuttle valve opens and the right hand side closes. The pressure reducing valve passage to the rear brake line is immediately cut off and the direct passage via the left hand shuttle valve is opened. Pressure from the master cylinder is therefore transmitted unrestricted directly to the rear brake pipe line. The effect of failure in the front brake circuit will be a considerable increase in foot pedal movement.

11.9 Pneumatic operated disc brakes (for trucks and trailers)

Heavy duty disc brake arrangements normally use a floating-caliper design which does not resort to hydraulic actuation, but instead relies on compressed air to supply the power source via a diaphragm operated air chamber actuator. The disc brake unit consists of a rotating disc attached to the road-wheel hub and a floating caliper supported on the caliper carrier which is itself bolted to the stub-axle or casing.

11.9.1 Floating caliper with integral half eccentric lever arm (Fig. 11.53(a and b))

When the brakes are applied air pressure pushes the actuator chamber diaphragm to the left hand side and so tilts the actuator lever about the two half needle roller bearing pivots (Fig. 11.53(a and b)). This results in the eccentric (off-set) bearing pin pushing the right hand friction pad towards the right hand side of the disc via the bridge block, see Fig.11.53(b). Simultaneously as the right hand friction pad bears against the right hand side of the disc, a reaction force now acts on the caliper and is transferred to the opposite friction pad so that both pads squeeze the disc with equal force. Thus the caliper in effect floats; this therefore centralizes the friction pads so that both pads apply equal pressure against their respective faces of the disc. The brake torque produced depends upon the air pressure relayed to the brake actuator chamber, the effective diaphragm area of the chamber and the leverage ratio created by the lever arm 'R' and eccentric off-set 'r', i.e. R/r. When the brakes are released the pull-off spring pushes the bridge block assembly back to the off position, thus producing a running clearance between the pads and disc, see Fig. 11.53(a).

Floating caliper with eccentric shaft and lever (Fig. 11.54(a)) With this type of heavy duty commercial vehicle disc brake a floating caliper is used in conjunction with an eccentric and lever to clamp the pads against the friction faces of the disc. The eccentric part of the eccentric shaft is surrounded

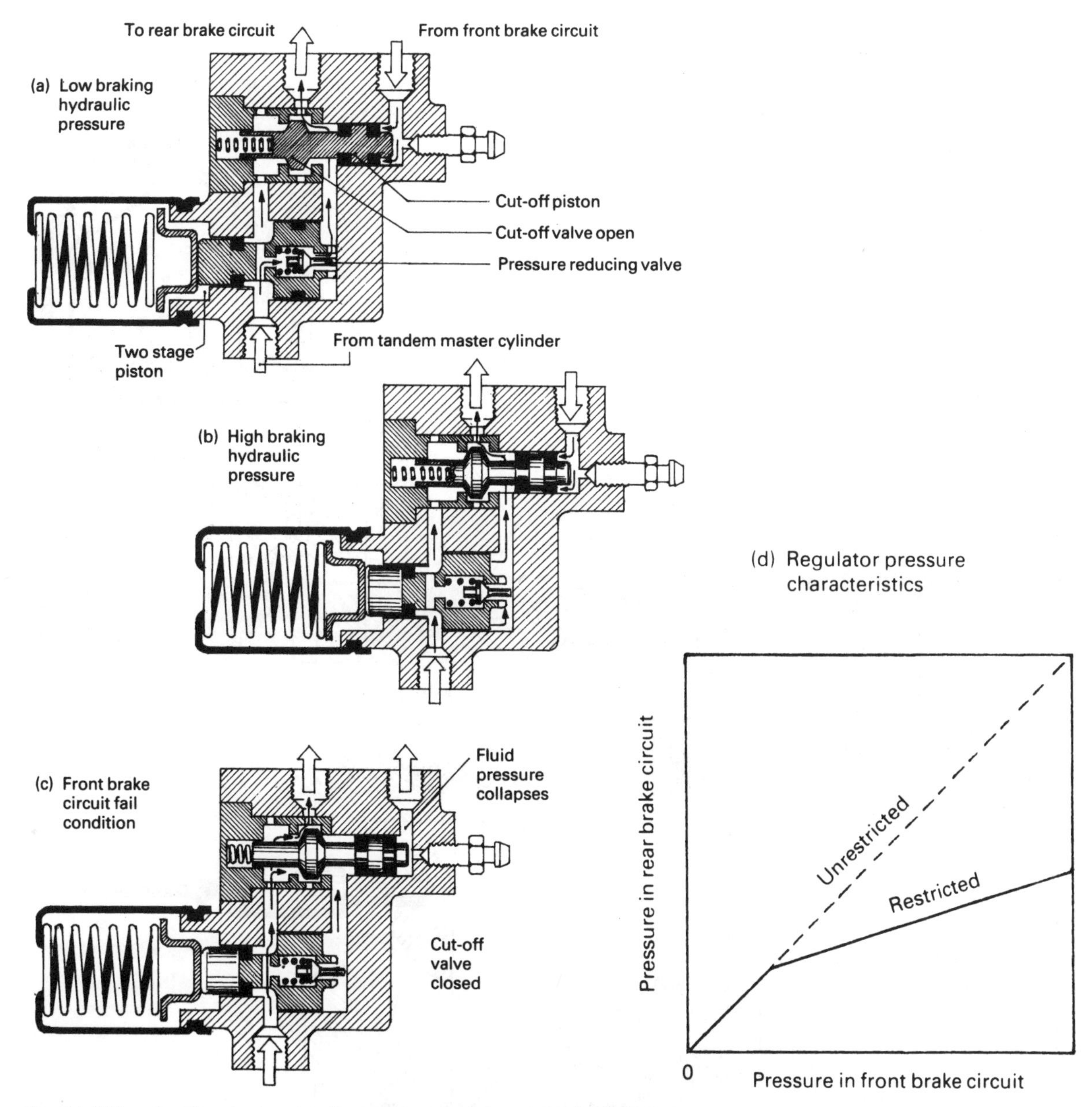

Fig. 11.52 (a–d) Rear brake circuit pressure regulator and cut-off device

by needle rollers positioned inside a bored hole in the bridge block and is connected to the inside pad via two threaded adjustment barrels and a load plate. On either side of the eccentric are stub shafts which are mounted via needle rollers in the caliper.

When the brakes are released the lever arm takes up a position in which the lobe side of the eccentric leans slightly to the right hand side of the vertical (Fig. 11.54(c)). As the brake is applied (Fig. 11.54(b)) the lever arm and eccentric swivels so that the lobe moves to the vertical position or just beyond, hence the bridge block will have moved to the right hand side (towards the disc face) by $x = x_2 - x_1$ where x equals pad take-up clearance. Thus when the brakes are applied compressed air is released into the actuator chamber; this pushes the diaphragm and push rod to the left hand side, causing the lever arm to rotate anticlockwise. As a result the eccentric lobe forces

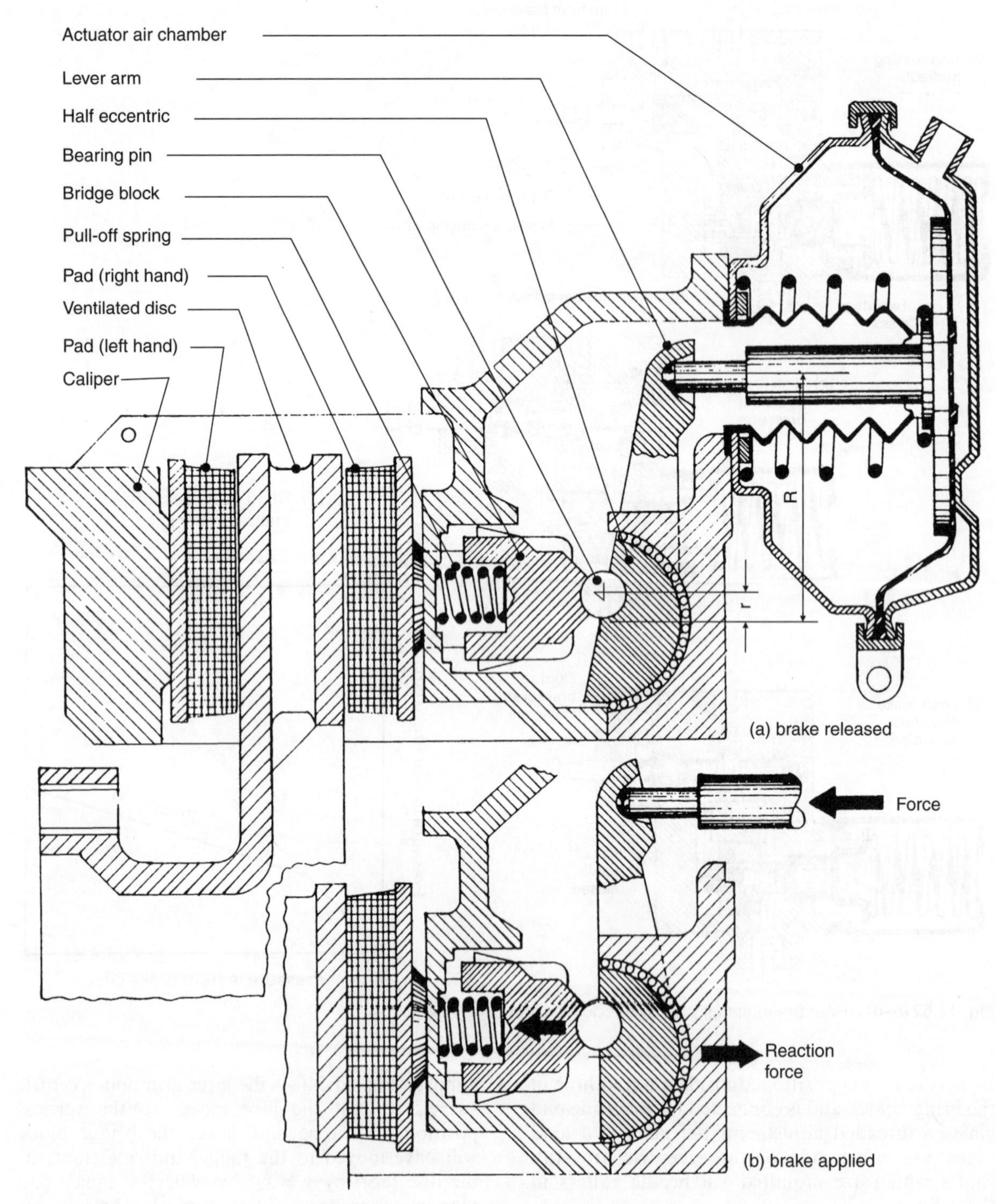

Fig. 11.53 (a and b) Pneumatic operated disc brake – floating caliper with integral half eccentric lever arm

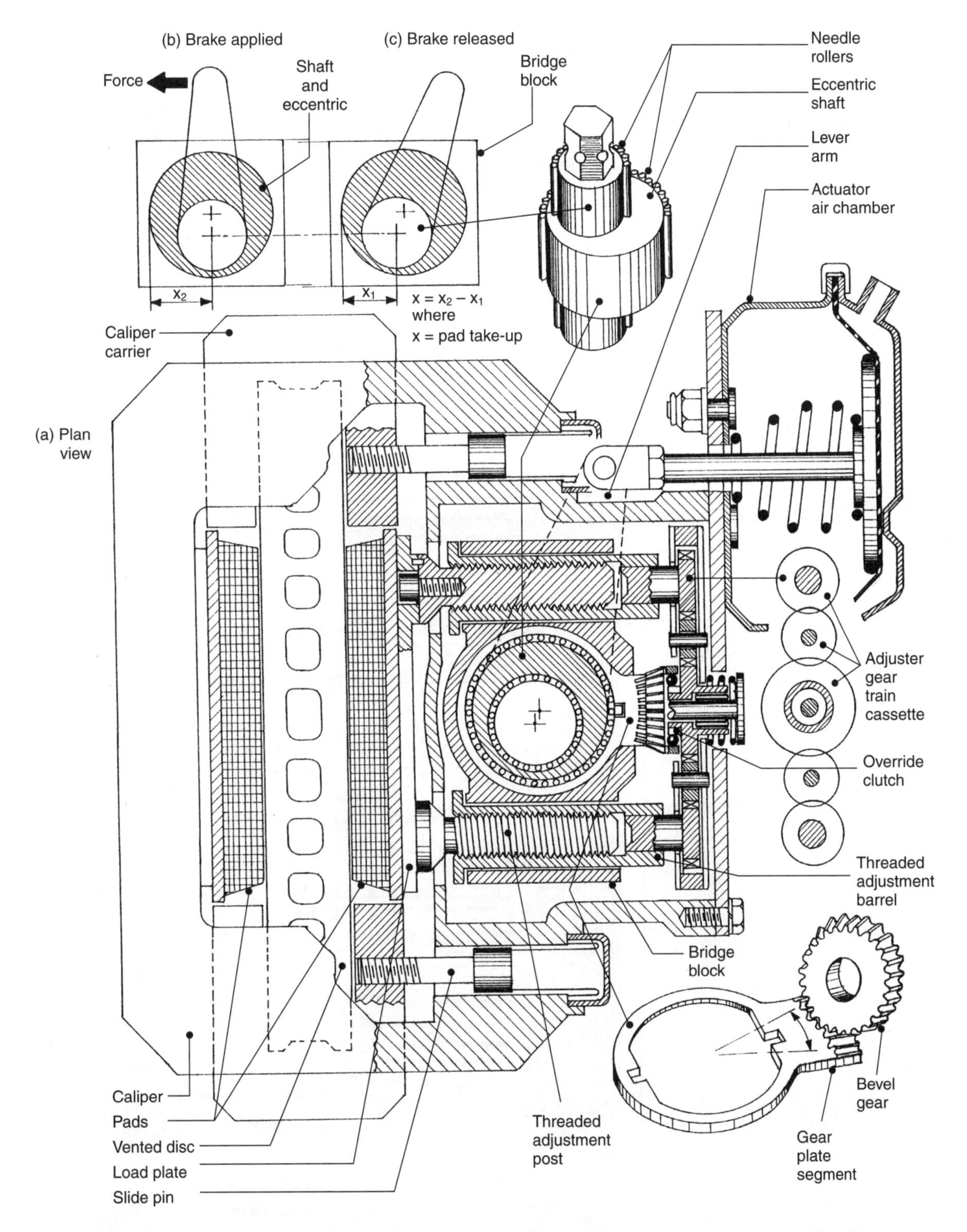

Fig. 11.54 (a–c) Pneumatic operated disc brake – eccentric shaft and lever with gear driven automatic adjustment mechanism

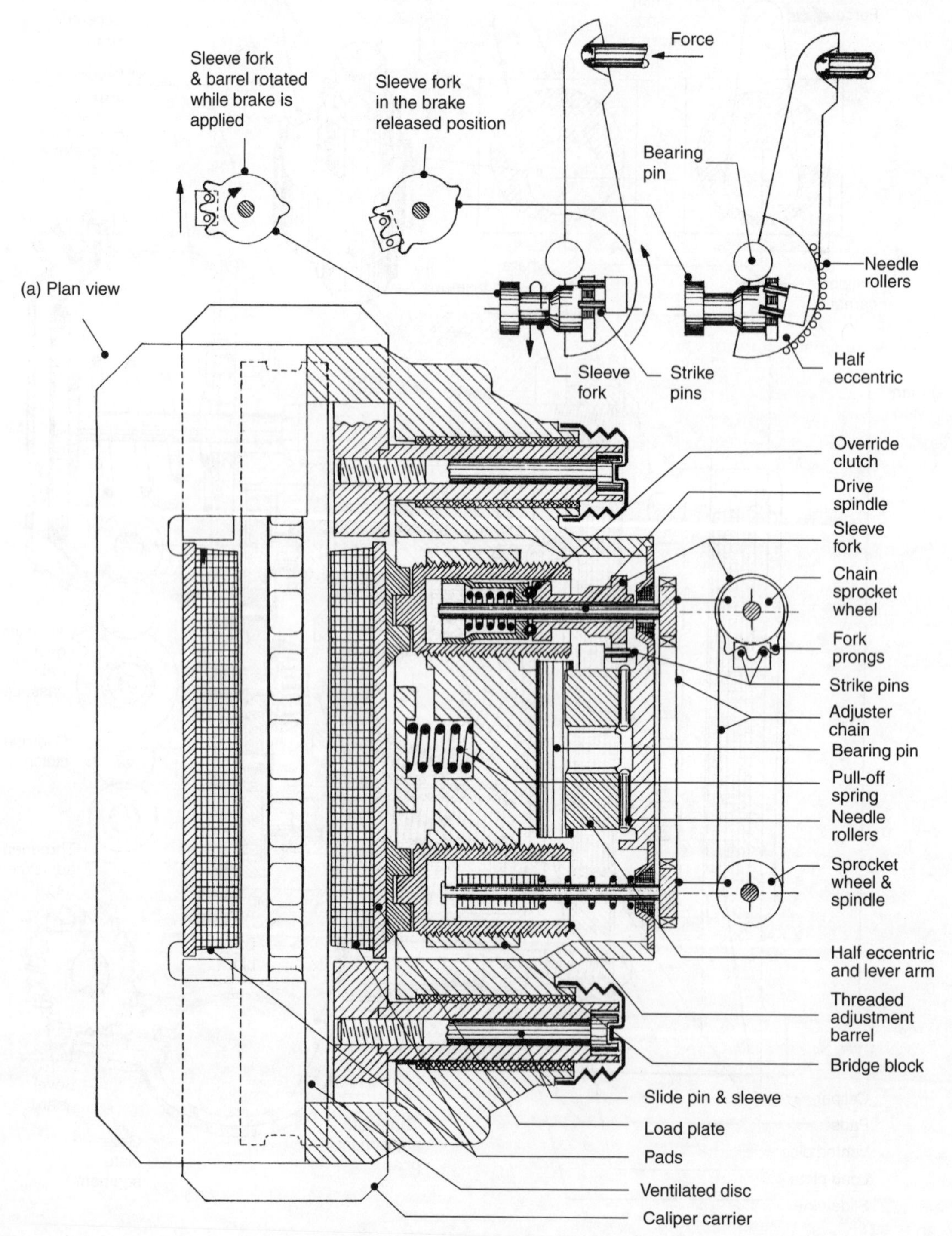

Fig. 11.55 (a–c) Pneumatic operated disc brake – half eccentric shaft and lever with gear driven automatic adjustment mechanism

the bridge block and consequently the inner pad towards the right hand face of the disc. Conversely a reaction force acting though the eccentric stub shaft and caliper pulls the whole caliper, and subsequently the outer pad, towards the left hand face of the disc until the desired amount of friction force is generated between the pads and disc to either slow down or park the vehicle.

Automatic pad clearance gear-driven type adjuster mechanism (Fig. 11.54(a, b and c)) A constant running clearance between the pad and disc is maintained with this mechanism; this device operates by the to and fro movement of the lever arm about the eccentric stub shafts every time the brakes are applied and released (Fig. 11.54(a)). Drawing together of the brake pads is achieved by partial rotation of the eccentric lobe within the bridge block, thus movement is transmitted to the pads via the two threaded adjustment barrels which are screwed either side of the eccentric onto the threaded adjuster posts which are rigidly attached to the inner brake pad load plate.

A gear-plate segment is attached to one side of the eccentric via a slot and tongue. The segment teeth mesh with a bevel gear which houses the override clutch (one-way clutch operating between balls rolling up and down inclined plains), see Fig. 11.54(a). Any partial rotation of the eccentric is transferred to the threaded adjustment barrels via the override clutch and the train of gears.

Thus every time the brake lever arm moves from the released to the applied position, the threaded adjustment barrels are partially screwed out from the threaded adjustment posts, thereby causing the load plate and pad to move further towards the inside face of the disc. Conversely each time the lever arm moves from the applied to the released position, the override one-way clutch disengages, so preventing the threaded adjustment barrels being screwed in again. Eventually after many braking applications, the threaded adjustment barrels will have screwed out the threaded adjustment posts sufficiently to cause the inner pad to touch the inner face of the disc.

Eventually after many braking applications, the threaded adjustment barrels will have screwed the threaded adjustment posts sufficiently out to cause the inner pad to touch the inner face of the disc. At this point, the slight tightening between the male and female threads generates sufficient friction in the screw threads and underneath the flange head of the threaded adjustment barrels to cause the override clutch to slip, hence further rotation of the threaded adjustment barrels ceases. As pad and disc wear occurs, the threaded adjustment barrels once again commence to turn; a constant running clearance is thus maintained during service.

Automatic pad clearance chain-driven type adjuster (Fig. 11.55(a, b and c)) This mechanism maintains a constant running clearance between the pads and disc; the adjuster is operated by the rocking movement of the lever arm each time the brakes are applied and released (Fig. 11.55(a and b)). Braking force is transferred from the lever arm (Fig. 11.55(a)) to the brake pad via the bridge block and the two threaded adjustment barrels which are screwed on either side of the lever arm into the bridge block; the shouldered blind ends of the threaded barrels fit into recesses formed in the load plate. On one side of the lever arm (Fig. 11.55(a, b and c)) are two fork pins which mesh with three prongs formed on the fork sleeve; this sleeve slides over a drive spindle situated inside one of the threaded adjustment barrels. An override clutch is formed at the opposite end to the pronged teeth of the fork sleeve (Fig. 11.55(a)), this consists of a race of balls rolling on ramps (inclined plains) formed by the ball-guide grooves.

Every time the brakes are applied the lever arm tilts causing the meshing strike pins to twist the fork sleeve clockwise and then back to its original position when the brakes are released (Fig. 11.55 (b and c)). Thus the clockwise movement of the sleeve is relayed to the threaded adjustment barrel via the override clutch, but the fork sleeve anticlockwise return movement of the override clutch releases the threaded adjustment barrel, thus the threaded barrel is progressively screwed towards the disc thereby taking up the running clearance caused by pad and disc wear. The running clearance is maintained by the slackness between the strike pin and prong teeth. Even take-up of the running clearance is obtained via the chain sprocket wheels and chain (Fig. 11.55(a)) which transfers the same rotary motion to the second threaded adjustment barrel. Any over-adjustment will cause the override clutch to slip. The sum clearance of both sides of the disc, that is, the total running clearance, should be within 0.6 and 0.9 mm. A larger clearance will cause a take-up clearance time delay whereas a very small clearance may lead to overheating of the discs and pads.

12 Air operated power brake equipment and vehicle retarders

12.1 Introduction to air powered brakes

As the size and weight of road vehicles increase there comes a time when not only are manual brakes inadequate, but there is no point in having power assistance because the amount of braking contributed by the driver's foot is insignificant relative to the principal source of power, be it vacuum or hydraulic energy, and therefore power operated brakes become essential. A further consideration is that the majority of heavy commercial or public service vehicles are propelled by diesel engines which do not have a natural source of vacuum and therefore require a vacuum pump (exhauster) driven from the engine to supply the vacuum energy. However, if a separate pump has to be incorporated to provide the necessary power transmitting media, a third energy source with definite advantages and few disadvantages can be used; that is compressed air.

Reciprocating compressors driven off the engine can operate efficiently and trouble-free at pressures in the region of 7–8 bar, whereas vacuum assisted brakes can only work at the most with depressions of 0.9 bar below atmospheric pressure. Consequently compressed air has a power factor advantage of between 7 and 8 times that for an equivalent vacuum source when used as a force transmitting media.

Conversely, hydraulic pumps are compelled to work at pressures of between 50 and 60 bar. The pressures generated in the pipe lines may reach values of 100 bar or even more. Consequently, because of these high pressures, small diameter servo cylinders and small bore pipes are utilized. This may appear to merit the use of hydraulic energy but, because of the very high working pressure in a hydraulic operated brake system, much more care has to be taken to avoid fluid leakage caused by wear or damage. Compressed air as a power transmitting media would operate at pressures of only one-tenth of an equivalent hydraulic source, but for large vehicles where there is more space, there is no real problem as much larger diameter cylinders can be used. In addition, if there is a leakage fault in a hydraulic layout it will eventually drain the supply fluid so that the brakes cannot continue to function, whereas small leakages of air in an air power operated braking system will not prevent the brakes operating even if this does take place at slightly reduced stopping efficiency.

12.2 Air operated power brake systems

12.2.1 Truck air over hydraulic brake system (Fig. 12.1)

Compressed air supply Air is drawn into the compressor and then discharged into and out of the wet tank where it is semi-dried; it then flows to the multi-circuit protection valve, here it divides to feed the two service reservoirs. At the same time, pressurized air from the reservoirs combine through internal passages in the multi-circuit protection valve to supply the remote spring brake actuator via the hand control valve.

Service line circuit (Fig. 12.1) There are two service lines feeding into a tandem power cylinder controlled by a dual foot valve, so that if a fault develops in one service line the air supply to the other circuit will not be interrupted. The air pressure is then converted to hydraulic pressure by the power piston push rod pushing the tandem master cylinder hydraulic piston forward. The hydraulic fluid supply is split into two circuits feeding the front and rear brake expander cylinders. To balance the proportion of braking provided by the rear axle according to the load carried, a hydraulic load sensing valve is installed on the tandem master cylinder rear axle output circuit. This therefore modifies the fluid pressure reaching the rear brake cylinders.

Secondary line circuit (Fig. 12.1) With the dual air and hydraulic lines, both systems operate independently and therefore provide a safeguard against failure of one or the other circuit. Thus the hand control valve is used only to park the vehicle.

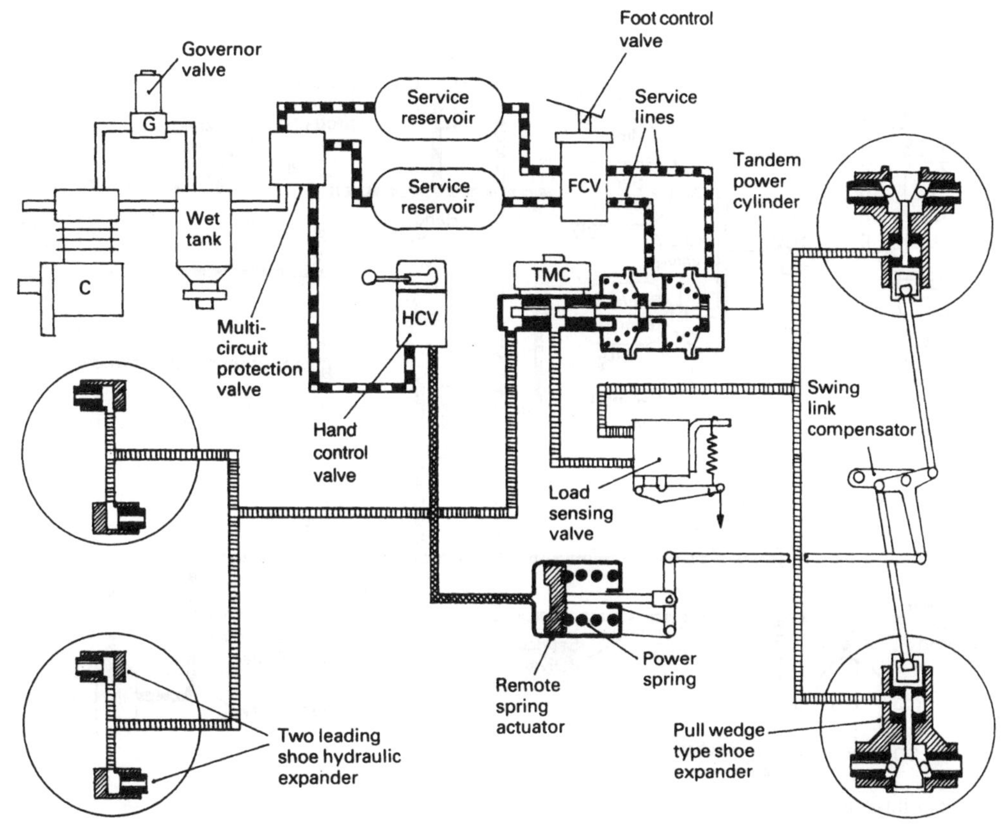

Fig. 12.1 Truck air over hydraulic system

Moving the hand valve lever from 'off' to 'park' position exhausts air from the remote spring actuator chamber. This permits the power spring within the actuator to expand and exert its full pull to the mechanical parking brake rod linkage.

12.2.2 Tractor three line brake system (Fig. 12.2)

Compressed air supply (Fig. 12.2) The compressor in this arrangement is controlled by a separate unloader valve. An alcohol evaporator is installed in the air intake, so that in cold weather alcohol can be introduced into the airstream to lower the freezing point of any water which may be present. When the compressor is running light, a check valve built into the evaporator prevents alcohol entering the air intake. Pressurized air from the compressor is then delivered to both the service and secondary park reservoirs via the check valves on the inlet side of each reservoir.

Service line circuit (Fig. 12.2) When the foot pedal is depressed, air from the service reservoir is permitted to flow directly to the tractor's front and rear service line chambers in each of the double diaphragm actuators which are mounted on the tractor axles. At the same time, a pressure signal is passed to the relay valve piston. This opens the valve so that the service storage line pressure flows from the service reservoir to the service line coupling (yellow) via the pressure protection valve. The pressure protection valve in the service storage (emergency) line and the relay valve in the service line safeguard the tractor's air supply, should a large air leak develop in the flexible tractor to trailer coupling hose or if any other fault causes a loss of air pressure.

Secondary line circuit (Fig. 12.2) Applying the hand control valve lever delivers a controlled air pressure from the secondary park reservoir to the front wheel secondary chambers, which form part of the double diaphragm actuators, and to the secondary line (red) coupling, which then delivers pressurized air to the trailer brakes via a flexible hose. Note that there is no secondary braking to

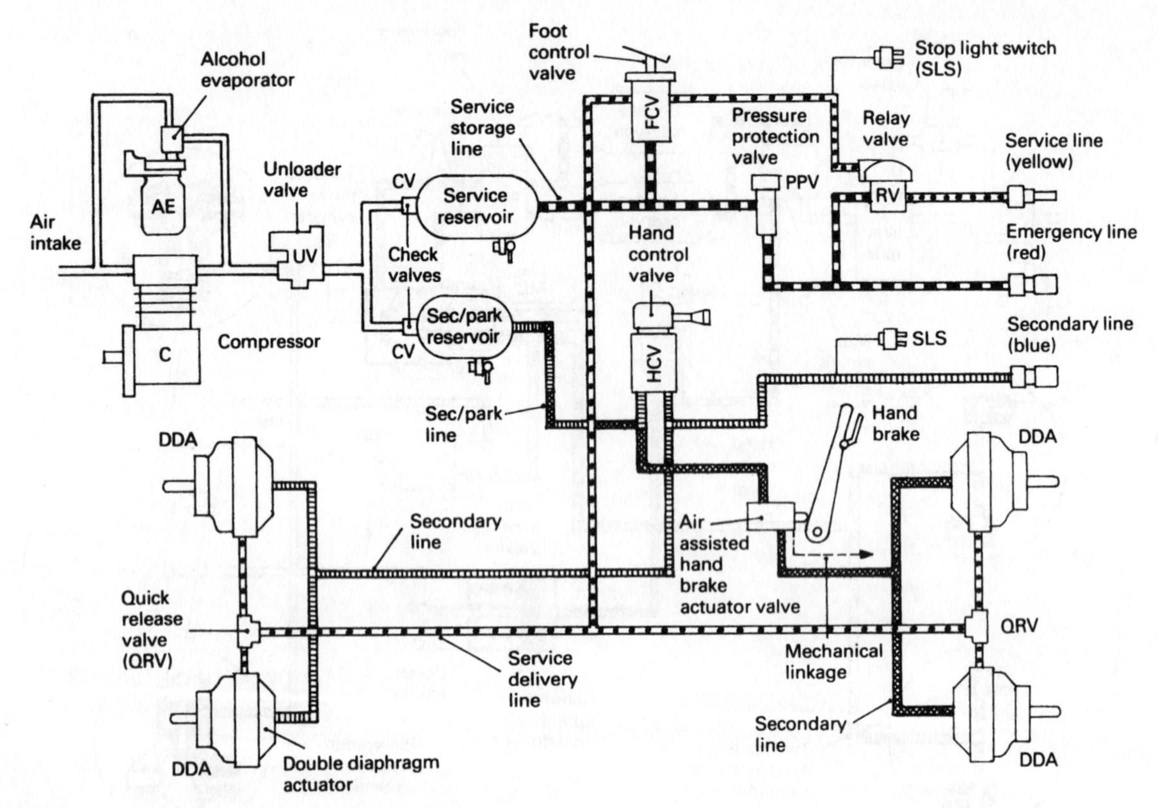

Fig. 12.2 Tractor three line brake system

the tractor's rear axle to reduce the risk of jack-knifing during an emergency application.

Parking circuit (Fig. 12.2) Applying the hand brake lever opens the hand brake valve so that pressurized air flows to the rear axle parking line chambers within the double diaphragm actuators to apply the brakes. At the same time, the mechanical parking linkage locks the brake shoes in the applied position and then releases the air from the parking actuator chambers. This parking brake is therefore mechanical with air assistance.

12.2.3 Trailer three line brake system (Fig. 12.3) All trailer air braking systems have their own reservoir which is supplied through the emergency line from the tractor's service reservoir.

Service line circuit (Fig. 12.3) When applying the brakes, air pressure from the tractor's relay valve signals the emergency relay valve to open and supply air pressure from the trailer's own reservoir to the trailer's service line brake actuator chambers relative to that applied to the tractor brakes. The

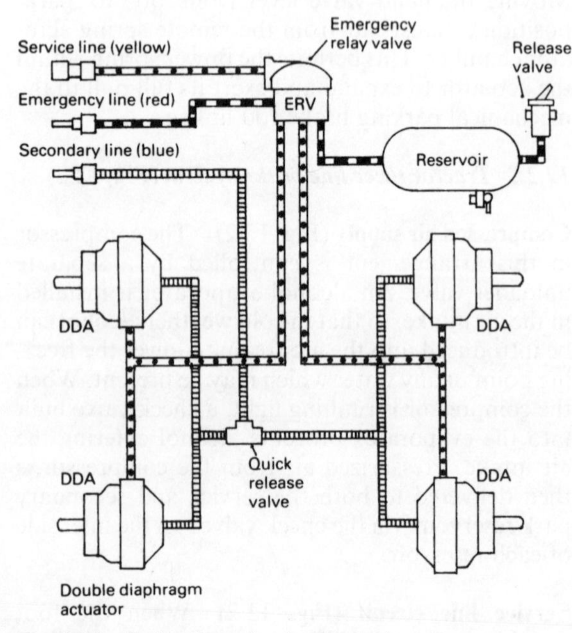

Fig. 12.3 Trailer three line brake system

object of the separate reservoir and relay valve installed on the trailer is to speed up the application and release of the trailer brakes, which are at some distance from the driver's foot control valve. Should there be a reduction in emergency line pressure below some predetermined minimum, the emergency relay valve will sense this condition and will automatically apply the trailer service brakes.

Secondary line circuit (Fig. 12.3) The secondary braking system of the trailer is controlled by the hand control valve mounted in front of the driver. Moving the hand control valve lever towards the applied position delivers a graduable air pressure via the secondary lines to the secondary chamber within each double diaphragm actuator. A quick release valve incorporated at the junction between the trailer's front and rear brakes speeds up the exhausting of the secondary chambers and, therefore, the release of the secondary brakes.

To release the trailer brakes when the trailer is detached from the tractor caused by the exhausting of the emergency line, a reservoir release valve is provided which should be moved to the 'open' piston to release the trailer brakes.

12.2.4 Towing truck or tractor spring brake three line system (Fig. 12.4)

Compressed air supply (Fig. 12.4) Air pressure is supplied by a compressor driven off the engine. Built into the compressor head is an unloaded mechanism which is controlled by a governor valve and which senses pressure change through the wet tank. Installed on the intake side of the compressor is an alcohol evaporator which feeds in very small quantities of alcohol spray when the compressor is pumping. As a result, it lowers the freezing temperature of the wet air induced into the compressor cylinder. When the compressor is running light, a check valve prevents alcohol spray entering the airstream, thereby reducing the alcohol consumption. The compressor supplies pressurized air to both service and secondary/park reservoirs via non-return check valves.

Service line circuit (Fig. 12.4) When the driver depresses the dual foot valve, air flows from the service reservoir through the service delivery line (yellow) directly to the front wheel service line actuator chamber, and indirectly via a variable load valve which regulates the air pressure,

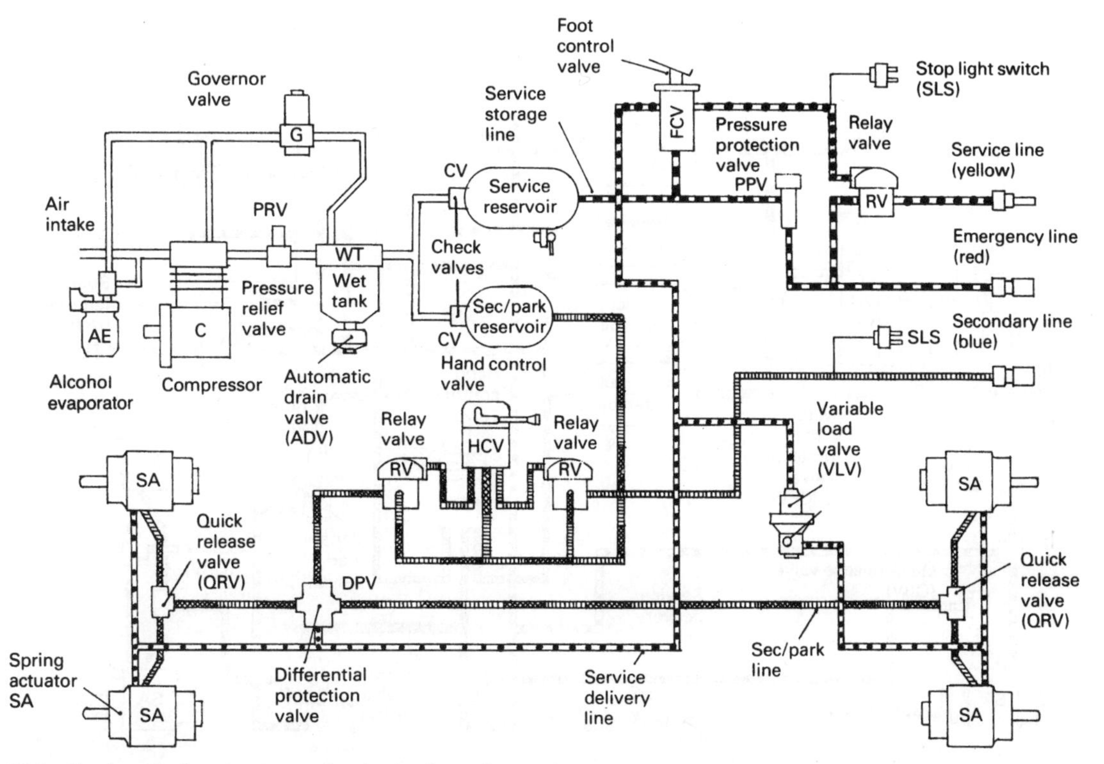

Fig. 12.4 Towing truck or tractor spring brake three line system

according to the loading imposed on the rear axle, to the rear wheel service chamber actuators. Compressed air is also delivered to both the service and the emergency line couplings via the relay valve and the pressure protection valve. This therefore safeguards the tractor air supply should there be a hose failure between the tractor and trailer. A differential protection valve is installed between the service line and the secondary/park line to prevent both systems operating simultaneously which would overload the foundation brakes.

Secondary/park line circuit (Fig. 12.4) Air is supplied from the secondary/park reservoir to the hand control valve and to a pair of relay valves. One relay valve controls the air delivered to the spring brake actuator, the other controls the service line air supply to the trailer brakes. With the hand control valve in the 'off' position, air is delivered through the secondary/park line relay valve to the spring brakes. The secondary/park spring brakes are held in the released position due to the compression of each power spring within the actuator. As the spring brakes are being released, the secondary line to the trailer is exhausted of compressed air via its relay valve. Moving the hand control valve lever to the 'on' position progressively reduces the secondary/park line pressure going to the spring brake. The secondary line pressure going to the trailer coupling increases, thereby providing a tractor to trailer brake match. Moving the hand control valve to the 'park' position exhausts the air from the trailer secondary line and the spring brake secondary/park line. The tractor foundation brakes are then applied by the thrust exerted by the power spring within the actuator alone. The release of the parking brake is achieved by delivering air to the spring brake when the hand control valve is moved to the 'off' position again.

12.2.5 Towing truck or tractor spring brake two line system (Fig. 12.5)

Compressed air supply (Fig. 12.5) The air supply from the compressor passes through the air dryer on its way to the multi-circuit protection. The output air supply is then shared between four reservoirs; two service, one trailer and one secondary/park reservoirs.

Service line circuit (Fig. 12.5) The air delivered to the service line wheel actuator chambers is

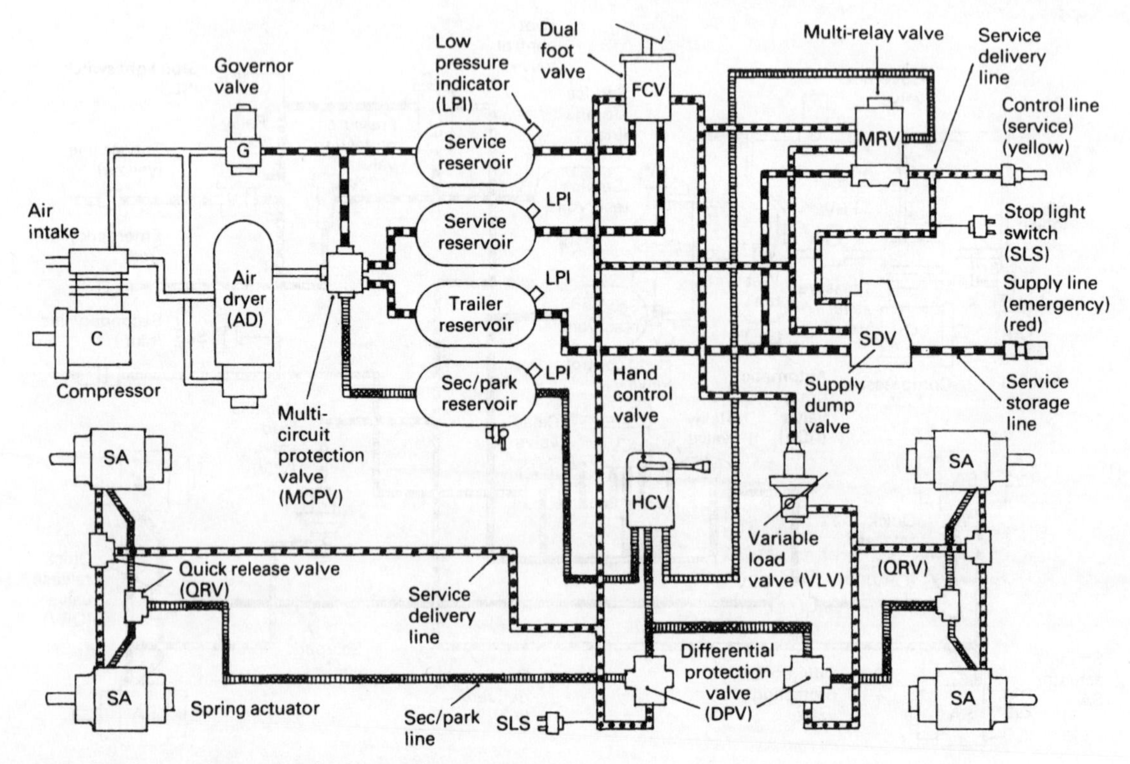

Fig. 12.5 Towing truck or spring brake two line system

provided by a dual foot valve which splits the service line circuits between the tractor's front and rear wheels. Therefore, if one or other service line circuit should develop a fault, the other circuit with its own reservoir will still function. At the same time as the tractor service brakes are applied, a signal pressure from the foot valve passes to the multi-relay valve. This opens an inlet valve which permits air from the trailer reservoir to flow to the control line (service line — yellow) trailer coupling.

To prevent both service line and secondary/park line supplies compounding, that is, operating at the same time, and overloading the foundation brakes, a differential protection valve is included for both the front and rear axle brakes.

Secondary/park line circuit (Fig. 12.5) A secondary braking system which incorporates a parking brake is provided by spring brakes which are installed on both front and rear axles. Control of the spring brakes is through a hand valve which provides an inverse signal to the multi-relay valve so that the trailer brakes can also be applied by the hand control valve.

With the hand control valve in the 'off' position the secondary line from the hand valve to the multi-relay valve, and the secondary/park line, also from the hand valve, going to the spring brake actuators via the differential protection valves, are both pressurized. This compresses the power springs, thereby releasing the spring brakes. During this period no secondary line pressure signal is passed to the trailer brakes via the multi-relay valve.

When the hand valve is moved towards the 'applied' position, the secondary line feeding the multi-relay valve and the secondary/park line going to the spring brakes reduces their pressures so that both the tractor's spring brakes and the trailer brakes are applied together in the required tractor to trailer proportions.

Moving the hand valve lever to the 'park' position exhausts the secondary/park line going to the spring brakes and pressurizes the secondary line going to the multi-relay valve. As a result, the power springs within the spring actuators exert their full thrust against the foundation brake cam lever and at the same time the trailer control line (service line) is exhausted of compressed air. Thus the vehicle is held stationary solely by the spring brakes.

Multi-relay valve (Fig. 12.25(a–d)) The purpose of the multi-relay valve is to enable each of the two service line circuits to operate independently should one malfunction, so that trailer braking is still provided. The multi-relay valve also enables the hand control valve to operate the trailer brakes so that the valve is designed to cope with three separate signals; the two service line pressure signals controlled by the dual foot valve and the hand valve secondary pressure signal.

Supply dump valve (Fig. 12.26(a, b and c)) The purpose of the supply dump valve is to automatically reduce the trailer emergency line pressure to 1.5 bar should the trailer service brake line fail after the next full service brake application within two seconds. This collapse of emergency line pressure signals to the trailer emergency valve to apply the trailer brakes from the trailer reservoir air supply, overriding the driver's response.

12.2.6 Trailer two line brake system (Fig. 12.6)
The difference with the two and three line trailer braking systems is that the two line only has a single control service line, whereas the three line has both a service line and a secondary line.

Control (service) line circuit (Fig. 12.6) On making a brake application, a pressure signal from the tractor control (service) line actuates the relay

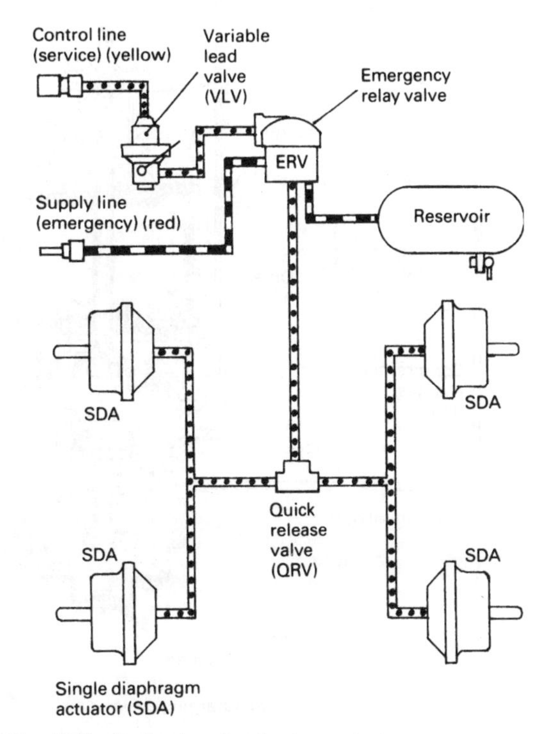

Fig. 12.6 Trailer two line brake system

portion of the emergency relay valve to deliver air pressure from the trailer reservoir to each of the single diaphragm actuator chambers. In order to provide the appropriate braking power according to the trailer payload, a variable load sensing valve is installed in the control line ahead of the emergency relay valve. This valve modifies the control line signal pressure so that the emergency relay valve only supplies the brake actuators with sufficient air pressure to retard the vehicle but not to lock the wheels. A quick-release valve may be included in the brake actuator feed line to speed up the emptying of the actuator chambers to release the brakes but usually the emergency relay valve exhaust valve provides this function adequately. If the supply (emergency) line pressure drops below a predetermined value, then the emergency portion of the emergency relay valve automatically passes air from the trailer reservoir to the brake actuators to stop the vehicle.

12.3 Air operated power brake equipment

12.3.1 Air dryer (Bendix) (Fig. 12.7(a and b))
Generally, atmospheric air contains water vapour which will precipitate if the temperature falls low enough. The amount of water vapour content of the air is measured in terms of relative humidity. A relative humidity of 100% implies that the air is saturated so that there will be a tendency for the air to condensate. The air temperature and pressure

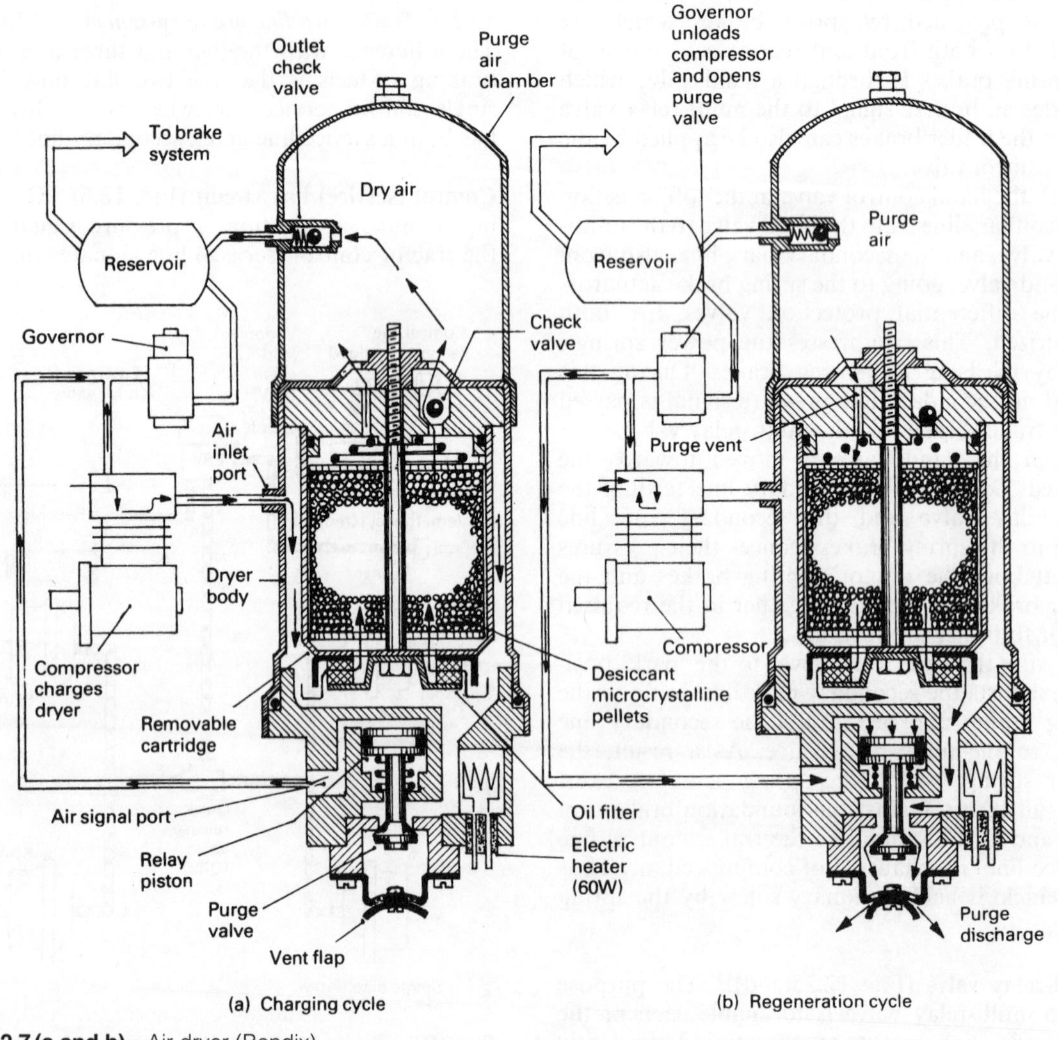

Fig. 12.7 (a and b) Air dryer (Bendix)

determines the proportion of water vapour retained in the air and the amount which condenses.

If the saturation of air at atmospheric pressure occurs when the relative humidity is 100% and the output air pressure from the compressor is 8 bar, that is eight times atmospheric pressure (a typical working pressure), then the compressed air will have a much lower saturation relative humidity equal to $\frac{100}{8} = 12.5\%$.

Comparing this 12.5% saturation relative humidity, when the air has been compressed, to the normal midday humidity, which can range from 60% in the summer to over 90% in the winter, it can be seen that the air will be in a state of permanent saturation.

However, the increase in air temperature which will take place when the air pressure rises will raise the relative humidity somewhat before the air actually becomes saturated, but not sufficiently to counteract the lowering of the saturation relative humidity when air is compressed.

The compressed air output from the compressor will always be saturated with water vapour. A safeguard against water condensate damaging the air brake equipment is obtained by installing an air dryer between the compressor and the first reservoir.

The air dryer unit cools, filters and dries all the air supplied to the braking system. The drying process takes place inside a desiccant cartridge which consists of many thousands of small microcrystalline pellets. The water vapour is collected in the pores of these pellets. This process is known as *absorption*. There is no chemical change as the pellets absorb and release water so that, provided that the pores do not become clogged with oil or other foreign matter, the pellets have an unlimited life. The total surface area of the pellets is about 464 000 m^2. This is because each pellet has many minute pores which considerably increase the total surface area of these pellets.

Dry, clean air is advantageous because:

1 the absence of moisture prevents any lubricant in the air valves and actuators from being washed away,
2 the absence of moisture reduces the risk of the brake system freezing,
3 the absence of oil vapour in the airstream caused by the compressor's pumping action extends the life of components such as rubber diaphragms, hoses and 'O' rings,
4 the absence of water and oil vapour prevents sludge forming and material accumulating in the pipe line and restricting the air flow.

Charge cycle (Fig. 12.7(a)) Air from the compressor is pumped to the air dryer inlet port where it flows downwards between the dryer body and the cartridge wall containing the desiccant. This cools the widely but thinly spread air, causing it to condense onto the steel walls and drip to the bottom of the dryer as a mixture of water and oil (lubricating oil from the compressor cylinder walls). Any carbon and foreign matter will also settle out in this phase. The cooled air charge now changes its direction and rises, passing through the oil filter and leaving behind most of the water droplets and oil which were still suspended in the air. Any carbon and dirt which has remained with the air is now separated by the filter.

The air will now pass through the desiccant so that any water vapour present in the air is progressively absorbed into the microcrystalline pellet matrix. The dried air then flows up through both the check valve and purge vent into the purge air chamber. The dryness of the air at this stage will permit the air to be cooled at least 17°C before any more condensation is produced. Finally the air now filling the purge chamber passes out to the check valve and outlet port on its way to the brake system's reservoirs.

Regeneration cycle (Fig. 12.7(b)) Eventually the accumulated moisture will saturate the desiccant, rendering it useless unless the microcrystalline pellets are dried. Therefore, to enable the pellets to be continuously regenerated, a reverse flow of dry air from the purge air chamber is made to occur periodically by the cut-out and cut-in pressure cycle provided by the governor action.

When the reservoir air pressure reaches the maximum cut-out pressure, the governor inlet valve opens, allowing pressurized air to be transferred to the unloader plunger in the compressor cylinder head. At the same time, this pressure signal is transmitted to the purge valve relay piston which immediately opens the purge valve. The accumulated condensation and dirt in the base of the dryer is then rapidly expelled due to the existing air pressure in the lower part of the dryer. The sudden drop in air pressure in the desiccant cartridge chamber allows the upper purge chamber to discharge dry air back through the purge vent into the desiccant cartridge, downwards through the oil filter, finally escaping through the open purge valve into the atmosphere.

During the reverse air flow process, the expanding dry air moves through the desiccant and effectively absorbs the moisture from the crystals on its way out into the atmosphere. Once the dryer has been purged of condensation and moisture, the purge valve will remain open until the cylinder head unloader air circuit is permitted to exhaust and the compressor begins to recharge the reservoir. At this point the trapped air above the purge relay piston also exhausts, allowing the purge valve to close. Thus with the continuous rise and fall of air pressure the charge and regeneration cycles will be similarly repeated.

A 60 W electric heater is installed in the base of the dryer to prevent the condensation freezing during cold weather.

12.3.2 Reciprocating air compressors

The source of air pressure energy for an air brake system is provided by a reciprocating compressor driven by the engine by either belt, gear or shaft-drive at half engine speed. The compressor is usually base- or flange-mounted to the engine.

To prevent an excessively high air working temperature, the cast iron cylinder barrel is normally air cooled via the enlarged external surface area provided by the integrally cast fins surrounding the upper region of the cylinder barrel. For low to moderate duty, the cylinder head may also be air cooled, but for moderate to heavy-duty high speed applications, liquid coolant is circulated through the internal passages cast in the aluminium alloy cylinder head. The heat absorbed by the coolant is then dissipated via a hose to the engine's own cooling system. The air delivery temperature should not exceed 220°C.

Lubrication of the crankshaft plain main and big-end bearings is through drillings in the crankshaft, the pressurized oil supply being provided by the engine's lubrication system, whereas the piston and rings and other internal surfaces are lubricated by splash and oil mist. Surplus oil is permitted to drain via the compressor's crankcase back to the engine's sump. The total cylinder swept volume capacity needed for an air brake system with possibly auxiliary equipment for light, medium and heavy commercial vehicles ranges from about 150 cm^3 to 500 cm^3, which is provided by either single or twin cylinder reciprocating compressor. The maximum crankshaft speed of these compressors is anything from 1500 to 3000 rev/min depending upon maximum delivery air pressure and application. The maximum air pressure a compressor can discharge continuously varies from 7 to 11 bar. A more typical maximum pressure value would be 9 bar.

The quantity of air which can be delivered at maximum speed by these compressors ranges from 150 L/min to 500 L/min for a small to large size compressor. This corresponds to a power loss of something like 1.5 kW to 6 kW respectively.

Compressor operation When the crankshaft rotates, the piston is displaced up and down causing air to be drawn through the inlet port into the cylinder on the down stroke and the same air to be pushed out on the upward stroke through the delivery port. The unidirectional flow of the air supply is provided by the inlet and delivery valves. The suction and delivery action of the compressor may be controlled by either spring loaded disc valves (Fig. 12.9) or leaf spring (reed) valves (Fig. 12.8). For high speed compressors the reed type valve arrangements tend to be more efficient.

On the downward piston stroke the delivery valve leaf flattens and closes, thus preventing the discharged air flow reversing back into the cylinder (Fig. 12.8). At the same time the inlet valve is drawn away from its seat so that fresh air flows through the valve passage in its endeavour to fill the expanding cylinder space.

On the upward piston stroke the inlet valve leaf is pushed up against the inlet passage exit closing the valve. Consequently the trapped pressurized air is forced to open the delivery valve so that the air charge is expelled through the delivery port to the reservoir.

The sequence of events is continuous with a corresponding increase in the quantity of air delivered and the pressure generated.

The working pressure range of a compressor may be regulated by either an air delivery line mounted unloader valve (Figs 12.10 and 12.11) or an integral compressor unloader mechanism controlled by an external governor valve (Fig. 12.9). A further feature which is offered for some applications is a multiplate clutch drive which reduces pumping and frictional losses when the compressor is running light (Fig. 12.8).

Clutch operation (Fig. 12.8) With the combined clutch drive compressor unit, the compressor's crankshaft can be disconnected from the engine drive once the primary reservoir has reached its maximum working pressure and the compressor is running light to reduce the wear of the rotary bearings and reciprocating piston and rings and to eliminate the power consumed in driving the compressor.

The clutch operates by compressed air and is automatically controlled by a governor valve similar to that shown in Fig. 12.9.

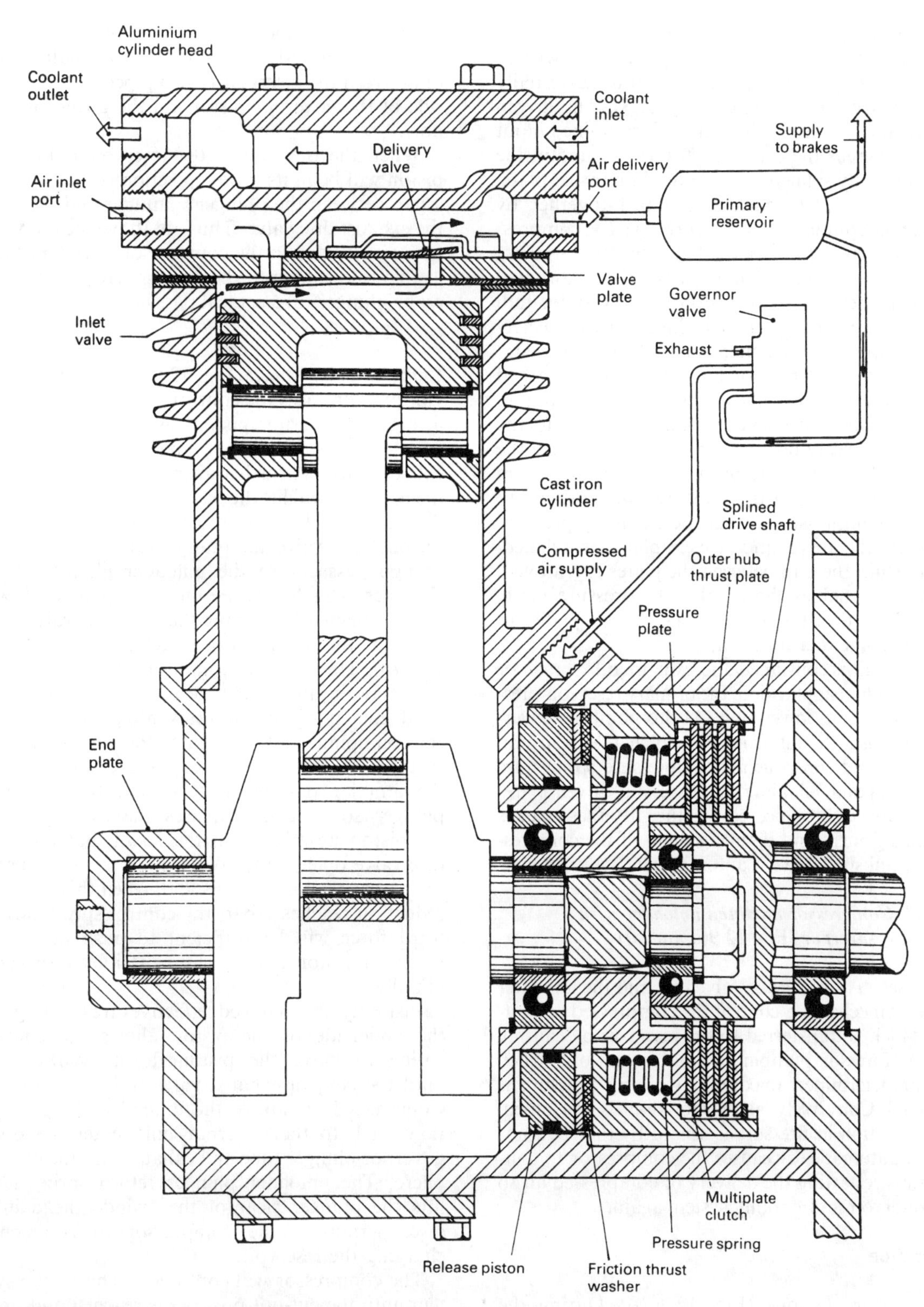

Fig. 12.8 Single cylinder air compressor with clutch drive

The multiplate clutch consists of four internally splined sintered bronze drive plates sandwiched between a pressure plate and four externally splined steel driven plates (Fig. 12.8). The driven plates fit over the enlarged end of the splined input shaft, whereas the driven plates are located inside the internally splined clutch outer hub thrust plate. The friction plate pack is clamped together by twelve circumferentially evenly spaced compression springs which react between the pressure plate and the outer hub thrust plate. Situated between the air release piston and the outer hub thrust plate are a pair of friction thrust washers which slip when the clutch is initially disengaged.

When the compressor air delivery has charged the primary reservoir to its preset maximum, the governor valve sends a pressure signal to the clutch air release piston chamber. Immediately the friction thrust washers push the clutch outer hub thrust plate outwards, causing the springs to become compressed so that the clamping pressure between the drive and driven plates is relaxed. As a result, the grip between the plates is removed. This then enables the crankshaft, pressure plate, outer hub thrust plate and the driven plates to rapidly come to a standstill.

As the air is consumed and exhausted by brake or air equipment application, the primary reservoir pressure drops to its lower limit. At this point the governor exhausts the air from the clutch release piston chamber and consequently the pressure springs are free to expand, enabling the drive and driven plates once again to be squeezed together. By these means the engagement and disengagement of the compressor's crankshaft drive is automatically achieved.

12.3.3 Compressor mounted unloader with separate governor (Fig. 12.9(a and b))

Purpose The governor valve unit and the unloader plunger mechanism control the compressed air output which is transferred to the reservoir by causing the compressor pumping action to 'cut-out' when the predetermined maximum working pressure is attained. Conversely, as the stored air is consumed, the reduction in pressure is sensed by the governor which automatically causes the compressor to 'cut-in', thus restarting the delivery of compressed air to the reservoir and braking system again.

Operation

Compressor charging (Fig. 12.9(a)) During the charging phase, air from the compressor enters the reservoir, builds up pressure and then passes to the braking system (Fig. 12.9(a)). A small sample of air from the reservoir is also piped to the underside of the governor piston via the governor inlet port.

When the pressure in the reservoir is low, the piston will be in its lowest position so that there is a gap between the plunger's annular end face and the exhaust disc valve. Thus air above the unloader plunger situated in the compressor's cylinder head is able to escape into the atmosphere via the governor plunger tube central passage.

Compressor unloaded (Fig. 12.9(b)) As the reservoir pressure rises the control spring is compressed lifting the governor piston until the exhaust disc valve contacts the plunger tube, thereby closing the exhaust valve. A further air pressure increase from the reservoir will lift the piston seat clear of the inlet disc valve. Air from the reservoir now flows around the inlet disc valve and plunger tube. It then passes though passages to the unloader plunger upper chamber. This forces the unloader plunger down, thus permanently opening the inlet disc valve situated in the compressor's cylinder head (Fig. 12.9(b)). Under these conditions the compressor will draw in and discharge air from the cylinder head inlet port, thereby preventing the compressor pumping and charging the reservoir any further. At the same time, air pressure acts on the annular passage area around the governor plunger stem. This increases the force pushing the piston upwards with the result that the inlet disc valve opens fully. When the brakes are used, the reservoir pressure falls and, when this pressure reduction reaches 1 bar, the control spring downward force will be sufficient to push down the governor piston and to close the inlet disc valve initially.

Instantly the reduced effective area acting on the underside of the piston allows the control spring to move the piston down even further until the control exhaust valve (tube/disc) opens. Compressed air above the unloader plunger will flow back to the governor unit, enter the open governor plunger tube and exhaust into the atmosphere. The unloader plunger return spring now lifts the plunger clear of the cylinder head inlet disc, permitting the compressor to commence charging the reservoir.

The compressor will continue to charge the system until the cut-out pressure is reached and once again the cycle will be repeated.

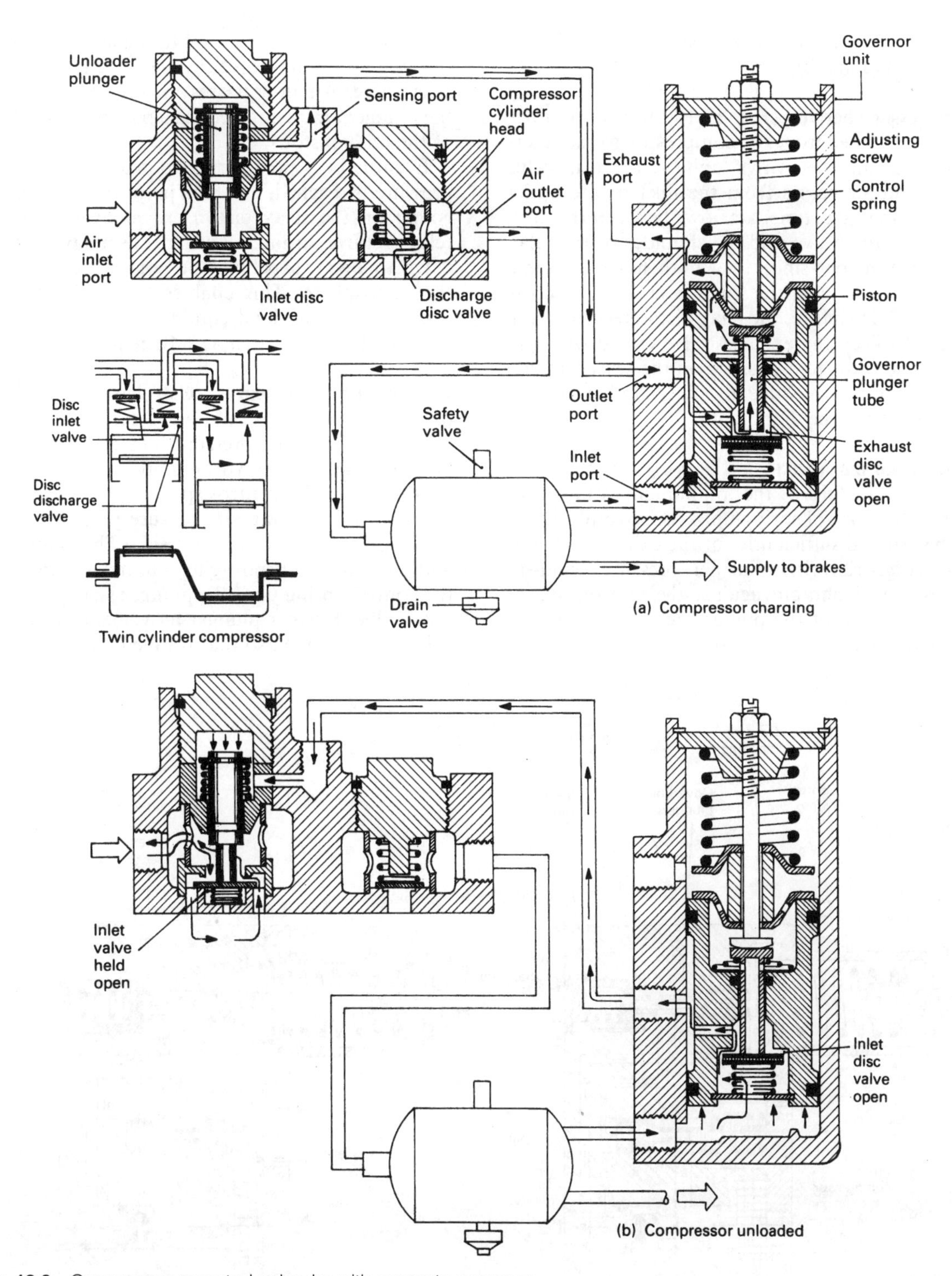

Fig. 12.9 Compressor mounted unloader with separate governor

12.3.4 Unloader valve (diaphragm type) (Fig. 12.10(a and b))

Compressor charging (Fig. 12.10(a)) When air is initially pumped from the compressor to the reservoir, the unloader valve unit non-return valve opens and air passes from the inlet to the outlet port. At the same time, air flows between the neck of the exhaust valve and the shoulder of the relay valve piston, but since they both have the same cross-sectional area, the force in each direction is equalized. Therefore, the relay piston return spring is able to keep the exhaust valve closed. Air will also move through a passage on the reservoir side of the non-return valve to the chamber on the plunger side of the diaphragm.

Compressor unloaded (Fig. 12.10(b)) As the reservoir pressure rises, the diaphragm will move against the control spring until the governor plunger has shifted sufficiently for the exhaust valve to close (Fig. 12.10(b)). Further pressure build-up moves the diaphragm against the control spring so that the end of the plunger enters its bore and opens the inlet valve. The annular end face of the plunger will also be exposed to the air pressure, so that the additional force produced fully opens the inlet valve. Air now passes through the centre of the plunger and is directed via a passage to the head of the relay piston.

Eventually a predetermined maximum cut-out pressure is reached, at which point the air pressure acting on the relay piston crown overcomes the relay return spring, causing the relay exhaust valve to open, expelling the compressed air into the atmosphere. This enables the compressor to operate under no-load conditions while the reservoir and braking system is sufficiently charged.

Compressor commences charging (Fig. 12.10 (a and b)) As the stored air is consumed during a braking cycle, the pressure falls until the cut-in point (minimum safe working pressure) is reached. At this point the control spring force equals and exceeds the opposing air pressure force acting on the diaphragm on the plunger side. The diaphragm and plunger will therefore tend to move away from the control spring until the plunger stem closes the inlet valve. Further plunger movement pushes the exhaust valve open so that trapped air in the relay

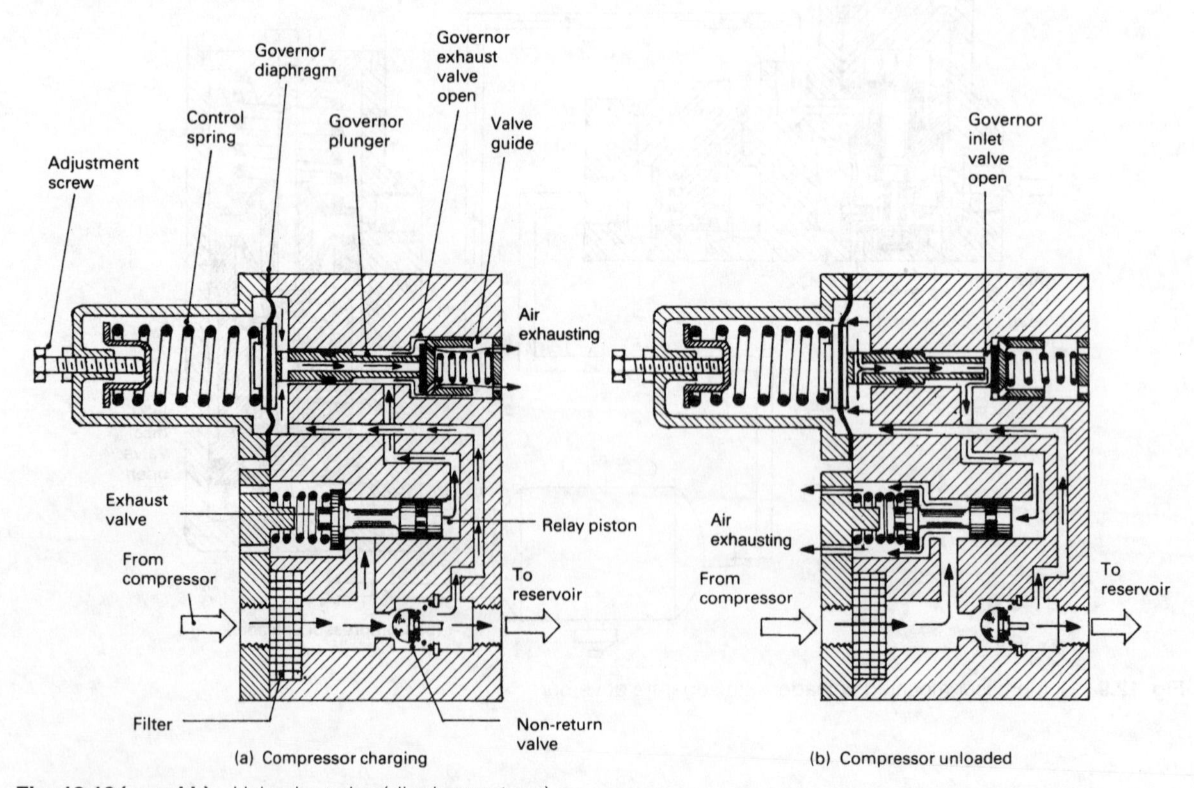

Fig. 12.10 (a and b) Unloader valve (diaphragm type)

piston crown chamber is able to escape to the atmosphere. The relay piston return spring closes the relay exhaust valve instantly so that compression of air again commences, permitting the reservoir to recharge to the pressure cut-out setting.

12.3.5 Unloader valve (piston type)
(Fig. 12.11(a and b))

Purpose The unloader valve enables the compressor to operate under no-load conditions, once the reservoir is fully charged, by automatically discharging the compressor's output into the atmosphere, and to reconnect the compressor output to the reservoir once the air pressure in the system drops to some minimum safe working value.

Operation

Compressor charging (Fig. 12.11(a)) When the compressor starts to charge, air will flow to the reservoir by way of the horizontal passage between the inlet and outlet ports.

The chamber above the relay piston is vented to the atmosphere via the open outlet pilot valve so that the return spring below the relay piston is able to keep the exhaust valve closed, thus permitting the reservoir to become charged.

Compressor unloaded (Fig. 12.11(b)) As the reservoir pressure acting on the right hand end face of the pilot piston reaches a maximum (cut-out setting), the pilot piston pushes away from its inlet seat. A larger piston area is immediately exposed to the air pressure, causing the pilot piston to rapidly move over to its outlet seat, thereby sealing the upper relay piston chamber atmospheric vent. Air will now flow along the space made between the pilot piston and its sleeve to act on the upper face of the relay piston. Consequently, the air pressure on both sides of the relay piston will be equalized momentarily. Air pressure acting down on the exhaust valve overcomes the relay piston return spring force and opens the compressor's discharge to the atmosphere. The exhaust valve will then be held fully open by the air pressure acting on the upper face of the relay piston. Compressed air from the compressor will be pumped directly to the atmosphere and so the higher pressure on the reservoir side of the non-return valve forces it to close, thereby preventing the stored air in the reservoir escaping.

Compressor commences charging (Fig. 12.11(a and b)) As the air pressure in the reservoir is discharged and lost to the atmosphere during brake applications the reservoir pressure drops. When the pressure has been reduced by approximately one bar below the cut-out setting (maximum pressure), the control spring overcomes the air pressure acting on the right hand face of the pilot piston, making it shift towards its inlet seat. The pilot piston outlet

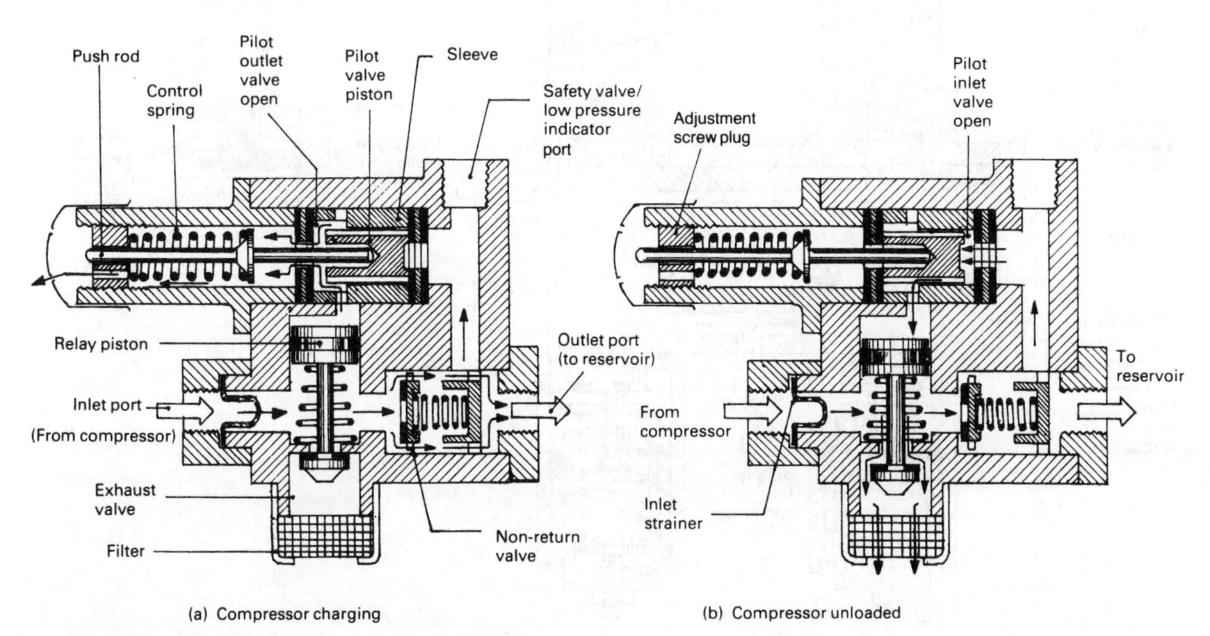

Fig. 12.11 (a and b) Unloader valve (piston type)

valve opens, causing the air pressure above the relay piston to escape to the atmosphere which allows the relay piston return spring to close the exhaust valve. The discharged air from the compressor will now be redirected to recharge the reservoir.

The difference between the cut-out and cut-in pressures is roughly one bar and it is not adjustable, but the maximum (cut-out) pressure can be varied over a wide pressure range by altering the adjustment screw setting.

12.3.6 Single- and multi-circuit protection valve (Fig. 12.12a)

Purpose Circuit protection valves are incorporated in the brake charging system to provide an independent method of charging a number of reservoirs to their operating minimum. Where there is a failure in one of the reservoir circuits, causing loss of air, they will isolate the affected circuit so that the remaining circuits continue to function.

Single element protection valve (Fig. 12.12(a)) When the compressor is charging, air pressure is delivered to the supply port where it increases until it is able to unseat the non-return disc valve against the closing force of the setting spring. Air will now pass between the valve disc and its seat before it enters the delivery port passage on its way to the reservoir. A larger area of the disc valve is now exposed to air pressure which forces the disc valve and piston to move further back against the already compressed setting spring. As the charging pressure in the reservoir increases, the air thrust on the disc and piston face also rises until it eventually pushes back the valve to its fully open position.

When the air pressure in the reservoir reaches its predetermined maximum, the governor or unloader valve cuts out the compressor. The light return spring around the valve stem, together with air pressure surrounding the disc, now closes the non-return valve, thereby preventing air escaping back through the valve. Under these conditions, the trapped air pressure keeps the disc valve on its seat and holds the setting spring and piston in the loaded position, away from the neck of the valve stem. As air is consumed from the reservoir, its pressure drops so that the compressor is signalled to cut in again (restarting pumping). The pressure on the compressor side of the non-return valve then builds up and opens the valve, enabling the reservoir to recharge.

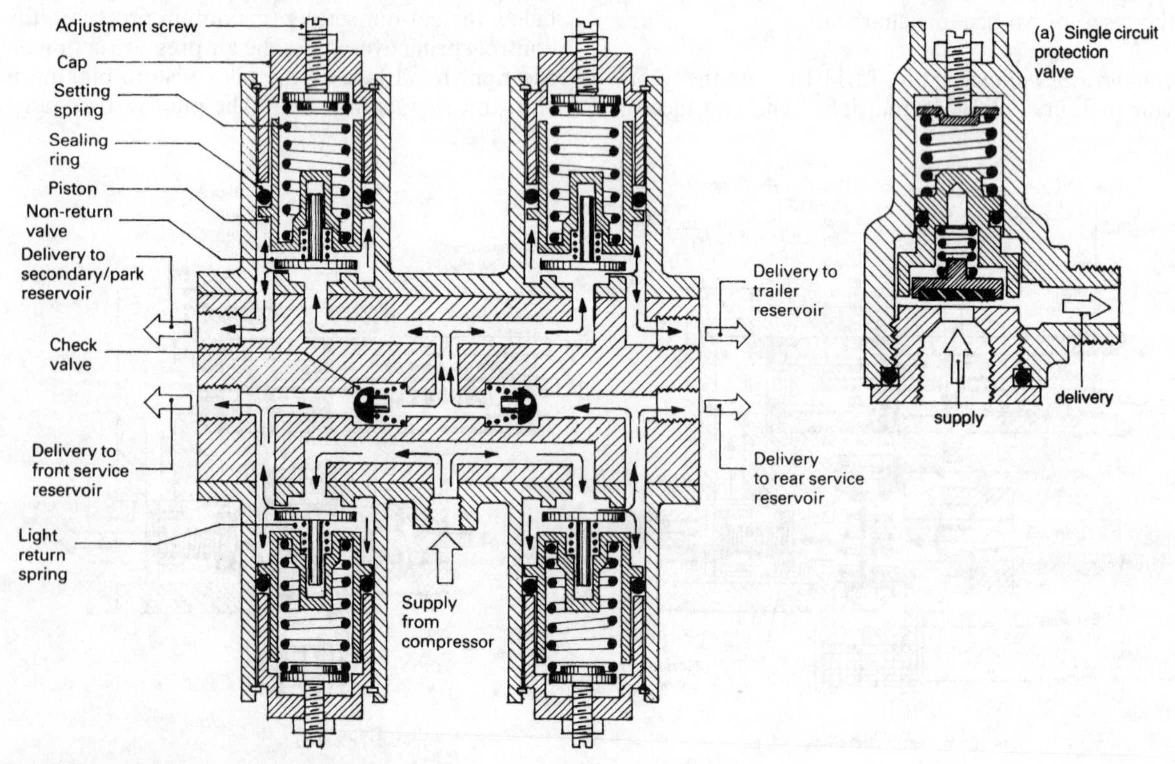

Fig. 12.12 Quadruple circuit protection valve

Should the air pressure in one of the reservoir systems drop roughly 2.1 bar or more, the setting spring stiffness overcomes the air pressure acting on the piston so that it moves against the disc valve to close the inlet passage. The existing air pressure stored in the reservoir will still impose a thrust against the piston, but because the valve face area exposed to the charge pressure is reduced by the annular seat area and is therefore much smaller, a pressure increase of up to 1.75 bar may be required to re-open the valve.

A total loss of air from one reservoir will automatically cause the setting spring of the respective protection valve to close the piston against the non-return valve.

Multi-element protection valve (Fig. 12.12) Multi-element protection valves are available in triple and quadruple element form. Each element contains the cap, piston, setting spring and non-return valve, similar to the single element protection valve.

Charging air from the compressor enters the supply port of the multi-element protection valve, increasing the pressure on the inlet face of the first and second valve element and controlling the delivery to the front and rear service reservoirs respectively. When the predetermined setting pressure is reached, both element non-return valves open, permitting air to pass through the valve to charge both service reservoirs.

The protection valves open and close according to the governor or unloader valve cutting in or cutting out the pumping operation of the compressor.

Internal passages within the multi-element valve body, protected by two non-return valves, connect the delivery from the first and second valve elements to the inlet of the third and fourth valve elements, which control the delivery to the secondary/park and the trailer reservoir supplies respectively. Delivery to the third and fourth valve elements is fed from the reservoir connected to the first and second valve element through passages within the body.

The additional check valves located in the body of the multi-protection valve act as a safeguard against cross-leakage between the front and rear service reservoirs. Failure of the front reservoir or circuit still permits the rear service reservoir to supply the third and fourth element valve. Alternatively, if the rear service reservoir should fail, the front service reservoir can cope adequately with delivering air charge to the third and fourth reservoir.

12.3.7 Pressure reducing valve (piston type) (Fig. 12.13(a, b and c))

Various parts of an air brake system may need to operate at lower pressures than the output pressure delivered to the reservoirs. It is therefore the function of the pressure reducing valve to decrease, adjust and maintain the air line pressure within some predetermined tolerance.

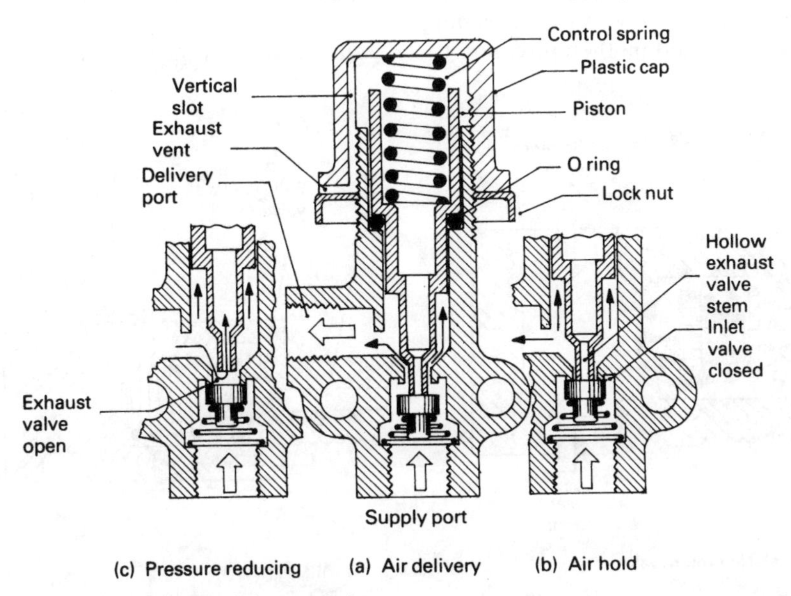

Fig. 12.13 (a–c) Pressure reducing valve (piston type)

Operation When the vehicle is about to start a journey, the compressor charges the reservoirs and air will flow through the system to the various components. Initially, air flows through to the pressure reducing valve supply port through the open inlet valve and out to the delivery port (Fig. 12.13(a)). As the air line pressure approaches its designed working value, the air pressure underneath the piston overcomes the stiffness of the control spring and lifts the piston sufficiently to close the inlet valve and cut off the supply of air passing to the brake circuit it supplies (Fig. 12.13(b)).

If the pressure in the delivery line exceeds the predetermined pressure setting of the valve spring, the extra pressure will lift the piston still further until the hollow exhaust stem tip is lifted clear of its seat. The surplus of air will now escape through the central exhaust valve stem into the hollow piston chamber where it passes out into the atmosphere via the vertical slot on the inside of the adjustable pressure cap (Fig. 12.13(c)). Delivery line air will continue to exhaust until it can no longer support the control spring. At this point, the spring pushes the piston down and closes the exhaust valve. After a few brake applications, the delivery line pressure will drop so that the control spring is able to expand further, thereby unseating the inlet valve. Hence the system is able to be recharged.

12.3.8 Non-return (check) valve (Fig. 12.14(a))

Purpose A non-return valve, sometimes known as a *check valve,* is situated in an air line system where it is necessary for the air to flow in one direction only. It is the valve's function therefore not to restrict the air flow in the forward direction, but to prevent any air movement in the reverse or opposite direction.

Operation (Fig. 12.14(a)) When compressed air is delivered to a part of the braking system via the non-return valve, the air pressure forces the spherical valve (sometimes disc) head of its seat against the resistance of the return spring. Air is then permitted to flow almost unrestricted through the valve. Should the air flow in the forward direction cease or even reverse, the return spring quickly closes to prevent air movement in the opposite direction occurring.

12.3.9 Safety valve (Fig. 12.14(b))

Purpose To protect the charging circuit of an air braking system from excessive air pressure, safety valves are incorporated and mounted at various positions in the system, such as on the compressor cylinder head, on the charging reservoir or in the pipe line between the compressor and reservoir.

Operation (Fig. 12.14(b)) If an abnormal pressure surge occurs in the charging system, the rise in air pressure will be sufficient to push the ball valve back against the regulating spring. The unseated ball now permits the excess air pressure to escape into the atmosphere. Air will exhaust to the atmosphere until the pressure in the charging system has been reduced to the blow-off setting determined by the initial spring adjustment. The regulating spring then forces the ball valve to re-seat so that no more air is lost from the charging system.

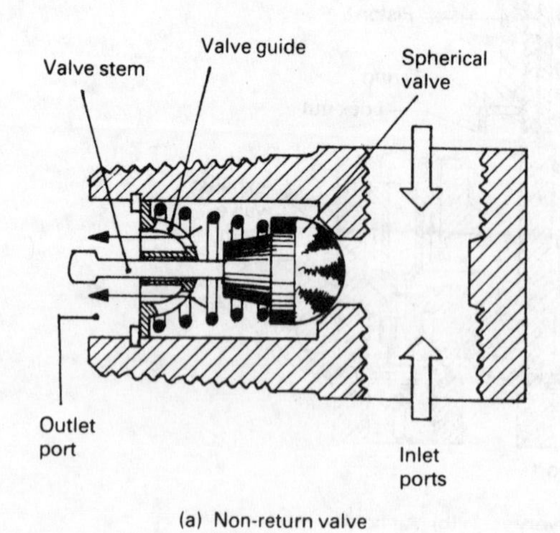

(a) Non-return valve

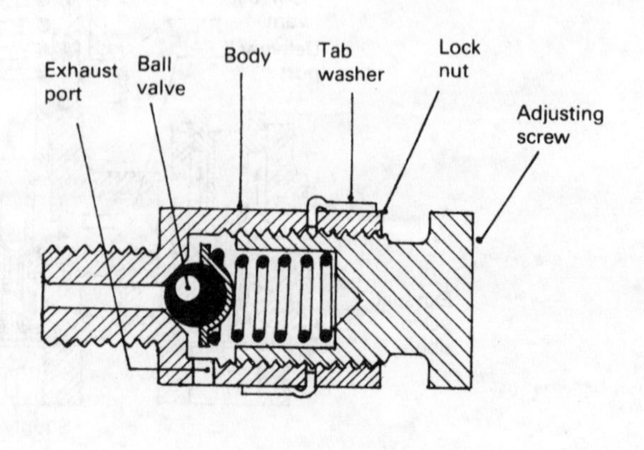

(b) Safety valve

Fig. 12.14 (a and b) Non-return and safety valves

12.3.10 Dual concentric foot control valve (Fig. 12.15(a and b))

Purpose The foot control valve regulates the air pressure passing to the brake system from the reservoir according to the amount the foot treadle is depressed. It also imparts a proportional reaction to the movement of the treadle so that the driver experiences a degree of brake application.

Operation

Applying brakes (Fig. 12.15(a)) Depressing the foot treadle applies a force through the graduating springs to the pistons, causing the exhaust hollow stem seats for both pistons to close the inlet/exhaust valves. With further depression of the foot pedal, the piston simultaneously unseats the inlet/exhaust valves and compressed air from the reservoirs passes through the upper and lower valves to the front and rear brake actuators respectively (or to the tractor and trailer brake actuators respectively).

Balancing (Fig. 12.15(a and b)) With the compressed air passing to the brake actuator chambers, air pressure is built up beneath the upper and lower pistons. Eventually the upthrust created by this air pressure equals the downward spring force; the pistons and valve carrier lift and the inlet valves close, thus interrupting the compressed air supply to the brake actuators. At the same time, the exhaust valves remain closed. The valves are then in a balanced condition with equal force above and

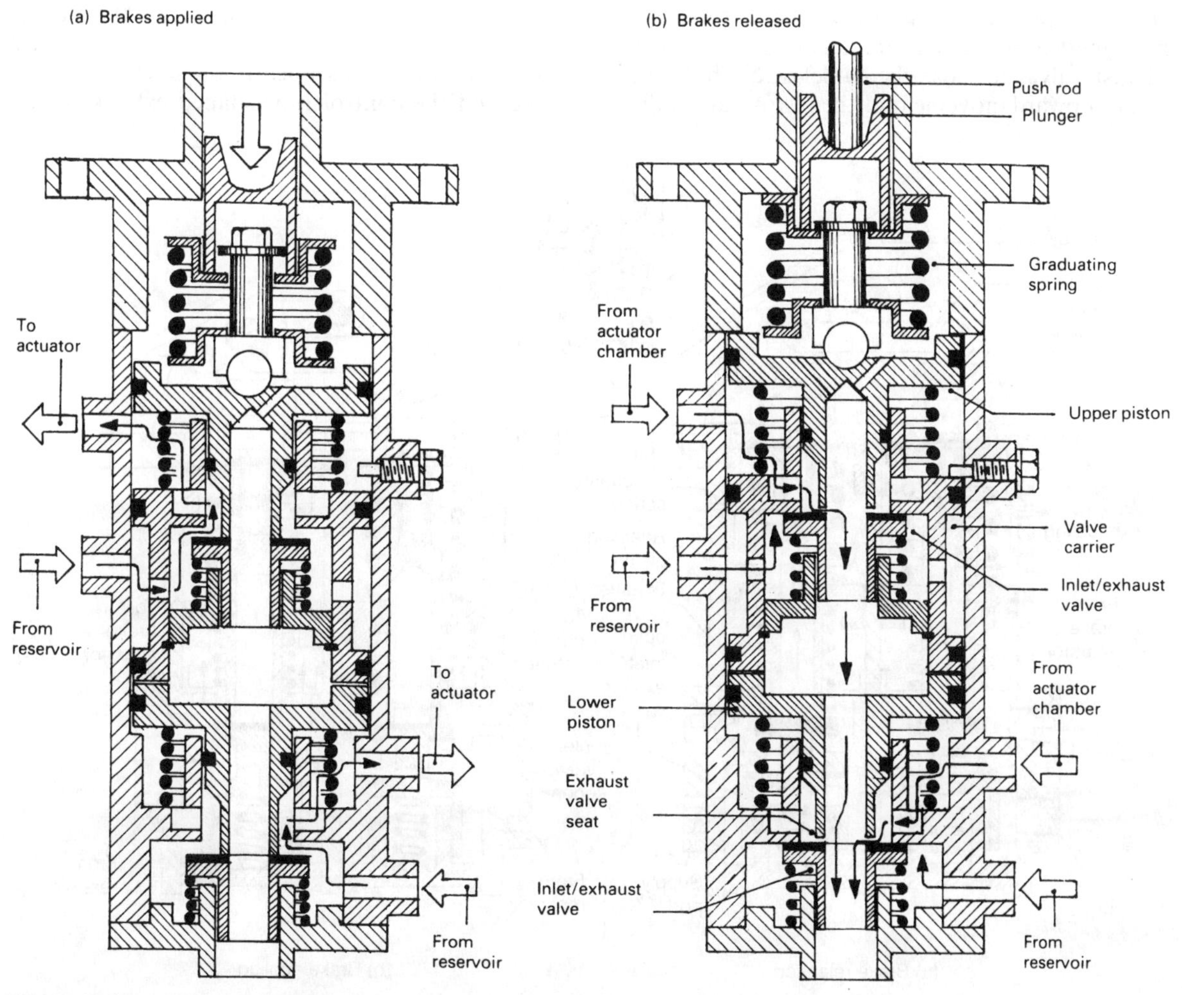

Fig. 12.15 (a and b) Dual concentric foot valve

below the upper piston and with equal air pressure being held in both halves of the brake line circuits.

Pushing the treadle down still further applies an additional force on top of the graduating spring. There will be a corresponding increase in the air pressure delivered and a new point of balance will be reached.

Removing some of the effort on the foot treadle reduces the force on top of the graduating spring. The pistons and valve carrier will then lift due to the air pressure and piston return springs. When this occurs the inlet valves remain closed and the exhaust valves open to exhausting air pressure from the brake actuators until a state of balance is obtained at lower pressure.

Releasing brakes (Fig. 12.15(b)) Removing the driver's force from the treadle allows the upper and lower piston and the valve carrier to rise to the highest position. This initially causes the inlet/exhaust valves to close their inlet seats, but with further upward movement of the pistons and valve assembly both exhaust valves open. Air from both brake circuits will therefore quickly escape to the atmosphere thus fully releasing the brakes.

12.3.11 Dual delta series foot control valve (Fig. 12.16)

Purpose The delta series of dual foot valves provide the braking system with two entirely separate foot controlled air valve circuits but which operate simultaneously with each other. Thus, if one half of the dual foot valve unit should develop a fault then the balance beam movement will automatically ensure that the other half of the twin valve unit continues to function.

Operation

Brakes released (12.16(a)) When the brakes are released, the return springs push up the piston, graduating spring and plunger assemblies for each half valve unit. Consequently the inlet disc valves close and the control tube shaped exhaust valves

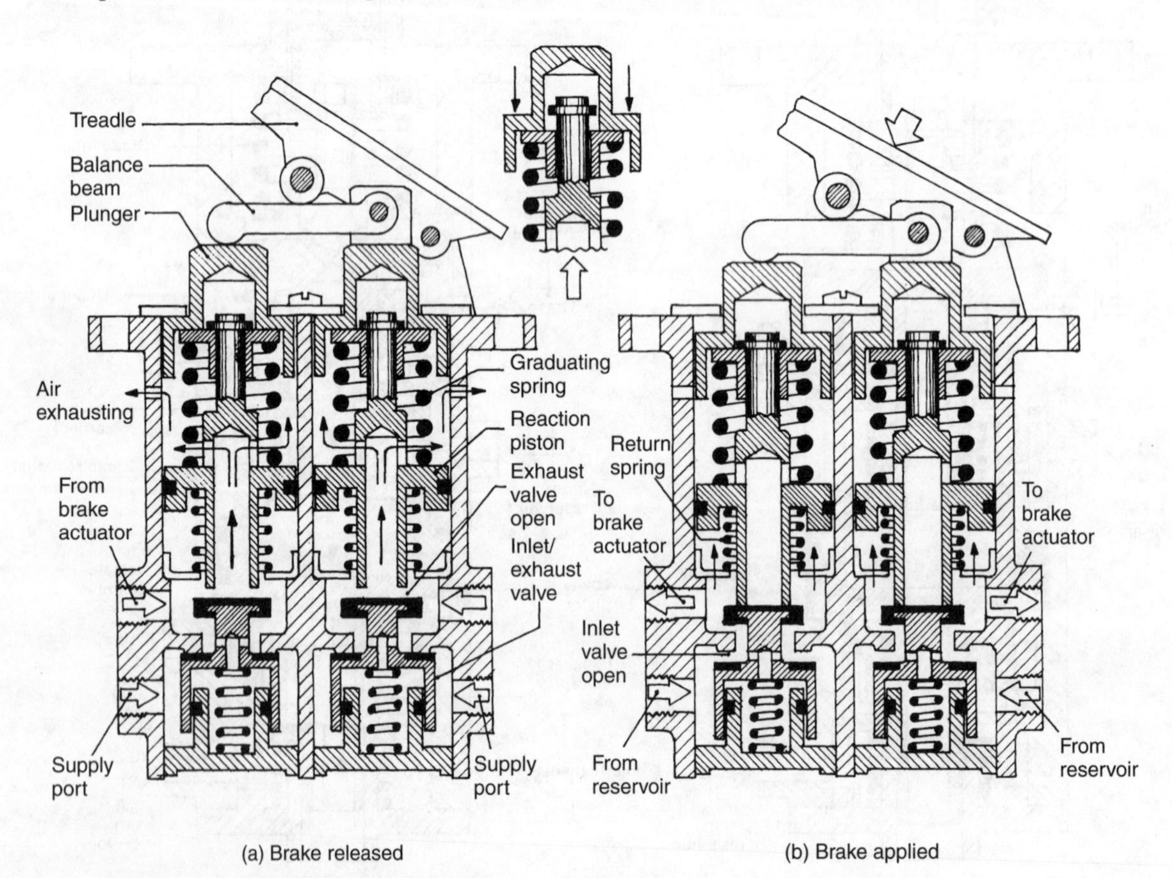

Fig. 12.16 (a and b) Dual delta foot control valve

open. This permits air to exhaust through the centre of the piston tube, upper piston chamber and out to the atmosphere.

Brakes applied (12.16(b)) When the foot treadle is depressed, a force is applied centrally to the balance beam which then shares the load between both plunger spring and piston assemblies. The downward plunger load initially pushes the piston tubular stem on its seat, closing the exhaust disc valve, and with further downward movement unseats and opens the inlet disc valve. Air from the reservoirs will now enter the lower piston chambers on its way to the brake actuators via the delivery ports.

As the air pressure builds up in the lower piston chambers it will oppose and compress the graduating springs until it eventually closes the inlet valve. The valve assembly is then in a lapped or balanced position where both exhaust and inlet valves are closed. Only when the driver applies an additional effort to the treadle will the inlet valve again open to allow a corresponding increase in pressure to pass through to the brake actuator.

The amount the inlet valve opens will be proportional to the graduating spring load, and the pressure reaching the brake actuator will likewise depend upon the effective opening area of the inlet valve. Immediately the braking effort to the foot treadle is charged, a new state of valve lap will exist so that the braking power caused by the air operating on the wheel brake actuator will be progressive and can be sensed by the driver by the amount of force being applied to the treadle. When the driver reduces the foot treadle load, the inlet valve closes and to some extent the exhaust valve will open, permitting some air to escape from the actuator to the atmosphere via central tube passages in the dual piston tubes. Thus the graduating spring driver-controlled downthrust and the reaction piston air-controlled upthrust will create a new state of valve lap and a corresponding charge to the braking power.

12.3.12 Hand control valve (Fig. 12.17(a and b))

Purpose These valves are used to regulate the secondary brake system on both the towing tractor

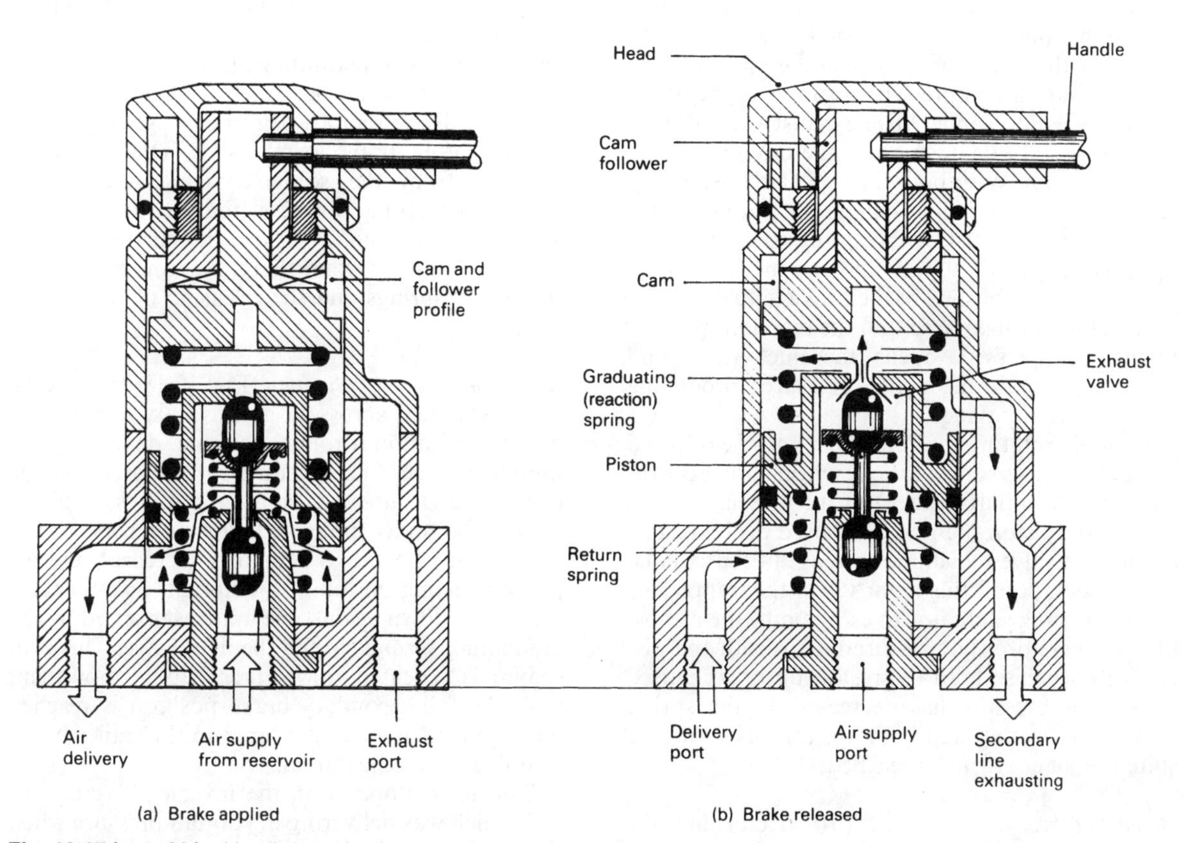

Fig. 12.17 (a and b) Hand control valve

and on the trailer. Usually only the tractor front axle has secondary braking to reduce the risk of a jack-knife during heavy emergency braking.

Operation

Applying brakes (Fig. 12.17(a)) Swivelling the handle from the released position enables the cam follower to slide over the matching inclined cam profile, thereby forcing the cam plate downwards against the graduating (reaction) spring. The stiffening of the reaction spring forces the piston to move downwards until the exhaust valve passage is closed. Further downward movement of the piston unseats the inlet valve, permitting compressed air from the reservoir to flow through the valve underneath the piston and out of the delivery port, to the front brake actuator and to the trailer brake actuator via the secondary line (blue) coupling to operate the brakes.

Balancing (Fig. 12.17(a and b)) The air supply passing through the valve gradually builds up an opposing upthrust on the underside of the piston until it eventually overcomes the downward force caused by the compressed reaction spring. Subsequently the piston lifts, causing the inlet valve to close so that the compressed air supply to the brake actuators is interrupted. The exhaust valve during this phase still remains seated, thereby preventing air exhaustion. With both inlet and exhaust valves closed, the system is in a balanced condition, thus the downward thrust of the spring is equal to the upthrust of the air supply and the predetermined air pressure established in the brake actuators.

Rotating the handle so that the reaction spring is further compressed, opens the inlet valve and admits more air at higher pressure, producing a new point of balance.

Partially rotating the handle back to the released position reduces some of the reaction spring downward thrust so that the existing air pressure is able to raise the piston slightly. The raised piston results in the inlet valve remaining seated, but the exhaust valve opens, permitting a portion of the trapped air inside the brake actuator to escape into the atmosphere. Therefore the pressure underneath the piston will decrease until the piston upthrust caused by the air pressure has decreased to the spring downthrust acting above the piston. Thus a new state of balance again is reached.

Releasing brakes (Fig. 12.17(b)) Returning the handle to the released position reduces the downward load of the reaction spring to fully raise the piston. As a result, the inlet valve closes and the exhaust valve is unseated, so that the air pressure in the brake actuator chambers collapses as the air is permitted to escape to the atmosphere.

12.3.13 Spring brake hand control valve (Fig. 12.18(a, b and c))

Purpose This hand control valve unit has two valve assemblies which, due to the cam profile design, is able to simultaneously deliver an 'upright' and an 'inverse' pressure. The valve unit is designed to provide pressure signals via the delivery of small volumes of air to the tractor spring brakes and the trailer's conventional diaphragm actuators. The required full volume of air is then able to pass from the secondary/park reservoir to the brake actuators via the relay valves to apply or release the brakes.

Operation

Spring brake release (Fig. 12.18(a)) When the vehicle is in motion with the brakes released, the upright valve assembly inlet valve is closed and the exhaust valve is unseated, permitting all the air in the trailer brake actuators to be expelled. Conversely the inverse valve assembly delivers a signal pressure to the spring brake relay valve. This results in the line from the secondary/park reservoir to the tractor spring brake actuators to be open. Thus a large volume of air will be delivered to the air chambers controlling the compression of the power springs and the releasing of the tractor brakes.

Secondary brake application (Fig. 12.18(b)) As the handle is moved across the gate to make a secondary brake application it rotates the cam, depressing the upright plunger. The exhaust valve closes and the inlet valve is unseated, causing compressed air to pass to the trailer brake actuator chambers. As the pressure in the brake actuators increases, the air pressure acting on top of the upright piston causes it to move down against the upthrust exerted by the graduating spring, closing the inlet valve. This procedure is repeated for further handle movement until the full secondary brake position is reached when the air pressure delivered to the trailer brake chamber is at a maximum.

During this operation the inverse valve assembly, which was delivering maximum pressure when the handle was in the 'off' position, is exhausting

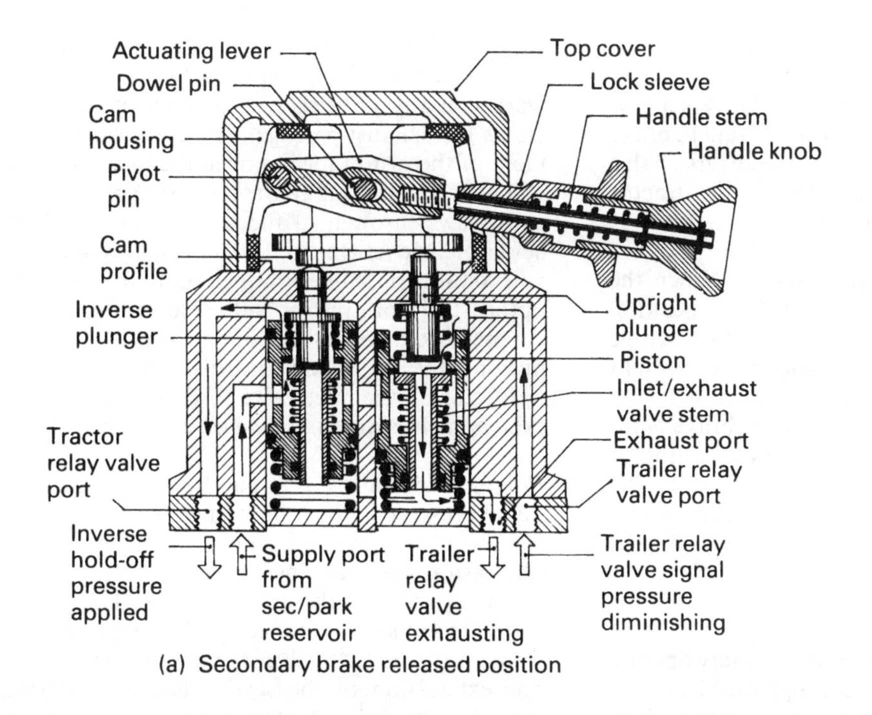

(a) Secondary brake released position

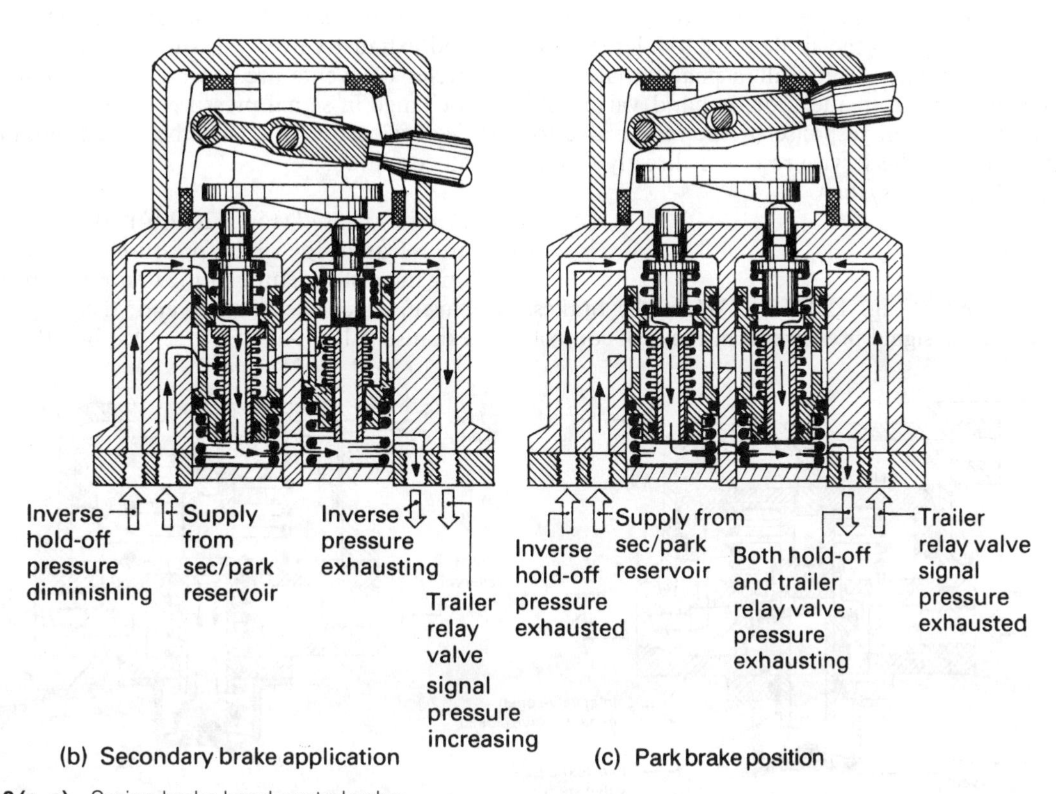

(b) Secondary brake application

(c) Park brake position

Fig. 12.18 (a–c) Spring brake hand control valve

until with the secondary brake position the delivered pressure is zero.

In other words, the upright valve delivers a gradually increasing pressure to the trailer brake actuators and, at the same time, the inverse valve assembly allows the air pressure on the tractor spring brake actuators to be gradually released.

Park brake application (Fig. 12.18(c)) When the handle is moved from the secondary brake position to the park position, the cam lifted by the leverage of the handle about its pivot allows the upright plunger and the inverse plunger to be raised. The air pressure in both tractor and trailer brake actuators then exhaust into the atmosphere. The tractor brakes are now applied in the park position by the mechanical force exerted by the spring actuators.

12.3.14 Relay valve (piston type) (Bendix) (Fig. 12.19(a and b))

Purpose The relay valve is used to rapidly operate a part of a braking system when signalled by either a foot or hand control valve. This is achieved by a small bore signal line feeding into the relay valve which then controls the air delivery to a large bore output service line. As a result, a small variation in signal pressure from the foot or hand valve will produce an instant response by the relay valve to admit air from the service reservoir directly to the service line brake system.

Operation

Brakes applied (Fig. 12.19(a)) When the brakes are applied, a signal pressure from the foot control valve (or hand control valve) reacts on the large control piston which responds by moving downwards rapidly until the centre stem of the piston closes the exhaust passage. The downwards movement of the piston pushes open the inlet valve. Air will now be admitted to the underside of the piston as it flows through to the service line and brake actuator. Movement of air from the service reservoir to the service line continues until the combined upthrust of both piston and valve springs and the air pressure balances the air signal pressure force, pushing the piston downwards. The piston now rises, closing the inlet valve so that both inlet and exhaust valves are in the *lapped* condition.

Brakes hold (Fig. 12.19(a and b)) A reduction in signal pressure now produces a greater force, pushing the piston upwards rather than downwards. The piston rises, closing the inlet valve, followed by the opening of the exhaust valve. The trapped air in the service line and actuator will now exhaust through the hollow valve stem to the atmosphere. The exhaustion of the service line air continues until the upward piston force balances the downward force caused by signal pressure. Both inlet and exhaust valves will subsequently close. These cycles of events are repeated the instant there is a change in signal pressure, be it an increasing or decreasing one, the valve being self-lapping under all conditions.

Brakes released (Fig. 12.19(b)) When the brakes are released, the signal pressure collapses, permitting the piston return spring to raise the piston; first closing the inlet valve, and then opening the exhaust valve. Air in the service line then escapes

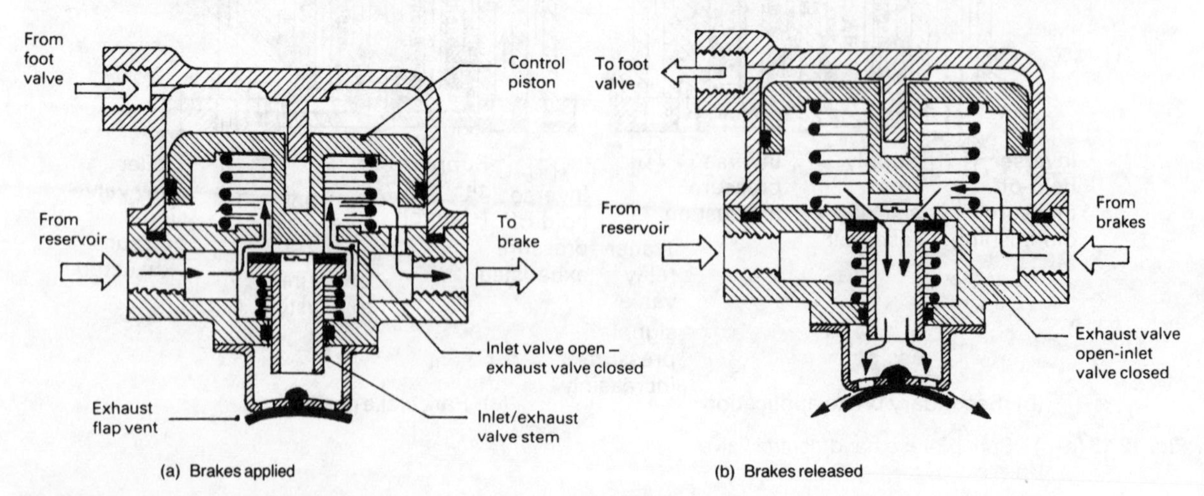

Fig. 12.19 (a and b) Relay valve

through the lower piston chamber and out into the atmosphere through the hollow valve stem.

12.3.15 Quick release valve
(Fig. 12.20(a, b and c))

Purpose The quick release valve (QRV) shortens the brake release time by speeding up the exhaustion of air from the brake actuator chambers, particularly if the actuators are some distance from the foot, hand or relay valve.

Operation

Applied position (Fig. 12.20(a)) When the brakes are applied, the air pressure from the foot or hand control valve enters the upper diaphragm chamber, forcing the diaphragm and its central stem down onto the exhaust port seat. The air pressure build-up then deflects downwards the circumferential diaphragm rim, thereby admitting air to the brake actuators via the pipe lines.

Hold position (Fig. 12.20(b)) Movement of air from the inlet port to the outlet ports permits air to occupy the underside of the diaphragm. Once the air pressure above and below the diaphragm has equalized, the diaphragm return spring upthrust pushes the outer diaphragm rim up onto its seat whilst the centre of the diaphragm and stem still seal off the exhaust port. Under these conditions, both inlet and exhaust passages are closed, preventing any additional air flow to occur to or from the brake actuators. The diaphragm is therefore in a state of 'hold'.

Released position (Fig. 12.20(c)) Releasing the air pressure above the diaphragm allows the trapped and pressurized air below the diaphragm to raise the central region of the diaphragm and stem. The trapped air in the brake lines and actuator chambers escape into this atmosphere.

Reducing the brake load slightly decreases the air pressure above the diaphragm, so that some of the air in the brake lines is allowed to escape before the pressure on both sides of the diaphragm balances again. The central region of the diaphragm moves down to close the exhaust port which moves the diaphragm into its 'hold' condition again.

The quick release valve therefore transfers any increased foot or hand valve control pressure through it to the brake actuators and quickly releases the air pressure from the brake actuators when the brake control valve pressure is reduced.

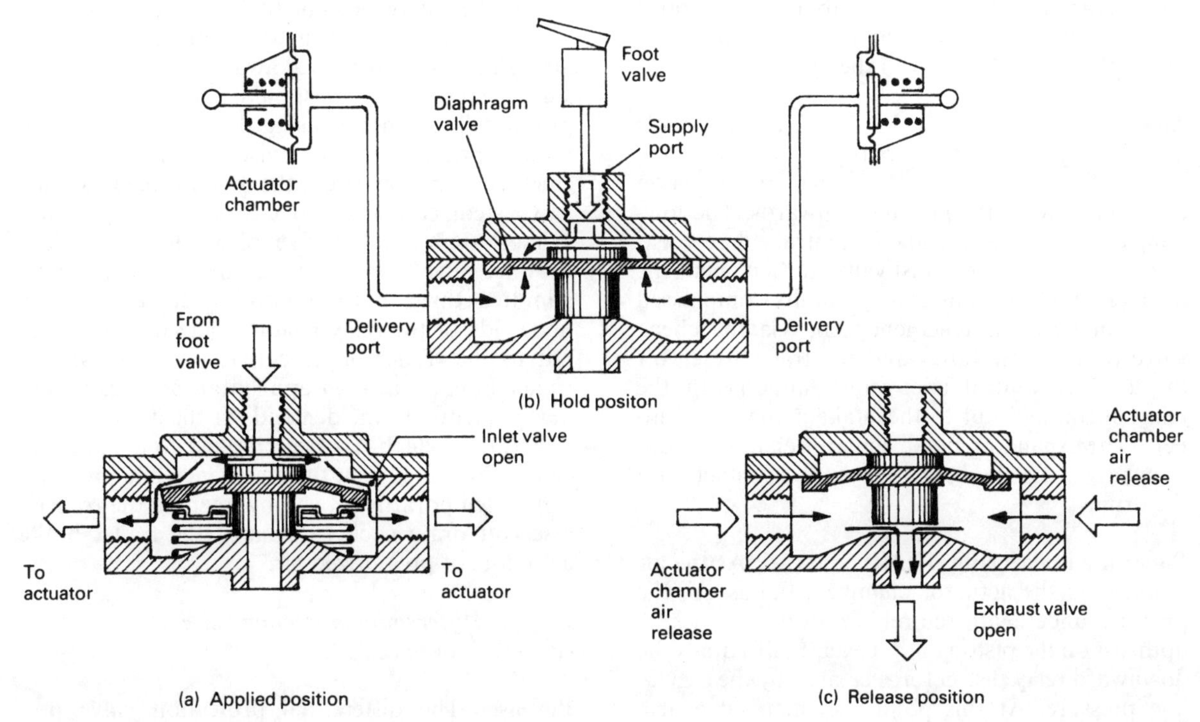

Fig. 12.20 (a–c) Quick release valve

By these means the air pressure in the brake actuators will always be similar to the delivery air pressure from the brake control valve.

12.3.16 Relay emergency valve
(Fig. 12.21(a–d))

Charging (Fig. 12.21(a)) Air delivery from the emergency line (red) enters the inlet port and strainer. The compressed air then opens the check valve, permitting air to flow across to and around the emergency piston, whence it passes to the outlet port leading to the trailer reservoir, enabling it to become charged.

If the reservoir is completely empty, both the relay piston and the emergency piston will be in their uppermost position. Under these conditions, the exhaust valve will be closed and the inlet valve open. Therefore some of the air flowing to the trailer reservoir will be diverted through the inlet valve to the brake actuator chambers, thereby operating the brakes. When the trailer reservoir charge pressure reaches 3.5 bar, air fed through a hole from the strainer pushes down on the annular area of the emergency piston causing the inlet to close. As the reservoir stored pressure rises to 4.2 bar, the downward air pressure force on the emergency piston moves the inlet/exhaust valve stem away from its exhaust seat, enabling the trapped air in the brake actuator chambers to escape to the atmosphere. The brakes will then be released.

Applying brakes (Fig. 12.21(b)) When the brakes are applied, a signal pressure is passed through the service line (yellow) to the upper relay piston chamber, forcing the piston downwards. The lowering of the relay piston and its central exhaust seat stem first closes the exhaust valve. It then opens the inlet valve which immediately admits compressed air from both the emergency line via the check valve (non-return valve) and the trailer reservoir through the central inlet valve, underneath the relay piston and out to the brake actuator chambers. The expanding brake actuator chambers subsequently press the brake shoes into contact with the drums.

Balancing brakes (Fig. 12.21(b and c)) As the air pressure in the actuator chambers builds up, the pressure underneath the relay piston increases its upthrust on the piston until it eventually equals the downward relay piston force created by the service line pressure. At this point the inlet valve also closes, so that both valves are now in a balanced state. Until a larger service line pressure is applied to the relay piston, the central stem will not move further down to open the inlet valve again and permit more air to pass to the brake actuator chambers. Conversely, if the foot brake is slightly released, initially the relay piston is permitted to rise, closing the inlet valve, followed by opening of the exhaust valve to release some of the air pressure acting on the brake actuator chambers.

Releasing brakes (Fig. 12.21(c)) Removing the load on the foot control valve first closes off the air supply to the service line and then releases the remaining air in the service line to the atmosphere. The collapse of service line pressure allows the relay piston to rise due to the existing brake actuator pressure acting upwards against the relay piston. The hollow valve stem immediately closes the inlet valve passage, followed by the relay piston centre stem exhaust seat lifting clear of the exhaust valve. Air is now free to escape underneath the relay piston through the central hollow inlet/exhaust valve inlet stem and out to the exhaust vent flap to the atmosphere. The brake actuators now move to the 'off' position, permitting the 'S' cam expanders to release the brake shoes from their drums.

Emergency position (Fig. 12.21(d)) If the air pressure in the emergency line (red) should drop below a predetermined minimum (normally 2 bar), due to air leakage or trailer breakaway, then the air pressure around the upper shoulder of the emergency piston will collapse, causing the emergency piston return spring to rapidly raise the piston. As the emergency piston rises, the hollow inlet/exhaust valve stem contacts and closes the relay piston exhaust stem seat. Further piston lift then opens the inlet valve. Air from the trailer reservoir is now admitted through the control inlet valve to the underside of the relay piston where it then passes out to the trailer brake actuator chambers. The trailer brakes are then applied automatically and independently to the demands of the driver.

A trailer which has been braked to a standstill, caused by a failure in the emergency line pressure, can be temporarily moved by opening the trailer's reservoir drain cock to exhaust the trailer brake actuators of pressurized air.

12.3.17 Differential protection valve
(Fig. 12.22(a, b and c))

Purpose The differential protection valve prevents both service brakes and secondary brakes applying their full braking force at any one time. The valve is designed to supply secondary line pressure to the spring brake release chambers when the service brakes are operating or to allow the service line pressure supplying the service brake chambers to decrease as the spring brakes are applied. By these means the spring and diaphragm actuator forces are prevented from compounding and overloading the combined spring and diaphragm actuator units and the foundation brakes which absorb the braking loads.

Operation

Brakes in off position (Fig. 12.22(a)) Releasing both the foot and hand brakes exhausts air from the service line. Air from the secondary line enters the secondary inlet port of the valve and flows between the outer piston and the casing to the spring brake output ports. It then passes to the actuator air chambers. The compressed air now holds the secondary springs in compression, thereby releasing the brake shoes from the drums.

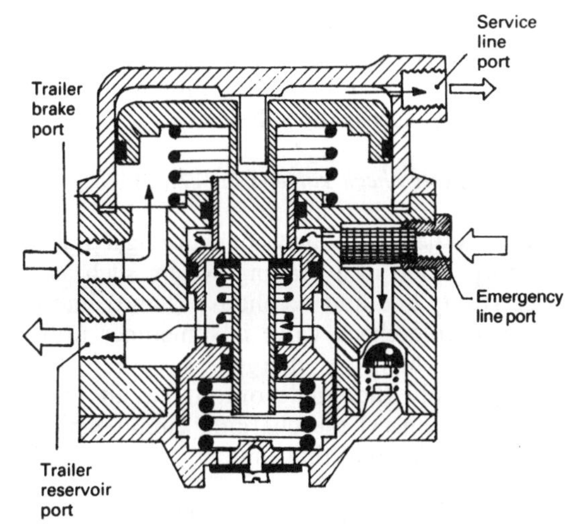

(a) Charging position

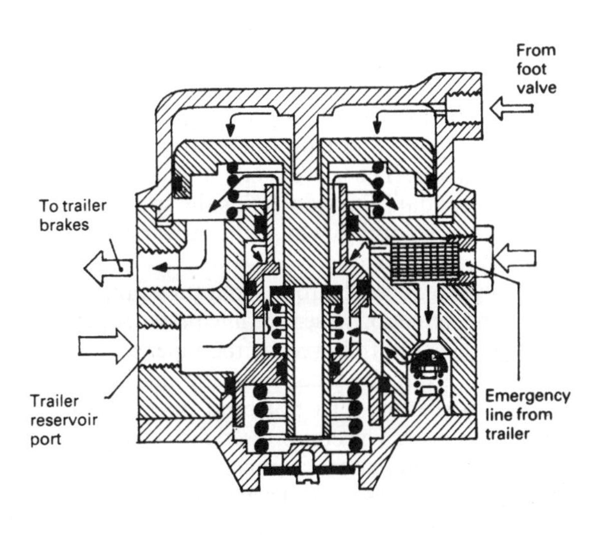

(b) Applying position

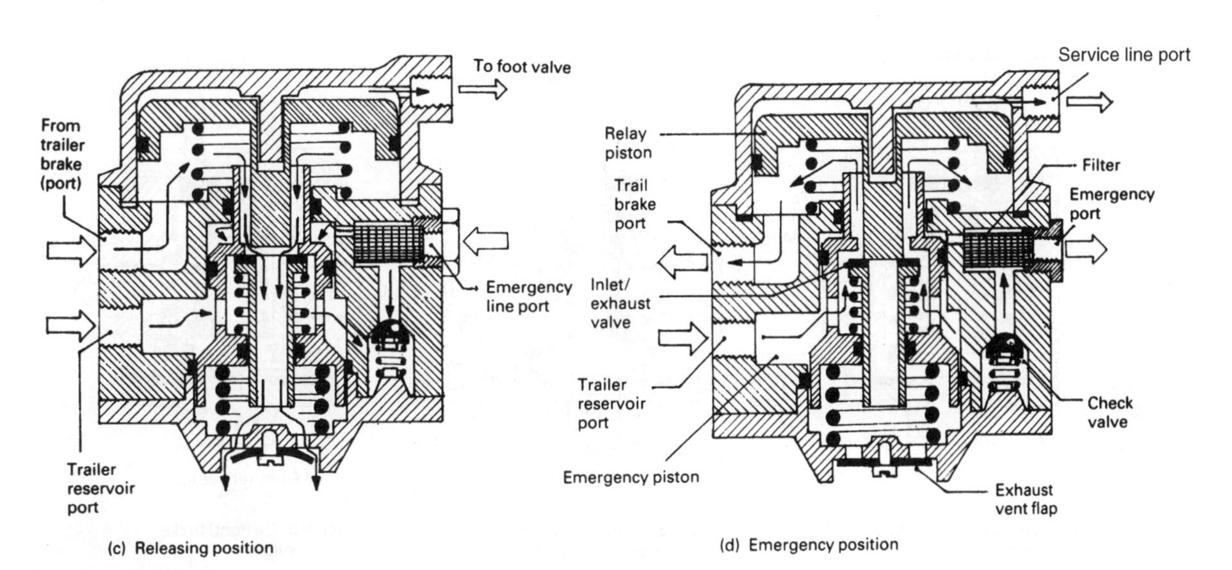

(c) Releasing position

(d) Emergency position

Fig. 12.21 (a–d) Relay emergency valve

Secondary (spring) brake application (Fig. 12.22(b)) When the secondary (spring) brakes are applied, following the initial application and holding of the service (foot) brakes, the compressed air in the spring actuator chambers and in the secondary line is exhausted via the differential protection valve to the atmosphere through the hand control valve. As the secondary line pressure reduces, the pressure trapped in the service line due to the previous foot brake application becomes greater than the decreasing pressure in the secondary line. It therefore causes the inner piston to be pushed across to block the secondary port air exit. Immediately afterwards, the outer piston is unseated so that service line air now flows through the valve from the service line inlet port to the spring delivery ports and from there to the spring actuator chambers. The service line air which has entered the secondary line now holds the springs so that they are not applied whilst the driver is still applying the foot brake.

As the driver reduces the foot pedal pressure, the corresponding reduction in service line pressure permits the outer piston, followed by the inner piston, to move away from the secondary line inlet port, closing the service line inlet port and opening the secondary inlet port. The compressed air occupying the spring brake actuator chambers is now permitted to fully exhaust so that the expanding springs re-apply the brakes simultaneously as the service (foot) brakes are being released.

Service (foot) brake application (Fig. 12.22(c)) When the service (foot) brakes are applied after a spring brake application, the secondary line will be exhausted of compressed air, which was essential for the spring brakes to operate. Therefore, as the service line pressure rises, it pushes the inner piston against its seat, closing the secondary line inlet port. With a further increase in service line pressure, the outer piston becomes unseated so that service line pressure can now flow through the valve and pass on to the spring brake actuators. This withdraws the spring brake force, thereby preventing the compounding of both spring and service chamber forces.

While the differential protection valve is in operation, an approximate 2.1 bar pressure differential between the service pressures and the delivered effective anti-compounding pressure will be maintained across the valve.

12.3.18 Double check valve (Fig. 12.23)

Purpose When two sources of charging a pipe line are incorporated in a braking system such as the service (foot) line and secondary (hand) line circuits, a double check valve is sometimes utilized to connect whichever charging system is being used to supply the single output circuit and to isolate (disconnect) the charging circuit which is not being operated at that time.

Operation (Fig. 12.23(a and b)) The two separate charging circuits (service and secondary lines) are joined together by the end inlet ports of the double check valve. When one of the brake systems is applied, air charge will be delivered to its double

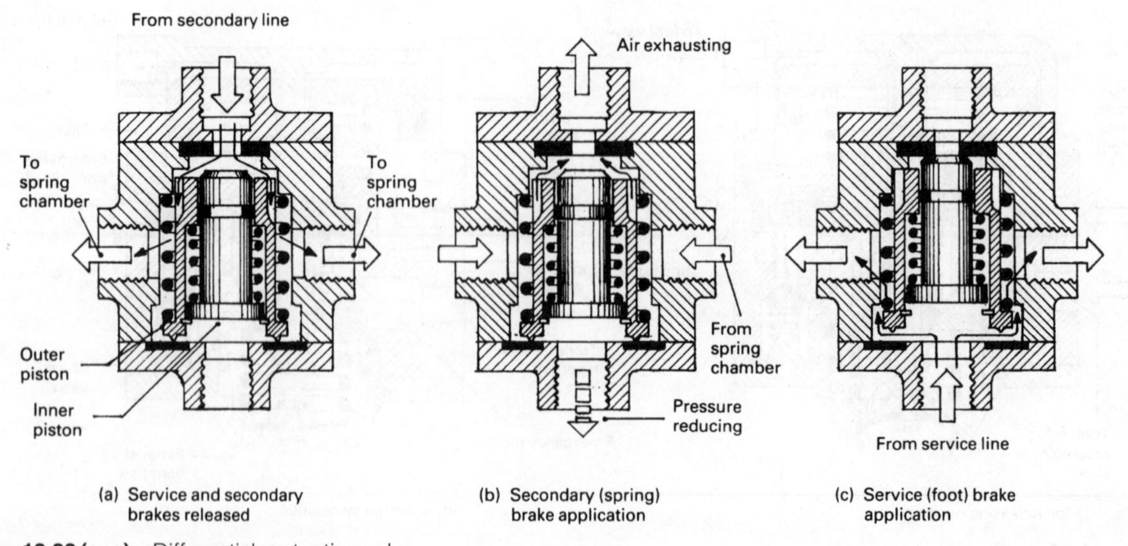

Fig. 12.22 (a–c) Differential protection valve

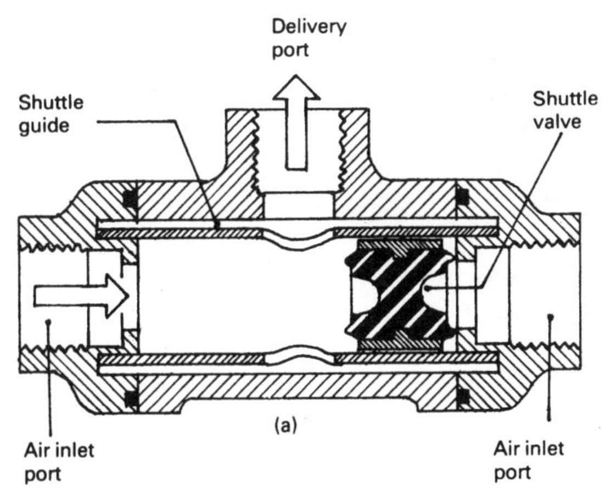

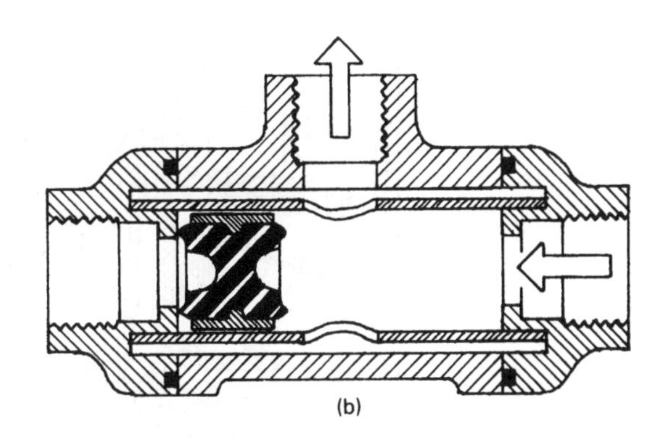

Fig. 12.23 (a and b) Double check valve

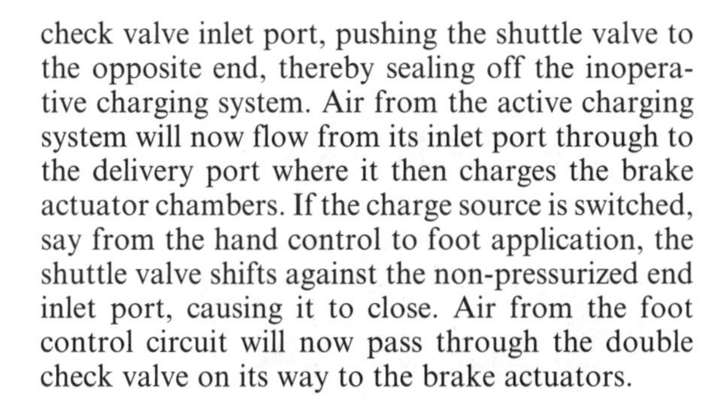

check valve inlet port, pushing the shuttle valve to the opposite end, thereby sealing off the inoperative charging system. Air from the active charging system will now flow from its inlet port through to the delivery port where it then charges the brake actuator chambers. If the charge source is switched, say from the hand control to foot application, the shuttle valve shifts against the non-pressurized end inlet port, causing it to close. Air from the foot control circuit will now pass through the double check valve on its way to the brake actuators.

12.3.19 Variable load valve (Fig. 12.24)

Purpose This valve is designed to sense the vertical load imposed on a particular axle by monitoring the charge in suspension height and to regulate the braking force applied to the axle's brakes in proportion to this loading. The valve therefore controls the brake actuator chamber air pressure in accordance with the load supported by the axle and the service line pressure.

Operation (Fig. 12.24(a, b and c)) The valve is mounted on the vehicle's chassis and its control lever is connected to the axle through a vertical adjustable link rod. The valve control lever is in its lowest position with the axle unladen, moving to its highest position as the axle load is increased to fully laden.

Brakes released (Fig. 12.24(a)) When the brakes are released, the service line pressure collapses, permitting the control piston to rise to its highest position. Because the valve stem rests on the ball pin, the inlet valve closes whereas the exhaust valve is unseated. Pressurized air in the brake actuator chambers and pipe line will subsequently flow underneath the diaphragm, up and around the hollow valve stem, past the exhaust valve and its seat into the atmosphere via the control exhaust passage.

Brakes applied (Fig. 12.24(b and c)) When the brakes are applied, service line pressure enters the upper piston chamber, pushing the control piston downwards. At the same time, some of the air is transferred through the external pipe to the lower clamp plunger, which is then forced upwards against the ball pin. As the control piston moves downwards, the exhaust/inlet valve stem closes the central exhaust passage and then uncovers the inlet valve passage. Air from the service line inlet port now passes through the inlet valve to the lower diaphragm chamber and from there it continues on its way to the brake actuator chambers.

If the axle is laden, the control lever ball pin will be in a high position so that the control piston does not move very far down before the exhaust valve is closed and the inlet valve is opened. Conversely, if the axle is unladen the control lever and ball pin will be in a much lower position so that the control piston has to move much further downwards.

When the brakes are released, the clamp plunger chamber is exhausted of air so that the valve stem assembly will not be rigidly attached to the ball pin and only becomes active during brake application. Hence unnecessary wear is avoided.

Brakes applied with heavy load (Fig. 12.24(b)) When the axle is laden, the ball pin will hold the valve stem in the high position, therefore the control piston will also be in the upper position. Under

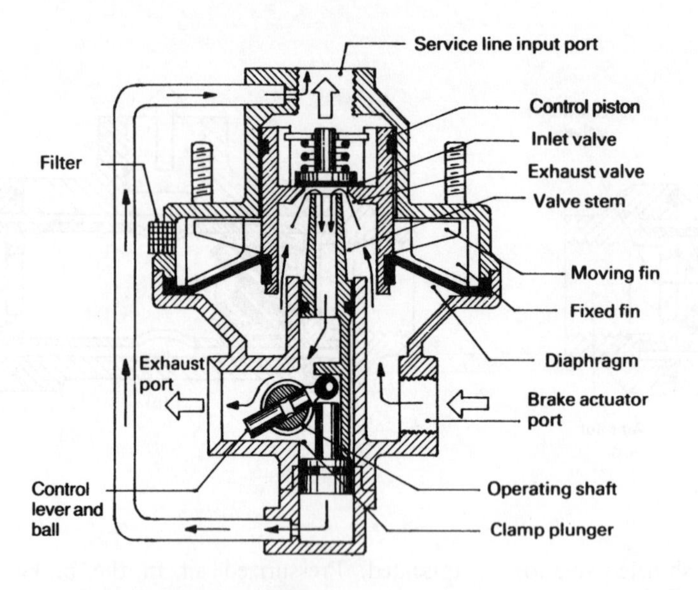

(a) Brakes released

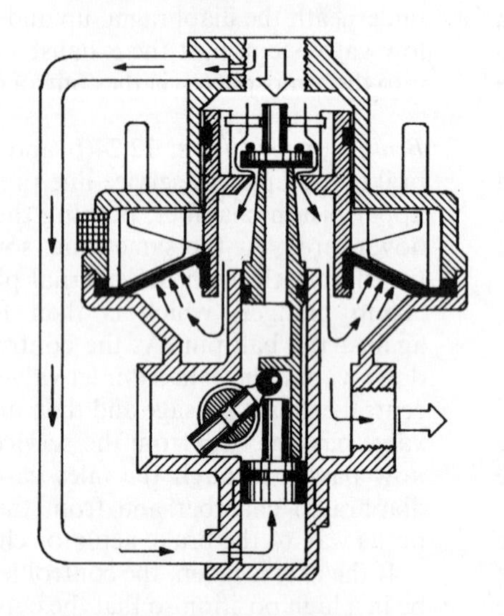

(b) Brakes applied with heavy load

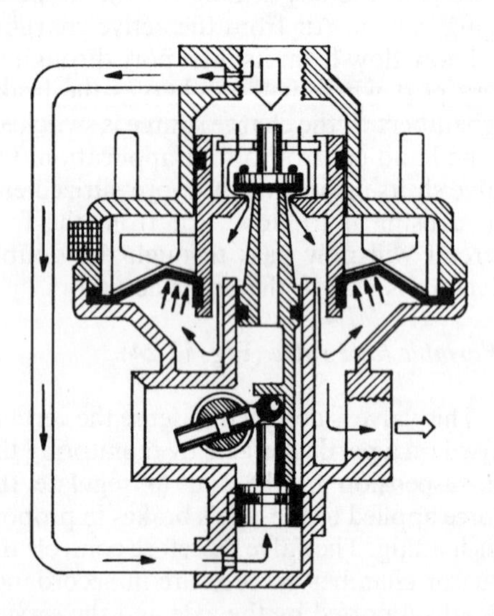

(c) Brakes applied with light load

Fig. 12.24 (a–c) Variable load valve

these conditions the underside of the diaphragm reacts against the fixed fins and only a small portion of the diaphragm area is supported by the moving fins attached to the piston. This means that very little piston upthrust is provided, which therefore permits the inlet valve to open wide and to admit a large air delivery pressure to the brake actuators. As the air supply flows through the valve, the pressure under the diaphragm increases until the upthrust acting on the varying effective area of the diaphragm equals that produced by the service line pressure acting on top of the control piston. The valve assembly now moves into a lapped condition whilst the forces imposed on the piston are in a state of balance.

Brakes applied with light load (Fig. 12.24(c)) When the axle is unladen, the ball pin will hold the valve stem in a lower position so that the control piston will be forced by the service line air pressure to move further down. Under these new conditions the underside of the diaphragm reacts against the moving fins more than the fixed ones. Consequently there will be a much larger diaphragm upthrust, tending to partially close the inlet valve whilst air pressure is being delivered to the brake actuator chambers. As a result, the piston will move to a new position of balance and the valve assembly again moves into a lapped condition.

It can be seen that the variable load valve automatically regulates the output air pressure delivered to the axle brake actuators in proportion to the laden weight imposed on the axle.

12.3.20 Multi-relay (triple) valve
(Fig. 12.25(a–d))

Purpose With a two line braking system the trailer has no secondary braking system. Therefore the tractor foot control valve and the hand control valve must each be able to apply the single trailer brake system independently. This is made possible by the utilization of a multi-relay (triple) valve which is very similar to the conventional single relay valve except that it has three signal sensing relay pistons placed one above the other.

Operation

Brakes released (Fig. 12.25(a)) If the brakes are released, all three relay pistons will rise to their uppermost positions due to the return spring upthrust. Consequently, the inlet valve closes and the exhaust valve will be unseated. This ensures that the trailer brake actuators are cleared of compressed air, so releasing the brakes.

Secondary line brake application (Fig. 12.25(b)) Applying the hand control valve handle sends a pressure signal to the lower relay piston (3). The lower relay piston will move downwards, initially closing the exhaust valve and then opening the inlet valve. A pressure signal will then pass from the trailer reservoir mounted on the tractor to the upper part of the trailer's emergency relay valve. As a result, air pressure from the supply line (red) now flows to the trailer brake actuators.

Service line brake application (Fig. 12.25(c and d)) When the foot control valve is depressed a signal pressure from the both halves of the foot valve is transmitted to the upper (1) and middle (2) relay valve pistons. Both relay pistons react immediately by moving down until the three relay pistons are pressed together. Further downward movement will close the exhaust valve and open the inlet valve. Air from the trailer reservoir mounted on the tractor will now pass to the emergency relay valve, permitting air from the supply line (red) to pass directly to the trailer brake actuators via the now opened passage passing through the emergency valve.

Should half of the dual foot valve service line circuit develop a fault, the other half service line circuit will still be effective and be able to operate the multi-relay valve.

12.3.21 Supply dump valve
(Fig. 12.26(a, b and c))

Purpose The supply dump valve has been designed to meet one of the *EEC Brake Safety Directive for Trailers,* which requires that if there is an imbalance of air pressure between the tractor service line and the trailer service line due to leakage or decoupling, then within two seconds of the next full service brake application the compressed air in the trailer supply (emergency) line will be dumped to the atmosphere, reducing the pressure to 1.5 bar. The result of the service line pressure collapse signals the trailer emergency valve to transfer compressed air stored in the trailer reservoir to the trailer brake actuators, so causing the brakes to be applied.

Operation

Brakes released (Fig. 12.26(a)) When the brake pedal is released, compressed air exhausts from the supply dump valve tractor and trailer service line sensing chambers. Under these conditions, the piston spring forces the piston and exhaust valve stem down and unseats the inlet valve. Air from the tractor emergency (supply) line is therefore free to flow through the supply dump valve to the trailer's emergency (supply) line to charge the trailer's reservoir.

Service brakes applied (Fig. 12.26(b)) When the foot pedal is depressed the tractor service line output from the foot control valve and the multi-relay valve output to the trailer service line both send a pressure signal. The air pressure in both the upper and lower chambers will therefore be approximately equal. Because the piston's upper surface area is greater than its underside area and the piston spring applies a downward thrust onto the piston, the piston will be forced to move to its lowest position. This lowering of the piston closes the exhaust valve and opens the inlet valve. Air is now able to flow from the tractor emergency

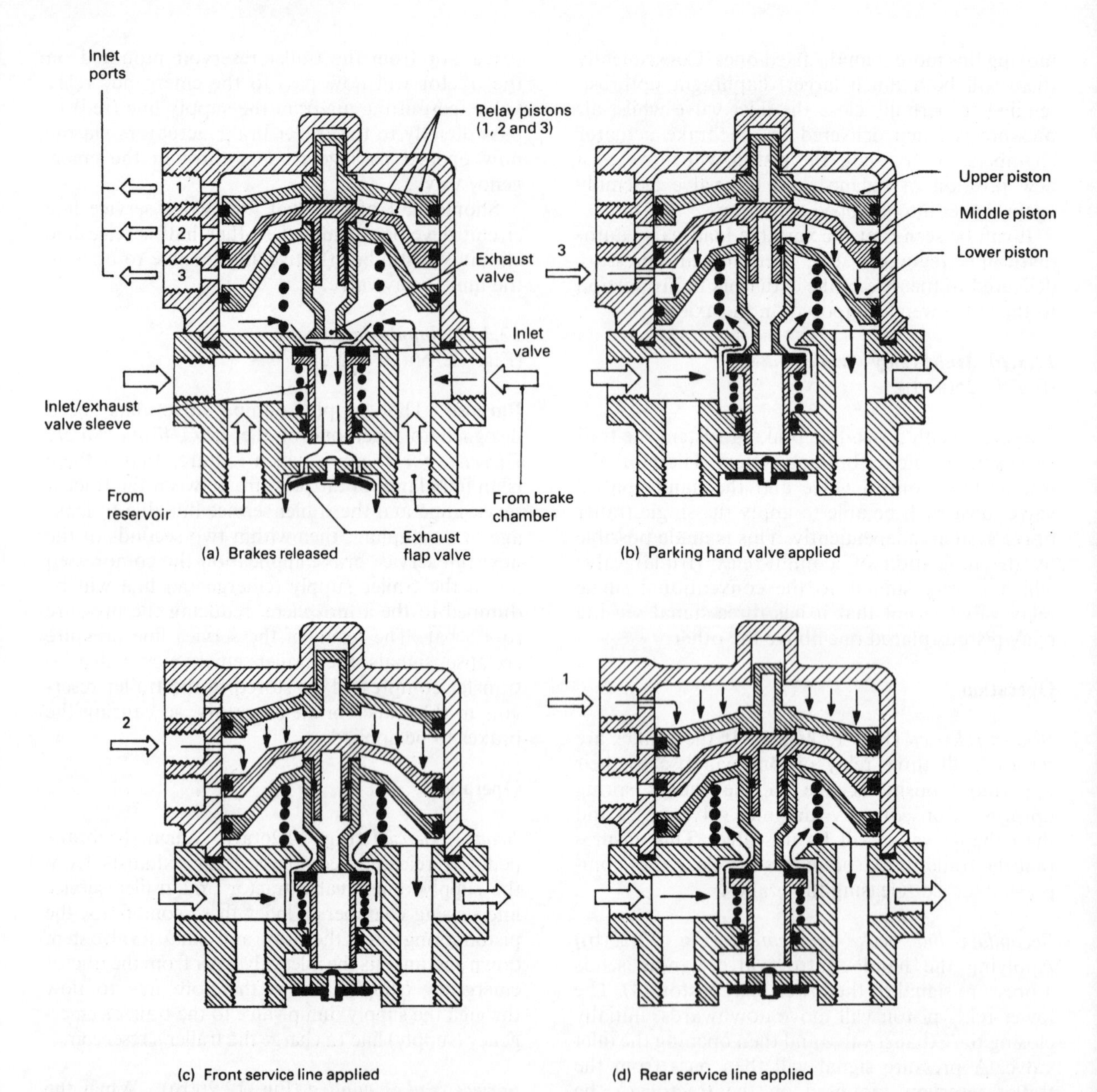

Fig. 12.25 (a–d) Multi-relay valve application

(supply) line to the trailer's emergency line via the open inlet valve mounted in the lower part of the dump valve. As a result, when the service line pressure signal is delivered to the emergency relay valve, the emergency (supply) line passes compressed air to the brake actuators on the trailer, thereby engaging the brakes.

Failure of service line pressure (Fig. 12.26(c)) Should the trailer service line be at fault, causing the piping or coupling to leak, the air pressure in the upper trailer service line sensing chamber will be lower than that in the tractor service line sensing chamber. Consequently the piston will lift, causing the inlet valve to close so that no more compressed air passes to the trailer emergency line and the exhaust valve becomes unseated. Air trapped in the trailer emergency line will immediately discharge through the centre hollow exhaust valve stem to the atmosphere. Once the trailer emergency

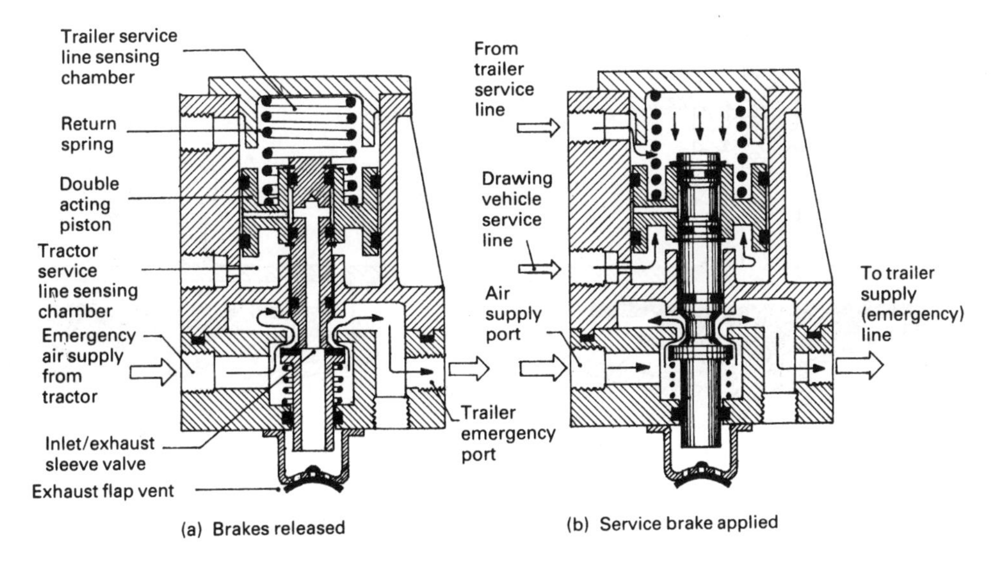

(a) Brakes released

(b) Service brake applied

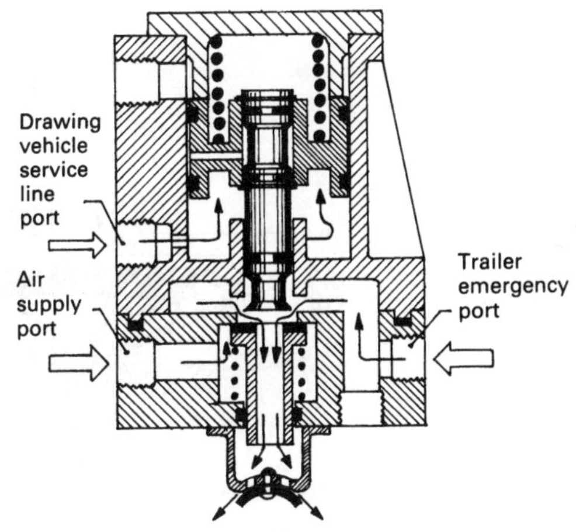

(c) Failure of service line pressure

Fig. 12.26 (a–c) Supply dump valve (Bendix)

line pressure has dropped below 2 bar, the emergency relay valve inlet passage opens, permitting the compressed air stored in the trailer reservoir to discharge into the trailer brake actuators. The towing and towed vehicles are therefore braked to a standstill.

12.3.22 Automatic reservoir drain valve
(Fig. 12.27(a–d))

Discharged air from the compressor entering the reservoir goes through a cycle of compression and expansion as it is exhausted during brake on/off applications. The consequence of the changing air density is the moisture, which is always present in the air, condenses against the cold walls of the reservoir, trickles down to the base of the chamber and thereby forms a common water pool. Permitting water to accumulate may result in the corroding of certain brake components and in cold weather this water may freeze thereby preventing the various braking valves from functioning correctly.

The object of the automatic reservoir drain valve is to constantly expel all the condensed unwanted water into the atmosphere from any container it is attached to.

Operation (Fig. 12.27(a–d)) If there is no air pressure in the braking system, both the inlet and exhaust valves will be in the closed position (Fig. 12.27(a)). Initially, as the compressor commences

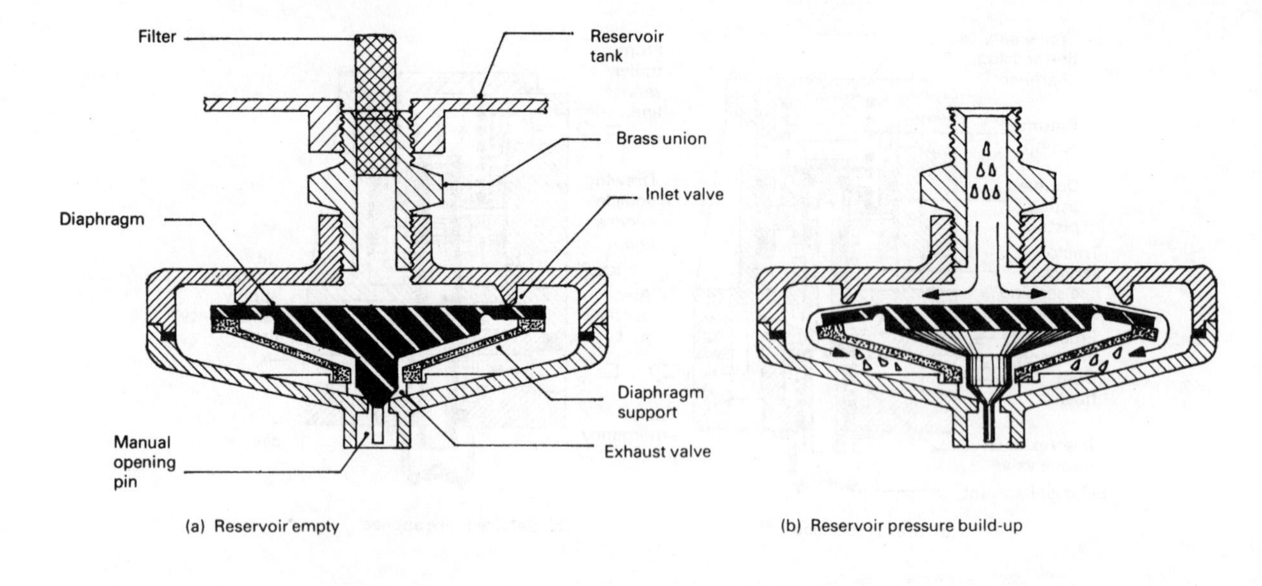

(a) Reservoir empty

(b) Reservoir pressure build-up

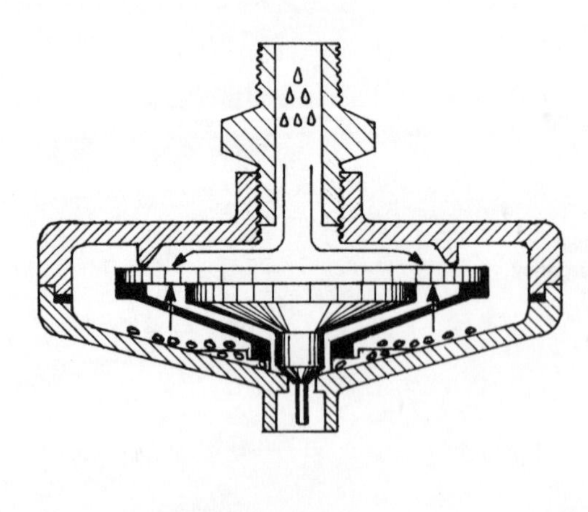

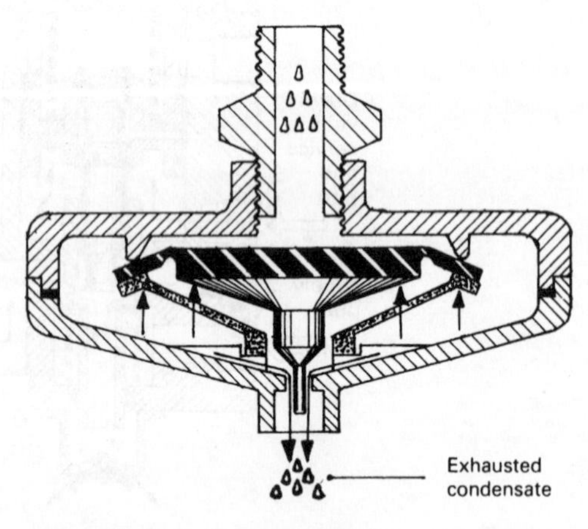

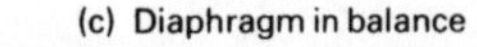

(c) Diaphragm in balance

(d) Reservoir pressure reduction

Fig. 12.27 (a–d) Automatic reservoir drain valve

to charge the reservoir, the air pressure rises and pushes open the inlet diaphragm valve. Condensed water collected above the diaphragm will now gravitate to the lower conical sump of the valve during the time the pressure is rising (Fig. 12.27(b)).

When peak governor pressure (cut-out pressure) is reached, the compressor is unloaded, cutting off further reservoir air supply so that the pressure above and below the valve diaphragm equalize. As a result, the diaphragm support below the diaphragm closes the inlet valve (Fig. 12.27(c)). Air is consumed due to the application of the brakes. There will be a reduction in reservoir air pressure so that the trapped air pressure below the diaphragm will be slightly higher than that above. Consequently the pressure difference between the two sides of the diaphragm will be sufficient to lift the middle portion of the diaphragm and the conical exhaust valve clear of its seat (Fig. 12.27(d)). Thus the air trapped underneath the diaphragm forcibly expels any condensate or foreign matter which has collected into the atmosphere.

Manual draining of the reservoir with the automatic valve is obtained by pushing up the vertical pin situated in the base of the exhaust port.

This type of automatic drain valve is designed for large reservoirs positioned away from the compressor. It is unsuitable for small sensing tanks or small volume condensers mounted near to the compressor.

12.3.23 Brake actuator chambers

Purpose Brake actuator chambers convert the energy of compressed air into the mechanical force and motion necessary to operate the foundation brakes which are attached to the ends of each axle casing or steering stub axles.

Brake actuators are designed for braking systems which must meet various application requirements. Broadly, brake actuators come into the following classifications:

1 single diaphragm chamber type,
2 double diaphragm chamber type,
3 triple diaphragm chamber type,
4 diaphragm/piston chamber push or pull type,
5 spring brake diaphragm/piston chamber type,
6 lock wedge actuator (now obsolete and will not therefore be described),
7 remote spring brakes.

Single diaphragm chamber type actuator (Fig. 12.28(a)) These actuators are normally used on single line trailers incorporating 'S' cam brake shoe expanders and slack adjuster levers.

The unit consists of two half chamber pressings supporting a flexible rubber diaphragm positioned inbetween. The sealed chamber subjected to compressed air is known as the *pressure plate chamber* whereas the chamber exposed to the atmosphere and housing the return spring is known as the *non-pressure plate chamber*. Holding the two half plates and diaphragm together is a clamp ring which, when tightened, wedges and seals the central diaphragm. The non-pressure plate also has studs attached, permitting the actuator to be bolted to the axle casing mounting bracket. A rigid disc and push rod transfer the air pressure thrust acting on the diaphragm when the brakes are applied to the slack adjuster lever.

Service line application When the brakes are applied, the foot brake valve is depressed, admitting compressed air into the pressure plate chamber. The build-up of pressure inside the chamber forces the diaphragm and push rod assembly outwards to apply the brakes. When the foot brake valve is released, the compressed air in the pressure plate chamber is exhausted to the atmosphere via the foot valve, permitting the return spring to move the diaphragm and push rod assembly back against the pressure plate wall in the off position.

Double diaphragm chamber type actuator (Fig. 12.28(b)) The double diaphragm chamber actuators are designed to be used when there are two separate air delivery circuits, known as the *service line (foot)* and the *secondary (hand) line*, systems operating on each foundation brake.

Service line application When the service (foot) brake is applied compressed air enters the service chamber via the intake ring. As the air pressure rises, the service diaphragm and push rod are forced outwards, applying the leverage to the slack adjuster.

Secondary application When the secondary (hand) brake is applied, compressed air enters the secondary chamber via the central pressure plate port. Rising air pressure forces both diaphragm and push rod outwards again, applying leverage to the slack adjuster and expanding the shoes against the brake drum.

Service and secondary brake line failure Should the service diaphragm puncture, the hand control valve secondary line can be used to operate the secondary diaphragm and push rod independently to apply the brakes.

Should the secondary diaphragm fail, the foot control valve service line will provide an alternative braking system. The air pressure in the service chamber automatically pushes the secondary diaphragm back against the central inlet port seal, preventing service line pressure escaping back through the secondary line due to a damaged or leaking secondary diaphragm.

If the secondary diaphragm and seal should fail together the service brake will still operate, provided the hand control valve is moved to the fully applied position because in this condition the exhaust valve exit for the secondary line is closed.

Triple diaphragm chamber type actuator (Fig. 12.28(c)) The triple diaphragm chamber actuator functions similarly to the double diaphragm actuator. Both types of actuators are designed to accommodate and merge both service and secondary braking systems into one integral wheel brake actuator, hence permitting both brake systems to operate independently of each other but applying their braking force and movement to one common slack adjuster lever.

Service line application When the service (foot) brake is depressed, compressed air passes through the inlet ring port between the central diaphragm and the service diaphragm nearest the push rod assembly. With increased air pressure, the secondary

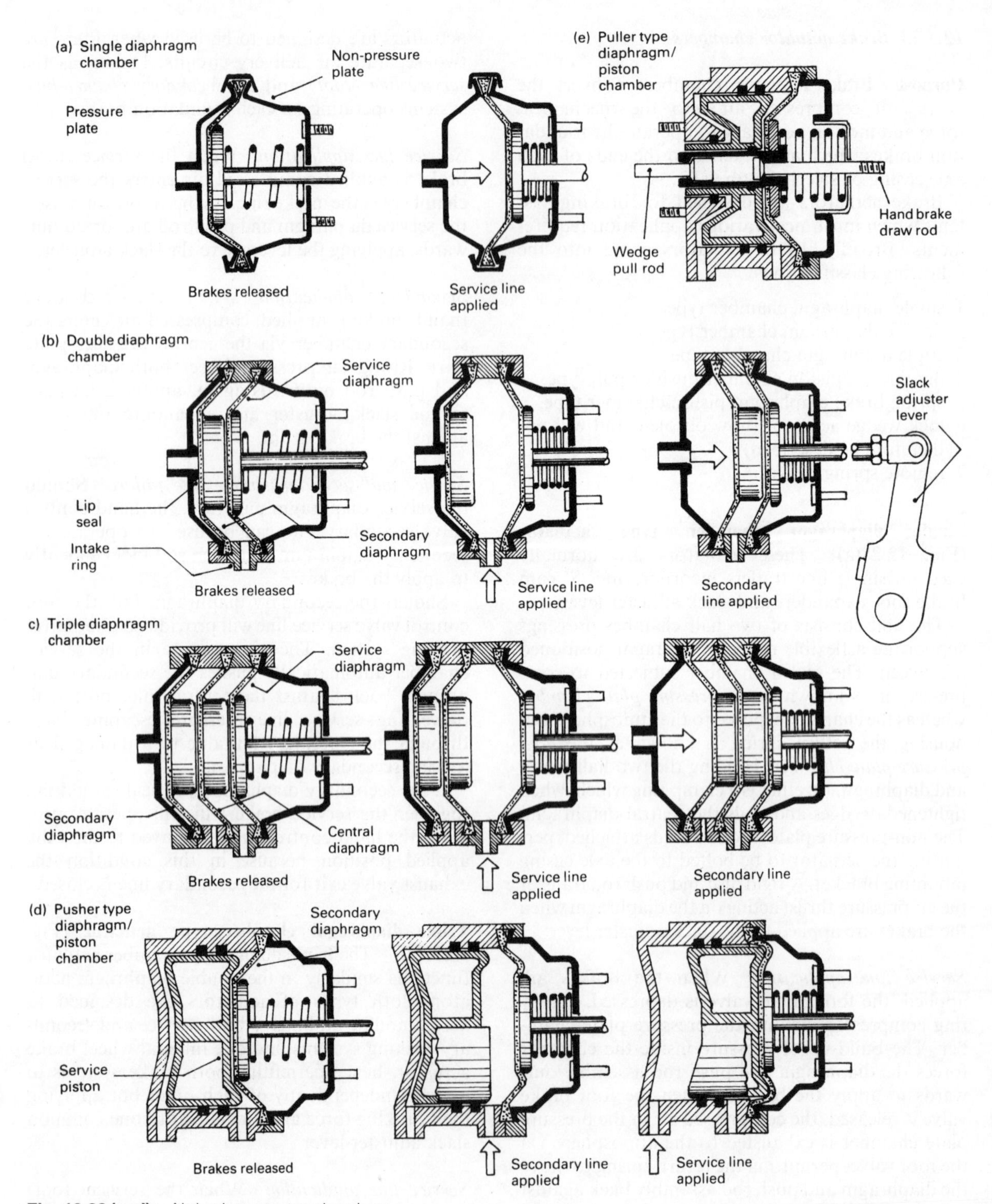

Fig. 12.28 (a–d) Air brake actuator chambers

and central diaphragm react against the pressure plate chamber and the service diaphragm forces the push rod assembly outwards. This results in the rotation of the slack adjuster lever, camshaft and cam expander, causing the brake shoes to grip the drums and apply the brakes.

Secondary line application When the secondary (hand) brake is operated, compressed air is admitted through the central end port between the pressure plate and the secondary diaphragm, causing the diaphragm and push rod assembly to expand outwards, again applying the brakes similarly to the service brake application. Releasing the hand operated lever permits the compressed air to exhaust out of the chamber through the hand control valve into the atmosphere. The return spring is now able to move all three diaphragms in the opposite direction towards the pressure plate chamber.

Service and secondary brake line failure Should the service diaphragm or central diaphragm rupture, there will be no interference with operation of the secondary diaphragm chamber. Likewise, if the secondary diaphragm fails, the service line air pressure operates only between the central and service diaphragm so that they will apply the brakes as normal.

Note that with the triple diaphragm arrangement, the central diaphragm performs the same function as the central inlet port seal in the secondary diaphragm chamber so that there is no possibility of a back leakage problem as with the double diaphragm actuator.

Diaphragm/piston chamber type actuator (Fig. 12.28) This diaphragm/piston type actuator operates similarly to the double and triple diaphragm type actuators, but because the whole cross-sectional area of the piston is effective in applying air pressure thrust (as opposed to a variable effective diaphragm area) the service chamber has a high mechanical efficiency.

Pusher type piston diaphragm chamber (Fig. 12.28(d))

Service line application When the brake pedal is depressed and the foot valve opens, air is supplied to the service port of the brake chamber. Air pressure is then applied between the rear of the chamber and the piston face. The thrust produced on the piston moves the piston, diaphragm and push rod outwards against the resistance of the return spring. The force and movement of the push rod actuates the wedge type of brake shoe expander, normally used with this type of brake actuator, to apply the brakes. Releasing the foot pedal permits the compressed air in the chambers to exhaust through the foot control valve into the atmosphere.

Secondary line application When the secondary hand control valve is operated, air is delivered to the brake chamber secondary port where it enters the space made between the piston and the diaphragm. As the air pressure increases, the piston reacts against the brake chamber's rear wall. The diaphragm and push rod are forced outwards again to force the wedge between the brake shoe expander to apply the brakes.

If air failure occurs in either the service or secondary chambers the other braking system can be operated independently to apply the brakes.

Puller type piston-diaphragm chamber (Fig. 12.28(e))

Service line application When air is delivered to the service chamber port, the piston, diaphragm, central sleeve and the wedge expander rod are all pulled outwards from the rear wall of the brake chamber, actuating the wedges of the brake shoe expander to apply the brakes.

Secondary line application Similarly when the secondary hand control is operated, air passes through the secondary chamber port to the space between the piston and diaphragm, so that the diaphragm and the central sleeve assembly are moved outwards from the chamber rear wall. As before, the brakes are applied.

When either the service or secondary line air pressure operates the brake chamber actuator, the hand brake draw rod remains stationary. When the hand brake is operated the draw rod connected to the hand brake cable (not shown) operates the sleeve and pull rod applying the brakes.

Spring brake actuator (Fig. 12.29(a–d)) Spring brake actuators are designed for use in air pressure braking systems to produce the force and movement necessary to operate the foundation brakes. The actuator uses air pressure to respond and control both service and secondary braking. The conversion of air pressure into mechanical force and travel for the service braking systems is obtained by a conventional diaphragm-operated servo chamber. In contrast, secondary braking and parking is achieved through the strain energy stored in a powerful coil spring which, when permitted to expand, applies mechanical effort and displacement to the foundation brake shoe expander.

Operation

All brakes released (Fig. 12.29(a)) Under normal driving conditions both service and secondary/park brake systems must be in the 'off' position. To release both braking systems, air pressure (above the low signal pressure) is supplied to the spring piston chamber which compresses the power spring and holds off the foundation brakes. At the same time air pressure is exhausted from the service line diaphragm chamber.

Service line applied (Fig. 12.29(b)) Air pressure controlled progressively by the foot valve is supplied to the service diaphragm chamber port. The cross-sectional air of the diaphragm exposed to the air pressure is subjected to a thrust forcing the push rod outwards. The combined movement and force is applied to the slack adjuster which then relays it to the camshaft, expander cam and shoes of the

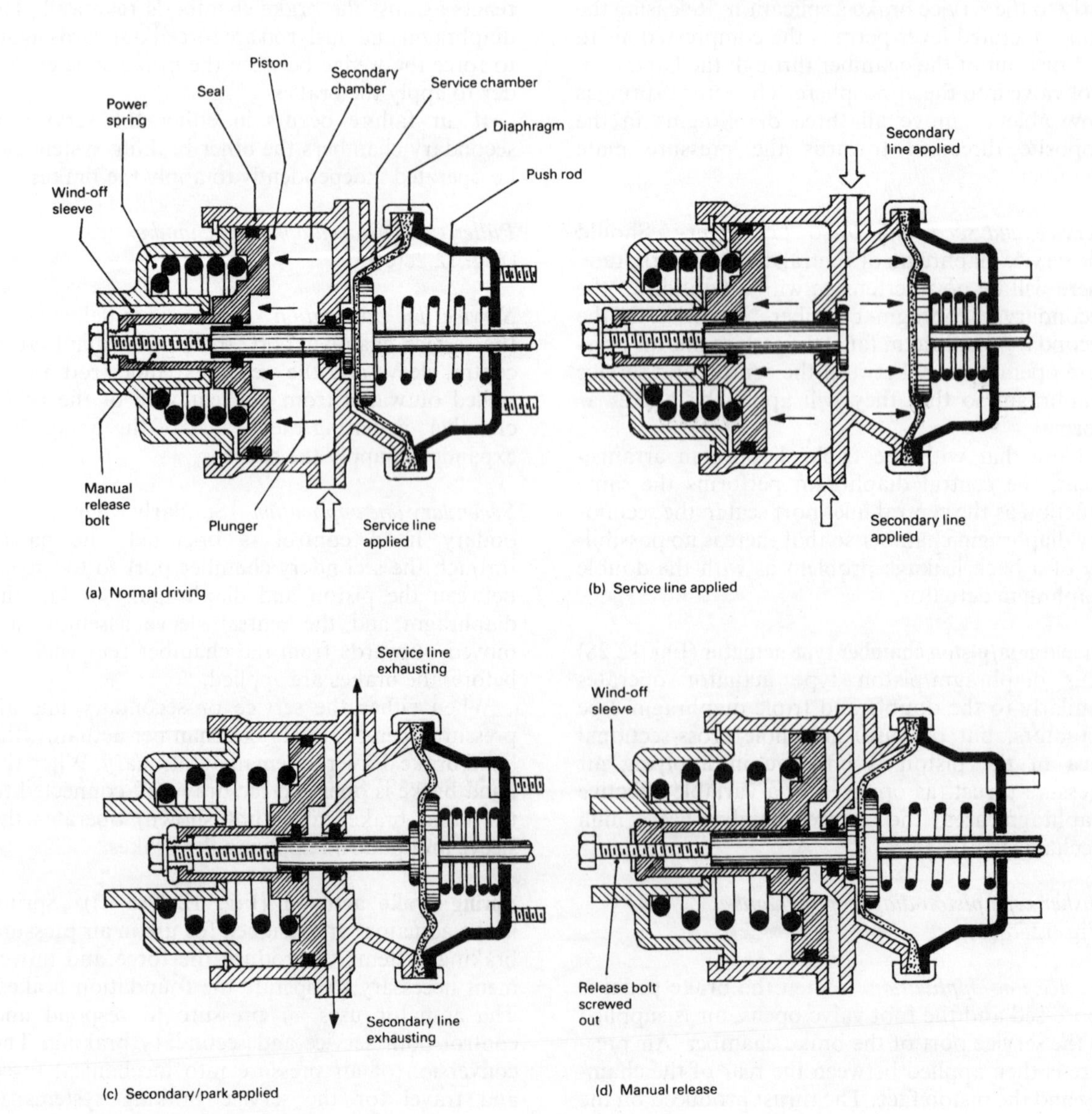

Fig. 12.29 (a–d) Spring brake actuator

foundation brakes. During service brake operation the power spring is held in compression by the secondary line air pressure so that it does not compound the service brake operational force.

Secondary/parking applied (Fig. 12.29(c)) Air pressure, controlled progressively by the hand valve, is released from the piston spring chamber, permitting the power spring to expand and push the plunger and push rod outwards against the resistance of the brake cam which is expanding the brake shoes up to the drum. When the vehicle is at a standstill, the hand lever is moved from secondary to park position, where any remaining air pressure in the secondary line and piston spring chamber is exhausted via the open hand control valve to the atmosphere. The brakes are then held on purely by the strain energy of the extended power spring and therefore are not dependent upon air pressure with its inherent leakage problem.

Manual release applied (Fig. 12.29(d)) Should there be a secondary line air supply failure, then the foundation brakes may be released for towing purposes or removal of the spring brake actuator by the readily accessible release bolt. Winding out the release bolt permits the plunger to move into the wind-off sleeve so that the push rod is able to return to the 'off' position, thereby releasing the brake shoes.

12.4 Vehicle retarders

12.4.1 Engine overrun vehicle retardation (Fig. 12.30)

When an engine is being overrun by the transmission system, the accelerator pedal is released and no fuel is available for combustion. Therefore the normal expansion in the power stroke by the burning products of combustion is replaced by the expansion of the compressed air alone. Consequently, the energy used in compressing the air on its upstroke is only partially given back on the downstroke, due to the friction and heat losses, so that additional work must be done to rotate the crankshaft (Fig. 12.30).

Thus an external energy source is necessary to keep this crankshaft rotating when no power is produced in the cylinder. This is conveniently utilized when the vehicle is being propelled by its own kinetic energy: the work put into overcoming the pumping, friction and heat losses while driving the transmission and engine reduces the vehicle kinetic energy, causing the vehicle to slow down. The retarding torque required on the road driving wheels to rotate the transmission and engine consists of the torque necessary to overcome the engine resistances when there is no power produced multiplied by the overall gear ratio of the transmission system.

12.4.2 Exhaust compression vehicle retardation (Fig. 12.30)

Principle of exhaust compression retarder The object of the exhaust brake (retarded) is to convert the exhaust gas expelling stroke into an air compressing power, absorbing upstroke when the accelerator pedal is released and the vehicle possesses a surplus of kinetic energy. This happens on overrun and tends to propel the vehicle forward, thus causing the transmission to rotate the engine. To achieve this negative work, a butterfly valve or slide valve is incorporated into the exhaust pipe as near to the manifold as possible, so that during the exhaust stroke the volume of air in the engine's cylinder will be displaced and contained within the exhaust system, causing its pressure to rise (Fig. 12.30). Just after the piston has reversed its direction and commences to move outwards on its induction stroke, the exhaust valve closes, trapping

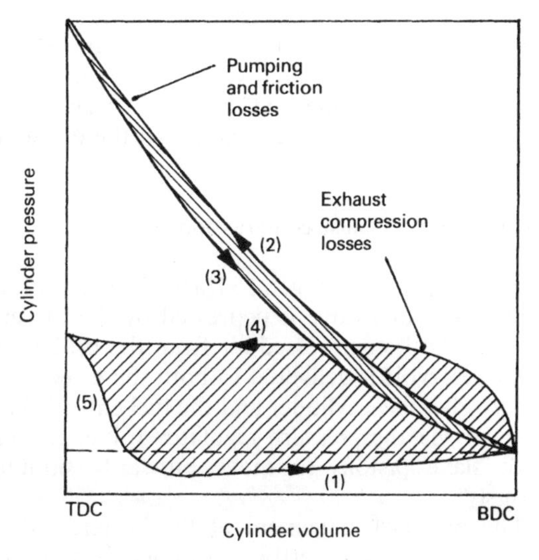

Fig. 12.30 Pressure/volume diagram for a four stroke engine installed with an exhaust compression

the air remaining between the exhaust port poppet valve and the exhaust butterfly brake valve in the exhaust system. Therefore approximately 90% of the exhaust upstroke is devoted in compressing air and absorbing energy (Fig. 12.30) before the inlet valve opens. This energy is then released to the atmosphere and cannot do useful work in pushing down the piston on the following induction stroke. Hence there is a net gain in negative work done by the engine whilst resisting the rotation of the crankshaft, which considerably helps to reduce the vehicle's speed.

Air will therefore be drawn into the cylinder on each induction stroke as normal, but on subsequent exhaust strokes the pressure in the exhaust manifold and pipe increases to a maximum of between 2.5 and 4.5 bar, depending upon the valve overlap compression ratio and engine speed.

With increased engine speed, the pressure build-up in the exhaust system eventually overcomes the exhaust valve spring closing force and unseats the exhaust valve. Pressurized air will then flow back into the cylinder and escape out of the induction port the next time the inlet valve opens. Consequently the upper limit to the pressure build-up on the exhaust upstroke is controlled by the amount of trapped compressed air stored in the exhaust system behind the exhaust valve when the exhaust valve opens and the piston commences on its modified exhaust upstroke.

With several cylinders feeding into a common exhaust manifold, pressure fluctuations are considerably reduced and a relatively high average pressure can be maintained. Subsequently a great deal of negative work can be done by all the cylinders collectively.

Operation of exhaust compression brake

Retarder operating (Fig. 12.31(a)) When the foot control (on/off) valve is depressed by the driver's left foot heel, compressed air from the brake system's reservoir is delivered to both the brake butterfly valve slave cylinder and the fuel cut-off slave cylinder via the pressure regulator valve, causing both slave pistons to move outwards simultaneously.

The outward movement of the butterfly slave piston causes the butterfly operating lever to rotate about its spindle to close the exhaust passage leading to the silencer. The engine then becomes a single stage low pressure compressor driven through the transmission by the road wheels. The air pressure established in the exhaust manifold and pipe due to blocking the exhaust exit reacts against the piston movement on most of the exhaust stroke, thus producing a retarding torque on the propeller shaft.

The outward movement of the fuel cut-off slave piston rotates the shift bell crank lever, causing the vertical link rod to pull down and fold the two governor link rods. As a result, the change speed lever of the injection pump is moved to the closed or full cut-off position and at the same time the accelerator pedal is drawn towards the floorboard.

The exhaust compression brake remains operative during the whole time the foot control valve is depressed.

Retarder inoperative (Fig. 12.31(b)) When the (on/off) foot valve is released during clutching or declutching, the compressed air in both slave cylinders is exhausted through the control valve. This permits the slave cylinders' respective return springs to open the butterfly valve, and to unfold the governor link rods so that their combined extended length moves the change speed lever to the full delivery position, and at the same time raises the accelerator pedal from the floorboards. The exhaust compression brake system is then inoperative and the engine can be driven normally once again.

12.4.3 Engine compressed air type retarder (Jacobs) (Fig. 12.33)

A cylinder compression retarder converts a power producing diesel engine into a power absorbing air compressor.

The compressed air engine retarder consists of a hydraulic circuit supplied by the engine oil pump which uses the existing injector rocker motion to open the exhaust valve at the end of the compression stroke via a pair of master and slave pistons actuated by a solenoid valve and a piston control valve.

These compressed air brake units are made to fit on top of the cylinder head and are designed for engines which incorporate combined pump and injector units such as the *Cummins* and *Detroit Diesel.*

Theory of operation With a conventional engine valve timing, work is done in compressing air on the inward compression stroke. The much reduced volume of air then gives out its energy by driving outwards the piston on its expansion stroke.

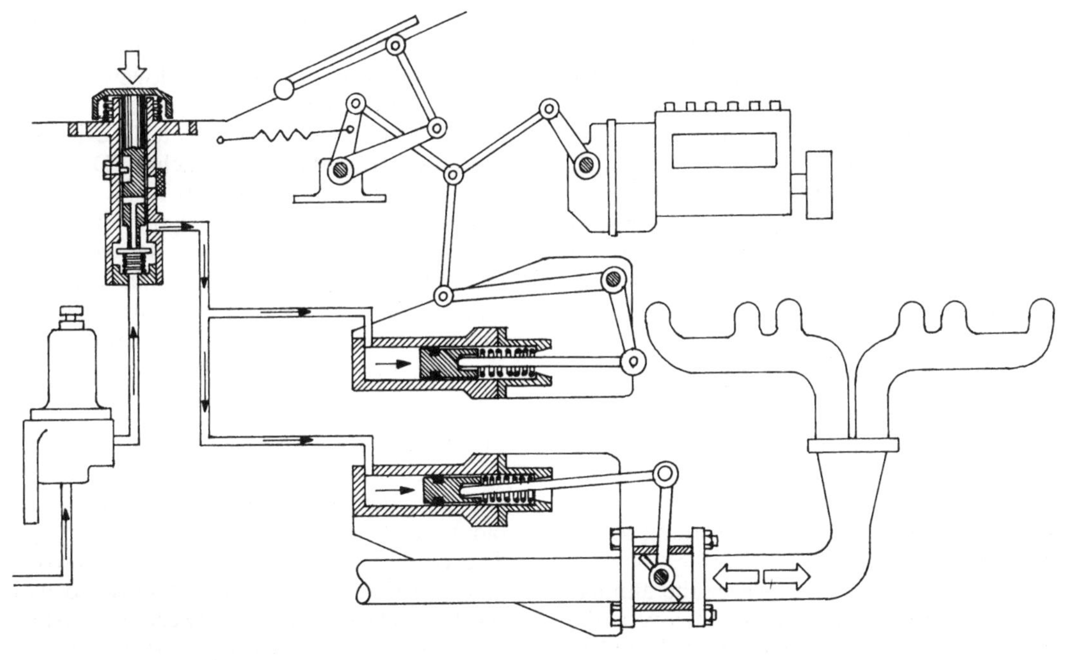

(a) Retarder operating

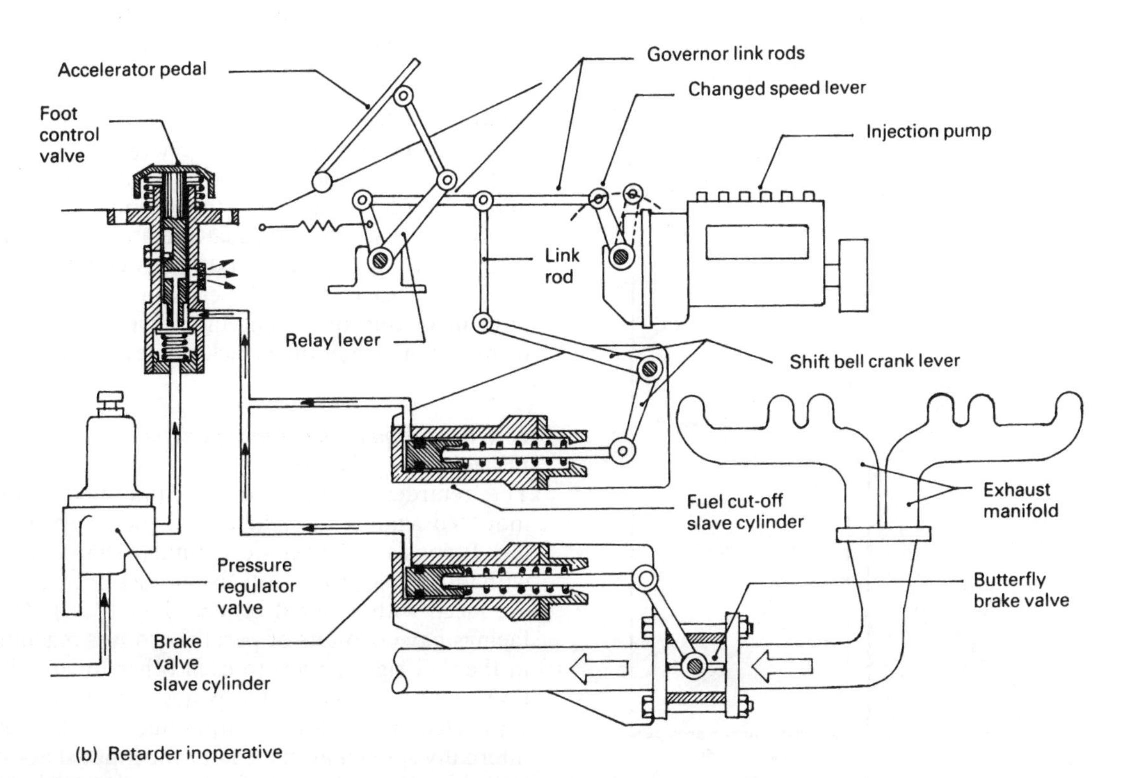

(b) Retarder inoperative

Fig. 12.31 (a and b) Exhaust compression (brake) type retarder

Therefore, except for frictional losses, there is very little energy lost in rotating the engine on overrun.

The adoption of a mechanism which modifies the exhaust valve timing to open the exhaust valves at the beginning of the expansion stroke causes the release of air into the atmosphere via the exhaust ports (*exhaust blowdown*) (Fig. 12.32), instead of making it do useful work in expanding the piston outwards. The effect of this is a net energy loss so that considerably more effort is required to crank the engine under these conditions.

Engine retarder engaged (exhaust blowdown) (Fig. 12.33(a)) When the accelerator pedal is released, the accelerator switch closes to complete the electrical circuit and thereby energizes the solenoid. The solenoid valve closes the oil return passage immediately and opens the passage leading to the control valve. Pressurized oil now pushes the control valve piston up to a point where the annular waste of the control valve piston uncovers the slave piston passage and unseats the ball. Oil at pump pressure therefore passes to the upper crown of the slave and master pistons. This forces the master piston downwards to take up the free movement between itself and the injector rocker adjustment screw. As the camshaft rotates, in the normal cycle of events, the cam lift forces the master piston upwards, causing the ball in the control valve to seat. The trapped contracting volume of oil

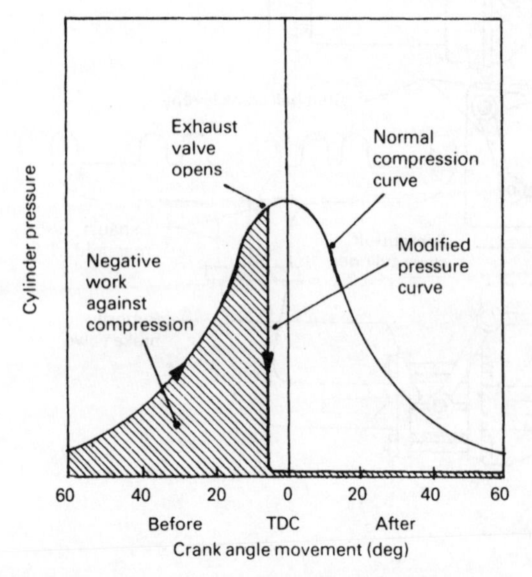

Fig. 12.32 Typical pressure/crank diagram for a four stroke engine installed with an engine compressor retarder

between the control valve and master piston now increases its pressure sufficiently to force the slave piston and valve cross head downwards. Consequently both exhaust valves are unseated approximately 1.5 mm and so as a result the exhaust valves open just before the piston reaches TDC position at the end of the compression stroke.

Whilst the engine is on exhaust blowdown, the vehicle speed is reducing and the engine is being overrun by the transmission, so that with the accelerator pedal released and the centrifugal governor weights thrown outwards, fuel is prevented from injecting into the engine cylinders.

Engine retarder disengaged (Fig. 12.33(b)) When the accelerator pedal is depressed, the solenoid circuit is interrupted, causing the de-energized solenoid valve to open the oil pump return passage. The oil pressure under the control valve piston collapses, causing the valve to move to its lowest position. The trapped oil between the master piston and slave piston therefore escapes through the passage opened by the control valve piston. The slave piston is then permitted to rise enabling, the cross head and exhaust valve to operate as normal.

Retarder control A master 'on/off' dash switch in combination with automatic accelerator and clutch switches allows the driver to operate the engine pump retarder under most conditions. Complete release of the accelerator pedal operates the retarder. Depression of the accelerator or clutch pedals opens the electrical circuit, permitting gear changes to be made during descent and prevents the engine from stalling when the vehicle comes to a halt.

12.4.4 Multiplate friction type retarder (Ferodo) (Fig. 12.34)

The retarder is an oil-cooled multiplate brake mounted against the rear end of the gearbox casing. It consists of four steel annular shaped plates with internal locating slots. Both sides of each plate are faced with sintered bronze (Fig. 12.34). These facings have two sets of parallel grooves machined in them at right angles to each other for distribution of the oil. The drive plates are aligned and supported on the slotted output hub, which is itself internally splined at one end to the input shaft and bolted to the output shaft at the other end. This provides a straight-through drive between the gearbox main shaft and the propeller shaft. Support is provided for the output hub and shaft by an inner

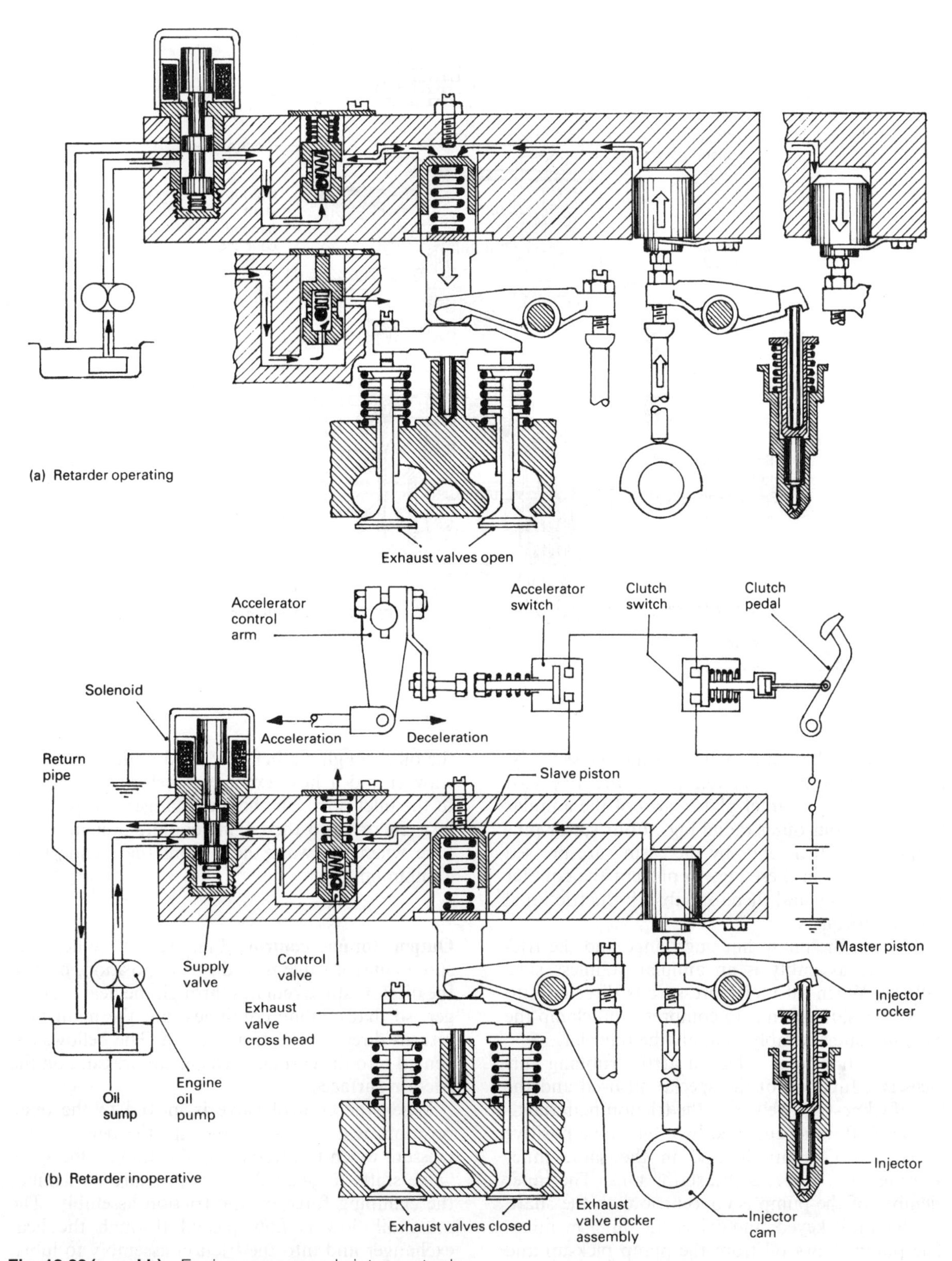

Fig. 12.33 (a and b) Engine compressed air type retarder

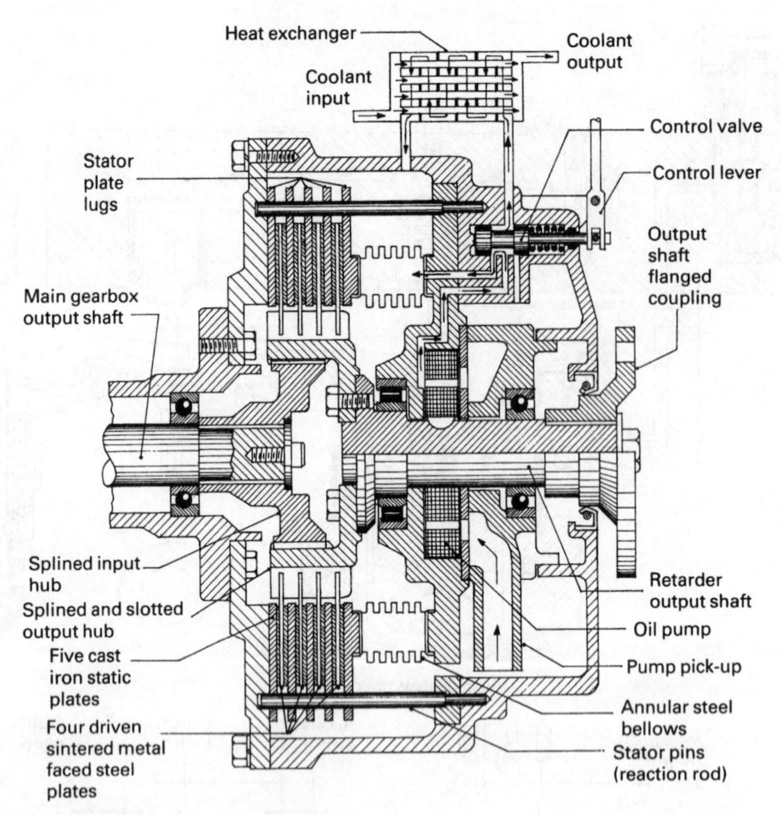

Fig. 12.34 Multiplate friction type retarder

and outer roller and ball bearing respectively. Interleaved with the driven plates are five cast iron stationary counter plates, also of the annular form, with four outer radial lugs. Four stator pins supported at their ends by the casing are pressed through holes in these lugs to prevent the counter plates rotating, and therefore absorb the frictional reaction torque.

Between the pump housing flange and the friction plate assembly is an annular stainless steel bellows. When oil under pressure is directed into the bellows, it expands to compress and clamp the friction plate assembly to apply the retarder.

The friction level achieved at the rubbing surfaces is a function of the special oil used and the film thickness, as well as of the friction materials.

The oil flow is generated by a lobe type positive displacement pump, housed in the same inner housing that supports the stator pins. The inner member of the pump is concentric with the shaft, to which it is keyed, and drives the outer member. The pump draws oil from the pump pick-up and circulates it through a control valve. It then passes the oil through a relief valve and a filter (both not shown) and a heat exchanger before returning it to the inlet port. The heat exchanger dissipates its heat energy into the engine cooling system at the time when the waste heat from the engine is at a minimum.

Output torque control (Fig. 12.34) When the spool control valve is in the 'off' position, part of the oil flow still circulates through the heat exchanger, so that cooling continues, but the main flow returns direct to the casing sump. The bellows are vented into the casing, releasing all pressure on the friction surfaces.

When the control valve is moved to the open position, it directs some oil into the bellows at a pressure which is governed by the amount the spool valve shifts to one side. This pressure determines the clamping force on the friction assembly. The main oil flow is now passed through the heat exchanger and into the friction assembly to lubricate and cool the friction plates.

12.4.5 Electro-magnetic eddy current type retarder (Telma) (Fig. 12.35(a, b and c))

The essential components are a stator, a support plate, which carries suitably arranged solenoids and is attached either to the chassis for mid-propeller shaft location or on the rear end of the gearbox (Fig. 12.35(a)), and a rotor assembly mounted on a flange hub. The stator consists of a steel dished plate mounted on a support bracket which is itself bolted to a rear gearbox flange. On the outward facing dished stator plate side are fixed eight solenoids with their axis parallel to that of the transmission. The rotor, made up of two soft steel discs facing the stator pole pieces, is bolted to a hub which is supported at the propeller shaft end by a ball bearing and at its other end by the gearbox output shaft. The drive from the gearbox output shaft is transferred to the propeller universal joint via the internally splined rotor hub sleeve. The rotor discs incorporate spiral shaped (turbine type blades) vanes to provide a large exposed area and to induce airflow sufficient to dissipate the heat generated by the current induced in the rotor and that produced in the stationary solenoid windings.

Four independent circuits are energized by the vehicle's battery through a relay box, itself controlled by a fingertip lever switch usually positioned under the steering wheel (Fig. 12.35(b and c)). These solenoid circuits are arranged in parallel as an added safety precaution because, in the event of failure of one circuit, the unit can still develop three-quarters of its normal power. The control lever has four positions beside 'off' which respectively energize two, four, six and eight poles. The solenoid circuit consumption is fairly heavy, ranging for a typical retarder from 40 to 180 amperes for a 12 volt system.

Operating principles (Fig. 12.36) If current is introduced to each pole piece winding, a magnetic flux is produced which interlinks each of the winding loops and extends across the air gap into the steel rotor disc, joining up with the flux created from adjacent windings (Fig. 12.36).

When the rotors revolve, a different section of the disc passes through the established flux so that in effect the flux in any part of the disc is continuously varying. As a result, the flux in any one segmental portion of the disc, as it sweeps across the faces of the pole pieces, increases and then decreases in strength as it moves towards and then away from the established flux field. The change in flux linkage with each segmental portion of the disc which passes an adjacent pole face induces an electromotive force (voltage) into the disc. Because the disc is an electrical conductor, these induced voltages will cause corresponding induced currents to flow in the rotor disc. These currents are termed *eddy currents* because of the way in which they whirl around within the metal.

Collectively the eddy currents produce an additional interlinking flux which opposes the motion of the rotor disc. This is really Lenz's Law which states that the direction of an induced voltage is such as to tend to set up a current flow, which in turn causes a force opposing the change which is producing the voltage. In other words, the eddy currents oppose the motion which produced them. Thus the magnetic field set up by these solenoids create eddy currents in the rotor discs as they revolve, and then eddy currents produce a magnetic drag force tending to slow down the rotors and consequently the propeller shaft (Fig. 12.36).

The induced eddy currents are created inside the steel discs in a perpendicular direction to the flux, and therefore heat (I^2Rt) is produced in the metal.

The retarding drag force or resisting torque varies with both the rotational speed of the rotor and propeller shaft and the strength of the electromagnetic field, which is itself controlled by the amount of current supplied.

12.4.6 Hydraulic type retarder (Voith) (Fig. 12.37)

The design of a hydraulic retarder is similar to that of a fluid coupling. Basically, the retarder consists of two saucer-shaped discs, a revolving rotor (or impellor) and a stationary stator (or reaction member) which are cast with a number of flat radial vanes or blades for directing the flowpath of the fluid. The rotor is bolted to the flange of the internally splined drive shaft hub, which is itself mounted over the external splines formed on both the gearbox mainshaft and the flanged output shaft, thereby coupling the two drive members together. Support to the drive shaft hub and rotor is given by a roller bearing recessed in the side of the stator, which is in turn housed firmly within the retarder casing.

Theory of operation (Fig. 12.37) The two half-saucer members are placed face to face so that fluid can rotate as a vortex within the cells created by the radial vanes (Fig. 12.37).

When the transmission drives the rotor on overrun and fluid (oil) is introduced into the spaces between the rotor and stator, the fluid is subject to centrifugal force causing it to be accelerated

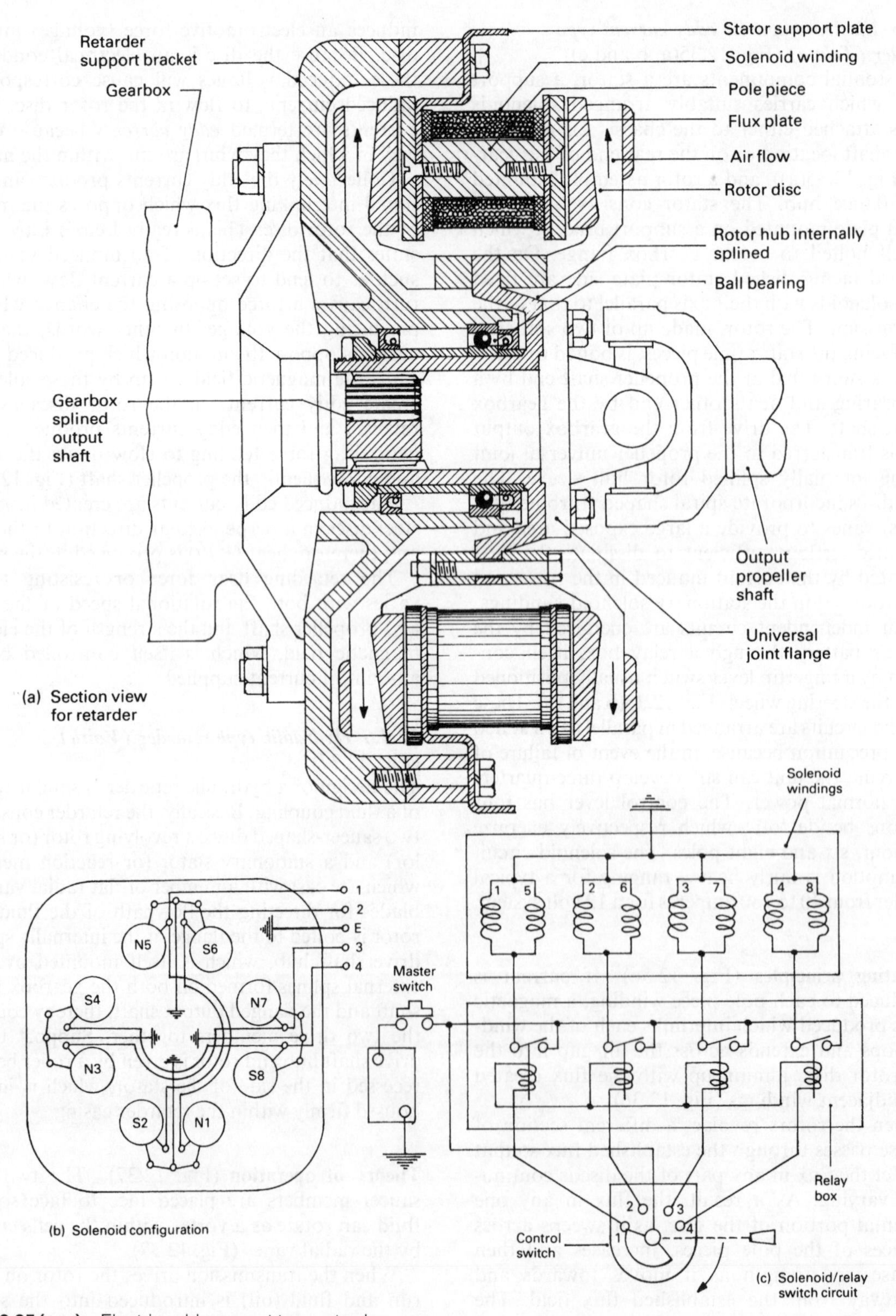

Fig. 12.35 (a–c) Electric eddy current type retarder

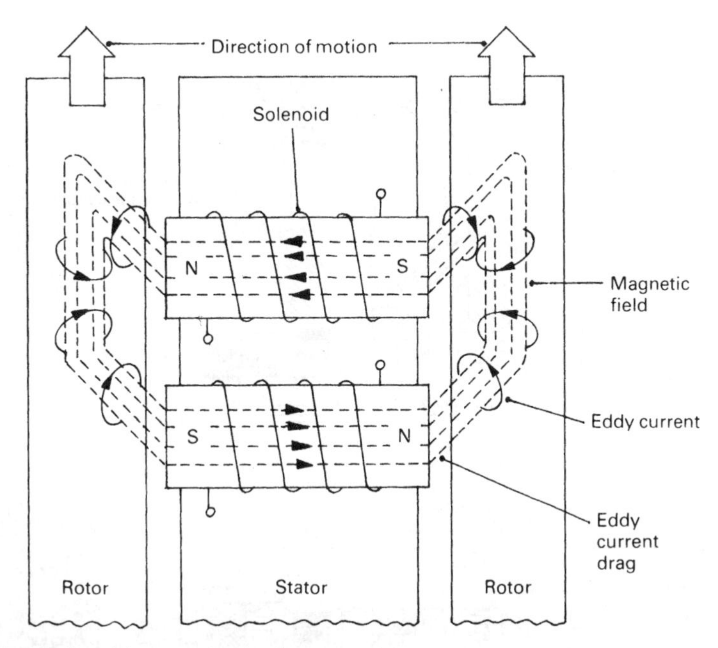

Fig. 12.36 Principle of electric eddy current retarder

radially outwards. As the fluid reaches the outmost periphery of the rotor cells, it is flung across the junction made between the rotor and stator faces. It then decelerates as it is guided towards the inner periphery of the rotor cells to where the cycle of events once again commences. The kinetic energy imparted to the fluid passing from the revolving rotor to the fixed stator produces a counter reaction against the driven rotor. This counter reaction therefore opposes the propelling energy at the road wheels developed by the momentum of the moving vehicle, causing the vehicle to reduce speed.

The kinetic energy produced by the rapidly moving fluid as it impinges onto the stator cells, and the turbulance created by the movement of the fluid between the cells is all converted into heat energy. Hence the kinetic energy of the vehicle is converted into heat which is absorbed by the fluid and then dissipated via a heat exchanger to the cooling system of the engine.

The poor absorption capacity of the hydraulic retarder increases almost with the cube of the propeller shaft speed for a given rotor diameter.

When the retarder is not in use the rotor rotates in air, generating a drag. In order to keep this drag as low as possible, a number of stator pins are mounted inside and around the stator cells. These disc-headed pins tend to interfere with the air circulating between the moving and stationary half-cells when they have been emptied of fluid, thereby considerably reducing the relatively large windage losses which normally exist.

Output torque control (Fig. 12.37) In order to provide good retardation at low speeds, the retarder is designed so that maximum braking torque is reached at approximately a quarter of the maximum rotor speed/propeller speed. However, the torque developed is proportional to the square of the speed, and when the vehicle speed increases, the braking torque becomes too great and must, therefore, be limited. This is simply achieved by means of a relief valve, controlling the fluid pressure which then limits the maximum torque.

The preloading of the relief valve spring is increased or reduced by means of an air pressure-regulated servo assisted piston (Fig. 12.37). The control valve can be operated either by a hand lever when the unit is used as a continuous retarder, or by the foot control valve when it is used for making frequent stops.

When the retarder foot control valve is depressed, air from the auxiliary air brake reservoir is permitted to flow to the servo cylinder and piston. The servo piston is pushed downwards relaying this movement to the relief spill valve via the inner spring. This causes the relief valve spool to partially close the return flow passage to the sump and to open the passage leading to the inner

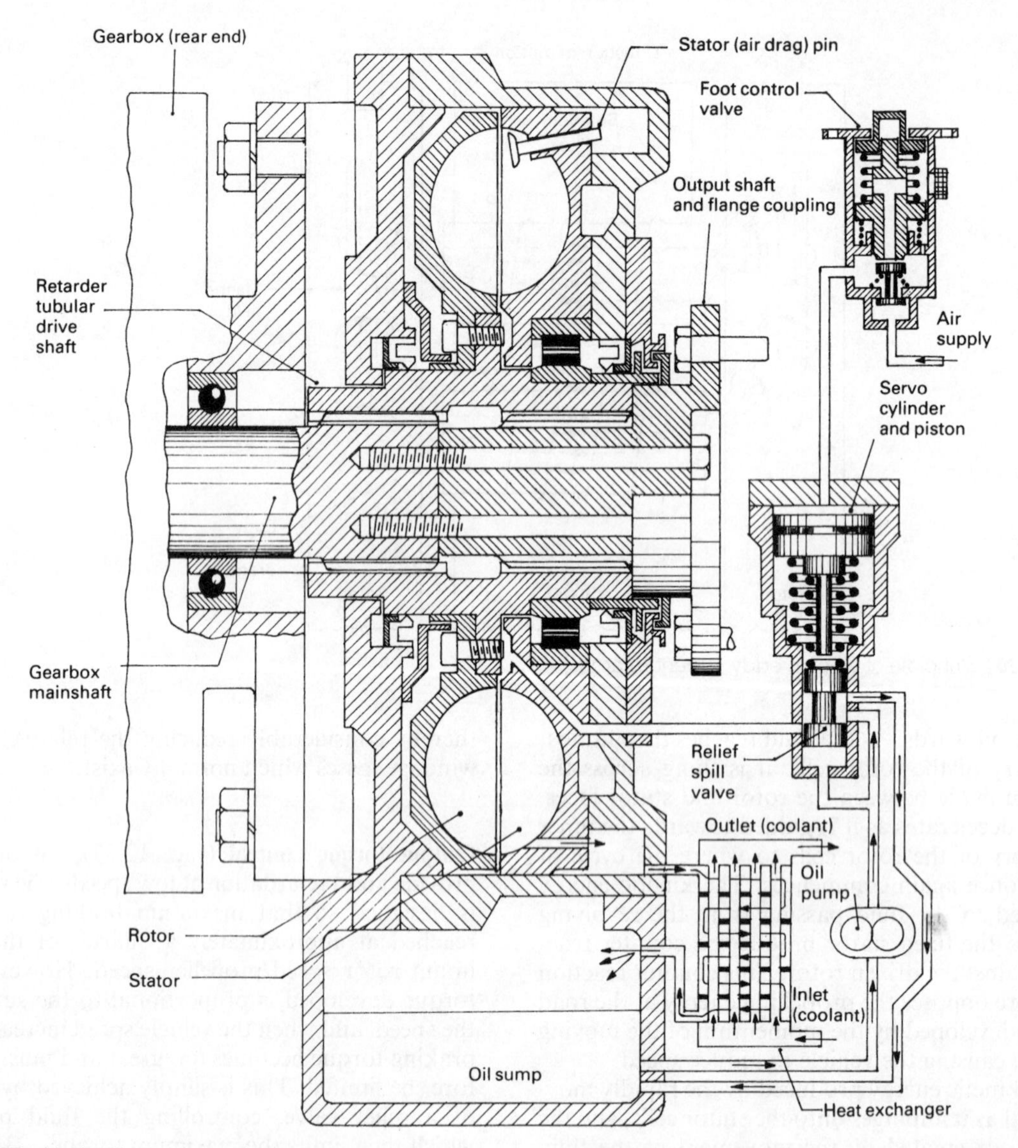

Fig. 12.37 Hydraulic type retarder

periphery of the stator. Fluid (oil) from the hydraulic pump now fills the rotor and stator cells according to the degree of retardation required, this being controlled by the foot valve movement. At any foot valve setting equilibrium is achieved between the air pressure acting on top of the servo piston and the opposing hydraulic pressure below the spool relief valve, which is itself controlled by the hydraulic pump speed and the amount of fluid escaping back to the engine's sump. The air feed pressure to the servo piston therefore permits the stepless and sensitive selection of any required retarding torque within the retarder's speed/torque characteristics.

Should the oil supply pressure become excessively high, the spool valve will lift against the control air pressure, causing the stator oil supply passage to partially close while opening the return flow passage so that fluid pressure inside the retarder casing is reduced.

12.4.7 A comparison of retarder power and torque absorbing characteristics

(Figs 12.38 and 12.39)

Retarders may be divided into those which utilize the engine in some way to produce a retarding effort and those which are mounted behind the

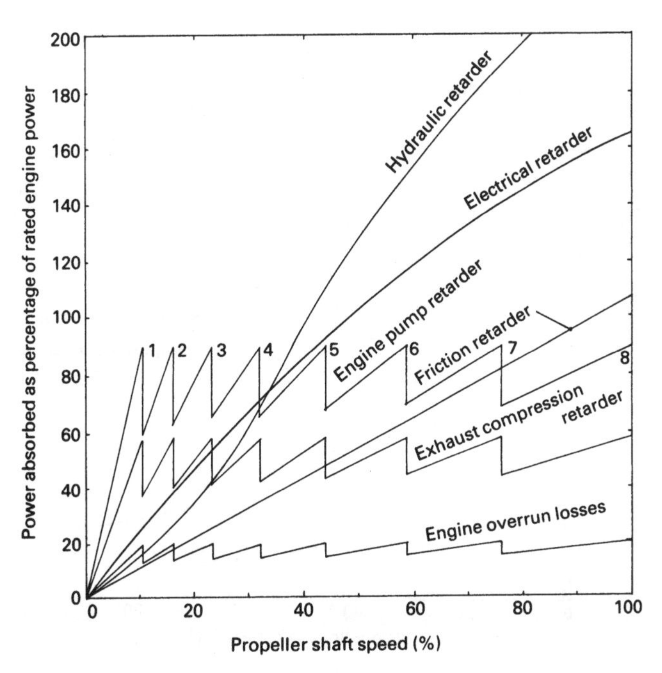

Fig. 12.38 Typical comparison of power absorption of various retarders relative to propellor shaft speed

gearbox, between the propeller shafts or in front of the final drive.

Retarders which convert the engine into a pump, such as the exhaust compression type or engine compressed air Jacobs type retarders, improve their performance in terms of power and torque absorption by using the gear ratios on overrun similarly to when the engine is used to propel the vehicle forwards. This is shown by the sawtooth power curve (Fig. 12.38) and the family of torque curves (Fig. 12.39) for the engine pump Jacobs type retarder. In the cases of the exhaust compression type retarder and engine overrun loss torque curves, the individual gear ratio torque curves are all shown merged into one for simplicity. Thus it can be seen that three methods, engine compressed air, exhaust compression and engine overrun losses, which use the engine to retard the vehicle, all depend for their effectiveness on the selection of the lowest possible gear ratio without over speeding the engine. As the gear ratio becomes more direct, the torque multiplication is reduced so that there is less turning resistance provided at the propeller shaft.

For retarders installed in the transmission after the gearbox there is only one speed range. It can be seen that retarders within this classification, such as the multi-friction plate, hydrokinetic and electrical eddy current type retarders all show an increase in power absorption in proportion to propeller shaft speed (Fig. 12.38). The slight deviation from a complete linear power rise for both hydraulic and electrical retarders is due to hydrodynamic and eddy current stabilizing conditions. It can be seen that in the lower speed range the hydraulic retarder absorbs slightly less power than the electrical retarder, but as the propeller shaft rises this is reversed and the hydraulic retarder absorbs proportionally more power, whereas the multiplate friction retarder produces a direct increase in power absorption throughout its speed range, but at a much lower rate compared to hydraulic and electrical retarders because of the difficulties in dissipating the generated heat.

When considering the torque absorption characteristics of these retarders (Fig. 12.39), the electrical retarder is capable of producing a high retarding torque when engaged almost immediately as the propeller shaft commences to rotate, reaching a peak at roughly 10% of its maximum speed range. It then declines somewhat, followed by a relatively constant output over the remainder of its speed range. However, the hydraulic retarder shows a slower resisting torque build-up which then gently exceeds that of the eddy current resisting torque curve, gradually reaching a peak followed by a very small decline as the propeller shaft speed approaches a maximum. In comparison to the

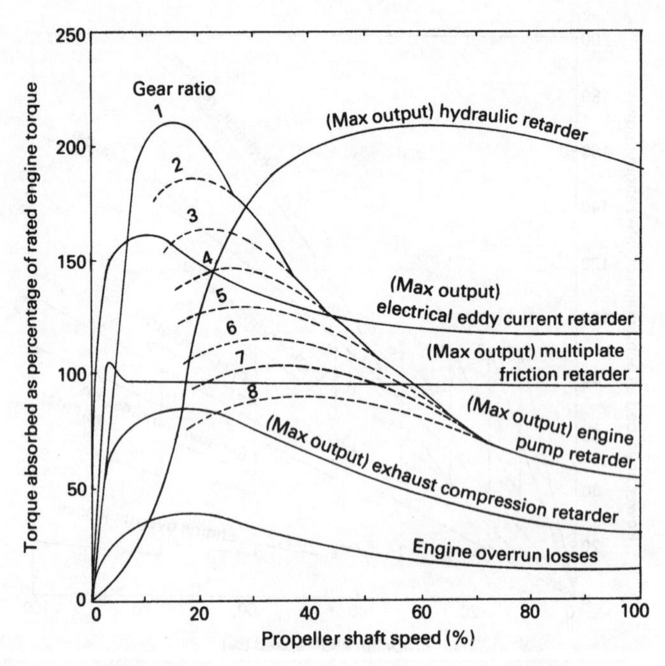

Fig. 12.39 Typical comparison of torque produced by various retarders relative to propeller shaft

other retarders, the multiplate friction retarder provides a resisting torque the instant the two sets of friction plates are pressed together. The relative slippage between plates provides the classical static high friction peak followed immediately by a much lower steady dynamic frictional torque which tends to be consistant throughout the retarder's operating speed range. What is not shown in Figs 12.38 and 12.39 is that the electrical, hydraulic and friction retarder outputs are controlled by the driver and are generally much reduced to suit the driving terrain of the vehicle.

12.5 Electronic-pneumatic brakes

12.5.1 Introduction to electronic-pneumatic brakes (Fig. 12.40)

The electronic-pneumatic brake (EPB) system controls the entire braking process; this includes ABS/TCS braking when conditions demand, and the layout consists of a single electronic-pneumatic brake circuit with an additional dual pneumatic circuit. The electronic-pneumatic part of the braking system is controlled via various electronic sensors: (1) brake pedal travel; (2) brake air pressure; (3) individual wheel speed; and (4) individual lining/pad wear. Electronic-pneumatic circuit braking does not rely on axle load sensing but relies entirely on the wheel speed and air pressure sensing.

The dual pneumatic brake system is split into three independent circuits known as the redundancy braking circuit, one for the front axle a second for the rear axle and a third circuit for trailer control. The dual circuit system is similar to that of a conventional dual line pneumatic braking system and takes over only if the electronic-pneumatic brake circuit should develop a fault. Hence the name redundancy circuit, since it is installed as a safety back-up system and may never be called upon to override the electronic-pneumatic circuit brakes. However, there will be no ABS/TCS function when the dual circuit redundancy back-up system takes over from the electronic-pneumatic circuit when braking.

The foot brake pedal movement corresponds to the driver's demand for braking and is monitored by the electronic control module (ECM) which then conveys this information to the various solenoid control valves and axle modules (AM); compressed air is subsequently delivered to each of the wheel brake actuators. Only a short application lag results from the instant reaction of the electronic-pneumatic circuit, and consequently it reduces the braking distance in comparison to a conventional pneumatic braking system.

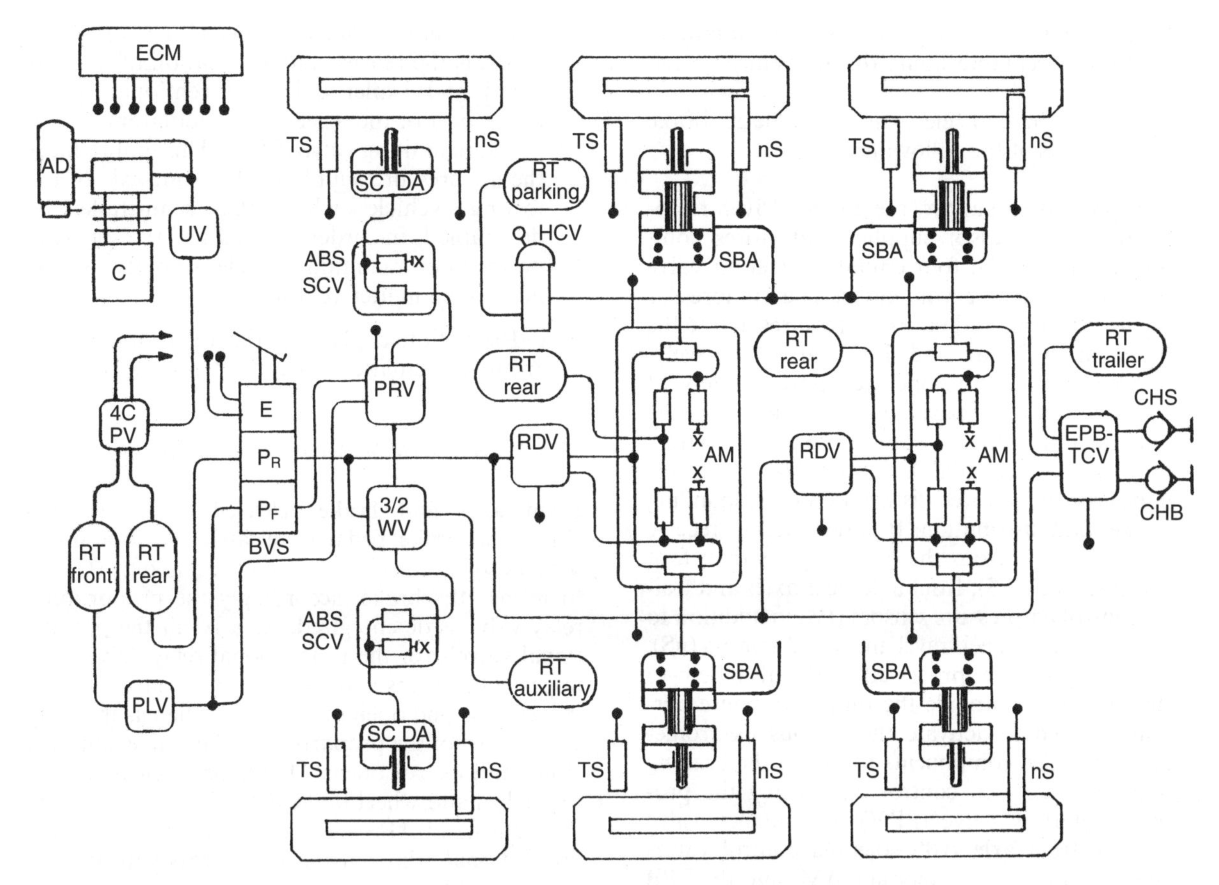

A list of key components and abbreviations used in the description of the electronic-pneumatic brake system is as follows:

No.	Component	Abbreviation
1	Electronic control module	ECM
2	Air dryer	AD
3	Compressor	C
4	Unloader valve	UV
5	Four circuit protection valve	4CPV
6	Reservoir tank (front/rear/trailer/auxiliary/parking)	RT etc
7	Brake value sensor	BVS
8	Proportional relay valve	PRV
9	3/2-way valve for auxiliary braking effect	3/2-WV-AB
10	ABS solenoid control valve	ABS-SCV
11	Single circuit diaphragm actuator	SCDA
12	Redundancy valve	RDV
13	Axle modulator	AM
14	Spring brake actuator	SBA
15	EPB trailer control valve	EPB-TCV
16	Park hand control valve	P-HCV
17	Coupling head for supply	CHS
18	Coupling head for brake	CHB
19	Travel sensor	TS
20	Speed sensor	nS
21	Pneumatic control front	P_F
22	Pneumatic control rear	P_R
23	Electrical sensors & switches	E
24	Air exit (exhaust)	x

Fig. 12.40 Electronic-pneumatic brake component layout

The electronic-pneumatic part of the braking system broadly divides the braking into three operation conditions:

1 Small differences between wheel speeds under part braking conditions; here the brake lining-disc wear is optimized between the front and rear axles.
2 Medium differences between wheel speeds; here the difference in wheel speed is signalled to the controls, causing wheel slip to be maintained similar on all axles. This form of brake control is known as adhesive adapted braking.
3 Large differences between wheel speeds and possibly a wheel locking tendency; here the magnitude of the spin-lock on each wheel is registered, triggering ABS/TCS intervention.

Note antilocking braking system (ABS) prevents the wheels from locking when the vehicle rapidly decelerates whereas a traction control system (TCS) prevents the wheels from spinning by maintaining slip within acceptable limits during vehicle acceleration.

The single circuit electronic-pneumatic brake circuit consists of the following:

1 Compressed air supply, the engine driven reciprocating compressor supplies and stores compressed air via the four circuit protection valve and numerous reservoir tanks. The compressor regulator cut-in and cut-out pressures are of the order of 10.2 bar and 12.3 bar respectively. Service foot circuits operate approximately at 10 bar whereas the parking and auxiliary circuits operate at a lower pressure of around 8.5 bar.
2 Electronic control module (ECM). This unit determines the brake force distribution corresponding to the load distribution. It is designed to receive signal currents from the following sources: foot travel sensors (TS), front axle, rear axles and trailer control air pressure sensors (PS) in addition to the individual wheel travel and speed sensors (nS). These inputs are processed and calculated to simultaneously provide the output response currents needed to activate the various electronically controlled components to match the braking requirements, such control units being the proportional relay valve (PRV), redundancy valve (RDV), front axle ABS solenoid control valves (ABS-SCV), rear axle module (AM) and the EPB trailer control valve (EPB-TCV).
3 Brake value sensor (BVS) unit which incorporates the pedal travel sensors (TS) and brake switches (BS) in addition to the dual circuit foot brake valve.
4 Redundancy valve (RDV): this valve switches into operating the rear axle dual circuit lines if a fault occurs in the electronic-pneumatic brake circuit.
5 Rear axle electronic-pneumatic axle module (AM) incorporating inlet and outlet solenoid valves used to control the application and release of the rear axle brakes.
6 Electronic-pneumatic proportional relay valve (PRV). This unit incorporates a solenoid relay valve which controls the amount of braking proportional to the needs of the front axle brakes.
7 Two front axle ABS solenoid control valves (ABS-SCV) which control the release and application of the front axle brakes.
8 Electronic-pneumatic brake-trailer control valve (EPB-TCV). This valve operates the trailer brakes via the trailer's conventional relay emergency valve during normal braking.
9 Parking hand control valve (P-HCV) which controls the release and application of the rear axle's and trailer axle's conventional spring brake part of the wheel brake actuators.
10 Pressure limiting valve (PLV). This unit reduces the air pressure supply to the front axle of the towing vehicle when the semi-trailer is de-coupled in order to reduce the braking power and maintain vehicle stability of the now much lighter vehicle.

A description explaining the operation of the electronic-pneumatic braking system now follows:

12.5.2 Front axle braking (Fig. 12.41(a–d))

Front axle foot brake released (Fig. 12.41(a)) When the brake pedal is released the foot travel sensors signal the electronic control module (ECM) to release the brake, accordingly the proportional relay valve is de-energized. As a result the proportional valve's (of the proportional relay valve unit) upper valve opens and its lower inlet valve and exit valves close and open respectively, whereas the relay valve's part of the proportional relay valve unit inlet closes and its exit opens. Hence air is released from the right hand wheel brake-diaphragm actuator via the right hand ABS solenoid control valve and the proportional relay valve exit, whereas with the left hand wheel brake-diaphragm actuator, compressed air is released via the left hand ABS solenoid control valve, 3/2-way valve and then out by the proportional relay valve exit.

Front axle foot brake applied (Fig. 12.41(b)) Air supply pressure from the front axle reservoir is directed to both the brake value sensor (BVS) and to the proportional relay valve (PRV).

When the driver pushes down the front brake pedal, the travel sensors incorporated within the brake value sensor (BVS) simultaneously measure the pedal movement and relay this information to the electronic control module (ECM). At the same time the brake switches close, thereby directing the electronic control module (ECM) to switch on the stop lights. Instantly the electronic control module (ECM) responds by sending a variable control current to the proportional valve situated in the proportional relay valve (PRV) unit. The energized solenoid allows the top valve to close whereas the lower control valve partially opens. Electronic-pneumatic control pressure now enters the relay valve's upper piston chamber, causing its piston to close the air exit and partially open the control valve, thereby permitting pre-calculated controlled brake pressure to be delivered to the wheel-diaphragm actuators via the ABS solenoid control valves for the right hand wheel and via the 3/2-way valve for auxiliary braking effect and the ABS solenoid control valve for the left hand wheel. For effective controlled braking the individual wheel speed sensors provide the electronic control module (ECM) with instant feed-back on wheel retardation and slip; this with the brake pedal movement sensors and pressure sensors enable accurate brake pressure control to be achieved at all times. Note the electronic-pneumatic brake (EPB) circuit has priority over the pneumatic modulated front pressure regulated by the brake value sensor (BVS) unit.

Front axle foot brake applied under ABS/TCS conditions (Fig. 12.41(c)) If the brakes are applied and the feed-back from the front axle speed sensors indicates excessive lock/slip the electronic control module will put the relevant ABS solenoid control valve into ABS mode. Immediately the ABS solenoid control valve attached to the wheel axle experiencing unstable braking energizes the solenoid valve, causing its inlet and exit valves to close and open respectively. Accordingly the wheel brake-diaphragm actuator will be depressurized thus avoiding wheel lock. The continuous monitoring of the wheel acceleration and deceleration by the electronic control module calculates current signal response to the ABS solenoid control valve to open and close respectively the inlet and exit valves, thus it controls the increase and decrease in braking pressure reaching the relevant wheel brake-diaphragm actuator; consequently the tendency of wheel skid is avoided.

Front axle foot brake applied with a fault in the electronic-pneumatics (Fig. 12.41(d)) If a fault develops in the electronic-pneumatic system the proportional relay valve shuts down, that is the solenoid proportional valve is de-energized causing its inlet valve to close and for its exit valve to open. Consequently when the brakes are applied the proportional relay valve's relay piston chamber is depressurized, making the relay valve's inlet and exit to close and open respectively. As a result, with the right hand ABS solenoid control valves de-energized air will exhaust from the right hand wheel brake-diaphragm actuator via the ABS solenoid control valve and the proportional relay valve. However, the collapse of the electro-pneumatic control pressure in the proportional relay valve causes the closure of the 3/2-way valve passage connecting the proportional relay valve to the left hand wheel brake actuator and opens the passages joining the auxiliary relay valve to the left hand wheel brake actuator via the left hand ABS solenoid control valve. Thus if the supply pressure from the front axle brake circuit is interrupted, the redundancy (pneumatic) rear axle brake pressure regulated by the brake valve sensor's foot control valve shifts over the 3/2-way valve into auxiliary braking effect position, that is, the 3/2-way valve blocks the passage between the proportional relay valve and the ABS solenoid control valve and then supplies modulated brake pressure from the 3/2-way valve to the left hand wheel brake-diaphragm actuator. Therefore the left hand front axle brake only, is designed to support the rear axle braking when the electronic-pneumatic brake circuit fails.

Front axle braking without trailer attached (Fig. 12.41(a–d)) When the semi-trailer is disconnected from its tractor the electronic control module responds by energizing the pressure-limiting valve solenoid. This results in the solenoid valve closing the direct by-pass passage leading to the proportional relay valve and opening the valve leading to the relay valve within the pressure-limiting valve unit (see Fig. 12.41(c)). This results in the solenoid valve shutting-off the front axle reservoir tank air supply from the proportional relay valve and at the same time re-routing the air supply via the pressure-limiting valve's relay valve which then reduces the maximum braking pressure reaching the proportional relay valve and hence the front axle brakes.

Limiting the air pressure reaching the front axle of the towing vehicle when the trailer is removed is essential in retaining the balance of front to rear axle braking power of the now much shorter overall vehicle base, thereby maintaining effective and stable vehicle retardation.

12.5.3 Rear axle braking (Fig. 12.42(a–d))

Rear axle — foot brake released (Fig. 12.42(a)) When the brake pedal is released the travel sensors within the brake value sensor (BVS) signal the electronic control module (ECM) which in turn informs the axle modulator to release the brake

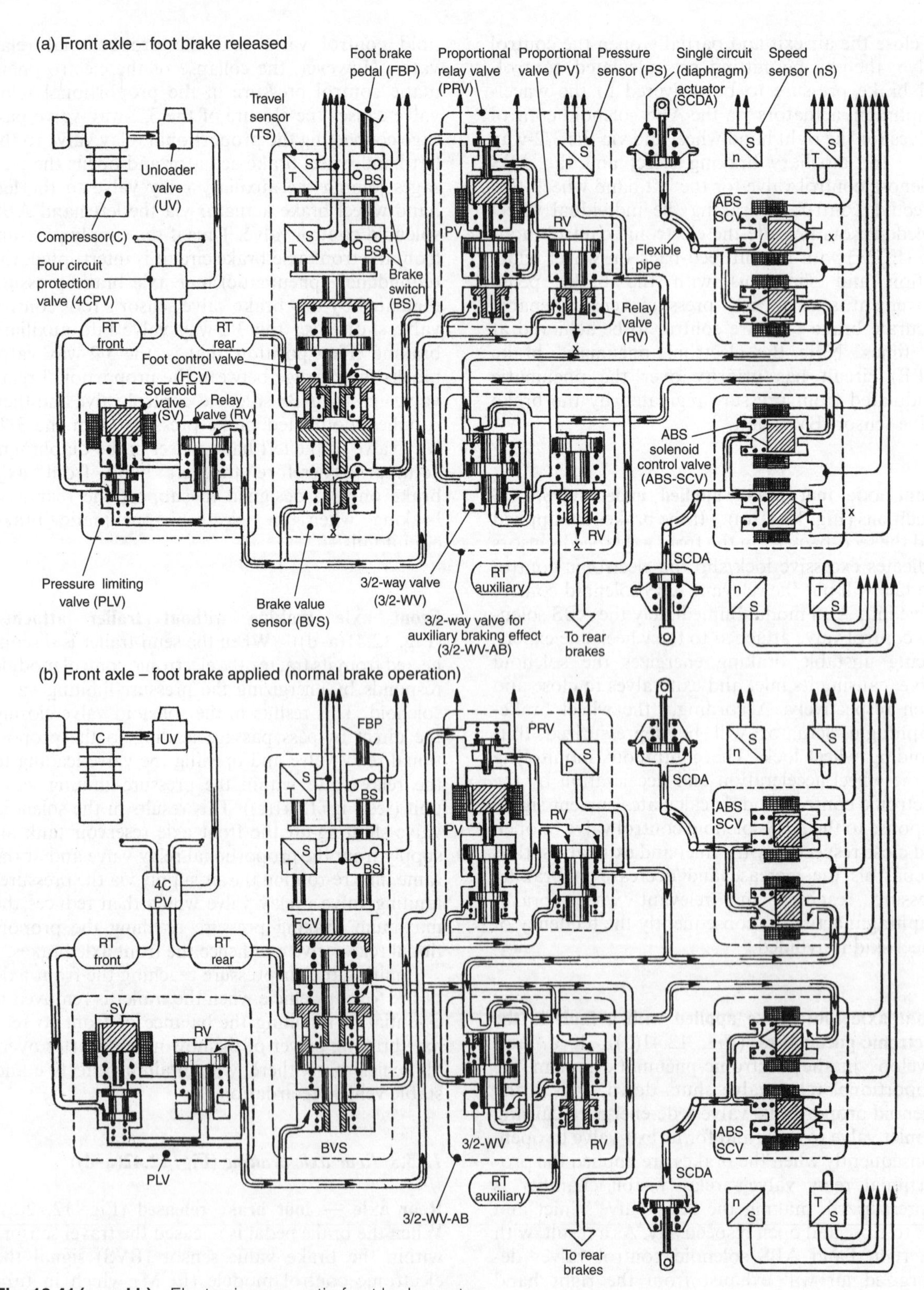

Fig. 12.41 (a and b) Electronic-pneumatic front brake system

(c) Front axle – foot brake applied under ABS/TCS conditions

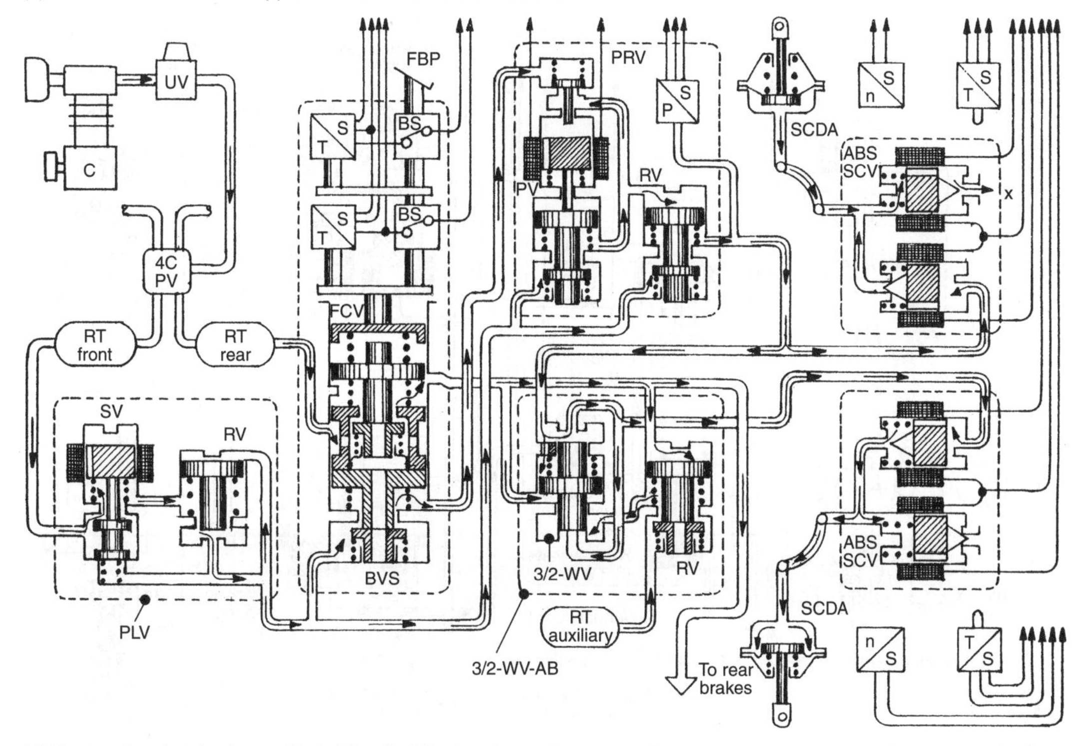

(d) Front axle – foot brake applied with a fault in the electronic-pneumatics

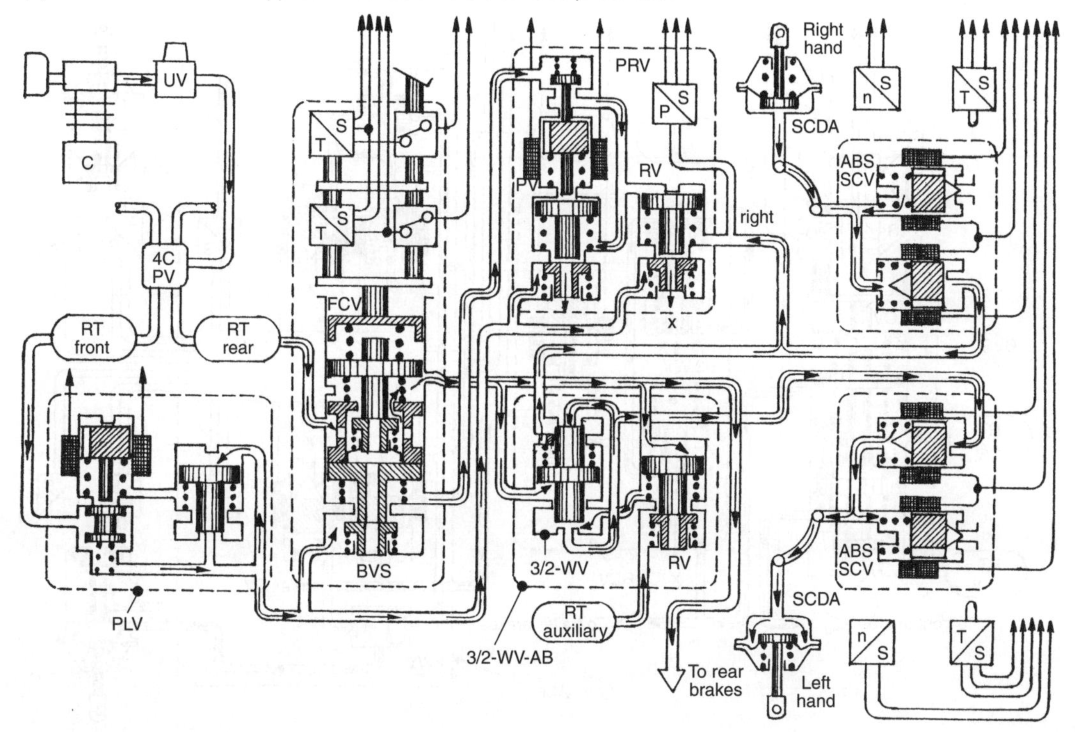

Fig. 12.41 (c and d) *Contd*

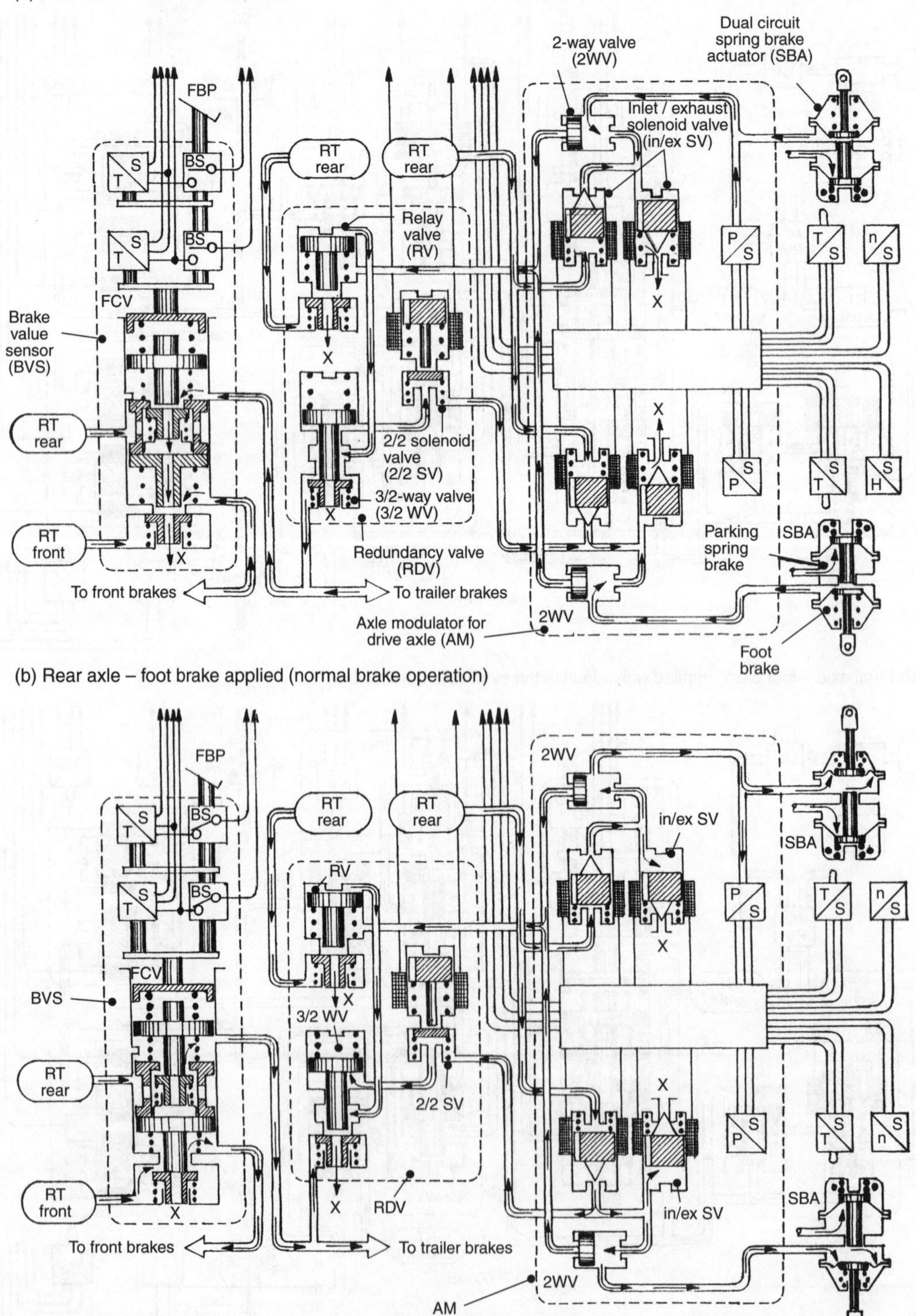

Fig. 12.42 (a and b) Electronic-pneumatic rear brake system

(c) Rear axle – foot brake applied under ABS/TCS conditions

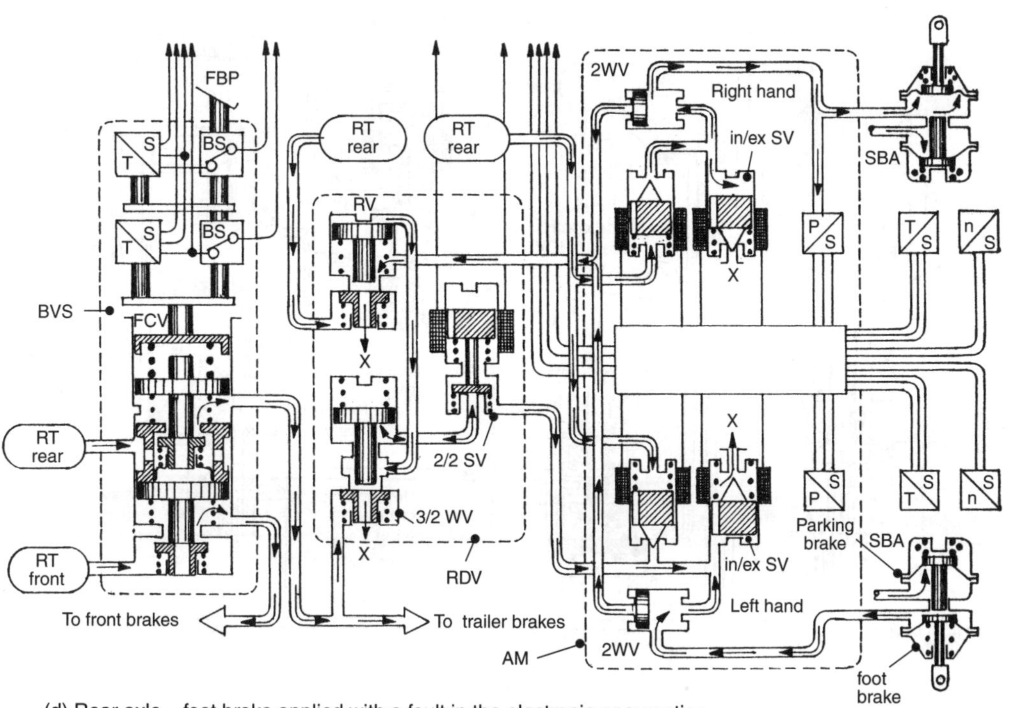

(d) Rear axle – foot brake applied with a fault in the electronic-pneumatics

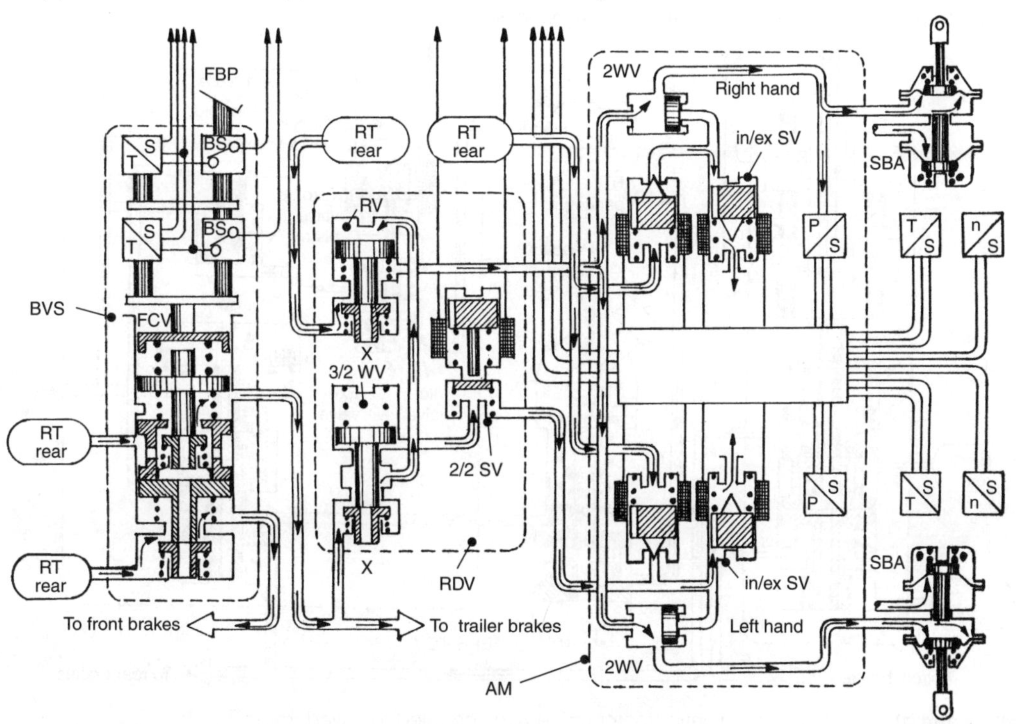

Fig. 12.42 (c and d) *Contd*

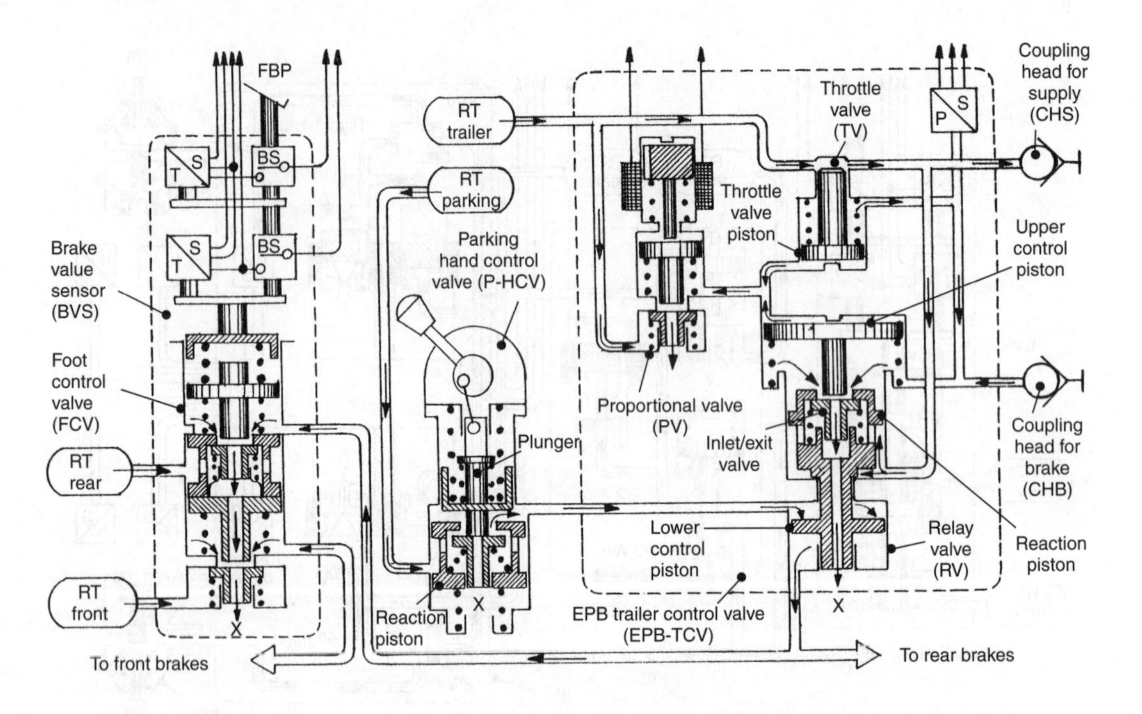

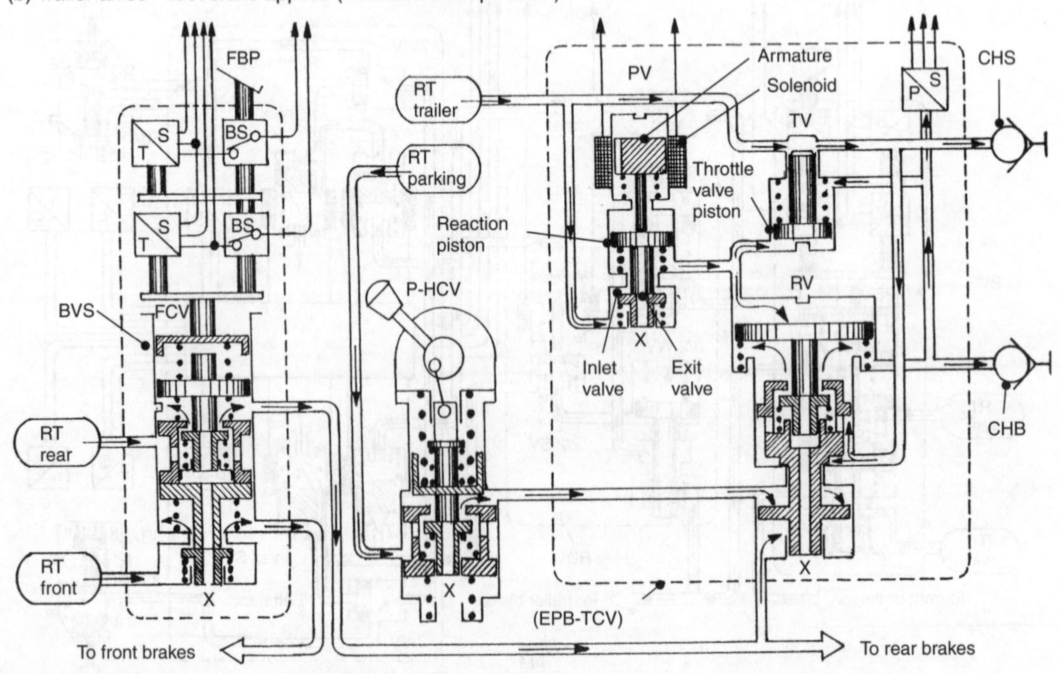

Fig. 12.43 (a and b) Electronic-pneumatic tractor unit brakes coupled to towed trailer

(c) Trailer axles – parking brake applied

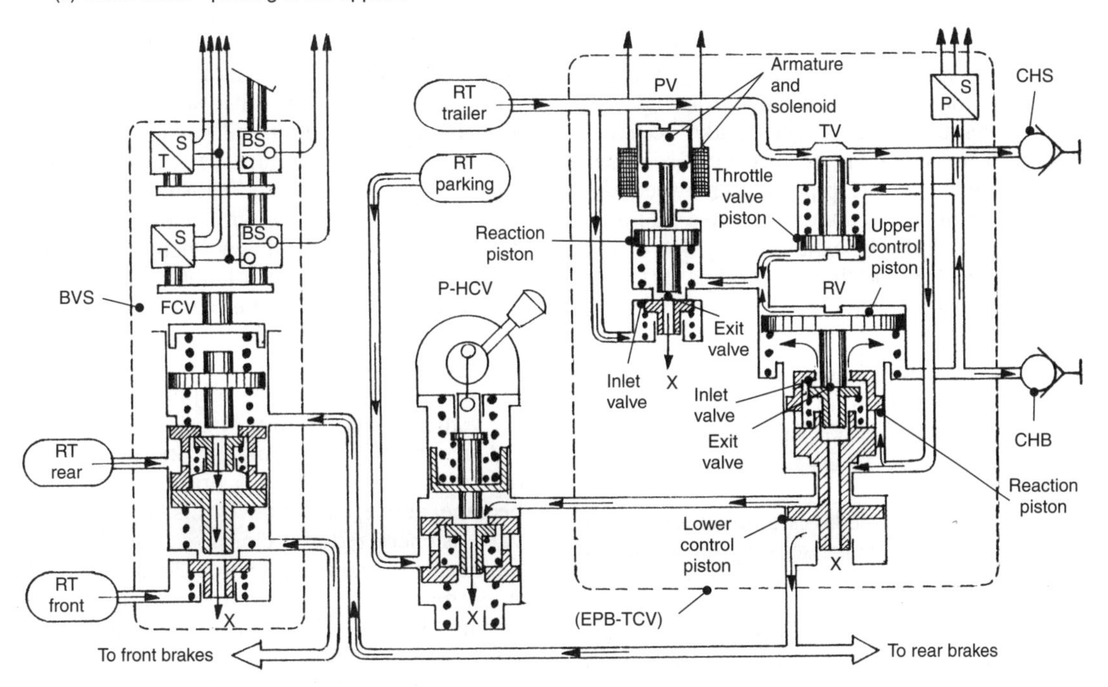

(d) Trailer axles – foot brake applied with a fault in the electronic-pneumatics

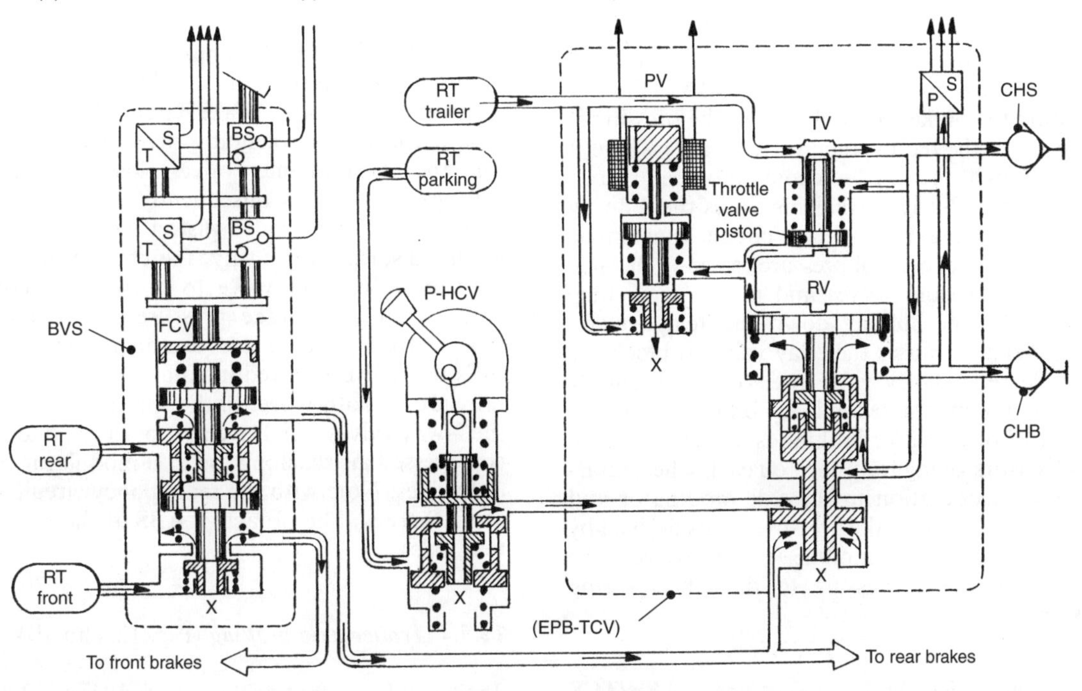

Fig. 12.43 (c and d) *Contd*

pressure in the wheel spring brake actuator brake lines. Consequently the axle modulator (AM) de-energizes the inlet and exhaust solenoid valves, causing the inlet valve to close and the exhaust valve to open, hence brake pressure will be prevented from reaching the spring brake actuator and any existing air pressure in the spring brake actuator (SBA) will be expelled via the exhaust solenoid valve. Air pressure in the pipe lines between the brake value sensor, the redundancy valve and axle modulator unit will also be exhausted by way of the foot control valve exit and the axle modulator (AM) exhaust solenoid valve exit.

Rear axle — foot brake applied (Fig. 12.42(b)) The rear axle reservoir tank delivers maximum supply pressure to the brake value sensor (BVS), the redundancy valve (RDV) and to the rear axle modulator (AM).

When the foot brake pedal is applied the travel sensor within the brake value sensor unit monitors the pedal movement and relays this information to the electronic control module. The brake switches will also close thereby informing the electronic control module (ECM) to operate the stop lights. The electronic control module (ECM) responds by signalling the axle modulators (AMs) to energize their corresponding inlet/exhaust solenoid valves. The exhaust valve will therefore close, whereas the inlet valves now open to permit rear reservoir tank supply pressure to flow via the 2-way valve to the dual circuit spring brake actuators (SBA) thereby operating the brakes. In addition, brake pressure is conveyed to the redundancy valve (RDV) where it flows though the 2/2 solenoid valve and then actuates the 3/2-way valve. This closes the 3/2-way valve, preventing redundancy circuit (pneumatic control pressure) control pressure reaching the axle modulator solenoid valves and at the same time exhausts the air holding down the relay valve's piston, hence it causes the relay valve to block the rear axle redundancy valve reservoir tank supply pressure entering the redundancy brake circuit.

Control of rear axle braking is achieved via the speed sensors giving feed-back on each wheel retardation or acceleration to the axle modulator and with the calculated brake pressure needs derived by the electronic control module (ECM) delivers the appropriate brake pressure to the wheel spring brake actuators.

Rear axles — foot brake applied under ABS/TCS conditions (Fig 12.42(c)) If the brakes are applied and the feed-back from one of the rear axle speed sensors registers excessive wheel lock/spin, then the electronic control module (ECM) will actuate the axle modulator (AM). Instantly the unstable breakaway wheel inlet/exhaust solenoid valves are energized, that is, the inlet solenoid valve closes and the corresponding exit valve opens, therefore blocking the air pressure passage leading to the wheel brake actuator and causing the air pressure in the actuator diaphragm chamber to collapse. The energizing and de-energizing (opening and closing) of each pair of solenoid valves is repeated continuously to adjust the magnitude of the wheel braking as demanded by the driver without the individual wheel locking when braking or for one of the wheels to spin due to poor road grip when the vehicle is accelerating.

During ABS/TCS braking conditions the redundancy (pneumatic circuit) brake system must not become active; to achieve this, the 2/2 solenoid valve is energized and closes so that air pressure is maintained underneath the 3/2-way valve piston. Accordingly the space above the relay valve remains open to the atmosphere and the 3/2-way valve inlet remains shut, hence locking out the redundancy brake circuit. Braking will revert to normal electronic-pneumatic control when the difference in wheel speeds is relatively small.

Rear axle — foot brake applied with a fault in the electronic-pneumatics (Fig. 12.42(d)) Should the electronic-pneumatic brake circuit fail, the axle modulator (AM) solenoid valve de-energizes so that the solenoid inlet valves close, whereas the exit valves open. Consequently compressed air under the 3/2-way valve piston escapes from the left hand solenoid exit valve thereby permitting the 3/2-way valve inlet valve to open. Foot control valve modulated brake pressure now enters the relay valve's upper piston chamber where it controls the delivery of redundancy circuit pneumatic pressure to both wheel brake spring actuators via the 2-way valves which are now positioned to block compressed air reaching the axle modulator solenoid valves. Note with the redundancy circuit operating there will be no active ABS at the front and rear axles.

12.5.4 Trailer axle braking (Fig. 12.43(a–d))

Trailer axle — foot brake released (Fig. 12.43(a)) When the brake pedal is released the foot travel sensor signals the electronic control module (ECM) to release the brakes by de-energizing the proportional valve (PV). As a result the proportional valve (PV) inlet closes and its exit opens so that air pressure in the throttle valve and relay valve piston chamber is exhausted. Accordingly the relay valve inlet closes and its exit opens, thus permitting the brake pressure leading to the coupling head to collapse and for the brakes to be released.

Trailer axles — foot brake applied (Fig. 12.43(b)) Maximum supply pressure from the rear axle and trailer reservoirs is routed to both the brake value sensor (BVS) and to the proportional relay valve (PRV) respectively.

When the driver operates the foot brake pedal, the travel sensors located inside the brake value sensor (BVS) unit measure the pedal downward movement and feed this information to the central electronic control module (ECM). At the same time the brake switch closes thereby instructing the electronic control module to switch on the stop light.

The electronic control module (ECM) responds by directing a calculated variable control current to the proportional valve (PV) which forms part of the EPB-trailer control valve unit. Energizing the proportional valve's solenoid, closes the exit valve and opens the inlet valve in proportion to the amount of braking requested. Controlled air pressure now enters the relay valve piston upper chamber; this closes the exit valve and opens the inlet valve in proportion to the degree of braking demanded. Modulated brake pressure will now pass to the coupling head brake circuit where it is then relayed to the trailer wheel brake actuators via the trailer-mounted relay emergency valve.

Trailer axle — parking brake applied (Fig. 12.43(c)) With the 'park' hand control valve in the 'off' position the hand control valve central plunger closes the exit and pushes open the inlet valve. Compressed air from the parking reservoir tank is therefore able to flow to the relay valve part of the EPB trailer control valve via the open inlet valve inside the hand control valve.

When the 'park' control valve lever moves towards 'park' position the central plunger rises, causing the exit to open and the inlet valve to close. Air pressure therefore exhausts from above the lower control piston within the relay valve. Supply pressure acting beneath the reaction piston will now be able to lift the reaction piston and inner valve assembly until the upper control piston plunger closes the exit. Further upward movement then opens the inlet valve, thus permitting supply air pressure to flow through the partially open inlet valve to the 'coupling head for brake', and hence to the trailer attached relay emergency valve where it modulates the supply pressure reaching the wheel brake actuators.

Trailer axle — foot brake applied with a fault in the electronic-pneumatics (Fig. 12.43(d)) If there is a fault in the electronic-pneumatic system the proportional valve (PV) is de-energized, causing its inlet valve to close and its exit to open; pressurized air is therefore able to exhaust from the relay valve's upper control piston chamber via the proportional valve exit. Note the relay valve consists of an upper control piston and a lower assembly with upper and lower piston regions and which incorporates an internal double seat inlet and exit valve. As a result the upper control piston moves to its uppermost position, the inlet valve initially closes and the trapped supply pressure and the redundancy brake pressure (foot control valve pneumatic pressure) acting underneath the inner assembly's upper and lower piston region, pushes up the assembly against the hand control valve park pressure sufficiently to close the exit valve and to open the inlet valve. Brake pressure will hence be delivered to the trailer wheel brake actuators via the 'coupling head for brakes'.

Brake line to trailer defective (Fig. 12.43(b)) If the brake line to the trailer fractures the output pressure from the relay valve drops, causing the pressure above the throttle valve piston to collapse; this forces the throttle valve piston to rise and partially close the throttle valve, thereby causing a rapid reduction in the supply pressure flowing to the coupling head supply line (diagram not shown with throttle valve in defective pipe line position). As a result the relay emergency valve mounted on the trailer switches into braking mode and hence overrides the electronic-pneumatic circuit brake control to bring the vehicle to rest.

13 Vehicle refrigeration

Refrigeration transport is much in demand to move frozen or chilled food from storage centres to shops and supermarkets. Thermally insulated body containers used for frozen and chilled food deliveries for both small rigid trucks and large articulated vehicles are shown in Figs 13.1 and 13.2 respectively. Refrigeration systems designed for motor vehicle trucks are basically made up of two parts supported on an aluminium alloy or steel frame. The condenser unit which is mounted outside the thermally insulated cold storage compartment, comprises a diesel engine (and optional electric motor) compressor, condenser coil, fan thermostat and accessories. The evaporator unit protruding inside the cold storage body contains the evaporator coil, evaporator fan expansion valve remote feeler bulb and any other accessories. For certain applications a standby electric motor is incorporated to drive the compressor when the truck is being parked at the loading or delivery site for a long period of time such as overnight, the electricity supply being provided by the local premises' mains power. Typical self-contained refrigeration unit arrangements incorporating an engine, compressor, evaporator, condenser, fans and any other accessories for small to medium and large frozen storage compartments are shown in Figs 13.3 and 13.4 respectively. Temperature control is fully automatic on a start–stop cycle. With small and medium size refrigeration systems the engine runs at full governed speed until the thermostat temperature setting is reached. It then automatically reduces speed and disconnects the magnetic or centrifugal clutch which stops the compressor. A slight increase of temperature will return the engine to full speed and again driving the compressor.

The cold storage compartment temperature for frozen food is usually set between −22°C and −25°C whereas the chilled compartment temperature is set between +3°C and +5°C.

13.1 Refrigeration terms (Fig. 13.5)

To understand the operating principles of a refrigeration system it is essential to appreciate the following terms:

Refrigerant This is the working fluid that circulates though a refrigeration system and produces both cooling and heating as it changes state. The desirable properties of a refrigerant fluid are such that it flows through the evaporator in its vapour state, absorbs heat from its surroundings, then transfers this heat via the flow of the refrigerant

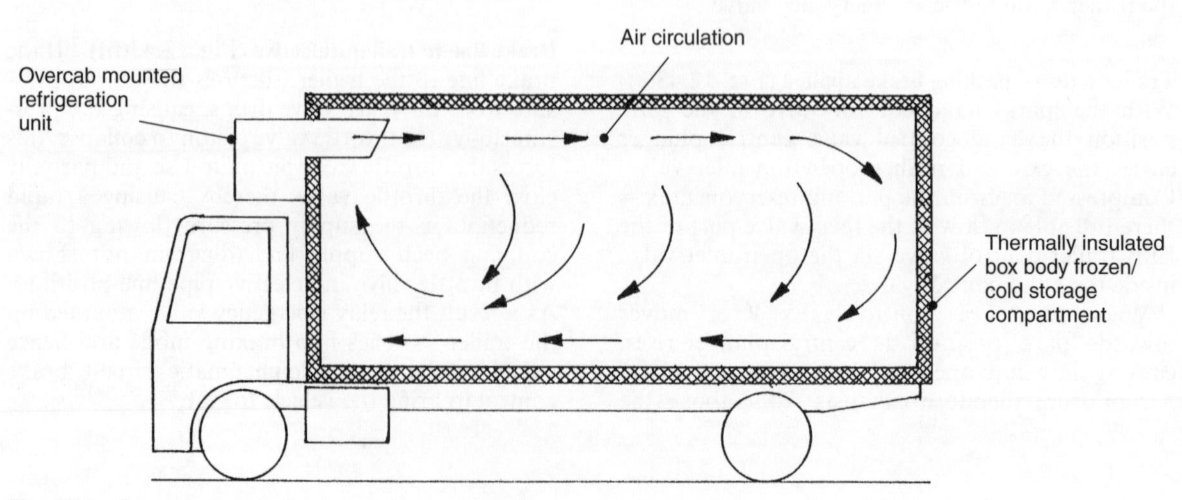

Fig. 13.1 Overcab mounted self-contained refrigeration system for small and medium rigid trucks

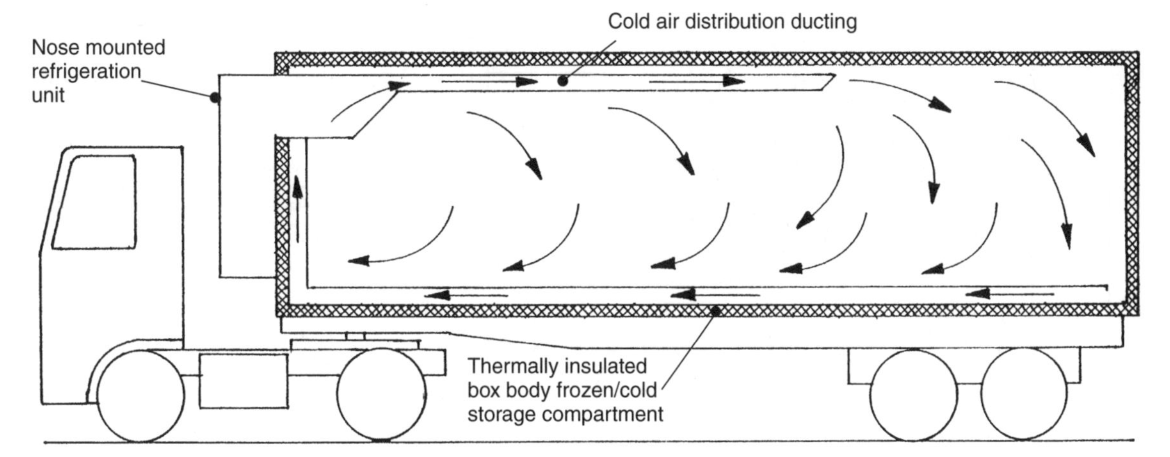

Fig. 13.2 Nose mounted self-contained refrigeration system for large articulated truck

to the condenser; the refrigerant then condenses to a liquid state and in the process dissipates heat taken in by the evaporator to the surrounding atmosphere. Refrigerants are normally in a vapour state at atmospheric pressure and at room temperature because they boil at temperatures below zero on the celsius scale; however, under pressure the refrigerant will convert to a liquid state.

Subcooled liquid (Fig. 13.5) This is a liquid at any temperature below its saturated (boiling) temperature.

Saturated temperature (Fig. 13.5) This is the temperature at which a liquid converts into vapour or a vapour converts into liquid, that is, the boiling point temperature.

Saturated liquid (Fig. 13.5) This is a liquid heated to its boiling point, that is, it is at the beginning of vaporization.

Saturated vapour (Fig. 13.5) This is the vapour which is formed above the surface of a liquid when heated to its boiling point.

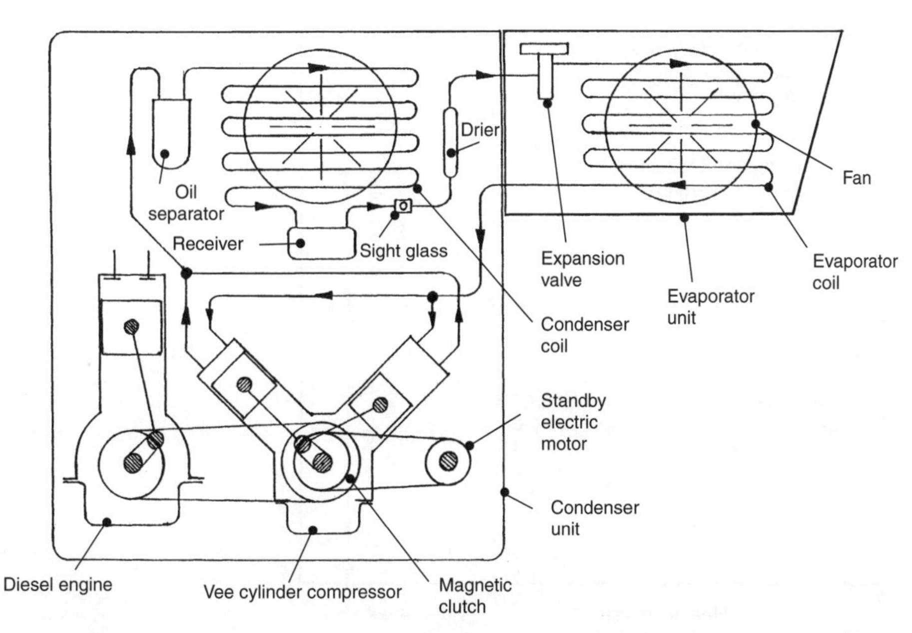

Fig. 13.3 Light to medium duty diesel engine and standby electric motor belt driven compressor refrigeration unit

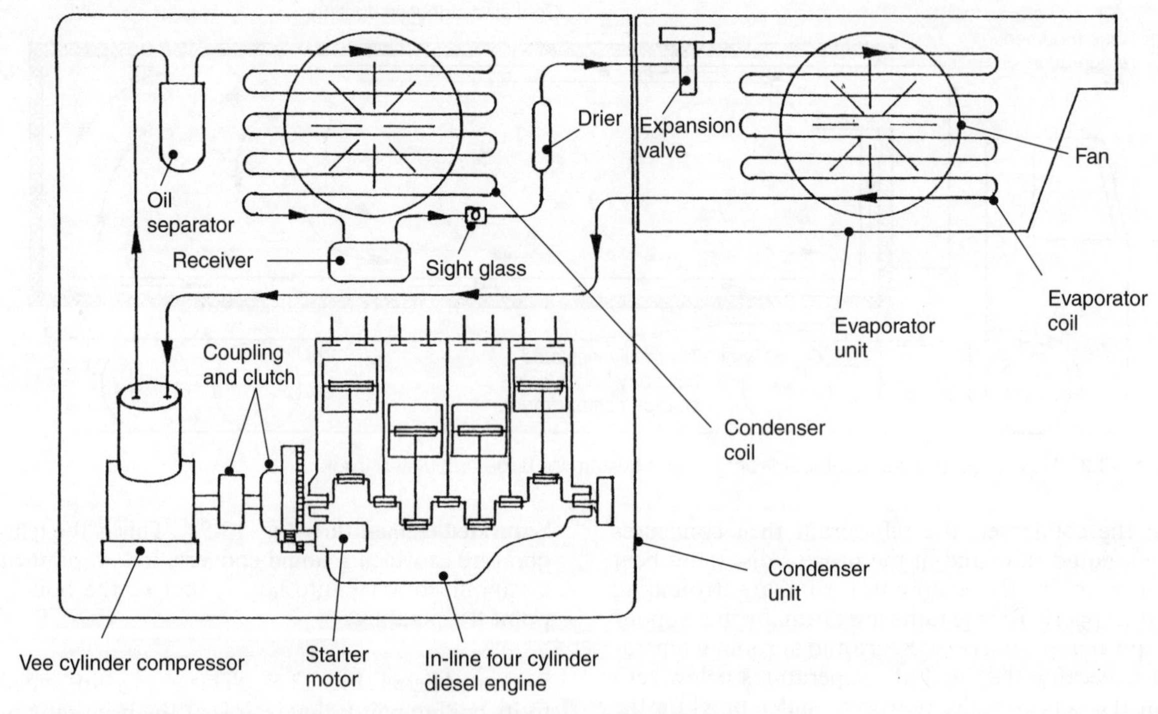

Fig. 13.4 Heavy duty diesel engine shaft driven compressor refrigeration unit

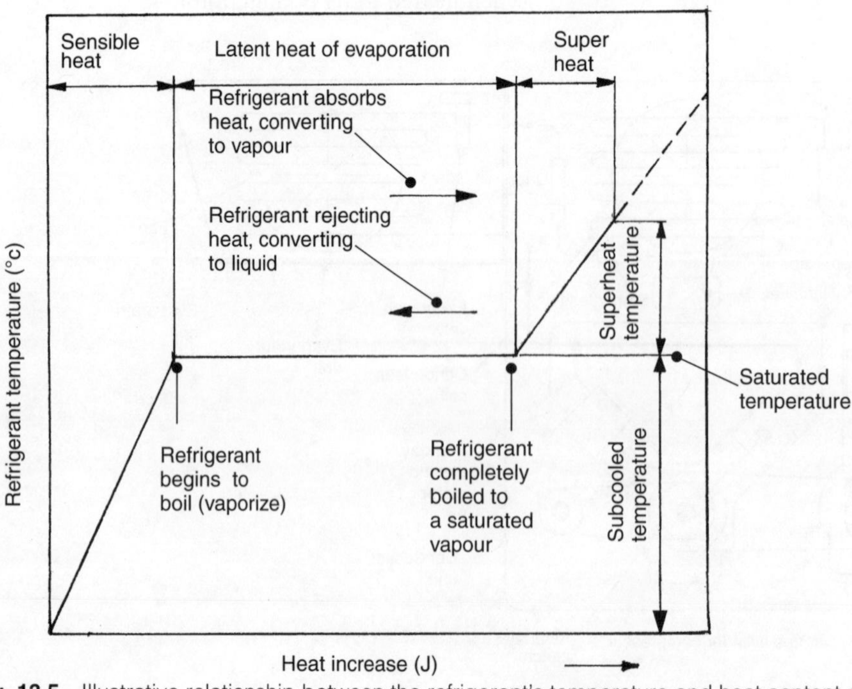

Fig. 13.5 Illustrative relationship between the refrigerant's temperature and heat content during a change of state

Latent heat of evaporation (Fig. 13.5) This is the heat needed to completely convert a liquid to a vapour and takes place without any temperature rise.

Superheated vapour (Fig. 13.5) This is a vapour heated to a temperature above the saturated temperature (boiling point); superheating can only occur once the liquid has been completely vaporized.

13.2 Principles of a vapour–compression cycle refrigeration system (Fig. 13.6)

1 High pressure subcooled liquid refrigerant at a typical temperature and pressure of 30 °C and 10 bar respectively flows from the receiver to the expansion valve via the sight glass and drier. The refrigerant then rapidly expands and reduces its pressure as it passes out from the valve restriction and in the process converts the liquid into a vapour flow.

2 The refrigerant now passes into the evaporator as a mixture of liquid and vapour, its temperature being lowered to something like −10 °C with a corresponding pressure of 2 bar (under these conditions the refrigerant will boil in the evaporator). The heat (latent heat of evaporation) necessary to cause this change of state will come from the surrounding frozen compartment in which the evaporator is exposed.

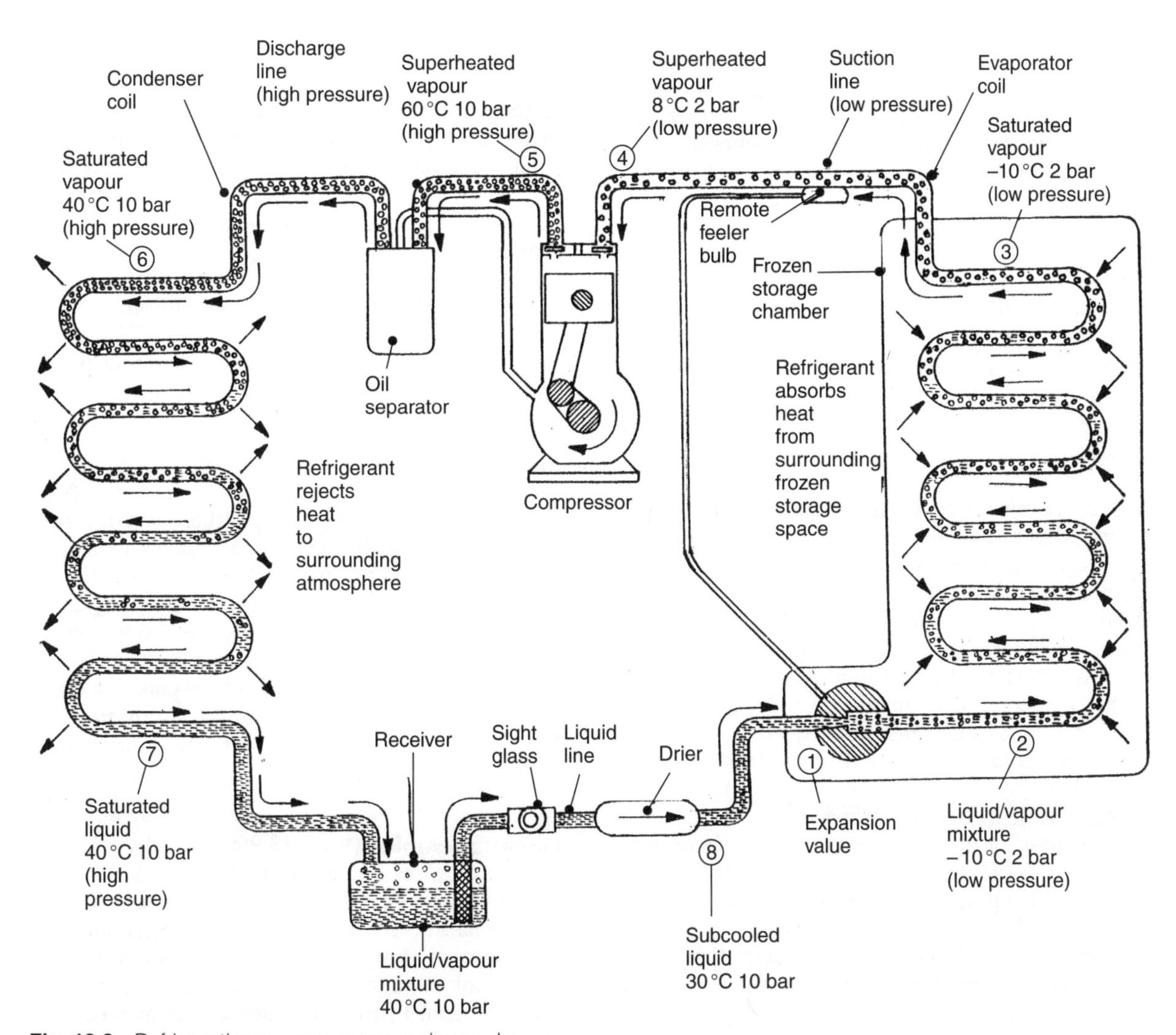

Fig. 13.6 Refrigeration vapour–compression cycle

3 As the refrigerant moves through the evaporator coil it absorbs heat and thus cools the space surrounding the coil. Heat will be extracted from the cold storage compartment until its pre-set working temperature is reached, at this point the compressor switches off. With further heat loss through the storage container insulation leakage, doors opening and closing and additional food products being stored, the compressor will automatically be activated to restore the desired degree of cooling. The refrigerant entering the evaporator tube completes the evaporation process as it travels through the evaporator coil so that the exit refrigerant from the evaporator will be in a saturated vapour state but still at the same temperature and pressure as at entry, that is, −10 °C and 2 bar respectively.
4 The refrigerant is now drawn towards the compressor via the suction line and this causes the heat from the surrounding air to superheat the refrigerant thus raising its temperature to something like 8 °C; however, there is no change in the refrigerant's pressure.
5 Once in the compressor the superheated vapour is rapidly compressed, consequently the superheated vapour discharge from the compressor is at a higher temperature and pressure in the order of 60 °C and 10 bar respectively.
6 Due to its high temperature at the exit from the compressor the refrigerant quickly loses heat to the surrounding air as it moves via the discharge line towards the condenser. Thus at the entry to the condenser the refrigerant will be in a saturated vapour state with its temperature now lowered to about 40 °C; however, there is no further change in pressure which is still therefore 10 bar.
7 On its way through the condenser the refrigerant saturated vapour condenses to a saturated liquid due to the stored latent heat in the refrigerant transferring to the surrounding atmosphere via the condenser coil metal walls. Note the heat dissipated to the surrounding atmosphere by the condenser coil is equal to the heat taken in by the evaporator coil from the cold storage compartment and the compressor.
8 After passing through the condenser where heat is given up to the surrounding atmosphere the saturated liquid refrigerant now flows into the receiver. Here the increased space permits a small amount of evaporation to occur, the refrigerant then completes the circuit to the expansion valve though the liquid line where again heat is lost to the atmosphere, and this brings the refrigerant's temperature down to something like 30 °C but without changing pressure which still remains at 10 bar.

13.3 Refrigeration system components

A description and function of the various components incorporated in a refrigeration system will be explained as follows:

13.3.1 Reciprocating compressor cycle of operation (Fig. 13.7(a–d))

Circulation of the refrigerant between the evaporator and the condenser is achieved by the pumping action of the compressor. The compressor draws in low pressure superheated refrigerant vapour from the evaporator and discharges it as high pressure superheated vapour to the condenser. After flowing through the condenser coil the high pressure refrigerant is now in a saturated liquid state; it then flows to the expansion valve losing heat on the way and thus causing the liquid to become subcooled. Finally the refrigerant expands on its way through the expansion valve causing it to convert into a liquid-vapour mix before re-entering the evaporator coil.

The reciprocating compressor completes a suction and discharge cycle every revolution; the outward moving piston from TDC to BDC forms the suction-stroke whereas the inward moving piston from BDC to TDC becomes the discharge stroke.

Suction stroke (Fig. 13.7(a and b)) As the crank shaft rotates past the TDC position the piston commences its suction stroke with the discharge reed valve closed and the suction reed valve open (Fig. 13.7(a and b)). The downward sweeping piston now reduces the cylinder pressure from P_1 to P_2 as its volume expands from V_1 to V_2, the vapour refrigerant in the suction line is now induced to enter the cylinder. The cylinder continues to expand and to be filled with vapour refrigerant at a constant pressure P_1 to the cylinder's largest volume of V_3, that is the piston's outermost position BDC, see Fig. 13.8.

Discharge stroke (Fig. 13.9(c and d)) As the crankshaft turns beyond BDC the piston begins its upward discharge stroke, the suction valve closes and the discharge valve opens (see Fig. 13.7(c and d)). The upward moving piston now compresses the refrigerant vapour thereby increasing the cylinder pressure from P_1 to P_2 through a volume reduction from V_3 to V_4 at which point the cylinder pressure

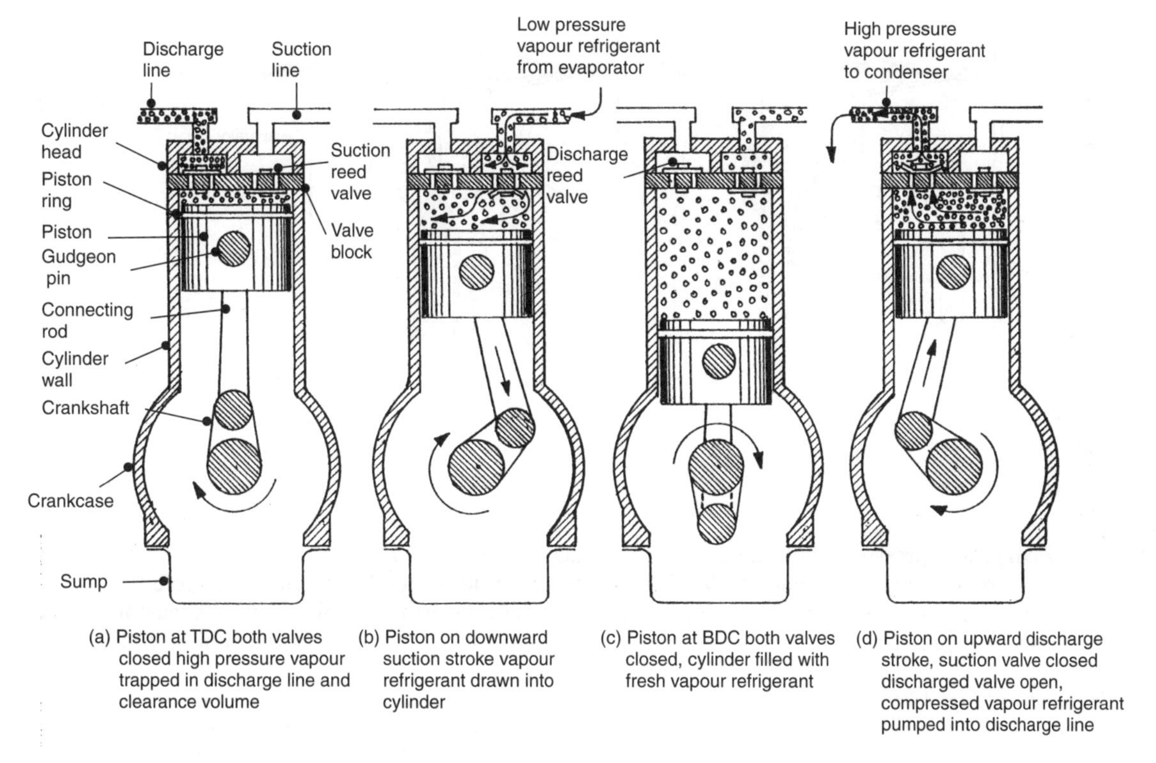

(a) Piston at TDC both valves closed high pressure vapour trapped in discharge line and clearance volume (b) Piston on downward suction stroke vapour refrigerant drawn into cylinder (c) Piston at BDC both valves closed, cylinder filled with fresh vapour refrigerant (d) Piston on upward discharge stroke, suction valve closed discharged valve open, compressed vapour refrigerant pumped into discharge line

Fig. 13.7 (a–d) Reciprocating compressor cycle of operation

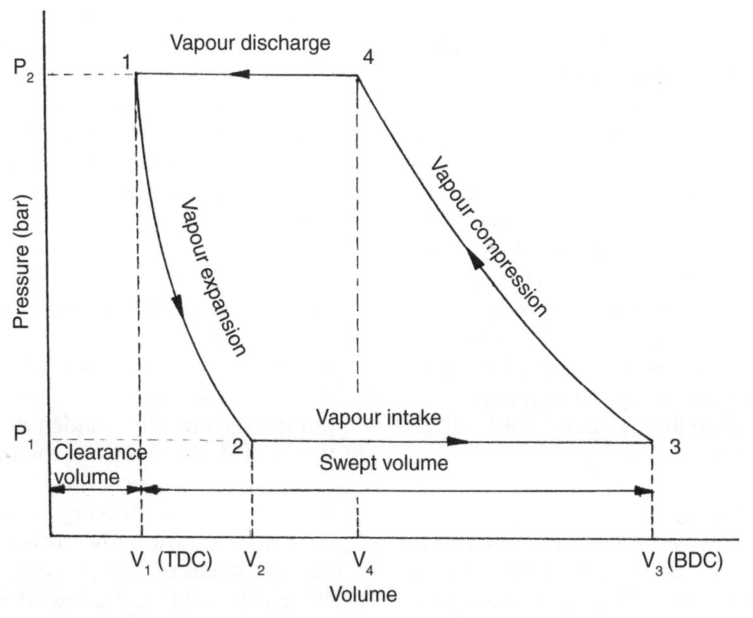

Fig. 13.8 Reciprocating compressor pressure-volume cycle

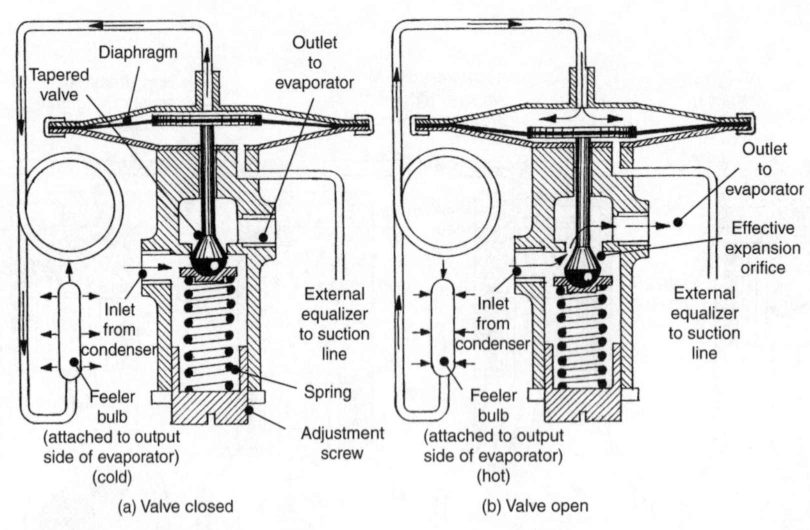

Fig. 13.9 (a and b) Thermostatic expansion valve

equals the discharge line pressure; the final cylinder volume reduction therefore from V_4 back to V_1 will be displaced into the high pressure discharge line at the constant discharge pressure of P_2 (see Fig. 13.8).

13.3.2 Evaporator (Fig. 13.6)

The evaporator's function is to transfer heat from the food being stored in the cold compartment into the circulating refrigerant vapour via the fins and metal walls of the evaporator coil tubing by convection and conduction respectively. The refrigerant entering the evaporator is nearly all liquid but as it moves through the tube coil, it quickly reaches its saturation temperature and is converted steadily into vapour. The heat necessary for this change of state comes via the latent heat of evaporation from the surrounding cold chamber atmosphere.

The evaporator consists of copper, steel or stainless steel tubing which for convenience is shaped in an almost zigzag fashion so that there are many parallel lengths bent round at their ends thus enabling the refrigerant to flow from side to side. To increase the heat transfer capacity copper fins are attached to the tubing so that relatively large quantities of heat surrounding the evaporator coil can be absorbed through the metal walls of the tubing, see Fig. 13.15(a and b).

13.3.3 Condenser (Fig. 13.6)

The condenser takes in saturated refrigerant vapour after it has passed though the evaporator and compressor, progressively cooling then takes place as it travels though the condenser coil, accordingly the refrigerant condenses and reverts to a liquid state. Heat will be rejected from the refrigerant during this phase change via conduction though the metal walls of the tubing and convection to the surrounding atmosphere.

A condenser consists of a single tube shaped so that there are many parallel lengths with semicircular ends which therefore form a continuous winding or coil. Evenly spaced cooling fins are normally fixed to the tubing, this greatly increases the surface area of the tubing exposed to the convection currents of the surrounding atmosphere, see Fig. 13.15(a and b).

Fans either belt driven or directly driven by an electric motor are used to increase the amount of air circulation around the condenser coil, this therefore improves the heat transfer taking place between the metal tube walls and fins to the surrounding atmosphere. This process is known as forced air convection.

13.3.4 Thermostatic expansion valve (Fig. 13.9(a and b))

An expansion valve is basically a small orifice which throttles the flow of liquid refrigerant being pumped from the condenser to the evaporator; the immediate exit from the orifice restriction will then be in the form of a rapidly expanding refrigerant, that is, the refrigerant coming out from the orifice is now a low pressure continuous liquid-vapour stream. The purpose of the thermostatic valve is to control the rate at which the refrigerant passes from the liquid line into the evaporator and to keep the pressure difference between the high and low pressure sides of the refrigeration system.

The thermostatic expansion valve consists of a diaphragm operated valve (see Fig. 13.9(a and b)). One side of the diaphragm is attached to a spring loaded tapered/ball valve, whereas the other side of the diaphragm is exposed to a refrigerant which also occupies the internal space of the remote feeler bulb which is itself attached to the suction line tube walls on the output side of the evaporator. If the suction line saturated/superheated temperature decreases, the pressure in the attached remote feeler bulb and in the outer diaphragm chamber also decreases. Accordingly the valve control spring thrust will partially close the taper/ball valve (see Fig. 13.9(a)). Consequently the reduced flow of refrigerant will easily now be superheated as it leaves the output from the evaporator. In contrast if the superheated temperature rises, the remote feeler bulb and outer diaphragm chamber pressure also increases, this therefore will push the valve further open so that a larger amount of refrigerant flows into the evaporator, see Fig. 13.9(b). The extra quantity of refrigerant in the evaporator means that less superheating takes place at the output from the evaporator. This cycle of events is a continuous process in which the constant superheated temperature control in the suction line maintains the desired refrigerant supply to the evaporator.

A simple type of thermostatic expansion valve assumes the input and output of an evaporator are both working at the same pressure; however, due to internal friction losses the output pressure will be slightly less than the input. Consequently the lower output pressure means a lower output saturated temperature so that the refrigerant will tend to vaporize completely before it reaches the end of the coil tubing. As a result this portion of tubing converted completely into vapour and which is in a state of superheat does not contribute to the heat extraction from the surrounding cold chamber so that the effective length of the evaporator coil is reduced. To overcome early vaporization and superheating, the diaphragm chamber on the valve-stem side is subjected to the output side of the evaporator down stream of the remote feeler bulb. This extra thrust opposing the remote feeler bulb pressure acting on the outer diaphragm chamber now requires a higher remote feeler bulb pressure to open the expansion valve.

13.3.5 Suction pressure valve (throttling valve) (Fig. 13.10(a and b))

This valve is incorporated in the compressor output suction line to limit the maximum suction

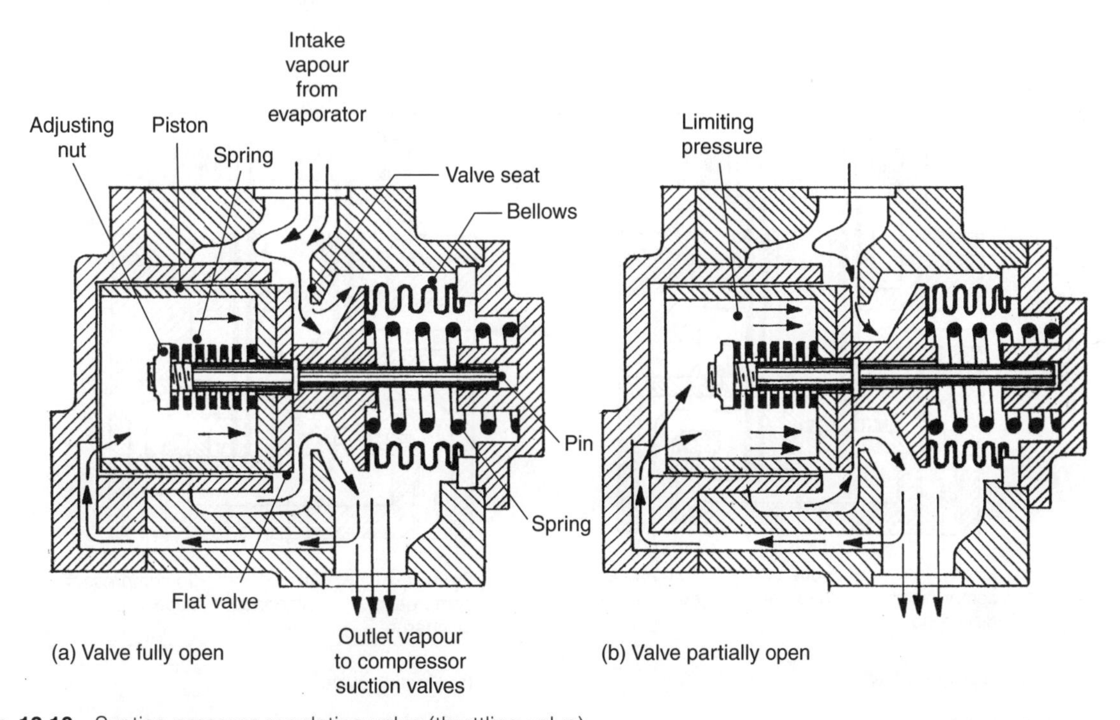

Fig. 13.10 Suction pressure regulating valve (throttling valve)

pressure generated by the compressor thereby safeguarding the compressor and drive engine/motor from overload. If the maximum suction pressure is exceeded when the refrigeration system is switched on and started up (pull down) excessive amounts of vapour or vapour/liquid or liquid refrigerant may enter the compressor cylinder, which could produce very high cylinder pressures; this would therefore cause severe strain and damage to the engine/electric motor components, conversely if the suction line pressure limit is set very low the evaporator may not operate efficiently.

The suction pressure valve consists of a combined piston and bellows controlled valve subjected to suction vapour pressure.

When the compressor is being driven by the engine/motor the output refrigerant vapour from the evaporator passes to the intake port of the suction pressure valve unit; this exposes the bellows to the refrigerant vapour pressure and temperature. Thus as the refrigerant pressure rises the bellows will contract against the force of the bellows spring; this restricts the flow of refrigerant to the compressor (see Fig. 13.10(a)). However, as the bellows temperature rises its internal pressure also increases and will therefore tend to oppose the contraction of the bellows. At the same time the piston will be subjected to the outlet vapour pressure from the suction pressure valve before entering the compressor cylinders, see Fig. 13.10(b). If this becomes excessive the piston and valve will move towards the closure position thus restricting the entry of refrigerant vapour or vapour/liquid to the compressor cylinders. Hence it can be seen that the suction pressure valve protects the compressor and drive against abnormally high suction line pressure.

13.3.6 Reverse cycle valve (Fig. 13.11(a and b))

The purpose of this valve is to direct the refrigerant flow so that the refrigerant system is in either a cooling or heating cycle mode.

Refrigerant cycle mode (Fig. 13.11(a)) With the pilot solenoid valve de-energized the suction passage to the slave cylinder of the reverse cycle valve is cut off whereas the discharge pressure supply from the compressor is directed to the slave piston. Accordingly the pressure build-up pushes the piston and both valve stems inwards; the left hand compressor discharge valve now closes the

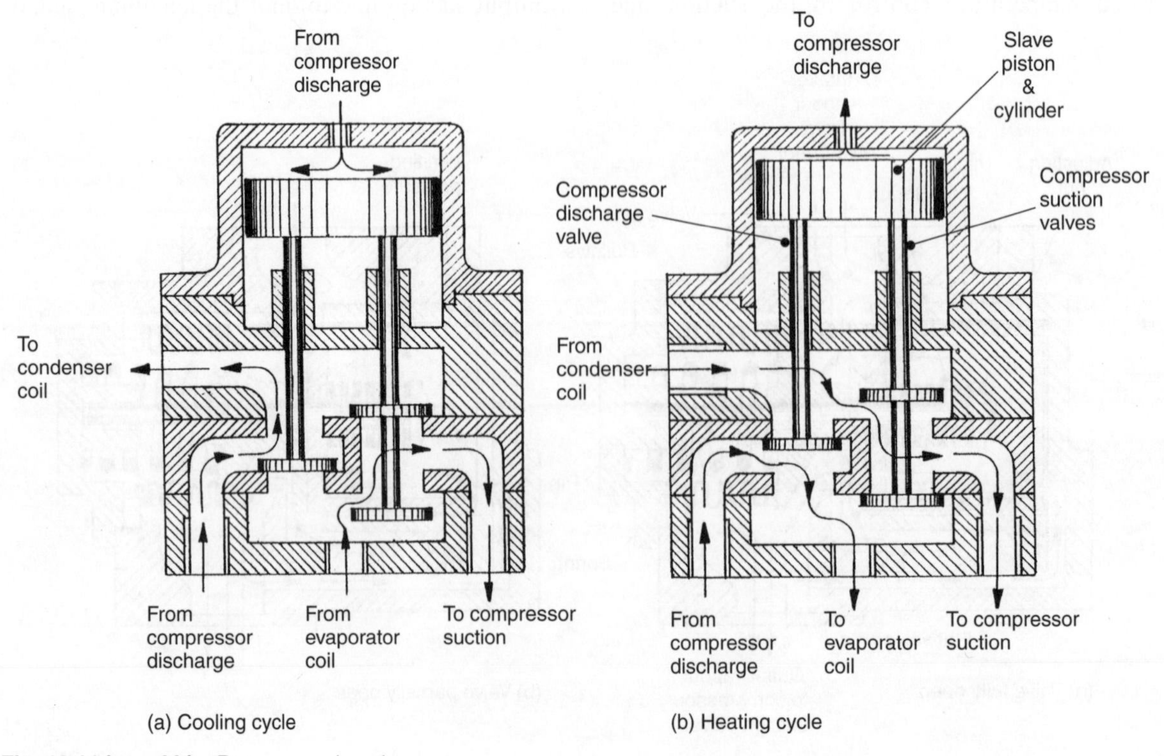

Fig. 13.11 (a and b) Reverse cycle valve

compressor discharge passage to the evaporator and opens the compressor discharge passage to the condenser whereas the right hand double compressor discharge valve closes the condenser to compressor suction passage and opens the evaporator to the compressor suction pressure.

Heat/defrost cycle mode (Fig. 13.11(b)) Energizing the pilot solenoid valve cuts off the compressor discharge pressure to the slave cylinder of the reverse cycle valve and opens it to the compressor suction line. As a result the trapped refrigerant vapour in the slave cylinder escapes to the compressor suction line thus permitting the slave piston and both valves to move to their outermost position. The left hand compressor discharge valve now closes the compressor discharge to the condenser passage and opens the compressor discharge to the evaporator passage whereas the right hand compressor suction double valve closes the evaporator to the compressor suction passage and opens the condenser to compressor suction pressure.

13.3.7 Drier (Fig. 13.12)

Refrigerant circulating the refrigerator system must be dry, that is, the fluid, be it a vapour or a liquid, should not contain water. Water in the form of moisture can promote the formation of acid which can attack the tubing walls and joints and cause the refrigerant to leak out. It may initiate the formation of sludge and restrict the circulation of the refrigerant. Moisture may also cause ice to form in the thermostatic expansion valve which again could reduce the flow of refrigerant. To overcome problems with water contamination driers are normally incorporated in the liquid line; these liquid line driers not only remove water contained in the refrigerant, they also remove sludge and other impurities. Liquid line driers are plumbed in on the output side of the receiver, this is because the moisture is concentrated in a relatively small space when the refrigerant is in a liquid state.

A liquid line drier usually takes the form of a cylindrical cartridge with threaded end connections so that the drier can be replaced easily when necessary. Filter material is usually packed in at both ends; in the example shown Fig. 13.12 there are layers, a coarse filter, felt pad and a fine filter. In between the filter media is a desiccant material, these are generally of the adsorption desiccant kind such as silica gel (silicon dioxide) or activated alumna (aluminium oxide). The desiccant substance has microscopic holes for the liquid refrigerant to pass through; however, water is attracted to the desiccant and therefore is prevented from moving on whereas the dry (free of water) clean refrigerant will readily flow through to the expansion valve.

13.3.8 Oil separator (Fig. 13.13)

Oil separators are used to collect any oil entering the refrigeration system through the compressor and to return it to the compressor crankcase and sump. The refrigerant may mix with the compressor's lubrication oil in the following way:

1 During the cycle of suction and discharge refrigerant vapour periodically enters and is displaced from the cylinders; however, if the refrigerant flow becomes excessive liquid will pass through the expansion valve and may eventually enter the suction line via the evaporator. The fluid may then drain down the cylinder walls to the crankcase and sump. Refrigerant mixing with oil dilutes it so that it loses its lubricating properties: the wear and tear of the various rubbing components in the compressor will therefore increase.

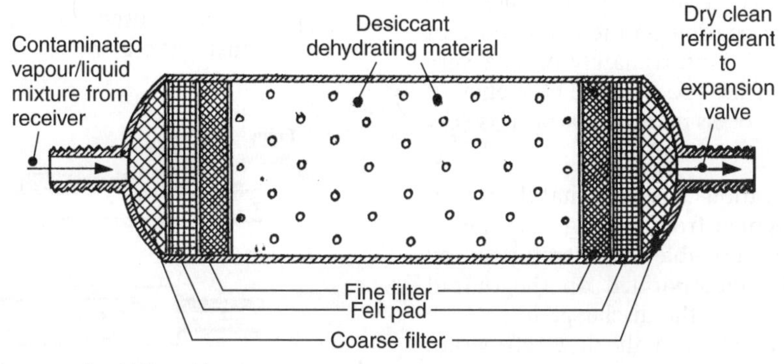

Fig. 13.12 Adsorption type liquid line drier

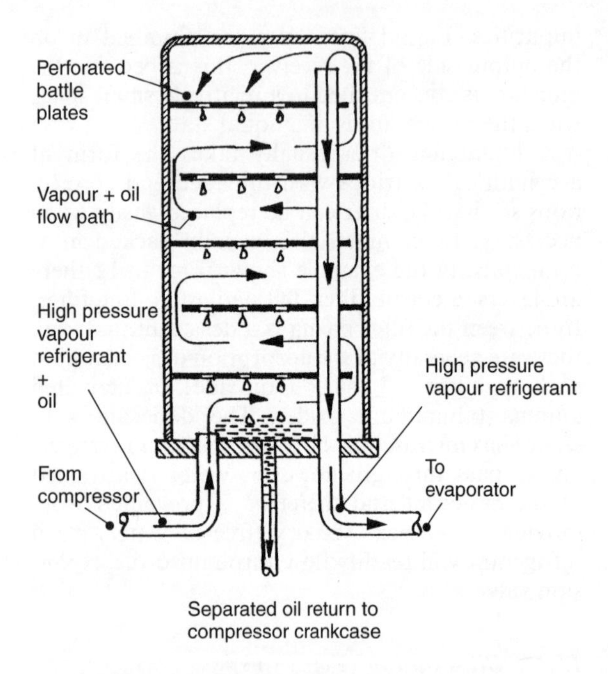

Fig. 13.13 Oil separator

2 When the refrigerator is switched off the now static refrigerant in the evaporator may condense and enter the suction line and hence the compressor cylinders where it drains over a period of time into the crankcase and sump.
3 Refrigerant mixing with the lubricant in the crankcase tends to produce oil frothing which finds its way past the pistons and piston rings into the cylinders; above each piston the oil will then be pumped out into the discharge line with the refrigerant where it then circulates. Oil does not cause a problem in the condenser as the temperature is fairly high so that the refrigerant remains suspended; however, in the evaporator the temperature is low so that the liquid oil separates from the refrigerant vapour, therefore tending to form a coating on the inside bore of the evaporator coil. Unfortunately oil is a very poor conductor of heat so that the efficiency of the heat transfer process in the evaporator is very much impaired.

After these observations it is clear that the refrigerant must be prevented from mixing with the oil but this is not always possible and therefore an oil separator is usually incorporated on the output side of the compressor in the discharge line which separates the liquid oil from the hot refrigerant vapour. An oil separator in canister form consists of a cylindrical chamber with a series of evenly spaced perforated baffle plates or wire mesh partitions attached to the container walls; each baffle plate has a small segment removed to permit the flow of refrigerant vapour (Fig. 13.13), the input from the compressor discharge being at the lowest point whereas the output is via the extended tube inside the container. A small bore pipe connects the base of the oil separator to the compressor crankcase to provide a return passage for the liquid oil accumulated. Thus when the refrigerator is operating, refrigerant will circulate and therefore passes though the oil separator. As the refrigerant/oil mix zigzags its way up the canister the heavier liquid oil tends to be attracted and attached to the baffle plates; the accumulating oil then spreads over the plates until it eventually drips down to the base of the canister, and then finally drains back to the compressor crankcase.

13.3.9 Receiver (Fig. 13.6)

The receiver is a container which collects the condensed liquid refrigerant and any remaining vapour from the condenser; this small amount of vapour will then have enough space to complete the condensation process before moving to the expansion valve.

13.3.10 Sight glass (Fig. 13.14)

This device is situated in the liquid line on the output side of the receiver; it is essentially a viewing port which enables the liquid refrigerant to be seen. Refrigerant movement or the lack of movement due to some kind of restriction, or bubbling caused by insufficient refrigerant, can be observed.

13.4 Vapour–compression cycle refrigeration system with reverse cycle defrosting

(Fig. 13.15(a and b))

A practical refrigeration system suitable for road transportation as used for rigid and articulated vehicles must have a means of both cooling and

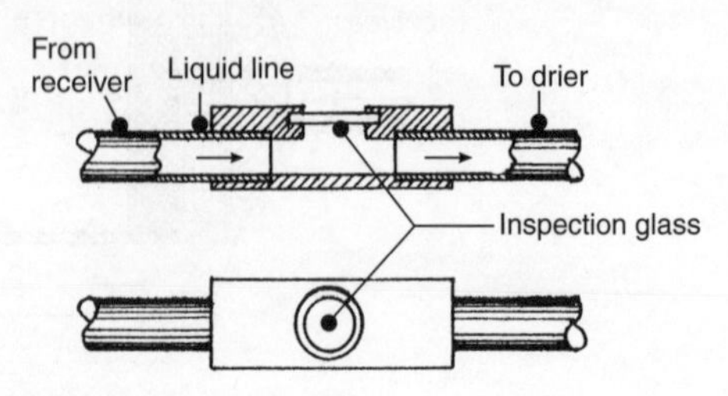

Fig. 13.14 Sight glass

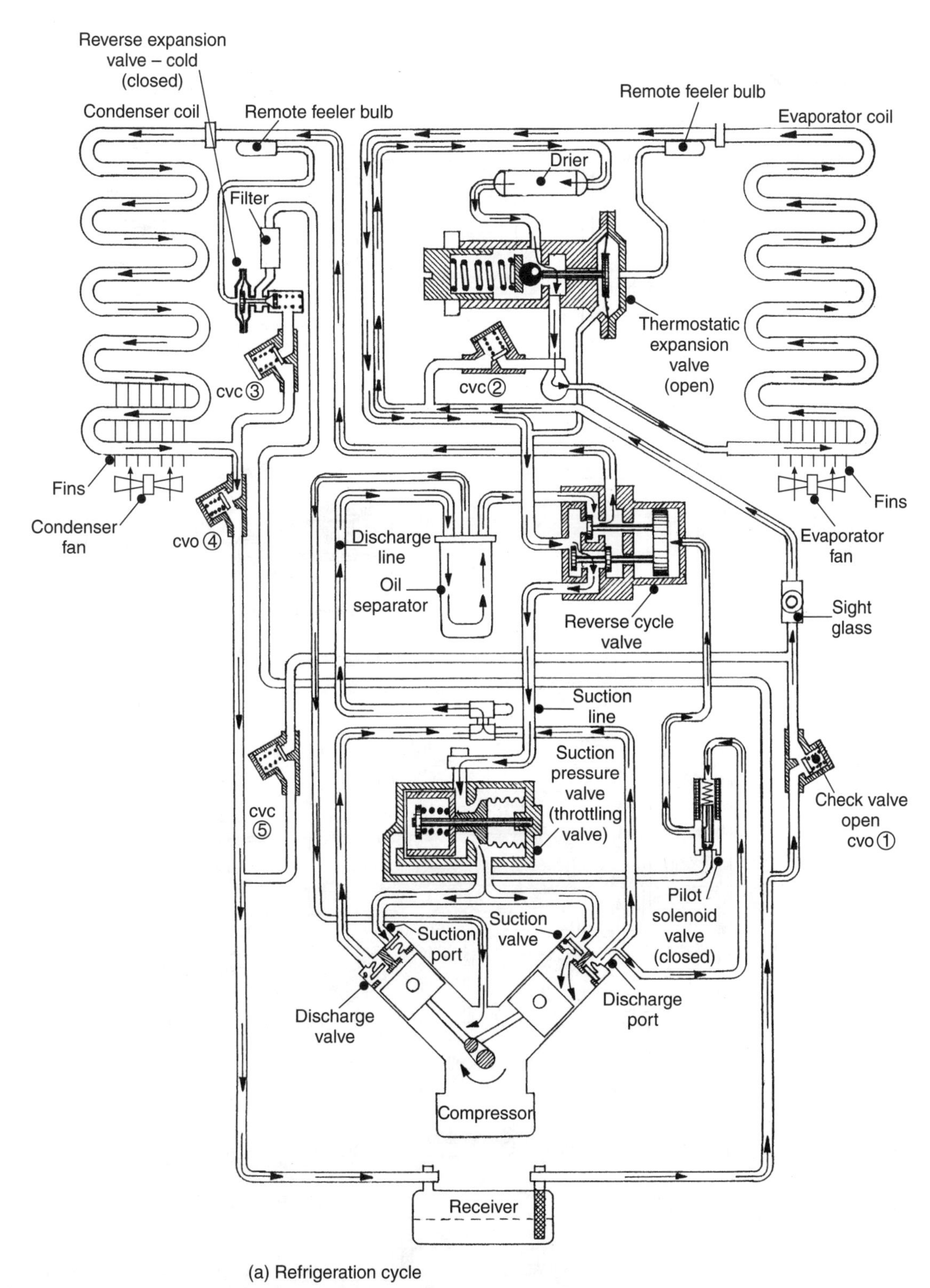

(a) Refrigeration cycle

Fig. 13.15 (a and b) Refrigeration system with reverse cycle defrosting

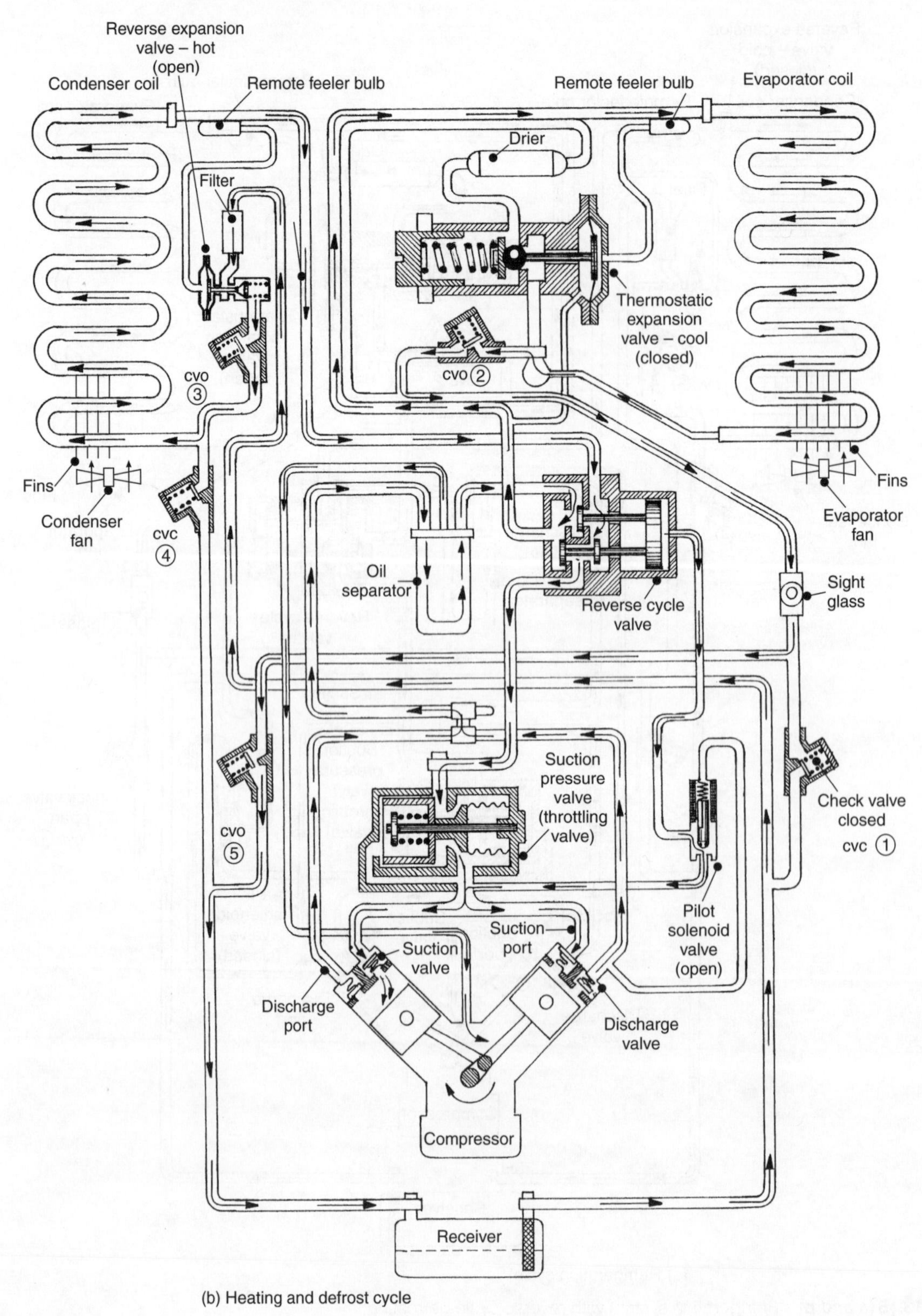

(b) Heating and defrost cycle

Fig. 13.15 *Contd*

defrosting the cold compartment. The operation of such a system involving additional valves enables the system to be switched between cooling and heat/defrosting, which will now be described.

13.4.1 Refrigeration cooling cycle (Fig. 13.15(a))

With the pilot solenoid valve de-energized and the compressor switched on and running the refrigerant commences to circulate through the system between the evaporator and condenser.

Discharge line pressure from the right hand compressor cylinder is transferred via the pilot valve to the reverse cycle valve; this pushes the slave piston and valves inwards to the left hand side into the 'cooling' position, see Fig. 13.15(a). Low pressure refrigerant from the receiver flows via the open check valve (1), sight glass and drier to the thermostatic expansion valve where rapid expansion in the valve converts the refrigerant to a liquid/vapour mixture. Low pressure refrigerant then passes through the evaporator coil where it absorbs heat from the cold storage compartment: the refrigerant then comes out from the evaporator as low pressure saturated vapour. Refrigerant now flows to the compressor suction port via the reverse cycle valve and suction pressure valve as superheated vapour. The compressor now converts the refrigerant to a high pressure superheated vapour before pumping it to the condenser inlet via the oil separator and reverse cycle valve; at this point the refrigerant will have lost heat to the surroundings so that it is now in a high pressure saturated vapour state. It now passes through the condenser where it gives out its heat to the surrounding atmosphere; during this process the high pressure refrigerant is transformed into a saturated liquid. Finally the main liquid refrigerant flows into the receiver via the open check valve (4) where there is enough space for the remaining vapour to condense. This cycle of events will be continuously repeated as the refrigerant is alternated between reducing pressure in the expansion valve before passing through the evaporator to take heat from the cold chamber, to increasing pressure in the compressor before passing through the condenser to give off its acquired heat to the surroundings. Note check valves (1) and (4) are open whereas check valves (2), (3) and (5) are closed for the cool cycle.

13.4.2 Heating and defrosting cycle (Fig. 13.15(b))

With constant use excessive ice may build up around the evaporator coil; this restricts the air movement so that the refrigerant in the evaporator is unable to absorb the heat from the surrounding atmosphere in the cold storage compartment, therefore a time will come when the evaporator must be defrosted.

Heating/defrosting is achieved by temporarily reversing the refrigerant flow circulation so that the evaporator becomes a heat dissipating coil and the condenser converts to a heat absorbing coil.

To switch to the heat/defrosting cycle the pilot solenoid valve is energized; this causes the solenoid valve to block the discharge pressure and connect the suction pressure to the servo cylinder reverse cycle valve, see Fig. 13.15(b). Subcooled high pressure liquid refrigerant is permitted to flow from the receiver directly to the now partially opened reverse thermostatic expansion valve (due to the now hot remote feeler bulb's increased pressure). The refrigerant expands in the reverse expansion valve and accordingly converts to a liquid/vapour; this then passes through the condenser via the open check valve (3) in the reverse direction to the normal refrigeration cycle and in the process absorbs heat from the surroundings so that it comes out as a low pressure saturated vapour. The refrigerant then flows to the compressor suction port via the reverse cycle valve and suction pressure valve but due to the high surrounding atmospheric temperature it is now superheated vapour. The compressor then transforms the low pressure superheated vapour into a high pressure superheated vapour and discharges it to the evaporator via the oil separator and reverse cycle valve. Hence the saturated vapour stream dissipates its heat through the tubing walls to the ice which is encasing the tubing coil until it has all melted. The refrigerant at the exit from the evaporator will now be in a saturated liquid state and is returned to the receiver via the open check valve (2), sight glass, and open check valve (5) for the heating/defrosting cycle to be repeated. Note during the refrigeration cycle the condenser's reverse expansion valve and remote feeler bulb sense the reduction in temperature at the exit from the condenser, thus the corresponding reduction in internal bulb pressure is relayed to the reverse expansion valve which therefore closes during the defrosting cycle. Defrosting is fully automatic. A differential air pressure switch which senses any air circulation restriction around the evaporator coil automatically triggers defrosting of the evaporator coil before ice formation can reduce its efficiency. A manual defrost switch is also provided.

14 Vehicle body aerodynamics

The constant need for better fuel economy, greater vehicle performance, reduction in wind noise level and improved road holding and stability for a vehicle on the move, has prompted vehicle manufacturers to investigate the nature of air resistance or drag for different body shapes under various operating conditions. Aerodynamics is the study of a solid body moving through the atmosphere and the interaction which takes place between the body surfaces and the surrounding air with varying relative speeds and wind direction. Aerodynamic drag is usually insignificant at low vehicle speed but the magnitude of air resistance becomes considerable with rising speed. This can be seen in Fig. 14.1 which compares the aerodynamic drag forces of a poorly streamlined, and a very highly streamlined medium sized car against its constant rolling resistance over a typical speed range. A vehicle with a high drag resistance tends only marginally to hinder its acceleration but it does inhibit its maximum speed and increases the fuel consumption with increasing speed.

Body styling has to accommodate passengers and luggage space, the functional power train, steering, suspension and wheels etc. thus vehicle design will conflict with minimizing the body surface drag so that the body shape finally accepted is nearly always a compromise.

An appreciation of the fundamentals of aerodynamics and the methods used to counteract high air resistance for both cars and commercial vehicles will now be explained.

14.1 Viscous air flow fundamentals

14.1.1 Boundary layer (Fig. 14.2)

Air has viscosity, that is, there is internal friction between adjacent layers of air, whenever there is relative air movement, consequently when there is sliding between adjacent layers of air, energy is dissipated. When air flows over a solid surface a thin boundary layer is formed between the main airstream and the surface. Any relative movement between the main airstream flow and the surface of

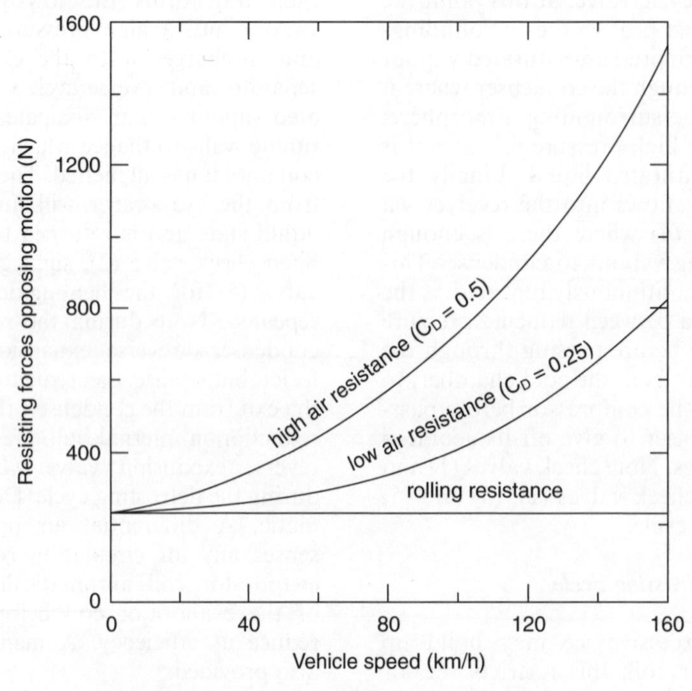

Fig. 14.1 Comparison of low and high aerodynamic drag forces with rolling resistance

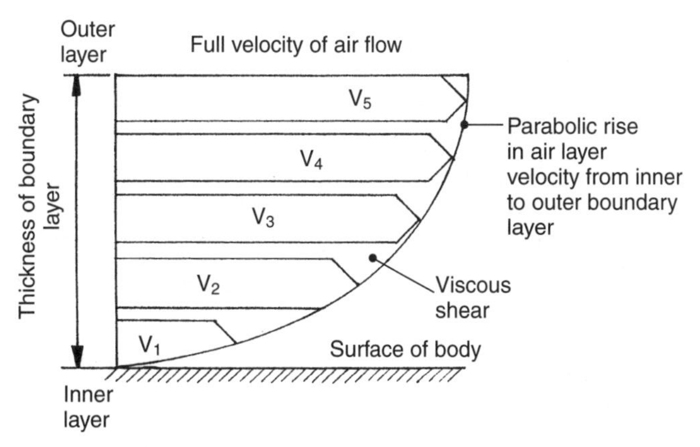

Fig. 14.2 Boundary layer velocity gradient

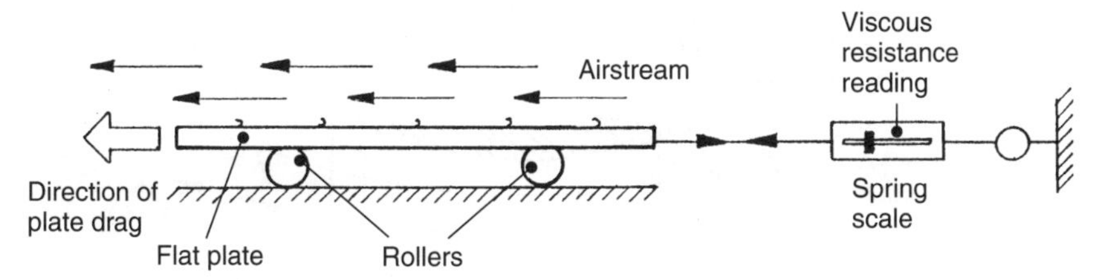

Fig. 14.3 Apparatus to demonstrate viscous drag

the body then takes place within this boundary layer via the process of shearing of adjacent layers of air. When air flows over any surface, air particles in intimate contact with the surface loosely attach themselves so that relative air velocity at the surface becomes zero, see Fig. 14.2. The relative speed of the air layers adjacent to the arrested air surface film will be very slow; however, the next adjacent layer will slide over an already moving layer so that its relative speed will be somewhat higher. Hence the relative air velocity further out from the surface rises progressively between air layers until it attains the unrestricted main airstream speed.

14.1.2 Skin friction (surface friction drag) (Fig. 14.3)

This is the restraining force preventing a thin flat plate placed edgewise to an oncoming airstream being dragged along with it (see Fig. 14.3), in other words, the skin friction is the viscous resistance generated within the boundary layer when air flows over a solid surface. Skin friction is dependent on the surface area over which the air flows, the degree of surface roughness or smoothness and the air speed.

14.1.3 Surface finish (Fig. 14.4(a and b))

Air particles in contact with a surface tend to be attracted to it, thus viscous drag will retard the layer of air moving near the surface. However, there will be a gradual increase in air speed from the inner to the outer boundary layer. The thickness of the boundary layer is influenced by the surface finish. A smooth surface, see Fig. 14.4(b), allows the free air flow velocity to be reached nearer the surface whereas a rough surface, see Fig. 14.4(a), widens the boundary so that the full air velocity will be reached further out from the surface. Hence the thicker boundary layer associated with a rough surface will cause more adjacent layers of air to be sheared, accordingly there will be more resistance to air movement compared with a smooth surface.

14.1.4 Venturi (Fig. 14.5)

When air flows through a diverging and converging section of a venturi the air pressure and its speed changes, see Fig. 14.5. Initially at entry the unrestricted air will be under atmospheric conditions where the molecules are relatively close together, consequently its pressure will be at its highest and its speed at its minimum.

As the air moves into the converging section the air molecules accelerate to maintain the volume flow. At the narrowest region in the venturi the random air molecules will be drawn

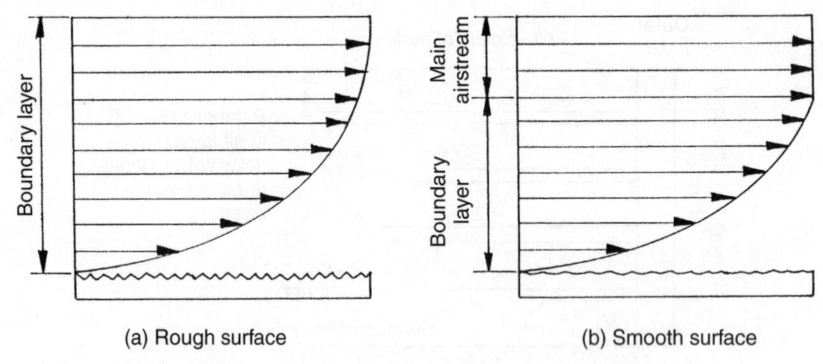

Fig. 14.4 (a and b) Influence of surface roughness on boundary layer velocity profile

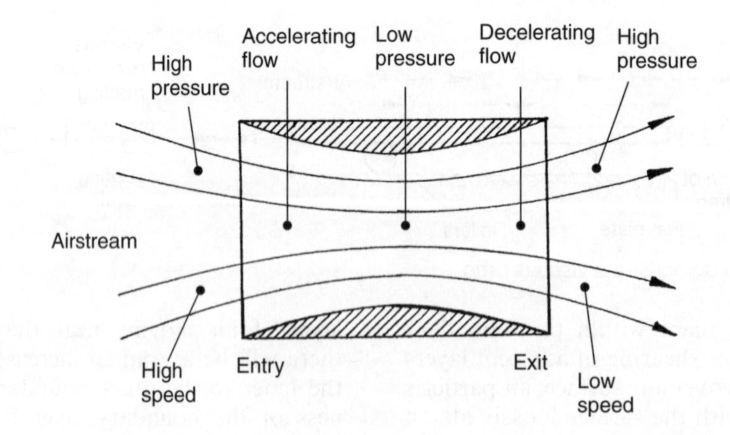

Fig. 14.5 Venturi

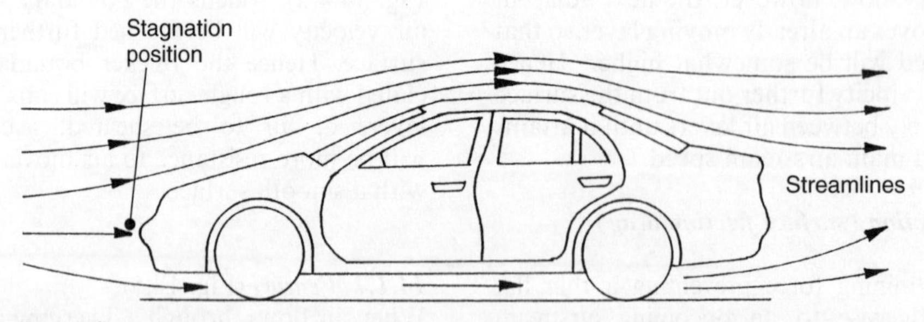

Fig. 14.6 Streamline air flow around car

apart thus creating a pressure drop and a faster movement of the air. Further downstream the air moves into the diverging or expanding section of the venturi where the air flow decelerates, the molecules therefore are able to move closer together and by the time the air reaches the exit its pressure will have risen again and its movement slows down.

*14.1.5 **Air streamlines*** (Fig. 14.6)

A moving car displaces the air ahead so that the air is forced to flow around and towards the rear. The pattern of air movement around the car can be visualized by airstreamlines which are imaginary lines across which there is no flow, see Fig. 14.6. These streamlines broadly follow the contour of the body but any sudden change in the car's shape

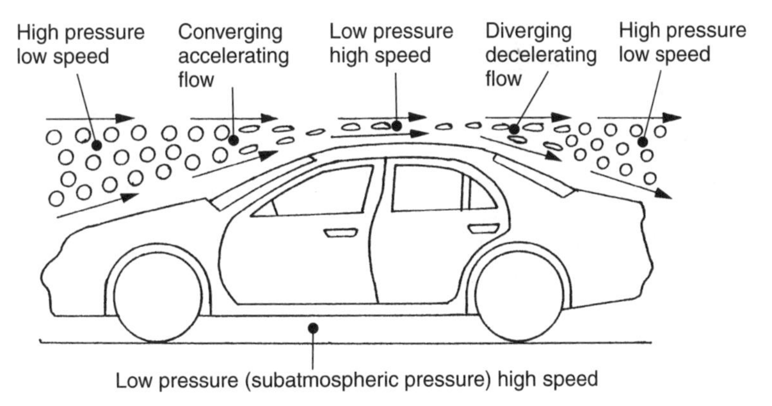

Fig. 14.7 Relative air speed and pressure conditions over the upper profile of a moving car

compels the streamlines to deviate, leaving zones of stagnant air pockets. The further these streamlines are from the body the more they will tend to straighten out.

*14.1.6 **Relative air speed and pressure conditions over the upper profile of a moving car*** (Figs 14.7 and 14.8)

The space between the upper profile of the horizontal outer streamlines relative to the road surface generated when the body is in motion can be considered to constitute a venturi effect, see Fig. 14.7. Note in effect it is the car that moves whereas air remains stationary; however, when wind-tunnel tests are carried out the reverse happens, air is drawn through the tunnel with the car positioned inside on a turntable so that the air passes over and around the parked vehicle. The air gap between the horizontal airstreamlines and front end bonnet (hood) and windscreen profile and the back end screen and boot (trunk) profile produces a diverging and converging air wedge, respectively. Thus the air scooped into the front wedge can be considered initially to be at atmospheric pressure and moving at car speed. As the air moves into the diverging wedge it has to accelerate to maintain the rate of volumetric displacement. Over the roof the venturi is at its narrowest, the air movement will be at its highest and the air molecules will be stretched further apart, consequently there will be a reduction in air pressure in this region. Finally the relative air movement at the rear of the boot will have slowed to car speed, conversely its pressure will have again risen to the surrounding atmospheric pressure conditions, thus allowing the random network of distorted molecules to move closer together to a more stable condition. As the air moves beyond the roof into the diverging wedge region it decelerates to cope with the enlarged flow space.

As can be seen in Fig. 14.8 the pressure conditions over and underneath the car's body can be plotted from the data; these graphs show typical pressure distribution trends only. The pressure over the rear half of the bonnet to the mid-front windscreen region where the airstream speed is slower is positive (positive pressure coefficient C_p), likewise the pressure over the mid-position of the rear windscreen and the rear end of the boot where the airstream speed has been reduced is also positive but of a lower magnitude. Conversely the pressure over the front region of the bonnet and particularly over the windscreen/roof leading edge and the horizontal roof area where the airstream speed is at its highest has a negative pressure (negative pressure coefficient C_p). When considering the air movement underneath the car body, the restricted airstream flow tends to speed up the air movement thereby producing a slight negative pressure distribution along the whole underside of the car. The actual pattern of pressure distribution above and below the body will be greatly influenced by the car's profile style, the vehicle's speed and the direction and intensity of the wind.

*14.1.7 **Lamina boundary layer*** (Fig. 14.9(a))

When the air flow velocity is low sublayers within the boundary layer are able to slide one over the other at different speeds with very little friction; this kind of uniform flow is known as lamina.

*14.1.8 **Turbulent boundary layer*** (Fig. 14.9(b))

At higher air flow velocity the sublayers within the boundary layer also increase their relative sliding speed until a corresponding increase in interlayer friction compels individual sublayers to randomly

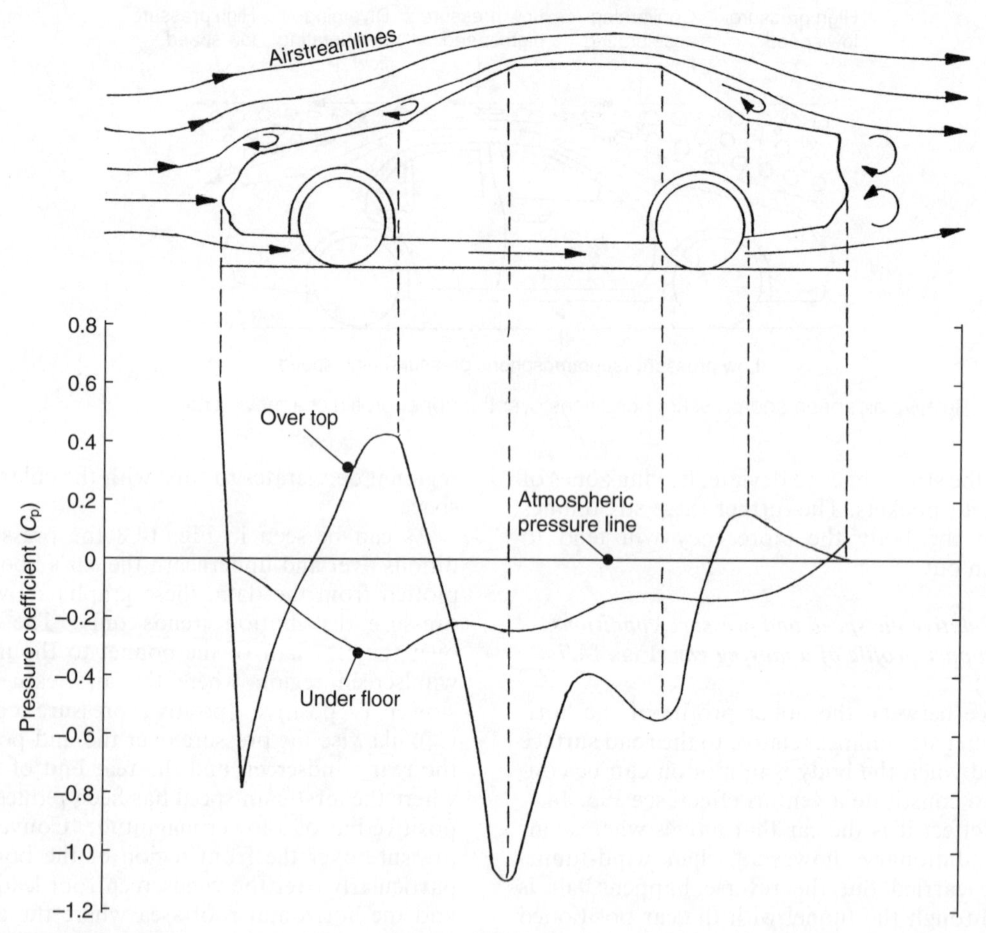

Fig. 14.8 Pressure distribution above and below the body structure

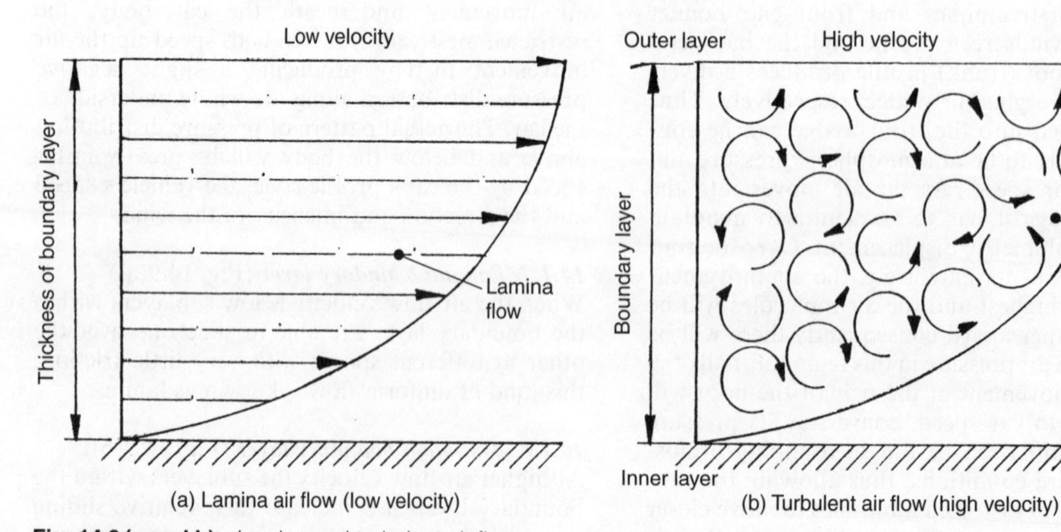

Fig. 14.9 (a and b) Lamina and turbulent air flow

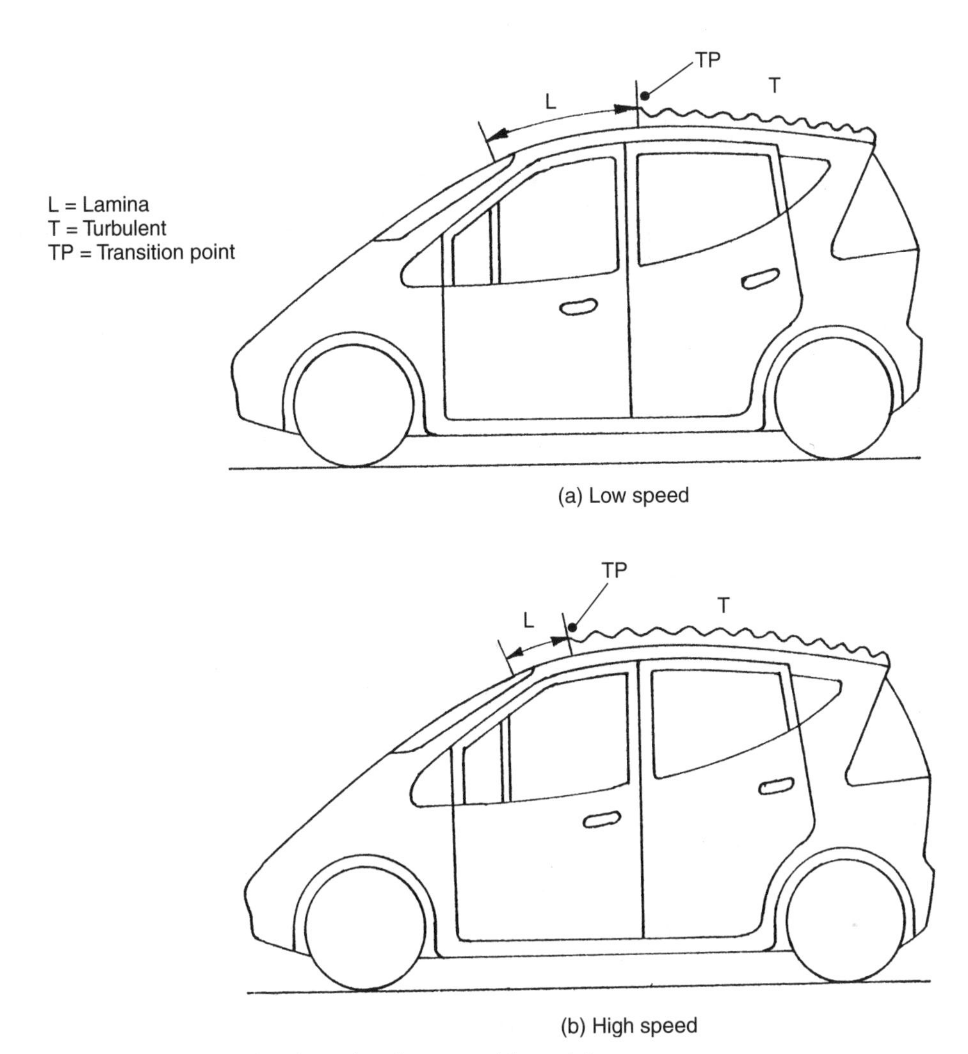

Fig. 14.10 (a and b) Lamina/turbulent boundary layer transition point

break away from the general direction of motion; they then whirl about in the form of eddies, but still move along with the air flow.

14.1.9 Lamina/turbulent boundary layer transition point (Fig. 14.10(a and b))

A boundary layer over the forward surface of a body, such as the roof, will generally be lamina, but further to the rear a point will be reached called the transition point when the boundary layer changes from a lamina to a turbulent one, see Fig. 14.10(a). As the speed of the vehicle rises the transition point tends to move further to the front, see Fig. 14.10(b), therefore less of the boundary layer will be lamina and more will become turbulent; accordingly this will correspond to a higher level of skin friction.

14.1.10 Flow separation and reattachment (Fig. 14.11(a and b))

The stream of air flowing over a car's body tends to follow closely to the contour of the body unless there is a sudden change in shape, see Fig. 14.11(a). The front bonnet (hood) is usually slightly curved and slopes up towards the front windscreen, from here there is an upward windscreen tilt (rake), followed by a curved but horizontal roof; the rear windscreen then tilts downwards where it either merges with the boot (trunk) or continues to slope gently downwards until it reaches the rear end of the car.

The air velocity and pressure therefore reaches its highest and lowest values, respectively, at the top of the front windscreen; however, towards the rear of the roof and when the screen tilts downwards

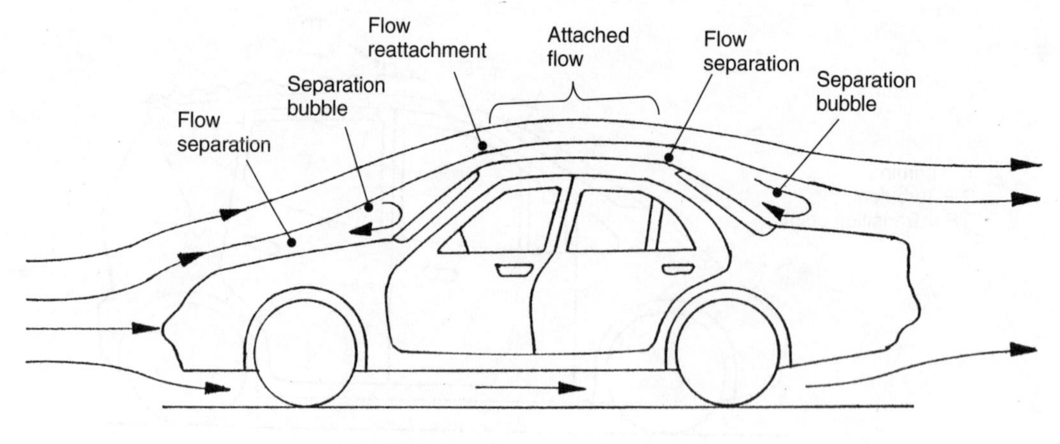

(a) Notch front and rear windscreens

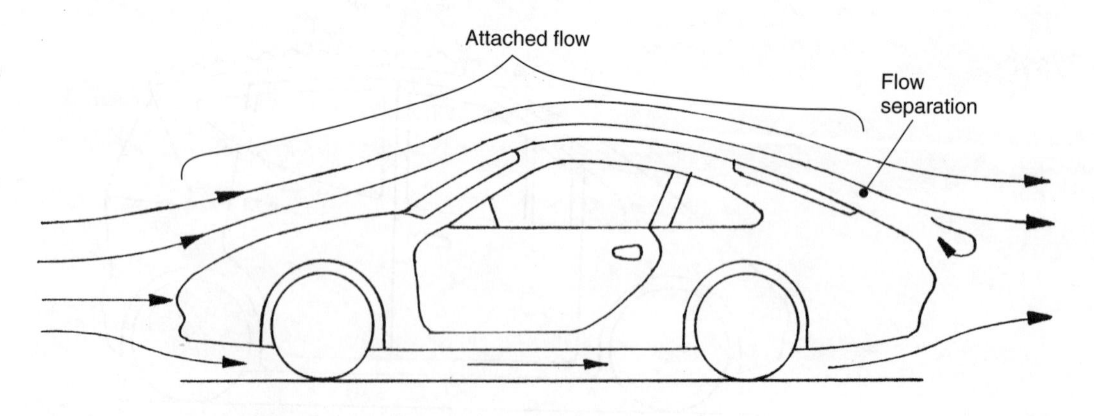

(b) Very streamlined shape

Fig. 14.11 (a and b) Flow separation and reattachment

there will be a reduction in air speed and a rise in pressure. If the rise in air pressure towards the rear of the car is very gradual then mixing of the airstream with the turbulent boundary layers will be relatively steady so that the outer layers will be drawn along with the main airstream, see Fig. 14.11(b). Conversely if the downward slope of the rear screen/boot is considerable, see Fig. 14.11(a), the pressure rise will be large so that the mixing rate of mainstream air with the boundary layers cannot keep the inner layers moving, consequently the slowed down boundary layers thicken. Under these conditions the mainstream air flow breaks away from the contour surface of the body, this being known as flow separation. An example of flow separation followed by reattachment can be visualized with air flowing over the bonnet and front windscreen; if the rake angle between the bonnet and windscreen is large, the streamline flow will separate from the bonnet and then reattach itself near the top of the windscreen or front end of the roof, see Fig. 14.11(a). The space between the separation and reattachment will then be occupied by circulating air which is referred to as a separation bubble, and if this rotary motion is vigorous a transverse vortex will be established.

14.2 Aerodynamic drag

14.2.1 Pressure (form) drag (Figs 14.12(a–e) and 14.13)

When viscous air flows over and past a solid form, vortices are created at the rear causing the flow

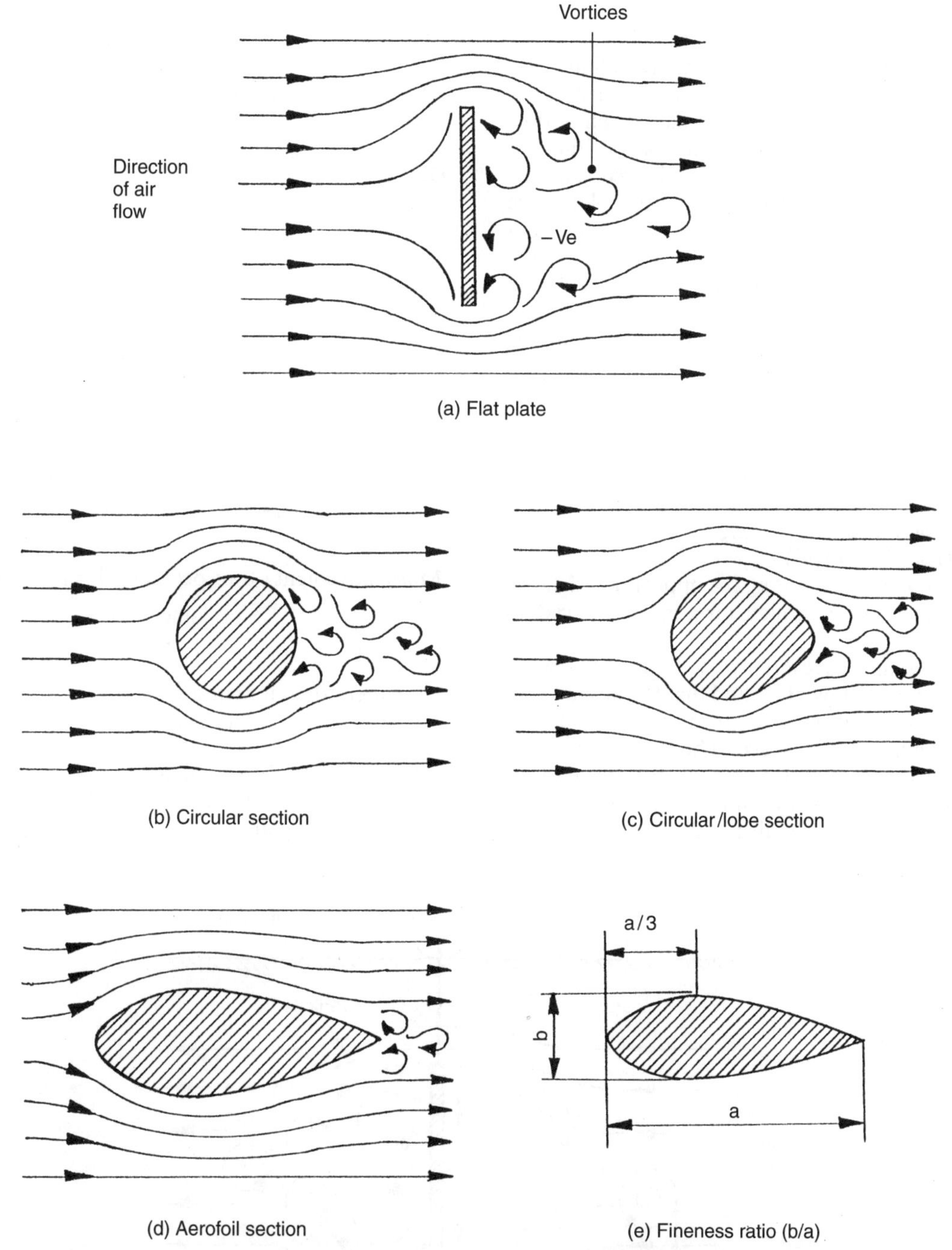

Fig. 14.12 (a–e) Air flow over various shaped sections

to deviate from the smooth streamline flow, see Fig. 14.12(a). Under these conditions the air flow pressure in front of the solid object will be higher than atmospheric pressure while the pressure behind will be lower than that of the atmosphere, consequently the solid body will be dragged (sucked) in the direction of air movement. Note that this effect is created in addition to the skin friction drag. An extreme example of pressure drag (sometimes known as form drag) can be seen in Fig. 14.13 where a flat plate placed at right angles to the air movement will experience a drag force in the direction of flow represented by the pulley weight which opposes the movement of the plate.

Pressure drag can be reduced by streamlining any solid form exposed to the air flow, for instance a round tube (Fig. 14.12(b)) encourages the air to flow smoothly around the front half and part of the rear before flow separation occurs thereby reducing the resistance by about half that of the flat plate. The resistance of a tube can be further reduced to about 15% of the flat plate by extending the rear of the circulating tube in the form of a curved tapering lobe, see Fig. 14.12(c). Even bigger reductions in resistance can be achieved by proportioning the tube section (see Fig. 14.12(d)) with a fineness ratio a/b of between 2 and 4 with the maximum thickness b set about one-third back from the nose, see Fig. 14.12(e). This gives a flow resistance of roughly one-tenth of a round tube or 5% of a flat plate.

14.2.2 Air resistance opposing the motion of a vehicle (Fig. 14.13)

The formula for calculating the opposing resistance of a body passing though air can be derived as follows:

Let us assume that a flat plate body (Fig. 14.13) is held against a flow of air and that the air particles are inelastic and simply drop away from the perpendicular plate surface. The density of air is the mass per unit volume and a cubic metre of air at sea level has an approximate mass of 1.225 kg, therefore the density of air is 1.225 kg/m^3.

Then let

$$\text{Mass} = m \text{ kg}$$
$$\text{Volume} = Q \text{ m}^3$$
$$\text{Density} = \rho \text{ kg/m}$$

Hence

$$\rho = \frac{m}{Q} \text{ kg/m}^{-3}$$

or

$$m = \rho Q \left(\frac{\text{kg}}{\text{m}^3}\text{m}^3\right)\text{kg}$$

Let

$$\text{Density of air flow} = \rho \text{ kg/m}^3$$
$$\text{Frontal area of plate} = A \text{ m}^2$$
$$\text{Velocity of air striking surface} = v \text{ m/s}$$
$$\text{Volume of air striking plate per second} = Q = vA \text{ m}^3$$
$$\text{Mass movement of air per second} = \rho Q = \rho \times vA$$
$$\text{since} \quad Q = vA$$
$$\text{Momentum of this air (mv)} = \rho vA \times v$$
$$\text{therefore momentum lost by air per second} = \rho A v^2$$

From Newton's second law the rate at which the movement of air is changed will give the force exerted on the plate.

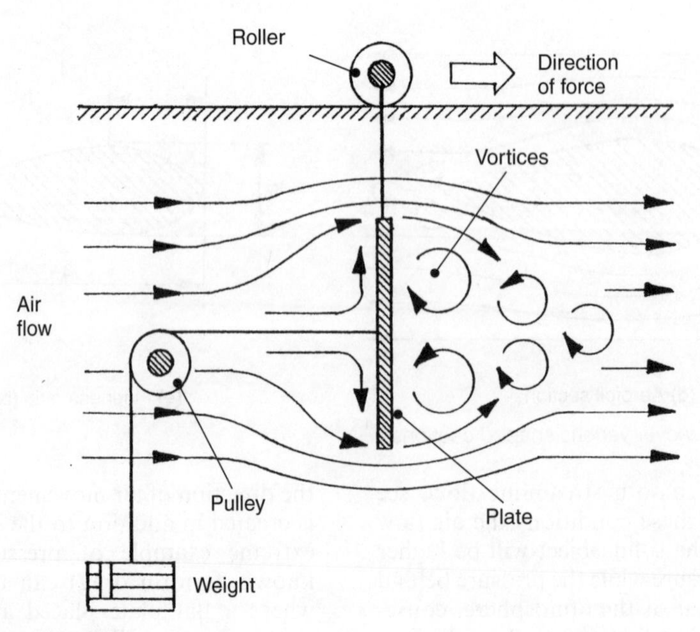

Fig. 14.13 Pressure drag apparatus

Hence

$$\text{force on plate} = \rho A v^2 \quad \text{Newton's}$$

However, the experimental air thrust against a flat plate is roughly 0.6 of the calculated $\rho A v^2$ force. This considerable 40% error is basically due to the assumption that the air striking the plate is brought to rest and falls away, where in fact most of the air escapes round the edges of the plate and the flow then becomes turbulent. In fact the theoretical air flow force does not agree with the actual experimental force (F) impinging on the plate, but it has been found to be proportional to $\rho A v^2$

hence

$$F \propto A v^2$$

therefore air resistance $F = C_D A v^2$ where C_D is the coefficient of proportionality.

The constant C_D is known as the coefficient of drag, it has no unity and its value will depend upon the shape of the body exposed to the airstream.

14.2.3 After flow wake (Fig. 14.14)

This is the turbulent volume of air produced at the rear end of a forward moving car and which tends to move with it, see Fig. 14.14. The wake has a cross-sectional area equal approximately to that of the rear vertical boot panel plus the rearward projected area formed between the level at which the air flow separates from the downward sloping rear window panel and the top edge of the boot.

14.2.4 Drag coefficient

The aerodynamic drag coefficient is a measure of the effectiveness of a streamline aerodynamic body shape in reducing the air resistance to the forward motion of a vehicle. A low drag coefficient implies that the streamline shape of the vehicle's body is such as to enable it to move easily through the surrounding viscous air with the minimum of resistance; conversely a high drag coefficient is caused by poor streamlining of the body profile so that there is a high air resistance when the vehicle is in motion. Typical drag coefficients for various classes of vehicles can be seen as follows:

Vehicle type	drag coefficient C_D
Saloon car	0.22–0.4
Sports car	0.28–0.4
Light van	0.35–0.5
Buses and coaches	0.4–0.8
Articulated trucks	0.55–0.8
Ridged truck and draw bar trailer	0.7–0.9

14.2.5 Drag coefficients and various body shapes (Fig. 14.15(a–f))

A comparison of the air flow resistance for different shapes in terms of drag coefficients is presented as follows:

(a) *Circular plate* (Fig. 14.15(a)) Air flow is head on, and there is an immediate end on pressure difference. Flow separation takes place at the rim; this provides a large vortex wake and a correspondingly high drag coefficient of 1.15.

(b) *Cube* (Fig. 14.15(b)) Air flow is head on but a boundary layer around the sides delays the flow separation; nevertheless there is still a large vortex wake and a high drag coefficient of 1.05.

(c) *Sixty degree cone* (Fig. 14.15(c)) With the piecing cone shape air flows towards the cone apex and then spreads outwards parallel to the shape of the cone surface. Flow separation however still takes place at the periphery thereby producing a wide vortex wake. This profile halves the drag coefficient to about 0.5 compared with the circular plate and the cube block.

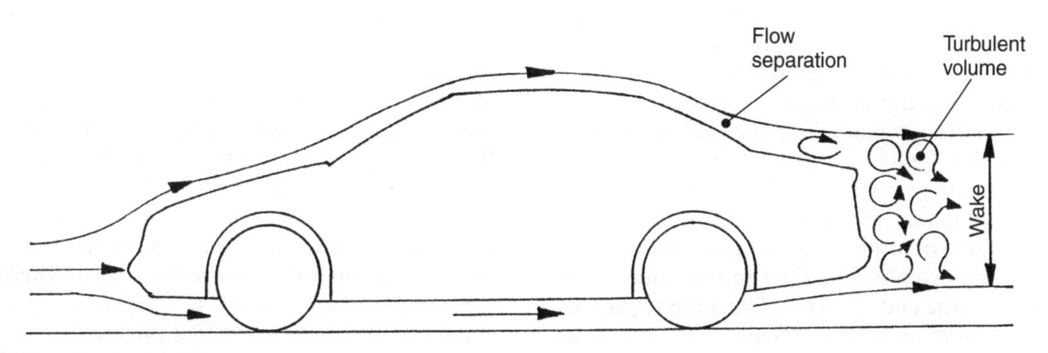

Fig. 14.14 After flow wake

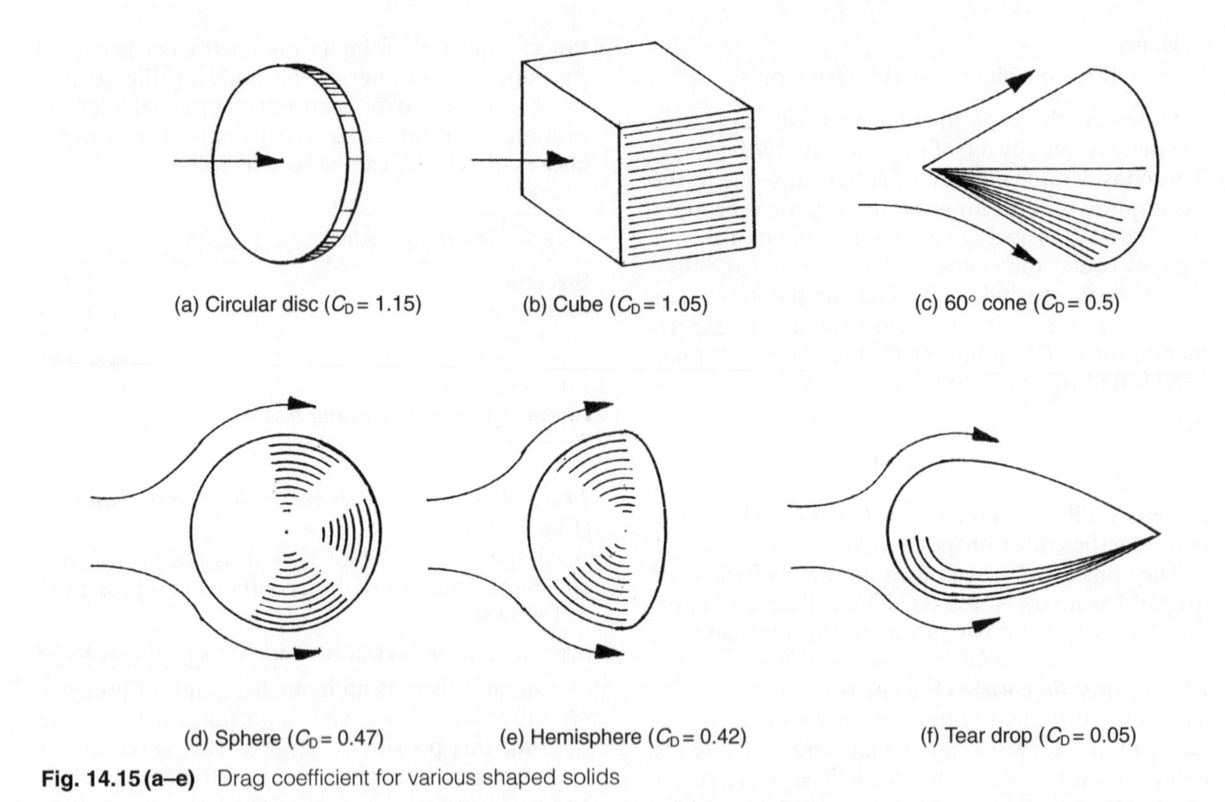

Fig. 14.15 (a–e) Drag coefficient for various shaped solids

(d) *Sphere* (Fig. 14.15(d)) Air flow towards the sphere, it is then diverted so that it flows outwards from the centre around the diverging surface and over a small portion of the converging rear half before flow separation occurs. There is therefore a slight reduction in the vortex wake and similarly a marginal decrease in the drag coefficient to 0.47 compared with the 60° cone.

(e) *Hemisphere* (Fig. 14.15(e)) Air flow towards and outwards from the centre of the hemisphere. The curvature of the hemisphere gradually aligns with the main direction of flow after which flow separation takes place on the periphery. For some unknown reason (possibly due to the very gradual alignment of surface curvature with the direction of air movement near the rim) the hemisphere provides a lower drag coefficient than the cone and the sphere shapes this, being of the order of 0.42.

(f) *Tear drop* (Fig. 14.15(f)) If the proportion of length to diameter is well chosen, for example 0.25, the streamline shape can maintain a boundary layer before flow separation occurs almost to the end of its tail. Thus the resistance to body movement will be mainly due to viscous air flow and little to do with vortex wake suction. With these contours the drag coefficient can be as low as 0.05.

14.2.6 Base drag (Fig. 14.16(a and b))

The shape of the car body largely influences the pressure drag. If the streamline contour of the body is such that the boundary layers cling to a converging rear end, then the vortex area is considerably reduced with a corresponding reduction in rear end suction and the resistance to motion. If the body was shaped in the form of a tear drop, the contour of the body would permit a boundary layer to continue a considerable way towards the tail before flow separation occurs, see Fig. 14.16(a), consequently the area heavily subjected to vortex swirl and negative pressure will be at a minimum. However, it is impractical to design a tear drop body with an extended tapering rear. end, but if the tail is cut off (bobtailed) at the point where the air flow separates from the contour of the body (see Fig. 14.16(b)), the same vortex (negative pressure) exists as if the tail was permitted to converge. The cut off cross-section area where flow separation would occur is known as the 'base area' and the negative vortex pressure produced is referred to as the 'base drag'. Thus there is a trend

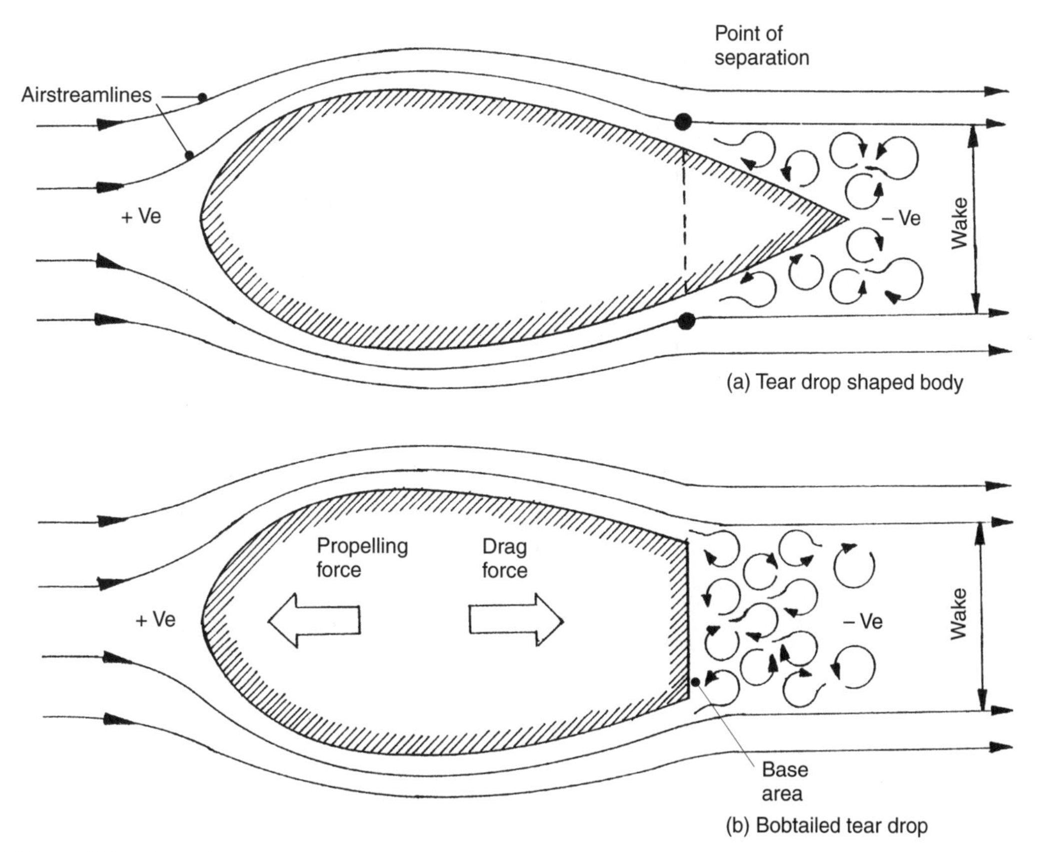

Fig. 14.16 (a and b) Base drag

for car manufacturers to design bodies that taper slightly towards the rear so that flow separation occurs just beyond the rear axle.

14.2.7 Vortices (Fig. 14.17)
Vortices are created around various regions of a vehicle when it is in motion. Vortices can be described as a swirling air mass with an annular cylindrical shape, see Fig. 14.17. The rotary speed at the periphery is at its minimal, but this increases inversely with the radius so that its speed near the centre is at a maximum. However, there is a central core where there is very little movement, consequently viscous shear takes place between adjacent layers of the static core and the fast moving air swirl; thus the pressure within the vortex will be below atmospheric pressure, this being much lower near the core than in the peripheral region.

14.2.8 Trailing vortex drag (Fig. 14.18(a and b))
Consider a car with a similar shape to a section of an aerofoil, see Fig. 14.18(a), when air flows from the front to the rear of the car, the air moves between the underside and ground, and over the raised upper body profile surfaces. Thus if the upper and lower airstreams are to meet at the rear at a common speed the air moving over the top must move further and therefore faster than the more direct underfloor airstream. The air pressure will therefore be higher in the slower underfloor airstream than that for the faster airstream moving over the top surface of the car. Now air moves from high to low pressure regions so that the high pressure airstream underneath the car will tend to move diagonally outwards and upwards towards the low pressure airstream flowing over the top of the body surface (see Fig. 14.18(b)). Both the lower and the upper airstreams eventually interact along the side-to-top profile edges on opposite sides of the body to form an inward rotary air motion that continues to whirl for some distance beyond the rear end of the forward moving car, see Fig. 14.18(a and b). The magnitude and intensity of these vortices will to a great extent depend upon the rear styling of the

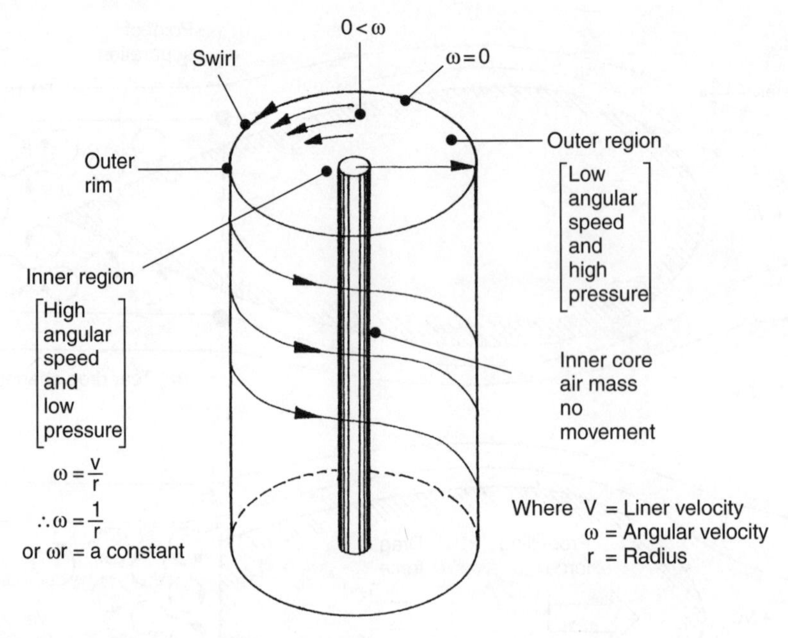

Fig. 14.17 The vortex

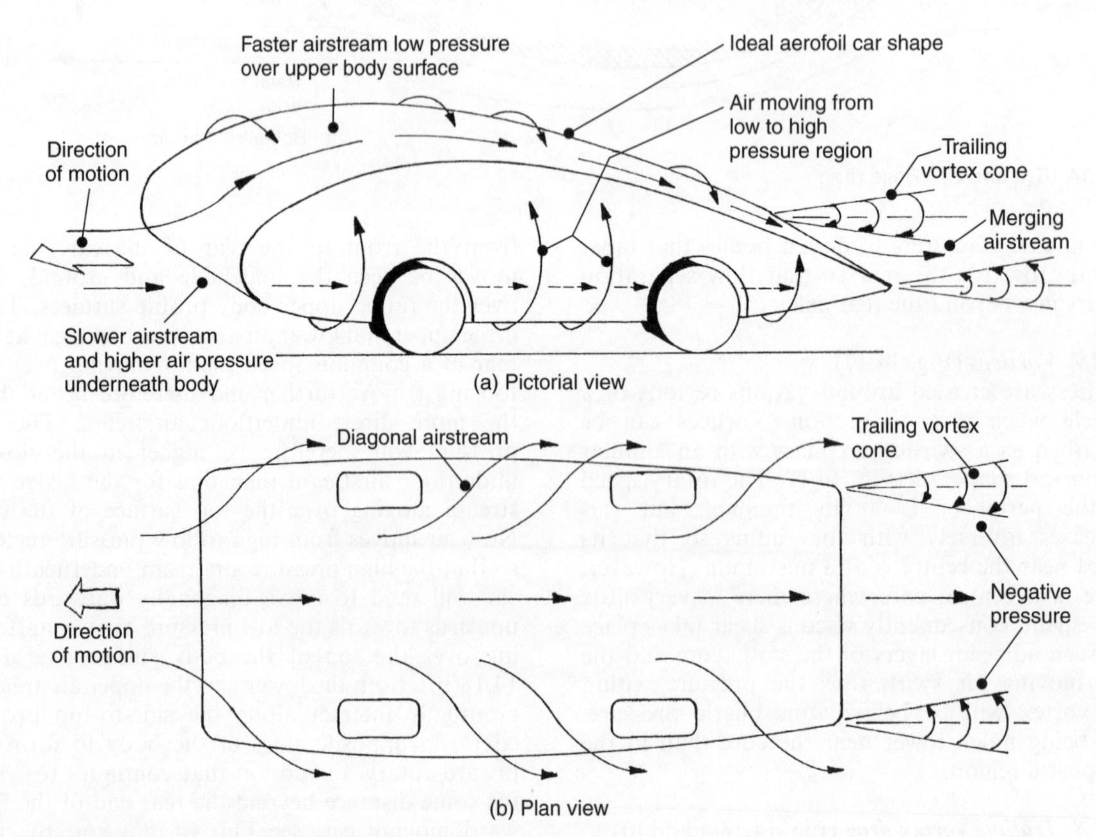

Fig. 14.18 (a and b) Establishment of trailing vortices

car. The negative (below atmospheric) pressure created in the wake of the trailing vortices at the rear of the car attempts to draw it back in the opposite direction to the forward propelling force; this resistance is therefore referred to as the 'trailing vortex drag'.

14.2.9 Attached transverse vortices

(Fig. 14.19(a and b))

Separation bobbles which form between the bonnet (hood) and front windscreen, the rear screen and boot (trunk) lid and the boot and rear light panel tend to generate attached transverse vortices (see Fig. 14.19(a and b)). The front attached vortices work their way around the 'A' post and then extend along the side windows to the rear of the car and beyond. Any overspill from the attached vortices in the rear window and rear light panel regions merges and strengthens the side panel vortices (see Fig. 14.19(b)); in turn the products of these secondary transverse vortices combine and enlarge the main trailing vortices.

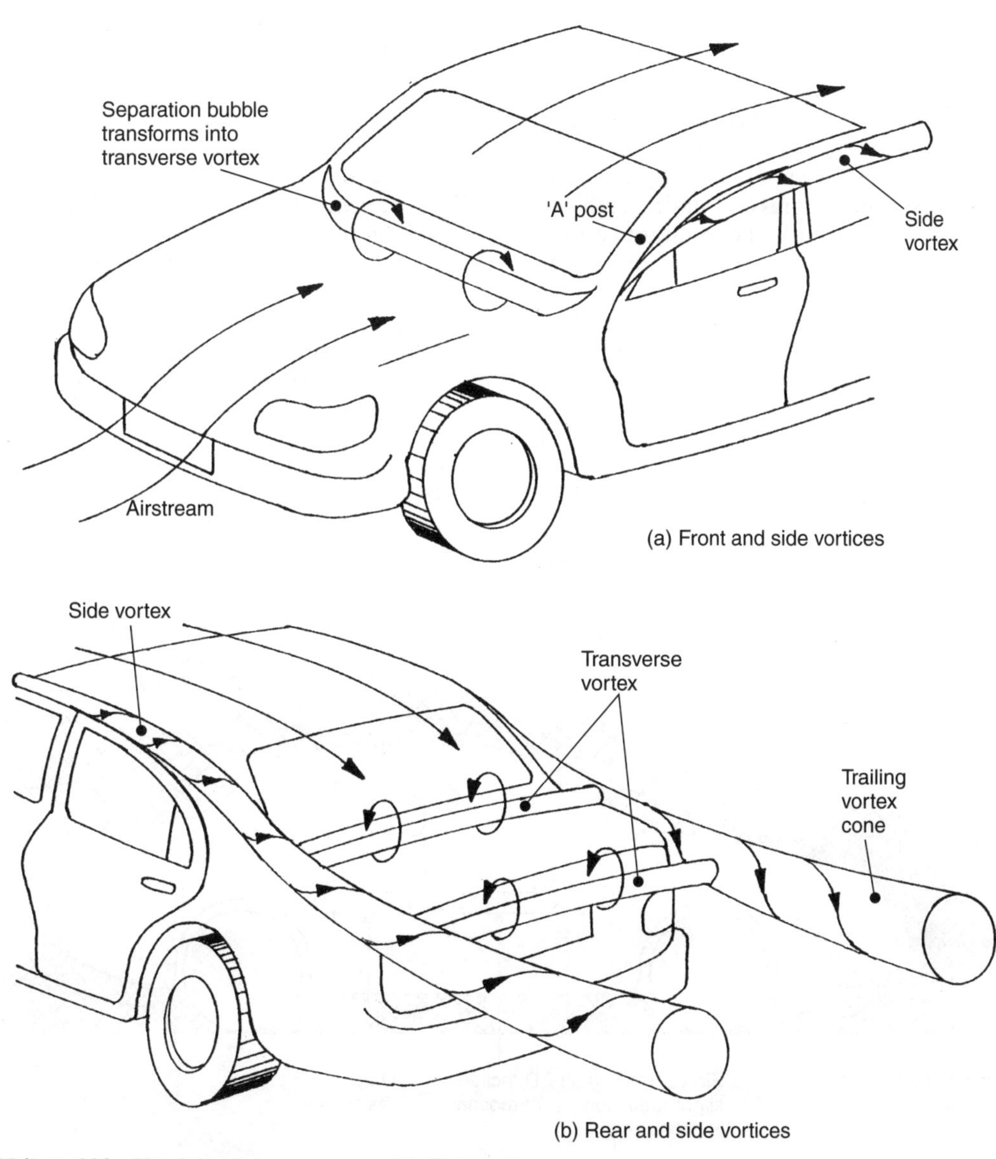

Fig. 14.19 (a and b) Notch back transverse and trailing vortices

14.3 Aerodynamic lift

14.3.1 Lift coefficients

The aerodynamic lift coefficient C_L is a measure of the difference in pressure created above and below a vehicle's body as it moves through the surrounding viscous air. A resultant upthrust or downthrust may be produced which mainly depend upon the body shape; however, an uplift known as positive lift is undesirable as it reduces the tyre to ground grip whereas a downforce referred to as negative lift enhances the tyre's road holding.

14.3.2 Vehicle lift (Fig. 14.20)

When a car travels along the road the airstream moving over the upper surface of the body from front to rear has to move further than the underside airstream which almost moves in a straight line (see Fig. 14.20). Thus the direct slower moving underside and the indirect faster moving top side airstream produces a higher pressure underneath the car than over it, consequently the resultant vertical pressures generated between the upper and under surfaces produce a net upthrust or lift. The magnitude of the lift depends mainly upon the styling profile of both over and under body surfaces, the distance of the underfloor above the ground, and the vehicle speed. Generally, the nearer the underfloor is to the ground the greater the positive lift (upward force); also the positive lift tends to increase with the square of the vehicle speed. Correspondingly a reduction in wheel load due to the lift upthrust counteracts the downward load; this therefore produces a reduction in the tyre to ground grip. If the uplift between the front and rear of the car is different, then the slip-angles generated by the front and rear tyres will not be equal; accordingly this will result in an under- or over-steer tendency instead of more neutral-steer characteristics. Thus uncontrolled lift will reduce the vehicle's road holding and may cause steering instability.

14.3.3 Underbody floor height versus aerodynamic lift and drag

(Figs 14.21(a and b) and 14.22)

With a large underfloor to ground clearance the car body is subjected to a slight negative lift force (downward thrust). As the underfloor surface moves closer to the ground the underfloor air space becomes a venturi, causing the air to move much faster underneath the body than over it, see Figs 14.21(a) and 14.22. Correspondingly with these changing conditions the air flow pressure on top of the body will be higher than for the underbody reduced venturi effect pressure, hence there will be a net down force (negative lift) tending to increase the contact pressure acting between the wheels and ground. Conversely a further reduction in underfloor to ground clearance makes it very restrictive for the underbody air flow (see Figs 14.21(b) and 14.22), so that much of the airstream is now compelled to flow over the body instead of underneath it, which results in an increase in air speed and a reduction in pressure over the top to cope with the reduction in the underfloor air

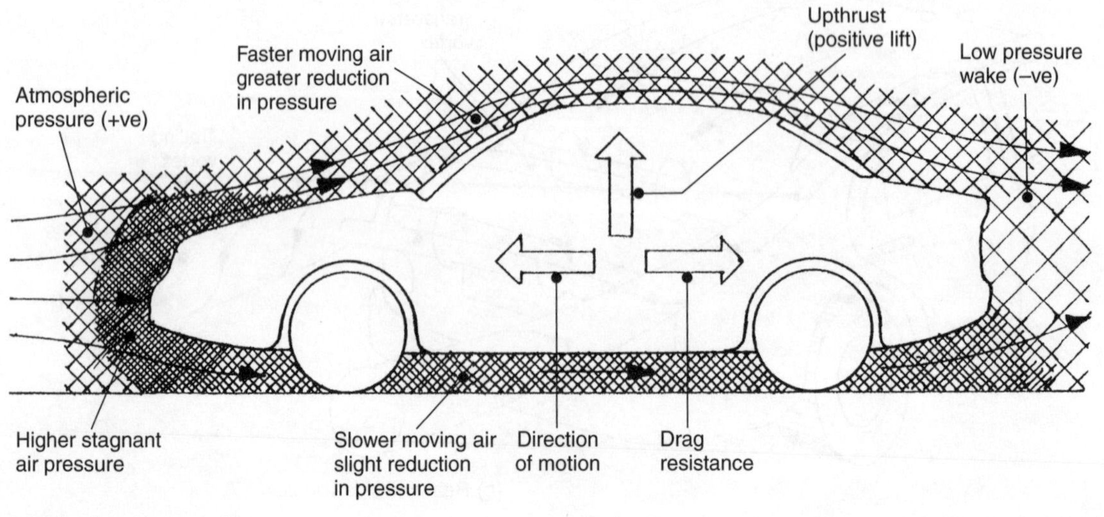

Fig. 14.20 Aerodynamic lift

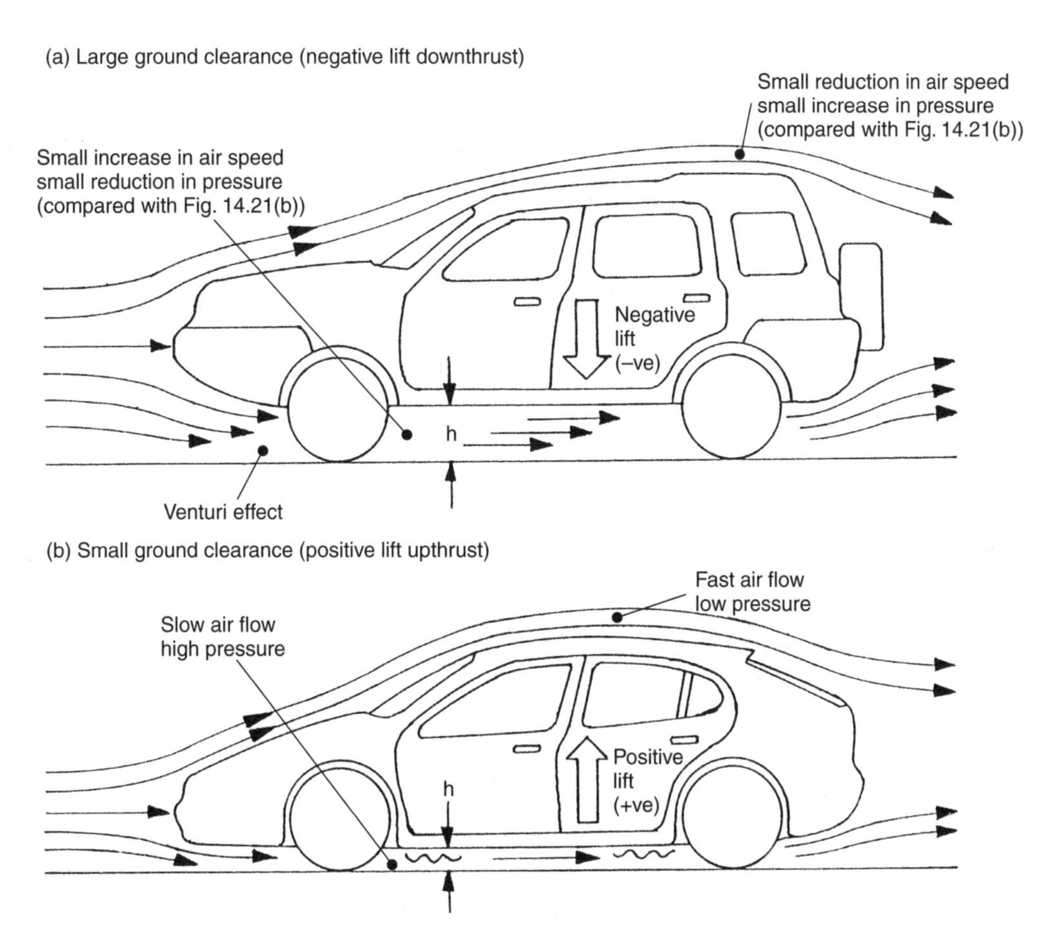

Fig. 14.21 (a and b) Effects of underfloor to ground clearance on the surrounding air speed, pressure and aerodynamic lift

movement. Thus the over and under pressure conditions have been reversed which subsequently now produces a net upward suction, that is, a tendency toward a positive lift.

14.3.4 Aerofoil lift and drag

(Figs 14.23(a–d), 14.24(a and b) and 14.25)

Almost any object moving through an airstream will be subjected to some form of lift and drag. Consider a flat plate inclined to the direction of air flow, the pressure of air above the surface of the plate is reduced while that underneath it is increased. As a result there will be a net pressure on the plate striving to force it both upwards and backwards, see Fig. 14.23(a). It will be seen that the vertical and horizontal components of the resultant reaction represents both lift and drag respectively, see Fig. 14.23(b). The greater the angle of inclination, the smaller will be the upward lift component, while the backward drag component will increase, see Fig. 14.23(c and d). Conversely as the angle of inclination decreases, the lift increases and the drag decreases; however, as the angle of inclination is reduced so does the resultant reaction force. If an aerofoil profile is used instead of the flat plate, (see Fig. 14.24(a and b)), the airstream over the top surface now has to move further and faster than the underneath air movement. This produces a greater pressure difference between the upper and lower surfaces and consequently greatly enhances the aerodynamic lift and promotes a smooth air flow over the upper profiled surface. A typical relationship between the C_L, C_D and angle of attack (inclination angle) is shown for an aerofoil section in Fig. 14.25.

14.3.5 Front end nose shape (Fig. 14.26(a–c))

Optimizing a protruding streamlined nose profile shape influences marginally the drag coefficient and to a greater extent the front end lift coefficient.

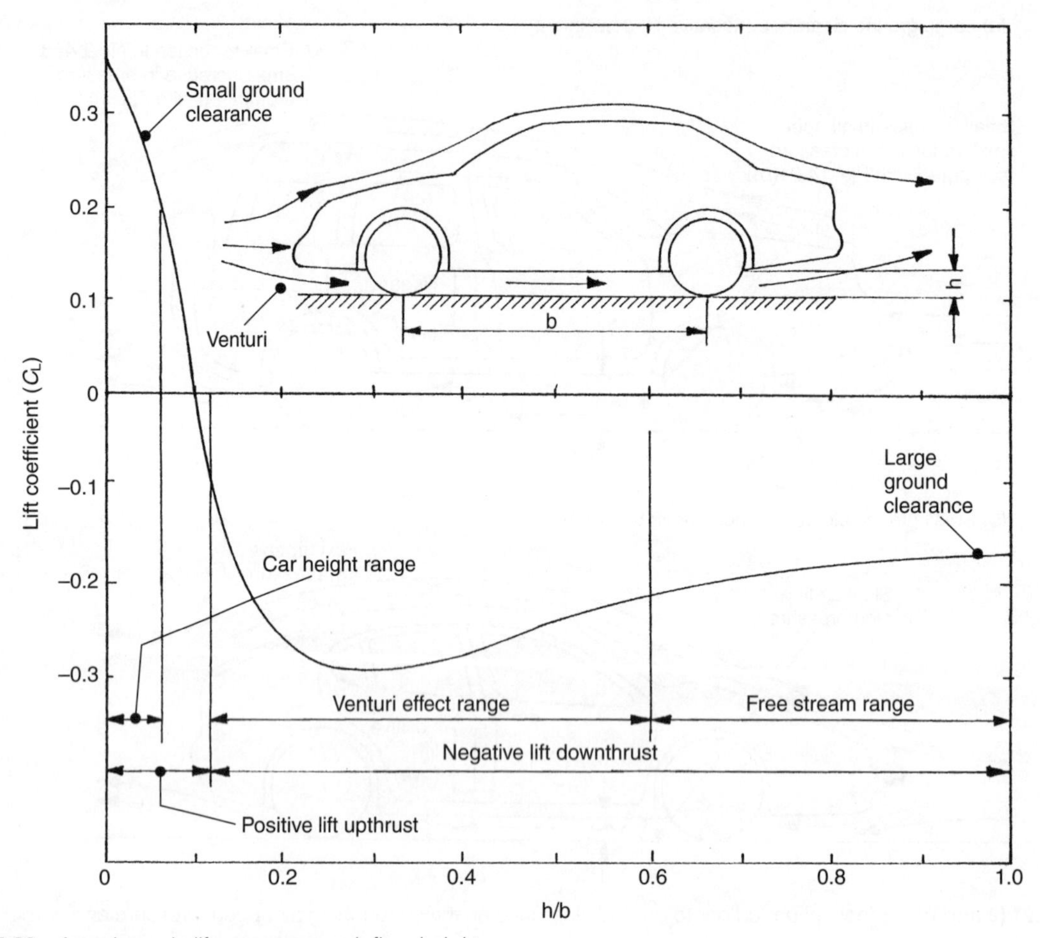

Fig. 14.22 Aerodynamic lift versus ground, floor height

With a downturned nose (see Fig. 14.26(a)) the streamlined nose profile directs the largest proportion of the air mass movement over the body, and only a relatively small amount of air flows underneath the body. If now a central nose profile is adopted (see Fig. 14.26(b)) the air mass movement is shared more evenly between the upper and lower body surfaces; however, the air viscous interference with the underfloor and ground still causes the larger proportion of air to flow above than below the car's body. Conversely a upturned nose (see Fig. 14.26(c)) induces still more air to flow beneath the body with the downward curving entry gap shape producing a venturi effect. Consequently the air movement will accelerate before reaching its highest speed further back at its narrowest body to ground clearance. Raising the mass airflow in the space between the body and ground increases the viscous interaction of the air with the under body surfaces and therefore forces the air flow to move diagonally out and upward from the sides of the car. It therefore strengthens the side and trailing vortices and as a result promotes an increase in front end aerodynamic lift force.

The three basic nose profiles discussed showed, under windtunnel tests, that the upturned nose had the highest drag coefficient C_D of 0.24 whereas there was very little difference between the central and downturned nose profiles which gave drag coefficients C_D of 0.223 and 0.224 respectively. However the front end lift coefficient C_L for the three shapes showed a marked difference, here the upturned nose profile gave a positive lift coefficient C_L of +0.2, the central nose profile provided an almost neutral lift coefficient C_L of +0.02, whereas the downturned nose profile generated a negative lift coefficient C_L of −0.1.

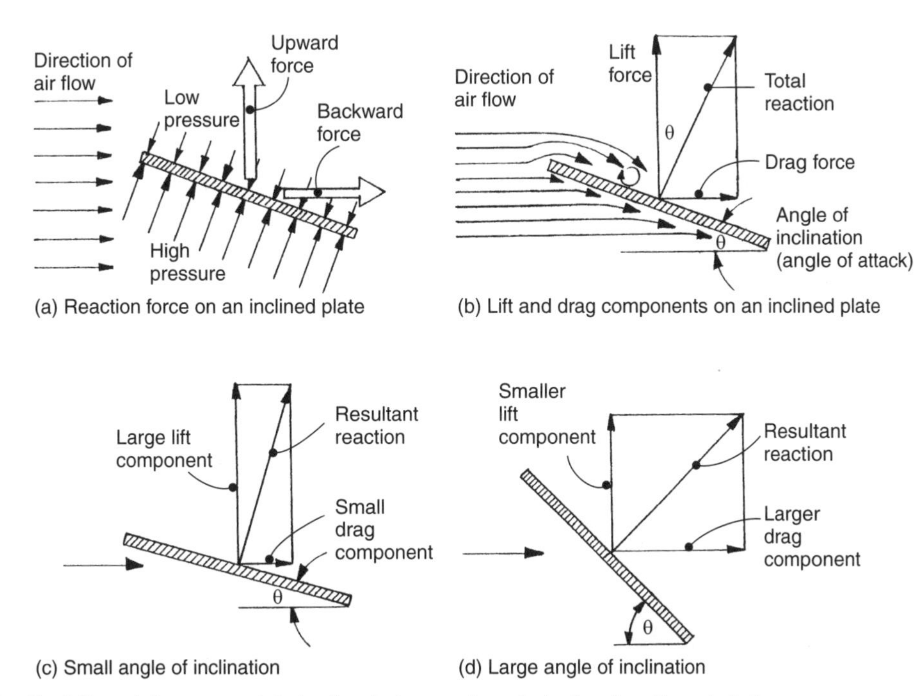

Fig. 14.23 (a–d) Lift and drag on a plate inclined at a small angle to the direction of air flow

14.4 Car body drag reduction

14.4.1 Profile edge rounding or chamfering (Fig. 14.27(a and b))

There is a general tendency for aerodynamic lift and drag coefficients to decrease with increased edge radius or chamfer: experiments carried out showed for a particular car shape (see Fig. 14.27(a)) how the drag coefficient was reduced from 0.43 to 0.40 with an edge radius/chamfer increasing from zero to 40 mm (see Fig. 14.27(b)), and there was a slightly greater reduction with chamfering than rounding the edges; however, beyond 40 mm radius there was no further advantage in increasing the edge radius or chamfer.

14.4.2 Bonnet slope and windscreen rake (Fig. 14.28(a–c))

Increasing the bonnet (hood) slope angle α from zero to roughly 10° reduces the drag coefficient, but beyond 10° the drag reduction is insignificant, see Fig. 14.28(b). Likewise, increasing the rake angle γ reduces the drag coefficient (see Fig. 14.28(c)) particularly when the rake angle becomes large;

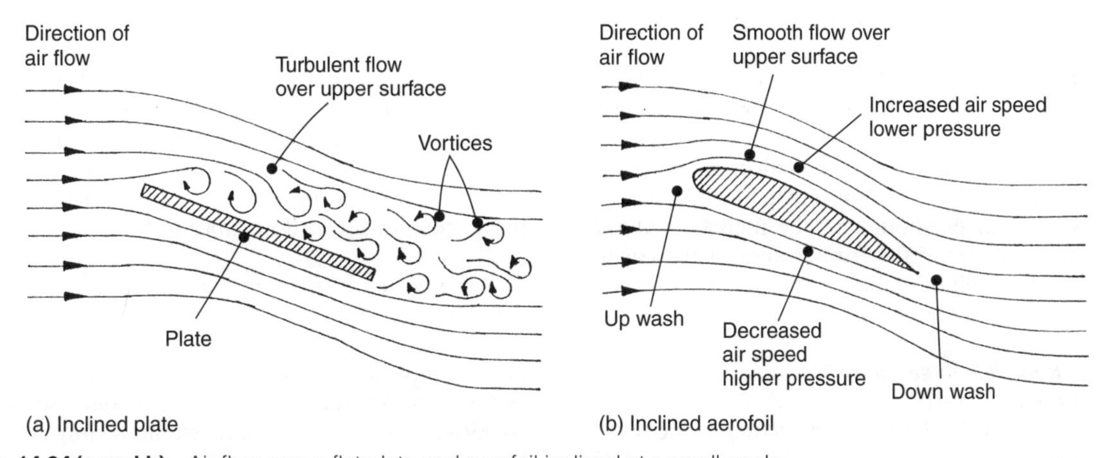

Fig. 14.24 (a and b) Air flow over a flat plate and aerofoil inclined at a small angle

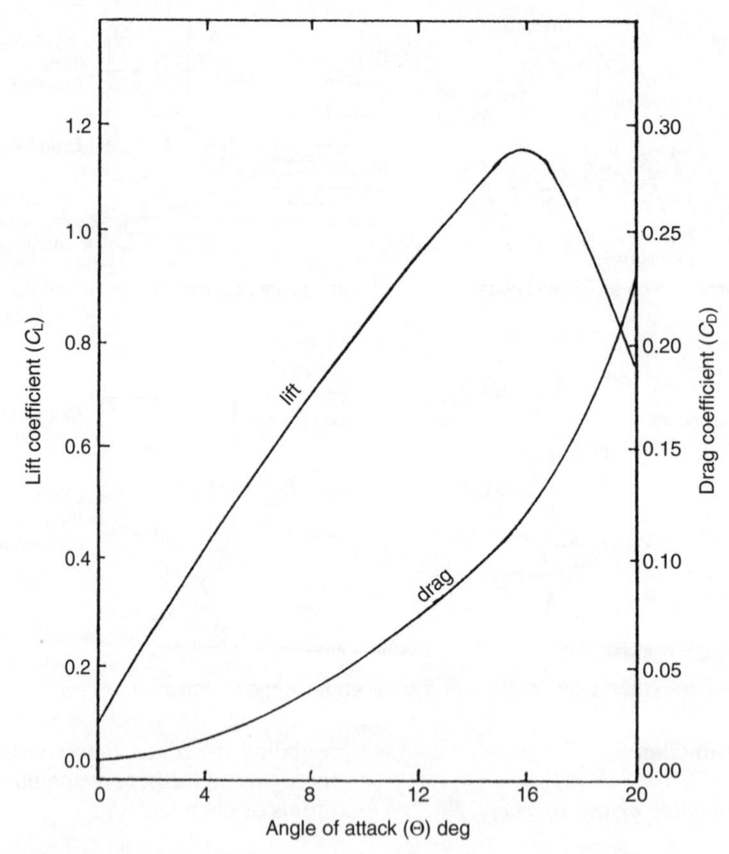

Fig. 14.25 Lift and drag coefficient versus angle of inclination (attack)

however, very large rake angles may conflict with the body styling.

14.4.3 Roof and side panel cambering

(Figs 14.29(a and b) and 14.30(a and b))
Cambering the roof (see Fig. 14.29(a and b)) and the side panels (see Fig. 14.30(a and b)) reduces the drag coefficient. However, if the roof camber curvature becomes excessive the drag coefficient commences to rise again (see Fig. 14.29(b)), whereas the reduction in drag coefficient with small amounts of side-panel cambering is marked (see Fig. 14.30(b)), but with excessive camber the reduction in the drag coefficient becomes only marginal. Both roof and side panel camber should not be increased at the expense of enlarging the frontal area of the car as this would in itself be counter-productive and would increase the drag coefficient.

14.4.4 Rear side panel taper

(Fig. 14.31(a and b))
Tapering inwards the rear side panel reduces the drag coefficient. This can be seen in Fig. 14.31 (a and b) which shows a marked reduction in the drag coefficient with both a 50 mm and then a 125 mm rear end contraction on either side of the car; however, there was no further reduction in the drag coefficient when the rear end contraction was increased to 200 mm.

14.4.5 Underbody rear end upward taper

(Fig. 14.32(a and b))
Tilting upwards the underfloor rear end produces a diffuser effect which shows a promising way to reduce the drag coefficient, see Fig. 14.32(a and b). However, it is important to select the optimum ratio of length of taper to overall car length and the angle β of upward inclination for best results.

14.4.6 Rear end tail extension

(Fig. 14.33(a and b))
Windtunnel investigation with different shaped tail models have shown that the minimal drag coefficients were produced with extended tails, see Fig. 14.33(a and b), but this shape would be impractical for design reasons. Conversely if the rear end tail is

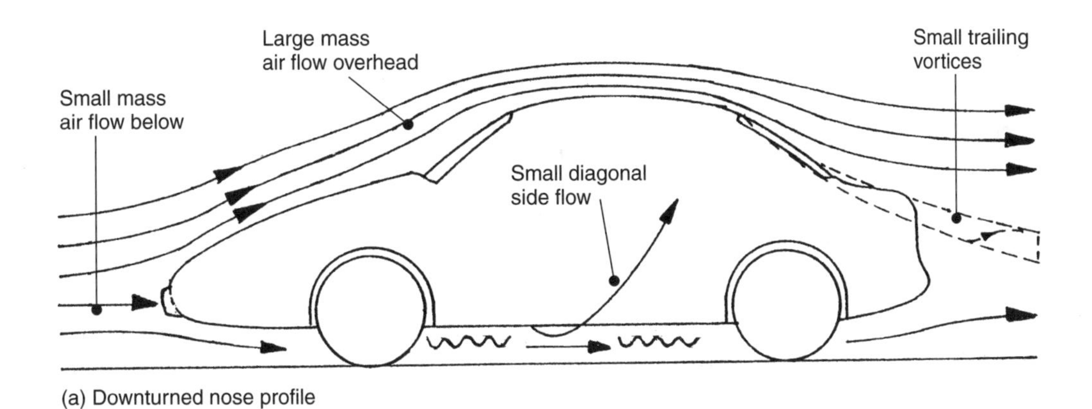

(a) Downturned nose profile

(b) Central nose profile

(c) Upturned nose profile

Fig. 14.26 (a–c) A greatly exaggerated air mass distribution around a car body for various nose profiles

cropped at various lengths and curved downwards there is an increase in the drag coefficient with each reduction in tail length beyond the rear wheels.

14.4.7 Underbody roughness

(Fig. 14.34(a and b))

The underbody surface finish influences the drag coefficient just as the overbody curvature, tapering, edge rounding and general shape dictates the drag resistance. Moulding in individual compartments in the underfloor pan to house the various components and if possible enclosing parts of the underside with plastic panels helps considerably to reduce the drag resistance. The underside of a body has built into it many cavities and protrusions to cater with the following structural requirements and operating

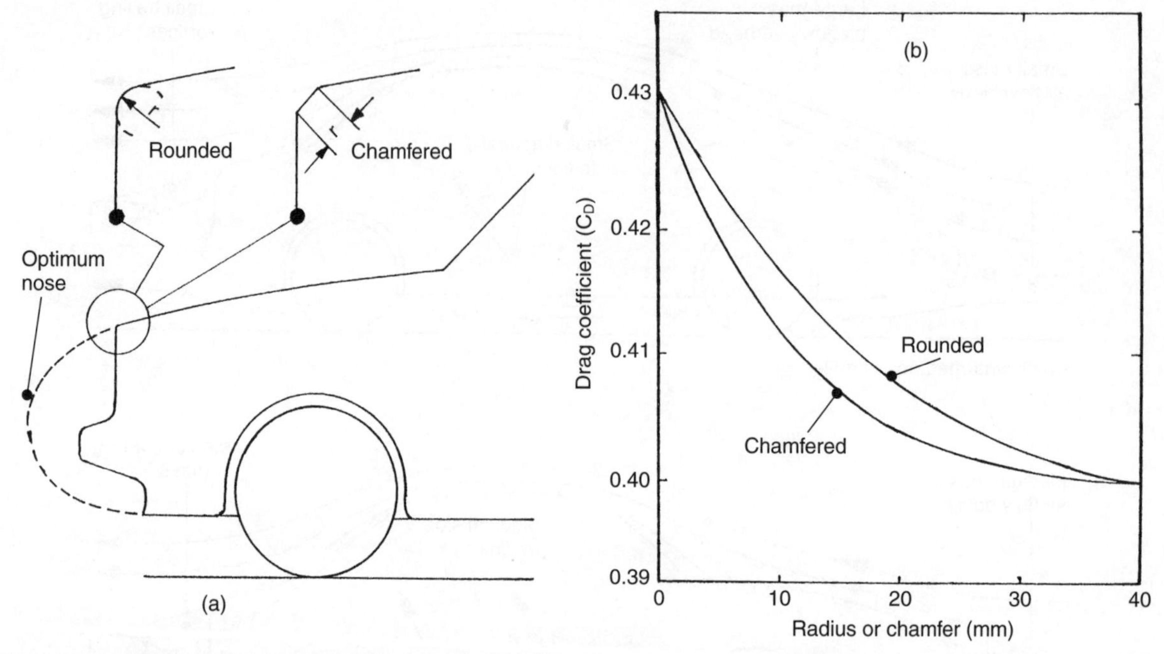

Fig. 14.27 (a and b) Influence of forebody bonnet (hood) edge shape on drag coefficient

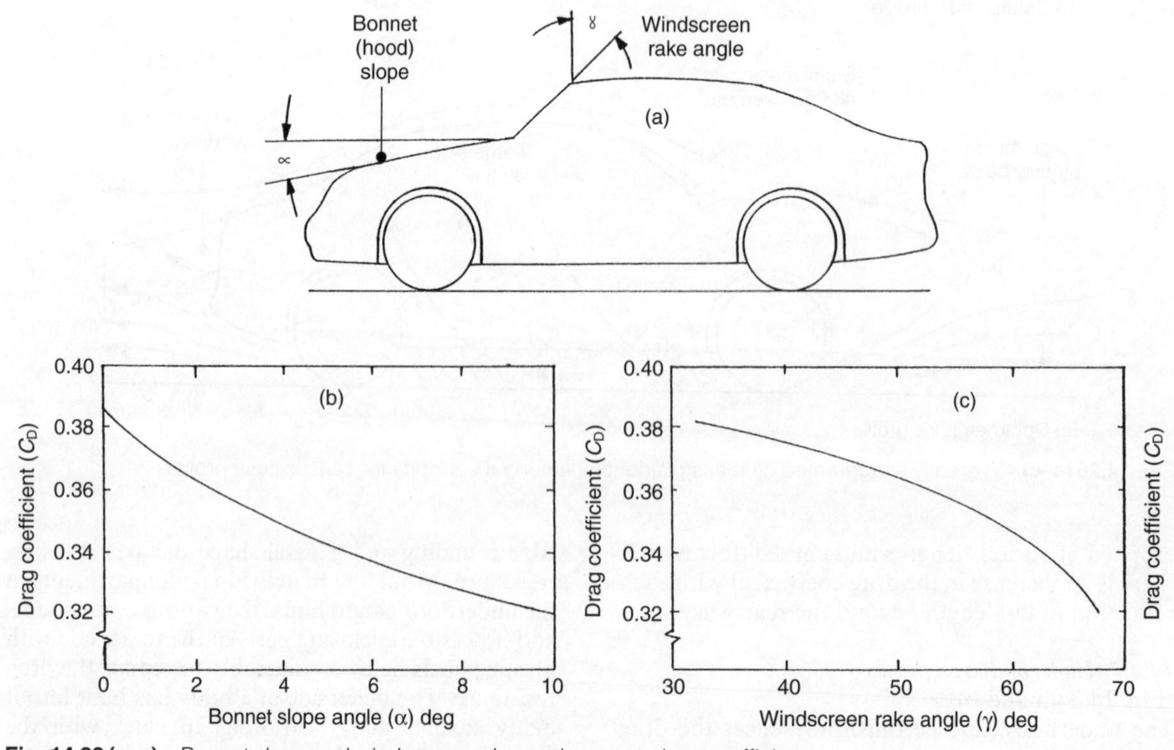

Fig. 14.28 (a–c) Bonnet slope and windscreen rake angle versus drag coefficient

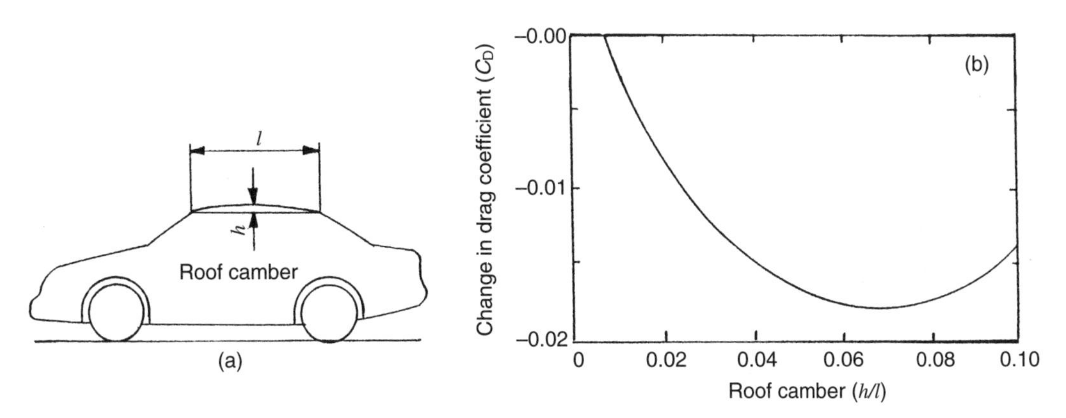

Fig. 14.29 (a and b) Effect of roof camber on drag coefficient

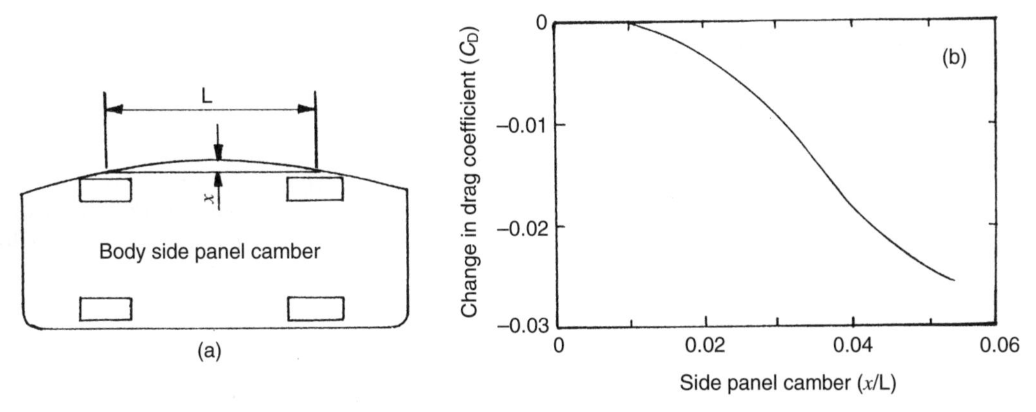

Fig. 14.30 (a and b) Effect of side panel camber on drag coefficient

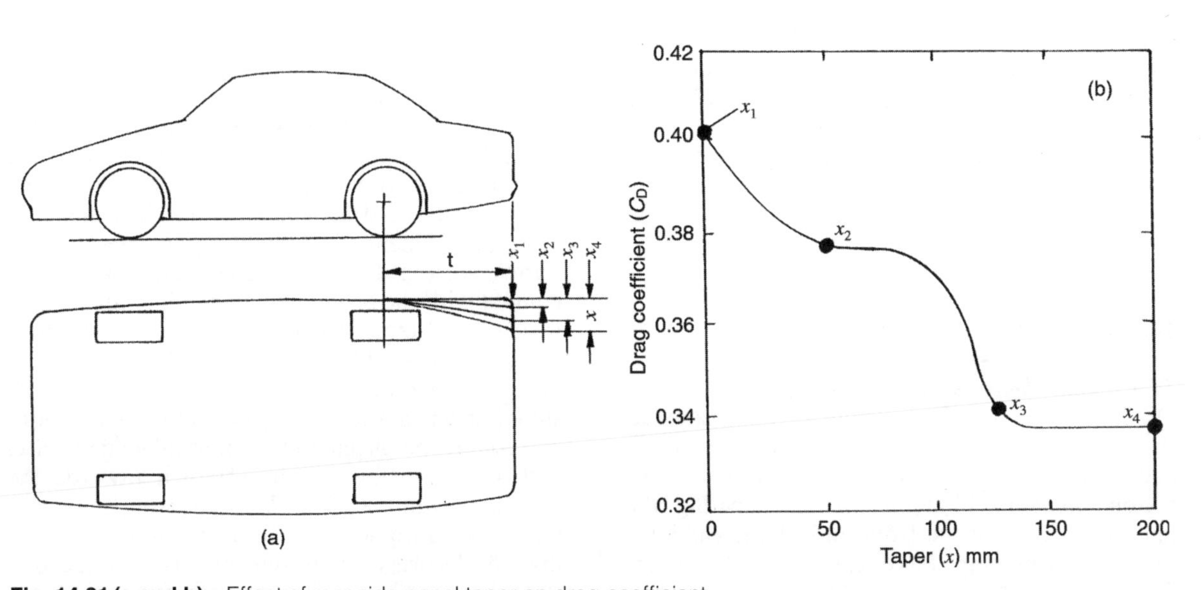

Fig. 14.31 (a and b) Effect of rear side panel taper on drag coefficient

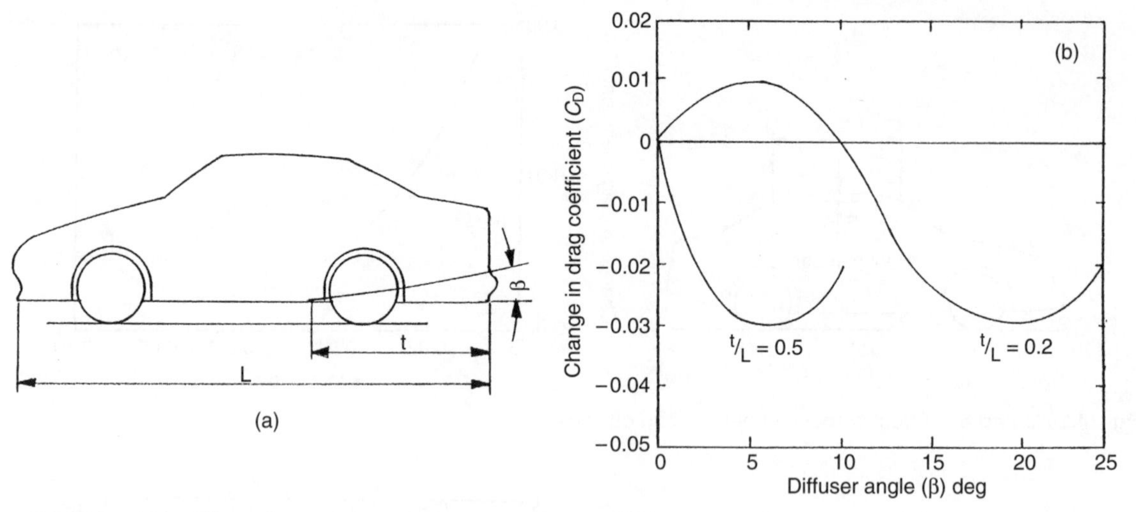

Fig. 14.32 (a and b) Effect of rear end upward taper on drag coefficient

components: front and rear wheel and suspension arch cavities, engine, transmission and steering compartment, side and cross member channelling, floor pan straightening ribs, jacking point straightening channel sections, structural central tunnel and rear wheel drive propeller shaft, exhaust system catalytic converter, silencer and piping, hand brake cable, and spare wheel compartment etc. A rough underbody produces excessive turbulence and friction losses and consequently raises the drag coefficient, whereas trapped air in the underside region slows down the air flow and tends to raise the underfloor pressure and therefore positive lift force. Vehicles with high drag coefficients gain least by smoothing the underside. The underfloor roughness or depth of irregularity defined as the centre line average peak to valley height for an average car is around +150 mm. A predictable relationship between the centre line average roughness and the drag coefficient for a given ground clearance and vehicle length is shown in Fig. 14.34(a and b).

14.5 Aerodynamic lift control

14.5.1 Underbody dams (Fig. 14.35(a–c))

Damming the underbody to ground clearance at the extreme rear blocks the underfloor airstream and causes a partial pressure build-up in this region, see Fig. 14.25(a), whereas locating the underbody dam in the front end of the car joins the rear low pressure wake region with the underfloor space, see Fig. 14.35(b). Thus with a rear end underfloor air dam the underfloor air flow pressure increase raises the aerodynamic upthrust, that is, it produces positive lift, see Fig. 14.35(a). Conversely a front end air dam reduces the underfloor air flow pressure thereby generating an aerodynamic downthrust, that is, it produces negative lift (see Fig. 14.35(b)). Experimental results show with a front end dam there is a decrease in front lift (negative lift) whereas there is a slight rise in rear end lift (positive lift) as the dam height is increased, and as would be expected, there is also a rise in drag as the frontal area of the dam is enlarged, see Fig. 14.35(c).

14.5.2 Exposed wheel air flow pattern (Fig. 14.36(a–c))

When a wheel rotates some distance from the ground air due to its viscosity attaches itself to the tread and in turn induces some of the surrounding air to be dragged around with it. Thus this concentric movement of air establishes in effect a weak vortex, see Fig. 14.36(a). If the rotating wheel is in contact with the ground it will roll forwards which makes windtunnel testing under these conditions difficult; this problem is overcome by using a supportive wheel and floor rig. The wheel is slightly submerged in a well opening equal to the tyre width and contact patch length for a normal loaded wheel and a steady flow of air is blown towards the frontal view of the wheel. With the wheel rig simulating a rotating wheel in contact with the ground, the wheel vortex air movement interacts and distorts the parallel main airstream.

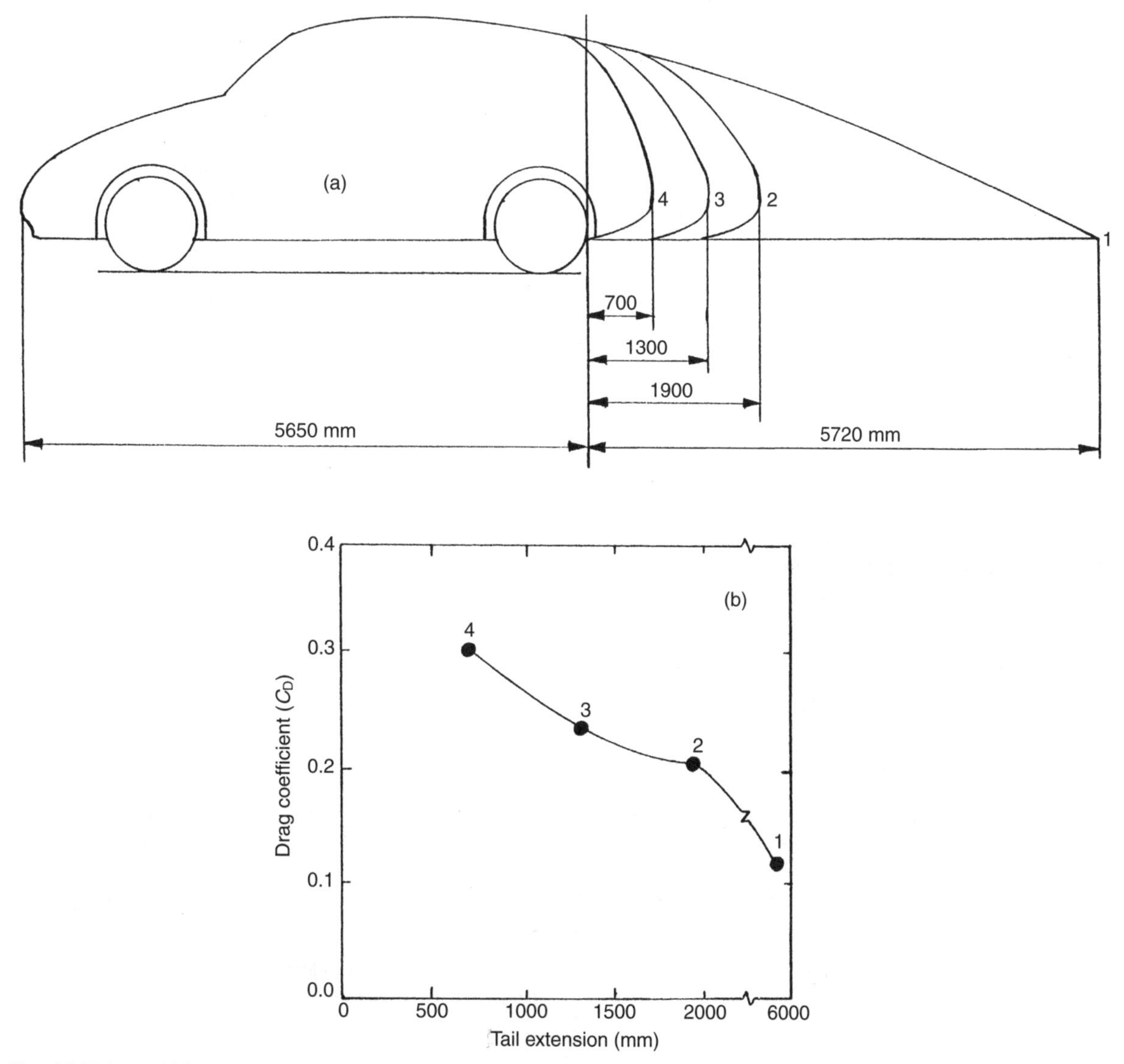

Fig. 14.33 (a and b) Effect of rear end tail extension on drag coefficient

The air flow pattern for an exposed wheel can be visualized and described in the following way. The air flow meeting the lower region of the wheel will be stagnant but the majority of the airstream will flow against the wheel rotation following the contour of the wheel until it reaches the top; it then separates from the vortex rim and continues to flow towards the rear but leaving underneath and in the wake of the wheel a series of turbulent vortices, see Fig. 14.36(b). The actual point of separation will creep forwards with increased rotational wheel speed. Air pressure distribution around the wheel will show a positive pressure build-up in the stagnant air flow front region of the wheel, but this changes rapidly to a high negative pressure where the main air flow breaks away from the wheel rim, see Fig. 14.36(c). It then declines to some extent beyond the highest point of the wheel, and then remains approximately constant around the rear wake region of the wheel. Under these described conditions, the exposed rotating wheel produces a resultant positive upward lift force which tends to reduce the adhesion between the tyre tread and ground.

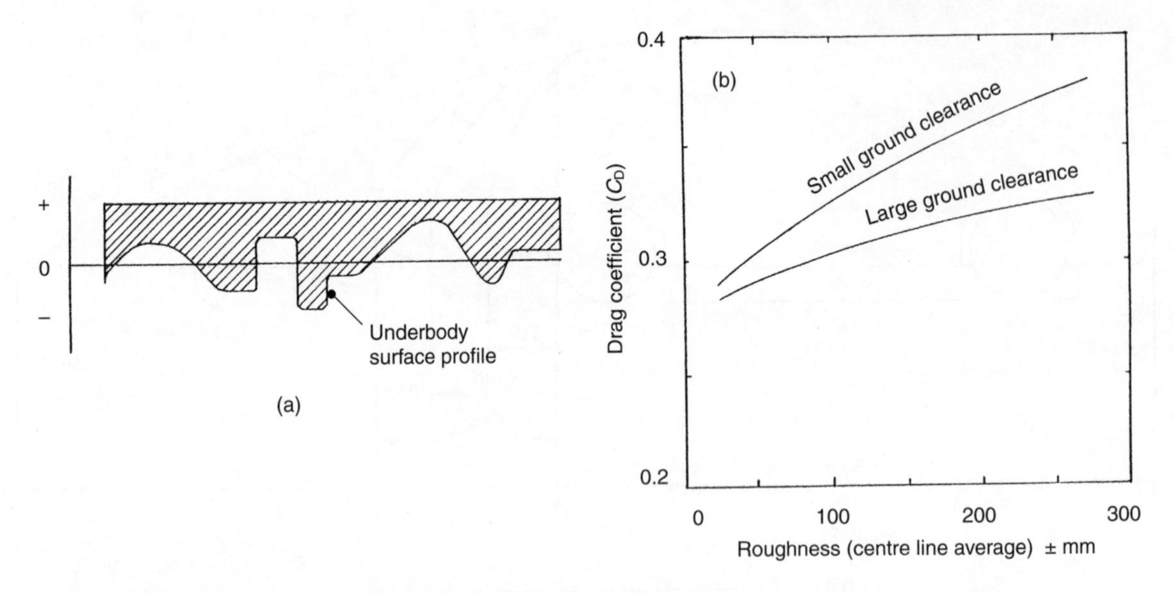

Fig. 14.34 (a and b) Effect of underbody roughness on drag coefficient

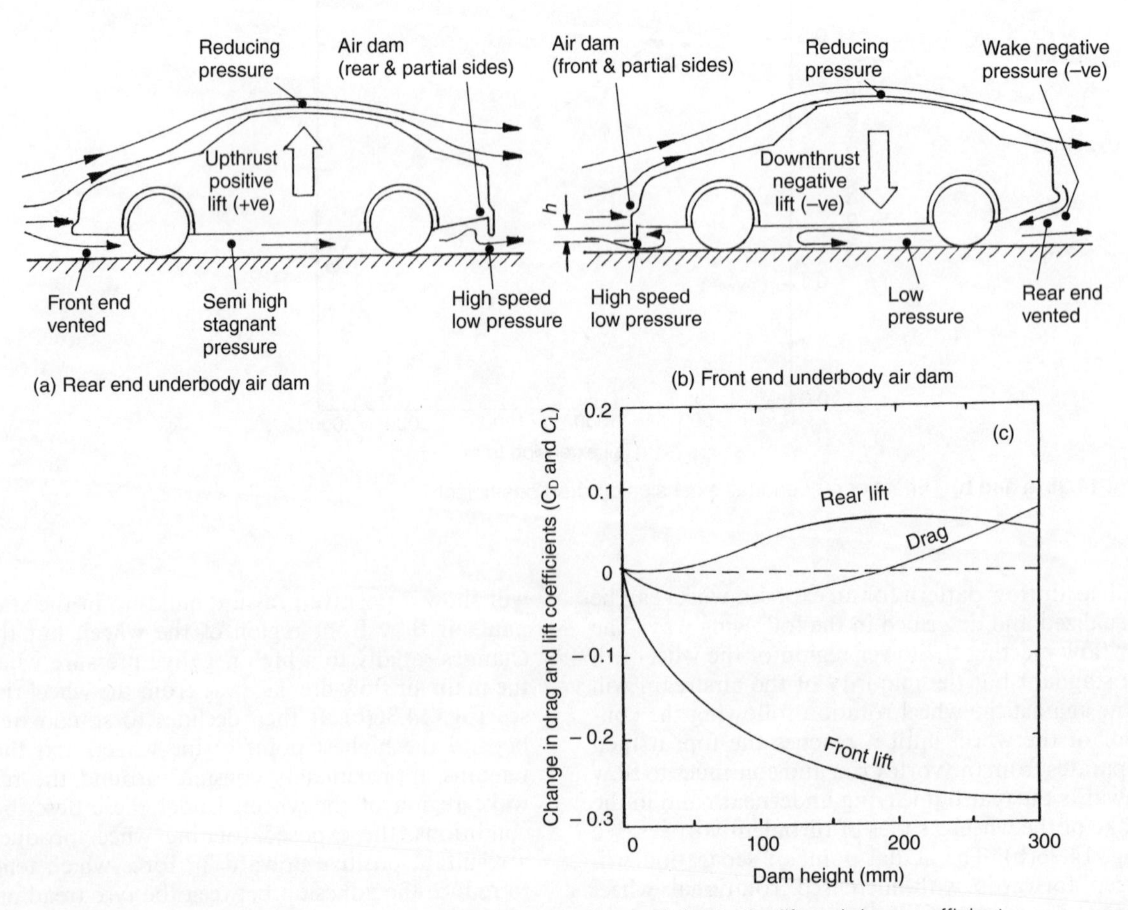

Fig. 14.35 (a–c) Effects of underbody front and rear end air dams relative to the lift and drag coefficient

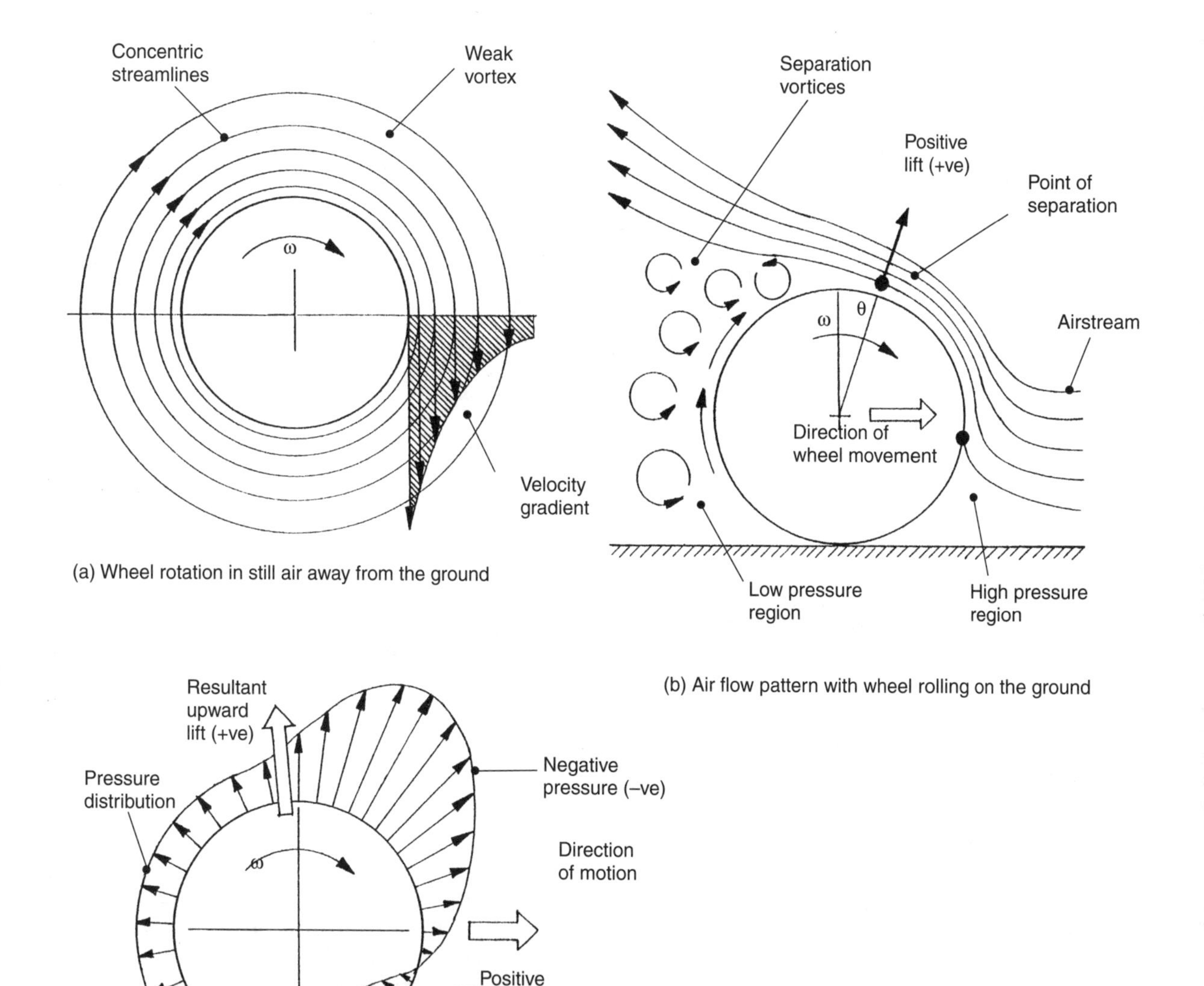

Fig. 14.36 (a–c) Exposed wheel air flow pattern and pressure distribution

14.5.3 Partial enclosed wheel air flow pattern

(Figs 14.37(a and b) and 14.38(a–c))

The air flow passing beneath the front of the car initially moves faster than the main airstream, this therefore causes a reduction in the local air pressure. At the rear of the rotating wheel due to viscous drag air will be scooped into the upper space formed between the wheel tyre and the wheel mudguard arch (see Fig. 14.37(a and b)). The air entrapped in the wheel arch cavity circulates towards the upper front of the wheel due to a slight pressure build-up and is then expelled through the front end wheel to the mudguard gap which is at a lower pressure in both a downward and sideward direction. Decreasing the clearance between the underside and the ground and shielding more of the wheel with the mudguard tends to produce a loss of momentum to the air so that both

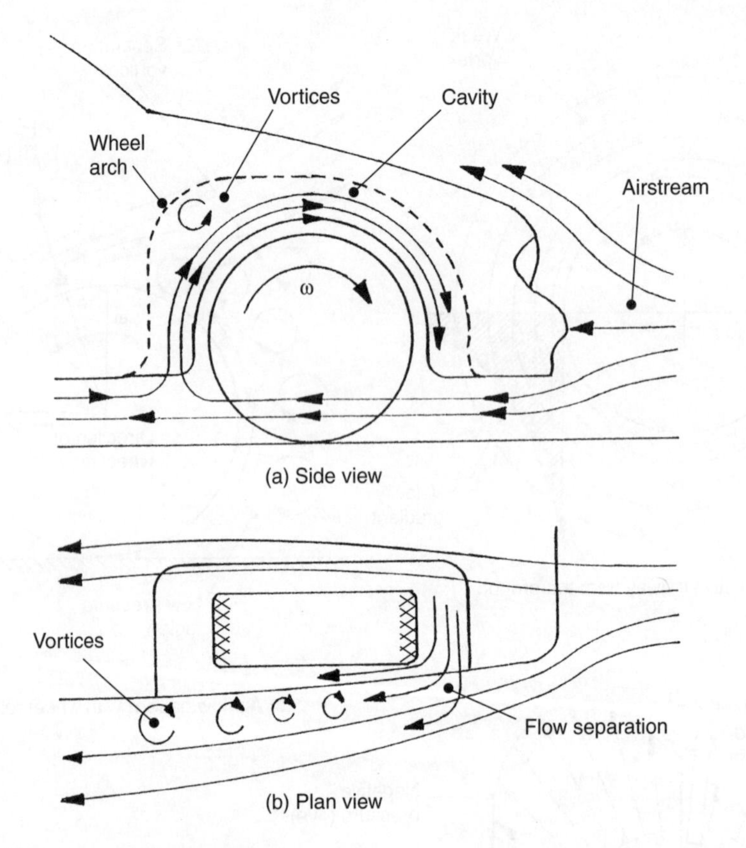

Fig. 14.37 (a and b) Wheel arch air flow pattern

aerodynamic lift C_{LW} and drag C_{DW} coefficients, and therefore forces, are considerably reduced Fig. 14.38(a–c).

14.5.4 Rear end spoiler (Fig. 14.39(a–c))

Generally when there is a gentle rear end body profile curvature change, it will be accompanied with a relatively fast but smooth streamline air flow over this region which does not separate from the upper surface. However, this results in lower local pressures which tend to exert a lift force (upward suction) at the rear end of the car. A lip, see Fig. 14.39(a), or small aerofoil spoiler, see Fig. 14.39(b), attached to the rear end of the car boot (trunk) interrupts the smooth streamline air flow thereby slowing down the air flow and correspondingly raising the upper surface local air pressure which effectively increase the downward force known as negative lift. A typical relationship between rear lift, front lift and drag coefficients relative to the spoiler lip height is shown in Fig. 14.39(c). The graph shows a general increase in negative lift (downward force) by increasing the spoiler lip height. However, this is at the expense of a slight rise in the front end lift coefficient, whereas the drag coefficient initially decreases and then marginally rises again with increased spoiler lip height. It should be appreciated that the break-up of the smooth streamline air flow and the increase in rear downward pressure should if possible be achieved without incurring too much, if any, increase in front end positive lift and aerodynamic drag.

14.5.5 Negative lift aerofoil wings (Fig. 14.40(a–c))

A negative lift wing is designed when attached to the rear end of the car to produce a downward thrust thereby enabling the traction generated by the rear driving wheels to be increased, or if a forward negative lift wing is fitted to improve the grip of both front steering wheels.

With the negative lift wing the aerofoil profile is tilted downward towards the front end with the negative and positive aerofoil section camber at the top and bottom respectively, see Fig. 14.40(a). The airstream therefore moving underneath the

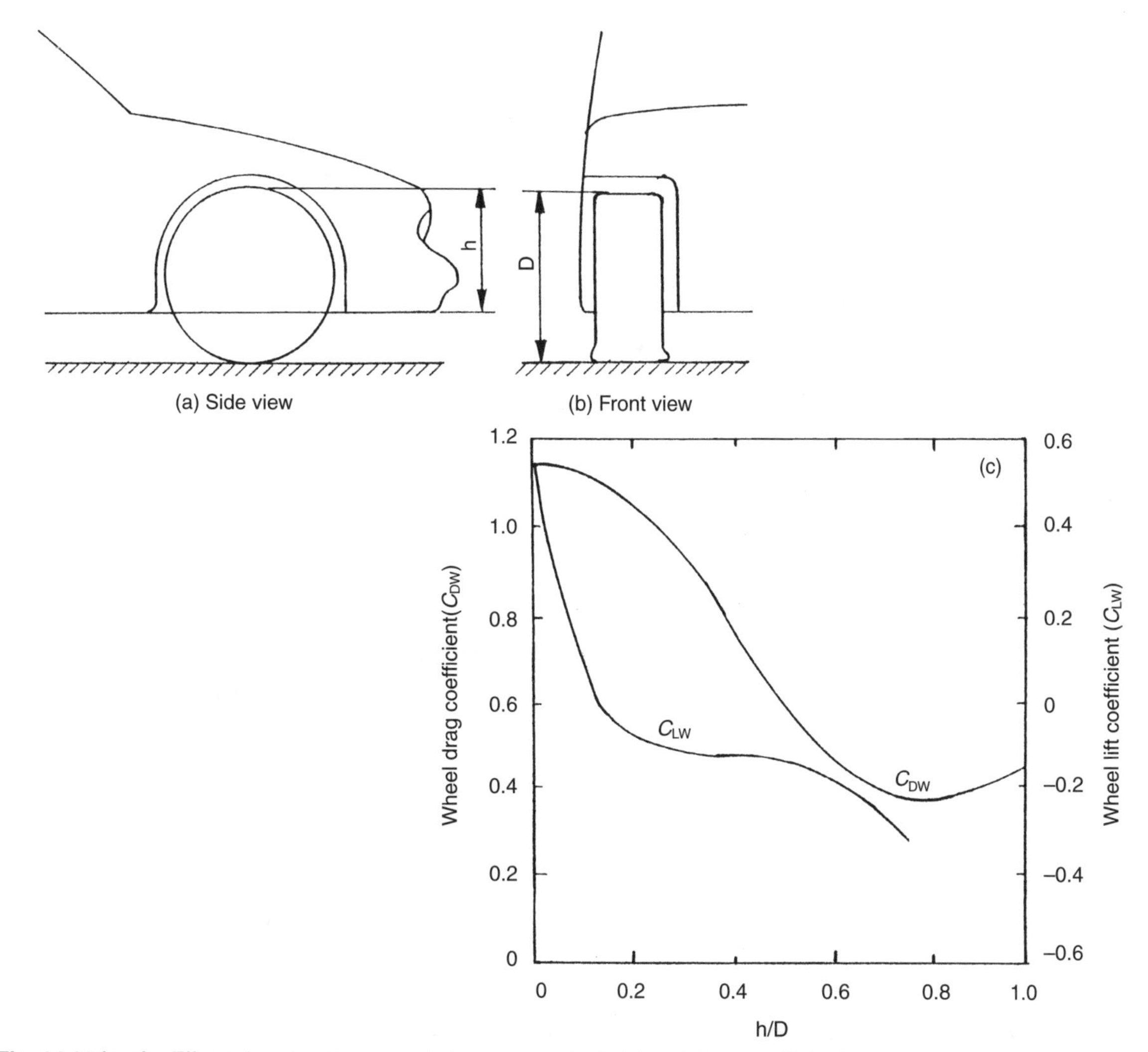

Fig. 14.38 (a–c) Effect of underside ground clearance on both lift and drag coefficients

aerofoil wing has to move further and faster than the airstream flowing over the upper surface; the pressure produced below the aerofoil wing is therefore lower than above. Consequently there will be a resultant downthrust perpendicular to the cord of the aerofoil (see Fig. 14.40(b)) which can be resolved into both a vertical downforce (negative lift) and a horizontal drag force. Enlarging the tilt angle of the wing promotes more negative lift (downthrust) but this is at the expense of increasing the drag force opposing the forward movement of the wing, thus a compromise must always be made between improving the downward wheel grip and the extra drag force opposing the motion of the car. Racing cars have the aerofoil wing over the rear wheel axles or just in front or behind them, see Fig. 14.40(c). However, the drag force produces a clockwise tilt which tends to lift the front wheels of the ground, therefore the front aerofoil wings (see Fig. 14.40(c)) are sometimes attached low down and slightly ahead of the front wheels to counteract the front end lift tendency.

14.6 Afterbody drag

14.6.1 Squareback drag (Figs 14.41 and 14.42)
Any car with a rear end (base) slope surface angle ranging from 90° to 50° is generally described as a squareback style (see Fig. 14.42). Between this angular surface inclination range for a squareback car there is very little change in the air flow pattern

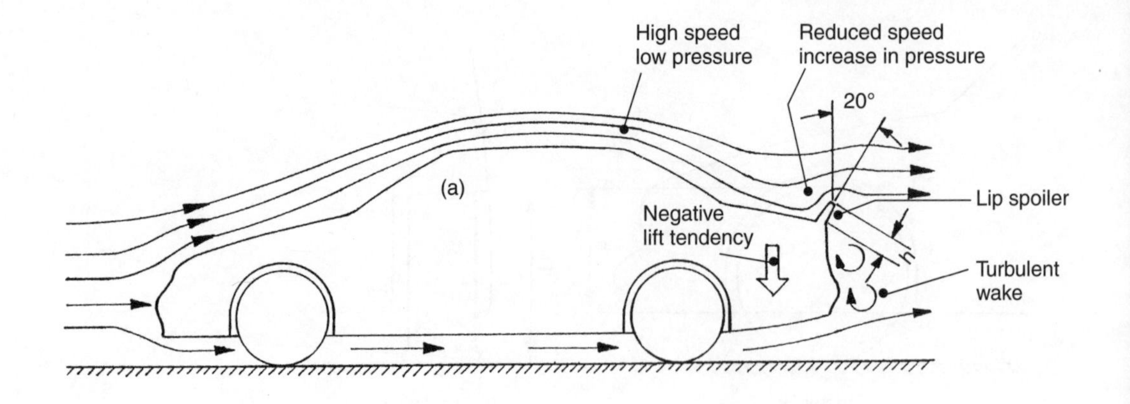

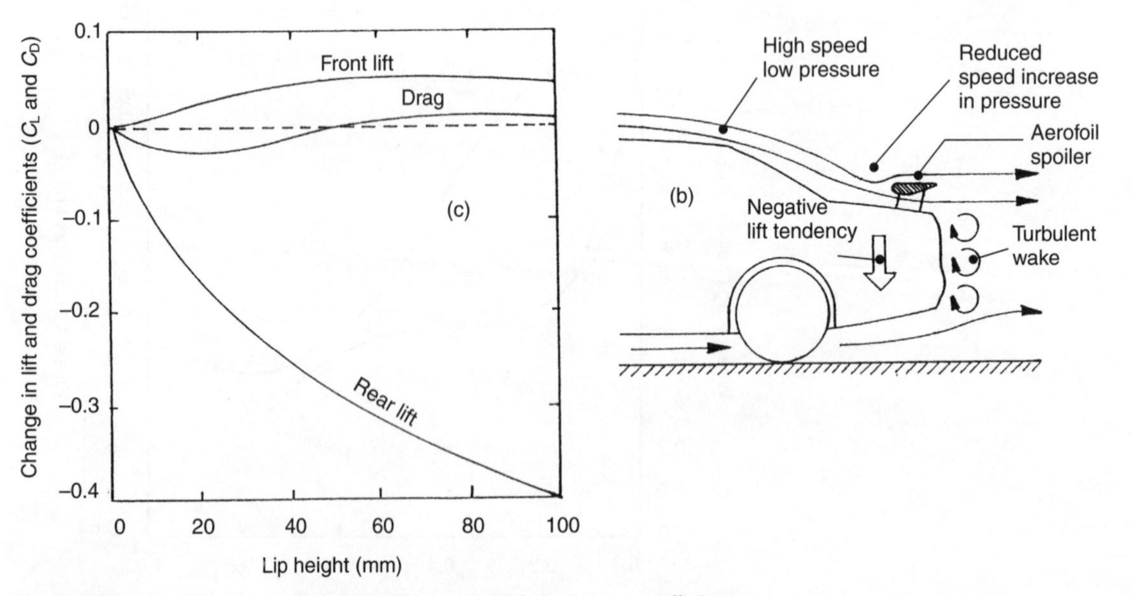

Fig. 14.39 (a–c) Effect of rear end spoiler on both lift and drag coefficients

and therefore there is virtually no variation in the afterbody drag (see Fig. 14.41). With a parallel sided squareback rear end configuration, the whole rear surface area (base area) becomes an almost constant low negative pressure wake region. Tapering the rear quarter side and roof of the body and rounding the rear end tends to lower the base pressure. In addition to the base drag, the afterbody drag will also include the negative drag due to the surrounding inclined surfaces.

14.6.2 Fastback drag (Figs 14.41 and 14.43)

When the rear slope angle is reduced to 25° or less the body profile style is known as a fastback, see Fig. 14.43. Within this much reduced rear end inclination the airstream flows over the roof and rear downward sloping surface, the airstream remaining attached to the body from the rear of the roof to the rear vertical light-plate and at the same time the condition which helps to generate attached and trailing vortices with the large sloping rear end is no longer there. Consequently the only rearward suction comes from the vertical rear end projected base area wake, thus as the rear end inclined angle diminishes, the drag coefficient decreases, see Fig. 14.41. However, as the angle approaches zero there is a slight rise in the drag coefficient again as the rear body profile virtually reverts to a squareback style car.

14.6.3 Hatchback drag (Figs 14.41, 14.44 and 14.45)

Cars with a rear sloping surface angle ranging from 50° to 25° are normally referred to as hatchback

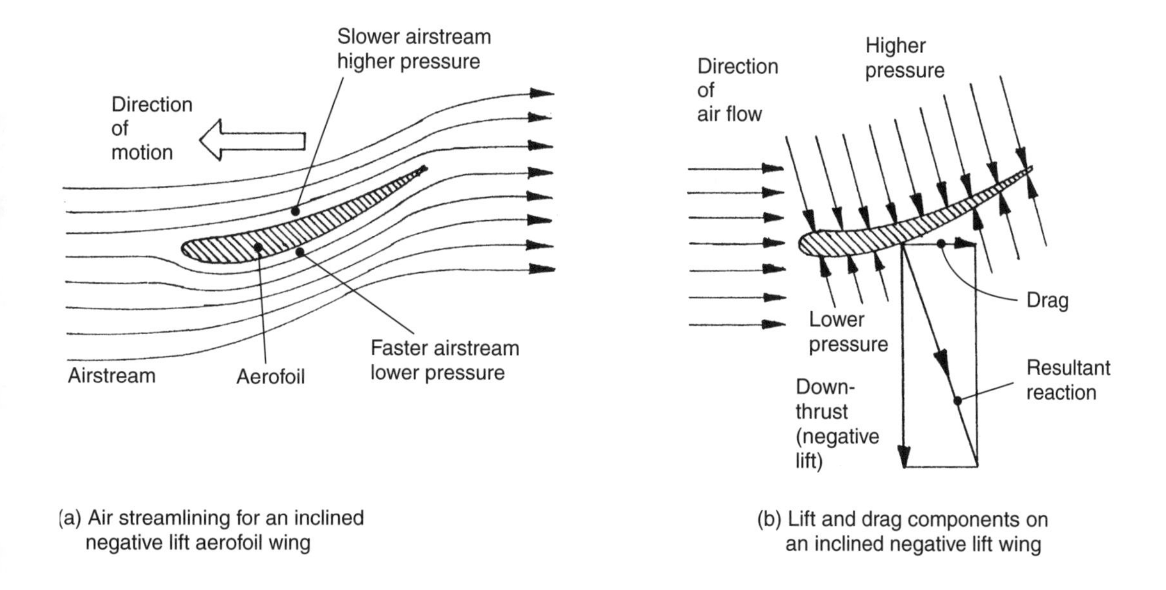

(a) Air streamlining for an inclined negative lift aerofoil wing

(b) Lift and drag components on an inclined negative lift wing

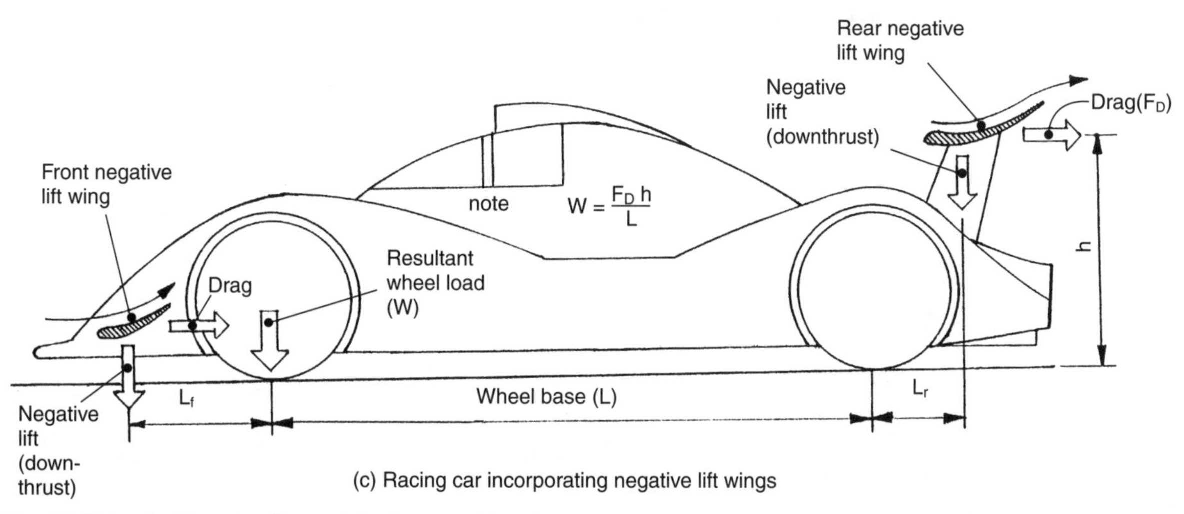

(c) Racing car incorporating negative lift wings

Fig. 14.40 (a–c) Negative lift aerofoil wing considerations

style, see Fig. 14.44. Within this rear end inclination range air flows over the rear edge of the roof and commences to follow the contour of the rear inclined surface; however, due to the steepness of the slope the air flow breaks away from the surface. At the same time some of the air flows from the higher pressure underfloor region to the lower pressure roof and rear sloping surface, then moves slightly inboard and rearward along the upper downward sloping surface. The intensity and direction of this air movement along both sides of the rear upper body edging causes the air to spiral into a pair of trailing vortices which are then pushed downward by the downwash of the airstream flowing over the rear edge of the roof, see Fig. 14.45. Subsequently these vortices re-attach themselves on each side of the body, and due to the air's momentum these vortices extend and trail well beyond the rear of the car. Hence not only does the rear negative wake base area include the vertical area and part of the rearward projected slope area where the airstream separates from the body profile, but it also includes the trailing conical vortices which also apply a strong suction pull against the forward motion of the car. As can be seen in Fig. 14.41 there is a critical slope angle range (20 to 35°) in which the drag coefficient rises steeply and should be avoided.

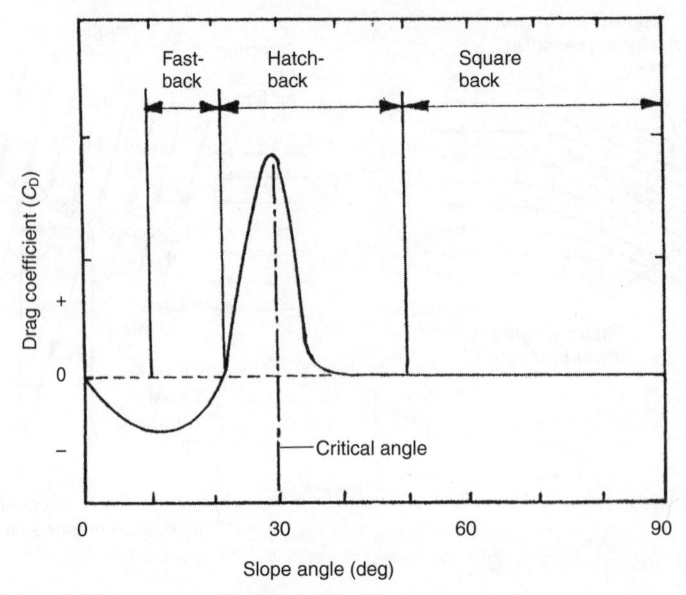

Fig. 14.41 Effect of rear panel slope angle on the afterbody drag

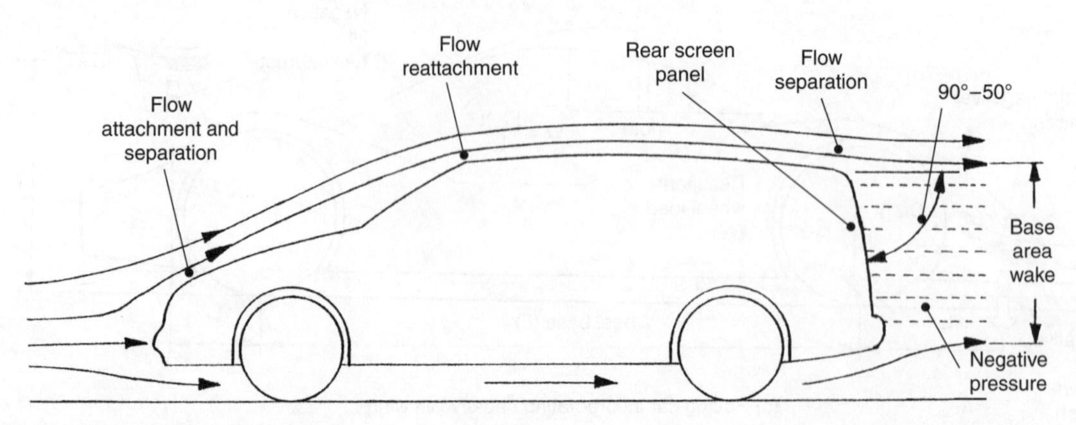

Fig. 14.42 Squareback configuration

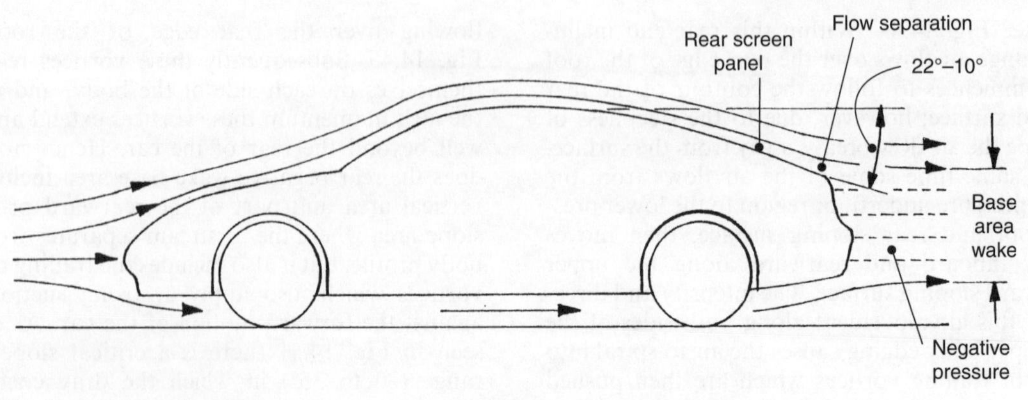

Fig. 14.43 Fastback configuration

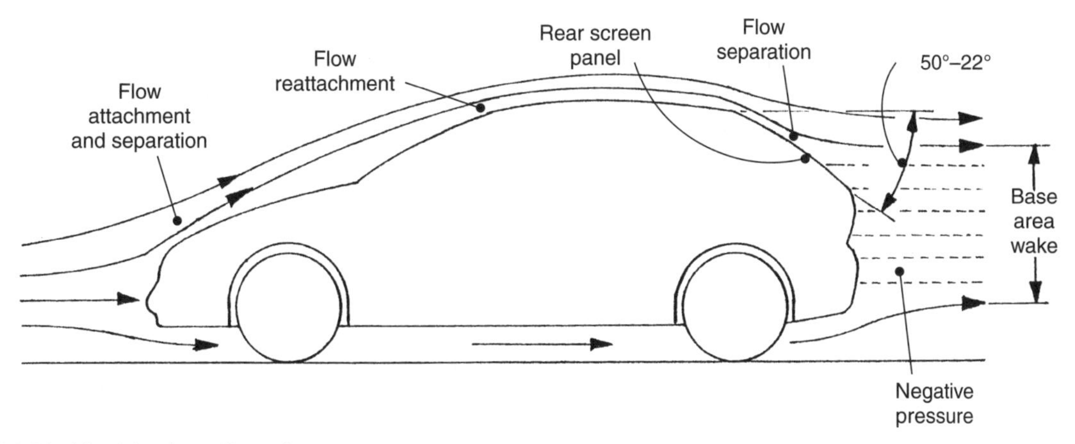

Fig. 14.44 Hatchback configuration

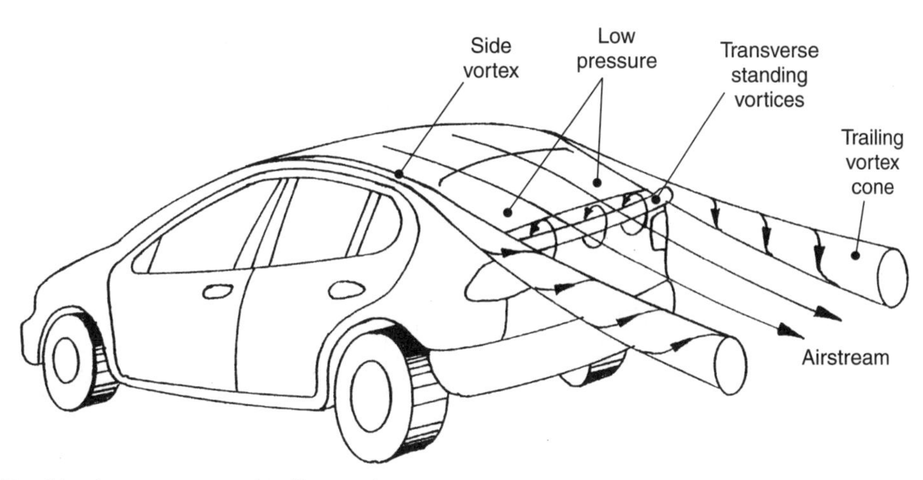

Fig. 14.45 Hatchback transverse and trailing vortices

14.6.4 Notchback drag (Figs 14.46, 14.47(a and b) and 14.48(a and b))

A notchback car style has a stepped rear end body profile in which the passenger compartment rear window is inclined downward to meet the horizontal rearward extending boot (trunk) lid (see Fig. 14.46). With this design, the air flows over the rear roof edge and follows the contour of the downward sloping rear screen for a short distance before separating from it; however, the downwash of the airstream causes it to re-attach itself to the body near the rear end extended boot lid. Thus the base-wake area will virtually be the vertical rear boot and light panel; however, standing vortices will be generated on each side of the body just inboard on the top surface of the rear window screen and boot lid, and will then be projected in the form of trailing conical vortices well beyond the rear end of the boot, see Fig. 14.19(b). Vortices will also be created along transverse rear screen to boot lid junction and across the rear of the panel light.

Experiments have shown (see Fig. 14.47(a)) that the angle made between the horizontal and the inclined line touching both the rear edges of the roof and the boot is an important factor in determining the afterbody drag. Fig. 14.47(b) illustrates the effect of the roof to boot line inclination; when this angle is increased from the horizontal the drag coefficient commences to rise until reaching a peak at an inclination of roughly 25°, after which the drag coefficient begins to decrease. From this it can be seen that raising the boot height or extending the boot length decreases the effective inclination angle Φ_e and therefore tends to reduce the drag coefficient. Conversely a very large effective inclination angle Φ_e will also cause a reduction in the

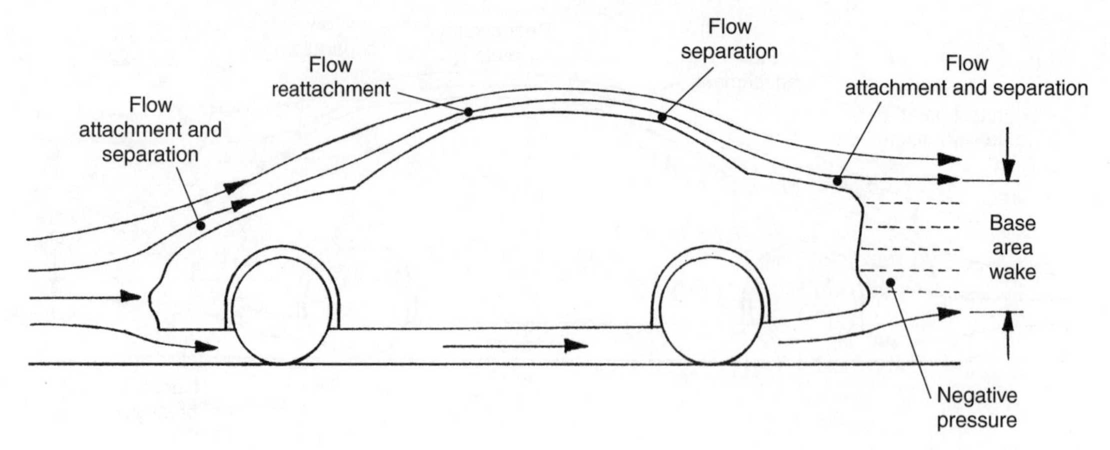

Fig. 14.46 Notchback configuration

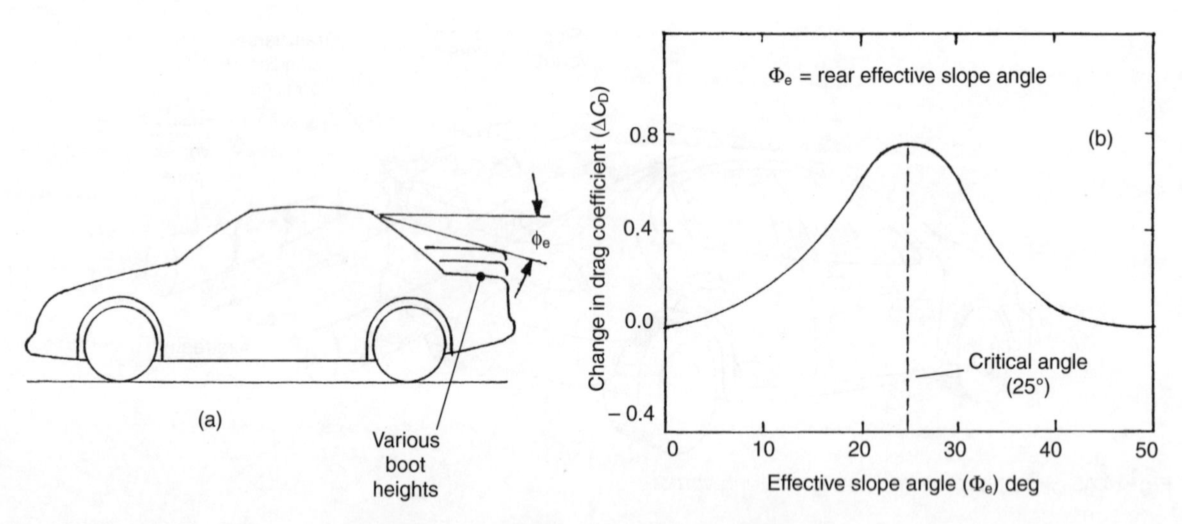

Fig. 14.47 (a and b) Influence of the effective slope angle on the drag coefficient

drag coefficient but at the expense of reducing the volume capacity of the boot. The drag coefficient relative to the rear boot profile can be clearly illustrated in a slightly different way, see Fig. 14.48(a). Here windtunnel tests show how the drag coefficient can be varied by altering the rear end profile from a downward sloping boot to a horizontal boot and then to a squareback estate shape. It will be observed (see Fig 14.48(b)) that there is a critical increase in boot height in this case from 50 to 150 mm when the drag coefficient rapidly decreases from 0.42 to 0.37.

14.6.5 Cabriolet cars (Fig. 14.49)

A cabriolet is a French noun and originally referred to a light two wheeled carriage drawn by one horse. Cabriolet these days describes a car with a folding roof such as a sports (two or four seater) or roadster (two seater) car. These cars may be driven with the folding roof enclosing the cockpit or with the soft roof lowered and the side screen windows up or down. Streamlining is such that the air flow follows closely to the contour of the nose and bonnet (hood), then moves up the windscreen before overshooting the screen's upper horizontal edge (see Fig. 14.49). If the rake angle of the windscreen is small (such as with a high mounted off road four wheel drive vehicle) the airstream will be deflected upward and rearward, but with a large rake angle windscreen the airstream will not rise much above the windscreen upper leading edge as the air flows over the open driver/passenger

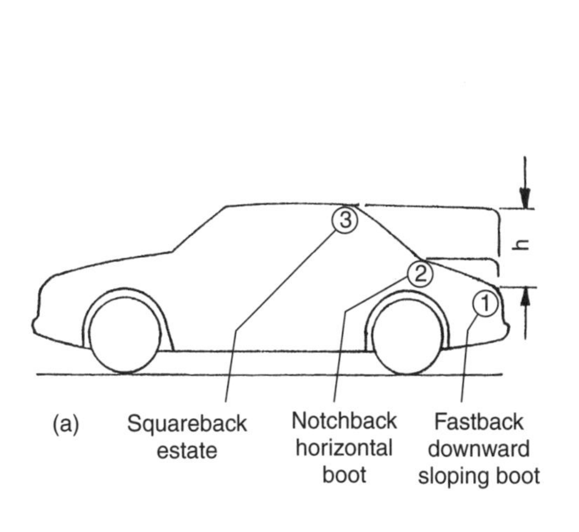

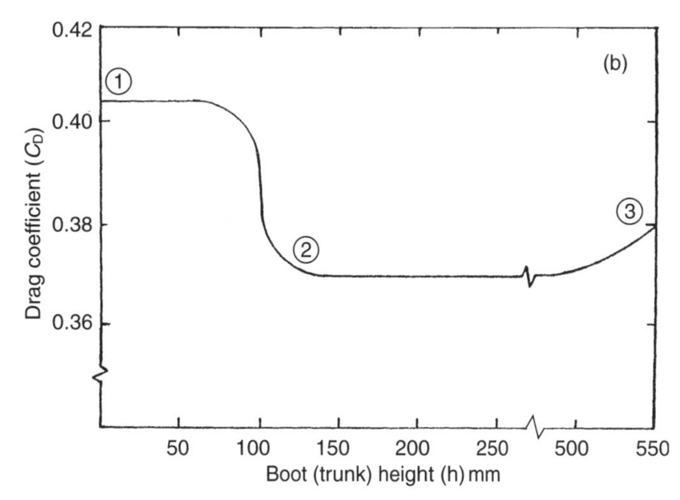

Fig. 14.48 (a and b) Effect of elevating the boot (trunk) height on the drag coefficient

compartment towards the rear of the car. A separation bubble forms between the airstream and the exposed and open seating compartment, the downstream air flow then re-attaches itself to the upper face of the boot (trunk). However, this bubble is unstable and tends to expand and burst in a cyclic fashion by the repetition of separation and re-attachment of the airstream on top of the boot (trunk), see Fig. 14.49. Thus the turbulent energy causes the bubble to expand and collapse and the fluctuating wake area (see Fig. 14.49), changing between h_1 and h_2 produces a relatively large drag resistance. With the side windscreens open air is drawn into the low pressure bubble region and in the process strong vortices are generated at the side entry to the seating compartment; this also therefore contributes to the car drag resistance. Typical drag coefficients for an open cabriolet car are given as follows: folding roof raised and side screens up C_D 0.35, folding roof down and side screens up C_D 0.38, and folding roof and side screens down C_D 0.41. Reductions in the drag coefficient can be made by attaching a header rail deflector, streamlining the roll over bar and by neatly storing or covering the folding roof, the most effective device to reduce drag being the header rail deflector.

14.7 Commercial vehicle aerodynamic fundamentals

14.7.1 The effects of rounding sharp front cab body edges (Fig. 14.50(a–d))

A reduction in the drag coefficient of large vehicles such as buses, coaches and trucks can be made by rounding the front leading edges of the vehicle.

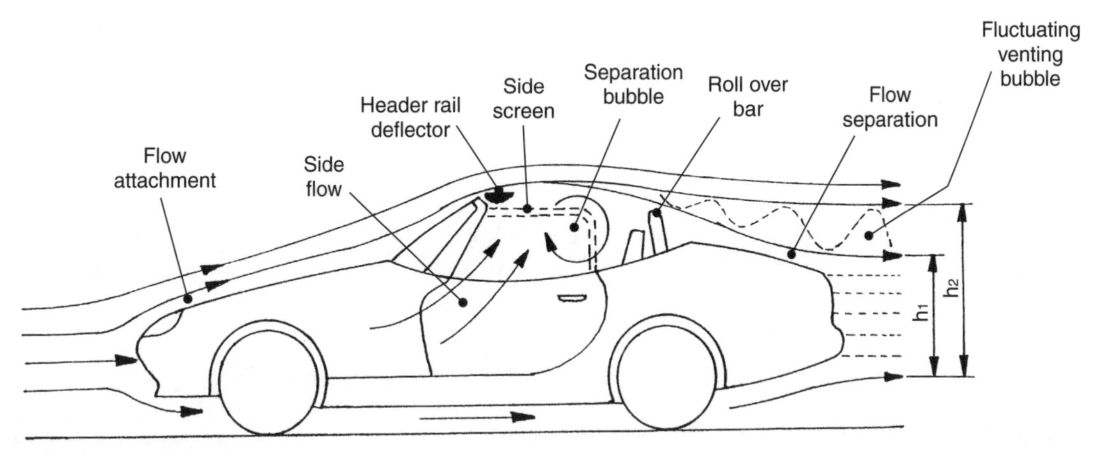

Fig. 14.49 Open cabriolet

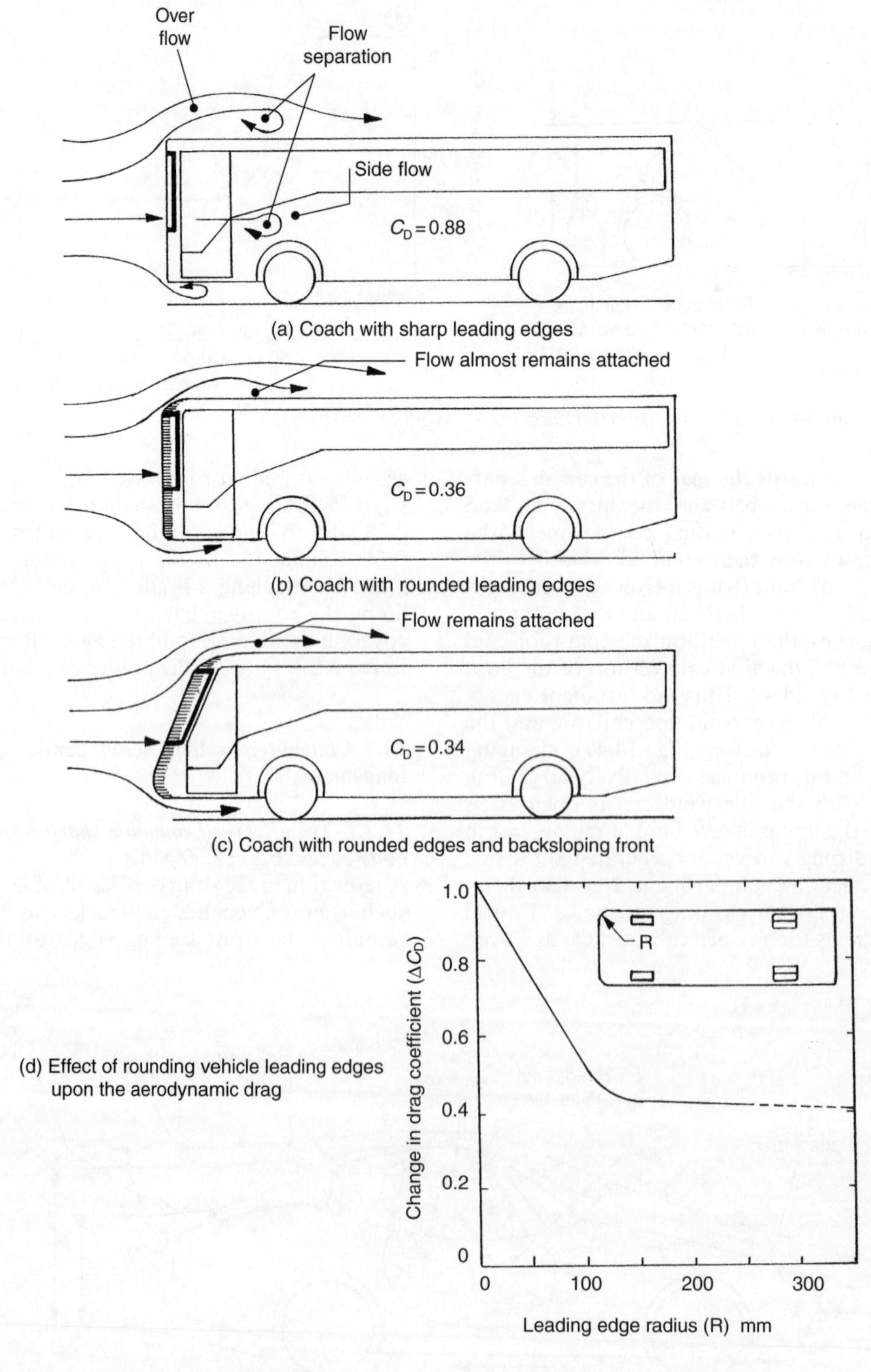

Fig. 14.50 (a–d) Forebody coach streamlining

Simulated investigations have shown a marked decrease in the drag coefficient from having sharp forebody edges (see Fig. 14.50(a)) to relatively large round leading edge radii, see Fig. 14.50(b). It can be seen from Fig. 14.50(d) that the drag coefficient progressively decreased as the round edge radius was increased to about 120 mm, but there was only a very small reduction in the drag coefficient with further increase in radii. Thus there is an optimum radius for the leading front edges, beyond this there is no advantage in increasing the rounding radius. The reduction in the drag coefficient due to rounding the edges is caused mainly by the change from flow separation to attached streamline flow for both cab roof and side panels, see Fig. 14.50(a and b). However, sloping back the front profile of the coach to provide further streamlining only made a marginal reduction in the drag coefficient, see Fig. 14.50(c).

14.7.2 The effects of different cab to trailer body heights with both sharp and rounded upper windscreen leading edges (Fig. 14.51(a–c))

A generalized understanding of the air flow over the upper surface of an articulated cab and trailer can be obtained by studying Fig. 14.51(a and b). Three different trailer heights are shown relative to one cab height for both a sharp upper windscreen leading edge (Fig. 14.51(a)) and for a rounded upper windscreen edge (Fig. 14.51(b)). It can be seen in the case of the sharp upper windscreen leading edge cab examples (Fig. 14.51(a)) that with the low trailer body the air flow cannot follow the contour of the cab and therefore overshoots both the cab roof and the front region of the trailer body roof thereby producing a relatively high coefficient of drag, see Fig. 14.51(c). With the medium height trailer body the air flow still overshoots (separates) the cab but tends to align and attach itself early to the trailer body roof thereby producing a relatively low coefficient of drag, see Fig. 14.51(c). However, with the high body the air flow again overshoots the cab roof; some of the air then hits the front of the trailer body, but the vast majority deflects off the trailer body leading edge before re-attaching itself further along the trailer body roof. Consequently the disrupted air flow produces a rise in the drag coefficient, see Fig. 14.51(c).

In the case of the rounded upper windscreen leading edge cab (see Fig. 14.51(b)), with a low trailer body the air flowing over the front windscreen remains attached to the cab roof, a small proportion will hit the front end of the trailer body then flow between the cab and trailer body, but the majority flows over the trailer roof leading edge and attaches itself only a short distance from the front edge of the trailer roof thereby producing a relatively low drag coefficient, see Fig. 14.51(c). With the medium height trailer body the air flow remains attached to the cab roof; some air flow again impinges on the front of the trailer body and is deflected between the cab and trailer body, but most of the air flow hits the trailer body leading edge and is deflected slightly upwards and only re-attaches itself to the upper surface some distance along the trailer roof. This combination therefore produces a moderate rise in the drag coefficient, see Fig. 14.51(c). In the extreme case of having a very high trailer body the air flow over the cab still remains attached and air still flows downwards into the gap made between the cab and trailer; however, more air impinges onto the vertical front face of the trailer body and the deflection of the air flow over the leading edge of the trailer body is even steeper than in the case of the medium height trailer body. Thus re-attachment of the air flow over the roof of the trailer body takes place much further along its length so that a much larger roof area is exposed to air turbulence; consequently there is a relatively high drag coefficient, see Fig. 14.51(c).

14.7.3 Forebody pressure distribution (Fig. 14.52(a and b))

With both the conventional cab behind the engine and the cab over or in front of the engine tractor unit arrangements there will be a cab to trailer gap to enable the trailer to be articulated when the vehicle is being manoeuvred. The cab roof to trailer body step, if large, will compel some of the air flow to impinge on the exposed front face of the trailer thereby producing a high pressure stagnation region while the majority of air flow will be deflected upwards. As it brushes against the upper leading edge of the trailer the air flow then separates from the forward region of the trailer roof before re-attaching itself further along the flat roof surface, see Fig. 14.52(a). As can be seen the pressure distribution shows a positive pressure (above atmospheric pressure) region air spread over the exposed front face of the trailer body with its maximum intensity (stagnant region) just above the level of the roof; this contrasts the negative pressure (below atmospheric pressure) generated air flow in the forward region of the trailer roof caused by the air flow separation turbulence. Note the negative pressure drops off towards the rear of the roof due to air flow re-attachment.

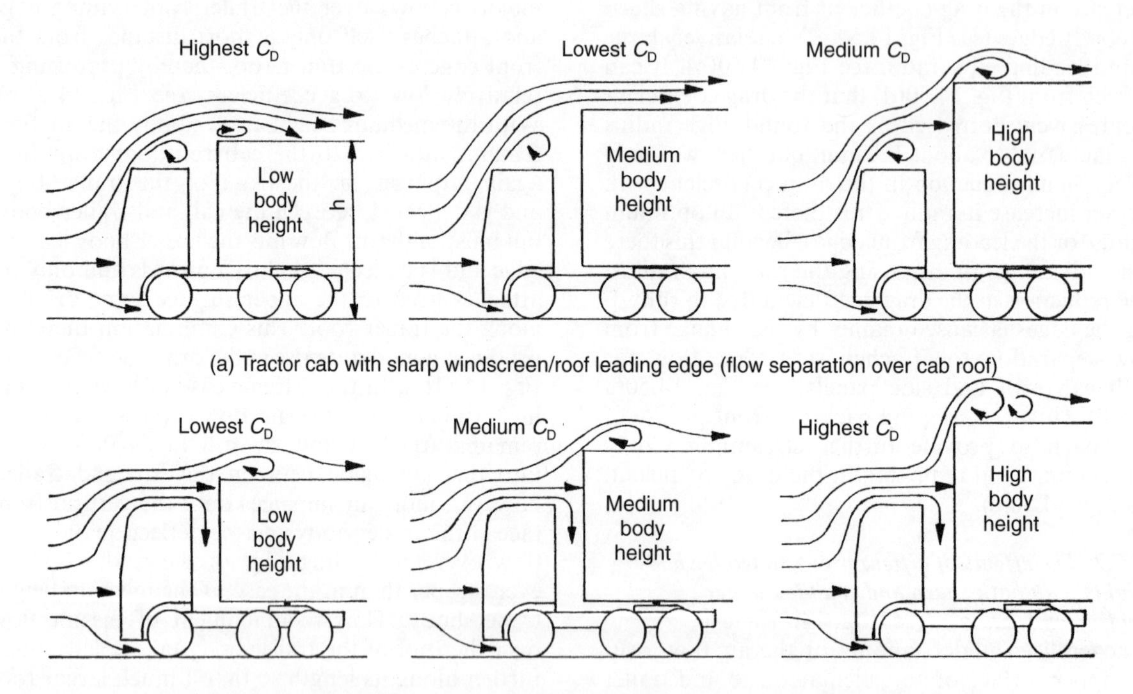

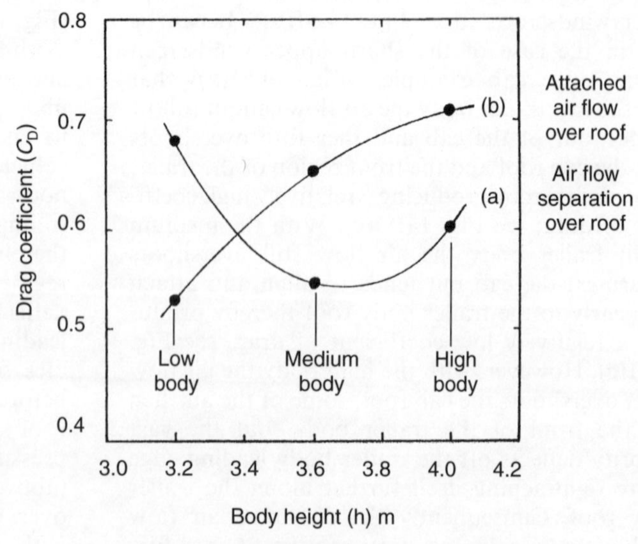

Fig. 14.51 (a–c) Comparison of air flow conditions with both sharp and rounded roof leading edge cab with various trailer body heights

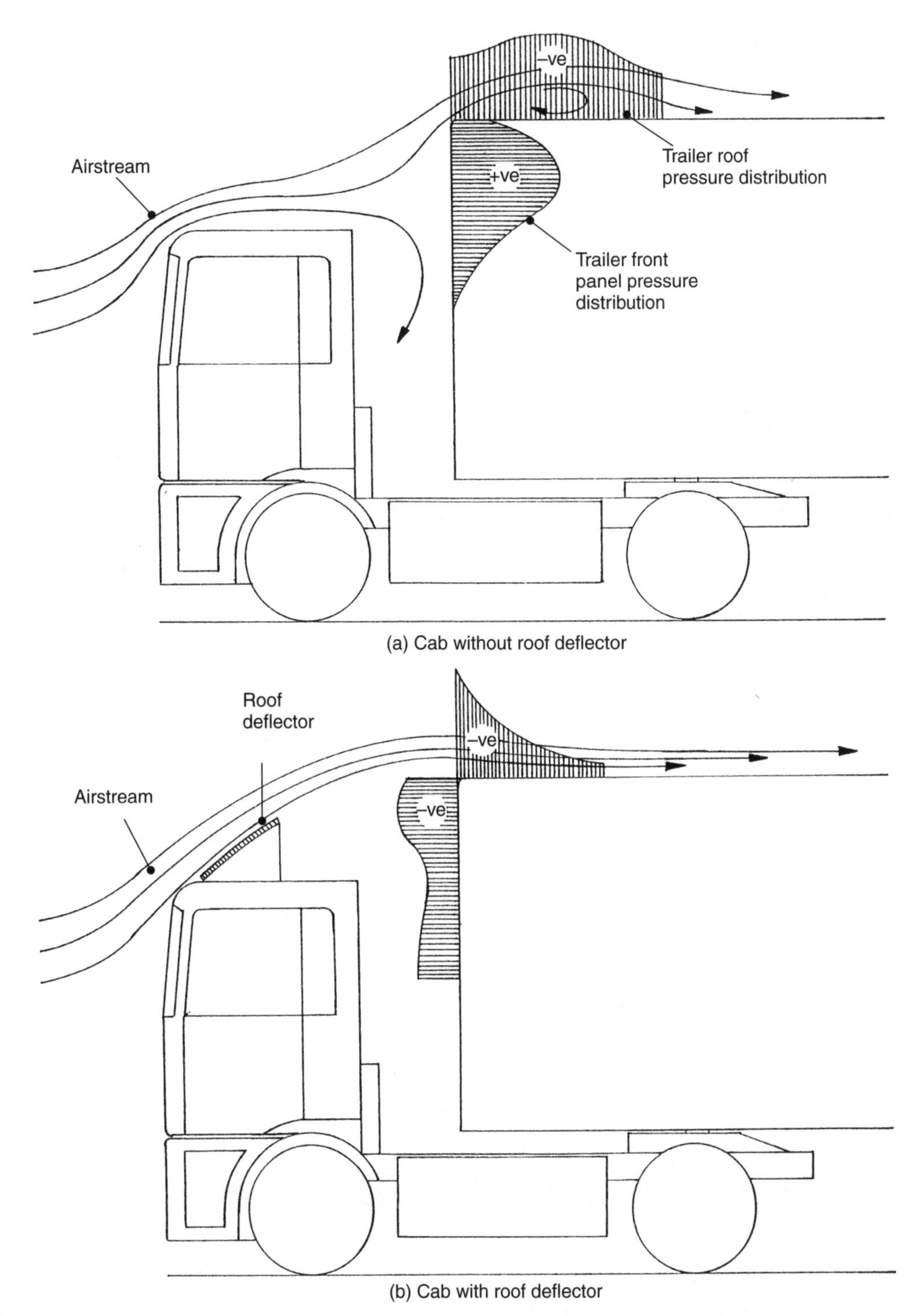

Fig. 14.52 (a and b) Trailer flow body pressure distribution with and without cab roof deflector

By fitting a cab roof deflector the pattern of air flow is diverted upwards and over the roof of the trailer body, there being only a slight degree of flow separation at the front end of the trailer body roof, see Fig. 14.52(b). Consequently the air flow moves directly between the cab roof deflector and the roof of the trailer body; it thus causes the air pressure in the cab to trailer gap to decrease, this negative pressure being more pronounced on the exposed upper vertical face of the trailer, hence the front face upper region of the trailer will actually reduce that portion of drag produced by the exposed frontal area of the trailer. Conversely the negative pressure created by the air flow over the leading edge of the roof falls rapidly, indicating early air flow re-attachment.

14.7.4 The effects of a cab to trailer body roof height step (Fig. 14.53(a and b))

Possibly the most important factor which contributes to a vehicle's drag resistance is the exposed area of the trailer body above the cab roof relative to the cab's frontal area (Fig. 14.53(a)). Investigation into the forebody drag of a truck in a windtunnel has been made where the trailer height is varied relative to a fixed cab height. The drag coefficient for different trailer body to cab height ratios (t/c) were then plotted as shown in Fig. 14.53(b). For this particular cab to trailer combination dimensions there was no noticeable change in the drag coefficient C of 0.63 with an increase in trailer body to cab height ratio until about 1.2, after which the drag coefficient commenced to rise in proportion to the increase in the trailer body to cab height ratio up to a t/c ratio of 1.5, which is equivalent to the maximum body height of 4.2 m; this corresponded to a maximum drag coefficient of 0.86. Hence increasing the trailer body step height ratio from 1.2 to 1.5 increases the step height from 0.56 m to 1.4 m and in turn raises the drag coefficient from 0.63 to 0.86. The rise in drag coefficient of 0.23 is considerable and therefore streamlining the air flow between the cab and trailer body roof is of great importance.

14.8 Commercial vehicle drag reducing devices

14.8.1 Cab roof deflectors (Figs 14.54(a and b), 14.55(a and b) and 14.56(a–c))

To partially overcome the large amount of extra drag experienced with a cab to trailer height mismatch a cab roof deflector is commonly used. This device prevents the air movement over the cab roof impinging on the upper front of the trailer body and then flowing between the cab and trailer gap, see Fig. 14.54(a). Instead the air flow is diverted by the uptilted deflector surface to pass directly between the cab to trailer gap and then to flow relatively smoothly over the surface of the trailer roof, see Fig. 14.54(b). These cab roof deflectors are beneficial in reducing the head on air flow but they do not perform so well when subjected to side winds. Slight improvements can be made to prevent air flowing underneath and across the deflector plate by enclosing the sides; this is usually achieved

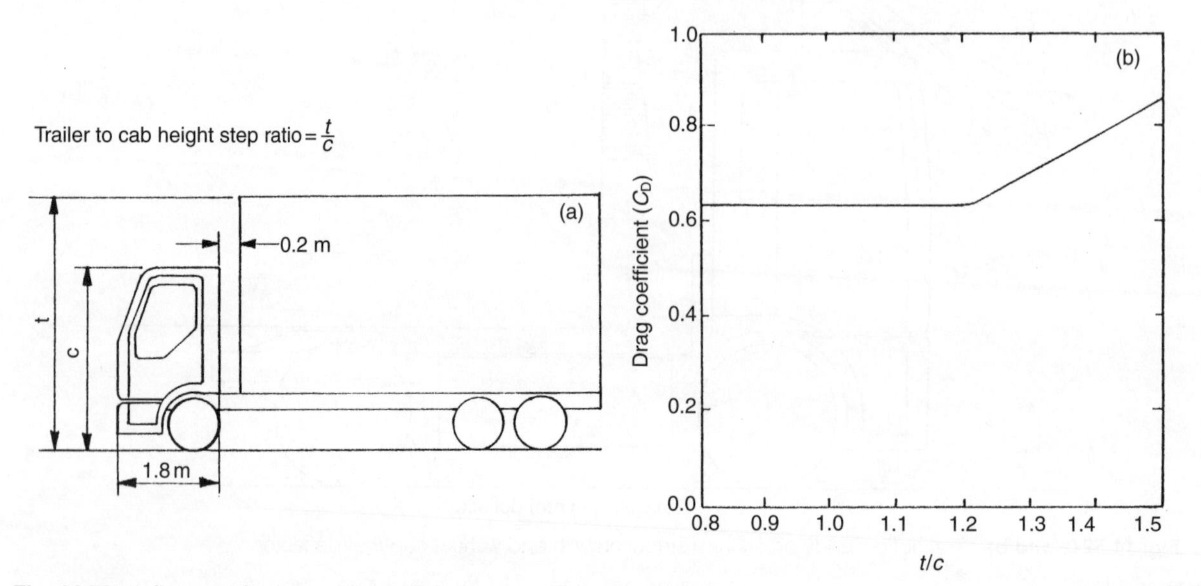

Fig. 14.53 Influence of cab to trailer body height upon the drag coefficient

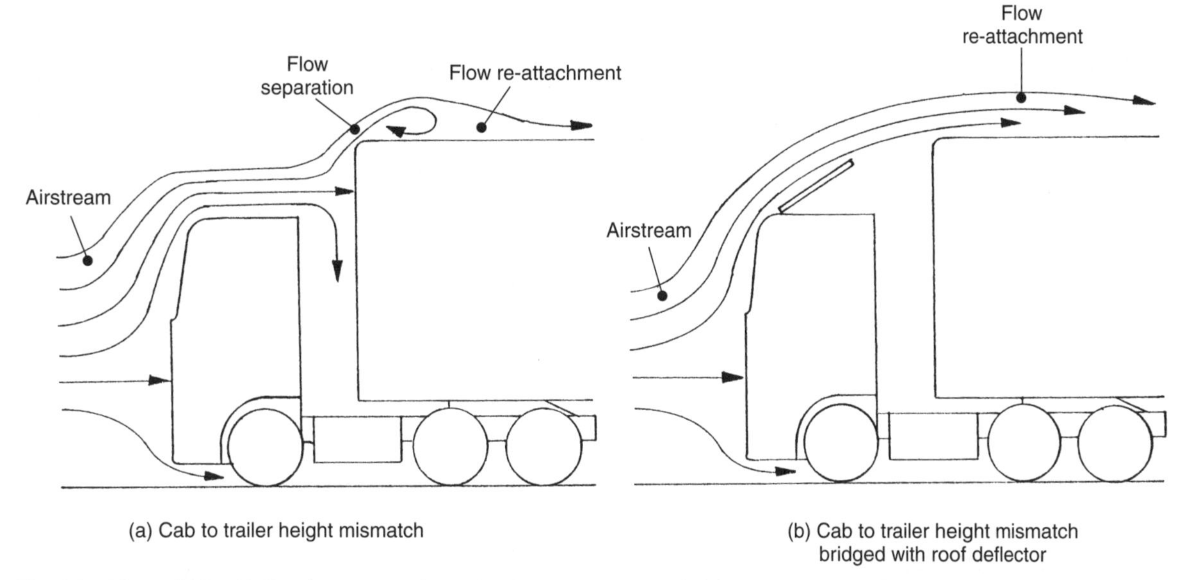

(a) Cab to trailer height mismatch

(b) Cab to trailer height mismatch bridged with roof deflector

Fig. 14.54 (a and b) Air flow between cab and trailer body with and without cab roof deflector

by using a fibre glass or plastic moulded deflector, see Fig. 14.55(b).

If trailers with different heights are to be coupled to the tractor unit while in service, then a mismatch of the deflector inclination may result which will again raise the aerodynamic drag. There are some cab deflector designs which can adjust the tilt of the cab deflector to optimize the cab to trailer air flow transition (see Fig. 14.55(a)), but in general altering the angle setting would be impractical. How the cab roof deflector effectiveness varies with deflector plate inclination is shown in Fig. 14.56(c) for both a narrow and a wide cab to trailer gap, representing a rigid truck and an articulated vehicle respectively (see Fig. 14.56(a and b)). These graphs illustrate the general trend and do not take into account the different cab to trailer heights, cab to trailer air gap width and the width to height ratio of the deflector plate. It can be seen that with a rigid truck having a small cab to trailer gap the

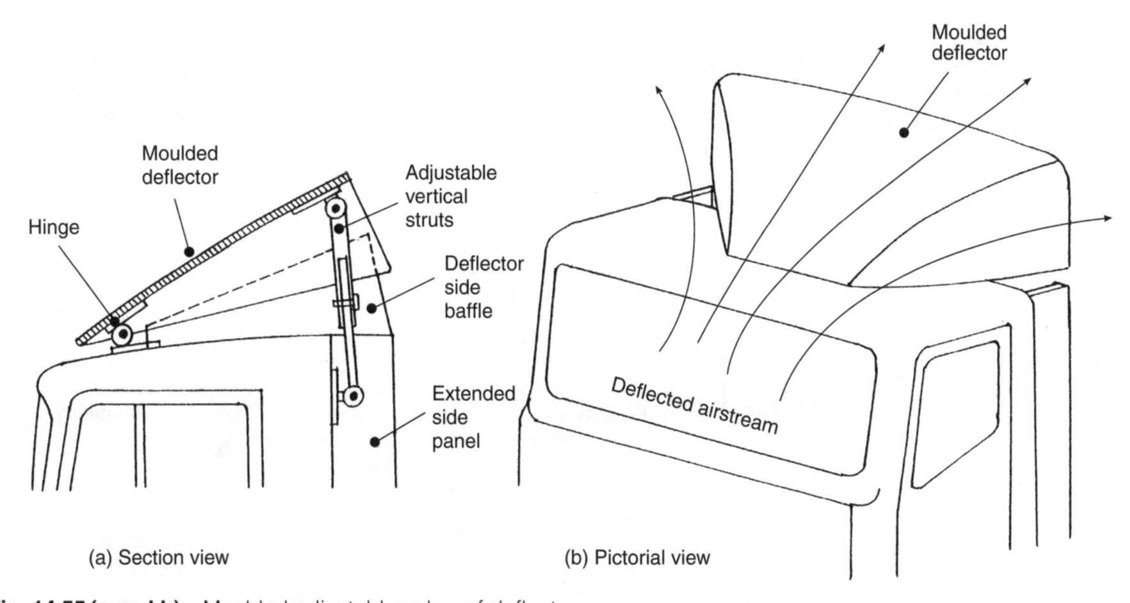

(a) Section view

(b) Pictorial view

Fig. 14.55 (a and b) Moulded adjustable cab roof deflector

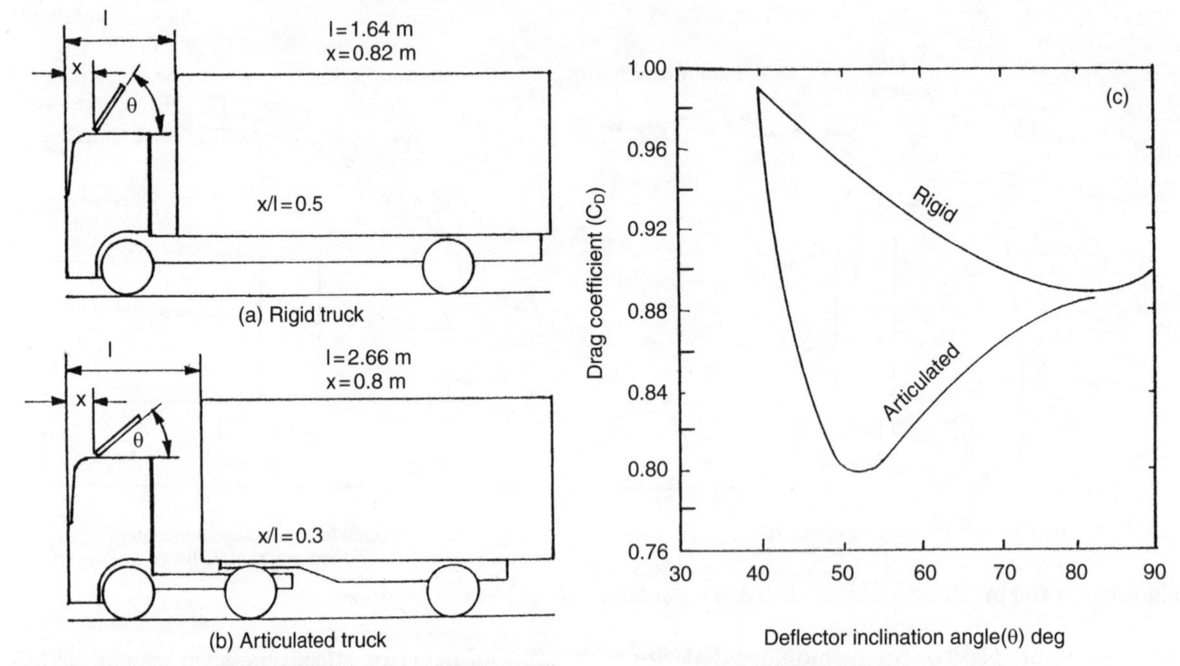

Fig. 14.56 (a–c) Optimizing roof deflector effectiveness for both rigid and articulated trucks

reduction in the drag coefficient with increased deflector plate inclination is gradual, reaching an optimum minimum at an inclination angle of 80° and then commencing to rise again, see Fig. 14.56(c). With the articulated vehicle having a large cab to trailer gap, the deflector plate effectiveness increases rapidly with an increase in the deflector inclination angle until the optimum angle of 50° is reached, see Fig. 14.56(c). Beyond this angle the drag coefficient begins to rise steadily again with further increase in the deflector plate angle; this indicates with the large gap of the articulated vehicle the change in drag coefficient is much more sensitive to deflector plate inclination.

14.8.2 Yaw angle (Figs 14.57 and 14.58)
With cars the influence of crosswinds on the drag coefficient is relatively small; however, with much larger vehicles a crosswind considerably raises the drag coefficient therefore not only does the direct air flow from the front but also the air movement from the side has to be considered. It is therefore necessary to study the effects crosswinds have on the vehicle's drag resistance, taking into account the velocity and angle of attack of the crosswind relative to the direction of motion of the vehicle and its road speed. This is achieved by drawing to scale a velocity vector triangle, see Fig. 14.57. The vehicle velocity vector line is drawn, then the crosswind

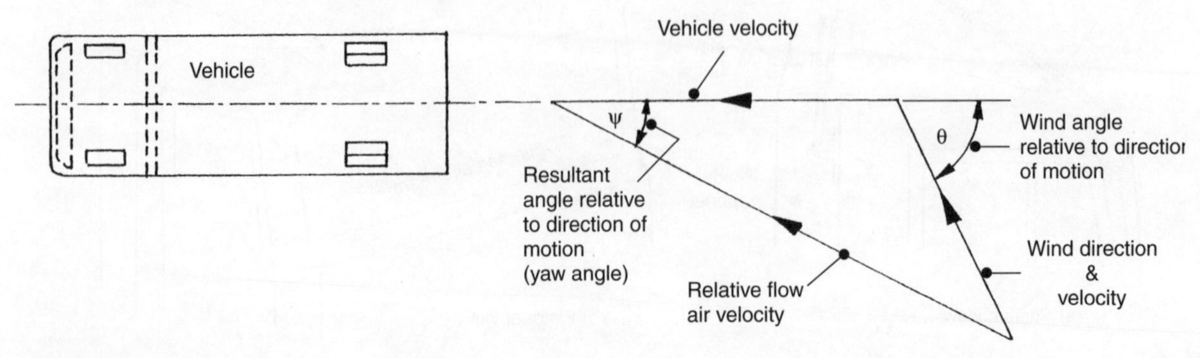

Fig. 14.57 The yaw angle

velocity vector at the crosswind angle to the direction of motion; a third line representing the relative air velocity then closes the triangle. The resultant angle made between the direction the vehicle is travelling and the resultant relative velocity is known as the yaw angle, and it is this angle which is used when investigating the effect of a crosswind on the drag coefficient.

In addition to head and tail winds vehicles are also subjected to crosswinds; crosswinds nearly always raise the drag coefficient, this being far more pronounced as the vehicle size becomes larger and the yaw angle (relative wind angle) is increased. The effect crosswinds have on the drag coefficient for various classes of vehicles expressed in terms of the yaw angle (relative wind angle) is shown in Fig. 14.58. Each class of vehicle with its own head on (zero yaw angle) air flow drag coefficient is given a drag coefficient of unity. It can be seen using a drag coefficient of 1.0 with zero yaw angle (no wind) that the drag coefficient for a car reaches a peak of 1.08 at a yaw angle of 20°, whereas for the van, coach, articulated vehicle and rigid truck and trailer the drag coefficient rose to 1.18, 1.35, 1.5 and 1.7 respectively for a similar yaw angle of 20°.

14.8.3 Cab roof deflector effectiveness versus yaw angle (Fig. 14.59(a and b))

The benefits of reducing the drag coefficient with a cab roof deflector are to some extent cancelled out when the vehicle is subjected to crosswinds. This can be demonstrated by studying data taken from

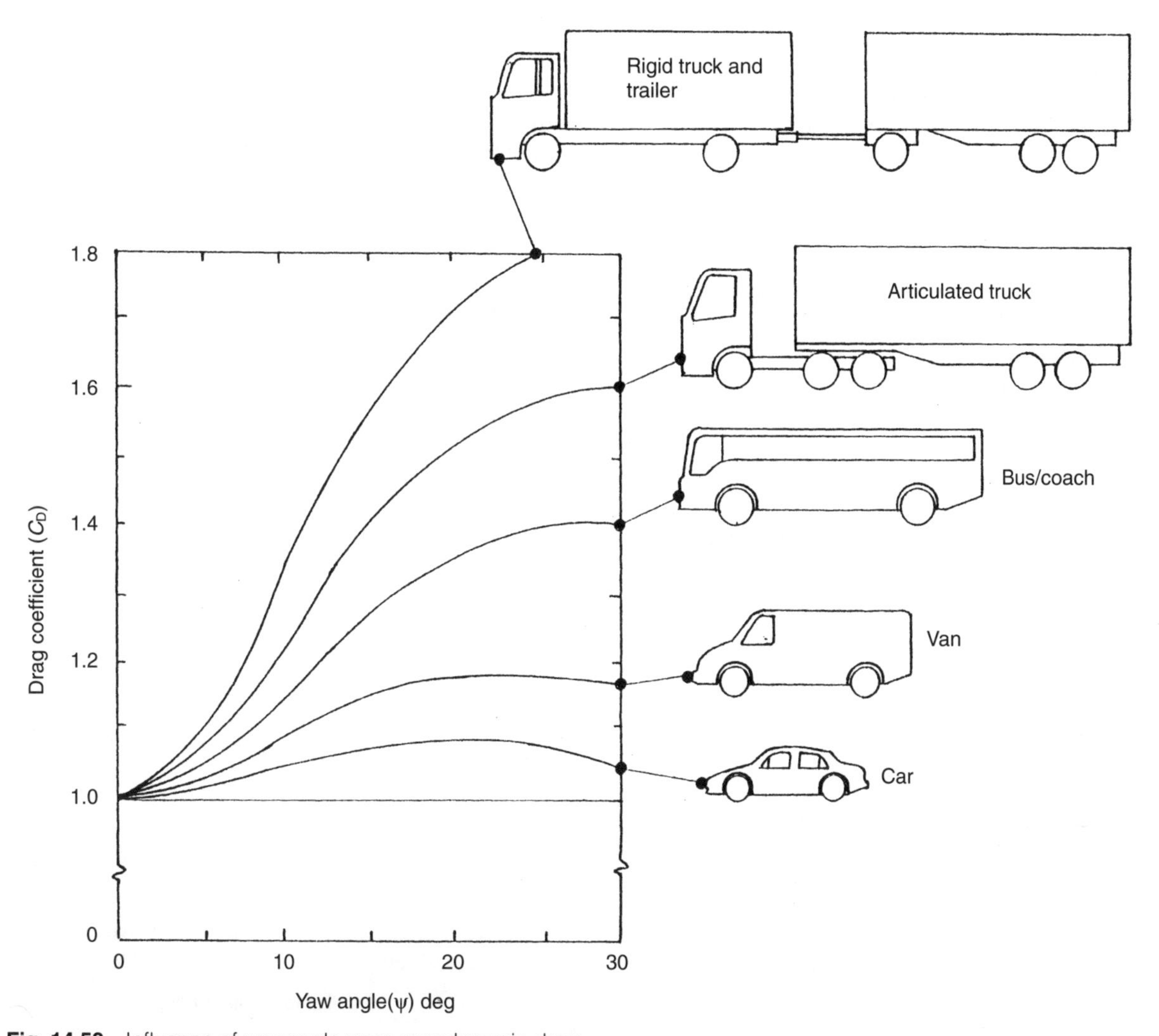

Fig. 14.58 Influence of yaw angle upon aerodynamic drag

one particular vehicle, see Fig. 14.59(a and b), which utilizes a cab roof deflector; here with zero yaw angle the drag coefficient reduces from 0.7 to 0.6 as the deflector inclination changes from 90° (vertical) to 50° respectively. With a 5° yaw angle (relative wind angle) the general trend of drag coefficient rises considerably to around 0.9 whereas the tilting of the deflector from the vertical over an angle of 40° only shows a marginal decrease in the drag coefficient of about 0.02; with a further 10° inclination decrease the drag coefficient then commenced to rise steeply to about 0.94. As the yaw angle is increased from 5 to 10° the drag coefficient rises even more to 1.03 with the deflector in the vertical position, however this increase in drag coefficient is not so much as from 0 to 5°. Hence the reduction in the drag coefficient from 1.03 to 0.98 as the deflector is tilted from the vertical to 40° is relatively small compared to the overall rise in drag coefficient due to crosswind effects. However, raising the yaw angle still further from 10 to 15° indicates on the graph that the yaw angle influence on the drag coefficient has peaked and is now beginning to decline: both the 10 and 15° yaw angle curves are similar in shape but the 15° yaw angle curve is now below that of the 10° yaw angle curve. Note the minimum drag coefficient deflection inclination angle is only relevant for the dimensions of this particular cab to trailer combination.

14.8.4 Comparison of drag resistance with various commercial vehicle cab arrangements relative to trailer body height (Fig. 14.60(a–e))

The drag coefficient of a tractor–trailer combination is influenced by the trailer body height and by different cab configurations such as a conventional low cab, low cab with roof deflector and high sleeper cab, see Fig. 14.60(a–c). Thus a high cab arrangement (see Fig. 14.60(c, d and e)) is shown to be more effective in reducing the drag coefficient than a low cab (see Fig. 14.60(a, d and e)) and therefore for long distance haulage the sleeper compartment above the driver cab has an advantage in having the sleeper area behind the driver's seat. Conversely with a low cab and a roof deflector which has an adjustable plate angle (see Fig. 14.60(b, d and e)), the drag coefficient can be kept almost constant for different trailer body heights. However, it is not always practical to adjust the deflector angle, but fortunately a great many commercial vehicle

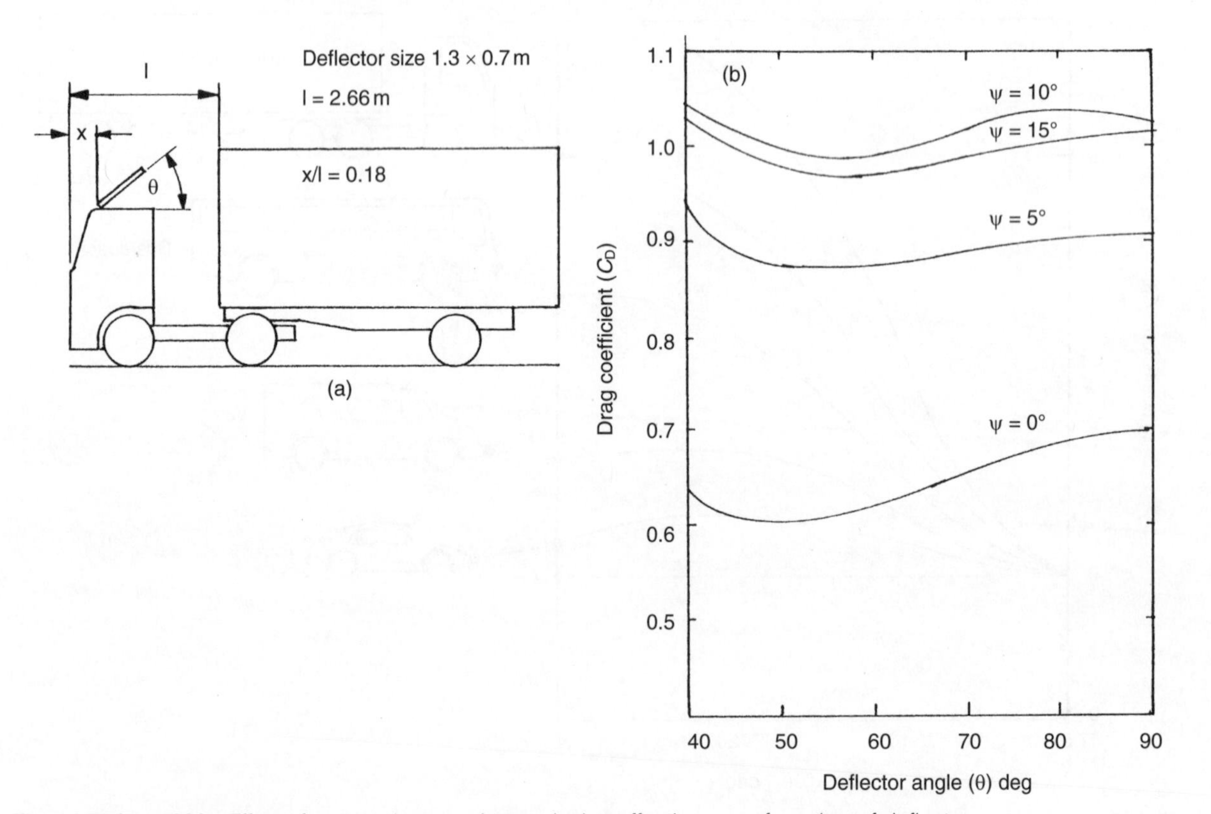

Fig. 14.59 (a and b) Effect of yaw angle upon drag reducing effectiveness of a cab roof deflector

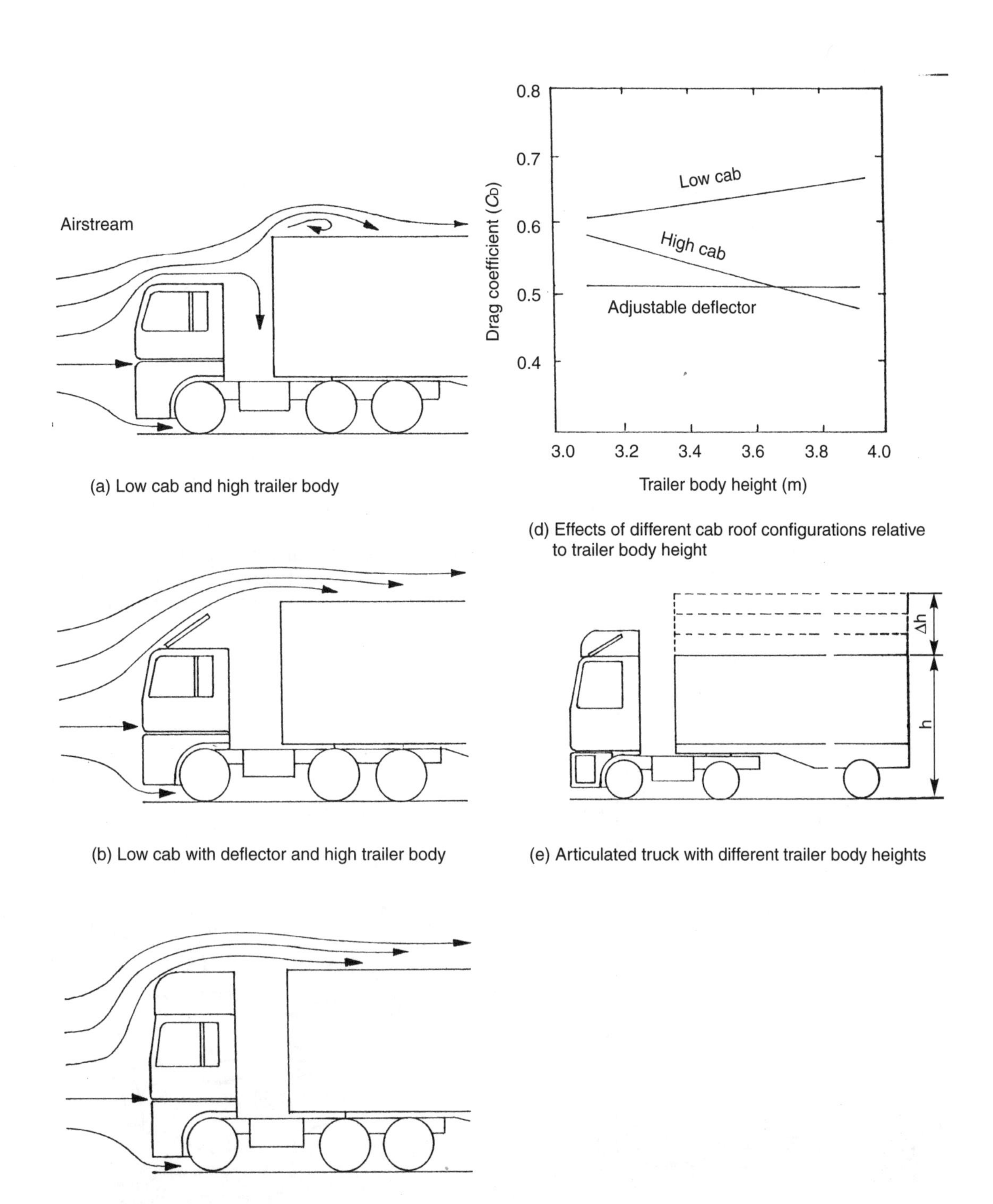

(a) Low cab and high trailer body

(b) Low cab with deflector and high trailer body

(c) High cab and trailer body

(d) Effects of different cab roof configurations relative to trailer body height

(e) Articulated truck with different trailer body heights

Fig. 14.60 (a–e) Methods of optimizing air flow conditions with different trailer body heights

cab–trailer combinations use the same size trailer bodies for one particular application so that the roof deflector angle can be pre-set to the minimum of drag resistance. With a low cab the drag coefficient tends to increase as the cab roof to body roof step height becomes larger whereas with a high cab the drag coefficient tends to decrease as trailer body height rises, see Fig. 14.60(d and e).

14.8.5 Corner vanes (Fig. 14.61(a–c))

The cab of a commercial vehicle resembles a cube with relatively flat upright front and side panels, thus with well rounded roof and side leading edges the cab still has a blunt front profile. When the vehicle moves forward the cab penetrates the surrounding air; however, the air flow passing over the top, underneath and around the sides will be far from being streamlined. Thus in particular the air flowing around the side leading edges of the cab may initially separate from the side panels, causing turbulence and a high resistance to air flow, see Fig. 14.61(a).

One method of reducing the forebody drag is to attach corner vanes on each side of the cab (Fig. 14.61(c)). The corner vane is set away from the rounded vertical edges and has several evenly spaced internal baffles which bridge the gap between the cab and corner vane walls. Air meeting the front face of the cab moves upwards and over the roof, while the rest flows to the left and right hand side leading edges. Some of this air also flows around the leading edge through the space formed between the cab and corner vanes (see Fig. 14.61 (b and c)); this then encourages the airstream to remain attached to the cab side panel surface. Air drag around the cab front and side panels is therefore kept to a minimum.

14.8.6 Cab to trailer body gap (Fig. 14.62)

Air passing between the cab and trailer body gap with an articulated vehicle due to crosswinds significantly increases the drag resistance. As the crosswind angle of attach is increased, the flow through the cab-trailer gap produces regions of flow separated on the sheltered side of the trailer body, see Fig. 14.62. This flow separation then tends to spread rearwards, eventually interacting with and enlarging the trailer wake, the net result being a rise in the rearward pull due to the enlarged negative pressure zone.

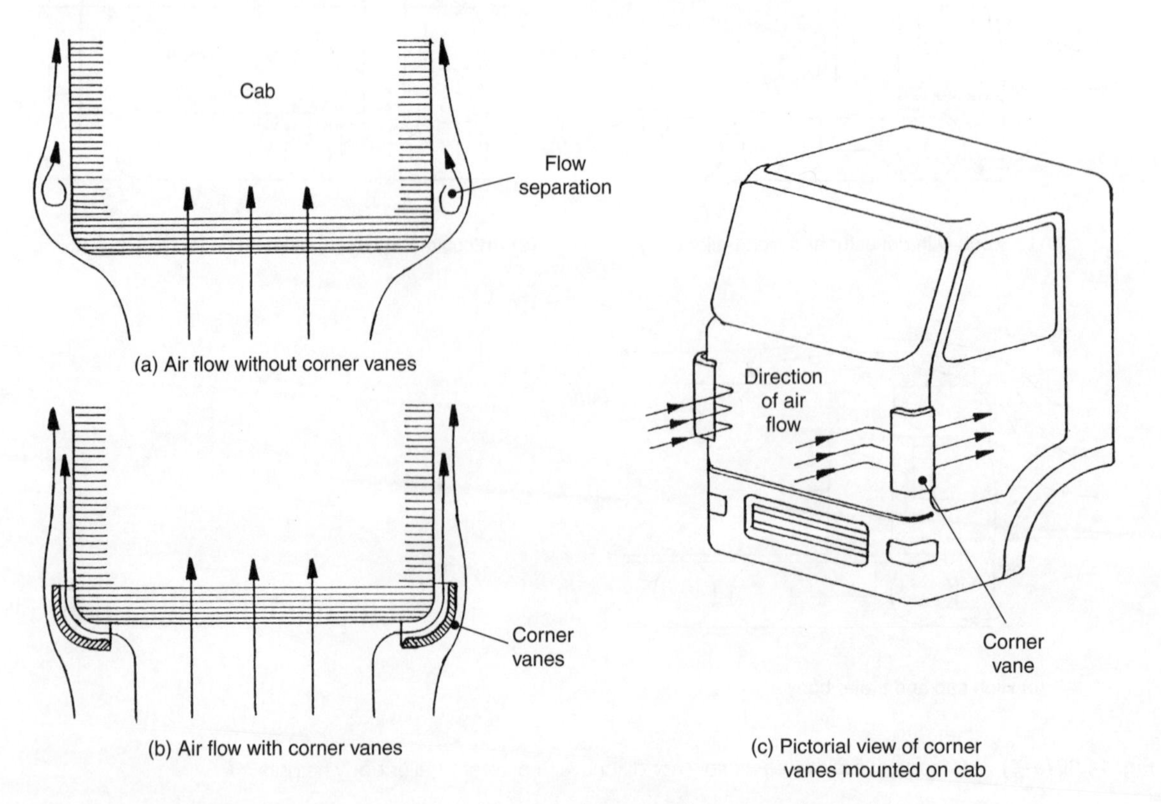

Fig. 14.61 (a–c) Influence of corner vanes in reducing cab side panel flow separation

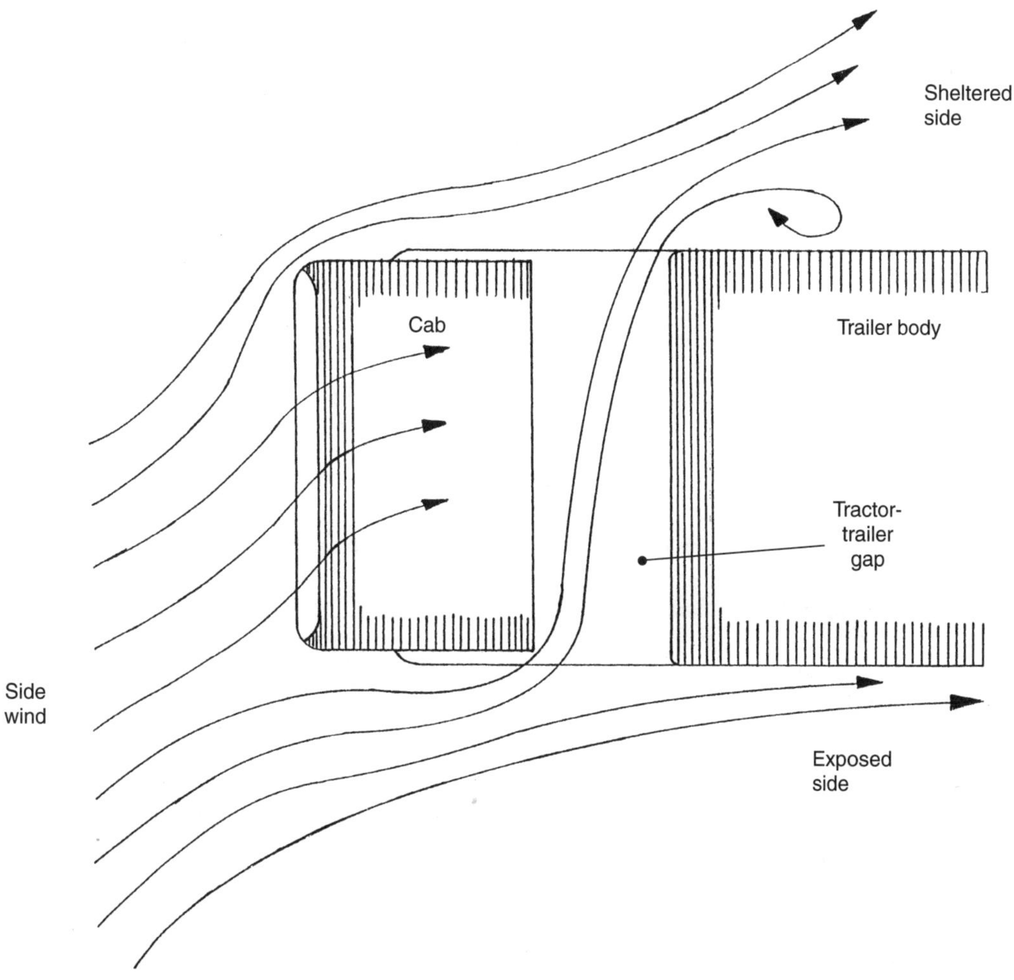

Fig. 14.62 Air flow through tractor-trailer gap with crosswind

14.8.7 Cab to trailer body gap seals (Fig. 14.63 (a and b), 14.64, 14.65(a and b) and 14.66(a–d)) Simple tilt plate cab roof deflectors when subjected to side winds tend to counteract the gain in head on airstream drag resistance unless the deflector sides are enclosed. With enclosed and streamlined cab roof deflector sides, see Fig. 14.55(a and b), improvements in the drag coefficient can be made with yaw angles up to about 20°, see Fig. 14.63(a and b). Further reductions in the drag coefficient are produced when the cab to trailer gap is sealed by some sort of partition which prevents air flowing through the cab to body gap, see Fig. 14.63 (a and b). The difficulty with using a cab to trailer air gap partition is designing some sort of curtain or plate which allows the trailer to articulate when manoeuvring the vehicle-trailer combination.

Cab to trailer gap seals can be divided into three basic designs:

1 Cab extended side panels
2 Centre line gap seals (splitter plate seal)
3 Windcheater roller edge device (forebody edge fairing).

Cab extended side panels (Fig. 14.64) These devices are basically rearward extended vertical panels attached to the rear edges of the cab which are angled towards the leading edges of the trailer body. This type of gap fairing (side streamlining) is effective in reducing the drag coefficient with increasing crosswind yaw angle. With zero and 10° yaw angles a drag coefficient reduction of roughly 0.05 and 0.22 respectively have been made possible.

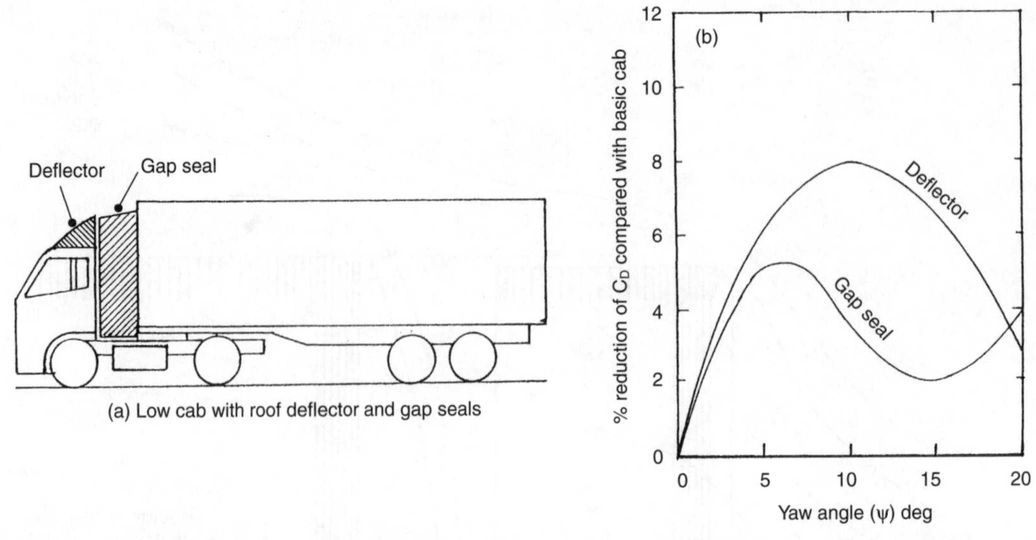

Fig. 14.63 (a and b) Drag reductions with crosswinds when incorporating a roof deflector and gap seal

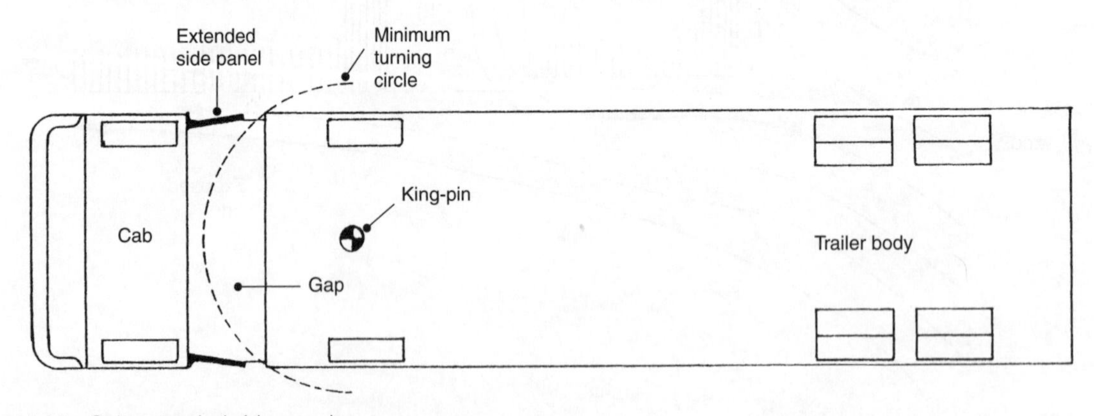

Fig. 14.64 Cab extended side panels

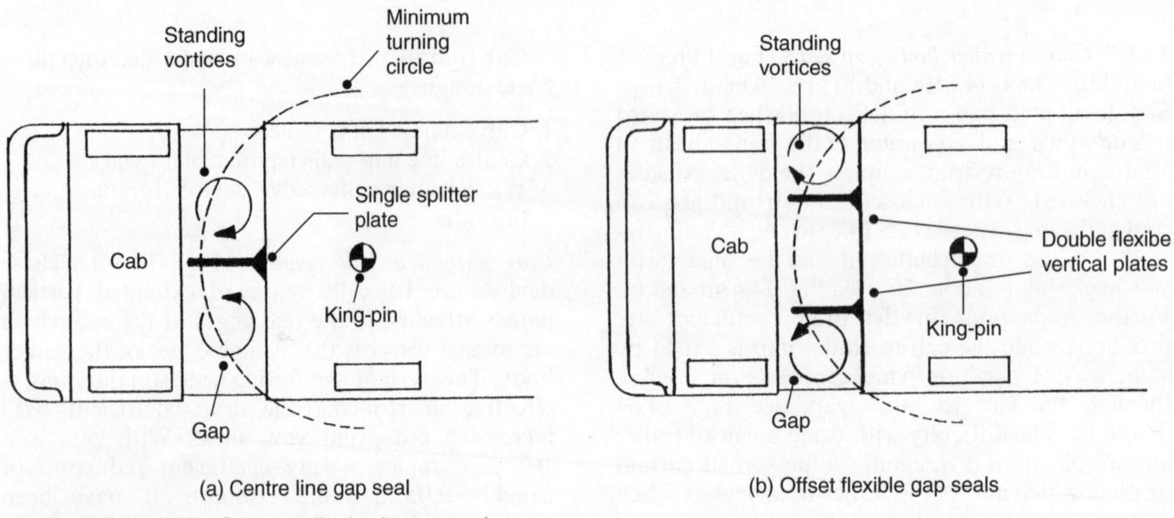

Fig. 14.65 (a and b) Cab to trailer body gap seals

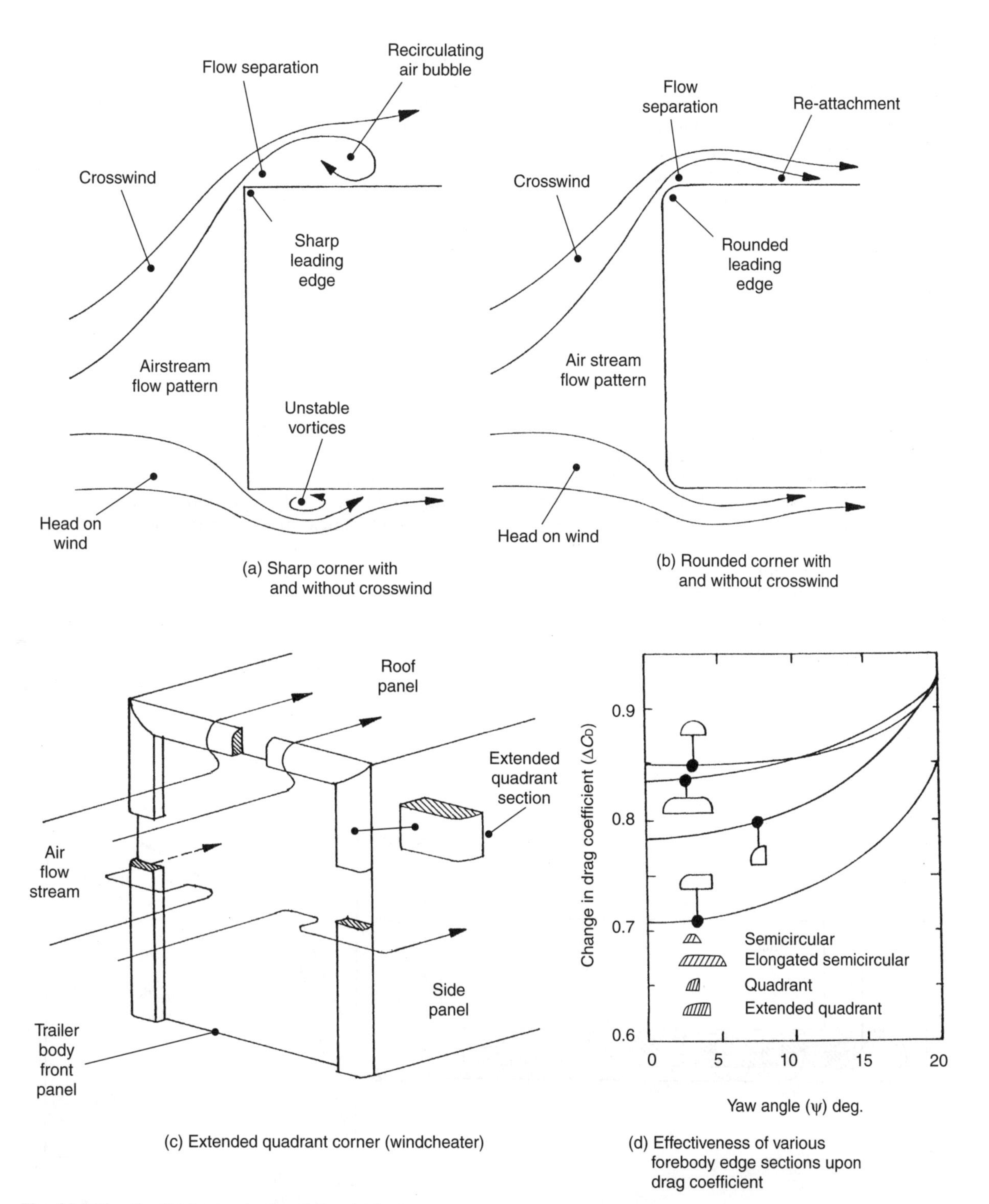

Fig. 14.66 (a–d) Trailer forebody edging fairings

Centre line and offset double gap seal (vortex stabilizer) (Fig. 14.65(a and b)) A central vertical partition or alternatively a pair of offset flexible vertical plates attached to the trailer body (see Fig. 14.65(a and b)) is effective in not only preventing crosswinds passing through the cab to trailer gap but also stabilizes the air flow entering the gap by generating a relatively stable vortex on either side of the plate or plates. The vortex stabilizer is slightly less effective than the extended side panel method in reducing the drag coefficient when side winds prevail.

Rolled edge windcheater (Fig. 14.66(a–d)) This device consists of an extended quadrant section moulding attached to the roof and both sides of the leading edges of the forebody trailer panel. When there are sharp leading edges around the trailer body air flowing through the cab to trailer space tends to overshoot and hence initially separate from the side panels of the trailer body (see Fig. 14.66(a)) and even with rounded edging there is still some overshoot and flow separation (Fig. 14.66(b)). The effectiveness of different sectioned forebody edge fairings are compared corresponding to a yaw angle (relative wide angle) variation from 0 to 20°, see Fig. 14.66(d). Here it can be seen that there is very little difference between the semi circular and elongated semi circular moulding but there is a moderate improvement in the drag coefficient at low yaw angles from 0 to 10° for the quadrant section; however, with the extended-quadrant moulding there is a considerable improvement as the yaw angle is increased from 0 to 20°. With the extended quadrant moulding (see Fig. 14.66(c)) the air flow tends to move tangentially between the cab to trailer air gap; some of the air then scrubs along the flat frontage of the trailer body until it reaches the extended-quadrant step, is then deflected slightly rearwards and then again forwards before closely following the contour of the curved corner. This makes it possible for the air flow to remain attached to the side panel surface of the trailer body, therefore keeping the drag resistance on the sheltered trailer panel side to the minimum.

14.8.8 Tractor and trailer skirting

(Fig. 14.67(a and b))

Crosswinds sweeping tangentially underneath the tractor and trailer chassis and between and around the road wheels and axles produce a rise in the drag coefficient. To partially counteract the increase in vehicle drag with increased yaw angle, side skirts can be attached either to the trailer or the tractor or both units. The effectiveness of both tractor and trailer skirting for one particular commercial vehicle is shown in Fig. 14.67(a and b); here it can be seen that with increased yaw angle (relative wind angle) the effectiveness of the trailer skirt peaked at a drag coefficient of 0.07 for a yaw

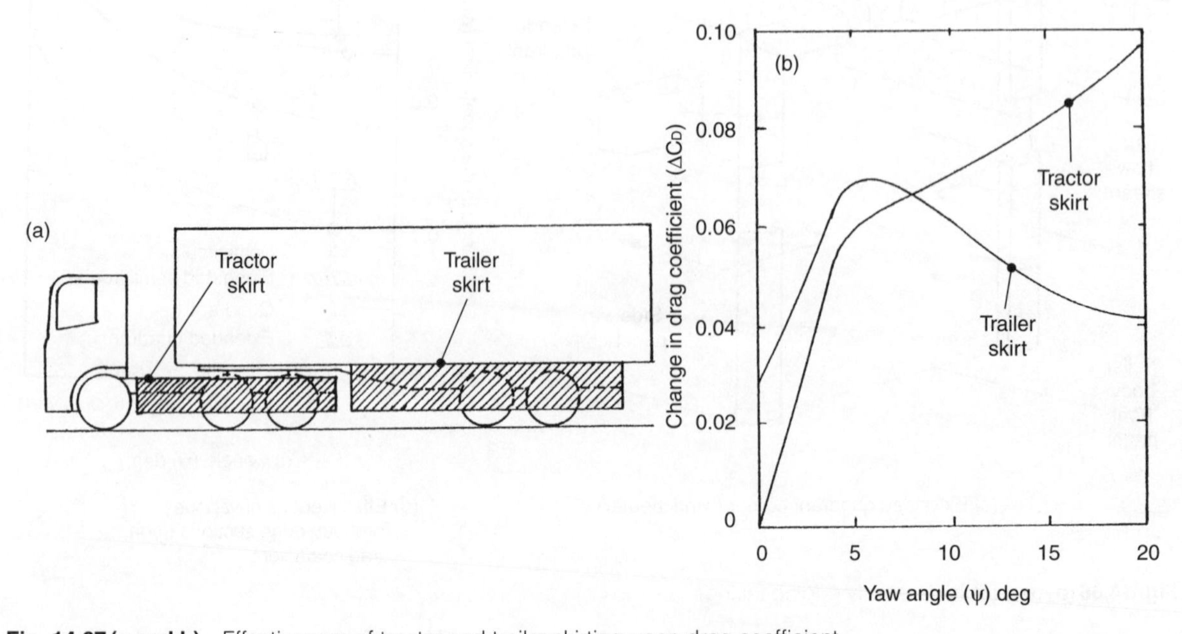

Fig. 14.67 (a and b) Effectiveness of tractor and trailer skirting upon drag coefficient

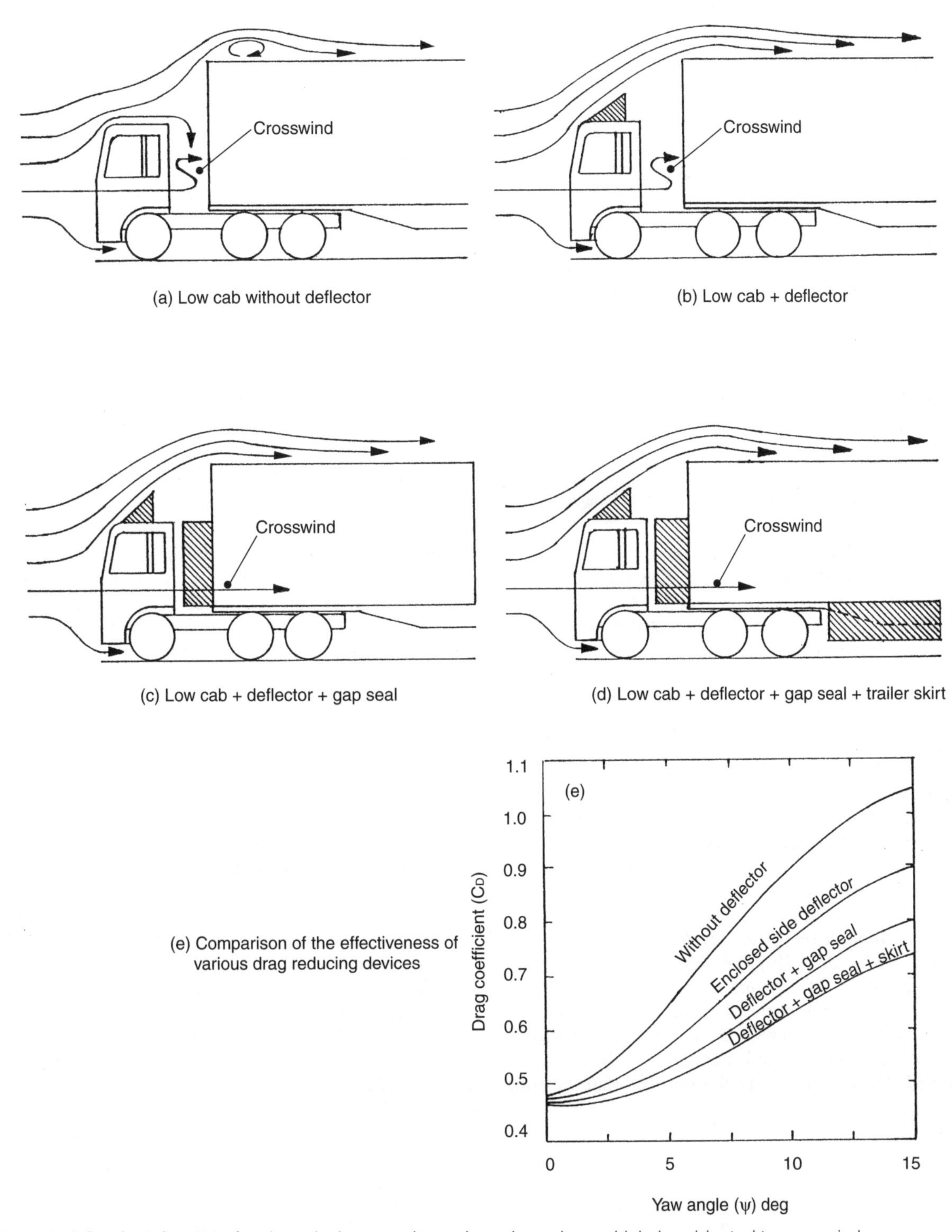

Fig. 14.68 (a–e) Influence of various devices used to reduce drag when vehicle is subjected to crosswinds

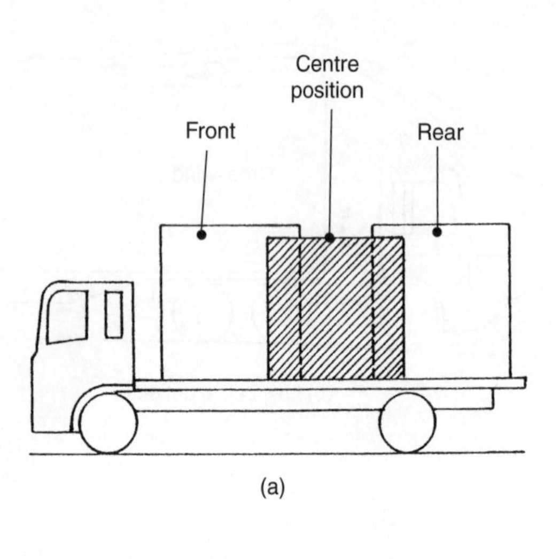

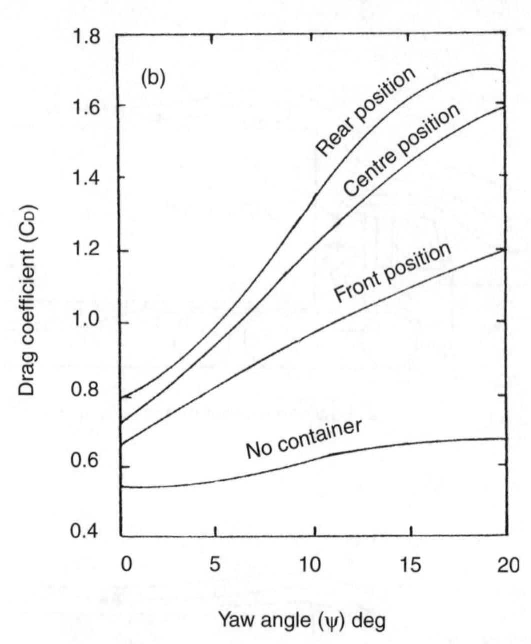

Fig. 14.69 (a and b) Effects of trailer load position upon drag coefficient

angle of 5°. The skirt effectiveness in reducing the drag coefficient then decreases steadily over an increasing yaw angle until it reaches its minimum of 0.04 at a yaw angle of 20°. Conversely the trailer skirt's effectiveness with respect to the yaw angle rose rapidly to 0.06 over a yaw angle range of 5°; the drag coefficient then continued to rise at a slower rate so that for a yaw angle of 20° the drag coefficient effectiveness reached a maximum of just over 0.09. However when considering attaching skirts to a vehicle there can be a problem with the accessibility for routine inspection and for maintenance of the steering, suspension, transmission and brakes; they can also restrict the cooling of the brake drums/discs.

14.8.9 Comparison of various devices used to reduce vehicle drag (Fig. 14.68(a–e))

A comparison of various devices used to reduce vehicle drag particularly when there are crosswinds can be seen in Fig. 14.68(a–e). The graph shows for a low cab and no roof deflector that the drag is at its highest due to the cab to trailer step and that the drag coefficient rises with increasing crosswind yaw angle (relative wide angle) from 0.48 to about 1.05 over a 15° increase in yaw angle (see Fig. 14.68(a and e)); a reduction in the drag coefficient occurs when a cab roof deflector is matched to the trailer body (Fig. 13.68(b and e)). When a cab to trailer gap seal is attached to the trailer there is a further reduction in the drag coefficient particularly with increasing yaw angles (Fig. 14.68(c and e)), and finally there is even a greater drag coefficient reduction obtained when fitting a trailer skirt (Fig. 14.68(d and e)).

14.8.10 Effects of trailer load position on a vehicle's drag resistance (Fig. 14.69(a and b))

The effects of positioning a container load on a platform container truck considerably influences the drag resistance. This becomes more noticeable with crosswinds (see Fig. 14.69(a and b)), and that with a yaw angle of 20° the drag coefficient without a container was only 0.7 whereas with the container mounted in front, centre and towards the rear, the drag coefficients reached 1.2, 1.6 and 1.7 respectively.

Index